GEOMETRIC
LINEAR ALGEBRA

 Volume 1

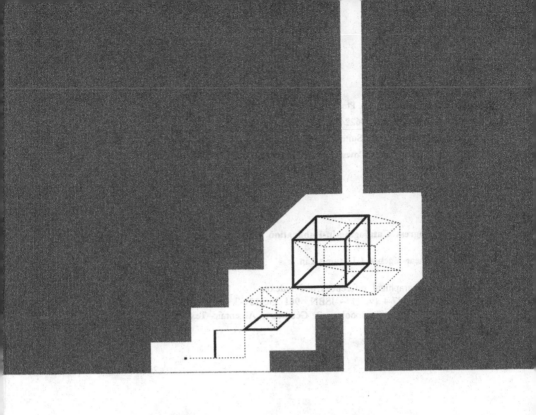

GEOMETRIC
LINEAR ALGEBRA

Volume 1

I-Hsiung Lin

National Taiwan Normal University, China

 World Scientific

NEW JERSEY • LONDON • SINGAPORE • BEIJING • SHANGHAI • HONG KONG • TAIPEI • CHENNAI

Published by

World Scientific Publishing Co. Pte. Ltd.

5 Toh Tuck Link, Singapore 596224

USA office: 27 Warren Street, Suite 401-402, Hackensack, NJ 07601

UK office: 57 Shelton Street, Covent Garden, London WC2H 9HE

Library of Congress Cataloging-in-Publication Data
Lin, Yixiong.
   Geometric linear algebra / I-hsiung Lin.
   p. cm.
   Includes bibliographical references and indexes.
   ISBN   981-256-087-4 (v. 1) -- ISBN   981-256-132-3 (v. 1 : pbk.)
     1. Algebras, Linear--Textbooks. 2. Geometry, Algebraic--Textbooks. I. Title.

QA184.2.L49 2005
512'.5--dc22

2005041722

British Library Cataloguing-in-Publication Data
A catalogue record for this book is available from the British Library.

Printed in Singapore by World Scientific Printers (S) Pte Ltd

To the memory of my
grandparents, parents
and to my family

PREFACE

What is linear algebra about?

Many objects such as buildings, furniture, etc. in our physical world are delicately constituted by counterparts of almost straight and flat shapes which, in geometrical terminology, are portions of straight lines or planes. A butcher's customers ordered the meat in various quantities, some by mass and some by price, so that he had to find answers to such questions as: What is the cost of 2/3 kg of the meat? What mass a customer should have if she only wants to spend 10 dollars? This can be solved by the linear relation $y = ax$ or $y = ax + b$ with b as a bonus. The same is, when traveling abroad, to know the value of a foreign currency in term of one's own. How many faces does a polyhedron with 30 vertices and 50 edges have? What is the Fahrenheit equivalent of 25°C? One experiences numerous phenomena in daily life, which can be put in the realms of straight lines or planes or linear relations of several unknowns.

To start with, the most fundamental and essential ideas needed in *geometry* are

1. (directed) line segment (including the extended line),
2. parallelogram (including the extended plane)

and the associated quantities such as length or signed length of line segment and angle between segments.

The *algebraic* equivalence, in global sense, is linear equations such as

$$a_{11}x_1 + a_{21}x_2 = b_1$$

or

$$a_{11}x_1 + a_{21}x_2 + a_{31}x_3 = b_1$$

and simultaneous equations composed of them. The core is how to determine whether such linear equations have a solution or solutions, and if so, how to find them in an effective way.

Algebra has operational priority over geometry, while the latter provides intuitively geometric motivation or interpretations to results of the former. Both play a role of head and tail of a coin in many situations.

Linear algebra is going to transform the afore-mentioned geometric ideas into two algebraic *operations* so that solving linear equations can be handled *linearly* and systematically. Its implication is far-reaching and its application is widely-open and touches almost every field in modern science. More precisely,

a directed line segment $\overrightarrow{AB} \rightarrow$ a vector $\vec{x}$;

ratio of signed lengths of (directed) line segments along the same line

$$\frac{\overrightarrow{PQ}}{\overrightarrow{AB}} = \alpha \rightarrow \vec{y} = \alpha\vec{x}, \text{ scalar multiplication of } \vec{x} \text{ by } \alpha.$$

See Fig. P.1.

Fig. P.1

Hence, the whole line can be described algebraically as $\alpha\vec{x}$ while α runs through the real numbers. While, the parallelogram in Fig. P.2 indicates that directed segments $\overrightarrow{OA}$ and $\overrightarrow{BC}$ represent the same vector $\vec{x}$, $\overrightarrow{OB}$ and $\overrightarrow{AC}$ represent the same vector $\vec{y}$ so that

the diagonal $\overrightarrow{OC} \rightarrow \vec{x} + \vec{y}$, the *addition* of vectors $\vec{x}$ and $\vec{y}$.

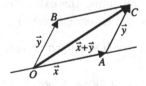

Fig. P.2

As a consequence, the whole plane can be described algebraically as the linear combinations $\alpha\vec{x} + \beta\vec{y}$ where α and β are taken from all the real numbers. In fact, parallelograms provide implicitly as an inductive process to construct and visualize higher dimensional spaces. One may imagine the line OA acting as an $(n-1)$-dimensional space, so that $\vec{x}$ is of the form $\alpha_1\vec{x}_1 + \cdots + \alpha_{n-1}\vec{x}_{n-1}$. In case the point C is outside the space, $\vec{y}$ cannot be expressed as such linear combinations. Then the addition $\alpha\vec{y} + \vec{x}$ will raise the space to the higher n-dimensional one.

As a whole, relying only on

$$\alpha\vec{x}, \quad \alpha \in \mathbb{R} \text{ (real field) and}$$
$$\vec{x} + \vec{y}$$

with appropriate operational properties and using the techniques:

linear combination,
linear dependence, and
linear independence

of vectors, plus deductive and inductive methods, one can develop and establish the whole theory of Linear Algebra, even formally and in a very abstract manner.

The main theme of the theory is about linear transformation which can be characterized as the mapping that *preserves* the ratio of the signed lengths of directed line segments along the same or parallel lines. Linear transformations between finite-dimensional vector spaces can be expressed as matrix equations $\vec{x}A = \vec{y}$, after choosing suitable coordinate systems as bases.

The matrix equation $\vec{x}A = \vec{y}$ has two main features. The *static structure* of it, when consider $\vec{y}$ as a constant vector $\vec{b}$, results from solving algebraically the system $\vec{x}A = \vec{b}$ of linear equations by the powerful and useful Gaussian elimination method. Rank of a matrix and its factorization as a product of simpler ones are the most important results among all. Rank provides insights into the geometric character of subspaces based on the concepts of linear combination, dependence and independence. While factorization makes the introduction of determinant easier and provides preparatory tools to understand another feature of matrices. The *dynamic structure*, when consider $\vec{y}$ as a varying vector, results from treating A as a linear transformation defined by $\vec{x} \to \vec{x}A = \vec{y}$. The kernel (for homogeneous linear equations) and range (for non-homogeneous linear equations) of a linear transformation, dimension theorem, invariant subspaces, diagonalizability, various decompositions of spaces or linear transformations and their canonical forms are the main topics among others.

When Euclidean concepts such as lengths and angles come into play, it is the inner product that combines both and the Pythagorean Theorem or orthogonality dominates everywhere. Therefore, linear operators $\vec{y} = \vec{x}A$

are much more specified and results concerned more fruitful, and provide wide and concrete applications in many fields.

Roughly speaking, using algebraic methods, linear algebra investigates the possibility and how of solving system of linear equations, or geometrically equivalent, studies the inner structures of spaces such as lines or planes and possible interactions between them. Nowadays, linear algebra turns out to be an indispensable shortcut from the global view to the local view of objects or phenomena in our universe.

The purpose of this introductory book

The teaching of linear algebra and its contents has become too algebraic and hence too abstract in the introduction of main concepts and the methods which are going to become formal and well established in the theory. Too fast abstraction of the theory definitely scares away many students whose majors are not in mathematics but need linear algebra very much in their careers.

For most beginners in a first course of linear algebra, the understanding of clearer pictures or the reasons why to do this and that seems more urgent and persuasive than the rigorousness of proofs and the completeness of the theory. Understanding cultivates interestingness to the subject and abilities of computation and abstraction.

To start from one's knowledge and experience does enhance the understanding of a new subject. As far as beginning linear algebra is concerned, I strongly believe that intuitive, even manipulatable, geometric objects or concepts are the best ways to open the gate of entrance. This is the momentum and the purpose behind the writing of this introductory book. I tried before (in Chinese), and I am trying to write this book in this manner, maybe not so successful as originally expected but away from the conventional style in quite a few places (refer to Appendix B).

This book is designed for beginners, like freshman and sophomore or honored high school students.

The general prerequisites to read this book are high-school algebra and geometry. Appendix A, which discuss sets, functions, fields, groups and polynomials, respectively, are intended to unify and review some basic ideas used throughout the book.

Features of the book

Most parts of the contents of this book are abridged briefly from my seven books on *The Introduction to Elementary Linear Algebra* (refer to [3–7], published in Chinese from 1982 to 1984, with the last two still unable to be published until now). I try to write the book in the following manner:

1. Use intuitive geometric concepts or methods to introduce or to motivate or to reinforce the creation of abstract or general theory in linear algebra.
2. Emphasize the geometric characterizations of results in linear algebra.
3. Apply known results in linear algebra to describe various geometries based on F. Klein's Erlanger' point of view.

Therefore, in order to vivify these connections of geometries with linear algebra in a convincing argument, I focus the discussion of the whole book on the real vector spaces $\mathbb{R}^1, \mathbb{R}^2$ and $\mathbb{R}^3$ endowed with more than 500 *graphic illustrations*. It is in this sense that I label this book the title as *Geometric Linear Algebra*. Almost each section is followed by a set of exercises.

4. Usually, each set of Exercises contains two parts: <A> and <B>. The former is designed to familiarize the readers with or to practice the established results in that section, while the latter contains challenging ones whose solutions, in many cases, need some knowledge to be exposed formally in sections that follow. In addition to these, some set of Exercises also contain parts <C> and <D>. <C> asks the readers to try to model after the content and to extend the process and results to vector spaces over arbitrary fields. <D> presents problems concerned with linear algebra, such as in real or complex calculus, differential equations and differential geometry, etc. Let such connections and applications of linear algebra say how important and useful it is.

The readers are asked to do all problems in <A> and are encouraged to try part in <B>, while <C> and <D> are optional and are left to more mature and serious students.

No applications outside pure mathematics are touched and the needed readers should consult books such as Gilbert Strang's *Linear Algebra and its Application*.

Finally, three points deviated from most existed conventional books on linear algebra should be cautioned. One is that chapters are divided according to affine, linear, and Euclidian structures of $\mathbb{R}^1, \mathbb{R}^2$ and $\mathbb{R}^3$, but not according to topics such as vectors spaces, determinants, etc. The other is that few definitions are formal and most of them are allowed to come to the surface in the middle of discussions, while main results obtained after a discussion are summarized and are numbered along with important formulas. The third one is that a point $\vec{x} = (x_1, x_2)$ is also treated as a position *vector* from the origin $\vec{0} = (0,0)$ to that point, when $\mathbb{R}^2$ is considered as a two-dimensional vector space, rather than the common used notation

$$\begin{pmatrix} x_1 \\ x_2 \end{pmatrix} \quad \text{or} \quad \begin{bmatrix} x_1 \\ x_2 \end{bmatrix}.$$

As a consequence of this convention, when a given 2×2 matrix A is considered to represent a linear transformation on $\mathbb{R}^2$ to act on the vector $\vec{x}$, we adopt $\vec{x}A$ and treat $\vec{x}$ as a 1×2 matrix but not $A\begin{bmatrix} x_1 \\ x_2 \end{bmatrix}$ to denote the image vector of $\vec{x}$ under A, unless otherwise stated. Similar explanation is valid for $\vec{x}A$ where $\vec{x} \in \mathbb{R}^m$ and A is an $m \times n$ matrix, etc.

In order to avoid getting lost and for a striking contrast, I compensate Appendix B and title it as *Fundamentals of Algebraic Linear Algebra* for the sake of reference and comparison.

Ways of writing and how to treat $\mathbb{R}^n$ for $n \geq 4$

The main contents are focused on the introduction of $\mathbb{R}^1, \mathbb{R}^2$ and $\mathbb{R}^3$, even though the results so obtained and the methods cultivated can almost be generalized verbatim to $\mathbb{R}^n$ for $n \geq 4$ or finite-dimension vector spaces over fields and, in many occasions, even to infinite-dimensional spaces.

As mentioned earlier, geometric motivation will lead the way of introduction to the well-established and formulated methods in the contents. So the general process of writing is as follows:

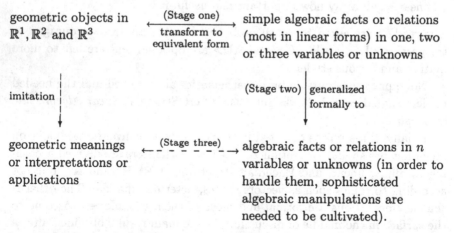

In most cases, we leave Stages two and three as Exercises <C> for mature students.

As a whole, we can use the following *problem*

Prove the identity $1^2 + 2^2 + 3^2 + \cdots + n^2 = \frac{1}{6}n(n+1)(2n+1)$, *where n is any natural number, by the mathematical induction.*

as a model to describe how we probe deeply into the construction and formulation of topics in *abstract* linear algebra. We proceed as follows. It is a dull business at the very beginning to prove this identity by testing both sides in cases $n = 1, n = 2, \ldots$ and then supposing both sides equal to each other in case $n = k$ and finally trying to show both sides equal when $n = k+1$. This is a well-established and sophiscated way of arguments, but it is not necessarily the best way to understand thoroughly the implications and the educational values this problem could provide. Instead, why not try the following steps:

1. How does one know beforehand that the sum of the left side is equal to $\frac{1}{6}n(n+1)(2n+1)$?
2. To pursue this answer, try trivial yet simpler cases when $n = 1, 2, 3$ and even $n = 4$, and then try to find out possible common rules owned by all of them.
3. Conjecture that the common rules found are still valid for general n.
4. Try to prove this conjecture formally by mathematical induction or some other methods.

Now, for $n = 1$, take a "shadow" unit square and a "white" unit square and put them side by side as Fig. P.3:

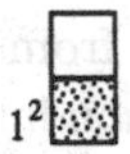

$$\frac{\text{area of shadow region}}{\text{area of the rectangle}} = \frac{1^2}{2 \cdot 1} = \frac{1}{2} = \frac{2 \cdot 1 + 1}{6 \cdot 1}.$$

Fig. P.3

For $n = 2$, use the same process and see Fig. P.4; for $n = 3$, see Fig. P.5.

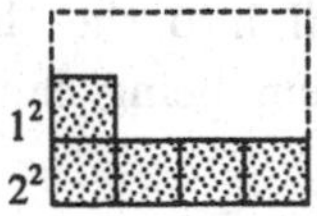

$$\frac{\text{area of shadow region}}{\text{area of the rectangle}} = \frac{1^2 + 2^2}{3 \cdot 4} = \frac{5}{12} = \frac{2 \cdot 2 + 1}{6 \cdot 2}.$$

Fig. P.4

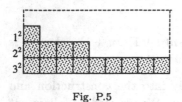

Fig. P.5

$$\frac{\text{area of shadow region}}{\text{area of the rectangle}} = \frac{1^2 + 2^2 + 3^2}{4 \cdot 9} = \frac{14}{36}$$
$$= \frac{7}{18} = \frac{2 \cdot 3 + 1}{6 \cdot 3}.$$

This suggests the conjecture

$$\frac{1^2 + 2^2 + 3^2 + \cdots + n^2}{(n+1)n^2} = \frac{2n+1}{6n}$$

$$\Rightarrow 1^2 + 2^2 + 3^2 + \cdots + n^2 = \frac{1}{6}n(n+1)(2n+1).$$

It is approximately in this manner that I wrote the contents of the book, in particular, in Chaps. 1, 2 and 4. Of course, this procedure is roundabout, overlapped badly in some cases and even makes one feel impatiently and sick. So I tried to summarize key points and main results on time. But I do strongly believe that it is a worthy way of educating beginners in a course of linear algebra.

Well, I am not able to realize physically the existence of four or higher dimensional spaces. Could you? How? It is algebraic method that convinces us properly the existence of higher dimensional spaces. Let me end up this puzzle with my own experience in the following story.

Some day in 1986, in a Taoism Temple in eastern Taiwan, I had a face-to-face dialogue with a person epiphanized (namely, making its presence or power felt) by the God Nuo Zha (also esteemed as the Third Prince in Chinese communities):

I asked: Does God exist?

Nuo Zha answered: Gods do exist and they live in spaces, from dimension seven to dimension thirteen. You common human being lives in dimension three, while dimensions four, five and six are buffer zones between human being and Gods. Also, there are "human being" in underearth, which are two-dimensional.

I asked: Does UFO (unfamiliar objects) really exist?

Nuo Zha answered: Yes. They steer improperly and fall into the three-dimensional space so that you human being can see them physically.

Believe it or not!

Sketch of the contents

Catch a quick glimpse of the **Contents** or **Sketch of the content** at the beginning of each chapter and one will have rough ideas about what might be going on inside the book.

Let us start from an example.

Fix a Cartesian coordinate system in space. Equation of a plane in space is $a_1x_1 + a_2x_2 + a_3x_3 = b$ with $b = 0$ if and only if the plane passes through the origin $(0,0,0)$. Geometrically, the planes $a_1x_1 + a_2x_2 + a_3x_3 = b$ and $a_1x_1 + a_2x_2 + a_3x_3 = 0$ are parallel to each other and they will be coincident by invoking a translation. Algebraically, the main advantage of a plane passing the origin over these that are not is that it can be vectorized as a two-dimensional *vector space* $\langle\langle \vec{v}_1, \vec{v}_2 \rangle\rangle = \{\alpha_1 \vec{v}_1 + \alpha_2 \vec{v}_2 \mid \alpha_1, \alpha_2 \in \mathbb{R}\}$, where $\vec{v}_1$ and $\vec{v}_2$ are linear independent *vectors* lying on $a_1x_1 + a_2x_2 + a_3x_3 = 0$, while $a_1x_1 + a_2x_2 + a_3x_3 = b$ is the image $\vec{x}_0 + \langle\langle \vec{v}_1, \vec{v}_2 \rangle\rangle$, called an *affine plane*, of $\langle\langle \vec{v}_1, \vec{v}_2 \rangle\rangle$ under a translation $\vec{x} \to \vec{x}_0 + \vec{x}$ where $\vec{x}_0$ is a point lying on the plane. Since any point in space can be chosen as the origin or as the zero vector, unnecessary distinction between vector and affine spaces, except possibly for pedagogic reasons, should be emphasized or exaggerated. This is the main reason why I put the affine and linear structures of $\mathbb{R}^1, \mathbb{R}^2$ and $\mathbb{R}^3$ together as **Part 1** which contains Chaps. 1–3.

When the concepts of length and angle come into our mind, we use inner product $\langle , \rangle$ to connect both. Then the plane $a_1x_1 + a_2x_2 + a_3x_3 = b$ can be characterized as $\langle \vec{x} - \vec{x}_0, \vec{a} \rangle = 0$ where $\vec{a} = (a_1, a_2, a_3)$ is the normal vector to the plane and $\vec{x} - \vec{x}_0$ is a *vector* lying on the plane which is determined by the *points* $\vec{x}_0$ and $\vec{x}$ in the plane. This is **Part 2**, the Euclidean structures of $\mathbb{R}^2$ and $\mathbb{R}^3$, which contains Chaps. 4 and 5.

In our vivid physical world, it is difficult to realize that the parallel planes $a_1x_1 + a_2x_2 + a_3x_3 = b$ $(b \neq 0)$ and $a_1x_1 + a_2x_2 + a_3x_3 = 0$ will intersect along a "line" within our sights. By central projection, it would be reasonable to imagine that they do intersect along an infinite or imaginary line l_∞. The adjoint of l_∞ to the plane $a_1x_1 + a_2x_2 + a_3x_3 = b$ constitutes a *projective plane*. This is briefly touched in Exs. <B> of Sec. 2.6 and Sec. 3.6, Ex. <B> of Sec. 2.8.5 and Sec. 3.8.4.

Changes of coordinates from $\vec{x} = (x_1, x_2)$ to $\vec{y} = (y_1, y_2)$ in $\mathbb{R}^2$:

$$y_1 = a_1 + a_{11}x_1 + a_{21}x_1$$
$$y_2 = a_2 + a_{12}x_1 + a_{22}x_2 \quad \text{or} \quad \vec{y} = \vec{x}_0 + \vec{x}A,$$

where $A = \begin{bmatrix} a_{11} & a_{12} \\ a_{21} & a_{22} \end{bmatrix}$ with $a_{11}a_{22} - a_{12}a_{21} \neq 0$ is called an *affine transformation* and, in particular, an invertible *linear transformation* if $\vec{x}_0 = (a_1, a_2) = \vec{0}$. This can be characterized as a one-to-one mapping from $\mathbb{R}^2$ onto $\mathbb{R}^2$ which preserves ratios of line segments along parallel lines (Secs. 1.3, 2.7, 2.8 and 3.8). If it preserves distances between any two points, then called a *rigid* or *Euclidean motion* (Secs. 4.8 and 5.8). While $\vec{y} = \sigma(\vec{x}A)$ for any scalar $\sigma \neq 0$ maps lines onto lines on the projective plane and is called a *projective transformation* (Sec. 3.8.4). The invariants under the group (Sec. A.4) of respective transformations constitute what F. Klein called *affine*, *Euclidean* and *projective* geometries (Secs. 2.8.4, 3.8.4, 4.9 and 5.9).

As important applications of exterior products (Sec. 5.1) in $\mathbb{R}^3$, *elliptic geometry* (Sec. 5.11) and *hyperbolic geometry* (Sec. 5.12) are introduced in the same manner as above. These two are independent of the others in the book.

Almost every text about linear algebra treats $\mathbb{R}^1$ trivially and obviously. Yes, really it is and hence some pieces of implicit information about $\mathbb{R}^1$ are usually ignored. Chapter 1 indicates that only *scalar multiplication* of a vector is just enough to describe a straight line and how the concept of *linear dependence* comes out of geometric intuition. Also, through vectorization and coordinatization of a straight line, one can realize why the abstract set $\mathbb{R}^1$ can be considered as standard representation of all straight lines. Changes of coordinates enable us to interpret the linear equation $y = ax + b$, $a \neq 0$, geometrically as an affine transformation preserving ratios of segment lengths. Above all, this chapter lays the foundation of inductive approach to the later chapters.

Ways of thinking and the methods adopted to realize them in Chap. 2 constitute a cornerstone for the development of the theory and a model to go after in Chap. 3 and even farther more. The fact that a point outside a given line is needed to construct a plane is algebraically equivalent to say that, in addition to scalar multiplication, the *addition of vectors* is needed in order, via concept of *linear independence* and method of *linear combination*, to go from a lower dimensional space (like straight line) to a higher one (like plane). Sections 2.2 up to 2.4 are counterparts of Secs. 1.1 up to 1.3 and they set up the abstract set $\mathbb{R}^2$ as the standard two-dimensional real *vector space* and changes of coordinates in $\mathbb{R}^2$. The existence of straight lines (Sec. 2.5) on $\mathbb{R}^2$ implicitly suggests that it is possible to discuss *vector* and *affine* subspaces in it. Section 2.6 formalizes affine coordinates and introduces another useful barycentric coordinates. The important concepts of *linear (affine) transformation* and its *matrix representation* related to *bases* are main theme in Secs. 2.7 and 2.8. The geometric behaviors of *elementary matrices* considered as linear transformations are investigated in Sec. 2.7.2 along with the *factorization of a matrix* in Sec. 2.7.5 as a product of elementary ones. While, Secs. 2.7.6–2.7.8 are concerned respectively with *diagonal, Jordan* and *rational canonical forms* of linear operators. Based on Sec. 2.7, Sec. 2.8.3 collects *invariants* under affine transformations and Sec. 2.8.4 introduces *affine geometry* in the plane. The last section Sec. 2.8.5 investigates affine invariants of *quadratic curves*.

Chapter 3, investigating $\mathbb{R}^3$, is nothing new by nature and in content from these in Chap. 2 but is more difficult in algebraic computations and in the manipulation of geometric intuition. What should be mentioned is that, basically, only middle-school algebra is enough to handle the whole Chap. 2 but I try to transform this classical form of algebra into rudimentary ones adopted in Linear Algebra which are going to become sophisticated and formally formulated in Chap. 3.

Chapters 4 and 5 use inner product $\langle , \rangle$ to connect concepts of length and angle. The whole theory concerned is based on the Pythagorean Theorem and *orthogonality* dominates everywhere. In addition to lines and planes, circles (Sec. 4.2), spheres (Sec. 5.2) and exterior product of vectors in $\mathbb{R}^3$ (Sec. 5.1) are discussed. One of the features here is that we use geometric intuition to define determinants of order 2 and 3 and to develop their algebraic operational properties (Secs. 4.3 and 5.3). An important by-product of nonnatural inner product (Secs. 4.4 and 5.4) is orthogonal matrix. Therefore, another feature is that we use geometric methods to prove *SVD for matrices* of order 2 and 3 (Secs. 4.5 and 5.5), and the *diagonalization of symmetric matrices* of order 2 and 3 (Secs. 4.7 and 5.7). Euclidean invariant and geometry are in Secs. 4.9 and 5.9. Euclidean invariants of quadratic curves and surfaces are in Secs. 4.10 and 5.10. As companions of Euclidean (also called parabolic) geometry, elliptic and hyperbolic geometries are sketched in Secs. 5.11 and 5.12, respectively.

Notations

Sections denoted by an asterisk (*) are optional and may be omitted.

[1] means the first book listed in the Reference, etc.

A.1 means the first section in Appendix A, etc.

Section 1.1 means the first section in Chap. 1. So Sec. 4.3 means the third section in Chap. 4, while Sec. 5.9.1 means the first subsection of Sec. 5.9, etc.

Exercise <A> 1 of Sec. 1.1 means the first problem in Exercises <A> of Sec. 1.1, etc.

(1.1.1) means the first numbered important or remarkable facts or summarized theorem in Sec. 1.1, etc.

Figure 3.6 means that the sixth figure in Chap. 3, etc. Fig. II.1 means the first figure in Part 2, etc. Figure A.1 means the first figure in Appendix A; similarly for Fig. B.1, etc.

The end of a proof or an Example is sometimes but not always marked by □ for attention.

For details, refer to **Index of Notations**.

Suggestions to the readers (how to use this book)

The materials covered in this book are rich and wide, especially in Exercises <C> and <D>. It is almost impossible to cover the whole book in a single course on linear algebra when being used as a textbook for beginners.

As a textbook, the depth and wideness of materials chosen, the degree of rigorousness in proofs and how many topics of applications to be covered depend, in my opinion, mainly on the purposes designed for the course and the students' mathematical sophistication and backgrounds. Certainly, there are various combinations of topics. The instructors always play a central role on many occasions. The following possible choices are suggested:

(1) For honored high school students: Chapters 1, 2 and 4 plus Exercises <A>.
(2) For freshman students: Chapters 1, 2 (up to Sec. 2.7), 3 (up to Sec. 3.7), 4 (up to Sec. 4.7 and Sec. 4.10) and/or 5 (up to Sec. 5.7 and Sec. 5.10) plus, at least, Ex. <A>, in a one-academic-year three-hour-per-week course. As far as teaching order, one can adopt this original arrangement in this book, or after finishing Chap. 1, try to combine Chaps. 2 and 3, 4 and 5 together according to the same titles of sections in each chapter.
(3) For sophomore students: Just like (2) but contains some selected problems from Ex. <B>.
(4) For a geometric course via linear algebra: Chapters 1, 2 (Sec. 2.8), 3 (Sec. 3.8), 4 (Sec. 4.8) and 5 (Secs. 5.8–5.12) in a one-academic-year three-hour-per-week course.

(5) For junior and senior students who have had some prior exposure to linear algebra: selective topics from the contents with emphasis on problem-solving from Exercises <C>, <D> and Appendix B.

Of course, there are other options up to one's taste.

In my opinion, this book might better be used as a reference book or a companion one to a formal course on linear algebra. In my experience of teaching linear algebra for many years, students often asked questions such as, among many others:

1. Why linear dependence and independence are defined in such way?
2. Why linear transformation $(f(\vec{x} + \vec{y}) = f(\vec{x}) + f(\vec{y}), f(\alpha, \vec{x}) = \alpha f(\vec{x}))$ is defined in such way? Is there any sense behind it?
3. Does the definition for eigenvalue seem so artificial and is its main purpose just for symmetric matrices?

Hence, all one needs to do is to cram up the algebraic rules of computation and the results concerned, pass the exams and get the credits. That is all. It is my hope that this book might provide a possible source of geometric explanation or introduction to abstract concept or results formulated in linear algebra (see **Features of the book**). But I am not sure that those geometric interpretations appeared in this book are the most suitable ones among all. Readers may try and provide a better one.

From Exercises <D>, readers can find possible connections and applications of linear algebra to other fields of pure mathematics or physics, which are mentioned briefly near the end of the **Sketch of the content** from Chap. 3 on.

Probably, Answers and Hints to problems in Exercises <A>, <B> and <C>, especially the latter two, should be attached near the end of the book. Anyway, I will prepare them but this takes time.

This book can be used in multiple ways.

Acknowledgements

I had the honor of receiving so much help as I prepared the manuscripts of this book.

Students listed below from my classes on linear algebra, advanced calculus and differential geometry typed my manuscripts:

1. Sophomore: Shu-li Hsieh, Ju-yu Lai, Kai-min Wang, Shih-hao Huang, Yu-ting Liao, Hung-ju Ko, Chih-chang Nien, Li-fang Pai;

2. Junior: S. D. Tom, Christina Chai, Sarah Cheng, I-ming Wu, Chih-chiang Huang, Chia-ling Chang, Shiu-ying Lin, Tzu-ping Chuang, Shih-hsun Chung, Wan-ju Liao, Siao-jyuan Wang;

3. Senior: Zheng-yue Chen, Kun-hong Xie, Shan-ying Chu, Hsiao-huei Wang, Bo-wen Hsu, Hsiao-huei Tseng, Ya-fan Yen, Bo-hua Chen, Wei-tzu Lu;

while

4. Bo-how Chen, Kai-min Wang, Sheng-fan Yang, Shih-hao Huang, Feng-sheng Tsai

graphed the figures, using GSP, WORD and FLASH; and

5. S. D. Tom, Siao-jyuan Wang, Chih-chiang Huang, Wan-ju Liao, Shih-hsun Chung, Chia-ling Chang

edited the initial typescript. They did these painstaking works voluntarily, patiently, dedicatedly, efficiently and unselfishly without any payment. Without their kind help, it is impossible to have this book coming into existence so soon. I'm especially grateful, with my best regards and wishes, to all of them.

And above all, special thanks should be given to Ms Shu-li Hsieh and Mr Chih-chiang Huang for their enthusiasm, carefulness, patience and constant assistance with trifles unexpected.

Teaching assistant Ching-yu Yang in the Mathematics Department, provided technical assistance with computer works occasionally. Prof. Shao-shiung Lin of National Taiwan University, Taipei reviewed the inital typescript and offered many valuable comments and suggestions for improving the text. Thank you both so much.

Also, thanks to Dr. K. K. Phua, Chairman and Editor-in-Chief, World Scientific, for his kind invitation to join this book in their publication, and to Ms Zhang Ji for her patience and carefulness in editing the book, and to these who might help correcting the English.

Of course, it is me who should take the responsibility of possible errata that remain. The author welcomes any positive and constructive comments and suggestions.

I-hsiung Lin
NTNU, Taipei, Taiwan
June 21, 2004

CONTENTS

PART 1

The Affine and Linear Structures of $\mathbb{R}^1$, $\mathbb{R}^2$ and $\mathbb{R}^3$

Introduction

Starting from intuitively geometric objects, we treat

1. a point as a zero vector,
2. a directed segment (along a line) as a vector, and
3. two directed segments along the same or parallel lines as the same vector if both have the same length and direction.

And hence, we define two vector operations: scalar multiplication $\alpha\vec{x}$ and addition $\vec{x} + \vec{y}$ and develop their operational properties. In the process, we single out the linear combination, dependence and independence among vectors as the main tools and establish the affine structures on a line, a plane and a space, respectively. Then, we extract the essence of concepts obtained and formulate, via rough ideas of linear isomorphism, the abstract sets $\mathbb{R}^1, \mathbb{R}^2$ and $\mathbb{R}^3$ as the standard one-dimensional, two-dimensional and three-dimensional vector spaces over the real field, respectively. So far, changes of coordinates in the same space are the most prominent results among all, which indicates implicitly the concepts of affine and linear transformations.

Then, we focus our attention to these mappings between spaces that preserve the ratios of signed lengths of segments along the same or parallel lines. They are affine transformations (see Secs. 1.4, 2.7 and 2.8), in particular, linear transformations if they map the zero vector into the zero vector when the spaces concerned are considered as vector spaces.

The main themes will be topics on linear transformations or, equivalently, real matrices of order $m \times n$, where $m, n = 1, 2, 3$, such as:

1. Eigenvalues and eigenvectors (Secs. 2.7.1, 2.7.2, 3.7.1 and 3.7.2).
2. Various decompositions, for example, elementary matrix factorizations, LU, LDU, LDL* and LPU, etc. (Secs. 2.7.5 and 3.7.5).
3. Rank, Sylvester's law of inertia (Secs. 2.7.1, 2.7.5, 3.7.1 and 3.7.5).

4. Diagonalizability (Secs. 2.7.6 and 3.7.6).
5. Jordan and rational canonical forms (Secs. 2.7.7, 2.7.8, 3.7.7 and 3.7.8).

A suitable choice of basis for the kernel or/and the image of a linear transformation will play a central role in handling these problems.

Based on results about linear transformations, affine transformations are composed of linear transformations followed by translations. We discuss topics such as:

1. Stretch, reflection, shearing, rotation and orthogonal reflection and their matrix representations and geometric mapping properties (Secs. 2.8.2 and 3.8.2).
2. Affine invariants (Secs. 2.8.3 and 3.8.3).
3. Affine geometries, including sketches of projective plane $P^2(\mathbb{R})$ and projective spaces $P^3(\mathbb{R})$ (Ex. <B> of Sec. 2.6, Sec. 2.8.4, Ex. <B> of Sec. 2.8.5 and Sec. 3.8.4).
4. Quadratic curves (Sec. 2.8.5) and quadrics (Sec. 3.8.5).

Chapter 1 is trivial in content but is necessary in the inductive process. Chapter 2 is crucial both in content and in method. Methods in Chap. 2 are essentially the extensions of geometric and algebraic ones which are learned from the middle school mathematical courses, in particular, the ways of solving simultaneous linear equations and their geometric interpretations. Hence, methods adopted in Chap. 2 play a transitive role from the middle school ones to the more sophisticated and well-established ones used in nowadays linear algebra, as will be seen and formulated in Chap. 3. In short, methods and contents in Chap. 2 can be considered as buffer zones between classical middle school algebra and modern linear algebra.

Based on our inductive construction and description about linear algebras on $\mathbb{R}^1$, $\mathbb{R}^2$ and $\mathbb{R}^3$, we hope that readers will possess enough solid foundations, both in geometric intuition and in algebraic manipulation. Thus, they can foresee, realize and construct what the n-dimensional vector space $\mathbb{R}^n$ ($n \geq 4$) and the linear algebras on it are, even on the more abstract vector spaces over fields. For this purpose, we have arranged intensively problems in Exercises <B> and <C> for minded readers to practice, and Appendix B for reference.

The use of matrices (of order $m \times n, m, n = 1, 2, 3$) and determinants (of order $m, m = 2, 3$) comes to surface naturally as we proceed without introducing them beforehand in a particularly selected section. Here in

Part 1, we emphasize the computational aspects of matrices and determinants. So, the needed readers should consult Sec. B.4 for matrices and Sec. B.6 for determinants. Sections 4.3 and 5.3 will formally introduce the theory of determinants of order 2 and 3, respectively, via geometric considerations.

CHAPTER 1

The One-Dimensional Real Vector Space $\mathbb{R}$ (or $\mathbb{R}^1$)

Introduction

Our theory starts from the following simple geometric

Postulate *A single point determines a unique zero-dimensional (vector) space.*

Usually, a little black point or spot is used as an intuitively geometric model of zero-dimensional space. Notice that "point" is an undefined term without length, width and height.

In the physical world, it is reasonable to imagine that there exits two different points. Hence, one has the

Postulate *Any two different points determine one and only one straight line.*

A straightened loop, extended beyond any finite limit in both directions, is a lively geometric model of a straight line. Mathematically, pick up two different points O and A on a flat paper, imagining extended beyond any limit in any direction, and then, connect O and A by a ruler. Now, we have a geometric model of an unlimited straight line L (see Fig. 1.1).

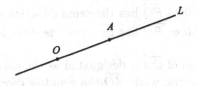

Fig. 1.1

As far as the basic concepts of straight lines are concerned, one should know the following facts (1)–(6).

5

(1) There are uncountably infinite points on L.

(2) The straight line determined by any two different points on L coincides with L.

(3) Any two points P and Q on L decide a *segment*, denoted by PQ:

    1. If $P = Q$ (i.e. P and Q coincide, and represent the same point), then the segment PQ degenerates into a single P (or Q);

    2. If $P \neq Q$ (i.e. P and Q are different points), then PQ consists of those points on L lying between P and Q (included).

(4) If one starts from point P, walking along L toward point Q, then one gets the *directed segment* $\overrightarrow{PQ}$; if from Q to P, reversing the direction, one has the *directed segment* $\overrightarrow{QP}$ (see Fig. 1.2).

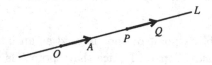

Fig. 1.2

(5) Arbitrarily fix two different points O and A on line L. Consider the segment OA as one unit in length. Then, one should be able to measure the *distance* between any two points P and Q on line L or the *length* of the segment PQ. As usual, distance and length are always non-negative real numbers.

In order to extend the mathematical knowledge we have up to now, here we introduce the *signed length* of a segment PQ as follows:

1. Let $P = Q$, then PQ has length zero;
2. Let $P \neq Q$,

    PQ has length $> 0 \Leftrightarrow \overrightarrow{PQ}$ has the same direction as $\overrightarrow{OA}$;
    PQ has length $< 0 \Leftrightarrow \overrightarrow{PQ}$ has the opposite direction as $\overrightarrow{OA}$.

Therefore, the direction of $\overrightarrow{OA}$ is designated as the *positive direction* of the line L with O as its origin, while $\overrightarrow{AO}$ the *negative direction*.

Remark For convenience, we endow PQ with two meanings: one represents the segment with endpoints P and Q, the other represents the length of that segment. Similarly, $\overrightarrow{PQ}$ has two meanings too: the directed segment from P to Q and the signed length of that segment.

Therefore, finally we have

(6) For any three points P, Q and R on line L, their signed lengths $\overrightarrow{PQ}, \overrightarrow{QR}$ and $\overrightarrow{PR}$ always satisfy the following identity:

$$\overrightarrow{PQ} + \overrightarrow{QR} = \overrightarrow{PR} \quad \text{(see Fig. 1.3)}.$$

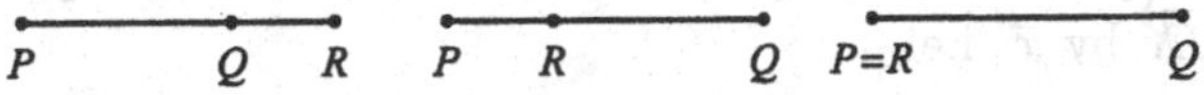

<table>
<tr><td>•——————• •</td><td>•——• •</td><td>•——————•</td></tr>
<tr><td>P Q R</td><td>P R Q</td><td>$P{=}R$ Q</td></tr>
</table>

Fig. 1.3

Sketch of the Content

Based on these facts, this chapter contains four sections, trying to vectorize (Sec. 1.1) and coordinatize (Sec. 1.2) the straight line, and studying the linear changes between different coordinate systems (Sec. 1.3). Invariants under affine transformation are discussed in Sec. 1.4.

The main result is that, under coordinatization, a straight line can be considered as a concrete geometric model of the real number system (field) $\mathbb{R}$. Hence, $\mathbb{R}$ is an abstract representation of the *one-dimensional vector space over the real field* $\mathbb{R}$.

Our introduction to two-dimensional (Chap. 2) and three-dimensional (Chap. 3) vector spaces will be modeled after the way we have treated here in Chap. 1.

1.1   Vectorization of a Straight Line: Affine Structure

Fix a straight line L.

We provide a directed segment $\overrightarrow{PQ}$ on line L as a (line) *vector*. If $P = Q, \overrightarrow{PQ}$ is called a *zero vector*, denoted by $\vec{0}$. On the contrary, $\overrightarrow{PQ}$ is a *nonzero vector* if $P \neq Q$.

Two vectors $\overrightarrow{PQ}$ and $\overrightarrow{P'Q'}$ are *identical*, i.e. $\overrightarrow{PQ} = \overrightarrow{P'Q'}$.

$\Leftrightarrow$ 1. $PQ = P'Q'$ (equal in length),

2. "the direction from P to Q (along L)" is the same as "the direction from P' to Q'".   (1.1.1)

We call properties 1 and 2 as the *parallel invariance of vector* (see Fig. 1.4).

Fig. 1.4

In particular, for any points P and Q on L, one has

$$\overrightarrow{PP} = \overrightarrow{QQ} = \vec{0}. \qquad (1.1.2)$$

Hence, zero vector is uniquely defined.

Now, fix any two different points O and X on L. For simplicity, denote the vector $\overrightarrow{OX}$ by $\vec{x}$, i.e.

$$\vec{x} = \overrightarrow{OX}.$$

Note that $\vec{x} \neq \vec{0}$.

For any fixed point P on L, the ratio of the signed length $\overrightarrow{OP}$ with respect to $\overrightarrow{OX}$ is

$$\frac{\overrightarrow{OP}}{\overrightarrow{OX}} = \alpha,$$

where the real number α has the following properties:

$\alpha = 0 \Leftrightarrow P = O$;
$0 < \alpha < 1 \Leftrightarrow P$ lies on the segment OX ($P \neq O, X$);
$\quad \alpha > 1 \Leftrightarrow P$ and X lie on the same side of O and $OP > OX$;
$\quad \alpha = 1 \Leftrightarrow P = X$; and
$\quad \alpha < 0 \Leftrightarrow P$ and X lie on the different sides of O.

In all cases, designate the vector

$$\overrightarrow{OP} = \alpha\overrightarrow{OX} = \alpha\vec{x}.$$

On the other hand, for any given $\alpha \in \mathbb{R}$, there corresponds one and only one point P on line L such that $\overrightarrow{OP} = \alpha\vec{x}$ holds (see Fig. 1.5).

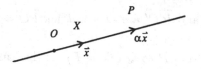

Fig. 1.5

Summarize as

The Vectorization of a straight line

Fix any two different points O and X on a straight line L and denote the vector $\overrightarrow{OX}$ by $\vec{x}$. Then the *scalar product* $\alpha\vec{x}$ of an arbitrary real number

$\alpha \in \mathbb{R}$ with the fixed vector $\vec{x}$ is suitable for describing any point P on L (i.e. the position vector $\overrightarrow{OP}$). Call the set

$$L(O; X) = \{\alpha\vec{x} \mid \alpha \in \mathbb{R}\}$$

the *vectorized space* of the line L with O as the *origin*, $\overrightarrow{OO} = \vec{0}$ as *zero vector* and $\vec{x}$ as the *base vector*. Elements in $L(O; X)$ are called (line) *vectors* which have the following algebraic operation properties: $\alpha, \beta \in \mathbb{R}$,

1. $(\alpha + \beta)\vec{x} = \alpha\vec{x} + \beta\vec{x} = \beta\vec{x} + \alpha\vec{x}$;
2. $(\alpha\beta)\vec{x} = \alpha(\beta\vec{x}) = \beta(\alpha\vec{x})$;
3. $1\vec{x} = \vec{x}$;
4. Let $\vec{0} = \overrightarrow{OO}$, then $\alpha\vec{x} + \vec{0} = \vec{0} + \alpha\vec{x} = \alpha\vec{x}$;
5. $(-\alpha)\vec{x} = -\alpha\vec{x}$; $\vec{x} + (-\vec{x}) = \vec{0} = \vec{x} - \vec{x}$;
6. $0\vec{x} = \alpha\vec{0} = \vec{0}$.

In short, via the position vector $\overrightarrow{OP} = \alpha\vec{x}$ for any α, points P on L have the above algebraic operation properties.   (1.1.3)

Using the concept of (1.1.3), one can establish the algebraic characterization for three points lying on the same line.

Suppose that points O, X and Y are collinear. Let $\vec{x} = \overrightarrow{OX}$ and $\vec{y} = \overrightarrow{OY}$.

In case $O = X = Y$: then $\vec{x} = \vec{y} = \vec{0}$, and hence $\vec{x} = \alpha\vec{y}$ or $\vec{y} = \alpha\vec{x}$ holds for any $\alpha \in \mathbb{R}$.

In case two of O, X and Y coincide, say $X \neq O = Y$: then $\vec{y} = \vec{0}$ and $\vec{y} = 0\vec{x}$ holds.

If O, X and Y are different from each other, owing to the fact that Y lies on the line determined by O and X, $\vec{y}$ belongs to $L(O; X)$. Hence, there exists $\alpha \in \mathbb{R}$ such that $\vec{y} = \alpha\vec{x}$.

We summarize these results as

Linear dependence of line vectors
Let $\vec{x} = \overrightarrow{OX}$ and $\vec{y} = \overrightarrow{OY}$. Then

(1) (geometric) Points O, X and Y are collinear.
$\Leftrightarrow$ (2) (algebraic) There exists $\alpha \in \mathbb{R}$ such that $\vec{y} = \alpha\vec{x}$ or $\vec{x} = \alpha\vec{y}$.
$\Leftrightarrow$ (3) (algebraic) There exist scalars $\alpha, \beta \in \mathbb{R}$, not all equal to zero, such that $\alpha\vec{x} + \beta\vec{y} = \vec{0}$.

In any one of these three cases, vectors $\vec{x}$ and $\vec{y}$ are said to be *linear dependent* (on each other).   (1.1.4)

As contrast to linear dependence, one has

$$\alpha\vec{x} = \vec{0} \Leftrightarrow 1^\circ\alpha = 0 \ (\vec{x} \text{ may not be } \vec{0}); \quad \text{or}$$
$$2^\circ\vec{x} = \vec{0} \ (\alpha \text{ may not be } 0).$$

Therefore, we have

Linear independence of a nonzero vector
Let $OX = \vec{x}$. Then

(1) (geometric) Points O and X are different.
$\Leftrightarrow$ (2) (algebraic) If there exists $\alpha \in \mathbb{R}$ such that $\alpha\vec{x} = \vec{0}$, then it is necessarily that $\alpha = 0$.

In any of these situations, vector $\vec{x}$ is said to be *linear independent* (from any other vector, whatsoever!).

(1.1.5)

That is to say, a single nonzero vector must be linearly independent.

Exercises
<A>

1. Interpret geometrically 1 to 6 in (1.1.3).
2. Finish the incomplete proofs of (1.1.4), for example, (3) $\Leftrightarrow$ (2) $\Rightarrow$ (1).

1.2   Coordinatization of a Straight Line: $\mathbb{R}^1$ (or $\mathbb{R}$)

Suppose the straight line L is provided with a fixed vectorized space $L(O; X)$, where $\vec{x} = \overrightarrow{OX}$ is the base vector.
   We call the set

$$\mathcal{B} = \{\vec{x}\}$$

a *basis* of $L(O; X)$.

For any given point $P \in L$, the fact that $\overrightarrow{OP} \in L(O; X)$ induces a unique $\alpha \in \mathbb{R}$ such that $\overrightarrow{OP} = \alpha\vec{x}$ holds. Then, this unique scalar, denoted by

$$\alpha = [\overrightarrow{OP}]_\mathcal{B} = [P]_\mathcal{B}$$

is defined to be the *coordinate* of the point P or the vector $\overrightarrow{OP}$ *with respect to the basis* $\mathcal{B}$. In particular,

$$[O]_\mathcal{B} = 0, \quad [X]_\mathcal{B} = 1.$$

For example:

$$[P]_\mathcal{B} = -2 \Leftrightarrow \overrightarrow{OP} = -2\vec{x};$$
$$[Q]_\mathcal{B} = \frac{3}{2} \Leftrightarrow \overrightarrow{OQ} = \frac{3}{2}\vec{x}.$$

See Fig. 1.6.

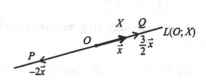

Fig. 1.6

Now we summarize as

The coordinatization of a straight line

Let $L(O; X)$ be an arbitrary vectorized space of line L, with $\mathcal{B} = \{\vec{x}\}$, $\vec{x} = \overrightarrow{OX}$, as a basis. The set

$$\mathbb{R}_{L(O;X)} = \{[P]_\mathcal{B} \mid P \in L\}$$

is called the *coordinatized space* of L with respect to $\mathcal{B}$. Explain further as follows.

(1) There is a one-to-one correspondence from any point P on line L onto corresponding number $[P]_\mathcal{B}$ of the real number system $\mathbb{R}$.

(2) Define a mapping $\Phi \colon L(O; X) \to \mathbb{R}$ by

$$\Phi(\alpha\vec{x}) = \alpha \quad (\text{or } \Phi(\overrightarrow{OP}) = [P]_\mathcal{B}, P \in L).$$

Then Φ is one-to-one, onto and preserves algebraic operations, i.e. for any $\alpha, \beta \in \mathbb{R}$,

  1. $\Phi(\beta(\alpha\vec{x})) = \beta\alpha$,

  2. $\Phi(\alpha\vec{x} + \beta\vec{x}) = \alpha + \beta$.   (1.2.1)

According to Sec. B.7 of Appendix B, mappings like Φ here are called *linear isomorphisms* and therefore, conceptually, $L(O; X)$, $\mathbb{R}_{L(O;X)}$ and $\mathbb{R}$ are considered being identical (see Fig. 1.7).

We have already known that the following are equivalent.

(1) Only two different points, needless a third one, are enough to determine a unique line.

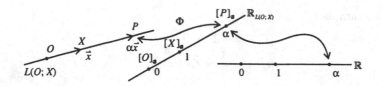

Fig. 1.7

(2) Only one nonzero vector is enough to generate the whole space $L(O; X)$ (refer to (1.1.4) and (1.1.5)).

Hence, we say that $L(O; X)$ is a *one-dimensional vector space* with zero vector $\vec{0}$. Accurately speaking, $L(O; X)$ is a *one-dimensional affine space* (see Sec. 2.8 or Fig. B.2) of the line L.

Owing to arbitrariness of O and X, the line L can be endowed with infinitely many vectorized spaces $L(O; X)$. But according to (1.2.1), no matter how O and X are chosen, we always have that

$$L(O; X) \underset{\Phi}{\simeq} \mathbb{R}_{L(O; X)} = \mathbb{R}. \qquad (1.2.2)$$

So, we assign $\mathbb{R}$ another role, representing the *standard one-dimensional real vector space* and denoted by

$$\mathbb{R}^1. \qquad (1.2.3)$$

A number α in $\mathbb{R}$ is identical with the position vector in $\mathbb{R}^1$, starting from 0 and pointing toward α, and is still denoted by α (see Fig. 1.8).

For the sake of reference and comparison, we summarize as

The real number system $\mathbb{R}$ and the standard one-dimensional vector space $\mathbb{R}^1$

(1) $\mathbb{R}$ (simply called the real field, refer to Sec. A.3)

    (a) *Addition* For any $x, y \in \mathbb{R}$,

$$x + y \in \mathbb{R}$$

satisfies the following properties.

1. (commutative) $x + y = y + x$.
2. (associative) $(x + y) + z = x + (y + z)$.
3. (zero element) 0: $0 + x = x$.
4. (inverse element) For each $x \in \mathbb{R}$, there exists a unique element in $\mathbb{R}$, denoted as $-x$, such that

$$x + (-x) = x - x = 0.$$

(b) *Multiplication* For any x, $y \in \mathbb{R}$,

$$xy \in \mathbb{R}$$

satisfies the following properties.

1. (commutative) $xy = yx$.
2. (associative) $(xy)z = x(yz)$.
3. (unit element) 1: $1x = x$.
4. (inverse element) For each nonzero $x \in \mathbb{R}$, there exists a unique element in $\mathbb{R}$, denoted by x^{-1} or $\frac{1}{x}$, such that

$$xx^{-1} = 1.$$

(c) The addition and multiplication satisfy the distributive law

$$x(y + z) = xy + xz.$$

(2) $\mathbb{R}^1$ (see (1.2.3) and refer to Sec. B.1)

(a) *Addition* To every pair of elements x and y in $\mathbb{R}^1$, there is a unique element

$$x + y \in \mathbb{R}^1$$

satisfying the following properties.

1. $x + y = y + x$.
2. $(x + y) + z = x + (y + z)$.
3. (zero vector 0) $x + 0 = x$.
4. (inverse vector) For each $x \in \mathbb{R}^1$, there exists a unique element in $\mathbb{R}^1$, denoted as $-x$, such that

$$x + (-x) = x - x = 0.$$

(b) *Scalar multiplication* To each $x \in \mathbb{R}^1$ and every real number $\alpha \in \mathbb{R}$, there exists a unique element

$$\alpha x \in \mathbb{R}^1$$

satisfying the following properties.

1. $1x = x$.
2. $\alpha(\beta x) = (\alpha\beta)x, \alpha, \beta \in \mathbb{R}$.

(c) The addition and scalar multiplication satisfy the distributive laws

$$(\alpha + \beta)x = \alpha x + \beta x;$$
$$\alpha(x + y) = \alpha x + \alpha y. \tag{1.2.4}$$

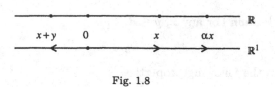

Fig. 1.8

See Fig. 1.8.

Remark On many occasions, no distinction between $\mathbb{R}$ and $\mathbb{R}^1$, in notation, will be specified and we simply denote $\mathbb{R}^1$ by $\mathbb{R}$.

In this sense, element in $\mathbb{R}$ has double meanings. One is to represent a number treated as a point on the real line. The other is to represent a position vector, pointing from 0 toward itself on the real line. One should remember that any element in $\mathbb{R}$, either as a number or as a vector, somewhat enjoys different algebraic operation properties as indicated in (1.2.4).

From now on, if necessarily or conveniently, we do not hesitate to use $\mathbb{R}$ to replace $\mathbb{R}^1$, both in notation and in meaning. Any straight line L, endowed with a vectorized space $L(O; X)$, is nothing but a concrete geometric model of $\mathbb{R}$ (i.e. $\mathbb{R}^1$).

1.3   Changes of Coordinates: Affine and Linear Transformations (or Mappings)

Let L be a straight line with two vectorized spaces $L(O; X)$ and $L(O'; Y)$ on it.

The same point P on L has different coordinates $[P]_\mathcal{B}$ and $[P]_{\mathcal{B}'}$, respectively, with respect to the different bases

$$\mathcal{B} = \{\vec{x}\}, \quad \vec{x} = \overrightarrow{OX}, \quad \text{and}$$
$$\mathcal{B}' = \{\vec{y}\}, \quad \vec{y} = \overrightarrow{O'Y}$$

(see Fig. 1.9).

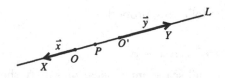

Fig. 1.9

Our purpose here is to find out the relation between $[P]_\mathcal{B}$ and $[P]_{\mathcal{B}'}$.

Suppose that, temporarily,

$$[P]_B = \mu \ (\Leftrightarrow \overrightarrow{OP} = \mu\vec{x}), \quad [O']_B = \alpha_0;$$
$$[P]_{B'} = \nu \ (\Leftrightarrow \overrightarrow{O'P} = \nu\vec{y}), \quad [O]_{B'} = \beta_0.$$

Since $\vec{x}$ and $\vec{y}$ are collinear, there exist constants α and β such that $\vec{y} = \alpha\vec{x}$ and $\vec{x} = \beta\vec{y}$. Hence $\vec{x} = \alpha\beta\vec{x}$ implies $(\alpha\beta - 1)\vec{x} = \vec{0}$. The linear independence of $\vec{x}$ shows that $\alpha\beta - 1 = 0$ should hold, i.e.

$$\alpha\beta = 1.$$

Now, owing to the fact that $\overrightarrow{OP} = \overrightarrow{OO'} + \overrightarrow{O'P}$,

$$\mu\vec{x} = \alpha_0\vec{x} + \nu\vec{y} = \alpha_0\vec{x} + \nu\alpha\vec{x} = (\alpha_0 + \nu\alpha)\vec{x}$$
$$\Rightarrow \mu = \alpha_0 + \nu\alpha, \quad \text{or}$$
$$[P]_B = [O']_B + \alpha[P]_{B'}. \tag{1.3.1}$$

Similarly, by using $\overrightarrow{O'P} = \overrightarrow{O'O} + \overrightarrow{OP}$, one has

$$\nu = \beta_0 + \beta\mu, \quad \text{or}$$
$$[P]_{B'} = [O]_{B'} + \beta[P]_B. \tag{1.3.2}$$

Remark (1.3.1) and (1.3.2) are reversible.

Suppose that (1.3.1) is true. Then one has (why $\alpha \neq 0$?)

$$\nu = -\frac{\alpha_0}{\alpha} + \frac{1}{\alpha}\mu.$$

But $\overrightarrow{OO'} = -\overrightarrow{O'O} \Rightarrow \alpha_0\vec{x} = -\beta_0\vec{y} = -\beta_0\alpha\vec{x} \Rightarrow \alpha_0 = -\beta_0\alpha$ or $\beta_0 = -\frac{\alpha_0}{\alpha}$. Since $\alpha\beta = 1$, therefore

$$\nu = \beta_0 + \beta\mu.$$

This is (1.3.2).
   Similarly, (1.3.1) is deduced from (1.3.2) with the same process.

Summarize the above results as

Coordinate changes of two vectorized spaces on the same line
Let

$$L(O; X) \text{ with basis } \mathcal{B} = \{\vec{x}\}, \vec{x} = \overrightarrow{OX}, \text{ and}$$
$$L(O'; Y) \text{ with basis } \mathcal{B}' = \{\vec{y}\}, \vec{y} = \overrightarrow{O'Y}$$

be two vectorized spaces of the line L. Suppose that

$$\vec{y} = \alpha\vec{x}, \quad \vec{x} = \beta\vec{y} \qquad (\text{therefore, } \alpha\beta = 1).$$

Then the coordinates $[P]_\mathcal{B}$ and $[P]_{\mathcal{B}'}$ with respect to bases $\mathcal{B}$ and $\mathcal{B}'$, of the same point P on L, satisfy the following reversible linear relations:

$$[P]_\mathcal{B} = [O']_\mathcal{B} + \alpha[P]_{\mathcal{B}'}, \quad \text{and}$$
$$[P]_{\mathcal{B}'} = [O]_{\mathcal{B}'} + \beta[P]_\mathcal{B}.$$

In particular, if $O = O'$, then $[O']_\mathcal{B} = [O]_{\mathcal{B}'} = 0.$ \hfill (1.3.3)

In specific terminology, Eqs. such as (1.3.1) and (1.3.2) are called *affine transformations* or *mappings* between affine spaces $L(O; X)$ and $L(O'; Y)$; in case $O = O'$, called (*invertible*) *linear transformation* (see Sec. B.7).

Finally, here is an example.

Example Determine the relation between the Centigrade (°C) and the Fahrenheit (°F) on the thermometer.

Solution (see Fig. 1.10) Suppose O and X are 0°C and 1°C respectively. Also, O' and Y are 0°F and 1°F. Then, we have

$$\overrightarrow{O'O} = 32\overrightarrow{O'Y} \quad \text{and} \quad \overrightarrow{OX} = \frac{9}{5}\overrightarrow{O'Y}.$$

Let $C = \{\overrightarrow{OX}\}$ and $F = \{\overrightarrow{O'Y}\}$. Now, for any given point P on the thermometer, we have

$$[P]_C = \text{the Centigrade degree of } P, \text{ and}$$
$$[P]_F = \text{the Fahrenheit degree of } P.$$

By using the fact that $\overrightarrow{O'P} = \overrightarrow{O'O} + \overrightarrow{OP}$, one has the following relation

$$[P]_F = [O]_F + \frac{9}{5}[P]_C = 32 + \frac{9}{5}[P]_F, \quad \text{or}$$
$$[P]_C = \frac{5}{9}\{[P]_F - 32\}$$

between $[P]_F$ and $[P]_C$. \hfill $\square$

Fig. 1.10

Exercises

<A> (Adopt the notations in (1.3.3).)

1. Suppose that $L(O;X)$ and $L(O';Y)$ are two vectorized spaces on the same line L. Let

$$[O]_{B'} = -5 \quad \text{and} \quad \vec{y} = \frac{1}{3}\vec{x}.$$

   (a) Locate the points O, X, O' and Y on the line L.
   (b) If a point $P \in L$ and $[P]_B = 0.2$, find $[P]_{B'}$? If $[P]_{B'} = 15$, what is $[P]_B$?

2. Construct two vectorized spaces $L(O;X)$ and $L(O';Y)$ on the same line L, and explain graphically the following equations as changes of coordinates with

$$[P]_B = x \quad \text{and} \quad [P]_{B'} = y, \quad P \in L.$$

   (a) $y = -2x$.
   (b) $y = \sqrt{3}x - \frac{5}{3}$.
   (c) $x = 6y$.
   (d) $x = -15y + 32$.

1.4 Affine Invariants

Construct a vectorized space $L(O;X)$ on the line L and a vectorized space $L'(O';X')$ on the line L', where L and L' may not be coincident. Let $\mathcal{B} = \{\overrightarrow{OX}\}$ and $\mathcal{B}' = \{\overrightarrow{O'X'}\}$ be the respective basis on L and L'. Also,

let

$$[P]_\mathcal{B} = x \quad \text{for } P \in L, \qquad \text{and}$$
$$[P']_{\mathcal{B}'} = y \quad \text{for } P' \in L'.$$

A mapping or transformation T from L onto L' (see Sec. A.2) is called an *affine mapping* or *transformation* if there exist constants a and $b \neq 0$ such that

$$T(x) = y = a + bx \qquad (1.4.1)$$

holds for all $P \in L$ and the corresponding $P' \in L'$. Note that $y = T(x)$ is one-to-one. In case $a = 0$, $y = T(x) = bx$ is called a *linear transformation* (isomorphism) from the vector space $L(O; X)$ onto the vector space $L'(O'; X')$. In this sense, change of coordinates on the same line as introduced in (1.3.3) is a special kind of affine mapping.

For any two fixed different points P_1 and P_2 with $[P_1]_\mathcal{B} = x_1$ and $[P_2]_\mathcal{B} = x_2$, the whole line L has *coordinate representation*

$$x = (1 - t)x_1 + tx_2, \quad t \in \mathbb{R} \qquad (1.4.2)$$

with respect to the basis $\mathcal{B}$. The *directed segment* $\overrightarrow{P_1P_2}$ or $\overrightarrow{x_1x_2}$ with P_1 as *initial point* and P_2 as *terminal point* is the set of points

$$x = (1 - t)x_1 + tx_2, \quad 0 \leq t \leq 1. \qquad (1.4.3)$$

In case $0 < t < 1$, the corresponding point x is called an *interior point* of $\overrightarrow{x_1x_2}$, otherwise (i.e. $t < 0$ or $t > 1$) an *exterior point*. See Fig. 1.11.

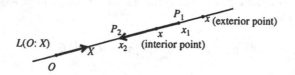

Fig. 1.11

Applying (1.4.1) and (1.4.3), we see that an affine mapping maps a (directed) segment $\overrightarrow{x_1x_2}$ onto a (directed) segment $\overrightarrow{y_1y_2}$, preserving end points, interior points and exterior points. In fact, a point $x = (1-t)x_1 + tx_2$ is mapped into the point

$$y = (1 - t)y_1 + ty_2 \qquad (1.4.4)$$

with $y_1 = a + bx_1$ and $y_2 = a + bx_2$.

Orient the line L by the basis vector $\overrightarrow{OX}$ in $L(O; X)$ and let $x_2 - x_1$ be the signed length of the segment $\overrightarrow{x_1 x_2}$ as we did in Sec. 1.1. For convenience, we also use $\overrightarrow{x_1 x_2}$ to denote its signed length. Then, by (1.4.2), we see that

$$(1 - t)(x - x_1) = t(x_2 - x)$$

$$\Rightarrow \frac{\overrightarrow{x_1 x}}{\overrightarrow{x x_2}} = \frac{t}{1 - t}, \quad t \neq 0, 1 \tag{1.4.5}$$

which is equal to $\frac{\overrightarrow{y_1 y}}{\overrightarrow{y y_2}}$, by using (1.4.4). This means that an affine mapping preserves the ratio of two line segments along the line (see also (1.4.6) and Ex. <A> 2).

Finally, $y_2 - y_1 = a + bx_2 - (a + bx_1) = b(x_2 - x_1)$ means that

$$\overrightarrow{y_1 y_2} = b\overrightarrow{x_1 x_2}. \tag{1.4.6}$$

Then, an affine mapping does not preserve the signed length except $b = 1$, and does not preserve the orientation of directed segment except $b > 0$.

We summarize as

Affine invariants
An affine transformation between straight lines preserves

1. (directed) line segment along with end points, interior points and exterior points, and
2. ratio of (signed) lengths of two line segments (along the same line) which are called *affine invariants*. It does not necessarily preserve
3. (signed) length, and
4. orientation. $\tag{1.4.7}$

Remark Affine coordinate system and affine (or barycentric) coordinate.

Let a straight line L be vectorized as $L(O; X)$, and P_1 and P_2 be two arbitrarily fixed different points. As usual, denote the basis $\mathcal{B} = \{\overrightarrow{OX}\}$ and $[P_1]_{\mathcal{B}} = x_1$, $[P_2]_{\mathcal{B}} = x_2$.

(1.4.2) can be rewritten as

$$x = \lambda_1 x_1 + \lambda_2 x_2, \quad \lambda_1, \lambda_2 \in \mathbb{R} \quad \text{and} \quad \lambda_1 + \lambda_2 = 1 \tag{1.4.8}$$

for any point P on L with $[P]_{\mathcal{B}} = x$. Then, we call the order pair

$$(\lambda_1, \lambda_2) \tag{1.4.9}$$

the *affine* or *barycentric coordinate* of the point P or x with respect to the *affine coordinate system* $\{P_1, P_2\}$ or $\{\overrightarrow{OP_1}, \overrightarrow{OP_2}\}$. In particular, $(\frac{1}{2}, \frac{1}{2})$ is the *barycenter* or the *middle point* of the segment $P_1 P_2$ or $x_1 x_2$.

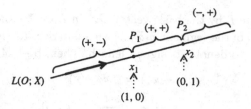

Fig. 1.12

x is an interior point of $P_1 P_2$ if and only if it has affine coordinate (λ_1, λ_2) with components $\lambda_1 > 0$ and $\lambda_2 > 0$. See Fig. 1.12. As a trivial consequence, the points P_1 and P_2 divide the whole line L into three different parts: $(+, -)$, $(+, +)$, and $(-, +)$.

Exercises

<A>

1. For each pair P_1, P_2 of different points on $L(O; X)$ and each pair P_1', P_2', of different points on $L'(O'; X')$, show that there exists a unique affine mapping T from $L(O; X)$ onto $L'(O'; X')$ such that

$$T(P_1) = P_1' \quad \text{and} \quad T(P_2) = P_2'.$$

2. A one-to-one and onto mapping $T: L(O; X) \rightarrow L'(O'; X')$ is affine if and only if T preserves ratio of signed lengths of any two segments.

CHAPTER 2

The Two-Dimensional Real Vector Space $\mathbb{R}^2$

Introduction

In our physical world, one can realize that there does exist a point, lying outside of a given straight line.

In Fig. 2.1, the point Q does not lie on the line L. This point Q and the moving point X on the line L generate infinitely many straight lines QX. We image that the collection of all such lines QX constitute a plane.

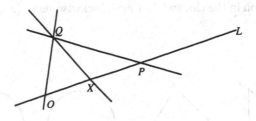

Fig. 2.1

Therefore, we formulate the

Postulate *Three noncollinear different points determine a unique plane.*

The face of a table or a piece of horizontally placed paper, imagined extended beyond limit in all directions, all can be considered as a geometric model of a plane. Usually, a parallelogram, including its interior, will act as a symbolic graph of a plane Σ (see Fig. 2.2).

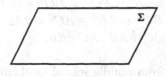

Fig. 2.2

21

About plane, one should know the following basic facts.

(1) A plane Σ contains uncountably many points.

(2) The straight line, determined by any pair of different points in a plane Σ, lies in Σ itself. Therefore, starting from a point and then a line on which the point does not lie, one can construct a plane.

(3) The plane, generated by any three points in a plane Σ, coincides with Σ.

(4) (Euclidean parallelism axiom) Let the point P and the line L be in the same plane Σ, with P lying outside of L. Then, there exists one and only one line on Σ, passing through the point P and parallel to the given line L (i.e. no point of intersection exists).

(5) A plane possesses the following Euclidean geometric concepts or quantities.

    1. Length (of a segment) or distance (between two points).
    2. Angle (between two intersection lines).
    3. Area (of a triangle).
    4. Rotation in the clockwise or anticlockwise sense (around a center).

See Fig. 2.3.

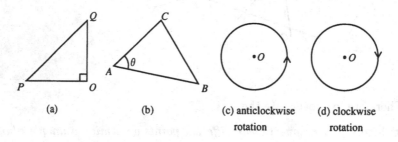

(a)   (b)   (c) anticlockwise   (d) clockwise
                                                      rotation   rotation

Fig. 2.3

(6) And the related basic formulas are

    1. (Pythagorean theorem) $PQ^2 = PO^2 + OQ^2$,
    2. (law of cosine) $BC^2 = AB^2 + AC^2 - 2AB \cdot AC \cos\theta$, and
    3. (area of a triangle) $\frac{1}{2}AB \cdot AC \sin\theta$.

All these can be learned from middle school mathematical courses.

Sketch of the Content

In this chapter, the development of our theory will be based on the Postulate mentioned on P.21. After introducing the concept of a plane vector (Sec. 2.1), we model after Sec. 1.1 to vectorize a plane (Sec. 2.2) and Sec. 1.2 to coordinatize it (Sec. 2.3). Therefore, we establish the abstract set

$$\mathbb{R}^2 = \{(x_1, x_2) \mid x_1, x_2 \in \mathbb{R}\}$$

with scalar multiplication and addition of vectors on it, which can be considered as a typical model of planes and is called the *standard two-dimensional real vector space*.

Changes of coordinates on the same plane (Sec. 2.4) result in the algebraic equations

$$y_1 = a_1 + a_{11}x_1 + a_{21}x_2, \quad y_2 = a_2 + a_{12}x_1 + a_{22}x_2$$

with $a_{11}a_{22} - a_{21}a_{12} \neq 0$. This is the *affine transformation* or *linear transformation* in case $a_1 = a_2 = 0$ to be formally discussed in Secs. 2.7 and 2.8.

Section 2.5 introduces straight lines in the affine plane $\mathbb{R}^2$ and its various equation representations in different coordinate systems.

Section 2.6 formally introduces the terminology, *affine basis* $\mathcal{B} = \{\vec{a}_0, \vec{a}_1, \vec{a}_2\}$, to replace the coordinate system $\Sigma(\vec{a}_0; \vec{a}_1, \vec{a}_2)$, and this will be used thereafter constantly. Hence, we also discuss *barycentric* or *affine coordinates* of a point with respect to an affine basis. Ex. <B> of Sec. 2.6 introduces primary ideas about projective plane.

The contents in Secs. 2.5 and 2.6 provide us the necessary backgrounds in geometric interpretation, which will be discussed in Secs. 2.7 and 2.8.

Just like Sec. 1.4, a one-to-one mapping T from $\mathbb{R}^2$ onto $\mathbb{R}^2$ that preserves signed ratios of line segments along the same line is characterized algebraically as

$$T(x) = \vec{x}_0 + \vec{x}A, \quad \text{where } A = [a_{ij}]_{2 \times 2} \text{ is invertible}, \ \vec{x} \in \mathbb{R}^2.$$

This is called *affine transformation*, and is called *linear* isomorphism, in the special case $\vec{x}_0 = \vec{0}$.

The main theme of this chapter lies in Sec. 2.7, concerning *linear transformations* and their various operations and properties, such as: geometric mapping properties, matrix representations, factorizations, *diagonal*, *Jordan* and *rational canonical forms*. Eigenvalues and eigenvectors as well as ranks of linear transformations are dominating, both conceptually and computationally.

Section 2.8 focuses on *affine transformations* and their geometric mapping properties, such as reflections, stretches, shearings and rotations. We single out *affine invariants* on the affine plane $\mathbb{R}^2$, talk about *affine geometry* illustrated by standard basic examples and apply both to the study of *quadratic curves*.

The primary connections among sections are listed as follows.

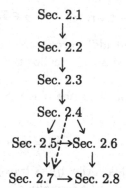

Sec. 2.1
↓
Sec. 2.2
↓
Sec. 2.3
↓
Sec. 2.4
↙ ↓ ↘
Sec. 2.5 ↦ Sec. 2.6
↓↓   ↓
Sec. 2.7 → Sec. 2.8

In the above diagram, one may go directly from Sec. 2.4 to Sec. 2.7 without seriously handicapping the understanding of the content in Sec. 2.7.

The manners presented, the techniques used and the results obtained in this chapter will be repeated and reconstructed for $\mathbb{R}^3$ in Chap. 3.

2.1   (Plane) Vector

Displace a point P on a fixed plane along a fixed direction to another point Q. The resulted directed segment $\overrightarrow{PQ}$ is called a *plane vector* (see Fig. 2.4). Therefore, the vector $\overrightarrow{PQ}$ should have two required properties, i.e.

1. length PQ, and
2. direction.   (2.1.1)

In case $Q = P$, call $\overrightarrow{PQ}$ a *zero vector*.

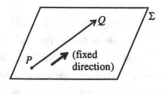

Fig. 2.4

In physics, a vector $\overrightarrow{PQ}$ can be interpreted as a force acting on the point P and moving in the fixed direction to the point Q (see Fig. 2.5). Under this circumstance, the vector $\overrightarrow{PQ}$ possesses both

1. the magnitude of acting force, and
2. the direction of movement, $\qquad$ (2.1.2)

as two main concepts.

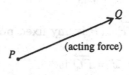

Fig. 2.5

Both displacement and acting force are concrete, vivid interpretation of vectors, which is going to be abstractly defined below.

In what follows, everything will happen on the same fixed plane.

Let the segment PQ be moved, while keeping parallel to each other, to the segment RS so that $PQSR$ forms a parallelogram (see Fig. 2.6). Then we designate that the vector $\overrightarrow{PQ}$ is *identical* with $\overrightarrow{RS}$. In other words,

$$\overrightarrow{PQ} = \overrightarrow{RS}$$

$\Leftrightarrow$ 1. $PQ = RS$ (in length), and
2. the direction from P to Q is the same as the direction from R to S.
$\qquad$ (2.1.3)

This is called the *parallel invariance* of vectors. As indicated in Fig. 2.6, $\overrightarrow{PR} = \overrightarrow{QS}$.

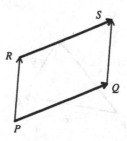

Fig. 2.6

Usually, we use $\vec{x}$ to represent a *vector*, that is to say, for any fixed point P, one can always find another point Q such that

$$\vec{x} = \overrightarrow{PQ}.$$

According to parallel invariance, there are infinitely many choices of $\overrightarrow{PQ}$ to represent the same $\vec{x}$, once these $\overrightarrow{PQ}$ have the same length and the same direction. Hence, $\vec{x}$ has a definite meaning, even though more abstract than $\overrightarrow{PQ}$. In particular, the *zero vector* is

$$\vec{0} = \overrightarrow{PP} \quad (P \text{ is any fixed point}), \tag{2.1.4}$$

and the *negative* of a vector $\vec{x} = \overrightarrow{PQ}$ is

$$-\vec{x} = (-1)\vec{x} = \overrightarrow{QP}. \tag{2.1.5}$$

Therefore, $\vec{x}$ and $-\vec{x}$ represent, respectively, two vectors equal in length but opposite in direction.

Now, two operations will be introduced among vectors.

Addition
Suppose $\vec{x} = \overrightarrow{PQ}$ and $\vec{y} = \overrightarrow{QS}$. Starting from P, along $\vec{x}$, to Q and then along $\vec{y}$, to S is the same as starting from P directly to S. Then, it is reasonable to define

$$\vec{x} + \vec{y} = \overrightarrow{PS} \tag{2.1.6}$$

(see Fig. 2.7). Physically, $\vec{x} + \vec{y}$ can be interpreted as the composite force of $\vec{x}$ and $\vec{y}$. The vector $\vec{x} + \vec{y}$ is called the *sum* of $\vec{x}$ and $\vec{y}$, and the binary operation of combining $\vec{x}$ and $\vec{y}$ into a single vector $\vec{x} + \vec{y}$ is called the *addition* of vectors.

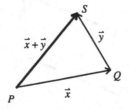

Fig. 2.7

Scalar multiplication

Suppose $\alpha \in \mathbb{R}$ and $\vec{x} = \overrightarrow{PQ}$. Keep P fixed. Stretch the segment PQ, α times, to a collinear segment PR. As for $\alpha > 0$ or $\alpha < 0$, the point R lies on the same side of P as Q or on the opposite of P, respectively (see Fig. 2.8). In all cases, we define the *scalar product* of the scalar α and the vector $\vec{x}$ by

$$\alpha\vec{x} = \overrightarrow{PR}, \tag{2.1.7}$$

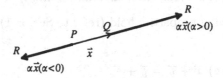

Fig. 2.8

and the whole process is called the *scalar multiplication* of vectors. In particular,

$$0\vec{x} = \overrightarrow{PP} = \vec{0}, \quad \text{and}$$
$$(-\alpha)\vec{x} = -(\alpha\vec{x}) = -\alpha\vec{x}. \tag{2.1.8}$$

Of course,

$$\alpha\vec{0} = \overrightarrow{PP} = \vec{0} \tag{2.1.9}$$

holds for all $\alpha \in \mathbb{R}$.

By the way, the *subtraction vector* of the vector $\vec{y}$ from the vector $\vec{x}$ is defined as

$$\vec{x} - \vec{y} = \vec{x} + (-\vec{y}). \tag{2.1.10}$$

See Fig. 2.9.

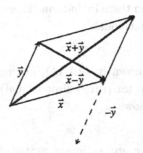

Fig. 2.9

We summarize what we have done, up to now, as

Algebraic operation properties of (plane) vectors

For each pair of vectors $\vec{x}$ and $\vec{y}$, there is a unique (plane) vector

$$\vec{x} + \vec{y} \quad \text{(called } \textit{addition}\text{),}$$

and for each scalar $\alpha \in \mathbb{R}$ and each vector $\vec{x}$, there is a unique (plane) vector

$$\alpha\vec{x} \quad \text{(called } \textit{scalar multiplication}\text{)}$$

such that the following properties hold (refer to Sec. B.1).

(a) Addition
   1. (commutative) $\vec{x} + \vec{y} = \vec{y} + \vec{x}$.
   2. (associative) $(\vec{x} + \vec{y}) + \vec{z} = \vec{x} + (\vec{y} + \vec{z})$.
   3. (zero vector) There is a vector, specifically denoted by $\vec{0}$, such that $\vec{x} + \vec{0} = \vec{x}$.
   4. (negative or inverse of a vector) For each (plane) vector $\vec{x}$, there exists another (plane) vector, denoted by $-\vec{x}$, such that $\vec{x} + (-\vec{x}) = \vec{0}$.

(b) Scalar multiplication
   1. $1\vec{x} = \vec{x}$.
   2. $\alpha(\beta\vec{x}) = (\alpha\beta)\vec{x}$.

(c) The addition and scalar multiplication satisfy the distributive laws

$$(\alpha + \beta)\vec{x} = \alpha\vec{x} + \beta\vec{x}, \quad \text{and}$$
$$\alpha(\vec{x} + \vec{y}) = \alpha\vec{x} + \alpha\vec{y}. \tag{2.1.11}$$

When comparing them with (1.2.4)(2), we can easily find that line vectors $\vec{x}$ and plane vectors $\vec{x}$ enjoy exactly the same operations and properties. The only difference between them is that, one is one-dimensional, by nature, while the other is two-dimensional.

Remark 2.1

Operation properties as shown in (2.1.11), not only hold for line vectors and plane vectors, but also for (three-dimensional) space vectors. Now, we will roughly explain as follows.

Line vectors

Suppose $\vec{x} \neq \vec{0}$. Then for any collinear vector $\vec{y}$, there exists a unique scalar $\alpha \in \mathbb{R}$ such that $\vec{y} = \alpha\vec{x}$. Therefore, in essence, the addition of $\vec{x}$

and $\vec{y}, \vec{x} + \vec{y}$ reduces to the scalar multiplication $\vec{x} + \vec{y} = \vec{x} + \alpha\vec{x} = (1 + \alpha)\vec{x}$.

Space vectors (see Chap. 3)

Pick up three arbitrary noncollinear points P, Q and R in space. Let $\vec{x} = \overrightarrow{PQ}$ and $\vec{y} = \overrightarrow{PR}$. These three points determine a unique plane Σ in space, which contains $\vec{x}$ and $\vec{y}$ as plane vectors (see Fig. 2.10). It follows that the space vectors $\vec{x}$, $\alpha\vec{x}$ (lying on a straight line in space) and $\vec{x} + \vec{y}$ (lying on the same plane extended by $\vec{x}$ and $\vec{y}$ in space, if $\vec{x}$ and $\vec{y}$ are linearly independent) will enjoy the operation properties listed in (2.1.11).

This suggests implicitly that only the addition and scalar multiplication of vectors are just good enough to describe (i.e. vectorize) the three-dimensional or even higher dimensional spaces. We will see it later in Chap. 3.

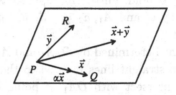

Fig. 2.10

Remark 2.2

It is appropriate, now, to say what is a vector space over a field.

For general definition of a field $\mathbb{F}$, please refer to Sec. A.3; and definition of a vector space V over a field $\mathbb{F}$, refer to Sec. B.1.

According to (1) in (1.2.4), $\mathbb{R}$ is a field and is called the real field. Therefore, (1.1.3) or (2) in (1.2.4) says that $\mathbb{R}^1$ or $\mathbb{R}$ is a vector space over the field $\mathbb{R}$, simply called a *real vector space*.

Similarly, (2.1.11) indicates that $\mathbb{R}^2$ is also a *real vector space*, but of two-dimensional (see Secs. 2.2 and 2.3).

The three-dimensional real vector spaces $\mathbb{R}^3$ will be defined in Secs. 3.1 and 3.2.

A vector space over the complex field $\mathbb{C}$ is specifically called a *complex vector space*.

The elements of the field $\mathbb{F}$ are called *scalars* and the elements of the vector space V are called *vectors*. The word "vector", without any practical meanings such as displacement or acting force, is now being used to describe any element of a vector space.

Exercises
<A>

1. Explain graphically the properties listed in (2.1.11).
2. Explain physically the properties listed in (2.1.11).

2.2   Vectorization of a Plane: Affine Structure

We take three noncollinear points O, A_1 and A_2, and then keep them fixed.
    Let

$$\vec{a}_1 = \overrightarrow{OA_1} \quad \text{and} \quad \vec{a}_2 = \overrightarrow{OA_2}.$$

Then $\vec{a}_1 \neq \vec{0}$ and $\vec{a}_2 \neq \vec{0}$ hold. Furthermore, since A_2 lies outside of the straight line generated by O and A_1, $\vec{a}_2 \neq \alpha\vec{a}_1$ for any $\alpha \in \mathbb{R}$. Similarly, $\vec{a}_1 \neq \alpha\vec{a}_2$ for any $\alpha \in \mathbb{R}$.

    Denote by Σ the plane determined by O, A_1 and A_2. Through any fixed point P on Σ, draw two straight lines parallel to the lines OA_1 and OA_2 respectively. And then intersect with OA_1 at point P_1 and with OA_2 at point P_2 (see Fig. 2.11). There exist two unique scalars $x_1, x_2 \in \mathbb{R}$ such that

$$\overrightarrow{OP_1} = x_1\vec{a}_1, \quad \text{and}$$
$$\overrightarrow{OP_2} = x_2\vec{a}_2.$$

Owing to the fact that OP_1PP_2 is a parallelogram, this implies that

$$\overrightarrow{OP} = \overrightarrow{OP_1} + \overrightarrow{P_1P} = \overrightarrow{OP_1} + \overrightarrow{OP_2} = x_1\vec{a}_1 + x_2\vec{a}_2.$$

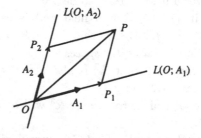

Fig. 2.11

Conversely, given any two scalars $x_1, x_2 \in \mathbb{R}$, we can find a unique point P_1 on the line OA_1 such that $\overrightarrow{OP_1} = x_1\vec{a}_1$, and a unique point P_2 on OA_2 such that $\overrightarrow{OP_2} = x_2\vec{a}_2$. Construct a parallelogram OP_1PP_2 with given sides OP_1 and OP_2. Then the position vector $\overrightarrow{OP}$ from O to its opposite vertex P satisfies $\overrightarrow{OP} = x_1\vec{a}_1 + x_2\vec{a}_2$.

Now fix x_1, and let x_2 run through all the real numbers in the equation $\overrightarrow{OP} = x_1\vec{a}_1 + x_2\vec{a}_2$. It geometrically results in the movement of the line $L(O; A_2)$ along the vector $x_1\vec{a}_1$ up to the parallel straight line

$$x_1\vec{a}_1 + L(O; A_2), \tag{2.2.1}$$

which passes through the point P_1 (see Fig. 2.12). Then, if x_1 starts to run through all the real numbers, the family of straight lines $x_1\vec{a}_1 + L(O; A_2)$, parallel to each other, will sweep through the whole plane Σ.

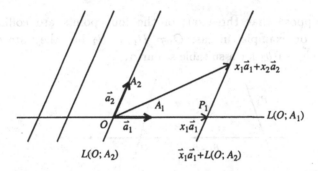

Fig. 2.12

Summarize the above results as

Algebraic vectorization of a plane

Let O, A_1 and A_2 be any fixed noncollinear points in a plane Σ. Let

$$\vec{a}_1 = \overrightarrow{OA_1} \quad \text{and} \quad \vec{a}_2 = \overrightarrow{OA_2}.$$

Then the *linear combination* $x_1\vec{a}_1 + x_2\vec{a}_2$, with *coefficients* x_1 and x_2, of the plane vectors $\vec{a}_1$ and $\vec{a}_2$ is suitable to describe any point P on Σ (i.e. the position vector $\overrightarrow{OP}$). Specifically, call the set

$$\Sigma(O; A_1, A_2) = \{x_1\vec{a}_1 + x_2\vec{a}_2 \mid x_1, x_2 \in \mathbb{R}\}$$

the *vectorized space* of the plane Σ with the point O as the *origin* or $\vec{0} = \overrightarrow{OO}$ as *zero vector* and $\vec{a}_1, \vec{a}_2$ as *basis vectors*. $\Sigma(O; A_1, A_2)$ indeed is a vector space over $\mathbb{R}$ (see (2.1.11) and Remark 2.2 there), with $\{\vec{a}_1, \vec{a}_2\}$ as a basis. (2.2.2)

Once we have these concepts, we are able to algebraically characterize the conditions for four points to be coplanar.

Suppose O, B_1, B_2 and B_3 are coplanar. Let $\vec{b}_i = \overrightarrow{OB_i}, i = 1, 2, 3$. The following three cases are considered.

Case 1 Suppose these four points are collinear (see Fig. 2.13). No matter what possibilities these four points might be situated, at least one of the vectors $\vec{b}_1, \vec{b}_2$ and $\vec{b}_3$ can be expressed as a linear combination of the other two. For example, in case $O = B_1 = B_2 \neq B_3$, we have $\vec{b}_1 = \vec{b}_2 = \vec{0}$ and $\vec{b}_3 \neq \vec{0}$, and hence $\vec{b}_1 = \alpha \vec{b}_2 + 0 \vec{b}_3$ for any scalar $\alpha \in \mathbb{R}$.

$$O = B_1 = B_2 = B_3 \qquad O = B_1 = B_2 \quad B_3 \qquad O = B_1 \quad B_2 = B_3 \qquad O \; B_1 \; B_2 \quad B_3$$

Fig. 2.13

Case 2 Suppose that three out of the four points are collinear (see Fig. 2.14). For example, in case $O \neq B_1 \neq B_2$ but they are collinear, then $\vec{b}_2 = \alpha \vec{b}_1 + 0 \vec{b}_3$ for a suitable scalar α.

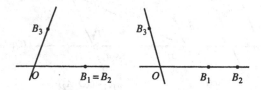

Fig. 2.14

Case 3 Suppose that any three of the four points are not collinear. As we already knew from Fig. 2.12, $\vec{b}_3 = \alpha \vec{b}_1 + \beta \vec{b}_2$ is true for some scalars α and β.

Conclusively, we have the

Linear dependence of plane vectors
The following statements hold and are equivalent.

(1) (geometric) Four points O, B_1, B_2 and B_3 are coplanar.
$\Leftrightarrow$ (2) (algebraic) Fix one of the points O, B_1, B_2 and B_3 as origin and hence produce three vectors, say O as origin and $\vec{b}_i = \overrightarrow{OB_i}$, $i = 1, 2, 3$. Then, at least one of the vectors $\vec{b}_1, \vec{b}_2$ and $\vec{b}_3$ is a linear combination of the other two vectors.
$\Leftrightarrow$ (3) (algebraic) There exist real numbers y_1, y_2 and y_3, not all equal to zero, such that $y_1 \vec{b}_1 + y_2 \vec{b}_2 + y_3 \vec{b}_3 = \vec{0}$.

Under these circumstances, the four points are called *affinely dependent* and the resulting plane vectors *linearly dependent.*   (2.2.3)

Therefore, any three coplanar vectors are linearly dependent.

From what we did at the beginning of this section, it follows immediately the

Linear independence of nonzero plane vectors

(1) (geometric) Three points O, B_1 and B_2 are not collinear.
$\Leftrightarrow$ (2) (algebraic) Fix one of the points O, B_1 and B_2 as origin and hence produce two vectors, say O as origin and $\vec{b}_i = \overrightarrow{OB_i}, i = 1, 2$. Then, for any $\alpha, \beta \in \mathbb{R}$, both $\vec{b}_1 \neq \alpha \vec{b}_2$ and $\vec{b}_2 \neq \beta \vec{b}_1$ are true.
$\Leftrightarrow$ (3) (algebraic) If there exist real numbers y_1 and y_2 such that $y_1 \vec{b}_1 + y_2 \vec{b}_2 = \vec{0}$ holds, then it is necessary that y_1 and y_2 should be equal to zero simultaneously, i.e. $y_1 = y_2 = 0$ is the only possibility.

Under these circumstances, three points are *affinely independent* and the resulting two plane vectors *linearly independent* (i.e. non-collinear).   (2.2.4)

As a consequence, two nonzero plane vectors can be either linearly dependent (i.e. collinear) or linearly independent (i.e. non-collinear).

Remark

The geometric fact that three points O, A_1 and A_2 are not collinear, is algebraically equivalent to the linear independence of plane vectors $\vec{a}_1 = \overrightarrow{OA_1}$ and $\vec{a}_2 = \overrightarrow{OA_2}$. Any common vector in $\Sigma(O; A_1, A_2)$ is produced, via linear combination $x_1 \vec{a}_1 + x_2 \vec{a}_2$, of a unique vector $x_1 \vec{a}_1$ in $L(O; A_1)$ and a unique vector $x_2 \vec{a}_2$ in $L(O; A_2)$. We combine these two facts together and write as

$$\Sigma(O; A_1, A_2) = L(O; A_1) \oplus L(O; A_2),   \text{(2.2.5)}$$

indicating that two intersecting (but not coincident) straight lines determine a unique plane.

Exercises

\<A\>

1. Prove (2.2.3) in detail.
2. Use notation in (2.2.2). For any three vectors $\vec{b}_1, \vec{b}_2, \vec{b}_3 \in \Sigma(O; A_1, A_2)$, prove that they are linearly dependent.
3. Prove (2.2.4) in detail.

2.3   Coordinatization of a Plane: $\mathbb{R}^2$

Provide the plane Σ a fixed vectorized space $\Sigma(O; A_1, A_2)$ with *basis*

$$\mathcal{B} = \{\vec{a}_1, \vec{a}_2\},$$

where $\vec{a}_1 = \overrightarrow{OA_1}$ and $\vec{a}_2 = \overrightarrow{OA_2}$ are called *basis vectors*. It should be mentioned that the order of appearance of $\vec{a}_1$ and $\vec{a}_2$ in $\mathcal{B}$ cannot be altered arbitrarily. Sometimes, it is just called an *ordered basis* to emphasize the order of the basis vectors. Hence, $\mathcal{B}$ is different from the ordered basis $\{\vec{a}_2, \vec{a}_1\}$.

Take any point $P \in \Sigma$. Then

$$\overrightarrow{OP} \in \Sigma(O; A_1, A_2)$$
$$\Leftrightarrow \overrightarrow{OP} = x_1 \vec{a}_1 + x_2 \vec{a}_2 \text{ for some } x_1, x_2 \in \mathbb{R}.$$

The respective coefficients x_1 and x_2 of $\vec{a}_1$ and $\vec{a}_2$ form a pair (x_1, x_2), denoted by

$$[\overrightarrow{OP}]_\mathcal{B} = [P]_\mathcal{B} = (x_1, x_2) \tag{2.3.1}$$

and called the *coordinate vector* of the point P or the vector $\overrightarrow{OP}$ *with respect to* the basis $\mathcal{B}$, where x_1 as the *first coordinate component* and x_2 the *second* one. In particular,

$$[O]_\mathcal{B} = (0,0),$$
$$[A_1]_\mathcal{B} = (1,0), \quad \text{and}$$
$$[A_2]_\mathcal{B} = (0,1).$$

See Fig. 2.15.

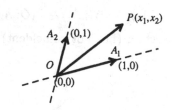

Fig. 2.15

They can be concluded as

The coordinatization of a plane

Choose a vectorized space $\Sigma(O; A_1, A_2)$ of a plane Σ, with basis $\mathcal{B} = \{\vec{a}_1, \vec{a}_2\}, \vec{a}_1 = \overrightarrow{OA_1}, \vec{a}_2 = \overrightarrow{OA_2}$. Point $P \in \Sigma$ has coordinate vector

$[P]_\mathcal{B} = (x_1, x_2) \Leftrightarrow \overrightarrow{OP} = x_1\vec{a}_1 + x_2\vec{a}_2$. The set of all these coordinate vectors

$$\mathbb{R}^2_{\Sigma(O;\,A_1,\,A_2)} = \{[P]_\mathcal{B} \mid P \in \Sigma\}$$

is called a *coordinatized space* of Σ, and is indeed a vector space over $\mathbb{R}$. Explain as follows.

(1) Points P in Σ are in one-to-one correspondence with the ordered pair $[P]_\mathcal{B} = (x_1, x_2)$ in $\mathbb{R}^2_{\Sigma(O;\,A_1,\,A_2)}$.
(2) Define the operations on the set $\mathbb{R}^2_{\Sigma(O;\,A_1,\,A_2)}$:

    1. addition $(x_1, x_2) + (y_1, y_2) = (x_1 + y_1, x_2 + y_2)$, and
    2. scalar multiplication $\alpha(x_1, x_2) = (\alpha x_1, \alpha x_2)$, where $\alpha \in \mathbb{R}$.

    They have all properties listed in (2.1.11), treating (x_1, x_2) as $\vec{x}$, etc. Hence, $\mathbb{R}^2_{\Sigma(O;\,A_1,\,A_2)}$ is a real vector space.
(3) Define a mapping $\Phi \colon \Sigma(O; A_1, A_2) \to \mathbb{R}^2_{\Sigma(O;\,A_1,\,A_2)}$ by

$$\Phi(x_1\vec{a}_1 + x_2\vec{a}_2) = (x_1, x_2) \quad \text{or} \quad \Phi(\overrightarrow{OP}) = [P]_\mathcal{B}.$$

    This Φ is one-to-one and onto, and it preserves vector operations, i.e.

    1. $\Phi(\overrightarrow{OP} + \overrightarrow{OQ}) = [P]_\mathcal{B} + [Q]_\mathcal{B} (= \Phi(\overrightarrow{OP}) + \Phi(\overrightarrow{OQ}))$, and
    2. $\Phi(\alpha\overrightarrow{OP}) = \alpha[P]_\mathcal{B} \ (= \alpha\Phi(\overrightarrow{OP})), \ \alpha \in \mathbb{R}$.

    Φ is called a *linear isomorphism* between $\Sigma(O; A_1, A_2)$ and $\mathbb{R}^2_{\Sigma(O;\,A_1,\,A_2)}$ (see Sec. B.7).

Hence, $\Sigma(O; A_1, A_2)$ and $\mathbb{R}^2_{\Sigma(O;\,A_1,\,A_2)}$ are considered identical.   (2.3.2)

See Fig. 2.16.

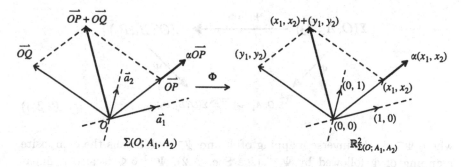

Fig. 2.16

Notice that the linear isomorphism Φ carries overall vector operations in $\Sigma(O; A_1, A_2)$ to the corresponding vector operations in $\mathbb{R}^2_{\Sigma(O; A_1, A_2)}$, a neat, easy-to-handle, but abstract space. In particular, Φ *preserves the linear dependence and independence of vectors.* To see these, suppose $\overrightarrow{b_1} = \alpha_{11}\overrightarrow{a_1} + \alpha_{12}\overrightarrow{a_2}$ and $\overrightarrow{b_2} = \alpha_{21}\overrightarrow{a_1} + \alpha_{22}\overrightarrow{a_2}$ are in $\Sigma(O; A_1, A_2)$, then

$$
\begin{aligned}
\Phi(y_1\overrightarrow{b_1} + y_2\overrightarrow{b_2}) &= \Phi((y_1\alpha_{11} + y_2\alpha_{21})\overrightarrow{a_1} + (y_1\alpha_{12} + y_2\alpha_{22})\overrightarrow{a_2}) \\
&= (y_1\alpha_{11} + y_2\alpha_{21}, y_1\alpha_{12} + y_2\alpha_{22}) \\
&= y_1(\alpha_{11}, \alpha_{12}) + y_2(\alpha_{21}, \alpha_{22}) \\
&= y_1\Phi(\overrightarrow{b_1}) + y_2\Phi(\overrightarrow{b_2})
\end{aligned}
$$

Since Φ is one-to-one and onto, what we claimed follows easily.

By (2.2.3) and (2.2.4), the following are equivalent.

1. (geometric) Three non-collinear points, needless the fourth one, are enough to determine a unique plane Σ.
2. (algebraic) Two linearly independent vectors, needless the third one, are able to generate, via linear combination, the whole space $\Sigma(O; A_1, A_2)$.

Owing to what we explained in the previous paragraph, vectorized space $\Sigma(O; A_1, A_2)$ and its isomorphic space $\mathbb{R}^2_{\Sigma(O; A_1, A_2)}$ are called *two-dimensional.*

Suppose that $\Sigma(O; A_1, A_2)$ and $\Sigma(O'; B_1, B_2)$ are two vectorized spaces of a plane Σ, and that $\Psi: \Sigma(O'; B_1, B_2) \to \mathbb{R}^2_{\Sigma(O'; B_1, B_2)}$ is the linear isomorphism as $\Phi: \Sigma(O; A_1, A_2) \to \mathbb{R}^2_{\Sigma(O'; A_1, A_2)}$ is the linear isomorphism, too. Then, we have the following diagram

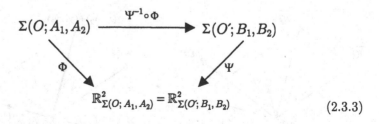

$$(2.3.3)$$

where Ψ^{-1} is the inverse mapping of Ψ and $\Psi^{-1} \circ \Phi$ means the composite mapping of Φ followed by Ψ^{-1} (see Sec. A.2). $\Psi^{-1} \circ \Phi$ is also a linear isomorphism. Observing this fact, we have the

Standard two-dimensional vector space $\mathbb{R}^2$ over $\mathbb{R}$

Let

$$\mathbb{R}^2 = \{(x_1, x_2) \,|\, x_1, x_2 \in \mathbb{R}\}$$

and designate $(x_1, x_2) = (y_1, y_2) \Leftrightarrow x_1 = y_1, x_2 = y_2$. Provide it with the following two operations,

(a) *addition* $(x_1, x_2) + (y_1, y_2) = (x_1 + y_1, x_2 + y_2)$,
(b) *scalar multiplication* $\alpha(x_1, x_2) = (\alpha x_1, \alpha x_2), \alpha \in \mathbb{R}$.

$\mathbb{R}^2$ is a two-dimensional real vector space, when treated (x_1, x_2) as vector $\vec{x}$ in (2.1.11). In particular,

$$\text{the zero vector } \vec{0} = (0, 0), \text{ and}$$

$$\text{the inverse vector } -\vec{x} = (-x_1, -x_2). \tag{2.3.4}$$

In this text, $\mathbb{R}^2$ is considered as an abstract model of a concrete plane Σ, in the sense that $\mathbb{R}^2$ is the universal representation of any vectorized space $\Sigma(O; A_1, A_2)$ or its coordinatized space $\mathbb{R}^2_{\Sigma(O; A_1, A_2)}$ of Σ.

Usually, element (x_1, x_2) in $\in \mathbb{R}^2$ is denoted by $\vec{x}$, i.e.

$$\vec{x} = (x_1, x_2), \quad \vec{y} = (y_1, y_2), \text{ etc.}$$

With this Convention, elements in $\mathbb{R}^2$ have double meanings:

1. (affine point of view) When $\mathbb{R}^2$ is considered as a plane, $\vec{x}$ is called a *point* and two points $\vec{x}, \vec{y}$ determine a vector $\vec{x} - \vec{y}$ or $\vec{y} - \vec{x}$ with $\vec{x} - \vec{x} = \vec{0}$.
2. (vector point of view) When considered as a vector space, $\vec{x}$ in $\mathbb{R}^2$ is called a *vector*, pointing from zero vector $\vec{0}$ toward the point $\vec{x}$ (for exact reason, see Definition 2.8.1 in Sec. 2.8).

If conveniently, we will feel free to use both of these two concepts. Refer to Fig. 2.17. Both traditionally and commonly used in existed textbooks of linear algebra, when the point (x_1, x_2) in $\mathbb{R}^2$ is considered as a position vector (as in 2), it is usually denoted by a column vector

$$\begin{pmatrix} x_1 \\ x_2 \end{pmatrix} \quad \text{or} \quad \begin{bmatrix} x_1 \\ x_2 \end{bmatrix}$$

and is treated as a 2×1 matrix. We would rather accept $\vec{x} = (x_1, x_2)$ *as a vector too than this traditional convention*, for simplicity and for later usage in connection with matrix computational works (see Sec. 2.4).

Remark 2.3
In the Euclidean sense, a segment has its length, and two lines from the angle between them.

For a coordinated space $\mathbb{R}^2_{\Sigma(O;A_1,A_2)}$ of a plane Σ, the basis vectors $\vec{e}_1 = (1,0)$ and $\vec{e}_2 = (0,1)$ together are usually said to form an *oblique* or *affine coordinate system* (see Fig. 2.17(a)). When the following additional requirements, that is,

1. the line OA_1 and OA_2 intersects orthogonally at O, and
2. the segments OA_1 and OA_2 have equal lengths,

are imposed, then $\vec{e}_1$ and $\vec{e}_2$ are said to form a *rectangular* or *Cartesian coordinate system* (see Fig. 2.17(b)).

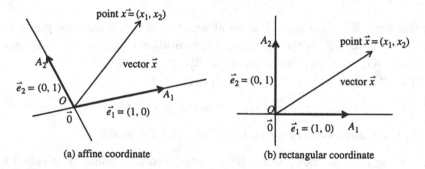

(a) affine coordinate   (b) rectangular coordinate

Fig. 2.17

From now on, if not specified, a plane endowed with a rectangular coordinate system will always be considered as a concrete geometric model of $\mathbb{R}^2$.

Remark 2.4 Reinterpretation of (2.2.5).
(2.2.5) is expected to be reinterpreted as

$$\mathbb{R}^2_{\Sigma(O;A_1,A_2)} = \mathbb{R}_{L(O;A_1)} \oplus \mathbb{R}_{L(O;A_2)} \quad \text{or} \tag{2.3.5}$$

$$\mathbb{R}^2 = \mathbb{R} \oplus \mathbb{R} \quad \text{(direct sum of } \mathbb{R} \text{ and } \mathbb{R}\text{)}. \tag{2.3.6}$$

Explain further as follows.

When considered being existed by itself, i.e. independent of being as a subset of the plane, we have already known from (1.2.2) that $\mathbb{R}_{L(O;A_1)} = \mathbb{R} = \mathbb{R}_{L(O;A_2)}$.

But, in reality, $L(O; A_1)$ is subordinated to the plane Σ, as part of it. For any given $P \in \Sigma$, then

$$P \text{ lies on the line } OA_1.$$
$$\Leftrightarrow \overrightarrow{OP} \in L(O; A_1), \text{ i.e. } \overrightarrow{OP} = x_1 \vec{a}_1 + 0\vec{a}_2.$$
$$\Leftrightarrow [P]_\mathcal{B} = (x_1, 0).$$

Therefore, we have

$$\mathbb{R}_{L(O; A_1)} = \{(x, 0) \mid x \in \mathbb{R}\} \subseteq \mathbb{R}^2_{\Sigma(O; A_1, A_2)}.$$

Similarly, the inclusive relation holds:

$$\mathbb{R}_{L(O; A_2)} = \{(0, y) \mid y \in \mathbb{R}\} \subseteq \mathbb{R}^2_{\Sigma(O; A_1, A_2)}.$$

It is easily seen that vectors in $\mathbb{R}_{L(O; A_1)}$ are closed under the addition and scalar multiplication which are inherited from those in $\mathbb{R}^2_{\Sigma(O; A_1, A_2)}$. This simply means that $\mathbb{R}_{L(O; A_1)}$ exists by itself as a vector space, and is called a one-dimensional *vector* (or *linear*) subspace of $\mathbb{R}^2_{\Sigma(O; A_1, A_2)}$. For the same reason, $\mathbb{R}_{L(O; A_2)}$ is also a one-dimensional vector subspace of $\mathbb{R}^2_{\Sigma(O; A_1, A_2)}$.

Observe the following two facts.

1. The vector subspaces $\mathbb{R}_{L(O; A_1)}$ and $\mathbb{R}_{L(O; A_2)}$ have only zero vector $\vec{0} = (0, 0)$ in common, i.e.

$$\mathbb{R}_{L(O; A_1)} \cap \mathbb{R}_{L(O; A_2)} = \{\vec{0}\}.$$

2. The sum vectors

$$(x_1, 0) + (0, x_2) = (x_1, x_2),$$

where $(x_1, 0) \in \mathbb{R}_{L(O; A_1)}$ and $(0, x_2) \in \mathbb{R}_{L(O; A_2)}$, generate the whole space $\mathbb{R}^2_{\Sigma(O; A_1, A_2)}$.   (2.3.7)

Just under these circumstances, we formulate (2.2.5) in the form of (2.3.5).

One can simplify (2.3.5) further.

For this purpose, define mappings

$$\Phi_1: \mathbb{R} \to \mathbb{R}_{L(O; A_1)} \quad \text{by } \Phi_1(x) = (x, 0), \quad \text{and}$$
$$\Phi_2: \mathbb{R} \to \mathbb{R}_{L(O; A_2)} \quad \text{by } \Phi_2(x) = (0, x). \qquad (2.3.8)$$

See Fig. 2.18.

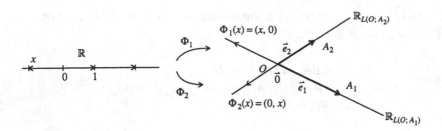

Fig. 2.18

It is obvious that Φ_1 and Φ_2 are linear isomorphisms (i.e. one-to-one, onto and preserving two vector operations) between vector spaces concerned. It is in this isomorphic sense that $\mathbb{R}_{L(O;\,A_1)}$ and $\mathbb{R}_{L(O;\,A_2)}$ are both considered identical with the standard one-dimensional real vector space $\mathbb{R}$, and we are able to rewrite (2.3.5) as (2.3.6). In other words, one just directly considers $\mathbb{R}$ as subspaces $\{(x,0)\,|\,x \in \mathbb{R}\}$ and $\{(0,x)\,|\,x \in \mathbb{R}\}$ and interpret $\mathbb{R}^2$ as (like (2.3.7))

$$\mathbb{R}^2 = \mathbb{R} \oplus \mathbb{R},$$

the (external) *direct sum* of its subspace $\mathbb{R}$.

Exercises

\<A\>

1. In the diagram (2.3.3), prove that $\Psi^{-1} \circ \Phi \colon \Sigma(O; A_1, A_2) \to \Sigma(O'; B_1, B_2)$ is a linear isomorphism. Then, modeling after Fig. 2.16, graphically explain the behavior of the mapping $\Psi^{-1} \circ \Phi$.
2. Prove that any three vectors in $\mathbb{R}^2$ must be *linearly dependent*, i.e. if $\vec{x}_1, \vec{x}_2, \vec{x}_3 \in \mathbb{R}^2$, then there exist scalars α_1, α_2 and α_3, not all equal to zero, such that $\alpha_1 \vec{x}_1 + \alpha_2 \vec{x}_2 + \alpha_3 \vec{x}_3 = \vec{0}$ (refer to (2.2.3)). Note that $\vec{x}_1, \vec{x}_2$ are *linearly independent* if $\alpha_1 \vec{x}_1 + \alpha_2 \vec{x}_2 = \vec{0}$ always implies $\alpha_1 = \alpha_2 = 0$.
3. A nonempty subset S of $\mathbb{R}^2$ is called a *vector* or *linear subspace* if the following properties holds:

   (1) $\vec{x}, \vec{y} \in S \Rightarrow \vec{x} + \vec{y} \in S$,
   (2) $\alpha \in \mathbb{R}$ and $\vec{x} \in S \Rightarrow \alpha\vec{x} \in S$ (in particular, $\vec{0} \in S$).

$\{\vec{0}\}$ and $\mathbb{R}^2$ itself are trivial subspaces of $\mathbb{R}^2$. A subspace which is not identical with $\mathbb{R}^2$ is called a *proper* subspace. Prove that the following are equivalent.

(a) S is a proper subspace which is not $\{\vec{0}\}$.

(b) There exists a vector $\vec{x}_0 \in S$ such that

$$S = \{\alpha\vec{x}_0 \,|\, \alpha \in \mathbb{R}\},$$

which is denoted by $\langle\langle\vec{x}_0\rangle\rangle$, called the subspace *generated* or *spanned* by $\vec{x}_0$. Note that $\vec{x}_0 \neq \vec{0}$ in this case.

(c) There exist constants a and b such that

$$S = \{\vec{x} = (x_1, x_2) \,|\, ax_1 + bx_2 = 0\}.$$

We simply call a straight line (refer to Sec. 2.5) passing $\vec{0} = (0,0)$ a one-dimensional subspace of $\mathbb{R}^2$ and denote S by $ax_1 + bx_2 = 0$.

Explain why a straight line $ax_1 + bx_2 = c$ with $c \neq 0$ is never a subspace.

4. A nonempty subset $\mathcal{B}$ of $\mathbb{R}^2$ is called a *basis* for $\mathbb{R}^2$ if

1. $\mathcal{B}$ is *linearly independent*, and
2. $\mathcal{B}$ *generates* or *spans* $\mathbb{R}^2$, namely, each vector in $\mathbb{R}^2$ can be expressed as a linear combination of vectors from $\mathcal{B}$ (see Sec. B.2).

Show that the number of basis vectors in any bases for $\mathbb{R}^2$ is equal to 2. This is the reason why we call $\mathbb{R}^2$ *two-dimensional*. Show that a proper nonzero subspace of $\mathbb{R}^2$ is *one-dimensional*. One calls $\{\vec{0}\}$ the *zero-dimensional* subspace.

5. Let S_1 and S_2 be two subspaces of $\mathbb{R}^2$. The *sum* of S_1 and S_2,

$$S_1 + S_2 = \{\vec{x}_1 + \vec{x}_2 \,|\, \vec{x}_1 \in S_1 \text{ and } \vec{x}_2 \in S_2\}$$

is still a subspace of $\mathbb{R}^2$. In case $S_1 \cap S_2 = \{\vec{0}\}$, write $S_1 + S_2$ as

$$S_1 \oplus S_2$$

and call the *direct sum* of S_1 and S_2. Suppose U is a subspace of $\mathbb{R}^2$, show that there exists a subspace V of $\mathbb{R}^2$ such that

$$\mathbb{R}^2 = U \oplus V.$$

Is V unique?

6. Let $\vec{x}_1, \ldots, \vec{x}_n$ be position vectors in $\mathbb{R}^2$ such that the terminal point of $\vec{x}_{j-1}$ is the initial point of $\vec{x}_j$ for $2 \leq j \leq n$ and the terminal point of

$\vec{x}_n$ is the initial point of $\vec{x}_1$. Then

$$\sum_{j=1}^{n} \vec{x}_j = \vec{x}_1 + \vec{x}_2 + \cdots + \vec{x}_n = \vec{0}.$$

<B>

1. Suppose $\vec{x}_1, \vec{x}_2, \vec{x}_3$ are in $\mathbb{R}^2$. For any other vector $\vec{x} \in \mathbb{R}^2$, show that there exist scalars $\alpha_1, \alpha_2, \alpha_3$, not all zero, such that

$$\vec{x}_1 + \alpha_1 \vec{x}, \quad \vec{x}_2 + \alpha_2 \vec{x}, \quad \vec{x}_3 + \alpha_3 \vec{x}$$

are linearly dependent.

2. Suppose $\vec{x}_1$ and $\vec{x}_2$ are linearly independent vectors in $\mathbb{R}^2$. For any $\vec{x} \in \mathbb{R}^2$, show that, among the vectors $\vec{x}, \vec{x}_1, \vec{x}_2$, at most one of these can be represented as a linear combination of the preceding ones.

3. Suppose $\vec{x}_1$ and $\vec{x}_2$ are linearly independent vectors in $\mathbb{R}^2$. If a vector $\vec{x} \in \mathbb{R}^2$ can be represented as linear combinations of $\vec{x}_1$ alone and $\vec{x}_2$ alone, then $\vec{x} = \vec{0}$.

4. Suppose any two vectors of the linearly dependent vectors $\vec{x}_1, \vec{x}_2$ and $\vec{x}_3$ are linearly independent. If the scalars a_1, a_2, a_3 satisfy $a_1 \vec{x}_1 + a_2 \vec{x}_2 + a_3 \vec{x}_3 = \vec{0}$, show that either $a_1 a_2 a_3 \neq 0$ or $a_1 = a_2 = a_3 = 0$. In the former case, if $b_1 \vec{x}_1 + b_2 \vec{x}_2 + b_3 \vec{x}_3 = \vec{0}$ also holds, then $a_1 : b_1 = a_2 : b_2 = a_3 : b_3$.

5. Suppose $\vec{x}_1, \vec{x}_2 \in \mathbb{R}^2$ satisfy the following:

   (1) $\{\vec{x}_1, \vec{x}_2\}$ *generates* $\mathbb{R}^2$, i.e. each $\vec{x} \in \mathbb{R}^2$ is a linear combination $a_1 \vec{x}_1 + a_2 \vec{x}_2$ of $\vec{x}_1$ and $\vec{x}_2$.
   (2) There exists a vector $\vec{x}_o \in \mathbb{R}^2$ which has a *unique* linear combination representation $\alpha \vec{x}_1 + \beta \vec{x}_2$.

   Then $\{\vec{x}_1, \vec{x}_2\}$ is a basis for $\mathbb{R}^2$. Is condition (2) necessary?

6. Let $\{\vec{x}_1, \vec{x}_2\}$ and $\{\vec{y}_1, \vec{y}_2\}$ be two bases for $\mathbb{R}^2$. Then at least one of $\{\vec{x}_1, \vec{y}_1\}$ and $\{\vec{x}_1, \vec{y}_2\}$ is a basis for $\mathbb{R}^2$. How about $\{\vec{x}_2, \vec{y}_1\}$ and $\{\vec{x}_2, \vec{y}_2\}$?

<C> Abstraction and generalization

Almost all the concepts we have introduced for $\mathbb{R}$ and $\mathbb{R}^2$ so far, can be generalized verbatim to abstract vector spaces over an arbitrary field. Such as:

Vector (or linear) space V over a field $\mathbb{F}$: $\mathbb{F}^n$.
Real or complex vector space: $\mathbb{R}^n, \mathbb{C}^n$.

Subspace (generated or spanned by a set S of vectors): $\langle\langle S \rangle\rangle$.

Zero subspace: $\{\vec{0}\}$.

Intersection subspace: $S_1 \cap S_2$.

Sum subspace: $S_1 + S_2$.

Direct sum of subspaces: $S_1 \oplus S_2$.

Linear combination (of vectors): $\sum_{i=1}^{n} \alpha_i \vec{x}_i$, $\vec{x}_i \in V$ and $\alpha_i \in \mathbb{F}$.

Linear dependence (of a set of vectors).

Linear independence (of a set of vectors).

Basis and basis vector: $\mathcal{B} = \{\vec{x}_1, \ldots, \vec{x}_n\}$.

Coordinate vector of a vector $\vec{x}$ with respect to a basis $\mathcal{B} = \{\vec{x}_1, \ldots, \vec{x}_n\}$:

$$[\vec{x}]_\mathcal{B} = (\alpha_1, \ldots, \alpha_n) \text{ if } \vec{x} = \sum_{i=1}^{n} \alpha_i \vec{x}_i.$$

Dimension (of a vector space V): $\dim V = n$.

Linear isomorphism.

Make sure you are able to achieve this degree of abstraction and generalization. If not, please refer to Secs. B.1, B.2, B.3 and B.7 for certainty. Then, try to extend these results stated in \<A\> and \<B\> as far as possible to a general vector space V over a field $\mathbb{F}$. For example, Ex. \<B\> 6 can be restated as

6'. Let $\{\vec{x}_1, \ldots, \vec{x}_n\}$ and $\{\vec{y}_1, \ldots, \vec{y}_n\}$ be two bases for an n-dimensional vector space V. Then there exists a permutation $j_1, j_2, \ldots, j_n$ of $1, 2, \ldots, n$ so that all

$$\{\vec{x}_1, \ldots, \vec{x}_{i-1}, \vec{y}_{j_i}, \vec{x}_{i+1}, \ldots, \vec{x}_n\}, \quad 1 \le i \le n$$

are bases for V.

Of course, better try to justify generalized results true by rigorous proofs or false by providing counterexamples in a project program or a seminar. During the proceeding, it is suggested to notice:

(1) Does the intuition experienced in the construction of the abstract spaces $\mathbb{R}$ and $\mathbb{R}^2$ help in solving the problems? In what way?

(2) Is more intuitive or algebraic experience, such as in $\mathbb{R}^3$ (see Chap. 3), needed?

(3) Does one have other sources of intuition concerning geometric concepts than those from $\mathbb{R}, \mathbb{R}^1$ and $\mathbb{R}^2$?

(4) Are the algebraic methods developed and used to solve problems in $\mathbb{R}$ and $\mathbb{R}^2$ still good? To what extend should it be generalized? Does the nature of a scalar field play an essential role in some problems?

(5) Need the algebraic methods be more unified and simplified? Need new methods such as matrix operations be introduced as early as possible and widely used?

(6) Are more mathematical backgrounds, sophistication or maturity needed?

Furthermore, try the following problems.

1. Model after (2) in (1.2.4) to explain that the set $\mathbb{C}$ of complex numbers is a one-dimensional vector space over the complex field $\mathbb{C}$.

   (a) Show that the complex number 1 itself constitutes a basis for $\mathbb{C}$. Try to find all other bases for $\mathbb{C}$.

   (b) Is there any intuitive interpretation for $\mathbb{C}$ as we did for $\mathbb{R}$ in Secs. 1.1 and 1.2?

   (c) Consider the set $\mathbb{C}$ of complex numbers as a vector space over the real field. Show that $\{1, i\}$, where $i = \sqrt{-1}$, is a basis for $\mathbb{C}$ and hence $\mathbb{C}$ is a two-dimensional real vector space. Find all other bases for this $\mathbb{C}$. Is there any possible relation between $\mathbb{R}^2$ and $\mathbb{C}$?

2. Consider the set $\mathbb{R}$ of real numbers as a vector space over the rational field $\mathbb{Q}$.

   (a) Make sure what scalar multiplication means in this case!

   (b) Two real numbers 1 and α are linearly independent if and only if α is an irrational number.

   (c) Is it possible for this $\mathbb{R}$ to be finite-dimensional?

3. Let the set

$$\mathbb{R}^+ = \{x \in \mathbb{R} \mid x > 0\}$$

be endowed with two operations:

(1) Addition $\oplus$: $x \oplus y = xy$ (multiplication of x and y), and

(2) Scalar multiplication $\odot$ of $x \in \mathbb{R}^+$ by $\alpha \in \mathbb{R}$: $\alpha \odot x = x^\alpha$.

   (a) Show that $\mathbb{R}^+$ is a real vector space with 1 as the zero vector and x^{-1} the inverse of x.

   (b) Show that each vector in $\mathbb{R}^+$ is linearly independent by itself and every two different vectors in $\mathbb{R}^+$ are linearly dependent.

   (c) Show that $\mathbb{R}^+$ is linear isomorphic to the real vector space $\mathbb{R}$.

4. Show that $B = \{\vec{x}_1, \vec{x}_2, \ldots, \vec{x}_n\}$, where

$$\vec{x}_j = \underbrace{(1, 1, \ldots, 1}_{j}, 0, \ldots, 0), \quad 1 \le j \le n$$

is a basis for $\mathbb{F}^n$ and find the coordinate $[\vec{x}]_B$ of $\vec{x} = (x_1, x_2, \ldots, x_n) \in \mathbb{F}^n$ with respect to B.

2.4 Changes of Coordinates: Affine and Linear Transformations (or Mappings)

Let $\Sigma(O; A_1, A_2)$ and $\Sigma(O'; B_1, B_2)$ be two geometric models of $\mathbb{R}^2$, with $\vec{a}_i = \overrightarrow{OA_i}$ and $\vec{b}_i = \overrightarrow{O'B_i}, i = 1, 2$ and the bases

$$B = \{\vec{a}_1, \vec{a}_2\},$$
$$B' = \{\vec{b}_1, \vec{b}_2\}.$$

Our main purpose here is to establish the relationship between the coordinate vectors $[P]_B$ and $[P]_{B'}$ of the same point $P \in \Sigma$ with respect to the bases B and B', respectively (see Fig. 2.19).

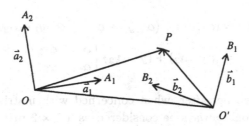

Fig. 2.19

The following notation will be adopted through this section:

$$[O']_B = [\overrightarrow{OO'}]_B = (\alpha_1, \alpha_2),$$
$$[O]_{B'} = [\overrightarrow{O'O}]_{B'} = (\beta_1, \beta_2),$$
$$[P]_B = [\overrightarrow{OP}]_B = (x_1, x_2),$$
$$[P]_{B'} = [\overrightarrow{O'P}]_{B'} = (y_1, y_2).$$

In view of parallel invariance of vectors, one may consider $\vec{b}_1$ and $\vec{b}_2$ as vectors in $\Sigma(O; A_1, A_2)$, and hence

$$[\vec{b}_i]_B = (\alpha_{i1}, \alpha_{i2}), \quad \text{i.e. } \vec{b}_i = \alpha_{i1}\vec{a}_1 + \alpha_{i2}\vec{a}_2, \ i = 1, 2.$$

Similarly, when considered $\vec{a}_1, \vec{a}_2 \in \Sigma(O'; B_1, B_2)$.

$$[\vec{a}_i]_{B'} = (\beta_{i1}, \beta_{i2}), \quad \text{i.e. } \vec{a}_i = \beta_{i1}\vec{b}_1 + \beta_{i2}\vec{b}_2, \ i = 1, 2.$$

As indicated in Fig. 2.19,

$$\overrightarrow{OP} = \overrightarrow{OO'} + \overrightarrow{O'P}$$

$$\Rightarrow x_1\vec{a}_1 + x_2\vec{a}_2 = (\alpha_1\vec{a}_1 + \alpha_2\vec{a}_2) + (y_1\vec{b}_1 + y_2\vec{b}_2)$$

$$= \alpha_1\vec{a}_1 + \alpha_2\vec{a}_2 + y_1(\alpha_{11}\vec{a}_1 + \alpha_{12}\vec{a}_2) + y_2(\alpha_{21}\vec{a}_1 + \alpha_{22}\vec{a}_2)$$

$$= (\alpha_1 + \alpha_{11}y_1 + \alpha_{21}y_2)\vec{a}_1 + (\alpha_2 + \alpha_{12}y_1 + \alpha_{22}y_2)\vec{a}_2.$$

$\Rightarrow$ (since $\vec{a}_1$ and $\vec{a}_2$ are linearly independent)

$$x_1 = \alpha_1 + \alpha_{11}y_1 + \alpha_{21}y_2,$$
$$x_2 = \alpha_2 + \alpha_{12}y_1 + \alpha_{22}y_2. \tag{2.4.1}$$

Suppose the reader is familiar with basic facts about matrix (if not, please refer to Sec. B.4). The above mentioned equations can be written into a single one like

$$(x_1, x_2) = (\alpha_1, \alpha_2) + (\alpha_{11}y_1 + \alpha_{21}y_2, \alpha_{12}y_1 + \alpha_{22}y_2)$$

$$= (\alpha_1 \quad \alpha_2) + (y_1 \quad y_2)\begin{bmatrix} \alpha_{11} & \alpha_{12} \\ \alpha_{21} & \alpha_{22} \end{bmatrix}. \tag{2.4.1'}$$

Note Hence and hereafter, when concerned with matrix computation, vector (x_1, x_2) would always be considered as a 1×2 matrix $(x_1 \ x_2)$ or $[x_1 \ x_2]$, a row matrix. So, occasionally, the notations (x_1, x_2) and $(x_1 \ x_2)$ might rise to some confusion.

Summarize the above result as part of the

Coordinate changes of two vectorized spaces of $\mathbb{R}^2$
Let $\Sigma(O; A_1, A_2)$ and $\Sigma(O'; B_1, B_2)$ be two vectorized spaces of $\mathbb{R}^2$ (also called *coordinate system* on $\mathbb{R}^2$), with bases

$$\mathcal{B} = \{\vec{a}_1, \vec{a}_2\}, \quad \vec{a}_i = \overrightarrow{OA_i}, \ i = 1, 2, \quad \text{and}$$
$$\mathcal{B}' = \{\vec{b}_1, \vec{b}_2\}, \quad \vec{b}_i = \overrightarrow{O'B_i}, \ i = 1, 2,$$

respectively. Then, the coordinate vectors $[P]_\mathcal{B}$ and $[P]_{\mathcal{B}'}$, of the same point $P \in \mathbb{R}^2$, satisfy the following two *formulas of coordinate changes*:

$$[P]_\mathcal{B} = [O']_\mathcal{B} + [P]_{\mathcal{B}'} A_\mathcal{B}^{\mathcal{B}'},$$
$$[P]_{\mathcal{B}'} = [O]_{\mathcal{B}'} + [P]_\mathcal{B} A_{\mathcal{B}'}^{\mathcal{B}},$$

(usually called *affine transformation* and *linear transformation* if $O = O'$) where

$$A_\mathcal{B}^{\mathcal{B}'} = \begin{bmatrix} [\vec{b}_1]_\mathcal{B} \\ [\vec{b}_2]_\mathcal{B} \end{bmatrix} = \begin{bmatrix} \alpha_{11} & \alpha_{12} \\ \alpha_{21} & \alpha_{22} \end{bmatrix}$$

is called the *transformation* or *transition matrix* of the basis $\mathcal{B}'$ with respect to basis $\mathcal{B}$;

$$A_{\mathcal{B}'}^{\mathcal{B}} = \begin{bmatrix} [\vec{a}_1]_{\mathcal{B}'} \\ [\vec{a}_2]_{\mathcal{B}'} \end{bmatrix} = \begin{bmatrix} \beta_{11} & \beta_{12} \\ \beta_{21} & \beta_{22} \end{bmatrix}$$

the transformation of $\mathcal{B}$ with respect to $\mathcal{B}'$, satisfy:

1. The determinants (see Sec. B.6 or Sec. 4.3)

$$\det A_\mathcal{B}^{\mathcal{B}'} = \begin{vmatrix} \alpha_{11} & \alpha_{12} \\ \alpha_{21} & \alpha_{22} \end{vmatrix} = \alpha_{11}\alpha_{22} - \alpha_{12}\alpha_{21} \neq 0,$$

$$\det A_{\mathcal{B}'}^{\mathcal{B}} = \begin{vmatrix} \beta_{11} & \beta_{12} \\ \beta_{21} & \beta_{22} \end{vmatrix} = \beta_{11}\beta_{22} - \beta_{12}\beta_{21} \neq 0.$$

2. The matrices $A_\mathcal{B}^{\mathcal{B}'}$ and $A_{\mathcal{B}'}^{\mathcal{B}}$ are invertible to each other, i.e.

$$A_\mathcal{B}^{\mathcal{B}'} A_{\mathcal{B}'}^{\mathcal{B}} = A_{\mathcal{B}'}^{\mathcal{B}} A_\mathcal{B}^{\mathcal{B}'} = I_2 = \begin{bmatrix} 1 & 0 \\ 0 & 1 \end{bmatrix},$$

and hence

$$\det A_\mathcal{B}^{\mathcal{B}'} = \frac{1}{\det A_{\mathcal{B}'}^{\mathcal{B}}};$$

$$(A_\mathcal{B}^{\mathcal{B}'})^{-1} = \frac{1}{\det A_\mathcal{B}^{\mathcal{B}'}} \begin{bmatrix} \alpha_{22} & -\alpha_{12} \\ -\alpha_{21} & \alpha_{11} \end{bmatrix} = A_{\mathcal{B}'}^{\mathcal{B}}.$$

3. Therefore, the two formulas are reversible, i.e.

$$[P]_{\mathcal{B}'} = -[O']_\mathcal{B}(A_\mathcal{B}^{\mathcal{B}'})^{-1} + [P]_\mathcal{B}(A_\mathcal{B}^{\mathcal{B}'})^{-1};$$
$$[O]_{\mathcal{B}'} = -[O']_\mathcal{B}(A_\mathcal{B}^{\mathcal{B}'})^{-1} = -[O']_\mathcal{B} A_{\mathcal{B}'}^{\mathcal{B}}.$$

In particular, if $O = O'$, then $[O]_{\mathcal{B}'} = [O']_\mathcal{B} = (0,0)$.   (2.4.2)

Proof Only 1, 2 and 3 are needed to be proved.

About 1 Note that $\vec{b}_i = \alpha_{i1}\vec{a}_1 + \alpha_{i2}\vec{a}_2$, one has:

$\vec{b}_1$ and $\vec{b}_2$ are linearly independent.

$\Leftrightarrow$ (see (2.2.4)) The only solution of y_1, y_2 in $y_1\vec{b}_1 + y_2\vec{b}_2 = \vec{0}$ is $y_1 = y_2 = 0$.

$\Leftrightarrow$ The equation

$$y_1(\alpha_{11}\vec{a}_1 + \alpha_{12}\vec{a}_2) + y_2(\alpha_{21}\vec{a}_1 + \alpha_{22}\vec{a}_2)$$
$$= (\alpha_{11}y_1 + \alpha_{21}y_2)\vec{a}_1 + (\alpha_{12}y_1 + \alpha_{22}y_2)\vec{a}_2 = \vec{0}$$

has zero solution only, i.e.

$$\alpha_{11}y_1 + \alpha_{21}y_2 = 0$$
$$\alpha_{12}y_1 + \alpha_{22}y_2 = 0$$

has only zero solution $y_1 = y_2 = 0$.

$\Leftrightarrow$ The coefficient matrix has the determinant

$$\det A_B^{\mathcal{B}'} = \begin{vmatrix} \alpha_{11} & \alpha_{12} \\ \alpha_{21} & \alpha_{22} \end{vmatrix} = \alpha_{11}\alpha_{22} - \alpha_{12}\alpha_{21} \neq 0.$$

This finishes the proof of 1.

About 2 By $\vec{a}_i = \beta_{i1}\vec{b}_1 + \beta_{i2}\vec{b}_2$, $i = 1, 2$,

$$\vec{b}_1 = \alpha_{11}\vec{a}_1 + \alpha_{12}\vec{a}_2$$
$$= \alpha_{11}(\beta_{11}\vec{b}_1 + \beta_{12}\vec{b}_2) + \alpha_{12}(\beta_{21}\vec{b}_1 + \beta_{22}\vec{b}_2)$$
$$= (\alpha_{11}\beta_{11} + \alpha_{12}\beta_{21})\vec{b}_1 + (\alpha_{11}\beta_{12} + \alpha_{12}\beta_{22})\vec{b}_2$$

$\Rightarrow$ (owing to linear independence of $\vec{b}_1$ and $\vec{b}_2$)

$$\alpha_{11}\beta_{11} + \alpha_{12}\beta_{21} = 1$$
$$\alpha_{11}\beta_{12} + \alpha_{12}\beta_{22} = 0.$$

Similarly, by $\vec{b}_2 = \alpha_{21}\vec{a}_1 + \alpha_{22}\vec{a}_2$,

$$\alpha_{21}\beta_{11} + \alpha_{22}\beta_{21} = 0$$
$$\alpha_{21}\beta_{12} + \alpha_{22}\beta_{22} = 1.$$

Put the above two sets of equations in the form of matrix product, and they can be simplified, in notation, as

$$\begin{bmatrix} \alpha_{11} & \alpha_{12} \\ \alpha_{21} & \alpha_{22} \end{bmatrix} \begin{bmatrix} \beta_{11} & \beta_{12} \\ \beta_{21} & \beta_{22} \end{bmatrix} = \begin{bmatrix} \beta_{11} & \beta_{12} \\ \beta_{21} & \beta_{22} \end{bmatrix} \begin{bmatrix} \alpha_{11} & \alpha_{12} \\ \alpha_{21} & \alpha_{22} \end{bmatrix} = \begin{bmatrix} 1 & 0 \\ 0 & 1 \end{bmatrix}$$

$$\Rightarrow A_B^{B'} A_{B'}^B = A_{B'}^B A_B^{B'} = I_2.$$

This means that $A_B^{B'}$ and $A_{B'}^B$ are invertible to each other. Actual computation in solving β_{ij} in terms of α_{ij} shows that

$$\left(A_B^{B'}\right)^{-1} = A_{B'}^B = \frac{1}{\alpha_{11}\alpha_{22} - \alpha_{12}\alpha_{21}} \begin{bmatrix} \alpha_{22} & -\alpha_{12} \\ -\alpha_{21} & \alpha_{11} \end{bmatrix}.$$

This is 2.

About 3 Multiply both sides of the first formula from the right by $\left(A_B^{B'}\right)^{-1}$, one has

$$[P]_B \left(A_B^{B'}\right)^{-1} = [O']_B \left(A_B^{B'}\right)^{-1} + [P]_{B'}$$

$$\Rightarrow [P]_{B'} = -[O']_B \left(A_B^{B'}\right)^{-1} + [P]_B A_{B'}^B$$

All we need to prove in the remaining is that $[O]_{B'} = -[O']_B \left(A_B^{B'}\right)^{-1}$. For this purpose, note first that $\overrightarrow{OO'} = -\overrightarrow{O'O}$. Therefore, remembering the notations we adopted at the beginning of this subsection,

$$\overrightarrow{OO'} = \alpha_1 \vec{a}_1 + \alpha_2 \vec{a}_2$$

$$= \alpha_1(\beta_{11}\vec{b}_1 + \beta_{12}\vec{b}_2) + \alpha_2(\beta_{21}\vec{b}_1 + \beta_{22}\vec{b}_2)$$

$$= (\alpha_1\beta_{11} + \alpha_2\beta_{21})\vec{b}_1 + (\alpha_1\beta_{12} + \alpha_2\beta_{22})\vec{b}_2$$

$$= -\overrightarrow{O'O}$$

$$= -(\beta_1\vec{b}_1 + \beta_2\vec{b}_2).$$

$$\Rightarrow \beta_1 = -(\alpha_1\beta_{11} + \alpha_2\beta_{21})$$

$$\beta_2 = -(\alpha_1\beta_{12} + \alpha_2\beta_{22}).$$

$$\Rightarrow (\beta_1\beta_2) = -(\alpha_1\beta_{11} + \alpha_2\beta_{21} \ \ \alpha_1\beta_{12} + \alpha_2\beta_{22})$$

$$= -(\alpha_1\alpha_2) \begin{bmatrix} \beta_{11} & \beta_{12} \\ \beta_{21} & \beta_{22} \end{bmatrix}.$$

This finishes 3. □

Example In $\mathbb{R}^2$, fix the following points

$$O = (1,0), \quad A_1 = (1,2), \quad A_2 = (0,1), \quad \text{and}$$
$$O' = (-1,-1), \quad B_1 = (0,0), \quad B_2 = (2,3).$$

Construct the vectorized spaces $\Sigma(O; A_1, A_2)$ and $\Sigma(O'; B_1, B_2)$, and then use them to justify the content of (2.4.2).

Solution Suppose

$$\vec{a}_1 = \overrightarrow{OA_1} = (1,2) - (1,0) = (0,2),$$
$$\vec{a}_2 = \overrightarrow{OA_2} = (0,1) - (1,0) = (-1,1);$$
$$\vec{b}_1 = \overrightarrow{O'B_1} = (0,0) - (-1,-1) = (1,1),$$
$$\vec{b}_2 = \overrightarrow{O'B_2} = (2,3) - (-1,-1) = (3,4),$$

and let

$$\mathcal{B} = \{\vec{a}_1, \vec{a}_2\}, \quad \mathcal{B}' = \{\vec{b}_1, \vec{b}_2\}$$

be bases of $\Sigma(O; A_1, A_2)$ and $\Sigma(O'; B_1, B_2)$ respectively (see Fig. 2.20).

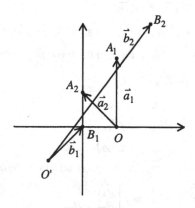

Fig. 2.20

To compute $A_B^{B'}$:

$$\vec{b}_1 = \alpha_{11}\vec{a}_1 + \alpha_{12}\vec{a}_2$$

$$\Rightarrow (1,1) = \alpha_{11}(0,2) + \alpha_{12}(-1,-1) = (-\alpha_{12}, 2\alpha_{11} + \alpha_{12})$$

$$\Rightarrow \begin{cases} \alpha_{12} = -1 \\ 2\alpha_{11} + \alpha_{12} = 1 \end{cases}$$

$$\Rightarrow \begin{cases} \alpha_{12} = -1 \\ \alpha_{11} = 1 \end{cases}$$

$$\Rightarrow [\vec{b}_1]_B = (1,-1).$$

Similarly,

$$\vec{b}_2 = \alpha_{21}\vec{a}_1 + \alpha_{22}\vec{a}_2$$

$$\Rightarrow (3,4) = \vec{b}_2 = \alpha_{21}(0,2) + \alpha_{22}(-1,1) = (-\alpha_{22}, 2\alpha_{21} + \alpha_{22})$$

$$\Rightarrow \begin{cases} \alpha_{22} = -3 \\ 2\alpha_{21} + \alpha_{22} = 4 \end{cases}$$

$$\Rightarrow \begin{cases} \alpha_{21} = \dfrac{7}{2} \\ \alpha_{22} = -3 \end{cases}$$

$$\Rightarrow [\vec{b}_2]_B = \left(\frac{7}{2}, -3\right).$$

Putting together, then

$$A_B^{B'} = \begin{bmatrix} [\vec{b}_1]_B \\ [\vec{b}_2]_B \end{bmatrix} = \begin{bmatrix} 1 & -1 \\ \frac{7}{2} & -3 \end{bmatrix}.$$

To compute $A_{B'}^{B}$:

$$\vec{a}_1 = \beta_{11}\vec{b}_1 + \beta_{12}\vec{b}_2$$

$$\Rightarrow (0,2) = \beta_{11}(1,1) + \beta_{12}(3,4) = (\beta_{11} + 3\beta_{12}, \beta_{11} + 4\beta_{12})$$

$$\Rightarrow \begin{cases} \beta_{11} + 3\beta_{12} = 0 \\ \beta_{11} + 4\beta_{12} = 2 \end{cases}$$

$$\Rightarrow \begin{cases} \beta_{11} = -6 \\ \beta_{12} = 2 \end{cases}$$

$$\Rightarrow [\vec{a}_1]_B = (-6, 2).$$

Also,

$$\vec{a}_2 = \beta_{21}\vec{b}_1 + \beta_{22}\vec{b}_2$$

$$\Rightarrow (-1, 1) = \beta_{21}(1, 1) + \beta_{22}(3, 4) = (\beta_{21} + 3\beta_{22}, \beta_{21} + 4\beta_{22})$$

$$\Rightarrow \begin{cases} \beta_{21} + 3\beta_{22} = -1 \\ \beta_{21} + 4\beta_{22} = 1 \end{cases}$$

$$\Rightarrow \begin{cases} \beta_{21} = -7 \\ \beta_{22} = 2 \end{cases}$$

$$\Rightarrow [\vec{a}_2]_{\mathcal{B}} = (-7, 2).$$

Hence,

$$A_{\mathcal{B}'}^{\mathcal{B}} = \begin{bmatrix} -6 & 2 \\ -7 & 2 \end{bmatrix}.$$

Through actual computation, one gets

$$A_{\mathcal{B}}^{\mathcal{B}'} A_{\mathcal{B}'}^{\mathcal{B}} = \begin{bmatrix} 1 & -1 \\ \frac{7}{2} & -3 \end{bmatrix} \begin{bmatrix} -6 & 2 \\ -7 & 2 \end{bmatrix} = \begin{bmatrix} 1 & 0 \\ 0 & 1 \end{bmatrix} = I_2, \quad \text{and}$$

$$A_{\mathcal{B}'}^{\mathcal{B}} A_{\mathcal{B}}^{\mathcal{B}'} = \begin{bmatrix} -6 & 2 \\ -7 & 2 \end{bmatrix} \begin{bmatrix} 1 & -1 \\ \frac{7}{2} & -3 \end{bmatrix} = \begin{bmatrix} 0 & 1 \\ 1 & 0 \end{bmatrix} = I_2,$$

which show that $A_{\mathcal{B}}^{\mathcal{B}'}$ and $A_{\mathcal{B}'}^{\mathcal{B}}$ are, in fact, invertible to each other. By the way,

$$\det A_{\mathcal{B}}^{\mathcal{B}'} = \begin{vmatrix} 1 & -1 \\ \frac{7}{2} & -3 \end{vmatrix} = -3 + \frac{7}{2} = \frac{1}{2},$$

$$\det A_{\mathcal{B}'}^{\mathcal{B}} = \begin{vmatrix} -6 & 2 \\ -7 & 2 \end{vmatrix} = -12 + 14 = 2 = \frac{1}{\det A_{\mathcal{B}}^{\mathcal{B}'}}$$

and

$$(A_{\mathcal{B}}^{\mathcal{B}'})^{-1} = \frac{1}{\det A_{\mathcal{B}}^{\mathcal{B}'}} \begin{bmatrix} -3 & 1 \\ -\frac{7}{2} & 1 \end{bmatrix} = 2 \begin{bmatrix} -3 & 1 \\ -\frac{7}{2} & 1 \end{bmatrix} = \begin{bmatrix} -6 & 2 \\ -7 & 2 \end{bmatrix} = A_{\mathcal{B}'}^{\mathcal{B}}$$

as expected from (2.4.2).

Finally,

$$\overrightarrow{OO'} = (-1,-1) - (1,0) = (-2,-1)$$
$$= \alpha_1 \vec{a}_1 + \alpha_2 \vec{a}_2 = \alpha_1(0,2) + \alpha_2(-1,1) = (-\alpha_1, 2\alpha_1 + \alpha_2)$$

$$\Rightarrow \begin{cases} \alpha_2 = 2 \\ 2\alpha_1 + \alpha_2 = -1 \end{cases}$$

$$\Rightarrow \begin{cases} \alpha_1 = -\dfrac{3}{2} \\ \alpha_2 = 2 \end{cases}$$

$$\Rightarrow [O']_\mathcal{B} = \left(-\frac{3}{2}, 2\right).$$

While,

$$\overrightarrow{O'O} = -\overrightarrow{OO'} = (2,1)$$
$$= \beta_1 \vec{b}_1 + \beta_2 \vec{b}_2 = \beta_1(1,1) + \beta_2(3,4) = (\beta_1 + 3\beta_2, \beta_1 + 4\beta_2)$$

$$\Rightarrow \begin{cases} \beta_1 + 3\beta_2 = 2 \\ \beta_1 + 4\beta_2 = 1 \end{cases}$$

$$\Rightarrow \begin{cases} \beta_1 = 5 \\ \beta_2 = -1 \end{cases}$$

$$\Rightarrow [O]_{\mathcal{B}'} = (5,-1).$$

Now, by actual computation, we do have

$$-[O']_\mathcal{B} A_{\mathcal{B}'}^\mathcal{B} = -\left(-\frac{3}{2} \quad 2\right) \begin{bmatrix} -6 & 2 \\ -7 & 2 \end{bmatrix} = -(9 - 14, -3 + 4)$$
$$= -(-5, 1) = (5, -1) = [O]_{\mathcal{B}'}.$$

The wanted formulas of changes of coordinates are

$$[P]_\mathcal{B} = \left(-\frac{3}{2} \quad 2\right) + [P]_{\mathcal{B}'} \begin{bmatrix} 1 & -1 \\ \frac{7}{2} & -3 \end{bmatrix} \quad \text{and}$$

$$[P]_{\mathcal{B}'} = (5 \quad -1) + [P]_\mathcal{B} \begin{bmatrix} -6 & 2 \\ -7 & 2 \end{bmatrix}.$$

For example, if $[P]_{\mathcal{B}'} = (5,2)$, i.e. $\overrightarrow{O'P} = 5\vec{b}_1 + 2\vec{b}_2$, then

$$[P]_\mathcal{B} = \left(-\frac{3}{2} \quad 2\right) + (5 \quad 2) \begin{bmatrix} 1 & -1 \\ \frac{7}{2} & -3 \end{bmatrix}$$
$$= \left(-\frac{3}{2}, 2\right) + (12, -11) = \left(\frac{21}{2}, -9\right)$$

which means $\overrightarrow{OP} = \frac{21}{2}\vec{a}_1 - 9\vec{a}_2$. $\square$

Remark The computations of $A_{\mathcal{B}}^{\mathcal{B}'}$ and $A_{\mathcal{B}'}^{\mathcal{B}}$.

Here, we suppose that the readers are familiar with basic matrix computation techniques, which may be obtained from Sec. B.4.

Suppose

$$\vec{a}_i = (a_{i1}, a_{i2}), \quad i = 1, 2, \quad \text{and} \quad \vec{b}_i = (b_{i1}, b_{i2}), \quad i = 1, 2$$

both are linearly independent. Let

$$A = \begin{bmatrix} \vec{a}_1 \\ \vec{a}_2 \end{bmatrix} = \begin{bmatrix} a_{11} & a_{12} \\ a_{21} & a_{22} \end{bmatrix} \quad \text{and} \quad B = \begin{bmatrix} \vec{b}_1 \\ \vec{b}_2 \end{bmatrix} = \begin{bmatrix} b_{11} & b_{12} \\ b_{21} & b_{22} \end{bmatrix}.$$

Then A and B are invertible matrices (see Ex. <A> 2).

Now,

$$\vec{b}_1 = \alpha_{11} \vec{a}_1 + \alpha_{12} \vec{a}_2$$
$$\vec{b}_2 = \alpha_{21} \vec{a}_1 + \alpha_{22} \vec{a}_2$$

$$\Rightarrow \begin{bmatrix} \vec{b}_1 \\ \vec{b}_2 \end{bmatrix} = \begin{bmatrix} \alpha_{11} \vec{a}_1 + \alpha_{12} \vec{a}_2 \\ \alpha_{21} \vec{a}_1 + \alpha_{22} \vec{a}_2 \end{bmatrix} = \begin{bmatrix} \alpha_{11} & \alpha_{12} \\ \alpha_{21} & \alpha_{22} \end{bmatrix} \begin{bmatrix} \vec{a}_1 \\ \vec{a}_2 \end{bmatrix}$$

$$\Rightarrow B = A_{\mathcal{B}}^{\mathcal{B}'} A, \text{ where } \mathcal{B} = \{\vec{a}_1, \vec{a}_2\} \text{ and } \mathcal{B}' = \{\vec{b}_1, \vec{b}_2\}$$

$$\Rightarrow A_{\mathcal{B}}^{\mathcal{B}'} = BA^{-1} = \begin{bmatrix} b_{11} & b_{12} \\ b_{21} & b_{22} \end{bmatrix} \begin{bmatrix} a_{11} & a_{12} \\ a_{21} & a_{22} \end{bmatrix}^{-1}. \tag{2.4.3}$$

Similarly,

$$A_{\mathcal{B}'}^{\mathcal{B}} = AB^{-1} = \begin{bmatrix} a_{11} & a_{12} \\ a_{21} & a_{22} \end{bmatrix} \begin{bmatrix} b_{11} & b_{12} \\ b_{21} & b_{22} \end{bmatrix}^{-1}. \tag{2.4.3'}$$

Exercises

<A>

1. Prove $[P]_{\mathcal{B}'} = [O]_{\mathcal{B}'} + [P]_{\mathcal{B}} A_{\mathcal{B}'}^{\mathcal{B}}$, without recourse to the established formula $[P]_{\mathcal{B}} = [O']_{\mathcal{B}} + [P]_{\mathcal{B}'} A_{\mathcal{B}}^{\mathcal{B}'}$.
2. For vectors in $\mathbb{R}^2$, prove that the following are equivalent.

   (a) $\vec{a}_1 = (a_{11}, a_{12})$ and $\vec{a}_2 = (a_{21}, a_{22})$ are linearly independent.
   (b) The determinant

$$\begin{vmatrix} \vec{a}_1 \\ \vec{a}_2 \end{vmatrix} = \begin{vmatrix} a_{11} & a_{12} \\ a_{21} & a_{22} \end{vmatrix} = a_{11}a_{22} - a_{12}a_{21} \neq 0.$$

(c) The square matrix

$$\begin{bmatrix} \vec{a_1} \\ \vec{a_2} \end{bmatrix} = \begin{bmatrix} a_{11} & a_{12} \\ a_{21} & a_{22} \end{bmatrix}$$

is invertible. In this case, prove that the inverse matrix is

$$\begin{bmatrix} a_{11} & a_{12} \\ a_{21} & a_{22} \end{bmatrix}^{-1} = \frac{1}{a_{11}a_{22} - a_{12}a_{21}} \begin{bmatrix} a_{22} & -a_{12} \\ -a_{21} & a_{11} \end{bmatrix}.$$

3. In $\mathbb{R}^2$, let $\vec{e_1} = (1,0)$ and $\vec{e_2} = (0,1)$. Then

$$\mathcal{N} = \{\vec{e_1}, \vec{e_2}\}$$

is a basis for the rectangular coordinate system $\Sigma(\vec{o}; \vec{e_1}, \vec{e_2})$. Owing to the fact that $\vec{x} = (x_1, x_2) = x_1\vec{e_1} + x_2\vec{e_2} = [\vec{x}]_\mathcal{N}$ for any $\vec{x} \in \mathbb{R}^2$, $\mathcal{N}$ is specifically called the *natural basis* of $\mathbb{R}^2$. Now, for any coordinate system $\Sigma(O; A_1, A_2)$ in $\mathbb{R}^2$ with basis $\mathcal{B} = \{\vec{a_1}, \vec{a_2}\}$, $\vec{a_i} = \overrightarrow{OA_i}$, $i = 1, 2$, prove that

$$[\vec{x}]_\mathcal{B} = \vec{x} \begin{bmatrix} \vec{a_1} \\ \vec{a_2} \end{bmatrix}^{-1},$$

where $\vec{x} \in \mathbb{R}^2$ is considered as a 1×2 matrix.

4. Take two sets of points in $\mathbb{R}^2$ as follows,

$$O = (-1,2), \quad A_1 = (4,-1), \quad A_2 = (-3,-4) \quad \text{and}$$
$$O' = (2,-3), \quad B_1 = (3,5), \quad B_2 = (-2,3).$$

Proceed as in the Example to justify (2.4.2) and use (2.4.3) to compute $A^\mathcal{B}_{\mathcal{B}'}$ and $A^{\mathcal{B}'}_\mathcal{B}$.

5. Construct two coordinate systems (i.e. vectorized spaces) $\Sigma(O; A_1, A_2)$ and $\Sigma(O'; B_1, B_2)$ in $\mathbb{R}^2$, and explain graphically the following formulas (refer to (2.4.1)) for changes of coordinates, where $[P]_\mathcal{B} = (x_1, x_2)$ and $[P]_{\mathcal{B}'} = (y_1, y_2)$.

(a) $x_1 = y_1 + y_2$, $x_2 = 2y_1 - 3y_2$.

(b) $x_1 = -\dfrac{1}{2} + 6y_1 - 3y_2$, $x_2 = 2 - 5y_1 + 4y_2$.

(c) $y_1 = \dfrac{3}{2}x_1 + \dfrac{1}{2}x_2$, $y_2 = \dfrac{1}{2}x_1 + \dfrac{3}{2}x_2$.

(d) $y_1 = 2 + 5x_1 + 6x_2$, $y_2 = 1 - 4x_1 + 7x_2$.

(e) $y_1 = 4 + \dfrac{1}{\sqrt{5}}x_1 - \dfrac{2}{\sqrt{5}}x_2$, $y_2 = -3 + \dfrac{2}{\sqrt{5}}x_1 + \dfrac{1}{\sqrt{5}}x_2$.

\<B\>

1. Consider the following system of equations

$$x_1 = \alpha_1 + \alpha_{11}y_1 + \alpha_{21}y_2$$
$$x_2 = \alpha_2 + \alpha_{12}y_1 + \alpha_{22}y_2,$$

where the *coefficient matrix*

$$\begin{bmatrix} \alpha_{11} & \alpha_{12} \\ \alpha_{21} & \alpha_{22} \end{bmatrix}$$

is invertible. Construct in $\mathbb{R}^2$ two coordinate systems so that the above prescribed equations will serve as changes of coordinates between them. How many pairs of such coordinate systems could exist?

2. Give a real 2×1 matrix

$$A = \begin{bmatrix} a_1 \\ a_2 \end{bmatrix}.$$

For vector $\vec{x} = (x_1, x_2) \in \mathbb{R}^2$, define

$$\vec{x}A = (x_1 x_2) \begin{bmatrix} a_1 \\ a_2 \end{bmatrix} = a_1 x_1 + a_2 x_2$$

and consider it as a vector in $\mathbb{R}$ (see the Note right after (2.4.1′)).

(a) Show that, for $\vec{x}, \vec{y} \in \mathbb{R}^2$ and $\alpha \in \mathbb{R}$,

$$(\alpha \vec{x} + \vec{y})A = \alpha(\vec{x}A) + \vec{y}A.$$

Hence A represents a *linear transformation* (see Secs. 2.7 and B.7) defined as $\vec{x} \to \vec{x}A$ from $\mathbb{R}^2$ to $\mathbb{R}$. The sets

$$\mathrm{Ker}(A) = \{\vec{x} \in \mathbb{R}^2 \mid \vec{x}A = 0\}, \text{ also denoted as } N(A);$$
$$\mathrm{Im}(A) = \{\vec{x}A \mid \vec{x} \in \mathbb{R}^2\}, \text{ also denoted as } R(A),$$

are respectively called the *kernel* and the *range* of A. Show that $\mathrm{Ker}(A)$ is a subspace of $\mathbb{R}^2$ while $\mathrm{Im}(A)$ a subspace of $\mathbb{R}$.

(b) Show that A, as a linear transformation, is onto $\mathbb{R}$, i.e. $\mathrm{Im}(A) = \mathbb{R}$, if and only if $\mathrm{Ker}(A) \neq \mathbb{R}^2$.

(c) Show that A is one-to-one if and only if $\mathrm{Ker}(A) = \{\vec{0}\}$. Is it possible that A is one-to-one?

(d) Suppose $f: \mathbb{R}^2 \to \mathbb{R}$ is a linear transformation. Show that there exist *unique* scalars a_1 and a_2 such that

$$f(\vec{x}) = (x_1 \quad x_2) \begin{bmatrix} a_1 \\ a_2 \end{bmatrix} = a_1 x_1 + a_2 x_2$$

where $\vec{x} = (x_1, x_2) \in \mathbb{R}^2$. In this case, call

$$[f]_{\mathcal{N}} = \begin{bmatrix} a_1 \\ a_2 \end{bmatrix}$$

the *matrix representation* of f related to the basis $\mathcal{N} = \{\vec{e}_1, \vec{e}_2\}$ for $\mathbb{R}^2$ and the basis $\{1\}$ for $\mathbb{R}$.

3. Give a real 1×2 matrix $A = [a_1 \quad a_2]$. Define the mapping $A: \mathbb{R} \to \mathbb{R}^2$ as

$$x \to xA = (a_1 x, a_2 x).$$

(a) Show that A, as a mapping, is a linear transformation from $\mathbb{R}$ to $\mathbb{R}^2$. What are $\mathrm{Ker}(A)$ and $\mathrm{Im}(A)$? Could A be both one-to-one and onto?

(b) Give a fixed subspace $a_1 x_1 + a_2 x_2 = 0$ of $\mathbb{R}^2$. Show that there exist infinitely many linear transformations f mapping $\mathbb{R}$ onto that subspace. Are such mappings one-to-one?

4. Let

$$A = \begin{bmatrix} a_{11} & a_{12} \\ a_{21} & a_{22} \end{bmatrix}$$

be a real 2×2 matrix. Define the mapping $A: \mathbb{R}^2 \to \mathbb{R}^2$ by

$$\vec{x} = (x_1, x_2) \to \vec{x}A = (x_1 \quad x_2) \begin{bmatrix} a_{11} & a_{12} \\ a_{21} & a_{22} \end{bmatrix}$$

$$= (a_{11} x_1 + a_{21} x_2, a_{12} x_1 + a_{22} x_2).$$

(a) Show that A is a linear transformation. Its kernel $\mathrm{Ker}(A)$ and range $\mathrm{Im}(A)$ are subspaces of $\mathbb{R}^2$.

(b) Show that $\dim \mathrm{Ker}(A) + \dim \mathrm{Im}(A) = \dim \mathbb{R}^2 = 2$ (see Sec. B.3).

(c) Show that the following are equivalent.

(1) A is one-to-one, i.e. $\mathrm{Ker}(A) = \{\vec{0}\}$.

(2) A is onto, i.e. $\mathrm{Im}(A) = \mathbb{R}^2$.

(3) A maps every basis $\mathcal{B} = \{\vec{x}_1, \vec{x}_2\}$ for $\mathbb{R}^2$ onto the basis $\{\vec{x}_1 A, \vec{x}_2 A\}$ for $\mathbb{R}^2$.

(4) A maps a basis $\mathcal{B} = \{\vec{x}_1, \vec{x}_2\}$ for $\mathbb{R}^2$ onto the basis $\{\vec{x}_1 A, \vec{x}_2 A\}$ for $\mathbb{R}^2$.

(5) A is invertible. In this case, A is called a *linear isomorphism*.

(d) Let $\mathcal{B} = \{\vec{x}_1, \vec{x}_2\}$ be any fixed basis for $\mathbb{R}^2$ and $\vec{y}_1, \vec{y}_2$ be any two vectors in $\mathbb{R}^2$, not necessarily linearly independent. Then, the mapping $f: \mathbb{R}^2 \to \mathbb{R}^2$ defined by

$$f(\alpha_1 \vec{x}_1 + \alpha_2 \vec{x}_2) = \alpha_1 \vec{y}_1 + \alpha_2 \vec{y}_2$$

is the *unique* linear transformation from $\mathbb{R}^2$ into $\mathbb{R}^2$ satisfying

$$f(\vec{x}_1) = \vec{y}_1, \quad f(\vec{x}_2) = \vec{y}_2.$$

Suppose $[f(\vec{x}_1)]_\mathcal{B} = (a_{11}, a_{12})$ and $[f(\vec{x}_2)]_\mathcal{B} = (a_{21}, a_{22})$. Then

$$[f(\vec{x})]_\mathcal{B} = [\vec{x}]_\mathcal{B}[f]_\mathcal{B}, \quad \text{where } [f]_\mathcal{B} = \begin{bmatrix} [f(\vec{x}_1)]_\mathcal{B} \\ [f(\vec{x}_2)]_\mathcal{B} \end{bmatrix} = \begin{bmatrix} a_{11} & a_{12} \\ a_{21} & a_{22} \end{bmatrix}.$$

This $[f]_\mathcal{B}$ is called the *matrix representation* of f related to the basis $\mathcal{B}$.

(e) Let $f: \mathbb{R}^2 \to \mathbb{R}^2$ be any linear transformation. Then (show that)

$$f(\vec{x}) = \vec{x} A,$$

where $A = [f]_\mathcal{N}$ and $\mathcal{N} = \{\vec{e}_1, \vec{e}_2\}$ is the natural basis for $\mathbb{R}^2$.

(f) Let $S: a_1 x_1 + a_2 x_2 = 0$ be a subspace of $\mathbb{R}^2$. Show that there are infinitely many linear transformations $f: \mathbb{R}^2 \to \mathbb{R}^2$ such that

$$S = \text{Ker}(f), \quad \text{and}$$
$$S = \text{Im}(f)$$

hold respectively.

(g) Let $S_1: a_{11} x_1 + a_{12} x_2 = 0$ and $S_2: a_{21} x_1 + a_{22} x_2 = 0$ be two subspaces of $\mathbb{R}^2$. Construct a linear transformation $f: \mathbb{R}^2 \to \mathbb{R}^2$ such that

$$f(S_1) = S_2.$$

How many such f are possible?

5. Let $\mathcal{B} = \{\vec{x}_1, \vec{x}_2\}$ and $\mathcal{B}' = \{\vec{y}_1, \vec{y}_2\}$ be two bases for $\mathbb{R}^2$ and $f: \mathbb{R}^2 \to \mathbb{R}^2$ be a linear transformation.

(a) Show that

$$[f]_{\mathcal{B}'} = A_\mathcal{B}^{\mathcal{B}'} [f]_\mathcal{B} A_{\mathcal{B}'}^\mathcal{B}.$$

(b) The *matrix representation* of f with respect to $\mathcal{B}$ and $\mathcal{B}'$ is defined as

$$[f]_{\mathcal{B}'}^{\mathcal{B}} = \begin{bmatrix} [f(\vec{x}_1)]_{\mathcal{B}'} \\ [f(\vec{x}_2)]_{\mathcal{B}'} \end{bmatrix} = \begin{bmatrix} a_{11} & a_{12} \\ a_{21} & a_{22} \end{bmatrix},$$

where $[f(\vec{x}_i)]_{\mathcal{B}'} = (a_{i1}, a_{i2})$, i.e. $f(\vec{x}_i) = a_{i1}\vec{y}_1 + a_{i2}\vec{y}_2$ for $i = 1, 2$. Show that

$$[f(\vec{x})]_{\mathcal{B}'} = [\vec{x}]_{\mathcal{B}}[f]_{\mathcal{B}'}^{\mathcal{B}};$$

in particular, $[f]_{\mathcal{B}}^{\mathcal{B}} = [f]_{\mathcal{B}}$ and $[\vec{x}]_{\mathcal{B}'} = [\vec{x}]_{\mathcal{B}}A_{\mathcal{B}'}^{\mathcal{B}}$. Show also that

$$[f]_{\mathcal{B}'}^{\mathcal{B}} = [f]_{\mathcal{B}}A_{\mathcal{B}'}^{\mathcal{B}} \quad \text{and} \quad [f]_{\mathcal{B}}^{\mathcal{B}'} = [f]_{\mathcal{B}'}A_{\mathcal{B}}^{\mathcal{B}'}.$$

\<C\> Abstraction and generalization

Let $\mathcal{B} = \{\vec{x}_1, \ldots, \vec{x}_n\}$ be a basis for an n-dimensional vector space V over $\mathbb{F}$. Then, the mapping $f: V \to \mathbb{F}^n$ defined by

$$f(\vec{x}) = [\vec{x}]_{\mathcal{B}}$$

is a linear isomorphism. Thus, V is isomorphic to $\mathbb{F}^n$.

Try to extend all the problems in Ex. \<B\> to $\mathbb{F}^n$. What is the counterpart of Ex. \<A\> 2 in $\mathbb{F}^n$? Please refer to Secs. B.4, B.5 and B.7, if necessary.

2.5 Straight Lines in a Plane

Let $\Sigma(O; A_1, A_2)$ be a fixed coordinate system (i.e. vectorized space) in $\mathbb{R}^2$, with basis $\mathcal{B} = \{\vec{a}_1, \vec{a}_2\}$ where $\vec{a}_i = \overrightarrow{OA_i}, i = 1, 2$.

Denote by L_i the straight line determined by O and $A_i, i = 1, 2$ (see Fig. 2.21).

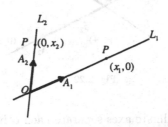

Fig. 2.21

Take a point $P \in \mathbb{R}^2$. Then

$$P \in L_1$$
$$\Leftrightarrow \overrightarrow{OP} \in L(O; A_1)$$
$$\Leftrightarrow [P]_{\mathcal{B}} = (x_1, 0).$$

This implies that, a point $P \in \mathbb{R}^2$ lies on the straight line L_1, if and only if the second component x_2 of the coordinate $[P]_{\mathcal{B}}$ of P with respect to the basis $\mathcal{B}$ is equal to zero, i.e. $x_2 = 0$. Hence, call

$$x_2 = 0 \tag{2.5.1}$$

the (*coordinate*) *equation* of L_1 with respect to $\mathcal{B}$, and L_1 the *first coordinate axis*. By exactly the same reason,

$$x_1 = 0 \tag{2.5.2}$$

is called the (*coordinate*) *equation* of the *second coordinate axis* L_2 with respect to $\mathcal{B}$.

These two coordinate axes intersect at the origin O and separate the whole plane $\Sigma(O; A_1, A_2)$ into four parts, called quadrants, according to the signs of x_1 and x_2 of the components of the coordinate $[P]_{\mathcal{B}} = (x_1, x_2)$, $P \in \mathbb{R}^2$, as follows.

> First quadrant: $x_1 > 0, x_2 > 0$.
> Second quadrant: $x_1 < 0, x_2 > 0$.
> Third quadrant: $x_1 < 0, x_2 < 0$.
> Fourth quadrant: $x_1 > 0, x_2 < 0$. $\tag{2.5.3}$

See Fig. 2.22.

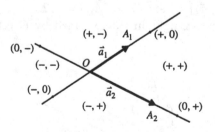

Fig. 2.22

Note that the two coordinate axes separate each other at O into four half-lines, symbolically denoted by $(+, 0), (0, +), (-, 0)$ and $(0, -)$.

Let X be a point lying on the straight line L, considered as a subset of $\Sigma(O; A_1, A_2)$, and generated by two different points A and B in $\mathbb{R}^2$.

Denote by

$$\vec{a} = \overrightarrow{OA} \quad \text{(viewed as a point on the line } L\text{)},$$
$$\vec{b} = \overrightarrow{AB} \quad \text{(viewed as a direction vector of the line } L\text{)}, \text{ and}$$
$$\vec{x} = \overrightarrow{OX} \quad \text{(viewed as a point on } L\text{)}.$$

See Fig. 2.23. Then, the position vector $\vec{x} - \vec{a} = \overrightarrow{AX}$ of the point $\vec{x}$ relative to the point $\vec{a}$ must be linearly dependent on the direction $\vec{b}$. Hence, there exists a scalar $t \in \mathbb{R}$ such that

$$\vec{x} - \vec{a} = t\vec{b}$$
$$\Rightarrow \vec{x} = \vec{a} + t\vec{b}, \quad t \in \mathbb{R} \tag{2.5.4}$$

called the *parametric equation* (in vector form) of the line L, passing through the point $\vec{a}$ with *direction vector* $\vec{b}$, in the coordinate system $\Sigma(O; A_1, A_2)$.

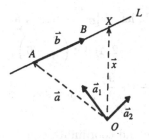

Fig. 2.23

Suppose

$$[A]_\mathcal{B} = [\vec{a}]_\mathcal{B} = (a_1, a_2),$$
$$[\vec{b}]_\mathcal{B} = (b_1, b_2),$$
$$[X]_\mathcal{B} = [\vec{x}]_\mathcal{B} = (x_1, x_2).$$

In terms with these coordinates, (2.5.4) can be rewritten as

$$[X]_\mathcal{B} = [\vec{a}]_\mathcal{B} + t[\vec{b}]_\mathcal{B} \tag{2.5.5}$$

$$\Rightarrow (x_1, x_2) = (a_1, a_2) + t(b_1, b_2) = (a_1 + tb_1, a_2 + tb_2)$$

$$\Rightarrow \begin{cases} x_1 = a_1 + tb_1 \\ x_2 = a_2 + tb_2 \end{cases}, \quad t \in \mathbb{R} \tag{2.5.6}$$

also called *the parametric equation* of L. Furthermore, eliminating the parameter t from (2.5.6), one gets

$$\frac{x_1 - a_1}{b_1} = \frac{x_2 - a_2}{b_2}, \quad \text{or}$$

$$\alpha x_1 + \beta x_2 + \gamma = 0 \tag{2.5.7}$$

called the *coordinate equation* of the line L in the coordinate system $\Sigma(O; A_1, A_2)$, with coordinate equations (2.5.1) and (2.5.2) as special cases.

Example (continued from the Example in Sec. 2.4) Find the equation of the line determined by the points $A = (-3, 0)$ and $B = (3, 3)$ in $\mathbb{R}^2$, respectively, in $\Sigma(O; A_1, A_2)$ and $\Sigma(O'; B_1, B_2)$.

Solution In the coordinate system $\Sigma(O; A_1, A_2)$, let

$$\vec{a} = \overrightarrow{OA} = (-3, 0) - (1, 0) = (-4, 0) \Rightarrow [\vec{a}]_B = (-2, 4),$$

$$\vec{b} = \overrightarrow{AB} = (3, 3) - (-3, 0) = (6, 3) \Rightarrow [\vec{b}]_B = \left(\frac{9}{2}, -6\right).$$

For arbitrary point $X \in L$, let $[X]_B = (x_1, x_2)$ and (2.5.5) shows that

$$(x_1, x_2) = (-2, 4) + t\left(\frac{9}{2}, -6\right), \quad t \in \mathbb{R}$$

$$\Rightarrow \begin{cases} x_1 = -2 + \frac{9}{2}t \\ x_2 = 4 - 6t \end{cases}, \quad \text{or} \quad 4x_1 + 3x_2 - 4 = 0$$

is the required parametric and coordinate equations of L in $\Sigma(O; A_1, A_2)$.

In the coordinate system $\Sigma(O'; B_1, B_2)$, let

$$\vec{a} = \overrightarrow{O'A} = (-3, 0) - (-1, -1) = (-2, 1) \Rightarrow [\vec{a}]_{B'} = (-11, 3),$$

$$\vec{b} = \overrightarrow{AB} = (6, 3) \Rightarrow [\vec{b}]_{B'} = (15, -3).$$

For point X in L, let $[X]_{B'} = (y_1, y_2)$. Then

$$(y_1, y_2) = (-11, 3) + t(15, -3), \quad t \in \mathbb{R}$$

$$\Rightarrow \begin{cases} y_1 = -11 + 15t \\ y_2 = 3 - 3t \end{cases}, \quad \text{or} \quad y_1 + 5y_2 - 4 = 0$$

is the required equations in $\Sigma(O'; B_1, B_2)$. $\qquad \square$

Remark Change formulas of the equations of the same line in different coordinate systems.

Adopt the notations and results in (2.4.2). From (2.5.5), the equation of the line L in $\Sigma(O; A_1, A_2)$ is

$$[X]_{\mathcal{B}} = [\vec{a}]_{\mathcal{B}} + t[\vec{b}]_{\mathcal{B}}, \quad X \in L.$$

Via the change of coordinate formula $[X]_{\mathcal{B}'} = [O]_{\mathcal{B}'} + [X]_{\mathcal{B}} A_{\mathcal{B}'}^{\mathcal{B}}$, one obtains

$$[X]_{\mathcal{B}'} = [O]_{\mathcal{B}'} + \{[\vec{a}]_{\mathcal{B}} + t[\vec{b}]_{\mathcal{B}}\} A_{\mathcal{B}'}^{\mathcal{B}}, \tag{2.5.8}$$

which is the equation of the same line L in another coordinate system $\Sigma(O'; B_1, B_2)$.

For example, we use the above example and compute as follows. As we already knew that

$$[O]_{\mathcal{B}'} = (5, -1) \quad \text{and} \quad A_{\mathcal{B}'}^{\mathcal{B}} = \begin{bmatrix} -6 & 2 \\ -7 & 2 \end{bmatrix},$$

using $[X]_{\mathcal{B}} = (x_1, x_2)$ and $[X]_{\mathcal{B}'} = (y_1, y_2)$, we have

$$
\begin{aligned}
(y_1, y_2) &= (5, -1) + \left\{ (-2, 4) + t\left(\frac{9}{2} - 6\right) \begin{bmatrix} -6 & 2 \\ -7 & 2 \end{bmatrix} \right\} \\
&= (5, -1) + (-16 + 15t, 4 - 3t) \\
&= (-11 + 15t, 3 - 3t).
\end{aligned}
$$

Hence $y_1 = -11 + 15t, y_2 = 3 - 3t$ as shown in the example.

Finally, we discuss

The relative positions of two lines in a plane
Let

L_1: $\vec{x} = \vec{a}_1 + t\vec{b}_1$ (passing the point $\vec{a}_1$ with direction $\vec{b}_1$),
L_2: $\vec{x} = \vec{a}_2 + t\vec{b}_2$ (passing the point $\vec{a}_2$ with direction $\vec{b}_2$)

be two given lines in $\mathbb{R}^2$. Then L_1 and L_2 may have the following three relative positions. They are,

1. *coincident* $(L_1 = L_2)$ $\Leftrightarrow$ the vectors $\vec{a}_1 - \vec{a}_2$, $\vec{b}_1$ and $\vec{b}_2$ are linearly dependent,
2. *parallel* $(L_1 /\!/ L_2)$ $\Leftrightarrow$ $\vec{b}_1$ and $\vec{b}_2$ are linearly dependent, but linearly independent of $\vec{a}_2 - \vec{a}_1$, and
3. *intersecting* (in a unique single point) $\Leftrightarrow$ $\vec{b}_1$ and $\vec{b}_2$ are linearly independent. $\tag{2.5.9}$

See Fig. 2.24.

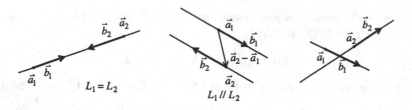

Fig. 2.24

Proof Since $\vec{b}_1$ and $\vec{b}_2$ are nonzero vectors, only these three cases on the right sides of 1, 2 and 3 need to be considered. Only sufficiencies are proved and the necessities are left to the readers.

Case 1 Let $\vec{b}_2 = \alpha \vec{b}_1$ and $\vec{a}_2 - \vec{a}_1 = \beta \vec{b}_1$. For point $\vec{x} \in L_1$, then

$$\vec{x} = a_1 + t\vec{b}_1 = \vec{a}_2 - \beta\vec{b}_1 + t\vec{b}_1 = \vec{a}_2 + \frac{t - \beta}{\alpha}\vec{b}_2,$$

which means $\vec{x} \in L_2$. Conversely, if $\vec{x} \in L_2$, then

$$\vec{x} = a_2 + t\vec{b}_2 = \vec{a}_1 + \beta\vec{b}_1 + t\alpha\vec{b}_1 = \vec{a}_1 + (\beta + t\alpha)\vec{b}_1$$

indicates that $\vec{x} \in L_1$. Therefore, $L_1 = L_2$ holds.

Case 2 Suppose $\vec{b}_2 = \alpha\vec{b}_1$, but $\vec{a}_2 - \vec{a}_1 \neq \beta\vec{b}_1$ for any scalar $\beta \in \mathbb{R}$. If there exists a point $\vec{x}$ common to both L_1 and L_2, then two scalars t_1 and t_2 can be found so that

$$\vec{a}_1 + t_1\vec{b}_1 = a_2 + t_2\vec{b}_2 = \vec{a}_2 + t_2\alpha\vec{b}_1$$
$$\Rightarrow \vec{a}_2 - \vec{a}_1 = (t_1 - t_2\alpha)\vec{b}_1$$

contradicting to the hypotheses. Hence, being no point of intersection, L_1 and L_2 should be parallel to each other.

Case 3 Since the vector $\vec{a}_2 - \vec{a}_1$ is coplanar with linearly independent vectors $\vec{b}_1$ and $\vec{b}_2$, there exist unique scalars t_1 and t_2 such that

$$\vec{a}_2 - \vec{a}_1 = t_1\vec{b}_1 + t_2\vec{b}_2$$
$$\Rightarrow \vec{a}_1 + t_1\vec{b}_1 = \vec{a}_2 + (-t_2)\vec{b}_2,$$

which means that L_1 and L_2 have only one point of intersection. $\quad\square$

According to (2.5.4), the line determined by two distinct points $\vec{a}_1$ and $\vec{a}_2$ on the plane $\mathbb{R}^2$ has the parametric equation

$$\vec{x} = \vec{a}_1 + t(\vec{a}_2 - \vec{a}_1) = (1 - t)\vec{a}_1 + t\vec{a}_2, \quad t \in \mathbb{R} \qquad (2.5.10)$$

Just like (1.4.2) and (1.4.3), the *directed segment* $\vec{a}_1\vec{a}_2$ with the *initial point* $\vec{a}_1$ and *terminal point* $\vec{a}_2$ is the set of points

$$\vec{x} = (1 - t)\vec{a}_1 + t\vec{a}_2, \quad 0 \le t \le 1. \qquad (2.5.11)$$

If $0 < t < 1$, the point $\vec{x}$ is called an *interior point* of $\vec{a}_1\vec{a}_2$ with *end points* $\vec{a}_1$ and $\vec{a}_2$; if $t < 0$ or $t > 1$, $\vec{x}$ is called an *exterior point*. See Fig. 2.25 (compare with Fig. 1.11). $\frac{1}{2}(\vec{a}_1 + \vec{a}_2)$ is called the *middle point* of $\vec{a}_1\vec{a}_2$.

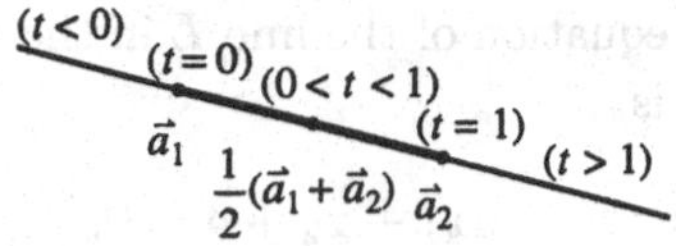

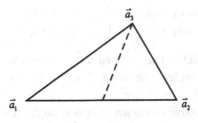

Fig. 2.25

By a *triangle*

$$\triangle \vec{a}_1\vec{a}_2\vec{a}_3 \qquad (2.5.12)$$

with three noncollinear points $\vec{a}_1$, $\vec{a}_2$ and $\vec{a}_3$ as *vertices*, we mean the plane figure formed by three consecutive segments $\vec{a}_1\vec{a}_2$, $\vec{a}_2\vec{a}_3$ and $\vec{a}_3\vec{a}_1$, which are called *sides*. See Fig. 2.26.

Fig. 2.26

A line joining a vertex to the midpoint of the opposite side is called a *median* of a triangle. Note that

$$\left[\vec{a}_3 - \frac{1}{2}(\vec{a}_1 + \vec{a}_2)\right] + \left[\vec{a}_1 - \frac{1}{2}(\vec{a}_2 + \vec{a}_3)\right] + \left[\vec{a}_2 - \frac{1}{2}(\vec{a}_3 + \vec{a}_1)\right] = \vec{0}$$

and the three medians of $\triangle \vec{a}_1\vec{a}_2\vec{a}_3$ meet at its *centroid* (refer to Ex. <B> 7 of Sec. 3.5).

For a somewhat different treatment of line segments and triangles, please refer to Sec. 2.6.

Exercises

<A>

1. Use the notations and results from Ex. <A> 4 in Sec. 2.4. Denote by L the line determined by the points $A = (1,3)$ and $B = (-6,1)$ in $\mathbb{R}^2$.

   (a) Find the parametric and coordinate equations of L in $\Sigma(O; A_1, A_2)$ and $\Sigma(O'; B_1, B_2)$, respectively.
   (b) Check the answers in (a), by using (2.5.8).

2. It is known that the equation of the line L in the rectangular coordinate system $\Sigma(\vec{o}; \vec{e}_1, \vec{e}_2)$ is

$$-3x_1 + x_2 + 4 = 0.$$

Find the equation of L in the coordinate system $\Sigma(O; A_1, A_2)$, where $O = (1,2)$, $A_1 = (-2,-3)$ and $A_2 = (-4,5)$. Conversely, if the line L' has equation $y_1 + 6y_2 - 3 = 0$ in $\Sigma(O; A_1, A_2)$, what is it in $\Sigma(\vec{o}; \vec{e}_1, \vec{e}_2)$?

<B>

1. Give scalars $a_i, b_i, i = 1,2$ and $c_i, d_i, i = 1,2$ with the requirement that b_1 and b_2, d_1 and d_2, are not equal to zero simultaneously, and

$$\begin{cases} x_1 = a_1 + tb_1 \\ x_2 = a_2 + tb_2 \end{cases}; \quad \begin{cases} y_1 = c_1 + td_1 \\ y_2 = c_2 + td_2 \end{cases}, \quad t \in \mathbb{R}.$$

Construct, in $\mathbb{R}^2$, two coordinate systems $\Sigma(O; A_1, A_2)$ and $\Sigma(O'; B_1, B_2)$ on which the same line L has the equations described above, respectively.

2. Prove that the relative positions of two straight lines are independent of the choices of coordinate systems in $\mathbb{R}^2$.

3. Let

$$L_i: \vec{x} = \vec{a}_i + t\vec{b}_i, \quad t \in \mathbb{R}, \quad i = 1,2,3$$

be three given lines in $\mathbb{R}^2$. Describe all possible relative positions among them and characterize it in each case.

4. Show that the whole plane $\mathbb{R}^2$ cannot be represented as a countable union of distinct straight lines, all passing through a fixed point.

5. The image $\vec{x}_0 + S$ of a subspace S of $\mathbb{R}^2$ under the *translation* $\vec{x} \rightarrow \vec{x}_0 + \vec{x}$ is called an *affine subspace* of $\mathbb{R}^2$ (refer to Fig. B.2). The dimension of $\vec{x}_0 + S$ is defined to be the dimension of S, i.e. $\dim(\vec{x}_0 + S) = \dim S$. Then, $\vec{x}_0 + S = \vec{y}_0 + S$ if and only if $\vec{x}_0 - \vec{y}_0 \in S$; in particular, $\vec{x}_0 + S = S$ if and only if $\vec{x}_0 \in S$. Show that:

   (a) Each point in $\mathbb{R}^2$ is a zero-dimensional affine subspace of $\mathbb{R}^2$.

   (b) Each straight line in $\mathbb{R}^2$ is a one-dimensional affine subspace.

   (c) $\mathbb{R}^2$ itself is the only two-dimensional affine subspace.

6. The composite $T(\vec{x}) = \vec{x}_0 + f(\vec{x})$ of a linear isomorphism $f: \mathbb{R}^2 \rightarrow \mathbb{R}^2$ followed by a translation $\vec{x} \rightarrow \vec{x}_0 + \vec{x}$ is called an *affine transformation* or *mapping*. Therefore, related to the natural basis $\mathcal{N} = \{\vec{e}_1, \vec{e}_2\}$ of $\mathbb{R}^2$, T can be represented as

$$T(\vec{x}) = \vec{x}_0 + \vec{x}A \quad \text{or} \quad y_1 = \alpha_1 + a_{11}x_1 + a_{21}x_2$$
$$y_2 = \alpha_2 + a_{12}x_1 + a_{22}x_2,$$

where $A = [a_{ij}]$ is an invertible 2×2 matrix .

   (a) For any two line segments $\vec{a}_1\vec{a}_2$ and $\vec{b}_1\vec{b}_2$, there exist infinitely many affine transformations T such that

$$T(\vec{a}_i) = \vec{b}_i, \quad i = 1, 2 \quad \text{and}$$
$$T(\vec{a}_1\vec{a}_2) = \vec{b}_1\vec{b}_2.$$

   (b) For any two given $\Delta\vec{a}_1\vec{a}_2\vec{a}_3$ and $\Delta\vec{b}_1\vec{b}_2\vec{b}_3$, there exists a unique affine transformation T such that

$$T(\vec{a}_i) = \vec{b}_i, \quad i = 1, 2, 3 \quad \text{and}$$
$$T(\Delta\vec{a}_1\vec{a}_2\vec{a}_3) = \Delta\vec{b}_1\vec{b}_2\vec{b}_3.$$

In the following Exs. 7 and 8, readers should have basic knowledge about the Euclidean plane (refer to Chap. 4, if necessary).

7. Let $\vec{a}_1 = (-1, -2)$, $\vec{a}_2 = (-2, 1)$ and $\vec{a}_3 = (1, 3)$.

   (a) Compute the three interior angles, side lengths and the area of $\Delta\vec{a}_1\vec{a}_2\vec{a}_3$.

   (b) Give affine transformation

$$T(\vec{x}) = (2, 3) + \vec{x}\begin{bmatrix} 2 & -1 \\ -6 & -5 \end{bmatrix}.$$

   Let $\vec{b}_i = T(\vec{a}_i)$, $i = 1, 2, 3$. Graph $\Delta\vec{b}_1\vec{b}_2\vec{b}_3$ and $\Delta\vec{a}_1\vec{a}_2\vec{a}_3$. Then, compute the three interior angles, side lengths and the area of $\Delta \vec{b_1} \vec{b_2} \vec{b_3}$. Compare these quantities with (a). What can you find?

(c) Find the fourth point $\vec{a_4}$ in $\mathbb{R}^2$ so that $\vec{a_1}, \vec{a_2}, \vec{a_3}, \vec{a_4}$ (in this order) constitute the vertices of a *parallelogram*, denoted by $\square \vec{a_1} \vec{a_2} \vec{a_3} \vec{a_4}$. Compute the area of $\square \vec{a_1} \vec{a_2} \vec{a_3} \vec{a_4}$. Do the same questions as in (b). In particular, do the image points $\vec{b_1}, \vec{b_2}, \vec{b_3}, \vec{b_4}$ form a parallelogram? What is its area?

8. In $\mathbb{R}^2$ (of course, with a rectangular coordinate system $\Sigma(\vec{o}; \vec{e_1}, \vec{e_2})$), the length of a vector $\vec{x} = (x_1, x_2)$ is denoted by

$$|\vec{x}| = (x_1^2 + x_2^2)^{1/2}.$$

The unit circle in $\mathbb{R}^2$ is

$$|\vec{x}|^2 = 1 \quad \text{or} \quad x_1^2 + x_2^2 = 1.$$

See Fig. 2.27.

(a) Find the equation of the unit circle in the coordinate system $\Sigma(O; A_1, A_2)$, where $O = (0,0)$, $A_1 = (1,2)$, $A_2 = (-3,1)$.

(b) If the equation obtained in (a) is observed in the coordinate system $\Sigma(\vec{o}; \vec{e_1}, \vec{e_2})$, then what curve does it represent? Try to graph this curve and find the area enclosed by it. What is the ratio of this area with respect to that of unit disk?

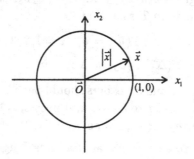

Fig. 2.27

<C> Abstraction and generalization

Let $\mathbb{I}_2 = \{0, 1\}$ be the residue classes of modulus 2 (see Sec. A.3). $\mathbb{I}_2$ is a finite field with two elements. Then, $\mathbb{I}_2^2 = \{(0,0), (0,1), (1,0), (1,1)\}$ is a two-dimensional vector space over $\mathbb{I}_2$ having subspaces $\{(0,0)\}, \{(0,0),$

$(1, 0)\}, \{(0, 0),(0, 1)\}$ and $\{(0, 0),(1, 1)\}$. Moreover,

$$\mathbb{I}_2^2 = \{(0,0)\} \cup \{(0,0),(1,0)\} \cup \{(0,0),(0,1)\} \cup \{(0,0),(1,1)\}.$$

See Fig. 2.28. Since

$$\{(0,1),(1,1)\} = (0,1) + \{(0,0),(1,0)\},$$

$\{(0,1),(1,1)\}$ is an affine straight line not passing through $(0, 0)$. The other one is $\{(1, 0), (1, 1)\}$.

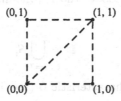

Fig. 2.28

1. Let $\mathbb{I}_3 = \{0,1,2\}$ and construct the vector space $\mathbb{I}_3^3$ over $\mathbb{I}_3$. See Fig. 2.29 and try to spot the vectors $(2, 1, 1)$, $(1, 2, 1)$, $(1, 1, 2)$.

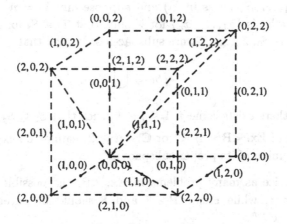

Fig. 2.29   (Dotted lines do not contain points except $\cdot$.)

(a) There are 27 vectors in $\mathbb{I}_3^3$. List them all.
(b) List all 13 one-dimensional subspaces. How many two-dimensional subspaces are there? List 10 of them. How many vectors are there in each subspace?

(c) How many affine straight lines not passing through (0, 0, 0) are there?

(d) How many different ordered bases for $\mathbb{I}_3^3$ are there?

(e) Is

$$\begin{bmatrix} 1 & 0 & 0 \\ 0 & 1 & 0 \\ 0 & 0 & 1 \end{bmatrix} = \begin{bmatrix} 1 & 2 & 1 \\ 2 & 0 & 1 \\ 1 & 1 & 1 \end{bmatrix} \begin{bmatrix} 1 & 1 & 1 \\ 1 & 0 & 2 \\ 1 & 2 & 1 \end{bmatrix}$$

true? Anything to do with changes of coordinates (or bases)?

(f) Let S_i, $1 \leq i \leq 13$, denote the 13 one-dimensional subspaces of $\mathbb{I}_3^3$. Then

$$\mathbb{I}_3^3 = \bigcup_{i=1}^{13} S_i.$$

2. Suppose $\mathbb{F}$ is a field of characteristic 0 (see Sec. A.3) and V is a vector space over $\mathbb{F}$.

   (a) (extension of Ex. <B> 4) V cannot be covered by any *finite* number of *proper* subspaces of V. That is, suppose $S_1, \ldots, S_k$ are any finite number of proper subspaces of V, then there exists $\vec{x} \in V$ such that $\vec{x} \notin \bigcup_{i=1}^{k} S_i$.

   (b) Let $S_1, \ldots, S_k$ be as in (a) and suppose dim $V = n$. Then, there exists a basis $\{\vec{x}_1, \ldots, \vec{x}_n\}$ for V such that $\vec{x}_j \notin S_j$ for $1 \leq j \leq k$.

   (c) Suppose $S_0, S_1, \ldots, S_k$ are subspaces of V such that

$$S_0 \subseteq \bigcup_{j=1}^{k} S_j.$$

   Then there exists some j, $1 \leq j \leq k$, such that $S_0 \subseteq S_j$ holds.

3. (extension of Ex. <B> 4) $\mathbb{R}^n$ or $\mathbb{C}^n$ ($n \geq 2$) cannot be expressed as a countable union of its proper subspaces.

Try to generalize as many problems in Ex. <B> as possible to abstract space V over $\mathbb{F}$, while Exs. <B> 7 and 8 should be extended to the Euclidean space $\mathbb{R}^n$.

2.6   Affine and Barycentric Coordinates

Recall the convention stated right before Remark 2.3 in Sec. 2.3: when $\mathbb{R}^2$ is considered as an affine plane, its element $\vec{x} = (x_1, x_2)$ is treated as a point, while if considered as a vector space, $\vec{x} = (x_1, x_2)$ represents a position vector from $\vec{o}$ to the point $\vec{x}$ itself. Sometimes, points in $\mathbb{R}^2$ are also denoted by capital letters such as A, X, etc.

Suppose $\vec{a}_0, \vec{a}_1$ and $\vec{a}_2$ are non-collinear points in $\mathbb{R}^2$. Then, the vectors $\vec{v}_1 = \vec{a}_1 - \vec{a}_0$ and $\vec{v}_2 = \vec{a}_2 - \vec{a}_0$ are linear independent, and we call the points $\vec{a}_0, \vec{a}_1, \vec{a}_2$ *affinely independent* (refer to (2.2.4)) and the ordered set

$$\mathcal{B} = \{\vec{a}_0, \vec{a}_1, \vec{a}_2\} \quad \text{or} \quad \{\vec{v}_1, \vec{v}_2\}$$

an *affine basis* with $\vec{a}_0$ as the *base point*. See Fig. 2.30. If so, the vectorized space $\Sigma(\vec{a}_0; \vec{a}_1, \vec{a}_2)$ is a geometric model for $\mathbb{R}^2$.

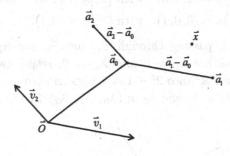

Fig. 2.30

Note that the affine basis $\mathcal{N} = \{\vec{o}, \vec{e}_1, \vec{e}_2\}$ is the standard basis $\{\vec{e}_1, \vec{e}_2\}$ for $\mathbb{R}^2$ when considered as a vector space.

Here in this section, we are going to introduce affine and barycentric coordinates for points in the affine plane $\mathbb{R}^2$ and changes of their coordinates. Conceptually, these are somewhat different versions of results stated in (2.2.2)–(2.2.4), so it could be skipped if one wants.

Suppose $\Sigma(\vec{a}_0; \vec{a}_1, \vec{a}_2)$ is a vectorized space of $\mathbb{R}^2$ with $\mathcal{B} = \{\vec{a}_0, \vec{a}_1, \vec{a}_2\}$ as an affine basis.

For any given $\vec{x} \in \mathbb{R}^2$, there exist unique constants x_1 and x_2 such that

$$\vec{x} - \vec{a}_0 = x_1(\vec{a}_1 - \vec{a}_0) + x_2(\vec{a}_2 - \vec{a}_0)$$

$$\begin{aligned}
\Rightarrow \vec{x} &= \vec{a}_0 + x_1(\vec{a}_1 - \vec{a}_0) + x_2(\vec{a}_2 - \vec{a}_0) \\
&= (1 - x_1 - x_2)\vec{a}_0 + x_1\vec{a}_1 + x_2\vec{a}_2 \\
&= \lambda_0\vec{a}_0 + \lambda_1\vec{a}_1 + \lambda_2\vec{a}_2, \quad \lambda_0 + \lambda_1 + \lambda_2 = 1, \quad (2.6.1)
\end{aligned}$$

where $\lambda_0 = 1 - x_1 - x$, $\lambda_1 = x_1$, $\lambda_2 = x_2$. The ordered triple

$$(\vec{x})_{\mathcal{B}} = (\lambda_0, \lambda_1, \lambda_2), \quad \lambda_0 + \lambda_1 + \lambda_2 = 1 \quad (2.6.2)$$

is called the (normalized) *barycentric coordinate* of the point $\vec{x}$ with respect to the affine basis $\mathcal{B}$. Once λ_1 and λ_2 are known, $\lambda_0 = 1 - \lambda_1 - \lambda_2$ is then uniquely determined. Hence, occasionally, we just call

$$[\vec{x}]_{\mathcal{B}} = [\vec{x} - \vec{a}_0]_{\mathcal{B}} = (x_1, x_2) = (\lambda_1, \lambda_2), \qquad (2.6.3)$$

the coordinate vector of the vector $\vec{x} - \vec{a}_0$ with respect to the basis $\{\vec{a}_1 - \vec{a}_0, \vec{a}_2 - \vec{a}_0\}$ as introduced in (2.3.1), the *affine coordinate* of $\vec{x}$, for simplicity.

In particular,

$$(\vec{a}_0)_{\mathcal{B}} = (1,0,0) \quad \text{with } [\vec{a}_0]_{\mathcal{B}} = (0,0),$$
$$(\vec{a}_1)_{\mathcal{B}} = (0,1,0) \quad \text{with } [\vec{a}_1]_{\mathcal{B}} = (1,0), \quad \text{and}$$
$$(\vec{a}_2)_{\mathcal{B}} = (0,0,1) \quad \text{with } [\vec{a}_2]_{\mathcal{B}} = (0,1).$$

The *coordinate axis* passing through $\vec{a}_1$ and $\vec{a}_2$ has equation $\lambda_0 = 0$, while the other two are $\lambda_1 = 0$ and $\lambda_2 = 0$, respectively. These three axes divide the plane $\mathbb{R}^2$ into $2^3 - 1 = 7$ regions according to the positive or negative signs of λ_0, λ_1 and λ_2 in $(\lambda_0, \lambda_1, \lambda_2)$. See Fig. 2.31.

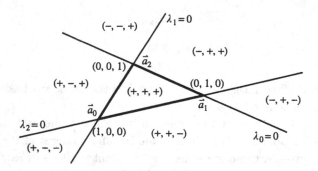

Fig. 2.31

Note that $\left(\frac{1}{3}, \frac{1}{3}, \frac{1}{3}\right)$ is the *barycenter* of the *base triangle* $\triangle \vec{a}_0 \vec{a}_1 \vec{a}_2$.

Let $\Sigma(\vec{b}_0; \vec{b}_1, \vec{b}_2)$ be another vectorized space of $\mathbb{R}^2$ with $\mathcal{B}' = \{\vec{b}_0, \vec{b}_1, \vec{b}_2\}$ the corresponding affine basis. For any $\vec{x} \in \mathbb{R}^2$, denote

$$(\vec{x})_{\mathcal{B}'} = (\mu_0, \mu_1, \mu_2), \quad \mu_0 + \mu_1 + \mu_2 = 1 \quad \text{or}$$
$$[\vec{x} - \vec{b}_0]_{\mathcal{B}'} = (y_1, y_2) = (\mu_1, \mu_2).$$

Then, by (2.4.2), the change of coordinates from $\Sigma(\vec{a}_0; \vec{a}_1, \vec{a}_2)$ to $\Sigma(\vec{b}_0; \vec{b}_1, \vec{b}_2)$ is

$$[\vec{x} - \vec{b}_0]_{\mathcal{B}'} = [\vec{a}_0 - \vec{b}_0]_{\mathcal{B}'} + [\vec{x} - \vec{a}_0]_{\mathcal{B}} A_{\mathcal{B}'}^{\mathcal{B}}, \qquad (2.6.4)$$

where

$$A_{B'}^B = \begin{bmatrix} [\vec{a}_1 - \vec{a}_0]_{B'} \\ [\vec{a}_2 - \vec{a}_0]_{B'} \end{bmatrix} = \begin{bmatrix} \beta_{11} & \beta_{12} \\ \beta_{21} & \beta_{22} \end{bmatrix}$$

is the transition matrix, which is invertible. Suppose

$$[\vec{a}_0 - \vec{b}_0]_{B'} = (p_1, p_2),$$

i.e. $\vec{a}_0 - \vec{b}_0 = p_1(\vec{b}_1 - \vec{b}_0) + p_2(\vec{b}_2 - \vec{b}_0)$. Then, (2.6.4) can be rewritten as

$$(y_1 \quad y_2) = (p_1 \quad p_2) + (x_1 \quad x_2) \begin{bmatrix} \beta_{11} & \beta_{12} \\ \beta_{21} & \beta_{22} \end{bmatrix} \qquad (2.6.5)$$

or

$$(y_1 \quad y_2 \quad 1) = (x_1 \quad x_2 \quad 1) \begin{bmatrix} \beta_{11} & \beta_{12} & 0 \\ \beta_{21} & \beta_{22} & 0 \\ p_1 & p_2 & 1 \end{bmatrix}. \qquad (2.6.5')$$

In particular, if $\vec{a}_0 = \vec{b}_0$ holds, then $(p_1, p_2) = (0, 0)$ and (2.6.5) reduces to

$$(y_1 \quad y_2) = (x_1 \quad x_2) \begin{bmatrix} \beta_{11} & \beta_{12} \\ \beta_{21} & \beta_{22} \end{bmatrix} \qquad (2.6.7)$$

or (2.6.4) reduces to $[\vec{x} - \vec{a}_0]_{B'} = [\vec{x} - \vec{a}_0]_B A_{B'}^B$, which is a linear transformation. In case $\vec{b}_i - \vec{b}_0 = \vec{a}_i - \vec{a}_0$, $i = 1, 2$, $A_{B'}^B = I_2$ is the 2×2 identity matrix and (2.6.5) reduces to

$$(y_1, y_2) = (p_1, p_2) + (x_1, x_2), \qquad (2.6.8)$$

or (2.6.4) reduces to $[\vec{x} - \vec{b}_0]_{B'} = [\vec{a}_0 - \vec{b}_0]_{B'} + [\vec{x} - \vec{a}_0]_B$, which represents a translation. Therefore, *a change of coordinates is a composite mapping of a linear transformation followed by a translation.*

Exercises

<A>

1. Suppose $\vec{a}_i = (a_{i1}, a_{i2}) \in \mathbb{R}^2$, $i = 1, 2, 3$.

   (a) Prove that $\{\vec{a}_1, \vec{a}_2, \vec{a}_3\}$ is an affine basis for $\mathbb{R}^2$ if and only if the matrix

   $$\begin{bmatrix} a_{11} & a_{12} & 1 \\ a_{21} & a_{22} & 1 \\ a_{31} & a_{32} & 1 \end{bmatrix}$$

   is invertible.

(b) Prove that $\{\vec{a}_1, \vec{a}_2, \vec{a}_3\}$ is an affine basis if and only if, for any permutation $\sigma\colon \{1,2,3\} \to \{1,2,3\}$, $\{\vec{a}_{\sigma(1)}, \vec{a}_{\sigma(2)}, \vec{a}_{\sigma(3)}\}$ is an affine basis.

2. Let $\vec{a}_i = (a_{i1}, a_{i2}) \in \mathbb{R}^2$, $i = 0, 1, 2$ and suppose $\mathcal{B} = \{\vec{a}_0, \vec{a}_1, \vec{a}_2\}$ forms an affine basis. For any point $\vec{x} = (x_1, x_2) \in \mathbb{R}^2$, let $(\vec{x})_{\mathcal{B}} = (\lambda_0, \lambda_1, \lambda_2)$ be as in (2.6.2) with $\lambda_0 = 1 - \lambda_1 - \lambda_2$. Show that

$$(\lambda_1, \lambda_2) = (\vec{x} - \vec{a}_0) \begin{bmatrix} \vec{a}_1 - \vec{a}_0 \\ \vec{a}_2 - \vec{a}_0 \end{bmatrix}^{-1}$$

with

$$\lambda_1 = \frac{1}{\det \begin{bmatrix} \vec{a}_1 - \vec{a}_0 \\ \vec{a}_2 - \vec{a}_0 \end{bmatrix}} \begin{vmatrix} a_{01} & a_{02} & 1 \\ x_1 & x_2 & 1 \\ a_{21} & a_{22} & 1 \end{vmatrix},$$

$$\lambda_2 = \frac{1}{\det \begin{bmatrix} \vec{a}_1 - \vec{a}_0 \\ \vec{a}_2 - \vec{a}_0 \end{bmatrix}} \begin{vmatrix} a_{01} & a_{02} & 1 \\ a_{11} & a_{12} & 1 \\ x_1 & x_2 & 1 \end{vmatrix},$$

and, conversely,

$$\vec{x} - \vec{a}_0 = (\lambda_1 \quad \lambda_2) \begin{bmatrix} \vec{a}_1 - \vec{a}_0 \\ \vec{a}_2 - \vec{a}_0 \end{bmatrix}$$

with the expected results (see (2.6.1))

$$x_i = \lambda_0 a_{0i} + \lambda_1 a_{1i} + \lambda_2 a_{2i}, \quad i = 1, 2.$$

3. Let $\vec{a}_i = (a_{i1}, a_{i2})$, $i = 0, 1, 2$ and $\vec{b}_j = (b_{j1}, b_{j2})$, $j = 1, 2$ be points in $\mathbb{R}^2$ with natural basis $\{\vec{0}, \vec{e}_1, \vec{e}_2\}$. Let $\vec{x} = (x_1, x_2) \in \mathbb{R}^2$. Suppose $\vec{b}_1 \ne \vec{b}_2$. Denote by L the straight line determined by $\vec{b}_1$ and $\vec{b}_2$. See Fig. 2.32.

(a) Show that the equation of L in $\{\vec{0}, \vec{e}_1, \vec{e}_2\}$ is

$$\frac{x_1 - b_{11}}{b_{21} - b_{11}} = \frac{x_2 - b_{12}}{b_{22} - b_{12}}, \quad \text{or}$$

$$\begin{vmatrix} x_1 & x_2 & 1 \\ b_{11} & b_{12} & 1 \\ b_{21} & b_{22} & 1 \end{vmatrix} = 0.$$

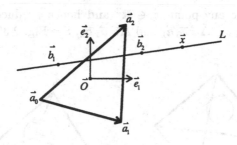

Fig. 2.32

(b) Suppose $\mathcal{B} = \{\vec{a}_0, \vec{a}_1, \vec{a}_2\}$ is an affine basis for $\mathbb{R}^2$. Let $[\vec{x} - \vec{a}_0]_{\mathcal{B}} = (\lambda_1, \lambda_2)$, $[\vec{b}_1 - \vec{a}_0]_{\mathcal{B}} = (\alpha_{11}, \alpha_{12})$ and $[\vec{b}_2 - \vec{a}_0]_{\mathcal{B}} = (\alpha_{21}, \alpha_{22})$ according to (2.6.3). Show that L has equation

$$\frac{\lambda_1 - \alpha_{11}}{\alpha_{21} - \alpha_{11}} = \frac{\lambda_2 - \alpha_{12}}{\alpha_{22} - \alpha_{12}} \quad \text{or}$$

$$\begin{vmatrix} \lambda_1 & \lambda_2 & 1 \\ \alpha_{11} & \alpha_{12} & 1 \\ \alpha_{21} & \alpha_{22} & 1 \end{vmatrix} = 0.$$

(c) Let $(\vec{x})_{\mathcal{B}} = (\lambda_0, \lambda_1, \lambda_2)$, $(\vec{b}_1)_{\mathcal{B}} = (\alpha_{10}, \alpha_{11}, \alpha_{12})$ and $(\vec{b}_2)_{\mathcal{B}} = (\alpha_{20}, \alpha_{21}, \alpha_{22})$ be as in (2.6.2). Show that L has equation

$$\frac{\lambda_0 - \alpha_{10}}{\alpha_{20} - \alpha_{10}} = \frac{\lambda_1 - \alpha_{11}}{\alpha_{21} - \alpha_{11}} = \frac{\lambda_2 - \alpha_{12}}{\alpha_{22} - \alpha_{12}}, \quad \text{or}$$

$$\begin{vmatrix} \lambda_0 & \lambda_1 & \lambda_2 \\ \alpha_{10} & \alpha_{11} & \alpha_{12} \\ \alpha_{20} & \alpha_{21} & \alpha_{22} \end{vmatrix} = 0.$$

(d) Try to deduce equations in (a)–(c) among themselves.

<B>

1. Suppose $\triangle \vec{a}_0 \vec{a}_1 \vec{a}_2$ is a base triangle. Let $\vec{p}$ and $\vec{q}$ be points inside and outside of $\triangle \vec{a}_0 \vec{a}_1 \vec{a}_2$, respectively. Prove that the line segment connecting $\vec{p}$ and $\vec{q}$ will intersect $\triangle \vec{a}_0 \vec{a}_1 \vec{a}_2$ at one point.

2. *Oriented triangle and signed areas as coordinates*
   Suppose $\vec{a}_1$, $\vec{a}_2$ and $\vec{a}_3$ are non-collinear points in $\mathbb{R}^2$. The ordered triples $\{\vec{a}_1, \vec{a}_2, \vec{a}_3\}$ is said to determine an *oriented triangle* denoted by $\triangle \vec{a}_1 \vec{a}_2 \vec{a}_3$ which is also used to represent its *signed area*, considered positive if the ordered triple is in anticlockwise sense and negative

otherwise. Take any point $\vec{x} \in \mathbb{R}^2$ and hence produce three oriented
triangles $\triangle \vec{x}\, \vec{a}_2\, \vec{a}_3$, $\triangle \vec{x}\, \vec{a}_3\, \vec{a}_1$ and $\triangle \vec{x}\, \vec{a}_1\, \vec{a}_2$. See Fig. 2.33.

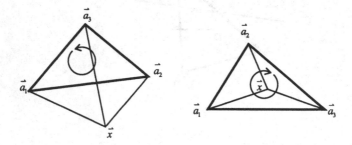

Fig. 2.33

Let $S = \triangle \vec{a}_1\, \vec{a}_2\, \vec{a}_3$, $S_1 = \triangle \vec{x}\, \vec{a}_2\, \vec{a}_3$, $S_2 = \triangle \vec{x}\, \vec{a}_3\, \vec{a}_1$ and $S_3 = \triangle \vec{x}\, \vec{a}_1\, \vec{a}_2$.
Then $S = S_1 + S_2 + S_3$ and the triple (S_1, S_2, S_3) is called the *area
coordinate* of the point $\vec{x}$ with respect to *coordinate* or *base triangle*
$\triangle \vec{a}_1\, \vec{a}_2\, \vec{a}_3$ with $\vec{a}_1, \vec{a}_2$ and $\vec{a}_3$ as *base points* and S_1, S_2, S_3 as its *coordinate components*.

(a) Let $S_1 : S_2 : S_3 = \lambda_1 : \lambda_2 : \lambda_3$ and call $(\lambda_1 : \lambda_2 : \lambda_3)$ a *homogenous
    area* or *barycentric coordinate*. In case $\lambda_1 + \lambda_2 + \lambda_3 = 1$, such λ_1,
    λ_2 and λ_3 are uniquely determined and denote $(\lambda_1 : \lambda_2 : \lambda_3)$ simply
    by $(\lambda_1, \lambda_2, \lambda_3)$ as usual and call the coordinate *normalized*. Given
    area coordinate (S_1, S_2, S_3) with $S = S_1 + S_2 + S_3$, then $(\lambda S_1 :
    \lambda S_2 : \lambda(S - S_1 - S_2))$ is a barycentric coordinate for any scalar
    $\lambda \neq 0$. Conversely, given barycentric coordinate $(\lambda_1 : \lambda_2 : \lambda_3)$ with
    $\lambda_1 + \lambda_2 + \lambda_3 \neq 0$,
    $$\left(\frac{\lambda_1 S}{\lambda_1 + \lambda_2 + \lambda_3}, \frac{\lambda_2 S}{\lambda_1 + \lambda_2 + \lambda_3}, \frac{\lambda_3 S}{\lambda_1 + \lambda_2 + \lambda_3} \right)$$
    is the corresponding area coordinate. In case $\lambda_1 + \lambda_2 + \lambda_3 = 0$,
    $(\lambda_1 : \lambda_2 : \lambda_3)$ is said to represent an *ideal* or *infinite point* in the realm
    of projective geometry (see Ex. <B> of Sec. 2.8.5 and Sec. 3.8.4).
(b) Owing to the fact that $S = S_1 + S_2 + S_3$, the third quantity is
    uniquely determined once two of S_1, S_2 and S_3 have been decided.
    Therefore, one may use (S_1, S_2) or $\left(\frac{S_1}{S}, \frac{S_2}{S} \right)$ to represent the point
    $\vec{x}$ and is called the *affine coordinate* of $\vec{x}$ with respect to *affine basis*
    $\{\vec{a}_3, \vec{a}_1, \vec{a}_2\}$ or $\{\vec{a}_1 - \vec{a}_3, \vec{a}_2 - \vec{a}_3\}$ with $\vec{a}_3$ as a *base point* or *vertex* (see
    Fig. 2.34(a)). In case the line segments $\vec{a}_3\, \vec{a}_1$ and $\vec{a}_3\, \vec{a}_2$ have equal
    length, say one unit and the angle $\angle \vec{a}_1\, \vec{a}_3\, \vec{a}_2 = 90°$, then $\{\vec{a}_3, \vec{a}_1, \vec{a}_2\}$

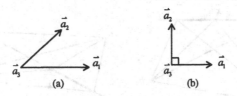

Fig. 2.34

is called an *orthogonal* or *rectangular* or *Cartesian coordinate system* (see Fig. 2.34(b), compare with Fig. 2.17). For generalization in $\mathbb{R}^n$, see Sec. 5.9.5.

(c) Suppose the vertices $\vec{a}_1, \vec{a}_2$ and $\vec{a}_3$ of the coordinate triangle $\triangle \vec{a}_1 \vec{a}_2 \vec{a}_3$ have respective coordinates

$$(x_1, y_1), \quad (x_2, y_2) \text{ and } (x_3, y_3)$$

in a certain orthogonal or affine coordinate system. Let the point $\vec{x} = (x, y)$ have barycentric coordinate $(\lambda_1 : \lambda_2 : \lambda_3)$. Show that

$$x = \frac{\lambda_1 x_1 + \lambda_2 x_2 + \lambda_3 x_3}{\lambda_1 + \lambda_2 + \lambda_3}, \quad y = \frac{\lambda_1 y_1 + \lambda_2 y_2 + \lambda_3 y_3}{\lambda_1 + \lambda_2 + \lambda_3}.$$

Then, compare these relations with those stated in (2.6.1) and Ex. <A> 2 both in content and in proof.

3. Suppose $\vec{x}_1$ and $\vec{x}_2$ have respective normalized barycentric coordinates $(\lambda_{11}, \lambda_{12}, \lambda_{13})$ and $(\lambda_{21}, \lambda_{22}, \lambda_{23})$. A point $\vec{x}$ on the line $\vec{x}_1 \vec{x}_2$ divides the segment $\vec{x}_1 \vec{x}_2$ into the ratio $\vec{x}_1 \vec{x} : \vec{x} \vec{x}_2 = k$ which is positive if $\vec{x}$ is an interior point of $\vec{x}_1 \vec{x}_2$ and negative if exterior. Show that $\vec{x}$ has normalized barycentric coordinate

$$\left(\frac{\lambda_{11} + k\lambda_{21}}{1+k}, \frac{\lambda_{12} + k\lambda_{22}}{1+k}, \frac{\lambda_{13} + k\lambda_{23}}{1+k} \right)$$

by the following two methods.

(a) Use (2.5.10).

(b) Let $\vec{a}$ and $\vec{b}$ be two points other than $\vec{x}_1$, $\vec{x}_2$ and $\vec{x}$. Show that

$$\triangle \vec{x} \vec{a} \vec{b} = \frac{1}{1+k} \triangle \vec{x}_1 \vec{a} \vec{b} + \frac{k}{1+k} \triangle \vec{x}_2 \vec{a} \vec{b}.$$

See Fig. 2.35.

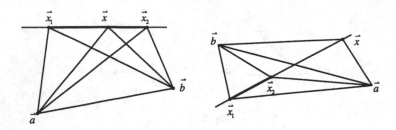

Fig. 2.35

4. Fix a coordinate triangle $\triangle \vec{a}_1 \vec{a}_2 \vec{a}_3$. Show that the equation of the line passing through the point $\vec{a}$ with (a_1, a_2, a_3) and the point $\vec{b}$ with (b_1, b_2, b_3) is

$$\begin{vmatrix} \lambda_1 & \lambda_2 & \lambda_3 \\ a_1 & a_2 & a_3 \\ b_1 & b_2 & b_3 \end{vmatrix} = 0, \quad \text{or}$$

$$\begin{vmatrix} a_2 & a_3 \\ b_2 & b_3 \end{vmatrix} \lambda_1 + \begin{vmatrix} a_3 & a_1 \\ b_3 & b_1 \end{vmatrix} \lambda_2 + \begin{vmatrix} a_1 & a_2 \\ b_1 & b_2 \end{vmatrix} \lambda_3 = 0,$$

where the coefficients satisfy

$$\begin{vmatrix} a_2 & a_3 \\ b_2 & b_3 \end{vmatrix} : \begin{vmatrix} a_3 & a_1 \\ b_3 & b_1 \end{vmatrix} : \begin{vmatrix} a_1 & a_2 \\ b_1 & b_2 \end{vmatrix} = h_1 : h_2 : h_3$$

with h_1, h_2 and h_3 the respective *signed distances* from the vertices $\vec{a}_1$, $\vec{a}_2$ and $\vec{a}_3$ to the line. Hence, designate h_i and h_j to have the same sign if both $\vec{a}_i$ and $\vec{a}_j$ lie on the same side of the line, the opposite sign if on the opposite sides (see Fig. 2.36). Use the following two methods.

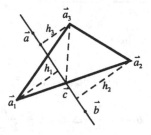

Fig. 2.36

(a) Try to use Ex. 3 and note that, in Fig. 2.36,

$$\frac{\begin{vmatrix} a_3 & a_1 \\ b_3 & b_1 \end{vmatrix}}{\begin{vmatrix} a_2 & a_3 \\ b_2 & b_3 \end{vmatrix}} = -\frac{\Delta \vec{c}\,\vec{a}_2\,\vec{a}_3}{\Delta \vec{a}_1\,\vec{c}\,\vec{a}_3} = \frac{\vec{a}_2\,\vec{c}}{\vec{a}_1\,\vec{c}} = \frac{h_1}{h_2}, \text{ etc.}$$

(b) Treat the line as the y-axis in a rectangular coordinate system (see Fig. 2.37). Then, a point $\vec{x}$ on the line has its abscissa (via Ex. 2(c))

$$x = \frac{\lambda_1 x_1 + \lambda_2 x_2 + \lambda_3 x_3}{\lambda_1 + \lambda_2 + \lambda_3} = 0$$

with $h_1 = x_1$, $h_2 = x_2$ and $h_3 = x_3$.

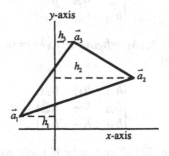

Fig. 2.37

Furthermore, deduce the fact that the equation of the line in an affine coordinate system is

$$\begin{vmatrix} x_1 & x_2 & 1 \\ a_1 & a_2 & 1 \\ b_1 & b_2 & 1 \end{vmatrix} = 0,$$

where $x_1 = \lambda_1$, $x_2 = \lambda_2$. For concrete examples, please see Ex. 6.

5. Fix a coordinate triangle $\Delta \vec{a}_1\,\vec{a}_2\,\vec{a}_3$.

   (a) Show that the point of intersection of the two lines

   $$h_{11}\lambda_1 + h_{12}\lambda_2 + h_{13}\lambda_3 = 0,$$
   $$h_{21}\lambda_1 + h_{22}\lambda_2 + h_{23}\lambda_3 = 0,$$

   has the barycentric coordinate

   $$\left(\begin{vmatrix} h_{12} & h_{13} \\ h_{22} & h_{23} \end{vmatrix} : \begin{vmatrix} h_{13} & h_{11} \\ h_{23} & h_{21} \end{vmatrix} : \begin{vmatrix} h_{11} & h_{12} \\ h_{21} & h_{22} \end{vmatrix} \right).$$

(b) Show that the line determined by the points (h_{11}, h_{12}, h_{13}) and (h_{21}, h_{22}, h_{23}) has the equation

$$\begin{vmatrix} \lambda_1 & \lambda_2 & \lambda_3 \\ h_{11} & h_{12} & h_{13} \\ h_{21} & h_{22} & h_{23} \end{vmatrix} = 0, \quad \text{or}$$

$$\begin{vmatrix} h_{12} & h_{13} \\ h_{22} & h_{23} \end{vmatrix} \lambda_1 + \begin{vmatrix} h_{13} & h_{11} \\ h_{23} & h_{21} \end{vmatrix} \lambda_2 + \begin{vmatrix} h_{11} & h_{12} \\ h_{21} & h_{22} \end{vmatrix} \lambda_3 = 0.$$

(c) Three points (h_{i1}, h_{i2}, h_{i3}), $i = 1, 2, 3$ are collinear if and only if

$$\begin{vmatrix} h_{11} & h_{12} & h_{13} \\ h_{21} & h_{22} & h_{23} \\ h_{31} & h_{32} & h_{33} \end{vmatrix} = 0.$$

While, three lines $h_{i1}\lambda_1 + h_{i2}\lambda_2 + h_{i3}\lambda_3 = 0$, $i = 1, 2, 3$, are concurrent if and only if

$$\begin{vmatrix} h_{11} & h_{12} & h_{13} \\ h_{21} & h_{22} & h_{23} \\ h_{31} & h_{32} & h_{33} \end{vmatrix} = 0.$$

(d) Two lines are parallel if and only if their equations in barycentric coordinates are

$$h_1\lambda_1 + h_2\lambda_2 + h_3\lambda_3 = 0,$$
$$(h_0 + h_1)\lambda_1 + (h_0 + h_2)\lambda_2 + (h_0 + h_3)\lambda_3 = 0,$$

where h_0 is a constant.

6. Give a coordinate triangle $\triangle \vec{a}_1 \vec{a}_2 \vec{a}_3$ and use $|\vec{a}_1 \vec{a}_2| = |\vec{a}_1 - \vec{a}_2|$ to denote the length of the side $\vec{a}_1 \vec{a}_2$, etc. See Fig. 2.38.

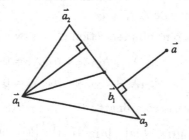

Fig. 2.38

(a) Show that the three sides have the respective equations:

$$\vec{a}_1\,\vec{a}_2: \lambda_3 = 0,$$
$$\vec{a}_2\,\vec{a}_3: \lambda_1 = 0,$$
$$\vec{a}_3\,\vec{a}_1: \lambda_2 = 0.$$

(b) The three medians have the equations:

$$\text{median on } \vec{a}_1\,\vec{a}_2: \lambda_1 - \lambda_2 = 0,$$
$$\text{median on } \vec{a}_2\,\vec{a}_3: \lambda_2 - \lambda_3 = 0,$$
$$\text{median on } \vec{a}_3\,\vec{a}_1: \lambda_3 - \lambda_1 = 0.$$

(c) The three altitudes have the equations:

$$\text{altitude on } \vec{a}_1\,\vec{a}_2: |\vec{a}_3\,\vec{a}_1| \cos\angle\vec{a}_1 \cdot \lambda_1 - |\vec{a}_2\,\vec{a}_3| \cos\angle\vec{a}_2 \cdot \lambda_2 = 0,$$
$$\text{altitude on } \vec{a}_2\,\vec{a}_3: |\vec{a}_1\,\vec{a}_2| \cos\angle\vec{a}_2 \cdot \lambda_2 - |\vec{a}_3\,\vec{a}_1| \cos\angle\vec{a}_3 \cdot \lambda_3 = 0,$$
$$\text{altitude on } \vec{a}_3\,\vec{a}_1: |\vec{a}_2\,\vec{a}_3| \cos\angle\vec{a}_3 \cdot \lambda_3 - |\vec{a}_1\,\vec{a}_2| \cos\angle\vec{a}_1 \cdot \lambda_1 = 0.$$

(d) From the given point $\vec{a}$ with $(\alpha_1 : \alpha_2 : \alpha_3)$ draw three perpendicular lines $\vec{a}\,\vec{b}_1$, $\vec{a}\,\vec{b}_2$ and $\vec{a}\,\vec{b}_3$ to the three sides $\vec{a}_2\,\vec{a}_3$, $\vec{a}_3\,\vec{a}_1$ and $\vec{a}_1\,\vec{a}_2$ respectively. Show that $\vec{a}\,\vec{b}_1$ has equation

$$|\vec{a}_1\,\vec{a}_2| \cos\angle\vec{a}_2 \cdot \lambda_2 - |\vec{a}_3\,\vec{a}_1| \cos\angle\vec{a}_3 \cdot \lambda_3$$
$$- \frac{|\vec{a}_1\,\vec{a}_2| \cos\angle\vec{a}_2 \cdot \lambda_2 - |\vec{a}_3\,\vec{a}_1| \cos\angle\vec{a}_3 \cdot \lambda_3}{\alpha_1 + \alpha_2 + \alpha_3}(\lambda_1 + \lambda_2 + \lambda_3) = 0.$$

What are equations for $\vec{a}\,\vec{b}_2$ and $\vec{a}\,\vec{b}_3$?

2.7   Linear Transformations (Operators)

As known already in Sec. 1.4, an affine transformation between two lines is characterized as a one-to-one and onto mapping which preserves ratios of signed lengths of two segments. With exactly the same manner, we define affine transformation between planes.

Consider $\mathbb{R}^2$ as an affine plane.

A one-to-one mapping T from $\mathbb{R}^2$ onto $\mathbb{R}^2$ is called an *affine transformation* or *mapping* if T preserves ratios of signed lengths of two segments along the same line or parallel lines. That is, for any points $\vec{x}_1$, $\vec{x}_2$ and $\vec{x}_3$

and scalar $\alpha \in \mathbb{R}$,

$$\vec{x}_3 - \vec{x}_1 = \alpha(\vec{x}_2 - \vec{x}_1)$$
$$\Rightarrow T(\vec{x}_3) - T(\vec{x}_1) = \alpha[T(\vec{x}_2) - T(\vec{x}_1)] \qquad (2.7.1)$$

always holds; in short, $\vec{x}_1\vec{x}_3 = \alpha\vec{x}_1\vec{x}_2$ always implies $T(\vec{x}_1)T(\vec{x}_3) = \alpha T(\vec{x}_1)T(\vec{x}_2)$. See Fig. 2.39.

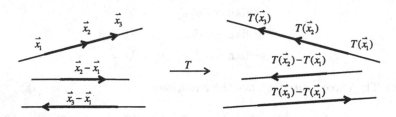

Fig. 2.39

In particular, an affine transformation maps lines onto lines and segments onto segments.

What follows immediately is to describe an affine transformation with linear structure of $\mathbb{R}^2$. For this purpose, fix an arbitrary point $\vec{x}_0 \in \mathbb{R}^2$ and define a mapping $f\colon \mathbb{R}^2 \to \mathbb{R}^2$ by

$$f(\vec{x} - \vec{x}_0) = T(\vec{x}) - T(\vec{x}_0), \quad \vec{x} \in \mathbb{R}^2.$$

This f is one-to-one, onto and $f(\vec{0}) = \vec{0}$ holds. Also, for any scalar $\alpha \in \mathbb{R}$, we have by definition (2.7.1)

$$f(\alpha(\vec{x} - \vec{x}_0)) = \alpha(T(\vec{x}) - T(\vec{x}_0)) = \alpha f(\vec{x} - \vec{x}_0), \quad \vec{x} \in \mathbb{R}^2. \qquad (*_1)$$

Furthermore, take any two other points $\vec{x}_1$ and $\vec{x}_2$ and construct a parallelogram as in Fig. 2.40. Let $\vec{x}_3$ be the middle point of the line segments $\vec{x}_1\vec{x}_2$, i.e. $\vec{x}_3 = \frac{1}{2}(\vec{x}_1 + \vec{x}_2)$. Then $\vec{x}_3$ is also the middle point of the line segment connecting $\vec{x}_0$ and $\vec{x}_1 + \vec{x}_2 - \vec{x}_0$. Now

$$\vec{x}_2 - \vec{x}_1 = 2(\vec{x}_3 - \vec{x}_1) \Rightarrow T(\vec{x}_2) - T(\vec{x}_1) = 2(T(\vec{x}_3) - T(\vec{x}_1));$$
$$(\vec{x}_1 + \vec{x}_2 - \vec{x}_0) - \vec{x}_0 = 2(\vec{x}_3 - \vec{x}_0) \Rightarrow T(\vec{x}_1 + \vec{x}_2 - \vec{x}_0) - T(\vec{x}_0)$$
$$= 2(T(\vec{x}_3) - T(\vec{x}_0)).$$

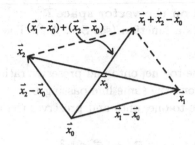

Fig. 2.40

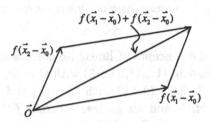

Fig. 2.41

Then

$$f((\vec{x}_1 - \vec{x}_0) + (\vec{x}_2 - \vec{x}_0)) = f((\vec{x}_1 + \vec{x}_2 - \vec{x}_0) - \vec{x}_0)$$
$$= T(\vec{x}_1 + \vec{x}_2 - \vec{x}_0) - T(\vec{x}_0) = 2(T(x_3) - T(\vec{x}_0))$$
$$= 2[(T(\vec{x}_3) - T(\vec{x}_1)) + (T(\vec{x}_1) - T(\vec{x}_0))]$$
$$= T(\vec{x}_2) - T(\vec{x}_1) + T(\vec{x}_1) - T(\vec{x}_0) + (T(\vec{x}_1) - T(\vec{x}_0))$$
$$= (T(\vec{x}_2) - T(\vec{x}_0)) + (T(\vec{x}_1) - T(\vec{x}_0))$$
$$= f(\vec{x}_1 - \vec{x}_0) + f(\vec{x}_2 - \vec{x}_0). \tag{$*_2$}$$

Therefore, the original *affine transformation* can be described by

$$T(\vec{x}) = T(\vec{x}_0) + f(\vec{x} - \vec{x}_0), \quad \vec{x} \in \mathbb{R}^2, \tag{2.7.2}$$

where f has the operational properties $(*_1)$ and $(*_2)$.

In this section, we focus our attention on mappings from $\mathbb{R}^2$ into $\mathbb{R}^2$ having properties $(*_1)$ and $(*_2)$, the so-called linear transformation, and leave the discussion of affine transformation (2.7.1) or (2.7.2) to Sec. 2.8.

In short, an affine transformation (2.7.2) keeping $\vec{0}$ fixed, i.e. $T(\vec{0}) = \vec{0}$, is a linear isomorphism. We summarize the above result as

Linear isomorphism on the vector space $\mathbb{R}^2$

Suppose $f\colon \mathbb{R}^2 \to \mathbb{R}^2$ is a function (see Sec. A.2). Then the following are equivalent.

(1) (geometric) f is one-to-one, onto and preserves ratios of signed lengths of line segments along the same line passing $\vec{0}$.

(2) (algebraic) f is one-to-one, onto and preserves the linear structures of $\mathbb{R}^2$, i.e.

    1. $f(\alpha\vec{x}) = \alpha f(\vec{x})$ for $\alpha \in \mathbb{R}$ and $\vec{x} \in \mathbb{R}^2$,

    2. $f(\vec{x} + \vec{y}) = f(\vec{x}) + f(\vec{y})$ for $\vec{x}, \vec{y} \in \mathbb{R}^2$.

Such a f is called a *linear isomorphism* or *invertible linear operator* on $\mathbb{R}^2$.

$$(2.7.3)$$

We have encountered the concepts of linear isomorphisms on quite a few occasions, for examples: Φ in (1.2.1), (1.3.3) with $O = O'$, and (1.4.1) with $a = 0$ and $b \neq 0$; Φ in (2.3.2), (2.4.2) with $O = O'$, and (2.6.7). *A linear isomorphism f is invertible and its inverse function* $f^{-1}\colon \mathbb{R}^2 \to \mathbb{R}^2$ *is a linear isomorphism too* (refer to Sec. A.2) and

$$f \circ f^{-1} = f^{-1} \circ f = 1_{\mathbb{R}^2} \quad \text{(the identity mapping on } \mathbb{R}^2\text{).}$$

The proof of this easy fact is left to the readers.

    If the "one-to-one" and "onto" conditions are dropped in (2.7.3), the resulting function f is called a linear transformation. Easy examples are touched on Exs. <B> 2–5 in Sec. 2.4. It seems timely now to give a formal definition for it.

Linear transformation

Let V and W be either the vector space $\mathbb{R}$ or $\mathbb{R}^2$. A function $f\colon V \to W$ is called a *linear transformation* if

    1. f preserves scalar multiplication of vectors, i.e.

$$f(\alpha\vec{x}) = \alpha f(\vec{x}) \quad \text{for } \alpha \in \mathbb{R} \text{ and } \vec{x} \in V,$$

    2. f preserves addition of vectors, i.e.

$$f(\vec{x} + \vec{y}) = f(\vec{x}) + f(\vec{y}) \quad \text{for } \vec{x}, \vec{y} \in V.$$

In case $W = V$, f is called a *linear operator* and if $W = \mathbb{R}$, a *linear functional*. In particular, $f(\vec{0}) = \vec{0}$.

$$(2.7.4)$$

Hence, a linear transformation (from V to V) which is both one-to-one and onto is a linear isomorphism. This definition is still suitable for arbitrary vector spaces over the same field (see Secs. B.1 and B.7).

We call the readers attention, once and for all, to the following concepts and notations concerned:

A *vector* (or *linear*) *subspace* (see Sec. B.1), simply called a *subspace*.

Hom(V, W) or L(V, W): the vector space of linear transformations from V to W (see Sec. B.7).

Hom(V, V) or L(V, V): the vector space of linear operators on V.

Ker(f) or N(f): the *kernel*$\{\vec{x} \in V \mid f(\vec{x}) = \vec{0}\}$ of an $f \in$ H(V, W) (see Sec. B.7).

Im(f) or R(f): the *range*$\{f(\vec{x}) \mid \vec{x} \in V\}$ of an $f \in$ H(V, W) (see Sec. B.7).

Note that Ker(f) is a subspace of V and Im(f) a subspace of W. If f is a linear operator on V, a subspace U of V is called *invariant* under f or simply *f-invariant* if

$$f(U) \subseteq U$$

i.e. for any $\vec{x} \in U, f(\vec{x}) \in U$ always holds. For examples,

$$\{\vec{0}\},\ V,\ \text{Ker}(f)\ \text{and Im}(f)$$

are trivial invariant subspaces of V for any linear operator f.

This section is divided into eight subsections.

Section 2.7.1 formulates what a linear operator looks like in the Cartesian coordinate system $\mathcal{N} = \{\vec{e}_1, \vec{e}_2\}$, and then Sec. 2.7.2 presents some basic but important elementary operators with their *eigenvalues* and *eigenvectors*.

Section 2.7.3 discusses various matrix representations of a linear operator related to different bases for $\mathbb{R}^2$ and the relations among them. The *rank* of a linear operator or a matrix is an important topic here.

Some theoretical treatment, independent of particular choice of a basis for $\mathbb{R}^2$, about linear operators will be given in Sec. 2.7.4.

From Sec. 2.7.5 to Sec. 2.7.8, we will investigate various decompositions of a linear operator or matrix.

Geometric mapping properties of elementary operators or matrices are discussed in Sec. 2.7.5. Therefore, algebraically, a square matrix can be expressed as a product of elementary matrices. And hence, geometrically, its mapping behaviors can be tracked.

Section 2.7.6 deepens the important concepts of eigenvalues and eigenvectors introduced in Sec. 2.7.2. If a linear operator or matrix has two distinct eigenvalues, then it is diagonalizable as a diagonal matrix as its canonical form.

In case a linear operator or matrix has two coincident eigenvalues and is not diagonalizable, Sec. 2.7.7 investigates its *Jordan canonical form* in a suitable basis for $\mathbb{R}^2$.

Finally, Sec. 2.7.8 discusses how to get the *rational canonical form* for linear operators or matrices which do not have real eigenvalues.

2.7.1   *Linear operators in the Cartesian coordinate system*

Fix $\mathbb{R}^2$ with the Cartesian coordinate system $\mathcal{N} = \{\vec{e}_1, \vec{e}_2\}$, see Fig. 2.17(b). Here, we will formally do the exercise Ex. <B> 4 of Sec. 2.4.

Let $f \colon \mathbb{R}^2 \to \mathbb{R}^2$ be a linear operator. For any $\vec{x} = (x_1, x_2) \in \mathbb{R}^2$, since

$$\vec{x} = x_1 \vec{e}_1 + x_2 \vec{e}_2 \quad \text{and}$$
$$f(\vec{x}) = x_1 f(\vec{e}_1) + x_2 f(\vec{e}_2),$$

f is *completely determined by the vectors* $f(\vec{e}_1)$ *and* $f(\vec{e}_2)$. Suppose

$$f(\vec{e}_i) = (a_{i1}, a_{i2}) = a_{i1} \vec{e}_1 + a_{i2} \vec{e}_2, \quad i = 1, 2.$$

Then

$$f(\vec{x}) = x_1(a_{11}\vec{e}_1 + a_{12}\vec{e}_2) + x_2(a_{21}\vec{e}_1 + a_{22}\vec{e}_2)$$
$$= (x_1 a_{11} + x_2 a_{21})\vec{e}_1 + (x_1 a_{12} + x_2 a_{22})\vec{e}_2$$
$$= (x_1 a_{11} + x_2 a_{21}, \; x_1 a_{12} + x_2 a_{22})$$
$$= (x_1 \quad x_2) \begin{bmatrix} a_{11} & a_{12} \\ a_{21} & a_{22} \end{bmatrix} = \vec{x}[f]_{\mathcal{N}}, \quad \text{with } [f]_{\mathcal{N}} = \begin{bmatrix} a_{11} & a_{12} \\ a_{21} & a_{22} \end{bmatrix},$$

$$(2.7.5)$$

where the matrix $[f]_{\mathcal{N}} = [a_{ij}]_{2 \times 2}$ is called the *matrix representation* of f with respect to the basis $\mathcal{N}$.

Conversely, for a given real 2×2 matrix $A = [a_{ij}]$, define the mapping $f \colon \mathbb{R}^2 \to \mathbb{R}^2$ by

$$f(\vec{x}) = \vec{x}A, \quad x \in \mathbb{R}^2. \qquad (2.7.6)$$

Here in $\vec{x}A$, we consider the vector $\vec{x} = (x_1, x_2)$ as a 1×2 matrix $[x_1 \quad x_2]$. By $(\alpha\vec{x}_1 + \vec{x}_2)A = (\alpha\vec{x}_1)A + \vec{x}_2 A = \alpha(\vec{x}_1 A) + \vec{x}_2 A$, it follows that such a

f is linear and $[f]_\mathcal{N} = A$. Henceforth, we do not make distinction between f and A in (2.7.6) and adopt the

Convention
For a given real 2×2 matrix $A = [a_{ij}]$, when considered as a linear operator $A: \mathbb{R}^2 \to \mathbb{R}^2$, it is defined as

$$\vec{x} \to \vec{x}A \quad \text{in } \mathcal{N} = \{\vec{e}_1, \vec{e}_2\}. \tag{2.7.7}$$

Therefore, a matrix A is an algebraic symbol as well as a linear operator on $\mathbb{R}^2$ if necessary. It is this point of view that matrices play the core of the theory of linear algebra on finite dimensional vector spaces.

It is well-known that $\{\vec{0}\}$ is the only zero-dimensional subspace of $\mathbb{R}^2$, $\mathbb{R}^2$ is the only two-dimensional subspace while one-dimensional subspaces are nothing but straight lines $ax_1 + bx_2 = 0$ passing $(0, 0)$ or $\langle\langle\vec{x}\rangle\rangle$ generated by nonzero vectors $\vec{x}$. See Exs. <A> 3–5 of Sec. 2.3.

For a nonzero matrix $A = [a_{ij}]_{2 \times 2}$, considered as a linear operator on $\mathbb{R}^2$, then

$\vec{x} = (x_1, x_2) \in \text{Ker}(A)$, the kernel space of A
$\Leftrightarrow \vec{x}A = 0$
$\Leftrightarrow a_{11}x_1 + a_{21}x_2 = 0$
$\quad a_{12}x_1 + a_{22}x_2 = 0$
$\Leftrightarrow$ (see Sec. 2.5) $\vec{x}$ lies on the lines $\vec{x} = t(-a_{21}, a_{11})$ and $\vec{x} = t(-a_{22}, a_{12})$.

In case $(a_{11}, a_{21}) = (0, 0) = \vec{0}$, $a_{11}x_1 + a_{22}x_2 = 0$ is satisfied by all vectors $\vec{x} \in \mathbb{R}^2$ and should be considered as the equation for the whole space $\mathbb{R}^2$. Suppose $(-a_{21}, a_{11}) \neq \vec{0}$ and $(-a_{22}, a_{12}) \neq \vec{0}$. According to (2.5.9),

The lines $\vec{x} = t(-a_{21}, a_{11})$ and $\vec{x} = t(-a_{22}, a_{12})$ coincide.
$\Leftrightarrow$ The direction vectors $(-a_{21}, a_{11})$ and $(-a_{22}, a_{12})$ are linearly dependent.
$\Leftrightarrow \dfrac{-a_{21}}{-a_{22}} = \dfrac{a_{11}}{a_{12}}$ or $\dfrac{a_{11}}{a_{21}} = \dfrac{a_{12}}{a_{22}}$ or

$\det A = \det \begin{bmatrix} a_{11} & a_{12} \\ a_{21} & a_{22} \end{bmatrix} = \begin{vmatrix} a_{11} & a_{12} \\ a_{21} & a_{22} \end{vmatrix} = a_{11}a_{22} - a_{12}a_{21} = 0.$

In this case, the kernel space $\mathrm{Ker}(A)$ is the one-dimensional subspace, say $a_{11}x_1 + a_{21}x_2 = 0$. On the other hand,

The lines $\vec{x} = t(-a_{21}, a_{11})$ and $\vec{x} = t(-a_{22}, a_{12})$ intersect at $\vec{0}$.

$$\Leftrightarrow \frac{-a_{21}}{-a_{22}} \neq \frac{a_{11}}{a_{12}} \quad \text{or} \quad \frac{a_{11}}{a_{21}} \neq \frac{a_{12}}{a_{22}} \quad \text{or}$$

$$\det A = \begin{vmatrix} a_{11} & a_{12} \\ a_{21} & a_{22} \end{vmatrix} = a_{11}a_{22} - a_{12}a_{21} \neq 0.$$

In this case, $\mathrm{Ker}(A) = \{\vec{0}\}$ holds.

What are the corresponding range spaces of A?
Suppose $\mathrm{Ker}(A)$ is $a_{11}x_1 + a_{21}x_2 = 0$. Then

$$\vec{y} = (y_1, y_2) \in \mathrm{Im}(A), \text{ the range space of } A$$
$$\Leftrightarrow y_1 = a_{11}x_1 + a_{21}x_2$$
$$y_2 = a_{12}x_1 + a_{22}x_2 \quad \text{for some } \vec{x} = (x_1, x_2) \in \mathbb{R}^2.$$
$$\Leftrightarrow \left(\text{remember that } \frac{a_{12}}{a_{11}} = \frac{a_{22}}{a_{21}} = \lambda \right)$$
$$\vec{y} = (a_{11}x_1 + a_{21}x_2)(1, \lambda) \quad \text{for } \vec{x} = (x_1, x_2) \in \mathbb{R}^2.$$

This means that the range space $\mathrm{Im}(A)$ is a straight line passing $\vec{0}$.
If $\mathrm{Ker}(A) = \{\vec{0}\}$, then

$$\vec{y} = (y_1, y_2) \in \mathrm{Im}(A)$$
$$\Leftrightarrow y_1 = a_{11}x_1 + a_{21}x_2$$
$$y_2 = a_{12}x_1 + a_{22}x_2 \quad \text{for some } \vec{x} = (x_1, x_2) \in \mathbb{R}^2.$$
$$\Leftrightarrow (\text{solve simultaneous equations with } x_1 \text{ and } x_2 \text{ as unknowns})$$
$$x_1 = \frac{1}{\det A}(a_{22}y_1 - a_{21}y_2) = \frac{1}{\det A} \begin{vmatrix} y_1 & y_2 \\ a_{21} & a_{22} \end{vmatrix},$$
$$x_2 = \frac{1}{\det A}(-a_{12}y_1 + a_{11}y_2) = \frac{1}{\det A} \begin{vmatrix} a_{11} & a_{12} \\ y_1 & y_2 \end{vmatrix}.$$

This is the *Cramer rule* (formulas).
Therefore, the range space $\mathrm{Im}(A) = \mathbb{R}^2$.
We summarize as

The kernel and the range of a linear operator
Let $A = [a_{ij}]_{2\times 2}$ be a real matrix, considered as a linear operator on $\mathbb{R}^2$ (see (2.7.7)).

(1) $\mathrm{Ker}(A)$ and $\mathrm{Im}(A)$ are subspaces of $\mathbb{R}^2$ and the *dimension theorem* is

$$\dim \mathrm{Ker}(A) + \dim \mathrm{Im}(A) = \dim \mathbb{R}^2 = 2,$$

where dim Ker(A) is called the *nullity* of A and dim Im(A) the *rank* of A (see (2.7.44)), particularly denoted as r(A).

(2) If Ker(A) = $\mathbb{R}^2$, then $A = O_{2\times2}$ is the zero matrix or zero linear operator and Im(A) = $\{\vec{0}\}$. Note the rank r(O) = 0.

(3) The following are equivalent.

 1. The rank r(A) = 1.
 2. $A \neq O_{2\times2}$, det $A = 0$.
 3. $A \neq O_{2\times2}$, the row (or column) vectors are linearly dependent.
 4. The nullity dim Ker(A) = $2 - 1 = 1 < 2$.

Both Ker(A) and Im(A) are straight lines passing $\vec{0}$. Suppose Ker(A) = $\langle\langle\vec{x}_1\rangle\rangle$ and take any vector $\vec{x}_2 \in \mathbb{R}^2$ such that $\mathbb{R}^2 = \langle\langle\vec{x}_1\rangle\rangle \oplus \langle\langle\vec{x}_2\rangle\rangle$, then

$$Im(A) = \langle\langle\vec{x}_2 A\rangle\rangle.$$

Hence, the *restriction operator* $A|_{\langle\langle\vec{x}_2\rangle\rangle}: \langle\langle\vec{x}_2\rangle\rangle \to \langle\langle\vec{x}_2 A\rangle\rangle$ is one-to-one and onto (see Sec. A.2).

(4) The following are equivalent (see Exs. <A> 2 and <B> 4 of Sec. 2.4).

 1. Ker(A) = $\{\vec{0}\}$, i.e. $\vec{x}A = \vec{0}$ if and only if $\vec{x} = \vec{0}$.
 2. A is one-to-one.
 3. A is onto, i.e. Im(A) = $\mathbb{R}^2$.
 4. The rank r(A) = 2.
 5. Two row vectors of A are linearly independent. So are the column vectors.
 6. det $A \neq 0$.
 7. A is invertible and hence is a linear isomorphism.
 8. A maps every or a basis $\{\vec{x}_1, \vec{x}_2\}$ for $\mathbb{R}^2$ onto a basis $\{\vec{x}_1 A, \vec{x}_2 A\}$ for $\mathbb{R}^2$.

For a given $\vec{y} \in \mathbb{R}^2$, the equation $\vec{x}A = \vec{y}$ has the unique solution $\vec{x} = \vec{y}A^{-1}$, where

$$A^{-1} = \frac{1}{\det A}\begin{bmatrix} a_{22} & -a_{12} \\ -a_{21} & a_{11} \end{bmatrix}. \tag{2.7.8}$$

These results provide us insights needed to construct various linear operators with different mapping properties.

For reminiscence and reference, we combine (2.5.9)–(2.5.12), Ex. <B> 5 of Sec. 2.5, (2.7.3) and (2.7.8) together to obtain

The general geometric mapping properties of a linear operator
Suppose $A = [a_{ij}]_{2\times 2}$ is a nonzero real matrix.

(1) Suppose r(A) = 2. Hence $A: \mathbb{R}^2 \to \mathbb{R}^2$, as a linear isomorphism, preserves:

1. The relative positions (i.e. coincidence, parallelism and intersection) of two straight lines.
2. Line segment (interior points to interior points).
3. The ratio of signed lengths of line segments along the same or parallel lines (in particular, midpoint to midpoint).
4. Triangle (interior points to interior points).
5. Parallelogram (interior points to interior points).
6. Bounded set (i.e. a set that lies inside the interior of a triangle).
7. The ratio of areas.

It does not necessarily preserve:

(a) length,
(b) angle,
(c) area, and
(d) directions(anticlockwise or clockwise) (see Ex. <A> 4 of Sec. 2.8.1 for formal definition) or orientations.

(2) Suppose r(A) = 1. Suppose Ker(A) = $\langle\langle \vec{x}_1 \rangle\rangle$ and take any $\vec{x}_2 \in \mathbb{R}^2$ such that $\mathbb{R}^2 = \langle\langle \vec{x}_1 \rangle\rangle \oplus \langle\langle \vec{x}_2 \rangle\rangle$, then Im($A$) = $\langle\langle \vec{x}_2 A \rangle\rangle$. In particular,

1. For any $\vec{v} \in$ Ker(A), A maps the straight line $\vec{v} + \langle\langle \vec{x}_2 \rangle\rangle$ one-to-one and onto $\langle\langle \vec{x}_2 A \rangle\rangle$ and hence preserves ratio of signed lengths of segments along such parallel lines.
2. For any $\vec{x}_0 \in \mathbb{R}^2$, A maps the whole line $\vec{x}_0 + $ Ker(A) into a single point $\vec{x}_0 A$.   (2.7.9)

See Fig. 2.42.

Readers can easily prove these results. So the proofs are left to the readers. If any difficulty arises, you just accept these facts or turn to Sec. 2.8.3 for detailed proofs. They are also true for affine transformations (see (2.7.1) and (2.7.2)) and are called *affine invariants*.

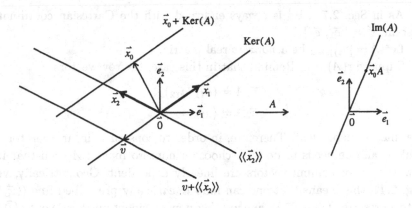

Fig. 2.42

Exercises

\<A\>

1. Prove (2.7.9) in detail, if possible.
2. Endow $\mathbb{R}$ with natural basis $\mathcal{N}_0 = \{1\}$ and $\mathbb{R}^2$ with $\mathcal{N} = \{\vec{e}_1, \vec{e}_2\}$.

   (a) Try to formulate all linear transformations from $\mathbb{R}$ to $\mathbb{R}^2$ in terms of $\mathcal{N}_0$ and $\mathcal{N}$.
   (b) Same as (a) for linear transformations from $\mathbb{R}^2$ to $\mathbb{R}$.

3. Which properties stated in (2.7.8) and (2.7.9) are still true for linear transformations from $\mathbb{R}$ to $\mathbb{R}^2$? State and prove them.
4. Do the same problem as in 3 for linear transformations from $\mathbb{R}^2$ to $\mathbb{R}$.

\<C\> Abstraction and generalization

Read Secs. B.4, B.6 and B.7 if necessary, and try to do the following problems.

1. Extend (2.7.8) and (2.7.9) to linear operators on $\mathbb{C}$ or $\mathbb{C}^2$.
2. Extend (2.7.8) and (2.7.9) to linear transformation from $\mathbb{F}^m$ to $\mathbb{F}^n$.
3. Extend (2.7.8) and (2.7.9) to linear transformation from a finite-dimensional vector space V to a vector space W.

2.7.2 *Examples*

This subsection will be concentrated on constructing some basic linear operators, and we will emphasize their geometric mapping properties.

As in Sec. 2.7.1, $\mathbb{R}^2$ is always endowed with the Cartesian coordinate system $\mathcal{N} = \{\vec{e}_1, \vec{e}_2\}$.

Let $A = [a_{ij}]_{2\times 2}$ be a *nonzero* real matrix.

Suppose $r(A) = 1$. Remind that in this case, the row vectors

$$\vec{e}_1 A = (a_{11}, a_{12}),$$
$$\vec{e}_2 A = (a_{21}, a_{22})$$

are linearly dependent. Therefore, in order to construct linear operator of rank 1, all one needs to do is to choose a nonzero matrix A such that its row vectors or column vectors are linearly dependent. Geometrically, via Fig. 2.42, this means that one can map *linearly* any prescribed line $\langle\langle \vec{x}_1 \rangle\rangle$ onto a prescribed line $\langle\langle \vec{v} \rangle\rangle$, and another noncoincident line $\langle\langle \vec{x}_2 \rangle\rangle$ onto $\{\vec{0}\}$.

Here, we list some standard operators of this feature.

Example 1 The linear operator

$$A = \begin{bmatrix} \lambda & 0 \\ 0 & 0 \end{bmatrix} \quad \text{with } \lambda \neq 0$$

has the properties:

1. $\vec{e}_1 A = \lambda \vec{e}_1$ and $\vec{e}_2 A = \vec{0} = 0 \cdot \vec{e}_2$. λ and 0 are called *eigenvalues* of A with corresponding *eigenvectors* $\vec{e}_1$ and $\vec{e}_2$, respectively.
2. $\operatorname{Ker}(A) = \langle\langle \vec{e}_2 \rangle\rangle$, while $\operatorname{Im}(A) = \langle\langle \vec{e}_1 \rangle\rangle$ is an *invariant* line (subspace) of A, i.e. A maps $\langle\langle \vec{e}_1 \rangle\rangle$ into itself.
3. A maps every line $\vec{x}_0 + \langle\langle \vec{x} \rangle\rangle$, where $\vec{x}$ is linearly independent of $\vec{e}_2$, one-to-one and onto the line $\langle\langle \vec{e}_1 \rangle\rangle$.

See Fig. 2.43(a) and (b). In Fig. 2.43(b), we put the two plane coincide and the arrow signs indicate how A preserves ratios of signed lengths of segments.

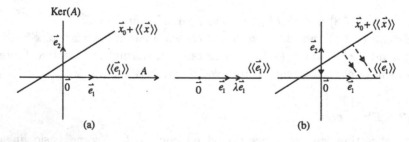

Fig. 2.43

Operators

$$\begin{bmatrix} 0 & \lambda \\ 0 & 0 \end{bmatrix}, \quad \begin{bmatrix} 0 & 0 \\ \lambda & 0 \end{bmatrix} \quad \text{and} \quad \begin{bmatrix} 0 & 0 \\ 0 & \lambda \end{bmatrix}, \quad \text{where } \lambda \neq 0$$

are all of this type.

Example 2 The linear operator

$$A = \begin{bmatrix} a & b \\ 0 & 0 \end{bmatrix} \quad \text{with } ab \neq 0,$$

which is equivalent to $(x_1, x_2)A = (ax_1, bx_1)$, has the properties:

1. Let $\vec{v} = (a, b)$. Then

$$\vec{v}A = a\vec{v} \quad \text{and} \quad \vec{e}_2 A = \vec{0} = 0 \cdot \vec{e}_2.$$

   a and 0 are called *eigenvalues* of A corresponding to *eigenvectors* $\vec{v}$ and $\vec{e}_2$.

2. $\text{Ker}(A) = \langle\langle \vec{e}_2 \rangle\rangle$, while $\text{Im}(A) = \langle\langle \vec{v} \rangle\rangle$ is an *invariant* line (subspace).

3. For any vector $\vec{x}$, linear independent of $\vec{e}_2$, A maps the line $\vec{x}_0 + \langle\langle \vec{x} \rangle\rangle$ one-to-one and onto the line $\langle\langle \vec{v} \rangle\rangle$.

See Fig. 2.44(a) and (b).

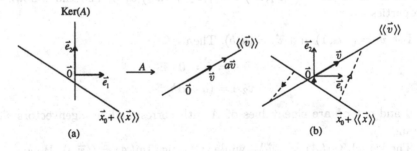

Fig. 2.44

Operators

$$\begin{bmatrix} a & 0 \\ b & 0 \end{bmatrix}, \quad \begin{bmatrix} 0 & b \\ 0 & a \end{bmatrix} \quad \text{and} \quad \begin{bmatrix} 0 & 0 \\ a & b \end{bmatrix}, \quad \text{where } ab \neq 0$$

are all of this type.

Suppose $ab\alpha \neq 0$. Consider the linear operator

$$f(x_1, x_2) = (a(x_1 + \alpha x_2), b(x_1 + \alpha x_2)) = (x_1 + \alpha x_2)(a, b)$$

and see if it has any invariant subspace. Firstly,

$$f(x_1, x_2) = \vec{0}$$
$$\Leftrightarrow x_1 + \alpha x_2 = 0.$$

Hence, the kernel space $\mathrm{Ker}(f)$ is the line $x_1 + \alpha x_2 = 0$. Suppose $\vec{x} = (x_1, x_2)$ does not lie on $x_1 + \alpha x_2 = 0$ and $f(\vec{x}) = \lambda \vec{x}$ for some scalar λ, then

$$f(\vec{x}) = \lambda \vec{x}$$
$$\Leftrightarrow (x_1 + \alpha x_2)(a, b) = \lambda(x_1, x_2)$$
$$\Leftrightarrow bx_1 - ax_2 = 0 \ (\text{and } \lambda = a + \alpha b).$$

This means that the operator f keeps the line $bx_1 - ax_2 = 0$ invariant, and maps a point $\vec{x} = (x_1, x_2)$ on it into another point $f(\vec{x}) = (a + \alpha b)\vec{x}$, still on that line. Summarize as part of

Example 3 The operator

$$A = \begin{bmatrix} a & b \\ \alpha a & \alpha b \end{bmatrix}, \quad \text{where } \alpha\, ab \neq 0,$$

which is equivalent to $(x_1, x_2)A = (x_1 + \alpha x_2)(a, b)$, has the following properties:

1. Let $\vec{v}_1 = (-\alpha, 1)$ and $\vec{v}_2 = (a, b)$. Then

$$\vec{v}_1 A = \vec{0} = 0 \cdot \vec{v}_1$$
$$\vec{v}_2 A = (a + \alpha b)\vec{v}_2.$$

0 and $a + \alpha b$ are eigenvalues of A with corresponding eigenvectors $\vec{v}_1$ and $\vec{v}_2$.

2. The kernel $\mathrm{Ker}(A) = \langle\langle \vec{v}_1 \rangle\rangle$, while the range $\mathrm{Im}(A) = \langle\langle \vec{v}_2 \rangle\rangle$. Hence

$$\mathrm{Ker}(A) = \mathrm{Im}(A)$$
$$\Leftrightarrow \vec{v}_1 = (-\alpha, 1) \text{ and } \vec{v}_2 = (a, b) \text{ are linearly dependent.}$$
$$\Leftrightarrow a + \alpha b = 0.$$

In this case, A does not have invariant lines. If $\mathrm{Ker}(A) \neq \mathrm{Im}(A)$, then $\mathbb{R}^2 = \mathrm{Ker}(A) \oplus \mathrm{Im}(A)$ and $\mathrm{Im}(A)$ is an invariant line (subspace) of A.

3. For any vector $\vec{x} \in \mathbb{R}^2$ which is linearly independent of $\vec{v}_1, A$ maps the straight line $\vec{x}_0 + \langle\langle \vec{x} \rangle\rangle$ one-to-one and onto the line Im(A).

See Fig. 2.45.

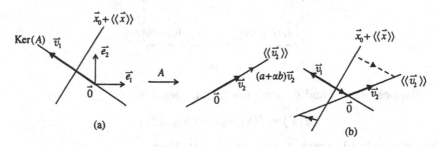

Fig. 2.45

From now on, suppose the rank r(A) = 2.

Example 4 The operator

$$A = \begin{bmatrix} \lambda_1 & 0 \\ 0 & \lambda_2 \end{bmatrix}, \quad \text{where } \lambda_1\lambda_2 \neq 0$$

or $\vec{x}A = (x_1, x_2)A = (\lambda_1 x_1, \lambda_2 x_2)$ has the properties:

1. $\vec{e}_1 A = \lambda_1 \vec{e}_1$ and $\vec{e}_2 A = \lambda_2 \vec{e}_2$. λ_1 and λ_2 are eigenvalues of A with corresponding eigenvectors $\vec{e}_1$ and $\vec{e}_2$.
2. In case $\lambda_1 = \lambda_2 = \lambda$, then $A = \lambda I_2$ and is called an *enlargement* with scale λ. Consequently, A keeps every line passing $\vec{0}$ invariant and has only $\vec{0}$ as an *invariant point*, if $\lambda \neq 1$.
3. In case $\lambda_1 \neq \lambda_2$, A has two invariant lines (subspaces) $\langle\langle \vec{e}_1 \rangle\rangle$ and $\langle\langle \vec{e}_2 \rangle\rangle$, and is called a *stretching*.
4. Stretching keeps the following invariants:

   a. *Length* if $\lambda_1 = \lambda_2 = 1$.
   b. *Angle* if $\lambda_1 = \lambda_2$.
   c. *Area* if $\lambda_1\lambda_2 = 1$.
   d. *Sense (direction)* if $\lambda_1\lambda_2 > 0$.

See Fig. 2.46.

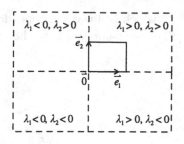

Fig. 2.46

Suppose $ab \neq 0$ and consider the linear operator

$$f(\vec{x}) = f(x_1, x_2) = (bx_2, ax_1).$$

No point is fixed except $\vec{0}$. Suppose $\vec{x} \neq \vec{0}$, then

$\langle\langle\vec{x}\rangle\rangle$ is an invariant line (subspace).

$\Leftrightarrow$ There exists a constant $\lambda \neq 0$ such that $f(\vec{x}) = \lambda\vec{x}$, or

$$bx_2 = \lambda x_1,$$
$$ax_1 = \lambda x_2.$$

In this case, $x_1 \neq 0$ and $x_2 \neq 0$ hold and consequently, $\lambda^2 = ab$. Therefore, in case $ab < 0$, there does not exist any invariant line. Now, suppose $ab > 0$. Then $\lambda = \pm\sqrt{ab}$ and there exists two invariant lines $\sqrt{|a|}x_1 = \pm\sqrt{|b|}x_2$.

In order to get insight into the case $ab < 0$, let us consider firstly a special case: $a = 1$ and $b = -1$. Note that

$$\vec{x} = (x_1, x_2) \xrightarrow[\begin{bmatrix} 0 & 1 \\ 1 & 0 \end{bmatrix}]{} (x_2, x_1) \xrightarrow[\begin{bmatrix} -1 & 0 \\ 0 & 1 \end{bmatrix}]{} (-x_2, x_1) = \vec{x}\begin{bmatrix} 0 & 1 \\ -1 & 0 \end{bmatrix}.$$

This means that the point (x_1, x_2) *reflects* along the line $x_1 = x_2$ into the point (x_2, x_1), and then reflects along x_2-axis into the point $(-x_2, x_1)$, the image point of (x_1, x_2) under f. See Fig. 2.47.

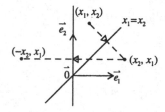

Fig. 2.47

We summarize these results in

Example 5 The linear operator

$$A = \begin{bmatrix} 0 & a \\ b & 0 \end{bmatrix}, \quad \text{where } ab \neq 0$$

has the following properties:

(1) In case $ab > 0$. Let $\vec{v}_1 = (\sqrt{|b|}, \sqrt{|a|})$ and $\vec{v}_2 = (-\sqrt{|b|}, \sqrt{|a|})$.
   1. $\vec{v}_1 A = \sqrt{ab}\,\vec{v}_1$ and $\vec{v}_2 A = -\sqrt{ab}\,\vec{v}_2$. Thus, $\sqrt{ab}$ and $-\sqrt{ab}$ are eigenvalues of A with corresponding eigenvectors $\vec{v}_1$ and $\vec{v}_2$.
   2. The invariant lines (subspaces) are $\langle\langle \vec{v}_1 \rangle\rangle$ and $\langle\langle \vec{v}_2 \rangle\rangle$.

Read Explanation below and see Fig. 2.48.

(2) In case $ab < 0$. A does not have (non-zero) eigenvectors and hence A does not have any invariant line. Suppose $a > 0$ and $b < 0$, then $-ab > 0$ and

$$A = \begin{bmatrix} 0 & a \\ -b & 0 \end{bmatrix} \begin{bmatrix} -1 & 0 \\ 0 & 1 \end{bmatrix} = \begin{bmatrix} 1 & 0 \\ 0 & -1 \end{bmatrix} \begin{bmatrix} 0 & a \\ -b & 0 \end{bmatrix}.$$

The mapping $\vec{x} \to \vec{x}A$ can be decomposed as

$$\vec{x} = (x_1, x_2) \xrightarrow[\begin{bmatrix} 0 & a \\ -b & 0 \end{bmatrix}]{} (-bx_2, ax_1) \xrightarrow[\begin{bmatrix} -1 & 0 \\ 0 & 1 \end{bmatrix}]{} (bx_2, ax_1) = \vec{x}A$$

See Fig. 2.49.

Explanation For a given $\vec{x} \in \mathbb{R}^2$, it might be difficult to pinpoint the position of $\vec{x}A$ in the coordinate system $\mathcal{N} = \{\vec{e}_1, \vec{e}_2\}$. Since $\vec{v}_1$ and $\vec{v}_2$ are linearly independent and thus $\mathcal{B} = \{\vec{v}_1, \vec{v}_2\}$ is a basis for $\mathbb{R}^2$. Any $\vec{x} \in \mathbb{R}^2$ can be uniquely expressed as

$$\vec{x} = \alpha_1 \vec{v}_1 + \alpha_2 \vec{v}_2, \quad \text{i.e. } [\vec{x}]_{\mathcal{B}} = (\alpha_1, \alpha_2)$$
$$\Rightarrow \vec{x}A = \alpha_1 \vec{v}_1 A + \alpha_2 \vec{v}_2 A$$
$$= \sqrt{ab}\,(\alpha_1 \vec{v}_1 - \alpha_2 \vec{v}_2), \quad \text{i.e. } [\vec{x}A]_{\mathcal{B}} = \sqrt{ab}\,(\alpha_1, -\alpha_2).$$

This suggests how easily it is in the new coordinate system $\mathcal{B}$ to determine $\vec{x}A$:

$$\vec{x} \to [\vec{x}]_{\mathcal{B}} = (\alpha_1, \alpha_2) \to (\alpha_1, -\alpha_2) \to \sqrt{ab}\,(\alpha_1, -\alpha_2) = [\vec{x}A]_{\mathcal{B}} \to \vec{x}A.$$

This is the essence of the *spectral decomposition* of A (refer to Sec. 2.7.6).
See Figs. 2.48 and 2.49 where $\vec{u}_1 = (\sqrt{-b}, \sqrt{a})$, $\vec{u}_2 = (-\sqrt{-b}, \sqrt{a})$.

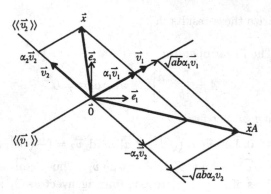

Fig. 2.48

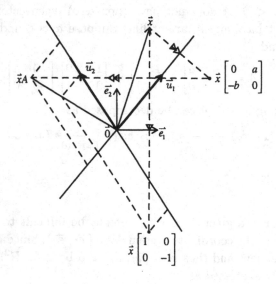

Fig. 2.49

If $a = b = 1$ or $a = -b = 1$, the readers are urged to simplify results stated in Example 5 and Figs. 2.48 and 2.49.

Suppose $abc \neq 0$. Since

$$\begin{bmatrix} a & 0 \\ b & c \end{bmatrix} = b \begin{bmatrix} \frac{a}{b} & 0 \\ 1 & \frac{c}{b} \end{bmatrix},$$

for simplicity, we may suppose that $b = 1$ and let $a = \lambda_1$, $c = \lambda_2$. Consider the linear operator

$$f(x_1, x_2) = (\lambda_1 x_1 + x_2, \lambda_2 x_2), \quad \text{where } \lambda_1 \lambda_2 \neq 0.$$

Let $\vec{x} = (x_1, x_2) \neq \vec{0}$. Then

$\langle\langle \vec{x} \rangle\rangle$ is an invariant line (subspace) of f.

$\Leftrightarrow$ There exists constant $\lambda \neq 0$ such that $f(\vec{x}) = \lambda \vec{x}$, i.e.

$$\lambda_1 x_1 + x_2 = \lambda x_1$$
$$\lambda_2 x_2 = \lambda x_2.$$

In case $x_2 \neq 0, \lambda = \lambda_2$ will give $\lambda_1 x_1 + x_2 = \lambda_2 x_1$, and hence $(\lambda_1 - \lambda_2) x_1 = -x_2$ implies that $\lambda_1 \neq \lambda_2$. So $\vec{x} = (1, \lambda_2 - \lambda_1)$ is a required vector. If $x_2 = 0$, then $\lambda_1 x_1 = \lambda x_1$ would give $\lambda = \lambda_1$ and $\vec{x} = (1, 0)$ is another choice. We put these in

Example 6 The linear operator

$$A = \begin{bmatrix} \lambda_1 & 0 \\ 1 & \lambda_2 \end{bmatrix}, \quad \text{where } \lambda_1 \lambda_2 \neq 0$$

has the following properties:

(1) In case $\lambda_1 \neq \lambda_2$. Let $\vec{v}_1 = \vec{e}_1, \vec{v}_2 = (1, \lambda_2 - \lambda_1)$.

    1. $\vec{v}_1 A = \lambda_1 \vec{v}_1$ and $\vec{v}_2 A = \lambda_2 \vec{v}_2$. Thus, λ_1 and λ_2 are eigenvalues of A with corresponding eigenvectors $\vec{v}_1$ and $\vec{v}_2$.

    2. $\langle\langle \vec{v}_1 \rangle\rangle$ and $\langle\langle \vec{v}_2 \rangle\rangle$ are invariant lines of A.

In fact, $\mathcal{B} = \{\vec{v}_1, \vec{v}_2\}$ is a basis for $\mathbb{R}^2$ and

$$[A]_{\mathcal{B}} = PAP^{-1} = \begin{bmatrix} \lambda_1 & 0 \\ 0 & \lambda_2 \end{bmatrix}, \quad \text{where } P = \begin{bmatrix} \vec{v}_1 \\ \vec{v}_2 \end{bmatrix} = \begin{bmatrix} 1 & 0 \\ 1 & \lambda_1 - \lambda_2 \end{bmatrix}$$

is the *matrix representation* of A with respect to $\mathcal{B}$ (see Exs. <B> 4, 5 of Sec. 2.4 and Sec. 2.7.3). See Fig. 2.50.

(2) In case $\lambda_1 = \lambda_2 = \lambda$. Then

$$\vec{e}_1 A = \lambda \vec{e}_1$$

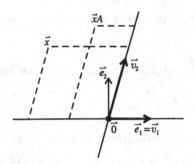

Fig. 2.50

and $\langle\langle\vec{e}_1\rangle\rangle$ is the only invariant subspace of A. Also

$$A = \begin{bmatrix} \lambda & 0 \\ 1 & \lambda \end{bmatrix} = \lambda I_2 + \begin{bmatrix} 0 & 0 \\ 1 & 0 \end{bmatrix} = \lambda \begin{bmatrix} 1 & 0 \\ \frac{1}{\lambda} & 1 \end{bmatrix}$$

shows that A has the mapping behavior as

$$\vec{x} = (x_1, x_2) \xrightarrow[\text{enlargement } \lambda I_2]{} (\lambda x_1, \lambda x_2)$$

$A \searrow \qquad\qquad \Big\downarrow \text{translation along } (x_2, 0)$

$$\vec{x}A = (\lambda x_1 + x_2, \lambda x_2)$$

See Fig. 2.51.

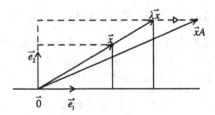

Fig. 2.51

A special operator of this type is

$$S = \begin{bmatrix} 1 & 0 \\ a & 1 \end{bmatrix}, \quad \text{where } a \neq 0,$$

which is called a *shearing*. S maps each point $\vec{x} = (x_1, x_2)$ to the point $(x_1 + ax_2, x_2)$ along the line passing $\vec{x}$ and parallel to the x_1-axis, to the right if $a > 0$ with a distance ax_2, proportional to the x_2-coordinate x_2 of $\vec{x}$ by a fixed constant a, and to the left if $a < 0$ by the same manner. Therefore,

1. S keeps every point on the x_1-axis fixed which is the only invariant subspace of it.
2. S moves every point (x_1, x_2) with $x_2 \ne 0$ along the line parallel to the x_1-axis, through a distance with a constant proportion a to its distance to the x_1-axis, to the point $(x_1 + ax_2, x_2)$. Thus, each line parallel to x_1-axis is an invariant line.

See Fig. 2.52.

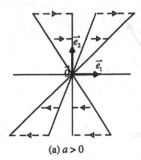 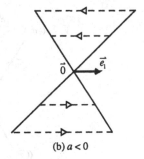

(a) $a > 0$　　　　　　　　　(b) $a < 0$

Fig. 2.52

Note that linear operators

$$\begin{bmatrix} \lambda_1 & 1 \\ 0 & \lambda_2 \end{bmatrix}, \text{ where } \lambda_1\lambda_2 \ne 0; \quad \begin{bmatrix} 1 & a \\ 0 & 1 \end{bmatrix}, \text{ where } a \ne 0,$$

can be investigated in a similar manner.

Suppose $ab \ne 0$. Consider the linear operator

$$f(\vec{x}) = f(x_1, x_2) = (bx_2, x_1 + ax_2).$$

Let $\vec{x} = (x_1, x_2) \ne \vec{0}$. Then,

$\langle\langle \vec{x} \rangle\rangle$ is an invariant line (subspace).

$\Leftrightarrow$ There exists a constant $\lambda \ne 0$ such that $f(\vec{x}) = \lambda\vec{x}$, i.e.

$$bx_2 = \lambda x_1$$
$$x_1 + ax_2 = \lambda x_2.$$

Since $b \neq 0$, $x_1 \neq 0$ should hold. Now, put $x_2 = \frac{\lambda}{b} x_1$ into the second equation and we get, after eliminating x_1,

$$\lambda^2 - a\lambda - b = 0.$$

The existence of such a nonzero λ depends on the discriminant $a^2 + 4b \geq 0$.

Case 1 $a^2 + 4b > 0$. Then $\lambda^2 - a\lambda - b = 0$ has two real roots

$$\lambda_1 = \frac{a + \sqrt{a^2 + 4b}}{2}, \quad \lambda_2 = \frac{a - \sqrt{a^2 + 4b}}{2}.$$

Solve $f(\vec{x}) = \lambda_i \vec{x}$, $i = 1, 2$, and we have the corresponding vector $\vec{v}_i = (b, \lambda_i)$. It is easy to see that $\vec{v}_1$ and $\vec{v}_2$ are linearly independent. Therefore, f has two different invariant lines $\langle\langle \vec{v}_1 \rangle\rangle$ and $\langle\langle \vec{v}_2 \rangle\rangle$. See Fig. 2.53.

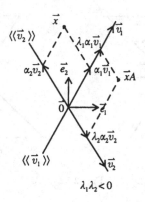

Fig. 2.53

Case 2 $a^2 + 4b = 0$. $\lambda^2 - a\lambda - b = 0$ has two equal real roots

$$\lambda = \frac{a}{2}, \frac{a}{2}.$$

Solve $f(\vec{x}) = \lambda \vec{x}$, and the corresponding vector is $\vec{v} = (-a, 2)$. Then $\langle\langle \vec{v} \rangle\rangle$ is the only invariant line. See Fig. 2.54.

Case 3 $a^2 + 4b < 0$. f does not have any invariant line. See Fig. 2.55.

Summarize these results in

Example 7 The linear operator

$$A = \begin{bmatrix} 0 & 1 \\ b & a \end{bmatrix}, \quad \text{where } ab \neq 0$$

has the following properties:

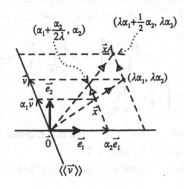

Fig. 2.54

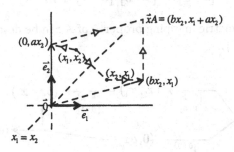

Fig. 2.55

(1) $a^2 + 4b > 0$. Let $\lambda_1 = \frac{a+\sqrt{a^2+4b}}{2}$ and $\lambda_2 = \frac{a-\sqrt{a^2+4b}}{2}$ and $\vec{v}_i = (b, \lambda_i)$ for $i = 1, 2$.

   1. $\vec{v}_i A = \lambda_i \vec{v}_i$, $i = 1, 2$. Thus, λ_1 and λ_2 are eigenvalues of A with corresponding eigenvectors $\vec{v}_1$ and $\vec{v}_2$.

   2. $\langle\langle \vec{v}_1 \rangle\rangle$ and $\langle\langle \vec{v}_2 \rangle\rangle$ are invariant lines (subspaces) of $\mathbb{R}^2$ under A.

In the basis $\mathcal{B} = \{\vec{v}_1, \vec{v}_2\}$, A can be represented as

$$[A]_\mathcal{B} = PAP^{-1} = \begin{bmatrix} \lambda_1 & 0 \\ 0 & \lambda_2 \end{bmatrix}, \quad \text{where } P = \begin{bmatrix} \vec{v}_1 \\ \vec{v}_2 \end{bmatrix} = \begin{bmatrix} b & \lambda_1 \\ b & \lambda_2 \end{bmatrix}.$$

(see Exs. <B> 4, 5 of Sec. 2.4 and Sec. 2.7.3.) See Fig. 2.53.

(2) $a^2 + 4b = 0$. Let $\lambda = \frac{a}{2}, \frac{a}{2}$ and $\vec{v} = (-a, 2)$.

   1. $\vec{v}A = \lambda\vec{v}$. λ is an eigenvalue of multiplicity 2 of A with corresponding eigenvector $\vec{v}$.

   2. $\langle\langle \vec{v} \rangle\rangle$ is an invariant line (subspace) of $\mathbb{R}^2$ under A.

In the basis $\mathcal{B} = \{\vec{v}, \vec{e}_1\}$, A can be represented as

$$[A]_{\mathcal{B}} = PAP^{-1} = \begin{bmatrix} \lambda & 0 \\ \frac{1}{2} & \lambda \end{bmatrix} = \lambda \begin{bmatrix} 1 & 0 \\ \frac{1}{2\lambda} & 1 \end{bmatrix} = \lambda I_2 + \begin{bmatrix} 0 & 0 \\ \frac{1}{2} & 0 \end{bmatrix},$$

where $P = \begin{bmatrix} \vec{v} \\ \vec{e}_1 \end{bmatrix} = \begin{bmatrix} -a & 2 \\ 1 & 0 \end{bmatrix}$.

This is the composite map of a shearing followed by an enlargement with scale λ. See Figs. 2.52 and 2.54.

(3) $a^2 + 4b < 0$. A does not have any invariant line. Notice that

$$\begin{bmatrix} 0 & 1 \\ b & a \end{bmatrix} = \begin{bmatrix} 0 & 1 \\ 1 & 0 \end{bmatrix} \begin{bmatrix} b & 0 \\ 0 & 1 \end{bmatrix} + \begin{bmatrix} 0 & 0 \\ 0 & a \end{bmatrix}.$$

Hence, the geometric mapping property of A can be described as follows in $\mathcal{N} = \{\vec{e}_1, \vec{e}_2\}$.

See Fig. 2.55.

The operator

$$\begin{bmatrix} a & 1 \\ b & 0 \end{bmatrix}, \quad \text{where } ab \neq 0$$

is of the same type.

Is there any more basic operator than these mentioned from Example 1 to Example 7? No more! It will turn eventually out that these operators are sufficiently enough to describe any operators on $\mathbb{R}^2$ in various ways, both algebraically and geometrically. Please refer to Secs. 2.7.5 to 2.7.8.

Before we are able to do so, we have to realize, as we learned from these examples, that the natural Cartesian coordinate system $\mathcal{N} = \{\vec{e}_1, \vec{e}_2\}$ is not always the best way to describe all possible linear operators whose definitions are independent of any particular choice of basis for $\mathbb{R}^2$. A suitable choice of a basis $\mathcal{B} = \{\vec{x}_1, \vec{x}_2\}$ for $\mathbb{R}^2$, according to the *features* of a given linear operator or a matrix (acts, as a linear operator in (2.7.7))

$$A = \begin{bmatrix} a_{11} & a_{12} \\ a_{21} & a_{22} \end{bmatrix},$$

will reduce as many entries a_{ij} to zero as possible. That is, after a change of coordinate from $\mathcal{N}$ to $\mathcal{B}$, and in the eyes of $\mathcal{B}, A$ becomes

$$[A]_{\mathcal{B}} = PAP^{-1} = \begin{bmatrix} \lambda_1 & 0 \\ 0 & \lambda_2 \end{bmatrix} \text{ or } \begin{bmatrix} \lambda & 0 \\ 1 & \lambda \end{bmatrix} \text{ or } \begin{bmatrix} 0 & 1 \\ b & a \end{bmatrix}. \qquad (2.7.10)$$

It is $[A]_{\mathcal{B}}$ that makes algebraic computations easier and geometric mapping properties clearer and our life happier.

Through these examples, for a given real 2×2 matrix $A = [a_{ij}]$, its main *features* are of two aspects:

1. the rank r(A),
2. the existence of an invariant line (subspace), if any.

We treated these concepts in examples via classical algebraic methods. Now, we reformulate them formally in the language, commonly used in linear algebra, which will be adopted throughout the book.

Definition Let $A = [a_{ij}]_{2\times2}$ be a real matrix and considered as a linear operator as in (2.7.7). If there exist a scalar $\lambda \in \mathbb{R}$ and a *nonzero* vector $\vec{x} \in \mathbb{R}^2$ such that

$$\vec{x}A = \lambda\vec{x},$$

then λ is called an *eigenvalue* (or *characteristic root*) of A and $\vec{x}$ an associated or corresponding *eigenvector* (or *characteristic vector*). (2.7.11)

In case $\lambda \neq 0$ and $\vec{x} \neq \vec{0}$ is an associated eigenvector, then this geometrically means that the line $\langle\langle\vec{x}\rangle\rangle$ is an *invariant line (subspace)* of $\mathbb{R}^2$ under A, and if $\lambda = 1, A$ keeps each vector along $\langle\langle\vec{x}\rangle\rangle$ (or point on $\langle\langle\vec{x}\rangle\rangle$) fixed, which is called a *line of invariant points*. If $\lambda = -1$, reverse it.

How to determine if A has eigenvalues and how to find them if they do exist? Suppose there does exist a nonzero vector $\vec{x}$ such that

$\vec{x}A = \lambda\vec{x}$ for some scalar $\lambda \in \mathbb{R}$

$\Leftrightarrow \vec{x}(A - \lambda I_2) = \vec{0}$

$\Leftrightarrow$ (by (3) in (2.7.8))

$$\det(A - \lambda I_2) = \det\begin{bmatrix} a_{11} - \lambda & a_{12} \\ a_{21} & a_{22} - \lambda \end{bmatrix} = \begin{vmatrix} a_{11} - \lambda & a_{12} \\ a_{21} & a_{22} - \lambda \end{vmatrix}$$

$$= \lambda^2 - (a_{11} + a_{22})\lambda + (a_{11}a_{22} - a_{12}a_{21}) = 0.$$

This means that any eigenvalue of A, if exists, is a root of the quadratic equation $\lambda^2 - (a_{11} + a_{22})\lambda + a_{11}a_{22} - a_{12}a_{21} = 0$. Solve this equation for real root λ and reverse the processes mentioned above to obtain the corresponding eigenvectors $\vec{x}$.

Summarize the above as

The procedure of computing eigenvalues and eigenvectors
Let $A = [a_{ij}]_{2 \times 2}$ be a real matrix.

1. Compute the *characteristic polynomial* of A,

$$\det(A - tI_2) = t^2 - (a_{11} + a_{22})t + a_{11}a_{22} - a_{12}a_{21} = t^2 - (\operatorname{tr} A)t + \det A,$$

where $\operatorname{tr}A = a_{11} + a_{22}$ is called the *trace* of A and $\det A$ the *determinant* of A.

2. Solve the *characteristic equation* of A,

$$\det(A - tI_2) = 0$$

for real roots λ, which are *real* eigenvalues of A.

3. Solve the simultaneous homogenous linear equations in two unknowns

$$\vec{x}(A - \lambda I_2) = 0, \quad \text{or}$$
$$(a_{11} - \lambda)x_1 + a_{21}x_2 = 0$$
$$a_{12}x_1 + (a_{22} - \lambda)\vec{x}_2 = 0, \quad \text{where } \vec{x} = (x_1, x_2).$$

Any *nonzero* vector solution $\vec{x} = (x_1, x_2)$ is an associated eigenvector of λ.

A has two distinct, coincident or no real eigenvalues according to the discriminant $(\operatorname{tr} A)^2 - 4\det A > 0, = 0$ or < 0, respectively.   (2.7.12)

It should be mentioned that definition (2.7.11) and terminologies in (2.7.12) are still valid for $n \times n$ matrix $A = [a_{ij}]$ over a field $\mathbb{F}$. For examples,

eigenvalue λ: $\vec{x}A = \lambda\vec{x}$ for $\vec{x} \neq 0$ in $\mathbb{F}^n$ and $\lambda \in \mathbb{F}$.
eigenvector $\vec{x}$: $\vec{x} \neq \vec{0}$ such that $\vec{x}A = \lambda\vec{x}$ for some $\lambda \in \mathbb{F}$.
Characteristic polynomial:

$$\det(A - tI_n) = (-1)^n t^n + a_{n-1}t^{n-1} + \cdots + a_1 t + a_0,$$
$$\text{where } a_{n-1} = (-1)^{n-1}\operatorname{tr} A \text{ and}$$
$$\operatorname{tr}A = \sum_{i=1}^{n} a_{ii} \text{ is the } trace \text{ of } A \text{ and } a_0 = \det A. \quad (2.7.13)$$

Please refer to Sec. B.10 if necessary. We will feel free to use them if needed.

The most simple and idealistic one is that A has two *distinct* real eigenvalues λ_1 and λ_2 with respective eigenvectors $\vec{x}_1$ and $\vec{x}_2$. In this case, $\vec{x}_1$ and $\vec{x}_2$ are *linearly independent*. To see this, suppose there exist constants α_1 and α_2 such that

$$\alpha_1\vec{x}_1 + \alpha_2\vec{x}_2 = \vec{0}$$

$\Rightarrow$ (perform A to both sides) $\alpha_1\lambda_1\vec{x}_1 + \alpha_2\lambda_2\vec{x}_2 = \vec{0}$

$\Rightarrow$ (by eliminating $\vec{x}_2$ from the above two equations)

$$\alpha_1(\lambda_1 - \lambda_2)\vec{x}_1 = \vec{0}$$

$\Rightarrow$ (since $\lambda_1 \neq \lambda_2$ and $\vec{x}_1 \neq \vec{0}$) $\alpha_1 = 0$ and hence $\alpha_2 = 0$. $\quad$ (2.7.14)

Thus, $\mathcal{B} = \{\vec{x}_1, \vec{x}_2\}$ is a new basis for $\mathbb{R}^2$. Since

$$\vec{x}_1 A = \lambda_1\vec{x}_1$$
$$\vec{x}_2 A = \lambda_2\vec{x}_2$$

$$\Rightarrow \begin{bmatrix} \vec{x}_1 \\ \vec{x}_2 \end{bmatrix} A = \begin{bmatrix} \lambda_1 & 0 \\ 0 & \lambda_2 \end{bmatrix}\begin{bmatrix} \vec{x}_1 \\ \vec{x}_2 \end{bmatrix}$$

$$\Rightarrow [A]_\mathcal{B} = PAP^{-1} = \begin{bmatrix} \lambda_1 & 0 \\ 0 & \lambda_2 \end{bmatrix}, \quad \text{where } P = \begin{bmatrix} \vec{x}_1 \\ \vec{x}_2 \end{bmatrix}. \quad (2.7.15)$$

This is the *matrix representation* of the linear operator A with respect to $\mathcal{B}$. We encountered this in Example 2 through Example 7. For further discussion about matrix representation, see Sec. 2.7.3. Note that

$$\text{tr}\,[A]_\mathcal{B} = \text{tr}\,A = \lambda_1 + \lambda_2 \quad \text{and} \quad \det[A]_\mathcal{B} = \det A = \lambda_1\lambda_2. \quad (2.7.16)$$

From (2.7.15), we have

$$\vec{x}_1 A^2 = (\vec{x}_1 A)A = (\lambda_1\vec{x}_1)A = \lambda_1(\vec{x}_1 A) = \lambda_1(\lambda_1\vec{x}_1) = \lambda_1^2\vec{x}_1,$$
$$\vec{x}_2 A^2 = \lambda_2^2\vec{x}_2$$

$\Rightarrow$ (for any scalars α_1 and α_2)

$$(\alpha_1\vec{x}_1 + \alpha_2\vec{x}_2)A^2 = \lambda_1^2\alpha_1\vec{x}_1 + \lambda_2^2\alpha_2\vec{x}_2$$

$$= (\lambda_1 + \lambda_2)(\alpha_1\lambda_1\vec{x}_1 + \alpha_2\lambda_2\vec{x}_2) - \lambda_1\lambda_2(\alpha_1\vec{x}_1 + \alpha_2\vec{x}_2)$$

$$= (\lambda_1 + \lambda_2)(\alpha_1\vec{x}_1 + \alpha_2\vec{x}_2)A - \lambda_1\lambda_2(\alpha_1\vec{x}_1 + \alpha_2\vec{x}_2)I_2$$

$\Rightarrow$ (since $\mathcal{B} = \{\vec{x}_1, \vec{x}_2\}$ is a basis for $\mathbb{R}^2$)

$$\vec{x}[A^2 - (\lambda_1 + \lambda_2)A + \lambda_1\lambda_2 I_2] = \vec{0} \quad \text{for all } \vec{x} \in \mathbb{R}^2$$

$$\Rightarrow A^2 - (\text{tr}\,A)A + (\det A)I_2 = O_{2\times 2}. \quad (2.7.17)$$

This matrix identity holds for any 2×2 real matrix A even if A has coincident eigenvalues or A does not have real eigenvalues. Since a direct computation shows that

$$A^2 = \begin{bmatrix} a_{11}^2 + a_{12}a_{21} & a_{12}(a_{11} + a_{22}) \\ a_{21}(a_{11} + a_{22}) & a_{12}a_{21} + a_{22}^2 \end{bmatrix}$$

$$= (a_{11} + a_{22}) \begin{bmatrix} a_{11} & a_{12} \\ a_{21} & a_{22} \end{bmatrix} - (a_{11}a_{22} - a_{12}a_{21}) \begin{bmatrix} 1 & 0 \\ 0 & 1 \end{bmatrix}$$

$$= (\operatorname{tr} A)A - (\det A)\mathrm{I}_2, \qquad\qquad (2.7.18)$$

the result (2.7.17) follows. For other proofs, see Exs.<A> 7 and <B> 5.

Summarize the above as

Cayley–Hamilton Theorem (Formula)

Let $A = [a_{ij}]_{2 \times 2}$ be a real matrix.

(1) A satisfies its characteristic polynomial $\det(A - t\mathrm{I}_2)$, i.e.

$$A^2 - (\operatorname{tr} A)A + (\det A)\mathrm{I}_2 = O.$$

Geometrically, this is equivalent to that

$$\vec{x}A^2 = (\operatorname{tr} A)(\vec{x}A) + (\det A)\vec{x}, \quad \text{for any } \vec{x} \in \mathbb{R}^2,$$

and hence $\vec{x}A^2 \in \langle\langle \vec{x}, \vec{x}A \rangle\rangle$ for any $\vec{x} \in \mathbb{R}^2$.

(2)

$$A((\operatorname{tr} A)\mathrm{I}_2 - A) = (\det A)\mathrm{I}_2$$

shows that

$$A \text{ is invertible} \Leftrightarrow \det A \neq 0,$$

and, in this case,

$$A^{-1} = \frac{1}{\det A}((\operatorname{tr} A)\mathrm{I}_2 - A). \qquad\qquad (2.7.19)$$

Notice that (2.7.19) still holds for any $n \times n$ matrix A over a field $\mathbb{F}$ (see (2.7.13)). It is the geometric equivalence of the Cayley–Hamilton formula that enables us to choose suitable basis $\mathcal{B}$ for $\mathbb{R}^2$ so that, in the eyes of $\mathcal{B}$, A becomes $[A]_{\mathcal{B}}$ as in (2.7.10). For details, see Secs. 2.7.6–2.7.8.

Exercises

<A>

1. Model after Example 1 and Fig. 2.43 to investigate the mapping
   properties of each of the following operators:

$$\begin{bmatrix} 0 & \lambda \\ 0 & 0 \end{bmatrix}, \quad \begin{bmatrix} 0 & 0 \\ \lambda & 0 \end{bmatrix} \quad \text{and} \quad \begin{bmatrix} 0 & 0 \\ 0 & \lambda \end{bmatrix},$$

   where $\lambda \neq 0$.

2. Model after Example 2 and Fig. 2.44, do the same problem for operators:

$$\begin{bmatrix} a & 0 \\ b & 0 \end{bmatrix}, \quad \begin{bmatrix} 0 & b \\ 0 & a \end{bmatrix} \quad \text{and} \quad \begin{bmatrix} 0 & 0 \\ b & a \end{bmatrix},$$

   where $ab \neq 0$.

3. Model after Example 6 and Figs. 2.50–2.52, do the same problem for
   operators:

$$\begin{bmatrix} \lambda_1 & 1 \\ 0 & \lambda_2 \end{bmatrix} \quad \text{and} \quad \begin{bmatrix} 1 & a \\ 0 & 1 \end{bmatrix},$$

   where $\lambda_1\lambda_2 \neq 0$ and $a \neq 0$.

4. Model after Example 7 and Figs. 2.53–2.55, do the same problem for the
   operator

$$\begin{bmatrix} a & 1 \\ b & 0 \end{bmatrix},$$

   where $ab \neq 0$.

5. Give two sets, $\{\vec{x}_1, \vec{x}_2\}$ and $\{\vec{y}_1, \vec{y}_2\}$, of vectors in $\mathbb{R}^2$. Does there always
   exist a linear operator $f\colon \mathbb{R}^2 \to \mathbb{R}^2$ such that f maps one set of vectors
   onto another set of vectors? How many such f can be found?

6. Let $\vec{x}_1 = (2,3)$, $\vec{x}_2 = (-3,1)$, $\vec{x}_3 = (-2,-1)$ and $\vec{x}_4 = (1,-4)$ be given
   vectors in $\mathbb{R}^2$. Do the following problems.

   (a) Find a linear operator f on $\mathbb{R}^2$ keeping $\vec{x}_1$ and $\vec{x}_2$ fixed, i.e.
       $f(\vec{x}_i) = \vec{x}_i$ for $i = 1, 2$. Is such a f unique? Does the same f keep
       $\vec{x}_3$ and $\vec{x}_4$ fixed?
   (b) Find a linear operator f such that $f(\vec{x}_1) = f(\vec{x}_3) = \vec{0}$. What
       happens to $f(\vec{x}_2)$ and $f(\vec{x}_4)$?
   (c) Find a linear operator f such that $\mathrm{Ker}(f) = \mathrm{Im}(f) = \langle\langle \vec{x}_1 \rangle\rangle$. How
       many of f are possible?
   (d) Find a linear operator f such that $\mathrm{Ker}(f) = \langle\langle \vec{x}_2 \rangle\rangle$ while
       $\mathrm{Im}(f) = \langle\langle \vec{x}_3 \rangle\rangle$. How many of f are possible?

(e) Find a linear operator f such that $f(\vec{x}_1) = \vec{x}_4$ and $f(\vec{x}_2) = \vec{x}_3$. Is such a f unique?

(f) Does there exist a linear operator f such that $f(\vec{x}_1) = \vec{x}_3$ and $f(-2\vec{x}_1) = f(\vec{x}_4)$? Why?

(g) Find a linear operator f such that $f(\vec{x}_1) = \vec{x}_3 + \vec{x}_4$ and $f(\vec{x}_2) = \vec{x}_3 - \vec{x}_4$.

(h) Does there exist a linear operator f such that $f(\vec{x}_i) = i\vec{x}_i$ for $i = 1, 2, 3, 4$?

(i) Find all possible linear operators mapping $\triangle \vec{0}\, \vec{x}_2 \vec{x}_4$ onto $\triangle \vec{0}\, \vec{x}_1 \vec{x}_3$.

(j) Find all possible affine transformations mapping $\triangle \vec{x}_2 \vec{x}_3 \vec{x}_4$ onto $\triangle \vec{x}_3 \vec{x}_4 \vec{x}_1$.

7. Suppose $A = [a_{ij}]_{2\times 2}$ is a real matrix, which is not a *scalar* matrix. Show that these exists at least one nonzero vector $\vec{x}_0$ such that $\mathcal{B} = \{\vec{x}_0, \vec{x}_0 A\}$ is a basis for $\mathbb{R}^2$. Try to model after (2.7.17) to show that there exist constants α and β such that $A^2 + \alpha A + \beta I_2 = 0$.

8. Let

$$A = \begin{bmatrix} -1 & 5 \\ -6 & 7 \end{bmatrix}.$$

(a) Show that the characteristic polynomial of A is $t^2 - 6t + 23$.

(b) Justify Cayley–Hamilton formula for A.

(c) Use (b) to compute A^{-1}.

(d) Try to find a basis $\mathcal{B} = \{\vec{x}_1, \vec{x}_2\}$ for $\mathbb{R}^2$, so that

$$[A]_{\mathcal{B}} = PAP^{-1} = \begin{bmatrix} 0 & 1 \\ -23 & 6 \end{bmatrix}, \quad \text{where } P = \begin{bmatrix} \vec{x}_1 \\ \vec{x}_2 \end{bmatrix}.$$

How many such $\mathcal{B}$ are there?

<B>

1. Let $A = [a_{ij}]_{2\times 2}$ be a matrix such that

$$A^2 = O_{2\times 2},$$

and we call A a *nilpotent* matrix of index 2 if $A \neq O$ (refer to Ex. 7 of Sec. B.4).

(a) By algebraic computation, show that

$$A^2 = O$$

$$\Leftrightarrow \operatorname{tr} A = a_{11} + a_{22} = 0 \quad \text{and} \quad a_{12}a_{21} = -a_{11}^2.$$

Give some numerical examples for such A.

(b) By geometric consideration, show that
$$A^2 = O$$
$$\Leftrightarrow \operatorname{Im}(A) \subseteq \operatorname{Ker}(A).$$

Then, consider the following three cases:

(1) $\operatorname{Ker}(A) = \mathbb{R}^2$, thus $\operatorname{Im}(A) = \{\vec{0}\}$ and $A = O_{2\times 2}$.

(2) $\operatorname{Ker}(A) = \langle\langle \vec{x}_1 \rangle\rangle$ where $\vec{x}_1 \neq 0$, thus $\operatorname{Im}(A) = \langle\langle \vec{x}_1 \rangle\rangle$. Take any fixed vector $\vec{x}_2$ which is linearly independent of $\vec{x}_1$, then $\vec{x}_2 A = \lambda \vec{x}_1$ for some scalar $\lambda \neq 0$. Show that
$$A = \begin{bmatrix} \vec{x}_1 \\ \vec{x}_2 \end{bmatrix}^{-1} \begin{bmatrix} 0 & 0 \\ \lambda & 0 \end{bmatrix} \begin{bmatrix} \vec{x}_1 \\ \vec{x}_2 \end{bmatrix}.$$

(3) Is it possible that $\operatorname{Ker}(A) = \{\vec{0}\}$?

(c) What are $\operatorname{tr} A$ and $\det A$? How much can $A^2 - (\operatorname{tr} A)A + (\det A)I_2 = O$ help in determining A?

2. A matrix $A = [a_{ij}]_{2\times 2}$ is called *idempotent* if
$$A^2 = A$$
(refer to Ex. 6 of Sec. B.4 and Ex. 7 of Sec. B.7).

(a) Using purely algebraic method, try to determine all idempotent 2×2 real matrices.

(b) Via geometric consideration, show that
$$A^2 = A$$

$\Leftrightarrow$ Each nonzero vector in $\operatorname{Im}(A)$ is an eigenvector of A associated to the eigenvalue 1.

Then,

(1) $\operatorname{Im}(A) = \{\vec{0}\}$ implies $A = O_{2\times 2}$.

(2) $\operatorname{Im}(A) = \langle\langle \vec{x}_1 \rangle\rangle$, where $\vec{x}_1 \neq \vec{0}$ implies that $\vec{x}_1 A = \vec{x}_1$ and there exists an $\vec{x}_2$ which is linearly independent of $\vec{x}_1$ with $\operatorname{Ker}(A) = \langle\langle \vec{x}_2 \rangle\rangle$. Hence $\vec{x}_2 A = \vec{0}$. Therefore,
$$A = \begin{bmatrix} \vec{x}_1 \\ \vec{x}_2 \end{bmatrix}^{-1} \begin{bmatrix} 1 & 0 \\ 0 & 0 \end{bmatrix} \begin{bmatrix} \vec{x}_1 \\ \vec{x}_2 \end{bmatrix}.$$

(3) $\operatorname{Im}(A) = \mathbb{R}^2$ implies that $\operatorname{Ker}(A) = \{\vec{0}\}$. For any two linearly independent vectors $\vec{x}_1$ and $\vec{x}_2$, $\vec{x}_1 A = \vec{x}_1$ and $\vec{x}_2 A = \vec{x}_2$ implies $A = I_2$.

(c) Does $A^2 - (\operatorname{tr} A)A + (\det A)I_2 = O$ help in determining such A?

How?

3. A matrix $A = [a_{ij}]_{2\times 2}$ is called *involutory* if

$$A^2 = I_2$$

(refer to Ex. 9 of Sec. B.4 and Ex. 8 of Sec. B.7).

(a) By purely algebraic method, show that

    (1) $A^2 = I_2$

$\Leftrightarrow$ (2) $A = \pm I_2$ or $a_{11} + a_{22} = 0$ and $a_{11}^2 = 1 - a_{12}a_{21}$.

Give some numerical examples for such A.

(b) By geometric consideration, show that

    (1) $A^2 = I_2$

$\Leftrightarrow$ (2) $(A - I_2)(A + I_2) = O$

$\Leftrightarrow$ (3) $\text{Im}(A - I_2) \subseteq \text{Ker}(A + I_2)$ and $\text{Im}(A + I_2) \subseteq \text{Ker}(A - I_2)$.

Then,

(1) If $r(A - I_2) = 0$, then $A = I_2$.

(2) If $r(A - I_2) = 1$, then there exists a basis $\{\vec{x}_1, \vec{x}_2\}$ for $\mathbb{R}^2$ so that

$$A = \begin{bmatrix} \vec{x}_1 \\ \vec{x}_2 \end{bmatrix}^{-1} \begin{bmatrix} 1 & 0 \\ 0 & -1 \end{bmatrix} \begin{bmatrix} \vec{x}_1 \\ \vec{x}_2 \end{bmatrix}.$$

(3) If $r(A - I_2) = 2$, then $A = -I_2$.

(c) Show that

$$\mathbb{R}^2 = \text{Im}(A - I_2) \oplus \text{Im}(A + I_2),$$
$$\text{Im}(A + I_2) = \text{Ker}(A - I_2), \quad \text{Im}(A - I_2) = \text{Ker}(A + I_2).$$

Then, try to find all such A.

(d) Try to use $A^2 - (\text{tr } A)A + (\det A)I_2 = O$ to determine A as far as possible.

4. Let $A = [a_{ij}]_{2\times 2}$ be a real matrix such that

$$A^2 = -I_2$$

(refer to Ex. 9 of Sec. B.7).

(a) Show that there does not exist a nonzero vector $\vec{x}_0$ such that $\vec{x}_0 A = \lambda \vec{x}_0$ for any $\lambda \neq 0$. That is, A does not have any invariant line.

(b) Take any nonzero vector $\vec{x}_1$ and show that

$$A = \begin{bmatrix} \vec{x}_1 \\ \vec{x}_1 A \end{bmatrix}^{-1} \begin{bmatrix} 0 & 1 \\ -1 & 0 \end{bmatrix} \begin{bmatrix} \vec{x}_1 \\ \vec{x}_1 A \end{bmatrix}.$$

(c) Is it possible to choose a basis $\mathcal{B}$ for $\mathbb{R}^2$ such that $[A]_{\mathcal{B}}$ is a diagonal matrix? Why?

5. Let $A = [a_{ij}]_{2 \times 2}$ be a *real* matrix without real eigenvalues, i.e. $(\operatorname{tr} A)^2 - 4 \det A < 0$ holds.

(a) Let $\lambda = \lambda_1 + i\lambda_2$ be a *complex* eigenvalue of A, considered as a linear operator defined by $\vec{x} \to \vec{x}A$ where $\vec{x} \in \mathbb{C}^2$. That is to say, there exists a nonzero $\vec{x} = \vec{x}_1 + i\vec{x}_2 \in \mathbb{C}^2$ where $\vec{x}_1, \vec{x}_2 \in \mathbb{R}^2$ such that

$$\vec{x}A = \lambda\vec{x}.$$

Show that $\vec{y}A = \bar{\lambda}\vec{y}$ where $\vec{y} = \vec{x}_1 - i\vec{x}_2$ is the conjugate vector of $\vec{x}$.

(b) Show that

$$\vec{x}_1 A = \lambda_1 \vec{x}_1 - \lambda_2 \vec{x}_2,$$
$$\vec{x}_2 A = \lambda_2 \vec{x}_1 + \lambda_1 \vec{x}_2.$$

(c) Show that $\vec{x}_1$ and $\vec{x}_2$ are linearly independent in $\mathbb{R}^2$ and hence $\mathcal{B} = \{\vec{x}_1, \vec{x}_2\}$ is a basis for $\mathbb{R}^2$. Then, in $\mathcal{B}$,

$$[A]_{\mathcal{B}} = PAP^{-1} = \begin{bmatrix} \lambda_1 & -\lambda_2 \\ \lambda_2 & \lambda_1 \end{bmatrix}, \quad \text{where } P = \begin{bmatrix} \vec{x}_1 \\ \vec{x}_2 \end{bmatrix}.$$

See Example 10 in Sec. 5.7.

(d) Try to use (b) or (c) to show that $A^2 - (\operatorname{tr} A)A + (\det A)I_2 = O$.

(e) Suppose the readers are familiar with basic knowledge concerned plane trigonometry. Try to show that

$$\begin{bmatrix} \lambda_1 & \lambda_2 \\ -\lambda_2 & \lambda_1 \end{bmatrix} = r \begin{bmatrix} \cos\theta & \sin\theta \\ -\sin\theta & \cos\theta \end{bmatrix},$$

where $r = \sqrt{\lambda_1^2 + \lambda_2^2}$ and $\theta \in \mathbb{R}$. The right side matrix can be considered (under field isomorphism) as the complex number $z = re^{i\theta}$. Try to explain its geometric mapping properties.

<C> Abstraction and generalization

Read Secs. B.4, B.6 and B.7 if necessary and try to do the following problems.

1. Extend Example 1 through Example 7 to linear operators on $\mathbb{C}^2$, endowed with the natural basis $\mathcal{N} = \{\vec{e}_1, \vec{e}_2\}$. Emphasize the similarities and differences between linear operators on $\mathbb{R}^2$ and those on $\mathbb{C}^2$, both algebraically and geometrically. For example, what happens to (3) in Example 7?

2. Try to find as many "basic" linear operators on $\mathbb{F}^n$, endowed with the natural basis $\mathcal{N} = \{\vec{e}_1, \ldots, \vec{e}_n\}$, as possible. Is this an easy job if one has only mean knowledge about the ranks and invariant subspaces of linear operators on $\mathbb{F}^n$? Why not read Secs. B.10, B.11 and B.12 if necessary.

2.7.3   *Matrix representations of a linear operator in various bases*

Remind that the definition (2.7.4) for a linear operator on $\mathbb{R}^2$ is independent of any particular choice of bases for $\mathbb{R}^2$. In Sec. 2.7.1 we considered a linear operator on $\mathbb{R}^2$ as a real 2×2 matrix in the Cartesian basis $\mathcal{N} = \{\vec{e}_1, \vec{e}_2\}$. But various concrete examples in Sec. 2.7.2 indicated that $\mathcal{N}$ is not always the best choice of bases in investigating both algebraic and geometric properties of a linear operator. It is in this background that, here in this subsection, we formally introduce the matrix representations of a linear operator with respect to various bases for $\mathbb{R}^2$.

Two matrix representations A and B of a linear operator with respect to two bases are *similar*, i.e. there exists an invertible matrix P representing change of coordinates (see Sec. 2.4) such that

$$B = PAP^{-1}. \tag{2.7.20}$$

Similarity among matrices is an equivalent relation (see Sec. A.1) and it *preserves*

1. the rank of a linear operator or a square matrix,
2. the characteristic polynomial and hence eigenvalues of a linear operator or a matrix,
3. the determinant of a matrix, and
4. the trace of a matrix. $\hspace{3cm}$ (2.7.21)

Refer to Exs. 25, 27, 31 and 32 of Sec. B.4 for more information concerned.

Also, reminding readers for linear operators f, g on $\mathbb{R}^2$, there are several relational linear *operations* and are mentioned as follows.

1. The addition $g + f$: $(g + f)(\vec{x}) = g(\vec{x}) + f(\vec{x}), \vec{x} \in \mathbb{R}^2$.
2. The scalar multiplication αf: $(\alpha f)(\vec{x}) = \alpha f(\vec{x}), \alpha \in \mathbb{R}$.
3. The composite $g \circ f$: $(g \circ f)(\vec{x}) = g(f(\vec{x})), \vec{x} \in \mathbb{R}^2$.
4. The invertible operator f^{-1} (if f is one-to-one and onto).

For general reference, see Sec. A.2. Addition and scalar multiplication are also defined for linear transformations to form a vector space V to a vector space W over the same field. Therefore, the set $\text{Hom}(V, W)$ or $\text{L}(V, W)$ of all such linear transformations form a *vector space* over $\mathbb{F}$ (see Sec. B.7). As a consequence, the set $\text{Hom}(V, V)$ of operators on V forms an *associative algebra with identity* and has the set $\text{GL}(V, V)$ of invertible linear operators become a *group* under the composite operation. If $\dim V = m < \infty$ and $\dim W = n < \infty$, $\text{Hom}(V, W)$ can be realized (under isomorphism) as $\text{M}(m, n; \mathbb{F})$, the set of all $m \times n$ matrices over $\mathbb{F}$ and $\text{GL}(V, V)$ as $\text{GL}(n; \mathbb{F})$, the set of all invertible matrices in $\text{M}(n; \mathbb{F}) = \text{M}(n, n; \mathbb{F})$. See Secs. B.4 and B.7.

Let us return to linear operators on $\mathbb{R}^2$ and proceed.

Choose a pair of bases $\mathcal{B} = \{\vec{a}_1, \vec{a}_2\}$ and $\mathcal{C} = \{\vec{b}_1, \vec{b}_2\}$ for the vector space $\mathbb{R}^2$. Suppose $f \colon \mathbb{R}^2 \to \mathbb{R}^2$ is a linear operator.

Let

$$f(\vec{a}_i) = \sum_{j=1}^{2} a_{ij} \vec{b}_j, \quad i = 1, 2$$

i.e. $[f(\vec{a}_i)]_{\mathcal{C}} = (a_{i1}, a_{i2}), i = 1, 2$. Therefore, for any vector $x \in \mathbb{R}^2$,

$$\vec{x} = \sum_{i=1}^{2} x_i \vec{a}_i$$

$$\Rightarrow f(\vec{x}) = \sum_{i=1}^{2} x_i f(\vec{a}_i) = \sum_{i=1}^{2} x_i \left(\sum_{j=1}^{2} a_{ij} \vec{b}_j \right) = \sum_{j=1}^{2} \left(\sum_{i=1}^{2} x_i a_{ij} \right) \vec{b}_j$$

$\Rightarrow$ (by (2.3.1) or (2.6.3) and Note in (2.4.1'))

$$[f(\vec{x})]_{\mathcal{C}} = \left(\sum_{i=1}^{2} x_i a_{i1}, \sum_{i=1}^{2} x_i a_{i2} \right)$$

$$= (x_1 \quad x_2) \begin{bmatrix} a_{11} & a_{12} \\ a_{21} & a_{22} \end{bmatrix}$$

$$= [\vec{x}]_{\mathcal{B}} [f]_{\mathcal{C}}^{\mathcal{B}},$$

where

$$[f]_{\mathcal{C}}^{\mathcal{B}} = \begin{bmatrix} [f(\vec{a}_1)]_{\mathcal{C}} \\ [f(\vec{a}_2)]_{\mathcal{C}} \end{bmatrix} = \begin{bmatrix} a_{11} & a_{12} \\ a_{21} & a_{22} \end{bmatrix} \tag{2.7.22}$$

is called the *matrix representation* of the linear operator f with respect to bases $\mathcal{B}$ and $\mathcal{C}$.

Suppose $g: \mathbb{R}^2 \to \mathbb{R}^2$ is another linear operator and $\mathcal{D}$ is a third basis for $\mathbb{R}^2$. Then, according to (2.7.22),

$$[g(\vec{x})]_\mathcal{D} = [\vec{x}]_C[g]_\mathcal{D}^C.$$

Therefore, the composite linear operator $g \circ f: \mathbb{R}^2 \to \mathbb{R}^2$ satisfies

$$[(g \circ f)(\vec{x})]_\mathcal{D} = [f(\vec{x})]_C[g]_\mathcal{D}^C = [\vec{x}]_B[f]_C^B[g]_\mathcal{D}^C, \quad x \in \mathbb{R}^2$$

which implies

$$[(g \circ f)]_\mathcal{D}^B = [f]_C^B[g]_\mathcal{D}^C. \tag{2.7.22'}$$

In case f is a linear isomorphism and $g = f^{-1}: \mathbb{R}^2 \to \mathbb{R}^2$ is the inverse isomorphism, take $\mathcal{D} = B$ in (2.7.22'), then

$$[f]_C^B[f^{-1}]_B^C = [1_{\mathbb{R}^2}]_B^B = I_2 = \begin{bmatrix} 1 & 0 \\ 0 & 1 \end{bmatrix},$$

$$[f^{-1}]_B^C[f]_C^B = [1_{\mathbb{R}^2}]_C^C = I_2,$$

which means that $[f]_C^B$ is invertible and

$$[f]_C^{B-1} = [f^{-1}]_B^C.$$

Conversely, given a real matrix $A = [a_{ij}]_{2\times2}$, there exists a unique linear transformation f such that $[f]_C^B = A$. In fact, define $f: \mathbb{R}^2 \to \mathbb{R}^2$ by

$$f(\vec{a}_i) = \sum_{j=1}^2 a_{ij}\vec{b}_j, \quad i = 1, 2$$

and, then, for $\vec{x} = \sum_{i=1}^2 x_i\vec{a}_i$, linearly by

$$f(\vec{x}) = \sum_{i=1}^2 x_i f(\vec{a}_i)$$

It is easy to check that f is linear and $[f]_C^B = A$ holds.

Summarize as (refer to Sec. B.7 for more detailed and generalized results)

The matrix representation of linear operator with respect to bases
Suppose B, C and $\mathcal{D}$ are bases for $\mathbb{R}^2$, and f and g are linear operators on $\mathbb{R}^2$. Then, the matrix representation (2.7.22) has the following properties:

(a) $[f(\vec{x})]_C = [\vec{x}]_B[f]_C^B, \vec{x} \in \mathbb{R}^2$.

(b) *Linearity*:

$$[f + g]_C^B = [f]_C^B + [g]_C^B,$$
$$[\alpha f]_C^B = \alpha[f]_C^B, \quad \alpha \in \mathbb{R}^2.$$

(c) *Composite transformation*:

$$[g \circ f]_D^B = [f]_C^B[g]_D^C.$$

(d) *Inverse transformation*: f is a linear isomorphism, i.e. *invertible* if and only if for any bases B and C, the matrix $[f]_C^B$ is invertible and

$$[f]_C^{B^{-1}} = [f^{-1}]_B^C.$$

(e) *Change of coordinates*: Suppose B' and C' are bases for $\mathbb{R}^2$. Then (refer to (2.4.2))

$$[f]_{C'}^{B'} = A_B^{B'}[f]_C^B A_{C'}^C,$$

where $A_B^{B'}$ is the transition matrix from the basis B' to the basis B and similarly for $A_C^{C'}$. So the following diagram is commutative.

$$
\begin{array}{ccc}
\mathbb{R}^2 & \xrightarrow{[f]_C^B} & \mathbb{R}^2 \\
(B) & & (C) \\
A_B^{B'} \uparrow & & \downarrow A_{C'}^C = A_C^{C'-1} \\
(B') & & (C) \\
\mathbb{R}^2 & \xrightarrow[{[f]_{C'}^{B'}}]{} & \mathbb{R}^2
\end{array}
\tag{2.7.23}
$$

In case $C = B$, we write

$$[f]_B = [f]_B^B \tag{2.7.24}$$

in short. f is called *diagonalizable* if $[f]_B$ is a diagonal matrix for some basis B for $\mathbb{R}^2$. Then, it follows from (e) that, if $C = B$ and $C' = B'$,

$$[f]_{B'} = A_B^{B'}[f]_B A_B^{B'-1} \tag{2.7.25}$$

holds. Then, $[f]_{B'}$ is said to be *similar* to $[f]_B$. Similarity among matrices is an important concept in the theory of linear algebra. It enables us to investigate the geometric behaviors of linear or affine transformations from different choices of bases. For concrete examples, see Sec. 2.7.2.

Actually, (b) and (d) in (2.7.23) tell us implicitly more information, but in an abstract setting.

Let the set of linear operators on $\mathbb{R}^2$ be denoted by

$$\mathrm{Hom}(\mathbb{R}^2, \mathbb{R}^2) \text{ or } L(\mathbb{R}^2, \mathbb{R}^2) = \{f \colon \mathbb{R}^2 \to \mathbb{R}^2 \text{ is a linear operator}\}, \quad (2.7.26)$$

which is a vector space over $\mathbb{R}$. As usual, let $\mathcal{N} = \{\vec{e}_1, \vec{e}_2\}$ be the natural basis for $\mathbb{R}^2$.

Four specified linear operators $f_{ij} \colon \mathbb{R}^2 \to \mathbb{R}^2, 1 \le i, j \le 2$ are defined as follows:

$$f_{ij}(\vec{e}_k) = \begin{cases} \vec{e}_j, & \text{if } k = i, \\ \vec{0}, & \text{if } k \ne i, \end{cases} \qquad (2.7.27)$$

and $f_{ij}(\vec{x}) = f_{ij}(x_1, x_2) = x_1 f_{ij}(\vec{e}_1) + x_2 f_{ij}(\vec{e}_2)$ linearly. If $f \in L(\mathbb{R}^2, \mathbb{R}^2)$, let $f(\vec{e}_k) = \sum_{j=1}^{2} a_{kj} \vec{e}_j$. Then, for $\vec{x} = (x_1, x_2) = \sum_{k=1}^{2} x_k \vec{e}_k$,

$$f(\vec{x}) = \sum_{k=1}^{2} x_k f(\vec{e}_k) = \sum_{k=1}^{2} x_k \left(\sum_{j=1}^{2} a_{kj} \vec{e}_j \right) = \sum_{k=1}^{2} \sum_{j=1}^{2} a_{kj} x_k \vec{e}_j$$

$$= \sum_{k=1}^{2} \sum_{j=1}^{2} a_{kj} x_k f_{kj}(\vec{e}_k) = \sum_{k=1}^{2} \sum_{j=1}^{2} a_{kj} f_{kj}(x_k \vec{e}_k)$$

$$= \sum_{k=1}^{2} \sum_{j=1}^{2} a_{kj} f_{kj}(\vec{x})$$

$$\Rightarrow f = \sum_{k=1}^{2} \sum_{j=1}^{2} a_{kj} f_{kj}. \qquad (2.7.28)$$

This means that any operator f is a *linear combination* of $f_{kj}, 1 \le k, j \le 2$, where f_{kj} are *linear independent*, i.e.

$$\sum_{k=1}^{2} \sum_{j=1}^{2} a_{kj} f_{kj} = 0 \text{ (zero operator)} \Leftrightarrow a_{kj} = 0, \quad 1 \le k, \; j \le 2.$$

Thus, $\{f_{11}, f_{12}, f_{21}, f_{22}\}$ forms a *basis* for $\mathrm{Hom}(\mathbb{R}^2, \mathbb{R}^2)$ and therefore, $\mathrm{Hom}(\mathbb{R}^2, \mathbb{R}^2)$ is a *4-dimensional real vector space* (refer to Sec. B.3).

Corresponding to (2.7.26), let the set of real 2×2 matrices be denoted by

$$M(2; \mathbb{R}) = \{A = [a_{ij}]_{2 \times 2} \mid a_{ij} \in \mathbb{R}, 1 \le i, j \le 2\}, \qquad (2.7.29)$$

which is also a real vector space with matrix addition and scalar multiplication: $B = [b_{ij}]_{2 \times 2}$ and $\alpha \in \mathbb{R}$,

$$A + B = [a_{ij} + b_{ij}];$$
$$\alpha A = [\alpha a_{ij}].$$

Let

$$E_{11} = \begin{bmatrix} 1 & 0 \\ 0 & 0 \end{bmatrix}, \quad E_{12} = \begin{bmatrix} 0 & 1 \\ 0 & 0 \end{bmatrix}, \quad E_{21} = \begin{bmatrix} 0 & 0 \\ 1 & 0 \end{bmatrix}, \quad E_{22} = \begin{bmatrix} 0 & 0 \\ 0 & 1 \end{bmatrix}. \qquad (2.7.30)$$

Note that f_{ij} defined in (2.7.27) has matrix representation

$$[f_{ij}]_{\mathcal{N}} = E_{ij}, \quad 1 \leq i, \ j \leq 2.$$

For any $A = [a_{ij}] \in M(2; \mathbb{R})$, we have

$$A = \sum_{i=1}^{2} \sum_{j=1}^{2} a_{ij} E_{ij} \qquad (2.7.31)$$

and $A = O$ (zero matrix) if and only if $a_{ij} = 0, 1 \leq i, j \leq 2$. Thus, $\{E_{11}, E_{12}, E_{21}, E_{22}\}$ forms a *basis* for $M(2; \mathbb{R})$ which hence is a 4-dimensional real vector space (refer to Sec. B.3).

Now, (b)–(d) in (2.7.23) guarantee

The isomorphism between $\mathrm{Hom}(\mathbb{R}^2, \mathbb{R}^2)$ and $M(2; \mathbb{R})$

In natural basis $\mathcal{N} = \{\vec{e}_1, \vec{e}_2\}$ for $\mathbb{R}^2$, define mapping $\Phi \colon \mathrm{Hom}(\mathbb{R}^2, \mathbb{R}^2) \to M(2; \mathbb{R})$ by

$$\Phi(f) = [f]_{\mathcal{N}}.$$

Then Φ is a *linear isomorphism*, i.e. Φ is one-to-one and onto and preserves linear operations:

1. $\Phi(f + g) = \Phi(f) + \Phi(g)$,
2. $\Phi(\alpha f) = \alpha \Phi(f)$, for $\alpha \in \mathbb{R}$ and $f, g \in \mathrm{Hom}(\mathbb{R}^2, \mathbb{R}^2)$.

Moreover, this Φ induces an *algebra isomorphism* between the associative algebra $\mathrm{Hom}(\mathbb{R}^2, \mathbb{R}^2)$ and the associative algebra $M(2; \mathbb{R})$, i.e.

1. $\Phi(g \circ f) = \Phi(f)\Phi(g)$,
2. $\Phi(1_{\mathbb{R}^2}) = I_2$,

where $1_{\mathbb{R}^2} \colon \mathbb{R}^2 \to \mathbb{R}^2$ is the identity operator. $\qquad (2.7.32)$

The vector space $\mathrm{Hom}(\mathbb{R}^2, \mathbb{R}^2)$ has an important subset:

$$GL(\mathbb{R}^2, \mathbb{R}^2) \qquad (2.7.33)$$

the set of all *invertible* linear operators on $\mathbb{R}^2$. Under the composition of functions, $GL(\mathbb{R}^2, \mathbb{R}^2)$ forms a *group* (see Sec. A.4), i.e. for each pair $f, g \in GL(\mathbb{R}^2, \mathbb{R}^2)$, $f \circ g \in GL(\mathbb{R}^2, \mathbb{R}^2)$ always holds and satisfies the following properties: for all $f, g, h \in GL(\mathbb{R}^2, \mathbb{R}^2)$

1. $h \circ (g \circ f) = (h \circ g) \circ f$.
2. $1_{\mathbb{R}^2} \circ f = f \circ 1_{\mathbb{R}^2}$, where $1_{\mathbb{R}^2}$ is the identity operator on $\mathbb{R}^2$.

3. There exists a unique element $f^{-1} \in \mathrm{GL}(\mathbb{R}^2, \mathbb{R}^2)$ such that

$$f \circ f^{-1} = f^{-1} \circ f = 1_{\mathbb{R}^2}.$$

Similarly, in vector space $\mathrm{M}(2; \mathbb{R})$, the set

$$\mathrm{GL}(2; \mathbb{R}) \tag{2.7.34}$$

of all invertible matrices forms a *group* under the operation of matrix multiplication. These two groups are essentially the same one in the following sense.

The isomorphism between $\mathrm{GL}(\mathbb{R}^2, \mathbb{R}^2)$ and $\mathrm{GL}(2; \mathbb{R})$
The linear (or algebra) isomorphism Φ in (2.7.32) induces a *group* isomorphism between $\mathrm{GL}(\mathbb{R}^2, \mathbb{R}^2)$ and $\mathrm{GL}(2; \mathbb{R})$, i.e. for each $f, g \in \mathrm{GL}(\mathbb{R}^2; \mathbb{R}^2)$,

1. $\Phi(g \circ f) = \Phi(f)\Phi(g)$,
2. $\Phi(1_{\mathbb{R}^2}) = I_2$,
3. $\Phi(f^{-1}) = \Phi(f)^{-1}$.

Hence, both are called the *real general linear group* on $\mathbb{R}^2$. (2.7.35)

Notice that, in terms of another basis $\mathcal{B}$ for $\mathbb{R}^2, \mathrm{GL}(\mathbb{R}^2, \mathbb{R}^2)$ is group isomorphic to the *conjugate* group

$$PGL(2; \mathbb{R})P^{-1} = \{PAP^{-1} \mid A \in \mathrm{GL}(2; \mathbb{R})\} \tag{2.7.36}$$

of $\mathrm{GL}(2; \mathbb{R})$, where P is the transition matrix from $\mathcal{B}$ to $\mathcal{N}$.

Here comes some notations to be used throughout the whole text. For $f \in \mathrm{Hom}(\mathbb{R}^2, \mathbb{R}^2)$ and $A \in \mathrm{M}(2; \mathbb{R})$ and n a positive integer,

$$\begin{aligned}
f^0 &= 1_{\mathbb{R}^2} \\
f^2 &= f \circ f \quad \text{(composite of function)}, \\
f^n &= f \circ f^{n-1} \quad \text{if } n \geq 2, \\
f^{-n} &= (f^{-1})^n \quad \text{if } f \text{ is invertible},
\end{aligned}$$

and correspondingly,

$$\begin{aligned}
A^0 &= 1_{\mathbb{R}^2}, \\
A^2 &= A \cdot A \quad \text{(matrix multiplication)}, \\
A^n &= A \cdot A^{n-1} \quad \text{if } n \geq 2, \\
A^{-n} &= (A^{-1})^n \quad \text{if } A \text{ is invertible}. \tag{2.7.37}
\end{aligned}$$

These are still valid for $\mathrm{Hom}(V; V)$ and $\mathrm{M}(n; \mathbb{F})$.

As a conclusion up to this point, (2.7.23), (2.7.32) and (2.7.35) all together indicate that the study of linear operators on $\mathbb{R}^2$ can be reduced to the study of real 2×2 matrices in a fixed basis for $\mathbb{R}^2$, as we did in Secs. 2.7.1 and 2.7.2, but the emphasis should be paid on the possible connections among various representations of a single linear operator owing to different choice of bases. That is, similarity of matrices plays an essential role in the interchange of a linear operator and its various matrix representations.

In what remains in this subsection, we are going to find out some invariants under the operation of similarity of matrices.

The easiest one among them is the determinant of a square matrix. Suppose $A \in M(2; \mathbb{R})$ and $P \in GL(2; \mathbb{R})$. Then (2.4.2) shows that

The invariance of the determinant of a square matrix under similarity

$$\det PAP^{-1} = \det A.$$

Hence, the *determinant* of a linear operator f on $\mathbb{R}^2$ is well-defined as

$$\det f = \det[f]_{\mathcal{B}},$$

where $\mathcal{B}$ is any fixed basis for $\mathbb{R}^2$ (for geometric meaning, see (2.8.44)).

(2.7.38)

Owing to

$$PAP^{-1} - tI_2 = P(A - tI_2)P^{-1},$$

it comes immediately

The invariance of characteristic polynomial and eigenvalues of a square matrix under similarity

$$\det(PAP^{-1} - tI_2) = \det(A - tI_2)$$

and hence PAP^{-1} and A have the same set of eigenvalues. Hence, the *characteristic polynomial* of a linear operator f on $\mathbb{R}^2$ is

$$\det(f - tI_{\mathbb{R}^2}) = \det([f]_{\mathcal{B}} - tI_2),$$

where $\mathcal{B}$ is any fixed basis for $\mathbb{R}^2$. (2.7.39)

Notice that

$$\vec{x} A = \lambda \vec{x} \quad \text{for } \vec{x} \in \mathbb{R}^2$$
$$\Leftrightarrow (\vec{x} P^{-1})(PAP^{-1}) = \lambda(\vec{x} P^{-1}). \tag{2.7.40}$$

This means that if $\vec{x}$ is an eigenvector of A (or an operator) associated with the eigenvalue λ in the natural coordinate system $\mathcal{N}$, then $\vec{x} P^{-1}$ is the corresponding eigenvector of PAP^{-1} associated with the same eigenvalue λ in any fixed basis $\mathcal{B}$ for $\mathbb{R}^2$, where P is the transition matrix from $\mathcal{B}$ to $\mathcal{N}$ and $\vec{x} P^{-1} = [\vec{x}]_{\mathcal{B}}$. See the diagram.

$$
\begin{array}{ccc}
\mathbb{R}^2 & \xrightarrow{\ \ A\ \ } & \mathbb{R}^2 \\
(\mathcal{N}) & & (\mathcal{N}) \\
P \uparrow & & \downarrow P^{-1} \\
(\mathcal{B}) & & (\mathcal{B}) \\
\mathbb{R}^2 & \xrightarrow[PAP^{-1}]{} & \mathbb{R}^2
\end{array}
$$

How about the trace of a square matrix?

Suppose $A = [a_{ij}]_{2\times 2}$ and $B = [b_{ij}]_{2\times 2}$. Actual computation shows that

$$
\begin{aligned}
\text{tr}(AB) &= (a_{11}b_{11} + a_{12}b_{21}) + (a_{21}b_{12} + a_{22}b_{22}) \\
&= (b_{11}a_{11} + b_{12}a_{21}) + (b_{21}a_{12} + b_{22}a_{22}) \\
&= \text{tr}(BA). \tag{2.7.41}
\end{aligned}
$$

Suppose $P_{2\times 2}$ is invertible. Let PA replace A and P^{-1} replace B in the above equality, we obtain

The invariance of trace of a square matrix under similarity

$$\text{tr}(PAP^{-1}) = \text{tr}\,A.$$

Hence, the *trace* of a linear operator f on $\mathbb{R}^2$ is well-defined as

$$\text{tr}\,f = \text{tr}[f]_{\mathcal{B}},$$

where $\mathcal{B}$ is any fixed basis for $\mathbb{R}^2$. $\tag{2.7.42}$

For more properties concerned with trace, refer to Exs. 25–30 of Sec. B.4. The invariance of trace can be proved indirectly from the invariance of characteristic polynomial as (2.7.12) shows.

Finally, how about the rank of an operator or square matrix? The rank we defined for a linear operator or matrix in Sec. 2.7.1 is, strictly speaking, not well-defined, since we still do not know if the nonnegative integer is unchangeable subject to changes of bases.

Let A and B be two real 2×2 matrices treated as linear operators on $\mathbb{R}^2$ (see (2.7.7)). Without loss of generality, both may be supposed to be nonzero matrices.

In case $r(A) = 1$, i.e. dim $\mathrm{Im}(A) = 1$: Two separate cases are considered as follows.

Case 1 Let $r(B) = 1$, then $\mathrm{Im}(AB) = \{\vec{0}\}$ or $\mathrm{Im}(AB) = \langle\langle \vec{v} \rangle\rangle$ where $\vec{v} \neq \vec{0}$ according to the image $\mathrm{Im}(A) = \mathrm{Ker}(B)$ or not. Hence,

$$r(AB) = \begin{cases} 0 \Leftrightarrow \mathrm{Im}(A) = \mathrm{Ker}(B) \\ r(A) = r(B) = 1 \Leftrightarrow \mathrm{Im}(A) \cap \mathrm{Ker}(B) = \{\vec{0}\}. \end{cases}$$

Case 2 Let $r(B) = 2$, then

$$r(AB) = r(A) = 1 < 2 = r(B).$$

In case $r(A) = 2$, still two separate cases are considered as follows.

Case 1 If $r(B) = 1$, then $r(AB) = 1$ holds.
Case 2 If $r(B) = 2$, then $r(AB) = 2$ holds.

Summarize the above as

The lower and upper bounds for the rank of the product matrix of two matrices
Let $A_{2\times 2}$ and $B_{2\times 2}$ be two real matrices.

(1) If either A or B is zero, the

$$r(AB) = 0 \leq \min\{r(A), r(B)\}.$$

(2) If both A and B are nonzero,

$$r(A) + r(B) - 2 \leq r(AB) \leq \min\{r(A), r(B)\},$$

where

$$r(AB) = r(A) + r(B) - 2 \Leftrightarrow \mathrm{Ker}(B) \subseteq \mathrm{Im}(A),$$
$$r(AB) = r(A) \Leftrightarrow \mathrm{Im}(A) \cap \mathrm{Ker}(B) = \{\vec{0}\},$$
$$r(AB) = r(B) \Leftrightarrow \mathbb{R}^2 = \mathrm{Im}(A) + \mathrm{Ker}(B). \tag{2.7.43}$$

These results are still suitable for matrices $A_{k\times m}$ and $B_{m\times n}$ over fields, with m replacing 2 (see Ex. 3 of Sec. B.7). Try to reprove (2.7.43) by using the dimension theorem stated in (2.7.8) and then try to prove these results for general matrices (see Ex. <C>).

As an easy consequence of (2.7.43), we have

The invariance of the rank of a matrix under similarity

Let $A_{2\times 2}$ be a real matrix.

(1) If $P_{2\times 2}$ is invertible, then P preserves the rank of A, i.e.

$$\mathrm{r}(PA) = \mathrm{r}(AP) = \mathrm{r}(A).$$

(2) Hence, if P is invertible,

$$\mathrm{r}(PAP^{-1}) = \mathrm{r}(A).$$

The *rank* of a linear operator is well-defined as

$$\mathrm{r}[f] = \mathrm{r}([f]_{\mathcal{B}}),$$

where $\mathcal{B}$ is any fixed basis for $\mathbb{R}^2$.   (2.7.44)

These results are still usable for arbitrary matrices over fields. Readers are urged to prove (2.7.44) directly without recourse to (2.7.43).

For the sake of reference and generalization, we introduce four closely related subspaces associated with a given matrix.

Let $A = [a_{ij}]_{2\times 2}$ be a real matrix. By interchange of rows and columns of A, the resulted matrix, denoted as

$$A^*, \qquad\qquad (2.7.45)$$

is called the *transpose* of A (see Sec. B.4).

Then, associated with A the following four subspaces:

$$\mathrm{Im}(A) \text{ or } \mathrm{R}(A) = \{\vec{x}A \mid \vec{x} \in \mathbb{R}^2\}$$
$$= \text{the subspace of } \mathbb{R}^2 \text{ generated by the row vectors}$$
$$A_{1*} = (a_{11}, a_{12}) \text{ and } A_{2*} = (a_{21}, a_{22}) \text{ of } A,$$
$$\mathrm{Ker}(A) \text{ or } \mathrm{N}(A) = \{\vec{x} \in \mathbb{R}^2 \mid \vec{x}A = \vec{0}\} \qquad (2.7.46)$$

are called respectively the *row space* and the *left kernel* space of A, while

$$\mathrm{Im}(A^*) \text{ or } \mathrm{R}(A^*) = \{\vec{x}A^* \mid \vec{x} \in \mathbb{R}^2\}$$
$$= \text{the subspace of } \mathbb{R}^2 \text{ generated by the column}$$
$$\text{vectors } A_{*1} = \begin{bmatrix} a_{11} \\ a_{21} \end{bmatrix} \text{ and } A_{*2} = \begin{bmatrix} a_{12} \\ a_{22} \end{bmatrix} \text{ of } A,$$
$$\mathrm{Ker}(A^*) \text{ or } \mathrm{N}(A^*) = \{\vec{x} \in \mathbb{R}^2 \mid \vec{x}A^* = \vec{0}\} \qquad (2.7.47)$$

are called respectively the *column space* and the *right kernel* space of A.

Also, define the

row rank of $A = \dim \text{Im}(A)$

= the maximal number of linearly independent row vectors of A,

column rank of $A = \dim \text{Im}(A^*)$

= the maximal number of linearly independent column vectors of A. (2.7.48)

Note that the row rank of A is the rank of A we defined in Sec. 2.7.1 and (2.7.44) and is sometimes called the *geometric rank* of A. As a contrast, the *algebraic rank* of A is defined to be the largest integer r such that some $r \times r$ *submatrix* of A has a nonzero determinant (see Sec. B.6) if A itself is a nonzero matrix.

Then (2.7.8) says

The equalities of three ranks of a matrix
Let $A_{2\times 2}$ be a real matrix.

(1) If $A = O_{2\times 2}$, the zero matrix, then the rank of A is defined as

$$r(A) = 0.$$

(2) If $A \neq O_{2\times 2}$, then the rank $r(A)$ satisfies

the row rank of A = the column rank of A

= the algebraic rank of A

= $r(A)$.

Notice that $1 \leq r(A) \leq 2$. (2.7.49)

What we have said here from (2.7.46) to (2.7.49) are still true for arbitrary $m \times n$ matrices A over a field $\mathbb{F}$ (see Ex. 15 of Sec. B.4, Secs. B.5 and B.6, Ex. 2 of Sec. B.7 and Sec. B.8): $0 \leq r(A) \leq \min(m, n)$.

Exercises

<A>

1. Let the linear operator f on $\mathbb{R}^2$ be defined as

$$f(\vec{x}) = \vec{x} \begin{bmatrix} -6 & 5 \\ 1 & 3 \end{bmatrix}$$

in $\mathcal{N} = \{e_1, e_2\}$. Let $\mathcal{B} = \{(-1, 1), (1, 1)\}$ and $\mathcal{C} = \{(2, -3), (4, -2)\}$.

(a) Compute the transition matrices $A_{\mathcal{B}}^{\mathcal{C}}$ and $A_{\mathcal{C}}^{\mathcal{B}}$, and justify that

$$(A_{\mathcal{B}}^{\mathcal{C}})^{-1} = A_{\mathcal{C}}^{\mathcal{B}}$$

by direct computation. Also $A_{\mathcal{B}}^{\mathcal{C}} = A_{\mathcal{N}}^{\mathcal{C}} A_{\mathcal{B}}^{\mathcal{N}}$.

(b) Compute $[f]_{\mathcal{B}}$ and $[f]_{\mathcal{C}}$ and show that they are similar by finding an invertible $P_{2\times2}$ such that $[f]_{\mathcal{C}} = P[f]_{\mathcal{B}}P^{-1}$.

(c) Compute $[f]_{\mathcal{B}}^{\mathcal{N}}$ and $[f]_{\mathcal{C}}^{\mathcal{N}}$, and show that $[f]_{\mathcal{B}}^{\mathcal{N}} = [f]_{\mathcal{C}}^{\mathcal{N}} A_{\mathcal{B}}^{\mathcal{C}}$.

(d) Compute $[f]_{\mathcal{C}}^{\mathcal{B}}$ and $[f]_{\mathcal{B}}^{\mathcal{C}}$, and show that $[f]_{\mathcal{B}}^{\mathcal{C}} = A_{\mathcal{B}}^{\mathcal{C}} [f]_{\mathcal{C}}^{\mathcal{B}} A_{\mathcal{B}}^{\mathcal{C}}$.

(e) Show that $[f]_{\mathcal{C}}^{\mathcal{B}} = A_{\mathcal{N}}^{\mathcal{B}} [f]_{\mathcal{N}} A_{\mathcal{C}}^{\mathcal{N}}$.

(f) Show that $[f]_{\mathcal{C}}^{\mathcal{B}} = A_{\mathcal{N}}^{\mathcal{B}} [f]_{\mathcal{C}}^{\mathcal{N}} = [f]_{\mathcal{N}}^{\mathcal{B}} A_{\mathcal{C}}^{\mathcal{N}}$.

2. Let

$$f(\vec{x}) = \vec{x} \begin{bmatrix} 2 & 3 \\ -4 & -6 \end{bmatrix}$$

and $\mathcal{B} = \{(1,-1),(-2,1)\}$ and $\mathcal{C} = \{(-1,-2),(3,-4)\}$. Do the same questions as in Ex. 1.

3. Find a linear operator f on $\mathbb{R}^2$ and a basis $\mathcal{B}$ for $\mathbb{R}^2$ such that

$$[f(\vec{x})]_{\mathcal{B}} = [\vec{x}]_{\mathcal{N}} \begin{bmatrix} -1 & 1 \\ 1 & 1 \end{bmatrix}$$

for $\vec{x} \in \mathbb{R}^2$. How many such f and $\mathcal{B}$ could we find?

4. Find the linear operator f on $\mathbb{R}^2$ and a basis $\mathcal{B}$ for $\mathbb{R}^2$ such that

$$[f]_{\mathcal{B}} = \begin{bmatrix} \vec{v}_1 \\ \vec{v}_2 \end{bmatrix} \begin{bmatrix} 3 & 2 \\ -1 & 4 \end{bmatrix} \begin{bmatrix} \vec{v}_1 \\ \vec{v}_2 \end{bmatrix}^{-1},$$

where $\vec{v}_1 = (1,-1)$ and $\vec{v}_2 = (-5,6)$. How many possible choices for $\mathcal{B}$ are there?

5. Let

$$A = \begin{bmatrix} 1 & 0 \\ 1 & 1 \end{bmatrix}.$$

Does there exist a linear operator f on $\mathbb{R}^2$ and two bases $\mathcal{B}$ and $\mathcal{C}$ for $\mathbb{R}^2$ such that $[f]_{\mathcal{B}} = I_2$ and $[f]_{\mathcal{C}} = A$? Note that both I_2 and A has the same characteristic polynomial $(t-1)^2$, and hence the same set of eigenvalues.

6. Let

$$A = \begin{bmatrix} 1 & 1 \\ -1 & 1 \end{bmatrix} \quad \text{and} \quad B = \begin{bmatrix} 1 & 2 \\ -1 & 0 \end{bmatrix}.$$

(a) Give as many different reasons as possible to justify that there does not exist a linear operator f on $\mathbb{R}^2$ and two bases B and C for $\mathbb{R}^2$ such that $[f]_B = A$ and $[f]_C = B$.

(b) Find some invertible matrices $P_{2\times2}$ and $Q_{2\times2}$ such that $PAQ = B$.

7. For any linear operator f on $\mathbb{R}^2$, there exist a basis $B = \{\vec{x}_1, \vec{x}_2\}$ and another basis $C = \{\vec{y}_1, \vec{y}_2\}$ for $\mathbb{R}^2$ such that

$$\begin{bmatrix} \vec{x}_1 \\ \vec{x}_2 \end{bmatrix} [f]_N \begin{bmatrix} \vec{y}_1 \\ \vec{y}_2 \end{bmatrix}^{-1} = [f]_C^B = \begin{bmatrix} 0 & 0 \\ 0 & 0 \end{bmatrix} \text{ or } \begin{bmatrix} 1 & 0 \\ 0 & 0 \end{bmatrix} \text{ or } \begin{bmatrix} 1 & 0 \\ 0 & 1 \end{bmatrix}.$$

8. Find nonzero matrices $A_{2\times2}$ and $B_{2\times2}$ such that AB has each possible rank.

9. Suppose $A_{2\times2}$ and $B_{2\times2}$ are matrices (or $n \times n$ matrices). Then

$$r(AB) = 2 \Leftrightarrow r(A) = r(B) = 2.$$

10. Let A be a 2×2 real matrix (or any $n \times n$ matrix). Then $r(A) = 1$ if and only if there exist matrices $B_{2\times2}$ and $C_{2\times2}$, each of rank 1, such that $A = BC$.

11. Let A be a 2×2 real matrix (or any $n \times n$ matrix). Prove that the following are equivalent.

 (1) A is invertible (and $(A^{-1})^{-1} = A$).
 (2) A^* is invertible (and hence, $(A^*)^{-1} = (A^{-1})^*$).
 (3) There exists a real 2×2 matrix B such that $AB = I_2$ (and hence $B = A^{-1}$).
 (4) There exists a real 2×2 matrix B such that $BA = I_2$ (and hence $B = A^{-1}$).
 (5) The matrix equation $AX = O_{2\times2}$ always implies that $X = O$.

 Compare with (4) in (2.7.8) and refer to Ex. 16 of Sec. B.4.

12. Let A and B be 2×2 real matrices (or any $n \times n$ matrices). Prove that AB is invertible if and only if A and B are invertible, and

$$(AB)^{-1} = B^{-1}A^{-1}.$$

 Compare with Ex. 9. What happens if $A_{m\times n}$ and $B_{n\times m}$ with $m \neq n$?

13. Let A be a 2×2 real matrix (or any $n \times n$ matrix).

 (a) Show that A is invertible if and only if A does not have zero eigenvalues.
 (b) A and A^* have the same characteristic polynomial and hence same eigenvalues.

(c) Suppose $\vec{x}A = \lambda\vec{x}$ and A is invertible, then $\vec{x}A^{-1} = \frac{1}{\lambda}\vec{x}$, i.e. $\frac{1}{\lambda}$ is an eigenvalue of A^{-1} associated with the same eigenvector $\vec{x}$.

(d) Suppose $\vec{x}A = \lambda\vec{x}$, then $\vec{x}A^p = \lambda^p\vec{x}$ for integer $p \geq 1$.

(e) Let $g(t) = a_k t^k + \cdots + a_1 t + a_0$ be any polynomial with real coefficients. Define

$$g(A) = a_k A^k + \cdots + a_1 A + a_0 I_2.$$

Suppose $\vec{x}A = \lambda\vec{x}$, then

$$\vec{x}\, g(A) = g(\lambda)\vec{x}.$$

Refer to Exercises of Sec. B.10.

14. Let A and B be 2×2 real matrices (or any $n \times n$ matrices). Show that

$$r(A + B) \leq r(A) + r(B)$$

with equality if and only if $\text{Im}(A+B) = \text{Im}(A)\oplus\text{Im}(B)$. Give examples of 2×2 nonzero matrices A and B so that $A + B$ has each possible rank.

15. Let A be a 2×2 real matrix (or any $n \times n$ matrix). Show that there exists another 2×2 real matrix B such that

(1) $BA = O$, and

(2) $r(A) + r(B) = \dim \mathbb{R}^2 = 2$.

16. Suppose A is a 2×2 real matrix (or any $n \times n$ matrix) such that $r(A^2) = r(A)$. Show that

$$\text{Im}(A) \cap \text{Ker}(A) = \{\vec{0}\}.$$

17. Let A be a 2×2 real matrix (or any $n \times n$ matrix) such that $r(A) = 1$. Show that there exists a unique scalar $\lambda \in \mathbb{R}$ such that $A^2 = \lambda A$, and $I_2 - A$ is invertible if $\lambda \neq 1$.

18. Let A be a 2×2 real matrix. Show that

$$r(A) = r(A^*A) = r(AA^*)$$

by testing case by case according to the rank of A. But, this equality still holds for any $m \times n$ matrix A over a field $\mathbb{F}$ with characteristic zero. Do you have any idea to prove this? Geometrically or algebraically?

19. Let A, B and C be 2×2 real matrices. Show that

$$r(AB) + r(BC) \leq r(B) + r(ABC)$$

by testing case by case according to the ranks of A, B and C. Frobenius obtained this inequality at 1911, and it is still true for arbitrary matrices $A_{p\times m}, B_{m\times n}$ and $C_{n\times q}$. Any ideas to prove this? See Ex. <C> 11.

20. We say two vectors $\vec{x}$ and $\vec{y}$ in $\mathbb{R}^2$ are *orthogonal* if $\vec{x}\,\vec{y}^* = 0$ and is denoted as $\vec{x}\perp\vec{y}$. For a vector subspace S of $\mathbb{R}^2$, let

$$S^\perp = \{\vec{x} \in \mathbb{R}^2 \mid \vec{x}\perp\vec{y} \text{ for each } \vec{y} \in S\}.$$

Then $S^\perp$ is a subspace of $\mathbb{R}^2$ and is called the *orthogonal complement* of S in $\mathbb{R}^2$. For any nonzero real 2×2 matrix A, show that, both geometrically and algebraically,

(1) $\mathrm{Im}(A)^\perp = \mathrm{Ker}(A^*)$,
(2) $\mathrm{Ker}(A^*)^\perp = \mathrm{Im}(A)$ and $\mathbb{R}^2 = \mathrm{Ker}(A^*) \oplus \mathrm{Im}(A)$,
(3) $\mathrm{Im}(A^*)^\perp = \mathrm{Ker}(A)$,
(4) $\mathrm{Ker}(A)^\perp = \mathrm{Im}(A^*)$ and $\mathbb{R}^2 = \mathrm{Ker}(A) \oplus \mathrm{Im}(A^*)$. Use the following matrices

$$\begin{bmatrix} -5 & 2 \\ 15 & -6 \end{bmatrix}, \quad \begin{bmatrix} 3 & -2 \\ 4 & 6 \end{bmatrix}$$

to justify the above results. Also, prove that
(5) $\mathrm{Ker}(AA^*) = \mathrm{Ker}(A)$ and hence, $\mathrm{r}(A) = \mathrm{r}(AA^*)$ (see Ex. 18).

All these results are true for arbitrary $m \times n$ matrix A over the real field $\mathbb{R}$ (for details, see Sec. B.8). Any ideas to prove these?

21. Let A be a real 2×2 matrix (or any $n \times n$ matrix over a field $\mathbb{F}$, not of characteristic equal to 2). Show that

(a) A is idempotent (see Ex. <B> 2 of Sec. 2.7.2), i.e. $A^2 = A$ if and only if $\mathrm{r}(A) + \mathrm{r}(I_2 - A) = 2$.
(b) A is involutory (see Ex. <B> 3 of Sec. 2.7.2), i.e. $A^2 = I_2$ if and only if $\mathrm{r}(I_2 + A) + \mathrm{r}(I_2 - A) = 2$.

<B>

1. Both geometrically and algebraically, do the following problems:

(a) $\begin{bmatrix} \lambda & 0 \\ 0 & 0 \end{bmatrix}$ and $\begin{bmatrix} 0 & 0 \\ 0 & \lambda \end{bmatrix}$, where $\lambda \neq 0$, are similar.

(b) What happens to $\begin{bmatrix} \lambda_1 & 0 \\ 0 & 0 \end{bmatrix}$ and $\begin{bmatrix} 0 & 0 \\ 0 & \lambda_2 \end{bmatrix}$ for similarity if λ_1 and λ_2 are different?

(c) $\begin{bmatrix} 0 & \lambda \\ 0 & 0 \end{bmatrix}$ and $\begin{bmatrix} 0 & 0 \\ \lambda & 0 \end{bmatrix}$ are similar if $\lambda \neq 0$.

(d) $\begin{bmatrix} \lambda & 0 \\ 0 & 0 \end{bmatrix}$ and $\begin{bmatrix} 0 & \lambda \\ 0 & 0 \end{bmatrix}$ are not similar if $\lambda \neq 0$.

2. Let

$$I_2 = \begin{bmatrix} 1 & 0 \\ 0 & 1 \end{bmatrix}, \quad J = \begin{bmatrix} 0 & 1 \\ 1 & 0 \end{bmatrix}, \quad E = \begin{bmatrix} 1 & 1 \\ 0 & 0 \end{bmatrix},$$

$$F = \begin{bmatrix} 0 & 0 \\ 1 & 1 \end{bmatrix}, \quad G = \begin{bmatrix} 1 & 0 \\ 1 & 0 \end{bmatrix}, \quad H = \begin{bmatrix} 0 & 1 \\ 0 & 1 \end{bmatrix}.$$

Adopt the Cartesian coordinate system $\mathcal{N} = \{\vec{e}_1, \vec{e}_2\}$ in $\mathbb{R}^2$.

(a) Explain geometrically and algebraically that it is impossible for I_2 and J to be similar. Instead of this, is it possible that

$$I_2 = PJP^*$$

for invertible P? How about $I_2 = PJQ$ for invertible P and Q?

(b) Show that E, F, G and H are never similar to I_2.

(c) Figure 2.56(a) and (b) explain graphically the mapping properties of E and F, respectively. Both are projections (see Sec. 2.7.4).

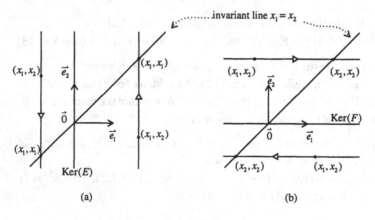

(a)   (b)

Fig. 2.56

Do you have a strong feeling that they should be similar? Yes, they are! Could you see directly from Fig. 2.56 that

$$JFJ = JFJ^{-1} = E?$$

Try to find invertible matrices P and Q such that

$$PEP^{-1} = QFQ^{-1} = \begin{bmatrix} 1 & 0 \\ 0 & 0 \end{bmatrix}.$$

Is it necessary that $J = P^{-1}Q$?

(d) Look at Fig. 2.57(a) and (b). Explain geometrically and algebraically that G and H are similar. Also,

$$JHJ = JHJ^{-1} = G.$$

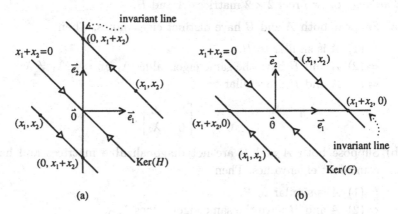

Fig. 2.57

Find invertible matrices R and S such that

$$RGR^{-1} = SHS^{-1} = \begin{bmatrix} 1 & 0 \\ 0 & 0 \end{bmatrix}.$$

(e) Therefore, E, F, G and H are similar to each other. Explain these both geometrically and algebraically.

(f) Determine if the following matrices

$$\begin{bmatrix} -1 & 1 \\ 0 & 0 \end{bmatrix}, \quad \begin{bmatrix} 1 & -1 \\ 0 & 0 \end{bmatrix}, \quad \begin{bmatrix} 1 & 1 \\ 0 & 0 \end{bmatrix}$$

are similar. Justify your answers graphically.

(g) Determine if the following matrices

$$\begin{bmatrix} a & b \\ 0 & 0 \end{bmatrix}, \quad \begin{bmatrix} a & 0 \\ b & 0 \end{bmatrix}, \quad \begin{bmatrix} 0 & b \\ 0 & a \end{bmatrix}, \quad \begin{bmatrix} 0 & 0 \\ b & a \end{bmatrix}, \quad \text{where } ab \neq 0$$

are similar.

3. Let

$$M = \begin{bmatrix} 1 & 1 \\ 1 & 0 \end{bmatrix}, \quad N = \begin{bmatrix} 0 & 1 \\ 1 & 1 \end{bmatrix}, \quad U = \begin{bmatrix} 1 & 1 \\ 0 & 1 \end{bmatrix}, \quad V = \begin{bmatrix} 1 & 0 \\ 1 & 1 \end{bmatrix}.$$

Explain geometrically and algebraically the following statements.

(a) M and N are similar.

(b) U and V are similar.

(c) M and U are not similar.

4. *Similarity* of two real 2×2 matrices A and B.

(a) Suppose both A and B have distinct eigenvalues. Then

(1) A is similar to B.

$\Leftrightarrow$ (2) A and B have the same eigenvalues λ_1 and $\lambda_2, \lambda_1 \neq \lambda_2$.

$\Leftrightarrow$ (3) A and B are similar to

$$\begin{bmatrix} \lambda_1 & 0 \\ 0 & \lambda_2 \end{bmatrix}.$$

(b) Suppose both A and B are not diagonalizable matrices and have coincident eigenvalues. Then

(1) A is similar to B.

$\Leftrightarrow$ (2) A and B have the same eigenvalues λ, λ.

$\Leftrightarrow$ (3) A and B are similar to or have the same Jordan canonical form (see Sec. 2.7.7, if necessary)

$$\begin{bmatrix} \lambda & 0 \\ 1 & \lambda \end{bmatrix}.$$

(c) Suppose A and B do not have real eigenvalues. Then

(1) A is similar to B.

$\Leftrightarrow$ (2) A and B have the same characteristic polynomial

$$t^2 + a_1 t + a_0 \quad \text{with} \quad a_1^2 - 4a_0 < 0.$$

$\Leftrightarrow$ (3) A and B have the same rational canonical form (see Sec. 2.7.8, if necessary)

$$\begin{bmatrix} 0 & 1 \\ -a_0 & -a_1 \end{bmatrix}.$$

Try to explain these results graphically (refer to Figs. 2.46–2.55).

5. Let A and B be two real 2×2 matrices.

(a) Show that AB and BA have the same characteristic polynomial

$$\det(AB - tI_2) = \det(BA - tI_2)$$

and hence the same eigenvalues, if any.

(b) Let

$$A = \begin{bmatrix} 0 & 0 \\ 1 & 1 \end{bmatrix}, \quad B = \begin{bmatrix} 0 & -1 \\ 0 & 1 \end{bmatrix}.$$

Show that AB and BA are not similar.

(c) If either A or B is invertible, the AB and BA are similar.

\<C\> Abstraction and generalization

All the materials in this section can be generalized verbatim or under minor but suitable changes to abstract finite-dimensional vector spaces over a field or $m \times n$ matrices over a field. Minded readers are able to find out their counterparts widely scattered throughout the Appendix B. The main difficulties we encountered are that the methods adopted here are, in many occasions, far insufficiently and ineffectively to provide proofs for these generalized and abstract extensions. More trainings such as experience with linear operators on $\mathbb{R}^3$ and 3×3 matrices (see Chap. 3) and more sophisticated backgrounds are indispensable.

For example, the case-by-case examination in the proof of (2.7.43) is hardly suitable to prove the generalized results for $A_{k \times m}$ and $B_{m \times n}$. On the other hand, the dimension theorem stated in (1) of (2.7.8) might be a hopeful alternative, a method seemed so roundabout and tedious in treating such a simple case as $A_{2 \times 2}$ and $B_{2 \times 2}$. Let us try it now for matrices $A_{k \times m}$ and $B_{m \times n}$ over an arbitrary field $\mathbb{F}$. Consider A and B as linear transformations defined as

$$\vec{x} \in \mathbb{F}^k \xrightarrow{A} \vec{x}A \in \mathbb{F}^m \xrightarrow{B} (\vec{x}A)B = \vec{x}(AB) \in \mathbb{F}^n.$$

Then, watch the steps (and better think geometrically at the same time):

$$\begin{aligned}
r(AB) &= \dim \operatorname{Im}(AB) = \dim(B(\operatorname{Im}(A))) \\
&= \dim \operatorname{Im}(A) - \dim(\operatorname{Im}(A) \cap \operatorname{Ker}(B)) \\
&\geq \dim \operatorname{Im}(A) - \dim \operatorname{Ker}(B) \\
&= \dim \operatorname{Im}(A) - (-\dim \operatorname{Im}(B) + m) \\
&= \dim \operatorname{Im}(A) + \dim \operatorname{Im}(B) - m \\
&= r(A) + r(B) - m.
\end{aligned}$$

Can you follow it and give precise reasons for each step? Or, do you get lost? Where? How? If the answer is positive, try to do the following problems and you will have a good chance to be successful.

1. Generalize (2.7.23) through (2.7.37) to linear transformations from a finite-dimensional vector space V to another finite-dimensional vector space W and $M(m, n; \mathbb{F})$.

2. Generalize (2.7.38) through (2.7.49) to suitable linear operators on $\mathbb{F}^n$ or $n \times n$ matrices over a field $\mathbb{F}$ and prove them.

3. Do Exs. <A> 9 through 21, except 19, for general cases stated in the parentheses.

4. Let $A_{k \times m}$ and $B_{m \times n}$ be such that $r(AB) = m$. Show that $r(A) = r(B) = m$. Is the converse true?

5. Let $B_{k \times n}$ be a $k \times n$ sub-matrix of $A_{m \times n}$ obtained from A by deleting $(m - k)$ row vectors. Show that

$$r(B) \geq r(A) + k - m.$$

6. Suppose $A_{n \times n}$ has at least $n^2 - n + 1$ entries equal to zero. Show that $r(A) \leq n - 1$ and show an example that $r(A) = n - 1$ is possible.

7. Suppose $A_{m \times n}$ and $B_{m \times p}$ are combined to form a partitioned matrix $[A \quad B]_{m \times (n+p)}$ (see Ex. <C> of Sec. 2.7.5). Show that

$$r([A \quad B]) \leq r(A) + r(B)$$

   and the equality is possible.

8. Let $\text{adj}\, A$ denote the adjoint matrix of $A_{n \times n}$ (see (3.3.2) or Sec. B.6). Show that

   (1) $r(A) = n \Leftrightarrow r(\text{adj}\, A) = n$. In this case, $\det(\text{adj}\, A) = (\det A)^{n-1}$.
   (2) $r(A) = n - 1 \Leftrightarrow r(\text{adj}\, A) = 1$.
   (3) $r(A) \leq n - 2 \Leftrightarrow r(\text{adj}\, A) = 0$.
   (4) $\text{adj}(\text{adj}\, A) = (\text{adj}\, A)^{n-2} A$ if $n > 2$; $\text{adj}(\text{adj}\, A) = A$ if $n = 2$.

9. For $A_{k \times m}$ and $B_{m \times n}$, show that $r(AB) = r(A)$ if and only if $\vec{x}(AB) = \vec{0}$ always implies $\vec{x}A = \vec{0}$ where $\vec{x} \in \mathbb{F}^k$ (refer to Ex. 11).

10. Let $A_{k \times m}, B_{m \times n}$, and $C_{l \times k}$ be such that $r(A) = r(AB)$. Show that $r(CA) = r(CAB)$ (refer to Ex. 9).

11. For $A_{k \times m}, B_{m \times n}$ and $C_{n \times p}$.

   (a) Show that, if $x \in \mathbb{F}^k$,

$$\dim\{\vec{x}A \in \mathbb{F}^m \mid \vec{x}AB = \vec{0}\} = k - r(AB) - [k - r(A)]$$
$$= r(A) - r(AB).$$

   Then, try to deduce that $r(A) + r(B) - m \leq r(AB)$. When does equality hold?

   (b) Show that $\text{Im}(AB) \cap \text{Ker}(C) \subseteq \text{Im}(B) \cap \text{Ker}(C)$.

(c) Hence, show that

$$r(AB) + r(BC) \le r(B) + r(ABC).$$

When will equality hold?

12. Let $A \in M(m, n; \mathbb{F})$ and $B \in M(m, p; \mathbb{F})$.

(a) Construct a homogeneous system of linear equations whose solution space is $\operatorname{Ker}(A) \cap \operatorname{Ker}(B)$.

(b) Construct a homogeneous system of linear equations whose solution space is $\operatorname{Ker}(A) + \operatorname{Ker}(B)$.

13. Let $A_{m \times n}, B_{p \times m}$ and $C_{q \times m}$ satisfy

$$r(A) + r(B) = m, \quad BA = O_{p \times n}, \quad CA = O_{q \times n}.$$

Then $\operatorname{Im}(B) = \operatorname{Ker}(A)$. Show that there exists $D_{q \times p}$ such that $C = DB$, and such a D is unique if and only if $r(B) = p$.

14. Try to do Ex. 10 of Sec. B.7.

15. Try to do Ex. 11 of Sec. B.7.

16. Try to do Ex. 12 of Sec. B.7.

2.7.4 *Linear transformations (operators)*

This subsection will devote to the study of some specified linear transformations between $\mathbb{R}$ and $\mathbb{R}^2$ and specially, linear operators on $\mathbb{R}^2$. All the results obtained can be easily extended, both in contents and in proofs, to linear transformations or operators from a finite-dimensional vector space V to another finite-dimensional vector space W over the same field $\mathbb{F}$, but not necessarily of the same dimension.

The feature is that we will try our best to obtain these results without recourse to any particular choices of the bases for the space, i.e. the matrix representations of the linear transformations (or operators). Then, for comparison, we will write these results in their corresponding matrix forms.

Suppose V_1 and V_2 are subspaces of $\mathbb{R}^2$ such that $\mathbb{R}^2 = V_1 \oplus V_2$ holds and hence, each $\vec{x} \in \mathbb{R}^2$ can be uniquely expressed as $\vec{x} = \vec{x}_1 + \vec{x}_2$ where $\vec{x}_1 \in V_1$ and $\vec{x}_2 \in V_2$. A linear operator $f \colon \mathbb{R}^2 \to \mathbb{R}^2$ is called a *projection* of $\mathbb{R}^2$ onto V_1 along V_2 if

$$f(\vec{x}) = \vec{x}_1 \qquad (2.7.50)$$

for each $\vec{x} \in \mathbb{R}^2$ (refer to Ex. 13 and Fig. B.3 of Sec. B.7 for definition in abstract space).

Of course, the zero linear operator is the projection of $\mathbb{R}^2$ onto $\{\vec{0}\}$ along the whole space $\mathbb{R}^2$. The identity operator $1_{\mathbb{R}^2}$ is the projection of $\mathbb{R}^2$ onto itself along $\{\vec{0}\}$. Now, we have

The equivalent criteria of a projection on $\mathbb{R}^2$

Let f be nonzero linear operator on $\mathbb{R}^2$ with rank 1.

- (1) f is a projection of $\mathbb{R}^2$ onto V_1 along V_2.
- $\Leftrightarrow$ (2) $r(f) + r(1_{\mathbb{R}^2} - f) = 2$.
- $\Leftrightarrow$ (3) $1_{\mathbb{R}^2} - f$ is a projection of $\mathbb{R}^2$ onto V_2 along V_1.
- $\Leftrightarrow$ (4) $f^2 = f$ (and hence f is the projection of $\mathbb{R}^2$ onto $\operatorname{Im}(f)$ along $\operatorname{Ker}(f)$), i.e. f is idempotent.
- $\Leftrightarrow$ (5) f has eigenvalues 1 and 0.
- $\Leftrightarrow$ (6) There exists a basis $\mathcal{B} = \{\vec{x}_1, \vec{x}_2\}$ such that

$$[f]_{\mathcal{B}} = P[f]_{\mathcal{N}}P^{-1} = \begin{bmatrix} 1 & 0 \\ 0 & 0 \end{bmatrix}, \quad P = \begin{bmatrix} \vec{x}_1 \\ \vec{x}_2 \end{bmatrix},$$

where, as usual, $\mathcal{N} = \{\vec{e}_1, \vec{e}_2\}$ is the natural basis for $\mathbb{R}^2$.

See Fig. 2.58.   (2.7.51)

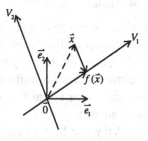

Fig. 2.58

These results still hold for projections on a finite dimensional vector space V, and (1), (3) and (4) are also true even for infinite-dimensional space.

Proof (1) $\Leftrightarrow$ (2) $\Leftrightarrow$ (3) are trivial.

(1) $\Leftrightarrow$ (4): For each $\vec{x} \in \mathbb{R}^2, f(\vec{x}) \in \operatorname{Im}(f) = V_1$ and hence $f(\vec{x}) = f(\vec{x}) + \vec{0}$ shows that $f(f(\vec{x})) = f(\vec{x})$. This means that $f^2 = f$ holds. Conversely, $f^2 = f$ implies that $f(f(\vec{x})) = f(\vec{x})$. Since $r(f) = 1$, let

$\text{Im}(f) = \langle\langle \vec{x}_1 \rangle\rangle$ and $\text{Ker}(f) = \langle\langle \vec{x}_2 \rangle\rangle$. Then f is the projection of $\mathbb{R}^2$ onto $\text{Im}(f)$ along $\text{Ker}(f)$.

(1) $\Rightarrow$ (5): Since for $\vec{x}_1 \in V_1, f(\vec{x}_1) = \vec{x}_1$ and for $\vec{x}_2 \in V_2, f(\vec{x}_2) = \vec{0} = 0 \cdot \vec{x}_2, f$ has only eigenvalues 1 and 0.

(5) $\Leftrightarrow$ (6): obviously.

(6) $\Leftrightarrow$ (4): Notice that $[f]_\mathcal{B}^2 = [f]_\mathcal{B} \Leftrightarrow [f]_\mathcal{N}^2 = [f]_\mathcal{N} \Leftrightarrow f^2 = f$. □

Suppose f is a linear operator on $\mathbb{R}^2$.

If the rank $\text{r}(f) = 2$, then f is invertible and $f^{-1} \circ f = 1_{\mathbb{R}^2}$, the identity operator on $\mathbb{R}^2$ which is a trivial projection.

In case $\text{r}(f) = 1$, let $\mathcal{B} = \{\vec{x}_1, \vec{x}_2\}$ be a basis for $\mathbb{R}^2$ such that $f(\vec{x}_1) \neq \vec{0}$ and $f(\vec{x}_2) = \vec{0}$. Let $\gamma = \{\vec{y}_1, \vec{y}_2\}$ be a basis for $\mathbb{R}^2$ where $\vec{y}_1 = f(\vec{x}_1)$. Define a linear operator $g: \mathbb{R}^2 \to \mathbb{R}^2$ such that

$$g(\vec{y}_i) = \vec{x}_i, \quad i = 1, 2.$$

Therefore,

$$(g \circ f)(\vec{x}_1) = g(f(\vec{x}_1)) = g(\vec{y}_1) = \vec{x}_1$$
$$\Rightarrow (g \circ f)^2(\vec{x}_1) = (g \circ f)(\vec{x}_1) = g(\vec{y}_1) = \vec{x}_1;$$

and

$$(g \circ f)(\vec{x}_2) = g(f(\vec{x}_2)) = g(\vec{0}) = \vec{0}$$
$$\Rightarrow (g \circ f)^2(\vec{x}_2) = (g \circ f)(\vec{x}_2) = \vec{0}.$$

Hence $(g \circ f)^2(\vec{x}) = (g \circ f)(\vec{x})$ for all $\vec{x} \in \mathbb{R}^2$.

We summarize as

The projectionalization of an operator
Let f be a linear operator on $\mathbb{R}^2$. Then there exists an invertible linear operator g on $\mathbb{R}^2$ such that

$$(g \circ f)^2 = g \circ f,$$

i.e. $g \circ f$ is a projection on $\mathbb{R}^2$ (see Fig. 2.59). Equivalently, for any real 2×2 matrix A, there exists an invertible matrix $P_{2 \times 2}$ such that

$$(AP)^2 = AP. \tag{2.7.52}$$

The above result still holds for linear operators on finite-dimensional vector space or any $n \times n$ matrix over a field.

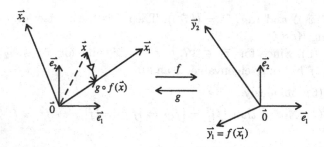

Fig. 2.59

For example, let

$$A = \begin{bmatrix} 1 & 1 \\ -2 & -2 \end{bmatrix}.$$

Then $\vec{e}_1 A = (1,1) = \vec{y}_1$ and $\vec{x}_2 = (2,1)$ satisfying $\vec{x}_2 A = \vec{0}$. Take arbitrary $\vec{y}_2$ linearly independent of $\vec{y}_1$, say $\vec{y}_2 = \vec{e}_2$ for simplicity. Define a square matrix $P_{2\times 2}$ such that

$$\vec{y}_1 P = \vec{e}_1$$
$$\vec{y}_2 P = \vec{x}_2$$

$$\Rightarrow P = \begin{bmatrix} \vec{y}_1 \\ \vec{y}_2 \end{bmatrix}^{-1} \begin{bmatrix} \vec{e}_1 \\ \vec{x}_2 \end{bmatrix} = \begin{bmatrix} 1 & 1 \\ 0 & 1 \end{bmatrix}^{-1} \begin{bmatrix} 1 & 0 \\ 2 & 1 \end{bmatrix} = \begin{bmatrix} 1 & -1 \\ 0 & 1 \end{bmatrix} \begin{bmatrix} 1 & 0 \\ 2 & 1 \end{bmatrix}$$

$$= \begin{bmatrix} -1 & -1 \\ 2 & 1 \end{bmatrix}.$$

Then,

$$AP = \begin{bmatrix} 1 & 1 \\ -2 & -2 \end{bmatrix} \begin{bmatrix} -1 & -1 \\ 2 & 1 \end{bmatrix} = \begin{bmatrix} 1 & 0 \\ -2 & 0 \end{bmatrix}$$

does satisfy $(AP)^2 = AP$.

Suppose f is a nonzero linear operator on $\mathbb{R}^2$ and has rank equal to 1.

Let $\text{Ker}(f) = \langle\langle \vec{x}_2 \rangle\rangle$ with $\vec{x}_2 \neq 0$. Take an arbitrary vector $\vec{x}_1$, linearly independent of $\vec{x}_2$, so that $\mathcal{B} = \{\vec{x}_1, \vec{x}_2\}$ forms a basis for $\mathbb{R}^2$. Thus, $\vec{y}_1 = f(\vec{x}_1) \neq \vec{0}$. Extend $\{\vec{y}_1\}$ to a basis $\mathcal{C} = \{\vec{y}_1, \vec{y}_2\}$ for $\mathbb{R}^2$.

Define linear operators g and h on $\mathbb{R}^2$ respectively as follows.

$$g(\vec{e}_i) = \vec{x}_i,$$
$$h(\vec{y}_i) = \vec{e}_i, \quad \text{for } i = 1, 2.$$

Both g and h are isomorphisms. Hence, the composite $h \circ f \circ g$ is a linear operator with the properties that

$$h \circ f \circ g(\vec{e}_1) = (h \circ f)(\vec{x}_1) = h(\vec{y}_1) = \vec{e}_1,$$
$$h \circ f \circ g(\vec{e}_2) = (h \circ f)(\vec{x}_2) = h(\vec{0}) = \vec{0}$$
$$\Rightarrow \text{if } \vec{x} = (x_1, x_2) = x_1 \vec{e}_1 + x_2 \vec{e}_2,$$
$$h \circ f \circ g(x_1, x_2) = (x_1, 0).$$

This means $h \circ f \circ g$ is the projection of $\mathbb{R}^2$ onto the axis $\langle\langle \vec{e}_1 \rangle\rangle$ along the axis $\langle\langle \vec{e}_2 \rangle\rangle$. See Fig. 2.60.

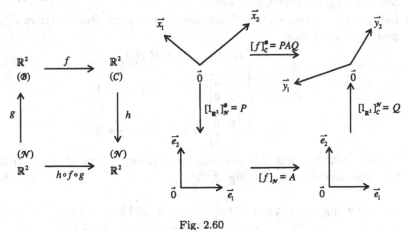

Fig. 2.60

In terms of natural basis $\mathcal{N} = \{\vec{e}_1, \vec{e}_2\}$ and $\mathcal{B}$ and $\mathcal{C}$, we have (see (2.7.23))

$$[h \circ f \circ g]_{\mathcal{N}} = [g]_{\mathcal{B}}^{\mathcal{N}} [f]_{\mathcal{C}}^{\mathcal{B}} [h]_{\mathcal{N}}^{\mathcal{C}} = \begin{bmatrix} 1 & 0 \\ 0 & 0 \end{bmatrix}$$

$$\Rightarrow [f]_{\mathcal{C}}^{\mathcal{B}} = I_2^{-1} \begin{bmatrix} 1 & 0 \\ 0 & 0 \end{bmatrix} I_2^{-1} = \begin{bmatrix} 1 & 0 \\ 0 & 0 \end{bmatrix}, \qquad (2.7.53)$$

which means that, the matrix representation of f with respect to the bases $\mathcal{B}$ and $\mathcal{C}$ is the matrix $\begin{bmatrix} 1 & 0 \\ 0 & 0 \end{bmatrix}$. Equivalently,

$$[f]_{\mathcal{N}} = [1_{\mathbb{R}^2}]_{\mathcal{B}}^{\mathcal{N}} [f]_{\mathcal{C}}^{\mathcal{B}} [1_{\mathbb{R}^2}]_{\mathcal{N}}^{\mathcal{C}}$$

$$\Rightarrow [f]_{\mathcal{C}}^{\mathcal{B}} = [1_{\mathbb{R}^2}]_{\mathcal{N}}^{\mathcal{B}} [f]_{\mathcal{N}} [1_{\mathbb{R}^2}]_{\mathcal{C}}^{\mathcal{N}}$$

$$= \begin{bmatrix} \vec{x}_1 \\ \vec{x}_2 \end{bmatrix} [f]_{\mathcal{N}} \begin{bmatrix} \vec{y}_1 \\ \vec{y}_2 \end{bmatrix}^{-1} = \begin{bmatrix} 1 & 0 \\ 0 & 0 \end{bmatrix}. \qquad (2.7.54)$$

See Fig. 2.60.

We summarize as

The rank theorem of a linear operator on $\mathbb{R}^2$
Let f be a linear operator on $\mathbb{R}^2$.

(1) There exist linear isomorphisms g and h on $\mathbb{R}^2$ such that, for any $\vec{x} = (\vec{x}_1, \vec{x}_2) \in \mathbb{R}^2$,

$$h \circ f \circ g(x_1, x_2) = \begin{cases} (0,0), & \text{if } r(f) = 0, \\ (x_1, 0), & \text{if } r(f) = 1, \\ (x_1, x_2), & \text{if } r(f) = 2. \end{cases}$$

See the diagram beside Fig. 2.60.

(2) Let $A_{2\times 2}$ be a real matrix. Then there exist invertible matrices $P_{2\times 2}$ and $Q_{2\times 2}$ such that

$$PAQ = \begin{cases} O_{2\times 2}, & \text{if } r(A) = 0, \\ \begin{bmatrix} 1 & 0 \\ 0 & 0 \end{bmatrix}, & \text{if } r(A) = 1, \\ \begin{bmatrix} 1 & 0 \\ 0 & 1 \end{bmatrix} = I_2, & \text{if } r(A) = 2. \end{cases}$$

Matrices on the right side are called the *normal form* of A according to its respective rank (see Fig. 2.60). $\hspace{3em}$ (2.7.55)

In general setting, for a $m \times n$ matrix A over a field $\mathbb{F}$, there exist invertible matrices $P_{m\times m}$ and $Q_{n\times n}$ such that the *normal form* of A is

$$PAQ = \begin{bmatrix} I_r & O \\ O & O \end{bmatrix}_{m\times n}, \quad r = r(A), \hspace{2em} (2.7.56)$$

where $I_0 = O$ in case $r = 0$.

Suppose again f is a nonzero linear operator on $\mathbb{R}^2$.

If f has rank 2, so are f^n for any $n \geq 2$ and hence $r(f^n) = 2, n \geq 2$.

In case f has rank 1, let $\text{Ker}(f) = \langle\langle \vec{x}_2 \rangle\rangle$ and $\vec{x}_1$ be linearly independent of $\vec{x}_2$ so that $\text{Im}(f) = \langle\langle f(\vec{x}_1) \rangle\rangle$. If $\langle\langle f(\vec{x}_1) \rangle\rangle = \langle\langle \vec{x}_2 \rangle\rangle$, then

$$f^2(\vec{x}_1) = f(f(\vec{x}_1)) = f(\lambda \vec{x}_2) = \lambda f(\vec{x}_2) = \vec{0}$$

for some $\lambda \in \mathbb{R}$. Together with $f^2(\vec{x}_2) = \vec{0}$, this means that $f^2 = 0$, the zero linear operator. Therefore, $r(f^2) = r(f^3) = \cdots = 0$. If $\langle\langle f(\vec{x}_1) \rangle\rangle \cap \langle\langle \vec{x}_2 \rangle\rangle = \{\vec{0}\}$, then $f^2(\vec{x}_1) = f(f(\vec{x}_1)) \neq \vec{0}$, otherwise $f(\vec{x}_1) \in \langle\langle \vec{x}_2 \rangle\rangle$, a contradiction. Hence, $\text{Im}(f) = \langle\langle f(\vec{x}_1) \rangle\rangle = \langle\langle f^2(\vec{x}_1) \rangle\rangle$ which, in turn, is equal to $\langle\langle f^n(\vec{x}_1) \rangle\rangle$ for any integer $n \geq 3$. This means $r(f^n) = 1$ for $n \geq 1$ in this case.

As a conclusion, we summarize as

The ranks of iterative linear operators

Let f be a linear operator on $\mathbb{R}^2$.

(1) Then

$$r(f^2) = r(f^3) = r(f^4) = \cdots .$$

In fact,

1. If $r(f) = 0$, then $r(f^n) = 0$ for $n \geq 1$.
2. If $r(f) = 1$ and $\text{Ker}(f) = \text{Im}(f)$, then $r(f^n) = 0$ for $n \geq 2$, while if $r(f) = 1$ and $\text{Ker}(f) \cap \text{Im}(f) = \{\vec{0}\}$, then $r(f^n) = 1$ for $n \geq 1$.
3. If $r(f) = 2$, then $r(f^n) = 2$ for $n \geq 1$,

(2) For any real 2×2 matrix A,

$$r(A^2) = r(A^3) = r(A^4) = \cdots . \tag{2.7.57}$$

The aforementioned method can be modified a little bit to prove the general result: for any $n \times n$ matrix A over a field,

$$r(A^n) = r(A^{n+1}) = \cdots . \tag{2.7.58}$$

always holds.

Exercises

<A>

1. Notice that $\mathbb{R}^2 = \langle\langle \vec{e}_1 + \vec{e}_2 \rangle\rangle \oplus \langle\langle \vec{e}_1 \rangle\rangle = \langle\langle \vec{e}_1 + \vec{e}_2 \rangle\rangle \oplus \langle\langle \vec{e}_2 \rangle\rangle$. Suppose f and g are respectively, the projections of $\mathbb{R}^2$ onto $\langle\langle \vec{e}_1 + \vec{e}_2 \rangle\rangle$ along $\langle\langle \vec{e}_2 \rangle\rangle$ and along $\langle\langle \vec{e}_1 \rangle\rangle$. Show that $f \circ g = g$ and $g \circ f = f$.
2. Suppose f and g are idempotent linear operators, i.e. $f^2 = f$ and $g^2 = g$ hold.

    (a) Show that $\text{Im}(f) = \text{Im}(g) \Leftrightarrow g \circ f = f$ and $f \circ g = g$.
    (b) Show that $\text{Ker}(f) = \text{Ker}(g) \Leftrightarrow g \circ f = g$ and $f \circ g = f$.

    (*Note* These results still hold for idempotent linear operators or projections on a finite-dimensional vector space.)
3. Let $A_{2\times 2}$ be

$$\begin{bmatrix} -3 & 4 \\ 1 & -2 \end{bmatrix} \quad \text{or} \quad \begin{bmatrix} -3 & 4 \\ -9 & 12 \end{bmatrix}.$$

    Try to find respective invertible matrix $P_{2\times 2}$ such that $(AP)^2 = AP$. How many such P are possible?

4. Let the linear operator f be defined as

$$f(\vec{x}) = \vec{x} \begin{bmatrix} 6 & 5 \\ -12 & -10 \end{bmatrix}$$

in $\mathcal{N} = \{\vec{e}_1, \vec{e}_2\}$. Try to find linear isomorphisms g and h on $\mathbb{R}^2$ such that $f \circ g \circ h(x_1, x_2) = (x_1, 0)$ for all $\vec{x} = (x_1, x_2) \in \mathbb{R}^2$. How many such g and h are there? Also, find invertible matrices $P_{2\times 2}$ and $Q_{2\times 2}$ such that

$$P \begin{bmatrix} 6 & 5 \\ -12 & -10 \end{bmatrix} Q = \begin{bmatrix} 1 & 0 \\ 0 & 0 \end{bmatrix}$$

and explain it graphically (see Fig. 2.60).

5. Let f be a linear operator on $\mathbb{R}^2$ and k a positive integer. Show that

$$\operatorname{Im}(f^k) = \operatorname{Im}(f^{2k}) \Leftrightarrow \mathbb{R}^2 = \operatorname{Im}(f^k) \oplus \operatorname{Ker}(f^k).$$

(*Note* This result still holds for linear operators on a finite-dimensional vector space.)

6. Let $f: \mathbb{R}^2 \to \mathbb{R}$ and $g: \mathbb{R} \to \mathbb{R}^2$ be linear transformations. Show that $g \circ f: \mathbb{R}^2 \to \mathbb{R}^2$ is never invertible. How about $f \circ g$?
(*Note* Suppose $m > n$ and $f \in \operatorname{Hom}(\mathbb{F}^m, \mathbb{F}^n)$ and $g \in \operatorname{Hom}(\mathbb{F}^n, \mathbb{F}^m)$. Then the composite $g \circ f$ is not invertible. What is its counterpart in matrices?)

7. Let $f \in \operatorname{Hom}(\mathbb{R}^2, \mathbb{R})$ and $g \in \operatorname{Hom}(\mathbb{R}, \mathbb{R}^2)$.

   (a) Show that f is onto if and only if there exists $h \in \operatorname{Hom}(\mathbb{R}, \mathbb{R}^2)$ such that

$$f \circ h = 1_{\mathbb{R}} \text{ (the identity operator on } \mathbb{R}).$$

   In this case, f is called *right invertible* and h a *right inverse* of f.

   (b) Show that g is one-to-one if and only if there exists $h \in \operatorname{Hom}(\mathbb{R}^2, \mathbb{R})$ such that

$$h \circ g = 1_{\mathbb{R}}.$$

   Then, g is called *left-invertible* and h a *left inverse* of g.

   (*Note* In general, let $f \in \operatorname{Hom}(V, W)$. Then

   (1) f is onto $\Leftrightarrow$ there exists $h \in \operatorname{Hom}(W, V)$ such that $f \circ h = 1_W$.
   (2) f is one-to-one $\Leftrightarrow$ there exists $h \in \operatorname{Hom}(W, V)$ such that $h \circ f = 1_V$.

   What are the counterparts for matrices (see Ex. 5 of Sec. B.7)? Try to give suitable geometric interpretations. See also Ex. <B> 5(d), (e) of Sec. 3.3 and Ex. <B> 6(e), (f) of Sec. 3.7.3.)

8. Let $A_{2\times2}$ and $B_{2\times2}$ be real matrices. Show that

$$\det(AB - tI_2) = \det(BA - tI_2)$$

by the following steps (for other proof, try to use (2.7.12) and see Ex. <B> 5 of Sec. 2.7.3).

(a) We may suppose $r(A) = 1$ and $A^2 = A$. By (2.7.51), there exists invertible $P_{2\times2}$ such that

$$A = P^{-1} \begin{bmatrix} 1 & 0 \\ 0 & 0 \end{bmatrix} P.$$

Then

$$\det(AB - tI_2) = \det\left(\begin{bmatrix} 1 & 0 \\ 0 & 0 \end{bmatrix} PBP^{-1} - tI_2\right), \quad \text{and}$$

$$\det(BA - tI_2) = \det\left(PBP^{-1} \begin{bmatrix} 1 & 0 \\ 0 & 0 \end{bmatrix} - tI_2\right).$$

Let $PBP^{-1} = \begin{bmatrix} b_{11} & b_{12} \\ b_{21} & b_{22} \end{bmatrix}$. Then

$$\det(AB - tI_2) = \det(BA - tI_2) = t(t - b_{11}).$$

(b) Suppose $r(A) = 1$ only. By (2.7.52), there exists invertible $P_{2\times2}$ such that $(AP)^2 = AP$ holds. Then, use (a) to finish the proof.

(*Note*   For general matrices $A_{m\times n}$ and $B_{n\times m}$ over a field, exactly the same method guarantees that

$$(-1)^{n-m}t^n\det(AB - tI_m) = t^m\det(BA - tI_n).$$

Do you know how to do this? Refer to Ex. <C> 9 of Sec. 2.7.5.)

9. Suppose $f, g \in \operatorname{Hom}(\mathbb{R}^2, \mathbb{R}^2)$ and $f \circ g = 0$. Show that

$$r(f) + r(g) \leq 2.$$

When will the equality hold?

(*Note*   This result still holds for $f, g \in \operatorname{Hom}(V, V)$, where $\dim V = n < \infty$, and $f \circ g = 0$.)

10. Suppose $f, g, h \in \operatorname{Hom}(\mathbb{R}^2, \mathbb{R}^2)$ and $f \circ g \circ h = 0$. Show that

$$r(f) + r(g) + r(h) \leq 2 \cdot 2 = 4.$$

When will equality hold?

(*Note*   For $f, g, h \in \operatorname{Hom}(V, V)$, where $\dim V = n < \infty$, and $f \circ g \circ h = 0$, then $r(f) + r(g) + r(h) \leq 2n$ holds.)

11. Suppose $f, g \in \text{Hom}(\mathbb{R}^2, \mathbb{R}^2)$ and $f + g = 1_{\mathbb{R}^2}$. Show that

$$\text{r}(f) + \text{r}(g) \geq 2$$

and the equality holds if and only if $f^2 = f$, $g^2 = g$ and $f \circ g = g \circ f = 0$. (*Note* This result still holds for $f, g \in \text{Hom}(V, V)$, where $\dim V = n$ and $f + g = 1_V$. Then $r(f) + r(g) \geq n$ and the equality holds if and only if $f^2 = f$, $g^2 = g$ and $f \circ g = g \circ f = 0$.)

12. Suppose $f \in \text{Hom}(\mathbb{R}^2, \mathbb{R}^2)$.

(a) If there exists $g \in \text{Hom}(\mathbb{R}^2, \mathbb{R}^2)$ so that $g \circ f = 1_{\mathbb{R}^2}$ or $f \circ g = 1_{\mathbb{R}^2}$ holds, then f is invertible and $g = f^{-1}$.

(b) If there exists a *unique* $g \in \text{Hom}(\mathbb{R}^2, \mathbb{R}^2)$ so that $f \circ g = 1_{\mathbb{R}^2}$, then $g \circ f = 1_{\mathbb{R}^2}$ holds and this f is invertible.

(c) There exists a $g \in \text{Hom}(\mathbb{R}^2, \mathbb{R}^2)$ such that

$$f \circ g \circ f = f.$$

In this case, $(f \circ g)^2 = f \circ g$ and $(g \circ f)^2 = g \circ f$ hold simultaneously.

(*Note* Let V be a vector space over a field $\mathbb{F}$. Then (a) still holds for $f \in \text{Hom}(V, V)$ in case $\dim V < \infty$, while (b) and (c) hold for arbitrary V. Moreover, if there exist two different g and h in $\text{Hom}(V, V)$ so that $f \circ g = f \circ h = 1_V$, then there are infinitely many $k \in \text{Hom}(V, V)$ so that $f \circ k = 1_V$ in case the underlying field $\mathbb{F}$ is of characteristic zero.)

13. Suppose $f, g \in \text{Hom}(\mathbb{R}^2, \mathbb{R}^2)$ such that $\text{r}(f) \leq \text{r}(g)$ holds. Then, there exist $h_1, h_2 \in \text{Hom}(\mathbb{R}^2, \mathbb{R}^2)$ such that

$$f = h_2 \circ g \circ h_1$$

and h_1 and h_2 can be chosen as invertible linear operators in case $\text{r}(f) = \text{r}(g)$.

(*Note* For $f, g \in \text{Hom}(V, V)$, where $\dim V < \infty$, the result still holds.)

14. Suppose $f \in \text{Hom}(\mathbb{R}^2, \mathbb{R}^2)$ commutes with any $g \in \text{Hom}(\mathbb{R}^2, \mathbb{R}^2)$, i.e. $f \circ g = g \circ f$ for all such g. Show that, it is necessary and sufficient that there exists scalar $\alpha \in \mathbb{R}$ such that

$$f = \alpha 1_{\mathbb{R}^2}.$$

Try the following steps.

(1) For any $\vec{x} \in \mathbb{R}^2$, $\{\vec{x}, f(\vec{x})\}$ should be linearly dependent.

(2) Take any *nonzero* $\vec{x}_0 \in \mathbb{R}^2$, then $f(\vec{x}_0) = \alpha \vec{x}_0$ for some scalar α. Thus, $f(\vec{x}) = \alpha \vec{x}$ should hold for any $\vec{x} \in \mathbb{R}^2$.

(*Note* The result still holds for $f \in \text{Hom}(V, V)$ where V is arbitrary.)

15. Suppose $f, g \in \text{Hom}(\mathbb{R}^2, \mathbb{R}^2)$.

    (a) There exist two bases $\mathcal{B}$ and $\mathcal{C}$ for $\mathbb{R}^2$ so that $[f]_\mathcal{B} = [g]_\mathcal{C}$ if and only if

    $$g = h \circ f \circ h^{-1}$$

    where $h \in \text{Hom}(\mathbb{R}^2, \mathbb{R}^2)$ is an invertible operator.

    (b) For any basis $\mathcal{B}$ for $\mathbb{R}^2$, $[f]_\mathcal{B}$ is the same matrix if and only if $f = \alpha 1_{\mathbb{R}^2}$ for some scalar $\alpha \in \mathbb{R}$.

    (*Note* The above result still holds for any $f \in \text{Hom}(V, V)$, where $\dim V = n < \infty$.)

<B>

1. Let $f, g \in \text{Hom}(\mathbb{R}^2, \mathbb{R}^2)$ and f be diagonalizable but not a scalar matrix.

    (a) Suppose $f \circ g = g \circ f$. Then f and g have a common eigenvector, i.e. there exists a nonzero vector $\vec{x}$ such that

    $$f(\vec{x}) = \lambda \vec{x} \quad \text{and} \quad g(\vec{x}) = \mu \vec{x}$$

    for some scalars λ and μ. Therefore, g is diagonalizable.

    (b) Suppose $f \circ g = g \circ f$. Then, there exists a basis $\mathcal{B}$ for $\mathbb{R}^2$ such that $[f]_\mathcal{B}$ and $[g]_\mathcal{B}$ are both diagonal matrices.

    (*Note* Suppose $A_{n \times n}$ and $B_{n \times n}$ are complex matrices such that $AB = BA$. Then there exists an invertible matrix P such that PAP^{-1} and PBP^{-1} are both upper (or lower) triangular matrices. See Ex. <C> 10 of Sec. 2.7.6.)

2. Let $f_i \in \text{Hom}(\mathbb{R}^2, \mathbb{R}^2)$ for $1 \le i \le 2$ satisfy

    $$f_i \circ f_j = \delta_{ij} f_i, \quad 1 \le i, j \le 2.$$

    (a) Show that either each $\text{r}(f_i) = 0$ for $1 \le i \le 2$ or each $\text{r}(f_i) = 1$ for $1 \le i \le 2$.

    (b) Suppose $g_i \in \text{Hom}(\mathbb{R}^2, \mathbb{R}^2)$ for $1 \le i \le 2$ satisfy $g_i \circ g_j = \delta_{ij} g_i$, $1 \le i, j \le 2$.

    Show that, there exists a linear isomorphism φ on $\mathbb{R}^2$ such that

    $$g_i = \varphi \circ f_i \circ \varphi^{-1}, \quad 1 \le i \le 2.$$

    (*Note* Let $\dim V = n$ and $f_i \in \text{Hom}(V, V)$ for $1 \le i \le n$ satisfy $f_i \circ f_j = \delta_{ij} f_i$, $1 \le i, j \le n$. Then the above results still hold.)

3. For $f, g \in \text{Hom}(\mathbb{R}^2, \mathbb{R}^2)$, define the *bracket operation* $[,]$ of f and g as

$$[f, g] = f \circ g - g \circ f.$$

Show that, for any $f, g, h \in \text{Hom}(\mathbb{R}^2, \mathbb{R}^2)$,

(1) (bilinearity) $[,]$ is bilinear, i.e. $[f, \cdot]: g \to [f, g]$ and $[\cdot, g]: f \to [f, g]$ are linear operators on $\text{Hom}(\mathbb{R}^2, \mathbb{R}^2)$,
(2) $[f, f] = 0$, and
(3) (Jacobi identity)

$$[f, [g, h]] + [g, [h, f]] + [h, [f, g]] = 0.$$

Then, the algebra $\text{Hom}(\mathbb{R}^2, \mathbb{R}^2)$ or the vector space $\mathbb{R}^2$, is called a *Lie algebra* according to the bracket operation $[,]$ and the properties 1–3. (*Note* Let $\dim V < \infty$. Then $\text{Hom}(V, V)$ or the vector space V is a Lie algebra with the bracket operation $[,]$. Equivalently, $\text{M}(n; \mathbb{F})$ with $[A, B] = AB - BA$ is also a Lie algebra.)

4. (a) Suppose $f \in \text{Hom}(\mathbb{R}^2, \mathbb{R}^2)$. Show that (see (2.7.39))

$$\det(f - t1_{\mathbb{R}^2}) = t^2 - (\text{tr } f)t + \det f$$

and hence, f satisfies its characteristic polynomial (see (2.7.19))

$$f^2 - (\text{tr } f)f + (\det f)1_{\mathbb{R}^2} = 0.$$

(b) Show that $f^2 = -\lambda 1_{\mathbb{R}^2}$ with $\lambda > 0$ if and only if

$$\det f > 0 \quad \text{and} \quad \text{tr } f = 0.$$

(c) For $f, g \in \text{Hom}(\mathbb{R}^2, \mathbb{R}^2)$, show that

$$f \circ g + g \circ f = (\text{tr } f)g + (\text{tr } g)f + [\text{tr }(f \circ g) - (\text{tr} f) \cdot (\text{tr } g)]1_{\mathbb{R}^2}.$$

(*Note* (a) and (c) are still true for any vector space V with $\dim V = 2$, while (b) is true only for real vector space.)

<C>

1. Do (2.7.51) for finite-dimensional vector space V.
2. Do (2.7.52) for finite-dimensional vector space V.
3. Do (2.7.56). Prove that a matrix $A_{n \times n}$ of rank r can be written as a sum of r matrices of rank 1.

4. Do (2.7.58).
5. Prove the statements inside the so many Notes in Exs. <A> and <B>.
6. Try your best to do as many problems as possible from Ex. 13 through Ex. 24 in Sec. B.7.
7. For any $A, B \in M(n; \mathbb{C})$ such that B is invertible, show that there exists a scalar $\lambda \in \mathbb{C}$ such that $A + \lambda B$ is not invertible. How many such different λ could be chosen?
8. Show that $T: M(n; \mathbb{F}) \to M(n; \mathbb{F})$ is a linear operator if and only if, there exists $A \in M(n; \mathbb{F})$ such that

$$T(X) = XA, \quad X \in M(n; \mathbb{F}).$$

Therefore, show that $T: M(n; \mathbb{F}) \to M(n; \mathbb{F})$ is a linear operator if and only if, there exist a positive integer k and matrices $Q_j, R_j \in M(n; \mathbb{R})$ for $1 \le j \le k$ such that

$$T(X) = \sum_{j=1}^{k} Q_j X R_j, \quad X \in M(n; \mathbb{F}).$$

9. Suppose $\dim V < \infty$ and $T: \mathrm{Hom}(V, V) \to \mathbb{F}$ is a linear functional.

   (a) There exist a linear operator $f_0 \in \mathrm{Hom}(V, V)$, uniquely determined by T, such that

   $$T(f) = \mathrm{tr}\,(f \circ f_0), \quad f \in \mathrm{Hom}(V, V).$$

   (b) Suppose T satisfies $T(g \circ f) = T(f \circ g)$ for any $f, g \in \mathrm{Hom}(V, V)$. Show that there exists a scalar λ such that

   $$T(f) = \lambda \, \mathrm{tr}\,(f), \quad f \in \mathrm{Hom}(V, V).$$

   (c) Suppose $T: \mathrm{Hom}(V, V) \to \mathrm{Hom}(V, V)$ is a linear transformation satisfying that $T(g \circ f) = T(g) \circ T(f)$ for any $f, g \in \mathrm{Hom}(V, V)$ and $T(1_V) = 1_V$. Show that

   $$\mathrm{tr}\, T(f) = \mathrm{tr}\, f, \quad f \in \mathrm{Hom}(V, V).$$

10. For $P \in M(m; \mathbb{R})$ and $Q \in M(n; \mathbb{R})$, define $\sigma(P, Q)$: $M(m, n; \mathbb{R}) \rightarrow M(m, n; \mathbb{R})$ as

$$\sigma(P, Q)X = PXQ^*, \quad X \in M(m, n; \mathbb{R}).$$

Show that $\sigma(P, Q)$ is linear and

$$\det \sigma(P, Q) = (\det P)^n (\det Q)^m.$$

(*Note*   In tensor algebra, $\sigma(P, Q)$ is denoted as

$$P \otimes Q$$

and is called the *tensor product* of P and Q. Results obtained here can be used to discuss the orientations of Grassmann manifolds and projective spaces.)

\<D\> Applications

(2.7.55) or (2.7.56) can be localized to obtain *the rank theorem* for continuously differentiable mapping from open set in $\mathbb{R}^m$ to $\mathbb{R}^n$. We will mention this along with other local versions of results from linear algebra in Chaps. 4 and 5.

2.7.5   *Elementary matrices and matrix factorizations*

The essence of *elimination method* in solving systems of linear equations lies on the following three basic operations:

1. Type 1: Interchange any pair of equations.
2. Type 2: Multiply any equation by a nonzero scalar.
3. Type 3: Replace any equation by its sum with a multiple of any other equation.

These three operations on equations are, respectively, the same operations on the corresponding *coefficient vectors* without changing the order of unknowns. Therefore, there correspond three types of operations on row vectors of the *coefficient matrix* of the equations.

For example, let us put the system of equations

$$\begin{cases} 5x_1 - 3x_2 = 6 \\ x_1 + 4x_2 = 2 \end{cases}$$

with its matrix form

$$\begin{bmatrix} 5 & -3 \\ 1 & 4 \end{bmatrix} \begin{bmatrix} x_1 \\ x_2 \end{bmatrix} = \begin{bmatrix} 6 \\ 2 \end{bmatrix}$$

side by side for comparison when solving the equations as follows. $(1) \leftrightarrow (2)$ means type 1, $(2) + (-5) \times (1)$ means type 3, and $-\frac{1}{23}(2)$ means type 2, etc.

$$
\begin{array}{ll}
5x_1 - 3x_2 = 6 & (1) \\
x_1 + 4x_2 = 2 & (2)
\end{array}
\qquad
\begin{bmatrix} 5 & -3 \\ 1 & 4 \end{bmatrix} \begin{bmatrix} x_1 \\ x_2 \end{bmatrix} = \begin{bmatrix} 6 \\ 2 \end{bmatrix}
$$

$$\downarrow (1) \leftrightarrow (2) \qquad\qquad \downarrow E_{(1)(2)}$$

$$
\begin{array}{ll}
x_1 + 4x_2 = 2 & (1) \\
5x_1 - 3x_2 = 6 & (2)
\end{array}
\qquad
\begin{bmatrix} 1 & 4 \\ 5 & -3 \end{bmatrix} \begin{bmatrix} x_1 \\ x_2 \end{bmatrix} = \begin{bmatrix} 2 \\ 6 \end{bmatrix}
$$

$$\downarrow (2) + (-5) \times (1) \qquad\qquad \downarrow E_{(2)-5(1)}$$

$$
\begin{array}{ll}
x_1 + 4x_2 = 2 & (1) \\
-23x_2 = -4 & (2)
\end{array}
\qquad
\begin{bmatrix} 1 & 4 \\ 0 & -23 \end{bmatrix} \begin{bmatrix} x_1 \\ x_2 \end{bmatrix} = \begin{bmatrix} 2 \\ -4 \end{bmatrix}
$$

$$\downarrow -\frac{1}{23}(2) \qquad\qquad \downarrow E_{-\frac{1}{23}(2)}$$

$$
\begin{array}{ll}
x_1 + 4x_2 = 2 & (1) \\
x_2 = \dfrac{4}{23} & (2)
\end{array}
\qquad
\begin{bmatrix} 1 & 4 \\ 0 & 1 \end{bmatrix} \begin{bmatrix} x_1 \\ x_2 \end{bmatrix} = \begin{bmatrix} 2 \\ \frac{4}{23} \end{bmatrix}
$$

$$\downarrow (1) + (-4) \times (2) \qquad\qquad \downarrow E_{(1)-4(2)}$$

$$
\begin{array}{ll}
x_1 = \dfrac{30}{23} & (1) \\
x_2 = \dfrac{4}{23} & (2)
\end{array}
\qquad
\begin{bmatrix} 1 & 0 \\ 0 & 1 \end{bmatrix} \begin{bmatrix} x_1 \\ x_2 \end{bmatrix} = \begin{bmatrix} \frac{30}{23} \\ \frac{4}{23} \end{bmatrix}.
$$

Thus, we introduce three types of *elementary row (or column) operations* on a matrix $A_{2\times 2}$ or any $A_{m\times n}$ as follows:

Type 1: interchange of ith row and jth row, $i \neq j$,

Type 2: multiply ith row by a scalar $\alpha \neq 0$,

Type 3: add a multiple of ith row to jth row, $\qquad$ (2.7.59)

and the corresponding column operations. The matrices obtained by performing these elementary row or column operations on the *identity matrix*

I_2 or I_n are called *elementary matrices* and are denoted respectively as

row operations: $E_{(i)(j)}$, $E_{\alpha(i)}$ and $E_{(j)+\alpha(i)}$,

column operations: $F_{(i)(j)}$, $F_{\alpha(i)}$ and $F_{(j)+\alpha(i)}$.   (2.7.60)

For example, 2×2 elementary matrices are:

$$E_{(1)(2)} = \begin{bmatrix} 0 & 1 \\ 1 & 0 \end{bmatrix} = F_{(1)(2)},$$

$$E_{\alpha(1)} = \begin{bmatrix} \alpha & 0 \\ 0 & 1 \end{bmatrix} = F_{\alpha(1)}; \quad E_{\alpha(2)} = \begin{bmatrix} 1 & 0 \\ 0 & \alpha \end{bmatrix} = F_{\alpha(2)},$$

$$E_{(2)+\alpha(1)} = \begin{bmatrix} 1 & 0 \\ \alpha & 1 \end{bmatrix} = F_{(1)+\alpha(2)},$$

$$E_{(1)+\alpha(2)} = \begin{bmatrix} 1 & \alpha \\ 0 & 1 \end{bmatrix} = F_{(2)+\alpha(1)}.$$

Also, refer to Sec. 2.7.2 for geometric mapping properties of these elementary matrices. For more theoretical information about or obtained by elementary matrices, please refer to Sec. B.5.

To see the advantages of the introduction of elementary matrices, let us start from concrete examples.

Example 1 Let

$$A = \begin{bmatrix} 1 & 2 \\ 4 & 10 \end{bmatrix}.$$

(1) Solve the equation $\vec{x}A = \vec{b}$ where $\vec{x} = (x_1, x_2) \in \mathbb{R}^2$ and $\vec{b} = (b_1, b_2)$ is a constant vector.

(2) Investigate the geometric mapping properties of A.

Solution Where written out, $\vec{x}A = \vec{b}$ is equivalent to

$$(x_1, x_2) \begin{bmatrix} 1 & 2 \\ 4 & 10 \end{bmatrix} = (b_1, b_2) \quad \text{or} \quad \begin{cases} x_1 + 4x_2 = b_1 \\ 2x_1 + 10x_2 = b_2 \end{cases}.$$

A is called the *coefficient matrix* of the equations and the 3×2 matrix $\begin{bmatrix} A \\ \vec{b} \end{bmatrix}_{3 \times 2}$ its *augmented matrix*. Apply consecutive column operations to the augmented matrix as follows.

$$\begin{bmatrix} A \\ -- \\ \vec{b} \end{bmatrix} = \begin{bmatrix} 1 & 2 \\ 4 & 10 \\ ----- \\ b_1 & b_2 \end{bmatrix} \xrightarrow{F_{(2)-2(1)}} \begin{bmatrix} 1 & 0 \\ 4 & 2 \\ ----- \\ b_1 & b_2-2b_1 \end{bmatrix} = \begin{bmatrix} A \\ -- \\ \vec{b} \end{bmatrix} F_{(2)-2(1)} \qquad (*1)$$

$$\xrightarrow{F_{\frac{1}{2}(2)}} \begin{bmatrix} 1 & 0 \\ 4 & 1 \\ ------ \\ b_1 & \frac{b_2-2b_1}{2} \end{bmatrix} = \begin{bmatrix} A \\ -- \\ \vec{b} \end{bmatrix} F_{(2)-2(1)}F_{\frac{1}{2}(2)} \qquad (*2)$$

$$\xrightarrow{F_{(1)-4(2)}} \begin{bmatrix} 1 & 0 \\ 0 & 1 \\ ------------ \\ 5b_1 - 2b_2 & \frac{b_2-2b_1}{2} \end{bmatrix}$$

$$= \begin{bmatrix} A \\ -- \\ \vec{b} \end{bmatrix} F_{(2)-2(1)}F_{\frac{1}{2}(2)}F_{(1)-4(2)}.$$

We can deduce some valuable information from the above process and the final result follows.

(a) *The solution of* $\vec{x}A = \vec{b}$

$$\begin{cases} x_1 = 5b_1 - 2b_2 \\ x_2 = -b_1 + \dfrac{1}{2}b_2 \end{cases}$$

is the solution of the equations. It is better to put in matrix form as

$$\vec{x} = (x_1 \quad x_2) = (b_1 \quad b_2) \begin{bmatrix} 5 & -1 \\ -2 & \frac{1}{2} \end{bmatrix} = \vec{b}A^{-1}.$$

In particular, we obtain the inverse A^{-1} of A.

(b) *The invertibility of A and its inverse* A^{-1}

$$I_2 = \begin{bmatrix} 1 & 0 \\ 0 & 1 \end{bmatrix} = AF_{(2)-2(1)}F_{\frac{1}{2}(2)}F_{(1)-4(2)}$$

$$\Rightarrow A^{-1} = F_{(2)-2(1)}F_{\frac{1}{2}(2)}F_{(1)-4(2)} = \begin{bmatrix} 1 & -2 \\ 0 & 1 \end{bmatrix}\begin{bmatrix} 1 & 0 \\ 0 & \frac{1}{2} \end{bmatrix}\begin{bmatrix} 1 & 0 \\ -4 & 1 \end{bmatrix}$$

$$= \frac{1}{2}\begin{bmatrix} 10 & -2 \\ -4 & 1 \end{bmatrix} = \begin{bmatrix} 5 & -1 \\ -2 & \frac{1}{2} \end{bmatrix}.$$

By the way, A^{-1} can be written as a product of elementary matrices.

(c) *A as a product of elementary matrices*

$$A = F^{-1}_{(1)-4(2)} F^{-1}_{\frac{1}{2}(2)} F^{-1}_{(2)-2(1)} = F_{(1)+4(2)} F_{2(2)} F_{(2)+2(1)}$$

$$= \begin{bmatrix} 1 & 0 \\ 4 & 1 \end{bmatrix} \begin{bmatrix} 1 & 0 \\ 0 & 2 \end{bmatrix} \begin{bmatrix} 1 & 2 \\ 0 & 1 \end{bmatrix}. \tag{2.7.61}$$

This factorization provides another way to investigate the geometric mapping properties of A than those presented in Sec. 2.7.4.

(d) *The computations of* $\det A$ *and* $\det A^{-1}$

$$\det A = \det \begin{bmatrix} 1 & 0 \\ 4 & 1 \end{bmatrix} \cdot \det \begin{bmatrix} 1 & 0 \\ 0 & 2 \end{bmatrix} \cdot \det \begin{bmatrix} 1 & 2 \\ 0 & 1 \end{bmatrix} = 1 \cdot 2 \cdot 1 = 2,$$

$$\det A^{-1} = \det \begin{bmatrix} 1 & -2 \\ 0 & 1 \end{bmatrix} \cdot \det \begin{bmatrix} 1 & 0 \\ 0 & \frac{1}{2} \end{bmatrix} \cdot \det \begin{bmatrix} 1 & 0 \\ -4 & 1 \end{bmatrix} = 1 \cdot \frac{1}{2} \cdot 1 = \frac{1}{2}.$$

Note When written in the *transpose form* $A^* \vec{x}^* = \vec{b}^*$, we can apply elementary row operations to the *augmented matrix* $[A^* \mid \vec{b}^*]_{2 \times 3}$ and obtain exactly the same factorizations for A and A^{-1} and solve the equations. The only difference is that, when applying row operations to A^*, the corresponding elementary matrices should multiply A in the front stepwise, i.e.

$$E_{(1)-4(2)} E_{\frac{1}{2}(2)} E_{(2)-2(1)} A^* = I_2,$$

which is the same as $A F_{(2)-2(1)} F_{\frac{1}{2}(2)} F_{(1)-4(2)} = I_2$ obtained when applying column operations.

Adopt the factorization in (2.7.61) and see Fig. 2.61 for geometric mapping properties step by step.

Note that another way to study the geometric mapping properties is to find the eigenvalues $\frac{11 \pm \sqrt{113}}{2}$ of A and model after Fig. 2.50 and Example 6 ($\lambda_1 \neq \lambda_2$) in Sec. 2.7.2.

Note that (2.7.61) can also be written as

$$A = \begin{bmatrix} 1 & 0 \\ 4 & 2 \end{bmatrix} \begin{bmatrix} 1 & 2 \\ 0 & 1 \end{bmatrix} = \begin{bmatrix} 1 & 0 \\ 4 & 1 \end{bmatrix} \begin{bmatrix} 1 & 2 \\ 0 & 2 \end{bmatrix}, \tag{2.7.62}$$

which is a product of a lower triangular matrix before an upper triangular matrix. Readers are urged to explain how these factorizations affect the mapping properties in Fig. 2.61.

It is worthy to notice that the appearance of lower and upper triangular matrices factorization in (2.7.62) is not accidental and is not necessarily a consequence of the previous elementary matrices factorization in (2.7.61).

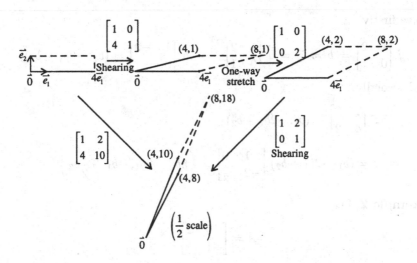

Fig. 2.61

We get it in the algebraic operation process up to the steps $(*_1)$ and $(*_2)$. Stop there and we deduce that

$$AF_{(2)-2(1)} = \begin{bmatrix} 1 & 0 \\ 4 & 2 \end{bmatrix}$$

$$\Rightarrow A = \begin{bmatrix} 1 & 0 \\ 4 & 2 \end{bmatrix} F^{-1}_{(2)-2(1)} = \begin{bmatrix} 1 & 0 \\ 4 & 2 \end{bmatrix} F_{(2)+2(1)} = \begin{bmatrix} 1 & 0 \\ 4 & 2 \end{bmatrix} \begin{bmatrix} 1 & 2 \\ 0 & 1 \end{bmatrix},$$

and

$$AF_{(2)-2(1)} F_{\frac{1}{2}(2)} = \begin{bmatrix} 1 & 0 \\ 4 & 1 \end{bmatrix}$$

$$\Rightarrow A = \begin{bmatrix} 1 & 0 \\ 4 & 1 \end{bmatrix} F^{-1}_{\frac{1}{2}(2)} F^{-1}_{(2)-2(1)} = \begin{bmatrix} 1 & 0 \\ 4 & 1 \end{bmatrix} \begin{bmatrix} 1 & 2 \\ 0 & 2 \end{bmatrix}.$$

Does factorization like (2.7.62) help in solving the equations $\vec{x}A = \vec{b}$? Yes, it does. Take the first factorization for example,

$$\vec{x} \begin{bmatrix} 1 & 2 \\ 4 & 10 \end{bmatrix} = \vec{b}$$

$$\Leftrightarrow \vec{x} \begin{bmatrix} 1 & 0 \\ 4 & 2 \end{bmatrix} = \vec{y} \text{ and } \vec{y} \begin{bmatrix} 1 & 2 \\ 0 & 1 \end{bmatrix} = \vec{b} \text{ where } \vec{y} = (y_1, y_2).$$

Solve firstly

$$\vec{y}\begin{bmatrix} 1 & 2 \\ 0 & 1 \end{bmatrix} = \vec{b} \Rightarrow \vec{y} = \vec{b}\begin{bmatrix} 1 & 2 \\ 0 & 1 \end{bmatrix}^{-1} = \vec{b}\begin{bmatrix} 1 & -2 \\ 0 & 1 \end{bmatrix} = (b_1, -2b_1 + b_2),$$

and secondly

$$\vec{x}\begin{bmatrix} 1 & 0 \\ 4 & 2 \end{bmatrix} = (b_1, -2b_1 + b_2)$$

$$\Rightarrow \vec{x} = (b_1, -2b_1 + b_2)\begin{bmatrix} 1 & 0 \\ -2 & \frac{1}{2} \end{bmatrix} = \left(5b_1 - 2b_1, -b_1 + \frac{1}{2}b_2\right).$$

Example 2 Let

$$A = \begin{bmatrix} 0 & 2 \\ -1 & 1 \end{bmatrix}.$$

Do the same problems as in Example 1.

Solution For $\vec{x} = (x_1, x_2)$ and $\vec{b} = (b_1, b_2)$, $\vec{x}A = \vec{b}$ is equivalent to

$$\begin{cases} 0 \cdot x_1 - x_2 = b_1 \\ 2x_1 + x_2 = b_2 \end{cases}$$

which can be easily solved as $x_1 = \frac{1}{2}(b_1 + b_2)$ and $x_2 = -b_1$.

The shortcoming of A, for the purpose of a general theory to be established later in this subsection, is that its leading diagonal entry is zero. To avoid this happening, we exchange the first row and the second row of A and the resulted matrix amounts to

$$B = \begin{bmatrix} -1 & 1 \\ 0 & 2 \end{bmatrix} = \begin{bmatrix} 0 & 1 \\ 1 & 0 \end{bmatrix}\begin{bmatrix} 0 & 2 \\ -1 & 1 \end{bmatrix} = E_{(1)(2)}A.$$

Then, perform column operations to

$$\begin{bmatrix} E_{(1)(2)}A \\ \hline \vec{b} \end{bmatrix} = \begin{bmatrix} B \\ \hline \vec{b} \end{bmatrix} = \begin{matrix} & x_2 & x_1 \\ & \begin{bmatrix} -1 & 1 \\ 0 & 2 \\ \hline b_1 & b_2 \end{bmatrix} \end{matrix}$$

$$\xrightarrow[F_{-(1)}]{} \begin{bmatrix} 1 & 1 \\ 0 & 2 \\ \hline -b_1 & b_2 \end{bmatrix} = \begin{bmatrix} B \\ \hline \vec{b} \end{bmatrix} F_{-(1)}$$

$$\xrightarrow{F_{(2)-(1)}} \left[\begin{array}{cc|} 1 & 0 \\ 0 & 2 \\ \hline -b_1 & b_1+b_2 \end{array}\right] = \left[\begin{array}{c} B \\ \hline \vec{b} \end{array}\right] F_{-(1)}F_{(2)-(1)}$$

$$\xrightarrow{F_{\frac{1}{2}(2)}} \left[\begin{array}{cc|} 1 & 0 \\ 0 & 1 \\ \hline -b_1 & \frac{b_1+b_2}{2} \end{array}\right] = \left[\begin{array}{c} B \\ \hline \vec{b} \end{array}\right] F_{-(1)}F_{(2)-(1)}F_{\frac{1}{2}(2)}.$$

Note that $\vec{x}A = (\vec{x}E_{(1)(2)})(E_{(1)(2)}A) = (\vec{x}E_{(1)(2)})B = (x_2 \; x_1)B = \vec{b}$ so that the first column corresponds to x_2 while the second one to x_1. Equivalently, we can perform row operations to

$$[(E_{(1)(2)}A)^*|\vec{b}^*] = [A^*F_{(1)(2)}|\vec{b}^*] = [B^*|\vec{b}^*] = \begin{bmatrix} -1 & 0 & : & b_1 \\ 1 & 2 & : & b_2 \end{bmatrix} \begin{matrix} x_2 \\ x_1 \end{matrix}$$

$$\xrightarrow{E_{-(1)}} \begin{bmatrix} 1 & 0 & : & -b_1 \\ 1 & 2 & : & b_2 \end{bmatrix} \xrightarrow{E_{(2)-(1)}} \begin{bmatrix} 1 & 0 & : & -b_1 \\ 0 & 2 & : & b_1+b_2 \end{bmatrix}$$

$$\xrightarrow{E_{\frac{1}{2}(2)}} \begin{bmatrix} 1 & 0 & : & -b_1 \\ 0 & 1 & : & \frac{b_1+b_2}{2} \end{bmatrix}.$$

In this case, the first row corresponds to x_2 while the second one to x_1.

(a) *The solution of* $\vec{x}A = \vec{b}$

$$\begin{cases} x_1 = \dfrac{b_1+b_2}{2} \\ x_2 = -b_1 \end{cases}, \quad \text{or}$$

$$\vec{x} = (x_1, \; x_2) = (b_1 \; b_2) \begin{bmatrix} \frac{1}{2} & -1 \\ \frac{1}{2} & 0 \end{bmatrix} = \vec{b}A^{-1}.$$

(b) *The invertibility of A and its inverse* A^{-1}

$$I_2 = \begin{bmatrix} 1 & 0 \\ 0 & 1 \end{bmatrix} = BF_{-(1)}F_{(2)-(1)}F_{\frac{1}{2}(2)}$$

$$\Rightarrow B^{-1} = (E_{(1)(2)}A)^{-1} = A^{-1}E_{(1)(2)}^{-1} = F_{-(1)}F_{(2)-(1)}F_{\frac{1}{2}(2)}$$

$$\Rightarrow A^{-1} = F_{-(1)}F_{(2)-(1)}F_{\frac{1}{2}(2)}F_{(1)(2)}$$

$$= \begin{bmatrix} -1 & 0 \\ 0 & 1 \end{bmatrix}\begin{bmatrix} 1 & -1 \\ 0 & 1 \end{bmatrix}\begin{bmatrix} 1 & 0 \\ 0 & \frac{1}{2} \end{bmatrix}\begin{bmatrix} 0 & 1 \\ 1 & 0 \end{bmatrix} = \begin{bmatrix} \frac{1}{2} & -1 \\ \frac{1}{2} & 0 \end{bmatrix}.$$

(c) *The elementary factorization of A*

$$A = F^{-1}_{(1)(2)} F^{-1}_{\frac{1}{2}(2)} F^{-1}_{(2)-(1)} F^{-1}_{-(1)}$$

$$= F_{(1)(2)} F_{2(2)} F_{(2)+(1)} F_{-(1)}$$

$$= \begin{bmatrix} 0 & 1 \\ 1 & 0 \end{bmatrix} \begin{bmatrix} 1 & 0 \\ 0 & 2 \end{bmatrix} \begin{bmatrix} 1 & 1 \\ 0 & 1 \end{bmatrix} \begin{bmatrix} -1 & 0 \\ 0 & 1 \end{bmatrix}$$

$$= \begin{bmatrix} 0 & 1 \\ 1 & 0 \end{bmatrix} \begin{bmatrix} -1 & 0 \\ 0 & 2 \end{bmatrix} \begin{bmatrix} 1 & -1 \\ 0 & 1 \end{bmatrix} \qquad (2.7.63)$$

$$\Rightarrow \begin{bmatrix} 0 & 1 \\ 1 & 0 \end{bmatrix} A = \begin{bmatrix} -1 & 1 \\ 0 & 2 \end{bmatrix} = \begin{bmatrix} 1 & 0 \\ 0 & 1 \end{bmatrix} \begin{bmatrix} -1 & 1 \\ 0 & 2 \end{bmatrix} = \begin{bmatrix} -1 & 0 \\ 0 & 2 \end{bmatrix} \begin{bmatrix} 1 & -1 \\ 0 & 1 \end{bmatrix}.$$

$$(2.7.64)$$

(d) *The determinants* $\det A$ *and* $\det A^{-1}$

$$\det A = \det \begin{bmatrix} 0 & 1 \\ 1 & 0 \end{bmatrix} \cdot \det \begin{bmatrix} 1 & 0 \\ 0 & 2 \end{bmatrix} \cdot \det \begin{bmatrix} 1 & 1 \\ 0 & 1 \end{bmatrix} \cdot \det \begin{bmatrix} -1 & 0 \\ 0 & 1 \end{bmatrix}$$

$$= (-1) \cdot 2 \cdot 1 \cdot (-1) = 2;$$

$$\det A^{-1} = \det \begin{bmatrix} -1 & 0 \\ 0 & 1 \end{bmatrix} \cdot \det \begin{bmatrix} 1 & -1 \\ 0 & 1 \end{bmatrix} \cdot \det \begin{bmatrix} 1 & 0 \\ 0 & \frac{1}{2} \end{bmatrix} \cdot \det \begin{bmatrix} 0 & 1 \\ 1 & 0 \end{bmatrix}$$

$$= (-1) \cdot 1 \cdot \left(\frac{1}{2} \right) \cdot (-1) = \frac{1}{2}.$$

See Fig. 2.62. Note that A does not have real eigenvalues.

Example 3 Let

$$A = \begin{bmatrix} 1 & 2 \\ 2 & -7 \end{bmatrix}.$$

(1) Solve the equation $A\vec{x}^* = \vec{b}^*$ where $\vec{x} = (x_1, x_2)$ and $\vec{b} = (b_1, b_2)$.
(2) Investigate the geometric mapping properties of A.

Solution As against $\vec{x}A = \vec{b}$ in Examples 1 and 2, here we use column vector $\vec{x}^* = \begin{bmatrix} x_1 \\ x_2 \end{bmatrix}$ as unknown vector and $\vec{b}^* = \begin{bmatrix} b_1 \\ b_2 \end{bmatrix}$ as constant vector in $A\vec{x}^* = \vec{b}^*$ with A as *coefficient matrix* and $[A \,|\, \vec{b}^*]_{2 \times 3}$ as *augmented* matrix.

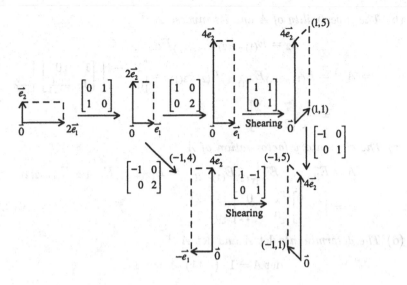

Fig. 2.62

Now, apply consecutive row operations to

$$[A|\vec{b}^*] = \begin{bmatrix} 1 & 2 & \vdots & b_1 \\ 2 & -7 & \vdots & b_2 \end{bmatrix}$$

$$\xrightarrow[E_{(2)-2(1)}]{} \begin{bmatrix} 1 & 2 & \vdots & b_1 \\ 0 & -11 & \vdots & b_2 - 2b_1 \end{bmatrix} = E_{(2)-2(1)}[A|\vec{b}^*]$$

$$\xrightarrow[E_{-\frac{1}{11}(2)}]{} \begin{bmatrix} 1 & 2 & \vdots & b_1 \\ 0 & 1 & \vdots & \frac{2b_1 - b_2}{11} \end{bmatrix} = E_{-\frac{1}{11}(2)}E_{(2)-2(1)}[A|\vec{b}^*]$$

$$\xrightarrow[E_{(1)-2(2)}]{} \begin{bmatrix} 1 & 0 & \vdots & \frac{7b_1 + 2b_2}{11} \\ 0 & 1 & \vdots & \frac{2b_1 - b_2}{11} \end{bmatrix} = E_{(1)-2(2)}E_{-\frac{1}{11}(2)}E_{(2)-2(1)}[A|\vec{b}^*].$$

(a) *The solution of $A\vec{x}^* = \vec{b}^*$*

$$\begin{cases} x_1 = \dfrac{1}{11}(7b_1 + 2b_2) \\ x_2 = \dfrac{1}{11}(2b_1 - b_2) \end{cases} , \quad \text{or}$$

$$\vec{x}^* = \begin{bmatrix} x_1 \\ x_2 \end{bmatrix} = \frac{1}{11} \begin{bmatrix} 7 & 2 \\ 2 & -1 \end{bmatrix} \begin{bmatrix} b_1 \\ b_2 \end{bmatrix} = A^{-1}\vec{b}^*.$$

(b) *The invertibility of A and its inverse* A^{-1}

$$I_2 = E_{(1)-2(2)} E_{-\frac{1}{11}(2)} E_{(2)-2(1)} A$$

$$\Rightarrow A^{-1} = E_{(1)-2(2)} E_{-\frac{1}{11}(2)} E_{(2)-2(1)} = \begin{bmatrix} 1 & -2 \\ 0 & 1 \end{bmatrix} \begin{bmatrix} 1 & 0 \\ 0 & -\frac{1}{11} \end{bmatrix} \begin{bmatrix} 1 & 0 \\ -2 & 1 \end{bmatrix}$$

$$= \frac{1}{11} \begin{bmatrix} 7 & 2 \\ 2 & -1 \end{bmatrix}.$$

(c) *The elementary factorization of A*

$$A = E^{-1}_{(2)-2(1)} E^{-1}_{-\frac{1}{11}(2)} E^{-1}_{(1)-2(2)} = E_{(2)+2(1)} E_{-11(2)} E_{(1)+2(2)}$$

$$= \begin{bmatrix} 1 & 0 \\ 2 & 1 \end{bmatrix} \begin{bmatrix} 1 & 0 \\ 0 & -11 \end{bmatrix} \begin{bmatrix} 1 & 2 \\ 0 & 1 \end{bmatrix}. \tag{2.7.65}$$

(d) *The determinants* det A *and* det A^{-1}

$$\det A = 1 \cdot (-11) \cdot 1 = -11,$$

$$\det A^{-1} = 1 \cdot \left(-\frac{1}{11}\right) \cdot 1 = -\frac{1}{11}.$$

See Fig. 2.63.

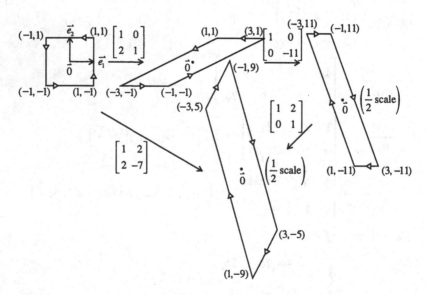

Fig. 2.63

Note that A has two distinct real eigenvalues $-3 \pm \sqrt{20}$.

Example 4 Let

$$A = \begin{bmatrix} 2 & 3 \\ -4 & -6 \end{bmatrix}.$$

(1) Solve the equation $A\vec{x}^* = \vec{b}^*$.
(2) Try to investigate the geometric mapping properties of A.

Solution Write $A\vec{x}^* = \vec{b}^*$ out as

$$\begin{cases} 2x_1 + 3x_2 = b_1 \\ -4x_1 - 6x_2 = b_2 \end{cases}.$$

The equations have solutions if and only if $2b_1 + b_2 = 0$. In this case, the solutions are $x_1 = \frac{1}{2}(b_1 - 3x_2)$ with x_2 arbitrary scalars.

Apply row operations to

$$[A \mid \vec{b}^*] = \begin{bmatrix} 2 & 3 & \vdots & b_1 \\ -4 & -6 & \vdots & b_2 \end{bmatrix}$$

$$\xrightarrow[E_{\frac{1}{2}(1)}]{} \begin{bmatrix} 1 & \frac{3}{2} & \vdots & \frac{b_1}{2} \\ -4 & -6 & \vdots & b_2 \end{bmatrix} = E_{\frac{1}{2}(1)}[A \mid \vec{b}^*]$$

$$\xrightarrow[E_{(2)+4(1)}]{} \begin{bmatrix} 1 & \frac{3}{2} & \vdots & \frac{b_1}{2} \\ 0 & 0 & \vdots & 2b_1 + b_2 \end{bmatrix} = E_{(2)+4(1)}E_{\frac{1}{2}(1)}[A \mid \vec{b}^*].$$

Since $r(A) = 1$ and elementary matrices preserve ranks, then

$$A\vec{x}^* = \vec{b}^* \text{ has a solution.}$$

$$\Leftrightarrow r(A) = r([A \mid \vec{b}^*]) \Leftrightarrow 2b_1 + b_2 = 0.$$

This constrained condition coincides with what we obtained via traditional method. On the other hand,

$$\begin{bmatrix} 1 & \frac{3}{2} \\ 0 & 0 \end{bmatrix} = E_{(2)+4(1)}E_{\frac{1}{2}(1)}A$$

$$\Rightarrow A = E_{\frac{1}{2}(1)}^{-1}E_{(2)+4(1)}^{-1}\begin{bmatrix} 1 & \frac{3}{2} \\ 0 & 0 \end{bmatrix} = \begin{bmatrix} 2 & 0 \\ 0 & 1 \end{bmatrix}\begin{bmatrix} 1 & 0 \\ -4 & 1 \end{bmatrix}\begin{bmatrix} 1 & \frac{3}{2} \\ 0 & 0 \end{bmatrix}.$$

See Fig. 2.64.

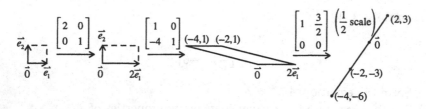

Fig. 2.64

We summarize what we obtained in Examples 1–4 in a general setting.

About elementary matrices

1. $E_{(i)(j)} = F_{(i)(j)}$; $E_{\alpha(i)} = F_{\alpha(i)}$; $E_{(j)+\alpha(i)} = F_{(i)+\alpha(j)}$.
2. $\det E_{(i)(j)} = -1$, $\det E_{\alpha(i)} = \alpha$; $\det E_{(j)+\alpha(i)} = 1$.
3. $E^*_{(i)(j)} = F_{(i)(j)}$; $E^*_{\alpha(i)} = F_{\alpha(i)}$; $E^*_{(j)+\alpha(i)} = F_{(j)+\alpha(i)}$.
4. $E^{-1}_{(i)(j)} = E_{(i)(j)}$; $E^{-1}_{\alpha(i)} = E_{\frac{1}{\alpha}(i)}$; $E^{-1}_{(j)+\alpha(i)} = E_{(j)-\alpha(i)}$.
5. The matrix obtained by performing an elementary row or column operation on a given matrix $A_{m \times n}$ of the type E or F is

$$EA \text{ or } AF,$$

where $E_{m \times m}$ and $F_{n \times n}$ are the corresponding elementary matrices.

(2.7.66)

Proofs are easy and are left to the readers.

By a *permutation matrix* P of *order* n is a matrix obtained from I_n by arbitrary exchange of orders of rows or columns of I_n. That is, for any permutation $\sigma \colon \{1, 2, \ldots, n\} \to \{1, 2, \ldots, n\}$, the matrix

$$P = \begin{bmatrix} \vec{e}_{\sigma(1)} \\ \vdots \\ \vec{e}_{\sigma(n)} \end{bmatrix}$$

(2.7.67)

is a permutation matrix. There are $n!$ permutation matrices of order n. Since P is a product of row exchanges of I_n,

1. $\det P = \begin{cases} 1, \text{even row exchanges from } I_n \\ -1, \text{odd row exchanges from } I_n. \end{cases}$
2. P is invertible and

$$P^{-1} = P^*.$$

Details are left to the readers.

Now, here are some of the main summaries.

How to determine the invertibility of a square matrix?
Let $A_{n \times n}$ be a real matrix (or any square matrix over a field).

(1) Perform a sequence of elementary row operations $E_1, \ldots, E_k$ needed to see if it is possible that

$$E_k E_{k-1} \cdots E_2 E_1 A = I_n.$$

(2) If it is, then A is invertible and its inverse is

$$A^{-1} = E_k E_{k-1} \cdots E_2 E_1.$$

(3) *Elementary matrix factorization of an invertible matrix* is

$$A = E_1^{-1} E_2^{-1} \cdots E_{k-1}^{-1} E_k^{-1}.$$

Hence, a square matrix is invertible if and only if it can be expressed as a product of finitely many elementary matrices.   (2.7.68)

Therefore, elementary row operation provides a method for computing the inverse of an invertible matrix, especially of large order.

The lower and upper triangular decompositions of a square matrix
Let $A_{n \times n}$ be a matrix (over $\mathbb{R}$ or any field $\mathbb{F}$).

(1) If only elementary row operations $E_1, \ldots, E_k$ of types 2 and 3 are needed (*no need of type* 1, exchange of rows) to transform A into an upper triangular matrix U'', i.e.

$$E_k E_{k-1} \cdots E_2 E_1 A = U''.$$

Then,

$$A = (E_1^{-1} E_2^{-1} \cdots E_{k-1}^{-1} E_k^{-1}) U''$$

$$= LU' = \begin{bmatrix} 1 & & 0 \\ \vdots & \ddots & \\ \cdots & \cdots & 1 \end{bmatrix} \begin{bmatrix} d_1 & \cdots & \cdots \\ & \ddots & \vdots \\ 0 & & d_n \end{bmatrix},$$

where L is lower triangular with diagonal entries all equal to 1 while U' is upper triangular with diagonal entries $d_1, d_2, \ldots, d_n$ (the nonzeros among them are called the *pivots* of A). Therefore, the following hold:

1. A is invertible if and only if $d_1, \ldots, d_n$ are nonzero. In this case, A can be factored as

$$A = \begin{bmatrix} 1 & & 0 \\ \vdots & \ddots & \\ \cdots & \cdots & 1 \end{bmatrix} \begin{bmatrix} d_1 & & 0 \\ & \ddots & \\ 0 & & d_n \end{bmatrix} \begin{bmatrix} 1 & \cdots & \cdots \\ & \ddots & \vdots \\ 0 & & 1 \end{bmatrix}$$

$$= LDU,$$

where the middle diagonal matrix P is called the *pivot matrix* of A, the lower triangular L is the same as before while this U is obtained from the original U' by dividing its ith row by d_i and hence, has all its diagonal entries equal to 1 now. Note that $\det A = \det D = d_1 \cdots d_n$.

2. As in 1,

$$A^* = U^* D L^*,$$

where U^* is lower triangular and L^* is upper and $D^* = D$.

3. Suppose A is also *symmetric* and invertible such that $A = LDU$, then $U = L^*$, the transpose of L and

$$A = LDL^*.$$

(2) If row exchanges are needed in transforming A into an upper triangular, there are two possibilities to handle this situation.

1. Do row exchanges before elimination. This means that there exists a permutation matrix P such that no row exchanges are needed now to transform PA into an upper triangular form. Then

$$PA = LU'$$

as in (1).

2. Hold all row exchanges until all nonzero entries below the diagonal are eliminated as possible. Then, try to permutate the order of rows of such a resulting matrix in order to produce an upper triangular form. Thus,

$$A = LPU,$$

where L is lower triangular with diagonal entries all equal to 1, P is a permutation matrix while U is an upper triangular with nonzero pivots along its diagonal.   (2.7.69)

As an example for Case (2), let

$$A = \begin{bmatrix} 0 & 1 & 3 \\ 2 & 4 & 0 \\ -1 & 0 & 5 \end{bmatrix}.$$

For $PA = LU$:

$$A \xrightarrow{P=E_{(1)(3)}} PA = \begin{bmatrix} -1 & 0 & 5 \\ 2 & 4 & 0 \\ 0 & 1 & 3 \end{bmatrix} \xrightarrow{E_{(2)+2(1)}} \begin{bmatrix} -1 & 0 & 5 \\ 0 & 4 & 10 \\ 0 & 1 & 3 \end{bmatrix}$$

$$\xrightarrow{E_{\frac{1}{4}(2)}} \begin{bmatrix} -1 & 0 & 5 \\ 0 & 1 & \frac{5}{2} \\ 0 & 1 & 3 \end{bmatrix} \xrightarrow{E_{(3)-(2)}} \begin{bmatrix} -1 & 0 & 5 \\ 0 & 1 & \frac{5}{2} \\ 0 & 0 & \frac{1}{2} \end{bmatrix}$$

$$\Rightarrow PA = E_{(2)+2(1)}^{-1} E_{\frac{1}{4}(2)}^{-1} E_{(3)-(2)}^{-1} \begin{bmatrix} -1 & 0 & 5 \\ 0 & 1 & \frac{5}{2} \\ 0 & 0 & \frac{1}{2} \end{bmatrix}$$

$$= \begin{bmatrix} 1 & 0 & 0 \\ -2 & 4 & 0 \\ 0 & 1 & 1 \end{bmatrix} \begin{bmatrix} -1 & 0 & 5 \\ 0 & 1 & \frac{5}{2} \\ 0 & 0 & \frac{1}{2} \end{bmatrix}$$

$$= \begin{bmatrix} -1 & 0 & 0 \\ 2 & 1 & 0 \\ 0 & 1 & 1 \end{bmatrix} \begin{bmatrix} -1 & 0 & 0 \\ 0 & 4 & 0 \\ 0 & 0 & \frac{1}{2} \end{bmatrix} \begin{bmatrix} 1 & 0 & -5 \\ 0 & 1 & \frac{5}{2} \\ 0 & 0 & 1 \end{bmatrix}.$$

For $A = LPU$:

$$A \xrightarrow{E_{(3)+\frac{1}{2}(2)}} \begin{bmatrix} 0 & 1 & 3 \\ 2 & 4 & 0 \\ 0 & 2 & 5 \end{bmatrix} \xrightarrow{E_{(3)-2(1)}} \begin{bmatrix} 0 & 1 & 3 \\ 2 & 4 & 0 \\ 0 & 0 & -1 \end{bmatrix} \xrightarrow{E_{(1)(2)}=P} \begin{bmatrix} 2 & 4 & 0 \\ 0 & 1 & 3 \\ 0 & 0 & -1 \end{bmatrix}$$

$$\Rightarrow A = E_{(3)+\frac{1}{2}(2)}^{-1} E_{(3)-2(1)}^{-1} E_{(1)(2)}^{-1} \begin{bmatrix} 2 & 4 & 0 \\ 0 & 1 & 3 \\ 0 & 0 & -1 \end{bmatrix}$$

$$= \begin{bmatrix} 1 & 0 & 0 \\ 0 & 1 & 0 \\ 1 & \frac{-1}{2} & 1 \end{bmatrix} \begin{bmatrix} 0 & 1 & 0 \\ 1 & 0 & 0 \\ 0 & 0 & 1 \end{bmatrix} \begin{bmatrix} 2 & 4 & 0 \\ 0 & 1 & 3 \\ 0 & 0 & -1 \end{bmatrix}.$$

The geometric mapping properties of elementary matrices of order 3 are deferred until Sec. 3.7.2 in Chap. 3.

Similar process guarantees

The factorizations of an $m \times n$ matrix
Let $A_{m \times n}$ be nonzero matrix over any field.

(1) There exists a lower triangular matrix $L_{m \times m}$ such that

$$A = LU_{m \times n},$$

where U is an *echelon* matrix with r pivot rows and r pivot columns, having zeros *below* pivots (see Sec. B.5), where r is the rank of A.

(2) There exists an invertible matrix $P_{m \times m}$ such that

$$PA$$

is the *row-reduced echelon* matrix of A (see Sec. B.5).

(3) There exist invertible matrices $P_{m \times m}$ and $Q_{n \times n}$ such that A has the *normal* form

$$PAQ = \begin{bmatrix} I_r & 0 \\ 0 & 0 \end{bmatrix}_{m \times n},$$

where $r = \mathrm{r}(A)$ (see (2.7.56) and Fig. 2.60). $\hspace{2cm}$ (2.7.70)

For more details and various applications of this kind of factorizations, see Sec. B.5 for a general reference.

In fact, this process of elimination by using elementary row and column operations provides us a far insight into the symmetric matrices.

Let us start from simple examples.

Let

$$A = \begin{bmatrix} 1 & -3 \\ -3 & 9 \end{bmatrix}.$$

Then

$$A \xrightarrow[E_{(2)+3(1)}]{} \begin{bmatrix} 1 & -3 \\ 0 & 0 \end{bmatrix} \xrightarrow[F_{(2)+3(1)}]{} \begin{bmatrix} 1 & 0 \\ 0 & 0 \end{bmatrix}$$

$$\Rightarrow A = E^{-1}_{(2)+3(1)} \begin{bmatrix} 1 & 0 \\ 0 & 0 \end{bmatrix} F^{-1}_{(2)+3(1)} = \begin{bmatrix} 1 & 0 \\ -3 & 1 \end{bmatrix} \begin{bmatrix} 1 & 0 \\ 0 & 0 \end{bmatrix} \begin{bmatrix} 1 & -3 \\ 0 & 1 \end{bmatrix} = LDL^*.$$

Let

$$B = \begin{bmatrix} 0 & 4 \\ 4 & 5 \end{bmatrix}.$$

Then,

$$B \xrightarrow[\substack{E_{(1)(2)} \\ F_{(1)(2)}}]{} E_{(1)(2)} B F_{(1)(2)} = E_{(1)(2)} B E^*_{(1)(2)} = \begin{bmatrix} 5 & 4 \\ 4 & 0 \end{bmatrix}$$

$$\xrightarrow[E_{(2)-\frac{4}{5}(1)}, F_{(2)-\frac{4}{5}(1)}]{} E_{(2)-\frac{4}{5}(1)} E_{(1)(2)} B E^*_{(1)(2)} E^*_{(2)-\frac{4}{5}(1)} = \begin{bmatrix} 5 & 0 \\ 0 & -\frac{16}{5} \end{bmatrix}$$

$$\Rightarrow B = E^{-1}_{(1)(2)} E^{-1}_{(2)-\frac{4}{5}(1)} \begin{bmatrix} 5 & 0 \\ 0 & \frac{-16}{5} \end{bmatrix} \left(E^{-1}_{(1)(2)} E^{-1}_{(2)-\frac{4}{5}(1)} \right)^*$$

$$= \begin{bmatrix} \frac{4}{5} & 1 \\ 1 & 0 \end{bmatrix} \begin{bmatrix} 5 & 0 \\ 0 & \frac{-16}{5} \end{bmatrix} \begin{bmatrix} \frac{4}{5} & 1 \\ 1 & 0 \end{bmatrix}$$

$$= PDP^*,$$

where $P = \begin{bmatrix} \frac{4}{5} & 1 \\ 1 & 0 \end{bmatrix}$ is not necessarily lower triangular and $D = \begin{bmatrix} 5 & 0 \\ 0 & -\frac{16}{5} \end{bmatrix}$ is diagonal. To simplify D one step further, let

$$Q = \begin{bmatrix} \frac{1}{\sqrt{5}} & 0 \\ 0 & \frac{\sqrt{5}}{4} \end{bmatrix}.$$

Then,

$$QDQ^* = \begin{bmatrix} 1 & 0 \\ 0 & -1 \end{bmatrix},$$

$$B = PDP^* = PQ^{-1}(QDQ^*)(Q^{-1}P)^*$$

$$= \begin{bmatrix} 4\sqrt{5} & \frac{4}{\sqrt{5}} \\ \sqrt{5} & 0 \end{bmatrix} \begin{bmatrix} 1 & 0 \\ 0 & -1 \end{bmatrix} \begin{bmatrix} 4\sqrt{5} & \sqrt{5} \\ \frac{4}{\sqrt{5}} & 0 \end{bmatrix}.$$

Or, equivalently,

$$RBR^* = \begin{bmatrix} 1 & 0 \\ 0 & -1 \end{bmatrix},$$

where

$$R = (PQ^{-1})^{-1} = QP^{-1} = \begin{bmatrix} 0 & \frac{1}{\sqrt{5}} \\ \frac{\sqrt{5}}{4} & -\sqrt{5} \end{bmatrix}.$$

These two examples tell us explicitly and inductively the following important results in the study of quadratic forms.

Congruence of real symmetric matrices

(1) A real matrix $A_{n \times n}$ is *congruent* to a diagonal matrix, i.e. there exists an invertible matrix $P_{n \times n}$ such that

$$PAP^* = \begin{bmatrix} d_1 & & & & 0 \\ & d_2 & & & \\ & & \ddots & & \\ 0 & & & & d_n \end{bmatrix}$$

if and only if A is a symmetric matrix.

(2) In case A is real symmetric, then there exists an invertible matrix $P_{n \times n}$ such that

$$PAP^* = \begin{bmatrix} 1 & & & & & & & & \\ & \ddots & & & & 0 & & \\ & & 1 & & & & & \\ & & & -1 & & & & \\ & & & & \ddots & & & \\ & & & & & -1 & & \\ & & & & & & 0 & \\ & 0 & & & & & & \ddots \\ & & & & & & & & 0 \end{bmatrix}_{n \times n} = \begin{bmatrix} I_k & 0 \\ & -I_l \\ 0 & 0 \end{bmatrix}_{n \times n}$$

where

the *index* k of A = the number of positive 1 in the diagonal,
the *signature* $s = k - l$ of A = the number k of positive 1 in the diagonal minus the number l of negative -1 in the diagonal,
the *rank* r of $A = k + l$.

(3) *Sylvester's law of inertia*
The index, signature and rank of a real symmetric matrix are *invariants* under congruence.

(4) Therefore, two real symmetric matrices of the same order are congruent if and only if they have the same invariants.   (2.7.71)

The result (1) still holds for symmetric matrices over a field of characteristic other than two (see Sec. A.3).

Exercises

<A>

1. Let

$$A = \begin{bmatrix} 4 & 3 \\ 5 & 2 \end{bmatrix}.$$

(a) Find the following factorizations of A:

(1) $A = P^{-1} \begin{bmatrix} \lambda_1 & 0 \\ 0 & \lambda_2 \end{bmatrix} P$, where P is invertible.

(2) $A = LU$ with pivots on the diagonal of U.

(3) $A = LDU$, where D is the pivot matrix of A.

(4) A as a product of elementary matrices.

(5) $A = LS$, where L is lower triangular and S is symmetric.

(b) Compute $\det A$ according to each factorization in (a).

(c) Solve the equation $A\vec{x}^* = \vec{b}^*$, where $\vec{x} = (x_1, x_2)$ and $\vec{b} = (b_1, b_2)$, by using the following methods:

(1) Apply elementary operations to the augmented matrix $[A \mid \vec{b}^*]$.

(2) Apply $A = LU$ in (a).

(3) Apply $A = LDU$ in (a).

(d) Compute A^{-1} by as many ways as possible, including the former four ones in (a).

(e) Find the image of the square with vertices at $\pm\vec{e}_1 \pm \vec{e}_2$ under the mapping $\vec{x} \rightarrow \vec{x}A$ by direct computation and by the former four factorizations shown in (a). Illustrate graphically at each step.

2. Let

$$A = \begin{bmatrix} 0 & 5 \\ 3 & -2 \end{bmatrix}.$$

Do the same problems as in Ex. 1, except (a)(2) is replaced by $E_{(1)(2)}A = LU$ and (a)(3) is replaced by $A = LPU$, where P is a permutation matrix.

3. Let

$$A = \begin{bmatrix} 1 & 2 \\ 2 & 7 \end{bmatrix}.$$

Do the same problems as in Ex. 1, except (a)(3) is replaced by $A = LDL^*$.

4. Let

$$A = \begin{bmatrix} 1 & 3 \\ -3 & 7 \end{bmatrix}.$$

Do the same problems as in Ex. 1 but pay attention to (a)(1). What change is needed in (a)(1)? And how?

5. Let

$$A = \begin{bmatrix} 1 & 4 \\ -3 & 7 \end{bmatrix}.$$

Do the same problems as in Ex. 1. What happens to (a)(1)?

6. Let

$$A = \begin{bmatrix} 2 & -4 \\ -6 & 12 \end{bmatrix}.$$

(a) Find the following factorizations of A:

   (1) $A = P^{-1} \begin{bmatrix} \lambda_1 & 0 \\ 0 & \lambda_2 \end{bmatrix} P$.

   (2) $A = E_1 E_2 \begin{bmatrix} 1 & -2 \\ 0 & 0 \end{bmatrix}$, where E_1 and E_2 are elementary matrices.

   (3) $A = LU$.

   (4) $A = LS$, where L is lower triangular and S is symmetric.

(b) Determine when the equation $A\vec{x}^* = \vec{b}^*$, where $\vec{x} = (x_1, x_2)$ and $\vec{b} = (b_1, b_2)$, has a solution and then, solve the equation by using the following methods.

   (1) Apply elementary row operations to $[A \mid \vec{b}^*]$.

   (2) Apply $A = LU$ in (a).

(c) Find the image of the triangle $\triangle \vec{a}_1 \vec{a}_2 \vec{a}_3$, where $\vec{a}_1 = (3,1)$, $\vec{a}_2 = (-2,3)$ and $\vec{a}_3 = (2,-4)$, under the mapping $\vec{x} \to \vec{x}A$ by direct computation and by the former three factorizations in (a). Illustrate graphically at each step.

7. Suppose $A = LDU$, where L, D, U are invertible.

(a) In case A has another factorization as $A = L_1 D_1 U_1$, show that L_1, D_1, U_1 are invertible and

$$L = L_1, \quad D = D_1 \quad \text{and} \quad U = U_1$$

by considering $L_1^{-1} LD = D_1 U_1 U^{-1}$.

(b) Rewrite A as $L(U^*)^{-1}(U^* DU)$. Show that there exists a lower triangular matrix L and a symmetric matrix S such that

$$A = LS.$$

Notice that the diagonal entries along L are all equal to 1.

(*Note* These results hold not just for $A_{2\times 2}$ over real but also for any invertible matrix $A_{n\times n}$ over a field $\mathbb{F}$.)

8. Let

$$A = \begin{bmatrix} 6 & 2 \\ 2 & 9 \end{bmatrix}.$$

(a) Show that A has eigenvalues $\lambda_1 = 10$ and $\lambda_2 = 5$. Find the respective eigenvectors $\vec{u}_1$, $\vec{u}_2$, and justify that

$$PAP^{-1} = \begin{bmatrix} 10 & 0 \\ 0 & 5 \end{bmatrix}, \quad P = \begin{bmatrix} \vec{u}_1 \\ \vec{u}_2 \end{bmatrix}.$$

In what follows, suppose you have concepts from plane Euclidean geometry (refer to Chap. 4, if necessary). Show that $\vec{u}_1$ is perpendicular to $\vec{u}_2$, i.e. their *inner product* $\langle \vec{u}_1, \vec{u}_2 \rangle = \vec{u}_1 \vec{u}_2^* = 0$. Divide $\vec{u}_i$ by its *length* $|\vec{u}_i|$, the resulting vector $\frac{\vec{u}_i}{|\vec{u}_i|}$ is of unit length for $i = 1, 2$. Let

$$Q = \begin{bmatrix} \frac{\vec{u}_1}{|\vec{u}_1|} \\ \frac{\vec{u}_2}{|\vec{u}_2|} \end{bmatrix}.$$

Justify that $Q^* = Q^{-1}$, i.e. Q is an *orthogonal matrix* and

$$QAQ^* = \begin{bmatrix} 10 & 0 \\ 0 & 5 \end{bmatrix}$$

still holds. Try to adjust Q to a new matrix R such that

$$RAR^* = \begin{bmatrix} 1 & 0 \\ 0 & 1 \end{bmatrix} = I_2$$

so that $A = R^{-1}(R^{-1})^*$.

(b) Try to graph the quadratic curve

$$6x_1^2 + 4x_1x_2 + 9x_2^2 = \langle \vec{x}A, \vec{x} \rangle = 1$$

in the Cartesian coordinate system $\mathcal{N} = \{\vec{e}_1, \vec{e}_2\}$, where $\vec{x} = (x_1, x_2)$, by any method available (for example, a computer). What does it look like? An ellipse, a parabola, a hyperbola or anything else? Determine its equations in the respective bases $\mathcal{B} = \{\vec{u}_1, \vec{u}_2\}$, $\mathcal{C} = \{\vec{v}_1, \vec{v}_2\}$ and $\mathcal{D} = \{\vec{w}_1, \vec{w}_2\}$, where $\vec{v}_1, \vec{v}_2$ are row vectors of Q and $\vec{w}_1, \vec{w}_2$ are row vectors of $\mathbb{R}$. In the eyes of $\mathcal{D}$, what does the quadratic curve look like?

(c) Applying elementary row operations to A, try to find an invertible matrix S such that

$$SAS^* = \begin{bmatrix} 6 & 0 \\ 0 & \frac{25}{3} \end{bmatrix}.$$

and another invertible matrix T such that

$$TAT^* = \begin{bmatrix} 1 & 0 \\ 0 & 1 \end{bmatrix} = I_2$$

so that $A = T^{-1}(T^{-1})^*$. Compare this T with $\mathbb{R}$ in (a). Are they the same?

(d) What are the equations of the quadratic curve in (b) in the respective bases $S = \{\vec{s}_1, \vec{s}_2\}$ and $T = \vec{t}_1, \vec{t}_2\}$ where $\vec{s}_1$, $\vec{s}_2$ are row vectors of S and $\vec{t}_1$, $\vec{t}_2$ are row vectors of T. Graph it in both bases.

(e) After your experience with (a)–(d), which of these bases for $\mathbb{R}^2$ is your favor in the understanding of that quadratic curve? Give some convincing reasons to support your choice.

9. Let

$$A = \begin{bmatrix} 7 & 4 \\ 4 & -8 \end{bmatrix}.$$

Do the same problems as in Ex. 8. Now, the corresponding quadratic curve is

$$\langle \vec{x} A, \vec{x} \rangle = 7x_1^2 + 8x_1 x_2 - 8x_2^2 = 1,$$

where $\vec{x} = (x_1, x_2)$ in $\mathcal{N} = \{\vec{e}_1, \vec{e}_2\}$.

10. Let

$$A = \begin{bmatrix} 2 & 4 \\ 4 & 8 \end{bmatrix}.$$

Do the same problems as in Ex. 8.

11. Let $A_{2 \times 2}$ be a nonzero real symmetric matrix. Suppose $P_{2 \times 2}$ and $Q_{2 \times 2}$ are two invertible matrices such that

$$PAP^* = \begin{bmatrix} \lambda_1 & 0 \\ 0 & \lambda_2 \end{bmatrix} \quad \text{and} \quad QAQ^* = \begin{bmatrix} \mu_1 & 0 \\ 0 & \mu_2 \end{bmatrix}.$$

Show that

(1) $\lambda_1 > 0, \lambda_2 > 0 \Leftrightarrow \mu_1 > 0, \mu_2 > 0$,

(2) $\lambda_1 < 0, \lambda_2 < 0 \Leftrightarrow \mu_1 < 0, \mu_2 < 0$,

(3) $\lambda_1 \lambda_2 < 0 \Leftrightarrow \mu_1 \mu_2 < 0$, and

(4) one of λ_1 and λ_2 is zero and the nonzero one is positive (or negative) $\Leftrightarrow$ one of μ_1 and μ_2 is zero and the nonzero one is positive (or negative).

(*Note* This is the Sylvester's law in (2.7.71) for 2×2 symmetric matrix. Ponder if your proof is still good for 3×3 or $n \times n$ symmetric matrix. If yes, try it; if no, is there any other way to attack this problem? See Ex. <B> 3 of Sec. 3.7.5.)

12. Factorize the nonzero matrix

$$A = \begin{bmatrix} a_{11} & a_{12} \\ a_{21} & a_{22} \end{bmatrix}$$

into products of elementary matrices and hence interpret its mapping properties (refer to (2.7.9)) directly and step by step by using the factorization.

<B>

Caution the following problems are combined and more challenging and basic knowledge about inner products (see Chap. 4) are needed.

1. Let

$$A = \begin{bmatrix} 0 & 1 & 2 \\ -1 & 0 & 1 \end{bmatrix}.$$

(a) A is an echelon matrix. Then

$$\begin{bmatrix} 0 & 1 \\ 1 & 0 \end{bmatrix} A = \begin{bmatrix} -1 & 0 & 1 \\ 0 & 1 & 2 \end{bmatrix}.$$

Also,

$$PA = \begin{bmatrix} 1 & 0 & -1 \\ 0 & 1 & 2 \end{bmatrix}, \quad \text{where } P = \begin{bmatrix} 0 & -1 \\ 1 & 0 \end{bmatrix}$$

is the row-reduced echelon matrix of A and

$$PAQ = \begin{bmatrix} 1 & 0 & 0 \\ 0 & 1 & 0 \end{bmatrix}, \quad \text{where } Q = \begin{bmatrix} 1 & 0 & 1 \\ 0 & 1 & -2 \\ 0 & 0 & 1 \end{bmatrix}$$

is the normal form of A. Justify such P and Q. Hence

$$A = P^{-1} \begin{bmatrix} 1 & 0 & 0 \\ 0 & 1 & 0 \end{bmatrix} Q^{-1}$$

$$= \begin{bmatrix} 0 & 1 \\ -1 & 0 \end{bmatrix} \begin{bmatrix} 1 & 0 & 0 \\ 0 & 1 & 0 \end{bmatrix} \begin{bmatrix} 1 & 0 & -1 \\ 0 & 1 & 2 \\ 0 & 0 & 1 \end{bmatrix}.$$

There are two ways to interpret geometrically PAQ or $A = P^{-1}(PAQ)Q^{-1}$.

(b) Fix the Cartesian's coordinate system $\mathcal{N} = \{\vec{e}_1, \vec{e}_2\}$ in $\mathbb{R}^2$. A, considered as the mapping $\vec{x} \to \vec{x}A$, is the composite of the following three consecutive maps:

$$\vec{x} \to \vec{x}P^{-1} \to xP^{-1}(PAQ) \to xP^{-1}(PAQ)Q^{-1} = \vec{x}A.$$

See the upper part of Fig. 2.65. The other way is to take basis $\mathcal{B} = \{\vec{u}_1, \vec{u}_2\}$ for $\mathbb{R}^2$, where $\vec{u}_1 = (0, -1)$ and $\vec{u}_2 = (1, 0)$ are row vectors of P, and the basis $\mathcal{C} = \{\vec{v}_1, \vec{v}_2, \vec{v}_3\}$ for $\mathbb{R}^3$, where $\vec{v}_1 = (1, 0, -1)$, $\vec{v}_2 = (0, 1, 2)$ and $\vec{v}_3 = (0, 0, 1)$ are row vectors of Q^{-1}. Let $\mathcal{N} = \{\vec{e}_1', \vec{e}_2', \vec{e}_3'\}$ be the Cartesian coordinate system for $\mathbb{R}^3$. Then (refer to (2.7.23)),

$$P = P_{\mathcal{N}}^{\mathcal{B}}, \quad Q = Q_{\mathcal{C}}^{\mathcal{N}'}$$

$$\Rightarrow PAQ = P_{\mathcal{N}}^{\mathcal{B}} A_{\mathcal{N}'}^{\mathcal{N}} Q_{\mathcal{C}}^{\mathcal{N}'} = A_{\mathcal{C}}^{\mathcal{B}} = \begin{bmatrix} 1 & 0 & 0 \\ 0 & 1 & 0 \end{bmatrix},$$

which is the *matrix representation* of $A = A_{\mathcal{N}'}^{\mathcal{N}}$, in terms of $\mathcal{B}$ and $\mathcal{C}$. Thus, $\vec{x} \to \vec{x}A = \vec{y}$ is equivalent to $[\vec{x}]_{\mathcal{B}} \to [\vec{x}]_{\mathcal{B}} A_{\mathcal{C}}^{\mathcal{B}} = [\vec{y}]_{\mathcal{C}}$. See the lower part of Fig. 2.65.

(c) The image of the unit circle $x_1^2 + x_2^2 = 1$, where $\vec{x} = (x_1, x_2)$, under the mapping A, is the intersection of the hyperplane $y_1 - 2y_2 + y_3 = 0$ with the cylinder $y_1^2 + y_2^2 = 1$ which is an ellipse, where $\vec{y} = \vec{x}A = (y_1, y_2, y_3)$. See Fig. 2.66.

   An interesting problem concerned is that where are the semiaxes of the ellipse and what are their lengths. Also, is it possible to choose a basis for $\mathbb{R}^3$ on which the ellipse has equation like $a_1\beta_1^2 + a_2\beta_2^2 = 1$? Do you have any idea in solving these two challenging problems?

(d) Actual computation shows that

$$AA^* = \begin{bmatrix} 5 & 2 \\ 2 & 2 \end{bmatrix}.$$

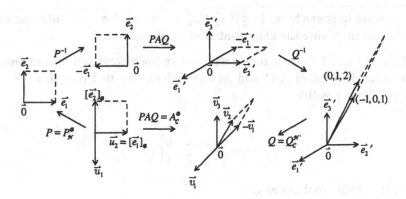

Fig. 2.65

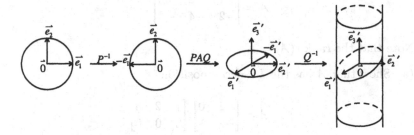

Fig. 2.66

Show that AA^* has eigenvalues 1 and 6 with corresponding eigenvectors $\vec{v}_1 = \frac{1}{\sqrt{5}}(-1, 2)$ and $\vec{v}_2 = \frac{1}{\sqrt{5}}(2, 1)$. Thus,

$$R(AA^*)R^{-1} = \begin{bmatrix} 1 & 0 \\ 0 & 6 \end{bmatrix}, \quad \text{where } R = \begin{bmatrix} \vec{v}_1 \\ \vec{v}_2 \end{bmatrix} = \frac{1}{\sqrt{5}} \begin{bmatrix} -1 & 2 \\ 2 & 1 \end{bmatrix}.$$

Note that $R^* = R^{-1}$. Take the new basis $\mathcal{B} = \{\vec{v}_1, \vec{v}_2\}$ for $\mathbb{R}^2$ and remember that, for $\vec{x} = (x_1, x_2) \in \mathbb{R}^2$,

$$\vec{x}R^{-1} = [\vec{x}]_{\mathcal{B}} = (\alpha_1, \alpha_2).$$

Therefore, for all $\vec{x} \in \mathbb{R}^2$ and hence all $[\vec{x}]_{\mathcal{B}} = (\alpha_1, \alpha_2)$,

$$\vec{x}AA^*\vec{x}^* = \vec{x}R^{-1}(RAA^*R^{-1})(\vec{x}R^{-1})^* = (\alpha_1 \quad \alpha_2) \begin{bmatrix} 1 & 0 \\ 0 & 6 \end{bmatrix} \begin{bmatrix} \alpha_1 \\ \alpha_2 \end{bmatrix}$$

$$= \alpha_1^2 + 6\alpha_2^2 \leq 6(\alpha_1^2 + \alpha_2^2), \quad \text{and}$$

$$\geq \alpha_1^2 + \alpha_2^2.$$

What happens to $\alpha_1^2 + \alpha_2^2$ if $x_1^2 + x_2^2 = 1$ holds? Do these information help in solving the aforementioned problems?

It might be still a puzzle to you why one knows beforehand to start from the consideration of AA^* and its eigenvalues. Try to figure this out as precisely as possible.

2. Let

$$A = \begin{bmatrix} 1 & 1 & 0 \\ -1 & 2 & 1 \end{bmatrix}.$$

Do the same problems as in 1.

3. Let

$$A = \begin{bmatrix} 1 & 2 & 3 \\ -2 & -4 & -6 \end{bmatrix}.$$

Note that the rank $r(A) = 1$.

(a) Show that A has the LU-decomposition

$$A = \begin{bmatrix} 1 & 0 \\ -2 & 1 \end{bmatrix} \begin{bmatrix} 1 & 2 & 3 \\ 0 & 0 & 0 \end{bmatrix}.$$

Meanwhile,

$$PA = \begin{bmatrix} 1 & 2 & 3 \\ 0 & 0 & 0 \end{bmatrix}, \quad \text{where } P = \begin{bmatrix} 1 & 0 \\ 2 & 1 \end{bmatrix}$$

is the row-reduced echelon matrix and

$$PAQ = \begin{bmatrix} 1 & 0 & 0 \\ 0 & 0 & 0 \end{bmatrix}, \quad \text{where } Q = \begin{bmatrix} 1 & -2 & -3 \\ 0 & 1 & 0 \\ 0 & 0 & 1 \end{bmatrix}$$

is the normal form of A.

(b) (Refer to Ex. <A> 20 of Sec. 2.7.3. The natural inner products both in $\mathbb{R}^2$ and in $\mathbb{R}^3$ are needed and are urged to refer to the Introduction in Chap. 4.) Show that, for $\vec{x} = (x_1, x_2)$ and $\vec{y} = (y_1, y_2, y_3) = \vec{x}A$,

$$\text{Ker}(A) = \{\vec{x} \in \mathbb{R}^2 \mid x_1 - 2x_2 = 0\} = \langle\langle(2, 1)\rangle\rangle,$$
$$\text{Im}(A) = \{\vec{y} \in \mathbb{R}^3 \mid 2y_1 - y_2 = 3y_1 - y_3 = 0\} = \langle\langle(1, 2, 3)\rangle\rangle;$$

for $\vec{y} = (y_1, y_2, y_3) \in \mathbb{R}^3$ and $\vec{x} = (x_1, x_2) = \vec{y}A^*$ or $\vec{x}^* = A\vec{y}^*$,

$$\mathrm{Ker}(A^*) = \{\vec{y} \in \mathbb{R}^3 \mid y_1 + 2y_2 + 3y_3 = 0\}$$
$$= \langle\langle(-2, 1, 0), (-3, 0, 1)\rangle\rangle,$$
$$\mathrm{Im}(A^*) = \{\vec{x} \in \mathbb{R}^2 \mid 2x_1 + x_2 = 0\} = \langle\langle(1, -2)\rangle\rangle.$$

Justify that

$$\mathrm{Im}(A)^{\perp} = \mathrm{Ker}(A^*), \quad \mathrm{Ker}(A)^{\perp} = \mathrm{Im}(A^*), \quad \text{and}$$
$$\mathrm{Im}(A^*)^{\perp} = \mathrm{Ker}(A), \quad \mathrm{Ker}(A^*)^{\perp} = \mathrm{Im}(A),$$

also $\mathbb{R}^3 = \mathrm{Ker}(A^*) \oplus \mathrm{Im}(A)$ and $\mathbb{R}^2 = \mathrm{Ker}(A) \oplus \mathrm{Im}(A^*)$. See Fig. 2.67.

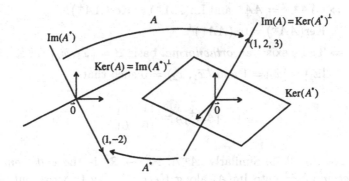

Fig. 2.67

(c) From our experience in (2.7.8) and Fig. 2.67, it is easily known that the line $\langle\langle(1, -2)\rangle\rangle = \mathrm{Im}(A^*)$ is an invariant line of AA^*, i.e. there exists a scalar λ such that

$$(1, -2)AA^* = \lambda(1, -2).$$

Actual computation shows that $\lambda = 70$. Hence, even though A, as a whole, does not have an inverse, *when restricted to the subspace* $\mathrm{Im}(A^*)$, it does satisfy

$$A\left(\frac{1}{70}A^*\right) = 1_{\mathrm{Im}(A^*)}.$$

Therefore, specifically denoted,

$$A^+ = \frac{1}{70}A^*$$

serves as a *right inverse* of the restriction $A|_{\mathrm{Im}(A^*)}$.

Since $\mathrm{Ker}(A) = \mathrm{Im}(A^*)^\perp$, for any $\vec{x} \in \mathbb{R}^2$, there exist unique scalars α and β such that $\vec{x} = \alpha(1,-2) + \beta(2,1)$ and hence

$$\vec{x}(AA^+) = \alpha(1,-2)$$
$$= \text{the orthogonal projection of } \vec{x} \text{ onto}$$
$$\mathrm{Im}(A^*) \text{ along } \mathrm{Ker}(A).$$

Therefore, AA^+ is the *orthogonal projection* of $\mathbb{R}^2$ onto $\mathrm{Im}(A^*)$ along $\mathrm{Ker}(A)$. Show that it can be characterized as:

AA^+ is symmetric and $(AA^+)^2 = AA^+$.

$\Leftrightarrow (AA^+)^2 = AA^+$ and $\mathrm{Im}(AA^+) = \mathrm{Ker}(AA^+)^\perp$,

$\mathrm{Ker}(AA^+) = \mathrm{Im}(AA^+)^\perp$.

$\Leftrightarrow$ There exists an *orthonormal* basis $\mathcal{B} = \{\vec{x}_1, \vec{x}_2\}$ for $\mathbb{R}^2$, i.e.

$|\vec{x}_1| = |\vec{x}_2| = 1$ and $\langle \vec{x}_1, \vec{x}_2 \rangle = 0$ such that

$$[AA^+]_B = \begin{bmatrix} 1 & 0 \\ 0 & 0 \end{bmatrix}.$$

See Fig. 2.68. Similarly, $A^+A\colon \mathbb{R}^3 \to \mathbb{R}^3$ is the *orthogonal projection* of $\mathbb{R}^3$ onto $\mathrm{Im}(A)$ along $\mathrm{Ker}(A^*)$. Try to write out similar characteristic properties for A^+A as for AA^+. See Fig. 2.69. A^+ is called the *generalized* or *pseudoinverse* of A. For details, see Note in Ex. <B> 3 of Sec. 3.7.3, Example 4 in Sec. 3.7.5, Secs. 4.5, 5.5 and B.8.

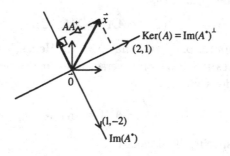

Fig. 2.68

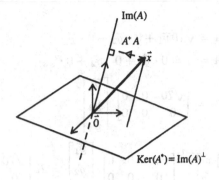

Fig. 2.69

(d) Since $r(A) = 1$, try to show that

$$A = \begin{bmatrix} 1 & 2 & 3 \\ -2 & -4 & -6 \end{bmatrix} = \begin{bmatrix} 1 \\ -2 \end{bmatrix} \begin{bmatrix} 1 & 2 & 3 \end{bmatrix}$$

$$= \sqrt{70} \begin{bmatrix} \frac{1}{\sqrt{5}} \\ \frac{-2}{\sqrt{5}} \end{bmatrix} \begin{bmatrix} \frac{1}{\sqrt{14}} & \frac{2}{\sqrt{14}} & \frac{3}{\sqrt{14}} \end{bmatrix}$$

$$\Rightarrow A^+ = \frac{1}{\sqrt{70}} \begin{bmatrix} \frac{1}{\sqrt{14}} \\ \frac{2}{\sqrt{14}} \\ \frac{3}{\sqrt{14}} \end{bmatrix} \begin{bmatrix} \frac{1}{\sqrt{5}} & \frac{-2}{\sqrt{5}} \end{bmatrix} = \frac{1}{70} A^*.$$

Here $\sqrt{70}$ is called the *singular value* of A (see (e)) and $(\sqrt{70})^2 = 70$ is the nonzero eigenvalue of AA^* as shown in (c).

(e) Suppose $\vec{x}_1 = \frac{1}{\sqrt{5}}(1, -2)$ which is intentionally normalized from $(1, -2)$ as a unit vector. Also, let $\vec{x} = \frac{1}{\sqrt{5}}(2, 1)$. Then

$$\begin{cases} \vec{x}_1 A = \dfrac{5}{\sqrt{5}}(1, 2, 3) = \sqrt{70}\left(\dfrac{1}{\sqrt{14}}, \dfrac{2}{\sqrt{14}}, \dfrac{3}{\sqrt{14}}\right), \\ \vec{x}_2 A = \vec{0}. \end{cases}$$

Let $\vec{y}_1 = \left(\frac{1}{\sqrt{14}}, \frac{2}{\sqrt{14}}, \frac{3}{\sqrt{14}}\right)$. Take any two orthogonal unit vectors $\vec{y}_2$ and $\vec{y}_3$ in $\mathrm{Ker}(A^*)$, say $\vec{y}_2 = \left(\frac{-2}{\sqrt{5}}, \frac{1}{\sqrt{5}}, 0\right)$ and $\vec{y}_3 = \left(\frac{-3}{\sqrt{70}}, \frac{-6}{\sqrt{70}}, \frac{5}{\sqrt{70}}\right)$.

Then

$$\begin{cases} \vec{x}_1 A = \sqrt{70}\,\vec{y}_1 + 0 \cdot \vec{y}_2 + 0 \cdot \vec{y}_3, \\ \vec{x}_2 A = \vec{0} = 0 \cdot \vec{y}_1 + 0 \cdot \vec{y}_2 + 0 \cdot \vec{y}_3, \end{cases}$$

$$\Rightarrow \begin{bmatrix} \vec{x}_1 \\ \vec{x}_2 \end{bmatrix} A = \begin{bmatrix} \sqrt{70} & 0 & 0 \\ 0 & 0 & 0 \end{bmatrix} \begin{bmatrix} \vec{y}_1 \\ \vec{y}_2 \\ \vec{y}_3 \end{bmatrix}$$

$$\Rightarrow A = \begin{bmatrix} \vec{x}_1 \\ \vec{x}_2 \end{bmatrix}^{-1} \begin{bmatrix} \sqrt{70} & 0 & 0 \\ 0 & 0 & 0 \end{bmatrix} \begin{bmatrix} \vec{y}_1 \\ \vec{y}_2 \\ \vec{y}_3 \end{bmatrix} = R \begin{bmatrix} \sqrt{70} & 0 & 0 \\ 0 & 0 & 0 \end{bmatrix} S, \qquad (*)$$

where

$$R = \begin{bmatrix} \frac{1}{\sqrt{5}} & \frac{-2}{\sqrt{5}} \\ \frac{2}{\sqrt{5}} & \frac{1}{\sqrt{5}} \end{bmatrix}^{-1} = \begin{bmatrix} \frac{1}{\sqrt{5}} & \frac{2}{\sqrt{5}} \\ \frac{-2}{\sqrt{5}} & \frac{1}{\sqrt{5}} \end{bmatrix} \quad \text{and}$$

$$S = \begin{bmatrix} \frac{1}{\sqrt{14}} & \frac{2}{\sqrt{14}} & \frac{3}{\sqrt{14}} \\ \frac{-2}{\sqrt{5}} & \frac{1}{\sqrt{5}} & 0 \\ \frac{-3}{\sqrt{70}} & \frac{-6}{\sqrt{70}} & \frac{5}{\sqrt{70}} \end{bmatrix}$$

are both orthogonal, i.e. $R^* = R^{-1}$ and $S^* = S^{-1}$. $(*)$ is called the *singular value decomposition* of A. See Fig. 2.70. Notice that the generalized inverse

$$\begin{bmatrix} \sqrt{70} & 0 & 0 \\ 0 & 0 & 0 \end{bmatrix}^+ = \begin{bmatrix} \frac{1}{\sqrt{70}} & 0 \\ 0 & 0 \\ 0 & 0 \end{bmatrix}$$

and hence, the generalized inverse of A is

$$A^+ = S^* \begin{bmatrix} \frac{1}{\sqrt{70}} & 0 \\ 0 & 0 \\ 0 & 0 \end{bmatrix} R^* = \frac{1}{\sqrt{70}} A^*.$$

(f) For $\vec{b} \in \mathbb{R}^3$ but $\vec{b} \notin \text{Im}(A)$, there is no solution for the equation $\vec{x} A = \vec{b}$. But, for any $\vec{b} \in \text{Im}(A)$, there are infinitely many solutions for $\vec{x} A = \vec{b}$. In fact, any point $\vec{x}$ along the line $\vec{b} A^+ + \text{Ker}(A)$ is a

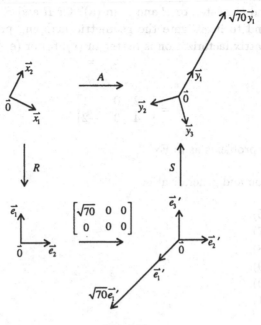

Fig. 2.70

solution and $\vec{b}A^+$ is the shortest solution (in distance from $\vec{0}$ or its length) among all, i.e.

$$|\vec{b}A^+| = \min_{\vec{x}A=\vec{b}} |\vec{x}|.$$

Equivalently, it is the vector $\vec{b}A^+$ that minimizes $|\vec{b} - \vec{x}A|$ for all $\vec{x} \in \mathbb{R}^2$, i.e.

$$|\vec{b} - (\vec{b}A^+)A| = \min_{\vec{x} \in \mathbb{R}^2} |\vec{b} - \vec{x}A|.$$

This is the theoretical foundation for the least square method in applications. Both explanations above can be used as definitions for generalized inverse of a matrix (see Sec. B.8).

(g) Find orthogonal matrices $T_{2\times 2}$ and $U_{3\times 3}$ such that

$$T(AA^*)T^{-1} = \begin{bmatrix} 70 & 0 \\ 0 & 0 \end{bmatrix}, \quad \text{and}$$

$$U(A^*A)U^{-1} = \begin{bmatrix} 70 & 0 & 0 \\ 0 & 0 & 0 \\ 0 & 0 & 0 \end{bmatrix}$$

Are T and U related to P and Q in (a)? Or R and S in (e)? In case you intend to investigate the geometric mapping properties of A, which matrix factorization is better in (a), (c) or (e)?

4. Let

$$A = \begin{bmatrix} -2 & 0 & 1 \\ 4 & 0 & -2 \end{bmatrix}.$$

Do the same problems as in Ex. 3

<C> Abstraction and generalization

1. Prove (2.7.66).
2. Prove (2.7.67).
3. Prove (2.7.68).
4. Prove (2.7.69).
5. Prove (2.7.70).
6. Prove (2.7.71).

Something more should be said about the *block matrix* or *partitioned matrix*. The augmented matrix $\left[\frac{A}{b}\right]$ or $[A \,|\, \vec{b}^*]$ is the simplest of such matrices. Let $A_{m \times n}$ be a matrix (over any field $\mathbb{F}$). Take arbitrarily positive integers $m_1, \ldots, m_p$ and $n_1, \ldots, n_q$ such that

$$\sum_{i=1}^{p} m_i = m, \quad \sum_{j=1}^{q} n_j = n.$$

Partition the rows of A according to m_1 rows, $\ldots, m_p$ rows and the columns of A according to n_1 columns, $\ldots, n_q$ columns. Then A is partitioned into $p \times q$ blocks as

$$A = \begin{bmatrix} A_{11} & A_{12} & \cdots & A_{1q} \\ A_{21} & A_{22} & \cdots & A_{2q} \\ \vdots & \vdots & & \vdots \\ A_{p1} & A_{p2} & \cdots & A_{pq} \end{bmatrix} = [A_{ij}],$$

where each $A_{ij}, 1 \le i \le p, 1 \le j \le q$, is an $m_i \times n_j$ matrix and is called a *submatrix* of A. A matrix of this type is called a *block* or *partitioned* matrix. Two block matrices $A_{m \times n} = [A_{ij}]$ and $B_{m \times n} = [B_{ij}]$ are said to be of the *same type* if $A_{ij} = B_{ij}$ for $1 \leq i \leq p, 1 \leq j \leq q$. Block matrices have the following operations as the usual matrices do (see Sec. B.4).

(1) **Addition** If $A = [A_{ij}]$ and $B = [B_{ij}]$ are of the same type, then

$$A + B = [A_{ij} + B_{ij}].$$

(2) **Scalar multiplication** For $A = [A_{ij}]$ and $\lambda \in \mathbb{F}$,

$$\lambda A = [\lambda A_{ij}].$$

(3) **Product** Suppose $A_{m \times n} = [A_{ij}]$ and $B_{n \times l} = [B_{jk}]$ where A_{ij} is $m_i \times n_j$ submatrix of A and B_{jk} is $n_j \times l_k$ submatrix of B for $1 \leq i \leq p$, $1 \leq j \leq q, 1 \leq k \leq t$, i.e. *the column numbers of each A_{ij} is equal to the row numbers of the corresponding B_{jk}*, then

$$AB = [C_{ik}], \quad C_{ik} = \sum_{j=1}^{q} A_{ij} B_{jk} \quad \text{for } 1 \leq i \leq p, 1 \leq k \leq t.$$

(4) **Transpose** If $A = [A_{ij}]$, then

$$A^* = [A_{ij}^*]^*.$$

(5) **Conjugate transpose** If $A = [A_{ij}]$ is a complex matrix, then

$$\bar{A}^* = [\bar{A}_{ij}^*]^*.$$

Block matrix of the following type

$$\begin{bmatrix} A_{11} & & & 0 \\ & A_{22} & & \\ & & \ddots & \\ 0 & & & A_{pp} \end{bmatrix} \text{ or } \begin{bmatrix} A_{11} & & & 0 \\ A_{21} & A_{22} & & \\ \vdots & & \ddots & \\ A_{p1} & A_{p2} & \cdots & A_{pp} \end{bmatrix} \text{ or }$$

$$\begin{bmatrix} A_{11} & A_{12} & \cdots & A_{1p} \\ & A_{22} & \cdots & A_{2p} \\ & & \ddots & \vdots \\ 0 & & & A_{pp} \end{bmatrix}$$

is called *pseudo-diagonal* or *pseudo-lower triangular* or *pseudo-upper triangular*, respectively.

Do the following problems.

7. Model after elementary row and column operations on matrices to prove the following *Schur's formulas*. Let

$$A = \begin{bmatrix} A_{11} & A_{12} \\ A_{21} & A_{22} \end{bmatrix}_{n \times n},$$

where A_{11} is an invertible $r \times r$ submatrix of A.

(a) $\begin{bmatrix} I_r & O \\ -A_{21}A_{11}^{-1} & I_{n-r} \end{bmatrix} \begin{bmatrix} A_{11} & A_{12} \\ A_{21} & A_{22} \end{bmatrix} = \begin{bmatrix} A_{11} & A_{12} \\ O & A_{22} - A_{21}A_{11}^{-1}A_{12} \end{bmatrix}.$

(b) $\begin{bmatrix} A_{11} & A_{12} \\ A_{21} & A_{22} \end{bmatrix} \begin{bmatrix} I_r & -A_{11}^{-1}A_{12} \\ O & I_{n-r} \end{bmatrix} = \begin{bmatrix} A_{11} & O \\ A_{21} & A_{22} - A_{21}A_{11}^{-1}A_{12} \end{bmatrix}.$

(c) $\begin{bmatrix} I_r & O \\ -A_{21}A_{11}^{-1} & I_{n-r} \end{bmatrix} \begin{bmatrix} A_{11} & A_{12} \\ A_{21} & A_{22} \end{bmatrix} \begin{bmatrix} I_r & -A_{11}^{-1}A_{12} \\ O & I_{n-r} \end{bmatrix}$

$$= \begin{bmatrix} A_{11} & O \\ O & A_{22} - A_{21}A_{11}^{-1}A_{12} \end{bmatrix}.$$

Hence, prove that

$$\mathrm{r}(A) = \mathrm{r}(A_{11}) \Leftrightarrow A_{22} = A_{21}A_{11}^{-1}A_{12}.$$

Also, prove that if A and A_{11} are invertible, then $A_{22} - A_{21}A_{11}^{-1}A_{12}$ is also invertible and

$$A^{-1} = \begin{bmatrix} I_r & -A_{11}^{-1}A_{12} \\ O & I_{n-r} \end{bmatrix} \begin{bmatrix} A_{11}^{-1} & O \\ O & (A_{22} - A_{21}A_{11}^{-1}A_{12})^{-1} \end{bmatrix} \begin{bmatrix} I_r & O \\ -A_{21}A_{11}^{-1} & I_{n-r} \end{bmatrix}.$$

8. Let A_{ij} be $n \times n$ matrices for $1 \le i, j \le 2$ and $A_{11}A_{21} = A_{21}A_{11}$.

(a) If A_{11} is invertible, then

$$\det \begin{bmatrix} A_{11} & A_{12} \\ A_{21} & A_{22} \end{bmatrix} = \det(A_{11}) \cdot \det(A_{22} - A_{21}A_{11}^{-1}A_{12})$$

$$= \det(A_{11}A_{22} - A_{21}A_{12}).$$

(b) If A_{ij} are complex matrices for $1 \le i, j \le 2$, then the invertibility of A_{11} may be dropped in (a) and the result still holds.

(c) In particular,

$$\det \begin{bmatrix} A_{11} & A_{12} \\ -A_{12}^* & A_{11}^* \end{bmatrix} = \det(A_{11}A_{11}^* + A_{12}^*A_{12}) \quad \text{if } A_{12}^*A_{11} = A_{11}A_{12}^*;$$

$$\det \begin{bmatrix} O & A_{12} \\ A_{12}^* & O \end{bmatrix} = (-1)^n |\det A_{12}|^2.$$

9. Let $A_{m\times n}$ and $B_{n\times m}$ be matrices over $\mathbb{F}$.

   (a) Show that

   $$\begin{bmatrix} I_m & O \\ -B & I_n \end{bmatrix} \begin{bmatrix} I_m & A \\ B & I_n \end{bmatrix} = \begin{bmatrix} I_m & A \\ O & I_n - BA \end{bmatrix}, \quad \text{and}$$

   $$\begin{bmatrix} I_m & -A \\ O & I_n \end{bmatrix} \begin{bmatrix} I_m & A \\ B & I_n \end{bmatrix} = \begin{bmatrix} I_m - AB & O \\ B & I_n \end{bmatrix}.$$

   (b) Show that $\det(I_n - BA) = \det(I_m - AB)$.

   (c) Show that $I_n - BA$ is invertible if and only if $I_m - AB$ is invertible. Try to use A, B and $(I_m - AB)^{-1}$ to represent $(I_n - BA)^{-1}$.

   (d) Use (b) to show that

   $$\det \begin{bmatrix} 1 + x_1 y_1 & x_1 y_2 & \cdots & x_1 y_n \\ x_2 y_1 & 1 + x_2 y_2 & \cdots & x_2 y_n \\ \vdots & \vdots & & \vdots \\ x_n y_1 & x_n y_2 & \cdots & 1 + x_n y_n \end{bmatrix} = 1 + x_1 y_1 + \cdots + x_n y_n.$$

   In general, for $A_{n\times 1}$ and $B_{n\times 1}$ and $\lambda \in \mathbb{F}$,

   $$\det(I_n - \lambda AB^*) = 1 - \lambda B^* A.$$

   Hence $I_n - \lambda AB^*$ is invertible if and only if $1 - \lambda B^* A \neq O$ and

   $$(I_n - \lambda AB^*)^{-1} = I_n + \frac{\lambda}{1 - \lambda B^* A} AB^*.$$

   (e) Show that, for $\lambda \in \mathbb{F}$,

   $$\lambda^n \det(\lambda I_m - AB) = \lambda^m \det(\lambda I_n - BA).$$

10. (a) Suppose $A \in M(n, \mathbb{F})$ satisfying $A^2 = A$, i.e. A is idempotent. Show that $r(A) = \text{tr}(A)$.

    (b) Let $A_i \in M(n, \mathbb{F})$ for $1 \leq i \leq k$ be such that $A_1 + \cdots + A_k = I_n$. Show that $A_i^2 = A_i$, $1 \leq i \leq k$, if and only if $r(A_1) + \cdots + r(A_k) = n$.

11. Let

$$A = \begin{bmatrix} B & C \\ O & I_r \end{bmatrix}$$

be a square matrix (and hence B is a square submatrix, say of order m).

   (a) Suppose $\det(B - I_m) \neq 0$, then

   $$A^n = \begin{bmatrix} B^n & (B^n - I_m)(B - I_m)^{-1} C \\ O & I_r \end{bmatrix}, \quad n = 0, 1, 2, \ldots.$$

   (b) In case $\det B \neq 0$, then (a) still holds for $n = -1, -2, \ldots$.

12. Show that

$$\begin{bmatrix} A_{m\times n} & I_m \\ O & B_{n\times m} \end{bmatrix}_{(m+n)\times(m+n)}$$

is invertible if and only if BA is invertible.

13. Let $A_{n\times n}$ and $B_{n\times n}$ be complex matrices. Show that

$$\det \begin{bmatrix} A & -B \\ B & A \end{bmatrix} = \det(A + iB)\det(A - iB).$$

In particular,

$$\det \begin{bmatrix} A & O \\ O & A \end{bmatrix} = (\det A)^2; \quad \det \begin{bmatrix} O & -B \\ B & O \end{bmatrix} = (\det B)^2.$$

14. Adopt notations in (2.7.66).

(a) Show that

$$E_{(i)+\alpha(j)} = E_{\alpha^{-1}(j)}E_{(i)+(j)}E_{\alpha(j)}, \quad \alpha \neq 0, \ i \neq j;$$
$$E_{(i)(j)} = E_{-(j)}E_{(j)+(i)}E_{-(i)}E_{(i)+(j)}E_{-(j)}E_{(j)+(i)}.$$

(b) Show that an invertible matrix $A_{n\times n}$ can always be expressed as a product of square matrices of the form $I_n + \alpha E_{ij}$.

15. Use (2.7.70) to show that a matrix $A_{m\times n}$ of rank $r \geq 1$ is a sum of r matrices of order $m \times n$, each of rank 1.

16. Let $A \in M(n, \mathbb{R})$ be a symmetric matrix with rank r. Show that at least one of its principal subdeterminants (see Sec. B.6) of order r is not equal to zero, and all nonzero such subdeterminants are of the same sign.

17. Let A be a $m \times n$ matrix. Show that $r(A) = n$ if and only if there exists an invertible matrix $P_{m\times m}$ such that

$$A = P \begin{bmatrix} I_n \\ O \end{bmatrix}_{m\times n}$$

by using the normal form of A in (2.7.70).

18. *The rank decomposition theorem of a matrix*
    Let $A_{m\times n}$ be a matrix. Show that

$$r(A) = r \geq 1$$

$\Leftrightarrow$ There exist a matrix $B_{m\times r}$ of rank r and a matrix $C_{r\times n}$ of rank r such that $A = BC$.

Then, show that, in case A is a real or complex matrix, the generalized inverse of A is

$$A^+ = C^*(CC^*)^{-1}(B^*B)^{-1}B^*.$$

19. By performing elementary row operations and elementary column operations of type 1, any $m \times n$ matrix can be reduced to the form

$$\begin{bmatrix} I_r & B \\ O & O \end{bmatrix},$$

where $r = r(A)$.

20. Elementary matrix operations shown in (2.7.59) can be extended to block matrices as follows. Let

$$A = \begin{bmatrix} A_{11} & A_{12} \\ A_{21} & A_{22} \end{bmatrix}_{m \times n},$$

where A_{11} is a $p \times q$ submatrix.

(1) Type 1

$$\begin{bmatrix} O & I_{m-p} \\ I_p & O \end{bmatrix} \begin{bmatrix} A_{11} & A_{12} \\ A_{21} & A_{22} \end{bmatrix} = \begin{bmatrix} A_{21} & A_{22} \\ A_{11} & A_{12} \end{bmatrix},$$

$$\begin{bmatrix} A_{11} & A_{12} \\ A_{21} & A_{22} \end{bmatrix} \begin{bmatrix} O & I_q \\ I_{n-q} & O \end{bmatrix} = \begin{bmatrix} A_{12} & A_{11} \\ A_{22} & A_{21} \end{bmatrix}.$$

(2) Type 2

$$\begin{bmatrix} B_{p \times p} & O \\ O & I_{m-p} \end{bmatrix} \begin{bmatrix} A_{11} & A_{12} \\ A_{21} & A_{22} \end{bmatrix} = \begin{bmatrix} BA_{11} & BA_{12} \\ A_{21} & A_{22} \end{bmatrix}, \text{ etc.;}$$

$$\begin{bmatrix} A_{11} & A_{12} \\ A_{21} & A_{22} \end{bmatrix} \begin{bmatrix} I_q & O \\ O & B_{(n-q) \times (n-q)} \end{bmatrix} = \begin{bmatrix} A_{11} & A_{12}B \\ A_{21} & A_{22}B \end{bmatrix}, \text{ etc.}$$

(3) Type 3

$$\begin{bmatrix} I_p & B \\ O & I_{m-p} \end{bmatrix} \begin{bmatrix} A_{11} & A_{12} \\ A_{21} & A_{22} \end{bmatrix} = \begin{bmatrix} A_{11} + BA_{21} & A_{12} + BA_{22} \\ A_{21} & A_{22} \end{bmatrix}, \text{ etc.;}$$

$$\begin{bmatrix} A_{11} & A_{12} \\ A_{21} & A_{22} \end{bmatrix} \begin{bmatrix} I_q & O \\ B & I_{n-q} \end{bmatrix} = \begin{bmatrix} A_{11} + A_{12}B & A_{12} \\ A_{21} + A_{22}B & A_{22} \end{bmatrix}, \text{ etc.}$$

Note that Schur's formulas in Ex. 7 are special cases.

21. Prove the following results for rank.

(a) If $A = [A_{ij}]$ is a block matrix, the $r(A) \geq r(A_{ij})$ for each i and j.

(b) $r\left(\begin{bmatrix} A & O \\ O & B \end{bmatrix}\right) = r(A) + r(B)$.

(c) $r\left(\begin{bmatrix} A & C \\ O & B \end{bmatrix}\right) \geq r\left(\begin{bmatrix} A & O \\ O & B \end{bmatrix}\right)$.

(d) Use (c) to prove that $r(A) + r(B) - n \leq r(AB)$, where $A_{m \times n}$ and $B_{n \times m}$.

22. Suppose $A_{m \times n}, B_{n \times p}$ and $C_{p \times q}$.

(a) Show that

$$\begin{bmatrix} I_m & A \\ O & I_n \end{bmatrix} \begin{bmatrix} ABC & O \\ O & B \end{bmatrix} \begin{bmatrix} I_q & O \\ -C & I_p \end{bmatrix} \begin{bmatrix} O & -I_q \\ I_p & O \end{bmatrix} = \begin{bmatrix} AB & O \\ B & BC \end{bmatrix}.$$

(b) Show that

$$r\begin{bmatrix} AB & O \\ O & BC \end{bmatrix} \leq r\begin{bmatrix} AB & O \\ B & BC \end{bmatrix} = r\begin{bmatrix} ABC & O \\ O & B \end{bmatrix}$$

and hence, deduce the *Frobenius inequality* (see Ex. <C> 11 of Sec. 2.7.3)

$$r(AB) + r(BC) \leq r(B) + r(ABC).$$

(c) Taking $B = I_n$ and using $B_{n \times p}$ to replace C, show the *Sylvester inequality* (see (2.7.43) and Ex. <C> of Sec. 2.7.3)

$$r(A) + r(B) - n \leq r(AB).$$

23. (a) Let $A_{m \times n}$ and $B_{m \times n}$ be real matrices. Show that

$$(\det AB^*)^2 \leq (\det AA^*)^2 (\det BB^*)^2.$$

(b) Let $A_{m \times n}$ be a complex matrix. Show that any principal subdeterminant of order r of $A\bar{A}^*$

$$A\bar{A}^* \begin{pmatrix} i_1 \dots i_r \\ i_1 \dots i_r \end{pmatrix} \geq 0, \quad 1 \leq i_1 < \dots < i_r \leq m$$

(see Sec. B.6).

2.7.6   *Diagonal canonical form*

The preliminary concepts about eigenvalues and eigenvectors of a linear operator or a real 2 × 2 matrix were introduced in Example 1 through Example 7 in Sec. 2.7.2, via the geometric terminology of invariant lines or subspaces. Formal definitions for them are in (2.7.11) and (2.7.13). See also (2.7.12), (2.7.14), (2.7.19) and (2.7.38), (2.7.39), (2.7.42).

From Secs. 2.7.2 to 2.7.5, we repeat again and again in examples and exercises to familiarize readers with these two concepts and methods to compute them, and to make readers realize implicitly and purposely how many advantages they might have in the investigation of geometric mapping properties of a linear operator.

Here in this subsection, we give two examples to end the study of the diagonalizability of a linear operator (see (2.7.24)) and to summarize formally their results in (2.7.72) and (2.7.73). *Examples presented here might make you feel cumbersome, boring, and sick. If so, please just skip this content and go directly to Exercises or Sec. 2.7.7.*

Example 1 Let $\mathbb{R}^2$ be endowed with Cartesian coordinate system $\mathcal{N} = \{\vec{e}_1, \vec{e}_2\}$ as Fig. 2.17(b) indicated. Investigate the geometric mapping properties of the linear transformation

$$f(x_1, x_2) = (2x_1 - x_2, -3x_1 + 4x_2)$$

$$= \vec{x} A, \quad \text{where } \vec{x} = (x_1, x_2) \text{ and } A = \begin{bmatrix} 2 & -3 \\ -1 & 4 \end{bmatrix}.$$

Solution It is easy to check that f is indeed a *linear transformation*.

Moreover, f is *isomorphic*. To see this, suppose $f(x_1, x_2) = 0$. This is equivalent to say that

$$2x_1 - x_2 = 0, \quad -3x_1 + 4x_2 = 0,$$

whose only solution is $\vec{x} = (x_1, x_2) = \vec{0}$. Hence f is one-to-one. Next, for any given $\vec{y} = (y_1, y_2)$, solve $f(x_1, x_2) = (y_1, y_2)$, i.e.

$$\begin{cases} 2x_1 - x_2 = y_1 \\ -3x_1 + 4x_2 = y_2 \end{cases}$$

$$\Rightarrow \begin{cases} x_1 = \dfrac{4}{5}y_1 + \dfrac{1}{5}y_2 \\ x_2 = \dfrac{3}{5}y_1 + \dfrac{2}{5}y_2 \end{cases}, \quad \text{with } A^{-1} = \frac{1}{5}\begin{bmatrix} 4 & 3 \\ 1 & 2 \end{bmatrix}.$$

The resulting vector $\vec{x} = \left(\frac{4}{5}y_1 + \frac{1}{5}y_2, \frac{3}{5}y_1 + \frac{2}{5}y_2\right)$ is then the (unique) solution to $f(\vec{x}) = \vec{y}$. Thus f is onto (see (2.7.8)). It is worth to notice that the above algebraic computation can be simplified as the computation of the inverse A^{-1} of the matrix A: $\vec{y} = \vec{x} A \Leftrightarrow \vec{x} = \vec{y} A^{-1}$.

f maps straight lines into straight lines and preserves ratios of lengths of line segments along the same line. The equation

$$a_1 x_1 + a_2 x_2 + b = 0 \tag{$*_1$}$$

of a line can be written in matrix form as

$$\vec{x} \begin{bmatrix} a_1 \\ a_2 \end{bmatrix} + b = 0.$$

Hence, the image of this line under f has the equation

$$\vec{y} A^{-1} \begin{bmatrix} a_1 \\ a_2 \end{bmatrix} + b = 0, \quad \text{or}$$

$$(4a_1 + 3a_2)y_1 + (a_1 + 2a_2)y_2 + 5b = 0, \tag{$*_2$}$$

which represents a straight line. The property that f preserves ratios of lengths segments is contained in the definition of f, a linear transformation. Do you see why? Of course, one can use parametric equation $\vec{x} = \vec{a} + t\vec{b}$ for the line (see (2.5.4)) to prove these results.

f preserves the relative positions of two lines (see (2.5.9)). Suppose the two lines have respective equation $\vec{x} = \vec{a}_1 + t\vec{b}_1$ and $\vec{x} = \vec{a}_2 + t\vec{b}_2$ where $t \in \mathbb{R}$. Note that the image of the line $\vec{x} = \vec{a}_i + t\vec{b}_i$ under f has the equation $\vec{y} = \vec{a}_i A + t\vec{b}_i A$ for $i = 1, 2$. Therefore, for example,

> two lines intersect at a point $\vec{x}_0$.
> $\Leftrightarrow$ $\vec{b}_1$ and $\vec{b}_2$ are linearly independent.
> $\Leftrightarrow$ $\vec{b}_1 A$ and $\vec{b}_2 A$ are linearly independent (because A is invertible).
> $\Leftrightarrow$ The two image lines intersect at a point $\vec{x}_0 A$.

Again, one should refer to Ex. <B> 4 of Sec. 2.4 or (2.7.8).

Suppose the $x_1 x_2$-plane and $y_1 y_2$-plane are deposited on the same plane. We are posed with the problem: *Is there any line coincident with its image under f? If yes, how to find it and how many can we find?*

For simplicity, suppose firstly the line passes through the origin $\vec{0} = (0, 0)$. Then in $(*_1)$ and in $(*_2)$, $b = 0$ should hold and both lines are coincident if and only if

$$\frac{4a_1 + 3a_2}{a_1} = \frac{a_1 + 2a_2}{a_2}.$$

There are two algebraic ways to treat the above equation.

(1) One is

$$a_2(4a_1 + 3a_2) = a_1(a_1 + 2a_2)$$
$$\Leftrightarrow a_1^2 - 2a_1a_2 - 3a_2^2 = (a_1 - 3a_2)(a_1 + a_2) = 0$$
$$\Leftrightarrow a_1 + a_2 = 0 \text{ which results in } a_1 : a_2 = 1 : -1, \text{ or}$$
$$a_1 - 3a_2 = 0 \text{ which results in } a_1 : a_2 = 3 : 1.$$

This means that the *line* $x_1 - x_2 = 0$ *and* $3x_1 + x_2 = 0$ *are kept invariant under the mapping f.*

(2) The other is

$$\frac{4a_1 + 3a_2}{a_1} = \frac{a_1 + 2a_2}{a_2} = \alpha$$

$$\Leftrightarrow \begin{cases} (4 - \alpha)a_1 + 3a_2 = 0 \\ a_1 + (2 - \alpha)a_2 = 0 \end{cases} \text{ or } \begin{bmatrix} 4 - \alpha & 3 \\ 1 & 2 - \alpha \end{bmatrix} \begin{bmatrix} a_1 \\ a_2 \end{bmatrix} = \begin{bmatrix} 0 \\ 0 \end{bmatrix}$$

$\Leftrightarrow$(Suppose there does exist $(a_1, a_2) \neq (0,0)$, and it solves equations.

Refer to Exs. <A> 2 and <B> 4 of Sec. 2.4.)

$$\det \begin{bmatrix} 4 - \alpha & 3 \\ 1 & 2 - \alpha \end{bmatrix} = \begin{vmatrix} 4 - \alpha & 3 \\ 1 & 2 - \alpha \end{vmatrix}$$

$$= (4 - \alpha)(2 - \alpha) - 3 = \alpha^2 - 6\alpha + 5 = 0$$

$\Leftrightarrow \alpha = 1, 5.$

The case $\alpha = 1$ will result in $a_1 + a_2 = 0$ which, in turn, implies that $x_1 - x_2 = 0$ is an invariant line. While $\alpha = 5$ results in $a_1 - 3a_2 = 0$ and hence $3x_1 + x_2 = 0$ is another invariant line. We obtain the same results as in (1).

Note 1 Method (2) above can be replaced by the following process. By the *vector forms* of $(*_1)$ and $(*_2)$, we have:

The line $\vec{x} \begin{bmatrix} a_1 \\ a_2 \end{bmatrix} = 0$ coincides with the line $\vec{x} A^{-1} \begin{bmatrix} a_1 \\ a_2 \end{bmatrix} = 0.$

$\Leftrightarrow$ There exists constant μ such that

$$A^{-1} \begin{bmatrix} a_1 \\ a_2 \end{bmatrix} = \mu \begin{bmatrix} a_1 \\ a_2 \end{bmatrix}.$$

$$\Leftrightarrow (A^{-1} - \mu I_2) \begin{bmatrix} a_1 \\ a_2 \end{bmatrix} = \begin{bmatrix} 0 \\ 0 \end{bmatrix}. \tag{$*_3$}$$

$$\Leftrightarrow \det(A^{-1} - \mu I_2) = \det\left[\frac{1}{5}\begin{bmatrix} 4 & 3 \\ 1 & 2 \end{bmatrix} - \mu\begin{bmatrix} 1 & 0 \\ 0 & 1 \end{bmatrix}\right]$$

$$= \frac{1}{25}\begin{vmatrix} 4 - 5\mu & 3 \\ 1 & 2 - 5\mu \end{vmatrix} = \frac{1}{5}(5\mu^2 - 6\mu + 1) = 0.$$

$$\Rightarrow \mu = 1 \text{ and } \mu = \frac{1}{5}.$$

If $\mu = 1$, by (∗$_3$), we have $a_1 - 3a_2 = 0$, and if $\mu = \frac{1}{5}$, then $a_1 - a_2 = 0$.

Note 2 Traditionally, especially in Cartesian coordinate system, we adopt (∗$_1$) as starting point for almost every computation work about straight lines. Why not use *parametric equation in vector form* as (2.5.4)? If so, then:

The line $\vec{x} = t\vec{b}$ coincides with its image $\vec{x} = t\vec{b}A$ under f.

$\Leftrightarrow$ There exists constant λ such that $\vec{b}A = \lambda\vec{b}$.

$\Leftrightarrow \vec{b}(A - \lambda I_2) = \vec{0}$.

$\Leftrightarrow$ (since $\vec{b} \neq \vec{0}$)

$$\det(A - \lambda I_2) = \begin{vmatrix} 2 - \lambda & -3 \\ -1 & 4 - \lambda \end{vmatrix} = \lambda^2 - 6\lambda + 5 = 0.$$

$$\Rightarrow \lambda = 1, 5.$$

In case $\lambda = 1$, to solve $\vec{b}(A - \lambda I_2) = \vec{b}(A - I_2) = \vec{0}$ is equivalent to solve

$$(b_1 \quad b_2)\begin{bmatrix} 2 - 1 & -3 \\ -1 & 4 - 1 \end{bmatrix} = 0 \text{ or } b_1 - b_2 = 0, \quad \text{where } \vec{b} = (b_1, b_2).$$

Hence $b_1 : b_2 = 1 : 1$ and the line $\vec{x} = t(1, 1)$ is an invariant line. In case $\lambda = 5$, to solve $\vec{b}(A - \lambda I_2) = \vec{b}(A - 5I_2) = 0$ is to solve

$$(b_1 \quad b_2)\begin{bmatrix} -3 & -3 \\ -1 & -1 \end{bmatrix} = 0 \quad \text{or} \quad 3b_1 + b_2 = 0.$$

Hence $b_1 : b_2 = 1 : -3$ and thus the line $\vec{x} = t(1, -3)$ is another invariant line.

We call, in Note 2 above, $\lambda_1 = 1$ and $\lambda_2 = 5$ the *eigenvalues* of the square matrix A, and $\vec{x}_1 = t(1, 1)$ for $t \neq 0$ *eigenvectors* of A related to λ_1 and $\vec{x}_2 = t(1, -3)$ for $t \neq 0$ *eigenvectors* of A related to $\lambda_2 = 5$. When comparing various algebraic methods mentioned above, the method in Note 2 is the simplest one to be generalized.

We will see its advantages over the others as we proceed. Notice that $\vec{x}_1 A = \vec{x}_1$ and $\vec{x}_2 A = 5\vec{x}_2$. See Fig. 2.71.

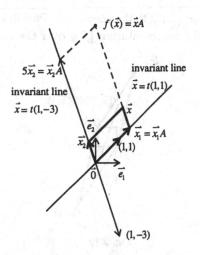

Fig. 2.71

An immediate advantage is as follows. Let $\vec{v}_1 = (1,1)$ and $\vec{v}_2 = (1,-3)$. Since

$$\det \begin{bmatrix} \vec{v}_1 \\ \vec{v}_2 \end{bmatrix} = \det \begin{bmatrix} 1 & 1 \\ 1 & -3 \end{bmatrix} = \begin{vmatrix} 1 & 1 \\ 1 & -3 \end{vmatrix} = -4 \neq 0,$$

so $\vec{v}_1$ and $\vec{v}_2$ are linearly independent. Thus $\mathcal{B} = \{\vec{v}_1, \vec{v}_2\}$ is a basis for $\mathbb{R}^2$. *What does the linear isomorphism f look like in $\mathcal{B}$?* Now,

$$f(\vec{v}_1) = \vec{v}_1 A = \vec{v}_1 = \vec{v}_1 + 0 \cdot \vec{v}_2$$
$$\Rightarrow [f(\vec{v}_1)]_{\mathcal{B}} = (1,0), \quad \text{and}$$
$$f(\vec{v}_2) = \vec{v}_2 A = 5\vec{v}_2 = 0 \cdot \vec{v}_1 + 5\vec{v}_2$$
$$\Rightarrow [f(\vec{v}_2)]_{\mathcal{B}} = (0,5).$$

Therefore, the matrix representation of f with respect to $\mathcal{B}$ is

$$[f]_{\mathcal{B}} = \begin{bmatrix} [f(\vec{v}_1)]_{\mathcal{B}} \\ [f(\vec{v}_2)]_{\mathcal{B}} \end{bmatrix} = \begin{bmatrix} 1 & 0 \\ 0 & 5 \end{bmatrix} = PAP^{-1}, \tag{$*_3$}$$

where

$$P = \begin{bmatrix} \vec{v}_1 \\ \vec{v}_2 \end{bmatrix} = \begin{bmatrix} 1 & 1 \\ 1 & -3 \end{bmatrix}.$$

Hence, for any $\vec{x} = (x_1, x_2) \in \mathbb{R}^2$, let $[\vec{x}]_{\mathcal{B}} = (\alpha_1, \alpha_2) = \vec{x}P^{-1}$, then

$$[f(\vec{x})]_{\mathcal{B}} = [\vec{x}]_{\mathcal{B}}[f]_{\mathcal{B}} = (\alpha_1 \quad \alpha_2) \begin{bmatrix} 1 & 0 \\ 0 & 5 \end{bmatrix} = (\alpha_1, 5\alpha_2) \tag{$*_4$}$$

(see (2.4.2), (2.6.4) or (2.7.23) if necessary). Parallelograms in Fig. 2.72 show how goodness the representation $[f]_\mathcal{B}$ over the original $[f]_\mathcal{N} = A$.

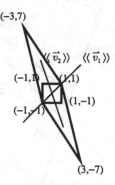

Fig. 2.72

A useful and formal reinterpretation of $(*_3)$ and $(*_4)$ is as follows. Since

$$PAP^{-1} = \begin{bmatrix} 1 & 0 \\ 0 & 5 \end{bmatrix} = \begin{bmatrix} 1 & 0 \\ 0 & 0 \end{bmatrix} + \begin{bmatrix} 0 & 0 \\ 0 & 5 \end{bmatrix} = \begin{bmatrix} 1 & 0 \\ 0 & 0 \end{bmatrix} + 5\begin{bmatrix} 0 & 0 \\ 0 & 1 \end{bmatrix}$$

$$\Rightarrow A = P^{-1}\begin{bmatrix} 1 & 0 \\ 0 & 0 \end{bmatrix}P + 5P^{-1}\begin{bmatrix} 0 & 0 \\ 0 & 1 \end{bmatrix}P = A_1 + 5A_2,$$

where

$$A_1 = P^{-1}\begin{bmatrix} 1 & 0 \\ 0 & 0 \end{bmatrix}P = \frac{1}{4}\begin{bmatrix} 3 & 1 \\ 1 & -1 \end{bmatrix}\begin{bmatrix} 1 & 0 \\ 0 & 0 \end{bmatrix}\begin{bmatrix} 1 & 1 \\ 1 & -3 \end{bmatrix} = \begin{bmatrix} \frac{3}{4} & \frac{3}{4} \\ \frac{1}{4} & \frac{1}{4} \end{bmatrix},$$

$$A_2 = P^{-1}\begin{bmatrix} 0 & 0 \\ 0 & 1 \end{bmatrix}P = \begin{bmatrix} \frac{1}{4} & -\frac{3}{4} \\ -\frac{1}{4} & \frac{3}{4} \end{bmatrix},$$

the claimed advantage will come to surface once we can handle the geometric mapping properties of both A_1 and A_2. Note that

$$\vec{x} = (x_1, x_2) \in \mathbb{R}^2 \quad \text{using } \mathcal{N} = \{\vec{e}_1, \vec{e}_2\}$$
$$\downarrow$$
$$\vec{x}P^{-1} = [\vec{x}]_\mathcal{B} = (\alpha_1, \alpha_2) \in \mathbb{R}^2 \quad \text{using } \mathcal{B} = \{\vec{v}_1, \vec{v}_2\}$$
$$\downarrow$$
$$\vec{x}P^{-1}\begin{bmatrix} 1 & 0 \\ 0 & 0 \end{bmatrix} = (\alpha_1, 0), \quad \text{the projection of } (\alpha_1, \alpha_2) \text{ onto } (\alpha_1, 0) \text{ in } \mathcal{B}$$
$$\downarrow$$
$$\vec{x}P^{-1}\begin{bmatrix} 1 & 0 \\ 0 & 0 \end{bmatrix}P = (\alpha_1 \quad 0)\begin{bmatrix} \vec{v}_1 \\ \vec{v}_2 \end{bmatrix} = \alpha_1\vec{v}_1 = \vec{x}A_1.$$

Hence A_1 defines a linear transformation $\vec{x} \in \mathbb{R}^2 \to \vec{x}A_1 \in \langle\langle \vec{v}_1 \rangle\rangle$, called the *eigenspace* of A related to the eigenvalue λ_1. Also,

$$\vec{x} \in \langle\langle \vec{v}_1 \rangle\rangle \Leftrightarrow \vec{x}A_1 = \vec{x}$$

and thus $A_1^2 = A_1$ holds. A_1 is called the *projection* of $\mathbb{R}^2$ onto $\langle\langle \vec{v}_1 \rangle\rangle$ along $\langle\langle \vec{v}_2 \rangle\rangle$. Similarly, A_2 defined by $\vec{x} \in \mathbb{R}^2 \to \vec{x}A_2 \in \langle\langle \vec{v}_2 \rangle\rangle$ is called the *projection* of $\mathbb{R}^2$ onto $\langle\langle \vec{v}_2 \rangle\rangle$ along $\langle\langle \vec{v}_1 \rangle\rangle$. For details, see (2.7.51).

We summarize as an abstract result.

The diagonal canonical form of a linear operator
Suppose $f(\vec{x}) = \vec{x}A$: $\mathbb{R}^2 \to \mathbb{R}^2$ is a linear operator where $A = [a_{ij}]_{2\times2}$ is a real matrix. Suppose the *characteristic polynomial*

$$\det(A - tI_2) = (t - \lambda_1)(t - \lambda_2), \quad \lambda_1 \neq \lambda_2.$$

Let $\vec{v}_i$ ($\neq \vec{0}$) be an *eigenvector* of f related to the *eigenvalue* λ_i, i.e. a *nonzero* solution vector of $\vec{x}A = \lambda_i\vec{x}$ for $i = 1, 2$. Then

$$\langle\langle \vec{v}_i \rangle\rangle = \{\vec{x} \in \mathbb{R}^2 \mid \vec{x}A = \lambda_i\vec{x}\}, \quad i = 1, 2$$

is an *invariant subspace* of $\mathbb{R}^2$, i.e. $f(\langle\langle \vec{v}_i \rangle\rangle) \subseteq \langle\langle \vec{v}_i \rangle\rangle$, and is called the *eigenspace* of f related to λ_i.

(1) $\mathcal{B} = \{\vec{v}_1, \vec{v}_2\}$ is a basis for $\mathbb{R}^2$ and

$$[f]_\mathcal{B} = PAP^{-1} = \begin{bmatrix} \lambda_1 & 0 \\ 0 & \lambda_2 \end{bmatrix}, \quad P = \begin{bmatrix} \vec{v}_1 \\ \vec{v}_2 \end{bmatrix}.$$

In this case, call f or A *diagonalizable*.
(2) Let

$$A_1 = P^{-1} \begin{bmatrix} 1 & 0 \\ 0 & 0 \end{bmatrix} P, \quad \text{and} \quad A_2 = P^{-1} \begin{bmatrix} 0 & 0 \\ 0 & 1 \end{bmatrix} P.$$

Then,

1. $\mathbb{R}^2 = \langle\langle \vec{v}_1 \rangle\rangle \oplus \langle\langle \vec{v}_2 \rangle\rangle$.
2. A_i: $\mathbb{R}^2 \to \mathbb{R}^2$ is the *projection* of $\mathbb{R}^2$ onto $\langle\langle \vec{v}_i \rangle\rangle$ along $\langle\langle \vec{v}_j \rangle\rangle$, $i \neq j$, i.e.

$$A_i^2 = A_i, \quad i = 1, 2.$$

3. $A_1A_2 = A_2A_1 = O$.
4. $I_2 = A_1 + A_2$.
5. $A = \lambda_1A_1 + \lambda_2A_2$.

This is called the *diagonal canonical decomposition* of A.   (2.7.72)

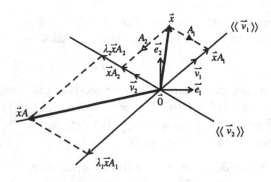

Fig. 2.73

Refer to Exs. 2 and 4 of Sec. B.11. See Fig. 2.73.

Let us return to the original mapping f in Example 1. *Try to find the image of the square with vertices at* $(1, 1)$, $(-1, 1)$, $(-1, -1)$ and $(1, -1)$ under the mapping f. Direct computation shows that

$$f(1, 1) = (1 \quad 1) \begin{bmatrix} 2 & -3 \\ -1 & 4 \end{bmatrix} = (1, 1),$$

$$f(-1, 1) = (-1 \quad 1) \begin{bmatrix} 2 & -3 \\ -1 & 4 \end{bmatrix} = (-3, 7),$$

$$f(-1, -1) = (-1 \quad -1) \begin{bmatrix} 2 & -3 \\ -1 & 4 \end{bmatrix} = (-1, -1) = -f(1, 1),$$

$$f(1, -1) = (1 \quad -1) \begin{bmatrix} 2 & -3 \\ -1 & 4 \end{bmatrix} = (3, -7) = -f(-1, 1).$$

Connect the image points $(1, 1)$, $(-3, 7)$, $(-1, -1)$ and $(3, -7)$ by consecutive line segments and the resulting parallelogram is the required one. See Fig. 2.72. Try to determine this parallelogram by using the diagonal canonical decomposition of A. By the way, the original square has area equal to 4 units. Do you know what is the area of the resulting parallelogram? It is 20 units. Why?   □

In general, a linear operator on $\mathbb{R}^2$ may not be one-to-one, i.e. not an isomorphism.

Example 2 Using $\mathcal{N} = \{\vec{e}_1, \vec{e}_2\}$, the Cartesian coordinate system, let the linear operator $f: \mathbb{R}^2 \to \mathbb{R}^2$ be defined as

$$f(x_1, x_2) = (2x_1 - 3x_2, -4x_1 + 6x_2)$$

$$= \vec{x}A, \quad \text{where } \vec{x} = (x_1, x_2) \text{ and } A = \begin{bmatrix} 2 & -4 \\ -3 & 6 \end{bmatrix}.$$

Try to investigate its geometric mapping properties.

Solution f is obviously linear. Now, the kernel of f is

$$\text{Ker}(f) = \{(x_1, x_2) \mid 2x_1 - 3x_2 = 0\} = \langle\langle(3, 2)\rangle\rangle,$$

which is a straight line passing through the origin. Hence f is not one-to-one and thus is not onto.

What is the range of f? Let $\vec{y} = (y_1, y_2) = \vec{x}A$, then

$$\begin{cases} 2x_1 - 3x_2 = y_1 \\ -4x_1 + 6x_2 = y_2 \end{cases}$$

$$\Rightarrow 2y_1 + y_2 = 0.$$

Therefore,

$$\text{Im}(f) = \{\vec{y} = (y_1, y_2) \in \mathbb{R}^2 \mid 2y_1 + y_2 = 0\} = \langle\langle(1, -2)\rangle\rangle.$$

Deposit the x_1x_2-plane and y_1y_2-plane on the same plane.
The characteristic polynomial of A is

$$\det(A - tI_2) = \begin{vmatrix} 2 - t & -4 \\ -3 & 6 - t \end{vmatrix} = t^2 - 8t.$$

Hence, f or A has two eigenvalues $\lambda_1 = 8$ and $\lambda_2 = 0$. Solve $\vec{x}A = 8\vec{x}$ and get the corresponding eigenvectors $\vec{x}_1 = t(1, -2)$ for $t \neq 0$. Similarly, solve $\vec{x}A = 0\vec{x} = \vec{0}$, and the corresponding eigenvectors $\vec{x}_2 = t(3, 2)$ for $t \neq 0$. Therefore, $\vec{v}_1 = (1, -2)$ and $\vec{v}_2 = (3, 2)$ are linearly independent and the eigenspaces are

$$\langle\langle\vec{v}_1\rangle\rangle = \{\vec{x} \in \mathbb{R}^2 \mid \vec{x}A = 8\vec{x}\} = \text{Im}(f),$$
$$\langle\langle\vec{v}_2\rangle\rangle = \{\vec{x} \in \mathbb{R}^2 \mid \vec{x}A = \vec{0}\} = \text{Ker}(f),$$

which are invariant subspaces of f. Also,

$$\mathbb{R}^2 = \text{Im}(f) \oplus \text{Ker}(f).$$

What follows can be easily handled by using classical high school algebra, but we prefer to adopt the vector method, because we are studying how to get into the realm of linear algebra.

How to find the images of lines parallel to the kernel $\langle\langle\vec{v}_2\rangle\rangle$? Let $\vec{x} = \vec{x}_0 + t\vec{v}_2$ be such a line and suppose it intersects the line $\langle\langle\vec{v}_1\rangle\rangle$ at the point $t_0\vec{v}_1$. Then

$$f(\vec{x}) = f(\vec{x}_0 + t\vec{v}_2) = f(\vec{x}_0) + tf(\vec{v}_2) = f(\vec{x}_0)$$
$$= f(t_0\vec{v}_1) = t_0 f(\vec{v}_1) = 8t_0\vec{v}_1.$$

This means that f maps the whole line $\vec{x} = \vec{x}_0 + t\vec{v}_2$ into a single point, 7 units away from the point of intersection of it with the line $\langle\langle\vec{v}_1\rangle\rangle$, along the line $\langle\langle\vec{v}_1\rangle\rangle$ and in the direction of the point of intersection. See Fig. 2.74.

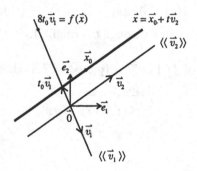

Fig. 2.74

What is the image of a line not parallel to the kernel $\langle\langle\vec{v}_2\rangle\rangle$? Let $\vec{x} = \vec{x}_0 + t\vec{u}$ be such a line where $\vec{u}$ and $\vec{v}_2$ are linearly independent. Suppose $\vec{u} = \vec{u}_1 + \vec{u}_2$ with $\vec{u}_1 \in \text{Im}(f)$ and $\vec{u}_2 \in \text{Ker}(f)$. Then

$$f(\vec{x}) = f(\vec{x}_0) + tf(\vec{u})$$
$$= f(\vec{x}_0) + tf(\vec{u}_1 + \vec{u}_2) = f(\vec{x}_0) + t(f(\vec{u}_1) + f(\vec{u}_2))$$
$$= f(\vec{x}_0) + tf(\vec{u}_1) = f(\vec{x}_0) + 8t\vec{u}_1,$$

where $f(\vec{x}_0) = 8t_0\vec{v}_1$ for some t_0. Since $\vec{u}_1 \neq 0$, the image is a line and coincides with the range line $\langle\langle\vec{v}_1\rangle\rangle$. Also, f maps the line $\vec{x} = \vec{x}_0 + t\vec{u}$ one-to-one and onto the line $\langle\langle\vec{v}_1\rangle\rangle$ and preserves ratios of signed lengths of segments along the line. See Fig. 2.75.

Now, we have a clearer picture about the mapping properties of f. For any $\vec{x} = (x_1, x_2)$, then

$$\vec{x} = \left(\frac{1}{4}x_1 - \frac{3}{8}x_2\right)\vec{v}_1 + \left(\frac{1}{4}x_1 + \frac{1}{8}x_2\right)\vec{v}_2$$
$$\Rightarrow f(\vec{x}) = \left(\frac{1}{4}x_1 - \frac{3}{8}x_2\right)f(\vec{v}_1) = \left(\frac{1}{4}x_1 - \frac{3}{8}x_2\right)\cdot 8\vec{v}_1 = (2x_1 - 3x_2)\vec{v}_1.$$

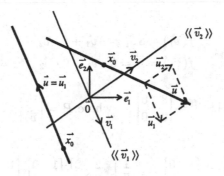

Fig. 2.75

Define the mapping $p: \mathbb{R}^2 \to \mathbb{R}^2$ by

$$p(\vec{x}) = \left(\frac{1}{4}x_1 - \frac{3}{8}x_2\right)\vec{v}_1.$$

p is linear and has the same kernel and range as f does. But p keeps every $\vec{x}$ in $\langle\langle\vec{v}_1\rangle\rangle$ fixed, i.e. $p^2 = p \circ p = p$ holds. This p is called the *projection* of $\mathbb{R}^2$ onto $\langle\langle\vec{v}_1\rangle\rangle$ along the kernel $\langle\langle\vec{v}_2\rangle\rangle$. Therefore,

$$\vec{x} \to p(\vec{x}) \to 8p(\vec{x}) = f(\vec{x}).$$

All one needs to do is to project $\vec{x}$ onto $\langle\langle\vec{v}_1\rangle\rangle$ along $\langle\langle\vec{v}_2\rangle\rangle$, and then to scalarly multiply the projected vector by 8. Finally, the resulting vector is $f(\vec{x})$. See Fig. 2.76 which is a special case of Fig. 2.73.

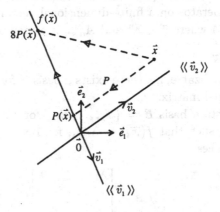

Fig. 2.76

We can use (2.7.72) to reinterpret the content of the above paragraph. To set up the canonical decomposition of f, let $\mathcal{B} = \{\vec{v}_1, \vec{v}_2\}$ which is a basis for $\mathbb{R}^2$. Then

$$f(\vec{v}_1) = 8\vec{v}_1 = 8\vec{v}_1 + 0 \cdot \vec{v}_2$$
$$f(\vec{v}_1) = \vec{0} = 0 \cdot \vec{v}_1 + 0 \cdot \vec{v}_2$$

$$\Rightarrow [f]_\mathcal{B} = PAP^{-1} = \begin{bmatrix} 8 & 0 \\ 0 & 0 \end{bmatrix}, \quad \text{where } P = \begin{bmatrix} \vec{v}_1 \\ \vec{v}_2 \end{bmatrix} = \begin{bmatrix} 1 & -2 \\ 3 & 2 \end{bmatrix}.$$

Let

$$A_1 = P^{-1} \begin{bmatrix} 1 & 0 \\ 0 & 0 \end{bmatrix} P = \frac{1}{8} \begin{bmatrix} 2 & 2 \\ -3 & 1 \end{bmatrix} \begin{bmatrix} 1 & 0 \\ 0 & 0 \end{bmatrix} \begin{bmatrix} 1 & -2 \\ 3 & 2 \end{bmatrix}$$

$$= \frac{1}{8} \begin{bmatrix} 2 & -4 \\ -3 & 6 \end{bmatrix} = \begin{bmatrix} \frac{1}{4} & -\frac{1}{2} \\ -\frac{3}{8} & \frac{3}{4} \end{bmatrix} = \frac{1}{8} A$$

as we might expect beforehand, i.e. $A = 8A_1$. Note that

$$p(\vec{x}) = \vec{x} A_1,$$

$$[p]_\mathcal{N} = A_1 \quad \text{and} \quad [p]_\mathcal{B} = \begin{bmatrix} 1 & 0 \\ 0 & 0 \end{bmatrix}. \qquad\qquad \square$$

For the sake of later reference, we generalize (2.7.72) to linear operators on $\mathbb{F}^n$ or $n \times n$ matrix over a field $\mathbb{F}$ as follows.

Diagonalization of linear operators or square matrices
Let f be a linear operator on a finite-dimensional vector space V (say, in the form $f(\vec{x}) = \vec{x} A$ where $\vec{x} \in \mathbb{F}^n$ and $A_{n \times n}$ is a matrix over $\mathbb{F}$).

(1) *Diagonalizability*

1. f is diagonalizable, i.e. there exists a basis $\mathcal{B}$ for V such that $[f]_\mathcal{B}$ is a diagonal matrix.
$\Leftrightarrow$ 2. There exists a basis $\mathcal{B} = \{\vec{x}_1, \ldots, \vec{x}_n\}$ for V and some scalars $\lambda_1, \ldots, \lambda_n$ such that $f(\vec{x}_i) = \lambda_i \vec{x}_i$, for $1 \le i \le n$. Under these circumstances,

$$[f]_\mathcal{B} = \begin{bmatrix} \lambda_1 & & 0 \\ & \ddots & \\ 0 & & \lambda_n \end{bmatrix}.$$

$\Leftrightarrow$ 3. *In case the characteristic polynomial of f is*

$$\det(f - t1_V) = (-1)^n (t - \lambda_1)^{r_1} \cdots (t - \lambda_k)^{r_k},$$

where $n = \dim V$ and $\lambda_1, \ldots, \lambda_k$ are distinct with $r_i \geq 1$ for $1 \leq i \leq k$ and $r_1 + \cdots + r_k = n$. Then, the *algebraic multiplicity* r_i of λ_i is equal to the dimension $\dim E_{\lambda_i}$, the *geometric multiplicity*, of the *eigenspace*

$$E_{\lambda_i} = \{\vec{x} \in V \mid f(\vec{x}) = \lambda_i \vec{x}\}$$

corresponding to the *eigenvalue* λ_i, for $1 \leq i \leq k$. *Nonzero* vectors in E_{λ_i} are called *eigenvectors* corresponding to λ_i.

⇔ 4. Let $\lambda_1, \ldots, \lambda_k$ be as in 3. $\varphi(t) = (t - \lambda_1) \cdots (t - \lambda_k)$ *annihilates* A, i.e.

$$\varphi(A) = (A - \lambda_1 I_n) \cdots (A - \lambda_k I_n) = O_{n \times n}$$

(see Ex. \<C\> 9(e)). This $\varphi(t)$ is called the *minimal polynomial* of A.

⇔ 5. $V = E_{\lambda_1} \oplus \cdots \oplus E_{\lambda_k}$, where $E_{\lambda_1}, \ldots, E_{\lambda_k}$ are as in 3.

⇐ 6. f has n *distinct* eigenvalues $\lambda_1, \ldots, \lambda_n$, where $n = \dim V$.

See Ex. \<C\> 4(e) for another diagonalizability criterion.

(2) *Diagonal canonical form or decomposition*

For simplicity, via a linear isomorphism, we may suppose that $V = \mathbb{F}^n$ and $f(\vec{x}) = \vec{x} A$ where A is an $n \times n$ matrix. Adopt notations in (1)3. Let $\mathcal{B} = \{\vec{x}_1, \ldots, \vec{x}_n\}$ be a basis for $\mathbb{F}^n$ consisting of eigenvectors of A such that

$$PAP^{-1} = [f]_a = \begin{bmatrix} \lambda_1 & & & & & 0 \\ & \ddots & & & & \\ & & \lambda_1 & & & \\ & & & \lambda_k & & \\ & & & & \ddots & \\ 0 & & & & & \lambda_k \end{bmatrix}, \quad P = \begin{bmatrix} \vec{x}_1 \\ \vdots \\ \vec{x}_i \\ \vdots \\ \vec{x}_n \end{bmatrix}.$$

Defined the following matrices or linear operators:

$$A_i = P^{-1} \begin{bmatrix} 0 & & & & & & \\ & \ddots & & & & & \\ & & 0 & & & 0 & \\ & & & 1 & & & \\ & & & & \ddots & & \\ & & & & & 1 & \\ & 0 & & & & & 0 \\ & & & & & & \ddots \\ & & & & & & & 0 \end{bmatrix} P, \quad 1 \leq i \leq k.$$

Then,

1. $\mathbb{F}^n = E_{\lambda_1} \oplus \cdots \oplus E_{\lambda_k}$.
2. Each $A_i \colon \mathbb{F}^n \to \mathbb{F}^n$ is a projection of $\mathbb{F}^n$ onto E_{λ_i} along $E_{\lambda_1} \oplus \cdots \oplus E_{\lambda_{i-1}} \oplus E_{\lambda_{i+1}} \oplus \cdots \oplus E_{\lambda_k}$, i.e.

$$A_i^2 = A_i, \quad 1 \le i \le k.$$

3. $A_i A_j = O_{n \times n}$ if $i \ne j, 1 \le i, j \le k$.
4. $I_n = A_1 + \cdots + A_k$.
5. $A = \lambda_1 A_1 + \cdots + \lambda_k A_k$. \hfill (2.7.73)

For more details, refer to Secs. B.11 and B.12. Also, note that, if $\dim(\mathrm{Ker}(f)) \ge 1$, then each nonzero vector in $\mathrm{Ker}(f)$ is an eigenvector corresponding to the eigenvalue 0.

Exercises

<A>

1. In Example 1, does linear isomorphism f have other invariant lines besides $x_1 - x_2 = 0$ and $3x_1 + x_2 = 0$? Find invariant lines of the affine transformation

$$f(\vec{x}) = \vec{x}_0 + \vec{x} A,$$

where $\vec{x}_0 \ne \vec{0}$, if any.

2. In Example 2, find all possible one-dimensional subspaces S of $\mathbb{R}^2$ such that

$$\mathbb{R}^2 = S \oplus \mathrm{Ker}(f).$$

For each $\vec{x} \in \mathbb{R}^2$, let

$$\vec{x} + \mathrm{Ker}(f) = \{ \vec{x} + \vec{v} \mid \vec{v} \in \mathrm{Ker}(f) \}$$

be the image of $\mathrm{Ker}(f)$ under the translation $\vec{v} \to \vec{x} + \vec{v}$ which is the line $\vec{x} + \langle\langle \vec{v}_2 \rangle\rangle$ parallel to the line $\langle\langle \vec{v}_2 \rangle\rangle$. Show that

$$\vec{x}_1 + \mathrm{Ker}(f) = \vec{x}_2 + \mathrm{Ker}(f) \Leftrightarrow \vec{x}_1 - \vec{x}_2 \in \mathrm{Ker}(f).$$

Denote the quotient set

$$\mathbb{R}^2 / \mathrm{Ker}(f) = \{ \vec{x} + \mathrm{Ker}(f) \mid \vec{x} \in \mathbb{R}^2 \}$$

and introduce two operations on it as follows:

(1) $\alpha(\vec{x} + \mathrm{Ker}(f)) = \alpha\vec{x} + \mathrm{Ker}(f), \alpha \in \mathbb{R}$,
(2) $(\vec{x}_1 + \mathrm{Ker}(f)) + (\vec{x}_2 + \mathrm{Ker}(f)) = (\vec{x}_1 + \vec{x}_2) + \mathrm{Ker}(f)$,

which are well-defined (refer to Sec. A.1). Then, show that $\mathbb{R}^2/\text{Ker}(f)$ is a vector space isomorphic to S mentioned above (see Sec. B.1). $\mathbb{R}^2/\text{Ker}(f)$ is called the *quotient space* of $\mathbb{R}^2$ modulus $\text{Ker}(f)$. What is $\mathbb{R}^2/S$?

3. In $\mathcal{N} = \{\vec{e}_1, \vec{e}_2\}$, define $f\colon \mathbb{R}^2 \to \mathbb{R}^2$ by

$$f(\vec{x}) = \vec{x}A, \quad \text{where } A = \begin{bmatrix} 5 & 1 \\ -7 & -3 \end{bmatrix} \text{ and } \vec{x} = (x_1, x_2).$$

(a) Model after the purely algebraic method in Example 1 to answer the following questions about f:

   (1) f is one-to-one and onto.
   (2) f maps straight lines into straight lines and preserves their relative positions.
   (3) f preserves ratios of signed lengths of line segments along the same or parallel lines.
   (4) f maps triangles into triangles with vertices, sides and interior to vertices, sides and interior, respectively.
   (5) f maps parallelograms into parallelograms.
   (6) Does f preserve orientations (counterclockwise or clockwise)? Why?
   (7) Does f preserve areas of triangles? Why?
   (8) How many invariant lines do f have? If any, find out.

   Now, for any fixed $\vec{x}_0 \in \mathbb{R}^2$ and $\vec{x}_0 \neq \vec{0}$, answer the same questions for the affine transformation

   $$T(\vec{x}) = \vec{x}_0 + f(\vec{x})$$

   as (1)–(8).

(b) Model after the linearly algebraic method in Example 1 to answer the same questions (1)–(8) as in (a). Notice the following process:

   (1) Compute the characteristic polynomial $\det(A - tI_2)$.
   (2) Solve $\det(A - tI_2) = 0$ to determine the eigenvalues λ_1 and λ_2.
   (3) Solve the equation $\vec{x}A = \lambda_i \vec{x}$ or $\vec{x}(A - \lambda_i I_2) = \vec{0}$ to determine the corresponding eigenvectors $\vec{x}_i \neq 0, i = 1, 2$.
   (4) Make sure if $\vec{x}_1$ and $\vec{x}_2$ are linearly independent.
   (5) Let $\mathcal{B} = \{\vec{x}_1, \vec{x}_2\}$, a basis for $\mathbb{R}^2$. Then set up

   $$[f]_\mathcal{B} = PAP^{-1} = \begin{bmatrix} \lambda_1 & 0 \\ 0 & \lambda_2 \end{bmatrix}, \quad \text{where } P = \begin{bmatrix} \vec{x}_1 \\ \vec{x}_2 \end{bmatrix}.$$

   (6) Justify (2) in (2.7.72) for this f or A.

   Then, try to use these information to do the problem.

(c) Compare the advantages of both methods with each other!

4. In $\mathcal{N} = \{\vec{e}_1, \vec{e}_2\}$, let $f\colon \mathbb{R}^2 \to \mathbb{R}^2$ be defined by

$$f(\vec{x}) = \vec{x}A, \quad \text{where } A = \begin{bmatrix} 6 & 2 \\ 2 & 9 \end{bmatrix}.$$

Do the same problems as in (a) of Ex. 3 via the linearly algebraic method. The symmetry in entries of A induces that the eigenvectors $\vec{x}_1$ and $\vec{x}_2$ are perpendicular to each other and could be chosen to be of unit length. In this case, say something about the geometric mapping properties of

$$P = \begin{bmatrix} \vec{x}_1 \\ \vec{x}_2 \end{bmatrix}.$$

5. Do the same problems as Ex. 4 for

$$f(\vec{x}) = \vec{x}A, \quad \text{where } \vec{x} = (x_1, x_2) \text{ and } A = \begin{bmatrix} 2 & 3 \\ 3 & -2 \end{bmatrix}.$$

6. In $\mathcal{N} = \{\vec{e}_1, \vec{e}_2\}$, let $f\colon \mathbb{R}^2 \to \mathbb{R}^2$ be defined by

$$f(\vec{x}) = \vec{x}A, \quad \text{where } \vec{x} = (x_1, x_2) \text{ and } A = \begin{bmatrix} 3 & -2 \\ -6 & 4 \end{bmatrix}.$$

(a) Use this f to justify (2.7.72), model after Example 2.
(b) Do the same problems as in (a) of Ex. 3. Be careful that some of these problems may not be true or have to be readjusted properly.

7. A square matrix of order 2 with coincident eigenvalues is diagonalizable if and only if the matrix is a scalar matrix.

8. $A_{2\times2}$ is diagonalizable $\Leftrightarrow A^*$ is diagonalizable.

9. If $A_{2\times2}$ is invertible, then A is diagonalizable $\Leftrightarrow A^{-1}$ is diagonalizable.

10. *Constructive* linear algebra

Fix the Cartesian coordinate system $\mathcal{N} = \{\vec{e}_1, \vec{e}_2\}$ on the plane $\mathbb{R}^2$. Where is the square with vertices $\vec{0} = (0,0)$, $\vec{e}_1 = (1,0)$, $\vec{e}_1 + \vec{e}_2 = (1,1)$ and $\vec{e}_2 = (0,1)$ in the coordinate system $\mathcal{B} = \{\vec{v}_1, \vec{v}_2\}$ for $\vec{v}_1 = \left(-\frac{1}{2}, -1\right)$ and $\vec{v}_2 = (-1, 2)$? This means that we have to find out $[\vec{0}]_{\mathcal{B}}, [\vec{e}_1]_{\mathcal{B}}, [\vec{e}_1 + \vec{e}_2]_{\mathcal{B}}$ and $[\vec{e}_2]_{\mathcal{B}}$ and reconsider these as points in the original system $\mathcal{N}$. Let

$$\vec{e}_1 = a_{11}\vec{v}_1 + a_{12}\vec{v}_2$$
$$\vec{e}_2 = a_{21}\vec{v}_1 + a_{22}\vec{v}_2$$

$$\Rightarrow \begin{bmatrix} \vec{e}_1 \\ \vec{e}_2 \end{bmatrix} = \begin{bmatrix} a_{11} & a_{12} \\ a_{21} & a_{22} \end{bmatrix} \begin{bmatrix} \vec{v}_1 \\ \vec{v}_2 \end{bmatrix}, \quad \text{where } \begin{bmatrix} \vec{v}_1 \\ \vec{v}_2 \end{bmatrix} = \begin{bmatrix} -\frac{1}{2} & -1 \\ -1 & 2 \end{bmatrix}$$

$$\Rightarrow \begin{bmatrix} a_{11} & a_{12} \\ a_{21} & a_{22} \end{bmatrix} = \begin{bmatrix} \vec{e}_1 \\ \vec{e}_2 \end{bmatrix} \begin{bmatrix} \vec{v}_1 \\ \vec{v}_2 \end{bmatrix}^{-1} = \begin{bmatrix} 1 & 0 \\ 0 & 1 \end{bmatrix} \cdot \frac{-1}{2} \begin{bmatrix} 2 & 1 \\ 1 & -\frac{1}{2} \end{bmatrix} = \begin{bmatrix} -1 & -\frac{1}{2} \\ -\frac{1}{2} & \frac{1}{4} \end{bmatrix}.$$

Hence $[\vec{e}_1]_\mathcal{B} = (-1, -\frac{1}{2})$, $[\vec{e}_2]_\mathcal{B} = (-\frac{1}{2}, \frac{1}{4})$ and $[\vec{e}_1 + \vec{e}_2]_\mathcal{B} = (-\frac{3}{2}, -\frac{1}{4})$. See Fig. 2.77.

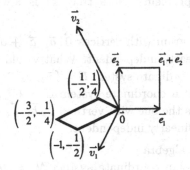

Fig. 2.77

Conversely, give a parallelogram with vertices $(0,0), (1,2), (-2,1)$ and $(-3, -1)$. Under what coordinate system $\mathcal{B} = \{\vec{v}_1, \vec{v}_2\}$ does the square look like the given parallelogram? This means that we have to find out $\vec{v}_1$ and $\vec{v}_2$ so that

$$[\vec{e}_1]_\mathcal{B} = (1,2), [\vec{e}_2]_\mathcal{B} = (-3,-1) \quad \text{and} \quad [\vec{e}_1 + \vec{e}_2]_\mathcal{B} = (-2,1).$$

Now

$$\vec{e}_1 = 1 \cdot \vec{v}_1 + 2 \cdot \vec{v}_2$$
$$\vec{e}_2 = (-3) \cdot \vec{v}_1 + (-1) \cdot \vec{v}_2$$

$$\Rightarrow \begin{bmatrix} \vec{e}_1 \\ \vec{e}_2 \end{bmatrix} = \begin{bmatrix} 1 & 2 \\ -3 & -1 \end{bmatrix} \begin{bmatrix} \vec{v}_1 \\ \vec{v}_2 \end{bmatrix}$$

$$\Rightarrow \begin{bmatrix} \vec{v}_1 \\ \vec{v}_2 \end{bmatrix} = \begin{bmatrix} 1 & 2 \\ -3 & -1 \end{bmatrix}^{-1} \begin{bmatrix} 1 & 0 \\ 0 & 1 \end{bmatrix} = \frac{1}{5} \begin{bmatrix} -1 & -2 \\ 3 & 1 \end{bmatrix}.$$

Thus $\vec{v}_1 = (-\frac{1}{5}, -\frac{2}{5})$, $\vec{v}_2 = (\frac{3}{5}, \frac{1}{5})$. See Fig. 2.78.

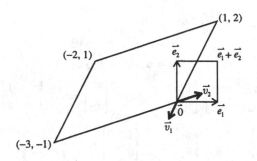

Fig. 2.78

Do the following problems. Note that $\mathbb{R}^2$ is always endowed with $\mathcal{N} = \{\vec{e}_1, \vec{e}_2\}$.

(a) Give a parallelogram with vertices $\vec{0}, \vec{a}_1, \vec{a}_1 + \vec{a}_2$ and $\vec{a}_2$ where $\vec{a}_1$ and $\vec{a}_2$ are linearly independent. What would this parallelogram look like in a coordinate system $\mathcal{B} = \{\vec{v}_2, \vec{v}_2\}$?

(b) Conversely, find a coordinate system $\mathcal{B} = \{\vec{v}_1, \vec{v}_2\}$ on which the parallelogram is the one with vertices $\vec{0}, \vec{b}_1, \vec{b}_1 + \vec{b}_2$ and $\vec{b}_2$ where $\vec{b}_1$ and $\vec{b}_2$ are linearly independent.

11. *Constructive* linear algebra

Adopt the Cartesian coordinate system $\mathcal{N} = \{\vec{e}_1, \vec{e}_2\}$ on $\mathbb{R}^2$. Let $\vec{a}_1 = (2,1), \vec{a}_2 = (-4,-2)$ and $\vec{b}_1 = (4,6), \vec{b}_2 = (-2,-3)$. See Fig. 2.79. Try to construct a linear operator, in $\mathcal{N}$, to map $\triangle \vec{0}\,\vec{a}_1\,\vec{b}_1$

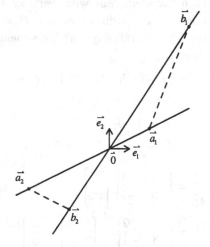

Fig. 2.79

onto $\Delta \vec{0}\,\vec{a}_2\,\vec{b}_2$. In case $\vec{a}_1$ and $\vec{b}_1$ are corresponding to $\vec{a}_2$ and $\vec{b}_2$ respectively, let linear operator f be defined as

$$f(\vec{a}_1) = \vec{a}_2 = -2\vec{a}_1,$$
$$f(\vec{b}_1) = \vec{b}_2 = -\frac{1}{2}\vec{b}_1$$

$$\Rightarrow \begin{bmatrix} \vec{a}_1 \\ \vec{b}_1 \end{bmatrix} [f]_N = \begin{bmatrix} -2 & 0 \\ 0 & -\frac{1}{2} \end{bmatrix} \begin{bmatrix} \vec{a}_1 \\ \vec{b}_1 \end{bmatrix}$$

$$\Rightarrow f(\vec{x}) = [\vec{x}]_N [f]_N = \vec{x} \begin{bmatrix} 2 & 1 \\ 4 & 6 \end{bmatrix}^{-1} \begin{bmatrix} -2 & 0 \\ 0 & -\frac{1}{2} \end{bmatrix} \begin{bmatrix} 2 & 1 \\ 4 & 6 \end{bmatrix} = \vec{x} \begin{bmatrix} -\frac{11}{4} & -\frac{9}{8} \\ \frac{3}{2} & \frac{1}{4} \end{bmatrix}.$$

This f is the required one. In case $\vec{a}_1$ and $\vec{b}_1$ are corresponding to $\vec{b}_2$ and $\vec{a}_2$ respectively, define linear operator g as

$$g(\vec{a}_2) = \vec{b}_2,$$
$$g(\vec{b}_2) = \vec{a}_2$$

$$\Rightarrow \begin{bmatrix} \vec{a}_2 \\ \vec{b}_2 \end{bmatrix} [g]_N = \begin{bmatrix} 0 & 1 \\ 1 & 0 \end{bmatrix} \begin{bmatrix} \vec{a}_2 \\ \vec{b}_2 \end{bmatrix}$$

$$\Rightarrow g(\vec{x}) = [\vec{x}]_N [g]_N = \vec{x} \begin{bmatrix} -4 & -2 \\ -2 & -3 \end{bmatrix}^{-1} \begin{bmatrix} 0 & 1 \\ 1 & 0 \end{bmatrix} \begin{bmatrix} -4 & -2 \\ -2 & -3 \end{bmatrix} = \vec{x} \begin{bmatrix} -\frac{1}{4} & \frac{5}{8} \\ \frac{3}{2} & \frac{1}{4} \end{bmatrix}.$$

Then the composite map

$$(g \circ f)(\vec{x}) = \vec{x} \cdot \frac{1}{64} \begin{bmatrix} -22 & -9 \\ 12 & 2 \end{bmatrix} \begin{bmatrix} -2 & 5 \\ 12 & 2 \end{bmatrix} = \vec{x} \begin{bmatrix} -1 & -2 \\ 0 & 1 \end{bmatrix}$$

is the required one. A direct computation for $g \circ f$ is easier and is shown as follows.

$$(g \circ f)(\vec{a}_1) = \vec{b}_2,$$
$$(g \circ f)(\vec{b}_1) = \vec{a}_2$$

$$\Rightarrow (g \circ f)(\vec{x}) = [\vec{x}]_N [g \circ f]_N = \vec{x} \begin{bmatrix} \vec{a}_1 \\ \vec{b}_1 \end{bmatrix}^{-1} \begin{bmatrix} \vec{b}_2 \\ \vec{a}_2 \end{bmatrix}$$

$$= \vec{x} \begin{bmatrix} 2 & 1 \\ 4 & 6 \end{bmatrix}^{-1} \begin{bmatrix} -2 & -3 \\ -4 & -2 \end{bmatrix} = \vec{x} \begin{bmatrix} -1 & -2 \\ 0 & 1 \end{bmatrix}.$$

Do the following problems.

(a) Show that

$$\begin{bmatrix} -1 & -2 \\ 0 & 1 \end{bmatrix} = \begin{bmatrix} 1 & -1 \\ 0 & 1 \end{bmatrix} \begin{bmatrix} -1 & 0 \\ 0 & 1 \end{bmatrix} \begin{bmatrix} 1 & 1 \\ 0 & 1 \end{bmatrix}.$$

Consider $\mathcal{B} = \{(1,1),(0,1)\}$ as a basis for $\mathbb{R}^2$. Try to construct graphically, in $\mathcal{B}$, the image $\triangle \vec{0}\,\vec{a_2}\,\vec{b_2}$ of $\triangle \vec{0}\,\vec{a_1}\,\vec{b_1}$.

(b) Find all possible linear operators that map $\triangle \vec{0}\,\vec{a_2}\,\vec{b_1}$ onto $\triangle \vec{0}\,\vec{a_1}\,\vec{b_2}$.

(c) Find all possible linear operators that map $\triangle \vec{a_1}\,\vec{a_2}\,\vec{b_1}$ onto $\triangle \vec{a_1}\,\vec{a_2}\,\vec{b_2}$.

(d) Find all possible linear operators that map $\triangle \vec{a_1}\,\vec{a_2}\,\vec{b_1}$ onto $\triangle \vec{b_1}\,\vec{b_2}\,\vec{a_1}$.

(e) Show that the linear operator mapping $\triangle \vec{0}\,\vec{a_1}\,\vec{b_1}$ onto $\triangle \vec{0}\,\vec{a_2}\,\vec{b_1}$ but keeping the side $\vec{0}\,\vec{b_1}$ fixed is

$$\begin{bmatrix} \vec{a_1} \\ \vec{b_1} \end{bmatrix}^{-1} \begin{bmatrix} -2 & 0 \\ 0 & 1 \end{bmatrix} \begin{bmatrix} \vec{a_1} \\ \vec{b_1} \end{bmatrix} = \begin{bmatrix} \vec{a_1} \\ \vec{b_1} \end{bmatrix}^{-1} \begin{bmatrix} \vec{a_2} \\ \vec{b_1} \end{bmatrix} = \frac{1}{4}\begin{bmatrix} -14 & -9 \\ 12 & 10 \end{bmatrix}.$$

Denote the above matrix by A. Hence, in $\mathcal{C} = \{\vec{a_1},\,\vec{b_1}\}$,

$$[A]_C = \begin{bmatrix} -2 & 0 \\ 0 & 1 \end{bmatrix}.$$

Try to explain geometrically by graphs the following two sequences of linear operators

$$A, \quad A\begin{bmatrix} -1 & 0 \\ 0 & 1 \end{bmatrix}, \quad A\begin{bmatrix} -1 & 0 \\ 0 & -1 \end{bmatrix}, \quad A\begin{bmatrix} 1 & 0 \\ 0 & -1 \end{bmatrix} \quad \text{and}$$

$$[A]_C, \quad [A]_C\begin{bmatrix} -1 & 0 \\ 0 & 1 \end{bmatrix}, \quad [A]_C\begin{bmatrix} -1 & 0 \\ 0 & -1 \end{bmatrix}, \quad [A]_C\begin{bmatrix} 1 & 0 \\ 0 & -1 \end{bmatrix}$$

in $\mathcal{N}$ and in $\mathcal{C}$ respectively. How about

$$A, \quad A\begin{bmatrix} 0 & -1 \\ 1 & 0 \end{bmatrix}, \quad A\begin{bmatrix} 0 & -1 \\ -1 & 0 \end{bmatrix}, \quad A\begin{bmatrix} 0 & 1 \\ -1 & 0 \end{bmatrix} \quad \text{and}$$

$$[A]_C, \quad [A]_C\begin{bmatrix} 0 & -1 \\ 1 & 0 \end{bmatrix}, \quad [A]_C\begin{bmatrix} 0 & -1 \\ -1 & 0 \end{bmatrix}, \quad [A]_C\begin{bmatrix} 0 & 1 \\ -1 & 0 \end{bmatrix}?$$

12. *Constructive* linear algebra

Adopt the Cartesian coordinate system $\mathcal{N} = \{\vec{e_1},\vec{e_2}\}$ on $\mathbb{R}^2$. Try to map the square with vertices $\vec{0}, \vec{e_1}, \vec{e_1} + \vec{e_2}$ and $\vec{e_2}$ onto a non-degenerated segment along the line $x_1 + x_2 = 0$ where $\vec{x} = (x_1, x_2)$. There are various ways to achieve this. The easiest way might be that each of the family of lines parallel to the x_1-axis will map its points or

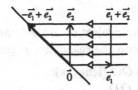

Fig. 2.80

points of intersection with the square into its points of intersection with the line $x_1 + x_2 = 0$. See Fig. 2.80. In terms of linear algebra, this is the projection of $\mathbb{R}^2$ along the x_1-axis, and in turn, eventually, becomes an easy problem concerning eigenvectors. Define $f : \mathbb{R}^2 \to \mathbb{R}^2$ by

$$f(\vec{e}_1) = \vec{0},$$
$$f(\vec{e}_2) = -\vec{e}_1 + \vec{e}_2.$$

Then

$$f(\vec{x}) = \vec{x} \begin{bmatrix} 0 & 0 \\ -1 & 1 \end{bmatrix}$$

is a required one. See Fig. 2.80. Note that $f(-\vec{e}_1 + \vec{e}_2) = -\vec{e}_1 + \vec{e}_2$ and the image segment is covered twice by the square under f. Actually, take any vector $\vec{v}$ which is linearly independent from $-\vec{e}_1 + \vec{e}_2$. Then a linear operator of $\mathbb{R}^2$ onto the line $x_1 + x_2 = 0$, defined as

$$f(\vec{v}) = \vec{0},$$
$$f(-\vec{e}_1 + \vec{e}_2) = \lambda(-\vec{e}_1 + \vec{e}_2) \quad \text{for scalar } \lambda \neq 0$$

will serve our purpose. See Fig. 2.81.

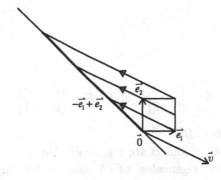

Fig. 2.81

Do the following problems.

(a) Does there exist any projection on $\mathbb{R}^2$ mapping the square with vertices $\pm \vec{e}_1 \pm \vec{e}_2$ onto the respective segment

    (1) BC with $OB = OC$ (lengths),
    (2) BD with $OB \neq OD$,
    (3) AE with $OA = OE$, or
    (4) OA on the line $x_1 = x_2$?

(see Fig. 2.82). If yes, try to find all such projections; if no, give the exact reason why and try to find some other kind of mappings that will do.

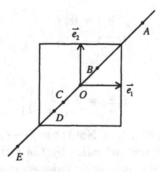

Fig. 2.82

(b) Via a linear operator followed by a translation, it is possible to map a triangle onto a non-degenerated line segment. Try to do this in all possible different ways.

<B>

1. Prove (2.7.72).
2. Let

$$A = \begin{bmatrix} 1 & 2 \\ 0 & 2 \end{bmatrix} \quad \text{and} \quad B = \begin{bmatrix} 3 & -8 \\ 0 & -1 \end{bmatrix}.$$

(a) Show that $AB = BA$.
(b) Show that both A and B are diagonalizable.
(c) Let λ be any eigenvalue of A and $\vec{v}$ be any corresponding eigenvector. Show that $\vec{v}B$ is also an eigenvector of A.

(d) Use (c) to show that there exists a basis $\mathcal{B} = \{\vec{v}_1, \vec{v}_2\}$ for $\mathbb{R}^2$ such that both $[A]_\mathcal{B}$ and $[B]_\mathcal{B}$ are diagonal matrices. In this case, A and B are called *simultaneously diagonalizable*.

3. There exist matrices $A_{2\times2}$ and $B_{2\times2}$ such that A is diagonalizable and $AB = BA$ holds, but B is not diagonalizable. For example,

$$A = I_2 \quad \text{and} \quad B = \begin{bmatrix} 1 & 0 \\ 1 & 1 \end{bmatrix}.$$

Try to find some other examples, if any.

4. Suppose A and B are similar 2×2 real matrices. Show that there exist bases $\mathcal{B}$ and $\mathcal{C}$ for $\mathbb{R}^2$ and a linear operator f on $\mathbb{R}^2$ such that

$$[f]_\mathcal{B} = A \quad \text{and} \quad [f]_\mathcal{C} = B$$

(try to refer to (2.7.25)).

(*Note* This result still holds for $n \times n$ matrices over a field.)

5. Let f be a nonzero linear operator on $\mathbb{R}^2$ (or any two-dimensional vector space). For any nonzero vector $\vec{x}$ in $\mathbb{R}^2$, the subspace

$$\langle\langle \vec{x}, f(\vec{x}), f^2(\vec{x}), \dots, \rangle\rangle, \quad \text{denoted as } C_f(\vec{x})$$

is called the *f-cycle subspace of $\mathbb{R}^2$ generated by* $\vec{x}$.

(a) Show that $C_f(\vec{x})$ is f-invariant, i.e. $f(\vec{y}) \in C_f(\vec{x})$ for each $\vec{y} \in C_f(\vec{x})$.

(b) Show that either $\mathbb{R}^2 = C_f(\vec{x})$ for some $\vec{x} \in \mathbb{R}^2$ or $f = \lambda 1_{\mathbb{R}^2}$ for some scalar λ.

(c) In case $f \neq \lambda 1_{\mathbb{R}^2}$ for any scalar λ, let g be a linear operator on $\mathbb{R}^2$ such that $g \circ f = f \circ g$. Then, there exists a polynomial $p(t)$ such that $g = p(f)$. Refer to Ex. <C> 6.

<C> Abstraction and generalization

1. Prove (2.7.73).

2. Suppose $A_{n\times n}$ is diagonalizable and PAP^{-1} is as in (2.7.73). Show that

$$(A - \lambda_1 I_n)(A - \lambda_2 I_n) \cdots (A - \lambda_k I_n) = O.$$

And hence, A satisfies its *minimal polynomial*

$$\varphi(t) = (t - \lambda_1)(t - \lambda_2) \cdots (t - \lambda_k)$$

and its characteristic polynomial $\det(A - tI_n)$, which is the Cayley–Hamilton theorem (see (2.7.13), (2.7.19), Exs. 5 and 9 below and Ex. 4 of Sec. B.10).

3. Let $f: M(n; \mathbb{F}) \to M(n; \mathbb{F})$ be defined by

$$f(A) = A^* \quad \text{(the transpose of } A\text{)}.$$

(a) Show that f is a linear operator on $M(n; \mathbb{F})$.
(b) Let $\mathcal{B} = \{E_{11}, E_{12}, \ldots, E_{1n}, E_{21}, \ldots, E_{n1}, \ldots, E_{nn}\}$ be the ordered basis for $M(n; \mathbb{F})$ (refer to (2.7.30) and Sec. B.4). Find $[f]_\mathcal{B}$.
(c) Show that ± 1 are the only eigenvalues of f.
(d) Determine all eigenvectors of f corresponding to ± 1.
(e) Is f diagonalizable? If yes, prove it; if not, say reason why.

4. Let f be a linear operator on a vector space V over a field $\mathbb{F}$. Remind that a subspace S of V is called f-*invariant* if $f(S) \subseteq S$ holds.

(a) *Examples* for f-invariant subspaces are:

(1) $\{\vec{0}\}, V, \text{Ker}(f)$ and $\text{Im}(f)$.
(2) For any nonzero $\vec{x} \in V$, the f-*cycle subspace generated by* $\vec{x}$:

$$C_f(\vec{x}) = \langle\langle \vec{x}, f(\vec{x}), f^2(\vec{x}), \ldots \rangle\rangle.$$

(3) For any $\lambda \in \mathbb{F}$, the subspace

$$G_\lambda(f) = \{\vec{x} \in V \mid (f - \lambda I_V)^k(\vec{x}) = \vec{0}$$

for some positive integer $k\}$.

In particular, if λ is an eigenvalue of $f, G_\lambda(f)$ is called the *generalized eigenspace* of f corresponding to λ, which contains the *eigenspace* $E_\lambda(f) = \{\vec{x} \in V \mid f(\vec{x}) = \lambda \vec{x}\}$ as a subspace.

(b) *Properties*

(1) Finite sum of f-invariant subspaces is f-invariant.
(2) Arbitrary intersection of f-invariant subspaces is f-invariant.
(3) If S is f-invariant and f is invertible, then S is also f^{-1}-invariant.
(4) If S is f-invariant and $V = S \oplus T$, then T is not necessarily f-invariant.
(5) If S is f-invariant, then the *restriction* $f|_S: S \to S$ is a linear operator.
(6) If S is f-invariant and $\vec{x}$ is an eigenvector of $f|_S$ associated with eigenvalue λ, then the same is true for f.
(7) Suppose $V = \text{Im}(f) \oplus S$ and S is f-invariant. Then $S \subseteq \text{Ker}(f)$ with equality if $\dim(V) < \infty$, but $S \subsetneq \text{Ker}(f)$ could happen if V is not finite-dimensional.

(c) *Matrix representation*

(1) A subspace S of V, where dim $V = n$, is f-invariant if and only if there exists a basis $\mathcal{B} = \{\vec{x}_1, \ldots, \vec{x}_k, \vec{x}_{k+1}, \ldots, \vec{x}_n\}$ for V such that $\mathcal{B}_1 = \{\vec{x}_1, \ldots, \vec{x}_k\}$ is a basis for S and

$$[f]_\mathcal{B} = \begin{bmatrix} A_{11} & O \\ A_{21} & A_{22} \end{bmatrix}_{n \times n},$$

where the restriction $f|_S$ has matrix representation A_{11} with respect to $\mathcal{B}_1$.

(2) In particular, $V = S_1 \oplus S_2$ where S_1 and S_2 are f-invariant subspaces if and only if, there exists a basis $\mathcal{B} = \{\vec{x}_1, \ldots, \vec{x}_k, \vec{x}_{k+1}, \ldots, \vec{x}_n\}$ for V such that

$$[f]_\mathcal{B} = \begin{bmatrix} A_{11} & O \\ O & A_{22} \end{bmatrix}_{n \times n},$$

where $\{\vec{x}_1, \ldots, \vec{x}_k\} = \mathcal{B}_1$ is a basis for S_1 with $[f|_{S_1}]_{\mathcal{B}_1} = A_{11}$ and $\{\vec{x}_{k+1}, \ldots, \vec{x}_n\} = \mathcal{B}_2$ is a basis for S_2 with $[f|_{S_2}]_{\mathcal{B}_2} = A_{22}$.

In Case (2), $f = f|_{S_1} \oplus f|_{S_2}$ is called the *direct sum* of $f|_{S_1}$ and $f|_{S_2}$. Another interpretation of A_{22} is as follows. Suppose S is a f-invariant subspace of V. Consider the *quotient space* of V modulus S (see Sec. B.1)

$$V/S = \{\vec{x} + S \mid \vec{x} \in V\}$$

and the *induced quotient operator* $\tilde{f}: V/S \to V/S$ defined by

$$\tilde{f}(\vec{x} + S) = f(\vec{x}) + S.$$

This $\tilde{f}$ is well-defined and is linear. Let $\pi: V \to V/S$ denote the *natural projection* defined by $\pi(\vec{x}) = \vec{x} + S$. Then the following diagram is commutative: $\pi \circ f = \tilde{f} \circ \pi$.

$$\begin{array}{ccc} V & \xrightarrow{f} & V \\ \pi \downarrow & & \downarrow \pi \\ V/S & \xrightarrow{\tilde{f}} & V/S \end{array}$$

(3) Suppose dim $V = n$. Adopt notations as in (1). Then $\mathcal{B}_2 = \{\vec{x}_{k+1} + S, \ldots, \vec{x}_n + S\}$ is a basis for V/S and

$$[f]_\mathcal{B} = \begin{bmatrix} A_{11} & 0 \\ A_{21} & A_{22} \end{bmatrix},$$

where $A_{11} = [f|_S]_{\mathcal{B}_1} = A_{11}$ and $A_{22} = [\tilde{f}]_{\mathcal{B}_2}$.

(d) *Characteristic polynomials of $f, f|_S$, and $\tilde{f}$*

Suppose $\dim V = n$ and S is a f-invariant subspace of V.

(1) The characteristic polynomial of $f|_S$ divides that of f, so does for $\tilde{f}$. Therefore, if the characteristic polynomial $\det(f - tI_V) = (-1)^n (t - \lambda_1)^{r_1} \cdots (t - \lambda_k)^{r_k}$, i.e. *splits* as linear factors, then so do the characteristic polynomials of $f|_S$ and $\tilde{f}$. This implies that any nonzero f-invariant subspace of V contains an eigenvector of f if $\det(f - t1_v)$ splits.

(2) Adopt notations in (c)(2). Then

$$\det(f - t1_V) = \det(f|_{S_1} - t1_{S_1}) \cdot \det(f|_{S_2} - t1_{S_2}).$$

(3) Adopt notation in (c)(3). Then

$$\det(f - t1_V) = \det(f|_S - t1_S) \cdot \det(\tilde{f} - t1_{V/S}).$$

Both (2) and (3) provide other ways to compute the characteristic polynomial of f.

(e) *Diagonalizability of $f, f|_S$, and $\tilde{f}$*

Suppose $\dim V = n$ and S is a f-invariant subspace of V.

(1) If f is diagonalizable, then so is $f|_S$.
(2) If f is diagonalizable, then so is $\tilde{f}$.
(3) If both $f|_S$ and $\tilde{f}$ are diagonalizable, then so is f.

5. Suppose f is a linear operator on a finite-dimensional vector space V.

(a) If $C_f(\vec{x})$ is the f-cycle subspace generated by $\vec{x} \neq \vec{0}$ and $\dim C_f(\vec{x}) = k$, then

(1) $\mathcal{B} = \{\vec{x}, f(\vec{x}), f^2(\vec{x}), \ldots, f^{(k-1)}(\vec{x})\}$ is a basis for $C_f(\vec{x})$. Hence there exist unique scalars $a_0, a_1, \ldots, a_{k-1}$ such that

$$f^{(k)}(\vec{x}) = -a_0\vec{x} - a_1 f(\vec{x}) - \cdots - a_{k-1} f^{(k-1)}(\vec{x}).$$

(2) Therefore,

$$[f|_{C_f(\vec{x})}]_{\mathcal{B}} = \begin{bmatrix} 0 & 1 & & & & \\ & 0 & 1 & & 0 & \\ & & 0 & & & \\ & 0 & & \ddots & & 1 \\ & & & & 0 & 1 \\ -a_0 & -a_1 & -a_2 & \cdots & -a_{k-2} & -a_{k-1} \end{bmatrix}_{k \times k}$$

and the characteristic polynomial $\varphi(t)$ of $f|_{C_f(\vec{x})}$ is

$$\det\left(f|_{C_f(\vec{x})} - t1_{C_f(\vec{x})}\right) = (-1)^k(t^k + a_{k-1}t^{k-1} + \cdots + a_1t + a_0).$$

with $t^k + a_{k-1}t^{k-1} + \cdots + a_1t + a_0$ as its minimal polynomial (see Ex. 9 below).

(3) Also, $\varphi(t)$ *annihilates* f, i.e. $\varphi(f|_{C_f(\vec{x})}) = 0$ or

$$f^k + a_{k-1}f^{k-1} + \cdots + a_1f + a_01_{C_f(\vec{x})} = 0$$

on $C_f(\vec{x})$.

(b) Use Ex. 4(d)(1) and (a) to prove the *Cayley–Hamilton theorem*, i.e. the characteristic polynomial of f annihilates itself.

6. Let f be a linear operator on a vector space V with dim $V < \infty$.

   (a) For $\vec{y} \in V$, then $\vec{y} \in C_f(\vec{x})$ if and only if there exists a polynomial $\varphi(t)$, whose degree can always be chosen to be not larger than dim $C_f(\vec{x})$, such that $\vec{y} = \varphi(f)\vec{x}$.

   (b) Suppose $V = C_f(\vec{x})$ for some $\vec{x} \in V$. If g is a linear operator on V, then $f \circ g = g \circ f$ if and only if $g = \varphi(f)$ for some polynomial $\varphi(t)$.

7. (continued from Exs. <B> 2 and 3) Two linear operators f and g on a finite dimensional vector space V are called *simultaneously diagonalizable* if there exists a basis $\mathcal{B}$ for V such that both $[f]_\mathcal{B}$ and $[g]_\mathcal{C}$ are diagonal matrices.

   (a) f and g are simultaneously diagonalizable if and only if, for any basis $\mathcal{C}$ for V, the matrices $[f]_\mathcal{C}$ and $[g]_\mathcal{C}$ are *simultaneously diagonalizable*.

   (b) If f and g are simultaneously diagonalizable, then f and g commutes, i.e. $f \circ g = g \circ f$.

   (c) Suppose $f \circ g = g \circ f$, then $\mathrm{Ker}(g)$ and $\mathrm{Im}(g)$ are f-invariant.

   (d) Suppose $f \circ g = g \circ f$ and f and g are diagonalizable, then f and g are simultaneously diagonalizable.

8. Let $A_{n \times n}$ and $B_{n \times n}$ be matrices over $\mathbb{C}$ and the characteristic polynomial of A be denoted by $\varphi(t)$. Then

   (1) A and B do not have common eigenvalues.

   $\Leftrightarrow$ (2) $\varphi(B)$ is invertible.

   $\Leftrightarrow$ (3) The matrix equation $AX = XB$ has only zero solution $X = O$.

   $\Leftrightarrow$ (4) The linear operator $f_{A,B}: M(n; \mathbb{F}) \to M(n; \mathbb{F})$ defined by

   $$f_{A,B}(X) = AX - XB$$

   is invertible.

9. A nonzero polynomial $p(t) \in P(\mathbb{F})$ (see Sec. A.5) is said to *annihilate* a matrix $A \in M(n; \mathbb{F})$ if $p(A) = O_{n \times n}$ and is called an *A-annihilator*. Therefore, the characteristic polynomial $f(t) = \det(A - tI_n)$ annihilates A. Let $A \in M(n; \mathbb{F})$ be a nonzero matrix.

(a) There exists a unique monic polynomial $\varphi(t)$ of the least degree that annihilates A. Such a $\varphi(t)$ is called the *minimal polynomial* of A. (Refer to Sec. B.11.)

(b) The minimal polynomial $\varphi(t)$ divides any A-annihilator, in particular, the characteristic polynomial $f(t)$ of A.

(c) Similar matrices have the same characteristic and minimal polynomial, but not the converse. For example,

$$A = \left[\begin{array}{cc|cc} 2 & 1 & & \\ 0 & 2 & & 0 \\ \hline & & 2 & 1 \\ & 0 & 0 & 2 \end{array}\right] \quad \text{and} \quad B = \left[\begin{array}{ccc|c} 2 & 1 & 0 & \\ 0 & 2 & 1 & 0 \\ 0 & 0 & 2 & \\ \hline & 0 & & 2 \end{array}\right].$$

Both have characteristic polynomial $f(t) = (t-2)^4$ and minimal polynomial $\varphi(t) = (t-2)^2$, yet they are not similar. Prove this.

(d) $\lambda \in \mathbb{F}$ is a zero of $\varphi(t) \Leftrightarrow \lambda$ is a zero of $f(t)$. Hence, each root of $\varphi(t) = 0$ is an eigenvalue of A, and vice versa.

(e) A is diagonalizable $\Leftrightarrow$ its minimal polynomial is

$$\varphi(t) = (t - \lambda_1)(t - \lambda_2) \cdots (t - \lambda_k),$$

where $\lambda_1, \ldots, \lambda_k$ are *all distinct* eigenvalues of A. Prove this by the following steps.

(1) Let $E_{\lambda_i} = \{\vec{x} \in \mathbb{F}^n \mid \vec{x}A = \lambda_i \vec{x}\}$ for $1 \le i \le k$. Then each $E_{\lambda_i} \neq \{\vec{0}\}$ and

$$E_{\lambda_i} \cap (E_{\lambda_1} + \cdots + E_{\lambda_{i-1}} + E_{\lambda_{i+1}} + \cdots + E_{\lambda_k}) = \{\vec{0}\}, \quad 1 \le i \le k.$$

(2) Let $p_i(t) = \prod_{j \neq i}^{k}(t - \lambda_j), 1 \le i \le k$. Show that there exist polynomials $\varphi_1(t), \ldots, \varphi_k(t)$ which are relatively prime such that

$$\varphi_1(t)p_1(t) + \cdots + \varphi_k(t)p_k(t) = 1.$$

Try to show that $\vec{x}\varphi_i(A)p_i(A) \in E_{\lambda_i}$ for each $\vec{x} \in \mathbb{F}^n$ and $1 \le i \le k$.

(3) Show that $\mathbb{F}^n = E_{\lambda_1} \oplus \cdots \oplus E_{\lambda_k}$.

Then, use (1)4 in (2.7.73) to finish the proof (refer to Sec. B.11).

(f) A is diagonalizable $\Leftrightarrow A^*$ is diagonalizable.

10. Let $A, B \in M(n; \mathbb{C})$ be nonzero matrices.

(a) (Schur, 1909) Use the fact that A has eigenvalues and the mathematical induction to show that A is similar to a lower triangular matrix, i.e. there exists an invertible $P_{n \times n}$, even unitarily (i.e. $\bar{P}^* = P^{-1}$), so that

$$PAP^{-1} = \begin{bmatrix} \lambda_1 & & & \\ & \lambda_2 & & 0 \\ & & \ddots & \\ & & & \lambda_n \end{bmatrix},$$

where the diagonal entries are eigenvalues of A.

(b) Suppose $AB = BA$, show that A and B have a common eigenvector by the following steps.

(1) Let λ be an eigenvalue of A and $E_\lambda = \{\vec{x} \in \mathbb{F}^n \mid \vec{x}A = \lambda\vec{x}\}$ have basis $\{\vec{x}_1, \ldots, \vec{x}_k\}$. Show that each $\vec{x}_j B \in E_\lambda$ for $1 \le j \le k$.

(2) Let $\vec{x}_j B = \sum_{i=1}^{k} b_{ij} \vec{x}_i$ for $1 \le j \le k$ and $Q = [b_{ij}]_{k \times k}$. Let $\vec{y} = (\alpha_1, \ldots, \alpha_k) \in \mathbb{C}^k$ be an eigenvector of Q, say $\vec{y}Q = \mu\vec{y}$ for $\mu \in \mathbb{C}$.

(3) Try to show that $\vec{x}_0 = \sum_{j=1}^{k} \alpha_j \vec{x}_j$ is a common eigenvector of A and B.

(c) Suppose $AB = BA$. Show that there exists an invertible $P_{n \times n}$ such that both PAP^{-1} and PBP^{-1} are lower triangular, i.e. A and B are *simultaneously lower triangularizable*.

(d) Give examples to show that the converse of (b) is not true, in general. Then, use

$$A = \begin{bmatrix} 1 & 0 & -1 & 0 \\ 0 & 1 & 0 & -1 \\ 1 & 0 & -1 & 0 \\ 0 & 1 & 0 & -1 \end{bmatrix} \quad \text{and} \quad B = \begin{bmatrix} 4 & -1 & -1 & 0 \\ -1 & 4 & 0 & -1 \\ 1 & 0 & 2 & -1 \\ 0 & 1 & -1 & 2 \end{bmatrix}$$

to justify (b).

11. Let $A_{n \times n}$ be a complex matrix. Then

(1) A is *nilpotent* (i.e. there exists some positive integer k such that $A^k = O_{n \times n}$, and the smallest such k is called the *index of nilpotency* of A).

$\Leftrightarrow$ (2) All eigenvalues of A are zeros.

$\Leftrightarrow$ (3) $A^n = O_{n \times n}$.

$\Leftrightarrow$ (4) $\operatorname{tr} A^l = 0$ for $l = 1, 2, 3, \ldots$.

$\Leftrightarrow$ (5) In case A has index n of nilpotency, then A is similar to

$$
\begin{bmatrix}
0 & & & & & \\
1 & 0 & & & 0 & \\
 & 1 & 0 & & & \\
 & & & \ddots & & \\
 & 0 & & 1 & 0 & \\
 & & & & 1 & 0
\end{bmatrix}_{n \times n}.
$$

12. Let $A \in M(m, n; \mathbb{C})$. Show that there exist unitary matrices $P_{m \times m}$ and $Q_{n \times n}$ so that

$$
PAQ =
\begin{bmatrix}
\lambda_1 & & & & & \\
 & \ddots & & & 0 & \\
 & & \lambda_r & & & \\
 & & & 0 & & \\
 & 0 & & & \ddots & \\
 & & & & & 0
\end{bmatrix}, \quad \lambda_1 \geq \lambda_2 \geq \cdots \geq \lambda_r > 0,
$$

where $r = \mathrm{r}(A) \geq 1$. Then, try to prove that

$$
\det(I_m - A\bar{A}^*) = \det(I_n - \bar{A}^* A).
$$

13. Let $A_{n \times n}$ be a complex matrix with trace $\operatorname{tr} A = 0$. Show that A is similar to a matrix whose diagonal entries are all equal to zeros, i.e. there exists an invertible $P_{n \times n}$ such that

$$
PAP^{-1} =
\begin{bmatrix}
0 & & & \\
 & 0 & & \\
 & & \ddots & \\
 & & & 0
\end{bmatrix}_{n \times n}.
$$

Try the following steps:

(1) Suppose A is similar to a lower triangular matrix such as in (a) of Ex. 10,

$$PAP^{-1} = \begin{bmatrix} \lambda_1 & & & & \\ b_{21} & \lambda_2 & & 0 & \\ b_{31} & b_{32} & \lambda_3 & & \\ \vdots & \vdots & \vdots & \ddots & \\ b_{n1} & b_{n2} & b_{n3} & \cdots & \lambda_n \end{bmatrix}, \quad P = \begin{bmatrix} \vec{x}_1 \\ \vec{x}_2 \\ \vdots \\ \vec{x}_n \end{bmatrix},$$

where $\operatorname{tr} A = \lambda_1 + \lambda_2 + \cdots + \lambda_n = 0$. Suppose $\lambda_1 \neq 0$. Let $\vec{v}_1 = \vec{x}_2 - \frac{b_{21}}{\lambda_1}\vec{x}_1$. Then $\mathcal{B} = \{\vec{v}_1, \vec{x}_2, \ldots, \vec{x}_n\}$ is a basis for $\mathbb{C}^n$. Now,

$$\vec{v}_1 A = \vec{x}_2 A - \frac{b_{21}}{\lambda_1}\vec{x}_1 A = b_{21}\vec{x}_1 + \lambda_2\vec{x}_2 - \frac{b_{21}}{\lambda_1}\lambda_1\vec{x}_1 = \lambda_2\vec{x}_2,$$

$$\vec{x}_2 A = b_{21}\vec{x}_1 + \lambda_2\vec{x}_2 = -\lambda_1\vec{v}_1 + \lambda_1\vec{x}_2 + \lambda_2\vec{x}_2$$
$$= -\lambda_1\vec{v}_1 + (\lambda_1 + \lambda_2)\vec{x}_2$$

$$\Rightarrow QAQ^{-1} = \begin{bmatrix} 0 & \lambda_2 & & & \\ -\lambda_1 & \lambda_1+\lambda_2 & & 0 & \\ b_{31} & b_{32} & \lambda_3 & & \\ \vdots & \vdots & \vdots & \ddots & \\ b_{n1} & b_{n2} & b_{n3} & \cdots & \lambda_n \end{bmatrix}, \quad Q = \begin{bmatrix} \vec{v}_1 \\ \vec{x}_2 \\ \vec{x}_3 \\ \vdots \\ \vec{x}_n \end{bmatrix}.$$

(2) Then, use the same process or mathematical induction on n to finish the proof.

For another proof (here, A can be an arbitrary $n \times n$ matrix over a field $\mathbb{F}$ of characteristic greater than n), try the following steps:

(1) A cannot have the form λI_n unless $\lambda = 0$.
(2) There exists a vector $\vec{u}_1 \in \mathbb{F}^n$ such that $\{\vec{u}_1, \vec{u}_1 A\}$ is linearly independent and $\{\vec{u}_1, \vec{u}_1 A, \vec{u}_3, \ldots, \vec{u}_n\} = \mathcal{B}$ is a basis for $\mathbb{F}^n$. Then

$$RAR^{-1} = \begin{bmatrix} 0 & 1 & 0 & \cdots & 0 \\ b_{21} & b_{22} & b_{23} & \cdots & b_{2n} \\ b_{31} & b_{32} & b_{33} & \cdots & b_{3n} \\ \vdots & \vdots & \vdots & \ddots & \vdots \\ b_{n1} & b_{n2} & b_{n3} & \cdots & b_{nn} \end{bmatrix} = \begin{bmatrix} O & A_{12} \\ A_{21} & A_{22} \end{bmatrix}$$

and $\operatorname{tr} A = \operatorname{tr} A_{22} = 0$. What is $R_{n \times n}$?

(3) By mathematical induction, there exists an $(n-1)\times(n-1)$ matrix S such that $SA_{22}S^{-1}$ is a matrix with zero diagonal entries.

(4) Then

$$\begin{bmatrix} 1 & O \\ O & S \end{bmatrix} RAR^{-1} \begin{bmatrix} 1 & O \\ O & S \end{bmatrix}^{-1} = \begin{bmatrix} O & A_{12}S^{-1} \\ SA_{21} & SA_{22}S^{-1} \end{bmatrix}$$

is a required one.

14. Show that

$$\{A \in \mathrm{M}(n;\ \mathbb{F}) \mid \mathrm{tr}\, A = 0\} = \{XY - YX \mid X, Y \in \mathrm{M}(n;\ \mathbb{F})\}$$

as subspaces of $\mathrm{M}(n;\ \mathbb{F})$ (refer to Ex. 30 in Sec. B.4), i.e. every square matrix A with $\mathrm{tr}\, A = 0$ can be expressed as $A = XY - YX$ for some square matrices X and Y. Try the following steps:

(1) Use Ex. 13.

(2) May suppose

$$A = \begin{bmatrix} 0 & a_{12} & \cdots & a_{1n} \\ a_{21} & 0 & \cdots & a_{2n} \\ \vdots & \vdots & & \vdots \\ a_{n1} & a_{n2} & \cdots & 0 \end{bmatrix}.$$

Take any diagonal matrix $X = \mathrm{diag}(\lambda_1, \lambda_2, \ldots, \lambda_n)$, where $\lambda_1, \ldots, \lambda_n$ are all distinct. Suppose $Y = [b_{ij}]_{n \times n}$. Then

$$XY - YX = \begin{bmatrix} 0 & b_{12}(\lambda_2 - \lambda_1) & \cdots & b_{1n}(\lambda_n - \lambda_1) \\ b_{21}(\lambda_1 - \lambda_2) & 0 & \cdots & b_{2n}(\lambda_n - \lambda_2) \\ \vdots & \vdots & & \vdots \\ b_{n1}(\lambda_1 - \lambda_n) & b_{n2}(\lambda_2 - \lambda_n) & \cdots & 0 \end{bmatrix}$$

and $XY - YX = A$ holds if and only if $b_{ij} = \frac{a_{ij}}{\lambda_j - \lambda_i}$ for $i \neq j$.

<D> Applications

For possible applications of diagonalizable linear operators to differential equations, we postpone the discussion to Sec. 3.7.6 in Chap. 3.

2.7.7 *Jordan canonical form*

In this subsection, we study the canonical form of these linear operators which have coincident eigenvalues but are not diagonalizable.

We start from two concrete examples and end up with a summarized result in general setting.

Example 1 In $\mathcal{N} = \{\vec{e}_1, \vec{e}_2\}$, let linear operator $f\colon \mathbb{R}^2 \to \mathbb{R}^2$ be defined as

$$f(x_1, x_2) = (x_1, -2x_1 + x_2)$$

$$= \vec{x}A, \quad \text{where } \vec{x} = (x_1, x_2) \text{ and } A = \begin{bmatrix} 1 & -2 \\ 0 & 1 \end{bmatrix}.$$

Try to investigate geometric mapping properties of f.

Solution The characteristic polynomial of f or A is

$$\det(A - tI_2) = \begin{vmatrix} 1 - t & -2 \\ 0 & 1 - t \end{vmatrix} = (t - 1)^2.$$

Thus, f has eigenvalues 1, 1. Suppose A is *diagonalizable*, i.e. there exists an invertible 2×2 matrix P such that

$$PAP^{-1} = \begin{bmatrix} \lambda_1 & 0 \\ 0 & \lambda_2 \end{bmatrix}$$

is a diagonal matrix. Since

$$\det(PAP^{-1} - tI_2) = \det \begin{bmatrix} \lambda_1 - t & 0 \\ 0 & \lambda_2 - t \end{bmatrix} = (t - \lambda_1)(t - \lambda_2)$$

$$= \det P(A - tI_2)P^{-1} = \det P \cdot \det(A - tI_2) \cdot \det P^{-1}$$

$$= \det(A - tI_2) = (t - 1)^2,$$

hence $\lambda_1 = \lambda_2 = 1$ should hold. Then

$$PAP^{-1} = \begin{bmatrix} 1 & 0 \\ 0 & 1 \end{bmatrix} = I_2$$

$$\Rightarrow A = P^{-1}IP = P^{-1}P = I_2,$$

which is not the original one. This indicates that A is *not* diagonalizable. It is reasonable to expect that the mapping properties of f would be more complicated.

Since $\det A = 1$, f is one-to-one and hence onto.

Just like Example 1 in Sec. 2.7.6. This f maps lines into lines and preserves their relative positions and ratios of signed lengths of line segments along the same line or parallel lines.

Let $\vec{y} = (y_1, y_2) = f(\vec{x}) = \vec{x}A.$ Then

$$y_1 = x_1,$$
$$y_2 = -2x_1 + x_2.$$

When solving $x_1 = x_1$ and $x_2 = -2x_1 + x_2$, it is easy to see that $x_1 = 0$ is the only invariant line passing through $\vec{0}$ under f. For a point $\vec{x} = (x_1, x_2)$ with a nonzero first component $x_1 \neq 0$, then

$$y_2 - x_2 = -2x_1$$

shows that the point $\vec{x}$ is moved parallel to the line $x_1 = 0$ and downward the distance $2x_1$ if $x_1 > 0$ and upward the distance $-2x_1$ if $x_1 < 0$. This is equivalent to say that

$$\frac{y_2 - x_2}{x_1} = -2,$$

which shows that $\vec{x}$ is moved along a line parallel to the line $x_1 = 0$ and the distance moved is proportional to its distance from $x_1 = 0$ by a constant scalar -2. Therefore, the line

$$x_1 = c \ (\text{constant})$$

is mapped onto itself with its point (c, x_2) into $(c, -2c + x_2)$. Such a line is also called an *invariant line* of f but is *not* an invariant subspace except $c = 0$. See Fig. 2.83. In Fig. 2.84, does the quadrilateral with vertices $(3, 1), (1, 2), (-2, 1)$ and $(-4, -2)$ have the same area and the same orientation as its image quadrilateral? Why?

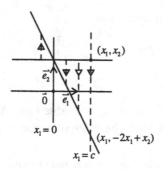

Fig. 2.83

But the concepts of eigenvalues and eigenvectors still can be used to investigate mapping properties of f.

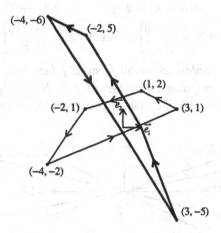

Fig. 2.84

We have already known that $\vec{x} = (x_1, x_2)$ is an eigenvector of f related to 1, i.e. $\vec{x}(A - I_2) = \vec{0}$ if and only if $x_1 = 0$ and $x_2 \neq 0$. What could $\vec{x}(A - I_2)^2 = \vec{0}$ be? Actual computation shows that

$$(A - I_2)^2 = (A - I_2)(A - I_2) = A^2 - 2A + I_2^2$$

$$= \begin{bmatrix} 1 & -4 \\ 0 & 1 \end{bmatrix} - 2 \begin{bmatrix} 1 & -2 \\ 0 & 1 \end{bmatrix} + \begin{bmatrix} 1 & 0 \\ 0 & 1 \end{bmatrix} = \begin{bmatrix} 0 & 0 \\ 0 & 0 \end{bmatrix} = O_{2 \times 2},$$

i.e. A satisfies its characteristic polynomial $(t - 1)^2 = t^2 - 2t + 1$ which illustrates the *Cayley–Hamilton theorem* (see (2.7.19)). Hence

$$G = \{\vec{x} \in \mathbb{R}^2 \mid \vec{x}(A - I_2)^2 = \vec{0}\} = \mathbb{R}^2$$

holds. There does exist vector $\vec{x} \in \mathbb{R}^2$ such that

$$\vec{x}(A - I_2) \neq \vec{0} \quad \text{but} \quad (\vec{x}(A - I_2))(A - I_2) = \vec{x}(A - I_2)^2 = \vec{0}.$$

For example, take $\vec{e}_1 = (1, 0)$ as our sample vector and consider

$$\vec{e}_1(A - I_2) = (1 \quad 0) \begin{bmatrix} 0 & -2 \\ 0 & 0 \end{bmatrix} = (0 \quad -2) = -2\vec{e}_2.$$

Thus $\{-2\vec{e}_2,\ \vec{e}_1\}$ is a basis $\mathcal{B}$ for $G = \mathbb{R}^2$. According to $\mathcal{B}$,

$$(-2\vec{e}_2)A = -2\vec{e}_2 = 1 \cdot (-2\vec{e}_2) + 0 \cdot \vec{e}_1,$$
$$\vec{e}_1 A = \vec{e}_1 - 2\vec{e}_2 = 1 \cdot (-2\vec{e}_2) + \vec{e}_1$$
$$\Rightarrow [f]_{\mathcal{B}} = PAP^{-1} = \begin{bmatrix} 1 & 0 \\ 1 & 1 \end{bmatrix}, \quad \text{where } P = \begin{bmatrix} -2\vec{e}_2 \\ \vec{e}_1 \end{bmatrix} = \begin{bmatrix} 0 & -2 \\ 1 & 0 \end{bmatrix},$$

which is called the *Jordan canonical form* of f (see Sec. B.12). This means that the geometric mapping properties of f in $\mathcal{N} = \{\vec{e}_1,\ \vec{e}_2\}$ is the same as $[f]_{\mathcal{B}}$ in $\mathcal{B} = \{-2\vec{e}_2, \vec{e}_1\}$. The latter is better illustrated in Fig. 2.85.

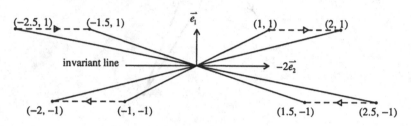

Fig. 2.85

One more example to practice methods in Example 1.

Example 2 In $\mathcal{N} = \{\vec{e}_1,\ \vec{e}_2\}$, let $f \colon \mathbb{R}^2 \to \mathbb{R}^2$ be defined as

$$f(x_1,\ x_2) = \left(x_1 + \frac{1}{2}x_2, -\frac{1}{2}x_1 + 2x_2 \right)$$
$$= \vec{x}A, \quad \text{where } \vec{x} = (x_1,\ x_2) \text{ and } A = \begin{bmatrix} 1 & -\frac{1}{2} \\ \frac{1}{2} & 2 \end{bmatrix}.$$

Investigate mapping properties of A.

Solution f is isomorphic. Its characteristic polynomial is

$$\det(A - tI_2) = \begin{vmatrix} 1-t & -\frac{1}{2} \\ \frac{1}{2} & 2-t \end{vmatrix} = \left(t - \frac{3}{2} \right)^2.$$

Therefore, f or A has eigenvalues $\frac{3}{2}, \frac{3}{2}$ and A is not diagonalizable. Solve $\vec{x}(A - \frac{3}{2}I_2) = 0$ and we get the corresponding eigenvectors $t(1, 1)$ for $t \neq 0$. Hence, it is expected that $x_1 - x_2 = 0$ is an invariant line (subspace) of A.

Of course, A satisfies its characteristic polynomial, i.e.

$$A^2 - 3A + \frac{9}{4}I_2 = \left(A - \frac{3}{2}I_2\right)^2 = O.$$

Thus

$$G = \left\{ \vec{x} \in \mathbb{R}^2 \;\middle|\; \vec{x}\left(A - \frac{3}{2}I_2\right)^2 = \vec{0} \right\} = \mathbb{R}^2.$$

Take a vector, say $\vec{e}_1 = (1,\ 0)$, so that

$$\vec{e}_1\left(A - \frac{3}{2}I_2\right) = \left(-\frac{1}{2},\ -\frac{1}{2}\right) = -\frac{1}{2}(\vec{e}_1 + \vec{e}_2) \neq \vec{0},$$

$$\vec{e}_1\left(A - \frac{3}{2}I_2\right)^2 = \vec{0}.$$

Then $\mathcal{B} = \{\vec{v}_1,\ \vec{v}_2\}$, where $\vec{v}_1 = -\frac{1}{2}(\vec{e}_1 + \vec{e}_2)$ and $\vec{v}_2 = \vec{e}_1$, is a basis for $\mathbb{R}^2$. Related to $\mathcal{B}$,

$$\vec{v}_1 A = \frac{3}{2}\vec{v}_1 = \frac{3}{2}\vec{v}_1 + 0 \cdot \vec{v}_2,$$

$$\vec{v}_2 A = \frac{3}{2}\vec{e}_1 - \frac{1}{2}(\vec{e}_1 + \vec{e}_2) = \vec{v}_1 + \frac{3}{2}\vec{v}_2$$

$$\Rightarrow [f]_{\mathcal{B}} = PAP^{-1} = \begin{bmatrix} \frac{3}{2} & 0 \\ 1 & \frac{3}{2} \end{bmatrix} = \frac{3}{2}I_2 + \begin{bmatrix} 0 & 0 \\ 1 & 0 \end{bmatrix},$$

where

$$P = \begin{bmatrix} \vec{v}_1 \\ \vec{v}_2 \end{bmatrix} = \begin{bmatrix} -\frac{1}{2} & -\frac{1}{2} \\ 1 & 0 \end{bmatrix}.$$

For any $\vec{x} \in \mathbb{R}^2$,

$\vec{x} = (x_1,\ x_2)$ in $\mathcal{N} = \{\vec{e}_1,\ \vec{e}_2\}$

$\downarrow$

$[\vec{x}]_{\mathcal{B}} = (\alpha_1,\ \alpha_2)$ in $\mathcal{B} = \{\vec{v}_1,\ \vec{v}_2\}$

$\downarrow$

$$[f(\vec{x})]_{\mathcal{B}} = [\vec{x}]_{\mathcal{B}}[f]_{\mathcal{B}} = \frac{3}{2}(\alpha_1,\ \alpha_2) + (\alpha_2,\ 0)$$

$$= \left(\frac{3}{2}\alpha_1 + \alpha_2,\ \frac{3}{2}\alpha_2\right) \quad \text{in } \mathcal{B}$$

$\downarrow$

$$f(\vec{x}) = \vec{x}A = \vec{x}P^{-1}[f]_{\mathcal{B}}P = \left(x_1 + \frac{1}{2}x_2,\ -\frac{1}{2}x_1 + 2x_2\right) \in \mathbb{R}^2 \quad \text{in } \mathcal{N}.$$

This means that $\langle\langle \vec{v}_1 \rangle\rangle$, i.e. $x_1 - x_2 = 0$, is the only invariant line (subspace) of f, on which each point $\vec{x}$ is moved to $f(\vec{x}) = \frac{3}{2}\vec{x}$. See Fig. 2.86.

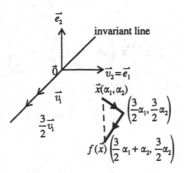

Fig. 2.86

We summarize Examples 1 and 2 as the following abstract result.

The Jordan canonical form of a linear operator

Suppose $f(\vec{x}) = \vec{x}A$: $\mathbb{R}^2 \to \mathbb{R}^2$ is a linear operator where $A = [a_{ij}]_{2\times2}$ is a real matrix. Suppose the characteristic polynomial is

$$\det(A - t\mathrm{I}_2) = (t - \lambda)^2,$$

so that A has eigenvalues λ, λ and A is *not* diagonalizable. Then $(A - \lambda\mathrm{I}_2)^2 = A^2 - 2\lambda A + \lambda^2\mathrm{I}_2 = O$. Let the *generalized eigenspace*

$$G_\lambda = \{\vec{x} \in \mathbb{R}^2 \mid \vec{x}(A - \lambda\mathrm{I}_2)^2 = \vec{0}\} = \mathbb{R}^2.$$

Take a vector $\vec{v}_2 \in G$ so that $\vec{v}_1 = \vec{v}_2(A - \lambda\mathrm{I}_2) \neq \vec{0}$. Then $\vec{v}_1$ is an eigenvector of A related to λ.

(1) $\mathcal{B} = \{\vec{v}_1, \vec{v}_2\}$ is a basis for $\mathbb{R}^2$ and

$$[f]_\mathcal{B} = PAP^{-1} = \begin{bmatrix} \lambda & 0 \\ 1 & \lambda \end{bmatrix} = \lambda\mathrm{I}_2 + \begin{bmatrix} 0 & 0 \\ 1 & 0 \end{bmatrix},$$

where

$$P = \begin{bmatrix} \vec{v}_1 \\ \vec{v}_2 \end{bmatrix}.$$

(2) The geometric mapping $\vec{x} \rightarrow f(\vec{x})$ in $\mathcal{N} = \{\vec{e}_1, \vec{e}_2\}$ is equivalent to the following mapping $[f]_{\mathcal{B}}$, in $\mathcal{B} = \{\vec{v}_1, \vec{v}_2\}$:

$$[\vec{x}]_{\mathcal{B}} = (\alpha_1, \alpha_2) \xrightarrow[\text{by scalar } \lambda]{\text{enlargement}} (\lambda\alpha_1, \lambda\alpha_2)$$

$$\downarrow \qquad\qquad\qquad \downarrow \text{translation along } (\alpha_2, 0)$$

$$[f(\vec{x})]_{\mathcal{B}} = [\vec{x}]_{\mathcal{B}}[f]_{\mathcal{B}} = \lambda(\alpha_1, \alpha_2) + (\alpha_2, 0) = (\lambda\alpha_1 + \alpha_2, \lambda\alpha_2)$$

This means that the one-dimensional subspace $\langle\langle\vec{v}_1\rangle\rangle$, i.e. $\alpha_2 = 0$, is the only invariant subspace of f except $\lambda = 1$. In case $\lambda = 1$, each line parallel to $\langle\langle\vec{v}_1\rangle\rangle$ is invariant under f. $\qquad\qquad$ (2.7.74)

For general results, please refer to Sec. B.11, in particular, Ex. 2 and Sec. B.12. Remind readers to review Example 6 and the explanations after it in Sec. 2.7.2, including Figs. 2.51 and 2.52.

Exercises

<A>

1. Model after Example 1 to investigate the geometric mapping proper- ties of

$$f(\vec{x}) = \vec{x}A, \quad \text{where } \vec{x} = (x_1, x_2) \quad \text{and} \quad A = \begin{bmatrix} -2 & 0 \\ 3 & -2 \end{bmatrix}.$$

2. Do the same problem as in Ex. 1 and Ex. <A> 3(a) of Sec. 2.7.6 if

   (a) $f(\vec{x}) = \vec{x}A$, where $\vec{x} = (x_1, x_2)$ and $A = \begin{bmatrix} 2 & 1 \\ -1 & 4 \end{bmatrix}$,

   (b) $f(\vec{x}) = \vec{x}A$, where $\vec{x} = (x_1, x_2)$ and $A = \begin{bmatrix} \lambda & 0 \\ b & \lambda \end{bmatrix}$, $b\lambda \neq 0$.

3. *Constructive* linear algebra

   Fix the Cartesian coordinate system $\mathcal{N} = \{\vec{e}_1, \vec{e}_2\}$ on $\mathbb{R}^2$. Let $\vec{v}_1 = (3, 1)$ and $\vec{v}_2 = (1, -2)$. Try to map the triangle $\triangle\vec{0}\,\vec{v}_1\vec{v}_2$ respec- tively onto $\triangle\vec{0}\,\vec{v}_1(-\vec{v}_2)$, $\triangle\vec{0}(-\vec{v}_2)(-\vec{v}_1)$ and $\triangle\vec{0}(-\vec{v}_1)\vec{v}_2$. See Fig. 2.87. For example, map $\triangle\vec{0}\,\vec{v}_1\vec{v}_2$ onto $\triangle\vec{0}\,\vec{v}_1(-\vec{v}_2)$ but keep $\vec{0}$ fixed. One way to do this is to find a linear isomorphism $f: \mathbb{R}^2 \rightarrow \mathbb{R}^2$ such that

$$f(\vec{v}_1) = \vec{v}_1 = 1 \cdot \vec{v}_1 + 0 \cdot \vec{v}_2$$
$$f(\vec{v}_2) = -\vec{v}_2 = 0 \cdot \vec{v}_1 + (-1) \cdot \vec{v}_2$$

$$\Rightarrow [f]_{\mathcal{B}} = \begin{bmatrix} 1 & 0 \\ 0 & -1 \end{bmatrix}, \quad \text{where } \mathcal{B} = \{\vec{v}_1, \vec{v}_2\}.$$

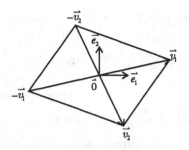

Fig. 2.87

In terms of the natural coordinate system $\mathcal{N} = \{\vec{e}_1,\ \vec{e}_2\}$,

$$[f]_\mathcal{N} = P^{-1}\begin{bmatrix} 1 & 0 \\ 0 & -1 \end{bmatrix}P, \quad \text{where } P = \begin{bmatrix} \vec{v}_1 \\ \vec{v}_2 \end{bmatrix} = \begin{bmatrix} 3 & 1 \\ 1 & -2 \end{bmatrix}$$

$$= -\frac{1}{7}\begin{bmatrix} -2 & -1 \\ -1 & 3 \end{bmatrix}\begin{bmatrix} 1 & 0 \\ 0 & -1 \end{bmatrix}\begin{bmatrix} 3 & 1 \\ 1 & -2 \end{bmatrix}$$

$$= \begin{bmatrix} \frac{5}{7} & \frac{4}{7} \\ \frac{6}{7} & -\frac{5}{7} \end{bmatrix}$$

$$\Rightarrow f(\vec{x}) = \vec{x}[f]_\mathcal{N} = \frac{1}{7}(5x_1 + 6x_2,\ 4x_1 - 5x_2).$$

The other way is to define $g\colon \mathbb{R}^2 \to \mathbb{R}^2$ as

$$g(\vec{v}_1) = -\vec{v}_2 = 0 \cdot \vec{v}_1 + (-1)\vec{v}_2$$
$$g(\vec{v}_2) = \vec{v}_1 = 1 \cdot \vec{v}_1 + 0 \cdot \vec{v}_2$$

$$\Rightarrow [g]_B = \begin{bmatrix} 0 & -1 \\ 1 & 0 \end{bmatrix} = \begin{bmatrix} 0 & 1 \\ 1 & 0 \end{bmatrix}\begin{bmatrix} 1 & 0 \\ 0 & -1 \end{bmatrix}$$

$$\Rightarrow [g]_\mathcal{N} = P^{-1}\begin{bmatrix} 0 & -1 \\ 1 & 0 \end{bmatrix}P = \begin{bmatrix} \frac{1}{7} & \frac{5}{7} \\ -\frac{10}{7} & -\frac{1}{7} \end{bmatrix}$$

$$\Rightarrow g(\vec{x}) = \vec{x}[g]_\mathcal{N} = \frac{1}{7}(x_1 - 10x_2,\ 5x_1 - x_2).$$

Note the role of the matrix $\begin{bmatrix} 0 & 1 \\ 1 & 0 \end{bmatrix}$ in the derivation of g.

(a) Model after f and g respectively and try to map $\triangle \vec{0}\,\vec{v}_1\vec{v}_2$ onto $\triangle \vec{0}(-\vec{v}_2)(-\vec{v}_1)$.

(b) Same as (a) but map $\triangle \vec{0}\,\vec{v}_1\vec{v}_2$ onto $\triangle \vec{0}(-\vec{v}_1)\vec{v}_2$.

The next problem is to map $\Delta \vec{0}\,\vec{v_1}\vec{v_2}$ onto $\Delta \vec{0}\,\vec{v_2}(\vec{v_1}+\vec{v_2})$, half of the parallelogram with vertices $\vec{0}$, $\vec{v_2}$, $\vec{v_1}+\vec{v_2}$ and $\vec{v_1}$. The mapping $h\colon \mathbb{R}^2 \to \mathbb{R}^2$ defined linearly by

$$h(\vec{v_1}) = \vec{v_1} + \vec{v_2},$$
$$h(\vec{v_2}) = \vec{v_2}$$

is probably the simplest choice among others. See Fig. 2.88. Geometrically, h represents the *shearing* with $\langle\langle \vec{v_2} \rangle\rangle$ as its invariant line (for details, refer to Example 6 in Sec. 2.7.2) and it moves a point $[\vec{x}]_B = (\alpha_1,\ \alpha_2)$ to the point $[h(\vec{x})]_B = (\alpha_1,\ \alpha_1+\alpha_2)$ along the line parallel to $\langle\langle \vec{v_2} \rangle\rangle$. Now

$$[h]_B = \begin{bmatrix} 1 & 1 \\ 0 & 1 \end{bmatrix}$$

$$\Rightarrow [h]_{\mathcal{N}} = P^{-1} \begin{bmatrix} 1 & 1 \\ 0 & 1 \end{bmatrix} P = \begin{bmatrix} \frac{9}{7} & -\frac{4}{7} \\ \frac{1}{7} & \frac{5}{7} \end{bmatrix} \overset{\text{def}}{=} A$$

$$\Rightarrow h(\vec{x}) = \vec{x}[h]_{\mathcal{N}} = \frac{1}{7}(9x_1 + x_2,\ -4x_1 + 5x_2).$$

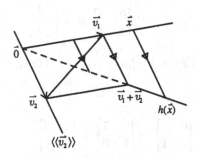

Fig. 2.88

By the way, we can use this h to justify (2.7.74) as follows. Firstly, the characteristic polynomial of h is

$$\det([h]_{\mathcal{N}} - tI_2) = \det \begin{bmatrix} 1-t & 1 \\ 0 & 1-t \end{bmatrix} = (t-1)^2$$

and hence h has eigenvalues 1, 1 but h is not diagonalizable. Take $\vec{y_2} = \frac{7}{2}\vec{e_1}$, then $\vec{y_1} = \vec{y_2}(A - I_2) = \vec{v_2}$ is an eigenvector of h and $\gamma = \{\vec{y_1},\ \vec{y_2}\}$ is a basis for $\mathbb{R}^2$. As might be expected, actual computation shows that

$$h(\vec{y}_1) = \vec{y}_1$$
$$h(\vec{y}_2) = \vec{y}_1 + \vec{y}_2$$

$$\Rightarrow [h]_\gamma = \begin{bmatrix} 1 & 0 \\ 1 & 1 \end{bmatrix}$$

$$\Rightarrow [h]_\mathcal{N} = Q^{-1} \begin{bmatrix} 1 & 0 \\ 1 & 1 \end{bmatrix} Q, \quad \text{where } Q = \begin{bmatrix} \vec{y}_1 \\ \vec{y}_2 \end{bmatrix} = \begin{bmatrix} 1 & -2 \\ \frac{7}{2} & 0 \end{bmatrix}$$

$$= \frac{1}{7} \begin{bmatrix} 0 & 2 \\ -\frac{7}{2} & 1 \end{bmatrix} \begin{bmatrix} 1 & 0 \\ 1 & 1 \end{bmatrix} \begin{bmatrix} 1 & -2 \\ \frac{7}{2} & 0 \end{bmatrix} = \begin{bmatrix} \frac{9}{7} & -\frac{4}{7} \\ \frac{1}{7} & \frac{5}{7} \end{bmatrix},$$

which coincides with the original one mentioned above. The other choice is to define $k: \mathbb{R}^2 \to \mathbb{R}^2$ by

$$k(\vec{v}_1) = \vec{v}_2,$$
$$k(\vec{v}_2) = \vec{v}_1 + \vec{v}_2$$

$$\Rightarrow [k]_\mathcal{B} = \begin{bmatrix} 0 & 1 \\ 1 & 1 \end{bmatrix}$$

$$\Rightarrow [k]_\mathcal{N} = P^{-1} \begin{bmatrix} 0 & 1 \\ 1 & 1 \end{bmatrix} P = \begin{bmatrix} \frac{6}{7} & -\frac{5}{7} \\ -\frac{11}{7} & \frac{1}{7} \end{bmatrix} \overset{\text{def}}{=} B$$

$$\Rightarrow k(\vec{x}) = \vec{x}[k]_\mathcal{N} = \frac{1}{7}(6x_1 - 11x_2, -5x_1 + x_2).$$

Note that

$$\begin{bmatrix} 0 & 1 \\ 1 & 1 \end{bmatrix} = \begin{bmatrix} 0 & 1 \\ 1 & 0 \end{bmatrix} + \begin{bmatrix} 0 & 0 \\ 0 & 1 \end{bmatrix}$$

$$= \begin{bmatrix} 0 & 1 \\ -1 & 0 \end{bmatrix} \begin{bmatrix} -1 & 0 \\ 0 & 1 \end{bmatrix} + \begin{bmatrix} 0 & 0 \\ 0 & 1 \end{bmatrix}.$$

Try to use these two algebraic representations on the right to explain the mapping properties of k in the basis $\mathcal{B}$. Now, k has the characteristic polynomial

$$\det([k]_\mathcal{N} - tI_2) = \det([k]_\mathcal{B} - tI_2) = \det \begin{bmatrix} -t & 1 \\ 1 & 1-t \end{bmatrix},$$

$$= t^2 - t - 1,$$

so that k has eigenvalues $\lambda_1 = \frac{1+\sqrt{5}}{2}$ and $\lambda_2 = \frac{1-\sqrt{5}}{2}$. Solve
$k(\vec{x}) = \lambda_i \vec{x}$, i.e.

$$(x_1 \quad x_2) \begin{bmatrix} \frac{6}{7} - \lambda_i & -\frac{5}{7} \\ -\frac{11}{7} & \frac{1}{7} - \lambda_i \end{bmatrix} = \vec{0}, \quad i = 1, 2$$

and we obtain the corresponding eigenvectors $\vec{x}_1 = (22, \ 5 + 7\sqrt{5})$
and $\vec{x}_2 = (22, \ 5 - 7\sqrt{5})$. Then $C = \{\vec{x}_1, \ \vec{x}_2\}$ is a basis for $\mathbb{R}^2$ and

$$R[k]_{\mathcal{N}}R^{-1} = \begin{bmatrix} \frac{1+\sqrt{5}}{2} & 0 \\ 0 & \frac{1-\sqrt{5}}{2} \end{bmatrix} = [k]_C,$$

where

$$R = \begin{bmatrix} \vec{x}_1 \\ \vec{x}_2 \end{bmatrix} = \begin{bmatrix} 22 & 5 + 7\sqrt{5} \\ 22 & 5 - 7\sqrt{5} \end{bmatrix}.$$

Therefore, $[k]_{\mathcal{N}} = R^{-1}[k]_C \, R$, i.e.

$$k(\vec{x}) = \vec{x}[k]_{\mathcal{N}} = \vec{x}R^{-1}[k]_C \, R = [\vec{x}]_C[k]_C R.$$

This means that, for a given $\vec{x} \in \mathbb{R}^2$, we can follow the following
steps to pinpoint $k(\vec{x})$:

$$\vec{x} \to [\vec{x}]_C \to [\vec{x}]_C[k]_C \to [\vec{x}]_C[k]_C R = k(\vec{x}).$$

Equivalently, by using (2.7.72), compute

$$B_1 = R^{-1} \begin{bmatrix} 1 & 0 \\ 0 & 0 \end{bmatrix} R$$

$$= \frac{-1}{308} \begin{bmatrix} 5 - 7\sqrt{5} & -5 - 7\sqrt{5} \\ -22 & 22 \end{bmatrix} \begin{bmatrix} 1 & 0 \\ 0 & 0 \end{bmatrix} \begin{bmatrix} 22 & 5 + 7\sqrt{5} \\ 22 & 5 - 7\sqrt{5} \end{bmatrix}$$

$$= \frac{-1}{14\sqrt{5}} \begin{bmatrix} 5 - 7\sqrt{5} & -10 \\ -22 & -5 - 7\sqrt{5} \end{bmatrix},$$

$$B_2 = R^{-1} \begin{bmatrix} 0 & 0 \\ 0 & 1 \end{bmatrix} R = \frac{-1}{14\sqrt{5}} \begin{bmatrix} -5 - 7\sqrt{5} & 10 \\ 22 & 5 - 7\sqrt{5} \end{bmatrix}$$

and we get the canonical decomposition of $[k]_{\mathcal{N}} = B$ as

$$I_2 = B_1 + B_2$$

$$[k]_{\mathcal{N}} = \frac{1 + \sqrt{5}}{2}B_1 + \frac{1 - \sqrt{5}}{2}B_2.$$

Refer to Fig. 2.73 and try to use the above decomposition to explain
geometrically how k maps $\triangle\vec{0}\,\vec{v}_1\vec{v}_2$ onto $\triangle\vec{0}\,\vec{v}_2(\vec{v}_1 + \vec{v}_2)$.

(c) Model after h and k to map $\Delta \vec{0}\,\vec{v}_1\,\vec{v}_2$ onto $\Delta \vec{0}\,(-\vec{v}_1)(-\vec{v}_1 - \vec{v}_2)$.

(d) In $\mathcal{B} = \{\vec{v}_1,\ \vec{v}_2\}$, a linear transformation $p\colon \mathbb{R}^2 \to \mathbb{R}^2$ has the representation

$$[p]_{\mathcal{B}} = \begin{bmatrix} -3 & 0 \\ 1 & -3 \end{bmatrix}.$$

Try to explain the geometric mapping properties of p in $\mathcal{B}$ and find the image of $\Delta \vec{0}\,\vec{v}_1\,\vec{v}_2$ under p. What is the representation of p in $\mathcal{N} = \{\vec{e}_1,\ \vec{e}_2\}$?

(e) In $\mathcal{B} = \{\vec{v}_1, \vec{v}_2\}$, a linear transformation $q\colon \mathbb{R}^2 \to \mathbb{R}^2$ has the representation

$$[q]_{\mathcal{B}} = \begin{bmatrix} 0 & 1 \\ -3 & -2 \end{bmatrix}.$$

Do the same question as (d). Refer to Sec. 2.7.8 if necessary.

(f) In $\mathcal{B} = \{\vec{v}_1, \vec{v}_2\}$, a linear transformation $r\colon \mathbb{R}^2 \to \mathbb{R}^2$ has the representation

$$[r]_{\mathcal{B}} = \begin{bmatrix} 0 & 1 \\ -2 & -3 \end{bmatrix}.$$

Do the same problem as (d). Refer to Sec. 2.7.8 if necessary.

<B>
1. Prove (2.7.74) and interpret its geometric mapping properties graphically.

<C> Abstraction and generalization

Read Secs. 3.7.7 and B.12 and try your best to do the exercises there.

2.7.8　Rational canonical form

Some linear operator may not have real eigenvalues.

Example 1 In $\mathcal{N} = \{\vec{e}_1,\ \vec{e}_2\}$, let linear operator $f\colon \mathbb{R}^2 \to \mathbb{R}^2$ be defined as

$$f(x_1,\ x_2) = (x_1 - 2x_2,\ x_1 - x_2)$$

$$= \vec{x}A, \quad \text{where } \vec{x} = (x_1,\ x_2) \text{ and } A = \begin{bmatrix} 1 & 1 \\ -2 & -1 \end{bmatrix}.$$

Try to investigate geometric mapping properties of f.

Solution f is one-to-one and hence is onto. The characteristic polynomial of f is

$$\det(A - tI_2) = \begin{vmatrix} 1-t & 1 \\ -2 & -1-t \end{vmatrix} = t^2 - 1 + 2 = t^2 + 1.$$

Thus, f does not have any real eigenvalues. This is equivalent to say that the simultaneous equations

$$\vec{x}A = \lambda\vec{x}, \quad \text{or}$$

$$(1 - \lambda)x_1 - 2x_2 = 0$$

$$x_1 - (1 + \lambda)x_2 = 0$$

do not have nonzero solution $\vec{x} = (x_1, x_2)$ for *any* real number λ. Hence, no line is invariant under f and only the point $\vec{0}$ is fixed by f.

A satisfies its characteristic polynomial $t^2 + 1$, i.e.

$$A^2 + I_2 = \begin{bmatrix} -1 & 0 \\ 0 & -1 \end{bmatrix} + \begin{bmatrix} 1 & 0 \\ 0 & 1 \end{bmatrix} = O.$$

Thus $A^2 + I_2$, as a linear operator on $\mathbb{R}^2$, annihilates all the vectors in $\mathbb{R}^2$, i.e.

$$G = \{\vec{x} \in \mathbb{R}^2 \mid \vec{x}(A^2 + I_2) = \vec{0}\} = \mathbb{R}^2.$$

Does this help in the investigation of choosing a better coordinate system for $\mathbb{R}^2$ so that the mapping properties of f becomes clearer?

Let us try and find some clue, if any.

In $\mathcal{N} = \{\vec{e}_1, \vec{e}_2\}$, consider the square with consecutive vertices $\vec{e}_1, \vec{e}_2, -\vec{e}_1$ and $-\vec{e}_2$. See Fig. 2.89. If we rotate the whole plane, with center at $\vec{0}$, through 90° in the counterclockwise direction, the resulting image of the square coincides with itself while its vertices permutate according to the ordering $\vec{e}_1 \rightarrow \vec{e}_2 \rightarrow -\vec{e}_1 \rightarrow -\vec{e}_2 \rightarrow \vec{e}_1$. Four such consecutive rotations will bring each vertex back to its original position (see Fig. 2.89). These four rotations together form a cyclic group of order 4 (refer to Sec. A.4). Such a rotation through 90° carries a point $\vec{x} = (x_1, x_2)$ into a point with coordinate $(-x_2, x_1)$ and, in $\mathcal{N} = \{\vec{e}_1, \vec{e}_2\}$, can be represented as

$$\vec{y} = \vec{x}J, \quad \text{where } J = \begin{bmatrix} 0 & 1 \\ -1 & 0 \end{bmatrix}.$$

See Fig. 2.90, where J is the composite mapping of the reflection (x_2, x_1) of the point (x_1, x_2) on the line $x_1 = x_2$ following by the reflection $(-x_2, x_1)$

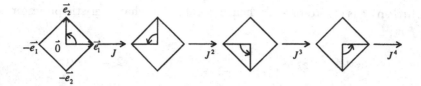

Fig. 2.89

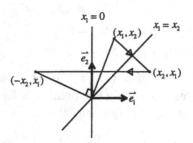

Fig. 2.90

of $(x_2,\ x_1)$ on the line $x_1 = 0$. Note that

$$\vec{e}_1 J = \vec{e}_2$$
$$\vec{e}_1 J^2 = \vec{e}_2 J = -\vec{e}_1$$
$$\vec{e}_1 J^3 = (-\vec{e}_1)J = -\vec{e}_2$$
$$\vec{e}_1 J^4 = (-\vec{e}_2)J = \vec{e}_1.$$

In particular, $J^2 + I_2 = O$ and $\{\vec{e}_1,\ \vec{e}_1 J\} = \{\vec{e}_1,\ \vec{e}_2\}$ are linearly independent and thus is a basis for $\mathbb{R}^2$.

When comparing $A^2 + I_2 = O$ and $J^2 + I_2 = O$, naturally we will have strong confidence in handling A by a similar way. For this purpose, take any fixed *nonzero* vector $\vec{v} \in G = \mathbb{R}^2$. Since A does not have any real eigenvalues, $\vec{v}$ and $\vec{v}A$ should be linearly independent and hence

$$\mathcal{B} = \{\vec{v},\ \vec{v}A\}$$

is a basis for $\mathbb{R}^2$. Now, since $\vec{v}(A^2 + I_2) = \vec{v}A^2 + \vec{v} = \vec{0}$,

$$\vec{v}A = 0 \cdot \vec{v} + 1 \cdot \vec{v}A$$
$$(\vec{v}A)A = \vec{v}A^2 = -\vec{v} = (-1)\vec{v} + 0 \cdot \vec{v}A$$
$$\Rightarrow [f]_{\mathcal{B}} = PAP^{-1} = \begin{bmatrix} 0 & 1 \\ -1 & 0 \end{bmatrix}, \quad \text{where } P = \begin{bmatrix} \vec{v} \\ \vec{v}A \end{bmatrix}.$$

This basis $\mathcal{B}$ and the resulting *rational canonical form* PAP^{-1} are exactly what we want (refer to Sec. B.12). Figure 2.91 illustrates mapping properties of f or A in the language of the basis $\mathcal{B}$. Note that

$$\{I_2, A, A^2 = -I_2, A^3 = -A\}$$

forms a cyclic group of order 4.

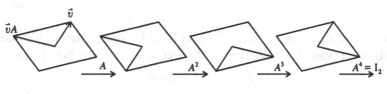

Fig. 2.91

Summarize Example 1 as the following abstract result.

The rational canonical form of a linear operator
Suppose $f(\vec{x}) = \vec{x}A \colon \mathbb{R}^2 \to \mathbb{R}^2$ is a linear operator where $A = [a_{ij}]_{2\times 2}$ is a real matrix. Suppose the characteristic polynomial is

$$\det(A - tI_2) = t^2 + a_1 t + a_0,$$

where $a_1^2 - 4a_0 < 0$, so that A does not have real eigenvalues (see Remark on the next page). Then $A^2 + a_1 A + a_0 I_2 = O$. Let the *generalized eigenspace*

$$G = \{\vec{x} \in \mathbb{R}^2 \mid \vec{x}(A^2 + a_1 A + a_0 I_2) = \vec{0}\} = \mathbb{R}^2.$$

Take any *nonzero* vector $\vec{v} \in \mathbb{R}^2$.

(1) $\mathcal{B} = \{\vec{v}, \vec{v}A\}$ is a basis for $\mathbb{R}^2$ and

$$[f]_{\mathcal{B}} = PAP^{-1} = \begin{bmatrix} 0 & 1 \\ -a_0 & -a_1 \end{bmatrix} = \begin{bmatrix} 0 & 1 \\ -1 & 0 \end{bmatrix}\begin{bmatrix} a_0 & 0 \\ 0 & 1 \end{bmatrix} + \begin{bmatrix} 0 & 0 \\ 0 & -a_1 \end{bmatrix},$$

where

$$P = \begin{bmatrix} \vec{v} \\ \vec{v}A \end{bmatrix}.$$

(2) The geometric mapping $\vec{x} \to f(\vec{x}) = \vec{x}A$ in $\mathcal{N} = \{\vec{e}_1, \vec{e}_2\}$ is equivalent to the mapping $[\vec{x}]_B \to [f(\vec{x})]_B = [\vec{x}]_B[f]_B$ in $B = \{\vec{v}, \vec{v}A\}$ as follows.

$$[\vec{x}]_B = (\alpha_1,\ \alpha_2) \xrightarrow[\begin{bmatrix} 0 & 1 \\ -1 & 0 \end{bmatrix}]{} (-\alpha_2,\ \alpha_1) \xrightarrow[\begin{bmatrix} a_0 & 0 \\ 0 & 1 \end{bmatrix}]{} (-a_0\alpha_2,\ \alpha_1)$$

$$\downarrow \qquad\qquad\qquad\qquad\qquad\qquad \Bigg\downarrow \begin{array}{l} \text{translation} \\ \text{along } (0,\ -a_1\alpha_2) \end{array}$$

$$[f(\vec{x})]_B = [\vec{x}]_B[f]_B = (-a_0\alpha_2,\ \alpha_1 - a_1\alpha_2) = (-a_0\alpha_2,\ \alpha_1) + (0,\ -a_1\alpha_2)$$

$$(2.7.75)$$

See Fig. 2.92 and refer to Secs. 3.7.8 and B.12 for generalized results. Readers should review Example 1 and the explanations associated, including Fig. 2.91.

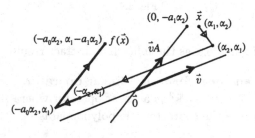

Fig. 2.92

Remark

Even if $a_1^2 - 4a_0 \geq 0$ so that A has real eigenvalues, (2.7.75) is still valid but not for every nonzero vector $\vec{v}$ in $\mathbb{R}^2$. All one needs to do is to choose a vector $\vec{v}$ in $\mathbb{R}^2$, which is *not* an eigenvector of A. Then, $\{\vec{v}, \vec{v}A\}$ is linear independent and hence forms a basis B for $\mathbb{R}^2$. In this case, (1) and (2) hold too.

Exercises

<A>

1. In $\mathcal{N} = \{\vec{e}_1, \vec{e}_2\}$, let $f \colon \mathbb{R}^2 \to \mathbb{R}^2$ be defined by

$$f(\vec{x}) = \vec{x}A \quad \text{where } \vec{x} = (x_1, x_2) \text{ and } A = \begin{bmatrix} -2 & 5 \\ 4 & -3 \end{bmatrix}.$$

(a) Model after Example 1 to justify (2.7.75).
(b) Do the same problem as in Ex. <A> 3(a) of Sec. 2.7.6.

2. Do the same problem as in 1 if

$$f(\vec{x}) = \vec{x}A, \quad \text{where } \vec{x} = (x_1, x_2) \text{ and } A = \begin{bmatrix} a & 1 \\ b & 0 \end{bmatrix}, \quad ab \neq 0.$$

<B>

1. Prove (2.7.75) and interpret it graphically.

<C> Abstraction and generalization

Read Secs. 3.7.8 and B.12 and try to do exercises there.

2.8 Affine Transformations

Topics here inherit directly from Sec. 2.7, in particular, (2.7.2).

In Exs. <B> 5 and 6 of Sec. 2.5, we gave definitions for affine subspaces and affine transformations on the plane $\mathbb{R}^2$. Now, we formally give definitions as follows.

Definition 2.8.1 Let X be a nonempty set and V an n-dimensional vector space over a field $\mathbb{F}$. Elements of X are called *points* which are temporarily denoted by capital letters $P, Q, R, \ldots$. If there exists a function: $X \times X \to V$ defined by

$$(P, Q) \in X \times X \to \overrightarrow{PQ} \in V$$

with the following properties:

1. For arbitrary three points P, Q, R on X,

$$\overrightarrow{PQ} + \overrightarrow{QR} = \overrightarrow{PR}.$$

2. For any point P in X and vector $\vec{x} \in V$, there corresponds a unique point Q in X such that

$$\overrightarrow{PQ} = \vec{x}.$$

Then, call the set X an *n-dimensional affine space* with the vector space V as its *difference space*. (2.8.1)

Since $\overrightarrow{PP} = \vec{0}$, hence any single point in X can be considered as the *zero vector* or *base point*. Also, call $\overrightarrow{PQ}$ a *position vector* with *initial point P* and *terminal point Q* or a *free vector* since law of parallel invariance holds for position vectors. Moreover, fix a point O in X as a base point, the correspondence

$$P \in X \leftrightarrow \overrightarrow{OP} \in V$$

is one-to-one and onto. Hence, one can treat X as V but the latter V has a specific $\vec{0}$, called the *zero vector* while the former X does not have any specific point in the sense that there is no difference between any two of its points. Conversely, if consider each vector in V as a point, then any two points $\vec{x}$ and $\vec{y}$ in V determine a vector

$$\vec{y} - \vec{x} \quad \text{(the *difference vector* from the point } \vec{x} \text{ to the point } \vec{y} \text{)}.$$

In this case, V is an *affine space with itself as difference space*. *We will adopt this Convention throughout the book.* This is the exact reason why we consider $\mathbb{R}$ and $\mathbb{R}^2$ both as affine spaces and as vector spaces (refer to the Convention before Remark 2.3 in Sec. 2.3).

Definition 2.8.2 Let S be a subspace of an n-dimensional vector space V over a field $\mathbb{F}$ and $\vec{x}_0$ be a fixed vector in V. The image

$$\vec{x}_0 + S = \{\vec{x}_0 + \vec{x} \mid \vec{x} \in S\}$$

of S under the translation $\vec{x} \to \vec{x}_0 + \vec{x}$ is called an *affine subspace* of V associated with the subspace S. Define the *dimension* of $\vec{x}_0 + S$ as dim S.

$$(2.8.2)$$

Note that $\vec{x}_0 + S = \vec{y}_0 + S$ if and only if $\vec{x}_0 - \vec{y}_0 \in S$. Also, $\vec{x}_0 + S = S$ if and only if $\vec{x}_0 \in S$.

Definition 2.8.3 Let $f\colon V \to V$ be a linear isomorphism of V onto itself and $\vec{x}_0$ be a fixed vector in V. The composite mapping

$$T(\vec{x}) = \vec{x}_0 + f(\vec{x})$$

of the linear isomorphism f followed by the translation $\vec{x} \to \vec{x}_0 + \vec{x}$ is called an *affine transformation* or *mapping* of the vector or affine space V.

$$(2.8.3)$$

This is the reason why we called such changes of coordinates in (1.3.2), (2.4.2) and (2.6.5) as affine transformations. Note that the above three definitions are still good for infinite-dimensional vector space.

Now we can formally characterize affine transformations on the plane in terms of linear transformations (see also Ex. <B> 1 of Sec. 2.8.3).

Affine transformation or mapping on the plane

Suppose $T: \mathbb{R}^2$ (affine space) $\to \mathbb{R}^2$ is a transformation. Then the following are equivalent.

(1) T is one-to-one, onto and preserves ratios of signed lengths of line segments along the same or parallel lines.
(2) For any fixed point $\vec{x}_0 \in \mathbb{R}^2$, there exists a unique linear isomorphism $f: \mathbb{R}^2$ (with $\vec{x}_0$ as base point) $\to \mathbb{R}^2$ such that

$$T(\vec{x}) = T(\vec{x}_0) + f(\vec{x} - \vec{x}_0), \quad \vec{x} \in \mathbb{R}^2.$$

Such a T is called an *affine transformation* or *mapping*. (2.8.4)

What should be mentioned is the expression of T is independent of the choice of the point $\vec{x}_0$. To see this, take any other point $\vec{y}_0$. Then,

$$T(\vec{y}_0) = T(\vec{x}_0) + f(\vec{y}_0 - \vec{x}_0) \quad \text{and}$$
$$T(\vec{x}) = T(\vec{x}_0) + f(\vec{x} - \vec{y}_0 + \vec{y}_0 - \vec{x}_0)$$
$$= T(\vec{x}_0) + f(\vec{y}_0 - \vec{x}_0) + f(\vec{x} - \vec{y}_0)$$
$$= T(\vec{y}_0) + f(\vec{x} - \vec{y}_0), \quad \vec{x} \in \mathbb{R}^2.$$

In particular,

$$T(\vec{x}) = T(\vec{0}) + f(\vec{x}), \quad \vec{x} \in \mathbb{R}^2. \tag{2.8.5}$$

In case $T(\vec{0}) = \vec{0}$, the affine transformation reduces to a linear isomorphism. In order to emphasize the "one-to-one and onto" properties, an affine transformation is usually called an *affine motion* in geometry.

The composite function of two affine transformations is again affine. To see this, let $T_1(\vec{x}) = T_1(\vec{0}) + f_1(\vec{x})$ and $T_2(\vec{x}) = T_2(\vec{0}) + f_2(\vec{x})$ be two affine transformations. Thus

$$(T_2 \circ T_1)(\vec{x}) = T_2(T_1(\vec{x}))$$
$$= T_2(\vec{0}) + f_2(T_1(\vec{0}) + f_1(\vec{x}))$$
$$= T_2(\vec{0}) + f_2(T_1(\vec{0})) + f_2(f_1(\vec{x}))$$
$$= (T_2 \circ T_1)(\vec{0}) + (f_2 \circ f_1)(\vec{x}), \quad \vec{x} \in \mathbb{R}^2 \tag{2.8.6}$$

with the prerequisite that the composite function of two linear isomorphisms is isomorphic, which can be easily verified.

By the very definitions, the inverse transformation $f^{-1}: \mathbb{R}^2 \to \mathbb{R}^2$ of f is isomorphic and the inverse transformation $T^{-1}: \mathbb{R}^2 \to \mathbb{R}^2$ is affine too.

In fact, we have

$$T^{-1}(\vec{x}) = -f^{-1}(T(\vec{0})) + f^{-1}(\vec{x}), \quad \vec{x} \in \mathbb{R}^2 \qquad (2.8.7)$$

and $T^{-1} \circ T(\vec{x}) = T \circ T^{-1}(\vec{x}) = \vec{x}$ for any $\vec{x} \in \mathbb{R}^2$.

Summarize as the

Affine group on the plane
The set of all affine transformations on the plane

$$G_a(2; \mathbb{R}) = \{\vec{x}_0 + f(\vec{x}) \mid \vec{x}_0 \in \mathbb{R}^2 \text{ and } f: \mathbb{R}^2 \to \mathbb{R}^2 \text{ is a linear isomorphism}\}$$

forms a group under the composite operation (see Sec. A.2) with

$$I: \mathbb{R}^2 \to \mathbb{R}^2 \quad \text{defined by } I(\vec{x}) = \vec{x}$$

as *identity element* and

$$-f^{-1}(\vec{x}_0) + f^{-1}(\vec{x}), \quad \vec{x} \in \mathbb{R}^2$$

as the *inverse element* of $\vec{x}_0 + f(\vec{x})$. Call $G_a(2; \mathbb{R})$ the *affine group* on the plane $\mathbb{R}^2$ (see Sec. A.4). $\qquad (2.8.8)$

This is the counterpart of (2.7.34) for affine transformations.

This section is divided into five subsections.

As a counterpart of Secs. 2.7.1 and 2.7.3, Sec. 2.8.1 introduces the matrix representations of an affine transformation in various affine bases. Concrete and basic affine transformations are illustrated graphically in Sec. 2.8.2 which can be compared with Sec. 2.7.2 in content.

From what we have experienced through examples in Sec. 2.8.2, we formally formulate these geometric quantities or properties that are invariant under the action of the affine group $G_a(2; \mathbb{R})$ as mentioned in (2.8.8). Section 2.8.3 studies the *affine invariants* on the affine plane and proves part (1) in (2.7.9). Results here are the plane version of those in Sec. 1.4, and they will form a model for later development in the affine space $\mathbb{R}^3$ (see Sec. 3.8.3).

The study of these geometric properties that are invariant under the affine group $G_a(2; \mathbb{R})$ constitutes basically what the so-called *affine geometry* on the plane. Section 2.8.4 will present some affine geometric problems as examples.

In particular, Sec. 2.8.5 will focus our attention on the affine invariants concerning the quadratic curve

$$\langle \vec{x}, \vec{x}B \rangle + 2\langle \vec{b}, \vec{x} \rangle + c = 0, \quad B = \begin{bmatrix} b_{11} & b_{12} \\ b_{12} & b_{22} \end{bmatrix}, \quad \vec{b} = (b_1, b_2), \quad c \in \mathbb{R},$$

where $\vec{x} = (x_1, x_2)$ and $\langle \vec{x}, \vec{x}B \rangle$ is treated as $\vec{x}(\vec{x}B)^* = \vec{x}B^*\vec{x}^* = \vec{x}B\vec{x}^*$.

2.8.1 *Matrix representations*

We consider the vector space $\mathbb{R}^2$ as an affine plane at the same time.

Remember (see Sec. 2.6) that three points $\vec{a}_0$, $\vec{a}_1$, $\vec{a}_2$ are said to constitute an *affine basis* $\mathcal{B} = \{\vec{a}_0, \vec{a}_1, \vec{a}_2\}$ with *base point* $\vec{a}_0$ if the vectors $\vec{a}_1 - \vec{a}_0$ and $\vec{a}_2 - \vec{a}_0$ are linearly independent.

Let $\mathcal{B} = \{\vec{a}_0, \vec{a}_1, \vec{a}_2\}$ and $\mathcal{C} = \{\vec{b}_0, \vec{b}_1, \vec{b}_2\}$ be two affine bases for the plane $\mathbb{R}^2$ and $T(\vec{x}) = \vec{x}_0 + f(\vec{x})$ an affine transformation on $\mathbb{R}^2$.

T may be expressed as

$$T(\vec{x}) = \vec{y}_0 + f(\vec{x} - \vec{a}_0),$$

where $\vec{y}_0 = T(\vec{a}_0) = \vec{x}_0 + f(\vec{a}_0)$. Therefore,

$$T(\vec{x}) - \vec{b}_0 = \vec{y}_0 - \vec{b}_0 + f(\vec{x} - \vec{a}_0)$$
$$\Rightarrow \text{ (by use of (2.6.3) and (a) in (2.7.23))}$$
$$[T(\vec{x}) - \vec{b}_0]_{\mathcal{C}} = [\vec{y}_0 - \vec{b}_0]_{\mathcal{C}} + [f(\vec{x} - \vec{a}_0)]_{\mathcal{C}}$$
$$= [\vec{y}_0 - \vec{b}_0]_{\mathcal{C}} + [\vec{x} - \vec{a}_0]_{\mathcal{B}}[f]_{\mathcal{C}}^{\mathcal{B}}$$

or, in short (see (2.3.1)),

$$[T(\vec{x})]_{\mathcal{C}} = [\vec{y}_0]_{\mathcal{C}} + [\vec{x}]_{\mathcal{B}}[f]_{\mathcal{C}}^{\mathcal{B}}, \tag{2.8.9}$$

which is called the *matrix representation* of the affine mapping T with respect to the affine bases $\mathcal{B} = \{\vec{a}_1 - \vec{a}_0, \vec{a}_2 - \vec{a}_0\}$ and $\mathcal{C} = \{\vec{b}_1 - \vec{b}_0, \vec{b}_2 - \vec{b}_0\}$. (2.8.9) can be written in matrix form

$$([T(\vec{x})]_{\mathcal{C}} \quad 1) = ([\vec{x}]_{\mathcal{B}} \quad 1) \begin{bmatrix} [f]_{\mathcal{C}}^{\mathcal{B}} & 0 \\ [T(\vec{a}_0)]_{\mathcal{C}} & 1 \end{bmatrix}. \tag{2.8.10}$$

Combining (2.8.8), (2.7.23) and (2.8.10), we have

The matrix representations of affine transformations (motions) with respect to affine bases

(1) Let $\mathcal{N} = \{\vec{0}, \vec{e}_1, \vec{e}_2\}$ be the *natural affine basis* for $\mathbb{R}^2$. Then, the affine group $G_a(2; \mathbb{R})$ in (2.8.8) is group isomorphic to the *matrix group*, also denoted as

$$G_a(2; \mathbb{R}) = \left\{ \begin{bmatrix} A & 0 \\ \vec{x}_0 & 1 \end{bmatrix} \,\middle|\, A \in GL(2; \mathbb{R}) \text{ and } \vec{x}_0 \in \mathbb{R}^2 \right\}$$

with matrix multiplication

$$\begin{bmatrix} A_1 & 0 \\ \vec{x}_1 & 1 \end{bmatrix} \begin{bmatrix} A_2 & 0 \\ \vec{x}_2 & 1 \end{bmatrix} = \begin{bmatrix} A_1 A_2 & 0 \\ \vec{x}_1 A_2 + \vec{x}_2 & 1 \end{bmatrix},$$

where the *identity motion* is

$$\begin{bmatrix} I_2 & 0 \\ 0 & 1 \end{bmatrix}$$

and the *inverse motion* is

$$\begin{bmatrix} A & 0 \\ \vec{x}_0 & 1 \end{bmatrix}^{-1} = \begin{bmatrix} A^{-1} & 0 \\ -\vec{x}_0 A^{-1} & 1 \end{bmatrix}.$$

$G_a(2;\ \mathbb{R})$ is called the *affine group* or *group of affine motions* on the plane with respect to the natural basis $\mathcal{N}$.

(2) Let $\mathcal{B} = \{\vec{a}_0,\ \vec{a}_1,\ \vec{a}_2\}$ be another affine basis for $\mathbb{R}^2$, the affine group with respect to $\mathcal{B}$ is the conjugate group

$$\begin{bmatrix} A_0 & 0 \\ \vec{a}_0 & 1 \end{bmatrix} G_a(2;\ \mathbb{R}) \begin{bmatrix} A_0 & 0 \\ \vec{a}_0 & 1 \end{bmatrix}^{-1}$$

$$= \left\{ \begin{bmatrix} A_0 A A_0^{-1} & 0 \\ (\vec{x}_0 + \vec{a}_0 A)A_0^{-1} - \vec{a}_0 A_0^{-1} & 1 \end{bmatrix} \middle| A \in \mathrm{GL}(2;\ \mathbb{R}) \text{ and } \vec{x}_0 \in \mathbb{R} \right\}$$

of the group $G_a(2;\ \mathbb{R})$, where A_0 is the transition matrix from $\mathcal{B} = \{\vec{a}_1 - \vec{a}_0,\ \vec{a}_2 - \vec{a}_0\}$ to $\mathcal{N} = \{\vec{e}_1,\ \vec{e}_2\}$ as bases for the vector space $\mathbb{R}^2$.   (2.8.11)

In case $\mathcal{B} = \mathcal{C}$, the matrix in (2.8.10) coincides with that in part (2) of (2.8.11). This is because

$$[f]_\mathcal{B} = A_\mathcal{N}^\mathcal{B}[f]_\mathcal{N}(A_\mathcal{N}^\mathcal{B})^{-1} = A_0[f]_\mathcal{N}A_0^{-1} = A_0 A A_0^{-1}$$

and

$$[T(\vec{a}_0)]_\mathcal{B} = [\vec{x}_0 + f(\vec{a}_0)]_\mathcal{N}A_\mathcal{B}^\mathcal{N} = (\vec{x}_0 + \vec{a}_0 A)A_0^{-1}$$
$$= (\vec{x}_0 + \vec{a}_0 A)A_0^{-1} - \vec{a}_0 A_0^{-1},$$

where $\vec{a}_0 A_0^{-1} = [\vec{a}_0]_\mathcal{N}A_\mathcal{B}^\mathcal{N} = [\vec{a}_0]_\mathcal{B} = \vec{0}$.

Notice that change of coordinates stated in (2.4.2) is a special kind of affine motions.

Using the natural affine basis $\mathcal{N} = \{\vec{0},\ \vec{e}_1,\ \vec{e}_2\}$, an affine transformation $T(\vec{x}) = \vec{x}_0 + \vec{x}A$ (see (2.8.5)) is decomposed as

$$\vec{x} \to \vec{x}A \text{ (keeping } \vec{0} \text{ fixed)} = \vec{y} \to \vec{x}_0 + \vec{y},   (2.8.12)$$

where $A \in \mathrm{GL}(2; \mathbb{R})$. In matrix terminology, it is nothing new but

$$\begin{bmatrix} A & 0 \\ \vec{x}_0 & 1 \end{bmatrix} = \begin{bmatrix} A & 0 \\ 0 & 1 \end{bmatrix} \begin{bmatrix} I_2 & 0 \\ \vec{x}_0 & 1 \end{bmatrix}, \quad A \in \mathrm{GL}(2; \mathbb{R}). \tag{2.8.13}$$

As a consequence, we can consider the real general group $\mathrm{GL}(2; \mathbb{R})$ in (2.7.34) as a *subgroup* of the affine group $\mathrm{G}_a(2; \mathbb{R})$ (for definition of a subgroup, see Ex. <A> 2).

A linear isomorphism on $\mathbb{R}^2$ is a special kind of affine transformation that keeps the zero vector $\vec{0}$ fixed. (2.8.12) says that an affine transformation is the composite of an affine transformation keeping $\vec{0}$ fixed and a translation along $T(\vec{0}) - \vec{0} = \vec{x}_0$. This fact is universally true for any point in $\mathbb{R}^2$.

To see this, let $\vec{x}_0$ be any fixed point in $\mathbb{R}^2$ and $T: \mathbb{R}^2 \to \mathbb{R}^2$ be an affine transformation. Then

$$\begin{aligned} T(\vec{x}) &= T(\vec{x}_0) + f(\vec{x} - \vec{x}_0) \\ &= T(\vec{x}_0) - \vec{x}_0 + (\vec{x}_0 + f(\vec{x} - \vec{x}_0)) \\ &= T(\vec{x}_0) - \vec{x}_0 + T_{\vec{x}_0}(\vec{x}), \end{aligned} \tag{2.8.14}$$

where $T_{\vec{x}_0}(\vec{x}) = \vec{x}_0 + f(\vec{x} - \vec{x}_0)$ is an affine transformation keeping the point $\vec{x}_0$ fixed. Therefore, we obtain the following result.

The decomposition of an affine transformation as an affine transformation keeping a point fixed and followed by a translation
Let $T: \mathbb{R}^2 \to \mathbb{R}^2$ be an affine transformation and $\vec{x}_0$ be any point in $\mathbb{R}^2$. Then T can be uniquely expressed as

$$T = f_2 \circ f_1,$$

where $f_2(\vec{x}) = [T(\vec{x}_0) - \vec{x}_0] + \vec{x}$ is the translation along $T(\vec{x}_0) - \vec{x}_0$, while $f_1(\vec{x}) = \vec{x}_0 + (T(\vec{x}) - T(\vec{x}_0))$: $\mathbb{R}^2 \to \mathbb{R}^2$ is an affine transformation keeping $\vec{x}_0$ fixed. (2.8.15)

Finally, we state

The fundamental theorem of affine transformations (motions)
For two arbitrary affine bases $\mathcal{B} = \{\vec{a}_0, \vec{a}_1, \vec{a}_2\}$ and $\mathcal{C} = \{\vec{b}_0, \vec{b}_1, \vec{b}_2\}$ for the plane, there exists a unique affine transformation $T: \mathbb{R}^2 \to \mathbb{R}^2$ satisfying

$$T(\vec{a}_i) = \vec{b}_i, \quad i = 0, 1, 2. \tag{2.8.16}$$

This is equivalent to say that there exists a unique linear isomorphism $f\colon \mathbb{R}^2 \to \mathbb{R}^2$ so that

$$f(\vec{a_i} - \vec{a_0}) = \vec{b_i} - \vec{b_0}, \quad i = 1, 2. \tag{2.8.17}$$

Then $T(\vec{x}) = \vec{b_0} + f(\vec{x} - \vec{a_0})$ is the required one.

We give an example to illustrate (2.8.15) and (2.8.16).

Example In $\mathbb{R}^2$, let

$$\vec{a_0} = (1,\ 2), \quad \vec{a_1} = (1,\ -1), \quad \vec{a_2} = (0,\ 1) \quad \text{and}$$
$$\vec{b_0} = (-2,\ -3), \quad \vec{b_1} = (3,\ -4), \quad \vec{b_2} = (-5,\ 1).$$

(a) Construct affine mappings T_1, T_2 and T such that

$$T_1(\vec{0}) = \vec{a_0}, \quad T_1(\vec{e_1}) = \vec{a_1}, \quad T_1(\vec{e_2}) = \vec{a_2};$$
$$T_2(\vec{0}) = \vec{b_0}, \quad T_2(\vec{e_1}) = \vec{b_1}, \quad T_2(\vec{e_2}) = \vec{b_2}, \quad \text{and}$$
$$T(\vec{a_i}) = \vec{b_i}, \quad i = 0,\ 1,\ 2.$$

(b) Show that $T = T_2 \circ T_1^{-1}$.

(c) Let $\vec{x_0} = (-2,\ -2)$. Express T as $f_2 \circ f_1$ where f_1 is a linear isomorphism keeping $\vec{x_0}$ fixed while f_2 is a suitable translation.

Solution See Fig. 2.93.

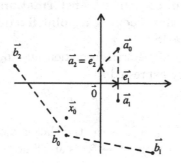

Fig. 2.93

(a) By computation,

$$\vec{a_1} - \vec{a_0} = (1,-1) - (1,2) = (0,-3) = 0\vec{e_1} - 3\vec{e_2},$$
$$\vec{a_2} - \vec{a_0} = (0,1) - (1,2) = (-1,-1) = -\vec{e_1} - \vec{e_2}, \quad \text{and}$$
$$\vec{a_0} = T_1(\vec{0}) = (1,2) = \vec{e_1} + 2\vec{e_2}.$$

Then T_1 has the equation

$$T_1(\vec{x}) = (1 \quad 2) + \vec{x} \begin{bmatrix} 0 & -3 \\ -1 & -1 \end{bmatrix}$$

or in coordinate form, if $\vec{x} = (x_1, x_2)$ and $\vec{y} = T_1(\vec{x}) = (y_1, y_2)$,

$$y_1 = 1 - x_2,$$
$$y_2 = 2 - 3x_1 - x_2.$$

Similarly,

$$\vec{b_1} - \vec{b_0} = (3, -4) - (-2, -3) = (5, -1),$$
$$\vec{b_2} - \vec{b_0} = (-5, 1) - (-2, -3) = (-3, 4) \quad \text{and}$$
$$\vec{b_0} = T_2(\vec{0}) = (-2, -3).$$

Then T_2 has equation

$$T_2(\vec{x}) = (-2 \quad -3) + \vec{x} \begin{bmatrix} 5 & -1 \\ -3 & 4 \end{bmatrix}, \quad \text{or}$$

$$y_1 = -2 + 5x_1 - 3x_2,$$
$$y_2 = -3 - x_1 + 4x_2.$$

Suppose $T(\vec{x}) = \vec{b_0} + f(\vec{x} - \vec{a_0})$ where f is a linear isomorphism. Thus

$$f(\vec{a_1} - \vec{a_0}) = f(0, -3) = f(0\vec{e_1} - 3\vec{e_2}) = 0f(\vec{e_1}) - 3f(\vec{e_2})$$
$$= -3f(\vec{e_2}) = \vec{b_1} - \vec{b_0} = (5, -1)$$
$$\Rightarrow f(\vec{e_2}) = \left(-\frac{5}{3}, \frac{1}{3}\right);$$

$$f(\vec{a_2} - \vec{a_0}) = f(-1, -1) = -f(\vec{e_1}) - f(\vec{e_2}) = \vec{b_2} - \vec{b_0} = (-3, 4)$$
$$\Rightarrow f(\vec{e_1}) = -f(\vec{e_2}) - (-3, 4) = \left(\frac{5}{3}, -\frac{1}{3}\right) - (-3, 4) = \left(\frac{14}{3}, -\frac{13}{3}\right).$$

Therefore,

$$T(\vec{x}) = (-2 \quad -3) + [\vec{x} - (1 \quad 2)] \begin{bmatrix} \frac{14}{3} & -\frac{13}{3} \\ -\frac{5}{3} & \frac{1}{3} \end{bmatrix}, \quad \text{or}$$

$$y_1 = -2 + \frac{14}{3}(x_1 - 1) - \frac{5}{3}(x_2 - 2) = -\frac{10}{3} + \frac{14}{3}x_1 - \frac{5}{3}x_2,$$
$$y_2 = -3 - \frac{13}{3}(x_1 - 1) + \frac{1}{3}(x_2 - 2) = \frac{2}{3} - \frac{13}{3}x_1 + \frac{1}{3}x_2.$$

(b) T_1^{-1} has the equation

$$x_1 = \frac{1}{3}(1 + y_1 - y_2), \quad x_2 = 1 - y_1.$$

Substituting these into the equation for T_2 (and change y_1, y_2 on the right side back to x_1, x_2 respectively), we have

$$y_1 = -2 + 5 \cdot \frac{1}{3}(1 + x_1 - x_2) - 3(1 - x_1) = -\frac{10}{3} + \frac{14}{3}x_1 - \frac{5}{3}x_2,$$

$$y_2 = -3 - \frac{1}{3}(1 + x_1 - x_2) + 4(1 - x_1) = \frac{2}{3} - \frac{13}{3}x_1 + \frac{1}{3}x_2,$$

which is the equation for T. Hence $T = T_2 \circ T_1^{-1}$. Readers are urged to verify this by matrix computation.

(c) By (2.8.15),

$$T(\vec{x}_0) = T(-2, -2) = (-2 \quad -3) + [(-2 \quad -2) - (1 \quad 2)]\begin{bmatrix} \frac{14}{3} & -\frac{13}{3} \\ -\frac{5}{3} & \frac{1}{3} \end{bmatrix}$$

$$= (-2 \quad -3) + (-3 \quad -4)\begin{bmatrix} \frac{14}{3} & -\frac{13}{3} \\ -\frac{5}{3} & \frac{1}{3} \end{bmatrix}$$

$$= (-2, -3) + \frac{1}{3}(-22, 35) = \left(-\frac{28}{3}, \frac{26}{3}\right)$$

and then

$$f_2(\vec{x}) = T(\vec{x}_0) - \vec{x}_0 + \vec{x}$$

$$= \left(-\frac{28}{3}, \frac{26}{3}\right) - (-2, -2) + \vec{x} = \left(-\frac{22}{3}, \frac{32}{3}\right) + \vec{x},$$

$$f_1(\vec{x}) = f_2^{-1}(T(\vec{x})) = \vec{x}_0 - T(\vec{x}_0) + T(\vec{x})$$

$$= (-2, -2) + (\vec{x} - (-2, -2))\begin{bmatrix} \frac{14}{3} & -\frac{13}{3} \\ -\frac{5}{3} & \frac{1}{3} \end{bmatrix}.$$

Hence, f_1 is an affine transformation keeping $(-2, -2)$ fixed, and f_2 is a translation. Of course, $T = f_2 \circ f_1$ holds.

Exercises

<A>

1. In $\mathbb{R}^2$, let

$$\vec{a}_0 = (-1, 2), \quad \vec{a}_1 = (5, -3), \quad \vec{a}_2 = (2, 1);$$
$$\vec{b}_0 = (-3, -2), \quad \vec{b}_1 = (4, -1), \quad \vec{b}_2 = (-2, 6).$$

(a) Show that $\mathcal{B} = \{\vec{a}_0, \vec{a}_1, \vec{a}_2\}$ and $\mathcal{C} = \{\vec{b}_0, \vec{b}_1, \vec{b}_2\}$ are affine bases for $\mathbb{R}^2$.

(b) Find the affine transformation T mapping $\vec{a}_i$ onto $\vec{b}_i$, $i = 0, 1, 2$. Express T in matrix forms with respect to $\mathcal{B}$ and $\mathcal{C}$, and with respect to the natural affine basis $\mathcal{N}$, respectively.

(c) Suppose T' is the affine transformation with matrix representation

$$T'(x_1, x_2) = \left(-\frac{1}{2} \quad \frac{5}{4}\right) + (x_1 \quad x_2) \begin{bmatrix} 2 & -1 \\ 5 & 6 \end{bmatrix}$$

with respect to natural basis $\mathcal{N}$. Find the equation of T' with respect to $\mathcal{B}$ and $\mathcal{C}$.

2. A nonempty subset S of a group (see Sec. A.4) is called a *subgroup* if $x, y \in S$ implies $x \circ y \in S$ and $x \in S$ implies $x^{-1} \in S$.

(a) Show that

$$S_+ = \{A \in \mathrm{GL}(2; \mathbb{R}) \,|\, \text{the determinant } \det A > 0\}$$

is a subgroup of $\mathrm{GL}(2; \mathbb{R})$.

(b) Show that

$$S' = \{A \in \mathrm{GL}(2; \mathbb{R}) \,|\, |\det A| = 1\}$$

is a subgroup of $\mathrm{GL}(2; \mathbb{R})$ and

$$S_1' = \{A \in S' \,|\, \det A = 1\}$$

is again a subgroup of S' and, of course, a subgroup of S_+ in (a). But the set $S_{-1}' = \{A \in S' \,|\, \det A = -1\}$ is not a subgroup of S'.

3. Let $\mathcal{B} = \{\vec{a}_1, \vec{a}_2\}$ be a basis for the vector space $\mathbb{R}^2$ and S be a subgroup of $\mathrm{GL}(2; \mathbb{R})$.

(a) For each $A \in S$, show that $\{\vec{a}_1 A, \vec{a}_2 A\}$ forms a basis for $\mathbb{R}^2$. Denote the set of all such bases by

$$S(\mathcal{B}) = \{\{\vec{a}_1 A, \vec{a}_2 A\} \,|\, A \in S\},$$

which is said to be *generated by the basis* $\mathcal{B}$ *with respect to subgroup* S.

(b) Suppose a basis $\mathcal{C} = \{\vec{b}_1, \vec{b}_2\} \in S(\mathcal{B})$. Show that

$$S(\mathcal{B}) = S(\mathcal{C}).$$

4. Two bases $\mathcal{B} = \{\vec{a}_1, \vec{a}_2\}$ and $\mathcal{C} = \{\vec{b}_1, \vec{b}_2\}$ for $\mathbb{R}^2$ are said to belong to the same *class* if there exists an $A \in S_+$ (defined in Ex. 2(a)) such that $\vec{b}_i = \vec{a}_i A$ or $\vec{a}_i = \vec{b}_i A, i = 1, 2$. Thus, all the bases for $\mathbb{R}^2$ are divided, with respect to the subgroup S_+, into two classes. Two bases are said to have *same orientation* if they belong to the same class, *opposite orientations* otherwise. The bases of one of the two classes are said to be *positively orientated* or *right-handed*; then, the bases of the other class are said to be *negatively oriented* or *left-handed*. $\mathbb{R}^2$ together with a definite class of bases is said to be *oriented*.

(*Note*   This acts as formal definitions for the so-called anticlockwise direction and clockwise direction mentioned occasionally in the text, say in Fig. 2.3 and (2.7.9).)

5. *Subgroup of affine transformations keeping a point fixed*
   Let $\vec{x}_0 \in \mathbb{R}^2$ be a point. An affine transformation of the form

$$T(\vec{x}) = \vec{x}_0 + (\vec{x} - \vec{x}_0)A,$$

where $A \in \mathrm{GL}(2; \mathbb{R})$, keeps $\vec{x}_0$ fixed, i.e.

$$T(\vec{x}_0) = \vec{x}_0.$$

See Fig. 2.94. All such transformations form a subgroup of $G_a(2; \mathbb{R})$, and is group isomorphic to

$$\left\{ \begin{bmatrix} A & 0 \\ 0 & 1 \end{bmatrix} \middle| A \in \mathrm{GL}(2; \mathbb{R}) \right\}$$

if the matrix representation of T with respect to an affine basis with $\vec{x}_0$ as base point is used.

Fig. 2.94

(*Note*   This is the reason why the real general linear group $\mathrm{GL}(2; \mathbb{R})$ can be treated as a subgroup of the affine group $G_a(2; \mathbb{R})$.)

6. *Subgroup of enlargement affine transformations keeping a point fixed*
   Affine transformation

   $$T(\vec{x}) = \vec{x}_0 + \alpha(\vec{x} - \vec{x}_0), \quad \alpha \in \mathbb{R}$$

   is an *enlargement* with *scale* α, keeping $\vec{x}_0$ fixed. See Fig. 2.95. All of
   them form a group which is a subgroup of that group mentioned in
   Ex. 5 and is group isomorphic to the group

   $$\left\{ \begin{bmatrix} \alpha I_2 & 0 \\ 0 & 1 \end{bmatrix} \middle| \alpha \in \mathbb{R} \text{ and } \alpha \neq 0 \right\}.$$

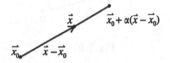

Fig. 2.95

7. *Subgroup of translations*
   The set of translations

   $$T(\vec{x}) = \vec{x}_0 + \vec{x}, \quad \vec{x}_0 \in \mathbb{R}^2$$

   forms a subgroup of $G_a(2;\mathbb{R})$. See Fig. 2.96. This subgroup is group
   isomorphic to the group

   $$\left\{ \begin{bmatrix} I_2 & 0 \\ \vec{x}_0 & 1 \end{bmatrix} \middle| \vec{x}_0 \in \mathbb{R}^2 \right\}.$$

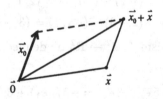

Fig. 2.96

8. *Subgroup of similarity transformations*
   Let $\vec{y}_0$ be a fixed point in $\mathbb{R}^2$. The affine transformation

   $$\begin{aligned} T(\vec{x}) &= \vec{y}_0 + \alpha(\vec{x} - \vec{x}_0) \\ &= (\vec{y}_0 - \vec{x}_0) + \vec{x}_0 + \alpha(\vec{x} - \vec{x}_0) \end{aligned}$$

   is the composite mapping of an enlargement followed by a translation
   and is called a *similarity transformation* or *mapping*. See Fig. 2.97.

They form a group which is group isomorphic to the group

$$\left\{ \begin{bmatrix} \alpha I_2 & 0 \\ \vec{y}_0 - \vec{x}_0 & 1 \end{bmatrix} \middle| \alpha \in \mathbb{R} \text{ and } \alpha \neq 0, \vec{x}_0 \in \mathbb{R}^2 \text{ with } \vec{y}_0 \text{ fixed} \right\}.$$

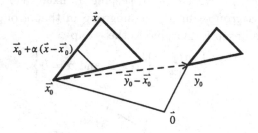

Fig. 2.97

9. *Constructive* linear algebra

Adopt the Cartesian coordinate system $\mathcal{N} = \{\vec{e}_1, \vec{e}_2\}$ on $\mathbb{R}^2$. Is it possible to construct an affine transformation mapping the triangle $\triangle\vec{a}_1\vec{a}_2\vec{a}_3$ with $\vec{a}_1 = (3,-1), \vec{a}_2 = (-2,2)$ and $\vec{a}_3 = (-4,-3)$, onto the triangle $\triangle\vec{b}_1\vec{b}_2\vec{b}_3$ with $\vec{b}_1 = (-1,1), \vec{b}_2 = (1,2)$ and $\vec{b}_3 = (2,-3)$? If yes, how? (2.8.16) guarantees the possibility, and offers a way how to find it. Construct side vectors

$$\vec{x}_1 = (-2,2) - (3,-1) = (-5,3),$$
$$\vec{x}_2 = (-4,-3) - (3,-1) = (-7,-2);$$
$$\vec{y}_1 = (1,2) - (-1,1) = (2,1),$$
$$\vec{y}_2 = (2,-3) - (-1,1) = (3,-4).$$

Let linear isomorphism $f \colon \mathbb{R}^2 \to \mathbb{R}^2$ be defined as

$$f(\vec{x}_i) = \vec{y}_i, \quad i = 1,2$$
$$\Rightarrow f(\vec{x}_1) = f(-5\vec{e}_1 + 3\vec{e}_2) = -5f(\vec{e}_1) + 3f(\vec{e}_2) = (2,1)$$
$$f(\vec{x}_2) = f(-7\vec{e}_1 - 2\vec{e}_2) = -7f(\vec{e}_1) - 2f(\vec{e}_2) = (3,-4)$$
$$\Rightarrow \begin{bmatrix} -5 & 3 \\ -7 & -2 \end{bmatrix} \begin{bmatrix} f(\vec{e}_1) \\ f(\vec{e}_2) \end{bmatrix} = \begin{bmatrix} 2 & 1 \\ 3 & -4 \end{bmatrix}$$
$$\Rightarrow \begin{bmatrix} f(\vec{e}_1) \\ f(\vec{e}_2) \end{bmatrix} = \begin{bmatrix} -5 & 3 \\ -7 & -2 \end{bmatrix}^{-1} \begin{bmatrix} 2 & 1 \\ 3 & -4 \end{bmatrix} = \frac{1}{31} \begin{bmatrix} -2 & -3 \\ 7 & -5 \end{bmatrix} \begin{bmatrix} 2 & 1 \\ 3 & -4 \end{bmatrix}$$
$$= \frac{1}{31} \begin{bmatrix} -13 & 10 \\ -1 & 27 \end{bmatrix}.$$

Hence, a required affine transformation is

$$f(\vec{x}) = (-1,1) + \frac{1}{31}(\vec{x} - (3,-1))\begin{bmatrix} -13 & 10 \\ -1 & 27 \end{bmatrix},$$

which is characterized by assigning the ordered vertices $\vec{a}_1, \vec{a}_2, \vec{a}_3$ to the respective ordered vertices $\vec{b}_1, \vec{b}_2, \vec{b}_3$. See Fig. 2.98.

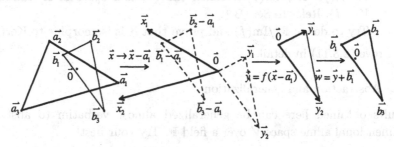

Fig. 2.98

If the ordered vertices $\vec{a}_1, \vec{a}_2, \vec{a}_3$ are mapped into ordered vertices $\vec{b}_1, \vec{b}_3, \vec{b}_2$, i.e. $\vec{x}_1 \rightarrow \vec{y}_2$ and $\vec{x}_2 \rightarrow \vec{y}_1$, try to see the reason why the required one is

$$f(\vec{x}) = (-1,1) + (\vec{x} - (3,-1)) \cdot \frac{1}{31}\begin{bmatrix} -2 & -3 \\ 7 & -5 \end{bmatrix}\begin{bmatrix} 0 & 1 \\ 1 & 0 \end{bmatrix}\begin{bmatrix} 2 & 1 \\ 3 & -4 \end{bmatrix}$$

$$= (-1,1) + \frac{1}{31}(\vec{x} - (3,-1))\begin{bmatrix} -12 & 5 \\ 11 & -33 \end{bmatrix}.$$

Do the following problems.

(a) Find all possible affine transformations mapping $\triangle \vec{a}_1 \vec{a}_2 \vec{a}_3$ onto $\triangle \vec{b}_1 \vec{b}_2 \vec{b}_3$ where $\triangle \vec{a}_1 \vec{a}_2 \vec{a}_3$ and $\triangle \vec{b}_1 \vec{b}_2 \vec{b}_3$ are as in Fig. 2.98.

(b) Do the same problem as (a), but for arbitrary triangles $\triangle \vec{a}_1 \vec{a}_2 \vec{a}_3$ and $\vec{b}_1 \vec{b}_2 \vec{b}_3$.

(c) Does there exist an affine transformation mapping the interior of a triangle onto the exterior of another triangle?

(d) Find all possible affine transformations mapping a parallelogram $\square \vec{a}_1 \vec{a}_2 \vec{a}_3 \vec{a}_4$ onto another parallelogram $\square \vec{b}_1 \vec{b}_2 \vec{b}_3 \vec{b}_4$.

10. Let $f : \mathbb{R}^2 \rightarrow \mathbb{R}^2$ be a linear transformation and its $\mathrm{Ker}(f)$ be a nonzero proper subspace of $\mathbb{R}^2$.

(a) For each $\vec{x} \in \mathbb{R}^2$, show that the preimage of $f(\vec{x})$ is the affine subspace

$$f^{-1}(f(\vec{x})) = \vec{x} + \mathrm{Ker}(f).$$

(b) Show that the quotient set

$$\mathbb{R}^2/\mathrm{Ker}(f) = \{f^{-1}(f(\vec{x})) \mid \vec{x} \in \mathbb{R}^2\}$$

is a vector space under $(\vec{x} + \mathrm{Ker}(f)) + (\vec{y} + \mathrm{Ker}(f)) = (\vec{x} + \vec{y}) + \mathrm{Ker}(f)$ and $\alpha(\vec{x} + \mathrm{Ker}(f)) = \alpha\vec{x} + \mathrm{Ker}(f)$ and is isomorphic to $\mathrm{Im}(f)$. $\mathbb{R}^2/\mathrm{Ker}(f)$ is called the *quotient space* of $\mathbb{R}^2$ modules $\mathrm{Ker}(f)$. Refer to Sec. B.1.

(c) Try to define $\mathbb{R}^2/\mathrm{Im}(f)$ and prove that it is isomorphic to $\mathrm{Ker}(f)$.

11. Prove (2.8.11) in detail.

<C> Abstraction and generalization

Results obtained here can be generalized almost verbatim to abstract n-dimensional affine space V over a field $\mathbb{F}$. Try your best!

2.8.2 *Examples*

This subsection will concentrate on how to construct some elementary affine transformations and their matrix representations in a suitable basis. It will be beneficial to compare the content with those in Sec. 2.7.2.

$\mathbb{R}^2$ is a vector space, and is also an affine plane as well. Occasionally, we need planar Euclidean concepts such as lengths, angles and areas as learned in high school courses. Please refer to the Introduction and Natural Inner Product in Part Two, if needed.

An *affine transformation* T that maps a point (x_1, x_2) onto a point (y_1, y_2) is of the form,

$$\begin{aligned} y_1 &= a_{11}x_1 + a_{21}x_2 + b_1 \\ y_2 &= a_{12}x_1 + a_{22}x_2 + b_2 \end{aligned} \qquad (2.8.18)$$

with the coefficient determinant

$$\Delta = \begin{vmatrix} a_{11} & a_{12} \\ a_{21} & a_{22} \end{vmatrix} = a_{11}a_{22} - a_{12}a_{21} \neq 0;$$

while, in the present matrix notation,

$$\vec{y} = \vec{x}_0 + \vec{x}A, \quad \vec{x}_0 = (b_1, b_2) \quad \text{and} \quad A = \begin{bmatrix} a_{11} & a_{12} \\ a_{21} & a_{22} \end{bmatrix} \qquad (2.8.19)$$

with $\det A \neq 0$, where $\vec{x} = (x_1, x_2)$ and $\vec{y} = (y_1, y_2)$ and $\vec{y} = T(\vec{x})$. The latter is the one we obtained in (2.8.9) or (2.8.10), where $\mathcal{B} = \mathcal{C}$ could be any affine basis for $\mathbb{R}^2$, via the geometric method stated in (2.8.4).

Remark In case

$$\Delta = \det A = 0, \tag{2.8.20}$$

the associated transformation is called a *singular* affine transformation, otherwise *nonsingular*. The affine transformations throughout the text will always mean nonsingular ones unless specified otherwise.

The following are important special cases of affine transformations.

Case 1 Translation

Fix a plane vector $\vec{x}_0$ (it could be the zero vector $\vec{0}$) and move all the point $\vec{x}$ in $\mathbb{R}^2$ along the direction $\vec{x}_0$. The resulting motion

$$T(\vec{x}) = \vec{x}_0 + \vec{x}, \tag{2.8.21}$$

with $A = I_2$, is called a *translation* of $\mathbb{R}^2$ along $\vec{x}_0$. See Fig. 2.96.

A translation preserves all the geometric mapping properties stated in (1) of (2.7.9). Every line parallel to $\vec{x}_0$ is an invariant line.

The set of all translations constitutes a subgroup of $G_a(2; \mathbb{R})$. Refer to Ex. <A> 7 of Sec. 2.8.1.

Case 2 Reflection

Take two intersecting but non-coincident straight lines OA_1 and OA_2 in $\mathbb{R}^2$, with O as point of intersection. For any point $X \in \mathbb{R}^2$, draw a line through X, parallel to OA_2, such that the line intersects OA_1 at the point P. Extend XP to X' such that $XP = PX'$. See Fig. 2.99(a). Then, the mapping $T: \mathbb{R}^2 \to \mathbb{R}^2$ defined by

$$T(X) = X'$$

is an affine transformation which can be easily verified by using (1) in (2.8.4) and is called the (*skew*) *reflection along the direction* $\overrightarrow{OA_2}$ of the plane $\mathbb{R}^2$ *with respect to the axis* OA_1. *This reflection keeps each point on OA_1 fixed and the axis OA_1 is then a line of invariant points. While, any line parallel to the direction* $\overrightarrow{OA_2}$ *is an invariant line.*

In order to linearize T, let the points O, A_1 and A_2 have respective coordinate $\vec{a}_0, \vec{a}_1$ and $\vec{a}_2$ in the Cartesian coordinate system $\mathcal{N} = \{\vec{e}_1, \vec{e}_2\}$. Then $\mathcal{B} = \{\vec{a}_0, \vec{a}_1, \vec{a}_2\}$ is an affine basis for $\mathbb{R}^2$. In terms of $\mathcal{B}$,

$$[T(\vec{x})]_{\mathcal{B}} = [\vec{x}]_{\mathcal{B}}[T]_{\mathcal{B}}, \quad [T]_{\mathcal{B}} = \begin{bmatrix} 1 & 0 \\ 0 & -1 \end{bmatrix}, \quad \vec{x} \in \mathbb{R}^2. \tag{2.8.22}$$

Notice that $[T(\vec{a}_0)]_{\mathcal{B}} = [\vec{a}_0]_{\mathcal{B}} = \vec{0}$. What is the equation of T in $\mathcal{N} = \{\vec{0}, \vec{e}_1, \vec{e}_2\}$? There are two ways to obtain this equation.

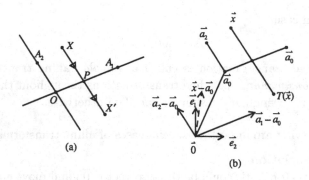

Fig. 2.99

One is to adopt the change of coordinates in (2.4.2). Now

$$[\vec{x}]_B = [\vec{0}]_B + [\vec{x}]_N A_B^N$$
$$[T(\vec{x})]_B = [\vec{0}]_B + [T(\vec{x})]_N A_B^N$$
$$\Rightarrow T(\vec{x}) = [T(\vec{x})]_N = (-[\vec{0}]_B + [T(\vec{x})]_B) A_N^B$$
$$= -[\vec{0}]_B A_N^B + \{[\vec{0}]_B + [\vec{x}]_N A_B^N\}[T]_B A_N^B$$
$$= -[\vec{0}]_B A_N^B + [\vec{0}]_B [T]_B A_N^B + [\vec{x}]_N A_B^N [T]_B A_N^B.$$

But $[\vec{a}_0]_B = [\vec{0}]_B + [\vec{a}_0]_N A_B^N = \vec{0}$ implies that

$$-[\vec{0}]_B A_N^B = [\vec{a}_0]_N A_B^N A_N^B = [\vec{a}_0]_N = \vec{a}_0.$$

Therefore,

$$T(\vec{x}) = \vec{a}_0 - [\vec{a}_0]_N A_B^N [T]_B A_N^B + [\vec{x}]_N A_B^N [T]_B A_N^B$$
$$= \vec{a}_0 + (\vec{x} - \vec{a}_0) A_B^N [T]_B A_N^B \tag{2.8.23}$$

is the required equation of T in terms of N.

The other way is to displace the vectors $\vec{a}_1 - \vec{a}_0$, $\vec{a}_2 - \vec{a}_0$ and $\vec{x} - \vec{a}_0$ to the origin $\vec{0}$. See Fig. 2.99(b). Then, observe each of the following steps:

$$\vec{x} - \vec{a}_0 \quad \text{(a vector in } N = \{\vec{e}_1, \vec{e}_2\})$$
$$\Rightarrow (\vec{x} - \vec{a}_0) A_B^N = [\vec{x}]_B \quad \text{(the coordinate of } \vec{x} - \vec{a}_0 \text{ in}$$
$$B = \{\vec{a}_1 - \vec{a}_0, \vec{a}_2 - \vec{a}_0\})$$
$$\Rightarrow (\vec{x} - \vec{a}_0) A_B^N [T]_B \quad \text{(the reflection in } B = \{\vec{a}_1 - \vec{a}_0, \vec{a}_2 - \vec{a}_0\})$$
$$\Rightarrow (\vec{x} - \vec{a}_0) A_B^N [T]_B A_N^B \quad \text{(the coordinate in } N = \{\vec{e}_1, \vec{e}_2\})$$
$$\Rightarrow \vec{a}_0 + (\vec{x} - \vec{a}_0) A_B^N [T]_B A_N^B \quad \text{(the reflection in } affine \text{ basis}$$
$$B = \{\vec{a}_0, \vec{a}_1, \vec{a}_2\}) \tag{2.8.24}$$

and this is the required $T(\vec{x})$.

We summarize the above result in

The reflection

Let $\vec{a}_0, \vec{a}_1$ and $\vec{a}_2$ be three distinct non-collinear points in $\mathbb{R}^2$. Then the *(skew) reflection* T along the *direction* $\vec{a}_2 - \vec{a}_0$ of $\mathbb{R}^2$ with respect to the *axis* $\vec{a}_0 + \langle\langle \vec{a}_1 - \vec{a}_0 \rangle\rangle$ has the following representations:

1. In the affine basis $\mathcal{B} = \{\vec{a}_0, \vec{a}_1, \vec{a}_2\}$,

$$[T(\vec{x})]_\mathcal{B} = [\vec{x}]_\mathcal{B}[T]_\mathcal{B}, \quad \text{where } [T]_\mathcal{B} = \begin{bmatrix} 1 & 0 \\ 0 & -1 \end{bmatrix}.$$

2. In the Cartesian coordinate system $\mathcal{N} = \{\vec{0}, \vec{e}_1, \vec{e}_2\}$,

$$T(\vec{x}) = \vec{a}_0 + (\vec{x} - \vec{a}_0)A^{-1}[T]_\mathcal{B}A, \quad \text{where } A = \begin{bmatrix} \vec{a}_1 - \vec{a}_0 \\ \vec{a}_2 - \vec{a}_0 \end{bmatrix}_{2 \times 2}$$

In case the direction $\vec{a}_2 - \vec{a}_0$ is perpendicular to the axis $\vec{a}_0 + \langle\langle \vec{a}_1 - \vec{a}_0 \rangle\rangle$, T is called an *orthogonal reflection* or simply *a symmetric motion* with respect to the *axis* $\vec{a}_0 + \langle\langle \vec{a}_1 - \vec{a}_0 \rangle\rangle$.

(1) A skew reflection preserves all the properties listed in (1) of (2.7.9) and
    (c) area.
(2) An orthogonal reflection, in addition to (1), also preserves

    (a) length,
    (b) angle,

but reverses the direction. $\hfill (2.8.25)$

Notice that the affine transformation T mentioned in (2.8.25) can be reduced to the form $(A^{-1}[T]_\mathcal{B}A$ there is now denoted as A below)

$$T(\vec{x}) = \vec{x}_0 + \vec{x}A, \quad A = \begin{bmatrix} a_{11} & a_{12} \\ a_{21} & a_{22} \end{bmatrix}, \qquad (2.8.26)$$

where $\det A \neq 0$. So we pose a *converse problem*: For a given affine transformation, how can we determine if it is a reflection and, if certainty, how to determine its direction and line of invariant points? The following are the steps:

1. Compute the eigenvalues of A. If A has eigenvalues 1 and -1, then T represents a reflection in case $\vec{x}(I_2 - A) = \vec{x}_0$ has a solution.
2. Compute an eigenvector $\vec{v}_1$ corresponding to 1, then $\vec{x}(I_2 - A) = \vec{x}_0$ or

$$\frac{1}{2}(\vec{0} + T(\vec{0})) + \langle\langle \vec{v}_1 \rangle\rangle = \frac{1}{2}\vec{x}_0 + \langle\langle \vec{v}_1 \rangle\rangle$$

is the line of invariant points of the reflection T.

3. Compute an eigenvector $\vec{v}_2$ corresponding to -1, then this $\vec{v}_2$ or $-\vec{v}_2$ is the direction of the reflection T. In fact, $\vec{x}_0$ is a required direction (why?) if $\vec{x}_0 \neq \vec{0}$.   (2.8.27)

Example 1 Let $\vec{a}_0 = (2,2), \vec{a}_1 = (4,1)$ and $\vec{a}_2 = (1,3)$. Determine the reflection along the direction $\vec{a}_2 - \vec{a}_0 = (-1,1)$ with $\vec{a}_0 + \langle\langle \vec{a}_1 - \vec{a}_0 \rangle\rangle = \vec{a}_0 + \langle\langle(2,-1)\rangle\rangle$ as the line of invariant points.

Solution In the affine basis $\mathcal{B} = \{\vec{a}_0, \vec{a}_1, \vec{a}_2\}$, the affine transformation T has the representation

$$[T(\vec{x})]_{\mathcal{B}} = [\vec{x}]_{\mathcal{B}} \begin{bmatrix} 1 & 0 \\ 0 & -1 \end{bmatrix}.$$

while, in $\mathcal{N} = \{\vec{0}, \vec{e}_1, \vec{e}_2\}$,

$$T(\vec{x}) = (2,2) + [\vec{x} - (2,2)] \begin{bmatrix} 2 & -1 \\ -1 & 1 \end{bmatrix}^{-1} \begin{bmatrix} 1 & 0 \\ 0 & -1 \end{bmatrix} \begin{bmatrix} 2 & -1 \\ -1 & 1 \end{bmatrix}$$

$$= (2,2) + [\vec{x} - (2,2)] \begin{bmatrix} 3 & -2 \\ 4 & -3 \end{bmatrix}$$

which can be simplified as

$$T(\vec{x}) = \vec{x}_0 + \vec{x}A, \quad \text{where } \vec{x}_0 = (-12,12) \text{ and } A = \begin{bmatrix} 3 & -2 \\ 4 & -3 \end{bmatrix}.$$

Now, suppose, on the contrary, that an affine transformation on $\mathbb{R}^2$ is given as above. We try to determine if it is a reflection.

Follow the steps in (2.8.27).

By computing

$$\det(A - tI_2) = \begin{vmatrix} 3-t & -2 \\ 4 & -3-t \end{vmatrix} = t^2 - 9 + 8 = t^2 - 1,$$

A has eigenvalues 1 and -1. Solve

$$(x_1 \quad x_2) \begin{bmatrix} 3-1 & -2 \\ 4 & -3-1 \end{bmatrix} = (0 \quad 0)$$

and get a corresponding eigenvector $\vec{v}_1 = (2,-1)$. Then the line

$$\frac{1}{2}\vec{x}_0 + \langle\langle\vec{v}_1\rangle\rangle = \frac{1}{2}(-12,12) + \langle\langle(2,-1)\rangle\rangle = (-6,6) + \langle\langle(2,-1)\rangle\rangle$$

is the line of invariant points. Note that this line is coincident with the line with equation $\vec{a}_0 + \langle\langle(2,-1)\rangle\rangle$ which is $x_1 + 2x_2 = 6$ in $\mathcal{N} = \{\vec{0}, \vec{e}_1, \vec{e}_2\}$.

On the other hand, solve

$$(x_1 \quad x_2) \begin{bmatrix} 3+1 & -2 \\ 4 & -3+1 \end{bmatrix} = (0 \quad 0)$$

and get a corresponding eigenvector $\vec{v}_2 = (-1, 1)$. Note that $\vec{x}_0 = (-12, 12)$ is an eigenvector too. Either $\vec{v}_2$ or $\vec{x}_0$, or any other corresponding eigenvector can be treated as the direction of the reflection. Readers are urged to draw a graph to explain everything above.

Case 3 Simple elongation and compression or strain or stretching
Let OA_1 and OA_2 be two non-coincident straight lines which interest at the point O. Take any fixed scalar $k \in \mathbb{R}$ (k could be positive or negative). For any point X on $\mathbb{R}^2$, draw a line through X and parallel to OA_2 so that the line interests the line OA_1 at the point P. Pick up the unique point X' on the line XP such that $X'P = kXP$. See Fig. 2.100. The mapping $T: \mathbb{R}^2 \rightarrow \mathbb{R}^2$ defined by

$$T(X) = X'$$

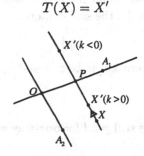

Fig. 2.100

is an affine transformation and is called a *simple elongation* and *compression* or *strain* or *one-way stretch* in the *direction* OA_2 of $\mathbb{R}^2$ with *axis* the line OA_1, and k is called the corresponding *scale factor*. A one-way stretching has the following features:

1. there is a line of invariant points (i.e. the axis),
2. all other points move on lines parallel to the direction and there is a stretch in this direction with scale factor k, and
3. all the lines parallel to the direction are invariant lines.

Notice the case $k = -1$ is the reflection mentioned in Case 2.

The processes to obtain (2.8.23), (2.8.24) and Fig. 2.99(b) are universal and thus are still suitable for stretching. Hence, we have results corresponding to (2.8.25) for

The one-way stretching

Let $\vec{a}_0$, $\vec{a}_1$ and $\vec{a}_2$ be three distinct non-collinear points in $\mathbb{R}^2$. The *one-way (skew) stretching* T in the *direction* $\vec{a}_2 - \vec{a}_0$ with $\vec{a}_0 + \langle\langle \vec{a}_1 - \vec{a}_0 \rangle\rangle$ as *axis* and *scale factor* $k \neq 0$ has the following representations:

1. In the affine basis $\mathcal{B} = \{\vec{a}_0, \vec{a}_1, \vec{a}_2\}$,

$$[T(\vec{x})]_\mathcal{B} = [\vec{x}]_\mathcal{B}[T]_\mathcal{B}, \quad \text{where } [T]_\mathcal{B} = \begin{bmatrix} 1 & 0 \\ 0 & k \end{bmatrix}.$$

2. In the Cartesian coordinate system $\mathcal{N} = \{\vec{0}, \vec{e}_1, \vec{e}_2\}$,

$$T(\vec{x}) = \vec{a}_0 + (\vec{x} - \vec{a}_0)A^{-1}[T]_\mathcal{B}A, \quad \text{where } A = \begin{bmatrix} \vec{a}_1 - \vec{a}_0 \\ \vec{a}_2 - \vec{a}_0 \end{bmatrix}_{2 \times 2}.$$

In case $\vec{a}_2 - \vec{a}_0$ is perpendicular to $\vec{a}_1 - \vec{a}_0$, T is called an *orthogonal one-way stretch* with *axis* $\vec{a}_0 + \langle\langle \vec{a}_1 - \vec{a}_0 \rangle\rangle$.

(1) A one-way stretch preserves all the properties listed in (1) of (2.7.9),
(2) but it enlarges the area by the scalar factor

$$|k|$$

and preserves the orientation if $k < 0$ and reverses the orientation if $k < 0$.   (2.8.28)

(2.8.26) and (2.8.27) are still good for one-way stretch, of course, subject to some minor charges.

To test the affine transformation (2.8.26) to see if it is a one-way stretch, follow the following steps:

1. Compute the eigenvalues of A. If A has eigenvalues 1 and $k \neq 1$, then T represents a one-way stretch if $\vec{x}(I_2 - A) = \vec{x}_0$ has a solution.
2. Compute an eigenvector $\vec{v}_1$ corresponding to 1, then $\vec{x}(I_2 - A) = \vec{x}_0$ or

$$\frac{1}{1-k}\vec{x}_0 + \langle\langle \vec{v}_1 \rangle\rangle$$

is the line of invariant points (i.e. the axis).
3. Compute an eigenvector $\vec{v}_2$ corresponding to k, and then $\vec{v}_2$ is the direction of the stretch. In particular,

$$\frac{1}{1-k}\vec{x}_0 \quad \text{or} \quad \vec{x}_0$$

is a direction (up to a nonzero scalar) if $\vec{x}_0 \neq \vec{0}$.   (2.8.29)

Example 2 Let $\vec{a}_0 = (-2, 1)$, $\vec{a}_1 = (1, -2)$ and $\vec{a}_2 = (3, 2)$. Determine the one-way stretch in the direction $\vec{a}_2 - \vec{a}_1 = (5, 1)$ with $\vec{a}_0 + \langle\langle \vec{a}_1 - \vec{a}_0 \rangle\rangle = \vec{a}_0 + \langle\langle (3, -3) \rangle\rangle$ as axis and respective scale factors $k = 2$ and $k = -3$.

Solution In the affine basis $\mathcal{B} = \{\vec{a}_0, \vec{a}_1, \vec{a}_2\}$, the one-way stretch T is

$$[T(\vec{x})]_{\mathcal{B}} = [\vec{x}]_{\mathcal{B}} \begin{bmatrix} 1 & 0 \\ 0 & k \end{bmatrix},$$

where $k = 2$ or -3, while in $\mathcal{N} = \{\vec{0}, \vec{e}_1, \vec{e}_2\}$,

$$T(\vec{x}) = (-2, 1) + [\vec{x} - (-2, 1)] \begin{bmatrix} 3 & -3 \\ 5 & 1 \end{bmatrix}^{-1} \begin{bmatrix} 1 & 0 \\ 0 & k \end{bmatrix} \begin{bmatrix} 3 & -3 \\ 5 & 1 \end{bmatrix}$$

$$= (-2, 1) + [\vec{x} - (-2, 1)] \cdot \frac{1}{18} \begin{bmatrix} 3 + 15k & -3 + 3k \\ -15 + 15k & 15 + 3k \end{bmatrix},$$

which can be simplified as

$$T(\vec{x}) = \vec{x}_0 + \vec{x} A,$$

where

$$\vec{x}_0 = (-2\ 1) - (-2\ 1) \cdot \frac{1}{6} \begin{bmatrix} 1 + 5k & -1 + k \\ -5 + 5k & 5 + k \end{bmatrix}$$

$$= \begin{cases} \frac{1}{6}(5, 1), & \text{if } k = 2, \\ -\frac{2}{3}(5, 1), & \text{if } k = -3; \end{cases}$$

$$A = \frac{1}{6} \begin{bmatrix} 11 & 1 \\ 5 & 7 \end{bmatrix} \text{ if } k = 2; \quad \frac{1}{3} \begin{bmatrix} -7 & -2 \\ -10 & 1 \end{bmatrix} \text{ if } k = -3.$$

Now, we consider the converse problems separately.

Suppose we have an affine transformation

$$T(\vec{x}) = \frac{1}{6}(5, 1) + \vec{x} A \quad \text{where } A = \frac{1}{6} \begin{bmatrix} 11 & 1 \\ 5 & 7 \end{bmatrix}$$

and we want to decide if this T is a one-way stretch. If yes, where is its axis and in what direction. Follow the steps suggested in (2.8.29) and proceed as follows.

1. A has characteristic polynomial

$$\det(A - tI_2) = \begin{vmatrix} \frac{11}{6} - t & \frac{1}{6} \\ \frac{5}{6} & \frac{7}{6} - t \end{vmatrix} = t^2 - 3t + 2 = (t - 1)(t - 2).$$

Hence A has two distinct eigenvalues 1 and 2 and it is a one-way stretch with scale factor $k = 2$.

2. Solve

$$\vec{x}(A - I_2) = (x_1 \quad x_2)\begin{bmatrix} \frac{5}{6} & \frac{1}{6} \\ \frac{5}{6} & \frac{1}{6} \end{bmatrix} = (0 \quad 0),$$

and get the eigenvectors $\vec{v} = t(1, -1)$ for $t \in \mathbb{R}$ and $t \neq 0$. Since $T(\vec{0}) = \frac{1}{6}(5, 1)$, the line

$$-\frac{1}{6}(5, 1) + \langle\langle(1, -1)\rangle\rangle = (-2, 1) + \langle\langle(3, -3)\rangle\rangle$$

is the axis as expected.

3. Solve

$$\vec{x}(A - 2I_2) = (x_1 \quad x_2)\begin{bmatrix} -\frac{1}{6} & \frac{1}{6} \\ \frac{5}{6} & -\frac{5}{6} \end{bmatrix} = (0 \quad 0)$$

and get the eigenvectors $\vec{u} = t(5, 1)$ for $t \in \mathbb{R}$ and $t \neq 0$. Then, any such $\vec{u}$ and, in particular, $\vec{x}_0 = \frac{1}{6}(5, 1)$ is the direction of T.

Next, we pose the following questions:

Q1. *What is the image of the line connecting* $(0, 3)$ *and* $(1, 0)$ *under* T? *Where do these two lines intersect?*

Q2. *What is the image of the triangle* $\triangle\vec{b}_1\vec{b}_2\vec{b}_3$, *where* $\vec{b}_1 = (0, 3)$, $\vec{b}_2 = (-4, 0)$ *and* $\vec{b}_3 = (1, -1)$, *under* T? *What are the areas of these two triangles?*

For Q1, compute the image points of $(0, 3)$ and $(1, 0)$ as

$$T(0, 3) = \frac{1}{6}(5, 1) + (0, 3) \cdot \frac{1}{6}\begin{bmatrix} 11 & 1 \\ 5 & 7 \end{bmatrix} = \frac{1}{3}(10, 11),$$

$$T(1, 0) = \frac{1}{6}(5, 1) + (1, 0) \cdot \frac{1}{6}\begin{bmatrix} 11 & 1 \\ 5 & 7 \end{bmatrix} = \frac{1}{3}(8, 1).$$

Thus, the image line of the original line $3x_1 + x_2 = 3$ has the equation, in $\mathcal{N}$,

$$5x_1 - x_2 - 13 = 0.$$

These two lines intersect at the point $(2, -3)$ which *lies on* the axis $(-2, 1) + \langle\langle(3, -3)\rangle\rangle$. Is this fact accidental or universally true for any line and its image line under T?

For Q2, compute

$$T(\vec{b}_2) = T(-4,0) = \frac{1}{6}(5,1) + (-4,0) \cdot \frac{1}{6}\begin{bmatrix} 11 & 1 \\ 5 & 7 \end{bmatrix} = \frac{1}{2}(-13,-1) = \vec{b}_2',$$

$$T(\vec{b}_3) = T(-1,1) = \frac{1}{6}(5,1) + (1,-1) \cdot \frac{1}{6}\begin{bmatrix} 11 & 1 \\ 5 & 7 \end{bmatrix} = \frac{1}{6}(11,-5) = \vec{b}_3'.$$

Let $\vec{b}_1' = T(\vec{b}_1) = \frac{1}{3}(10,11)$. Then the image of $\Delta\,\vec{b}_1\,\vec{b}_2\,\vec{b}_3$ is the triangle $\Delta\vec{b}_1'\vec{b}_2'\vec{b}_3'$. Note that both $\Delta\,\vec{b}_1\,\vec{b}_2\,\vec{b}_3$ and $\Delta\vec{b}_1'\vec{b}_2'\vec{b}_3'$ have the same orientation. As far as triangle areas are concerned, we have the following results (refer to Sec. 4.3, Ex. <A> of Sec. 2.6, or (2.8.44), if necessary):

$$\Delta\,\vec{b}_1\,\vec{b}_2\,\vec{b}_3 \text{ has area} = \frac{1}{2}\det\begin{bmatrix} \vec{b}_2 - \vec{b}_1 \\ \vec{b}_3 - \vec{b}_1 \end{bmatrix} = \frac{1}{2}\begin{vmatrix} -4 & -3 \\ 1 & -4 \end{vmatrix} = \frac{19}{2},$$

$$\Delta\vec{b}_1'\vec{b}_2'\vec{b}_3' \text{ has area} = \frac{1}{2}\det\begin{bmatrix} \vec{b}_2' - \vec{b}_1' \\ \vec{b}_3' - \vec{b}_1' \end{bmatrix} = \frac{1}{2}\cdot\frac{1}{36}\begin{vmatrix} -59 & -25 \\ -9 & -27 \end{vmatrix} = \frac{1368}{72} = 19.$$

$$\Rightarrow \frac{\text{The area of } \Delta\vec{b}_1'\vec{b}_2'\vec{b}_3'}{\text{The area of } \Delta\,\vec{b}_1\,\vec{b}_2\,\vec{b}_3} = 2 = \det A.$$

See Fig. 2.101.

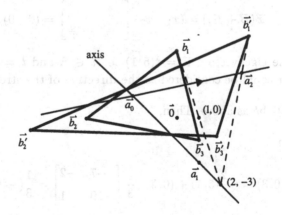

Fig. 2.101

On the other hand, suppose the affine transformation, for $k = -3$, is

$$T(\vec{x}) = -\frac{2}{3}(5,1) + \vec{x}B, \quad \text{where } B = \frac{1}{3}\begin{bmatrix} -7 & -2 \\ -10 & 1 \end{bmatrix}.$$

We want to consider the same problems as in the case $k = 2$.

1. The characteristic polynomial

$$\det(B - tI_2) = \begin{vmatrix} -\frac{7}{3} - t & -\frac{2}{3} \\ -\frac{10}{3} & \frac{1}{3} - t \end{vmatrix} = t^2 + 2t - 3 = (t + 3)(t - 1).$$

Hence, B has eigenvalues 1 and -3 and the corresponding T is a one-way stretching with scale factor -3.

2. Solve

$$\vec{x}(B - I_2) = (x_1 \quad x_2) \begin{bmatrix} -\frac{10}{3} & -\frac{2}{3} \\ -\frac{10}{3} & -\frac{2}{3} \end{bmatrix} = (0 \quad 0)$$

and get the eigenvectors $\vec{v} = t(1, -1)$ for $t \in \mathbb{R}$ and $t \neq 0$. The axis of stretching is

$$\frac{1}{1 - (-3)} \cdot \left(-\frac{2}{3}\right)(5, 1) + \langle\langle(1, -1)\rangle\rangle = -\frac{1}{6}(5, 1) + \langle\langle(1, -1)\rangle\rangle$$

$$= (-2, 1) + \langle\langle(3, -3)\rangle\rangle.$$

3. Solve

$$\vec{x}(B + 3I_2) = (x_1 \quad x_2) \begin{bmatrix} \frac{2}{3} & -\frac{2}{3} \\ -\frac{10}{3} & \frac{10}{3} \end{bmatrix} = (0 \quad 0)$$

and get the eigenvectors $\vec{u} = t(5, 1)$ for $t \in \mathbb{R}$ and $t = 0$. Any such eigenvector could be considered as the direction of the stretching.

Let Q1 and Q2 be as above. Then,

Q1. Compute

$$T(0, 3) = -\frac{2}{3}(5, 1) + (0, 3) \cdot \frac{1}{3} \begin{bmatrix} -7 & -2 \\ -10 & 1 \end{bmatrix} = \frac{1}{3}(-40, 1),$$

$$T(0, 1) = -\frac{2}{3}(5, 1) + (1, 0) \cdot \frac{1}{3} \begin{bmatrix} -7 & -2 \\ -10 & 1 \end{bmatrix} = \frac{1}{3}(-17, -4).$$

Hence, the image line of the original line $3x_1 + x_2 = 3$ is, in $\mathcal{N}$, $15x_1 + 69x_2 + 177 = 0$. They intersect at the point $(2, -3)$.

Q2. It is known that $T(\vec{b}_1) = \vec{b'}_1 = \frac{1}{3}(-40, 1)$. Compute

$$T(\vec{b}_2) = -\frac{2}{3}(5, 1) + (-4, 0) \cdot \frac{1}{3} \begin{bmatrix} -7 & -2 \\ -10 & 1 \end{bmatrix} = (6, 2) = \vec{b'}_2,$$

$$T(\vec{b}_3) = -\frac{2}{3}(5, 1) + (1, -1) \cdot \frac{1}{3} \begin{bmatrix} -7 & -2 \\ -10 & 1 \end{bmatrix} = \frac{1}{3}(-7, -5) = \vec{b'}_3.$$

Then $\Delta \vec{b'}_1 \vec{b'}_2 \vec{b'}_3$ has the *signed* area

$$\frac{1}{2} \det \begin{bmatrix} \vec{b'}_2 - \vec{b'}_1 \\ \vec{b'}_3 - \vec{b'}_1 \end{bmatrix} = \frac{1}{2} \cdot \frac{1}{9} \begin{vmatrix} 58 & 5 \\ 33 & -6 \end{vmatrix} = -\frac{57}{2}$$

$$\Rightarrow \frac{\text{the signed area of } \Delta \vec{b'}_1 \vec{b'}_2 \vec{b'}_3}{\text{the area of } \Delta \vec{b}_1 \vec{b}_2 \vec{b}_3} = -3 = \det B.$$

Notice that T reverses the orientation of $\Delta \vec{b}_1 \vec{b}_2 \vec{b}_3$. Hope that the readers will be able to give a graphical illustration just like Fig. 2.101.

Case 4 Two-way stretch

This is a combination of two one-way stretches whose lines of invariant points intersect at one point, the only invariant point. There are no invariant lines at all if the scale factors are distinct.

As an easy consequence of (2.8.28), we have

The two-way stretch

Let $\vec{a}_0$, $\vec{a}_1$ and $\vec{a}_2$ be three distinct non-collinear points in $\mathbb{R}^2$. The *two-way stretch* T, with $\vec{a}_0$ the only invariant point, which has scale factor k_1 along $\vec{a}_1 - \vec{a}_0$ and scale k_2 along $\vec{a}_2 - \vec{a}_0$, has the following representations:

1. In the affine basis $\mathcal{B} = \{\vec{a}_0, \vec{a}_1, \vec{a}_2\}$,

$$[T(\vec{x})]_{\mathcal{B}} = [\vec{x}]_{\mathcal{B}}[T]_{\mathcal{B}}, \quad \text{where } [T]_{\mathcal{B}} = \begin{bmatrix} k_1 & 0 \\ 0 & k_2 \end{bmatrix} = \begin{bmatrix} k_1 & 0 \\ 0 & 1 \end{bmatrix} \begin{bmatrix} 1 & 0 \\ 0 & k_2 \end{bmatrix},$$

where $k_1 \neq k_2$ and $k_1 k_2 \neq 0$.

2. In $\mathcal{N} = \{\vec{0}, \vec{e}_1, \vec{e}_2\}$,

$$T(\vec{x}) = \vec{a}_0 + (\vec{x} - \vec{a}_0)A^{-1}[T]_{\mathcal{B}}A, \quad \text{where } A = \begin{bmatrix} \vec{a}_1 - \vec{a}_0 \\ \vec{a}_2 - \vec{a}_0 \end{bmatrix}_{2 \times 2}.$$

In case $\vec{a}_1 - \vec{a}_0$ is perpendicular to $\vec{a}_2 - \vec{a}_0$, T is called an *orthogonal two-way stretch*.

(1) A two-way stretch preserves all the properties listed in (1) of (2.7.9),

(2) but it enlarges the area by the scale factor

$$|k_1 k_2|$$

and preserves the orientation if $k_1 k_2 > 0$ and reverses it if $k_1 k_2 < 0$.

(2.8.30)

If $k_1 = k_2 = k \neq 0, 1$, then

$$T(\vec{x}) = \vec{a}_0 + k(\vec{x} - \vec{a}_0) \qquad (2.8.31)$$

is called an *enlargement* with scale factor k (refer to Ex. <A> 6 of Sec. 2.8.1). See Fig. 2.102.

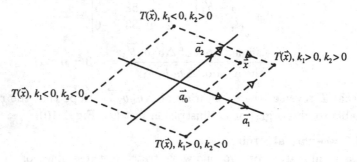

Fig. 2.102

(2.8.29) has a counterpart for a two-way stretch. The details are left to the readers. We go directly to an example.

Example 3 Let $\vec{a}_0 = (1,1), \vec{a}_1 = (2,2)$ and $\vec{a}_2 = (3,0)$. Determine the two-way stretch T, with $\vec{a}_0$ as the only invariant point, which has scale 3 along $\vec{a}_1 - \vec{a}_0$ and scale -2 along $\vec{a}_2 - \vec{a}_0$.

Solution In the basis $\mathcal{B} = \{\vec{a}_0, \vec{a}_1, \vec{a}_2\}$,

$$[T(\vec{x})]_\mathcal{B} = [\vec{x}]_\mathcal{B}[T]_\mathcal{B} \quad \text{where } [T]_\mathcal{B} = \begin{bmatrix} 3 & 0 \\ 0 & -2 \end{bmatrix}.$$

In $\mathcal{N} = \{\vec{0}, \vec{e}_1, \vec{e}_2\}$,

$$T(\vec{x}) = (1,1) + [\vec{x} - (1,1)] \begin{bmatrix} 1 & 1 \\ 2 & -1 \end{bmatrix}^{-1} \begin{bmatrix} 3 & 0 \\ 0 & -2 \end{bmatrix} \begin{bmatrix} 1 & 1 \\ 2 & -1 \end{bmatrix}$$

$$= (1,1) + [\vec{x} - (1,1)] \cdot \frac{-1}{3} \begin{bmatrix} 1 & -5 \\ -10 & -4 \end{bmatrix}$$

$$= (-2,-2) + \vec{x}A, \quad \text{where } A = -\frac{1}{3} \begin{bmatrix} 1 & -5 \\ -10 & -4 \end{bmatrix}.$$

Conversely, if T is given as above, how can one determine if it is a stretch? And, if yes, where are the directions of stretching, and what are the scale factors?

Watch the following steps.

1. The characteristic polynomial is

$$\det(A - tI_2) = \begin{vmatrix} -\frac{1}{3} - t & \frac{5}{3} \\ \frac{10}{3} & \frac{4}{3} - t \end{vmatrix} = t^2 - t - 6 = (t - 3)(t + 2).$$

Hence, A has eigenvalues 3 and -2, and the associated T is a two-way stretch.

2. Solve

$$\vec{x}(A - 3I_2) = (x_1 \quad x_2) \begin{bmatrix} -\frac{10}{3} & \frac{5}{3} \\ \frac{10}{3} & -\frac{5}{3} \end{bmatrix} = (0 \quad 0)$$

and get the eigenvectors $\vec{v}_1 = t(1, 1)\, t \in \mathbb{R}$ and $t \neq 0$. Any such $\vec{v}_1$ is the direction of one-way stretch with scale factor 3.

3. Solve

$$\vec{x}(A + 2I_2) = (\vec{x}_1 \quad \vec{x}_2) \begin{bmatrix} \frac{5}{3} & \frac{5}{3} \\ \frac{10}{3} & \frac{10}{3} \end{bmatrix} = (0 \quad 0)$$

and get the eigenvectors $\vec{v}_2 = t(2, -1)$ for $t \in \mathbb{R}$ and $t \neq 0$. Then, any such vector $\vec{v}_2$ is another direction of stretch with scale factor -2.

4. To find the only invariant point, solve

$$\vec{x} = (-2, -2) + \vec{x}A$$
$$\Rightarrow \vec{x}(A - I_2) = (2, 2)$$
$$\Rightarrow \vec{x} = (2 \quad 2)(A - I_2)^{-1} = (2 \quad 2) \cdot \left(-\frac{1}{6}\right) \begin{bmatrix} \frac{1}{3} & -\frac{5}{3} \\ -\frac{16}{3} & -\frac{4}{3} \end{bmatrix} = (1 \quad 1)$$

and the point $(1, 1)$ is the required one.

The square with vertices $\vec{a}_0 = (1, 1), (-1, 1), (-1, -1),$ and $(1, -1),$ in counterclockwise ordering, is mapped onto the parallelogram with vertices $\vec{a}_0 = (1, 1), T(-1, 1) = \frac{1}{3}(5, -7), T(-1, -1) = (-5, -5)$ and $T(1, -1) = -\frac{1}{3}(17, 5)$ in clockwise ordering. See Fig. 2.103.

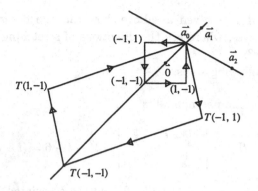

Fig. 2.103

Notice that the image parallelogram has

$$\text{the signed area} = \det \begin{bmatrix} T(-1,1) - \vec{a}_0 \\ T(1,-1) - \vec{a}_0 \end{bmatrix} = \begin{vmatrix} \frac{2}{3} & -\frac{10}{3} \\ -\frac{20}{3} & -\frac{8}{3} \end{vmatrix}$$

$$= -\frac{216}{9} = -24$$

$$\Rightarrow \frac{\text{the signed area of the parallelogram}}{\text{the area of the square}} = \frac{-24}{4} = -6 = \det A.$$

Case 5 Shearing (Euclidean notions are needed)
Fix a straight line OA on the plane $\mathbb{R}^2$ and pick up a line segment PQ on it. With PQ as one side, construct a rectangle $PQRS$ and then push the rectangle $PQRS$ in the direction parallel to OA, but keep points on OA fixed and the height of each point to OA invariant, to construct a new rectangle $PQR'S'$. See Fig. 2.104 (a). Of course, these two rectangles have the same area.

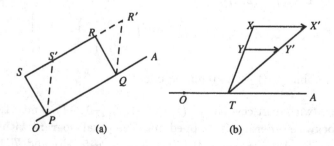

(a)   (b)

Fig. 2.104

What we care is that, how to determine the image Y' of a given point Y in the rectangle $PQRS$ under this mapping. Suppose Y does not lie on the line OA. Draw any line through Y which intersects PQ at T and RS at X. See Fig. 2.104 (b). The image X' of X must satisfy $XX' = SS' = RR'$. Draw a line through Y and parallel to OA so that it intersects TX' at the point Y'. This Y' is then the required one. Now

$$\Delta TXX' \text{ is similar to } \Delta TYY'.$$
$$\Rightarrow \frac{XX'}{TX} = \frac{YY'}{TY}$$
$$\Rightarrow \frac{XX'}{\text{the distance of } X \text{ to } OA} = \frac{YY'}{\text{the distance of } Y \text{ to } OA}.$$

This means that any point Y in the rectangle $PQRS$ moves, parallel to the line OA, a distance with *a fixed direct proportion* to its distance to the line OA.

Suppose every point X in the plane moves in this same manner, i.e.

$$\frac{\text{the distance } XX' \text{ moved from } X \text{ to } X' \text{ parallel to } OA}{\text{the (perpendicular) distance of } X \text{ to the line } OA} = k.$$

Then, the mapping $T\colon \mathbb{R}^2 \to \mathbb{R}^2$ defined as

$$T(X) = X'$$

is an affine transformation and is called a *shearing* with the line OA as *axis*, the only line of invariant points. Any line parallel to OA is invariant.

To linearize T, take two straight lines OA and OB, perpendicular to each other, so that $OA = OB = 1$ in length. Suppose in $\mathcal{N} = \{\vec{0}, \vec{e}_1, \vec{e}_2\}$, $O = \vec{a}_0$, $A = \vec{a}_1$ and $B = \vec{a}_2$. Then, $\mathcal{B} = \{\vec{a}_0, \vec{a}_1, \vec{a}_2\}$ is an orthonormal affine basis for $\mathbb{R}^2$ and for $\vec{x} \in \mathbb{R}^2$,

$$\vec{x} - \vec{a}_0 = \alpha_1(\vec{a}_1 - \vec{a}_0) + \alpha_2(\vec{a}_2 - \vec{a}_0), \quad \text{i.e. } [\vec{x}]_\mathcal{B} = (\alpha_1, \alpha_2)$$
$$\Rightarrow T(\vec{x}) - \vec{a}_0 = (\alpha_1 + k\alpha_2)(\vec{a}_1 - \vec{a}_0) + \alpha_2(\vec{a}_2 - \vec{a}_0)$$
$$\Rightarrow [T(\vec{x})]_\mathcal{B} = (\alpha_1 + k\alpha_2, \alpha_2) = (\alpha_1 \quad \alpha_2) \begin{bmatrix} 1 & 0 \\ k & 1 \end{bmatrix}$$
$$= [\vec{x}]_\mathcal{B} \begin{bmatrix} 1 & 0 \\ k & 1 \end{bmatrix}.$$

The afore-mentioned process for the definition of a shearing indicates that it preserves areas of a parallelogram and hence of a triangle. To see this analytically, let $\vec{x}_1$, $\vec{x}_2$ and $\vec{x}_3$ be three non-collinear points. Then,

$$\vec{x}_i = \vec{a}_0 + \alpha_{i1}(\vec{a}_1 - \vec{a}_0) + \alpha_{i2}(\vec{a}_2 - \vec{a}_0), \quad i = 1, 2, 3.$$

$$\Rightarrow T(\vec{x}_i) = \vec{a}_0 + (\alpha_{i1} + k\alpha_{i2})(\vec{a}_1 - \vec{a}_0) + \alpha_{i2}(\vec{a}_2 - \vec{a}_0), \quad i = 1, 2, 3.$$

$\Rightarrow$ The signed area of $\Delta T(\vec{x}_1)T(\vec{x}_2)T(\vec{x}_3)$

$$= \frac{1}{2}\begin{vmatrix} T(\vec{x}_2) - T(\vec{x}_1) \\ T(\vec{x}_3) - T(\vec{x}_1) \end{vmatrix}$$

$$= \frac{1}{2}\det \begin{bmatrix} [(\alpha_{21} + k\alpha_{22}) - (\alpha_{11} + k\alpha_{12})](\vec{a}_1 - \vec{a}_0) \\ + (\alpha_{22} - \alpha_{12})(\vec{a}_1 - \vec{a}_0) \\ [(\alpha_{31} + k\alpha_{32}) - (\alpha_{11} + k\alpha_{12})](\vec{a}_2 - \vec{a}_0) \\ + (\alpha_{32} - \alpha_{12})(\vec{a}_2 - \vec{a}_0) \end{bmatrix}$$

$$= \frac{1}{2}\det \begin{bmatrix} (\alpha_{21} - \alpha_{11}) + k(\alpha_{22} - \alpha_{12}) & \alpha_{22} - \alpha_{12} \\ (\alpha_{31} - \alpha_{11}) + k(\alpha_{32} - \alpha_{12}) & \alpha_{32} - \alpha_{12} \end{bmatrix} \begin{bmatrix} \vec{a}_1 - \vec{a}_0 \\ \vec{a}_2 - \vec{a}_0 \end{bmatrix}$$

$$= \frac{1}{2}\det \begin{bmatrix} \alpha_{21} - \alpha_{11} & \alpha_{22} - \alpha_{12} \\ \alpha_{31} - \alpha_{11} & \alpha_{32} - \alpha_{12} \end{bmatrix} \begin{bmatrix} \vec{a}_1 - \vec{a}_0 \\ \vec{a}_2 - \vec{a}_0 \end{bmatrix}$$

$$= \frac{1}{2}\begin{vmatrix} \vec{x}_2 - \vec{x}_1 \\ \vec{x}_3 - \vec{x}_1 \end{vmatrix}$$

$= $ The signed area of $\Delta \vec{x}_1 \vec{x}_2 \vec{x}_3$.

Therefore, a shearing preserves areas of plane domains bounded by non-self intersecting closed polygonal curves and hence any measurable plane set by a limit process.

We summarize as (one needs Euclidean concepts.)

The shearing

Let $\vec{a}_0$, $\vec{a}_1$ and $\vec{a}_2$ be three distinct points in $\mathbb{R}^2$ so that $|\vec{a}_1 - \vec{a}_0| = |\vec{a}_2 - \vec{a}_0| = 1$ and $(\vec{a}_1 - \vec{a}_0) \perp (\vec{a}_2 - \vec{a}_0)$, and $k \neq 0$ be a scalar. Then, the *shearing* T with *axis* $\vec{a}_0 + \langle\langle \vec{a}_1 - \vec{a}_0 \rangle\rangle$ and *coefficient* or *scale factor* k has the following representations.

1. In the affine basis $\mathcal{B} = \{\vec{a}_0, \vec{a}_1, \vec{a}_2\}$,

$$[T(\vec{x})]_{\mathcal{B}} = [\vec{x}]_{\mathcal{B}}[T]_{\mathcal{B}}, \quad \text{where } [T]_{\mathcal{B}} = \begin{bmatrix} 1 & 0 \\ k & 1 \end{bmatrix}.$$

2. In $\mathcal{N} = \{\vec{0}, \vec{e}_1, \vec{e}_2\}$,

$$T(\vec{x}) = \vec{a}_0 + (\vec{x} - \vec{a}_0)A^{-1}[T]_{\mathcal{B}}A, \quad \text{where } A = \begin{bmatrix} \vec{a}_1 - \vec{a}_0 \\ \vec{a}_2 - \vec{a}_0 \end{bmatrix}.$$

Here A is an *orthogonal matrix*, i.e. $A^{-1} = A^*$.

See Fig. 2.105.

(1) A shearing preserves all the properties listed in (1) of (2.7.9),
(2) and also preserves

    (c) area,
    (d) directions (i.e. orientations). (2.8.32)

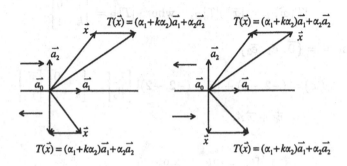

Fig. 2.105

For a generalization to *skew shearing*, see Ex. <A> 17.

As a counterpart of (2.8.27) for a shearing, we list the following steps:

1. Compute the eigenvalues of A. In case A has only one eigenvalue 1 with multiplicity 2 and A is *not* diagonalizable, then the associated T is a shearing.

2. The eigenspace

$$E = \{ \vec{x} \in \mathbb{R}^2 \,|\, \vec{x}A = \vec{x} \}$$

has dimension one (see (1)3. in (2.7.73)). Take any eigenvector $\vec{v}$ of unit length and take any $\vec{a_0}$ as a solution of $\vec{x}(I_2 - A) = \vec{x_0}$. Then $\vec{a_0} + \langle\langle\vec{v}\rangle\rangle$ or $\vec{x}(I_2 - A) = \vec{x_0}$ itself is the axis of the shearing.

3. Take any $\vec{u}$, of unit length and perpendicular to $\vec{v}$, and then $\vec{u}A - \vec{u} = k\vec{v}$ holds and the scalar k is the coefficient. (2.8.33)

We give an example to illustrate these concepts.

Example 4 Let $\vec{a_0} = (-2, -2)$ and $\vec{a_1} = (2, 1)$. Construct the shearings with axis $\vec{a_0} + \langle\langle\vec{a_1} - \vec{a_0}\rangle\rangle$ and coefficients $k = 2$ and $k = -1$, respectively.

Solution $\vec{a}_1 - \vec{a}_0 = (4,3)$ and its length $|\vec{a}_1 - \vec{a}_0| = 5$. Let

$$\vec{v}_1 = \frac{1}{5}(4,3),$$

$$\vec{v}_2 = \frac{1}{5}(-3,4).$$

Then $\mathcal{B} = \{\vec{a}_0, \vec{a}_0 + \vec{v}_1, \vec{a}_0 + \vec{v}_2\}$ is an orthonormal affine basis for $\mathbb{R}^2$.
In $\mathcal{B}$, the required affine transformation is

$$[T(\vec{x})]_{\mathcal{B}} = [\vec{x}]_{\mathcal{B}}[T]_{\mathcal{B}}, \quad \text{where } [T]_{\mathcal{B}} = \begin{bmatrix} 1 & 0 \\ k & 1 \end{bmatrix}.$$

While, in $\mathcal{N} = \{\vec{0}, \vec{e}_1, \vec{e}_2\}$,

$$T(\vec{x}) = (-2,-2) + [\vec{x} - (-2,-2)]\begin{bmatrix} \vec{v}_1 \\ \vec{v}_2 \end{bmatrix}^{-1} \begin{bmatrix} 1 & 0 \\ k & 1 \end{bmatrix} \begin{bmatrix} \vec{v}_1 \\ \vec{v}_2 \end{bmatrix}$$

$$= \vec{x}_0 + \vec{x}A,$$

where

$$A = \frac{1}{25}\begin{bmatrix} 25 - 12k & -9k \\ 16k & 25 + 12k \end{bmatrix}$$

$$= \begin{cases} \dfrac{1}{25}\begin{bmatrix} 1 & -18 \\ 32 & 49 \end{bmatrix} & \text{if } k = 2; \\[3mm] \dfrac{1}{25}\begin{bmatrix} 37 & 9 \\ -16 & 13 \end{bmatrix} & \text{if } k = -1, \end{cases}$$

$$\vec{x}_0 = (-2,-2) - (-2,-2)A = (-2,-2)(I_2 - A)$$

$$= \begin{cases} \dfrac{4}{25}(4,\ 3) & \text{if } k = 2; \\[3mm] -\dfrac{2}{25}(4,\ 3) & \text{if } k = -1. \end{cases}$$

Now, we treat the converse problems.

Suppose $T(\vec{x}) = \vec{x}_0 + \vec{x}A$ if $k = 2$. Notice the following steps:

1. The characteristic polynomial is

$$\det[A - tI_2] = \det \begin{bmatrix} \frac{1-25t}{5} & -\frac{18}{25} \\ \frac{32}{5} & \frac{49-25t}{25} \end{bmatrix} = t^2 - 2t + 1 = (t-1)^2.$$

Hence, A has coincident eigenvalues 1. The corresponding eigenspace is

$$E = \{\vec{x} \in \mathbb{R}^2 \mid \vec{x}(A - I_2) = \vec{0}\} = \langle\langle(4,3)\rangle\rangle = \langle\langle\vec{v}_1\rangle\rangle,$$

which is one-dimensional. Therefore A is not diagonalizable. See also Ex. <A> 7 of Sec. 2.7.6. Therefore, the associated T is a shearing.

2. Solve

$$\vec{x}(I_2 - A) = (x_1 \quad x_2) \begin{bmatrix} \frac{24}{25} & \frac{18}{25} \\ -\frac{32}{25} & -\frac{24}{25} \end{bmatrix} = \vec{x}_0 = \frac{4}{25}(4, \ 3),$$

which reduces to $3x_1 - 4x_2 = 2$. Then any point, say $(-2, -2)$, on the line $3x_1 - 4x_2 = 2$, is an invariant point. Therefore, $(-2, -2) + \langle\langle \vec{v}_1 \rangle\rangle$, which is $3x_1 - 4x_2 = 2$, is the axis.

3. Take a vector $\vec{v}_2$ such that $|\vec{v}_2| = 1$ and $\vec{v}_2 \perp \vec{v}_1$, say $\vec{v}_2 = \frac{1}{5}(-3, 4)$. Then

$$\vec{v}_2 A = \frac{1}{5}(-3, 4) \cdot \frac{1}{25} \begin{bmatrix} 1 & -18 \\ 32 & 49 \end{bmatrix} = (1, 2)$$

$$\Rightarrow \vec{v}_2 A - \vec{v}_2 = (1, 2) - \frac{1}{5}(-3, 4) = \frac{1}{5}(8, 6) = 2\vec{v}_1.$$

Hence, $k = 2$ is the coefficient.

We raise two questions as follows:

Q1. *What is the image of the straight line $x_1 + x_2 = 3$? Where do they intersect?*

Q2. *What is the image of the square with vertices $(1, 1), (-1, 1), (-1, -1)$ and $(1, -1)$? How much is its area?*

For Q1, one way to do this is to rewrite $x_1 + x_2 = 3$ as

$$(x_1 \quad x_2) \begin{bmatrix} 1 \\ 1 \end{bmatrix} = (x_1 \quad x_2) A \cdot A^{-1} \begin{bmatrix} 1 \\ 1 \end{bmatrix} = 3$$

$$\Rightarrow (\text{denote } \vec{y} = (y_1, y_2) = T(\vec{x}) \text{ temporarily})$$

$$\left[\vec{y} - \frac{4}{25}(4, 3) \right] \cdot \frac{1}{25} \begin{bmatrix} 49 & 18 \\ -32 & 1 \end{bmatrix} \begin{bmatrix} 1 \\ 1 \end{bmatrix} = 3$$

$$\Rightarrow (\text{change } y_1, y_2 \text{ back to } x_1, x_2 \text{ respectively})$$

$$67x_1 - 31x_2 = 103.$$

These two lines intersect at $(2, 1)$ which lies on axis $3x_1 - 4x_2 = 2$. This is absolutely not accidental, but why?

For Q2, by computation,

$$T(1,1) = \frac{4}{25}(4,3) + \frac{1}{25}(1,1)\begin{bmatrix} 1 & -18 \\ 32 & 49 \end{bmatrix} = \frac{1}{25}(49,43),$$

$$T(-1,1) = \frac{1}{25}(47,78),$$

$$T(-1,-1) = \frac{1}{25}(-17,-18),$$

$$T(1,-1) = \frac{1}{5}(-3,-11).$$

These four points form four vertices of a parallelogram whose area is equal to 4. Do you know why? See Fig. 2.106.

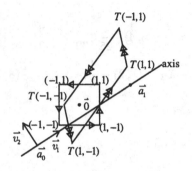

Fig. 2.106

Suppose $T(\vec{x}) = \vec{x}_0 + \vec{x}A$ if $k = -1$. We leave the corresponding details to the readers except Q2. For Q2, by computation,

$$T(1,1) = -\frac{2}{25}(4,3) + \frac{1}{25}(1,1)\begin{bmatrix} 37 & 9 \\ -16 & 13 \end{bmatrix} = \frac{1}{25}(13,16),$$

$$T(-1,1) = \frac{1}{25}(-61,-2),$$

$$T(-1,-1) = \frac{1}{25}(-29,-28),$$

$$T(1,-1) = \frac{1}{25}(45,-10).$$

They form a parallelogram of area equal to 4. See Fig. 2.107.

Case 6 Rotation (Euclidean notions are needed)

Take two perpendicular lines OA and OB as axes so that the line segments OA and OB are of unit length. Let θ be real such that $-2\pi < \theta \leq 2\pi$.

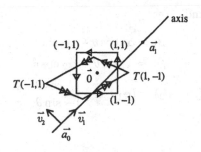

Fig. 2.107

For any point X in the plane, rotate the line segment OX, with center at O, through an angle θ to reach the line segment OX'. Define a mapping $T : \mathbb{R}^2 \to \mathbb{R}^2$ by

$$T(X) = X'.$$

This T is an affine transformation and is called the *rotation* with *center at* O of the plane *through angle* θ. It is *counterclockwise* if $\theta > 0$, *clockwise* if $\theta < 0$. See Fig. 2.108.

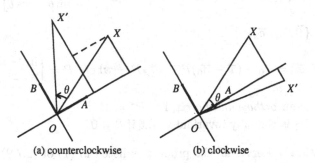

(a) counterclockwise   (b) clockwise

Fig. 2.108

To linearize T, let $O = \vec{a}_0$, $A = \vec{a}_1$ and $B = \vec{a}_2$ in $\mathcal{N} = \{\vec{0}, \vec{e}_1, \vec{e}_2\}$ and $\vec{v}_1 = \vec{a}_1 - \vec{a}_0 = (\alpha_{11}, \alpha_{12})$ and $\vec{v}_2 = \vec{a}_2 - \vec{a}_0 = (\alpha_{21}, \alpha_{22})$. Then

$$|\vec{v}_1| = |\vec{v}_2| = 1 \quad \text{and} \quad \vec{v}_1 \perp \vec{v}_2.$$

Now, $\mathcal{B} = \{\vec{a}_0, \vec{a}_1, \vec{a}_2\}$ is an affine orthonormal basis for $\mathbb{R}^2$. For $\vec{x} \in \mathbb{R}^2$, let

$$\vec{x} - \vec{a}_0 = \alpha_1 \vec{v}_1 + \alpha_2 \vec{v}_2 \quad \text{or} \quad [\vec{x}]_{\mathcal{B}} = (\alpha_1, \alpha_2),$$
$$T(\vec{x}) - \vec{a}_0 = \beta_1 \vec{v}_1 + \beta_2 \vec{v}_2 \quad \text{or} \quad [T(\vec{x})]_{\mathcal{B}} = (\beta_1, \beta_2).$$

Then Fig. 2.108(a) indicates

$$\beta_1 = \alpha_1 \cos\theta - \alpha_2 \sin\theta$$
$$\beta_2 = \alpha_1 \sin\theta + \alpha_2 \cos\theta$$
$$\Rightarrow (\beta_1 \quad \beta_2) = (\alpha_1 \quad \alpha_2) \begin{bmatrix} \cos\theta & \sin\theta \\ -\sin\theta & \cos\theta \end{bmatrix}.$$

We summarize as

The rotation
Let $\vec{a}_0$, $\vec{a}_1$ and $\vec{a}_2$ be three distinct non-collinear points so that

$$|\vec{a}_1 - \vec{a}_0| = |\vec{a}_2 - \vec{a}_0| = 1 \text{ (in length)}, \quad \text{and}$$
$$(\vec{a}_1 - \vec{a}_0) \perp (\vec{a}_2 - \vec{a}_0) \text{ (perpendicularity)}.$$

Then the *rotation* T with *center* at $\vec{a}_0$ and *through the angle* θ has the following representations.

1. In the orthonormal affine basis $\mathcal{B} = \{\vec{a}_0, \vec{a}_1, \vec{a}_2\}$,

$$[T(\vec{x})]_\mathcal{B} = [\vec{x}]_\mathcal{B}[T]_\mathcal{B}, \quad \text{where } [T]_\mathcal{B} = \begin{bmatrix} \cos\theta & \sin\theta \\ -\sin\theta & \cos\theta \end{bmatrix}.$$

2. In $\mathcal{N} = \{\vec{0}, \vec{e}_1, \vec{e}_2\}$,

$$T(\vec{x}) = \vec{a}_0 + (\vec{x} - \vec{a}_0)P^{-1}[T]_\mathcal{B}P, \quad \text{where } P = \begin{bmatrix} \vec{a}_1 - \vec{a}_0 \\ \vec{a}_2 - \vec{a}_0 \end{bmatrix}.$$

Here P is an *orthogonal matrix*, i.e. $P^* = P^{-1}$.
The center $\vec{a}_0$ is the only invariant point if $\theta \neq 0$.

(1) A rotation preserves all the properties listed in (1) of (2.7.9),
(2) and it also preserves

    (a) length,
    (b) angle,
    (c) area,
    (d) orientation.   (2.8.34)

In $\mathcal{N} = \{\vec{0}, \vec{e}_1, \vec{e}_2\}$, to test if a given affine transformation

$$T(\vec{x}) = \vec{x}_0 + \vec{x}A$$

is a rotation, try the following steps.

1. Justify that A is an orthogonal matrix, i.e. $A^* = A^{-1}$ and $\det A = 1$.
2. Put A in the form

$$\begin{bmatrix} \cos\theta & \sin\theta \\ -\sin\theta & \cos\theta \end{bmatrix}.$$

Then θ is the rotation angle.
3. The only solution of $\vec{x}A = \vec{x}$ (if $\theta \neq 0$) is $\vec{0}$. Hence $I_2 - A$ is invertible. Thus

$$\vec{x}_0(I_2 - A)^{-1}$$

is the center of rotation.   (2.8.35)

Notice that A does not have real eigenvalues if $\theta \neq 0, \pi$. The readers should consult Ex. <B> 5 of Sec. 2.7.2 for a more general setting.

Example 5 Determine whether

$$T(\vec{x}) = (1,1) + \vec{x}A, \quad \text{where } A = \begin{bmatrix} \frac{1}{\sqrt{10}} & \frac{3}{\sqrt{10}} \\ -\frac{3}{\sqrt{10}} & \frac{1}{\sqrt{10}} \end{bmatrix}$$

is a rotation. If yes, write down its rotation angle and center.

Solution It is easy to check that

$$A^* A = I_2$$

and $\det A = 1$. Put

$$\cos\theta = \frac{1}{\sqrt{10}} \quad \text{and} \quad \sin\theta = \frac{3}{\sqrt{10}},$$

then $\tan\theta = 3$ and $\theta = \tan^{-1} 3$ is the rotation angle. Also,

$$I_2 - A = \frac{1}{\sqrt{10}} \begin{bmatrix} \sqrt{10} - 1 & -3 \\ 3 & \sqrt{10} - 1 \end{bmatrix}$$

$$\Rightarrow \det(I_2 - A) = \frac{1}{10}(20 - 2\sqrt{10}) = \frac{2(\sqrt{10} - 1)}{\sqrt{10}}$$

$$\Rightarrow I_2 - A \text{ is invertible and}$$

$$(I_2 - A)^{-1} = \frac{\sqrt{10}}{2(\sqrt{10} - 1)} \cdot \frac{1}{\sqrt{10}} \begin{bmatrix} \sqrt{10} - 1 & 3 \\ -3 & \sqrt{10} - 1 \end{bmatrix}$$

$$= \frac{1}{2} \begin{bmatrix} 1 & \frac{\sqrt{10}+1}{3} \\ -\frac{\sqrt{10}+1}{3} & 1 \end{bmatrix}.$$

$$\Rightarrow \vec{x}_0(I_2 - A)^{-1} = (1 \quad 1) \cdot \frac{1}{2} \begin{bmatrix} 1 & \frac{\sqrt{10}+1}{3} \\ -\frac{\sqrt{10}+1}{3} & 1 \end{bmatrix} = \frac{1}{6}(2 - \sqrt{10}, 4 + \sqrt{10}),$$

which is the center of rotation.

Do you know how to find the equation of the image of the unit circle $x_1^2 + x_2^2 = 1$ under T?

Case 7 Reflection again (Euclidean notations are needed)

Example 6 Study the affine transformation

$$T(\vec{x}) = (-2, 2\sqrt{3}) + \vec{x}A, \quad \text{where } A = \begin{bmatrix} \frac{1}{2} & \frac{\sqrt{3}}{2} \\ \frac{\sqrt{3}}{2} & -\frac{1}{2} \end{bmatrix}.$$

Solution Although A is orthogonal, i.e. $A^* = A^{-1}$, but its determinant $\det A = -1$. Therefore, the associated T is *not* a rotation.

The characteristic polynomial of A is

$$\det(A - tI_2) = \begin{vmatrix} \frac{1}{2} - t & \frac{\sqrt{3}}{2} \\ \frac{\sqrt{3}}{2} & -\frac{1}{2} - t \end{vmatrix} = t^2 - 1,$$

so A has eigenvalues 1 and -1. According to 1 in (2.8.27), the associated T is a reflection.

Solve

$$\vec{x}(A - I_2) = (x_1 \quad x_2) \begin{bmatrix} -\frac{1}{2} & \frac{\sqrt{3}}{2} \\ \frac{\sqrt{3}}{2} & -\frac{3}{2} \end{bmatrix} = (0 \quad 0)$$

and get the associated eigenvectors $\vec{v} = t(\sqrt{3}, 1)$, for $t \in \mathbb{R}$ and $t \neq 0$. Pick up a unit vector $\vec{v}_1 = \frac{1}{2}(\sqrt{3}, 1)$. On the other hand, solve

$$\vec{x}(A + I_2) = (x_1 \quad x_2) \begin{bmatrix} \frac{3}{2} & \frac{\sqrt{3}}{2} \\ \frac{\sqrt{3}}{2} & \frac{1}{2} \end{bmatrix} = (0 \quad 0)$$

and get the associated eigenvectors $\vec{v} = t(-1, \sqrt{3})$. Pick up a unit vector $\vec{v}_2 = \frac{1}{2}(-1, \sqrt{3})$. This $\vec{v}_2$ is the direction of reflection. Note that $\vec{v}_2$ is perpendicular to $\vec{v}_1$, so that A is an *orthogonal* reflection with respect to the axis $\langle\langle \vec{v}_1 \rangle\rangle$.

To see if the original affine transformation $T(\vec{x}) = \vec{x}_0 + \vec{x}A$ is an orthogonal reflection with respect to a certain axis, this is our problem. Suppose the axis is of the form $\vec{a}_0 + \langle\langle \vec{v}_1 \rangle\rangle$, then

$$T(\vec{x}) = \vec{x}_0 + \vec{x}A = \vec{x}_0 + \vec{a}_0 A + (\vec{x} - \vec{a}_0)A$$

$$\Rightarrow \vec{a}_0 = \vec{x}_0 + \vec{a}_0 A$$

$$\Rightarrow \vec{x}_0 = \vec{a}_0(I_2 - A) \text{ which implies that } \vec{x}_0(I_2 + A) = \vec{0}, \text{ i.e. } \vec{x}_0 A = -\vec{x}_0$$

$$\text{or } \vec{x}_0 \text{ is an eigenvector of } A \text{ corresponding to } -1. \tag{2.8.36}$$

For the given $\vec{x}_0 = (-2, 2\sqrt{3})$, solve

$$\vec{x}(I_2 - A) = \vec{x} \begin{vmatrix} \frac{1}{2} & -\frac{\sqrt{3}}{2} \\ -\frac{\sqrt{3}}{2} & \frac{3}{2} \end{vmatrix} = (-2, 2\sqrt{3}), \quad \text{where } \vec{x} = (x_1, x_2)$$

and the axis is $x_1 - \sqrt{3}x_2 = -4$. Of course, any point on it, say $(-4, 0)$, can be used as $\vec{a}_0$ and the axis is the same as $\vec{a}_0 + \langle\langle \vec{v}_1 \rangle\rangle$. See Fig. 2.109.

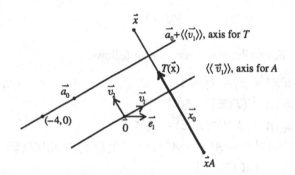

Fig. 2.109

We can reinterpret the transformation $\vec{x} \to T(\vec{x})$ as follows.

$$\vec{x} = (x_1, \ x_2) \xrightarrow[\text{with axis } \langle\langle \vec{e}_1 \rangle\rangle]{\text{orthogonal reflection}} \vec{x} \begin{bmatrix} 1 & 0 \\ 0 & -1 \end{bmatrix}$$

$$= (x_1, \ -x_2) \xrightarrow[\text{center at } \vec{0} \text{ and through } 60°]{\text{rotation with}} \vec{x} \begin{bmatrix} 1 & 0 \\ 0 & -1 \end{bmatrix} \begin{bmatrix} \frac{1}{2} & \frac{\sqrt{3}}{2} \\ -\frac{\sqrt{3}}{2} & \frac{1}{2} \end{bmatrix}$$

$$= \vec{x} \begin{bmatrix} \frac{1}{2} & \frac{\sqrt{3}}{2} \\ \frac{\sqrt{3}}{2} & -\frac{1}{2} \end{bmatrix}$$

$$= \vec{x} A \xrightarrow[\text{along } \vec{x}_0 = (-2, 2\sqrt{3})]{\text{translation}} \vec{x}_0 + \vec{x} A = T(\vec{x}).$$

Fig. 2.110 illustrates all the steps involved.

What is the image of the unit circle $x_1^2 + x_2^2 = 1$? As it might be expected beforehand, since A, as a mapping, does keep the circle invariant while the transformation $\vec{x} \to \vec{x}_0 + \vec{x}$ preserves everything, the image is the circle

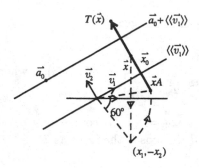

Fig. 2.110

$|\vec{x} - \vec{x}_0| = 1$. Formally, we prove this as follows.

$\langle \vec{x}, \vec{x} \rangle = \vec{x}\,\vec{x}^* = x_1^2 + x_2^2 = 1$

$\Rightarrow (T(\vec{x}) - \vec{x}_0)A^{-1}[(T(\vec{x}) - \vec{x}_0)A^{-1}]^* = 1$

$\Rightarrow (T(\vec{x}) - \vec{x}_0)A^{-1}(A^{-1})^*(T(\vec{x}) - \vec{x}_0)^* = 1$

$\Rightarrow (\text{Since } A^{-1}(A^{-1})^* = A^*A = A^{-1}A = I_2)(T(\vec{x}) - \vec{x}_0)(T(\vec{x}) - \vec{x}_0)^* = 1$

$\Rightarrow |T(\vec{x}) - \vec{x}_0| = 1$ if $|\vec{x}| = 1$.

In order to avoid getting confused, it should be mentioned and cautioned that the transformation $T(\vec{x}) = \vec{x}_0 + \vec{x}A$ in (2.8.26) is derived from $T(\vec{x}) = \vec{a}_0 + (\vec{x} - \vec{a}_0)B$, where $B = A^{-1}[T]_\mathcal{N}A$, so that

$$\vec{x}_0 = \vec{a}_0 - \vec{a}_0 B = \vec{a}_0(I_2 - B).$$

This implies explicitly why 3 in (2.8.27) and (2.8.29) should hold. But this is not the case in Example 6 where $T(\vec{x}) = \vec{x}_0 + \vec{x}A$ is given, independent of the form $\vec{a}_0 + (\vec{x} - \vec{a}_0)B$. That is why we have to recapture $\vec{a}_0$ from $\vec{x}_0$ in (2.8.36).

Therefore, we summarize the following steps to test if a given transformation, in $\mathcal{N} = \{\vec{0}, \vec{e}_1, \vec{e}_2\}$,

$$T(\vec{x}) = \vec{x}_0 + \vec{x}A$$

is an orthogonal reflection where A is an orthogonal matrix.

1. Justify $\det A = -1$ so that A has eigenvalues ± 1.
2. Put A in the form

$$A = \begin{bmatrix} \cos\theta & \sin\theta \\ \sin\theta & -\cos\theta \end{bmatrix} = \begin{bmatrix} 1 & 0 \\ 0 & -1 \end{bmatrix} \begin{bmatrix} \cos\theta & \sin\theta \\ -\sin\theta & \cos\theta \end{bmatrix}.$$

Then the eigenvectors corresponding to 1 are $\vec{v} = te^{i\frac{\theta}{2}} = t\left(\cos\frac{\theta}{2}, \sin\frac{\theta}{2}\right)$ for $t \in \mathbb{R}$ and $t \neq 0$, these to -1 are $\vec{u} = tie^{i\frac{\theta}{2}} = t\left(-\sin\frac{\theta}{2}, \cos\frac{\theta}{2}\right)$ for $t \in \mathbb{R}$ and $t \neq 0$.

3. To find the axis, suppose the axis passes the point $\vec{a}_0$. Then, solve

$$\vec{a}_0 = \vec{x}_0 + \vec{a}_0 A, \quad \text{or}$$
$$\vec{x}_0 = \vec{a}_0(I_2 - A)$$

to get $\vec{a}_0$ and $\vec{a}_0 + \langle\langle \vec{v} \rangle\rangle$, the *axis of the reflection*. In fact, this axis has the equation $\vec{x}_0 = \vec{x}(I_2 - A)$ in $\mathcal{N}$. Note that $\vec{x}_0$ should be an eigenvector corresponding to -1. $\hspace{2cm}$ (2.8.37)

Steps similar to these should replace those stated in (2.8.27) and (2.8.29) once $T(\vec{x}) = \vec{x}_0 + \vec{x}A$ is given beforehand, independent of being derived from $T(\vec{x}) = \vec{a}_0 + (\vec{x} - \vec{a}_0)B$.

Let $A = [a_{ij}]_{2\times 2}$ be an invertible matrix. We have the following elementary matrix factorization for A (see (2.7.68)).

Case 1 $a_{11} \neq 0$, and

$$A = \begin{bmatrix} 1 & 0 \\ \frac{a_{21}}{a_{11}} & 1 \end{bmatrix} \begin{bmatrix} 1 & \frac{a_{11}a_{12}}{\det A} \\ 0 & 1 \end{bmatrix} \begin{bmatrix} a_{11} & 0 \\ 0 & \frac{\det A}{a_{11}} \end{bmatrix}.$$

Case 2 $a_{11} = 0$, then $a_{12}a_{21} \neq 0$, and

$$A = \begin{bmatrix} 0 & 1 \\ 1 & 0 \end{bmatrix} \begin{bmatrix} 1 & \frac{a_{22}}{a_{12}} \\ 0 & 1 \end{bmatrix} \begin{bmatrix} a_{21} & 0 \\ 0 & a_{12} \end{bmatrix}.$$

Case 1 shows that an affine transformation $T(\vec{x}) = \vec{x}_0 + \vec{x}A$ in $\mathcal{N} = \{\vec{0}, \vec{e}_1, \vec{e}_2\}$ is the composite of

(a) a shearing: axis $\langle\langle \vec{e}_1 \rangle\rangle$ and coefficient $\frac{a_{21}}{a_{11}}$,

(b) a shearing: axis $\langle\langle \vec{e}_2 \rangle\rangle$ and coefficient $\frac{a_{11}a_{12}}{\det A}$,

(c) a two-way stretch: along $\langle\langle \vec{e}_1 \rangle\rangle$ with scale factor a_{11} and along $\langle\langle \vec{e}_2 \rangle\rangle$ with scale factor $\frac{\det A}{a_{11}}$, and

(d) a translation: along $\vec{x}_0$; $\hspace{4cm}$ (2.8.38)

while Case 2 shows that it is the composite of

(a) a reflection: axis $\langle\langle \vec{e}_1 + \vec{e}_2 \rangle\rangle$,

(b) a shearing: axis $\langle\langle \vec{e}_2 \rangle\rangle$ and coefficient $\frac{a_{22}}{a_{12}}$,

(c) a two-way stretch: along $\langle\langle \vec{e}_1 \rangle\rangle$ with scale factor a_{21} and along $\langle\langle \vec{e}_2 \rangle\rangle$ with scale factor a_{12}, and

(d) a translation: along $\vec{x}_0$. $\hspace{4cm}$ (2.8.39)

According to logical development arranged in this book, planar Euclidean notions will be formally introduced at the very beginning of Part 2 and Chap. 4. Unfortunately, we need them here in order to introduce shearings, rotations and orthogonal reflections. Hope it does not cause much trouble to the readers.

In Chap. 4, different decompositions for affine transformations from these in (2.8.38) and (2.8.39) will be formulated and proved. See Exs. <B> 5–9 of Sec. 4.8.

Exercises

<A>

1. Prove (2.8.27).
2. Let

$$T(\vec{x}) = \vec{x}_0 + \vec{x}A, \quad \text{where } A = \frac{1}{11}\begin{bmatrix} 13 & -8 \\ 6 & -13 \end{bmatrix}$$

   be an affine transformation in $\mathcal{N} = \{\vec{0}, \vec{e}_1, \vec{e}_2\}$.

   (a) Determine which $\vec{x}_0$ will guarantee that T is a reflection in a certain affine basis.
   (b) Find an affine basis in which T is a reflection and determine the direction of the reflection.
   (c) Find the image of the line $4x_1 + x_2 = c$, where c is a constant.
   (d) Find the image of the line $x_1 + x_2 = 1$. Where do they intersect?
   (e) Find the image of the unit circle $x_1^2 + x_2^2 = 1$ and illustrate it graphically.

3. Let

$$T(\vec{x}) = \vec{x}_0 + \vec{x}A, \quad \text{where } A = \begin{bmatrix} \frac{4}{5} & \frac{3}{5} \\ \frac{3}{5} & -\frac{4}{5} \end{bmatrix}.$$

   Do the same problems as in Ex. 2.

4. Prove (2.8.29).
5. Let k be a nonzero constant and $k \neq 1$. Let

$$T(\vec{x}) = \vec{x}_0 + \vec{x}A, \quad \text{where } A = \begin{bmatrix} -3k+4 & 6k-6 \\ -2k+2 & 4k-3 \end{bmatrix}$$

   be an affine transformation in $\mathcal{N} = \{\vec{0}, \vec{e}_1, \vec{e}_2\}$.

   (a) Determine all these $\vec{x}_0$ so that T is a one-way stretch in a certain affine basis.

(b) For these $\vec{x}_0$ in (a), determine the axis and the direction of the associated one-way stretch.

(c) Show that a line, not parallel to the direction, and its image under T will always intersect at a point lying on the axis.

(d) Find the image of the ellipse $x_1^2 + 2x_2^2 = 1$ under T if $k = 3$.

(e) Find the image of the hyperbola $x_1^2 - 2x_2^2 = 1$ under T if $k = -2$.

(f) Find the image of the parabola $x_1^2 = x_2$ under T if $k = \frac{1}{2}$.

6. In $\mathcal{N} = \{\vec{0}, \vec{e}_1, \vec{e}_2\}$, let

$$T(\vec{x}) = \vec{x}_0 + \vec{x}A,$$

where $A_{2\times2}$ is an invertible matrix with two distinct eigenvalues. Model after (2.8.27) and (2.8.29) to set up criteria for T to be a two-way stretch, including

(1) axes (i.e. directions of one-way stretches) and scale factors, and

(2) the only invariant point.

For $k_1 \neq k_2$ and $k_1 k_2 \neq 0$, let

$$A = \begin{bmatrix} 16k_1 - 15k_2 & 12(k_1 - k_2) \\ 20(-k_1 + k_2) & -15k_1 + 16k_2 \end{bmatrix}.$$

Justify your claims above for this $T(\vec{x}) = \vec{x}_0 + \vec{x}A$. Then, try to do the following problems.

(a) If $k_1 k_2 < 1$, find the image of the circle $x_1^2 + x_2^2 = 1$ under T.

(b) If $k_1 k_2 = 1$, find the image of the hyperbola $x_1^2 - x_2^2 = 1$ under T.

(c) If $k_1 k_2 > 1$, find the image of the parabola $x_2^2 = x_1$ under T.

Try to sketch these quadratic curves for some numerical k_1 and k_2.

7. Prove (2.8.33).

8. In $\mathcal{N} = \{\vec{0}, \vec{e}_1, \vec{e}_2\}$, let $k \neq 0$ be a scalar and

$$T(\vec{x}) = \vec{x}_0 + \vec{x}A, \quad \text{where } A = \frac{1}{4}\begin{bmatrix} 4 - \sqrt{3}k & -k \\ 3k & 4 + \sqrt{3}k \end{bmatrix}.$$

(a) Show that A has coincident eigenvalues 1 but A is not diagonalizable.

(b) Determine these $\vec{x}_0$ for which T is a shearing. In this case, where is the axis?

(c) Show that the coefficient of the shearing is k.

(d) Graph the image of the square with vertices at $(1, 0), (0, 1),$ $(-1, 0), (0, -1)$ under T, where $\vec{x}_0 = (\sqrt{3}, 1)$ and $k = 1$.

(e) Graph the image of the curve $x_1^2 - x_2^2 = 0$ under T, where $\vec{x}_0 = (-\sqrt{3}, -1)$ and $k = -2$.

(f) Graph the image of the curve $x_1^2 = a^2$, where a is a scalar, under T with $\vec{x}_0 = (3, \sqrt{3})$ and $k = 3$.

9. Prove (2.8.35).

10. Let $\alpha^2 + \beta^2 \neq 0$ and

$$T(\vec{x}) = \vec{x}_0 + \vec{x}A, \quad \text{where } A = \frac{1}{\sqrt{\alpha^2 + \beta^2}} \begin{bmatrix} \alpha & -\beta \\ \beta & \alpha \end{bmatrix}.$$

(a) Show that T is a rotation.

(b) Determine the center and the angle of rotation.

(c) Show that T has properties listed in (1) and (2) of (2.8.34).

11. Prove (2.8.37).

12. Let $\alpha^2 + \beta^2 \neq 0$ and

$$T(\vec{x}) = \vec{x}_0 + \vec{x}A, \quad \text{where } A = \frac{1}{\sqrt{\alpha^2 + \beta^2}} \begin{bmatrix} \alpha & \beta \\ \beta & -\alpha \end{bmatrix}.$$

Model after Example 6 to do all the problems within.

13. Let

$$A = \begin{bmatrix} -3 & 5 \\ 2 & 4 \end{bmatrix}.$$

Decompose A as $A = A_1 A_2 A_3$, where

$$A_1 = \begin{bmatrix} 1 & 0 \\ -\frac{2}{3} & 1 \end{bmatrix}, \quad A_2 = \begin{bmatrix} 1 & \frac{15}{22} \\ 0 & 1 \end{bmatrix}, \quad A_3 = \begin{bmatrix} -3 & 0 \\ 0 & \frac{22}{3} \end{bmatrix}.$$

and consider the affine transformation, in $\mathcal{N} = \{\vec{0}, \vec{e}_1, \vec{e}_2\}$,

$$T(\vec{x}) = (-2, 3) + \vec{x}A.$$

(a) Find the image of the straight line $x_1 + x_2 = 1$ under T via the following methods and graph it and its image.

(1) By direct computation.

(2) By steps indicated in (2.8.38).

(3) By diagonalization of A and try to use (2) of (2.7.72).

(b) Do the same problems as in (a) for the square $|x_1| + |x_2| = 2$.

(c) Do the same problems as in (a) for the circle $x_1^2 + x_2^2 = 1$.

(d) Justify (1) of (2.7.9).

14. Let

$$A = \begin{bmatrix} 0 & 5 \\ -3 & 4 \end{bmatrix}.$$

Decompose A as $A = A_1 A_2 A_3$, where

$$A_1 = \begin{bmatrix} 0 & 1 \\ 1 & 0 \end{bmatrix}, \quad A_2 = \begin{bmatrix} 1 & \frac{4}{5} \\ 0 & 1 \end{bmatrix}, \quad A_3 = \begin{bmatrix} -3 & 0 \\ 0 & 5 \end{bmatrix}$$

and consider the affine transformation, in $\mathcal{N} = \{\vec{0}, \vec{e}_1, \vec{e}_2\}$,

$$T(\vec{x}) = (3, -2) + \vec{x} A.$$

Do the same problems as in Ex. 13 but now use (2.8.39).

15. Fix an affine basis $\mathcal{B} = \{\vec{a}_0, \vec{a}_1, \vec{a}_2\}$, Remind that a one-way stretch in the direction $\vec{a}_2 - \vec{a}_0$ with $\vec{a}_0 + \langle\langle \vec{a}_1 - \vec{a}_0 \rangle\rangle$ as axes can be represented as

$$\begin{bmatrix} 1 & 0 \\ 0 & k \end{bmatrix},$$

where $k \neq 0$ is the scale factor.

(a) Show that the set of all such one-way stretches

$$\left\{ \begin{bmatrix} 1 & 0 \\ 0 & k \end{bmatrix} \,\middle|\, k \in \mathbb{R} \right\}$$

is a subgroup (see Ex. <A> 2 of Sec. 2.8.1) of $G_a(2; \mathbb{R})$. It is an abelian group.

(b) Show that

$$\left\{ \begin{bmatrix} 1 & 0 \\ 0 & k \end{bmatrix} \,\middle|\, k > 0 \right\}$$

forms an abelian subgroup of the group in (a).

(c) Suppose $\mathcal{B} = \mathcal{N} = \{\vec{0}, \vec{e}_1, \vec{e}_2\}$. Show that the unit circle can be transformed into an ellipse by an orthogonal one-way stretch and conversely. See Fig. 2.111(a), where the ellipse has equation $x_1^2 + \frac{1}{k^2} x_2^2 = 1$ for $0 < k < 1$.

(d) In $\mathcal{N}$, the unit circle can be transformed into a skew ellipse. See Fig. 2.111(b), where the ellipse has the equation $ax_1^2 + 2bx_1 x_2 + cx_2^2 = 1$ with suitable constants a, b and c. Try the following two ways.

(1) Use result in (c) and perform a suitable rotation to it.

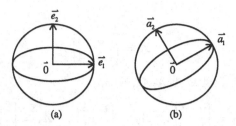

Fig. 2.111

(2) Let $\vec{a}_i = (a_{i1}, a_{i2})$ for $i = 1, 2$. Note that $|\vec{a}_1| = |\vec{a}_2| = 1$ and $\vec{a}_1 \perp \vec{a}_2$. Then $\mathcal{B} = \{\vec{0}, \vec{a}_1, \vec{a}_2\}$ is an orthonormal affine basis for $\mathbb{R}^2$. For some $0 < k < 1$, the ellipse in $\mathcal{B}$ is

$$[\vec{x}]_{\mathcal{B}} \begin{bmatrix} 1 & 0 \\ 0 & \frac{1}{k^2} \end{bmatrix} [\vec{x}]_{\mathcal{B}}^* = 1.$$

Then use the change of coordinates $[\vec{x}]_{\mathcal{B}} = [\vec{x}]_{\mathcal{N}} A_{\mathcal{B}}^{\mathcal{N}}$ to change it into the equation in $\mathcal{N}$.

16. Fix an affine basis $\mathcal{B} = \{\vec{a}_0, \vec{a}_1, \vec{a}_2\}$. Remind that a two-way stretch stated in 1 of (2.8.30).

   (a) Show that

   $$\left\{ \begin{bmatrix} k_1 & 0 \\ 0 & k_2 \end{bmatrix} \;\middle|\; k_1, k_2 \in \mathbb{R} \text{ and } k_1 k_2 \neq 0 \right\}$$

   is an abelian subgroup of $G_a(2; \mathbb{R})$.

   (b) In $\mathcal{N}$, a circle can be transformed into an ellipse of the form $a_{11}x_1^2 + 2a_{12}x_1x_2 + a_{22}x_2^2 = 1$ and conversely.

   (c) How a circle can be transformed into an ellipse of the form $a_{11}x_1^2 + 2a_{12}x_1x_2 + a_{22}x_2^2 + b_1x_1 + b_2x_2 + c = 0$? Try to give out the details.

17. Fix an affine basis $\mathcal{B} = \{\vec{a}_0, \vec{a}_1, \vec{a}_2\}$ for $\mathbb{R}^2$.

   (a) An affine transformation T having the equation, in $\mathcal{B}$,

   $$[T(\vec{x})]_{\mathcal{B}} = [\vec{x}]_{\mathcal{B}} \begin{bmatrix} 1 & 0 \\ k & 1 \end{bmatrix}, \quad k \neq 0.$$

   is called a *(skew) shearing*. Try to illustrate this T graphically, including its axis, i.e. the line of invariant points, and all its invariant lines.

(b) The set of all shearings

$$\left\{ \begin{bmatrix} 1 & 0 \\ k & 1 \end{bmatrix} \,\middle|\, k \in \mathbb{R} \right\}$$

forms an abelian subgroup of $G_a(2; \mathbb{R})$.

<B>

1. *Hyperbolic rotation*

Fix an affine basis $\mathcal{B} = \{\vec{a}_0, \vec{a}_1, \vec{a}_2\}$ for $\mathbb{R}^2$. Let f denote the one-way stretch in the direction $\vec{a}_2 - \vec{a}_0$, with $\vec{a}_0 + \langle\langle \vec{a}_1 - \vec{a}_0 \rangle\rangle$ as axis and scale factor k, while g denote the one-way stretch in the direction $\vec{a}_1 - \vec{a}_0$, with $\vec{a}_0 + \langle\langle \vec{a}_2 - \vec{a}_0 \rangle\rangle$ as axis and scale factor $\frac{1}{k}$. Then the composite

$$T = g \circ f = f \circ g$$

is called a *hyperbolic rotation*, See Fig. 2.112.

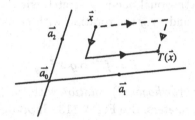

Fig. 2.112

(a) Show that

$$[T(\vec{x})]_\mathcal{B} = [\vec{x}]_\mathcal{B}[T]_\mathcal{B}, \quad \text{where}$$

$$[T]_\mathcal{B} = \begin{bmatrix} \frac{1}{k} & 0 \\ 0 & 1 \end{bmatrix} \begin{bmatrix} 1 & 0 \\ 0 & k \end{bmatrix} = \begin{bmatrix} \frac{1}{k} & 0 \\ 0 & k \end{bmatrix}.$$

Note that T is completely determined by $\vec{a}_0$, $\vec{a}_1$, a point $\vec{x}$ not lying on $\vec{a}_0 + \langle\langle \vec{a}_1 - \vec{a}_0 \rangle\rangle$ and its image point $\vec{x}' = T(\vec{x})$.

(b) The set of all hyperbolic rotations

$$\left\{ \begin{bmatrix} \frac{1}{k} & 0 \\ 0 & k \end{bmatrix} \,\middle|\, k \in \mathbb{R} \text{ and } k \neq 0 \right\}$$

forms an abelian subgroup of $G_a(2; \mathbb{R})$. This subgroup is useful in the study of hyperbolic (or Lobachevsky) geometry and theory of relativity.

(c) Suppose $|\vec{a}_1 - \vec{a}_0| = |\vec{a}_2 - \vec{a}_0| = 1$ and $(\vec{a}_1 - \vec{a}_0) \perp (\vec{a}_2 - \vec{a}_0)$, the resulting T is called a *hyperbolic (orthogonal) rotation*. Let $[\vec{x}]_\mathcal{B} = (\alpha_1, \alpha_2)$. Then a hyperbola with the lines $\vec{a}_0 + \langle\langle \vec{a}_1 - \vec{a}_0 \rangle\rangle$ and $\vec{a}_0 + \langle\langle \vec{a}_2 - \vec{a}_0 \rangle\rangle$ as asymptotes has equation $\alpha_1 \alpha_2 = c$, where c is a constant. Let $[T(\vec{x})]_\mathcal{B} = (\alpha'_1, \alpha'_2)$. Then $\alpha'_1 = \frac{1}{k}\alpha_1$ and $\alpha'_2 = k\alpha_2$ and hence

$$\alpha'_1 \alpha'_2 = \alpha_1 \alpha_2 = c,$$

which indicates that the image point $T(\vec{x})$ still lies on the same hyperbola $\alpha_1 \alpha_2 = c$. Therefore, such hyperbolas are invariant under the group mentioned in (b). In particular, the asymptotes $\vec{a}_0 + \langle\langle \vec{a}_1 - \vec{a}_0 \rangle\rangle$ and $\vec{a}_0 + \langle\langle \vec{a}_2 - \vec{a}_0 \rangle\rangle$ are invariant lines with $\vec{a}_0$ as the only point of invariant.

2. *Elliptic rotation*

Take two different points $\vec{a}_0$ and $\vec{a}_1$ in $\mathbb{R}^2$. Let k be a positive constant. Denote by f the orthogonal one-way stretch with axis $\vec{a}_0 + \langle\langle \vec{a}_1 - \vec{a}_0 \rangle\rangle$ and scale factor k, and by g the rotation with center at $\vec{a}_0$ and angle θ. The composite

$$T = f^{-1} \circ g \circ f$$

is called the *elliptic (orthogonal) rotation* with $\vec{a}_0 + \langle\langle \vec{a}_1 - \vec{a}_0 \rangle\rangle$ as axis and k and θ as parameters. See Fig. 2.113. Suppose $|\vec{a}_1 - \vec{a}_0| = 1$. Take another point $\vec{a}_2$ so that $|\vec{a}_2 - \vec{a}_0| = 1$ and $(\vec{a}_2 - \vec{a}_0) \perp (\vec{a}_1 - \vec{a}_0)$.

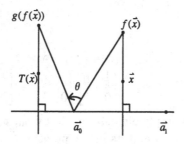

Fig. 2.113

Then $\mathcal{B} = \{\vec{a}_0, \vec{a}_1, \vec{a}_2\}$ is an affine orthonormal basis for $\mathbb{R}^2$.

(a) Show that

$$[T(\vec{x})]_\mathcal{B} = [\vec{x}]_\mathcal{B}[T]_\mathcal{B},$$

where

$$[T]_B = [f]_B[g]_B[f]_B^{-1} = \begin{bmatrix} 1 & 0 \\ 0 & k \end{bmatrix} \begin{bmatrix} \cos\theta & \sin\theta \\ -\sin\theta & \cos\theta \end{bmatrix} \begin{bmatrix} 1 & 0 \\ 0 & \frac{1}{k} \end{bmatrix}$$

$$= \begin{bmatrix} \cos\theta & \frac{1}{k}\sin\theta \\ -k\sin\theta & \cos\theta \end{bmatrix}.$$

(b) Suppose the rotation center $\vec{a}_0$, the axis $\vec{a}_0 + \langle\langle\vec{a}_1 - \vec{a}_0\rangle\rangle$ and the scale factor k are fixed. Then the set of all elliptic rotations

$$\left\{ \begin{bmatrix} \cos\theta & \frac{1}{k}\sin\theta \\ -k\sin\theta & \cos\theta \end{bmatrix} \,\middle|\, \theta \in \mathbb{R} \right\}$$

forms an abelian subgroup of $G_a(2;\mathbb{R})$. This group is of fundamental in the study of elliptic geometry.

(c) Let $0 < k < 1$. An ellipse with center at $\vec{a}_0$, the major axis $\vec{a}_0 + \langle\langle\vec{a}_1 - \vec{a}_0\rangle\rangle$ and the eccentricity $\sqrt{1-k^2}$ has the equation $k\alpha_1^2 + \alpha_2^2 = r > 0$, where $[\vec{x}]_B = (\alpha_1, \alpha_2)$. Its image under f is a circle $k\alpha_2^2 + k\alpha_2^2 = r$. Conversely, f^{-1} will recover the ellipses from the circles. Hence, T maps such a family of ellipses onto itself.

(d) An elliptic rotation is completely determined by its center $\vec{a}_0$, its major axis $\vec{a}_0 + \langle\langle\vec{a}_1 - \vec{a}_0\rangle\rangle$ and a pair $\vec{x}$, $\vec{x}'$ of points where $\vec{x} \neq \vec{a}_0$, $\vec{x}' \neq \vec{a}_0$.

3. An affine transformation with a unique line of invariant points (called the *axis*) is called an *affinity*. Suppose T is an affinity with axis L and T maps a point $\vec{x}_0$, not lying on L, into the point $\vec{x}_0'$.

   (a) In case the line $\vec{x}_0\vec{x}_0'$ is parallel to L, show that T is an orthogonal shearing with L as its axis (see (2.8.32)).

   (b) In case the line $\vec{x}_0\vec{x}_0'$ is not parallel to L, show that T is a one-way stretch in the direction $\vec{x}_0'$ and with axis L.

<C> Abstraction and generalization

For counterparts in $\mathbb{R}^3$, refer to Sec. 3.8.2.

Results detained here and Sec. 3.8.2 can be generalized to abstract n-dimensional affine space V over a field $\mathbb{F}$, in particular, the real field.

2.8.3 *Affine invariants*

We come back to (1) of (2.7.9) and prove all of them to be true in the content of affine transformations.

Affine invariants

Affine transformation $T(\vec{x}) = \vec{x}_0 + \vec{x}A$, where $A_{2\times2}$ is an invertible matrix, preserves 1–7 but not necessarily preserves a–d in (1) of (2.7.9).   (2.8.40)

Various examples in Sec. 2.8.2 convinced us that (2.8.40) indeed holds. Formal proof will be given as follows.

Proof For 1, adopt notations and results in (2.5.9). The image of the line L_i: $\vec{x} = \vec{a}_i + t\vec{b}_i$ for $i = 1, 2$ under $T(\vec{x}) = \vec{x}_0 + \vec{x}A$ is

$$T(L_i): T(\vec{x}) = \vec{x}_0 + \vec{a}_iA + t(\vec{b}_iA), \quad t \in \mathbb{R},$$

which is still a line passing the point $\vec{x}_0 + \vec{a}_iA$ with direction $\vec{b}_iA$. Therefore,

a. L_1 is coincident with L_2.

  $\Leftrightarrow$ The vectors $\vec{a}_2 - \vec{a}_1$, $\vec{b}_1$ and $\vec{b}_2$ are linearly dependent.
  $\Leftrightarrow$ (since A is invertible) $(\vec{x}_0 + \vec{a}_2A) - (\vec{x}_0 + \vec{a}_1A) = (\vec{a}_2 - \vec{a}_1)A$, $\vec{b}_1A$
    and $\vec{b}_2A$ are linearly dependent.
  $\Leftrightarrow$ $T(L_1)$ is coincident with $T(L_2)$.

b. L_1 is parallel to L_2.

  $\Leftrightarrow$ $\vec{b}_1$ and $\vec{b}_2$ are linearly dependent, but linearly independent of $\vec{a}_2 - \vec{a}_1$.
  $\Leftrightarrow$ $\vec{b}_1A$ and $\vec{b}_2A$ are linearly dependent, but linearly independent of

$$(\vec{x}_0 + \vec{a}_2A) - (\vec{x}_0 + \vec{a}_1A) = (\vec{a}_2 - \vec{a}_1)A.$$

  $\Leftrightarrow$ $T(L_1)$ is parallel to $T(L_2)$.

c. L_1 is intersecting with L_2 at a unique point.

  $\Leftrightarrow$ $\vec{b}_1$ and $\vec{b}_2$ are linearly independent.
  $\Leftrightarrow$ $\vec{b}_1A$ and $\vec{b}_2A$ are linearly independent.
  $\Leftrightarrow$ $T(C_1)$ intersects with $T(C_2)$ at a unique point.

Due to one-to-one and onto properties, point of intersection is mapped into point of intersection in Case c.

For 2, adopt notations and concepts concerned in (2.5.11). The image of the directed segment $\vec{a}_1\vec{a}_2$: $\vec{x} = (1-t)\vec{a}_1 + t\vec{a}_2, 0 \le t \le 1$, under T, is

$$T(\vec{x}) = \vec{x}_0 + (1-t)\vec{a}_1A + t\vec{a}_2A$$
$$= (1-t)(\vec{x}_0 + \vec{a}_1A) + t(\vec{x}_0 + \vec{a}_2A), \quad 0 \le t \le 1,$$

which is the directed line segment with the initial point $\vec{x}_0 + \vec{a}_1A$ and the terminal point $\vec{x}_0 + \vec{a}_2A$. It is obvious that interior points (i.e. $0 < t < 1$) are mapped into interior points and endpoints into endpoints.

For 3, adopt notations in (2.5.10). Take four points $\vec{x}_i$ on the line L: $\vec{x} = \vec{a}_1 + t(\vec{a}_2 - \vec{a}_1)$, say

$$\vec{x}_i = \vec{a}_1 + t_i(\vec{a}_2 - \vec{a}_1), \quad i = 1, 2, 3, 4.$$

The *directed* segments

$$\vec{x}_1\vec{x}_2 = \vec{x}_2 - \vec{x}_1 = (t_2 - t_1)(\vec{a}_2 - \vec{a}_1), \quad \text{and}$$
$$\vec{x}_3\vec{x}_4 = \vec{x}_4 - \vec{x}_3 = (t_4 - t_3)(\vec{a}_2 - \vec{a}_1)$$

have their respective image under T the *directed* segment

$$T(\vec{x}_1)T(\vec{x}_2) = T(\vec{x}_2) - T(\vec{x}_1) = (t_2 - t_1)(\vec{a}_2 - \vec{a}_1)A$$
$$T(\vec{x}_3)T(\vec{x}_4) = T(\vec{x}_4) - T(\vec{x}_3) = (t_4 - t_3)(\vec{a}_2 - \vec{a}_1)A.$$

Therefore (refer to (1.4.5)), in case $\vec{x}_3 \neq \vec{x}_4$, it follows that

$$\frac{T(\vec{x}_1)T(\vec{x}_2)}{T(\vec{x}_3)T(\vec{x}_4)} = \frac{t_2 - t_1}{t_4 - t_3} = \frac{\vec{x}_1\vec{x}_2}{\vec{x}_3\vec{x}_4} \quad \text{(ratios of signed lengths)},$$

which proves the claim. If $\vec{x}_1\vec{x}_2$ and $\vec{x}_3\vec{x}_4$ lie on different but parallel lines, the same proof will work by using 2 in (2.5.9).

For 4, refer to (2.5.12) and Fig. 2.26. By 2, it is easy to see that the image of $\triangle\vec{a}_1\vec{a}_2\vec{a}_3$ under T is the triangle $\triangle\vec{b}_1\vec{b}_2\vec{b}_3$ where $\vec{b}_i = T(\vec{a}_i)$ for $i = 1, 2, 3$. Let $\vec{a}$ be any fixed interior point of $\triangle\vec{a}_1\vec{a}_2\vec{a}_3$ (one might refer to Fig. 2.31 for a precise definition of interior points of a triangle). Extend the line segment $\vec{a}_3\vec{a}$ to meet the opposite side $\vec{a}_1\vec{a}_2$ at the point $\vec{a}_4$. Then, $\vec{a}_4$ is an interior point of $\vec{a}_1\vec{a}_2$. By 2 or 3, $T(\vec{a}_4)$ is an interior point of the side $\vec{b}_1\vec{b}_2$. Since the line segment $\vec{a}_1\vec{a}_4$ is mapped onto the line segment $\vec{b}_3T(\vec{a}_4)$, the interior of $\vec{a}$ implies that $T(\vec{a})$ is an interior point of $\vec{b}_3T(\vec{a}_4)$. See Fig. 2.114. Hence $T(\vec{a})$ is an interior point of the triangle $\triangle\vec{b}_1\vec{b}_2\vec{b}_3$.

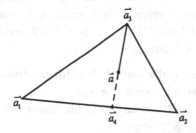

Fig. 2.114

For 5 and 6, both are easy consequences of 4.

For 7, let us start from a triangle $\triangle \vec{a}_1 \vec{a}_2 \vec{a}_3$ (see Fig. 2.114) and find the signed area of its image $\triangle \vec{b}_1 \vec{b}_2 \vec{b}_3$ under T. In what follows, it is supposed that the readers are familiar with the basic knowledge about the determinants of order 2 and it is suggested to refer to Sec. 4.3 or Ex. <B> of Sec. 2.6 if needed.

The image triangle $\triangle \vec{b}_1 \vec{b}_2 \vec{b}_3$, where $\vec{b}_i = T(\vec{a}_i) = \vec{x}_0 + \vec{a}_i A$ for $1 \le i \le 3$, has the

$$
\begin{aligned}
\text{signed area} &= \frac{1}{2} \det \begin{bmatrix} T(\vec{a}_2) - T(\vec{a}_1) \\ T(\vec{a}_3) - T(\vec{a}_1) \end{bmatrix} \\
&= \frac{1}{2} \det \begin{bmatrix} (\vec{a}_2 - \vec{a}_1)A \\ (\vec{a}_3 - \vec{a}_1)A \end{bmatrix} \\
&= \frac{1}{2} \det \begin{bmatrix} (\vec{a}_2 - \vec{a}_1) \\ (\vec{a}_3 - \vec{a}_1) \end{bmatrix} A \\
&= (\text{the signed area of } \triangle \vec{a}_1 \vec{a}_2 \vec{a}_3) \det A \qquad (2.8.41) \\
\Rightarrow \quad & \frac{\text{the area of } \triangle \vec{b}_1 \vec{b}_2 \vec{b}_3}{\text{the area of } \triangle \vec{a}_1 \vec{a}_2 \vec{a}_3} = |\det A|. \qquad (2.8.42)
\end{aligned}
$$

From this relation, it follows that an affine transformation preserves the ratio of (signed) areas of two triangles and hence, two non-self-intersecting polygons and finally, by limiting process, two measurable planar domains.

Examples in Sec. 2.8.2 show that a general affine transformation does not necessarily preserve length, angle and area except $\det A = 1$. It does

preserve the direction if $\det A > 0$, and
reverse the direction if $\det A < 0$. \qquad (2.8.43)

For formal definitions for directions or orientations for the plane, please refer to Ex. <A> 4 of Sec. 2.8.1. $\qquad \square$

As we have experienced so many numerical examples concerned with areas in the previous exercises and examples, such as examples in Exs. <B> 7 and 8 of Sec. 2.5 and in Sec. 2.8.2, we single out (2.8.41)–(2.8.43) as parts of the following summary.

The geometric interpretation of the determinant of a linear operator or a square matrix

Let $f(\vec{x}) = \vec{x}A$: $\mathbb{R}^2 \to \mathbb{R}^2$ be an invertible linear operator in the natural basis $\mathcal{N} = \{\vec{e}_1, \vec{e}_2\}$ for $\mathbb{R}^2$, where

$$
A = \begin{bmatrix} a_{11} & a_{12} \\ a_{21} & a_{22} \end{bmatrix}
$$

is an invertible real 2×2 matrix. The determinant

$$\det f = \det A = \begin{vmatrix} a_{11} & a_{12} \\ a_{21} & a_{22} \end{vmatrix}$$

= the signed area of the parallelogram $\square \vec{a}_1 \vec{a}_2$, where

$$\vec{a}_1 = f(\vec{e}_1) = (a_{11}, a_{12}) \text{ and } \vec{a}_2 = f(\vec{e}_2) = (a_{21}, a_{22})$$

$$= \frac{\text{the signed area of } \square \vec{a}_1 \vec{a}_2}{\text{the area of the square } \square \vec{e}_1 \vec{e}_2}.$$

See Fig. 2.115.

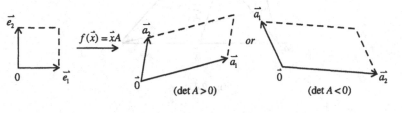

Fig. 2.115

Therefore, for any affine transformation $T(\vec{x}) = \vec{x}_0 + \vec{x} A$,

1. The signed area of the image

$$\Delta T(\vec{a}_1) T(\vec{a}_2) T(\vec{a}_3) = (\text{the signed area of } \Delta \vec{a}_1 \vec{a}_2 \vec{a}_3) \det A.$$

It is also true for any simple closed polygon and measurable planar domain.

2. T preserves orientation $\Leftrightarrow \det A > 0$; and

T reverses orientation $\Leftrightarrow \det A < 0$. $\qquad\qquad$ (2.8.44)

It is worthy of mentioning that $\det A = 0$ can be used to interpret the parallelogram $\square \vec{a}_1 \vec{a}_2$ is degenerated to a single point or a line segment in Fig. 2.115.

What kind of linear operator or associated affine transformation will preserve length, angle, and hence area? It is the orthogonal operator (matrix). This is one of the main topics to be touched in Part 2.

Exercises

<A>

1. Prove (2.8.40) by using (2.8.38) and (2.8.39).
2. Prove (2.8.44) by using (2.8.38) and (2.8.39).

3. On the sides $\vec{a}_1\vec{a}_2$, $\vec{a}_2\vec{a}_3$, and $\vec{a}_3\vec{a}_1$ of a triangle $\triangle\vec{a}_1\vec{a}_2\vec{a}_3$, pick three points $\vec{b}_3$, $\vec{b}_1$ and $\vec{b}_2$ respectively, so that the directed segments $\vec{a}_1\vec{b}_3 = 3\vec{b}_3\vec{a}_2$, $\vec{a}_2\vec{b}_1 = \vec{b}_1\vec{a}_3$ and $\vec{a}_3\vec{b}_2 = 2\vec{b}_2\vec{a}_1$.

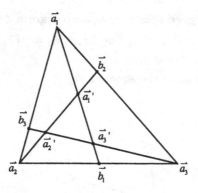

Fig. 2.116

See Fig. 2.116. Compute

$$\frac{\text{the area of } \triangle\vec{a}_1'\vec{a}_2'\vec{a}_3'}{\text{the area of } \triangle\vec{a}_1\vec{a}_2\vec{a}_3}$$

by the following methods.

(1) May suppose $\triangle\vec{a}_1\vec{a}_2\vec{a}_3$ as a right-angle triangle with its two legs lying on the axes and then use 7 in (2.8.40).

(2) Find an affine transformation mapping $\triangle\vec{a}_1\vec{a}_2\vec{a}_3$ onto $\triangle\vec{a}_1'\vec{a}_2'\vec{a}_3'$.

(3) Use Menelaus theorem (see Sec. 2.8.4 or Ex. <B> 3 below, if needed).

<B>

1. (Refer to [14]) Suppose $T: \mathbb{R}^2 \to \mathbb{R}^2$ has the properties:

(1) T is one-to-one and onto, and

(2) T maps any three collinear points into three collinear points.

Show that T is an affine transformation. Try the following steps:

(a) T maps straight lines into straight lines.

(b) T maps parallel lines into parallel lines.

(c) T maps intersecting lines into intersecting lines and preserves the point of intersection.

(d) T preserves the midpoint of a line segment.

(e) T maps n equidistant points on a line segment into n equidistant points on a line segment, for any natural number n.

(f) (**Darboux's Theorem**) T preserves interior points of a line segment.

(g) T preserves signed ratios of line segments along the same or parallel lines.

2. Give a triangle $\triangle \vec{a}_1 \vec{a}_2 \vec{a}_3$ and three positive number α, β, γ with $0 < \alpha < 1, 0 < \beta < 1$ and $0 < \gamma < 1$. Suppose the point $\vec{b}_1$ divides the side $\vec{a}_2 \vec{a}_3$ internally into the ratio $\alpha: 1 - \alpha$; $\vec{b}_2$ divides the side $\vec{a}_3 \vec{a}_1$ internally into the ratio $\beta: 1 - \beta$; and $\vec{b}_3$ divides the side $\vec{a}_1 \vec{a}_2$ internally into the ratio $\gamma: 1 - \gamma$. Also, suppose the line segments $\vec{a}_1 \vec{b}_1$, $\vec{a}_2 \vec{b}_2$ and $\vec{a}_3 \vec{b}_3$ intersect into a triangle $\triangle \vec{a}_1' \vec{a}_2' \vec{a}_3'$. (Refer to Fig. 2.116.)

(a) Compute

$$\frac{\text{the area of } \triangle \vec{a}_1' \vec{a}_2' \vec{a}_3'}{\text{the area of } \triangle \vec{a}_1 \vec{a}_2 \vec{a}_3}.$$

(b) (**Ceva Theorem**) The lines $\vec{a}_1 \vec{b}_1$, $\vec{a}_2 \vec{b}_2$ and $\vec{a}_3 \vec{b}_3$ are *concurrent* at a point if and only if

$$\frac{\alpha \beta \gamma}{(1 - \alpha)(1 - \beta)(1 - \gamma)} = 1.$$

3. Let $\triangle \vec{a}_1 \vec{a}_2 \vec{a}_3$ and $\vec{b}_1$, $\vec{b}_2$, $\vec{b}_3$ and α, β, γ be as in Ex. 3, except that now $\vec{b}_1$ divides $\vec{a}_2 \vec{a}_3$ externally into $\alpha: 1 - \alpha$. See Fig. 2.117.

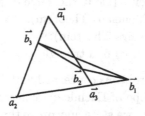

Fig. 2.117

(a) Compute

$$\frac{\text{the area of } \triangle \vec{b}_1 \vec{b}_2 \vec{b}_3}{\text{the area of } \triangle \vec{a}_1 \vec{a}_2 \vec{a}_3}.$$

(b) (**Menelaus Theorem**) The three points $\vec{b}_1$, $\vec{b}_2$ and $\vec{b}_3$ are collinear if and only if

$$\frac{\alpha \beta \gamma}{(1 - \alpha)(1 - \beta)(1 - \gamma)} = -1.$$

(c) What happens to (b) in case $\vec{b}_2 \vec{b}_3$ is parallel to $\vec{a}_2 \vec{a}_3$?

<C> Abstraction and generalization

Properties 1–6 in (2.8.40) are still true for affine transformation on an n-dimensional affine space V over a field $\mathbb{F}$, with suitable extensions as follows.

1. The relative positions of a r-dimensional affine subspace and a s-dimensional subspace, where $0 \leq r, s < n$.
4. n-dimensional simplex.
5. n-dimensional parallelepiped, etc.

While 6, 7 and a to d are still true for n-dimensional inner product space.

For a clearer picture about what we have claimed here, one might refer to Chap. 3 and Sec. 5.9.

2.8.4 *Affine geometry*

The study of these geometric properties invariant under the affine group $G_a(2; \mathbb{R})$constitutes the content of the so-called *affine geometry* on the affine plane $\mathbb{R}^2$. According to (2.8.40), roughly speaking, affine geometry investigates the following topics:

1. Parallelism.
2. The ratio of line segments (along the same or parallel lines).
3. Collinear points (e.g. Menelaus Theorem).
4. Concurrent lines (e.g. Ceva Theorem).
5. Centroid (or median point) of a triangle.

But not such geometric problems concerned with lengths, angles and hence areas which are in the scope of Euclidean geometry (refer to Sec. 4.9).

Here in this subsection, we state and prove three fundamental theorems on planar affine geometry: Menelaus, Ceva and Desargues theorems, in such a way that they can be extended verbatim to abstract affine plane over a field. One can refer to [19] for their historical backgrounds.

Menelaus Theorem
For any three non-collinear points $\vec{x}_1$, $\vec{x}_2$ and $\vec{x}_3$ in $\mathbb{R}^2$, pick up a point $\vec{y}_1$ on the line $\vec{x}_2 + \langle\langle \vec{x}_3 - \vec{x}_2 \rangle\rangle$, a point $\vec{y}_2$ on the line $\vec{x}_3 + \langle\langle \vec{x}_1 - \vec{x}_3 \rangle\rangle$ and a point $\vec{y}_3$ on the line $\vec{x}_1 + \langle\langle \vec{x}_1 - \vec{x}_2 \rangle\rangle$ so that $\vec{y}_1$, $\vec{y}_2$ and $\vec{y}_3$ do not coincide with the vertices of $\triangle \vec{x}_1 \vec{x}_2 \vec{x}_3$. Then

$$\vec{y}_1, \ \vec{y}_2 \ and \ \vec{y}_3 \ are \ collinear.$$
$$\Leftrightarrow (\vec{x}_2 \vec{y}_1 : \vec{y}_1 \vec{x}_3)(\vec{x}_3 \vec{y}_2 : \vec{y}_2 \vec{x}_1)(\vec{x}_1 \vec{y}_3 : \vec{y}_3 \vec{x}_2) = -1,$$

where $\vec{x}_2\vec{y}_1$ denotes the directed segment from $\vec{x}_2$ to $\vec{y}_1$, etc. Equivalently, if $\vec{y}_1$ divides $\vec{x}_2\vec{x}_3$ into $\alpha: 1-\alpha$, $\vec{y}_2$ divides $\vec{x}_3\vec{x}_1$ into $\beta: 1-\beta$ and $\vec{y}_3$ divides $\vec{x}_1\vec{x}_2$ into $\gamma: 1-\gamma$, then

$\vec{y}_1, \vec{y}_2$ and $\vec{y}_3$ are collinear.
$$\Leftrightarrow (1-\alpha)^{-1}\alpha(1-\beta)^{-1}\beta(1-\gamma)^{-1}\gamma = -1. \tag{2.8.45}$$

See Fig. 2.118.

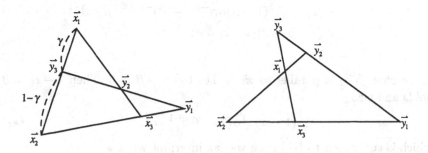

Fig. 2.118

Proof (For another proof, see Ex. <B> 3 of Sec. 2.8.3.) According to the assumptions, we have

$$\vec{y}_1 = (1-\alpha)\vec{x}_2 + \alpha\vec{x}_3, \quad \alpha \neq 0, 1$$
$$\vec{y}_2 = (1-\beta)\vec{x}_3 + \beta\vec{x}_1, \quad \beta \neq 0, 1$$
$$\vec{y}_3 = (1-\gamma)\vec{x}_1 + \gamma\vec{x}_2, \quad \gamma \neq 0, 1. \tag{$*_1$}$$

The necessity From the first two equations, it follows that

$$\vec{y}_2 - \vec{y}_1 = \beta\vec{x}_1 - (1-\alpha)\vec{x}_2 + (1-\alpha-\beta)\vec{x}_3.$$

Then $\vec{y}_1\vec{y}_2$ is parallel to $\vec{x}_1\vec{x}_2$ if and only if $1-\alpha-\beta = 0$ holds.

In case $\vec{y}_1\vec{y}_2$ is not parallel to $\vec{x}_1\vec{x}_2$, i.e. $1-\alpha-\beta \neq 0$. Eliminate $\vec{x}_3$ from the first two equations, one obtains the equation

$$(1-\beta)\vec{y}_1 - \alpha\vec{y}_2 = (1-\alpha)(1-\beta)\vec{x}_2 - \alpha\beta\vec{x}_1 \tag{$*_2$}$$
$\Rightarrow$ (multiply both sides by $(1-\alpha-\beta)^{-1}$)
$$(1-\alpha-\beta)^{-1}(1-\beta)\vec{y}_1 - (1-\alpha-\beta)^{-1}\alpha\vec{y}_2$$
$$= (1-\alpha-\beta)^{-1}(1-\alpha-\beta+\alpha\beta)\vec{x}_2 - (1-\alpha-\beta)^{-1}\alpha\beta\vec{x}_1,$$

which represents a point lying simultaneously on the segments $\vec{x}_1\vec{x}_2$ and $\vec{y}_1\vec{y}_2$ or their extended lines. Suppose this point is $\vec{y}_3$. Then, compare it with the third equation in $(*_1)$ and get the relations

$$1 - \gamma = -(1 - \alpha - \beta)^{-1}\alpha\beta$$
$$\gamma = (1 - \alpha - \beta)^{-1}(1 - \alpha)(1 - \beta)$$
$$\Rightarrow (1 - \alpha)^{-1}\alpha(1 - \beta)^{-1}\beta(1 - \gamma)^{-1}\gamma$$
$$= (1 - \alpha)^{-1}\alpha(1 - \beta)^{-1}\beta[-\beta^{-1}\alpha^{-1}(1 - \alpha - \beta)]$$
$$\times (1 - \alpha - \beta)^{-1}(1 - \alpha)(1 - \beta)$$
$$= -(1 - \alpha)^{-1}(1 - \alpha)\alpha\alpha^{-1}(1 - \beta)^{-1}(1 - \beta)\beta\beta^{-1}$$
$$\times (1 - \alpha - \beta)(1 - \alpha - \beta)^{-1}$$
$$= -1. \tag{$*_3$}$$

In case $\vec{y_1}\vec{y_2}$ is parallel to $\vec{x_1}\vec{x_2}$, i.e. $1 - \alpha - \beta = 0$. Then $1 - \alpha = \beta$ holds and also

$$(1 - \alpha)^{-1}\alpha(1 - \beta)\beta^{-1} = 1, \tag{$*_4$}$$

which is equivalent to $(*_3)$ once we can interpret what

$$\lim_{\gamma \to \infty} \frac{\gamma}{1 - \gamma} = -1 \tag{$*_5$}$$

means geometrically. To see this, notice that the lines $\vec{x_1} + \langle\langle \vec{x_2} - \vec{x_1}\rangle\rangle$ and $\vec{y_1} + \langle\langle \vec{y_2} - \vec{y_1}\rangle\rangle$ are parallel to each other and they never meet at a point in $\mathbb{R}^2$ within our sight. Suppose there were a point where these two lines meet. By comparing the third equation in $(*_1)$ to $(*_2)$, there exists some scalar $t \in \mathbb{R}$ such that

$$1 - \gamma = -t\alpha\beta$$
$$\gamma = t(1 - \alpha)(1 - \beta)$$
$$\Rightarrow (\text{using } (1 - \alpha)(1 - \beta) = 1 - \alpha - \beta + \alpha\beta = \alpha\beta)$$
$$1 - \gamma = -\gamma$$
$$\Rightarrow 1 = 0,$$

which is a contradiction. Now, by imagining the parallel lines $\vec{x_1} + \langle\langle \vec{x_2} - \vec{x_1}\rangle\rangle$ and $\vec{y_1} + \langle\langle \vec{y_2} - \vec{y_1}\rangle\rangle$ extended beyond any limit and met at a point, called the *infinite point* (refer to Ex. <B> 5 of Sec. 2.8.5), denoted as ∞. It is in this sense that $(*_5)$ means geometrically. See Fig. 2.119.

Refer to the Remark after this proof.

The sufficiency It is well-known that three points $\vec{y_1}, \vec{y_2}$ and $\vec{y_3}$ are collinear or affinely dependent (refer to Sec. 2.6) if and only if there exist

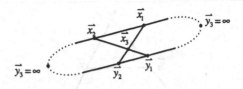

Fig. 2.119

scalars t_1, t_2, t_3, not all zeros, so that

$$t_1\vec{y}_1 + t_2\vec{y}_2 + t_3\vec{y}_3$$

$$= [t_2\beta + t_3(1-\gamma)]\vec{x}_1 + [t_3\gamma + t_1(1-\alpha)]\vec{x}_2 + [t_1\alpha + t_2(1-\beta)]\vec{x}_3 = \vec{0},$$

where $t_1 + t_2 + t_3 = 0$. By assumptions, $\vec{y}_1, \vec{y}_2$ and $\vec{y}_3$ do not coincide with any one of $\vec{x}_1, \vec{x}_2$ and $\vec{x}_3$. Therefore, $\vec{y}_1, \vec{y}_2$ and $\vec{y}_3$ are all distinct and t_1, t_2 and t_3 can be chosen, all not equal to zero, so that

$$t_2^{-1}t_3 = -(1-\gamma)^{-1}\beta,$$
$$t_3^{-1}t_1 = -(1-\alpha)^{-1}\gamma,$$
$$t_1^{-1}t_2 = -(1-\beta)^{-1}\alpha$$

hold. In this case,

$$t_1\vec{y}_1 + t_2\vec{y}_2 + t_3\vec{y}_3 = \vec{0}$$

and

$$t_1 + t_2 + t_3 = t_1(1-\alpha+\alpha) + t_2(1-\beta+\beta) + t_3(1-\gamma+\gamma)$$
$$= [t_2\beta + t_3(1-\gamma)] + [t_3\gamma + t_1(1-\alpha)] + [t_1\alpha + t_2(1-\beta)] = 0.$$

This proves that $\vec{y}_1, \vec{y}_2, \vec{y}_3$ are collinear.

In case, say $\vec{y}_3 = \infty$, the infinite point, then $t_3 = 0$ if and only if $\gamma = \infty$. At this time, the points $\vec{y}_1$ and $\vec{y}_2$ are still collinear. $\square$

Remark $(*_4)$ and $(*_5)$ are related to the well-known theorem: in a triangle $\triangle \vec{x}_1\vec{x}_2\vec{x}_3$, the line segments (see Fig. 2.120)

$$\vec{y}_1\vec{y}_2 \parallel \vec{x}_1\vec{x}_2$$

$$\Leftrightarrow \frac{\vec{x}_3\vec{y}_2}{\vec{y}_2\vec{x}_1} = \frac{\vec{x}_3\vec{y}_1}{\vec{y}_1\vec{x}_2} \quad \text{(ratio of signed lengths)},$$

which is a special case of Menelaus Theorem.

By the way, in case the points $\vec{y}_1, \vec{y}_2$ and $\vec{y}_3$ all do not lie between the vertices $\vec{x}_1, \vec{x}_2$ and $\vec{x}_3$ or just two of them lie between the vertices, the Menelaus Theorem is known also as *Pasch axiom*.

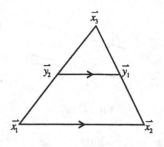

Fig. 2.120

Ceva Theorem *Adopt the notations and assumption for* $\vec{x}_1, \vec{x}_2, \vec{x}_3$ *and* $\vec{y}_1, \vec{y}_2, \vec{y}_3$ *and* α, β, γ *as in Menelaus Theorem. Then the segments*

$$\vec{x}_1\vec{y}_1, \ \vec{x}_2\vec{y}_2 \ \text{and} \ \vec{x}_3\vec{y}_3 \ \text{are concurrent.}$$

$$\Leftrightarrow (\vec{x}_2\vec{y}_1 : \vec{y}_1\vec{x}_3)(\vec{x}_3\vec{y}_2 : \vec{y}_2\vec{x}_1)(\vec{x}_1\vec{y}_3 : \vec{y}_3\vec{x}_2) = 1.$$

$$\Leftrightarrow (1-\alpha)^{-1}\alpha(1-\beta)^{-1}\beta(1-\gamma)^{-1}\gamma = 1. \tag{2.8.46}$$

See Fig. 2.121.

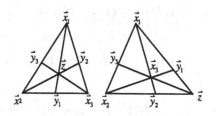

Fig. 2.121

Proof (For other proofs, see Ex. <B> 2 of Sec. 2.8.3 and Ex. <A> 1.)
Use Menelaus Theorem twice as follows. In $\triangle\vec{x}_1\vec{x}_2\vec{y}_2$,

$\vec{x}_3, \vec{z}, \vec{y}_3$ are collinear.

$$\Leftrightarrow (\vec{y}_2\vec{x}_3 : \vec{x}_3\vec{x}_1)(\vec{x}_1\vec{y}_3 : \vec{y}_3\vec{x}_2)(\vec{x}_2\vec{z} : \vec{z}\vec{y}_2) = -1;$$

while in $\triangle\vec{x}_2\vec{y}_2\vec{x}_3$,

$\vec{x}_1, \vec{z}, \vec{y}_1$ are collinear.

$$\Leftrightarrow (\vec{x}_2\vec{y}_1 : \vec{y}_1\vec{x}_3)(\vec{x}_3\vec{x}_1 : \vec{x}_1\vec{y}_2)(\vec{y}_2\vec{z} : \vec{z}\vec{x}_2) = -1.$$

By noting that $\vec{z}$ lies on the line segment $\vec{x}_2\vec{y}_2$, combine together these two results and finish the proof.   □

Desargues Theorem *Let* $\vec{x}_0, \vec{y}_1, \vec{z}_1; \vec{x}_0, \vec{y}_2, \vec{z}_2$ *and* $\vec{x}_0, \vec{y}_3, \vec{z}_3$ *be three sets of distinct collinear points in* $\mathbb{R}^2$, *lying on different lines. Suppose the line* *segments* $\vec{y}_2\vec{y}_3$ *and* $\vec{z}_2\vec{z}_3$ *meet at* $\vec{x}_1$, $\vec{y}_3\vec{y}_1$ *and* $\vec{z}_3\vec{z}_1$ *meet at* $\vec{x}_2$, $\vec{y}_1\vec{y}_2$ *and* $\vec{z}_1\vec{z}_2$ *meet at* $\vec{x}_3$. *Then*

$$\vec{x}_1, \vec{x}_2 \text{ and } \vec{x}_3 \text{ are collinear.} \qquad (2.8.47)$$

See Fig. 2.122.

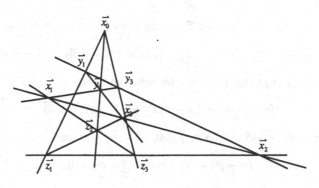

Fig. 2.122

Proof There exist α, β and γ in $\mathbb{R}$ such that

$$\vec{x}_0 = (1-\alpha)\vec{y}_1 + \alpha\vec{z}_1 = (1-\beta)\vec{y}_2 + \beta\vec{z}_2 = (1-\gamma)\vec{y}_3 + \gamma\vec{z}_3.$$

Hence

$$(1-\alpha)\vec{y}_1 - (1-\beta)\vec{y}_2 = -\alpha\vec{z}_1 + \beta\vec{z}_2. \qquad (*_1)$$

We claim that $\alpha \neq \beta$. To see this, suppose on the contrary that $\alpha = \beta$ does hold. Then $(*_1)$ reduces to

$$(1-\alpha)(\vec{y}_1 - \vec{y}_2) = \alpha(\vec{z}_2 - \vec{z}_1).$$

On the other hand, since $\vec{x}_3$ lies on $\vec{y}_1\vec{y}_2$ and $\vec{z}_1\vec{z}_2$, there exist t_1 and t_2 such that

$$\vec{x}_3 = (1-t_1)\vec{y}_1 + t_1\vec{y}_2 = (1-t_2)\vec{z}_1 + t_2\vec{z}_2, \quad t_1, t_2 \neq 0, 1$$
$$\Rightarrow \vec{z}_2 = (1-t_1)t_2^{-1}\vec{y}_1 + t_1 t_2^{-1}\vec{y}_2 - (1-t_2)t_2^{-1}\vec{z}_1$$
$$\Rightarrow (1-\alpha)(\vec{y}_1 - \vec{y}_2) = \alpha[(1-t_1)t_2^{-1}\vec{y}_1 + t_1 t_2^{-1}\vec{y}_2 - (1-t_2)t_2^{-1}\vec{z}_1 - \vec{z}_1]$$
$$\Rightarrow \vec{z}_1 = [(1-t_1) - t_2(1-\alpha)\alpha^{-1}]\vec{y}_1 + [t_1 + t_2(1-\alpha)\alpha^{-1}]\vec{y}_2.$$

Since the sum of the coefficients of $\vec{y}_1$ and $\vec{y}_2$ in the last relation is equal to 1, the point $\vec{z}_1$ lies on the segment $\vec{y}_1\vec{y}_2$. This will induce that $\vec{y}_2$ lies on the line $\vec{y}_1\vec{z}_1$ and hence $\vec{x}_0, \vec{y}_1, \vec{z}_1$, and $\vec{x}_0, \vec{y}_2, \vec{z}_2$ all lie on the same line, a contradiction to the assumption. Therefore, $\alpha \neq \beta$ is true.

Note that, in $(*_1)$, $(1 - \alpha) - (1 - \beta) = -\alpha + \beta$. This means that

$$\frac{1}{\beta - \alpha}[(1 - \alpha)\vec{y}_1 - (1 - \beta)\vec{y}_2] = \frac{1}{\beta - \alpha}[-\alpha\vec{z}_1 + \beta\vec{z}_2] = \vec{x}_3$$

$$\Rightarrow (1 - \alpha)\vec{y}_1 - (1 - \beta)\vec{y}_2 = -\alpha\vec{z}_1 + \beta\vec{z}_2 = (\beta - \alpha)\vec{x}_3, \quad \beta - \alpha \neq 0. \quad (*_2)$$

Similarly,

$$(1 - \beta)\vec{y}_2 - (1 - \gamma)\vec{y}_3 = -\beta\vec{z}_2 + \gamma\vec{z}_3 = (\gamma - \beta)\vec{x}_1, \quad \gamma - \beta \neq 0 \quad (*_3)$$

$$(1 - \gamma)\vec{y}_3 - (1 - \alpha)\vec{y}_1 = -\gamma\vec{z}_3 + \alpha\vec{z}_1 = (\alpha - \gamma)\vec{x}_2, \quad \alpha - \gamma \neq 0. \quad (*_4)$$

By adding $(*_2), (*_3)$ and $(*_4)$ side by side, we have

$$(\gamma - \beta)\vec{x}_1 + (\alpha - \gamma)\vec{x}_2 + (\beta - \alpha)\vec{x}_3 = \vec{0}$$

with coefficients $\gamma - \beta \neq 0$, $\alpha - \gamma \neq 0$, $(\beta - \alpha) \neq 0$ and

$$(\gamma - \beta) + (\alpha - \gamma) + (\beta - \alpha) = 0.$$

Thus, $\vec{x}_1, \vec{x}_2$ and $\vec{x}_3$ are collinear. $\qquad \square$

Exercises

<A>

1. Another proof of Ceva Theorem. Let $\vec{y}_1, \vec{y}_2$ and $\vec{y}_3$ be expressed as in $(*_1)$ in the proof of Menelaus Theorem.

   (a) Show that any point on the line $\vec{y}_1 + \langle\langle \vec{x}_1 - \vec{y}_1 \rangle\rangle$ can be expressed as

   $$(1 - t_1)\vec{y}_1 + t_1\vec{x}_1 = (1 - t_1)(1 - \alpha)\vec{x}_2 + (1 - t_1)\alpha\vec{x}_3 + t_1\vec{x}_1.$$

   Find similar expressions for points on the lines $\vec{y}_2 + \langle\langle \vec{x}_2 - \vec{y}_2 \rangle\rangle$ and $\vec{y}_3 + \langle\langle \vec{x}_3 - \vec{y}_3 \rangle\rangle$.

   (b) The three lines in (a) meet at a point $\vec{z}$ if and only if the three expressions in (a) for $\vec{z}$ will be coincident.

2. (One of the main features of affine geometry) In Fig. 2.123, show that $XY \| BC \Leftrightarrow V$ is the midpoint of BC, by the following two methods.

   (a) Use Ceva Theorem.

   (b) Take $\{A, B, C\}$ as an affine basis with A as base point (see Sec. 2.6). Let $A = (0, 0)$, $B = (1, 0)$, $C = (0, 1)$ and $X = (\alpha, 0)$, $0 < \alpha < 1$ and $Y = (0, \beta)$, $0 < \beta < 1$. Try to find out the affine coordinates of U and V and show that $\alpha = \beta \Leftrightarrow 2\overrightarrow{AV} = \overrightarrow{AB} + \overrightarrow{AC}$.

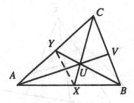

Fig. 2.123

3. Four straight lines, in the plane $\mathbb{R}^2$, meeting at six points $\vec{x}_1, \vec{x}_2, \vec{x}_3, \vec{x}_4$ and $\vec{y}_1, \vec{y}_2$ are said to form a *complete quadrilateral*. See Fig. 2.124. Let the segment $\vec{x}_1\vec{x}_3$ meet the segment $\vec{y}_1\vec{y}_2$ at the point $\vec{z}_1$, and $\vec{x}_2\vec{x}_4$ meet $\vec{y}_1\vec{y}_2$ at $\vec{z}_2$. Then $\vec{z}_1$ divides $\vec{y}_1\vec{y}_2$ into the *signed ratio* $\vec{y}_1\vec{z}_1 : \vec{z}_1\vec{y}_2$ and $\vec{z}_2$ divides $\vec{y}_1\vec{y}_2$ into the *signed ratio* $\vec{y}_1\vec{z}_2 : \vec{z}_2\vec{y}_2$. Show that the ratio of these two signed ratios is equal to -1, denoted as

$$(\vec{y}_1, \vec{y}_2; \ \vec{z}_1, \vec{z}_2) = (\vec{y}_1\vec{z}_1 : \vec{z}_1\vec{y}_2)(\vec{y}_1\vec{z}_2 : \vec{z}_2\vec{y}_2)^{-1} = -1$$

In this case, the four points $\vec{y}_1, \vec{y}_2, \vec{z}_1, \vec{z}_2$, in this ordering, are said to form *harmonic points*.

(*Note* In Fig. 2.124, $\vec{x}_2\vec{x}_4 \parallel \vec{y}_1\vec{y}_2 \Leftrightarrow \vec{z}_2 = \infty$, the infinite point. In this case, this problem reduces to Ex. 2 as a special case.)

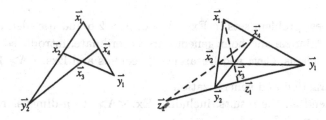

Fig. 2.124

4. In the plane $\mathbb{R}^2$, four different sets of three points $\vec{x}_0, \vec{y}_1, \vec{z}_1$; $\vec{x}_0, \vec{y}_2, \vec{z}_2$; $\vec{x}_0, \vec{y}_3, \vec{z}_3$ and $\vec{x}_0, \vec{y}_4, \vec{z}_4$ lie on different lines, all passing through the same point $\vec{x}_0$. Suppose both $\vec{y}_1, \vec{y}_2, \vec{y}_3, \vec{y}_4$ and $\vec{z}_1, \vec{z}_2, \vec{z}_3, \vec{z}_4$ are collinear. See Fig. 2.125. Now, let

$$\vec{y}_3 = (1 - t_1)\vec{y}_1 + t_1\vec{y}_2, \quad \vec{y}_4 = (1 - t_2)\vec{y}_1 + t_2\vec{y}_2,$$
$$\vec{z}_3 = (1 - s_1)\vec{z}_1 + s_1\vec{z}_2, \quad \vec{z}_4 = (1 - s_2)\vec{z}_1 + s_2\vec{z}_2.$$

Prove that

$$t_1(1 - t_1)^{-1}(1 - t_2)t_2^{-1} = s_1(1 - s_1)^{-1}(1 - s_2)s_2^{-1},$$

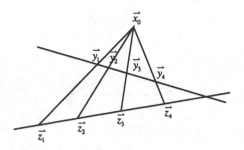

Fig. 2.125

which is called the *cross ratio* of the four points $\vec{y}_3, \vec{y}_4, \vec{y}_1, \vec{y}_2$, in this ordering, and is denoted as

$$(\vec{y}_3, \vec{y}_4;\ \vec{y}_1, \vec{y}_2) = t_1(1 - t_1)^{-1} : t_2(1 - t_2)^{-1} = t_1(1 - t_1)^{-1}(1 - t_2)t_2^{-1}.$$

In case $(\vec{y}_3, \vec{y}_4;\ \vec{y}_1, \vec{y}_2) = -1$, these four points form a set of harmonic points as in Ex. 3.

(*Note*　In Fig. 2.125, if the line passing $\vec{y}_i$'s is parallel to the line passing $\vec{z}_i$'s, to what extent can this problem be simplified both in statement and in proof?)

<B>
Try to review problem sets in Ex. <B> of Sec. 2.6 and use ideas, such as coordinate triangle and homogeneous area coordinate, introduced there to reprove Menelaus, Ceva and Desargues Theorems and Exs. <A> 1–4.

<C> Abstraction and generalization
Try to extend all the results, including Ex. <A>, to n-dimensional affine space over a field $\mathbb{F}$.

2.8.5　*Quadratic curves*

Review (from high-school mathematical courses)
In the Cartesian coordinate system $\mathcal{N} = \{\vec{e}_1, \vec{e}_2\}$, the graph of the equation of the second degree polynomial of two real variables x_1 and x_2 with real coefficients

$$b_{11}x_1^2 + 2b_{12}x_1x_2 + b_{22}x_2^2 + 2b_1x_1 + 2b_2x_2 + b = 0 \qquad (2.8.48)$$

is called a *quadratic curve*, where b_{11}, b_{12} and b_{22} are not all zeros. It is also called a *conic* in a traditional and more geometric feature.

To eliminate x_1 and x_2 terms, use a *translation* $x_1 = x_1' + h$, $x_2 = x_2' + k$ so that

$$b_{11}x_1'^2 + 2b_{12}x_1'x_2' + b_{22}x_2'^2 + 2(b_{11}h + b_{12}k + b_1)x_1' + 2(b_{12}h + b_{22}k + b_2)x_2'$$
$$+ b_{11}h^2 + 2b_{12}hk + b_{22}k^2 + 2b_1h + 2b_2k + b = 0. \tag{$*_1$}$$

Put the coefficients of x_1' and x_2' equal to zeros, i.e.

$$b_{11}h + b_{12}k + b_1 = 0$$
$$b_{12}h + b_{22}k + b_2 = 0.$$

Case 1 $b_{11}b_{22} - b_{12}^2 \neq 0$. The unique solution is

$$h = \frac{b_{12}b_2 - b_{22}b_1}{b_{11}b_{22} - b_{12}^2}, \qquad k = \frac{b_{12}b_1 - b_{11}b_1}{b_{11}b_{22} - b_{12}^2}. \tag{$*_2$}$$

The translation $x_1 = x_1' + h$, $x_2 = x_2' + k$ then transforms $(*_1)$ into

$$b_{11}x_1'^2 + 2b_{12}x_1'x_2' + b_{22}x_2'^2 + b' = 0, \tag{$*_3$}$$

where $b' = b_1h + b_2k + b = \dfrac{1}{b_{11}b_{22} - b_{12}^2} \begin{vmatrix} b_{11} & b_{12} & b_1 \\ b_{12} & b_{22} & b_2 \\ b_1 & b_2 & b \end{vmatrix}$.

To eliminate $x_1'x_2'$ term in $(*_3)$, use a *rotation*

$$x_1' = x_1'' \cos\theta - x_2'' \sin\theta$$
$$x_2' = x_1'' \sin\theta + x_2'' \cos\theta \tag{$*_4$}$$

so that

$$(b_{11}\cos^2\theta + 2b_{12}\cos\theta\sin\theta + b_{22}\sin^2\theta)x_1''^2$$
$$+ [(b_{22} - b_{11})\cos\theta\sin\theta + b_{12}(\cos^2\theta - \sin^2\theta)]x_1''x_2''$$
$$+ [b_{11}\sin^2\theta - 2b_{12}\cos\theta\sin\theta + b_{22}\cos^2\theta]x_2''^2 + b' = 0. \tag{$*_5$}$$

Put the coefficient of $x_1''x_2''$ equal to zero, i.e.

$$(b_{22} - b_{11})\cos\theta\sin\theta + b_{12}(\cos^2\theta - \sin^2\theta) = 0$$
$$\Rightarrow \quad \tan 2\theta = \frac{2b_{12}}{b_{11} - b_{22}}. \tag{$*_6$}$$

Choose the smallest positive θ so that $(*_6)$ holds. Then the rotation transforms $(*_3)$ into

$$b_{11}'x_1''^2 + b_{22}'x_2''^2 + b' = 0, \tag{$*_7$}$$

where $b'_{11} + b'_{22} = b_{11} + b_{12}$,

$$b'_{11} - b'_{22} = (b_{11} - b_{22}) \cos 2\theta + 2b_{12} \sin 2\theta,$$

$$b'^2_{12} - b'_{11}b'_{22} = b^2_{12} - b_{11}b_{22} \quad \text{(here } b'_{12} = 0\text{)},$$

$$b'_{11} = \frac{1}{2}\left(b_{11} + b_{22} \pm \sqrt{(b_{11} - b_{22})^2 + 4b^2_{12}}\,\right),$$

$$b'_{22} = \frac{1}{2}\left(b_{11} + b_{22} \mp \sqrt{(b_{11} - b_{22})^2 + 4b^2_{12}}\,\right).$$

Notice that the choice of plus and minus signs in b'_{11} and b'_{22} should obey the identity $b'_{11}b'_{22} = b_{11}b_{22} - b^2_{12}$.

Case 2 $b_{11}b_{22} - b^2_{12} = 0$. Note that, now, $b_{11} + b_{22} \neq 0$ and $b_{11}b_{22} \geq 0$ hold. From ($*_6$), let

$$\cos\theta = \sqrt{\frac{b_{11}}{b_{11} + b_{22}}} \quad \text{and} \quad \sin\theta = \pm\sqrt{\frac{b_{22}}{b_{11} + b_{22}}} \quad (\pm \text{ is the sign of } b_{11}b_{22}).$$

$$(*_8)$$

The rotation

$$x_1 = x'_1 \cos\theta - x'_2 \sin\theta$$
$$x_2 = x'_1 \sin\theta + x'_2 \cos\theta$$

will transform (2.8.48) into

$$b'_{22}x'^2_2 + 2b'_1 x'_1 + 2b'_2 x'_2 + b = 0, \qquad (*_9)$$

where $b'_{22} = b_{11} + b_{22}$,

$$b'_1 = b_1 \cos\theta + b_2 \sin\theta,$$
$$b'_2 = -b_1 \sin\theta + b_2 \cos\theta.$$

or $b'_{11}x'^2_1 + 2b'_1 x'_1 + 2b'_2 x'_2 + b = 0$ with $b'_{11} = b_{11} + b_{22}$. Since ($*_9$) can be rewritten as

$$\left(x'_2 + \frac{b'_2}{b'_{22}}\right)^2 + \frac{2b'_1}{b'_{22}}\left(x'_1 + \frac{bb'_{22} - b'^2_2}{2b'_{22}b'_1}\right) = 0,$$

the translation

$$x'_1 = x''_1 - \frac{bb'_{22} - b'^2_2}{2b'_{22}b'_1}, \qquad x'_2 = x''_2 - \frac{b'_2}{b'_{22}} \qquad (*_{10})$$

will transform ($*_9$) into

$$x''^2_2 = px''_1, \quad \text{where } p = -\frac{2b'_1}{b'_{22}}. \qquad (*_{11})$$

We summarize $(*_7)$, $(*_9)$ and $(*_{11})$ as

The standard forms of quadratic curves
In the Cartesian coordinate system $\mathcal{N} = \{\vec{e}_1, \vec{e}_2\}$, the quadratic curves are classified into the following nine standard forms, where $a_1 > 0, a_2 > 0$ and $a > 0$.

1. Ellipse $\dfrac{x_1^2}{a_1^2} + \dfrac{x_2^2}{a_2^2} = 1$.

2. Imaginary ellipse $\dfrac{x_1^2}{a_1^2} + \dfrac{x_2^2}{a_2^2} = -1$.

3. Two intersecting imaginary lines or point ellipse $\dfrac{x_1^2}{a_1^2} + \dfrac{x_2^2}{a_2^2} = 0$.

4. Hyperbola $\dfrac{x_1^2}{a_1^2} - \dfrac{x_2^2}{a_2^2} = 1$.

5. Two intersecting lines $\dfrac{x_1^2}{a_1^2} - \dfrac{x_2^2}{a_2^2} = 0$.

6. Parabola $x_2^2 = 2ax_1$.

7. Two parallel lines $x_1^2 = a^2$.

8. Two imaginary parallel lines $x_1^2 = -a^2$.

9. Two coincident lines $x_1^2 = 0$. $\hspace{2cm}$ (2.8.49)

See Fig. 2.126.

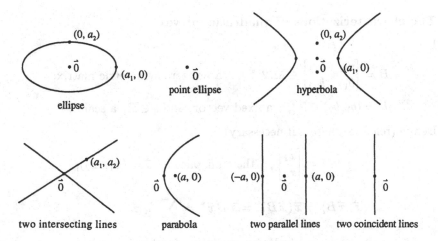

Fig. 2.126

These nine types of quadratic curves can be characterized respectively, via point sets, as follows.

1. A bounded set containing (at least) two distinct points.
2. Empty set.
3. A set containing only one point.
4. An unbounded set containing two non-intersecting branches.
5. Two distinct intersecting lines.
6. An unbounded set containing one (connected) branch.
7. Two distinct parallel lines.
8. Empty set.
9. Two coincident lines.   (2.8.50)

Unfortunately, we cannot distinguish type 2 from type 8 since both are empty set. But, in the eyes of the affine complex plane $\mathbb{C}^2$, both do exist and are of different types. In $\mathbb{C}^2$, the quadratic curves are simplified as five types only:

(a) Ellipse and hyperbola (combining 1, 2, 4).
(b) Parabola (only 6).
(c) Two intersecting lines (combining 3, 5).
(d) Two parallel lines (combining 7, 8).
(e) Two coincident lines (only 9).   (2.8.51)

On the other hand, a careful examination of the coefficients involved from $(*_1)$ to $(*_{11})$ will lead to

The characterizations of quadratic curves

Let

$$B = \begin{bmatrix} b_{11} & b_{12} \\ b_{12} & b_{22} \end{bmatrix} \in M(2;\mathbb{R}), \quad \text{a nonzero symmetric matrix;}$$

$$\vec{b} = (b_1, b_2) \in \mathbb{R}^2, \quad \text{a fixed vector; and } b \in \mathbb{R}, \text{ a scalar.}$$

Define (refer to Chap. 4 if necessary)

$$\vec{x}^* = \begin{bmatrix} x_1 \\ x_2 \end{bmatrix}, \quad \text{the transpace of } \vec{x} = (x_1,\ x_2);$$

$$\langle \vec{x}, \vec{x}B \rangle = \vec{x}(\vec{x}B)^* = \vec{x}B\vec{x}^* = \sum_{i,j=1}^{2} b_{ij}x_ix_j$$

(called a *quadratic form* in $x_1,\ x_2$);

and

$$\langle \vec{b}, \vec{x} \rangle = \vec{b} \, \vec{x}^* = \sum_{i=1}^{2} b_i x_i.$$

Also, let

$$\Delta = \begin{bmatrix} B & \vec{b}^* \\ \vec{b} & b \end{bmatrix} \quad \text{with } \det \Delta = \begin{vmatrix} b_{11} & b_{12} & b_1 \\ b_{12} & b_{22} & b_2 \\ b_1 & b_2 & b \end{vmatrix};$$

$$\det B = \begin{vmatrix} b_{11} & b_{12} \\ b_{12} & b_{22} \end{vmatrix} = b_{11}b_{22} - b_{12}^2; \quad \text{and}$$

$$\text{tr } B = b_{11} + b_{22}.$$

The quadratic curve

$$\langle \vec{x}, \vec{x}B \rangle + 2\langle \vec{b}, \vec{x} \rangle + b = 0$$

is, respectively,

1. Ellipse $\Leftrightarrow \det B > 0, \det \Delta < 0, \text{tr } B > 0$.
2. Imaginary ellipse $\Leftrightarrow \det B > 0, \det \Delta > 0, \text{tr } B > 0$.
3. Point ellipse $\Leftrightarrow \det B > 0, \det \Delta = 0$.
4. Hyperbola $\Leftrightarrow \det B < 0, \det \Delta \neq 0$.
5. Two intersecting lines $\Leftrightarrow \det B < 0, \det \Delta = 0$.
6. Parabola $\Leftrightarrow \det B = 0, \det \Delta \neq 0, \text{tr } B \neq 0$.
7. Two parallel lines $\Leftrightarrow$ In case $b_{11} \neq 0$ (or $b_{22} \neq 0$), $\det B = 0, \det \Delta = 0$
   and
   $$\begin{vmatrix} b_{11} & b_1 \\ b_1 & b \end{vmatrix} = b_{11}b - b_1^2 < 0 \quad \left(\text{or } \begin{vmatrix} b_{22} & b_2 \\ b_2 & b \end{vmatrix} < 0 \right).$$

8. Two imaginary parallel lines $\Leftrightarrow$ In case $b_{11} \neq 0$ (or $b_{22} \neq 0$), $\det B = 0$,
   $\det \Delta = 0$ and
   $$\begin{vmatrix} b_{11} & b_1 \\ b_1 & b \end{vmatrix} = b_{11}b - b_1^2 > 0 \quad \left(\text{or } \begin{vmatrix} b_{22} & b_2 \\ b_2 & b \end{vmatrix} > 0 \right).$$

9. Two coincident lines $\Leftrightarrow$ In case $b_{11} \neq 0$ (or $b_{22} \neq 0$), $\det B = 0$,
   $\det \Delta = 0$ and
   $$\begin{vmatrix} b_{11} & b_1 \\ b_1 & b \end{vmatrix} = b_{11}b - b_1^2 = 0 \quad \left(\text{or } \begin{vmatrix} b_{22} & b_2 \\ b_2 & b \end{vmatrix} = 0 \right). \qquad (2.8.52)$$

The quantity $\det B = b_{11}b_{22} - b_{12}^2$ is called the *discriminant* for quadratic curves. Sometimes, ellipse, hyperbola and parabola (note that in these cases, the rank of Δ is equal to 3) are called *non-degenerated conics* or *irreducible quadratic curves*, while the others are called *degenerated* or *reducible* (note that, the rank $r(\Delta) = 2$ for two intersecting lines and $r(\Delta) = 1$ for type 7 and type 9). Therefore, in short, for irreducible curves, it is

an ellipse $\Leftrightarrow \det B > 0$, or
a hyperbola $\Leftrightarrow \det B < 0$, or
a parabola $\Leftrightarrow \det B = 0$.

A point $\vec{c}$ is called a *center* of a quadratic curve γ if for each point $\vec{x}$ on γ, there is another point $\vec{y}$ on γ such that

$$\vec{c} = \frac{1}{2}(\vec{x} + \vec{y}). \tag{2.8.53}$$

According to this definition, among the standard forms in (2.8.49), ellipse, hyperbola and two intersecting lines all have center at $\vec{0}$, two parallel lines have every point on the line equidistant from both lines as its centers, two coincident lines have every point on it as its centers, while parabola does not have center. A non-degenerated conic with a (unique) center is called a *central-conic*. There are only two such curves: ellipse and hyperbola and the criterion for this is $\det B \neq 0$.

Remark
Instead of the conventional methods shown from $(*_1)$ to $(*_{11})$, one can employ effectively the techniques developed in linear algebra (even up to this point in this text) to give (2.8.49) and (2.8.52) a more concise, systematic proof that can be generalized easily to three-dimensional quadrics or even higher-dimensional ones. This *linearly algebraic method* mainly contains the following essentials:

(a) Homogeneous coordinates of points in affine plane $\mathbb{R}^2$ or spaces $\mathbb{R}^n$, $n \geq 2$ (refer to Ex. <B> of Sec. 2.6).
(b) The matrix representation of an affine transformation such as (2.8.10) or (2.8.11).
(c) Orthogonal matrix $P_{n \times n}$, i.e. $P^* = P^{-1}$ (refer to (2.8.32), (2.8.34) and (2.8.37)).
(d) The diagonalization of a symmetric matrix (refer to Sec. 2.7.6, Ex. 2 in Sec. B.11 and Ex. 7 in Sec. B.9). $\hspace{2cm}$ (2.8.54)

As a preview to this method, we appoint Ex. <B> 1 for the readers to practice around. In Secs. 3.8.5, 4.10 and 5.10, we will formulate this method to prove counterparts of (2.8.49) for quadratic curves and quadrics.

Affine point of view about quadratic curves

As we have known already in (2.4.2), a change of coordinates between affine bases is a special kind of affine transformations.

Conversely, any affine transformation on $\mathbb{R}^2$ can be treated as a change of coordinates between some affine bases. For example, for any fixed affine basis $\mathcal{B} = \{\vec{a}_0, \vec{a}_1, \vec{a}_2\}$ and a given affine transformation

$$T(\vec{x}) = \vec{x}_0 + \vec{x}A, \qquad (2.8.55)$$

where $A = [a_{ij}] \in GL(2; \mathbb{R})$, we

1. consider $\vec{x}_0$ as $[\vec{b}_0]_\mathcal{B}$ for some $\vec{b}_0 \in \mathbb{R}^2$, and
2. construct a new affine basis $\mathcal{B}' = \{\vec{b}_0, \vec{b}_1, \vec{b}_2\}$, where $\vec{b}_i = a_{i1}(\vec{a}_1 - \vec{a}_0) + a_{i2}(\vec{a}_2 - \vec{a}_0)$ for $i = 1, 2$ so that $(a_{i1}, a_{i2}) = [\vec{b}_i]_\mathcal{B}$ is the row vector of $A = A_\mathcal{B}^{\mathcal{B}'}$, the transition matrix from $\mathcal{B}'$ to $\mathcal{B}$, and
3. treat $\vec{x}$ as $[\vec{y}]_{\mathcal{B}'}$.

Then T is nothing but the change of coordinates from $\mathcal{B}'$ to $\mathcal{B}$: $[\vec{b}_0]_\mathcal{B} + [\vec{y}]_{\mathcal{B}'} A_\mathcal{B}^{\mathcal{B}'} = [\vec{y}]_\mathcal{B}$. Refer to (2.4.2) and Ex. <B> 1 of Sec. 2.4.

Under this convention, we notice the statement in the next paragraph.

Give an arbitrary affine transformation on $\mathbb{R}^2$

$$T(\vec{y}) = \vec{x} = \vec{y}_0 + \vec{y}A,$$

where $\vec{y}_0 \in \mathbb{R}^2$ is fixed and $A = [a_{ij}]_{2 \times 2} \in GL(2; \mathbb{R})$. The image of the quadratic curve (2.8.48), in $\mathcal{N} = \{\vec{0}, \vec{e}_1, \vec{e}_2\}$, under T is

$$(\vec{y}_0 + \vec{y}A)B(\vec{y}_0 + \vec{y}A)^* + 2\vec{b}(\vec{y}_0 + \vec{y}A)^* + b = 0$$
$$\Rightarrow \vec{y}ABA^*\vec{y}^* + 2\langle \vec{y}_0 BA^* + \vec{b}A^*, \vec{y}\rangle$$
$$+ (\vec{y}_0 B\vec{y}_0^* + 2\langle \vec{b}, \vec{y}_0\rangle + b) = 0, \qquad (2.8.56)$$

which is still a *quadratic curve* but in the *affine basis* $\mathcal{B} = \{\vec{y}_0, \vec{y}_0 + \vec{a}_1, \vec{y}_0 + \vec{a}_2\}$, where $\vec{a}_1 = (a_{11}, a_{12})$ and $\vec{a}_2 = (a_{21}, a_{22})$ are row vectors of A.

For two quadratic curves γ_1 and γ_2, no matter in what affine bases, if there exists an affine transformation T on $\mathbb{R}^2$ mapping γ_1 onto γ_2, i.e.

$$T(\gamma_1) = \gamma_2, \qquad (2.8.57)$$

then γ_1 and γ_2 are said to be *affinely equivalent* and are classified as the same type.

Note This definition is *not* good for types 2 and 8. We have compensated this deficiency by introducing the algebraic criteria in (2.8.52) instead of the geometric (and intuitive) criteria in (2.8.50).

As a consequence of (2.8.40), from (2.8.50) it follows easily that quadratic curves of different types in (2.8.49), except types 2 and 8, cannot be affinely equivalent.

However, quadratic curves of the same type in (2.8.49) indeed are affinely equivalent in the sense of (2.8.57).

For example, let γ_1 and γ_2 be two arbitrary ellipses on the plane $\mathbb{R}^2$. After suitable translation and rotation (see Sec. 2.8.2), one can transform γ_2 into a new location so that its center coincides with that of γ_1 and its major axis lies on that of γ_1. Use γ_2^* to denote this new-located ellipse. Choose the center as the origin of a Cartesian coordinate system and the common major axis as x_1-axis, then γ_1 and γ_2^* can be expressed as

$$\gamma_1: \frac{x_1^2}{a_1^2} + \frac{x_2^2}{b_1^2} = 1,$$

$$\gamma_2^*: \frac{x_1^2}{a_2^2} + \frac{x_2^2}{b_2^2} = 1.$$

Then, the affine transformation

$$y_1 = \frac{a_2}{a_1}x_1, \quad y_2 = \frac{b_2}{b_1}x_2 \tag{$*_{12}$}$$

will transform γ_1 onto γ_2^*. This means that γ_1 and γ_2^*, and hence γ_1 and γ_2 are affinely equivalent.

For two imaginary ellipses

$$\frac{x_1^2}{a_1^2} + \frac{x_2^2}{b_1^2} = -1, \quad \frac{x_1^2}{a_2^2} + \frac{x_2^2}{b_2^2} = -1$$

or point ellipses or hyperbolas, $(*_{12})$ still works.

For two parabolas

$$x_2^2 = 2a_1 x_1, \ a_1 \neq 0 \quad \text{and} \quad x_2^2 = 2a_2 x_2, \ a_2 \neq 0,$$

the affine transformation

$$y_1 = \frac{a_1}{a_2}x_1, \quad y_2 = x_2 \tag{$*_{13}$}$$

will do.

Readers definitely can handle the remaining types in (2.8.49), except types 2 and 8.

We summarize as

The classification of quadratic curves in affine geometry
The quadratic curves are classified into nine types stated in (2.8.49) under affine transformations (motions). (2.8.58)

We arrange Ex. <A> 3 for readers to prove this result by using (2.8.52) and by observing (2.8.59) below.

Let us come back to (2.8.56) and compute the following quantities (refer to (2.8.52)):

$$\det(ABA^*) = \det A \cdot \det B \cdot \det A^* = (\det A)^2 \det B;$$

$$
\det \Delta' = \det \begin{bmatrix} ABA^* & (\vec{y}_0 BA^* + bA^*)^* \\ \vec{y}_0 BA^* + bA^* & \vec{y}_0 B \vec{y}_0^* + 2\vec{b}\,\vec{y}_0^* + b \end{bmatrix}
$$

$$
= \det \begin{bmatrix} A & 0 \\ \vec{y}_0 & 1 \end{bmatrix} \begin{bmatrix} B & \vec{b}^* \\ \vec{b} & b \end{bmatrix} \begin{bmatrix} A & 0 \\ \vec{y}_0 & 1 \end{bmatrix}^*
$$

$$
= (\det A)^2 \det \begin{bmatrix} B & \vec{b}^* \\ \vec{b} & b \end{bmatrix} = (\det A)^2 \Delta;
$$

$$
\operatorname{tr}(ABA^*) = \operatorname{tr}(BA^*A)
$$
$$
= (a_{11}^2 + a_{21}^2)b_{11} + 2(a_{11}a_{12} + a_{21}a_{22})b_{12} + (a_{12}^2 + a_{22}^2)b_{22}
$$
$$(*_{14})$$

where $A = [a_{ij}]_{2 \times 2}$.

Since A is invertible and thus $\det A \neq 0$, we note that $\det ABA^*$ and $\det B, \det \Delta'$ and $\det \Delta$ all have the same signs.

The implication of $\operatorname{tr} B$ upon $\operatorname{tr}(ABA^*)$ is less obvious. A partial result is derived as follows. Suppose $\det B > 0$ and $\operatorname{tr} B > 0$ hold. Then

$$
b_{11}b_{22} > b_{12}^2 \quad \text{and} \quad b_{11} > 0, \quad b_{22} > 0
$$
$$
\Rightarrow -\sqrt{b_{11}b_{22}} < b_{12} < \sqrt{b_{11}b_{22}}
$$
$$
\Rightarrow \operatorname{tr}(ABA^*) > (a_{11}^2 + a_{21}^2)b_{11} - 2(a_{11}a_{12} + a_{21}a_{22})\sqrt{b_{11}b_{22}}
$$
$$
+ (a_{12}^2 + a_{22}^2)b_{22}
$$
$$
= (a_{11}\sqrt{b_{11}} - a_{12}\sqrt{b_{22}})^2 + (a_{21}\sqrt{b_{11}} - a_{22}\sqrt{b_{22}})^2 \geq 0.
$$

Since the inverse of an affine transformation is still affine, the assumptions that $\det(ABA^*) > 0$ and $\operatorname{tr}(ABA^*) > 0$ would imply that $\operatorname{tr} B > 0$ holds.

Summarize as

The affine invariants of quadratic curves
For a quadratic curve

$$\langle \vec{x}, \vec{x} B \rangle + 2\langle \vec{b}, \vec{x} \rangle + b = 0,$$

the signs or zeros of det B, and

$$\det \begin{bmatrix} B & \vec{b}^* \\ \vec{b} & b \end{bmatrix}$$

are *affine invariants*. In case det B and tr B are positive, then the positiveness of

$$\operatorname{tr} B$$

is an *affine invariant*.   (2.8.59)

Later in Sec. 4.10, we are going to prove that these three quantities are *Euclidean* invariants.

Exercises

<A>

1. For each of the following quadratic curves, do the following problems:

   (1) Determine what type of curve it is.
   (2) Determine the affine transformation needed to reduce it to its standard form.
   (3) Write out its standard form.
   (4) Sketch the curve in the original Cartesian coordinate system $\mathcal{N} = \{\vec{0}, \vec{e}_1, \vec{e}_2\}$ and in the affine orthonormal basis $\mathcal{B}$ where its standard form stands, of course, in the same affine plane $\mathbb{R}^2$.

   (a) $x_1^2 - 6x_1x_2 + x_2^2 - 4x_1 + 4x_2 + 3 = 0.$
   (b) $2x_1^2 + 4\sqrt{3}x_1x_2 + 6x_2^2 + (8 + \sqrt{3})x_1 + (8\sqrt{3} - 1)x_2 = 0.$
   (c) $2x_1^2 + 3x_1x_2 - 2x_2^2 + 5x_2 - 2 = 0.$
   (d) $2x_1^2 + 2x_1x_2 + 2x_2^2 - 4x_1 - 2x_2 - 1 = 0.$
   (e) $x_1^2 - 2x_1x_2 + x_2^2 + 2x_1 - 2x_2 + 1 = 0.$
   (f) $x_1^2 - 2x_1x_2 + x_2^2 - 3x_1 + 3x_2 + 2 = 0.$

2. Use $(*_1)$–$(*_{11})$ to prove (2.8.52) in detail.
3. Try to use (2.8.52) and (2.8.59) to prove (2.8.58) in detail.

4. (1994 Putnam Examination) Find the value of m so that the line $y = mx$ bisects the region

$$\left\{ (x, y) \in \mathbb{R}^2 \,\middle|\, \frac{x^2}{4} + y^2 \le 1, \ x \ge 0, \ y \ge 0 \right\}.$$

For such concept as center, tangent line, pole and polar, diameter and conjugate diameter for quadratic curves and the methods to derive them, please refer to Sec. 4.10.

<B>

1. In order to preview the linearly algebraic method indicated in (2.8.54) in the proof of (2.8.52), try the following steps.

   (a) Use $\frac{x_1}{x_3}$ and $\frac{x_2}{x_3}$, where $x_3 \ne 0$, to replace x_1 and x_2 respectively in (2.8.48), so that the quadratic curve has equation

   $$b_{11}x_1^2 + 2b_{12}x_1x_2 + b_{22}x_2^2 + 2b_1x_1x_3 + 2b_2x_2x_3 + bx_3^2 = 0$$

   in the homogeneous coordinate (x_1, x_2, x_3) for the affine plane $\mathbb{R}^2$. Note that from this equation (2.8.48) may be obtained by putting $x_3 = 1$. Rewrite the above equation in matrix form as

   $$\langle \vec{x}, \vec{x}\Delta \rangle = 0, \quad \text{where } \Delta = \begin{bmatrix} B & \vec{b}^* \\ \vec{b} & b \end{bmatrix} \text{ and } \vec{x} = (x_1, x_2, x_3).$$

   (b) Rewrite an affine transformation

   $$(y_1, y_2) \to \vec{v}_0 + (y_1, y_2)A, \quad \text{where } \vec{v}_0 \in \mathbb{R}^2$$

   in the homogeneous form

   $$\vec{x} = \vec{y} \begin{bmatrix} A & \vec{0}^* \\ \vec{v}_0 & 1 \end{bmatrix}, \quad \text{where } \vec{y} = (y_1, y_2, y_3).$$

   Note that this will reduce to the one stated in (2.8.10) or (2.8.11) by putting $x_3 = y_3 = 1$.

   (c) Then $\langle \vec{x}, \vec{x}\Delta \rangle = \vec{x}\Delta\vec{x}^* = 0$ under the affine transformation becomes

   $$\vec{y} \begin{bmatrix} A & \vec{0}^* \\ \vec{v}_0 & 1 \end{bmatrix} \Delta \begin{bmatrix} A & \vec{0}^* \\ \vec{v}_0 & 1 \end{bmatrix}^* \vec{y}^* = 0.$$

Now both B and ABA^* are symmetric matrices. Since B is diagonalizable, there exists an orthogonal matrix P so that

$$PBP^{-1} = \begin{bmatrix} \lambda_1 & 0 \\ 0 & \lambda_2 \end{bmatrix}$$

is a diagonal matrix. If chosen this P as A, then the equation becomes

$$\lambda_1 y_1^2 + \lambda_2 y_2^2 + 2b_1' y_1 y_3 + 2b_2' y_2 y_3 + b' y_3^2 = 0$$

or, by putting $y_3 = 1$,

$$\lambda_1 y_1^2 + \lambda_2 y_2^2 + 2b_1' y_1 + 2b_2' y_2 + b' = 0.$$

(d) Let

$$A = P = \begin{bmatrix} \cos\theta & \sin\theta \\ -\sin\theta & \cos\theta \end{bmatrix}, \quad \text{and}$$

$$\begin{bmatrix} A & \vec{0}^* \\ \vec{v}_0 & 1 \end{bmatrix} \Delta \begin{bmatrix} A & \vec{0}^* \\ \vec{v}_0 & 1 \end{bmatrix}^* = \begin{bmatrix} b_{11}' & b_{12}' & b_1' \\ b_{12}' & b_{22}' & b_2' \\ b_1' & b_2' & b' \end{bmatrix}.$$

Try to use b_{ij}, b_i, b to express b_{ij}', b_i' and b'.

(e) Use data obtained so far to prove (2.8.49) and (2.8.52).

The following three problems will introduce the classification of affine transformations according to the non-degenerated conics and the corresponding invariant subgroups of $G_a(2; \mathbb{R})$.

2. *Elliptic rotation* (refer to Ex. <B> 2 of Sec. 2.8.2)
Take the unit circle, in $\mathcal{N} = \{\vec{0}, \vec{e}_1, \vec{e}_2\}$,

$$x_1^2 + x_2^2 = 1$$

as the representative of the ellipses in the affine plane $\mathbb{R}^2$. The problem is to find all possible affine transformations on $\mathbb{R}^2$ that keep $x_1^2 + x_2^2 = 1$ invariant.

(a) Show that an affine transformation $\vec{x} = \vec{y}_0 + \vec{y}A$ keeps $x_1^2 + x_2^2 = 1$ invariant if and only if

$$\vec{y}AA^*\vec{y}^* + \vec{y}_0 A^*\vec{y}^* + \vec{y}A\vec{y}_0^* + \vec{y}_0\vec{y}_0^* = 1$$
$$\Leftrightarrow \vec{y}AA^*\vec{y}^* = 1 \quad \text{and} \quad \vec{y}_0 = \vec{0}$$
$$\Leftrightarrow AA^* = I_2 \quad \text{and} \quad \vec{y}_0 = \vec{0}$$

i.e. $\vec{y}_0 = \vec{0}$ and $A^* = A^{-1}$. Hence, A is an orthogonal matrix where

$$A = \begin{bmatrix} \cos\theta & \sin\theta \\ -\sin\theta & \cos\theta \end{bmatrix} \quad \text{or} \quad \begin{bmatrix} \cos\theta & \sin\theta \\ \sin\theta & -\cos\theta \end{bmatrix}$$

(see (2.8.34) and (2.8.37)).

(b) The set

$$\left\{ \begin{bmatrix} \cos\theta & \sin\theta \\ -\sin\theta & \cos\theta \end{bmatrix} \,\middle|\, \theta \in \mathbb{R} \right\}$$

forms a *transitive one-parameter subgroup* of $G_a(2;\mathbb{R})$, whose member is called an *elliptic rotation*. Transitivity means that for any two points on $x_1^2 + x_2^2 = 1$, there is a member of this group that transforms one of the points into another point.

3. *Hyperbolic rotation* (refer to Ex. <B> 1 of Sec. 2.8.2)
   Take the hyperbola, in $\mathcal{N} = \{\vec{0}, \vec{e}_1, \vec{e}_2\}$,

$$x_1 x_2 = 1$$

as the representative of hyperbolas. The problem is to find all possible affine transformations on $\mathbb{R}^2$ that keep $x_1 x_2 = 1$ invariant.

(a) $\vec{x} = \vec{y}_0 + \vec{x}A$ keeps $x_1 x_2 = 1$ invariant if and only if

$$\vec{y}A \begin{bmatrix} 0 & \frac{1}{2} \\ \frac{1}{2} & 0 \end{bmatrix} A^* \vec{y}^* + \vec{y}_0 \begin{bmatrix} 0 & \frac{1}{2} \\ \frac{1}{2} & 0 \end{bmatrix} A^* \vec{y}^* + \vec{y}A \begin{bmatrix} 0 & \frac{1}{2} \\ \frac{1}{2} & 0 \end{bmatrix} \vec{y}_0^*$$

$$+ \vec{y}_0 \begin{bmatrix} 0 & \frac{1}{2} \\ \frac{1}{2} & 0 \end{bmatrix} \vec{y}_0^* = 1$$

$$\Leftrightarrow A \begin{bmatrix} 0 & \frac{1}{2} \\ \frac{1}{2} & 0 \end{bmatrix} A^* = \begin{bmatrix} 0 & \frac{1}{2} \\ \frac{1}{2} & 0 \end{bmatrix} \quad \text{and} \quad \vec{y}_0 = \vec{0}$$

$\Leftrightarrow$ (let $A = [a_{ij}]_{2\times2}$)

$a_{11} = a_{22} = 0$ and $a_{12}a_{21} = 1$ or $a_{12} = a_{21} = 0$ and $a_{11}a_{22} = 1$.

(b) The set

$$\left\{ \begin{bmatrix} a & 0 \\ 0 & \frac{1}{a} \end{bmatrix} \,\middle|\, a \in \mathbb{R} \text{ and } a \neq 0 \right\}$$

forms a *transitive one-parameter subgroup* of $G_a(2;\mathbb{R})$, whose member is called a *hyperbolic rotation*.

(c) The asymptotes $x_1 = 0$ and $x_2 = 0$ of $x_1 x_2 = 1$ are the only invariant lines under this group. In fact, for each member A of this group,

$$(x_1, 0)A = a(x_1, 0) \quad \text{for each } x_1 \in \mathbb{R}, \quad \text{and}$$

$$(0, x_2)A = \frac{1}{a}(0, x_2) \quad \text{for each } x_2 \in \mathbb{R},$$

so A has eigenvectors $(x_1, 0)$ and $(0, x_2)$ with respective eigenvalues a and $\frac{1}{a}$.

4. *Parabolic translation*

Take the parabola, in $\mathcal{N} = \{\vec{0}, \vec{e}_1, \vec{e}_2\}$,

$$x_1 = x_2^2$$

as the representative of parabolas. The problem is to find all possible affine transformations that keep $x_1 = x_2^2$ invariant.

(a) Let $\vec{x} = \vec{y}_0 + \vec{y}A$ where $\vec{y}_0 = (b_1, b_2)$ and $A = [a_{ij}]_{2 \times 2} \in \mathrm{GL}(2; \mathbb{R})$. Then this affine transformation keeps $x_1 = x_2^2$ invariant if and only if, $\vec{y} = (y_1, y_2)$,

$$(b_1 + a_{11}y_1 + a_{21}y_2) - (b_2 + a_{12}y_1 + a_{22}y_2)^2 = 0$$
$$\Leftrightarrow a_{12} = 0, \; b_1 = b_2^2, \; a_{11} = 2a_{22}^2 \quad \text{and} \quad a_{21} = 2a_{22}b_2$$
$$\Leftrightarrow x_1 = a^2 y_1 + 2aby_2 + b^2$$
$$x_2 = ay_2 + b, \quad \text{or}$$

$$\vec{x} = (b^2, b) + \vec{y}\begin{bmatrix} a^2 & 0 \\ 2ab & a \end{bmatrix}, \quad a \neq 0, \; b \in \mathbb{R}.$$

All such transformations form a subgroup of two parameters a and b of $\mathrm{G}_a(2; \mathbb{R})$.

(b) Take $a = 1$. The set

$$\left\{ (b^2, b) + \vec{x}\begin{bmatrix} 1 & 0 \\ 2b & 1 \end{bmatrix} \; \middle| \; b \in \mathbb{R} \text{ and } \vec{x} \in \mathbb{R}^2 \right\}$$

forms a *transitive one-parameter subgroup* of $\mathrm{G}_a(2; \mathbb{R})$, whose member is called a *parabolic translation*.

(c) The linear part of a parabolic translation

$$\begin{bmatrix} 1 & 0 \\ 2b & 1 \end{bmatrix}$$

is a shearing (refer to Ex. 6 in Sec. 2.7.2 and (2.8.32)) and the x_1-axis $\langle\langle \vec{e}_1 \rangle\rangle$ is its line of invariant points.

In what follows, our purpose is to characterize non-degenerated conics by counting the number of infinite points lying on it. We adopt ideas and notations used in Ex. <B> of Sec. 2.6. Remind that any point in $\mathbb{R}^2$ is represented by a certain barycentric coordinate as $(\lambda_1: \lambda_2: \lambda_3)$ where $\lambda_1 + \lambda_2 + \lambda_3 \neq 0$. In case $\lambda_1 + \lambda_2 + \lambda_3 = 0, (\lambda_1: \lambda_2: \lambda_3)$ is said to represent an *infinite point* of the affine plane $\mathbb{R}^2$. The set of infinite points

$$l_\infty = \{(\lambda_1: \lambda_2: \lambda_3) | \lambda_1 + \lambda_2 + \lambda_3 = 0\}$$

is called the *infinite line* of $\mathbb{R}^2$ so that $\mathbb{R}^2 \cup l_\infty = P^2(\mathbb{R})$ is called the *projective plane* (refer to (3.8.60) and Ex. <A> 13 of Sec. 3.8.4.).

5. Any line in the affine plane $\mathbb{R}^2$ passes through a unique infinite point. Try the following steps.

   (a) By Ex. <B> 5 of Sec. 2.6, show that any ordinary line l has equation $c_1\lambda_1 + c_2\lambda_2 + c_3\lambda_3 = 0$ in barycentric coordinates, where c_1, c_2 and c_3 are not all zeros.

   (b) Let $(\lambda_1^*: \lambda_2^*: \lambda_3^*)$ be an infinite point on l, if any. Solve

   $$c_1\lambda_1^* + c_2\lambda_2^* + c_3\lambda_3^* = 0$$
   $$\lambda_1^* + \lambda_2^* + \lambda_3^* = 0$$

   and get $(\lambda_1^* : \lambda_2^* : \lambda_3^*) = (c_2 - c_3 : c_3 - c_1 : c_1 - c_2)$.

6. Parallel lines in $\mathbb{R}^2$ pass through the same infinite point. Try to use Ex. <B> 5(d) of Sec. 2.6 and Ex. 5.

7. Let $\triangle A_1 A_2 A_3$ be a coordinate triangle. Show that a quadratic curve Γ passing through the base points A_1, A_2 and A_3 has the equation

   $$c_1\lambda_2\lambda_3 + c_2\lambda_3\lambda_1 + c_3\lambda_1\lambda_2 = 0$$

in barycentric coordinates. Try to use Ex. <B> 2(c) of Sec. 2.6. Show that the number of infinite points lying on Γ has the following criteria: the discriminant

$$D = c_1^2 + c_2^2 + c_3^2 - 2c_2c_3 - 2c_3c_1 - 2c_1c_2 \begin{cases} > 0 \Leftrightarrow \text{two infinite points} \\ = 0 \Leftrightarrow \text{one infinite point} \\ < 0 \Leftrightarrow \text{none.} \end{cases}$$

8. Let Γ be a quadratic curve passing through the base points of a coordinate triangle. Show that

   (a) If Γ is a hyperbola, then there are two infinite points lying on Γ.
   (b) If Γ is a parabola, then there is only one infinite point lying on Γ.
   (c) If Γ is an ellipse, then there is no infinite point on Γ.

For (c), follow the steps (due to Professor Lu Yang):

(1) Suppose Γ is $\frac{x^2}{a^2} + \frac{y^2}{b^2} = 1$ in $\mathcal{N} = \{0, \vec{e}_1, \vec{e}_2\}$. Use Ex. <B> 2(c) of Sec. 2.6 to show that, in barycentric coordinates, Γ has equation

$$c_1\lambda_2\lambda_3 + c_2\lambda_3\lambda_1 + c_3\lambda_1\lambda_2 = 0,$$

where

$$c_1 = 2\left(\frac{x_2 x_3}{a^2} + \frac{y_2 y_3}{b^2} - 1\right) = -\left[\frac{(x_2 - x_3)^2}{a^2} + \frac{(y_2 - y_3)^2}{b^2}\right],$$

$$c_2 = 2\left(\frac{x_3 x_1}{a^2} + \frac{y_3 y_1}{b^2} - 1\right) = -\left[\frac{(x_3 - x_1)^2}{a^2} + \frac{(y_3 - y_1)^2}{b^2}\right],$$

$$c_3 = 2\left(\frac{x_1 x_2}{a^2} + \frac{y_1 y_2}{b^2} - 1\right) = -\left[\frac{(x_1 - x_2)^2}{a^2} + \frac{(y_1 - y_2)^2}{b^2}\right]$$

and $A_1 = (x_1, y_1), A_2 = (x_2, y_2), A_3 = (x_3, y_3)$ in $\mathcal{N} = \{0, \vec{e}_1, \vec{e}_2\}$.

(2) By Ex. 7, the discriminant is

$$D = \frac{Q_{xx}}{a^4} + \frac{Q_{xy}}{a^2 b^2} + \frac{Q_{yy}}{b^4}$$

where Q_{xx} is a polynomial in x_1, x_2, x_3; Q_{yy} is a polynomial in y_1, y_2, y_3; and Q_{xy} is a polynomial in $x_1, x_2, x_3, y_1, y_2, y_3$.

(3) Show that $Q_{xx} = Q_{yy} = 0$ and

$$2Q_{xy} = Q_{xx} + 2Q_{xy} + Q_{yy} = \cdots$$
$$= -16s(s - a_1)(s - a_2)(s - a_3)$$
$$= -16\Delta^2,$$

where $a_1 = A_2 A_3, a_2 = A_3 A_1, a_3 = A_1 A_2$ and $\Delta = \sqrt{s(s - a_1)(s - a_2)(s - a_3)}$ is the Herron's formula.

(4) Finally, $D = -\frac{16\Delta^2}{a^2 b^2} < 0$.

For (a), let Γ be $\frac{x^2}{a^2} - \frac{y^2}{b^2} = 1$, then

$$D = -\frac{2Q_{xy}}{a^2 b^2} = \frac{16\Delta^2}{a^2 b^2} > 0.$$

For (b), let Γ be $y^2 = 2px$, then

$$D = Q_{yy} = 0.$$

Remark As a whole, a quadratic curve

$$\langle \vec{x}, \vec{x}B \rangle + 2\langle \vec{b}, \vec{x} \rangle + b = 0$$

is classified to be

1. an *elliptic type* ⟺ (algebraic) det $B > 0$.

⟺ (geometric) containing no infinite point.

These contain types 1, 2, 3.

2. a *hyperbolic type* ⟺ det $B < 0$.

⟺ containing two infinite points.

These contain types 4, 5.

3. a *parabolic type* ⟺ det $B = 0$.

⟺ containing one infinite point.

These contain types 6, 7, 8, 9.

\<C\> Abstraction and generalization

See Secs. 3.8.5, 4.10 and 5.10.

CHAPTER 3

The Three-Dimensional Real Vector Space $\mathbb{R}^3$

Introduction

In our real world, there does exist a point lying outside a fixed given plane. For example, a lamp (considered as a point) hanging over a desk (considered as a plane) is such a case.

Figure 3.1 shows that one point R is not on the plane Σ. The family of the straight lines connecting R to all arbitrary points in Σ are considered, in imagination, to form a so-called three-dimensional space, physically inhabited by the human being.

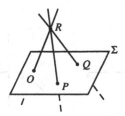

Fig. 3.1

Therefore, we have the

Postulate *Four different non-coplanar points determine a unique (three-dimensional) space.*

Here and only in this chapter, the term "space" will always mean three-dimensional as postulated above. Usually, a parallelepiped including its interior is considered as a symbolic graph of a space Γ (see Fig. 3.2).

One should be familiar with the following basic facts about space Γ.

(1) Γ contains uncountably many points.

(2) Γ contains the line generated by any two different points in it, the plane generated by any three non-collinear points in it, and coincides with the space determined by any four different non-coplanar points in it.

319

Fig. 3.2

(3) A space Γ possesses the following Euclidean geometric concepts or quantities.

1. Length.
2. Angle.
3. The area of a rectangle is equal to length times width, and therefore, the area of a parallelogram is equal to height times base (length).
4. The volume of a rectangular box is equal to length times width times height, and therefore, the volume of a parallelepiped is equal to height times base area (of a parallelogram).
5. The right-hand system and left-hand system.

See Fig. 3.3.

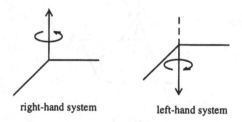

right-hand system   left-hand system

Fig. 3.3

A directed segment $\overrightarrow{PQ}$ in space Γ is called a (space) *vector*, considered identical when both have the same length and the same direction just as stated in Sec. 2.1. And the most important of all, space vectors satisfy all the properties as listed in (2.1.11) (refer to Remark 2.1 in Sec. 2.1). Hereafter, in this section, we will feel free to use these operational properties, if necessary. What we need is to note the following facts:

1. $\alpha \in \mathbb{R}$ and $\vec{x}$ is a space vector $\Rightarrow \alpha\vec{x}$ and $\vec{x}$ are collinear vectors in the space Γ.
2. $\vec{x}$ and $\vec{y}$ are space vectors $\Rightarrow \vec{x} + \vec{y}$ and $\vec{x}$, $\vec{y}$ are coplanar vectors in the space Γ.

3. $\vec{x}, \vec{y}$ and $\vec{z}$ are space vectors $\Rightarrow \vec{x} + \vec{y} + \vec{z}$ and $\vec{x}, \vec{y}, \vec{z}$ may be either coplanar or non-coplanar.

See Fig. 3.4.

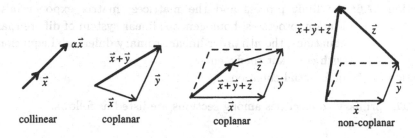

collinear   coplanar   coplanar   non-coplanar

Fig. 3.4

Sketch of the Content

Refer to the *Sketch of the Content* in the Introduction of Chap. 2.

The whole theoretical development will be based on the Postulate stated above.

Since the steps adopted are exactly the same like those in Chap. 2, only main results and difference between $\mathbb{R}^2$ and $\mathbb{R}^3$ will be mentioned and emphasized and the details of proofs will be left to the readers as good exercises.

Space vectors are already explained in Remark 2.1 of Sec. 2.1. We proceed at the very beginning to vectorize the space Γ (Sec. 3.1) and coordinatize it (Sec. 3.2), then discuss changes of coordinates (Sec. 3.3). The lines (Sec. 3.4) and the planes (Sec. 3.5) in $\mathbb{R}^3$ are introduced.

From Sec. 3.6 to Sec. 3.8, both titles and contents of the sections are parallel to those of the same numbered sections in Chap. 2.

Note that, in Sec. 3.6, we formally use affine basis $\mathcal{B} = \{\vec{a}_0, \vec{a}_1, \vec{a}_2, \vec{a}_3\}$ to replace the coordinate system Γ ($\vec{a}_0; \vec{a}_1, \vec{a}_2, \vec{a}_3$) introduced in Sec. 3.2 and is used thereafter.

Sections 3.4–3.6 lay the foundation for geometric interpretations of results to be developed in Secs. 3.7 and 3.8.

Linear operators (or transformations, Sec. 3.7) are the main theme in the whole Chap. 3. Except routine topics as in Sec. 2.7, exercises here contain some applications of the theory to the related fields, such as Markov processes and differential equations.

Section 3.8 investigates the affine transformations, affine invariants and geometry on $\mathbb{R}^3$.

Exercises <D> throughout the whole chapter contain the following topics:

Section 3.7.6: the limit process and the matrices; matrix exponentials; Markov processes; homogeneous linear system of differential equations; the nth order linear ordinary differential equation with constant coefficients.
Section 3.7.7: differential equations.

The primary connections among sections are listed as follows.

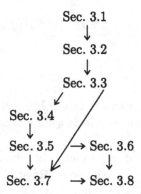

3.1 Vectorization of a Space: Affine Structure

Let O, A_1, A_2 and A_3 be four different non-coplanar points in a space Γ. Also, let the space vectors

$$\vec{a}_i = \overrightarrow{OA_i}, \quad i = 1, 2, 3.$$

Then $\vec{a}_i \neq \vec{0}$ for $i = 1, 2, 3$ and any one of the vectors $\vec{a}_1, \vec{a}_2, \vec{a}_3$ cannot be expressed as a linear combination of the other two, for example, $\vec{a}_1 \neq \alpha \vec{a}_2 + \beta \vec{a}_3$ for any scalars $\alpha, \beta \in \mathbb{R}$.

Take an arbitrarily fixed point $P \in \Gamma$. Notice that

Case 1 If O, A_1, A_2 and P are coplanar, then the vector $\overrightarrow{OP} = x_1 \vec{a}_1 + x_2 \vec{a}_2$ for some scalars x_1 and x_2.
Case 2 If O, A_3 and P are collinear, then $\overrightarrow{OP} = x_3 \vec{a}_3$ for some scalar x_3.

Therefore, one may suppose that P is positioned in space Γ so that Cases 1 and 2 cannot occur. Under this circumstance, points O, A_3 and P are coplanar. A unique line L, passing the point P and lying entirely in that plane, can be drawn parallel to the line generated by O and A_3 and intersects at a point Q with the plane generated by O, A_1 and A_2 (see Fig. 3.5).

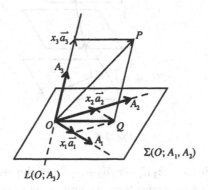

Fig. 3.5

Hence, there exist scalars x_1, x_2 and x_3 such that

$$\overrightarrow{OP} = \overrightarrow{OQ} + \overrightarrow{QP} \quad \text{with} \quad \overrightarrow{OQ} = x_1\vec{a}_1 + x_2\vec{a}_2 \quad \text{and} \quad \overrightarrow{QP} = x_3\vec{a}_3$$
$$\Rightarrow \overrightarrow{OP} = x_1\vec{a}_1 + x_2\vec{a}_2 + x_3\vec{a}_3.$$

Conversely, for any fixed scalars x_1, x_2 and x_3, there corresponds a unique point P in space Γ such that $\overrightarrow{OP} = x_1\vec{a}_1 + x_2\vec{a}_2 + x_3\vec{a}_3$ holds.

Fix a scalar $x_3 \in \mathbb{R}$. Move the plane $\Sigma(O; A_1, A_2)$ along the direction $x_3\vec{a}_3$ up to the parallel plane

$$x_3\vec{a}_3 + \Sigma(O; A_1, A_2) \tag{3.1.1}$$

(see Fig. 3.6). Then, let x_3 run through all the reals, the family of parallel planes (3.1.1) will fill the whole space Γ.

We summarize as (corresponding to (2.2.2))

Algebraic vectorization of a space
Let O, A_1, A_2 and A_3 be any fixed noncoplanar points in a space Γ. Let

$$\vec{a}_i = \overrightarrow{OA}_i, \quad i = 1, 2, 3$$

be space vectors.

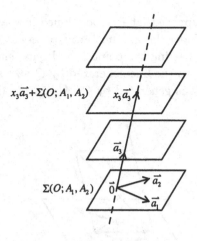

Fig. 3.6

(1) The *linear combination* $x_1\vec{a}_1 + x_2\vec{a}_2 + x_3\vec{a}_3$, of the vectors $\vec{a}_1$, $\vec{a}_2$ and $\vec{a}_3$ with corresponding *coefficients* x_1, x_2 and x_3, is suitable to describe any point P in Γ (i.e. the position vector $\overrightarrow{OP}$). Therefore, the set

$$\Gamma(O; A_1, A_2, A_3) = \{x_1\vec{a}_1 + x_2\vec{a}_2 + x_3\vec{a}_3 \mid x_1, x_2, x_3 \in \mathbb{R}\}$$

is called the *vectorized space* or a *coordinate system* with the point O as the *origin* (i.e. $\overrightarrow{OO} = \vec{0}$ as *zero vector*) and $\vec{a}_1$, $\vec{a}_2$ and $\vec{a}_3$ as *basis vectors*. It is indeed a *vector space* over $\mathbb{R}$ (see (2.1.11) and Remark 2.2 there), with $\{\vec{a}_1, \vec{a}_2, \vec{a}_3\}$ as an *ordered basis* (see (3.1.3) and (3.1.4)).

(2) Therefore,

$$\begin{aligned}
\Gamma(O; A_1, A_2, A_3) &= L(O; A_1) \oplus L(O; A_2) \oplus L(O; A_3) \\
&= L(O; A_3) \oplus \Sigma(O; A_1, A_2) \\
&= L(O; A_2) \oplus \Sigma(O; A_1, A_3) \\
&= L(O; A_1) \oplus \Sigma(O; A_2, A_3)
\end{aligned}$$

in the isomorphic sense.   (3.1.2)

Corresponding to (2.2.3), we have

Linear dependence of space vectors
The following are equivalent.

(1) (geometric) Five different points O, B_1, B_2, B_3 and B_4 lie in the same space Γ.

$\Leftrightarrow$ (2) (algebraic) Take any one of the five points as a base point and construct four vectors, for example, $\vec{b_i} = \overrightarrow{OB_i}$, $1 \leq i \leq 4$. Then, at least one of the four vectors $\vec{b_1}, \vec{b_2}, \vec{b_3}, \vec{b_4}$ can be expressed as a linear combination of the other three, i.e. $\vec{b_4} = y_1 \vec{b_1} + y_2 \vec{b_2} + y_3 \vec{b_3}$, etc.

$\Leftrightarrow$ (3) (algebraic) There exist scalars y_1, y_2, y_3, y_4, not all zero, such that $y_1 \vec{b_1} + y_2 \vec{b_2} + y_3 \vec{b_3} + y_4 \vec{b_4} = \vec{0}$.

In any of these cases, $\vec{b_1}, \vec{b_2}, \vec{b_3}$ and $\vec{b_4}$ are said to be *linearly dependent*.

$$(3.1.3)$$

Also, corresponding to (2.2.4), we have

Linear independence of nonzero vectors in space
The following are equivalent.

(1) (geometric) Four different points O, B_1, B_2 and B_3 are not coplanar (this implies implicitly that any three of them are non-collinear).

$\Leftrightarrow$ (2) (algebraic) Take any one of the four points as a base point and construct three vectors, for example, $\vec{b_i} = \overrightarrow{OB_i}$, $1 \leq i \leq 3$. Then, any one of the three vectors $\vec{b_1}, \vec{b_2}, \vec{b_3}$ cannot be expressed as a linear combination of the other two, i.e. $\vec{b_3} \neq y_1 \vec{b_1} + y_2 \vec{b_2}$ for any scalars y_1, y_2, etc.

$\Leftrightarrow$ (3) (algebraic) If there exist scalars y_1, y_2 and y_3, satisfying

$$y_1 \vec{b_1} + y_2 \vec{b_2} + y_3 \vec{b_3} = \vec{0},$$

then it is necessary that $y_1 = y_2 = y_3 = 0$.

In any of these cases, $\vec{b_1}, \vec{b_2}$ and $\vec{b_3}$ are said to be *linearly independent*.

$$(3.1.4)$$

It is observed that any one or any two vectors out of three linearly independent vectors must be linearly independent, too.

Exercises

<A>

1. Prove (3.1.2)–(3.1.4) in detail.
2. Suppose $\vec{b_1}, \vec{b_2}, \vec{b_3}, \vec{b_4}$ are elements of $\Gamma(O; A_1, A_2, A_3)$.

   (a) Prove that $\vec{b_1}, \vec{b_2}, \vec{b_3}$ and $\vec{b_4}$ should be linearly dependent.
   (b) If $\vec{b_1}$ and $\vec{b_2}$ are known linearly dependent, then so are $\vec{b_1}, \vec{b_2}$ and $\vec{b_3}$.

(c) If $\vec{b}_1, \vec{b}_2$ and $\vec{b}_3$ are linearly independent, is it true that $\vec{b}_1$ and $\vec{b}_2$ are linearly independent? Why?

(d) Is it possible that any three vectors out of $\vec{b}_1, \vec{b}_2, \vec{b}_3$ and $\vec{b}_4$ are linearly independent? Why?

<C> Abstraction and generalization

Try to describe $\Gamma(O; A_1, \ldots, A_n)$.

3.2   Coordinatization of a Space: $\mathbb{R}^3$

Just like (2.3.2), we have

The Coordinatization of a space

Fix an arbitrary vectorized space $\Gamma(O; A_1, A_2, A_3)$ of a space Γ, with ordered basis $\mathcal{B} = \{\vec{a}_1, \vec{a}_2, \vec{a}_3\}$, $\vec{a}_i = \overrightarrow{OA_i}$, $i = 1, 2, 3$. The *coordinate* of a point P in Γ with respect to $\mathcal{B}$ is defined and denoted by the ordered triple

$$[P]_\mathcal{B} = [\overrightarrow{OP}]_\mathcal{B} = (x_1, x_2, x_3) \Leftrightarrow \overrightarrow{OP} = x_1\vec{a}_1 + x_2\vec{a}_2 + x_3\vec{a}_3.$$

Call the set

$$\mathbb{R}^3_{\Gamma(O; A_1, A_2, A_3)} = \{(x_1, x_2, x_3) \mid x_1, x_2, x_3 \in \mathbb{R}\}$$

the *coordinatized space* of Γ with respect to $\mathcal{B}$. Explain as follows.

(1) Points in Γ are in one-to-one correspondence with the triples in $\mathbb{R}^3_{\Gamma(O; A_1, A_2, A_3)}$.

(2) Introduce into $\mathbb{R}^3_{\Gamma(O; A_1, A_2, A_3)}$ two operations as

1. *addition* $(x_1, x_2, x_3) + (y_1, y_2, y_3) = (x_1 + y_1, x_2 + y_2, x_3 + y_3)$,
2. *scalar multiplication* $\alpha(x_1, x_2, x_3) = (\alpha x_1, \alpha x_2, \alpha x_3)$, where $\alpha \in \mathbb{R}$,

which have all the properties listed in (2.1.11), treated (x_1, x_2, x_3) as $\vec{x}$, etc. Hence, $\mathbb{R}^3_{\Gamma(O; A_1, A_2, A_3)}$ is indeed a real *vector space*.

(3) Define a mapping $\Phi \colon \Gamma(O; A_1, A_2, A_3) \to \mathbb{R}^3_{\Gamma(O; A_1, A_2, A_3)}$ by

$$\Phi(x_1\vec{a}_1 + x_2\vec{a}_2 + x_3\vec{a}_3) = (x_1, x_2, x_3) \quad \text{or} \quad \Phi(\overrightarrow{OP}) = [P]_\mathcal{B},$$

then Φ is one-to-one, onto and preserves vector operations, i.e.

1. $\Phi(\overrightarrow{OP} + \overrightarrow{OQ}) = [P]_\mathcal{B} + [Q]_\mathcal{B}(= \Phi(\overrightarrow{OP}) + \Phi(\overrightarrow{OQ}))$,
2. $\Phi(\alpha\overrightarrow{OP}) = \alpha[P]_\mathcal{B}(= \alpha\Phi(\overrightarrow{OP}))$, $\alpha \in \mathbb{R}$.

Φ is called a *linear isomorphism* between the two vector spaces.

Therefore, conceptually, $\Gamma(O; A_1, A_2, A_3)$ and $\mathbb{R}^3_{\Gamma(O;A_1,A_2,A_3)}$ are considered identical.   (3.2.1)

The only main difference between $\Gamma(O; A_1, A_2, A_3)$ and $\mathbb{R}^3_{\Gamma(O;A_1,A_2,A_3)}$ is in notations used to represent them, even though, the former might be more concrete than the latter (see Fig. 3.7).

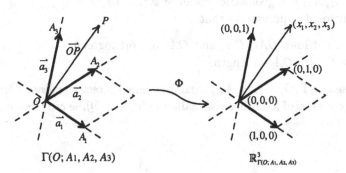

$$\Gamma(O; A_1, A_2, A_3) \qquad\qquad \mathbb{R}^3_{\Gamma(O; A_1, A_2, A_3)}$$

Fig. 3.7

The fact that four non-coplanar points determine a space is algebraically equivalent to three linearly independent vectors, no more or less, generate the whole space $\Gamma(O; A_1, A_2, A_3)$ or $\mathbb{R}^3_{\Gamma(O;A_1,A_2,A_3)}$. Therefore, they both are called three-dimensional vector spaces.

The diagram in (2.3.3) is still valid for $\Gamma(O; A_1, A_2, A_3)$ and $\mathbb{R}^3_{\Gamma(O;A_1,A_2,A_3)}$, replacing $\Sigma(O; A_1, A_2)$ and $\mathbb{R}^2_{\Sigma(O;A_1,A_2)}$ there, respectively.

Hence, we introduce the

Standard three-dimensional vector space $\mathbb{R}^3$ over $\mathbb{R}$
Let

$$\mathbb{R}^3 = \{(x_1, x_2, x_3) \mid x_1, x_2, x_3 \in \mathbb{R}\}$$

and $(x_1, x_2, x_3) = (y_1, y_2, y_3)$ mean that $x_i = y_i$, $1 \leq i \leq 3$. Define on $\mathbb{R}^3$ two operations as follows:

1. *addition* $(x_1, x_2, x_3) + (y_1, y_2, y_3) = (x_1 + y_1, x_2 + y_2, x_3 + y_3)$,
2. *scalar multiplication* $\alpha(x_1, x_2, x_3) = (\alpha x_1, \alpha x_2, \alpha x_3)$, where $\alpha \in \mathbb{R}$,

which have all the properties listed in (2.1.11), with (x_1, x_2, x_3) as $\vec{x}$ there, etc. Hence, $\mathbb{R}^3$ is a three-dimensional vector space over $\mathbb{R}$. In particular,

zero vector : $\vec{0} = (0, 0, 0)$;

the inverse vector of $\vec{x} = (x_1, x_2, x_3)$: $-\vec{x} = (-x_1, -x_2, -x_3)$, and the *natural basis* $\mathcal{N} = \{\vec{e}_1, \vec{e}_2, \vec{e}_3\}$ where

$$\vec{e}_1 = (1,0,0), \quad \vec{e}_2 = (0,1,0), \quad \vec{e}_3 = (0,0,1). \tag{3.2.2}$$

As a whole, $\mathbb{R}^3$ is the universal representative of any vectorized space $\Gamma(O; A_1, A_2, A_3)$ or coordinatized space $\mathbb{R}^3_{\Gamma(O;A_1,A_2,A_3)}$ of a space Γ, and $\Gamma(O; A_1, A_2, A_3)$ is a geometric model of $\mathbb{R}^3$.

If additional requirements that

1. the straight lines OA_1, OA_2 and OA_3 are orthogonal,
2. $OA_1 = OA_2 = OA_3$ in length. $\hspace{3cm}$ (3.2.3)

are imposed on $\Gamma(O; A_1, A_2, A_3)$, then it is called a *rectangular* or *Cartesian coordinate system* of $\mathbb{R}^3$; otherwise, called *oblique* or *affine coordinate system* (see Fig. 3.8).

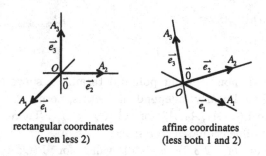

rectangular coordinates   affine coordinates
(even less 2)   (less both 1 and 2)

Fig. 3.8

Unless otherwise specified, from now on, $\mathbb{R}^3$ will always be endowed with rectangular coordinate system with $\mathcal{N} = \{\vec{e}_1, \vec{e}_2, \vec{e}_3\}$ as natural basis.

Elements in $\mathbb{R}^3$ are usually denoted by $\vec{x} = (x_1, x_2, x_3)$, $\vec{y} = (y_1, y_2, y_3)$, etc. They represent two kinds of concept as follows.

Case 1 (affine point of view) When $\mathbb{R}^3$ is considered as a space, element $\vec{x}$ is called a *point* and two points decide a unique vector $\vec{x} - \vec{y}$ or $\vec{y} - \vec{x}$ with $\vec{x} - \vec{x} = \vec{0}$.

Case 2 (vector point of view) When considered as a vector space, element $\vec{x}$ in $\mathbb{R}^3$ is called a *vector*, pointing from the zero vector $\vec{0}$ toward the point $\vec{x}$.

$$\hspace{10cm} (3.2.4)$$

See Fig. 3.9. In short, in Case 1, an arbitrary fixed point can be used as a base point in order to study the position vectors of the other points relative to the base point. If the base point is considered as zero vector $\vec{0}$, then Case 1 turns into Case 2.

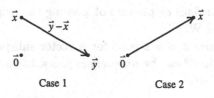

Case 1   Case 2

Fig. 3.9

Remark Reinterpretation of (2) in (3.1.2).

Corresponding to (2.3.5), one has the following

$$\mathbb{R}^3_{\Gamma(O;A_1,A_2,A_3)} = \mathbb{R}_{L(O;A_1)} \oplus \mathbb{R}_{L(O;A_2)} \oplus \mathbb{R}_{L(O;A_3)}$$
$$= \mathbb{R}_{L(O;A_1)} \oplus \mathbb{R}^2_{\Sigma(O;A_2,A_3)}$$
$$= \cdots \tag{3.2.5}$$

and hence, corresponding to (2.3.6), the following

$$\mathbb{R}^3 = \mathbb{R} \oplus \mathbb{R} \oplus \mathbb{R}$$
$$= \mathbb{R} \oplus \mathbb{R}^2, \tag{3.2.6}$$

where, in the sense of isomorphism,

$$\mathbb{R} \cong \{(x_1,0,0) \mid x_1 \in \mathbb{R}\}, \text{ etc.,} \quad \text{and}$$
$$\mathbb{R}^2 \cong \{(0,x_2,x_3) \mid x_2,x_3 \in \mathbb{R}\}, \text{ etc.}$$

Exercises

<A>

1. Explain (3.2.1) graphically and prove $1, 2$ in (3).
2. Any two vectorized spaces $\Gamma(O; A_1, A_2, A_3)$ and $\Gamma(O'; B_1, B_2, B_3)$ of a space Γ is isomorphic to each other. Prove this and try to explain it graphically. Indeed, there are infinitely many isomorphisms between them, why?
3. Explain and prove the equivalence of linear dependence as stated in (3.1.3) and linear independence as stated in (3.1.4), respectively, for $\mathbb{R}^3$.
4. Explain (3.2.5) just like (2.3.5).
5. Explain (3.2.6) just like (2.3.6).

6. *Vector* (or *linear*) *subspaces* of $\mathbb{R}^3$ are defined in $\mathbb{R}^3$ (instead of $\mathbb{R}^2$) exactly the same as in Ex. <A> 3 of Sec. 2.3. Try to find out graphically all vector subspaces of $\mathbb{R}^3$ (see Exs. 9 and 10 below).

7. Is it possible, for lines or planes not passing the origin $\vec{0}$ in $\mathbb{R}^3$, to be subspaces of $\mathbb{R}^3$? Why?

8. Define what it means $\mathcal{B}$ is a *basis* for a vector subspace S of $\mathbb{R}^3$. Prove that any vector in $\mathbb{R}^3$ can be *uniquely* expressed as a *linear combination* of vectors from basis.

   (a) Show that any nonzero vector subspace of $\mathbb{R}^3$ has a basis.

   (b) Show that the numbers of elements in any two bases for a vector subspace S are the same. This common number, denoted as

$$\dim S,$$

   is called the *dimension* of S. Note that $\dim\{\vec{0}\} = 0$.

9. Model after Ex. <A> 3 of Sec. 2.3 to characterize any one-dimensional vector subspace of $\mathbb{R}^3$.

10. Let S be a vector subspace of $\mathbb{R}^3$. Prove that the following are equivalent.

   (a) $\dim S = 2$.

   (b) There are two linearly independent vectors $\vec{x}_1$ and $\vec{x}_2$ in $\mathbb{R}^3$ so that

$$S = \langle\langle \vec{x}_1, \vec{x}_2 \rangle\rangle = \{\alpha_1\vec{x}_1 + \alpha_2\vec{x}_2 \mid \alpha_1, \alpha_2 \in \mathbb{R}\}.$$

   In this case, S is called *generated* or *spanned* by $\vec{x}_1$ and $\vec{x}_2$.

   (c) There exist scalars a_1, a_2 and a_3, not all zero, so that

$$S = \{\vec{x} = (x_1, x_2, x_3) \mid a_1x_1 + a_2x_2 + a_3x_3 = 0\}.$$

   In this case, simply call S a plane passing $\vec{0} = (0,0,0)$ and denote S by its equation $a_1x_1 + a_2x_2 + a_3x_3 = 0$.

11. Let S_1 and S_2 be subspaces of $\mathbb{R}^3$.

   (a) Show that the *intersection* $S_1 \cap S_2$ and the *sum* $S_1 + S_2$ (see Ex. <A> 5 of Sec. 2.3) are subspaces of $\mathbb{R}^3$. In case $S_1 \cap S_2 = \{\vec{0}\}$, denote $S_1 + S_2$ by

$$S_1 \oplus S_2 \quad \text{(the } \textit{direct sum} \text{ of } S_1 \text{ and } S_2\text{)}.$$

   (b) Prove that

$$\dim(S_1 \cap S_2) + \dim(S_1 + S_2) = \dim S_1 + \dim S_2$$

   so that $S_1 \cap S_2 = \{\vec{0}\}$ if and only if $\dim(S_1 + S_2) = \dim S_1 + \dim S_2$.

(c) For a given subspace S_1 of $\mathbb{R}^3$, try to find all possible subspaces S_2 of $\mathbb{R}^3$ so that

$$\mathbb{R}^3 = S_1 \oplus S_2.$$

(d) Suppose $S_1 \subseteq S_2$. Prove that the following are equivalent.
   (1) $S_1 = S_2$.
   (2) $S_1 \cap S_2 = S_2$.
   (3) $\dim S_1 = \dim S_2$.
   (4) $\dim(S_1 + S_2) = \dim S_1$.

12. Let $\vec{x}_1 = (1, 2, 1)$, $\vec{x}_2 = (-2, 1, 2)$, $\vec{x}_3 = (-1, 3, 3)$, $\vec{x}_4 = (0, -3, -5)$ and $\vec{x}_5 = (2, -9, -11)$ be vectors in $\mathbb{R}^3$.

(a) Show that $S = \{\vec{x}_1, \vec{x}_2, \vec{x}_3, \vec{x}_4, \vec{x}_5\}$ generates $\mathbb{R}^3$.
(b) Find all subsets of S that are bases for $\mathbb{R}^3$. How many *ordered* bases for $\mathbb{R}^3$ can be chosen from vectors in S? List them out.
(c) Find all linearly independent subsets of S and all linearly dependent subsets of S.
(d) Construct all possible but different subspaces of $\mathbb{R}^3$ from vectors in S. Among these subspaces, which are the intersections of the other two? Which are the sums of the other two?

13. Let $S_1 = \{(2, -3, 1), (1, 4, -2), (5, -2, 0), (1, -7, 3)\}$, and $S_2 = \{(4, -17, 7), (0, 6, 1)\}$.

(a) Determine the subspace $\langle\langle S_1 \rangle\rangle$ of $\mathbb{R}^3$ generated by S_1 and $\langle\langle S_2 \rangle\rangle$. Write out their equations in $\mathcal{N}$ (see Exs. 9 and 10).
(b) Determine $\langle\langle S_1 \rangle\rangle \cap \langle\langle S_2 \rangle\rangle$ and $\langle\langle S_1 \rangle\rangle + \langle\langle S_2 \rangle\rangle$ and write out their equations.

14. Give the system of homogeneous linear equations:

$$\frac{1}{3}x_1 + 2x_2 - 6x_3 = 0,$$

$$-4x_1 + 5x_3 = 0,$$

$$-3x_1 + 6x_2 - 13x_3 = 0,$$

$$-\frac{11}{3}x_1 + 2x_2 - x_3 = 0,$$

where $\vec{x} = (x_1, x_2, x_3)$ in $\mathcal{N}$.

(a) Solve the equations. Denote by S the set of all such solutions and try to explain S geometrically.
(b) Show that S is a subspace of $\mathbb{R}^3$ and find a basis for it.

15. Let $\vec{x}_1 = (3,1,1)$, $\vec{x}_2 = (2,5,-1)$, $\vec{x}_3 = (1,-4,2)$, $\vec{x}_4 = (4,-3,3)$ be vectors in $\mathbb{R}^3$.

   (a) Determine the subspace $\langle\langle \vec{x}_1, \vec{x}_2, \vec{x}_3, \vec{x}_4 \rangle\rangle$ and its dimension.

   (b) Let $\vec{y}_1 = (5,6,0)$, $\vec{y}_2 = (8,7,1)$. Show that $\vec{y}_1$ and $\vec{y}_2$ are linearly independent vectors in $\langle\langle \vec{x}_1, \vec{x}_2, \vec{x}_3, \vec{x}_4 \rangle\rangle$.

   (c) Which two vectors of $\vec{x}_1, \vec{x}_2, \vec{x}_3$ and $\vec{x}_4$ can be replaced by $\vec{y}_1$ and $\vec{y}_2$, say $\vec{x}_i$ and $\vec{x}_j$, so that $\langle\langle \vec{y}_1, \vec{y}_2, \vec{x}_i', \vec{x}_j' \rangle\rangle = \langle\langle \vec{x}_1, \vec{x}_2, \vec{x}_3, \vec{x}_4 \rangle\rangle$, where $\vec{x}_i'$ and $\vec{x}_j'$ are the remaining two vectors of $\vec{x}_1, \vec{x}_2, \vec{x}_3$ and $\vec{x}_4$? In how many ways?

   (*Note*   To do (b), you need algebraic computation to justify the claim there. Once this procedure has been finished, can you figure out any geometric intuition on which formal but algebraic proof for (c) will rely? For generalization, see Steinitz's Replacement Theorem in Sec. B.3.)

16. Find scalar k so that the vectors $(k,1,0)$, $(1,k,1)$ and $(0,1,k)$ are linearly dependent.

17. Find the necessary and sufficient conditions so that the vectors $(1,a_1,a_1^2)$, $(1,a_2,a_2^2)$ and $(1,a_3,a_3^2)$ are linearly dependent.

<B>

Review the *comments* in Ex. <C> of Sec. 2.3 and then try to extend problems in Ex. <B> there to counterparts in $\mathbb{R}^3$ (or $\mathbb{F}^n$ or more abstract vector space, if possible) and prove them true or false.

1. Suppose $\vec{x}_1, \vec{x}_2, \ldots, \vec{x}_k$ for $k \geq 2$ are linearly dependent vectors in $\mathbb{R}^3$. Let $\vec{x}$ be an arbitrary vector in $\mathbb{R}^3$.

   (a) The vectors $\vec{x}_1 + \vec{x}, \vec{x}_2 + \vec{x}, \ldots, \vec{x}_k + \vec{x}$ may be linearly dependent or independent. Find conditions that guarantee their dependence or independence and explain these conditions geometrically.

   (b) There do exist scalars $\alpha_1, \ldots, \alpha_k$, not all zeros, so that $\vec{x}_1 + \alpha_1\vec{x}$, $\vec{x}_2 + \alpha_2\vec{x}, \ldots, \vec{x}_k + \alpha_k\vec{x}$ are always linearly dependent. Any geometrical meaning?

2. Let $\{\vec{x}_1, \vec{x}_2, \vec{x}_3\}$ be a basis for $\mathbb{R}^3$. For any vector $\vec{x} \in \mathbb{R}^3$, at most one of the vectors $\vec{x}, \vec{x}_1, \vec{x}_2$ and $\vec{x}_3$ can be represented as a linear combination of the preceding ones.

3. Suppose $\{\vec{x}_1, \vec{x}_2, \vec{x}_3\}$ is a basis for $\mathbb{R}^3$. Let $\vec{x} \in \mathbb{R}^3$ be such that $\vec{x}$ can be expressed as linear combinations of any two of the vectors $\vec{x}_1, \vec{x}_2, \vec{x}_3$. Show that $\vec{x} = \vec{0}$ both algebraically and geometrically.

4. Let $\vec{x}_1, \vec{x}_2, \ldots, \vec{x}_k$ for $2 \leq k \leq 4$ be linearly dependent vectors in $\mathbb{R}^3$ such that any $(k-1)$ vectors of them are linearly independent.

(a) If $a_1\vec{x}_1 + a_2\vec{x}_2 + \cdots + a_k\vec{x}_k = \vec{0}$ for some scalars $a_1, a_2, \ldots, a_k$, show that either $a_1 a_2 \cdots a_k \neq 0$ or $a_1 = a_2 = \cdots = a_k = 0$.

(b) In case $a_1 a_2 \cdots a_k \neq 0$ and $b_1\vec{x}_1 + b_2\vec{x}_2 + \cdots + b_k\vec{x}_k = \vec{0}$ also holds, then $a_1 : b_1 = a_2 : b_2 = \cdots = a_k : b_k$.

5. Do Ex. <C>6' of Sec. 2.3 in case $n = 3$ and $V = \mathbb{R}^3$.

<C> Abstraction and generalization

Do problems in Ex. <C> of Sec. 2.3 if you missed them at that moment.

Problems that follow concern with vector spaces such as $\mathbb{F}^n$, $M(m, n; \mathbb{F})$, $P(\mathbb{F})$, $P_n(\mathbb{F})$ and $\mathcal{F}(X, \mathbb{F})$. For definitions, see Sec. B.1. Readers are required to be able to extend what have been learned in $\mathbb{R}$, $\mathbb{R}^2$ and $\mathbb{R}^3$ to abstract vector spaces over a field $\mathbb{F}$, not to say about geometric meanings, but at least about linearly algebraic computational techniques.

1. It is well-known that $P_2(\mathbb{R})$ has a natural basis $\{1, x, x^2\}$.

(a) Let a_0, a_1 and a_2 be distinct real scalars. Show that the unique polynomial $p_0(x) \in P_2(\mathbb{R})$ satisfying

$$p_0(a_j) = \delta_{0j}, \quad j = 0, 1, 2$$

is

$$p_0(x) = \frac{(x - a_1)(x - a_2)}{(a_0 - a_1)(a_0 - a_2)}.$$

Find other two polynomials $p_1(x), p_2(x) \in P_2(\mathbb{R})$ so that $p_i(a_j) = \delta_{ij}$ for $i = 1, 2$ and $j = 0, 1, 2$.

(b) Show that $\{p_0, p_1, p_2\}$ is a basis for $P_2(\mathbb{R})$.

(c) Construct three different bases for $P_2(\mathbb{R})$.

(d) Not every basis for $P_2(\mathbb{R})$ is derived as in (a) and (b). Show that $\{x^2 + 3x - 2, 2x^2 + 5x - 3, -x^2 - 4x + 4\}$ is a basis for $P_2(\mathbb{R})$.

(e) Show that $\{1 - 2x + 4x^2, 2 + x - 2x^2, -1 - x + 2x^2\}$ is not a basis for $P_2(\mathbb{R})$, but any two of them are linearly independent. Try to construct three different bases for $P_2(\mathbb{R})$ from this set of vectors.

(*Note*   For generalization, see Sec. B.3.)

2. Let $r_1, \ldots, r_n$ be distinct real numbers. Show that $\{e^{r_1 x}, e^{r_2 x}, \ldots, e^{r_n x}\}$ are linearly independent in $\mathcal{F}(\mathbb{R}, \mathbb{R})$.

3. Denote by N the set of positive integers. An element in $\mathcal{F}(\mathbb{N}, \mathbb{F})$ is called a *sequence* in $\mathbb{F}$ and is denoted by $\{a_n\}$. Let

$$V = \{\{a_n\} \in \mathcal{F}(\mathbb{N}, \mathbb{F}) \mid a_n \neq 0 \text{ for only a finite number of } n\}.$$

(a) Show that V is a proper subspace of $\mathcal{F}(\mathbb{N}, \mathbb{F})$.

(b) Find a basis for V. Is V finite-dimensional?

4. In M(2; $\mathbb{R}$), let

$$V = \left\{ \begin{bmatrix} a & -a \\ b & c \end{bmatrix} \middle| a, b, c \in \mathbb{R} \right\}, \quad \text{and}$$

$$W = \left\{ \begin{bmatrix} d & e \\ -d & f \end{bmatrix} \middle| d, e, f \in \mathbb{R} \right\}.$$

(a) Show that V and W are subspaces and $\dim V = \dim W = 3$. Find a basis for each of V and W.

(b) Find a basis for each of $V \cap W$ and $V + W$.

(c) Show that $\dim V + \dim W = \dim(V \cap W) + \dim(V + W)$.

5. Remind that M(n; $\mathbb{C}$) is the n-dimensional complex vector space, consisting of all complex matrices of order n, while M(n; $\mathbb{R}$) is the real one. Let

$$\mathrm{SL}(n; \mathbb{R}) = \{ A \in \mathrm{M}(n; \mathbb{R}) \mid \mathrm{tr}\, A = 0 \};$$

$$\mathrm{SL}(n; i\mathbb{R}) = \{ A \in \mathrm{M}(n; \mathbb{C}) \mid \text{entries of } A \text{ are all pure imaginaries and}$$
$$\mathrm{tr}\, A = 0 \};$$

$$\langle\langle E_{11}, iE_{11} \rangle\rangle = \text{the subspace of M}(n; \mathbb{C}), \text{ generated by } \{E_{11}, iE_{11}\},$$
$$\text{over the } \textit{reals}; \langle\langle E_{11} \rangle\rangle; \langle\langle iE_{11} \rangle\rangle;$$

$$\mathrm{M}(n; i\mathbb{R}) = \{ A \in \mathrm{M}(n; \mathbb{C}) \mid \text{entries of } A \text{ are all pure imaginaries} \};$$

$$S(n; \mathbb{R}) = \{ A \in \mathrm{M}(n; \mathbb{R}) \mid A^* = A \};$$

$$T(n; \mathbb{R}) = \{ A \in \mathrm{M}(n; \mathbb{R}) \mid A^* = -A \}.$$

(a) Show that M(n; $\mathbb{C}$) is $2n$-dimensional over $\mathbb{R}$. Find a basis for it.

(b) Show that both $\mathrm{SL}(n; \mathbb{R})$ and $\mathrm{SL}(n; i\mathbb{R})$ are $(n^2 - 1)$-dimensional over $\mathbb{R}$. Find a basis for each of them.

(c) Show that $S(n; \mathbb{R})$ is an $\frac{n(n+1)}{2}$-dimensional space over $\mathbb{R}$. Find a basis for it.

(d) Show that $T(n; \mathbb{R})$ is an $\frac{n(n-1)}{2}$-dimensional space over $\mathbb{R}$. Find a basis for it.

(e) What are about $\mathrm{SL}(n; i\mathbb{R})$ and $\mathrm{M}(n; i\mathbb{R})$?

(f) Show that

$$\mathrm{M}(n; \mathbb{C}) = \mathrm{SL}(n; \mathbb{R}) \oplus \langle\langle E_{11}, iE_{11} \rangle\rangle \oplus \mathrm{SL}(n; i\mathbb{R});$$
$$\mathrm{M}(n; \mathbb{R}) = \mathrm{SL}(n; \mathbb{R}) \oplus \langle\langle E_{11} \rangle\rangle$$
$$= S(n; \mathbb{R}) \oplus T(n; \mathbb{R})$$

etc. See the following diagram.

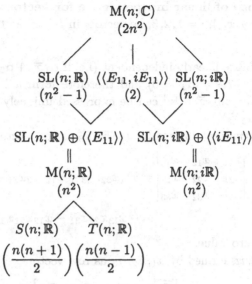

$$M(n; \mathbb{C})$$
$$(2n^2)$$

$$SL(n; \mathbb{R}) \quad \langle\langle E_{11}, iE_{11}\rangle\rangle \quad SL(n; i\mathbb{R})$$
$$(n^2 - 1) \qquad (2) \qquad (n^2 - 1)$$

$$SL(n; \mathbb{R}) \oplus \langle\langle E_{11}\rangle\rangle \quad SL(n; i\mathbb{R}) \oplus \langle\langle iE_{11}\rangle\rangle$$
$$\| \qquad\qquad\qquad \|$$
$$M(n; \mathbb{R}) \qquad\qquad M(n; i\mathbb{R})$$
$$(n^2) \qquad\qquad\quad (n^2)$$

$$S(n; \mathbb{R}) \qquad T(n; \mathbb{R})$$
$$\left(\frac{n(n+1)}{2}\right) \left(\frac{n(n-1)}{2}\right)$$

6. Find dimensions and bases for real vector spaces:

$$SU(n; \mathbb{C}) = \{A \in M(n; \mathbb{C}) \mid \bar{A}^* = -A \text{ (Skew-Hermitian) and tr } A = 0\};$$
$$SH(n; \mathbb{C}) = \{A \in M(n; \mathbb{C}) \mid \bar{A}^* = A \text{ (Hermitian) and tr } A = 0\}.$$

3.3 Changes of Coordinates: Affine Transformation (or Mapping)

Take points $\vec{0}$, $\vec{e}_1$, $\vec{e}_2$ and $\vec{e}_3$ in $\mathbb{R}^3$ and construct the (rectangular or affine) coordinate system $\Gamma(\vec{0}; \vec{e}_1, \vec{e}_2, \vec{e}_3)$ as in (3.2.3) with ordered basis

$$\mathcal{N} = \{\vec{e}_1, \vec{e}_2, \vec{e}_3\}.$$

Then, for point $\vec{x} = (x_1, x_2, x_3) \in \mathbb{R}^3$,

$$\vec{x} = x_1\vec{e}_1 + x_2\vec{e}_2 + x_3\vec{e}_3$$
$$\Rightarrow [\vec{x}]_{\mathcal{N}} = (x_1, x_2, x_3) = \vec{x} \qquad (3.3.1)$$

i.e. the coordinate of $\vec{x}$ with respect to the basis $\mathcal{N}$ is the *vector* $\vec{x}$. That is why $\Gamma(\vec{0}; \vec{e}_1, \vec{e}_2, \vec{e}_3)$ is called the *natural coordinate system* of $\mathbb{R}^3$ and $\mathcal{N}$ the *natural basis* of $\mathbb{R}^3$. This is the most commonly used coordinate system in $\mathbb{R}^3$, and unless specified, will be adopted throughout, without mentioning and indicating the notation $\Gamma(\vec{0}; \vec{e}_1, \vec{e}_2, \vec{e}_3)$.

The equivalent statements in Ex. <A> 2 of Sec. 2.4 can be, quite similarly, extended to $\mathbb{R}^3$. For convenience and later usage, we list them in

The equivalence of linear independence for vectors in $\mathbb{R}^3$

Let $\vec{x}_i = (x_{i1}, x_{i2}, x_{i3})$, $i = 1, 2, 3$, be vectors in $\mathbb{R}^3$. Then the following are equivalent.

(1) $\vec{x}_1, \vec{x}_2, \vec{x}_3$ are linearly independent (i.e. if $\alpha_1 \vec{x}_1 + \alpha_2 \vec{x}_2 + \alpha_3 \vec{x}_3 = \vec{0}$, then $\alpha_1 = \alpha_2 = \alpha_3 = 0$) and therefore form a *basis* $\{\vec{x}_1, \vec{x}_2, \vec{x}_3\}$ for $\mathbb{R}^3$ (i.e. any $\vec{x} \in \mathbb{R}^3$ can be expressed uniquely as $\vec{x} = \alpha_1 \vec{x}_1 + \alpha_2 \vec{x}_2 + \alpha_3 \vec{x}_3$).

$\Leftrightarrow$ (2) The *determinant*, formed by $\vec{x}_1, \vec{x}_2, \vec{x}_3$ as row vectors,

$$\begin{vmatrix} \vec{x}_1 \\ \vec{x}_2 \\ \vec{x}_3 \end{vmatrix} = \begin{vmatrix} x_{11} & x_{12} & x_{13} \\ x_{21} & x_{22} & x_{23} \\ x_{31} & x_{32} & x_{33} \end{vmatrix} = x_{11}x_{22}x_{33} + x_{12}x_{23}x_{31} + x_{13}x_{21}x_{32}$$

$$- x_{13}x_{22}x_{31} - x_{12}x_{21}x_{33} - x_{11}x_{23}x_{32}$$

has nonzero value.

$\Leftrightarrow$ (3) The *matrix*, formed by $\vec{x}_1, \vec{x}_2, \vec{x}_3$ as row vectors,

$$\begin{bmatrix} \vec{x}_1 \\ \vec{x}_2 \\ \vec{x}_3 \end{bmatrix} = \begin{bmatrix} x_{11} & x_{12} & x_{13} \\ x_{21} & x_{22} & x_{23} \\ x_{31} & x_{32} & x_{33} \end{bmatrix}$$

is *invertible* (i.e. if denoted by A, it means that there exists another 3×3 matrix B such that $AB = BA = I_3$ and B, denoted by A^{-1}, is called the *inverse matrix* of A). In this case, the inverse matrix is

$$A^{-1} = \frac{1}{\det A} \operatorname{adj} A \quad \text{with} \quad \operatorname{adj} A = \begin{bmatrix} A_{11} & A_{21} & A_{31} \\ A_{12} & A_{22} & A_{32} \\ A_{13} & A_{23} & A_{33} \end{bmatrix},$$

the *adjoint matrix* of A,

where $\det A$ is as in (2) and A_{ij} is the *cofactor* of x_{ij} in $\det A$, i.e. deleting the ith row and the jth column from $\det A$, and multiplying $(-1)^{i+j}$ to the remaining 2×2 determinant. For example,

$$A_{11} = (-1)^{1+1} \begin{vmatrix} x_{22} & x_{23} \\ x_{32} & x_{33} \end{vmatrix}, \quad A_{12} = (-1)^{1+2} \begin{vmatrix} x_{21} & x_{23} \\ x_{31} & x_{33} \end{vmatrix}, \text{ etc.}$$

$$(3.3.2)$$

For details about matrix and determinant, please refer to Secs. B.4–B.6. Perhaps, it might be easy for the readers to prove statements in Ex. <A> 2 of Sec. 2.4. But by exactly the same method you had experienced there, is it easy for you to prove the extended results in (3.3.2)? Where are the difficulties one might encounter? Can one find any easier way to prove them? Refer to Ex. <B> 3.

At least, for this moment, (3.3.2) is helpful in the computation of the

Coordinate changes of two coordinate systems in $\mathbb{R}^3$

Let

$\Gamma(O; A_1, A_2, A_3)$ with basis $\mathcal{B} = \{\vec{a}_1, \vec{a}_2, \vec{a}_3\}$, $\vec{a}_i = \overrightarrow{OA_i}$, $1 \leq i \leq 3$,   and

$\Gamma(O'; B_1, B_2, B_3)$ with basis $\mathcal{B}' = \{\vec{b}_1, \vec{b}_2, \vec{b}_3\}$, $\vec{b}_i = \overrightarrow{O'B_i}$, $1 \leq i \leq 3$

be two coordinate systems of $\mathbb{R}^3$. Then the coordinates $[P]_\mathcal{B}$ and $[P]_{\mathcal{B}'}$ of the point P in $\mathbb{R}^3$ have the following formulas of changes of coordinates (called *affine transformations* or *mappings*)

$$x_i = \alpha_i + \sum_{j=1}^{3} y_j \alpha_{ji}, \quad i = 1, 2, 3, \quad \text{and}$$

$$y_j = \beta_j + \sum_{i=1}^{3} x_i \beta_{ij}, \quad j = 1, 2, 3,$$

simply denoted by

$$[P]_\mathcal{B} = [O']_\mathcal{B} + [P]_{\mathcal{B}'} A_\mathcal{B}^{\mathcal{B}'}, \quad \text{and}$$

$$[P]_{\mathcal{B}'} = [O]_{\mathcal{B}'} + [P]_\mathcal{B} A_{\mathcal{B}'}^\mathcal{B},$$

where $[O']_\mathcal{B} = (\alpha_1, \alpha_2, \alpha_3)$, $[O]_{\mathcal{B}'} = (\beta_1, \beta_2, \beta_3)$, $[P]_\mathcal{B} = (x_1, x_2, x_3)$, $[P]_{\mathcal{B}'} = (y_1, y_2, y_3)$ and

$$A_\mathcal{B}^{\mathcal{B}'} = \begin{bmatrix} [\vec{b}_1]_\mathcal{B} \\ [\vec{b}_2]_\mathcal{B} \\ [\vec{b}_3]_\mathcal{B} \end{bmatrix} = \begin{bmatrix} \alpha_{11} & \alpha_{12} & \alpha_{13} \\ \alpha_{21} & \alpha_{22} & \alpha_{23} \\ \alpha_{31} & \alpha_{32} & \alpha_{33} \end{bmatrix} \text{ is called the } transition\ matrix \text{ of}$$

$$\mathcal{B}' \text{ with respect to } \mathcal{B}, \text{ and}$$

$$A_{\mathcal{B}'}^\mathcal{B} = \begin{bmatrix} [\vec{a}_1]_{\mathcal{B}'} \\ [\vec{a}_2]_{\mathcal{B}'} \\ [\vec{a}_3]_{\mathcal{B}'} \end{bmatrix} = \begin{bmatrix} \beta_{11} & \beta_{12} & \beta_{13} \\ \beta_{21} & \beta_{22} & \beta_{23} \\ \beta_{31} & \beta_{32} & \beta_{33} \end{bmatrix} \text{ is called the } transition\ matrix \text{ of}$$

$$\mathcal{B} \text{ with respect to } \mathcal{B}'$$

satisfying:

1. The determinants

$$\det A_\mathcal{B}^{\mathcal{B}'} = \begin{vmatrix} \alpha_{11} & \alpha_{12} & \alpha_{13} \\ \alpha_{21} & \alpha_{22} & \alpha_{23} \\ \alpha_{31} & \alpha_{32} & \alpha_{33} \end{vmatrix} \neq 0; \quad \det A_{\mathcal{B}'}^\mathcal{B} = \begin{vmatrix} \beta_{11} & \beta_{12} & \beta_{13} \\ \beta_{21} & \beta_{22} & \beta_{23} \\ \beta_{31} & \beta_{32} & \beta_{33} \end{vmatrix} \neq 0.$$

2. The matrices $A_B^{B'}$ and $A_{B'}^B$ are invertible to each other, i.e.

$$A_B^{B'} A_{B'}^B = A_{B'}^B A_B^{B'} = I_3 = \begin{bmatrix} 1 & 0 & 0 \\ 0 & 1 & 0 \\ 0 & 0 & 1 \end{bmatrix}$$

and therefore (see (3) in (3.3.2))

$$\left(A_B^{B'} \right)^{-1} = A_{B'}^B; \left(A_{B'}^B \right)^{-1} = A_B^{B'}.$$

3. Hence, these two formulas are reversible, i.e.

$$[P]_{B'} = -[O']_B \left(A_B^{B'} \right)^{-1} + [P]_B \left(A_B^{B'} \right)^{-1}, \quad \text{and}$$

$$[O]_{B'} = -[O']_B \left(A_B^{B'} \right)^{-1} = -[O']_B A_{B'}^B.$$

In particular, if $O = O'$, then $[O']_B = [O]_{B'} = (0,0,0)$ and the affine mapping is usually called a *linear mapping* or *transformation*.

$$(3.3.3)$$

The proofs (compare with those of (2.4.2)) are left to the readers. See Fig. 3.10 (compare with Fig. 2.19).

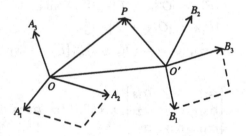

Fig. 3.10

Remark The computations of $A_B^{B'}$ and $A_{B'}^B$ (extending (2.4.3)).

Adopt the notations in (3.3.3) and use the results in (3.3.2) to help computation.

Let $\vec{a}_i = (a_{i1}, a_{i2}, a_{i3})$ and $\vec{b}_i = (b_{i1}, b_{i2}, b_{i3})$, $i = 1, 2, 3$, also

$$A = \begin{bmatrix} \vec{a}_1 \\ \vec{a}_2 \\ \vec{a}_3 \end{bmatrix} = \begin{bmatrix} a_{11} & a_{12} & a_{13} \\ a_{21} & a_{22} & a_{23} \\ a_{31} & a_{32} & a_{33} \end{bmatrix} \quad \text{and}$$

$$B = \begin{bmatrix} \vec{b}_1 \\ \vec{b}_2 \\ \vec{b}_3 \end{bmatrix} = \begin{bmatrix} b_{11} & b_{12} & b_{13} \\ b_{21} & b_{22} & b_{23} \\ b_{31} & b_{32} & b_{33} \end{bmatrix}.$$

Then A and B are invertible (see (3) in (3.3.2)).

By assumption, $[\vec{b}_i]_\mathcal{B} = (\alpha_{i1}, \alpha_{i2}, \alpha_{i3})$, $i = 1, 2, 3$, and then

$$\vec{b}_i = \alpha_{i1}\vec{a}_1 + \alpha_{i2}\vec{a}_2 + \alpha_{i3}\vec{a}_3$$

$$= (\alpha_{i1} \quad \alpha_{i2} \quad \alpha_{i3}) \begin{bmatrix} \vec{a}_1 \\ \vec{a}_2 \\ \vec{a}_3 \end{bmatrix}$$

$$= [\vec{b}_i]_\mathcal{B} A \quad \text{(remember that } [\vec{b}_i]_\mathcal{B} \text{ is viewed as a } 1 \times 3 \text{ matrix)}$$

$$\Rightarrow [\vec{b}_i]_\mathcal{B} = \vec{b}_i A^{-1}, \quad i = 1, 2, 3.$$

Similarly,

$$[\vec{a}_i]_{\mathcal{B}'} = \vec{a}_i B^{-1}, \quad i = 1, 2, 3.$$

Therefore,

$$A^{\mathcal{B}'}_{\mathcal{B}} = \begin{bmatrix} \vec{b}_1 A^{-1} \\ \vec{b}_2 A^{-1} \\ \vec{b}_3 A^{-1} \end{bmatrix} = \begin{bmatrix} \vec{b}_1 \\ \vec{b}_2 \\ \vec{b}_3 \end{bmatrix} A^{-1} = BA^{-1}, \quad \text{and}$$

$$A^{\mathcal{B}}_{\mathcal{B}'} = \begin{bmatrix} \vec{a}_1 B^{-1} \\ \vec{a}_2 B^{-1} \\ \vec{a}_3 B^{-1} \end{bmatrix} = \begin{bmatrix} \vec{a}_1 \\ \vec{a}_2 \\ \vec{a}_3 \end{bmatrix} B^{-1} = AB^{-1} \tag{3.3.4}$$

are the required formulas.

Example Give two sets of points

$$O = (1, 0, 0), \quad A_1 = (1, 1, 0), \quad A_2 = (0, 1, 1), \quad A_3 = (1, 0, 1), \quad \text{and}$$
$$O' = (-1, -1, -1), \quad B_1 = (1, -1, 1), \quad B_2 = (-1, 1, 1), \quad B_3 = (1, 1, 1)$$

in $\mathbb{R}^3$. Construct two coordinate systems $\Gamma(O; A_1, A_2, A_3)$ and $\Gamma(O'; B_1, B_2, B_3)$ and establish the formulas of changes of coordinates between them.

Solution Firstly, by simple computation, one has

$$\vec{a}_1 = \overrightarrow{OA_1} = (1, 1, 0) - (1, 0, 0) = (0, 1, 0),$$
$$\vec{a}_2 = \overrightarrow{OA_2} = (0, 1, 1) - (1, 0, 0) = (-1, 1, 1),$$
$$\vec{a}_3 = \overrightarrow{OA_3} = (1, 0, 1) - (1, 0, 0) = (0, 0, 1),$$
$$\vec{b}_1 = \overrightarrow{O'B_1} = (1, -1, 1) - (-1, -1, 1) = (2, 0, 2),$$
$$\vec{b}_2 = \overrightarrow{O'B_2} = (-1, 1, 1) - (-1, -1, -1) = (0, 2, 2),$$
$$\vec{b}_3 = \overrightarrow{O'B_3} = (1, 1, 1) - (-1, -1, -1) = (2, 2, 2), \quad \text{and}$$
$$\overrightarrow{OO'} = (-1, -1, -1) - (1, 0, 0) = (-2, -1, -1) = -\overrightarrow{O'O}.$$

Then, let

$$A = \begin{bmatrix} \vec{a_1} \\ \vec{a_2} \\ \vec{a_3} \end{bmatrix} = \begin{bmatrix} 0 & 1 & 0 \\ -1 & 1 & 1 \\ 0 & 0 & 1 \end{bmatrix} \quad \text{and} \quad B = \begin{bmatrix} \vec{b_1} \\ \vec{b_2} \\ \vec{b_3} \end{bmatrix} = \begin{bmatrix} 2 & 0 & 2 \\ 0 & 2 & 2 \\ 2 & 2 & 2 \end{bmatrix}$$

and compute, by the method indicated in (3) of (3.3.2),

$$A^{-1} = \begin{bmatrix} 1 & -1 & 1 \\ 1 & 0 & 0 \\ 0 & 0 & 1 \end{bmatrix},$$

$$B^{-1} = -\frac{1}{8} \begin{bmatrix} 0 & 4 & -4 \\ 4 & 0 & -4 \\ -4 & -4 & 4 \end{bmatrix} = \begin{bmatrix} 0 & -\frac{1}{2} & \frac{1}{2} \\ -\frac{1}{2} & 0 & \frac{1}{2} \\ \frac{1}{2} & \frac{1}{2} & -\frac{1}{2} \end{bmatrix}.$$

Then,

$$A_{\mathcal{B}}^{\mathcal{B}'} = BA^{-1} = \begin{bmatrix} 2 & 0 & 2 \\ 0 & 2 & 2 \\ 2 & 2 & 2 \end{bmatrix} \begin{bmatrix} 1 & -1 & 1 \\ 1 & 0 & 0 \\ 0 & 0 & 1 \end{bmatrix} = \begin{bmatrix} 2 & -2 & 4 \\ 2 & 0 & 2 \\ 4 & -2 & 4 \end{bmatrix},$$

$$A_{\mathcal{B}'}^{\mathcal{B}} = AB^{-1} = \begin{bmatrix} 0 & 1 & 0 \\ -1 & 1 & 1 \\ 0 & 0 & 1 \end{bmatrix} \begin{bmatrix} 0 & -\frac{1}{2} & \frac{1}{2} \\ -\frac{1}{2} & 0 & \frac{1}{2} \\ \frac{1}{2} & \frac{1}{2} & -\frac{1}{2} \end{bmatrix} = \begin{bmatrix} -\frac{1}{2} & 0 & \frac{1}{2} \\ 0 & 1 & -\frac{1}{2} \\ \frac{1}{2} & \frac{1}{2} & -\frac{1}{2} \end{bmatrix}.$$

Finally, using (3.3.3), for any point P in $\mathbb{R}^3$, we have the following formulas for changes of coordinates:

$$[P]_{\mathcal{B}} = (-3 \quad 2 \quad -3) + [P]_{\mathcal{B}'} \begin{bmatrix} 2 & -2 & 4 \\ 2 & 0 & 2 \\ 4 & -2 & 4 \end{bmatrix} \quad \text{and}$$

$$[P]_{\mathcal{B}'} = \left(0 \quad -\frac{1}{2} \quad 1 \right) + [P]_{\mathcal{B}} \begin{bmatrix} -\frac{1}{2} & 0 & \frac{1}{2} \\ 0 & 1 & -\frac{1}{2} \\ \frac{1}{2} & \frac{1}{2} & -\frac{1}{2} \end{bmatrix}.$$

Let $[P]_\mathcal{B} = (x_1, x_2, x_3)$ and $[P]_{\mathcal{B}'} = (y_1, y_2, y_3)$, the above two equations can be rewritten, respectively, as

$$\begin{cases} x_1 = -3 + 2y_1 + 2y_2 + 4y_3 \\ x_2 = 2 - 2y_1 - 2y_3 \\ x_3 = -3 + 4y_1 + 2y_2 + 4y_3, \quad \text{and} \end{cases}$$

$$\begin{cases} y_1 = -\dfrac{1}{2}x_1 + \dfrac{1}{2}x_3 \\ y_2 = \dfrac{1}{2} + x_2 + \dfrac{1}{2}x_3 \\ y_3 = 1 + \dfrac{1}{2}x_1 - \dfrac{1}{2}x_2 - \dfrac{1}{2}x_3. \end{cases}$$

Exercises

<A>

1. Prove (3.3.2) in detail, if possible (otherwise, see Ex. <B> 3).
2. Prove (3.3.3) in detail.
3. Let

   $$O = (1, -2, 0), \quad A_1 = (1, 1, 2), \quad A_2 = (-1, 2, -1), A_3 = (-2, -2, 3),$$

   and

   $$O' = (-1, -1, -2), \quad B_1 = (1, 1, 0), \quad B_2 = (0, 1, -1), \quad B_3 = (-1, 0, 1)$$

   be two sets of points in $\mathbb{R}^3$. Construct graphically two coordinate systems $\Gamma(O; A_1, A_2, A_3)$ and $\Gamma(O'; B_1, B_2, B_3)$ with $\vec{a}_i = \overrightarrow{OA_i}$ and $\vec{b}_i = \overrightarrow{O'B_i}$, $i = 1, 2, 3$.

   (a) Prove that the vectors $\vec{a}_1, \vec{a}_2, \vec{a}_3$ are linearly independent. Hence, $\mathcal{B} = \{\vec{a}_1, \vec{a}_2, \vec{a}_3\}$ is an ordered basis for $\Gamma(O; A_1, A_2, A_3)$.
   (b) Prove, similarly, that $\mathcal{B}' = \{\vec{b}_1, \vec{b}_2, \vec{b}_3\}$ indeed is an ordered basis for $\Gamma(O'; B_1, B_2, B_3)$.
   (c) Find formulas of changes of coordinates.

4. Let $\Gamma(O'; B_1, B_2, B_3)$ be a coordinate system in $\mathbb{R}^3$ with basis $\mathcal{B}' = \{\vec{b}_1, \vec{b}_2, \vec{b}_3\}$, where $\vec{b}_i = \overrightarrow{O'B_i}$, $i = 1, 2, 3$. Suppose a point $\vec{x} = (x_1, x_2, x_3) \in \mathbb{R}^3$ has coordinate $[\vec{x}]_{\mathcal{B}'} = (y_1, y_2, y_3)$ with respect

to $\mathcal{B}'$, satisfying

$$\begin{cases} x_1 = -6 + y_1 - y_2 + 2y_3, \\ x_2 = 5 + y_2 - y_3, \\ x_3 = 2 + y_1 + y_2 + y_3. \end{cases}$$

Determine the positions of the points O', B_1, B_2 and B_3; that is, the coordinates of these points with respect to the natural basis of $\mathbb{R}^3$.

5. Let the points O, A_1, A_2 and A_3 be as in Ex. 3, with $\mathcal{B} = \{\vec{a}_1, \vec{a}_2, \vec{a}_3\}$, where $\vec{a}_i = \overrightarrow{OA_i}$, $i = 1, 2, 3$, a basis for $\Gamma(O; A_1, A_2, A_3)$. If a point $\vec{x} \in \mathbb{R}^3$ has coordinate $[\vec{x}]_{\mathcal{B}} = (x_1, x_2, x_3)$, repeat Ex. 4.

6. Construct two coordinate systems $\Gamma(O; A_1, A_2, A_3)$ and $\Gamma(O'; B_1, B_2, B_3)$ with basis $\mathcal{B}$ and $\mathcal{B}'$, respectively, such that the coordinates $[\vec{x}]_{\mathcal{B}} = (x_1, x_2, x_3)$ and $[\vec{x}]_{\mathcal{B}'} = (y_1, y_2, y_3)$ of a point $\vec{x} \in \mathbb{R}^3$ satisfy the following equations, respectively.

(a) $x_1 = 1 + 2y_1 - 15y_2 + 22y_3,$
    $x_2 = -3 + y_1 - 7y_2 + 10y_3,$
    $x_3 = 4 - 4y_1 + 30y_2 - 43y_3;$

(b) $x_1 = 2y_1 + y_2 - y_3,$
    $x_2 = 2y_1 - y_2 + 2y_3,$
    $x_3 = 3y_1 + y_3;$

(c) $y_1 = -15 + x_1 + 2x_2 + x_3,$

    $y_2 = -7 + x_1 - x_2 + 2x_3,$

    $y_3 = 30 - x_1;$

(d) $y_1 = 4 + \dfrac{4}{3}x_1 - \dfrac{8}{3}x_2 - \dfrac{1}{3}x_3,$

    $y_2 = -3 + \dfrac{1}{3}x_1 + \dfrac{1}{3}x_2 - \dfrac{1}{3}x_3,$

    $y_3 = -6 - \dfrac{1}{3}x_1 + \dfrac{2}{3}x_2 + \dfrac{1}{3}x_3.$

<B>

1. Consider the following system of equations

$$\begin{aligned} x_1 &= \alpha_1 + \alpha_{11}y_1 + \alpha_{21}y_2 + \alpha_{31}y_3, \\ x_2 &= \alpha_2 + \alpha_{12}y_1 + \alpha_{22}y_2 + \alpha_{32}y_3, \\ x_3 &= \alpha_3 + \alpha_{13}y_1 + \alpha_{23}y_2 + \alpha_{33}y_3, \end{aligned}$$

where the *coefficient matrix*

$$\begin{bmatrix} \alpha_{11} & \alpha_{12} & \alpha_{13} \\ \alpha_{21} & \alpha_{22} & \alpha_{23} \\ \alpha_{31} & \alpha_{32} & \alpha_{33} \end{bmatrix}$$

is invertible. Try to construct, in $\mathbb{R}^3$, two coordinate systems, so that the above prescribed equations will serve as changes of coordinates between them. In fact, there are infinitely many such a pair of coordinate systems.

2. Let $\vec{x} = (x_1, x_2, x_3)$ and $\vec{y} = (y_1, y_2, y_3)$ be vectors in $\mathbb{R}^3$. Prove the following.

   (a) $\vec{x}$ is linearly independent by itself if and only if at least one of the components x_1, x_2, x_3 is not equal to zero.
   (b) $\vec{x}$ and $\vec{y}$ are linearly independent if and only if at least one of the following three determinants

   $$\begin{vmatrix} x_1 & x_2 \\ y_1 & y_2 \end{vmatrix}, \quad \begin{vmatrix} x_2 & x_3 \\ y_2 & y_3 \end{vmatrix}, \quad \begin{vmatrix} x_1 & x_3 \\ y_1 & y_3 \end{vmatrix}$$

   is not equal to zero. Try to explain this result geometrically.

   (*Note* Refer to the following Ex. 3 and try to think jointly. Are results here extendable to $\mathbb{R}^4$? $\mathbb{R}^n$?)

3. Prove (3.3.2), noting and answering the following questions.

   (1) Can the method used in proving Ex. <A>2 of Sec. 2.4 be adopted directly in order to prove (3.3.2)? If yes, try out all the details. Practically, is the method still valuable in proving similar results for 4×4 matrices, or even higher order matrices?
   (2) If not so easy, at which steps one might encounter difficulties? Can you overcome it by invoking some other methods which are powerful and still efficient in proving similar results for higher order matrices?
   (3) In the process of proof, are you able to give each algebraic quantity or step an exact geometric meaning or interpretation?

4. Give a 3×3 matrix

   $$A = \begin{bmatrix} a_{11} & a_{12} & a_{13} \\ a_{21} & a_{22} & a_{23} \\ a_{31} & a_{32} & a_{33} \end{bmatrix}.$$

   For a vector $\vec{x} = (x_1, x_2, x_3) \in \mathbb{R}^3$, designate

   $$\vec{x}A = (x_1 \quad x_2 \quad x_3) \begin{bmatrix} a_{11} & a_{12} & a_{13} \\ a_{21} & a_{22} & a_{23} \\ a_{31} & a_{32} & a_{33} \end{bmatrix}$$

   $$= \left(\sum_{i=1}^{3} a_{i1}x_i, \ \sum_{i=1}^{3} a_{i2}x_i, \ \sum_{i=1}^{3} a_{i3}x_i \right)$$

   and consider it as a vector in $\mathbb{R}^3$ (see the Convention before Remark 2.3 in Sec. 2.3). $\vec{x}A$ is called the *image vector* of $\vec{x}$ under the operation of A which represents a *linear transformation*, to be treated in detail in Sec. 3.7 (see Sec. B.7 for details).

(a) Show that the kernel $\text{Ker}(A)$ and the range $\text{Im}(A)$ are subspaces of $\mathbb{R}^3$.

(b) Show that $\dim \text{Ker}(A) + \dim \text{Im}(A) = \dim \mathbb{R}^3 = 3$.

(c) Show that the following are equivalent.

    (1) A is one-to-one, i.e. $\text{Ker}(A) = \{\vec{0}\}$.

    (2) A is onto, i.e. $\text{Im}(A) = \mathbb{R}^3$.

    (3) A maps every or a basis $\mathcal{B} = \{\vec{x}_1, \vec{x}_2, \vec{x}_3\}$ for $\mathbb{R}^3$ onto a basis $\{\vec{x}_1 A, \vec{x}_2 A, \vec{x}_3 A\}$ for $\mathbb{R}^3$.

    (4) A is invertible.

(d) Let $\mathcal{B} = \{\vec{x}_1, \vec{x}_2, \vec{x}_3\}$ be a basis for $\mathbb{R}^3$ and $\vec{y}_1, \vec{y}_2, \vec{y}_3$ be any vectors in $\mathbb{R}^3$. Show that there exists a unique linear transformation $f: \mathbb{R}^3 \to \mathbb{R}^3$ so that

$$f(\vec{x}_i) = \vec{y}_i \quad \text{for } 1 \le i \le 3.$$

(e) Let S be the subspace $x_1 + 2x_2 + 3x_3 = 0$ in $\mathbb{R}^3$. Show that there are infinitely many linear transformations $f: \mathbb{R}^3 \to \mathbb{R}^3$ with the following respective property:

    (1) $\text{Ker}(f) = S$.

    (2) $\text{Im}(f) = S$.

Is there any linear transformation $f: \mathbb{R}^3 \to \mathbb{R}^3$ such that $\text{Ker}(f) = \text{Im}(f) = S$? Try to explain your claim geometrically and analytically.

(f) Let S be the subspace in $\mathbb{R}^3$ defined by

$$-x_1 + 2x_2 + 3x_3 = 0, \quad 5x_1 + 2x_2 - x_3 = 0.$$

Do the same problem as in (e).

(g) Let S_1 be the subspace as in (e) and S_2 be the subspace as in (f). Show that there are infinitely many linear transformations $f: \mathbb{R}^3 \to \mathbb{R}^3$ with the following respective property:

    (1) $\text{Ker}(f) = S_1$ and $\text{Im}(f) = S_2$.

    (2) $\text{Ker}(f) = S_2$ and $\text{Im}(f) = S_1$.

5. Let $S_1 \subseteq \mathbb{R}^2$ be a subspace and $S_2 \subseteq \mathbb{R}^3$ be a subspace.

(a) Find all possible linear transformations $f: \mathbb{R}^2 \to \mathbb{R}^3$ with the following property, respectively:

    (1) f is one-to-one.

    (2) $\text{Ker}(f) = S_1$.

    (3) $\text{Im}(f) = S_2$. Be careful about the case that $\dim S_2 = 3$.

    (4) $\text{Ker}(f) = S_1$ and $\text{Im}(f) = S_2$.

(b) Find all possible linear transformations $g: \mathbb{R}^3 \to \mathbb{R}^2$ with the following property, respectively:

    (1) g is onto.
    (2) $\mathrm{Ker}(g) = S_2$.
    (3) $\mathrm{Im}(g) = S_1$.
    (4) $\mathrm{Ker}(g) = S_2$ and $\mathrm{Im}(g) = S_1$.

(c) Does there exist a linear transformation from $\mathbb{R}^2$ onto $\mathbb{R}^3$? How about a one-to-one linear transformation from $\mathbb{R}^3$ into $\mathbb{R}^2$? Why?

(d) Suppose $f: \mathbb{R}^2 \to \mathbb{R}^3$ is an one-to-one linear transformation. Find all possible linear transformations $g: \mathbb{R}^3 \to \mathbb{R}^2$ so that

$$g \circ f = 1_{\mathbb{R}^2} \quad \text{(the identity transformation on } \mathbb{R}^2\text{)}.$$

(e) Suppose $g: \mathbb{R}^3 \to \mathbb{R}^2$ is an onto linear transformation. Find all possible linear transformations $f: \mathbb{R}^2 \to \mathbb{R}^3$ so that

$$g \circ f = 1_{\mathbb{R}^2}.$$

6. A linear transformation f from $\mathbb{R}^3$ (or $\mathbb{R}^2$) to $\mathbb{R}$ is specially called a *linear functional*.

(a) Define the kernel $\mathrm{Ker}(f)$.

(b) Prove that there exist unique scalars a_1, a_2 and a_3 so that

$$f(\vec{x}) = a_1 x_1 + a_2 x_2 + a_3 x_3,$$

where $\vec{x} = (x_1, x_2, x_3)$ is in $\mathbb{R}^3$.

<C>Abstraction and generalization

Try to extend (3.3.3) to n-dimensional vector space over a field.

3.4 Lines in Space

Throughout this section, $\Gamma(O; A_1, A_2, A_3)$ is a fixed coordinate system with basis $\mathcal{B} = \{\vec{a}_1, \vec{a}_2, \vec{a}_3\}$, $\vec{a}_i = \overrightarrow{OA_i}$, $i = 1, 2, 3$.

The straight line determined by O and A_i is denoted by L_i, $i = 1, 2, 3$. Take a point P in $\mathbb{R}^3$. Then

$$P \in L_1$$
$$\Leftrightarrow \overrightarrow{OP} \in L(O; A_1)$$
$$\Leftrightarrow [P]_{\mathcal{B}} = (x_1, 0, 0).$$

This characterizes the fact that the second and the third components of the coordinate $[P]_{\mathcal{B}}$ for a point P lying on L_1 should be all equal to zero.

Therefore,

$$x_2 = 0, \quad x_3 = 0 \tag{3.4.1}$$

is called the *equation* of the *first coordinate axis* L_1 with respect to $\mathcal{B}$ (see Fig. 3.11). Similarly,

$$\begin{aligned} x_1 &= 0, \quad x_3 = 0, \\ x_1 &= 0, \quad x_2 = 0, \end{aligned} \tag{3.4.2}$$

are, respectively, called the equation of the *second* and the *third coordinate axes* L_2 and L_3 (see Fig. 3.11).

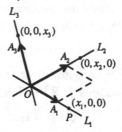

Fig. 3.11

These three coordinate axes, all together, separate the whole space $\mathbb{R}^3$ into $2^3 = 8$ quadrants, according to the positive and negative signs of components x_1, x_2, x_3 of the coordinate $[P]_\mathcal{B}$, $P \in \mathbb{R}^3$ (see Sec. 2.5).

By exactly the same way as we did in Sec. 2.5, one has

Equations of a line with respect to a fixed coordinate system in $\mathbb{R}^3$
The straight line L determined by two different points A and B in $\mathbb{R}^3$ has the following ways of representation in $\Gamma(O; A_1, A_2, A_3)$ with basis $\mathcal{B}$.

(1) *Parametric equation in vector form*
L passes the point $\vec{a} = \overrightarrow{OA}$ with the direction $\vec{b} = \overrightarrow{AB}$, and hence has the equation

$$\vec{x} = \vec{a} + t\vec{b}, \quad t \in \mathbb{R},$$

where $\vec{x} = \overrightarrow{OX}$ is the position vector of a point X on L with respect to O and is viewed as a point in $\mathbb{R}^3$.

(2) *Parametric equation with respect to basis $\mathcal{B}$*

$$[\vec{x}]_\mathcal{B} = [\vec{a}]_\mathcal{B} + t[\vec{b}]_\mathcal{B}, \quad t \in \mathbb{R}$$

or, let $[\vec{a}]_\mathcal{B} = (a_1, a_2, a_3), [\vec{b}]_\mathcal{B} = (b_1, b_2, b_3)$ and $[\vec{x}]_\mathcal{B} = (x_1, x_2, x_3)$,

$$\begin{aligned} x_1 &= a_1 + tb_1, \\ x_2 &= a_2 + tb_2, \\ x_3 &= a_3 + tb_3. \end{aligned}$$

(3) *Coordinate equation with respect to $\mathcal{B}$*

$$\frac{x_1 - a_1}{b_1} = \frac{x_2 - a_2}{b_2} = \frac{x_3 - a_3}{b_3}. \tag{3.4.3}$$

See Fig. 3.12 (compare with Fig. 2.23).

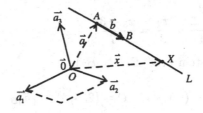

Fig. 3.12

Remark Changes of equations of the same line in different coordinate systems.

We adopt the notations and results from (3.3.3).

The line L in the coordinate system $\Gamma(O; A_1, A_2, A_3)$ has the equation

$$[X]_{\mathcal{B}} = [\vec{a}]_{\mathcal{B}} + t[\vec{b}]_{\mathcal{B}},$$

where $\vec{a} = \overrightarrow{OA}$ and $\vec{b} = \overrightarrow{AB}$ (direction) and X is a moving point on L. Via the change of coordinates $[X]_{\mathcal{B}'} = [O]_{\mathcal{B}'} + [X]_{\mathcal{B}} A_{\mathcal{B}'}^{\mathcal{B}}$, the equation of L in the coordinate system $\Gamma(O'; B_1, B_2, B_3)$ is

$$[X]_{\mathcal{B}'} = [O]_{\mathcal{B}'} + \{[\vec{a}]_{\mathcal{B}} + t[\vec{b}]_{\mathcal{B}}\} A_{\mathcal{B}'}^{\mathcal{B}}. \tag{3.4.4}$$

Example (continued from the example in Sec. 3.3) Find the equations of the straight line determined by the points $A = (1, 2, 3)$ and $B = (-2, 1, -1)$ in the coordinate systems $\Gamma(O; A_1, A_2, A_3)$ and $\Gamma(O'; B_1, B_2, B_3)$, respectively.

Solution In the coordinate system $\Gamma(O; A_1, A_2, A_3)$,

$$\vec{a} = \overrightarrow{OA} = (1, 2, 3) - (1, 0, 0) = (0, 2, 3), \quad [\vec{a}]_{\mathcal{B}} = (2, 0, 3), \quad \text{and}$$
$$\vec{b} = \overrightarrow{AB} = (-2, 1, -1) - (1, 2, 3) = (-3, -1, -4), \quad [\vec{b}]_{\mathcal{B}} = (-4, 3, -7).$$

Let X be a moving point on the line and $[X]_B = (x_1, x_2, x_3)$. Then, the equation of the line is

$$(x_1, x_2, x_3) = (2, 0, 3) + t(-4, 3, -7), \quad t \in \mathbb{R}$$

$$\Rightarrow \begin{cases} x_1 = 2 - 4t, \\ x_2 = 3t, \qquad t \in \mathbb{R} \qquad \text{or} \\ x_3 = 3 - 7t, \end{cases}$$

$$\frac{x_1 - 2}{-4} = \frac{x_2}{3} = \frac{x_3 - 3}{-7}.$$

In the coordinate system $\Gamma(O'; B_1, B_2, B_3)$,

$$\vec{a} = \overrightarrow{O'A} = (1, 2, 3) - (-1, -1, -1) = (2, 3, 4), \quad [\vec{a}]_{B'} = \left(\frac{1}{2}, 1, \frac{1}{2}\right) \quad \text{and}$$

$$\vec{b} = (-3, -1, -4), \quad [\vec{b}]_{B'} = \left(-\frac{3}{2}, -\frac{1}{2}, 0\right).$$

Let $[X]_{B'} = (y_1, y_2, y_3)$, the equation of the line is

$$(y_1, y_2, y_3) = \left(\frac{1}{2}, 1, \frac{1}{2}\right) + t\left(-\frac{3}{2}, -\frac{1}{2}, 0\right), \quad t \in \mathbb{R}$$

$$\Rightarrow \begin{cases} y_1 = \dfrac{1}{2} - \dfrac{3}{2}t, \\ y_2 = 1 - \dfrac{1}{2}t, \quad t \in \mathbb{R} \qquad \text{or} \\ y_3 = \dfrac{1}{2}, \end{cases}$$

$$\frac{y_1 - \frac{1}{2}}{-\frac{3}{2}} = \frac{y_2 - 1}{-\frac{1}{2}} = \frac{y_3 - \frac{1}{2}}{0}.$$

By using (3.4.4) and the results obtained in the example of Sec. 3.3, we are able to deduce the equation of the line in the coordinate system $\Gamma(O'; B_1, B_2, B_3)$ from that in $\Gamma(O; A_1, A_2, A_3)$ as follows.

$$(y_1, y_2, y_3) = \begin{pmatrix} 0 & -\frac{1}{2} & 1 \end{pmatrix} + \{(2 \quad 0 \quad 3) + t(-4 \quad 3 \quad -7)\} \begin{bmatrix} -\frac{1}{2} & 0 & \frac{1}{2} \\ 0 & 1 & -\frac{1}{2} \\ \frac{1}{2} & \frac{1}{2} & -\frac{1}{2} \end{bmatrix}$$

$$= \begin{pmatrix} 0 & -\frac{1}{2} & 1 \end{pmatrix} + \begin{pmatrix} \frac{1}{2} - \frac{3}{2}t & \frac{3}{2} - \frac{1}{2}t & -\frac{1}{2} \end{pmatrix}$$

$$= \begin{pmatrix} \frac{1}{2} - \frac{3}{2}t & 1 - \frac{1}{2}t & \frac{1}{2} \end{pmatrix},$$

which is identical with the above result.

Finally, we list the

Relative positions of two lines in space

Given two lines

$$L_1: \vec{x} = \vec{a}_1 + t\vec{b}_1, \ t \in \mathbb{R} \quad \text{and}$$
$$L_2: \vec{x} = \vec{a}_2 + t\vec{b}_2, \ t \in \mathbb{R}$$

in $\mathbb{R}^3$, they have the following four kinds of relative positions.

Case 1 Coincident $(L_1 = L_2) \Leftrightarrow$ the vectors $\vec{b}_1, \vec{b}_2$ and $\vec{a}_2 - \vec{a}_1$ are linearly dependent.

Case 2 Parallel $(L_1 \parallel L_2) \Leftrightarrow \vec{b}_1$ and $\vec{b}_2$ are linearly dependent, and $\vec{a}_2 - \vec{a}_1$ is linearly independent of $\vec{b}_1$ or $\vec{b}_2$.

Case 3 Intersecting (at a unique point) $\Leftrightarrow \vec{b}_1$ and $\vec{b}_2$ are linearly independent, and $\vec{a}_2 - \vec{a}_1$ is linearly dependent of $\vec{b}_1$ and $\vec{b}_2$.

Case 4 *Skew* (neither parallel nor intersecting) $\Leftrightarrow \vec{b}_1$ and $\vec{b}_2$ and $\vec{a}_2 - \vec{a}_1$ are linearly independent.

In Cases 1–3, the two lines are coplanar, while in Case 4, they are non-coplanar. (3.4.5)

Proofs are left to the readers (see Fig. 3.13).

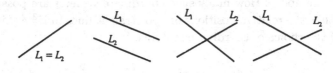

Fig. 3.13

Exercises

<A>

1. Prove (3.4.3) in detail.
2. Prove (3.4.5) in detail.
3. (continued from Ex. <A> 3 of Sec. 3.3) Let L be the line in $\mathbb{R}^3$ determined by the points $A = (-\frac{1}{2}, 0, 3)$ and $B = (\sqrt{2}, 1, 6)$.
   (a) Find the equations of L both in $\Gamma(O; A_1, A_2, A_3)$ and in $\Gamma(O'; B_1, B_2, B_3)$.
   (b) Use the results in (a) to justify (3.4.4).

4. Suppose a line in $\mathbb{R}^3$ has the equation

$$\frac{x_1 - 1}{\sqrt{2}} = \frac{x_2 + 3}{-2} = \frac{x_3 - 6}{4}$$

in the rectangular coordinate system $\Gamma(\vec{0}; \vec{e}_1, \vec{e}_2, \vec{e}_3)$. Find the equation of the line in another coordinate system $\Gamma(O; A_1, A_2, A_3)$ where $O = (0, \sqrt{3}, 1)$, $A_1 = (1, 0, 1)$, $A_2 = (1, -1, 0)$, $A_3 = (-\sqrt{2}, -1, -2)$.

5. (continued from Ex. 4) If the equation of a line in the coordinate system $\Gamma(O; A_1, A_2, A_3)$ is

$$\frac{y_1 + 2}{1} = \frac{y_2 - 1}{2} = \frac{y_3 + 5}{3},$$

find the equation of the line in $\Gamma(\vec{0}; \vec{e}_1, \vec{e}_2, \vec{e}_3)$.

<B>

1. Suppose that a_i, b_i and c_i, d_i, $i = 1, 2, 3$, are scalars, that b_1, b_2, b_3 and d_1, d_2, d_3, are not all equal to zero respectively. Let

$$\frac{x_1 - a_1}{b_1} = \frac{x_2 - a_2}{b_2} = \frac{x_3 - a_3}{b_3} \quad \text{and} \quad \frac{y_1 - c_1}{d_1} = \frac{y_2 - c_2}{d_2} = \frac{y_3 - c_3}{d_3}.$$

Try to construct two coordinate systems $\Gamma(O; A_1, A_2, A_3)$ and $\Gamma(O'; B_1, B_2, B_3)$ in $\mathbb{R}^3$ so that the respective equation of the same line with respect to either of the two coordinate systems is the one or the other given above. How many such coordinate systems are possible?

2. Prove that the relative positions of two straight lines in $\mathbb{R}^3$ are independent of the choice of coordinate systems.

3. Let

$$L_i: \vec{x} = \vec{a}_i + t\vec{b}_i, \quad t \in \mathbb{R}, \quad i = 1, 2, 3,$$

be three given lines in $\mathbb{R}^3$. Try to discuss all possible relative positions of them and use $\vec{a}_i, \vec{b}_i, i = 1, 2, 3$, to characterize each case.

4. Do there exist, in $\mathbb{R}^3$, infinitely many straight lines such that any two of them are skew to each other? If yes, try to construct one.

5. How to find the distance between two parallel lines or two skew lines in $\mathbb{R}^3$? Any formula for it?

3.5 Planes in Space

Suppose $\Gamma(O; A_1, A_2, A_3)$ with basis $\mathcal{B} = \{\vec{a}_1, \vec{a}_2, \vec{a}_3\}$, $\vec{a}_i = \overrightarrow{OA_i}$, $i = 1, 2, 3$, is a fixed coordinate system in $\mathbb{R}^3$, throughout this whole section.

Refer back to Fig. 3.11. Take any point $P \in \mathbb{R}^3$, then

$\overrightarrow{OP} \in \Sigma(O; A_1, A_2),$ the plane generated by O, A_1 and $A_2,$

$\Leftrightarrow [P]_{\mathcal{B}} = (x_1, x_2, 0),$

which characterizes the fact that the third component x_3 of the coordinate $[P]_{\mathcal{B}}$ with respect to $\mathcal{B}$ must be equal to zero. Hence we call

$$x_3 = 0 \tag{3.5.1}$$

the *equation* of the *coordinate plane* generated by O, A_1 and A_2. The other two *coordinate planes*, generated respectively by O, A_2, A_3 and O, A_1, A_3, have respective equation

$$x_1 = 0, \quad \text{and}$$
$$x_2 = 0. \tag{3.5.2}$$

Note that any two of these three coordinate planes intersect along a straight line which is a coordinate axis (see Sec. 3.4).

Three non-collinear points A, B and C in $\mathbb{R}^3$ determine a unique plane, still lying in $\mathbb{R}^3$. What is the equation of this plane in a fixed coordinate system?

In Fig. 3.14, let

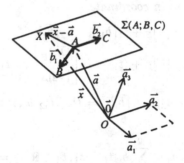

Fig. 3.14

$\vec{a} = \overrightarrow{OA}$ (viewed as a point in $\mathbb{R}^3$),

$\vec{b}_1 = \overrightarrow{AB},$

$\vec{b}_2 = \overrightarrow{AC}$ (both viewed as direction vectors in $\mathbb{R}^3$),

$\vec{x} = \overrightarrow{OX}$ (X, a moving point in $\mathbb{R}^3$ and $\vec{x}$ viewed as a point).

Notice that $\vec{b}_1$ and $\vec{b}_2$ are linearly independent.

For point $X \in \mathbb{R}^3$, then

$$\overrightarrow{OX} \in \Sigma(A; B, C), \quad \text{the plane generated by } A, B \text{ and } C$$
$$\Leftrightarrow \vec{x} - \vec{a} = t_1 \vec{b}_1 + t_2 \vec{b}_2, \quad t_1, t_2 \in \mathbb{R}$$
$$\Leftrightarrow \vec{x} = \vec{a} + t_1 \vec{b}_1 + t_2 \vec{b}_2, \quad t_1, t_2 \in \mathbb{R}.$$

This is the equation, in vector form, of the plane $\Sigma(A; B, C)$, passing through the *point* $\vec{a}$ with *directions* $\vec{b}_1$ and $\vec{b}_2$.

Just like what we did in Sec. 2.5, one may use coordinates of $\vec{x}$, $\vec{a}$, $\vec{b}$ and $\vec{c}$ with respect to $\mathcal{B}$, and eliminate the parameters t_1 and t_2 to obtain the coordinate equation of the plane in $\mathcal{B}$. Anyway, we summarize as

Equations of a plane with respect to a fixed coordinate system in $\mathbb{R}^3$

The plane Σ determined by non-collinear points A, B and C in $\mathbb{R}^3$ has the following representations in $\Gamma(O; A_1, A_2, A_3)$ with basis $\mathcal{B}$.

(1) *Parametric equation in vector form*

$$\vec{x} = \vec{a} + t_1 \vec{b}_1 + t_2 \vec{b}_2, \quad t_1, t_2 \in \mathbb{R}.$$

Σ passes through the point $\vec{a} = \overrightarrow{OA}$ with directions $\vec{b}_1 = \overrightarrow{AB}$ and $\vec{b}_2 = \overrightarrow{AC}$, while $\vec{x} = \overrightarrow{OX}$ is viewed as a point in Σ for $X \in \Sigma$.

(2) *Parametric equation in coordinates*

$$[\vec{x}]_{\mathcal{B}} = [\vec{a}]_{\mathcal{B}} + t_1 [\vec{b}_1]_{\mathcal{B}} + t_2 [\vec{b}_2]_{\mathcal{B}}$$
$$= [\vec{a}]_{\mathcal{B}} + (t_1 \quad t_2) \begin{bmatrix} [\vec{b}_1]_{\mathcal{B}} \\ [\vec{b}_2]_{\mathcal{B}} \end{bmatrix}, \quad t_1, t_2 \in \mathbb{R}.$$

Or, let $[\vec{a}]_{\mathcal{B}} = (a_1, a_2, a_3), [\vec{b}_i]_{\mathcal{B}} = (b_{i1}, b_{i2}, b_{i3})$, $i = 1, 2$, and $[\vec{x}]_{\mathcal{B}} = (x_1, x_2, x_3)$, then

$$x_j = a_j + \sum_{i=1}^{2} b_{ij} t_i, \quad t_1, t_2 \in \mathbb{R}, \quad j = 1, 2, 3.$$

(3) *Coordinate equation*

$$\begin{vmatrix} x_1 - a_1 & x_2 - a_2 & x_3 - a_3 \\ b_{11} & b_{12} & b_{13} \\ b_{21} & b_{22} & b_{23} \end{vmatrix} = 0,$$

or

$$\begin{vmatrix} b_{12} & b_{13} \\ b_{22} & b_{23} \end{vmatrix} (x_1 - a_1) + \begin{vmatrix} b_{13} & b_{11} \\ b_{23} & b_{21} \end{vmatrix} (x_2 - a_2) + \begin{vmatrix} b_{11} & b_{12} \\ b_{21} & b_{22} \end{vmatrix} (x_3 - a_3) = 0,$$

or simplified as

$$\alpha x_1 + \beta x_2 + \gamma x_3 + \delta = 0. \tag{3.5.3}$$

The details of the proof are left to the readers.

Remark Changes of equations of the same plane in different coordinate systems.

Notations and results in (3.3.3) are adopted directly in the following.

The plane Σ in another coordinate system $\Gamma(O'; B_1, B_2, B_3)$ with respect to $\mathcal{B}'$ has the equation

$$[X]_{\mathcal{B}'} = [O]_{\mathcal{B}'} + \left\{ [\vec{a}]_{\mathcal{B}} + (t_1 \quad t_2) \begin{bmatrix} [\vec{b}_1]_{\mathcal{B}} \\ [\vec{b}_2]_{\mathcal{B}} \end{bmatrix} \right\} A_{\mathcal{B}'}^{\mathcal{B}}, \tag{3.5.4}$$

where $X \in \Sigma$ and $[X]_{\mathcal{B}'} = [\overrightarrow{O'X}]_{\mathcal{B}'}$.

Example (continued from the example in Sec. 3.3) Find the equations of the plane determined by the points $A = (1, 2, 3)$, $B = (-2, 1, -1)$ and $C = (0, -1, 4)$, respectively, in $\Gamma(O; A_1, A_2, A_3)$ and $\Gamma(O'; B_1, B_2, B_3)$.

Solution One might refer to the example in Sec. 3.4 and results there.

In the coordinate system $\Gamma(O; A_1, A_2, A_3)$,

$$\vec{a} = \overrightarrow{OA} = (0, 2, 3), \quad [\vec{a}]_{\mathcal{B}} = (2, 0, 3);$$
$$\vec{b}_1 = \overrightarrow{AB} = (-3, -1, -4), \quad [\vec{b}_1]_{\mathcal{B}} = (-4, 3, -7);$$
$$\vec{b}_2 = \overrightarrow{AC} = (0, -1, 4) - (1, 2, 3) = (-1, -3, 1), \quad [\vec{b}_2]_{\mathcal{B}} = (-4, 1, 0).$$

For point X in the plane, let $[\vec{x}]_{\mathcal{B}} = [\overrightarrow{OX}]_{\mathcal{B}} = (x_1, x_2, x_3)$ and then, the equation is

$$(x_1, x_2, x_3) = (2, 0, 3) + t_1(-4, 3, -7) + t_2(-4, 1, 0), \quad t_1, t_2 \in \mathbb{R}$$

$$\Rightarrow \begin{cases} x_1 = 2 - 4t_1 - 4t_2, \\ x_2 = 3t_1 + t_2, \\ x_3 = 3 - 7t_1, \end{cases} \quad t_1, t_2 \in \mathbb{R}$$

$$\Rightarrow 7x_1 + 28x_2 - 8x_3 + 10 = 0.$$

In the coordinate system $\Gamma(O'; B_1, B_2, B_3)$,

$$\vec{a} = \overrightarrow{O'A} = (2, 3, 4), \quad [\vec{a}]_{\mathcal{B}'} = \left(\frac{1}{2}, 1, \frac{1}{2} \right);$$

$$\vec{b}_1 = \overrightarrow{AB} = (-3, -1, -4), \quad [\vec{b}_1]_{\mathcal{B}'} = \left(-\frac{3}{2}, -\frac{1}{2}, 0 \right);$$

$$\vec{b}_2 = \overrightarrow{AC} = (-1, -3, 1), \quad [\vec{b}_2]_{\mathcal{B}'} = \left(2, 1, -\frac{5}{2} \right).$$

Let $[X]_{B'} = (y_1, y_2, y_3)$, then the equation is

$$(y_1, y_2, y_3) = \left(\frac{1}{2}, 1, \frac{1}{2}\right) + t_1 \left(-\frac{3}{2}, -\frac{1}{2}, 0\right) + t_2 \left(2, 1, -\frac{5}{2}\right), \quad t_1, t_2 \in \mathbb{R}$$

$$\Rightarrow \begin{cases} y_1 = \dfrac{1}{2} - \dfrac{3}{2}t_1 + 2t_2, \\ y_2 = 1 - \dfrac{1}{2}t_1 + t_2, \quad t_1\, t_2 \in \mathbb{R} \\ y_3 = \dfrac{1}{2} - \dfrac{5}{2}t_2, \end{cases}$$

$$\Rightarrow 10y_1 - 30y_2 - 4y_3 + 27 = 0.$$

By the way, using (3.5.4) and results obtained in the example of Sec. 3.3, one has

$$(y_1, y_2, y_3) = \begin{pmatrix} 0 & -\frac{1}{2} & 1 \end{pmatrix} + \{(2 \quad 0 \quad 3) + t_1(-4 \quad 3 \quad -7)$$

$$+ t_2(-4 \quad 1 \quad 0)\} \begin{bmatrix} -\frac{1}{2} & 0 & \frac{1}{2} \\ 0 & 1 & -\frac{1}{2} \\ \frac{1}{2} & \frac{1}{2} & -\frac{1}{2} \end{bmatrix}$$

$$= \begin{pmatrix} 0 & -\frac{1}{2} & 1 \end{pmatrix} + \begin{pmatrix} \frac{1}{2} - \frac{3}{2}t_1 + 2t_2 & \frac{3}{2} - \frac{1}{2}t_1 + t_2 & -\frac{1}{2} - \frac{5}{2}t_2 \end{pmatrix}$$

$$= \begin{pmatrix} \frac{1}{2} - \frac{3}{2}t_1 + 2t_2 & 1 - \frac{1}{2}t_1 + t_2 & \frac{1}{2} - \frac{5}{2}t_2 \end{pmatrix}$$

just as established above.　　　　　　　　　　　　　　　　　　　　　□

Finally, we present the

Relative positions of a straight line and a plane in $\mathbb{R}^3$

Given a straight line L and a plane Σ in $\mathbb{R}^3$, respectively, as

$$L: \vec{x} = \vec{a} + t\vec{b}, \qquad\qquad t \in \mathbb{R}, \qquad \text{and}$$
$$\Sigma: \vec{x} = \vec{c} + t_1 \vec{d_1} + t_2 \vec{d_2}, \qquad t_1, t_2 \in \mathbb{R},$$

then L and Σ have the following relative positions:

1. Coincident $(L \subseteq \Sigma) \Leftrightarrow$ both $\vec{b}$ and $\vec{a} - \vec{c}$ are linearly dependent on $\vec{d_1}$ and $\vec{d_2}$.
2. Parallel $(L \parallel \Sigma) \Leftrightarrow \vec{b}$ is linearly dependent on $\vec{d_1}$ and $\vec{d_2}$ and $\vec{a} - \vec{c}$ is linearly independent of $\vec{d_1}$ and $\vec{d_2}$.
3. Intersecting (at a unique point) $\Leftrightarrow \vec{b}$, $\vec{d_1}$ and $\vec{d_2}$ are linearly independent.

$$(3.5.5)$$

See Fig. 3.15.

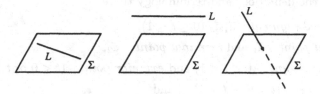

Fig. 3.15

Also is the

Relative positions of two planes in $\mathbb{R}^3$

Given two planes in $\mathbb{R}^3$ as

$$\Sigma_1: \vec{x} = \vec{a} + t_1 \vec{b_1} + t_2 \vec{b_2},$$
$$\Sigma_2: \vec{x} = \vec{c} + t_1 \vec{d_1} + t_2 \vec{d_2}, \quad t_1, t_2 \in \mathbb{R},$$

then, Σ_1 and Σ_2 have the following relative positions:

1. Coincident $(\Sigma_1 = \Sigma_2) \Leftrightarrow$ the vectors $\vec{a} - \vec{c}$, $\vec{d_1}$ and $\vec{d_2}$ are linearly dependent on $\vec{b_1}$ and $\vec{b_2}$.
2. Parallel $(\Sigma_1 \parallel \Sigma_2) \Leftrightarrow \vec{d_1}$ and $\vec{d_2}$ are linearly dependent on $\vec{b_1}$ and $\vec{b_2}$, but $\vec{a} - \vec{c}$ is linearly independent of $\vec{b_1}$ and $\vec{b_2}$.
3. Intersecting (along a straight line) $\Leftrightarrow$ at least three of the vectors $\vec{b_1}$, $\vec{b_2}$, $\vec{d_1}$ and $\vec{d_2}$ are linearly independent.

$$(3.5.6)$$

See Fig. 3.16.

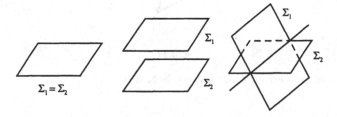

Fig. 3.16

According to (3.4.3), the line determined by the distinct points $\vec{a_1}$ and $\vec{a_2}$ in the space $\mathbb{R}^3$ has the parametric equation

$$\vec{x} = \vec{a_1} + t(\vec{a_2} - \vec{a_1}) = (1 - t)\vec{a_1} + t\vec{a_2}, \quad t \in \mathbb{R}. \qquad (3.5.7)$$

This is exactly the same as (2.5.10) where $\vec{a}_1$ and $\vec{a}_2$ are points in $\mathbb{R}^2$. Therefore the definitions and terminology there for

>*directed segment* $\vec{a}_1 \vec{a}_2$ ($0 \le t \le 1$),
>
>*initial point* $\vec{a}_1$ and *terminal point* $\vec{a}_2$,
>
>*interior point* ($0 < t < 1$) and *exterior point* ($t < 0$ or $t > 1$),
>
>*endpoints* ($t = 0$ or $t = 1$),   and
>
>*middle point* $\dfrac{1}{2}(\vec{a}_1 + \vec{a}_2)$,   (3.5.8)

all are still valid for space line (3.5.7). See also Fig. 2.25.

A *tetrahedron*

$$\triangle \vec{a}_1 \vec{a}_2 \vec{a}_3 \vec{a}_4 \tag{3.5.9}$$

with four non-coplanar points $\vec{a}_1, \vec{a}_2, \vec{a}_3$ and $\vec{a}_4$ as *vertices* is the space figure formed by six *edges* $\vec{a}_1 \vec{a}_2, \vec{a}_1 \vec{a}_3, \vec{a}_1 \vec{a}_4, \vec{a}_2 \vec{a}_3, \vec{a}_2 \vec{a}_4$ and $\vec{a}_3 \vec{a}_4$. See Fig. 3.17. It has four triangles $\triangle \vec{a}_1 \vec{a}_2 \vec{a}_3, \triangle \vec{a}_2 \vec{a}_3 \vec{a}_4, \triangle \vec{a}_3 \vec{a}_4 \vec{a}_1$ and $\triangle \vec{a}_4 \vec{a}_1 \vec{a}_2$ as *faces*. The plane determined by the points $\vec{a}_1, \vec{a}_2$ and $\frac{1}{2}(\vec{a}_3 + \vec{a}_4)$ is called the *median plane* along the edge $\vec{a}_1 \vec{a}_2$. The median planes along $\vec{a}_1 \vec{a}_2$, $\vec{a}_1 \vec{a}_3$ and $\vec{a}_1 \vec{a}_4$ will meet along a line passing through $\vec{a}_1$ and the centroid $\frac{1}{3}(\vec{a}_2 + \vec{a}_3 + \vec{a}_4)$ of $\triangle \vec{a}_2 \vec{a}_3 \vec{a}_4$. (See Ex. <B> 7.)

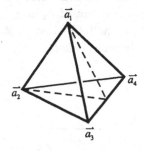

Fig. 3.17

For more related information, see Sec. 3.6.

Exercises

<A>

1. Prove (3.5.3) in detail.
2. Prove (3.5.5) in detail.
3. Prove (3.5.6) in detail.

4. (continued from Ex. <A> 3 of Sec. 3.3) Let $A = (-\frac{1}{2}, 0, 3)$, $B = (\sqrt{2}, 1, 6)$ and $C = (0, 1, 9)$ be three given points in $\mathbb{R}^3$.

   (a) Show that $\overrightarrow{OA}$, $\overrightarrow{OB}$ and $\overrightarrow{OC}$ are linearly independent, where $O = (0, 0, 0)$. Hence A, B and C determine a unique plane Σ.

   (b) Find equations of Σ in $\Gamma(O; A_1, A_2, A_3)$ and $\Gamma(O'; B_1, B_2, B_3)$ respectively.

   (c) Use (3.5.4) to check results obtained in (b).

5. In the natural rectangular coordinate system $\Gamma(\vec{0}; \vec{e}_1, \vec{e}_2, \vec{e}_3)$, the following equations

$$x_1 - x_2 + 2x_3 - 3 = 0,$$
$$-2x_1 + 3x_2 + 4x_3 + 12 = 0,$$
$$6x_1 - 8x_2 + 5x_3 - 7 = 0$$

represent, respectively, three different planes in $\mathbb{R}^3$.

   (a) Show that the planes intersect at a point. Find out the coordinate of the point.

   (b) Try to choose a new coordinate system $\Gamma(O; A_1, A_2, A_3)$, in which the above three planes are the coordinate planes.

6. According to natural coordinate system $\Gamma(\vec{0}; \vec{e}_1, \vec{e}_2, \vec{e}_3)$, a line and a plane are given, respectively, as

$$L: \frac{x_1 - 1}{2} = \frac{x_2 + 1}{-3} = \frac{x_3}{5}, \quad \text{and}$$
$$\Sigma: 2x_1 + x_2 + x_3 - 4 = 0.$$

   (a) Show that the line L and the plane Σ intersect at a point. Find out the coordinate of this point.

   (b) Construct another coordinate system $\Gamma(O; A_1, A_2, A_3)$ in $\mathbb{R}^3$ so that, according to this new system, the line L is a coordinate axis and the plane is a coordinate plane. How many choices of such a new system are possible?

7. Still in $\Gamma(\vec{0}; \vec{e}_1, \vec{e}_2, \vec{e}_3)$, a line

$$L: \frac{x_1 - 0}{1} = \frac{x_2 - 2}{1} = \frac{x_3 - 3}{2}$$

and two planes

$$\Sigma_1: x_1 - x_2 + x_3 = 1,$$
$$\Sigma_2: x_1 + 2x_2 + x_3 = 7$$

are given.

(a) Show that Σ_1 and Σ_2 intersect along a line L'. Find the equation of L'.

(b) Then, show that L and L' intersect at a point. Find the coordinate of this point.

(c) Construct a new coordinate system $\Gamma(O; A_1, A_2, A_3)$ in $\mathbb{R}^3$ such that Σ_1 and Σ_2 become coordinate planes of the new system which contains the given line L in its third coordinate plane.

8. Let

$$x_1 - 18x_2 + 6x_3 - 5 = 0$$

be the equation of a plane in $\Gamma(\vec{0}; \vec{e}_1, \vec{e}_2, \vec{e}_3)$. Find its equation in $\Gamma(O; A_1, A_2, A_3)$ as shown in Ex. <A> 4 of Sec. 3.4.

9. (continued from 8) Suppose

$$3y_1 + 4y_2 + 7y_3 = 1$$

is the equation of a plane in $\Gamma(O; A_1, A_2, A_3)$. Find its equation in $\Gamma(\vec{0}; \vec{e}_1, \vec{e}_2, \vec{e}_3)$.

10. Do the following questions in $\Gamma(\vec{0}; \vec{e}_1, \vec{e}_2, \vec{e}_3)$.

(a) Given a point $P_0 = (3, 4, -1)$ and two straight lines $\frac{x_1}{3} = x_2 - 1 = -x_3, x_1 + 1 = \frac{-x_2 - 1}{2} = x_3 + 1$, find the plane passing through the point P_0 and parallel to the two straight lines.

(b) Prove that the straight lines $\frac{x_1 - 1}{2} = \frac{x_2 + 2}{-3} = \frac{x_3 - 5}{5}$ and $x_1 = 7 + 3t$, $x_2 = 2 + 2t$, $x_3 = 1 - 2t$ lie on a plane simultaneously. Find the equation of the plane.

(c) Find a plane passing through the point $P_0 = (3, -2, 1)$ and containing the straight line $\frac{x_1 + 1}{2} = \frac{x_2 - 1}{3} = \frac{x_3}{4}$.

(d) Find a plane containing both the line $\frac{x_1 - 2}{3} = \frac{x_2 + 1}{2} = \frac{x_3 - 3}{-2}$ and the line $\frac{x_1 - 1}{3} = \frac{x_2 - 2}{2} = \frac{x_3 + 3}{-2}$.

<B>

1. Suppose

$$a_{ij}, \quad 1 \le i, \ j \le 3 \quad \text{and} \quad b_{ij}, \quad 1 \le i, \ j \le 3$$

are scalars such that the vectors (a_{12}, a_{22}, a_{32}) and (a_{13}, a_{23}, a_{33}), (b_{12}, b_{22}, b_{32}) and (b_{13}, b_{23}, b_{33}) are linearly independent, respectively.

Let

$$x_1 = a_{11} + a_{12}t_1 + a_{13}t_2, \quad y_1 = b_{11} + b_{12}t_1 + b_{13}t_2,$$
$$x_2 = a_{21} + a_{22}t_1 + a_{23}t_2, \quad y_2 = b_{21} + b_{22}t_1 + b_{23}t_2, \qquad t_1, t_2 \in \mathbb{R}.$$
$$x_3 = a_{31} + a_{32}t_1 + a_{33}t_2, \quad y_3 = b_{31} + b_{32}t_1 + b_{33}t_2,$$

Construct, in $\mathbb{R}^3$, two coordinate systems so that the equation of a plane with respect to either of the systems is the one or the other given above. How many such systems are possible?

2. Prove that the relative position of a straight line and a plane in $\mathbb{R}^3$ is independent of choices of coordinate systems concerned.

3. Do the same question as in Ex. 2 for two planes.

4. Give three planes in $\mathbb{R}^3$,

$$\Sigma_i: \vec{x} = \vec{a}_i + t_1 \vec{b}_{i1} + t_2 \vec{b}_{i2}, \quad i = 1, 2, 3, \ t_1, t_2 \in \mathbb{R}.$$

Explain graphically all possible relative positions of the three planes and characterize them by using the vectors $\vec{a}_i$, $\vec{b}_{i1}$, $\vec{b}_{i2}$, $i = 1, 2, 3$. Suppose equations of Σ_i are changed to coordinate forms as

$$\Sigma_i: \alpha_i x_1 + \beta_i x_2 + \gamma_i x_3 + \delta_i = 0, \quad i = 1, 2, 3.$$

Now, use the coefficients α_i, β_i, γ_i, δ_i, $i = 1, 2, 3$, to characterize the relative positions described above.

5. Do the same question as in Ex. 2 for three planes.

6. Is it possible to represent the whole space $\mathbb{R}^3$ as a union of finitely many or countably infinite planes? Why? Here, it is required that all the planes should pass through the origin $\vec{0}$.

7. In Fig. 3.17, show that the three median planes along the edges $\vec{a}_1 \vec{a}_2$, $\vec{a}_1 \vec{a}_3$ and $\vec{a}_1 \vec{a}_4$ meet along the line passing through $\vec{a}_1$ and $\frac{1}{3}(\vec{a}_2 + \vec{a}_3 + \vec{a}_4)$.

8. Let V be a vector subspace of $\mathbb{R}^3$ and $\vec{x}_0$ a fixed point in $\mathbb{R}^3$. Call the set

$$\vec{x}_0 + V = \{\vec{x}_0 + v \,|\, v \in V\}$$

an *affine subspace* of $\mathbb{R}^3$ passing through the point $\vec{x}_0$, obtained by moving the vector space V parallelly along the direction $\vec{x}_0$ until up to the point $\vec{x}_0$ (see Fig. 3.18).

(a) Determine all the affine subspaces of $\mathbb{R}^3$.

(b) What are the equations to represent affine subspaces?

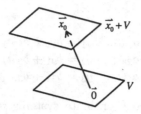

Fig. 3.18

9. In $\mathbb{R}^3$ with natural coordinate system $\Gamma(\vec{0}; \vec{e}_1, \vec{e}_2, \vec{e}_3)$, one uses

$$|\vec{x}| = \left(x_1^2 + x_2^2 + x_3^2\right)^{1/2}$$

to denote the *length* of a vector $\vec{x} = (x_1, x_2, x_3)$. Then

$$|\vec{x}|^2 = x_1^2 + x_2^2 + x_3^2 = 1$$

is the equation of *unit sphere* in $\mathbb{R}^3$ (see Fig. 3.19).

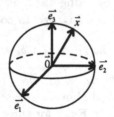

Fig. 3.19

(a) Try to find out all the coordinate systems $\Gamma(\vec{0}; \vec{a}_1, \vec{a}_2, \vec{a}_3)$ in $\mathbb{R}^3$, with basis $\mathcal{B} = \{\vec{a}_1, \vec{a}_2, \vec{a}_3\}$, so that the unit sphere still has the equation

$$y_1^2 + y_2^2 + y_3^2 = 1,$$

where $[\vec{x}]_\mathcal{B} = (y_1, y_2, y_3)$.

(b) Suppose the coordinate system $\Gamma(O; A_1, A_2, A_3)$ is the same as in Ex. <A> 3 of Sec. 3.3. Determine the equation of the unit sphere in $\Gamma(O; A_1, A_2, A_3)$. Then, try to construct a coordinate system $\Gamma(O; B_1, B_2, B_3)$ in which the equation is unchanged. How many such systems are there?

(c) One views the equation obtained in (b) in the eyes of $\Gamma(\vec{0}; \vec{e}_1, \vec{e}_2, \vec{e}_3)$. Then, what kind of surface does it represent? Try to graph the surface and compute the volume it encloses. Compare this volume with that of the unit sphere. What is the ratio? Refer to Secs. 5.4 and 5.10 if necessary.

10. Points

$$\vec{a}_1 = (1,1,0), \quad \vec{a}_2 = (0,1,1), \quad \vec{a}_3 = (1,0,1)$$

in $\mathbb{R}^3$ form a coordinate system $\Gamma(\vec{0}; \vec{a}_1, \vec{a}_2, \vec{a}_3)$ with basis $\mathcal{B} = \{\vec{a}_1, \vec{a}_2, \vec{a}_3\}$. For point $X \in \mathbb{R}^3$, let $[X]_\mathcal{B} = (y_1, y_2, y_3)$. Then, the equation

$$y_1^2 + y_2^2 + y_3^2 = 1$$

represents a certain kind of surface in $\mathbb{R}^3$ (see Fig. 3.20). When viewing it in $\Gamma(\vec{0}; \vec{e}_1, \vec{e}_2, \vec{e}_3)$, what is the surface? What are the equation of the surface and the volume it encloses? Refer to Secs. 5.4 and 5.10 if necessary.

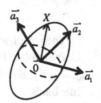

Fig. 3.20

<C> Abstraction and generalization

Read Ex. <C> of Sec. 2.5 and try to do problems there if you missed them at that time. Try to extend as many problems in Ex. <B> as possible to abstract spaces V over a field $\mathbb{F}$. While Exs. <B> 9 and 10 should be in $\mathbb{R}^n$ or finite-dimensional inner product spaces.

3.6 Affine and Barycentric Coordinates

Contents here are parallel to those in Sec. 2.6. So, only sketch is given as follows.

Let $\vec{a}_0, \vec{a}_1, \vec{a}_2$ and $\vec{a}_3$ be four points in $\mathbb{R}^3$. Then

$\vec{a}_0, \vec{a}_1, \vec{a}_2$ and $\vec{a}_3$ are non-coplanar in *affine* space $\mathbb{R}^3$.

$\Leftrightarrow$ The vectors $\vec{a}_1 - \vec{a}_0, \vec{a}_2 - \vec{a}_0, \vec{a}_3 - \vec{a}_0$ are linearly independent in the vector space $\mathbb{R}^3$.

In this case, $\vec{a}_0, \vec{a}_1, \vec{a}_2$ and $\vec{a}_3$ are said to be *affinely independent* and the ordered set

$$\mathcal{B} = \{\vec{a}_0, \vec{a}_1, \vec{a}_2, \vec{a}_3\} \quad \text{or} \quad \{\vec{a}_1 - \vec{a}_0, \vec{a}_2 - \vec{a}_0, \vec{a}_3 - \vec{a}_0\} \qquad (3.6.1)$$

an *affine basis* for $\mathbb{R}^3$ with $\vec{a}_0$ as the *base point*. See Fig. 3.21.

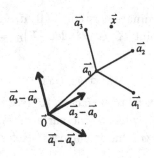

Fig. 3.21

For such an affine basis $\mathcal{B}$, the vectorized space $\Gamma(\vec{a}_0; \vec{a}_1, \vec{a}_2, \vec{a}_3)$ is a geometric model for $\mathbb{R}^3$. Conversely, for a given vectorized space $\Gamma(\vec{a}_0; \vec{a}_1, \vec{a}_2, \vec{a}_3)$, the set $\mathcal{B} = \{\vec{a}_0, \vec{a}_1, \vec{a}_2, \vec{a}_3\}$ is an affine basis for $\mathbb{R}^3$.

Note that the rectangular affine basis

$$\mathcal{N} = \{\vec{0}; \vec{e}_1, \vec{e}_2, \vec{e}_3\} \qquad (3.6.2)$$

is the standard basis for $\mathbb{R}^3$ when considered as a vector space.

Fix an affine basis $\mathcal{B} = \{\vec{a}_0, \vec{a}_1, \vec{a}_2, \vec{a}_3\}$ for $\mathbb{R}^3$. For any point $\vec{x} \in \mathbb{R}^3$, there exist unique scalars x_1, x_2 and x_3 so that

$$\vec{x} - \vec{a}_0 = \sum_{i=1}^{3} x_i(\vec{a}_i - \vec{a}_0)$$

$$\Rightarrow \vec{x} = \vec{a}_0 + \sum_{i=1}^{3} x_i(\vec{a}_i - \vec{a}_0)$$

$$= (1 - x_1 - x_2 - x_3)\vec{a}_0 + x_1\vec{a}_1 + x_2\vec{a}_2 + x_3\vec{a}_3$$

$$= \lambda_0 \vec{a}_0 + \lambda_1 \vec{a}_1 + \lambda_2 \vec{a}_2 + \lambda_3 \vec{a}_3, \qquad (3.6.3)$$

where $\lambda_0 = 1 - x_1 - x_2 - x_3, \lambda_1 = x_1, \lambda_2 = x_2$ and $\lambda_3 = x_3$. The ordered quadruple

$$(\vec{x})_{\mathcal{B}} = (\lambda_0, \lambda_1, \lambda_2, \lambda_3) \quad \text{where } \lambda_0 + \lambda_1 + \lambda_2 + \lambda_3 = 1 \qquad (3.6.4)$$

is called the (normalized) *barycentric coordinate* of the point $\vec{x}$ with respect to the affine basis $\mathcal{B}$. Occasionally, the associated

$$[\vec{x}]_\mathcal{B} = [\vec{x} - \vec{a_0}]_\mathcal{B} = (x_1, x_2, x_3) = (\lambda_1, \lambda_2, \lambda_3) \qquad (3.6.5)$$

is called the *affine coordinate* of $\vec{x}$, for simplicity.

In particular,

$$(\vec{a_0})_\mathcal{B} = (1, 0, 0, 0) \quad \text{with} \quad [\vec{a_0}]_\mathcal{B} = (0, 0, 0),$$
$$(\vec{a_1})_\mathcal{B} = (0, 1, 0, 0) \quad \text{with} \quad [\vec{a_1}]_\mathcal{B} = (1, 0, 0),$$
$$(\vec{a_2})_\mathcal{B} = (0, 0, 1, 0) \quad \text{with} \quad [\vec{a_2}]_\mathcal{B} = (0, 1, 0),$$
$$(\vec{a_3})_\mathcal{B} = (0, 0, 0, 1) \quad \text{with} \quad [\vec{a_3}]_\mathcal{B} = (0, 0, 1).$$

Hence,

the *coordinate axis* $\vec{a_0}\,\vec{a_1}$ has equation $\lambda_2 = \lambda_3 = 0$, and

the *coordinate plane* $\vec{a_0}\,\vec{a_1}\,\vec{a_2}$ has equation $\lambda_3 = 0$,

etc.

There are four coordinate planes. They all together divide the space $\mathbb{R}^3$ into $2^4 - 1 = 15$ regions. The one with $(+, +, +, +)$ is the interior of the *base tetrahedron* $\triangle\vec{a_0}\,\vec{a_1}\,\vec{a_2}\,\vec{a_3}$ whose *barycenter* is $\left(\frac{1}{4}, \frac{1}{4}, \frac{1}{4}, \frac{1}{4}\right)$. See Fig. 3.22.

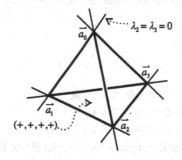

Fig. 3.22

Notice that $\triangle\vec{a_0}\,\vec{a_1}\,\vec{a_2}\,\vec{a_3}$ can be easily described as the set of the points $(\lambda_0, \lambda_1, \lambda_2, \lambda_3)$ where $\lambda_0 \geq 0, \lambda_1 \geq 0, \lambda_2 \geq 0, \lambda_3 \geq 0$ and $\lambda_0 + \lambda_1 + \lambda_2 + \lambda_3 = 1$. In this expression, what does $\lambda_3 = 0$ mean? How about $\lambda_1 = \lambda_3 = 0$?

Let $\Gamma(\vec{b_0}; \vec{b_1}, \vec{b_2}, \vec{b_3})$ be another vectorized space in $\mathbb{R}^3$ with affine basis $\mathcal{B}' = \{\vec{b_0}, \vec{b_1}, \vec{b_2}, \vec{b_3}\}$.

Then, the change of coordinates from $\Gamma(\vec{a_0}; \vec{a_1}, \vec{a_2}, \vec{a_3})$ to $\Gamma(\vec{b_0}; \vec{b_1}, \vec{b_2}, \vec{b_3})$ is

$$[\vec{x} - \vec{b_0}]_{\mathcal{B}'} = [\vec{a_0} - \vec{b_0}]_{\mathcal{B}'} + [\vec{x} - \vec{a_0}]_\mathcal{B} A_{\mathcal{B}'}^\mathcal{B}, \quad \vec{x} \in \mathbb{R}^3$$

or

$$(y_1, y_2, y_3) = (p_1, p_2, p_3) + (x_1, x_2, x_3)A_{B'}^B,$$

or

$$(y_1 \quad y_2 \quad y_3 \quad 1) = (x_1 \quad x_2 \quad x_3 \quad 1)\begin{bmatrix} & & & 0 \\ & A_{B'}^B & & 0 \\ & & & 0 \\ p_1 & p_2 & p_3 & 1 \end{bmatrix}_{4 \times 4}, \qquad (3.6.6)$$

where

$$[\vec{x} - \vec{a_0}]_B = (x_1, x_2, x_3), \quad [\vec{y} - \vec{b_0}]_{B'} = (y_1, y_2, y_3),$$
$$[\vec{a_0} - \vec{b_0}]_{B'} = (p_1, p_2, p_3) \quad \text{and}$$
$$A_{B'}^B = \begin{bmatrix} [\vec{a_1} - \vec{a_0}]_{B'} \\ [\vec{a_2} - \vec{a_0}]_{B'} \\ [\vec{a_3} - \vec{a_0}]_{B'} \end{bmatrix} = \begin{bmatrix} \beta_{11} & \beta_{12} & \beta_{13} \\ \beta_{21} & \beta_{22} & \beta_{23} \\ \beta_{31} & \beta_{32} & \beta_{33} \end{bmatrix} \quad \text{is the transition matrix.}$$

Just like what (2.6.7) and (2.6.8) indicated, *a change of coordinates is a composite of a linear isomorphism followed by a translation.* This is the typical form of an *affine transformation* (see Sec. 2.8).

Exercises

\<A>

1. Generalize Ex. \<A> 1 of Sec. 2.6 to $\mathbb{R}^3$.
2. Generalize Ex. \<A> 2 of Sec. 2.6 to $\mathbb{R}^3$.
3. Generalize Ex. \<A> 3 of Sec. 2.6 to planes in $\mathbb{R}^3$.
4. State and prove the counterparts for lines in $\mathbb{R}^3$ of Ex. \<A> 3 in Sec. 2.6.
5. Use the results obtained in Exs. 3 and 4 to prove (3.4.5), (3.5.5) and (3.5.6).

\<B>

Try to extend problems in Ex. \<B> of Sec. 2.6 to $\mathbb{R}^3$. Be careful that, in some cases, lines in $\mathbb{R}^2$ should be replaced by planes in $\mathbb{R}^3$ here.

\<C> Abstraction and generalization

Try to extend the contents of Ex. \<B> to abstract finite-dimensional affine space over a field $\mathbb{F}$. For partial extension to $\mathbb{R}^n$, see Sec. 5.9.5.

3.7 Linear Transformations (Operators)

All the concepts, definitions concerned, linearly algebraic expositions and results in Sec. 2.7 are still valid in $\mathbb{R}^3$. Readers are urged to check them carefully in the realm of $\mathbb{R}^3$, particularly from (2.7.1) to (2.7.4). We will feel free to use them, including notations such as $\text{Ker}(f), \text{Im}(f)$, in this section.

The section is also divided into eight subsections, each of which as a counterpart of the same numbered subsections in Sec. 2.7. Results obtained here are generalized ones of the corresponding results stated in Sec. 2.7, in the manner that they look much more like linearly algebraic theorems than those in Sec. 2.7.

Section 3.7.1: Discuss linear operators in $\mathcal{N} = \{\vec{e}_1, \vec{e}_2, \vec{e}_3\}$ from different algebraic and geometric points of view. Main topics are kernels, ranks and mapping prospects.

Section 3.7.2: Here illustrates basic essential linear operators and their eigenvalues and eigenvectors, if exist.

Section 3.7.3: Treat some special linear operators independent of bases for $\mathbb{R}^3$.

Section 3.7.4: Introduce matrix representations of a linear operator with respect to various bases for $\mathbb{R}^3$ and the relations among them.

Section 3.7.5: Various decompositions of a square matrix, such as elementary matrices, LU and LDU, etc., and right and left and generalized inverses are discussed by examples.

Section 3.7.6: Put special emphasis on diagonalizable operators and their characterizations and varieties of their usefulness.

Section 3.7.7: Discuss Jordan canonical forms for operators with coincident eigenvalues and their applications in solving special kinds of differential equations.

Section 3.7.8: Discuss the rational canonical forms for operators, especially these which do not have their eigenvalues all real.

3.7.1 *Linear operators in the Cartesian coordinate system*

We have previewed some concepts about linear operators in Ex. <B> 4 of Sec. 3.3 and linear transformations among $\mathbb{R}, \mathbb{R}^2$ and $\mathbb{R}^3$ there in Exs. <B> 5 and 6. Here in this section, we will treat them in great details. Results obtained are parallel to the contents of Sec. 2.7.1. In case the processes to obtain these results are similar to those counterparts in Sec. 2.7.1, the details will be left to the readers as exercises.

Fix the vector space $\mathbb{R}^3$ with its natural basis $\mathcal{N} = \{\vec{e}_1, \vec{e}_2, \vec{e}_3\}$ (see Fig. 3.8) throughout the whole section.

Just like (2.7.7), we have

The formula for linear operator on $\mathbb{R}^3$ (in $\mathcal{N}$)
In $\mathcal{N} = \{\vec{e}_1, \vec{e}_2, \vec{e}_3\}$,

1. $f \colon \mathbb{R}^3 \to \mathbb{R}^3$ is a linear operator.
⇔ 2. There exists a (unique) real 3×3 matrix $A = [a_{ij}]_{3 \times 3}$ such that, for $\vec{x} = (x_1, x_2, x_3)$,

$$f(\vec{x}) = \vec{x} A$$

$$= \left(\sum_{i=1}^{3} a_{i1} x_i, \ \sum_{i=1}^{3} a_{i2} x_i, \ \sum_{i=1}^{3} a_{i3} x_i \right)$$

where, for $1 \le i \le 3$, $f(\vec{e}_i) = (a_{i1}, a_{i2}, a_{i3})$ is the ith row vector of A, and

$$A = \begin{bmatrix} f(\vec{e}_1) \\ f(\vec{e}_2) \\ f(\vec{e}_3) \end{bmatrix}_{3 \times 3} = \begin{bmatrix} a_{11} & a_{12} & a_{13} \\ a_{21} & a_{22} & a_{23} \\ a_{31} & a_{32} & a_{33} \end{bmatrix}, \quad \text{which is } [f]_{\mathcal{N}}. \quad (3.7.1)$$

In case $\mathbb{R}^m$ for $m = 1, 2, 3$ or any positive integer and $\mathbb{R}^n$ for $n = 1, 2, 3$ or any positive integer, any *linear transformation* $f \colon \mathbb{R}^m \to \mathbb{R}^n$ can be expressed as, in terms of *natural bases* for $\mathbb{R}^m$ and $\mathbb{R}^n$,

$$f(\vec{x}) = \vec{x} A, \quad \vec{x} \in \mathbb{R}^m. \quad (3.7.2)$$

where A is a certain real $m \times n$ matrix.

In what follows, we adopt the *convention* that a real 3×3 matrix A will stand for a linear operator on $\mathbb{R}^3$ as

$$\vec{x} \to \vec{x} A. \quad (3.7.3)$$

So does $A_{m \times n}$ act as a linear transformation $\vec{x} \to \vec{x} A$ from $\mathbb{R}^m$ to $\mathbb{R}^n$.

Fix $A = [a_{ij}] \in M(3; \mathbb{R})$.

The kernel $\mathrm{Ker}(A)$ and the image $\mathrm{Im}(A)$ of A are subspaces of $\mathbb{R}^3$. They could be dimension zero, one, two or three. Moreover,

$$\dim \mathrm{Ker}(A) + \dim \mathrm{Im}(A) = \dim \mathbb{R}^3 = 3, \quad (3.7.4)$$

where $\dim \mathrm{Ker}(A)$ is called the *nullity* of A and $\dim \mathrm{Im}(A)$ is called the *rank* of A, denoted as $\mathrm{r}(A)$.

We separate into four cases in the following, according to dim Ker(A).

Case 1 dim Ker(A) = 3   Then Ker(A) = $\mathbb{R}^3$ and $A = O_{3\times3}$ is the only possibility. In this case, Im(A) = $\{\vec{0}\}$ holds and the rank $r(A) = 0$.

Case 2 dim Ker(A) = 2   Then $A \neq O_{3\times3}$. Also,

(1) Ker(A) is a plane through $\vec{0} = (0,0,0)$.

⇔ (2) (algebraic and geometric) The homogeneous equations $\vec{x}A = \vec{0}$, i.e. the three planes

$$a_{1j}x_1 + a_{2j}x_2 + a_{3j}x_3 = 0, \quad 1 \leq j \leq 3$$

coincide with a single plane in $\mathbb{R}^3$ (refer to (3.5.6)). See Fig. 3.23.

Fig. 3.23

⇔ (3) (algebraic) All subdeterminants of order 2 are equal to zero, namely,

$$\det \begin{bmatrix} a_{i_1j_1} & a_{i_1j_2} \\ a_{i_2j_1} & a_{i_2j_2} \end{bmatrix} = 0 \quad \text{for } 1 \leq i_1 < i_2 \leq 3 \text{ and } 1 \leq j_1 < j_2 \leq 3$$

and hence det $A = 0$ (refer to Ex. <B> 4 of Sec. 3.5). For example, the point $(a_{21}, -a_{11}, 0)$ lying on that plane will result in

$$a_{12}a_{21} - a_{22}a_{11} = - \begin{vmatrix} a_{11} & a_{12} \\ a_{21} & a_{22} \end{vmatrix} = 0, \quad \text{etc.}$$

⇔ (4) (linearly algebraic) A has at least one *nonzero* row vector, say $\vec{a}_1 = (a_{11}, a_{12}, a_{13})$ and there are scalars α and β such that the other two row vectors $\vec{a}_i = (a_{i1}, a_{i2}, a_{i3})$, $i = 2, 3$, satisfy

$$\vec{a}_2 = \alpha\vec{a}_1 \quad \text{and} \quad \vec{a}_3 = \beta\vec{a}_1.$$

So do the three column vectors of A.

⇔ (5) (linearly algebraic) The *row rank* and the *column rank* of A are equal to 1, i.e.

The maximal number of linearly independent row vectors of A
= the maximal number of linearly independent column vectors of A
= 1.

$\Leftrightarrow$ (6) (geometric) In case $A_{*2} = \alpha A_{*1}$ and $A_{*3} = \beta A_{*1}$ where $A_{*j} = \begin{bmatrix} a_{1j} \\ a_{2j} \\ a_{3j} \end{bmatrix}$

for $j = 1, 2, 3$, are column vectors, then the image $\operatorname{Im}(A)$ is the straight line

$$\langle\langle (1, \alpha, \beta) \rangle\rangle = \{ t(1, \alpha, \beta) \mid t \in \mathbb{R} \}$$

or $x_2 = \alpha x_1, x_3 = \beta x_1$.

$\Leftrightarrow$ (7) The rank $r(A) = 1$.   (3.7.5)

For a general setting, let the kernel

$$\operatorname{Ker}(A) = \langle\langle \vec{x}_1, \vec{x}_2 \rangle\rangle$$

and take *any* vector $\vec{x}_3 \in \mathbb{R}^3$ which is linearly independent of $\vec{x}_1$ and $\vec{x}_2$, so that $\mathcal{B} = \{ \vec{x}_1, \vec{x}_2, \vec{x}_3 \}$ forms a basis for $\mathbb{R}^3$. Then, the range

$$\operatorname{Im}(A) = \langle\langle \vec{x}_3 A \rangle\rangle$$

is a *fixed* line $\langle\langle \vec{v} \rangle\rangle$, $\vec{v} \neq 0$, through $\vec{0}$. It is easy to see that (refer to (3.5.6))

1. A maps each plane $\vec{x}_0 + \operatorname{Ker}(A)$, parallel to $\operatorname{Ker}(A)$, into a single point (or vector) $\vec{x}_0 A$ on $\langle\langle \vec{v} \rangle\rangle$.
2. A maps each plane $\vec{u} + \langle\langle \vec{u}_1, \vec{u}_2 \rangle\rangle$ where only one of $\vec{u}_1$ and $\vec{u}_2$ is linearly dependent on $\vec{x}_1$ and $\vec{x}_2$, intersecting $\operatorname{Ker}(A)$ along a line, onto the whole line $\langle\langle \vec{v} \rangle\rangle$.

                                                                                                                        (3.7.6)

See Fig. 3.24.

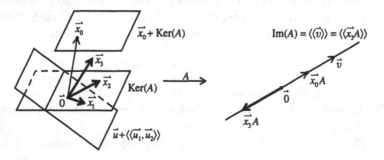

Fig. 3.24

Take any two linearly independent vectors $\vec{v}_1, \vec{v}_2$ in $\mathbb{R}^3$ so that $\mathcal{B}' = \{\vec{v}_1, \vec{v}_2, \vec{x}_3 A\}$ is a basis for $\mathbb{R}^3$. Then,

$$\vec{x}_1 A = \vec{0} = 0\vec{v}_1 + 0\vec{v}_2 + 0\vec{x}_3 A,$$
$$\vec{x}_2 A = \vec{0} = 0\vec{v}_1 + 0\vec{v}_2 + 0\vec{x}_3 A,$$
$$\vec{x}_3 A = 0\vec{v}_1 + 0\vec{v}_2 + 1\vec{x}_3 A,$$

$$\Rightarrow \begin{bmatrix} \vec{x}_1 \\ \vec{x}_2 \\ \vec{x}_3 \end{bmatrix} A = \begin{bmatrix} 0 & 0 & 0 \\ 0 & 0 & 0 \\ 0 & 0 & 1 \end{bmatrix} \begin{bmatrix} \vec{v}_1 \\ \vec{v}_2 \\ \vec{x}_3 A \end{bmatrix}$$

$$\Rightarrow [A]_{\mathcal{B}'}^{\mathcal{B}} = P_{\mathcal{N}}^{\mathcal{B}} A Q_{\mathcal{B}'}^{\mathcal{N}} = \begin{bmatrix} 0 & 0 & 0 \\ 0 & 0 & 0 \\ 0 & 0 & 1 \end{bmatrix}, \quad \text{where}$$

$$P_{\mathcal{N}}^{\mathcal{B}} = \begin{bmatrix} \vec{x}_1 \\ \vec{x}_2 \\ \vec{x}_3 \end{bmatrix} \quad \text{is the transition matrix from } \mathcal{B} \text{ to } \mathcal{N}, \text{ and}$$

$$Q_{\mathcal{B}'}^{\mathcal{N}} = \begin{bmatrix} \vec{v}_1 \\ \vec{v}_2 \\ \vec{x}_3 A \end{bmatrix}^{-1} \quad \text{is the transition matrix from } \mathcal{N} \text{ to } \mathcal{B}'. \quad (3.7.7)$$

This $[A]_{\mathcal{B}'}^{\mathcal{B}}$ is the *matrix representation* of A with respect to bases $\mathcal{B}$ and $\mathcal{B}'$. There are infinitely many such $\mathcal{B}$ and $\mathcal{B}'$ that put A into this *canonical* form.

The *quotient space* of all affine subspaces (refer to Ex. <B> 8 of Sec. 3.5) modulus $\text{Ker}(A)$

$$\mathbb{R}^3/\text{Ker}(A) = \{\vec{x} + \text{Ker}(A) \mid \vec{x} \in \mathbb{R}^3\} \quad (3.7.8)$$

is a one-dimensional real vector space which is linearly isomorphic to $\text{Im}(A)$. See Fig. 3.25. Therefore,

$$\dim \text{Ker}(A) + \dim \mathbb{R}^3/\text{Ker}(A) = \dim \mathbb{R}^3 = 3.$$

Case 3 $\dim \text{Ker}(A) = 1$ Of course, $A \neq O_{3\times 3}$. Also,

(1) (geometric) $\text{Ker}(A)$ is a line through $\vec{0} = (0, 0, 0)$.

$\Leftrightarrow$ (2) (algebraic and geometric) The homogeneous linear equations $\vec{x} A = \vec{0}$, i.e. the three planes

$$\sum_{i=1}^{3} a_{ij} x_i = 0 \quad \text{for } 1 \leq j \leq 3$$

intersect along a straight line in $\mathbb{R}^3$, passing through $\vec{0}$. See Fig. 3.26.

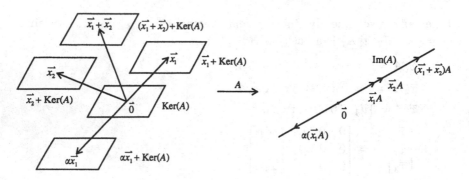

Fig. 3.25

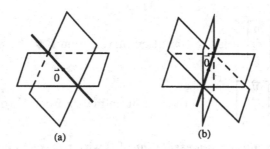

Fig. 3.26

$\Leftrightarrow$ (3) (algebraic) In Fig. 3.26(a), suppose the planes $\sum_{i=1}^{3} a_{ij}x_i = 0$ for $j = 1, 2$ coincide while the third plane $\sum_{i=1}^{3} a_{i3}x_i = 0$ intersects the former two along a line. Then,

$$\det \begin{bmatrix} a_{i1} & a_{i2} \\ a_{j1} & a_{j2} \end{bmatrix} = 0 \quad \text{for } 1 \leq i < j \leq 3 \quad \text{and } \textit{at least } \text{one of}$$

$$\det \begin{bmatrix} a_{i1} & a_{i3} \\ a_{j1} & a_{j3} \end{bmatrix}, \quad 1 \leq i < j \leq 3, \quad \text{is } \textit{not} \text{ equal to zero,}$$

and thus $\det A = 0$. In Fig. 3.26(b), at least three of

$$\det \begin{bmatrix} a_{i_1 j_1} & a_{i_1 j_2} \\ a_{i_2 j_1} & a_{i_2 j_2} \end{bmatrix}, \quad 1 \leq i_1 < i_2 \leq 3, \ 1 \leq j_1 < j_2 \leq 3$$

are *not* equal to zero. Since the intersecting line of the former two planes lies on the third one, then

$$\det A = \begin{vmatrix} a_{11} & a_{12} & a_{13} \\ a_{21} & a_{22} & a_{23} \\ a_{31} & a_{32} & a_{33} \end{vmatrix}$$

$$= a_{13} \begin{vmatrix} a_{21} & a_{22} \\ a_{31} & a_{32} \end{vmatrix} - a_{23} \begin{vmatrix} a_{11} & a_{12} \\ a_{31} & a_{32} \end{vmatrix} + a_{33} \begin{vmatrix} a_{11} & a_{12} \\ a_{21} & a_{22} \end{vmatrix} = 0.$$

In conclusion, $\det A = 0$ and at least one of the subdeterminents of order 2 is *not* equal to zero.

$\Leftrightarrow$ (4) (linearly algebraic) Two row (or column) vectors of A are linearly independent and the three row (or column) vectors are linearly dependent, say

$$\alpha_1 A_{*1} + \alpha_2 A_{*2} + \alpha_3 A_{*3} = \vec{0}^*$$

for scalars $\alpha_1, \alpha_2, \alpha_3$, not all zero, where A_{*j} for $j = 1, 2, 3$ are column vectors.

$\Leftrightarrow$ (5) (linearly algebraic)

The row rank of A = the column rank of A = 2.

$\Leftrightarrow$ (6) (geometric) Adopt notations in (4). The image Im(A) is the plane

$$\alpha_1 x_1 + \alpha_2 x_2 + \alpha_3 x_3 = 0.$$

$\Leftrightarrow$ (7) The rank r$(A) = 2$. $\hspace{3cm}$ (3.7.9)

Let

$$\text{Ker}(A) = \langle\langle \vec{x}_1 \rangle\rangle$$

and take two arbitrarily linearly independent vectors $\vec{x}_2$ and $\vec{x}_3$ in $\mathbb{R}^3$ so that $\mathcal{B} = \{\vec{x}_1, \vec{x}_2, \vec{x}_3\}$ forms a basis for $\mathbb{R}^3$. Then, the range

$$\text{Im}(A) = \langle\langle \vec{x}_2 A, \vec{x}_3 A \rangle\rangle$$

is a *fixed* plane through $\vec{0}$. It is easy to see that (refer to (3.5.5))

1. A maps each line $\vec{x}_0 + \text{Ker}(A)$, parallel to Ker$(A)$, into a single point $\vec{x}_0 A$ on Im(A).
2. A maps each line $\vec{v}_0 + \langle\langle \vec{u} \rangle\rangle$, where $\vec{u}$ is linearly independent of $\vec{x}_1$, not parallel to Ker(A), onto a line on Im(A).
3. A maps each plane, parallel to Ker(A), into a line on Im(A). When will the image line pass $\vec{0}$, i.e. a subspace?

4. A maps each plane, *not* parallel to Ker(A), onto Im(A) in a one-to-one manner.

(3.7.10)

See Fig. 3.27.

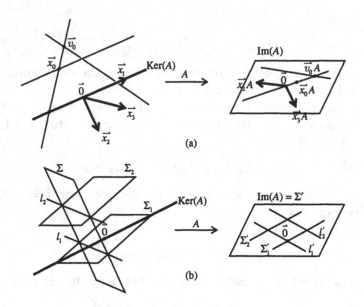

(a)

(b)

Fig. 3.27

Take any nonzero vector $\vec{v}$ in $\mathbb{R}^3$ so that $\mathcal{B}' = \{\vec{v}, \vec{x}_2 A, \vec{x}_3 A\}$ is a basis for $\mathbb{R}^3$. Then the *matrix representation* of A with respect to $\mathcal{B}$ and $\mathcal{B}'$ is

$$[A]_{\mathcal{B}'}^{\mathcal{B}} = P_{\mathcal{N}}^{\mathcal{B}} A Q_{\mathcal{B}'}^{\mathcal{N}} = \begin{bmatrix} 0 & 0 & 0 \\ 0 & 1 & 0 \\ 0 & 0 & 1 \end{bmatrix},$$

where $P_{\mathcal{N}}^{\mathcal{B}} = \begin{bmatrix} \vec{x}_1 \\ \vec{x}_2 \\ \vec{x}_3 \end{bmatrix}$ and $Q_{\mathcal{B}'}^{\mathcal{N}} = \begin{bmatrix} \vec{v} \\ \vec{x}_2 A \\ \vec{x}_3 A \end{bmatrix}^{-1}$. (3.7.11)

The quotient space of $\mathbb{R}^3$ modulus Ker(A),

$$\mathbb{R}^3/\text{Ker}(A) = \{\vec{x} + \text{Ker}(A) \mid \vec{x} \in \mathbb{R}^3\} \tag{3.7.12}$$

is a two-dimensional vector space which is isomorphic to $\mathbb{R}^2$. Also

$$\dim \text{Ker}(A) + \dim \mathbb{R}^3/\text{Ker}(A) = \dim \mathbb{R}^3 = 3.$$

See Fig. 3.28.

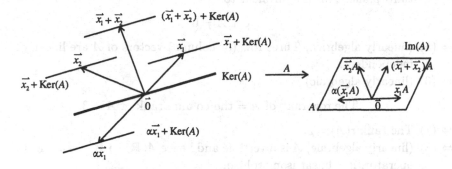

Fig. 3.28

Case 4 dim Ker$(A) = 0$ Then $A \neq O_{3\times3}$. Also,

(1) Ker$(A) = \{\vec{0}\}$, i.e. A is one-to-one.

⇔ (2) A is onto, i.e. Im$(A) = \mathbb{R}^3$.

⇔ (3) (algebraic and geometric) The homogeneous equation $\vec{x}A = \vec{0}$ has only zero solution $\vec{0}$, i.e. the three planes

$$\sum_{i=1}^{3} a_{ij}x_i = 0 \quad \text{for } 1 \leq j \leq 3$$

meet at only one point, namely, the origin $\vec{0}$. See Fig. 3.29.

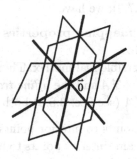

Fig. 3.29

⇔ (4) (algebraic) The (nonzero) direction

$$\left(\begin{vmatrix} a_{21} & a_{22} \\ a_{31} & a_{32} \end{vmatrix}, \begin{vmatrix} a_{31} & a_{32} \\ a_{11} & a_{12} \end{vmatrix}, \begin{vmatrix} a_{11} & a_{12} \\ a_{21} & a_{22} \end{vmatrix} \right)$$

of the intersection line of the first two planes does not lie on the third plane. This is equivalent to

$$\det A \neq 0.$$

$\Leftrightarrow$ (5) (linearly algebraic) Three row (or column) vectors of A are linearly independent.

$\Leftrightarrow$ (6) (linearly algebraic)

The row rank of A = the column rank of $A = 3$.

$\Leftrightarrow$ (7) The rank $r(A) = 3$.

$\Leftrightarrow$ (8) (linearly algebraic) A is invertible and hence $A \colon \mathbb{R}^3 \to \mathbb{R}^3$, as a linear operator, is a linear isomorphism.

$\Leftrightarrow$ (9) (linearly algebraic) A maps every or a basis $\{\vec{x}_1, \vec{x}_2, \vec{x}_3\}$ for $\mathbb{R}^3$ onto a basis $\{\vec{x}_1 A, \vec{x}_2 A, \vec{x}_3 A\}$ for $\mathbb{R}^3$.

$$(3.7.13)$$

In (9), let $\mathcal{B} = \{\vec{x}_1, \vec{x}_2, \vec{x}_3\}$ and $\mathcal{B}' = \{\vec{x}_1 A, \vec{x}_2 A, \vec{x}_3 A\}$ which are bases for $\mathbb{R}^3$. Then the *matrix representation* of A with respect to $\mathcal{B}$ and $\mathcal{B}'$ is

$$[A]^{\mathcal{B}}_{\mathcal{B}'} = P^{\mathcal{B}}_{\mathcal{N}} A Q^{\mathcal{N}}_{\mathcal{B}'} = \begin{bmatrix} 1 & 0 & 0 \\ 0 & 1 & 0 \\ 0 & 0 & 1 \end{bmatrix} = I_3,$$

$$\text{where} \quad P^{\mathcal{B}}_{\mathcal{N}} = \begin{bmatrix} \vec{x}_1 \\ \vec{x}_2 \\ \vec{x}_3 \end{bmatrix} \quad \text{and} \quad Q^{\mathcal{N}}_{\mathcal{B}'} = \begin{bmatrix} \vec{x}_1 A \\ \vec{x}_2 A \\ \vec{x}_3 A \end{bmatrix}^{-1}. \tag{3.7.14}$$

As a counterpart of (2.7.9), we have

The general geometric mapping properties of an invertible linear operator

Let $A = [a_{ij}]_{3 \times 3}$ be an invertible real matrix. Then, the *linear isomorphism* $A \colon \mathbb{R}^3 \to \mathbb{R}^3$ defined by $\vec{x} \to \vec{x} A$ and the *affine transformation* $T \colon \mathbb{R}^3 \to \mathbb{R}^3$ defined by $T(\vec{x}) = \vec{x}_0 + \vec{x} A$ (see Sec. 3.8.3) both preserve:

1. Line segment (interior points to interior points) and line.
2. Triangle and parallelogram (interior points to interior points) and plane.
3. Tetrahedron and parallelepiped (interior points to interior points).
4. The relative positions (see (3.4.5), (3.5.5) and (3.5.6)) of straight lines and planes.
5. Bounded set.
6. The ratio of signed lengths of line segments along the same or parallel lines.

7. The ratio of solid volumes.

$\bigg($ *Note* Let $\vec{a}_i = (a_{i1}, a_{i2}, a_{i3})$, $i = 1, 2, 3$, be row vectors of A. Denote by $\square\vec{a}_1\vec{a}_2\vec{a}_3$ the parallelepiped (see Fig. 3.2) with vertex at $\vec{0}$ and side vectors $\vec{a}_1, \vec{a}_2$ and $\vec{a}_3$, i.e.

$$\square\vec{a}_1\vec{a}_2\vec{a}_3 = \left\{\sum_{i=1}^{3} \lambda_i \vec{a}_i \mid 0 \le \lambda_i \le 1 \text{ for } 1 \le i \le 3\right\}.$$

Then (for details, see Sec. 5.3)

$$\frac{\text{the signed volume of } \square\vec{a}_1\vec{a}_2\vec{a}_3}{\text{the volume of } \vec{e}_1\vec{e}_2\vec{e}_3} = \det A.\bigg)$$

They do not necessarily preserve:

a. Angle.
b. (linear) Length.
c. (planar) Area.
d. (solid) Volume, except $\det A = \pm 1$.
e. Directions or orientations (see Fig. 3.3): preserving the orientation if $\det A > 0$ and reversing the orientation if $\det A < 0$.

$$(3.7.15)$$

Proofs are left to the readers. In case difficulties arise, just review Sec. 2.8.3 carefully and model after proofs there or see Sec. 3.8.3.

Exercises

<A>

1. Prove (3.7.5) and (3.7.6) in detail.
2. Prove (3.7.8) and (3.7.12).
3. Prove (3.7.9) and (3.7.10) in detail.
4. Prove (3.7.13) in detail.
5. Prove (3.7.15) in detail.
6. Let $A = [a_{ij}] \in M(3; \mathbb{R})$ be a nonzero matrix, considered as the linear operator $\vec{x} \to \vec{x}A$. Denote by A_{1*}, A_{2*}, A_{3*} the three row vectors of A and A_{*1}, A_{*2}, A_{*3} the three column vectors of A. Let $\vec{x} = (x_1, x_2, x_3)$, $\vec{y} = (y_1, y_2, y_3)$. Notice that

$$\vec{x}A = x_1 A_{1*} + x_2 A_{2*} + x_3 A_{3*}$$
$$= (\vec{x}A_{*1}, \vec{x}A_{*2}, \vec{x}A_{*3}) = \vec{y}.$$

The following provides a method of how to determine the kernel $\text{Ker}(A)$ and the range $\text{Im}(A)$ by intuition or inspection.

(a) Suppose $r(A) = 1$. We may suppose that $A_{1*} \neq \vec{0}$ and $A_{2*} = \alpha_2 A_{1*}$ and $A_{3*} = \alpha_3 A_{1*}$ for some scalars α_2 and α_3. Then

$$x_1 A_{1*} + x_2 A_{2*} + x_3 A_{3*} = (x_1 + \alpha_2 x_2 + \alpha_3 x_3) A_{1*} = \vec{0}$$
$$\Leftrightarrow x_1 + \alpha_2 x_2 + \alpha_3 x_3 = 0,$$

which is the kernel Ker(A). We may suppose that $A_{*1} \neq \vec{0}^*$ and $A_{*2} = \beta_2 A_{*1}$ and $A_{*3} = \beta_3 A_{*1}$ for some scalars β_2 and β_3. Then

$$A_{*2} = \beta_2 A_{*1}, \quad A_{*3} = \beta_3 A_{*1}$$
$$\Leftrightarrow \vec{x} A_{*2} = \beta_2 \vec{x} A_{*1}, \quad \vec{x} A_{*3} = \beta_3 \vec{x} A_{*1} \quad \text{for all } \vec{x} \in \mathbb{R}^3.$$
$$\Leftrightarrow y_2 = \beta_2 y_1, \quad y_3 = \beta_3 y_1,$$

which is the image Im(A).

(b) Suppose $r(A) = 2$. May suppose that A_{1*} and A_{2*} are linearly independent and $A_{3*} = \alpha_1 A_{1*} + \alpha_2 A_{2*}$. Then,

$$x_1 A_{1*} + x_2 A_{2*} + x_3 A_{3*} = (x_1 + \alpha_1 x_3) A_{1*} + (x_2 + \alpha_2 x_3) A_{2*} = \vec{0}.$$
$$\Leftrightarrow x_1 + \alpha_1 x_3 = 0, \quad x_2 + \alpha_2 x_3 = 0,$$

which is Ker(A). May suppose that A_{*1} and A_{*2} are linearly independent and $A_{*3} = \beta_1 A_{*1} + \beta_2 A_{*2}$. Then,

$$A_{*3} = \beta_1 A_{*1} + \beta_2 A_{*2}$$
$$\Leftrightarrow \vec{x} A_{*3} = \beta_1 \vec{x} A_{*1} + \beta_2 \vec{x} A_{*2} \quad \text{for all } \vec{x} \in \mathbb{R}^3.$$
$$\Leftrightarrow y_3 = \beta_1 y_1 + \beta_2 y_2,$$

which is Im(A).

Try to determine Ker(A) and Im(A) for each of the following matrices:

$$A = \begin{bmatrix} 6 & -10 & 2 \\ -3 & 5 & -1 \\ -9 & 15 & -3 \end{bmatrix}; \quad A = \begin{bmatrix} 12 & 0 & -2 \\ 0 & 4 & 6 \\ 3 & -1 & -2 \end{bmatrix}.$$

7. Formulate the features for Ker(A) and Im(A) and the geometric mapping properties for each of the following linear transformations, in the natural bases for spaces concerned.

(a) $f: \mathbb{R} \to \mathbb{R}^3$ defined by $f(\vec{x}) = \vec{x} A, A \in M(1; 3; \mathbb{R})$.
(b) $f: \mathbb{R}^2 \to \mathbb{R}^3$ defined by $f(\vec{x}) = \vec{x} A, A \in M(2; 3; \mathbb{R})$.
(c) $f: \mathbb{R}^3 \to \mathbb{R}$ defined by $f(\vec{x}) = \vec{x} A, A \in M(3; 1; \mathbb{R})$.
(d) $f: \mathbb{R}^3 \to \mathbb{R}^2$ defined by $f(\vec{x}) = \vec{x} A, A \in M(3; 2; \mathbb{R})$.

<B>

The system of non-homogeneous linear equations in three unknowns x_1, x_2, x_3

$$\sum_{i=1}^{3} a_{ij} x_i = b_j, \quad j = 1, 2, 3$$

has a solution if and only if the constant vector $\vec{b} = (b_1, b_2, b_3)$ lies on the range $\text{Im}(A)$ of the linear operator

$$f(\vec{x}) = \vec{x} A, \quad A = [a_{ij}]_{3 \times 3}.$$

This is equivalent to saying that the *coefficient matrix A* and its *augmented matrix* $[A \mid \vec{b}^*]$ have the same rank, i.e.

$$r(A) = r\left([A \mid \vec{b}^*]\right).$$

1. Try to use these concepts and methods to redo Ex. <B> 4 of Sec. 3.5 but emphasize the following aspects.

   (a) Find all relative positions of the three planes and characterize each case by using ranks and determinants.

   (b) What kinds of relative positions will guarantee the existence of a solution? And how many solutions are there?

2. Try to use Ex. 1 to solve the following sets of equations and graph these planes concerned, if possible.

   (a) $5x_1 + 3x_2 + 7x_3 = 4$;
   $3x_1 + 26x_2 + 2x_3 = 9$;
   $7x_1 + 2x_2 + 10x_3 = 5$.

   (b) $x_1 - 3x_2 + x_3 = 1$;
   $2x_1 - 3x_2 = 4$;
   $x_2 + 2x_3 = 0$.

   (c) $2x_1 + 6x_2 + 11 = 0$;
   $6x_1 + 20x_2 - 6x_3 + 3 = 0$;
   $6x_1 - 18x_3 + 1 = 0$.

   (d) $2x_1 + x_2 = 2$;
   $3x_2 - 2x_3 = 4$;
   $x_2 + 3x_3 = 1$.

   (e) $x_1 + x_2 + x_3 = 6$;
   $3x_1 - 2x_2 - x_3 = 7$;
   $x_1 - 4x_2 - 3x_3 = -5$.

   (f) $x_1 + 2x_2 - x_3 = 5$;
   $3x_1 - x_2 + 2x_3 = 2$;
   $2x_1 + 11x_2 - 7x_3 = -2$.

   (g) $3x_1 + x_2 - 3x_3 = 0$;
   $2x_1 + 2x_2 - 3x_3 = 0$;
   $x_1 - 5x_2 + 9x_3 = 0$.

3. Solve the equations:

$$(a+3)x_1 - 2x_2 + 3x_3 = 4;$$
$$3x_1 + (a-3)x_2 + 9x_3 = b;$$
$$4x_1 - 8x_2 + (a+14)x_3 = c,$$

where a, b, c are constants.

<C> Abstraction and generalization

Do problems in Ex. <C> of Sec. 2.7.1.

Results in (3.7.5), (3.7.9) and (3.7.13), and the methods to obtain them can be unified and generalized to linear transformation $f: \mathbb{F}^m \to \mathbb{F}^n$ or matrix $A[a_{ij}] \in M(m, n; \mathbb{F})$. We proceed as follows.

An $(m-1)$-dimensional subspace S of $\mathbb{F}^m$ is called a *hypersubspace* and its image $\vec{x}_0 + S$, under the translation $\vec{x} \to \vec{x}_0 + \vec{x}$, a *hyperplane* which is an $(m-1)$-dimensional affine subspace of $\mathbb{F}^m$. These two geometric objects can be characterized through the concept of linear functional on $\mathbb{F}^m$.

Suppose $\{\vec{x}_1, \ldots, \vec{x}_{m-1}\}$ is any fixed basis for the hypersubspace S and $\mathcal{B} = \{\vec{x}_1, \ldots, \vec{x}_{m-1}, \vec{x}_m\}$ is a basis for $\mathbb{F}^m$. Then the linear functional $f: \mathbb{F}^m \to \mathbb{F}$ defined by

$$f(\vec{x}_i) = \begin{cases} 0, & 1 \le i \le m-1 \\ 1, & i = m \end{cases}$$

has precisely its

$$\mathrm{Ker}(f) = S;$$
$$\mathrm{Im}(f) = \mathbb{F}.$$

To find the expression of f in terms of the natural basis $\mathcal{N} = \{\vec{e}_1, \ldots, \vec{e}_m\}$ for $\mathbb{F}^m$, let $\vec{e}_i = \sum_{j=1}^m a_{ij}\vec{x}_j$ for $1 \le i \le m$, namely, $[\vec{e}_i]_\mathcal{B} = (a_{i1}, \ldots, a_{im})$. Then

$$\vec{x} = (x_1, \ldots, x_m) = \sum_{i=1}^m x_i \vec{e}_i$$

$$\Rightarrow f(\vec{x}) = \sum_{i=1}^m x_i f(\vec{e}_i) = \sum_{i=1}^m x_i \left(\sum_{j=1}^m a_{ij} f(\vec{x}_j) \right) = \sum_{i=1}^m a_{im} x_i.$$

Note that at least one of $a_{1m}, a_{2m}, \ldots, a_{mm}$ is not equal to zero. As a consequence,

$$\text{the hypersubspace } S: \sum_{i=1}^{m} a_{im} x_i = 0, \quad \text{and}$$

$$\text{the hyperplane } \vec{x}_0 + S: \sum_{i=1}^{m} a_{im} x_i = b, \qquad (3.7.16)$$

where $b = \sum_{i=1}^{m} a_{im} x_{i0}$ if $\vec{x}_0 = (x_{10}, x_{20}, \ldots, x_{m0})$.

Give a linear functional $f: \mathbb{F}^m \to \mathbb{F}$. Since

$$\dim \text{Ker}(f) + \dim \text{Im}(f) = \dim \mathbb{F}^m = m,$$

it follows that if f is nonzero, then $\dim \text{Ker}(f) = m - 1$ and $\text{Ker}(f)$ is a hypersubspace. Also, for any (nonzero) subspace V of $\mathbb{F}^m$, the dimension identity

$$\dim V \cap \text{Ker}(f) + \dim(V + \text{Ker}(f)) = \dim V + \dim \text{Ker}(f)$$

$$\Rightarrow \dim V \cap \text{Ker}(f) = \begin{cases} \dim V, & \text{if } V \subseteq \text{Ker}(f) \\ \dim V - 1, & \text{if } f \text{ is nonzero and } V \not\subseteq \text{Ker}(f). \end{cases}$$
$$(3.7.17)$$

For a nonzero matrix $A = [a_{ij}] \in M(m, n; \mathbb{F})$, when considered as a linear transformation from $\mathbb{F}^m$ into $\mathbb{F}^n$ defined by $\vec{x} \to \vec{x} A$, then

$$\dim \text{Ker}(A) + \dim \text{Im}(A) = \dim \mathbb{F}^m = m, \quad \text{and}$$

$$\text{Ker}(A) = \bigcap_{j=1}^{n} \text{Ker}(f_j), \qquad (3.7.18)$$

where $f_j(\vec{x}) = \sum_{i=1}^{m} a_{ij} x_i$ are the linear functionals determined by the jth column vector of A, for each $j, 1 \le j \le n$.

Try to do the following problems.

1. What peculiar vectors $\text{Ker}(f_j), 1 \le j \le n$, might possess?

   (a) Suppose $\dim \text{Ker}(f_1) = m - 1$ and $a_{11} \ne 0$, for simplicity. Show that the following vectors

$$(a_{21}, -a_{11}, 0, \ldots, 0),$$
$$(a_{31}, 0, -a_{11}, \ldots, 0),$$
$$\vdots$$
$$(a_{m1}, 0, \ldots, 0, -a_{11})$$

   form a basis for $\text{Ker}(f_1)$. What happens if $a_{k1} \ne 0$ for some k, where $1 \le k \le m$? These facts are still valid for any $\text{Ker}(f_j), 1 \le j \le n$.

   (b) For $1 \le i_1 < i_2 < i_3 \le m, \text{Ker}(f_1) \cap \text{Ker}(f_2)$ contains the vector

$$\left(0, \ldots, 0, \det \begin{bmatrix} a_{i_2 1} & a_{i_2 2} \\ a_{i_3 1} & a_{i_3 2} \end{bmatrix}, 0, \ldots, 0, \det \begin{bmatrix} a_{i_3 1} & a_{i_3 2} \\ a_{i_1 1} & a_{i_1 2} \end{bmatrix}, \right.$$

$$\uparrow \qquad\qquad\qquad\qquad \uparrow$$
$$i_1\text{th coordinate} \qquad\qquad i_2\text{th coordinate}$$

$$\left. 0, \ldots, 0, \det \begin{bmatrix} a_{i_1 1} & a_{i_1 2} \\ a_{i_2 1} & a_{i_2 2} \end{bmatrix}, 0, \ldots, 0 \right)$$

$$\uparrow$$
$$i_3\text{th coordinate}$$

   which might be denoted as $\vec{v}_{i_1 i_2 i_3}$. Do you see why? Note that it could happen that $\vec{v}_{i_1 i_2 i_3} = \vec{0}$. But when? When all such $\vec{v}_{i_1 i_2 i_3} = \vec{0}$?

   (c) For each $1 \le i_1 < i_2 < \cdots < i_k < i_{k+1} \le m$, the subspace $\bigcap_{j=1}^k \text{Ker}(f_j)$ contains the vector $\vec{v}_{i_1 i_2 \cdots i_k i_{k+1}}$ whose

   $i_l\text{th coordinate}$

$$= (-1)^{l+1} \det \begin{bmatrix} a_{i_1 1} & a_{i_1 2} & \cdots & a_{i_1, l-1} & a_{i_1, l+1} & \cdots & a_{i_1, k+1} \\ \vdots & \vdots & & \vdots & \vdots & & \vdots \\ a_{i_{l-1}, 1} & a_{i_{l-1}, 2} & \cdots & a_{i_{l-1}, l-1} & a_{i_{l-1}, l+1} & \cdots & a_{i_{l-1}, k+1} \\ a_{i_{l+1}, 1} & a_{i_{l+1}, 2} & \cdots & a_{i_{l+1}, l-1} & a_{i_{l+1}, l+1} & \cdots & a_{i_{l+1}, k+1} \\ \vdots & \vdots & & \vdots & \vdots & & \vdots \\ a_{i_{k+1}, 1} & a_{i_{k+1}, 2} & \cdots & a_{i_{k+1}, l-1} & a_{i_{k+1}, l+1} & \cdots & a_{i_{k+1}, k+1} \end{bmatrix}$$

   for $1 \le l \le k+1$,   and

   $p\text{th coordinate} = 0$ for $p \ne i_l, 1 \le p \le m$ and $1 \le l \le k+1$.

Try to Figure out when $\vec{v}_{i_1 i_2 \cdots i_k i_{k+1}} \neq \vec{0}$ for some $1 \leq i_1 < i_2 < \cdots < i_k < i_{k+1} \leq m$.

(*Note* The inductive processes in (a), (b) and (c) all together suggest implicitly an *inductive definition* for determinant of order m over a field $\mathbb{F}$.)

2. What are the possible dimensions for $\bigcap_{j=1}^{n} \operatorname{Ker}(f_j)$ and how to characterize them in each case? Here we suppose that, for each $1 \leq j \leq n$, dim $\operatorname{Ker}(f_j) = m - 1$ holds.

   (a) In case $n = 2$, the following are equivalent:

   (1) The hypersubspaces $\operatorname{Ker}(f_1)$ and $\operatorname{Ker}(f_2)$ coincide, i.e. $\operatorname{Ker}(f_1) = \operatorname{Ker}(f_2)$ holds.

   (2) There exists a nonzero scalar α so that $f_2 = \alpha f_1$.

   (3) The ratios of the corresponding coefficients in $\operatorname{Ker}(f_1)$: $\sum_{i=1}^{m} a_{i1} x_i = 0$ and $\operatorname{Ker}(f_2)$: $\sum_{i=1}^{m} a_{i2} x_i = 0$ are equal, namely,

   $$\frac{a_{11}}{a_{12}} = \frac{a_{21}}{a_{22}} = \cdots = \frac{a_{m1}}{a_{m2}}.$$

   (4) All the submatrices of order 2 of the matrix

   $$A = \begin{bmatrix} a_{11} & a_{12} \\ a_{21} & a_{22} \\ \vdots & \vdots \\ a_{m1} & a_{m2} \end{bmatrix}$$

   have zero determinant, namely, for $1 \leq i_1 < i_2 \leq m$,

   $$\det \begin{bmatrix} a_{i_1 1} & a_{i_1 2} \\ a_{i_2 1} & a_{i_2 2} \end{bmatrix} = 0.$$

   (5) The two column vectors of A are linearly dependent, and the maximal number of linearly independent row vectors of A is 1.

   Then, what happens if $\operatorname{Ker}(f_1) \neq \operatorname{Ker}(f_2)$, i.e. dim$(\operatorname{Ker}(f_1) \cap \operatorname{Ker}(f_2)) = m - 2$?

   (b) In case $n = 3$, the dimensions for $\bigcap_{j=1}^{3} \operatorname{Ker}(f_j)$ could be $n-1, n-2$ or $n - 3$ which is, respectively, the case (3.7.5), (3.7.9) or (3.7.13) if $n = 3$. Try to figure out possible characteristic properties, both algebraic and geometric.

(c) Suppose the matrix $A = [a_{ij}]_{m \times n}$ has a submatrix of order k whose determinant is *not* equal to zero. Say, for simplicity,

$$\det \begin{bmatrix} a_{11} & a_{12} & \cdots & a_{1k} \\ a_{21} & a_{22} & \cdots & a_{2k} \\ \vdots & & & \vdots \\ a_{k1} & a_{k1} & \cdots & a_{kk} \end{bmatrix} \neq 0.$$

Then it follows easily that the first k column vectors $A_{*1}, \ldots, A_{*k}$ of A should be linearly independent. So are the first k row vectors $A_{1*}, \ldots, A_{k*}$. For

$$\alpha_1 A_{*1} + \cdots + \alpha_k A_{*k} = \vec{0}^*, \quad \text{where } \vec{0} \in \mathbb{F}^m.$$

$$\Rightarrow (\alpha_1 \ \cdots \ \alpha_k) \begin{bmatrix} a_{11} & \cdots & a_{k1} \\ \vdots & & \vdots \\ a_{1k} & \cdots & a_{kk} \end{bmatrix} = \vec{0}, \quad \text{where } \vec{0} \in \mathbb{F}^k.$$

$$\Rightarrow (\alpha_1, \ldots, \alpha_k) = \vec{0} \quad \text{in } \mathbb{F}^k.$$

Furthermore, in this case,

$$\dim \bigcap_{j=1}^{k} \text{Ker}(f_j) = m - k.$$

To see this, one might use (3.7.17) k times by noting that $A_{*1}, \ldots, A_{*k}$ are linearly independent. Or, we might observe that

$$\vec{x} = (x_1, \ldots, x_m) \in \bigcap_{j=1}^{k} \text{Ker}(f_j)$$

$$\Leftrightarrow \sum_{i=1}^{m} a_{ij} x_i = 0 \quad \text{for } 1 \le j \le k.$$

$$\Leftrightarrow \sum_{i=1}^{k} a_{ij} x_i = - \sum_{i=k+1}^{m} a_{ij} x_i \quad \text{for } 1 \le j \le k.$$

$$\Leftrightarrow (x_1 \ \cdots \ x_k)$$

$$= -(x_{k+1} \ \cdots \ x_m) \begin{bmatrix} a_{k+1,1} & \cdots & a_{k+1,k} \\ \vdots & & \vdots \\ a_{m1} & \cdots & a_{mk} \end{bmatrix} \begin{bmatrix} a_{11} & \cdots & a_{1k} \\ \vdots & & \vdots \\ a_{k1} & \cdots & a_{kk} \end{bmatrix}^{-1},$$

which indicates immediately that $\dim \bigcap_{j=1}^{k} \text{Ker}(f_j) = m - k$. Also, Ex. 1 might be helpful in handling this problem.

(d) (continuation of (c)) Conversely, suppose dim $\bigcap_{j=1}^{k} \mathrm{Ker}(f_j) = m-k$, then there exists at least one $1 \le i_1 < i_2 < \cdots < i_k \le m$ so that

$$\det \begin{bmatrix} a_{i_1 1} & \cdots & a_{i_1 k} \\ \vdots & & \vdots \\ a_{i_k 1} & \cdots & a_{i_k k} \end{bmatrix} \ne 0.$$

Suppose on the contrary that all such submatrices of order k have zero determinant. Let p be the largest integer so that there exists at least one submatrix of order p of the matrix $[A_{*1} \cdots A_{*k}]_{m \times k}$, having nonzero determinant. Then $p < k$ holds. And the argument in (c) shows that $\bigcap_{j=1}^{k} \mathrm{Ker}(f_j)$ should have dimension $m - p$ rather than $m - k$, a contradiction.

(e) Try to use or model after (c) and (d) to characterize the case that dim $\bigcap_{j=1}^{k} \mathrm{Ker}(f_j) = m - p$, where $1 \le p \le k$.

As a conclusion, we figure out our final result as (adopt notations in (3.7.18)):

1. The linear transformation $A \colon \mathbb{F}^m \to \mathbb{F}^n$ has its kernel the dimension dim $\mathrm{Ker}(A) = m - r$.

$\Leftrightarrow$ 2. (geometric) The n hypersubspaces $\mathrm{Ker}(f_j)$, $1 \le j \le n$, intersect into a $(m - r)$-dimensional subspace $\bigcap_{j=1}^{n} \mathrm{Ker}(f_j) = \mathrm{Ker}(A)$.

$\Leftrightarrow$ 3. (algebraic) r is the largest integer so that some $r \times r$ submatrix of A has a *nonzero* determinant.

$\Leftrightarrow$ 4. (geometric and linearly algebraic)

The maximal number of linearly independent row vectors of A
= The maximal number of linearly independent column vectors
of A

= r,

the former is called the *row rank* of A and the latter the *column rank* of A.

$\Leftrightarrow$ 5. (algebraic) row rank of A = column rank of A = r.

$\Leftrightarrow$ 6. (linearly algebraic) the rank of $A = r = \dim \mathrm{Im}(A)$, denoted as $\mathrm{r}(A)$; in case $\mathrm{r}(A) = m \le n$,

$\Leftrightarrow$ 7. A is one-to-one, i.e. $\mathrm{Ker}(A) = \{\vec{0}\}$; in case $\mathrm{r}(A) = n \leq m$,

$\Leftrightarrow$ 8. A is onto, i.e. $\mathrm{Im}(A) = \mathbb{F}^n$; in case $m = n$ and $\mathrm{r}(A) = n$,

$\Leftrightarrow$ 9. A is a linear isomorphism on $\mathbb{F}^n$.

$\Leftrightarrow$ 10. A maps every or a basis for $\mathbb{F}^n$ onto a basis for $\mathbb{F}^n$.

$$(3.7.19)$$

For other approaches to these results, refer to Exs. 2 and 24 of Sec. B.7, see also Secs. B.8 and 5.9.3.

Let V and W be finite-dimensional vector spaces over the same field. Suppose $\dim V = m$, and $\mathcal{B}$ is a basis for V, while $\dim W = n$ and C is a basis for W. The mapping $\Phi\colon V \to \mathbb{F}^m$ defined by $\Phi(\vec{x}) = [\vec{x}]_{\mathcal{B}}$, the coordinate vector of $\vec{x} \in V$ with respect to $\mathcal{B}$, is a linear isomorphism (refer to (2.3.2) and (3.2.1)). Similarly, $\Psi\colon W \to \mathbb{F}^n$ defined by $\Psi(\vec{y}) = [\vec{y}]_C$ is a linear isomorphism. Let $\mathcal{N}$ and $\mathcal{N}'$ be natural bases for $\mathbb{F}^m$ and $\mathbb{F}^n$, respectively. For each $f \in \mathrm{L}(V, W)$, define $g \in \mathrm{L}(\mathbb{F}^m, \mathbb{F}^n)$ by $g([\vec{x}]_{\mathcal{B}}) = [f(\vec{x})]_C$. Then $g \circ \Phi = \Psi \circ f$ reflects the commutative of the following diagram

$$
\begin{array}{ccc}
V & \overset{f}{\to} & W \\
\Phi \downarrow & & \downarrow \Psi \\
\mathbb{F}^m & \underset{g}{\to} & \mathbb{F}^n
\end{array}.
$$

Since $[\Phi]_{\mathcal{N}}^{\mathcal{B}}[g]_{\mathcal{N}'}^{\mathcal{N}} = [f]_C^{\mathcal{B}}[\Psi]_{\mathcal{N}'}^C$ (see Secs. 2.7.3 and 3.7.3), we can recapture various properties in (3.7.19) for f via $[g]_{\mathcal{N}'}^{\mathcal{N}}$, and hence, via $[f]_C^{\mathcal{B}}$.

3.7.2 *Examples*

It is supposed or suggested that readers are familiar with those basic examples presented in Sec. 2.7.2 and general terminologies, such as

1. eigenvalues, eigenvectors, characteristic polynomial,
2. line of invariant points, invariant line, and
3. Caylay–Hamilton theorem,

stated in the later part of that section.

Give a nonzero matrix $A = [a_{ij}] \in \mathrm{M}(3; \mathbb{R})$, considered as a linear operator $\vec{x} \to \vec{x}A$ on $\mathbb{R}^3$. According to the rank $\mathrm{r}(A) = 1, 2, 3$ and (3.7.7),

(3.7.11), (3.7.14), A has the following respective *canonical* form

$$\begin{bmatrix} 1 & 0 & 0 \\ 0 & 0 & 0 \\ 0 & 0 & 0 \end{bmatrix}, \quad \begin{bmatrix} 1 & 0 & 0 \\ 0 & 1 & 0 \\ 0 & 0 & 0 \end{bmatrix}, \quad \begin{bmatrix} 1 & 0 & 0 \\ 0 & 1 & 0 \\ 0 & 0 & 1 \end{bmatrix}.$$

For operator A with rank $r(A) = 1$ or $r(A) = 2$, its algebraic and geometric mapping properties are *essentially* the same as operators on $\mathbb{R}^2$, after choosing a fixed basis $\mathcal{B}$ for $\mathbb{R}^3$ where at least one of the basis vectors is from its kernel. Therefore, we will focus on examples here on operators of rank 3.

Example 1 The operator (compare with Example 4 in Sec. 2.7.2)

$$A = \begin{bmatrix} \lambda_1 & 0 & 0 \\ 0 & \lambda_2 & 0 \\ 0 & 0 & \lambda_3 \end{bmatrix}, \quad \text{where } \lambda_1 \lambda_2 \lambda_3 \neq 0$$

has, in $\mathcal{N} = \{\vec{e}_1, \vec{e}_2, \vec{e}_3\}$, the properties:

1. $\vec{e}_i A = \lambda_i \vec{e}_i$ for $i = 1, 2, 3$. So $\vec{e}_i$ is an eigenvector of A corresponding to the eigenvalue λ_i and $\langle\langle \vec{e}_i \rangle\rangle$ is an invariant line (subspace) of A.
2. In case $\lambda_1 = \lambda_2 = \lambda_3 = \lambda$, then

$$A = \lambda I_3$$

   is called an *enlargement* with *scale* λ. In this case, A keeps every line passing $\vec{0}$ invariant.
3. In case $\lambda_1 \neq \lambda_2 = \lambda_3 = 1$, A is called a *one-way stretch* along the invariant line $\langle\langle \vec{e}_1 \rangle\rangle$, while $\langle\langle \vec{e}_2, \vec{e}_3 \rangle\rangle$ is the *plane of invariant points*.
4. In case $\lambda_1 \neq \lambda_2 \neq \lambda_3 = 1$, A is called a *two-way stretch* along $\langle\langle \vec{e}_1 \rangle\rangle$ and $\langle\langle \vec{e}_2 \rangle\rangle$, while $\langle\langle \vec{e}_3 \rangle\rangle$ is the line of invariant points.
5. In case $\lambda_1 \neq \lambda_2 \neq \lambda_3 \neq 1$, A is simply called a *stretch* or a *three-way stretch* and $\vec{0}$ is the only invariant point.

This is the simplest linear operator among all. It just maps a vector $\vec{x} = (x_1, x_2, x_3)$ into the vector $\vec{x}A = (\lambda_1 x_1, \lambda_2 x_2, \lambda_3 x_3)$. See Fig. 3.30.

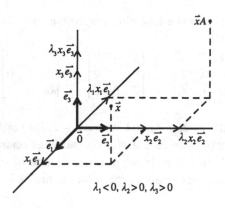

$$\lambda_1 < 0, \lambda_2 > 0, \lambda_3 > 0$$

Fig. 3.30

If A is the matrix representation of a linear operator B on $\mathbb{R}^3$ with respect to a certain basis $\mathcal{B} = \{\vec{x}_1, \vec{x}_2, \vec{x}_3\}$, namely,

$$[B]_{\mathcal{B}} = PBP^{-1} = A = \begin{bmatrix} \lambda_1 & 0 & 0 \\ 0 & \lambda_2 & 0 \\ 0 & 0 & \lambda_3 \end{bmatrix}, \quad \text{where } P = \begin{bmatrix} \vec{x}_1 \\ \vec{x}_2 \\ \vec{x}_3 \end{bmatrix}, \quad (3.7.20)$$

then the action of $A = [B]_{\mathcal{B}}$ on a vector $[\vec{x}]_{\mathcal{B}} = (\alpha_1, \alpha_2, \alpha_3)$ can be similarly illustrated as in Fig. 3.31. In this case, B is called *diagonalizable*.

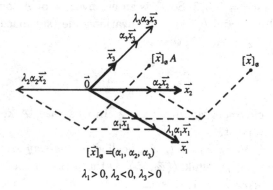

$$[\vec{x}]_a = (\alpha_1, \alpha_2, \alpha_3)$$
$$\lambda_1 > 0, \lambda_2 < 0, \lambda_3 > 0$$

Fig. 3.31

Example 2 The operator (compare with Example 5 in Sec. 2.7.2)

$$A = \begin{bmatrix} 0 & a & 0 \\ b & 0 & 0 \\ 0 & 0 & c \end{bmatrix}, \quad \text{with} \quad abc \neq 0$$

has the following properties: The characteristic polynomial is $-(t-c)(t^2-ab)$.

(1) **Case 1** $ab > 0$ and $c \neq \sqrt{ab}$ or $-\sqrt{ab}$

eigenvalues	eigenvectors
$\lambda_1 = \sqrt{ab}$	$\vec{v}_1 = (\sqrt{|b|}, \sqrt{|a|}, 0)$
$\lambda_2 = -\sqrt{ab}$	$\vec{v}_2 = (-\sqrt{|b|}, \sqrt{|a|}, 0)$
$\lambda_3 = c$	$\vec{v}_3 = \vec{e}_3 = (0, 0, 1)$

$\mathcal{B} = \{\vec{v}_1, \vec{v}_2, \vec{v}_3\}$ is a basis for $\mathbb{R}^3$ and the axes $\langle\langle \vec{v}_1 \rangle\rangle, \langle\langle \vec{v}_2 \rangle\rangle, \langle\langle \vec{v}_3 \rangle\rangle$ are invariant lines.

1. Since

$$A = \begin{bmatrix} 0 & 1 & 0 \\ 1 & 0 & 0 \\ 0 & 0 & 1 \end{bmatrix} \begin{bmatrix} b & 0 & 0 \\ 0 & a & 0 \\ 0 & 0 & c \end{bmatrix}$$

in $\mathcal{N} = \{\vec{e}_1, \vec{e}_2, \vec{e}_3\}$, A can be decomposed as

$$\vec{x} = (x_1, x_2, x_3) \to \vec{x} \begin{bmatrix} 0 & 1 & 0 \\ 1 & 0 & 0 \\ 0 & 0 & 1 \end{bmatrix} = (x_2, x_1, x_3)$$

$$\to (x_2, x_1, x_3) \begin{bmatrix} b & 0 & 0 \\ 0 & a & 0 \\ 0 & 0 & c \end{bmatrix} = (bx_2, ax_1, cx_3) = \vec{x}A$$

where the first mapping is a *reflection* with respect to the plane $x_1 = x_2$ while the second one is a *stretch* (see Example 1). See Fig. 3.32 (and Fig. 2.49).

2. In $\mathcal{B} = \{\vec{v}_1, \vec{v}_2, \vec{v}_3\}$,

$$[A]_{\mathcal{B}} = PAP^{-1} = \begin{bmatrix} \sqrt{ab} & & 0 \\ & -\sqrt{ab} & \\ 0 & & c \end{bmatrix}, \quad \text{where } P = \begin{bmatrix} \vec{v}_1 \\ \vec{v}_2 \\ \vec{v}_3 \end{bmatrix}.$$

Refer to Fig. 3.31 (and compare with Fig. 2.48). What happens if $c = \sqrt{ab}$ or $-\sqrt{ab}$.

(2) **Case 2** $ab < 0$

A has only one real eigenvalue c with associated eigenvector $\vec{e}_3$. Thus $\langle\langle \vec{e}_3 \rangle\rangle$ is the only invariant line (subspace). Suppose $a > 0$ and $b < 0$, then $-ab > 0$ and

$$A = \begin{bmatrix} 0 & a & 0 \\ -b & 0 & 0 \\ 0 & 0 & c \end{bmatrix} \begin{bmatrix} -1 & 0 & 0 \\ 0 & 1 & 0 \\ 0 & 0 & 1 \end{bmatrix} = \begin{bmatrix} 1 & 0 & 0 \\ 0 & -1 & 0 \\ 0 & 0 & 1 \end{bmatrix} \begin{bmatrix} 0 & a & 0 \\ -b & 0 & 0 \\ 0 & 0 & c \end{bmatrix}.$$

Therefore, in $\mathcal{N}$, A can be decomposed as

$$\vec{x} = (x_1, x_2, x_3) \to \vec{x} \begin{bmatrix} 0 & a & 0 \\ -b & 0 & 0 \\ 0 & 0 & c \end{bmatrix} \to \vec{x} \begin{bmatrix} 0 & a & 0 \\ -b & 0 & 0 \\ 0 & 0 & c \end{bmatrix} \begin{bmatrix} -1 & 0 & 0 \\ 0 & 1 & 0 \\ 0 & 0 & 1 \end{bmatrix} = \vec{x} A$$

where the first mapping is of type as in (1) while the second one is a *reflection* with respect to the coordinate plane $\langle\langle \vec{e}_2, \vec{e}_3 \rangle\rangle$. See Fig. 3.33 (and Fig. 2.49).

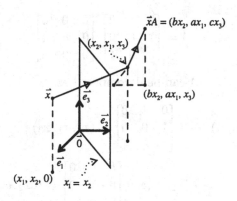

Fig. 3.32

(*Note* A satisfies its characteristic polynomial, namely

$$(A - cI_3)(A^2 - abI_3) = A^3 - cA^2 - abA + abcI_3 = O_{3\times3}.)$$

Remark   Other matrix representations for the case that $ab < 0$
     For simplicity, *replace b by $-b$ and then suppose $b > 0$ in A.* In this case, A satisfies $(A - cI_3)(A^2 + abI_3) = O$. By computation,

$$A^2 = \begin{bmatrix} -ab & 0 & 0 \\ 0 & -ab & 0 \\ 0 & 0 & c^2 \end{bmatrix} \Rightarrow (A^2 + abI_3) = \begin{bmatrix} 0 & 0 & 0 \\ 0 & 0 & 0 \\ 0 & 0 & c^2 + ab \end{bmatrix}.$$

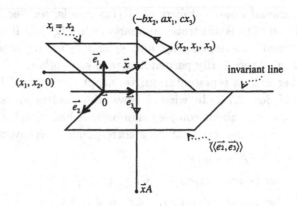

Fig. 3.33

Since $c^2 + ab > 0, r(A^2 + abI_3) = 1$. As we have experienced in the example in Sec. 2.7.8, let us consider vector $\vec{x}$ satisfying

$$\vec{x}A \neq \vec{0},$$
$$\vec{x}(A^2 + abI_3) = \vec{0}.$$

Such a $\vec{x}$ is not an eigenvector of A corresponding to c. Hence, $\vec{x} \in \langle\langle\vec{e_1}, \vec{e_2}\rangle\rangle$ should hold. Actual computation shows that any vector $\vec{x}$ in $\langle\langle\vec{e_1}, \vec{e_2}\rangle\rangle$ satisfies $\vec{x}(A^2 + abI_3) = \vec{0}$ and conversely. Choose an arbitrarily fixed nonzero $\vec{v} \in \langle\langle\vec{e_1}, \vec{e_2}\rangle\rangle$. Then $\vec{v}$ and $\vec{v}A$ are linearly independent, since

$$\alpha\vec{v} + \beta\vec{v}A = \vec{0}$$
$$\Rightarrow \alpha\vec{v}A + \beta\vec{v}A^2 = \alpha\vec{v}A - ab\beta\vec{v} = \vec{0}$$
$$\Rightarrow (\alpha^2 + ab\beta^2)\vec{v} = \vec{0}$$
$$\Rightarrow (\text{since } v \neq \vec{0} \text{ and } ab > 0) \ \alpha = \beta = 0.$$

Therefore, $\mathcal{B} = \{\vec{v}, \vec{v}A, \vec{e_3}\}$ forms a basis for $\mathbb{R}^3$. In $\mathcal{B}$,

$$\vec{v}A = 0 \cdot \vec{v} + 1 \cdot \vec{v}A + 0 \cdot \vec{e_3}$$
$$(\vec{v}A)A = \vec{v}A^2 = -ab\vec{v} = -ab\vec{v} + 0 \cdot \vec{v}A + 0 \cdot \vec{e_3}$$
$$\vec{e_3}A = c\vec{e_3}$$

$$\Rightarrow [A]_{\mathcal{B}} = P \begin{bmatrix} 0 & a & 0 \\ -b & 0 & 0 \\ 0 & 0 & c \end{bmatrix} P^{-1} = \begin{bmatrix} 0 & 1 & 0 \\ -ab & 0 & 0 \\ 0 & 0 & c \end{bmatrix}, \quad \text{where } P = \begin{bmatrix} \vec{v} \\ \vec{v}A \\ \vec{e_3} \end{bmatrix}.$$

$$(3.7.21)$$

This is the *rational canonical form* of A (for details, see Sec. 3.7.8). The action of $[A]_B$ on $[\vec{x}]_B$ is illustrated similarly as in Fig. 3.32 if a and $-b$ are replaced by 1 and $-ab$ respectively, and $\vec{e}_1$ and $\vec{e}_2$ are replaced by $\vec{v}$ and $\vec{v}A$ which are not necessarily perpendicular to each other.

For another matrix representation, we adopt the method indicated in Ex. <B> 5 of Sec. 2.7.2. In what fallows, the readers are supposed to have basic knowledge about complex numbers. A has *complex* eigenvalues $\pm\sqrt{ab}\,i$. Let $\vec{x} = (x_1, x_2, x_3) \in \mathbb{C}^3$ be a corresponding eigenvector. Then

$$\vec{x}A = \sqrt{ab}\,i\,\vec{x}$$
$$\Leftrightarrow (-bx_2, ax_1, cx_3) = \sqrt{ab}\,i\,(x_1, x_2, x_3)$$
$$\Leftrightarrow bx_2 = -\sqrt{ab}\,i\,x_1, \quad ax_1 = \sqrt{ab}\,i\,x_2,$$
$$cx_3 = \sqrt{ab}\,i\,x_3 \quad \text{(recall that } c^2 + ab > 0)$$
$$\Leftrightarrow \vec{x} = t(\sqrt{b}i, \sqrt{a}, 0) \quad \text{for } t \in \mathbb{C}.$$

Note that $(\sqrt{b}i, \sqrt{a}, 0) = (0, \sqrt{a}, 0) + i(\sqrt{b}, 0, 0)$. By equating the real and the imaginary parts of both sides of

$$\left[(0, \sqrt{a}, 0) + i(\sqrt{b}, 0, 0)\right] A = \sqrt{ab}\,i\left[(0, \sqrt{a}, 0) + i(\sqrt{b}, 0, 0)\right]$$
$$\Rightarrow (0, \sqrt{a}, 0)\,A = -\sqrt{ab}\,(\sqrt{b}, 0, 0),$$
$$(\sqrt{b}, 0, 0)A = \sqrt{ab}\,(0, \sqrt{a}, 0).$$

Combined with $\vec{e}_3A = c\vec{e}_3$, we obtain the matrix representation of A in the basis $\mathcal{C} = \{(\sqrt{b}, 0, 0), (0, \sqrt{a}, 0), \vec{e}_3\} = \{\sqrt{b}\,\vec{e}_1, \sqrt{a}\,\vec{e}_2, \vec{e}_3\}$

$$[A]_\mathcal{C} = Q \begin{bmatrix} 0 & a & 0 \\ -b & 0 & 0 \\ 0 & 0 & c \end{bmatrix} Q^{-1} = \begin{bmatrix} 0 & \sqrt{ab} & 0 \\ -\sqrt{ab} & 0 & 0 \\ 0 & 0 & c \end{bmatrix}$$

$$= \sqrt{ab} \begin{bmatrix} 0 & 1 & 0 \\ -1 & 0 & 0 \\ 0 & 0 & \frac{c}{\sqrt{ab}} \end{bmatrix}, \quad \text{where } Q = \begin{bmatrix} \sqrt{b} & 0 & 0 \\ 0 & \sqrt{a} & 0 \\ 0 & 0 & 1 \end{bmatrix}. \quad (3.7.22)$$

$[A]_\mathcal{C}$ can be decomposed as, in $\mathcal{C}$,

$$[\vec{x}]_\mathcal{C} \rightarrow [\vec{x}]_\mathcal{C} \begin{bmatrix} 0 & 1 & 0 \\ -1 & 0 & 0 \\ 0 & 0 & 1 \end{bmatrix} \rightarrow [\vec{x}]_\mathcal{C} \begin{bmatrix} 0 & 1 & 0 \\ -1 & 0 & 0 \\ 0 & 0 & 1 \end{bmatrix} \begin{bmatrix} 1 & 0 & 0 \\ 0 & 1 & 0 \\ 0 & 0 & \frac{c}{\sqrt{ab}} \end{bmatrix}$$

$$\rightarrow [\vec{x}]_\mathcal{C}\sqrt{ab} \begin{bmatrix} 0 & 1 & 0 \\ -1 & 0 & 0 \\ 0 & 0 & \frac{c}{\sqrt{ab}} \end{bmatrix} = [\vec{x}]_\mathcal{C}[A]_\mathcal{C} = [\vec{x}A]_\mathcal{C},$$

where the first mapping is a *rotation* through 90° in the coordinate plane $\langle\langle \sqrt{b}\,\vec{e}_1, \sqrt{a}\,\vec{e}_2\rangle\rangle$, the second one is a *one-way stretch* along $\langle\langle \vec{e}_3\rangle\rangle$ and the third one is an *enlargement* with scale $\sqrt{ab}$. The readers are urged to illustrate these mappings graphically.

Example 3 The operator (compare with Example 6 in Sec. 2.7.2)

$$A = \begin{bmatrix} a & 0 & 0 \\ b & c & 0 \\ 0 & 0 & 1 \end{bmatrix}, \quad \text{where } abc \neq 0$$

has the following properties:

Case 1 $a \neq c \neq 1$

eigenvalues	eigenvectors
$\lambda_1 = a$	$\vec{v}_1 = \vec{e}_1$
$\lambda_2 = c$	$\vec{v}_2 = (b, c - a, 0)$
$\lambda_3 = 1$	$\vec{v}_3 = \vec{e}_3$

In the basis $\mathcal{B} = \{\vec{v}_1, \vec{v}_2, \vec{v}_3\}$, A has the representation

$$[A]_{\mathcal{B}} = PAP^{-1} = \begin{bmatrix} a & 0 & 0 \\ 0 & c & 0 \\ 0 & 0 & 1 \end{bmatrix}, \quad \text{where } P = \begin{bmatrix} \vec{v}_1 \\ \vec{v}_2 \\ \vec{v}_3 \end{bmatrix}.$$

Refer to Fig. 3.31 for the mapping properties of A in $\mathcal{B}$.

Case 2 $a \neq c = 1$

eigenvalues	eigenvectors
$\lambda_1 = a$	$\vec{v}_1 = \vec{e}_1$
$\lambda_2 = 1, 1$	$\vec{v}_2 = (b, 1 - a, 0)$ and $\vec{v}_3 = \vec{e}_3$

In the basis $\mathcal{B} = \{\vec{v}_1, \vec{v}_2, \vec{v}_3\}$,

$$[A]_{\mathcal{B}} = P \begin{bmatrix} a & 0 & 0 \\ b & 1 & 0 \\ 0 & 0 & 1 \end{bmatrix} P^{-1} = \begin{bmatrix} a & 0 & 0 \\ 0 & 1 & 0 \\ 0 & 0 & 1 \end{bmatrix}, \quad \text{where } P = \begin{bmatrix} \vec{v}_1 \\ \vec{v}_2 \\ \vec{v}_3 \end{bmatrix}.$$

Refer to Fig. 3.31.

Case 3 $a = c = 1$

1 is the only eigenvalue of A with associated eigenvectors $\vec{e}_1$ and $\vec{e}_3$.

1. Since $b \neq 0$,

$$A = \begin{bmatrix} 1 & 0 & 0 \\ b & 1 & 0 \\ 0 & 0 & 1 \end{bmatrix}$$

is not diagonalizable.

2. The coordinate plane $\langle\langle \vec{e}_1, \vec{e}_3 \rangle\rangle$ is the plane (subspace) of invariant points of A.

3. Since $\vec{x} = (x_1, x_2, x_3) \to \vec{x}A = (x_1 + bx_2, x_2, x_3)$, A moves a point $\vec{x}$ along a line parallel to the axis $\langle\langle \vec{e}_1 \rangle\rangle$ through a distance with a constant proposition b to its distance x_2 from the $\langle\langle \vec{e}_1 \rangle\rangle$ axis, i.e.

$$\frac{(x_1 + bx_2) - x_1}{x_2} = b$$

to the point $\vec{x}A$. This A is called a *shearing* along the plane $\langle\langle \vec{e}_1, \vec{e}_3 \rangle\rangle$ with scale factor b. See Fig. 3.34.

4. Hence, every plane $\vec{v} + \langle\langle \vec{e}_1, \vec{e}_3 \rangle\rangle$ parallel to $\langle\langle \vec{e}_1, \vec{e}_3 \rangle\rangle$ is an invariant plane.

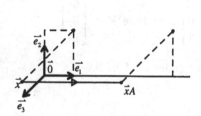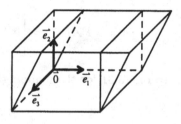

Fig. 3.34

Case 4 $a = c \neq 1$

a and 1 are eigenvalues of A with respective eigenvectors $\vec{e}_1$ and $\vec{e}_3$.

1. A satisfies its characteristic polynomial $-(t - a)^2(t - 1)$, i.e.

$$(A - aI_3)^2(A - I_3) = O_{3\times3}.$$

2. Since $(A - aI_3)(A - I_3) \neq O_{3\times3}$, A is not the diagonalizable (refer to Ex. <C> 9 of Sec. 2.7.6 or Secs. 3.7.6 and 3.7.7).

3. Since

$$A = \begin{bmatrix} a & 0 & 0 \\ b & a & 0 \\ 0 & 0 & 1 \end{bmatrix} = \begin{bmatrix} a & 0 & 0 \\ 0 & a & 0 \\ 0 & 0 & 1 \end{bmatrix} \begin{bmatrix} 1 & 0 & 0 \\ \frac{b}{a} & 1 & 0 \\ 0 & 0 & 1 \end{bmatrix} = \begin{bmatrix} a & 0 & 0 \\ 0 & a & 0 \\ 0 & 0 & 1 \end{bmatrix} + \begin{bmatrix} 0 & 0 & 0 \\ b & 0 & 0 \\ 0 & 0 & 0 \end{bmatrix},$$

therefore A can be decomposed, in $\mathcal{N}$, as an *enlargement* on the coordinate plane $\langle\langle \vec{e}_1, \vec{e}_2 \rangle\rangle$ with scale a followed by a *shearing* along $\langle\langle \vec{e}_1, \vec{e}_3 \rangle\rangle$ with scale $\frac{b}{a}$. See Fig. 3.34.

Remark Jordan canonical form of a matrix
    Let

$$B = \begin{bmatrix} \lambda_1 & 0 & 0 \\ b & \lambda_1 & 0 \\ a & c & \lambda_2 \end{bmatrix}$$

where $b \neq 0$ and $\lambda_1 \neq \lambda_2$. B has characteristic polynomial $-(t-\lambda_1)^2(t-\lambda_2)$ and hence has eigenvalues λ_1, λ_1 and λ_2. Since

$$(B - \lambda_1 I_3)(B - \lambda_2 I_3) = \begin{bmatrix} 0 & 0 & 0 \\ b & 0 & 0 \\ a & c & \lambda_2 - \lambda_1 \end{bmatrix} \begin{bmatrix} \lambda_1 - \lambda_2 & 0 & 0 \\ b & \lambda_1 - \lambda_2 & 0 \\ a & c & 0 \end{bmatrix}$$

$$= \begin{bmatrix} 0 & 0 & 0 \\ b(\lambda_1 - \lambda_2) & 0 & 0 \\ bc & 0 & 0 \end{bmatrix} \neq O_{3 \times 3},$$

B is not diagonalizable. As we have experienced in Sec. 2.7.7, consider

$$(B - \lambda_1 I_3)^2 = \begin{bmatrix} 0 & 0 & 0 \\ 0 & 0 & 0 \\ bc + a(\lambda_2 - \lambda_1) & c(\lambda_2 - \lambda_1) & (\lambda_2 - \lambda_1)^2 \end{bmatrix}$$

and take a vector $\vec{v}_2$ satisfying $\vec{v}_2(B - \lambda_1 I_3)^2 = \vec{0}$ but $\vec{v}_1 = \vec{v}_2(B - \lambda_1 I_3) \neq \vec{0}$. Say $\vec{v}_2 = \vec{e}_2$, then

$$\vec{v}_1 = \vec{v}_2(B - \lambda_1 I_3) = (b, 0, 0) = b\vec{e}_1.$$

Solve $\vec{x}B = \lambda_2\vec{x}$ and take an eigenvector $\vec{v}_3 = (bc - a(\lambda_1 - \lambda_2), -c(\lambda_1 - \lambda_2), (\lambda_1 - \lambda_2)^2)$. Then $C = \{\vec{v}_1, \vec{v}_2, \vec{v}_3\}$ is a basis for $\mathbb{R}^3$. Since

$$\vec{v}_1 B = b\vec{e}_1 B = b\lambda_1\vec{e}_1 = \lambda_1\vec{v}_1 = \lambda_1\vec{v}_1 + 0\cdot\vec{v}_2 + 0\cdot\vec{v}_3,$$
$$\vec{v}_2 B = \vec{e}_2 B = (b, \lambda_1, 0) = b\vec{e}_1 + \lambda_1\vec{e}_2 = \vec{v}_1 + \lambda_1\cdot\vec{v}_2 + 0\cdot\vec{v}_3,$$
$$\vec{v}_3 B = \lambda_2\vec{v}_3 = 0\cdot\vec{v}_1 + 0\cdot\vec{v}_2 + \lambda_2\cdot\vec{v}_3$$

$$\Rightarrow [B]_C = QBQ^{-1} = \begin{bmatrix} \lambda_1 & 0 & 0 \\ 1 & \lambda_1 & 0 \\ 0 & 0 & \lambda_2 \end{bmatrix}, \quad \text{where } Q = \begin{bmatrix} \vec{v}_1 \\ \vec{v}_2 \\ \vec{v}_3 \end{bmatrix}. \qquad (3.7.23)$$

This is called the *Jordan canonical form* of B (for details, see Sec. 3.7.7). $[B]_C$ belongs to Case 4 in Example 3 by noticing that

$$[B]_C = \lambda_2 \begin{bmatrix} \frac{\lambda_1}{\lambda_2} & 0 & 0 \\ \frac{1}{\lambda_2} & \frac{\lambda_1}{\lambda_2} & 0 \\ 0 & 0 & 1 \end{bmatrix}. \qquad (3.7.24)$$

Example 4 The linear operator

$$A = \begin{bmatrix} \lambda & 0 & 0 \\ b & \lambda & 0 \\ 0 & c & \lambda \end{bmatrix}, \quad \text{where } bc \neq 0$$

has the following properties:

1. A satisfies its characteristic polynomial $-(t - \lambda)^3$, i.e.

$$(A - \lambda I_3)^3 = O_{3\times 3}$$

but

$$A - \lambda I_3 = \begin{bmatrix} 0 & 0 & 0 \\ b & 0 & 0 \\ 0 & c & 0 \end{bmatrix} \neq O_{3\times 3}, \quad (A - \lambda I_3)^2 = \begin{bmatrix} 0 & 0 & 0 \\ 0 & 0 & 0 \\ bc & 0 & 0 \end{bmatrix} \neq O_{3\times 3}.$$

2. Hence, A has eigenvalues $\lambda, \lambda, \lambda$ with associated eigenvectors $t\vec{e}_1, t \in \mathbb{R}$ and $t \neq 0$ and A is not diagonalizable.
3. Notice that A can be written as the sum of the following linear operators:

$$A = \lambda I_3 + (A - \lambda I_3)$$

$$= \lambda \begin{bmatrix} 1 & 0 & 0 \\ \frac{b}{\lambda} & 1 & 0 \\ 0 & 0 & 1 \end{bmatrix} + \begin{bmatrix} 0 & 0 & 0 \\ 0 & 0 & 0 \\ 0 & c & 0 \end{bmatrix}.$$

Note that $\langle\langle\vec{e}_1\rangle\rangle$ is the only invariant line (subspace). See Fig. 3.35 for some geometric feeling.

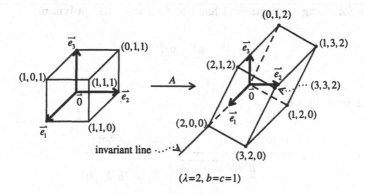

Fig. 3.35

Remark Jordan canonical form of a matrix

Let A be as in Example 4.

Try to choose a nonzero vector $\vec{v}_3$ so that

$$\vec{v}_3(A - \lambda I_3)^3 = \vec{0},$$
$$\vec{v}_2 = \vec{v}_3(A - \lambda I_3) \neq \vec{0}, \quad \text{and}$$
$$\vec{v}_1 = \vec{v}_2(A - \lambda I_3) = \vec{v}_3(A - \lambda I_3)^2 \text{ is an eigenvector of } A.$$

Take $\vec{v}_3 = \vec{e}_3$, then $\vec{v}_2 = (0, c, 0) = c\vec{e}_2$ and $\vec{v}_1 = (bc, 0, 0) = bc\vec{e}_1$. Now $\mathcal{B} = \{\vec{v}_1, \vec{v}_2, \vec{v}_3\}$ is a basis for $\mathbb{R}^3$. Thus

$$\vec{v}_1 A = bc\vec{e}_1 A = bc(\lambda, 0, 0) = \lambda\vec{v}_1 = \lambda\vec{v}_1 + 0 \cdot \vec{v}_2 + 0 \cdot \vec{v}_3,$$
$$\vec{v}_2 A = c\vec{e}_2 A = c(b, \lambda, 0) = bc\vec{e}_1 + \lambda c\vec{e}_2 = 1 \cdot \vec{v}_1 + \lambda \cdot \vec{v}_2 + 0 \cdot \vec{v}_3,$$
$$\vec{v}_3 A = \vec{e}_3 A = (0, c, \lambda) = c\vec{e}_2 + \lambda\vec{e}_3 = 0 \cdot \vec{v}_1 + 1 \cdot \vec{v}_2 + \lambda \cdot \vec{v}_3$$

$$\Rightarrow [A]_{\mathcal{B}} = PAP^{-1} = \begin{bmatrix} \lambda & 0 & 0 \\ 1 & \lambda & 0 \\ 0 & 1 & \lambda \end{bmatrix}, \quad \text{where } P = \begin{bmatrix} bc\vec{e}_1 \\ c\vec{e}_2 \\ \vec{e}_3 \end{bmatrix}. \qquad (3.7.25)$$

$[A]_{\mathcal{B}}$ is called the *Jordan canonical form* of A (for details, see Sec. 3.7.7). See Fig. 3.35 for $\lambda = 2$.

Example 5 The linear operator (compare with Example 7 in Sec. 2.7.2)

$$A = \begin{bmatrix} 0 & 1 & 0 \\ b & a & 0 \\ 0 & 0 & c \end{bmatrix}, \quad \text{where } bc \neq 0$$

has the following properties: A has the characteristic polynomial

$$-(t^2 - at - b)(t - c).$$

Case 1 $a^2 + 4b > 0$

eigenvalues	eigenvectors
$\lambda_1 = \dfrac{a + \sqrt{a^2 + 4b}}{2}$	$\vec{v}_1 = (b, \lambda_1, 0)$
$\lambda_2 = \dfrac{a - \sqrt{a^2 + 4b}}{2}$	$\vec{v}_2 = (b, \lambda_2, 0)$
$\lambda_3 = c$	$\vec{v}_3 = \vec{e}_3$

In the basis $\mathcal{B} = \{\vec{v}_1, \vec{v}_2, \vec{v}_3\}$, A is diagonalizable as

$$[A]_\mathcal{B} = PAP^{-1} = \begin{bmatrix} \lambda_1 & 0 & 0 \\ 0 & \lambda_2 & 0 \\ 0 & 0 & \lambda_3 \end{bmatrix}, \quad \text{where } P = \begin{bmatrix} \vec{v}_1 \\ \vec{v}_2 \\ \vec{v}_3 \end{bmatrix}.$$

See Fig. 3.31.

Case 2 $a^2 + 4b = 0$ and $a \neq 2c$

eigenvalues	eigenvectors
$\lambda = \dfrac{a}{2}, \dfrac{a}{2}$	$\vec{v} = (-a, 2, 0)$
$\lambda_3 = c$	$\vec{v}_3 = \vec{e}_3$

1. Since

$$A - \frac{a}{2}I_3 = \begin{bmatrix} -\frac{a}{2} & 1 & 0 \\ b & \frac{a}{2} & 0 \\ 0 & 0 & c - \frac{a}{2} \end{bmatrix},$$

$$\left(A - \frac{a}{2}I_3\right)^2 = \begin{bmatrix} 0 & 0 & 0 \\ 0 & 0 & 0 \\ 0 & 0 & \left(c - \frac{a}{2}\right)^2 \end{bmatrix},$$

$$A - cI_3 = \begin{bmatrix} -c & 1 & 0 \\ b & a-c & 0 \\ 0 & 0 & 0 \end{bmatrix}, \quad \text{and}$$

$$\left(A - \frac{a}{2}I_3\right)(A - cI_3) = \begin{bmatrix} \frac{ac}{2}+b & \frac{a}{2}-c & 0 \\ b(\frac{a}{2}-c) & -(\frac{ac}{2}+b) & 0 \\ 0 & 0 & 0 \end{bmatrix} \neq O_{3\times3},$$

thus A is not diagonalizable.

2. Now, $\mathcal{B} = \{\vec{v}, \vec{e}_1, \vec{e}_3\}$ is a basis for $\mathbb{R}^3$. Since

$$\vec{v}A = \lambda\vec{v},$$

$$\vec{e}_1A = (0,1,0) = \frac{1}{2}\vec{v} + \lambda\vec{e}_1,$$

$$\vec{e}_3A = c\vec{e}_3$$

$$\Rightarrow [A]_{\mathcal{B}} = P \begin{bmatrix} 0 & 1 & 0 \\ b & a & 0 \\ 0 & 0 & c \end{bmatrix} P^{-1} = \begin{bmatrix} \lambda & 0 & 0 \\ \frac{1}{2} & \lambda & 0 \\ 0 & 0 & c \end{bmatrix}$$

$$= c \begin{bmatrix} \frac{\lambda}{c} & 0 & 0 \\ \frac{1}{2c} & \frac{\lambda}{c} & 0 \\ 0 & 0 & 1 \end{bmatrix}, \quad \text{where } P = \begin{bmatrix} \vec{v} \\ \vec{e}_1 \\ \vec{e}_3 \end{bmatrix}.$$

For mapping properties of A in $\mathcal{B}$, please refer to Example 4. Note that $\langle\langle\vec{v}, \vec{e}_3\rangle\rangle$ is an invariant plane (subspace) of A.

See Fig. 3.36 (compare with Fig. 2.54).

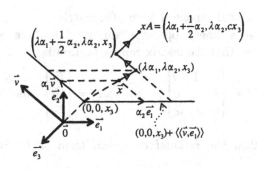

Fig. 3.36

Case 3 $a^2 + 4b < 0$

A has only one real eigenvalue c with associated eigenvector $\vec{e}_3$. Hence $\langle\langle \vec{e}_3 \rangle\rangle$ is an invariant line (subspace) of A. Since

$$A = \begin{bmatrix} 0 & 1 & 0 \\ 1 & 0 & 0 \\ 0 & 0 & 1 \end{bmatrix} \begin{bmatrix} b & 0 & 0 \\ 0 & 1 & 0 \\ 0 & 0 & 1 \end{bmatrix} + \begin{bmatrix} 0 & 0 & 0 \\ 0 & a & 0 \\ 0 & 0 & c-1 \end{bmatrix},$$

then, A is the *sum* of the following two linear operators.

1. The first one is the composite of the *reflection* with respect to the plane $x_1 = x_2$ (see Example 2) followed by a *one-way stretch* along $\langle\langle \vec{e}_1 \rangle\rangle$ with scale b (see Example 1).
2. The second one is a *mapping* of $\mathbb{R}^3$ onto the plane $\langle\langle \vec{e}_2, \vec{e}_3 \rangle\rangle$ if $a \neq 0, c \neq 1$; onto the axis $\langle\langle \vec{e}_2 \rangle\rangle$ if $a \neq 0$ and $c = 1$; and onto the axis $\langle\langle \vec{e}_3 \rangle\rangle$ if $a = 0$ and $c \neq 1$ (see (3.7.34)).

See Fig. 3.37 (compare with Fig. 2.55).

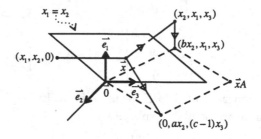

Fig. 3.37

Remark The rational canonical form of a matrix

Let $B_{3\times 3}$ be a linear operator on $\mathbb{R}^3$. If there exists a basis $\mathcal{B} = \{\vec{v}_1, \vec{v}_2, \vec{v}_3\}$ so that the matrix representation is

$$[B]_\mathcal{B} = PBP^{-1} = \begin{bmatrix} 0 & 1 & 0 \\ b & a & 0 \\ 0 & 0 & c \end{bmatrix}, \quad \text{where } P = \begin{bmatrix} \vec{v}_1 \\ \vec{v}_2 \\ \vec{v}_3 \end{bmatrix} \text{ and } a^2 + 4b < 0,$$

(3.7.26)

then $[B]_\mathcal{B}$ is called the *rational canonical form* of B. For details, see Sec. 3.7.8.

To sum up what we obtained so far in this subsection and to expect what we will do in Secs. 3.7.6–3.7.8, let $A_{3\times 3}$ be a nonzero real matrix whose characteristic polynomial is

$$\det(A - tI_3) = -(t^3 + b_2t^2 + b_1t + b_0) = -(t - \lambda_1)(t^2 + a_1t + a_0).$$

Cayley–Hamilton theorem says that characteristic polynomial annihilates A, i.e.

$$A^3 + b_2A^2 + b_1A + b_0I_3 = (A - \lambda_1I_3)(A^2 + a_1A + a_0I_3) = O_{3\times3}. \quad (3.7.27)$$

For a direct proof of this matrix identity, see Ex. <A> 3, also refer to (2.7.19), Ex. <C> 5 of Sec. 2.7.6 and Ex. 4 of Sec. B.10. Note that A is invertible if and only if

$$-b_0 = \det A \neq 0 \quad \text{and then} \quad A^{-1} = -\frac{1}{b_0}(A^2 + b_2A + b_1I_3). \quad (3.7.28)$$

The canonical forms of a nonzero matrix $A_{3\times3}$ under similarity

Case 1 $a_1^2 - 4a_0 > 0$
$t^2 + a_1t + a_0 = (t - \lambda_2)(t - \lambda_3)$ and A has real eigenvalues $\lambda_1, \lambda_2, \lambda_3$.

1. If $\lambda_1 \neq \lambda_2 \neq \lambda_3$, A is diagonalizable and is similar to

$$\begin{bmatrix} \lambda_1 & & 0 \\ & \lambda_2 & \\ 0 & & \lambda_3 \end{bmatrix}.$$

2. If at least two of $\lambda_1, \lambda_2, \lambda_3$ are equal, see Case 2.

Case 2 $a_1^2 - 4a_0 = 0$
$t^2 + a_1t + a_0 = (t - \lambda_2)^2$ and A has real eigenvalues $\lambda_1, \lambda_2, \lambda_2$.

1. If $\lambda_1 \neq \lambda_2$ and $(A - \lambda_1I_3)(A - \lambda_2I_3) = O_{3\times3}$, A is diagonalizable and is similar to

$$\begin{bmatrix} \lambda_1 & & 0 \\ & \lambda_2 & \\ 0 & & \lambda_2 \end{bmatrix}.$$

2. If $\lambda_1 \neq \lambda_2$ and $(A - \lambda_1I_3)(A - \lambda_2I_3) \neq O_{3\times3}$, A is *not* diagonalizable and is similar to the *Jordan canonical form* or the *rational canonical form*

$$\begin{bmatrix} \lambda_2 & 0 & 0 \\ 1 & \lambda_2 & 0 \\ 0 & 0 & \lambda_1 \end{bmatrix} \quad \text{or} \quad \begin{bmatrix} 0 & 1 & 0 \\ -\lambda_2^2 & 2\lambda_2 & 0 \\ 0 & 0 & \lambda_1 \end{bmatrix}.$$

3. If $\lambda_1 = \lambda_2 = \lambda_3 = \lambda$ and $A - \lambda I_3 = O_{3\times3}$, then A itself is λI_3.

4. If $\lambda_1 = \lambda_2 = \lambda_3 = \lambda$ but $A - \lambda I_3 \neq O_{3\times3}$ and $(A - \lambda I_3)^2 = O_{3\times3}$, then A is similar to the *Jordan canonical form* or the *rational canonical form*

$$\begin{bmatrix} \lambda & 0 & 0 \\ 1 & \lambda & 0 \\ 0 & 0 & \lambda \end{bmatrix} \quad \text{or} \quad \begin{bmatrix} 0 & 1 & 0 \\ -\lambda^2 & 2\lambda & 0 \\ 0 & 0 & \lambda \end{bmatrix}.$$

5. If $\lambda_1 = \lambda_2 = \lambda_3 = \lambda$ but $(A - \lambda I_3) \neq O_{3\times3}, (A - \lambda I_3)^2 \neq O_{3\times3}$ and $(A - \lambda I_3)^3 = O_{3\times3}$, then A is similar to the *Jordan canonical form* or the *rational canonical form*

$$\begin{bmatrix} \lambda & 0 & 0 \\ 1 & \lambda & 0 \\ 0 & 1 & \lambda \end{bmatrix} \quad \text{or} \quad \begin{bmatrix} 0 & 1 & 0 \\ 0 & 0 & 1 \\ \lambda^3 & -3\lambda^2 & 3\lambda \end{bmatrix}.$$

Case 3 $a_1^2 - 4a_0 < 0$

A is similar to the *rational canonical form*

$$\begin{bmatrix} 0 & 1 & 0 \\ -a_0 & -a_1 & 0 \\ 0 & 0 & \lambda_1 \end{bmatrix}. \tag{3.7.29}$$

Exercises

<A>

1. For each of the following matrices A (linear operators), do the following problems.

   (1) Determine the rank $r(A)$.
   (2) Compute the characteristic polynomial $\det(A - tI_3)$ and real eigenvalues and the associated eigenvectors.
   (3) Justify the Cayley–Hamilton theorem and use it to compute A^{-1} if A is invertible (see (3.7.27) and (3.7.28)).
   (4) Determine invariant lines or/and planes (subspaces and affine subspaces); also, lines or planes of invariant points, if any.
   (5) Illustrate graphically the essential mapping properties of A in $\mathcal{N} = \{\vec{e}_1, \vec{e}_2, \vec{e}_3\}$.
   (6) Try to find a basis $\mathcal{B} = \{\vec{v}_1, \vec{v}_2, \vec{v}_3\}$, if possible, so that $[A]_{\mathcal{B}}$ is in its canonical form (see (3.7.29)).
   (7) Redo (5) in $\mathcal{B}$.

(a) A is one of

$$\begin{bmatrix} 0 & 0 & 0 \\ 0 & a & 0 \\ 0 & 0 & 0 \end{bmatrix}, \begin{bmatrix} 0 & 0 & a \\ 0 & 0 & 0 \\ 0 & 0 & 0 \end{bmatrix} \text{ or } \begin{bmatrix} 0 & 0 & 0 \\ 0 & 0 & 0 \\ 0 & a & 0 \end{bmatrix}, \quad \text{where } a \neq 0.$$

(b) A is one of

$$\begin{bmatrix} 0 & a & 0 \\ 0 & 0 & 0 \\ b & 0 & 0 \end{bmatrix}, \begin{bmatrix} 0 & 0 & a \\ 0 & 0 & 0 \\ 0 & 0 & b \end{bmatrix} \text{ or } \begin{bmatrix} 0 & 0 & a \\ 0 & b & 0 \\ 0 & 0 & 0 \end{bmatrix}, \quad \text{where } ab \neq 0.$$

(c) A is one of

$$\begin{bmatrix} 0 & a & 0 \\ b & 0 & c \\ 0 & 0 & 0 \end{bmatrix}, \begin{bmatrix} 0 & 0 & 0 \\ a & 0 & 0 \\ b & c & 0 \end{bmatrix}, \begin{bmatrix} a & 0 & 0 \\ 0 & 0 & 0 \\ 0 & b & c \end{bmatrix} \text{ or } \begin{bmatrix} 0 & 0 & 0 \\ 0 & a & b \\ c & 0 & 0 \end{bmatrix},$$

where $abc \neq 0$.

(d) A is one of

$$\begin{bmatrix} 0 & 0 & a \\ b & 0 & 0 \\ 0 & c & 0 \end{bmatrix}, \begin{bmatrix} 0 & 0 & a \\ 0 & b & 0 \\ c & 0 & 0 \end{bmatrix} \text{ or } \begin{bmatrix} 0 & a & 0 \\ b & 0 & 0 \\ 0 & 0 & c \end{bmatrix}, \quad \text{where } abc \neq 0.$$

Note that one can assign a, b and c numerical numbers so that the characteristic polynomial can be handled easier.

(e) A is one of

$$\begin{bmatrix} 0 & -1 & -1 \\ -3 & -1 & -2 \\ 7 & 5 & 6 \end{bmatrix}, \begin{bmatrix} -3 & 3 & -2 \\ -7 & 6 & -3 \\ 1 & -1 & 2 \end{bmatrix} \text{ or } \begin{bmatrix} 0 & 1 & -1 \\ -4 & 4 & -2 \\ -2 & 1 & 1 \end{bmatrix}.$$

(f) $A = \begin{bmatrix} 0 & 1 & 0 \\ 0 & 0 & 1 \\ 8 & -12 & 6 \end{bmatrix}$ into Jordan canonical form.

(g) $A = \begin{bmatrix} -2 & 1 & 0 \\ 0 & -2 & 1 \\ 0 & 0 & -2 \end{bmatrix}$ into Jordan and rational canonical forms.

2. Recall that $\vec{e}_1 = (1,0,0)$, $\vec{e}_2 = (0,1,0)$, and $\vec{e}_3 = (0,0,1)$. Let $\vec{x}_1 = (0,1,1)$, $\vec{x}_2 = (1,0,1)$, $\vec{x}_3 = (1,1,0)$, and $\vec{x}_4 = (1,1,1)$. These seven points form vertices of a unit *cube*. See Fig. 3.38. This cube can be divided into six tetrahedra of equal volumes. Three of them are shown in the figure. Any such a tetrahedron is called *part* of the cube. Similar situation happens to a parallelepiped.

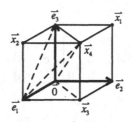

Fig. 3.38

(a) Find a *linear* transformation mapping the tetrahedron $\Delta \vec{o}\,\vec{e}_1\,\vec{e}_2\,\vec{e}_3$ one-to-one and *onto* the $\Delta \vec{o}\,\vec{e}_1\,\vec{x}_3\,\vec{x}_4$.

  (1) Determine the other vertices of the parallelepiped that contains $\Delta \vec{o}\,\vec{e}_1\,\vec{x}_3\,\vec{x}_4$ as part of it. Find the volume of it.

  (2) How many such linear transformations are there? Write them out in matrix forms. Are there any connections among them, say, via permutation matrices (see (2.7.67))?

  (3) If we replace "onto" by "*into* the tetrahedron $\Delta \vec{o}\,\vec{e}_1\,\vec{x}_3\,\vec{x}_4$" so that the range spaces contain a vertex, an edge or a face of it, find the corresponding linear transformations. Not just one only, but many of them!

(b) Do the same questions as in (a), but onto and into the tetrahedron $\Delta \vec{o}\,\vec{e}_1\,\vec{e}_3\,\vec{x}_4$.

(c) Do the same questions as in (a), but onto and into the tetrahedron $\Delta \vec{e}_1\,\vec{x}_2\,\vec{e}_3\,\vec{x}_4$.

  (*Note* Here, "linear" should be replaced by "affine". By an *affine* transformation, we mean a mapping of the form

$$T(\vec{x}) = \vec{x}_0 + \vec{x}A, \quad \vec{x}_0, \vec{x} \in \mathbb{R}^3,$$

where A is a linear operator (refer to (2.8.20) and Sec. 3.8).)

(d) Do the same questions as in (a), but from $\Delta \vec{e}_1\,\vec{x}_3\,\vec{e}_2\,\vec{x}_4$ onto or into $\Delta \vec{x}_1\,\vec{e}_2\,\vec{e}_3\,\vec{x}_4$.

3. Prove Cayley–Hamilton theorem (3.7.27) by the following methods.

  (a) Use canonical forms in (3.7.29).

  (b) $A_{3\times3}$ has at least one real eigenvalue λ so that there exists a basis $\mathcal{B} = \{\vec{v}_1, \vec{v}_2, \vec{v}_3\}$ for $\mathbb{R}^3$ in which

$$[A]_{\mathcal{B}} = PAP^{-1} = \begin{bmatrix} \lambda_1 & 0 & 0 \\ b_{21} & b_{22} & b_{23} \\ b_{31} & b_{32} & b_{33} \end{bmatrix}.$$

(c) Try to simplify for $A_{3\times3}$ what Exs. <B> 5 and <C> 5 of Sec. 2.7.6 say.

(d) Try to model after (2.7.18) for $A_{3\times3}$. Is this a beneficial way to prove this theorem?

(e) Let $\varphi(\lambda) = \det(A - \lambda I_3)$, the characteristic polynomial of A. Try the following steps.

(1) Take λ so that $\varphi(\lambda) \neq 0$. Then (see (3.3.2) or Sec. B.6)
$$(A - \lambda I_3)^{-1} = \tfrac{1}{\varphi(\lambda)} \operatorname{adj}(A - \lambda I_3).$$

(2) Each entry of the adjoint matrix $\operatorname{adj}(A - \lambda I_3)$ is a polynomial in λ of degree less than 3. Hence, there exist constant matrices $B_2, B_1, B_0 \in M(3; \mathbb{R})$ such that
$$\operatorname{adj}(A - \lambda I_3) = B_2\lambda^2 + B_1\lambda + B_0.$$

(3) Multiply both sides of $\operatorname{adj}(A - \lambda I_3) \cdot (A - \lambda I_3) = \varphi(\lambda)I_3$ out and equate the corresponding coefficient matrices of $\lambda^3, \lambda^2, \lambda$ and λ^0 to obtain
$$-B_2 = -I_3$$
$$B_2A - B_1 = -b_2I_3$$
$$B_1A - B_0 = -b_1I_3$$
$$B_0A = -b_0I_3.$$

Then try to eliminate B_2, B_1, B_0 from the right sides.

(*Note* Methods in (c) and (e) are still valid for $n \times n$ matrices over a field $\mathbb{F}$.)

4. Let $A = [a_{ij}]_{3\times3}$.

(a) Show that the characteristic polynomial
$$\det(A - tI_3) = -t^3 + \operatorname{tr}(A)t^2 + b_1t + \det A.$$

Try to use $a_{ij}, 1 \leq i, j \leq 3$, to express the coefficient b_1.

(b) Prove (3.7.28).

(c) Let
$$A = \begin{bmatrix} -9 & 4 & 4 \\ -8 & 3 & 4 \\ -16 & 8 & 7 \end{bmatrix}.$$

(1) Try to use (b) to compute A^{-1}.

(2) Show that A is diagonalizable and find an invertible matrix P such that PAP^{-1} is a diagonal matrix.

(3) Compute A^n for $n = \pm2, \pm3$.

(4) Compute $A^5 - 5A^3 - A^2 + 23A + 17I_3$.

\<B\>

1. Try to prove (3.7.29). For details, refer to Secs. 3.7.6–3.7.8.

2. (Refer to Ex. \<B\> 1 of Sec. 2.7.2, Ex. 7 of Sec. B.4 and Ex. 5 of Sec. B.12.) Let $A \in M(3; \mathbb{R})$ be a nilpotent matrix of index 2, namely $A \neq O_{3\times3}$ but $A^2 = O$.

(a) Show that A is not one-to-one, i.e. not invertible.

(b) If λ is an eigenvalue (it could be a complex number) of A, then it is necessarily that $\lambda = 0$.

(c) What is the characteristic polynomial of A? What its canonical form looks like?

(d) Show that

$$A \neq O, \quad A^2 = O \Leftrightarrow \operatorname{Im}(A) \subseteq \operatorname{Ker}(A).$$

Use this to determine all such A. For example, $\operatorname{Ker}(A) = \{\vec{0}\}$ or $\mathbb{R}^3$ is impossible if $A \neq O$ and hence, $\operatorname{Im}(A) \neq \{\vec{0}\}$ or $\mathbb{R}^3$. Consider the following cases.

(1) $\dim \operatorname{Ker}(A) = 2$ and $\operatorname{Ker}(A) = \langle\langle \vec{x}_1, \vec{x}_2 \rangle\rangle$. Choose $\vec{x}_3$ so that $\{\vec{x}_1, \vec{x}_2, \vec{x}_3\}$ is a basis for $\mathbb{R}^3$. Then $\vec{x}_3 A = \alpha_1 \vec{x}_1 + \alpha_2 \vec{x}_2$ and

$$PAP^{-1} = \begin{bmatrix} 0 & 0 & 0 \\ 0 & 0 & 0 \\ \alpha_1 & \alpha_2 & 0 \end{bmatrix}, \quad \text{where } P = \begin{bmatrix} \vec{x}_1 \\ \vec{x}_2 \\ \vec{x}_3 \end{bmatrix}.$$

Is $\operatorname{Im}(A) = \operatorname{Ker}(A) = \langle\langle \vec{x}_1, \vec{x}_2 \rangle\rangle$ a possibility?

(2) $\dim \operatorname{Ker}(A) = 1$ will eventually lead to

$$PAP^{-1} = \begin{bmatrix} 0 & 0 & 0 \\ \alpha_1 & 0 & 0 \\ \alpha_2 & 0 & 0 \end{bmatrix}.$$

3. Let $A \in M(3; \mathbb{R})$ be a nilpotent matrix of index 3, i.e. $A \neq O_{3\times3}$, $A^2 \neq O$ but $A^3 = O$.

(a) Show that all the eigenvalues of A are zeros. What is its canonical form?

(b) Show that $\operatorname{tr}(A^k) = 0$ for $k = 1, 2, 3, 4, \ldots$.

(c) Show that

(1) $A \neq O, A^2 \neq O$ but $A^3 = O$.

$\Leftrightarrow$ (2) $1 = \dim \operatorname{Ker}(A) < 2 = \dim \operatorname{Ker}(A^2) < 3 = \dim \operatorname{Ker}(A^3)$.

$\Leftrightarrow$ (3) $2 = r(A) > 1 = r(A^2) > 0 = r(A^3)$.

Hence, show that there exists a vector $\vec{x} \in \mathbb{R}^3$ so that $\vec{x}A \neq \vec{0}, \vec{x}A^2 \neq \vec{0}$ and $\vec{x}A^3 = \vec{0}$, and $\mathcal{B} = \{\vec{x}, \vec{x}A, \vec{x}A^2\}$ forms a basis for $\mathbb{R}^3$. What is $[A]_\mathcal{B}$? See Ex. <C> 3 of Sec. 3.7.7 and Ex. 5 of Sec. B.12 for general setting.

4. Does there exist a matrix $A \in M(3; \mathbb{R})$ such that $A^k \neq O$ for $k = 1, 2, 3$ but $A^4 = O$? If yes, give explicitly such an A; if not, give a precise reason (or proof).

5. (Refer to Ex. <B> 2 of Sec. 2.7.2, Ex. 6 of Sec. B.4 and Ex. 7 of Sec. B.7.) Let $A \in M(3; \mathbb{R})$ be idempotent, i.e. $A^2 = A$.

   (a) Show that each eigenvalue of A is either 0 or 1 .
   (b) Guess what are possible characteristic polynomials and canonical forms for such A. Could you justify true or false of your statements?
   (c) Show that

   $$A^2 = A.$$

   $\Leftrightarrow$ Each nonzero vector in Im (A) is an eigenvector of A associated to the eigenvalue 1.

   Try to use this to determine all such A up to similarity.

6. (Refer to Ex. <B> 3 of Sec. 2.7.2, Ex. 9 of Sec. B.4 and Ex. 8 of Sec. B.7.) Let $A \in M(3; \mathbb{R})$ be involutory, i.e. $A^2 = I_3$.

   (a) Show that each eigenvalue of A is either 1 or -1.
   (b) Consider the following cases:

       (1) $A - I_3 = O_{3 \times 3}$.
       (2) $A + I_3 = O_{3 \times 3}$.
       (3) $A - I_3 \neq O_{3 \times 3}, A + I_3 \neq O_{3 \times 3}$ but $(A - I_3)(A + I_3) = O$.

   Write out the canonical form of A for each case. Prove them!
   (c) Show that

   $$\text{Ker}(A - I_3) = \text{Im}(A + I_3),$$
   $$\text{Ker}(A + I_3) = \text{Im}(A - I_3), \quad \text{and}$$
   $$\mathbb{R}^3 = \text{Ker}(A - I_3) \oplus \text{Ker}(A + I_3).$$

   Thus, determine the canonical forms for all such A.

7. (Refer to Ex. <B> 4 of Sec. 2.7.2 and Ex. 9 of Sec. B.7.) Try to show that there does not exist any real 3×3 matrix A such that

   $$A^2 = -I_3.$$

8. (Refer to Ex. <B> 5 of Sec. 2.7.2.) Let $A_{3 \times 3}$ be a real matrix which has only one real eigenvalue λ. Show that there exists a basis

$B = \{\vec{x}_1, \vec{x}_2, \vec{x}_3\}$ such that

$$[A]_B = PAP^{-1} = \begin{bmatrix} \lambda & 0 & 0 \\ 0 & \lambda_1 & \lambda_2 \\ 0 & -\lambda_2 & \lambda_1 \end{bmatrix}, \quad \text{where } P = \begin{bmatrix} \vec{x}_1 \\ \vec{x}_2 \\ \vec{x}_3 \end{bmatrix} \text{ and } \lambda_2 \neq 0.$$

In particular, A has an invariant subspace $\langle\langle \vec{x}_2, \vec{x}_3 \rangle\rangle$ of dimension two.

<C> Abstraction and generalization.

Refer to Ex. <C> of Sec. 2.7.2.

3.7.3 *Matrix representations of a linear operator in various bases*

The process adopted, the definitions concerned and all the results obtained in Sec. 2.7.3 can be generalized verbatim to linear operators on $\mathbb{R}^3$. All we need to do is to change trivially from $\mathbb{R}^2$ to $\mathbb{R}^3$. For examples:

$\mathbb{R}^2$	$\mathbb{R}^3$
$f \colon \mathbb{R}^2 \to \mathbb{R}^2$ is a linear operator. $B = \{\vec{a}_1, \vec{a}_2\}$ is a basis for $\mathbb{R}^2$.	$f \colon \mathbb{R}^3 \to \mathbb{R}^3$ is a linear operator. $B = \{\vec{a}_1, \vec{a}_2, \vec{a}_3\}$ is a basis for $\mathbb{R}^3$.
$[f]_C^B = \begin{bmatrix} a_{11} & a_{12} \\ a_{21} & a_{22} \end{bmatrix}$, etc.	$[f]_C^B = \begin{bmatrix} a_{11} & a_{12} & a_{13} \\ a_{21} & a_{22} & a_{23} \\ a_{31} & a_{32} & a_{33} \end{bmatrix}$ $= \begin{bmatrix} [f(\vec{a}_1)]_C \\ [f(\vec{a}_2)]_C \\ [f(\vec{a}_3)]_C \end{bmatrix}$, etc.
$I_2 = \begin{bmatrix} 1 & 0 \\ 0 & 1 \end{bmatrix}$	$I_3 = \begin{bmatrix} 1 & 0 & 0 \\ 0 & 1 & 0 \\ 0 & 0 & 1 \end{bmatrix}$
$A = \sum_{i=1}^{2} \sum_{j=1}^{2} a_{ij} E_{ij}$	$A = \sum_{i=1}^{3} \sum_{j=1}^{3} a_{ij} E_{ij}$
$\text{Hom}(\mathbb{R}^2, \mathbb{R}^2)$ or $L(\mathbb{R}^2, \mathbb{R}^2)$	$\text{Hom}(\mathbb{R}^3, \mathbb{R}^3)$ or $L(\mathbb{R}^3, \mathbb{R}^3)$
$M(2; \mathbb{R})$	$M(3; \mathbb{R})$
$1_{\mathbb{R}^2}$	$1_{\mathbb{R}^3}$
$GL(2; \mathbb{R})$	$GL(3; \mathbb{R})$
$O_{2 \times 2}$	$O_{3 \times 3}$
$A_{2 \times 2}, B_{2 \times 2}$, etc.	$A_{3 \times 3}, B_{3 \times 3}$, etc.

We will feel no hesitation to use these converted results for $\mathbb{R}^3$. Recall that $f \in \text{Hom}(\mathbb{R}^3, \mathbb{R}^3)$ is called *diagonalizable* if there exists a basis $\mathcal{B}$ for $\mathbb{R}^3$ so that $[f]_\mathcal{B}$ is a diagonal matrix. In this case, $[f]_\mathcal{C}$ is similar to a diagonal matrix for any basis $\mathcal{C}$ for $\mathbb{R}^3$, and vice versa.

We list following results for reference.

The invariance of a square matrix or a linear operator under similarity

Let $A_{3 \times 3}$ (or $A_{n \times n}$) be a real matrix. Then the following are *invariants* under similarity:

1. The determinant $\det(PAP^{-1}) = \det(A)$.
2. The characteristic polynomial $\det(PAP^{-1} - tI_3) = \det(A - tI_3)$, and hence the set of eigenvalues.
3. The trace $\text{tr}(PAP^{-1}) = \text{tr}\,A$.
4. The rank $\text{r}(PAP^{-1}) = \text{r}(A)$.

Hence, for a linear operator $f \colon \mathbb{R}^3 \to \mathbb{R}^3$ (or $f \colon \mathbb{R}^n \to \mathbb{R}^n$) and any fixed basis $\mathcal{B}$ for $\mathbb{R}^3$, the following are *well-defined*:

1. $\det f = \det[f]_\mathcal{B}$.
2. $\det(f - t1_{\mathbb{R}^3}) = \det([f]_\mathcal{B} - tI_3)$.
3. $\text{tr}\,f = \text{tr}[f]_\mathcal{B}$.
4. $\text{r}(f) = \text{r}([f]_\mathcal{B})$. $\qquad\qquad\qquad\qquad\qquad\qquad\qquad$ (3.7.30)

In (2.7.43), the following change

$$\text{r}(A) + \text{r}(B) - 3 \leq \text{r}(AB) \leq \min\{\text{r}(A), \text{r}(B)\} \qquad (3.7.31)$$

is needed for $A_{3 \times 3}$ and $B_{3 \times 3}$. For general case, see Ex. <C> in Sec. 2.7.3.

For $A \in \text{M}(m, n; \mathbb{R})$ or $A \in \text{M}(m, n; \mathbb{F})$ where $m, n = 1, 2, 3, \ldots$, as before in (2.7.46) and (2.7.47), let the

row space:	$\text{Im}(A)$	or	$R(A) = \{\vec{x}A \mid \vec{x} \in \mathbb{R}^m\}$,
left kernel:	$\text{Ker}(A)$	or	$N(A) = \{\vec{x} \in \mathbb{R}^m \mid \vec{x}A = \vec{0}\}$,
column space:	$\text{Im}(A^*)$	or	$R(A^*) = \{\vec{x}A^* \mid \vec{x} \in \mathbb{R}^n\}$,
right kernel:	$\text{Ker}(A^*)$	or	$N(A^*) = \{\vec{x} \in \mathbb{R}^n \mid \vec{x}A^* = \vec{0}\}$.

$\qquad\qquad\qquad\qquad\qquad\qquad\qquad\qquad\qquad\qquad\qquad\qquad$ (3.7.32)

Then, we have (refer to Sec. 3.7.1 and, in particular, to Ex. <C> there)

The equalities of three ranks of a matrix $A_{m \times n}$**:**

(1) If $A = O_{m \times n}$, $r(A) = 0$.
(2) If $A \neq O_{m \times n}$, then

  the row rank of A (i.e. $\dim \text{Im}(A)$)
  $=$ the column rank of A (i.e. $\dim \text{Im}(A^*)$)
  $=$ the algebraic rank of A (how to define it?)
  $= r(A)$ (called the *rank* of A).
  Notice that $1 \leq r(A) \leq \min\{m, n\}$.

By the way, in case $P_{m \times m}$ and $Q_{n \times n}$ are invertible matrices, then

$$r(PA) = r(AQ) = r(A),$$

i.e. operations on A by invertible matrices preserve the rank of A.

$$(3.7.33)$$

We end up this subsection by the following

Example Let the linear operator f on $\mathbb{R}^3$ be defined as

$$f(\vec{x}) = \vec{x} A, \quad \text{where } A = \begin{bmatrix} -1 & 0 & 1 \\ 0 & 3 & -2 \\ -2 & 3 & 0 \end{bmatrix}$$

in $\mathcal{N} = \{\vec{e}_1, \vec{e}_2, \vec{e}_3\}$. Let $\mathcal{B} = \{\vec{a}_1, \vec{a}_2, \vec{a}_3\}$ where $\vec{a}_1 = (1, 1, 0)$, $\vec{a}_2 = (1, 0, 1)$, $\vec{a}_3 = (0, 1, 1)$ and $\mathcal{C} = \{\vec{b}_1, \vec{b}_2, \vec{b}_3\}$ where $\vec{b}_1 = (-1, -1, 1)$, $\vec{b}_2 = (-1, 1, -1)$, $\vec{b}_3 = (1, -1, -1)$. Do the same questions as in Ex. <A> 1 of Sec. 2.7.3.

Solution Denote by A_{i*} the ith row vector of A for $i = 1, 2, 3$ and A_{*j} the jth column vector for $j = 1, 2, 3$.

Since $A_{3*} = 2A_{1*} + A_{2*}$ or $A_{*3} = -A_{*1} - \frac{2}{3}A_{*2}$, then the rank $r(A) = 2$. By direct computation,

$$\det \begin{bmatrix} \vec{a}_1 \\ \vec{a}_2 \\ \vec{a}_3 \end{bmatrix} = \begin{vmatrix} 1 & 1 & 0 \\ 1 & 0 & 1 \\ 0 & 1 & 1 \end{vmatrix} = -2; \quad \det \begin{bmatrix} \vec{b}_1 \\ \vec{b}_2 \\ \vec{b}_3 \end{bmatrix} = \begin{vmatrix} -1 & -1 & 1 \\ -1 & 1 & -1 \\ 1 & -1 & -1 \end{vmatrix} = 4$$

therefore $\mathcal{B}$ and $\mathcal{C}$ are bases for $\mathbb{R}^3$.

 (a) By using (3.3.5),

$$A_\mathcal{B}^\mathcal{C} = \begin{bmatrix} \vec{b}_1 \\ \vec{b}_2 \\ \vec{b}_3 \end{bmatrix} \begin{bmatrix} \vec{a}_1 \\ \vec{a}_2 \\ \vec{a}_3 \end{bmatrix}^{-1} = \begin{bmatrix} -1 & -1 & 1 \\ -1 & 1 & -1 \\ 1 & -1 & -1 \end{bmatrix} \cdot \left(-\frac{1}{2}\right) \begin{bmatrix} -1 & -1 & 1 \\ -1 & 1 & -1 \\ 1 & -1 & -1 \end{bmatrix}$$

$$= -\frac{1}{2}\begin{bmatrix} 3 & -1 & -1 \\ -1 & 3 & -1 \\ -1 & -1 & 3 \end{bmatrix},$$

$$A_C^B = \begin{bmatrix} \vec{a_1} \\ \vec{a_2} \\ \vec{a_3} \end{bmatrix}\begin{bmatrix} \vec{b_1} \\ \vec{b_2} \\ \vec{b_3} \end{bmatrix}^{-1} = \begin{bmatrix} 1 & 1 & 0 \\ 1 & 0 & 1 \\ 0 & 1 & 1 \end{bmatrix}\cdot\left(-\frac{1}{2}\right)\begin{bmatrix} 1 & 1 & 0 \\ 1 & 0 & 1 \\ 0 & 1 & 1 \end{bmatrix} = -\frac{1}{2}\begin{bmatrix} 2 & 1 & 1 \\ 1 & 2 & 1 \\ 1 & 1 & 2 \end{bmatrix}.$$

It happens to be that

$$\begin{bmatrix} \vec{a_1} \\ \vec{a_2} \\ \vec{a_3} \end{bmatrix}^{-1} = -\frac{1}{2}\begin{bmatrix} \vec{b_1} \\ \vec{b_2} \\ \vec{b_3} \end{bmatrix} \quad \text{and hence,} \quad \begin{bmatrix} \vec{b_1} \\ \vec{b_2} \\ \vec{b_3} \end{bmatrix}^{-1} = -\frac{1}{2}\begin{bmatrix} \vec{a_1} \\ \vec{a_2} \\ \vec{a_3} \end{bmatrix}.$$

By above matrix relations or actual computation, we have

$$A_B^C A_C^B = I_3.$$

Similarly,

$$A_N^C = \begin{bmatrix} \vec{b_1} \\ \vec{b_2} \\ \vec{b_3} \end{bmatrix}\begin{bmatrix} \vec{e_1} \\ \vec{e_2} \\ \vec{e_3} \end{bmatrix}^{-1} = \begin{bmatrix} \vec{b_1} \\ \vec{b_2} \\ \vec{b_3} \end{bmatrix}I_3 = \begin{bmatrix} -1 & -1 & 1 \\ -1 & 1 & -1 \\ 1 & -1 & -1 \end{bmatrix},$$

$$A_B^N = \begin{bmatrix} \vec{e_1} \\ \vec{e_2} \\ \vec{e_3} \end{bmatrix}\begin{bmatrix} \vec{a_1} \\ \vec{a_2} \\ \vec{a_3} \end{bmatrix}^{-1} = I_3\begin{bmatrix} \vec{a_1} \\ \vec{a_2} \\ \vec{a_3} \end{bmatrix}^{-1} = -\frac{1}{2}\begin{bmatrix} -1 & -1 & 1 \\ -1 & 1 & -1 \\ 1 & -1 & -1 \end{bmatrix}$$

and hence, $A_B^C = A_N^C A_B^N$ holds.

(b) Note that $[f]_N = A$ as given above. Now, by (3.3.4) and (3.3.5),

$$[f]_B = [f]_B^B = \begin{bmatrix} [f(\vec{a_1})]_B \\ [f(\vec{a_2})]_B \\ [f(\vec{a_3})]_B \end{bmatrix} = \begin{bmatrix} [\vec{a_1}A]_B \\ [\vec{a_2}A]_B \\ [\vec{a_3}A]_B \end{bmatrix} = \begin{bmatrix} \vec{a_1}A \\ \vec{a_2}A \\ \vec{a_3}A \end{bmatrix}\begin{bmatrix} \vec{a_1} \\ \vec{a_2} \\ \vec{a_3} \end{bmatrix}^{-1}$$

$$= \begin{bmatrix} \vec{a_1} \\ \vec{a_2} \\ \vec{a_3} \end{bmatrix}A\begin{bmatrix} \vec{a_1} \\ \vec{a_2} \\ \vec{a_3} \end{bmatrix}^{-1} = A_N^B [f]_N^N A_B^N$$

$$= \begin{bmatrix} 1 & 1 & 0 \\ 1 & 0 & 1 \\ 0 & 1 & 1 \end{bmatrix}\begin{bmatrix} -1 & 0 & 1 \\ 0 & 3 & -2 \\ -2 & 3 & 0 \end{bmatrix}\cdot\left(-\frac{1}{2}\right)\begin{bmatrix} -1 & -1 & 1 \\ -1 & 1 & -1 \\ 1 & -1 & -1 \end{bmatrix}$$

$$= -\frac{1}{2}\begin{bmatrix} -3 & 5 & -3 \\ 1 & 5 & -7 \\ -6 & 10 & -6 \end{bmatrix}.$$

Similarly,

$$[f]_C = [f]_C^C = A_N^C [f]_N^N A_C^N$$

$$= \begin{bmatrix} -1 & -1 & 1 \\ -1 & 1 & -1 \\ 1 & -1 & -1 \end{bmatrix} \begin{bmatrix} -1 & 0 & 1 \\ 0 & 3 & -2 \\ -2 & 3 & 0 \end{bmatrix} \cdot -\frac{1}{2} \begin{bmatrix} 1 & 1 & 0 \\ 1 & 0 & 1 \\ 0 & 1 & 1 \end{bmatrix}$$

$$= -\frac{1}{2} \begin{bmatrix} -1 & 0 & 1 \\ 3 & 0 & -3 \\ -5 & 4 & -3 \end{bmatrix}.$$

Direct computation shows that both $[f]_B$ and $[f]_C$ are of rank 2. By what we obtained for $[f]_B$ and $[f]_C$,

$$[f]_N = [f]_N^N = A_B^N [f]_B A_N^B = A_C^N [f]_C A_N^C$$
$$\Rightarrow [f]_C = A_N^C A_B^N [f]_B A_N^B A_C^N = P[f]_B P^{-1},$$

where

$$P = A_N^C A_B^N = A_B^C$$

as shown above.

(c) By definition, (3.3.4) and (3.3.5),

$$[f]_B^N = \begin{bmatrix} [f(\vec{e}_1)]_B \\ [f(\vec{e}_2)]_B \\ [f(\vec{e}_3)]_B \end{bmatrix} = \begin{bmatrix} [A_{1*}]_B \\ [A_{2*}]_B \\ [A_{3*}]_B \end{bmatrix} = \begin{bmatrix} A_{1*} \\ A_{2*} \\ A_{3*} \end{bmatrix} \begin{bmatrix} \vec{a}_1 \\ \vec{a}_2 \\ \vec{a}_3 \end{bmatrix}^{-1} = A \begin{bmatrix} \vec{a}_1 \\ \vec{a}_2 \\ \vec{a}_3 \end{bmatrix}^{-1}$$

$$= \begin{bmatrix} -1 & 0 & 1 \\ 0 & 3 & -2 \\ -2 & 3 & 0 \end{bmatrix} \cdot \left(-\frac{1}{2}\right) \begin{bmatrix} -1 & -1 & 1 \\ -1 & 1 & -1 \\ 1 & -1 & -1 \end{bmatrix} = -\frac{1}{2} \begin{bmatrix} 2 & 0 & -2 \\ -5 & 5 & -1 \\ -1 & 5 & -5 \end{bmatrix};$$

$$[f]_C^N = A \begin{bmatrix} \vec{b}_1 \\ \vec{b}_2 \\ \vec{b}_3 \end{bmatrix}^{-1} = \begin{bmatrix} -1 & 0 & 1 \\ 0 & 3 & -2 \\ -2 & 3 & 0 \end{bmatrix} \cdot \left(-\frac{1}{2}\right) \begin{bmatrix} 1 & 1 & 0 \\ 1 & 0 & 1 \\ 0 & 1 & 1 \end{bmatrix}$$

$$= -\frac{1}{2} \begin{bmatrix} -1 & 0 & 1 \\ 3 & -2 & 1 \\ 1 & -2 & 3 \end{bmatrix}.$$

Note that both $[f]_B^N$ and $[f]_C^N$ have ranks equal to 2. Direct computation shows that

$$[f]_B^N = [f]_C^N A_B^C.$$

(d) Just like (c),

$$[f]_C^B = \begin{bmatrix} [f(\vec{a}_1)]_C \\ [f(\vec{a}_2)]_C \\ [f(\vec{a}_3)]_C \end{bmatrix} = \begin{bmatrix} [\vec{a}_1 A]_C \\ [\vec{a}_2 A]_C \\ [\vec{a}_3 A]_C \end{bmatrix} = \begin{bmatrix} \vec{a}_1 A \\ \vec{a}_2 A \\ \vec{a}_3 A \end{bmatrix} \begin{bmatrix} \vec{b}_1 \\ \vec{b}_2 \\ \vec{b}_3 \end{bmatrix}^{-1} = \begin{bmatrix} \vec{a}_1 \\ \vec{a}_2 \\ \vec{a}_3 \end{bmatrix} A \begin{bmatrix} \vec{b}_1 \\ \vec{b}_2 \\ \vec{b}_3 \end{bmatrix}^{-1}$$

$$= A_N^B [f]_N^N A_C^N$$

$$= \begin{bmatrix} 1 & 1 & 0 \\ 1 & 0 & 1 \\ 0 & 1 & 1 \end{bmatrix} \begin{bmatrix} -1 & 0 & 1 \\ 0 & 3 & -2 \\ -2 & 3 & 0 \end{bmatrix} \cdot -\frac{1}{2} \begin{bmatrix} 1 & 1 & 0 \\ 1 & 0 & 1 \\ 0 & 1 & 1 \end{bmatrix} = -\frac{1}{2} \begin{bmatrix} 2 & -2 & 2 \\ 0 & -2 & 4 \\ 4 & -4 & 4 \end{bmatrix}$$

$$= \begin{bmatrix} -1 & 1 & -1 \\ 0 & 1 & -2 \\ -2 & -2 & -2 \end{bmatrix};$$

$$[f]_B^C = A_N^C [f]_N^N A_B^N$$

$$= \begin{bmatrix} -1 & -1 & 1 \\ -1 & 1 & -1 \\ 1 & -1 & -1 \end{bmatrix} \begin{bmatrix} -1 & 0 & 1 \\ 0 & 3 & -2 \\ -2 & 3 & 0 \end{bmatrix} \cdot -\frac{1}{2} \begin{bmatrix} -1 & -1 & 1 \\ -1 & 1 & -1 \\ 1 & -1 & -1 \end{bmatrix}$$

$$= -\frac{1}{2} \begin{bmatrix} 2 & 0 & -2 \\ -6 & 0 & 6 \\ 6 & -12 & -6 \end{bmatrix} = \begin{bmatrix} -1 & 0 & 1 \\ 3 & 0 & -3 \\ -3 & 6 & 3 \end{bmatrix}.$$

Note that both $[f]_C^B$ and $[f]_B^C$ have ranks equal to 2. Also

$$[f]_N^N = A_B^N [f]_C^B A_N^C = A_C^N [f]_B^C A_N^B$$

$$\Rightarrow [f]_B^C = A_N^C A_B^N [f]_C^B A_N^C A_B^N = A_B^C [f]_C^B A_B^C.$$

(e) This is contained in (d).

(f) By direct computation or by using known results in (c) and (d), it follows that

$$[f]_C^B = A_N^B [f]_N^N A_C^N = A_N^B [f]_C^N$$

where, like (c), $[f]_C^N = [f]_N^N A_C^N$ holds. Similarly, $[f]_C^B = [f]_N^B A_C^N$. □

Exercises

<A>

1. In $\mathcal{N} = \{\vec{e}_1, \vec{e}_2, \vec{e}_3\}$, let

$$f(\vec{x}) = \vec{x} A, \quad \text{where } A = \begin{bmatrix} 1 & 1 & -1 \\ 1 & -1 & 0 \\ 2 & 0 & 1 \end{bmatrix}.$$

Let $\mathcal{B} = \{\vec{a}_1, \vec{a}_2, \vec{a}_3\}$ where $\vec{a}_1 = (-1, 2, 1)$, $\vec{a}_2 = (2, 4, 5)$, $\vec{a}_3 = (1, 2, -3)$ and $\mathcal{C} = \{\vec{b}_1, \vec{b}_2, \vec{b}_3\}$ where $\vec{b}_1 = (1, 0, 1)$, $\vec{b}_2 = (1, 2, 4)$, $\vec{b}_3 = (2, 2, -1)$. Model after the example and do the same questions as in Ex. $<A> 1$ of Sec. 2.7.3.

2. Find a linear operator f on $\mathbb{R}^3$ and a basis $\mathcal{B}$ for $\mathbb{R}^3$ so that

$$[f(\vec{x})]_{\mathcal{B}} = [\vec{x}]_{\mathcal{N}} \begin{bmatrix} -1 & 1 & 1 \\ 0 & 2 & 5 \\ -2 & 4 & -3 \end{bmatrix}, \quad \vec{x} \in \mathbb{R}^3.$$

Describe all possible such f and $\mathcal{B}$.

3. Let

$$A = \begin{bmatrix} 1 & 0 & 0 \\ 1 & 1 & 0 \\ 0 & 0 & 1 \end{bmatrix} \quad \text{and} \quad B = \begin{bmatrix} 1 & 0 & 0 \\ 1 & 1 & 0 \\ 0 & 1 & 1 \end{bmatrix}.$$

(a) Do there exist a linear operator f on $\mathbb{R}^3$ and two bases $\mathcal{B}$ and $\mathcal{C}$ for $\mathbb{R}^3$ so that $[f]_{\mathcal{B}} = A$ and $[f]_{\mathcal{C}} = B$? Give precise reason.

(b) Show that there exist invertible matrices P and Q so that

$$PAQ = B.$$

4. Find nonzero matrices $A_{3\times3}$ and $B_{3\times3}$ so that AB has each possible rank. Show that

$$r(AB) = 3 \Leftrightarrow r(A) = r(B) = 3.$$

5. Generalize Exs. $<A> 10$ through 21 of Sec. 2.7.3 to $\mathbb{R}^3$ or real 3×3 matrices and prove them. For example, in Ex. 20, the *orthogonal complement* of a subspace S in $\mathbb{R}^3$ is now defined as

$$S^{\perp} = \{\vec{x} \in \mathbb{R}^3 \mid \vec{x}\vec{y}^* = 0 \text{ for each } \vec{y} \in S\}.$$

For a nonzero 3×3 matrix A, then

(1) $\text{Im}(A)^{\perp} = \text{Ker}(A^*)$.
(2) $\text{Ker}(A^*)^{\perp} = \text{Im}(A)$, and $\mathbb{R}^3 = \text{Ker}(A^*) \oplus \text{Im}(A)$.
(3) $\text{Im}(A^*)^{\perp} = \text{Ker}(A)$.
(4) $\text{Ker}(A)^{\perp} = \text{Im}(A^*)$, and $\mathbb{R}^3 = \text{Ker}(A) \oplus \text{Im}(A^*)$.

For each of the following matrices A:

$$\begin{bmatrix} 2 & 5 & 3 \\ -6 & 1 & -2 \\ 2 & 21 & 10 \end{bmatrix}, \quad \begin{bmatrix} -1 & -2 & 3 \\ 3 & 6 & -9 \\ 2 & 4 & -6 \end{bmatrix} \quad \text{and} \quad \begin{bmatrix} 1 & 0 & -2 \\ 4 & 6 & -3 \\ -5 & -1 & 2 \end{bmatrix},$$

find a basis for each of the subspaces $\text{Im}(A)$, $\text{Ker}(A)$, $\text{Im}(A^*)$ and $\text{Ker}(A^*)$ and justify the above relations (1)–(4).

6. Extend (2.7.23), (2.7.28), (2.7.31), (2.7.32) and (2.7.35) to $\mathbb{R}^3$ and prove all of them.
7. Prove (3.7.30).
8. Prove (3.7.31).
9. Prove (3.7.33).

<B>

Read back Sec. 2.7.3 carefully and extend all possible definitions and results there to linear transformations from $\mathbb{R}^m$ to $\mathbb{R}^n$ where $m, n = 1, 2, 3$; in particular, the cases $m \neq n$. Please refer to Secs. B.4, B.6 and B.7 if necessary. We will be free to use them in what follows if needed.

1. Let $\mathcal{N} = \{1\}$ be the natural basis for $\mathbb{R}$ and $\mathcal{N}' = \{\vec{e}_1, \vec{e}_2, \vec{e}_3\}$ for $\mathbb{R}^3$.

   (a) A mapping $f \colon \mathbb{R} \to \mathbb{R}^3$ is a linear transformation if and only if there exists a vector $\vec{a} = (a_1, a_2, a_3)$ in $\mathbb{R}^3$ such that
   $$f(x) = x\vec{a} \quad \text{for } x \in \mathbb{R},$$
   namely, $[f(x)]_{\mathcal{N}'} = [x]_{\mathcal{N}}[f]_{\mathcal{N}'}^{\mathcal{N}}$, in which
   $$[f]_{\mathcal{N}'}^{\mathcal{N}} = [[f(1)]_{\mathcal{N}'}]_{1 \times 3}, \quad \text{where } [f(1)]_{\mathcal{N}'} = f(1) = \vec{a},$$
   is the *matrix representation* of f with respect to $\mathcal{N}$ and $\mathcal{N}'$.

   (b) Does there exist a basis $\mathcal{B}$ for $\mathbb{R}$ so that $[f]_{\mathcal{N}'}^{\mathcal{B}} = \vec{b} \in \mathbb{R}^3$ but $\vec{b}$ is linearly independent of $\vec{a}$? Notice that
   $$[f]_{\mathcal{N}'}^{\mathcal{B}} = [1_{\mathbb{R}}]_{\mathcal{N}}^{\mathcal{B}}[f]_{\mathcal{N}'}^{\mathcal{N}},$$
   where $1_{\mathbb{R}} \colon \mathbb{R} \to \mathbb{R}$ is the identity transformation and hence $[1_{\mathbb{R}}]_{\mathcal{N}}^{\mathcal{B}}$ is the transition matrix from the basis $\mathcal{B}$ to the basis $\mathcal{N}$.

   (c) Do there exist a basis $\mathcal{B}$ for $\mathbb{R}$ and a basis $\mathcal{B}'$ for $\mathbb{R}^3$ so that $[f]_{\mathcal{B}'}^{\mathcal{B}} = \vec{b}$ where $\vec{b} \in \mathbb{R}^3$ is a preassigned vector? If yes, prove it; if no, under what conditions that it will become true. Notice that
   $$[f]_{\mathcal{B}'}^{\mathcal{B}} = [1_{\mathbb{R}}]_{\mathcal{N}}^{\mathcal{B}}[f]_{\mathcal{N}'}^{\mathcal{N}}[1_{\mathbb{R}^3}]_{\mathcal{B}'}^{\mathcal{N}'}.$$

   (d) For any fixed straight line $\vec{x} = \vec{x}_0 + t\vec{v}$ for $t \in \mathbb{R}$, in $\mathbb{R}^3$, show that there exists an *affine* transformation mapping $\mathbb{R}$ one-to-one and onto that line. How many such transformations are there?

2. Let $\mathcal{N} = \{\vec{e}_1, \vec{e}_2\}$ be the natural basis for $\mathbb{R}^2$ and $\mathcal{N}' = \{\vec{e}_1', \vec{e}_2', \vec{e}_3'\}$ the natural basis for $\mathbb{R}^3$.

(a) A mapping $f \colon \mathbb{R}^2 \to \mathbb{R}^3$ is a linear transformation if and only if there exists a real 2×3 matrix $A = [a_{ij}]$ so that

$$f(\vec{x}) = \vec{x}A, \quad \vec{x} \in \mathbb{R}^2,$$

i.e.

$$[f(\vec{x})]_{\mathcal{N}'} = [\vec{x}]_{\mathcal{N}}A, \quad \text{where } A = [f]_{\mathcal{N}'}^{\mathcal{N}} = \begin{bmatrix} [f(\vec{e}_1)]_{\mathcal{N}'} \\ [f(\vec{e}_2)]_{\mathcal{N}'} \end{bmatrix}.$$

(b) Give a fixed real matrix $B_{2\times3}$. Do there exist a basis $\mathcal{B}$ for $\mathbb{R}^2$ and a basis $\mathcal{B}'$ for $\mathbb{R}^3$ so that $[f]_{\mathcal{B}'}^{\mathcal{B}} = B$ holds? Notice that $[f]_{\mathcal{B}'}^{\mathcal{B}} = [1_{\mathbb{R}^2}]_{\mathcal{N}}^{\mathcal{B}}[f]_{\mathcal{N}'}^{\mathcal{N}}[1_{\mathbb{R}^3}]_{\mathcal{B}'}^{\mathcal{N}'}$.

(c) Show that f can be one-to-one but never onto $\mathbb{R}^3$.

(d) Show that there are infinitely many *affine* transformations

$$T(\vec{x}) = \vec{y}_0 + \vec{x}A, \quad \vec{y}_0 \in \mathbb{R}^3 \text{ fixed and } \vec{x} \in \mathbb{R}^2,$$

where $A_{2\times3}$ is of rank 2, mapping $\mathbb{R}^2$ one-to-one and onto any preassigned two-dimensional plane in $\mathbb{R}^3$.

(e) Show that any affine transformation from $\mathbb{R}^2$ into $\mathbb{R}^3$ preserves relative positions of two straight lines in $\mathbb{R}^2$ (see (2.5.9)).

(f) Let $T(\vec{x}) = \vec{y}_0 + \vec{x}A$ be an affine transformation from $\mathbb{R}^2$ into $\mathbb{R}^3$ (see (d)).

    (1) Show that the image of the unit square with vertices $\vec{0}$, $\vec{e}_1$, $\vec{e}_1 + \vec{e}_2$ and $\vec{e}_2$ under T is a parallelogram (see Fig. 3.39). Compute the planar area of this parallelogram.

    (2) The image of a triangle in $\mathbb{R}^2$ under T is a triangle $\Delta T(\vec{a}_1)T(\vec{a}_2)T(\vec{a}_3)$ (see Fig. 3.39). Compute

$$\frac{\text{the area of } \Delta T(\vec{a}_1)T(\vec{a}_2)T(\vec{a}_3)}{\text{the area of } \Delta \vec{a}_1\vec{a}_2\vec{a}_3}.$$

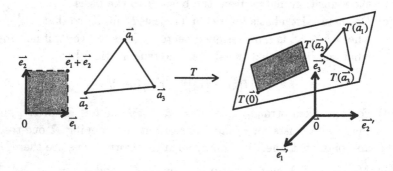

Fig. 3.39

(3) What is the image of the unit circle $x_1^2 + x_2^2 = 1$ in $\mathbb{R}^2$ under T? How about its area? Refer to Fig. 2.66.

3. The system of three non-homogeneous linear equations in two unknowns x_1 and x_2

$$\sum_{i=1}^{2} a_{ij}x_i = b_j, \quad j = 1, 2, 3$$

can be rewritten as

$$\vec{x}A = \vec{b}, \quad \text{where } A = \begin{bmatrix} a_{11} & a_{12} & a_{13} \\ a_{21} & a_{22} & a_{23} \end{bmatrix},$$
$$\vec{b} = (b_1, b_2, b_3) \text{ and } \vec{x} = (x_1, x_2) \in \mathbb{R}^2.$$

(a) Prove that the following are equivalent.

(1) $\vec{x}A = \vec{b}$ has a solution.

(2) (linearly algebraic) The vector $\vec{b}$ lies on the range space of A, i.e.

$$\vec{b} \in \text{Im}(A).$$

(3) (geometric) The three lines

$$l_1: a_{11}x_1 + a_{21}x_2 = b_1,$$
$$l_2: a_{12}x_1 + a_{22}x_2 = b_2,$$
$$l_3: a_{13}x_1 + a_{23}x_2 = b_3$$

in $\mathbb{R}^2$ are either all coincident, two coincident and one intersecting or three intersecting at a single point but never any two of them intersecting at different points. See Fig. 3.40 (refer to Fig. 2.24).

(4) (algebraic)

$$\text{Fig. 3.39(a)} \Leftrightarrow \frac{a_{11}}{a_{1i}} = \frac{a_{21}}{a_{2i}} = \frac{b_1}{b_i} \quad \text{for } i = 2, 3$$

(infinitely many solutions).

$$\text{Fig. 3.39(b)} \Leftrightarrow \frac{a_{11}}{a_{12}} = \frac{a_{21}}{a_{22}} = \frac{b_1}{b_2} \text{ but } \frac{a_{11}}{a_{13}} \neq \frac{a_{21}}{a_{23}}$$

(a unique solution).

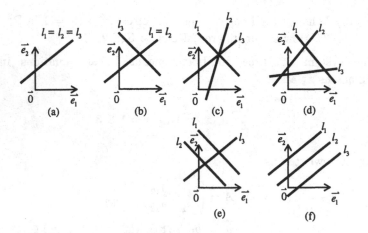

Fig. 3.40

Fig. 3.39(c) $\Leftrightarrow \dfrac{a_{11}}{a_{12}} \neq \dfrac{a_{21}}{a_{22}}, \dfrac{a_{12}}{a_{13}} \neq \dfrac{a_{22}}{a_{23}}, \dfrac{a_{13}}{a_{11}} \neq \dfrac{a_{23}}{a_{21}}$ but

$$\Delta = \begin{vmatrix} a_{11} & a_{21} & b_1 \\ a_{12} & a_{22} & b_2 \\ a_{13} & a_{23} & b_3 \end{vmatrix} = 0 \quad \text{(a unique solution)}.$$

Fig. 3.39(d) $\Leftrightarrow \dfrac{a_{11}}{a_{12}} \neq \dfrac{a_{21}}{a_{22}}, \dfrac{a_{12}}{a_{13}} \neq \dfrac{a_{22}}{a_{23}}, \dfrac{a_{13}}{a_{11}} \neq \dfrac{a_{23}}{a_{21}}$ but

$\Delta \neq 0$ (no solution).

Fig. 3.39(e) $\Leftrightarrow \dfrac{a_{11}}{a_{12}} = \dfrac{a_{21}}{a_{22}} \neq \dfrac{b_1}{b_2}$ but $\dfrac{a_{13}}{a_{11}} \neq \dfrac{a_{23}}{a_{21}}$ and

$\Delta \neq 0$ (no solution).

Fig. 3.39(f) $\Leftrightarrow \dfrac{a_{11}}{a_{12}} = \dfrac{a_{21}}{a_{22}} \neq \dfrac{b_1}{b_2}, \dfrac{a_{12}}{a_{13}} = \dfrac{a_{22}}{a_{23}} \neq \dfrac{b_2}{b_3},$

$\dfrac{a_{13}}{a_{11}} = \dfrac{a_{23}}{a_{21}} \neq \dfrac{b_3}{b_1}$ but $\Delta \neq 0$ (no solution).

What happens if $l_1 = l_2 \parallel l_3$?

(5) (linearly algebraic) The rank of the *augmented matrix* $\left[\dfrac{A}{b}\right]_{3\times 3}$ is equal to that of the *coefficient matrix* A, i.e.

$$\mathrm{r}\left(\left[\frac{A}{b}\right]\right) = \mathrm{r}(A).$$

(b) In case $\vec{x}A = \vec{b}$ has a solution where $A \neq O_{2 \times 3}$. Then

   (1) $\vec{x}A = \vec{b}$ has a unique solution.

$\Leftrightarrow$ (2) $\vec{x}A = \vec{0}$ has only one solution $\vec{x} = \vec{0}$, i.e. $\text{Ker}(A) = \{\vec{0}\}$.

$\Leftrightarrow$ (3) The linear transformation A: $\mathbb{R}^2 \to \mathbb{R}^3$ is one-to-one, i.e. $r(A) = 2$.

$\Leftrightarrow$ (4) AA^*, as a square matrix of order 2, is invertible. Thus, the unique solution (refer to Ex. <A> 5(1)) is

$$\vec{b}A^*(AA^*)^{-1}.$$

On the other hand,

   (1) $\vec{x}A = \vec{b}$ has infinitely many solutions.

$\Leftrightarrow$ (2) $\vec{x}A = \vec{0}$ has infinitely many solutions, i.e. the *solution space* $\text{Ker}(A)$ has $\dim \text{Ker}(A) = 1$.

$\Leftrightarrow$ (3) The linear transformation A: $\mathbb{R}^2 \to \mathbb{R}^3$ is not one-to-one, i.e. the rank $r(A) = 1$.

$\Leftrightarrow$ (4) AA^* is not invertible and $r(AA^*) = 1$.

If $\vec{x}_0$ is a solution of $\vec{x}A = \vec{b}$, then the *solution affine subspace*

$$\vec{x}_0 + \text{Ker}(A)$$

is the solution set of $\vec{x}A = \vec{b}$ (see Fig. 3.41). Among so many solutions, there exists a unique solution $\vec{v}$ whose distance to $\vec{0}$ is the smallest one, i.e.

$$|\vec{v}| = \min_{\vec{x}A=\vec{b}} |\vec{x}|.$$

The remaining question is that how to find $\vec{v}$, more explicitly, how to determine $\vec{v}$ via A. If $r(A) = 2$, then by the former part of (b), it is known that $\vec{v} = \vec{b}A^*(AA^*)^{-1}$.

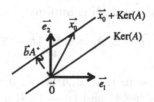

Fig. 3.41

(c) Suppose $r(A) = 1$. Then $r(AA^*) = 1$ and AA^* is then not invertible. We may suppose $a_{11} \neq 0$. Rewrite A as

$$A = \begin{bmatrix} a_{11} & A_{12} \\ a_{21} & A_{22} \end{bmatrix} = BC,$$

where

$$B = \begin{bmatrix} a_{11} \\ a_{21} \end{bmatrix}, C = \begin{bmatrix} 1 & a_{11}^{-1} A_{12} \end{bmatrix} \quad \text{and}$$

$$A_{12} = \begin{bmatrix} a_{12} & a_{13} \end{bmatrix}, \quad A_{22} = \begin{bmatrix} a_{22} & a_{23} \end{bmatrix}.$$

Therefore $r(B) = r(C) = 1$. Now

$$\vec{x} A = \vec{b}$$
$$\Rightarrow \vec{x} AA^* = \vec{b} A^*$$
$$\Rightarrow \vec{x}(BC)(BC)^* = \vec{x} B(CC^*)B^* = \vec{b} C^* B^*$$
$$\Rightarrow \vec{x} B(CC^*)(B^*B) = \vec{b} C^*(B^*B)$$
$$\Rightarrow (\text{since } B^*B \text{ and } CC^* \text{ are invertible})$$
$$\vec{x} B = \vec{b} C^*(CC^*)^{-1}$$
$$\Rightarrow (\text{since } r(B) = 1) \text{ The } required \text{ solution is}$$
$$\vec{v} = \vec{b} C^*(CC^*)^{-1}(B^*B)^{-1}B^* \quad (\text{Why?}).$$

(d) Let $A = \begin{bmatrix} 1 & 1 & 1 \\ -1 & -1 & -1 \end{bmatrix}$. Show that $r(A) = 1$ and

$$A^+ = C^*(CC^*)^{-1}(B^*B)^{-1}B^* = \frac{1}{6} \begin{bmatrix} 1 & -1 \\ 1 & -1 \\ 1 & -1 \end{bmatrix}.$$

4. (continued from Ex. 3) Given $A = [a_{ij}]_{2 \times 3}$ and $\vec{b} = (b_1, b_2, b_3) \in \mathbb{R}^3$, suppose $\vec{x} A = \vec{b}$ doesn't have any solution $\vec{x} \in \mathbb{R}^2$. The problem is to find $\vec{x}_0 \in \mathbb{R}^2$ so that $|\vec{b} - \vec{x}_0 A|$ is minimal, i.e.

$$|\vec{b} - \vec{x}_0 A| = \min_{\vec{x} \in \mathbb{R}^2} |\vec{b} - \vec{x} A|.$$

Geometrically, this means that $\vec{x}_0 A$ is the *orthogonal projection* of $\vec{b}$ onto the range space $\text{Im}(A)$ and $|\vec{b} - \vec{x}_0 A|$ is the *distance* from $\vec{b}$ to $\text{Im}(A)$. For details, see Chaps. 4 and 5. See also Fig. 3.42.

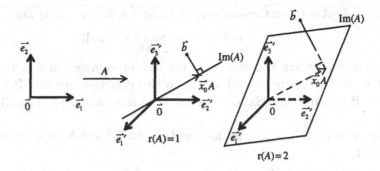

Fig. 3.42

Now,

$$(\vec{b} - \vec{x}_0 A) \perp \vec{x} A \quad \text{for all } \vec{x} \in \mathbb{R}^2$$
$$\Leftrightarrow (\vec{b} - \vec{x}_0 A)(\vec{x} A)^* = (\vec{b} - \vec{x}_0 A) A^* \vec{x}^* = 0 \quad \text{for all } \vec{x} \in \mathbb{R}^2$$
$$\Leftrightarrow (\vec{b} - \vec{x}_0 A) A^* = \vec{b} A^* - \vec{x}_0 A A^* = \vec{0}, \text{ i.e.}$$
$$\vec{x}_0 A A^* = \vec{b} A^*.$$

(a) In case $r(A) = 2$, then $r(AA^*) = 2$ and AA^* is invertible. Therefore,

$$\vec{x}_0 = \vec{b} A^* (AA^*)^{-1}$$

$\Rightarrow$ The projection vector of $\vec{b}$ on $\text{Im}(A)$ is $\vec{b} A^* (AA^*)^{-1} A$.

(b) In case $r(A) = 1$, then $r(AA^*) = 1$ and AA^* is not invertible. Decompose A as in Ex. 3(c) and show that

$$\vec{x}_0 = \vec{b} C^* (CC^*)^{-1} (B^* B)^{-1} B^*$$

$\Rightarrow$ The projection vector of $\vec{b}$ on $\text{Im}(A)$ is

$$\vec{b} C^* (CC^*)^{-1} (B^* B)^{-1} B^* A = \vec{b} C^* (CC^*)^{-1} C.$$

(*Note* Combining Ex. 3 and Ex. 4 together, for $A_{2 \times 3} \neq O_{2 \times 3}$ the 3×2 matrix

$$A^+ = \begin{cases} A^* (AA^*)^{-1}, & \text{if } r(A) = 2 \\ C^* (CC^*)^{-1} (B^* B)^{-1} B^*, & \text{if } r(A) = 1 \text{ and} \\ & \quad A = BC \text{ as in Ex. 3(c)} \end{cases}$$

is called the *generalized* or *pseudoinverse* of A. This A^+ has the following properties:

(1) $|\vec{b} A^+| = \min_{\vec{x} A = \vec{b}} |\vec{x}|$ if $\vec{x} A = \vec{b}$ has a solution.

(2) $\vec{b}A^+A$ is the *orthogonal projection* of $\vec{b} \in \mathbb{R}^3$ on $\text{Im}(A)$ and
$$|\vec{b} - \vec{b}A^+A| = \min_{\vec{x} \in \mathbb{R}^2} |\vec{b} - \vec{x}A|.$$

These results are also still valid for real or complex $m \times n$ matrix A with $m \le n$. For general setting in this direction, see Sec. B.8 and Fig. B.9, also refer to Ex. <B> of Sec. 2.7.5 and Example 4 in Sec. 3.7.5, Secs. 4.5, 5.5.)

5. Let $\mathcal{N} = \{\vec{e}_1, \vec{e}_2, \vec{e}_3\}$ be the natural basis for $\mathbb{R}^3$ and $\mathcal{N}' = \{1\}$ the one for $\mathbb{R}$.

   (a) A mapping $f: \mathbb{R}^3 \to \mathbb{R}$ is a linear transformation, specifically called a *linear functional*, if and only if there exist scalars a_1, a_2, a_3 so that
   $$f(\vec{x}) = a_1 x_1 + a_2 x_2 + a_3 x_3 = \vec{x} \begin{bmatrix} a_1 \\ a_2 \\ a_3 \end{bmatrix}, \quad \vec{x} \in \mathbb{R}^3.$$

   i.e.
   $$[f(\vec{x})]_{\mathcal{N}'} = [\vec{x}]_{\mathcal{N}}[f]_{\mathcal{N}'}^{\mathcal{N}}, \quad \text{where } [f]_{\mathcal{N}'}^{\mathcal{N}} = \begin{bmatrix} a_1 \\ a_2 \\ a_3 \end{bmatrix}.$$

   (b) Give a matrix $B = [b_i]_{3 \times 1}$. Find conditions so that there exist a basis $\mathcal{B}$ for $\mathbb{R}^3$ and a basis $\mathcal{B}'$ for $\mathbb{R}$ so that $[f]_{\mathcal{B}'}^{\mathcal{B}} = B$.

   (c) Show that f can be onto but never one-to-one.

   (d) Suppose $f: \mathbb{R}^3 \to \mathbb{R}$ is a linear functional such that $\text{Im}(f) = \mathbb{R}$ holds. Try to define the quotient space
   $$\mathbb{R}^3/\text{Ker}(f).$$
   Show that it is linear isomorphic to $\mathbb{R}$ (refer to Fig. 3.25).

   (e) Let $f_j: \mathbb{R}^3 \to \mathbb{R}$ be the linear functional satisfying
   $$f_j(\vec{e}_i) = \delta_{ij}, \quad 1 \le i, j \le 3.$$
   Then any linear functional $f: \mathbb{R}^3 \to \mathbb{R}$ can be uniquely expressed as
   $$f = f(\vec{e}_1)f_1 + f(\vec{e}_2)f_2 + f(\vec{e}_3)f_3.$$

   (f) The set of all linear functionals from $\mathbb{R}^3$ to $\mathbb{R}$, namely
   $$(\mathbb{R}^3)^* = \text{Hom}(\mathbb{R}^3, \mathbb{R})$$
   (see Ex. 19 of Sec. B.7) is a three-dimensional real vector space, called the *(first) dual space* of $\mathbb{R}^3$, with $\{f_1, f_2, f_3\}$ as a basis which is called the *dual basis* of $\mathcal{N}$ in $(\mathbb{R}^3)^*$. How to define the dual basis $\mathcal{B}^*$ for $(\mathbb{R}^3)^*$ of a basis $\mathcal{B}$ for $\mathbb{R}^3$?

(g) Let S be a subspace (or a nonempty subset) of $\mathbb{R}^3$. The set

$$S^0 = \{f \in (\mathbb{R}^3)^* \mid f(\vec{x}) = 0 \text{ for all } \vec{x} \in S\}$$

is a subspace of $(\mathbb{R}^3)^*$ and is called the *annihilator* of S in $(R^3)^*$. Show that

(1) $\dim S + \dim S^0 = \dim \mathbb{R}^3 = 3$.
(2) $S_1 \subseteq S_2 \leftrightarrow S_1^0 \supseteq S_2^0$.
(3) $(S_1 + S_2)^0 = S_1^0 \cap S_2^0$.
(4) $(S_1 \cap S_2)^0 = S_1^0 + S_2^0$.

Try to explain (1)–(4) geometrically. Therefore, $S \to S^0$ sets up a one-to-one correspondence between the family of subspaces of $\mathbb{R}^3$ and the family of subspaces of $(\mathbb{R}^3)^*$ but reverses the inclusion relation. For example,

Two planes intersecting along a line through $\vec{0}$ in $\mathbb{R}^3$.

$\leftrightarrow$ Two lines generating a plane through $\vec{0}$ in $(\mathbb{R}^3)^*$.

Thus, $S \to S^0$ reflects geometrically the *dual* properties between $\mathbb{R}^3$ and $(\mathbb{R}^3)^*$ and the latter is then called the *dual space* of the former.

(h) The dual space of $(\mathbb{R}^3)^*$

$$(\mathbb{R}^3)^{**} = \left((\mathbb{R}^3)^*\right)^* = \operatorname{Hom}\left((\mathbb{R}^3)^*, \mathbb{R}\right)$$

is called the *second dual space* of $\mathbb{R}^3$. For each $\vec{x} \in \mathbb{R}^3$, define a mapping $\vec{x}^{**} \colon (\mathbb{R}^3)^* \to \mathbb{R}$ by

$$\vec{x}^{**}(f) = f(\vec{x}), \quad f \in (\mathbb{R}^3)^*.$$

Show that

(1) $\vec{x}^{**} \in (\mathbb{R}^3)^{**}$.
(2) $\vec{x} \to \vec{x}^{**}$ sets up a linear isomorphism from $\mathbb{R}^3$ onto $(\mathbb{R}^3)^{**}$ in a natural way, i.e. independent of choices of bases for $\mathbb{R}^3$ and $(\mathbb{R}^3)^{**}$.

Therefore, each basis for $(\mathbb{R}^3)^*$ is the dual basis of some basis for $\mathbb{R}^3$. Occasionally, $\vec{x}^{**}(f) = f(\vec{x})$ is rewritten as

$$\langle \vec{x}, f \rangle = \langle f, \vec{x} \rangle, \quad \vec{x} \in \mathbb{R}^3 \text{ and } f \in (\mathbb{R}^3)^*.$$

which indicates implicitly the duality between $\mathbb{R}^3$ and $(\mathbb{R}^3)^*$. Also, show that for a nonempty set S in $\mathbb{R}^3$,

$$(S^0)^0 = S^{00} = \langle\langle S \rangle\rangle, \quad \text{the subspace generated by } S.$$

(i) Let $\varphi \colon \mathbb{R}^3 \to \mathbb{R}^3$ be a linear operator. Define a mapping $\varphi^* \colon (\mathbb{R}^3)^* \to (\mathbb{R}^3)^*$ by

$$\varphi^*(f)(\vec{x}) = f(\varphi(\vec{x})), \quad f \in (\mathbb{R}^3)^* \text{ and } \vec{x} \in \mathbb{R}^3; \quad \text{or}$$
$$\langle \vec{x}, \varphi^*(f) \rangle = \langle \varphi(\vec{x}), f \rangle.$$

Then, such a φ^* is linear and is unique and is called the *adjoint* or the *dual* of φ. Let $\mathcal{B}$ and $\mathcal{C}$ be the bases for $\mathbb{R}^3$ and $\mathcal{B}^*$ and $\mathcal{C}^*$ the corresponding dual bases for $(\mathbb{R}^3)^*$. Then

$$\left([\varphi]_{\mathcal{C}}^{\mathcal{B}} \right)^* = [\varphi^*]_{\mathcal{B}^*}^{\mathcal{C}^*},$$

namely, $[\varphi^*]_{\mathcal{B}^*}^{\mathcal{C}^*}$ is the transpose of $[\varphi]_{\mathcal{C}}^{\mathcal{B}}$.

(j) Discuss $\mathbb{R}^*$ and $(\mathbb{R}^2)^*$. How to define $\varphi^* \colon (\mathbb{R}^n)^* \to (\mathbb{R}^m)^*$ of a linear transformation $\varphi \colon \mathbb{R}^m \to \mathbb{R}^n$, where $m, n = 1, 2, 3$?

(*Note*   For a general setting, please refer to Ex. 19 through Ex. 24 of Sec. B.7.)

6. Let $\mathcal{N} = \{\vec{e}_1, \vec{e}_2, \vec{e}_3\}$ be the natural basis for $\mathbb{R}^3$ and $\mathcal{N}' = \{\vec{e}_1', \vec{e}_2'\}$ the one for $\mathbb{R}^2$.

(a) A mapping $f \colon \mathbb{R}^3 \to \mathbb{R}^2$ is a linear transformation if and only if there exists a matrix $A = [a_{ij}]_{3 \times 2}$ such that

$$f(\vec{x}) = \vec{x} A, \quad \vec{x} \in \mathbb{R}^3,$$

namely,

$$[f(\vec{x})]_{\mathcal{N}'} = [\vec{x}]_{\mathcal{N}} [f]_{\mathcal{N}'}^{\mathcal{N}}, \quad \text{where } [f]_{\mathcal{N}'}^{\mathcal{N}} = A_{3 \times 2}.$$

(b) For a matrix $B = [b_{ij}]_{3 \times 2}$, find conditions so that there exist a basis $\mathcal{B}$ for $\mathbb{R}^3$ and a basis $\mathcal{B}'$ for $\mathbb{R}^2$ so that $[f]_{\mathcal{B}'}^{\mathcal{B}} = B$.

(c) Show that f can be onto but never one-to-one.

(d) Show that $\mathbb{R}^3 / \mathrm{Ker}(f)$ is isomorphic to $\mathrm{Im}(f)$.

(e) Let $f \colon \mathbb{R}^3 \to \mathbb{R}^2$ be a linear transformation. Then

(1) f is onto.

$\Leftrightarrow$ (2) There exists a linear transformation $g \colon \mathbb{R}^2 \to \mathbb{R}^3$ so that $f \circ g = 1_{\mathbb{R}^2}$, the identity operator on $\mathbb{R}^2$.

$\Leftrightarrow$ (3) There exist a basis $\mathcal{B}$ for $\mathbb{R}^3$ and a basis $\mathcal{B}'$ for $\mathbb{R}^2$ so that $[f]_{\mathcal{B}'}^{\mathcal{B}}$ is left invertible, i.e. there is a matrix $B_{2 \times 3}$ so that $B[f]_{\mathcal{B}'}^{\mathcal{B}} = I_2$.

In this case, f is called *right invertible*. Use

$$f(\vec{x}) = \vec{x}A, \quad \text{where } A = \begin{bmatrix} -5 & 3 \\ 4 & -1 \\ 2 & 6 \end{bmatrix}$$

to justify (1)–(3).

(f) Let $g \colon \mathbb{R}^2 \to \mathbb{R}^3$ be a linear transformation. Then

    (1) g is one-to-one.

$\Leftrightarrow$ (2) There exists a linear transformation $f \colon \mathbb{R}^3 \to \mathbb{R}^2$ so that $f \circ g = 1_{\mathbb{R}^2}$, the identity operator on $\mathbb{R}^2$.

$\Leftrightarrow$ (3) There exist a basis $\mathcal{B}'$ for $\mathbb{R}^2$ and a basis $\mathcal{B}$ for $\mathbb{R}^3$ so that $[g]_{\mathcal{B}}^{\mathcal{B}'}$ is right invertible, i.e. there is a matrix $B_{3 \times 2}$ so that $[g]_{\mathcal{B}'}^{\mathcal{B}} B = I_2$.

In this case, g is called *left invertible*. Use

$$g(\vec{x}) = \vec{x} \begin{bmatrix} -5 & 4 & 2 \\ 3 & -1 & 6 \end{bmatrix}$$

to justify (1)–(3).

(g) Prove counterparts of (e) and (f) for linear transformations $f \colon \mathbb{R}^3 \to \mathbb{R}$ and $g \colon \mathbb{R} \to \mathbb{R}^3$.

(*Note* For (e), (f) and (g), general cases can be found in Ex. 5 of Sec. B.7 and Ex. 3 of Sec. B.8. Also, refer to Ex. <A> 7 of Sec. 2.7.4 and Ex. <B> 5 (d), (e) of Sec. 3.3.)

(h) Let $f \colon \mathbb{R}^3 \to \mathbb{R}^2$ be defined as

$$f(\vec{x}) = \vec{x} \begin{bmatrix} 1 & 0 \\ 0 & -1 \\ -1 & 1 \end{bmatrix}, \quad \vec{x} \in \mathbb{R}^3.$$

Find the image of the unit sphere $x_1^2 + x_2^2 + x_3^2 = 1$ under f. How about the image of the unit closed ball $x_1^2 + x_2^2 + x_3^2 \le 1$? Watch the following facts: Let $\vec{y} = (y_1, y_2) = f(\vec{x})$.

(1) $\mathrm{Ker}(f) = \langle\langle (1,1,1) \rangle\rangle$. Hence, f is one-to-one on the plane $x_1 + x_2 + x_3 = 0$.

(2) $y_1 = x_1 - x_3$, $y_2 = -x_2 + x_3 \Rightarrow x_1 = y_1 + x_3$ and $x_2 = -y_2 + x_3$. Substitute these two equations into $x_1 + x_2 + x_3 = 0$ and obtain $x_3 = \frac{1}{3}(y_2 - y_1)$ and hence $x_1 = \frac{1}{3}(2y_1 + y_2)$, $x_2 = -\frac{1}{3}(y_1 + 2y_2)$.

(3) Now, consider the image of the circle $x_1^2 + x_2^2 + x_3^2 = 1$, $x_1 + x_2 + x_3 = 0$ under f.

(i) Let $f(\vec{x}) = \vec{x}A$ be a linear transformation from $\mathbb{R}^3$ onto $\mathbb{R}^2$. Do the same question as in (h) by trying the following method:

    (1) Consider $\mathbb{R}^3 = \operatorname{Ker}(f) \oplus \operatorname{Ker}(f)^{\perp}$ so that $f|_{\operatorname{Ker}(f)^{\perp}}$ is one-to-one.

    (2) Use (e) to find a matrix $B_{2\times 3}$ so that $BA = I_2$. Then $\vec{y} = \vec{x}A$ implies that $\vec{y}B \in \vec{x} + \operatorname{Ker}(f)$ for $\vec{y} \in \mathbb{R}^2$ and vice versa. Indeed, $B = (A^*A)^{-1}A^*$.

See Fig. 3.43.

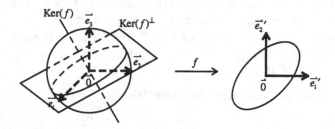

Fig. 3.43

7. The system of two non-homogeneous linear equations in three unknowns x_1, x_2 and x_3

$$\sum_{i=1}^{3} a_{ij}x_i = b_j, \quad j = 1, 2$$

can be rewritten as

$$\vec{x}A = \vec{b}, \quad \text{where } A = \begin{bmatrix} a_{11} & a_{12} \\ a_{21} & a_{22} \\ a_{31} & a_{32} \end{bmatrix} \neq O_{3\times 2}, \ \vec{b} = (b_1, b_2) \in \mathbb{R}^2 \text{ and }$$

$$\vec{x} = (x_1, x_2, x_3) \in \mathbb{R}^3.$$

(a) Prove that the following are equivalent.

    (1) $\vec{x}A = \vec{b}$ has a solution.

    (2) (linearly algebraic) The vector $\vec{b}$ lies on the range space of A, i.e. $\vec{b} \in \operatorname{Im}(A)$.

    (3) (geometric) The two planes

$$\Sigma_1 \colon a_{11}x_1 + a_{21}x_2 + a_{31}x_3 = b_1,$$
$$\Sigma_2 \colon a_{12}x_1 + a_{22}x_2 + a_{32}x_3 = b_2$$

in $\mathbb{R}^3$ are either coincident or intersecting along a line but never parallel (see (3.5.6) and Fig. 3.16).

(4) (algebraic)

$$\text{Coincident } \Sigma_1 = \Sigma_2 \Leftrightarrow \frac{a_{11}}{a_{12}} = \frac{a_{21}}{a_{22}} = \frac{a_{31}}{a_{32}} = \frac{b_1}{b_2};$$

intersecting along a line $\Leftrightarrow$ at least two of the ratios $\dfrac{a_{11}}{a_{12}}, \dfrac{a_{21}}{a_{22}}$

and $\dfrac{a_{31}}{a_{32}}$ are not equal.

(5) (linearly algebraic) The *coefficient* matrix A and the *augmented* *matrix* $\left[\frac{A}{\vec{b}}\right]$ have the same rank, i.e.

$$\mathrm{r}\left(\left[\frac{A}{\vec{b}}\right]\right) = \mathrm{r}(A) = \begin{cases} 1, & \text{if coincidence;} \\ 2, & \text{if intersection.} \end{cases}$$

Therefore, it is worth mentioned that

$$\vec{x}A = \vec{b} \text{ has no solution.}$$

$\Leftrightarrow$ The planes Σ_1 and Σ_2 are parallel.

$$\Leftrightarrow \frac{a_{11}}{a_{12}} = \frac{a_{21}}{a_{22}} = \frac{a_{31}}{a_{32}} \neq \frac{b_1}{b_2}.$$

$$\Leftrightarrow \mathrm{r}(A) = 1 < \mathrm{r}\left(\left[\frac{A}{\vec{b}}\right]\right) = 2.$$

(b) In case $\vec{x}A = \vec{b}$ has a solution. Then

(1) The linear transformation $A : \mathbb{R}^3 \to \mathbb{R}^2$ is onto, i.e. $\mathrm{r}(A) = 2$.

$\Leftrightarrow$ (2) The *solution space* $\mathrm{Ker}(A)$ of $\vec{x}A = \vec{0}$ is a one-dimensional subspace of $\mathbb{R}^3$.

$\Leftrightarrow$ (3) For each point $\vec{b} \in \mathbb{R}^2$, the solution set of $\vec{x}A = \vec{b}$ is a one-dimensional affine subspace of $\mathbb{R}^3$.

$\Leftrightarrow$ (4) A^*A is an invertible 2×2 matrix, i.e. $\mathrm{r}(A^*A) = 2$. Thus, a *particular solution* of $\vec{x}A = \vec{b}$ is

$$\vec{b}(A^*A)^{-1}A^*.$$

In this case, the *solution affine subspace* of $\vec{x}A = \vec{b}$ is

$$\vec{b}(A^*A)^{-1}A^* + \mathrm{Ker}(A),$$

which is perpendicular to the vector $\vec{b}(A^*A)^{-1}A^*$, namely, for any $\vec{x} \in \mathrm{Ker}(A)$,

$$\vec{x}(\vec{b}(A^*A)^{-1}A^*)^* = \vec{x}A(A^*A)^{-1}\vec{b}^*$$

$$= \vec{0}(A^*A)^{-1}\vec{b}^* = \vec{0}.$$

Hence, the point $\vec{b}(A^*A)^{-1}A^*$ has the shortest distance, among all the points lying on the solution affine subspace, to the origin $\vec{0}$

(see Fig. 3.44), i.e.

$$|\vec{b}\,(A^*A)^{-1}A^*| = \min_{\vec{x}\,A=\vec{b}} |\vec{x}|.$$

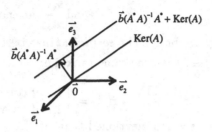

Fig. 3.44

(c) In case $\vec{x}A = \vec{b}$ has a solution. Then

   (1) The linear transformation $A\colon \mathbb{R}^3 \to \mathbb{R}^2$ is not onto, i.e. $r(A) = 1$.

   $\Leftrightarrow$ (2) The *solution space* $\mathrm{Ker}(A)$ of $\vec{x}A = \vec{0}$ is a two-dimensional subspace of $\mathbb{R}^3$.

   $\Leftrightarrow$ (3) For the point $\vec{b}$ for which $\vec{x}A = \vec{b}$ has a solution, the solution set is a two-dimensional affine subspace of $\mathbb{R}^3$.

   $\Leftrightarrow$ (4) For the point $\vec{b}$ for which $\vec{x}A = \vec{b}$ has a solution, A^*A is not invertible and

$$r(A^*A) = r(A^*A + \vec{b}^*\vec{b}) = 1.$$

If $\vec{x}_0$ is a particular solution of $\vec{x}A = \vec{b}$, then the affine plane

$$\vec{x}_0 + \mathrm{Ker}(A)$$

is the solution set of $\vec{x}A = \vec{b}$ (see Fig. 3.45). There exists a unique point $\vec{v}$ on the plane whose distance to $\vec{0}$ is the smallest one, i.e.

$$|\vec{v}| = \min_{\vec{x}\,A=\vec{b}} |\vec{x}|.$$

This $\vec{v}$ is going to be perpendicular to both $\mathrm{Ker}(A)$ and $\vec{x}_0 + \mathrm{Ker}(A)$.

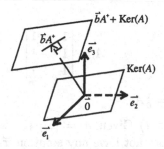

Fig. 3.45

To find such a $\vec{v}$, we model after Ex. 3(c) and proceed as follows. May suppose $a_{11} \neq 0$ and rewrite A as

$$A = BC, \quad \text{where } B = \begin{bmatrix} a_{11} \\ A_{21} \end{bmatrix}_{3 \times 1} \text{ and } C = \begin{bmatrix} 1 & a_{11}^{-1} A_{12} \end{bmatrix}_{1 \times 2} \text{ and}$$

$$A_{21} = \begin{bmatrix} a_{21} \\ a_{31} \end{bmatrix}, \quad A_{12} = [a_{12}]_{1 \times 1}$$

so that $\mathrm{r}(B) = \mathrm{r}(C) = 1$. Then

$\vec{x} A = \vec{b}$
$\Rightarrow \vec{x} A A^* = \vec{b} A^*$ or $\vec{x} B(CC^*)B^* = \vec{b} C^* B^*$
$\Rightarrow \vec{x} B(CC^*)(B^*B) = \vec{b} C^*(B^*B)$
$\Rightarrow \vec{x} B(CC^*) = \vec{b} C^*$ or $\vec{x} B = \vec{b} C^*(CC^*)^{-1}$
$\Rightarrow$ The required solution is
$\quad \vec{v} = \vec{b} C^*(CC^*)^{-1}(B^*B)^{-1}B^* = \vec{b} A^+$
where $A^+ = C^*(CC^*)^{-1}(B^*B)^{-1}B^*$.

Notice that, for any $\vec{x} \in \mathbb{R}^3$, $\vec{x} A = \vec{0}$ if and only if $\vec{x} B = \vec{0}$, and hence, for such $\vec{x}$,

$$\vec{x}(\vec{b} A^+)^* = \vec{x} B(B^*B)^{-1}(CC^*)^{-1}C = 0.$$

This means that the vector $\vec{b} A^+$ is indeed perpendicular to $\mathrm{Ker}(A)$ and it is a point lying on the affine plane $\vec{x}_0 + \mathrm{Ker}(A)$, since

$$(\vec{b} A^+)A = \vec{b} C^*(CC^*)^{-1}(B^*B)^{-1}B^* BC$$
$$= \vec{b} C^*(CC^*)^{-1}C = \vec{x} BC = \vec{x} A = \vec{b}.$$

(d) Let

$$A = \begin{bmatrix} 1 & -1 \\ 1 & -1 \\ 1 & -1 \end{bmatrix}.$$

Show that $A^+ = \frac{1}{6}A^*$.

8. (continued from Ex. 7) Given $A = [a_{ij}]_{3\times2}$ and $\vec{b} = (b_1, b_2) \in \mathbb{R}^2$, suppose $\vec{x}A = \vec{b}$ does not have any solution $\vec{x} \in \mathbb{R}^3$. As in Ex. 4, the problem is to find $\vec{x}_0 \in \mathbb{R}^3$ so that

$$|\vec{b} - \vec{x}_0 A| = \min_{\vec{x} \in \mathbb{R}^3} |\vec{b} - \vec{x}A|.$$

This means that $\vec{x}_0 A$ is the *orthogonal projection* of $\vec{b}$ onto the range space $\mathrm{Im}(A)$ and $|\vec{b} - \vec{x}_0 A|$ is the *distance* from $\vec{b}$ to $\mathrm{Im}(A)$. For details, see Chaps. 4 and 5. See also Fig. 3.46.

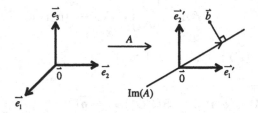

Fig. 3.46

According to Ex. 7, $r(A) = 1$ should hold in this case. Just like Ex. 4, we have

$$(\vec{b} - \vec{x}_0 A) \perp \vec{x}A \quad \text{for all } \vec{x} \in \mathbb{R}^3$$

$$\Leftrightarrow \vec{x}_0 AA^* = \vec{b}A^*.$$

Since $r(AA^*) = r(A) = 1$, model after Ex. 7(c) and $A = BC$ there, then

$$\vec{x}_0 = \vec{b}C^*(CC^*)^{-1}(B^*B)^{-1}B^* = \vec{b}A^*$$

$$\Rightarrow \text{The projection vector of } \vec{b} \text{ on } \mathrm{Im}(A) \text{ is}$$

$$\vec{b}C^*(CC^*)^{-1}(B^*B)^{-1}B^*A = \vec{b}C^*(CC^*)^{-1}C.$$

(*Note* Combing Exs. 7 and 8 together, for $A_{3\times2} \neq O_{3\times2}$, the 2×3 matrix

$$A^+ = \begin{cases} (A^*A)^{-1}A^*, & \text{if } r(A) = 2 \\ C^*(CC^*)^{-1}(B^*B)^{-1}B^*, & \text{if } r(A) = 1 \text{ and } A = BC \text{ as in Ex. 7(c)} \end{cases}$$

is called the *generalized* or *pseudoinverse* of A. This A^+ has the following properties:

(1) $|\vec{b}A^+| = \min_{\vec{x}A=\vec{b}} |\vec{x}|$ if $\vec{x}A = \vec{b}$ has a solution.

(2) $\vec{b}A^+A$ is the *orthogonal projection* of $\vec{b}$ on $\text{Im}(A)$ and

$$|\vec{b} - \vec{b}A^+A| = \min_{\vec{x}\in\mathbb{R}^3} |\vec{b} - \vec{x}A|.$$

Similar results hold for real or complex matrix $A_{m\times n}$ with $m \geq n$. See Sec. B.8, also refer to Exs. <B> and <C> 18 of Sec. 2.7.5 and compare with the note in Ex. 4.)

9. Do the same problems as in Ex. 4 and Ex. 8 for $A_{1\times2}, A_{2\times1}, A_{1\times3}$ and $A_{3\times1}$.

10. (a) N points (x_i, y_i), $1 \leq i \leq n$, in the plane $\mathbb{R}^2$ are collinear if and only if, for $\vec{a}_i = (x_i, y_i) - (x_1, y_1)$, $2 \leq i \leq n$, the matrix

$$\begin{bmatrix} \vec{a}_2 \\ \vec{a}_3 \\ \vdots \\ \vec{a}_n \end{bmatrix}$$

has rank equal to or less than 2.

(b) N straight lines $a_{i1}x_1 + a_{i2}x_2 + b_i = 0$, $1 \leq i \leq n$, in the plane $\mathbb{R}^2$ are concurrent at a point if and only if, for $\vec{a}_i = (a_{i1}, a_{i2})$, $1 \leq i \leq n$,

$$A = \begin{bmatrix} \vec{a}_1 \\ \vec{a}_2 \\ \vdots \\ \vec{a}_n \end{bmatrix} \quad \text{and} \quad [A \mid \vec{b}^*] = \begin{bmatrix} \vec{a}_1 & b_1 \\ \vec{a}_2 & b_2 \\ \vdots & \vdots \\ \vec{a}_n & b_n \end{bmatrix}, \quad \text{where } \vec{b} = (b_1, \dots, b_n)$$

have the same rank 2, these lines are coincident along a line if and only if $r(A) = r([A \mid \vec{b}^*]) = 1$.

11. In $\mathbb{R}^3$.

(a) Find necessary and sufficient conditions for n points (x_i, y_i, z_i), $1 \leq i \leq n$, to be coplanar or collinear.

(b) Find necessary and sufficient conditions for n planes $a_{i1}x_1 + a_{i2}x_2 + a_{i3}x_3 + b_i = 0$, $1 \leq i \leq n$, to be concurrent at a point, intersecting along a line or coincident along a plane.

<C>

Read Ex. <C> of Sec. 2.7.3 and do all the problems there if you missed them at that time.

Also, do the following problems. Refer to Exs. 19–24 of Sec. B.7, if necessary.

1. A mapping $f: \mathbb{C}^3 \to \mathbb{C}^3$ is defined by

$$f(x_1, x_2, x_3) = (3ix_1 - 2x_3 - ix_3, \; ix_2 + 2x_3, \; x_1 + 4x_3).$$

(a) In the natural basis $\mathcal{N} = \{\vec{e}_1, \vec{e}_2, \vec{e}_3\}$, f can be expressed as

$$f(\vec{x}) = [\vec{x}]_{\mathcal{N}}[f]_{\mathcal{N}}, \quad \text{where } [f]_{\mathcal{N}} = [f]_{\mathcal{N}}^{\mathcal{N}} = \begin{bmatrix} 3i & 0 & 1 \\ -2 & i & 0 \\ -i & 2 & 4 \end{bmatrix}.$$

(b) Let $\vec{x}_1 = (1, 0, i)$, $\vec{x}_2 = (-1, 2i, 1)$, $\vec{x}_3 = (2, 1, i)$. Show that both $\mathcal{B} = \{\vec{x}_1, \vec{x}_2, \vec{x}_3\}$ and $f(\mathcal{B}) = \mathcal{B}' = \{f(\vec{x}_1), f(\vec{x}_2), f(\vec{x}_3)\}$ are bases for $\mathbb{C}^3$. What is $[f]_{\mathcal{B}'}^{\mathcal{B}}$?

(c) By direct computation, show that

$$[f]_{\mathcal{B}}^{\mathcal{B}} = \begin{bmatrix} \frac{23-3i}{2} & \frac{1-5i}{2} & -5+i \\ \frac{-15+9i}{2} & \frac{5+5i}{2} & 5 - 5i \\ 19 - 5i & -5i & -10 + 3i \end{bmatrix}.$$

Compute the transition matrix $P_{\mathcal{N}}^{\mathcal{B}}$ and justify that $[f]_{\mathcal{B}}^{\mathcal{B}} = P_{\mathcal{N}}^{\mathcal{B}}[f]_{\mathcal{N}}^{\mathcal{N}}P_{\mathcal{B}}^{\mathcal{N}}$.

(d) Compute $[f]_{\mathcal{B}'}^{\mathcal{B}'}$ and $P_{\mathcal{B}}^{\mathcal{B}'}$ and justify that $[f]_{\mathcal{B}'}^{\mathcal{B}'} = P_{\mathcal{B}}^{\mathcal{B}'}[f]_{\mathcal{B}}^{\mathcal{B}}P_{\mathcal{B}'}^{\mathcal{B}}$.

(e) Let $g: \mathbb{C}^3 \to \mathbb{C}^3$ be defined by

$$g(x_1, x_2, x_3) = (x_1 + 2ix_2, x_2 + x_3, -ix_1 + ix_3).$$

Compute

(1) $g \circ f$ and $f \circ g$,

(2) $[g]_{\mathcal{B}'}^{\mathcal{B}}$ and $[g]_{\mathcal{B}}^{\mathcal{B}}$, $[g]_{\mathcal{B}'}^{\mathcal{B}'}$, and

(3) $[g \circ f]_{\mathcal{B}}^{\mathcal{B}}$ and $[f \circ g]_{\mathcal{B}'}^{\mathcal{B}'}$, $[f \circ g]_{\mathcal{B}'}^{\mathcal{B}}$.

2. In $P_n(\mathbb{R})$ (see Secs. A.5 and B.3), let

$$\mathcal{N} = \{1, x, x^2, \ldots, x^n\}, \text{ and}$$
$$\mathcal{B} = \{1, x + x_0, (x + x_0)^2, \ldots, (x + x_0)^n\},$$

where $x_0 \in \mathbb{R}$ is a fixed number. $\mathcal{N}$ is called the *natural basis* for $P_n(\mathbb{R})$.

(a) Show that $\mathcal{B}$ is a basis for $P_n(\mathbb{R})$.

(b) Let $D\colon P_n(\mathbb{R}) \to P_n(\mathbb{R})$ be the differential operator, i.e. $D(p) = p'$. Show that

$$[D]_{\mathcal{N}} = [D]_{\mathcal{N}}^{\mathcal{N}} = \begin{bmatrix} 0 & 0 & 0 & 0 & \cdots & 0 & 0 \\ 1 & 0 & 0 & 0 & \cdots & 0 & 0 \\ 0 & 2 & 0 & 0 & \cdots & 0 & 0 \\ 0 & 0 & 3 & 0 & \cdots & 0 & 0 \\ \vdots & \vdots & \vdots & \vdots & \ddots & \vdots & \vdots \\ 0 & 0 & 0 & 0 & \cdots & 0 & 0 \\ 0 & 0 & 0 & 0 & \cdots & n & 0 \end{bmatrix}_{(n+1)\times(n+1)}$$

(c) Show that $[D]_{\mathcal{B}} = [D]_{\mathcal{N}}$.

(d) Compute the transition matrices $P_{\mathcal{B}}^{\mathcal{N}}$ and $P_{\mathcal{N}}^{\mathcal{B}}$.

(e) Compute $P_{\mathcal{B}}^{\mathcal{N}}$ for $n = 2$ and justify that $[D]_{\mathcal{B}} = P_{\mathcal{N}}^{\mathcal{B}}[D]_{\mathcal{N}}P_{\mathcal{B}}^{\mathcal{N}}$.

(f) What happen to (b)–(e) if D is considered as a linear transformation from $P_n(\mathbb{R})$ to $P_{n-1}(\mathbb{R})$.

(g) Define $\Phi\colon P_n(\mathbb{R}) \to \mathbb{R}^{n+1}$ by

$$\Phi\left(\sum_{i=0}^{n} a_i x^i\right) = (a_0, a_1, \ldots, a_n).$$

Show that Φ is a linear isomorphism and

$$[\Phi]_{\mathcal{N}'}^{\mathcal{N}} = I_{n+1},$$

where $\mathcal{N}' = \{\vec{e}_1', \ldots, \vec{e}_{n+1}'\}$ is the natural basis for $\mathbb{R}^{n+1}$. What is

$$[\Phi \circ D \circ \Phi^{-1}]_{\mathcal{N}'}^{\mathcal{N}'}$$

where D is as in (b)?

3. In $P_2(\mathbb{R})$, let $\mathcal{N} = \{1, x, x^2\}$ be the natural basis and let

$$\mathcal{B} = \{x^2 - x + 1, x + 1, x^2 + 1\},$$
$$\mathcal{B}' = \{x^2 + x + 4, 4x^2 - 3x + 2, 2x^2 + 3\},$$
$$\mathcal{C} = \{x^2 - x, x^2 + 1, x - 1\},$$
$$\mathcal{C}' = \{2x^2 - x + 1, x^2 + 3x - 2, -x^2 + 2x + 1\}.$$

(a) Use Ex. 2(g) to show that $\mathcal{B}$, $\mathcal{B}'$, $\mathcal{C}$ and $\mathcal{C}'$ are bases for $P_2(\mathbb{R})$.

(b) Show that $\{5x^2 - 2x - 3, -2x^2 + 5x + 5\}$ is linear independent in $P_2(\mathbb{R})$ and extend it to form a basis for $P_2(\mathbb{R})$.

(c) Find a subset of $\{2x^2 - x, x^2 + 21x - 2, 3x^2 + 5x + 2, 9x - 9\}$ that is a basis for $P_2(\mathbb{R})$.

(d) Compute the transition matrices P_C^B and $P_{C'}^{B'}$. Let Φ be as in Ex. 2(g) and notice that

(1) Φ transforms a basis $\mathcal{B}$ for $P_2(\mathbb{R})$ onto a basis $\Phi(\mathcal{B})$ for $\mathbb{R}^3$.

(2) What is $[\Phi]_{\Phi(\mathcal{B})}^{\mathcal{B}}$?

(3) $P_C^B = [\Phi]_{\Phi(\mathcal{B})}^{\mathcal{B}} P_{\Phi(C)}^{\Phi(\mathcal{B})} [\Phi]_C^{\Phi(C)}$.

(e) Let $T\colon P_2(\mathbb{R}) \to P_2(\mathbb{R})$ be defined by

$$T(p)(x) = p'(x) \cdot (3 + x) + 2p(x).$$

Show that T is a linear operator. Compute $[T]_{\mathcal{N}}$ and $[T]_{\mathcal{B}}$ and justify that $[T(p)]_\mathcal{B} = [p]_\mathcal{B}[T]_\mathcal{B}$ by supposing $p(x) = 3 - 2x + x^2$.

(f) Compute $[T]_C^B$ and $[T]_{C'}^{B'}$ and justify that $[T]_{C'}^{B'} = P_B^{B'}[T]_C^B P_{C'}^C$.

(g) Let $U\colon P_2(\mathbb{R}) \to \mathbb{R}^3$ be defined by

$$U(a + bx + cx^2) = (a + b, c, a - b).$$

Show that U is a linear isomorphism. Use $\mathcal{N}' = \{\vec{e}_1, \vec{e}_2, \vec{e}_3\}$ to denote the natural basis for $\mathbb{R}^3$. Compute $[U]_{\mathcal{N}'}^{\mathcal{N}}$ and $[U]_{\mathcal{N}'}^{\mathcal{B}}$, $[U^{-1}]_{\mathcal{B}}^{\mathcal{N}'}$ and justify that

$$[U]_{\mathcal{N}'}^{\mathcal{B}}[U^{-1}]_{\mathcal{B}}^{\mathcal{N}'} = [U^{-1}]_{\mathcal{B}}^{\mathcal{N}'}[U]_{\mathcal{N}'}^{\mathcal{B}} = I_3,$$

i.e. $([U]_{\mathcal{N}'}^{\mathcal{B}})^{-1} = [U^{-1}]_{\mathcal{B}}^{\mathcal{N}'}$.

(h) Compute $[U]_{\mathcal{N}'}^{\mathcal{N}}$ and justify that $[U \circ T]_{\mathcal{N}'}^{\mathcal{N}} = [T]_{\mathcal{N}}[U]_{\mathcal{N}'}^{\mathcal{N}}$.

(i) Define $V\colon P_2(\mathbb{R}) \to M(2\colon \mathbb{R})$ by

$$V(p) = \begin{bmatrix} p'(0) & 0 \\ 2p(1) & p''(3) \end{bmatrix}.$$

Show that V is a linear transformation and compute $[V]_{\mathcal{N}''}^{\mathcal{N}}$ where $\mathcal{N}'' = \{E_{11}, E_{12}, E_{21}, E_{22}\}$ is the natural basis for $M(2\colon \mathbb{R})$ (see Sec. B.4). Verify that $[V(p)]_{\mathcal{N}''} = [p]_{\mathcal{N}}[V]_{\mathcal{N}''}^{\mathcal{N}}$ if $p(x) = 4 - 6x + 3x^2$.

(j) Define linear functionals $f_i\colon P_2(\mathbb{R}) \to \mathbb{R}$, for $i = 1, 2, 3$, by

$$f_1(a + bx + cx^2) = a,$$
$$f_2(a + bx + cx^2) = b,$$
$$f_3(a + bx + cx^2) = c.$$

Show that $\mathcal{N}^* = \{f_1, f_2, f_3\}$ is the dual basis of $\mathcal{N} = \{1, x, x^2\}$ in $P_2(\mathbb{R})^*$.

(k) It is easy to see that

$$ax^2 + bx + c = (c - a - b)(x^2 - x + 1) + (c - a)(x + 1) + (2a + b - c)(x^2 + 1).$$

Then, try to find the dual basis $\mathcal{B}^*$ of $\mathcal{B}$ in $P_2(\mathbb{R})^*$. How about C^*?

(l) Define $g_i\colon P_2(\mathbb{R}) \to \mathbb{R}$, for $i = 1, 2, 3$, by

$$g_1(p) = \int_0^1 p(t)\, dt,$$

$$g_2(p) = \int_0^2 p(t)\, dt,$$

$$g_3(p) = \int_0^3 p(t)\, dt.$$

Show that $\{g_1, g_2, g_3\}$ is a basis for $P_2(\mathbb{R})^*$. Try to find a basis for $P_2(\mathbb{R})$ so that its dual basis in $P_2(\mathbb{R})^*$ is $\{g_1, g_2, g_3\}$.

(m) Let T be as in (e). Compute $[T^*]_{\mathcal{B}^*}^{\mathcal{C}^*}$ and justify that it is equal to $([T]_{\mathcal{C}}^{\mathcal{B}})^*$. Try to find $T^*(f)$ if $f(ax^2 + bx + c) = a + b + c$ by the following methods:

  (1) $[T^*(f)]_{\mathcal{B}^*} = [f]_{\mathcal{C}^*} \cdot [T^*]_{\mathcal{B}^*}^{\mathcal{C}^*}$.
  (2) By definition of T^*, for $p \in P_2(\mathbb{R})$, then

$$T^*(f)(p) = f(T(p)).$$

(n) Let U be as in (g). Describe $U^*\colon (\mathbb{R}^3)^* \to (P_2(\mathbb{R}))^*$.
(o) Let V be as in (i). Describe $V^*\colon M(2\colon \mathbb{R})^* \to (P_2(\mathbb{R}))^*$.

4. In $M(2\colon \mathbb{R})$, let

$$\mathcal{N} = \{E_{11}, E_{12}, E_{21}, E_{22}\},$$

$$\mathcal{B} = \left\{ \begin{bmatrix} 1 & 0 \\ 0 & 1 \end{bmatrix}, \begin{bmatrix} 1 & 1 \\ 0 & 0 \end{bmatrix}, \begin{bmatrix} 1 & 0 \\ 1 & 0 \end{bmatrix}, \begin{bmatrix} 0 & 0 \\ 1 & 1 \end{bmatrix} \right\},$$

$$\mathcal{C} = \left\{ \begin{bmatrix} 0 & 1 \\ 1 & 1 \end{bmatrix}, \begin{bmatrix} 1 & 0 \\ 1 & 1 \end{bmatrix}, \begin{bmatrix} 1 & 1 \\ 0 & 1 \end{bmatrix}, \begin{bmatrix} 1 & 1 \\ 1 & 0 \end{bmatrix} \right\}.$$

(a) Show that $\mathcal{B}$ and $\mathcal{C}$ are bases for $M(2\colon \mathbb{R})$.
(b) Show that $\left\{ \begin{bmatrix} -2 & 5 \\ 1 & 3 \end{bmatrix}, \begin{bmatrix} 0 & 4 \\ 2 & -1 \end{bmatrix} \right\}$ is linearly independent in $M(2\colon \mathbb{R})$ and extend it to a basis for $M(2\colon \mathbb{R})$.
(c) How many bases for $M(2\colon \mathbb{R})$ can be selected from the set

$$\left\{ \begin{bmatrix} 1 & 1 \\ 0 & 0 \end{bmatrix}, \begin{bmatrix} 1 & 1 \\ 1 & 1 \end{bmatrix}, \begin{bmatrix} 0 & 1 \\ 1 & 0 \end{bmatrix}, \begin{bmatrix} -1 & 0 \\ 0 & 1 \end{bmatrix}, \begin{bmatrix} 0 & 1 \\ 0 & -1 \end{bmatrix}, \begin{bmatrix} -1 & 1 \\ 0 & 2 \end{bmatrix} \right\}?$$

(d) Let $\Phi\colon M(2\colon \mathbb{R}) \to \mathbb{R}^4$ be defined by

$$\Phi\left(\begin{bmatrix} a_{11} & a_{12} \\ a_{21} & a_{22} \end{bmatrix}\right) = (a_{11}, a_{12}, a_{21}, a_{22}).$$

Show that Φ is a linear isomorphism. Hence, $\Phi(\mathcal{N}) = \{\vec{e}_1, \vec{e}_2, \vec{e}_3, \vec{e}_4\}$ is the natural basis for $\mathbb{R}^4$. What are $\Phi(\mathcal{B})$ and $\Phi(\mathcal{C})$? Show that

$$[\Phi]_{\Phi(\mathcal{N})}^{\mathcal{N}} = I_4.$$

(e) Compute the transition matrices $P_{\mathcal{N}}^{\mathcal{B}}$, $P_{\mathcal{N}}^{\mathcal{C}}$ and $P_{\mathcal{C}}^{\mathcal{B}}$ by using the following methods:

(1) Direct computation via definitions.
(2) $P_{\mathcal{C}}^{\mathcal{B}} = [\Phi]_{\Phi(\mathcal{B})}^{\mathcal{B}} P_{\Phi(\mathcal{C})}^{\Phi(\mathcal{B})} [\Phi]_{\mathcal{C}}^{\Phi(\mathcal{C})}$, etc.

Verify that

$$P_{\mathcal{C}}^{\mathcal{B}} = P_{\mathcal{N}}^{\mathcal{B}} P_{\mathcal{C}}^{\mathcal{N}}.$$

(f) Define $T\colon M(2\colon \mathbb{R}) \to M(2\colon \mathbb{R})$ by

$$T(A) = A^* \quad \text{(the transpose of } A\text{)}.$$

Show that T is a linear isomorphism. Compute $[T]_{\mathcal{N}}$, $[T]_{\mathcal{B}}$, $[T]_{\mathcal{C}}$ and $[T]_{\mathcal{C}}^{\mathcal{B}}$ by using the methods indicated in (e):

(1) Direct computation.
(2) Via $\mathbb{R}^4$. Consider the following diagram

$$M(2\colon \mathbb{R}) \xrightarrow{T} M(2\colon \mathbb{R})$$
$$\Phi\downarrow \qquad\qquad \Phi\downarrow$$
$$\mathbb{R}^4 \xrightarrow{\tilde{T}} \mathbb{R}^4$$

where $T = \Phi^{-1} \circ \tilde{T} \circ \Phi$. What is $\tilde{T}$?

Verify that $[T]_{\mathcal{C}}^{\mathcal{B}} = P_{\mathcal{N}}^{\mathcal{B}} [T]_{\mathcal{N}} P_{\mathcal{C}}^{\mathcal{N}}$ and $[T^{-1}]_{\mathcal{B}}^{\mathcal{C}} = ([T]_{\mathcal{C}}^{\mathcal{B}})^{-1}$.

(g) Compute the dual bases $\mathcal{B}^*$ and $\mathcal{C}^*$ of $\mathcal{B}$ and $\mathcal{C}$ in $M(2\colon \mathbb{R})^*$. Show that

$$[T^*]_{\mathcal{B}^*}^{\mathcal{C}^*} = ([T]_{\mathcal{C}}^{\mathcal{B}})^*$$

and compute $T^*(\mathrm{tr})$ where $\mathrm{tr}\colon M(2\colon \mathbb{R}) \to \mathbb{R}$ defined by $\mathrm{tr}(A) = a_{11} + a_{22}$ is the trace of A.

(h) Show that ± 1 are the only eigenvalues of T. Find all eigenvectors of T corresponding to 1 and -1, respectively.

(i) Define $U\colon M(2\colon \mathbb{R}) \to P_2(\mathbb{R})$ by

$$U\left(\begin{bmatrix} a_{11} & a_{12} \\ a_{21} & a_{22} \end{bmatrix}\right) = (a_{11} + a_{12}) + 2a_{22}x + a_{12}x^2.$$

Determine $\mathrm{Ker}(U)$ and the rank $\mathrm{r}(U)$. Let $\mathcal{N}_0 = \{1, x, x^2\}$ be the natural basis for $P_2(\mathbb{R})$ and $\mathcal{B}_0$ and $\mathcal{C}_0$ be $\mathcal{B}$ and $\mathcal{C}$ in Ex. 3 respectively. Do the following problems.

(1) Compute $[U]_{\mathcal{N}_0}^{\mathcal{N}}$, $[U]_{\mathcal{B}_0}^{\mathcal{B}}$ and $[U]_{\mathcal{C}_0}^{\mathcal{C}}$.
(2) Verify that $[U(A)]_{\mathcal{B}_0} = [A]_{\mathcal{B}}[U]_{\mathcal{B}_0}^{\mathcal{B}}$.
(3) Verify that $[U]_{\mathcal{C}_0}^{\mathcal{C}} = P_{\mathcal{N}}^{\mathcal{C}}[U]_{\mathcal{N}_0}^{\mathcal{N}} P_{\mathcal{C}_0}^{\mathcal{N}_0}$.
(4) Describe U^* and verify that $[U^*]_{\mathcal{B}^*}^{\mathcal{B}_0^*} = ([U]_{\mathcal{B}_0}^{\mathcal{B}})^*$. Compute $U^*(p)$ if $p(x) = 1 + x + x^2$.
(5) Verify that $[U \circ T]_{\mathcal{C}_0}^{\mathcal{C}} = [T]_{\mathcal{B}}^{\mathcal{C}}[U]_{\mathcal{C}_0}^{\mathcal{B}} = [T]_{\mathcal{C}}[U]_{\mathcal{C}_0}^{\mathcal{C}}$.

3.7.4 *Linear transformations (operators)*

Just like Sec. 3.7.3 to Sec. 2.7.3, the definitions, the results and the proofs concerned in Sec. 2.7.4 can be generalized, as claimed there, to higher dimensional vector spaces or matrices, in particular to $\mathbb{R}^n$ for $n = 1, 2, 3$.

Here we list corresponding results in $\mathbb{R}^3$ and leave their proofs to readers. As a generalization of (2.7.50) and (2.7.51), we have the

Projection on $\mathbb{R}^3$

Let $f\colon \mathbb{R}^3 \to \mathbb{R}^3$ (or $\mathbb{R}^n \to \mathbb{R}^n$ for $n \geq 2$) be a *nonzero* linear operator with *rank* equal to r, $1 \leq r \leq 2$.

(1) f is a projection of $\mathbb{R}^3 = V_1 \oplus V_2$ onto V_1 along V_2, i.e. for each $\vec{x} = \vec{x}_1 + \vec{x}_2 \in \mathbb{R}^3$, where $\vec{x}_1 \in V_1$, and $\vec{x}_2 \in V_2$, then

$$f(\vec{x}) = \vec{x}_1.$$

See Fig. 3.47.

$\Leftrightarrow$ (2) $\mathrm{r}(f) + \mathrm{r}(1_{\mathbb{R}^3} - f) = \dim \mathbb{R}^3 = 3$.
$\Leftrightarrow$ (3) $1_{\mathbb{R}^3} - f$ is a projection of $\mathbb{R}^3$ onto V_2 along V_1.
$\Leftrightarrow$ (4) $f^2 = f$ (and hence f is the projection of $\mathbb{R}^3$ onto $\mathrm{Im}(f)$ along $\mathrm{Ker}(f)$).
$\Leftrightarrow$ (5) f has only eigenvalues 1 and 0.

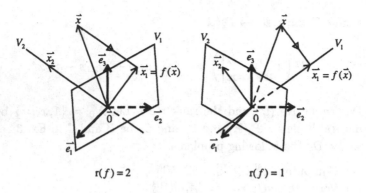

$$r(f) = 2 \qquad\qquad r(f) = 1$$

Fig. 3.47

⇔ (6) There exists a basis $\mathcal{B} = \{\vec{x}_1, \vec{x}_2, \vec{x}_3\}$ for $\mathbb{R}^3$ such that

$$[f]_{\mathcal{B}} = P[f]_N P^{-1} = \begin{bmatrix} I_r & 0 \\ 0 & 0 \end{bmatrix}_{3\times3}, \quad P = \begin{bmatrix} \vec{x}_1 \\ \vec{x}_2 \\ \vec{x}_3 \end{bmatrix},$$

where $N = \{\vec{e}_1, \vec{e}_2, \vec{e}_3\}$ is the natural basis for $\mathbb{R}^3$.   (3.7.34)

Of course, the zero linear operator is the projection of $\mathbb{R}^3$ onto $\{\vec{0}\}$ along the whole space $\mathbb{R}^3$, while the identity operator $1_{\mathbb{R}^3}$ is the only projection of $\mathbb{R}^3$ onto itself along $\{\vec{0}\}$.

As a counterpart of (2.7.52), it is

The projectionalization

(1) For each linear operator f on $\mathbb{R}^3$ (or on $\mathbb{R}^n$ for $n \geq 2$), there exists an invertible linear operator g on $\mathbb{R}^3$ such that

$$(g \circ f)^2 = g \circ f,$$

i.e. $g \circ f$ is a projection on $\mathbb{R}^3$.

(2) For any real 3×3 (or $n \times n$) matrix A, there exists an invertible matrix $P_{3\times3}$ such that

$$(AP)^2 = AP.   (3.7.35)$$

(2.7.55) and (2.7.56) become

The rank theorem

Let $f: \mathbb{R}^m \to \mathbb{R}^n$ be a linear transformation for $m, n = 1, 2, 3$ (and for any $m, n \geq 1$). Suppose the rank $r(f) = r \leq \min(m, n)$.

(1) There exist an invertible linear operator g on $\mathbb{R}^m$ and an invertible linear operator h on $\mathbb{R}^n$ so that

$$h \circ f \circ g(x_1, \ldots, x_m) = \begin{cases} (0, \ldots, 0), & \text{if } r(f) = 0 \\ (x_1, \ldots, x_r, \underbrace{0, \ldots, 0}_{n-r}), & \text{if } r(f) = r \geq 1. \end{cases}$$

See the diagram below and Fig. 3.48.

(2) Let $A_{m \times n}$ be a real matrix. Then there exist invertible matrices $P_{m \times m}$ and $Q_{n \times n}$ such that

$$PAQ = \begin{cases} O_{m \times n}, & \text{if } r(A) = 0, \\ \begin{bmatrix} I_{r \times r} & O_{r \times (n-r)} \\ O_{(m-r) \times r} & O_{(m-r) \times (n-r)} \end{bmatrix}_{m \times n}, & \text{if } r(A) = r \geq 1, \end{cases}$$

which is called the *normal form* of A. See Fig. 3.49. $\qquad$ (3.7.36)

To illustrate (1) graphically and, at the same time, to provide a sketch of its proof, let $\mathcal{N} = \{\vec{e}_1, \ldots, \vec{e}_m\}$ and $\mathcal{N}' = \{\vec{e}_1', \ldots, \vec{e}_n'\}$ be the respective natural bases for $\mathbb{R}^m$ and $\mathbb{R}^n$. Let $\mathcal{B} = \{\vec{x}_1, \ldots, \vec{x}_r, \vec{x}_{r+1}, \ldots, \vec{x}_m\}$ be a basis for $\mathbb{R}^m$ so that $\{f(\vec{x}_1), \ldots, f(\vec{x}_r)\}$ is a basis for $\text{Im}(f)$ which is extended to form a basis $\mathcal{C} = \{f(\vec{x}_1), \ldots, f(\vec{x}_r), \vec{y}_{r+1}, \ldots, \vec{y}_n\}$ for $\mathbb{R}^n$. Define linear operators $g \colon \mathbb{R}^m \to \mathbb{R}^m$ and $h \colon \mathbb{R}^n \to \mathbb{R}^n$ by

$$g(\vec{e}_i) = \vec{x}_i, \quad 1 \leq i \leq m,$$
$$h(f(\vec{x}_i)) = \vec{e}_i' \text{ for } 1 \leq i \leq r \quad \text{and} \quad h(\vec{y}_j) = \vec{e}_j' \text{ for } r+1 \leq j \leq n.$$

Then $h \circ f \circ g$ has the required property. See the following diagram and Fig. 3.48 for $m = 3$, $n = 2$ and $r(f) = 1$.

$$
\begin{array}{ccc}
\mathbb{R}^m & \overset{f}{\longrightarrow} & \mathbb{R}^n \\
(\mathcal{B}) & & (\mathcal{C}) \\
g\uparrow & & \downarrow h \\
(\mathcal{N}) & & (\mathcal{N}') \\
\mathbb{R}^m & \overset{h \circ f \circ g}{\longrightarrow} & \mathbb{R}^n
\end{array}
$$

In short, a linear transformation $f \colon \mathbb{R}^m \to \mathbb{R}^n$, of rank r, can be described as a projection $h \circ f \circ g$ mapping $\mathbb{R}^m$ onto the subspace generated by first r coordinate axes of $\mathbb{R}^n$, after suitably readjusting coordinate axes in $\mathbb{R}^m$

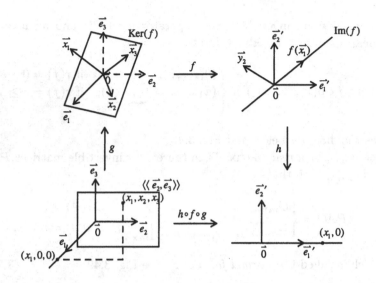

Fig. 3.48

and $\mathbb{R}^n$. Suppose

$$f(\vec{x}) = \vec{x}A,$$

$$P = [1_{\mathbb{R}^m}]_{\mathcal{N}}^{\mathcal{B}} = \begin{bmatrix} \vec{x}_1 \\ \vdots \\ \vec{x}_r \\ \vdots \\ \vec{x}_m \end{bmatrix}_{m \times m} \quad \text{and} \quad Q^{-1} = [1_{\mathbb{R}^n}]_{\mathcal{N}'}^{\mathcal{C}} = \begin{bmatrix} f(\vec{x}_1) \\ \vdots \\ f(\vec{x}_r) \\ \vec{y}_{r+1} \\ \vdots \\ \vec{y}_n \end{bmatrix}_{n \times n}.$$

Then

$$PAQ = \begin{bmatrix} I_r & 0 \\ 0 & 0 \end{bmatrix} = [f]_{\mathcal{C}}^{\mathcal{B}} = [g]_{\mathcal{N}}^{\mathcal{B}}[f]_{\mathcal{N}'}^{\mathcal{N}}[h]_{\mathcal{C}}^{\mathcal{N}'}. \qquad (3.7.37)$$

This is (2) in (3.7.36). See Fig. 3.49.

The counterpart of (2.7.57) and (2.7.58) is

The nullities and the ranks of iterative linear operators
Let f be a linear operator on $\mathbb{R}^3$.

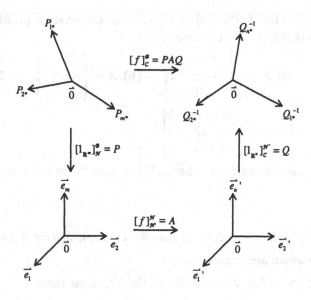

Fig. 3.49

(1) Then

$$\dim \operatorname{Ker}(f^3) = \dim \operatorname{Ker}(f^4) = \cdots = \dim \operatorname{Ker}(f^n) = \cdots, \quad \text{for } n \geq 3;$$
$$\operatorname{r}(f^3) = \operatorname{r}(f^4) = \cdots = \operatorname{r}(f^n) = \cdots, \quad \text{for } n \geq 3.$$

(2) For any real 3×3 matrix A,

$$\operatorname{r}(A^3) = \operatorname{r}(A^4) = \cdots = \operatorname{r}(A^n) = \cdots, \quad \text{for } n \geq 3.$$

In general, for a linear operator $f: \mathbb{R}^n \to \mathbb{R}^n$, there exists a least positive integer k so that

    1. $\operatorname{Ker}(f^k) = \operatorname{Ker}(f^{k+1}) = \cdots$ and $\operatorname{Im}(f^k) = \operatorname{Im}(f^{k+1}) = \cdots$, and
    2. $\mathbb{R}^n = \operatorname{Ker}(f^k) \oplus \operatorname{Im}(f^k)$.

                                                       (3.7.38)

Exercises

\<A\>

1. Prove (3.7.34), (3.7.35), (3.7.36) and (3.7.38).
2. For each of the following matrices A, do the following problems.

    (1) Find invertible matrix P such that AP is a projection on $\mathbb{R}^3$.
    (2) Find invertible matrices P and Q so that PAQ is in its normal form.

(3) Use A to justify (3.7.38) and determine the smallest positive integer k so that $\mathrm{r}(A^k) = \mathrm{r}(A^n)$ for $n \geq k$.

$$\text{(a) } A = \begin{bmatrix} 6 & 1 & -5 \\ 2 & -3 & 4 \\ 3 & 7 & -1 \end{bmatrix}. \quad \text{(b) } A = \begin{bmatrix} -3 & -6 & 15 \\ -1 & -2 & 5 \\ 2 & 4 & -10 \end{bmatrix}.$$

$$\text{(c) } A = \begin{bmatrix} -2 & 6 & 3 \\ 0 & 12 & 10 \\ 4 & 0 & 4 \end{bmatrix}.$$

3. Do Exs. \<A\> 2 and 5–15 of Sec. 2.7.4 in case $\mathbb{R}^n$ or $n \times n$ matrix A for $n \geq 3$.

\<B\>

Do Exs. \<B\> 1–3 of Sec. 2.7.4 in case $\mathbb{R}^n$ or $n \times n$ matrix A for $n \geq 3$.

\<C\> Abstraction and generalization

Read Ex. \<C\> of Sec. 2.7.4 and do all the problems there.

\<D\> Applications

Do the following problems

1. Remind that the $(n+1)$-dimensional vector space $P_n(\mathbb{R})$ has the natural basis $\mathcal{N} = \{1, x, \ldots, x^{n-1}, x^n\}$. Let $D: P_n(\mathbb{R}) \to P_{n-1}(\mathbb{R}) \subseteq P_n(\mathbb{R})$ be the *differential operator*

$$D(p(x)) = p'(x)$$

and $I: P_{n-1}(\mathbb{R}) \to P_n(\mathbb{R})$ be the *integral operator*

$$I(q(x)) = \int_0^x q(t)\, dt.$$

(a) Show that $D \circ I = 1_{P_{n-1}(\mathbb{R})}$, the identity operator on $P_{n-1}(\mathbb{R})$ and

$$[D \circ I]_{\mathcal{N}'} = I_{\mathcal{N}}^{\mathcal{N}'} D_{\mathcal{N}'}^{\mathcal{N}} = I_n,$$

where $\mathcal{N}' = \{1, x, \ldots, x^{n-1}\}$ is the natural basis for $P_{n-1}(\mathbb{R})$. Is this anything to do with the *Newton–Leibniz theorem*?

(b) Is $I \circ D = 1_{P_n(\mathbb{R})}$ true? Why? Any readjustment needed?

(c) Show that, for $1 \leq k < n$,

$$P_n(\mathbb{R}) = P_k(\mathbb{R}) \oplus \langle\langle x^{k+1}, \ldots, x^n \rangle\rangle,$$

where $P_k(\mathbb{R})$ is an invariant subspace of D, and $\langle\langle x^{k+1}, \ldots, x^n \rangle\rangle$ is not. Therefore

$$P_n(\mathbb{R})/P_k(\mathbb{R}) \text{ is isomorphic to } \langle\langle x^{k+1}, \ldots, x^n \rangle\rangle.$$

(d) Compute the matrix representations

$$[D \mid P_k(\mathbb{R})]_{\mathcal{N}_1} \quad \text{and} \quad [D|_{P_n(\mathbb{R})/P_k(\mathbb{R})}]_{\mathcal{N}_2},$$

where $\mathcal{N}_1 = \{1, x, \ldots, x^k\}$ and $D|_{P_n(\mathbb{R})/P_k(\mathbb{R})}$ is the *linear operator* mapping the coset $x^j + P_k(\mathbb{R})$ to $jx^{j-1} + P_k(\mathbb{R})$ for $k+1 \leq j \leq n$ and $\mathcal{N}_2 = \{x^{k+1} + P_k(\mathbb{R}), \ldots, x^n + P_k(\mathbb{R})\}$.

2. Let $a_0, a_1, \ldots, a_n$ be distinct real numbers.

(a) Define a linear functional $f_i \colon P_n(\mathbb{R}) \to \mathbb{R}$ by $f_i(p) = p(a_i)$ for $0 \leq i \leq n$. Show that $\mathcal{B}^* = \{f_0, f_1, \ldots, f_n\}$ is a basis for $P_n(\mathbb{R})^*$, the dual space of $P_n(\mathbb{R})$.

(b) There exists a (unique) basis $\mathcal{B} = \{p_0, p_1, \ldots, p_n\}$ for $P_n(\mathbb{R})$ so that $\mathcal{B}^*$ in (a) is the dual basis of $\mathcal{B}$ in $P_n(\mathbb{R})^*$, i.e.

$$f_i(p_j) = p_j(a_i) = \delta_{ij}, \quad 0 \leq i, j \leq n$$

(refer to Ex. <B> 5(h) of Sec. 3.7.3 and Exs. 19 and 21(c) of Sec. B.7). $p_0, p_1, \ldots, p_n$ are called the *Lagrange polynomials* associated with $a_0, a_1, \ldots, a_n$ (see Sec. B.3).

(c) For any scalars $\alpha_0, \alpha_1, \ldots, \alpha_n \in \mathbb{R}$, there exists a unique polynomial $p \in P_n(\mathbb{R})$ such that $p(a_i) = \alpha_i$ for $0 \leq i \leq n$. In fact,

$$p = \sum_{i=0}^{n} p(a_i) p_i = \sum_{i=0}^{n} \alpha_i p_i.$$

This is the *Lagrange interpolation formula*.

(d) Show that

$$\int_a^b p(t)\, dt = \sum_{i=0}^{n} p(a_i) \int_a^b p_i(t)\, dt.$$

Now, divide $[a, b]$ into n equal parts with $a_i = a + \frac{i(b-a)}{n}$ for $0 \leq i \leq n$.

(1) Take $n = 1$. Show that

$$\int_a^b p(t)\, dt = \frac{(b-a)}{2} [p(a) + p(b)].$$

This is the *trapezoidal rule* for polynomials.

(2) Take $n = 2$. Calculate

$$\int_a^b p(t)\, dt = \sum_{i=0}^{2} p(a_i) \int_a^b p_i(t)\, dt$$

which will yield the *Simpson's rule* for polynomials.

3.7.5   *Elementary matrices and matrix factorizations*

All the definitions and theoretical results concerned in Sec. 2.7.5 are still true for real matrices of order 3. Here in this subsection, we use concrete examples to illustrate these results stated there without rewriting or coping them.

Elementary matrices of order 3 are as follows.

Type 1:

$$E_{(1)(2)} = \begin{bmatrix} 0 & 1 & 0 \\ 1 & 0 & 0 \\ 0 & 0 & 1 \end{bmatrix} = F_{(1)(2)};$$

$$E_{(1)(3)} = F_{(1)(3)}; \quad E_{(2)(3)} = F_{(2)(3)}.$$

Type 2:

$$E_{\alpha(1)} = \begin{bmatrix} \alpha & 0 & 0 \\ 0 & 1 & 0 \\ 0 & 0 & 1 \end{bmatrix} = F_{\alpha(1)}, \quad \alpha \neq 0;$$

$$E_{\alpha(2)} = F_{\alpha(2)}; \quad E_{\alpha(3)} = F_{\alpha(3)}.$$

Type 3:

$$E_{(2)+\alpha(1)} = \begin{bmatrix} 1 & 0 & 0 \\ \alpha & 1 & 0 \\ 0 & 0 & 1 \end{bmatrix} = F_{(1)+\alpha(2)}; \quad E_{(3)+\alpha(1)} = F_{(1)+\alpha(3)};$$

$$E_{(1)+\alpha(2)} = F_{(2)+\alpha(1)}; \quad E_{(3)+\alpha(2)} = F_{(2)+\alpha(3)};$$

$$E_{(1)+\alpha(3)} = F_{(3)+\alpha(1)}; \quad E_{(2)+\alpha(3)} = F_{(3)+\alpha(2)}. \tag{3.7.39}$$

For geometric mapping properties of these elementary matrices, please refer to examples in Sec. 3.7.2.

In what follows, we list a series of examples to illustrate systematically the general results stated in (2.7.68)–(2.7.71) and more.

Example 1 Let

$$A = \begin{bmatrix} 1 & 1 & 0 \\ 4 & 3 & -5 \\ 1 & 1 & 5 \end{bmatrix}.$$

(1) Solve $A\vec{x}^* = \vec{b}^*$ where $\vec{x} = (x_1, x_2, x_3) \in \mathbb{R}^3$ and $\vec{b} = (b_1, b_2, b_3)$.
(2) Determine if A is invertible. If yes, compute A^{-1} and express A and A^{-1} as products of elementary matrices.

(3) Compute $\det A$ and $\det A^{-1}$.

(4) Find LU and LDU decompositions of A.

(5) Compute $\text{Im}(A)$, $\text{Ker}(A)$ and $\text{Im}(A^*)$, $\text{Ker}(A^*)$ (see (3.7.32)).

(6) Investigate the geometric mapping properties of A.

Solution Perform elementary row operations to

$$[A|\vec{b}^*|I_3] = \begin{bmatrix} 1 & 1 & 0 & | & b_1 & | & 1 & 0 & 0 \\ 4 & 3 & -5 & | & b_2 & | & 0 & 1 & 0 \\ 1 & 1 & 5 & | & b_3 & | & 0 & 0 & 1 \end{bmatrix}$$

$$\xrightarrow[\substack{E_{(2)-4(1)} \\ E_{(3)-(1)}}]{} \begin{bmatrix} 1 & 1 & 0 & | & b_1 & | & 1 & 0 & 0 \\ 0 & -1 & -5 & | & b_2 - 4b_1 & | & -4 & 1 & 0 \\ 0 & 0 & 5 & | & b_3 - b_1 & | & -1 & 0 & 1 \end{bmatrix} \qquad (*1)$$

$$\xrightarrow[\substack{E_{-(2)} \\ E_{\frac{1}{5}(3)}}]{} \begin{bmatrix} 1 & 1 & 0 & | & b_1 & | & 1 & 0 & 0 \\ 0 & 1 & 5 & | & 4b_1 - b_2 & | & 4 & -1 & 0 \\ 0 & 0 & 1 & | & \frac{1}{5}(b_3 - b_1) & | & -\frac{1}{5} & 0 & \frac{1}{5} \end{bmatrix}$$

$$\xrightarrow[\substack{E_{(1)-(2)} \\ E_{(2)-5(3)}}]{} \begin{bmatrix} 1 & 0 & -5 & | & -3b_1 + b_2 & | & -3 & 1 & 0 \\ 0 & 1 & 0 & | & 5b_1 - b_2 - b_3 & | & 5 & -1 & -1 \\ 0 & 0 & 1 & | & \frac{1}{5}(b_3 - b_1) & | & -\frac{1}{5} & 0 & \frac{1}{5} \end{bmatrix}$$

$$\xrightarrow[E_{(1)+5(3)}]{} \begin{bmatrix} 1 & 0 & 0 & | & -4b_1 + b_2 + b_3 & | & -4 & 1 & 1 \\ 0 & 1 & 0 & | & 5b_1 - b_2 - b_3 & | & 5 & -1 & -1 \\ 0 & 0 & 1 & | & \frac{1}{5}(b_3 - b_1) & | & -\frac{1}{5} & 0 & \frac{1}{5} \end{bmatrix}.$$
$$(*2)$$

Stop at (*1):

$$\begin{array}{ccc} x_1 & x_2 & x_3 \end{array}$$
$$\begin{bmatrix} 1 & 1 & 0 & | & b_1 \\ 0 & -1 & -5 & | & b_2 - 4b_1 \\ 0 & 0 & 5 & | & b_3 - b_1 \end{bmatrix} \Rightarrow \begin{cases} x_1 + x_2 = b_1 \\ -x_2 - 5x_3 = b_2 - 4b_1 \\ 5x_3 = b_3 - b_1 \end{cases}$$

$$\Rightarrow \begin{cases} x_1 = -4b_1 + b_2 + b_3 \\ x_2 = 5b_1 - b_2 - b_3 \\ x_3 = \dfrac{1}{5}(b_3 - b_1). \end{cases}$$

This is the solution of the equations $A\vec{x}^* = \vec{b}^*$. On the other hand,

$$E_{(3)-(1)}E_{(2)-4(1)}A = \begin{bmatrix} 1 & 1 & 0 \\ 0 & -1 & -5 \\ 0 & 0 & 5 \end{bmatrix}$$

$$\Rightarrow A = E_{(2)-4(1)}^{-1} E_{(3)-(1)}^{-1} \begin{bmatrix} 1 & 1 & 0 \\ 0 & -1 & -5 \\ 0 & 0 & 5 \end{bmatrix} = E_{(2)+4(1)} E_{(3)+(1)} \begin{bmatrix} 1 & 1 & 0 \\ 0 & -1 & -5 \\ 0 & 0 & 5 \end{bmatrix}$$

$$= \begin{bmatrix} 1 & 0 & 0 \\ 4 & 1 & 0 \\ 0 & 0 & 1 \end{bmatrix} \begin{bmatrix} 1 & 0 & 0 \\ 0 & 1 & 0 \\ 1 & 0 & 1 \end{bmatrix} \begin{bmatrix} 1 & 1 & 0 \\ 0 & -1 & -5 \\ 0 & 0 & 5 \end{bmatrix}$$

$$= \begin{bmatrix} 1 & 0 & 0 \\ 4 & 1 & 0 \\ 1 & 0 & 1 \end{bmatrix} \begin{bmatrix} 1 & 1 & 0 \\ 0 & -1 & -5 \\ 0 & 0 & 5 \end{bmatrix} \quad \text{(LU-decomposition)}$$

$$= \begin{bmatrix} 1 & 0 & 0 \\ 4 & 1 & 0 \\ 1 & 0 & 1 \end{bmatrix} \begin{bmatrix} 1 & 0 & 0 \\ 0 & -1 & 0 \\ 0 & 0 & 5 \end{bmatrix} \begin{bmatrix} 1 & 1 & 0 \\ 0 & 1 & 5 \\ 0 & 0 & 1 \end{bmatrix} \quad \text{(LDU-decomposition)}.$$

From here, it is easily seen that

$$\det A = \text{ the product of the pivots } 1, -1 \text{ and } 5 = -5.$$
$$\det A^{-1} = (\det A)^{-1} = -\frac{1}{5}.$$

Also, the four subspaces (see (3.7.32)) are:

$$\text{Im}(A) = \langle\langle (1,1,0), (0,-1,-5), (0,0,5)\rangle\rangle = \mathbb{R}^3;$$
$$\text{Ker}(A) = \{\vec{0}\};$$
$$\text{Im}(A^*) = \langle\langle (1,0,0), (1,-1,0), (0,-5,5)\rangle\rangle = \mathbb{R}^3;$$
$$\text{Ker}(A^*) = \{\vec{0}\};$$

which can also be obtained from (*2) (see Application 8 in Sec. B.5).

Stop at (*2):
A is invertible, since

$$E_{(1)+5(3)} E_{(2)-5(3)} E_{(1)-(2)} E_{\frac{1}{5}(3)} E_{-(2)} E_{(3)-(1)} E_{(2)-4(1)} A = I_3.$$

Therefore,

$$A^{-1} = \begin{bmatrix} -4 & 1 & 1 \\ 5 & -1 & -1 \\ -\frac{1}{5} & 0 & \frac{1}{5} \end{bmatrix}$$

$$= E_{(1)+5(3)} E_{(2)-5(3)} E_{(1)-(2)} E_{\frac{1}{5}(3)} E_{-(2)} E_{(3)-(1)} E_{(2)-4(1)}$$

$$\Rightarrow \det A^{-1} = \frac{1}{5} \cdot (-1) = -\frac{1}{5};$$

and

$$A = E^{-1}_{(2)-4(1)} E^{-1}_{(3)-(1)} E^{-1}_{-(2)} E^{-1}_{\frac{1}{5}(3)} E^{-1}_{(1)-(2)} E^{-1}_{(2)-5(3)} E^{-1}_{(1)+5(3)}$$

$$= E_{(2)+4(1)} E_{(3)+(1)} E_{-(2)} E_{5(3)} E_{(1)+(2)} E_{(2)+5(3)} E_{(1)-5(3)}$$

$$= \begin{bmatrix} 1 & 0 & 0 \\ 4 & 1 & 0 \\ 0 & 0 & 1 \end{bmatrix} \begin{bmatrix} 1 & 0 & 0 \\ 0 & 1 & 0 \\ 1 & 0 & 1 \end{bmatrix} \begin{bmatrix} 1 & 0 & 0 \\ 0 & -1 & 0 \\ 0 & 0 & 1 \end{bmatrix} \begin{bmatrix} 1 & 0 & 0 \\ 0 & 1 & 0 \\ 0 & 0 & 5 \end{bmatrix}$$

$$\begin{bmatrix} 1 & 1 & 0 \\ 0 & 1 & 0 \\ 0 & 0 & 1 \end{bmatrix} \begin{bmatrix} 1 & 0 & 0 \\ 0 & 1 & 5 \\ 0 & 0 & 1 \end{bmatrix} \begin{bmatrix} 1 & 0 & -5 \\ 0 & 1 & 0 \\ 0 & 0 & 1 \end{bmatrix}$$

$$\Rightarrow \det A = (-1) \cdot 5 = -5.$$

From the elementary matrix factorization of A, we can recapture the LDU and hence LU decomposition, since

$$E_{(2)+4(1)} E_{(3)+(1)} = \begin{bmatrix} 1 & 0 & 0 \\ 4 & 1 & 0 \\ 1 & 0 & 1 \end{bmatrix} = L,$$

$$E_{-(2)} E_{5(3)} = \begin{bmatrix} 1 & 0 & 0 \\ 0 & -1 & 0 \\ 0 & 0 & 5 \end{bmatrix} = D,$$

$$E_{(1)+(2)} E_{(2)+5(3)} E_{(1)-5(3)} = \begin{bmatrix} 1 & 1 & 0 \\ 0 & 1 & 5 \\ 0 & 0 & 1 \end{bmatrix} = U.$$

This is within our reasonable expectation, because in the process of obtaining (*2), we use $E_{(2)-4(1)}$ and $E_{(3)-(1)}$ to transform the original A into an upper triangle as shown in (*1) and the lower triangle L should be $\left(E_{(3)-(1)} E_{(2)-4(1)}\right)^{-1} = E_{(2)+4(1)} E_{(3)+(1)}$.

The LU-decomposition can help solving $A\vec{x}^* = \vec{b}^*$. Notice that

$$A\vec{x}^* = \vec{b}^*$$

$$\Leftrightarrow \begin{bmatrix} 1 & 1 & 0 \\ 0 & -1 & -5 \\ 0 & 0 & 5 \end{bmatrix} \vec{x}^* = \vec{y}^* \quad \text{and} \quad \begin{bmatrix} 1 & 0 & 0 \\ 4 & 1 & 0 \\ 1 & 0 & 1 \end{bmatrix} \vec{y}^* = \vec{b}^*$$

$$\Leftrightarrow \vec{x}^* = \begin{bmatrix} 1 & 1 & 0 \\ 0 & -1 & -5 \\ 0 & 0 & 5 \end{bmatrix}^{-1} \begin{bmatrix} 1 & 0 & 0 \\ 4 & 1 & 0 \\ 1 & 0 & 1 \end{bmatrix}^{-1} \vec{b}^*$$

$$\Leftrightarrow \vec{x}^* = A^{-1}\vec{b}^*.$$

Readers are urged to carry out actual computations to solve out the solution.

The elementary matrices, LU and LDU decompositions can be used to help investigating geometric mapping properties of A, better using GSP. For example, the image of the unit cube under A is the parallelepiped as shown in Fig. 3.50. This parallelepiped can be obtained by performing successive mappings $E_{(2)+4(1)}$ to the cube followed by $E_{(3)+(1)} \cdots$ then by $E_{(1)-5(3)}$. Also (see Sec. 5.3),

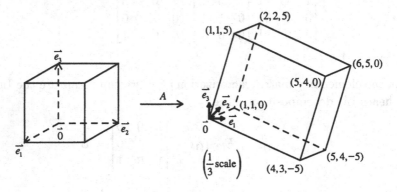

Fig. 3.50

the *signed* volume of the parallelepiped $= \det \begin{bmatrix} 1 & 1 & 0 \\ 4 & 3 & -5 \\ 1 & 1 & 5 \end{bmatrix} = -5$

$$\Rightarrow \frac{\text{the signed volume of the parallelepiped}}{\text{the volume of the unit cube}} = \det A.$$

Since $\det A = -5 < 0$, so A reverses the orientations in $\mathbb{R}^3$. $\square$

Example 2 Let

$$A = \begin{bmatrix} 0 & 1 & -1 \\ 2 & 4 & 6 \\ 2 & 6 & 4 \end{bmatrix}.$$

Do problems similar to Example 1.

Solution Perform elementary row operations to

$$[A|\vec{b}^*|I_3] = \begin{bmatrix} 0 & 1 & -1 & | & b_1 & | & 1 & 0 & 0 \\ 2 & 4 & 6 & | & b_2 & | & 0 & 1 & 0 \\ 2 & 6 & 4 & | & b_3 & | & 0 & 0 & 1 \end{bmatrix} \begin{matrix} (x_1) \\ (x_2) \\ (x_3) \end{matrix}$$

$$\xrightarrow{E_{(1)(2)}} \begin{bmatrix} 2 & 4 & 6 & | & b_2 & | & 0 & 1 & 0 \\ 0 & 1 & -1 & | & b_1 & | & 1 & 0 & 0 \\ 2 & 6 & 4 & | & b_3 & | & 0 & 0 & 1 \end{bmatrix} \begin{matrix} (x_1) \\ (x_2) \\ (x_3) \end{matrix}$$

$$\xrightarrow[E_{\frac{1}{2}(1)}]{E_{(3)-(1)}} \begin{bmatrix} 1 & 2 & 3 & | & \frac{b_2}{2} & | & 0 & \frac{1}{2} & 0 \\ 0 & 1 & -1 & | & b_1 & | & 1 & 0 & 0 \\ 0 & 2 & -2 & | & b_3 - b_2 & | & 0 & -1 & 1 \end{bmatrix}$$

$$\xrightarrow{E_{(3)-2(2)}} \begin{bmatrix} 1 & 2 & 3 & | & \frac{b_2}{2} & | & 0 & \frac{1}{2} & 0 \\ 0 & 1 & -1 & | & b_1 & | & 1 & 0 & 0 \\ 0 & 0 & 0 & | & b_3 - b_2 - 2b_1 & | & -2 & -1 & 1 \end{bmatrix} \quad (*3)$$

$$\xrightarrow{E_{(1)-2(2)}} \begin{bmatrix} 1 & 0 & 5 & | & \frac{b_2}{2} - 2b_1 & | & -2 & \frac{1}{2} & 0 \\ 0 & 1 & -1 & | & b_1 & | & 1 & 0 & 0 \\ 0 & 0 & 0 & | & b_3 - b_2 - 2b_1 & | & -2 & -1 & 1 \end{bmatrix}. \quad (*4)$$

Notice that, since the leading entry of the first row of A is zero, exchange of row 1 and row 2 is needed as the first row operation.

From (*3),

$$A\vec{x}^* = \vec{b}^* \text{ has a solution } \vec{x}.$$

$$\Leftrightarrow E_{(3)-2(2)} E_{\frac{1}{2}(1)} E_{(3)-(1)} E_{(1)(2)} A \begin{bmatrix} x_2 \\ x_1 \\ x_3 \end{bmatrix} = \begin{bmatrix} 1 & 2 & 3 \\ 0 & 1 & -1 \\ 0 & 0 & 0 \end{bmatrix} \begin{bmatrix} x_1 \\ x_2 \\ x_3 \end{bmatrix}$$

$$= \begin{bmatrix} \dfrac{b_2}{2} \\ b_1 \\ b_3 - b_2 - 2b_1 \end{bmatrix}$$

has a solution $\vec{x}$.

$$\Leftrightarrow b_3 - b_2 - 2b_1 = 0.$$

In this case,

$$\begin{cases} x_1 + 2x_2 + 3x_3 = \dfrac{b_2}{2} \\ x_2 - x_3 = b_1 \end{cases}$$

$$\Rightarrow \begin{cases} x_2 = b_1 + x_3 \\ x_1 = -2b_1 + \dfrac{b_2}{2} - 5x_3 \\ x_3 \in \mathbb{R} \text{ is arbitrary.} \end{cases}$$

$$\Rightarrow \vec{x} = \left(-2b_1 + \frac{b_2}{2} - 5x_3, b_1 + x_3, x_3\right)$$

$$= \left(-2b_1 + \frac{b_2}{2}, b_1, 0\right) + x_3(-5, 1, 1), \quad x_3 \in \mathbb{R}.$$

Hence, the solution set is the affine line $\left(-2b_1 + \frac{b_2}{2}, b_1, 0\right) + \langle\langle(-5, 1, 1)\rangle\rangle$ in $\mathbb{R}^3$ with $\left(-2b_1 + \frac{b_2}{2}, b_1, 0\right)$ as a particular solution. It is worth mentioning that, $A\vec{x}^* = \vec{b}^*$ is the system of equations

$$\begin{cases} x_2 - x_3 = b_1 \\ 2x_1 + 4x_2 + 6x_3 = b_2 \\ 2x_1 + 6x_2 + 4x_3 = b_3. \end{cases}$$

For this system of equations to have a solution, it is necessary and sufficient that, after eliminating x_1, x_2, x_3 from the equations,

$$b_2 - 4b_1 - 10x_3 + 6b_1 + 6x_3 + 4x_3 = b_3$$

which is $b_3 - b_2 - 2b_1 = 0$, as claimed above.

(*3) tells us that A is not invertible.

But (*3) does indicate that

$$E_{(3)-2(2)}E_{\frac{1}{2}(1)}E_{(3)-(1)}\left(E_{(1)(2)}A\right) = \begin{bmatrix} 1 & 2 & 3 \\ 0 & 1 & -1 \\ 0 & 0 & 0 \end{bmatrix} \text{ (upper triangle)}$$

$$\Rightarrow E_{(1)(2)}A = \begin{bmatrix} 2 & 4 & 6 \\ 0 & 1 & -1 \\ 2 & 6 & 4 \end{bmatrix} = E_{(3)-(1)}^{-1}E_{\frac{1}{2}(1)}^{-1}E_{(3)-2(2)}^{-1}\begin{bmatrix} 1 & 2 & 3 \\ 0 & 1 & -1 \\ 0 & 0 & 0 \end{bmatrix}$$

$$= E_{(3)+(1)}E_{2(1)}E_{(3)+2(2)}\begin{bmatrix} 1 & 2 & 3 \\ 0 & 1 & -1 \\ 0 & 0 & 0 \end{bmatrix}$$

$$= \begin{bmatrix} 2 & 0 & 0 \\ 0 & 1 & 0 \\ 2 & 2 & 1 \end{bmatrix}\begin{bmatrix} 1 & 2 & 3 \\ 0 & 1 & -1 \\ 0 & 0 & 0 \end{bmatrix}$$

$$= \begin{bmatrix} 1 & 0 & 0 \\ 0 & 1 & 0 \\ 2 & 2 & 1 \end{bmatrix}\begin{bmatrix} 2 & 4 & 6 \\ 0 & 1 & -1 \\ 0 & 0 & 0 \end{bmatrix} \text{ (LU-decomposition)}$$

$$= \begin{bmatrix} 1 & 0 & 0 \\ 0 & 1 & 0 \\ 2 & 2 & 1 \end{bmatrix}\begin{bmatrix} 2 & 0 & 0 \\ 0 & 1 & 0 \\ 0 & 0 & 0 \end{bmatrix}\begin{bmatrix} 1 & 2 & 3 \\ 0 & 1 & -1 \\ 0 & 0 & 0 \end{bmatrix} \text{ (LDU-decomposition)}.$$

Refer to (2) in (2.7.69) and the example after this. A can be decomposed as follows too.

$$A \xrightarrow[E_{(3)-(2)}]{} \begin{bmatrix} 0 & 1 & -1 \\ 2 & 4 & 6 \\ 0 & 2 & -2 \end{bmatrix} \xrightarrow[E_{(3)-2(1)}]{} \begin{bmatrix} 0 & 1 & -1 \\ 2 & 4 & 6 \\ 0 & 0 & 0 \end{bmatrix} \xrightarrow[E_{(1)(2)}]{} \begin{bmatrix} 2 & 4 & 6 \\ 0 & 1 & -1 \\ 0 & 0 & 0 \end{bmatrix}$$

$$\Rightarrow A = E_{(3)-(2)}^{-1} E_{(3)-2(1)}^{-1} E_{(1)(2)}^{-1} \begin{bmatrix} 2 & 4 & 6 \\ 0 & 1 & -1 \\ 0 & 0 & 0 \end{bmatrix}$$

$$= \begin{bmatrix} 1 & 0 & 0 \\ 0 & 1 & 0 \\ 2 & 1 & 1 \end{bmatrix} \begin{bmatrix} 0 & 1 & 0 \\ 1 & 0 & 0 \\ 0 & 0 & 1 \end{bmatrix} \begin{bmatrix} 2 & 4 & 6 \\ 0 & 1 & -1 \\ 0 & 0 & 0 \end{bmatrix} \quad \text{(LPU-decomposition)}.$$

From (*4), firstly, (*4) says that A is not invertible. Secondly, (*4) also says that $A\vec{x}^* = \vec{b}^*$ has a solution if and only if $b_3 - b_2 - 2b_1 = 0$, and the solutions are $x_1 = \frac{b_2}{2} - 2b_1 - 5x_3$, $x_2 = b_1 + x_3$, $x_3 \in \mathbb{R}$. Third, (*4) indicates that

$$E_{(1)-2(2)} E_{(3)-2(2)} E_{\frac{1}{2}(1)} E_{(3)-(1)} E_{(1)(2)} A = \begin{bmatrix} 1 & 0 & 5 \\ 0 & 1 & -1 \\ 0 & 0 & 0 \end{bmatrix}$$

$$\Rightarrow PA = \begin{bmatrix} 1 & 0 & 5 \\ 0 & 1 & -1 \\ 0 & 0 & 0 \end{bmatrix},$$

where $P = \begin{bmatrix} -2 & \frac{1}{2} & 0 \\ 1 & 0 & 0 \\ -2 & -1 & 1 \end{bmatrix}_{3\times 3} = E_{(1)-2(2)} E_{(3)-2(2)} E_{\frac{1}{2}(1)} E_{(3)-(1)} E_{(1)(2)}.$

(row-reduced echelon matrix of A)

Now perform elementary column operations to

$$\begin{bmatrix} PA \\ \hline I_3 \end{bmatrix} = \begin{bmatrix} 1 & 0 & 5 \\ 0 & 1 & -1 \\ 0 & 0 & 0 \\ \hline 1 & 0 & 0 \\ 0 & 1 & 0 \\ 0 & 0 & 1 \end{bmatrix} \xrightarrow[F_{(3)+(2)}]{F_{(3)-5(1)}} \begin{bmatrix} 1 & 0 & 0 \\ 0 & 1 & 0 \\ 0 & 0 & 0 \\ \hline 1 & 0 & -5 \\ 0 & 1 & 1 \\ 0 & 0 & 1 \end{bmatrix} = \begin{bmatrix} I_2 & 0 \\ 0 & 0 \\ \hline Q_{3\times 3} \end{bmatrix}$$

$$\Rightarrow PAQ = \begin{bmatrix} I_2 & 0 \\ 0 & 0 \end{bmatrix}_{3\times 3}, \quad \text{where } Q = \begin{bmatrix} 1 & 0 & -5 \\ 0 & 1 & 1 \\ 0 & 0 & 1 \end{bmatrix} = F_{(3)-5(1)} F_{(3)+(2)}.$$

(the normal form of A)

From $PA = R$, the row-reduced echelon matrix of A, it follows that

$\text{Im}(A) = \langle\langle(1,0,5),(0,1,-1)\rangle\rangle$, where the basis vectors are the first and the second row vectors of PA;

$\text{Ker}(A) = \langle\langle(-2,-1,1)\rangle\rangle$, where $(-2,-1,1)$ is the third row vector of P;

$\text{Im}(A^*) = \langle\langle(0,2,2),(1,4,6)\rangle\rangle$, generated by the first and the second column vectors of A;

$\text{Ker}(A^*) = \langle\langle(-5,1,1)\rangle\rangle$, generated by the fundamental solution $(-5,1,1)$ of $A\vec{x}^* = \vec{0}$.

Refer to Application 8 in Sec. B.5.

Various decompositions of A, such as LU, LDU, LPU, row-reduced echelon matrix and normal form, can be used to study the geometric mapping properties of A (refer to (3.7.9) and (3.7.10)). Of course, the application of GSP method in drawing graphs is strongly convincing, especially when applying stepwise to each member of an elementary matrix factorization of a matrix.

By computation,

$$\text{Ker}(A) = \langle\langle(-2,-1,1)\rangle\rangle,$$
$$\text{Im}(A) = \{(x_1,x_2,x_3) \mid 5x_1 - x_2 - x_3 = 0\} = \langle\langle\vec{e}_1 A, \vec{e}_2 A\rangle\rangle$$
$$= \langle\langle(0,1,-1),(2,4,6)\rangle\rangle.$$

Therefore, the parallelepiped with side vectors $\vec{e}_1, \vec{e}_2$ and $(-2,-1,1)$ has the image under A the parallelogram with side vector $\vec{e}_1 A$ and $\vec{e}_2 A$. See Fig. 3.51.

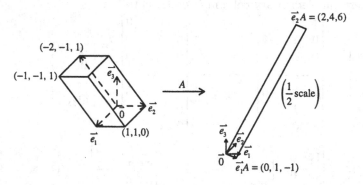

Fig. 3.51

By the way (why? See Sec. 5.3 if needed),

$$\text{the volume of the parallelepiped} = \begin{vmatrix} 1 & 0 & 0 \\ 0 & 1 & 0 \\ -2 & -1 & 1 \end{vmatrix} = 1,$$

$$\text{the area of the parallelogram} = \begin{vmatrix} \langle \vec{e}_1 A, \vec{e}_1 A \rangle & \langle \vec{e}_1 A, \vec{e}_2 A \rangle \\ \langle \vec{e}_1 A, \vec{e}_2 A \rangle & \langle \vec{e}_2 A, \vec{e}_2 A \rangle \end{vmatrix}^{\frac{1}{2}}$$

$$= \begin{vmatrix} 2 & -2 \\ -2 & 56 \end{vmatrix}^{\frac{1}{2}} = 6\sqrt{3}.$$

What is the preimage of the parallelogram $\square(\vec{e}_1 A)(\vec{e}_2 A)$? Readers are urged to find the image under A of the unit cube with side vectors $\vec{e}_1, \vec{e}_2$ and $\vec{e}_3$.

Yet the other way to visualize the mapping properties of A are to see if A is diagonalizable. By computation, we have

eigenvalues	eigenvectors
$\lambda_1 = 0$	$\vec{v}_1 = t(-2, -1, 1)$, for $t \in \mathbb{R}$ and $t \neq 0$
$\lambda_2 = 10$	$\vec{v}_2 = t(12, 31, 29)$
$\lambda_3 = -2$	$\vec{v}_3 = t(0, 1, -1)$.

So A is diagonalizable (see Case 1 in (3.7.29) or Sec. 3.7.6). Refer to Figs. 3.30 and 3.31. □

Example 3 Let

$$A = \begin{bmatrix} 1 & -3 & 0 \\ -3 & 2 & -1 \\ 0 & -1 & 4 \end{bmatrix}.$$

Do problems similar to Example 1.

Solution Perform elementary row operations to

$$\left[A \mid \vec{b}^* \mid I_3 \right] = \begin{bmatrix} 1 & -3 & 0 & \vline & b_1 & \vline & 1 & 0 & 0 \\ -3 & 2 & -1 & \vline & b_2 & \vline & 0 & 1 & 0 \\ 0 & -1 & 4 & \vline & b_3 & \vline & 0 & 0 & 1 \end{bmatrix}$$

$$\xrightarrow[E_{(2)+3(1)}]{} \begin{bmatrix} 1 & -3 & 0 & \vline & b_1 & \vline & 1 & 0 & 0 \\ 0 & -7 & -1 & \vline & b_2 + 3b_1 & \vline & 3 & 1 & 0 \\ 0 & -1 & 4 & \vline & b_3 & \vline & 0 & 0 & 1 \end{bmatrix}$$

$$\xrightarrow[E_{-\frac{1}{7}(2)}]{} \left[\begin{array}{ccc|c|ccc} 1 & -3 & 0 & b_1 & 1 & 0 & 0 \\ 0 & 1 & \frac{1}{7} & -\frac{b_2+3b_1}{7} & -\frac{3}{7} & -\frac{1}{7} & 0 \\ 0 & -1 & 4 & b_3 & 0 & 0 & 1 \end{array}\right]$$

$$\xrightarrow[E_{(3)+(2)}]{} \left[\begin{array}{ccc|c|ccc} 1 & -3 & 0 & b_1 & 1 & 0 & 0 \\ 0 & 1 & \frac{1}{7} & -\frac{b_2+3b_1}{7} & -\frac{3}{7} & -\frac{1}{7} & 0 \\ 0 & 0 & \frac{29}{7} & -\frac{b_2+3b_1}{7}+b_3 & -\frac{3}{7} & -\frac{1}{7} & 1 \end{array}\right] \qquad (*5)$$

$$\xrightarrow[\substack{E_{\frac{7}{29}(3)} \\ E_{(1)+3(2)}}]{} \left[\begin{array}{ccc|c|ccc} 1 & 0 & \frac{3}{7} & -\frac{3b_2+2b_1}{7} & -\frac{2}{7} & -\frac{3}{7} & 0 \\ 0 & 1 & \frac{1}{7} & -\frac{b_2+3b_1}{7} & -\frac{3}{7} & -\frac{1}{7} & 0 \\ 0 & 0 & 1 & -\frac{b_2+3b_1-7b_3}{29} & -\frac{3}{29} & -\frac{1}{29} & \frac{7}{29} \end{array}\right]$$

$$\xrightarrow[\substack{E_{(1)-\frac{3}{7}(3)} \\ E_{(2)-\frac{1}{7}(3)}}]{} \left[\begin{array}{ccc|c|ccc} 1 & 0 & 0 & -\frac{7b_1+12b_2+3b_3}{29} & -\frac{7}{29} & -\frac{12}{29} & -\frac{3}{29} \\ 0 & 1 & 0 & -\frac{12b_1+4b_2+b_3}{29} & -\frac{12}{29} & -\frac{4}{29} & -\frac{1}{29} \\ 0 & 0 & 1 & -\frac{3b_1+b_2-7b_3}{29} & -\frac{3}{29} & -\frac{1}{29} & \frac{7}{29} \end{array}\right]. \qquad (*6)$$

From (*5):

$$A\vec{x}^* = \vec{b}^* \text{ has a solution } \vec{x}.$$
$$\Leftrightarrow PA\vec{x}^* = P\vec{b}^* \text{ has a solution } \vec{x}, \quad \text{where}$$

$$P = E_{(3)+(2)}E_{-\frac{1}{7}(2)}E_{(2)+3(1)} = \left[\begin{array}{ccc} 1 & 0 & 0 \\ -\frac{3}{7} & -\frac{1}{7} & 0 \\ -\frac{3}{7} & -\frac{1}{7} & 1 \end{array}\right].$$

In this case,

$$\begin{cases} x_1 - 3x_2 = b_1 \\ x_2 + \frac{1}{7}x_3 = -\frac{1}{7}(b_2+3b_1) \\ \frac{29}{7}x_3 = -\frac{1}{7}(b_2+3b_1-7b_3) \end{cases} \Rightarrow \begin{cases} x_1 = -\frac{1}{29}(7b_1+12b_2+3b_3) \\ x_2 = -\frac{1}{29}(12b_1+4b_2+b_3) \\ x_3 = -\frac{1}{29}(3b_1+b_2-7b_3) \end{cases}.$$

On the other hand,

$$PA = \left[\begin{array}{ccc} 1 & -3 & 0 \\ 0 & 1 & \frac{1}{7} \\ 0 & 0 & \frac{29}{7} \end{array}\right]$$

implies that A is invertible and

$$\det(PA) = \det P \cdot \det A = -\frac{1}{7}\det A = \frac{29}{7}$$

$$\Rightarrow \det A = -29 \quad \text{and} \quad \det A^{-1} = -\frac{1}{29}.$$

Since

$$P^{-1} = E_{(2)-3(1)}E_{-7(2)}E_{(3)-(2)} = \begin{bmatrix} 1 & 0 & 0 \\ -3 & -7 & 0 \\ 0 & -1 & 1 \end{bmatrix}$$

$$\Rightarrow A = \begin{bmatrix} 1 & 0 & 0 \\ -3 & -7 & 0 \\ 0 & -1 & 1 \end{bmatrix}\begin{bmatrix} 1 & -3 & 0 \\ 0 & 1 & \frac{1}{7} \\ 0 & 0 & \frac{29}{7} \end{bmatrix}$$

$$= \begin{bmatrix} 1 & 0 & 0 \\ -3 & 1 & 0 \\ 0 & \frac{1}{7} & 1 \end{bmatrix}\begin{bmatrix} 1 & -3 & 0 \\ 0 & -7 & -1 \\ 0 & 0 & \frac{29}{7} \end{bmatrix} \quad \text{(LU-decomposition)}$$

$$= \begin{bmatrix} 1 & 0 & 0 \\ -3 & 1 & 0 \\ 0 & \frac{1}{7} & 1 \end{bmatrix}\begin{bmatrix} 1 & 0 & 0 \\ 0 & -7 & 0 \\ 0 & 0 & \frac{29}{7} \end{bmatrix}\begin{bmatrix} 1 & -3 & 0 \\ 0 & 1 & \frac{1}{7} \\ 0 & 0 & 1 \end{bmatrix} \quad \text{(LDL*-decomposition)}.$$

These decompositions can help in solving the equation $A\vec{x}^* = \vec{b}^*$.

Moreover,

$$PAP^* = \begin{bmatrix} 1 & -3 & 0 \\ 0 & 1 & \frac{1}{7} \\ 0 & 0 & \frac{29}{7} \end{bmatrix}\begin{bmatrix} 1 & -\frac{3}{7} & -\frac{3}{7} \\ 0 & -\frac{1}{7} & -\frac{1}{7} \\ 0 & 0 & 1 \end{bmatrix} = \begin{bmatrix} 1 & 0 & 0 \\ 0 & -\frac{1}{7} & 0 \\ 0 & 0 & \frac{29}{7} \end{bmatrix}. \quad (*7)$$

Notice that

$$P^* = E^*_{(2)+3(1)}E^*_{-\frac{1}{7}(2)}E^*_{(3)+(2)} = F_{(2)+3(1)}F_{-\frac{1}{7}(2)}F_{(3)+(2)}.$$

Therefore, (*7) might be expected beforehand since A is a *symmetric* matrix. (*7) means that, when performing at the same time elementary row operations and column operations of the same types, we will get a diagonal matrix. In fact

$$A \xrightarrow[F_{(2)+3(1)}]{E_{(2)+3(1)}} E_{(2)+3(1)}AF_{(2)+3(1)} = \begin{bmatrix} 1 & 0 & 0 \\ 0 & -7 & -1 \\ 0 & -1 & 4 \end{bmatrix}$$

$$\xrightarrow[F_{-\frac{1}{7}(2)}]{E_{-\frac{1}{7}(2)}} \begin{bmatrix} 1 & 0 & 0 \\ 0 & -\frac{1}{7} & \frac{1}{7} \\ 0 & \frac{1}{7} & 4 \end{bmatrix} \xrightarrow[F_{(3)+(2)}]{E_{(3)+(2)}} \begin{bmatrix} 1 & 0 & 0 \\ 0 & -\frac{1}{7} & 0 \\ 0 & 0 & \frac{29}{7} \end{bmatrix} \qquad (*8)$$

which is (*7). Interchange row 2 and row 3, and let

$$P_1 = E_{(2)(3)}P = \begin{bmatrix} 1 & 0 & 0 \\ -\frac{3}{7} & -\frac{1}{7} & 1 \\ -\frac{3}{7} & -\frac{1}{7} & 0 \end{bmatrix}.$$

Therefore

$$P_1 AP_1^* = \begin{bmatrix} 1 & 0 & 0 \\ 0 & \frac{29}{7} & 0 \\ 0 & 0 & -\frac{1}{7} \end{bmatrix} = \begin{bmatrix} 1 & 0 & 0 \\ 0 & \sqrt{\frac{29}{7}} & 0 \\ 0 & 0 & \frac{1}{\sqrt{7}} \end{bmatrix} \begin{bmatrix} 1 & 0 & 0 \\ 0 & 1 & 0 \\ 0 & 0 & -1 \end{bmatrix} \begin{bmatrix} 1 & 0 & 0 \\ 0 & \sqrt{\frac{29}{7}} & 0 \\ 0 & 0 & \frac{1}{\sqrt{7}} \end{bmatrix}$$

$$\Rightarrow QP_1 AP_1^* Q^* = \begin{bmatrix} 1 & 0 & 0 \\ 0 & 1 & 0 \\ 0 & 0 & -1 \end{bmatrix}, \quad \text{where } Q = \begin{bmatrix} 1 & 0 & 0 \\ 0 & \sqrt{\frac{7}{29}} & 0 \\ 0 & 0 & \sqrt{7} \end{bmatrix}.$$

Now, let

$$S = QP_1 = \begin{bmatrix} 1 & 0 & 0 \\ -\frac{3\sqrt{7}}{7\sqrt{29}} & -\frac{\sqrt{7}}{7\sqrt{29}} & \frac{\sqrt{7}}{\sqrt{29}} \\ -\frac{3\sqrt{7}}{7} & -\frac{\sqrt{7}}{7} & 0 \end{bmatrix}$$

$$\Rightarrow SAS^* = \begin{bmatrix} 1 & 0 & 0 \\ 0 & 1 & 0 \\ 0 & 0 & -1 \end{bmatrix} \quad (A \text{ is } congruent \text{ to a diagonal matrix}). \quad (*9)$$

Hence, the *index* of A is equal to 2, the *signature* is equal to $2 - 1 = 1$, and the rank of A is $2 + 1 = 3$ (see (2.7.71)).

The invertible matrix S in (*9) is not necessarily unique. A is diagonalizable (see Secs. 3.7.6 and 5.7). A has characteristic polynomial $\det(A - tI_3) = -t^3 + 7t^2 - 4t - 29$ and has two positive eigenvalues λ_1, λ_2

and one negative eigenvalue λ_3. Let $\vec{v}_1, \vec{v}_2$ and $\vec{v}_3$ be corresponding eigenvectors so that

$$R = \begin{bmatrix} \vec{v}_1 \\ \vec{v}_2 \\ \vec{v}_3 \end{bmatrix}$$

is an *orthogonal* matrix, i.e. $R^* = R^{-1}$. Then

$$RAR^{-1} = RAR^*$$

$$= \begin{bmatrix} \lambda_1 & 0 & 0 \\ 0 & \lambda_2 & 0 \\ 0 & 0 & \lambda_3 \end{bmatrix}$$

$$= \begin{bmatrix} \sqrt{\lambda_1} & 0 & 0 \\ 0 & \sqrt{\lambda_2} & 0 \\ 0 & 0 & \sqrt{-\lambda_3} \end{bmatrix} \begin{bmatrix} 1 & 0 & 0 \\ 0 & 1 & 0 \\ 0 & 0 & -1 \end{bmatrix} \begin{bmatrix} \sqrt{\lambda_1} & 0 & 0 \\ 0 & \sqrt{\lambda_2} & 0 \\ 0 & 0 & \sqrt{-\lambda_3} \end{bmatrix}$$

$$\Rightarrow S_1 A S_1^* = \begin{bmatrix} 1 & 0 & 0 \\ 0 & 1 & 0 \\ 0 & 0 & -1 \end{bmatrix}, \quad \text{where } S_1 = \begin{bmatrix} \frac{1}{\sqrt{\lambda_1}} & 0 & 0 \\ 0 & \frac{1}{\sqrt{\lambda_2}} & 0 \\ 0 & 0 & \frac{1}{\sqrt{-\lambda_3}} \end{bmatrix} \begin{bmatrix} \vec{v}_1 \\ \vec{v}_2 \\ \vec{v}_3 \end{bmatrix}.$$

$$(*10)$$

This S_1 is not equal to S in (*9) in general.

From (*6):

(*6) solves $A\vec{x}^* = \vec{b}^*$ immediately, just by inspection.

Also, A is invertible and

$$A^{-1} = E_{(2)-\frac{1}{7}(3)} E_{(1)-\frac{3}{7}(3)} E_{(1)+3(2)} E_{\frac{7}{29}(3)} E_{(3)+(2)} E_{-\frac{1}{7}(2)} E_{(2)+3(1)}$$

$$= -\frac{1}{29} \begin{bmatrix} 7 & 12 & 3 \\ 12 & 4 & 1 \\ 3 & 1 & -7 \end{bmatrix},$$

$$A = E_{(2)-3(1)} E_{-7(2)} E_{(3)-(2)} E_{\frac{29}{7}(3)} E_{(1)-3(2)} E_{(1)+\frac{3}{7}(3)} E_{(2)+\frac{1}{7}(3)}.$$

Note that A^{-1} is also a symmetric matrix. Again, GSP will provide a stepwise graphing of the mapping properties of A according to the above elementary matrix factorization.

To the end, let us *preview* one of the main applications of the concept
of congruence among matrices. Give the *quadric* from

$$\langle \vec{x}, \vec{x}A \rangle = x_1^2 + 2x_2^2 + 4x_3^2 - 6x_1x_2 - 2x_2x_3 = 1, \quad \vec{x} = (x_1, x_2, x_3)$$

in $\mathbb{R}^3$. Let

$$\vec{y} = (y_1, y_2, y_3) = \vec{x} S^{-1}$$

= the coordinate vector of $\vec{x}$ in the coordinate system $\mathcal{B}$ formed by three row vectors of S.

Then

$$\langle x, \vec{x} A \rangle = \vec{x} A \vec{x}^* = (\vec{x} S^{-1})(SAS^*)(\vec{x} S^{-1})^*$$
$$= \vec{y}(SAS^*)\vec{y}^*$$
$$= y_1^2 + y_2^2 - y_3^2 = 1.$$

This means that, in $\mathcal{B}$, the quadric looks like a *hyperboloid of one sheet* (refer to Fig. 3.90) and can be used as a model for *hyperbolic geometry* (see Sec. 5.12).

The following examples are concerned with matrices of order $m \times n$ where $m \neq n$.

Example 4 Let

$$A = \begin{bmatrix} 2 & 1 & 0 \\ -1 & 0 & 1 \end{bmatrix}.$$

(1) Find LU decomposition and the normal form of A (see (2.7.70)).
(2) Find a matrix $B_{3 \times 2}$ so that $AB = I_2$ (refer to Ex. <A> 7 of Sec. 2.7.4, Ex. 5 of Sec. B.7 and Exs. 3–5 of Sec. B.8).
(3) Try to investigate the geometric mapping properties of AA^* and A^*A (refer to Ex. <B> of Sec. 2.7.5 and Ex. <B> of Sec. 3.7.4).

Caution that, for the understanding of (3), one needs basic knowledge about Euclidean structures of $\mathbb{R}^2$ and $\mathbb{R}^3$, and one can refer to Part 2 if necessary.

Solution Perform elementary column operations to

$$\begin{bmatrix} A \\ \hline I_3 \end{bmatrix} = \begin{bmatrix} 2 & 1 & 0 \\ -1 & 0 & 1 \\ \hline 1 & 0 & 0 \\ 0 & 1 & 0 \\ 0 & 0 & 1 \end{bmatrix} \xrightarrow[F_{(1)+\frac{1}{2}(3)}]{F_{\frac{1}{2}(1)}} \begin{bmatrix} 1 & 1 & 0 \\ 0 & 0 & 1 \\ \hline \frac{1}{2} & 0 & 0 \\ 0 & 1 & 0 \\ \frac{1}{2} & 0 & 1 \end{bmatrix} \xrightarrow[F_{(2)(3)}]{F_{(2)-(1)}} \begin{bmatrix} 1 & 0 & 0 \\ 0 & 1 & 0 \\ \hline \frac{1}{2} & 0 & -\frac{1}{2} \\ 0 & 0 & 1 \\ \frac{1}{2} & 1 & -\frac{1}{2} \end{bmatrix}.$$

This indicates that

$$A \begin{bmatrix} \frac{1}{2} & 0 & -\frac{1}{2} \\ 0 & 0 & 1 \\ \frac{1}{2} & 1 & -\frac{1}{2} \end{bmatrix} = \begin{bmatrix} 1 & 0 & 0 \\ 0 & 1 & 0 \end{bmatrix} = [I_2 \quad 0],$$

which is the *normal form* of A. From this, we extract a submatrix $B_{3\times 2}$ so that

$$AB = I_2, \quad \text{where } B = \begin{bmatrix} \frac{1}{2} & 0 \\ 0 & 0 \\ \frac{1}{2} & 1 \end{bmatrix}.$$

Hence, A is *right invertible* and B is one of its right inverses. In general, the right inverses B can be found in the following way. Let $B = [\vec{v}_1^* \quad \vec{v}_2^*]_{3\times 2}$. Then

$$AB = A \begin{bmatrix} \vec{v}_1^* & \vec{v}_2^* \end{bmatrix} = \begin{bmatrix} A\vec{v}_1^* & A\vec{v}_2^* \end{bmatrix} = I_2$$
$$\Leftrightarrow A\vec{v}_1^* = \vec{e}_1^*$$
$$A\vec{v}_2^* = \vec{e}_2^*.$$

Suppose $\vec{v}_1 = (x_1, x_2, x_3)$, then

$$A\vec{v}_1^* = \vec{e}_1^* \Leftrightarrow 2x_1 + x_2 = 1$$
$$- x_1 + x_3 = 0$$
$$\Leftrightarrow \vec{v}_1 = (0,1,0) + t_1(1,-2,1) \quad \text{for } t_1 \in \mathbb{R}.$$

Similarly, if $\vec{v}_2 = (x_1, x_2, x_3)$, then

$$A\vec{v}_2^* = \vec{e}_2^* \Leftrightarrow 2x_1 + x_2 = 0$$
$$- x_1 + x_3 = 1$$
$$\Leftrightarrow \vec{v}_2 = (0,0,1) + t_2(1,-2,1) \quad \text{for } t_2 \in \mathbb{R}.$$

Therefore, the *right inverses* are

$$B = \begin{bmatrix} t_1 & t_2 \\ 1 - 2t_1 & -2t_2 \\ t_1 & 1 + t_2 \end{bmatrix}_{3\times 2} \quad \text{for } t_1, t_2 \in \mathbb{R}. \tag{*11}$$

On the other hand, perform elementary row operations to

$$\left[A \mid \vec{b}^* \mid I_2\right] = \begin{bmatrix} 2 & 1 & 0 \mid b_1 \mid 1 & 0 \\ -1 & 0 & 1 \mid b_2 \mid 0 & 1 \end{bmatrix}$$

$$\xrightarrow[E_{\frac{1}{2}(1)},\ E_{(2)+(1)}]{} \begin{bmatrix} 1 & \frac{1}{2} & 0 \mid \frac{b_1}{2} \mid \frac{1}{2} & 0 \\ 0 & \frac{1}{2} & 1 \mid b_2 + \frac{b_1}{2} \mid \frac{1}{2} & 1 \end{bmatrix} \qquad (*12)$$

$$\xrightarrow[E_{2(2)},\ E_{(1)-\frac{1}{2}(2)}]{} \begin{bmatrix} 1 & 0 & -1 \mid -b_2 \mid 0 & -1 \\ 0 & 1 & 2 \mid 2b_2 + b_1 \mid 1 & 2 \end{bmatrix}. \qquad (*13)$$

From (*12),

$$E_{(2)+(1)} E_{\frac{1}{2}(2)} A = \begin{bmatrix} 1 & \frac{1}{2} & 0 \\ 0 & \frac{1}{2} & 1 \end{bmatrix} \quad \text{(an echelon matrix)}$$

$$\Rightarrow A = E_{2(1)} E_{(2)-(1)} \begin{bmatrix} 1 & \frac{1}{2} & 0 \\ 0 & \frac{1}{2} & 1 \end{bmatrix}$$

$$= \begin{bmatrix} 2 & 0 \\ -1 & 1 \end{bmatrix} \begin{bmatrix} 1 & \frac{1}{2} & 0 \\ 0 & \frac{1}{2} & 1 \end{bmatrix} \quad \text{(LU-decomposition)}.$$

Refer to (2.7.70).

From (*13),

$$A\vec{x}^* = \vec{b}^* \text{ has a solution } \vec{x} = (x_1, x_2, x_3).$$

$$\Leftrightarrow \begin{bmatrix} 1 & 0 & -1 \\ 0 & 1 & 2 \end{bmatrix} \begin{bmatrix} x_1 \\ x_2 \\ x_3 \end{bmatrix} = \begin{bmatrix} -b_2 \\ 2b_2 + b_1 \end{bmatrix} \text{ has a solution } \vec{x}.$$

$$\Rightarrow x_1 = -b_2 + x_3$$
$$x_2 = b_1 + 2b_2 - 2x_3, \quad x_3 \in \mathbb{R}$$
$$\Rightarrow \vec{x} = (-b_2, b_1 + 2b_2, 0) + t(1, -2, 1), \quad t \in \mathbb{R}.$$

Thus the solution set is a one-dimensional affine subspace in $\mathbb{R}^3$.

Also, (*13) indicates that

$$E_{(1)-\frac{1}{2}(2)} E_{2(2)} E_{(2)+(1)} E_{\frac{1}{2}(1)} A$$

$$= \begin{bmatrix} 1 & 0 & -1 \\ 0 & 1 & 2 \end{bmatrix} \quad \text{(row-reduced echelon matrix)}$$

$$\Rightarrow PA = \begin{bmatrix} 1 & 0 & -1 \\ 0 & 1 & 2 \end{bmatrix}, \quad \text{where } P = E_{(1)-\frac{1}{2}(2)} E_{2(2)} E_{(2)+(1)} E_{\frac{1}{2}(1)} = \begin{bmatrix} 0 & -1 \\ 1 & 2 \end{bmatrix}.$$

Then, perform elementary column operations to

$$
\begin{bmatrix} PA \\ \hline I_3 \end{bmatrix} = \left[\begin{array}{ccc} 1 & 0 & -1 \\ 0 & 1 & 2 \\ \hline 1 & 0 & 0 \\ 0 & 1 & 0 \\ 0 & 0 & 1 \end{array} \right] \xrightarrow[F_{(3)-2(2)}]{F_{(3)+(1)}} \left[\begin{array}{ccc} 1 & 0 & 0 \\ 0 & 1 & 0 \\ \hline 1 & 0 & 1 \\ 0 & 1 & -2 \\ 0 & 0 & 1 \end{array} \right]
$$

$$
\Rightarrow PAQ = \begin{bmatrix} I_2 & O_{1\times1} \end{bmatrix}_{2\times3}, \text{ where } Q = \begin{bmatrix} 1 & 0 & 1 \\ 0 & 1 & -2 \\ 0 & 0 & 1 \end{bmatrix}.
$$

This is the *normal form* of A.

For part (3), we consider the solvability of the system of equation $\vec{x}A = \vec{b}$ where $\vec{x} = (x_1, x_2) \in \mathbb{R}^2$ and $\vec{b} = (b_1, b_2, b_3) \in \mathbb{R}^3$.

Simple calculation or from (*13) shows that

$$
\mathrm{Ker}(A) = \{\vec{0}\} = \mathrm{Im}(A^*)^{\perp},
$$
$$
\mathrm{Im}(A) = \{(y_1, y_2, y_3) \in \mathbb{R}^3 \mid y_1 - 2y_2 + y_3 = 0\} = \langle\langle(2,1,0),(0,1,2)\rangle\rangle
$$
$$
= \langle\langle(1,0,-1),(0,1,2)\rangle\rangle = \mathrm{Ker}(A^*)^{\perp},
$$
$$
\mathrm{Ker}(A^*) = \langle\langle(1,-2,1)\rangle\rangle = \mathrm{Im}(A)^{\perp},
$$
$$
\mathrm{Im}(A^*) = \langle\langle(2,-2),(1,2)\rangle\rangle = \langle\langle(2,-1),(1,0)\rangle\rangle = \mathbb{R}^2 = \mathrm{Ker}(A)^{\perp},
$$

and

$$
AA^* = \begin{bmatrix} 5 & -2 \\ -2 & 2 \end{bmatrix},
$$
$$
(AA^*)^{-1} = \frac{1}{6}\begin{bmatrix} 2 & 2 \\ 2 & 5 \end{bmatrix}.
$$

For any fixed $\vec{b} \in \mathbb{R}^3$, $\vec{x}A = \vec{b}$ may have a solution or not if and only if the distance from $\vec{b}$ to the range space $\mathrm{Im}(A)$ is zero or not (see Ex. <B> 4 of Sec. 3.7.3 and Fig. 3.42). Suppose $\vec{x}_0 \in \mathbb{R}^2$ so that

$$
|\vec{b} - \vec{x}_0 A| = \min_{\vec{x} \in \mathbb{R}^2} |\vec{b} - \vec{x}A|
$$
$$
\Leftrightarrow (\vec{b} - \vec{x}_0 A) \perp \vec{x}A \quad \text{for all } \vec{x} \in \mathbb{R}^2
$$
$$
\Leftrightarrow (\vec{b} - \vec{x}_0 A)(\vec{x}A)^* = 0 \quad \text{for all } \vec{x} \in \mathbb{R}^2
$$
$$
\Leftrightarrow \vec{x}_0 AA^* = \vec{b}A^*
$$
$$
\Leftrightarrow \vec{x}_0 = \vec{b}A^*(AA^*)^{-1}. \tag{*14}
$$

and $\vec{x}_0 A = \vec{b} A^*(AA^*)^{-1}A$ is the *orthogonal projection* of $\vec{b}$ on $\operatorname{Im}(A)$.The operator

$$A^*(AA^*)^{-1}A = \begin{bmatrix} 2 & -1 \\ 1 & 0 \\ 0 & 1 \end{bmatrix} \cdot \frac{1}{6}\begin{bmatrix} 2 & 2 \\ 2 & 5 \end{bmatrix} \cdot \begin{bmatrix} 2 & 1 & 0 \\ -1 & 1 & 0 \end{bmatrix}$$

$$= \frac{1}{6}\begin{bmatrix} 5 & 2 & -1 \\ 2 & 2 & 2 \\ -1 & 2 & 5 \end{bmatrix} \qquad (*15)$$

is the *projection* of $\mathbb{R}^3$ onto $\operatorname{Im}(A)$ along $\operatorname{Ker}(A^*)$ (see (3.7.34)). In fact,

$$A^*(AA^*)^{-1}A \text{ is symmetric and}$$
$$(A^*(AA^*)^{-1}A)^2 = A^*(AA^*)^{-1}A.$$

Moreover, it is *orthogonal* in the sense that

$$\operatorname{Im}(A^*(AA^*)^{-1}A)^{\perp} = \operatorname{Ker}(A^*(AA^*)^{-1}A) = \operatorname{Ker}(A^*) = \operatorname{Im}(A)^{\perp},$$
$$\operatorname{Ker}(A^*(AA^*)^{-1}A)^{\perp} = \operatorname{Im}(A^*(AA^*)^{-1}A) = \operatorname{Im}(A).$$

See Fig. 3.52. Such an operation is simply called an *orthogonal projection* of $\mathbb{R}^3$ onto $\operatorname{Im}(A)$ along $\operatorname{Ker}(A^*)$. We have defined $A^*(AA^*)^{-1}$ as the *generalized inverse* A^+ of A (see the Note in Ex. <B> 4 of Sec. 3.7.3 or Sec. B.8). Try to find A^+ out of (*11).

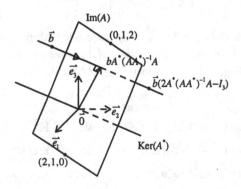

Fig. 3.52

A subsequent problem is to find the *reflection* or *symmetric point* of $\vec{b}$ with respect to the plane $\text{Im}(A)$ (see Fig. 3.52). It is the point

$$\vec{b}\,A^*(AA^*)^{-1}A + (\vec{b}\,A^*(AA^*)^{-1}A - \vec{b}) = \vec{b}\,(2A^*(AA^*)^{-1}A - I_3). \quad (*16)$$

Hence, denote the operator

$$P_A = 2A^*(AA^*)^{-1}A - I_3 = 2A^+A - I_3$$

$$= 2 \cdot \frac{1}{6} \begin{bmatrix} 5 & 2 & -1 \\ 2 & 2 & 2 \\ -1 & 2 & 5 \end{bmatrix} - I_3 = \begin{bmatrix} \frac{2}{3} & \frac{2}{3} & -\frac{1}{3} \\ \frac{2}{3} & -\frac{1}{3} & \frac{2}{3} \\ -\frac{1}{3} & \frac{2}{3} & \frac{2}{3} \end{bmatrix}. \quad (*17)$$

P_A is an *orthogonal* matrix, i.e. $P_A^* = P_A^{-1}$ and P_A is *symmetric*.

By simple computation or even by geometric intuition, we have:

eigenvalues of A^+A	eigenvalues of P_A	eigenvectors
1	1	$\vec{v}_1 = \left(\dfrac{2}{\sqrt{5}}, \dfrac{1}{\sqrt{5}}, 0 \right)$
1	1	$\vec{v}_2 = \left(\dfrac{1}{\sqrt{30}}, -\dfrac{2}{\sqrt{30}}, -\dfrac{5}{\sqrt{30}} \right)$
0	-1	$\vec{v}_3 = \left(\dfrac{1}{\sqrt{6}}, -\dfrac{2}{\sqrt{6}}, \dfrac{1}{\sqrt{6}} \right)$

$\mathcal{B} = \{\vec{v}_1, \vec{v}_2, \vec{v}_3\}$ is an *orthonormal basis* for $\mathbb{R}^3$. In $\mathcal{B}$,

$$[A^+A]_{\mathcal{B}} = RA^+R^{-1} = \begin{bmatrix} 1 & 0 & 0 \\ 0 & 1 & 0 \\ 0 & 0 & 0 \end{bmatrix}, \text{ where } R = \begin{bmatrix} \vec{v}_1 \\ \vec{v}_2 \\ \vec{v}_3 \end{bmatrix} \text{ is orthogonal;}$$

$$[P_A]_{\mathcal{B}} = RP_AR^{-1} = \begin{bmatrix} 1 & 0 & 0 \\ 0 & 1 & 0 \\ 0 & 0 & -1 \end{bmatrix}.$$

Try to explain $[A^+A]_{\mathcal{B}}$ and $[P_A]_{\mathcal{B}}$ graphically.

As compared with A^+A,

$$A^*A = \begin{bmatrix} 2 & -1 \\ 1 & 0 \\ 0 & 1 \end{bmatrix} \begin{bmatrix} 2 & 1 & 0 \\ -1 & 0 & 1 \end{bmatrix} = \begin{bmatrix} 5 & 2 & -1 \\ 2 & 1 & 0 \\ -1 & 0 & 1 \end{bmatrix} : \mathbb{R}^3 \to \mathbb{R}^3$$

is not a projection, even though $\text{Im}(A)$ is an invariant subspace of A^*A. A^*A has eigenvalues $0, 1$ and 6 with corresponding eigenvectors $(1, -2, 1), (0, 1, 2)$

and $(5, 2, -1)$. Readers are urged to model after Fig. 3.31 to explain graphically the mapping properties of A^*A and A^+A.

What we obtained for this particular A is universally true for any real 2×3 matrix of rank 2. We summarize them in

A real matrix $A_{2\times 3}$ of rank 2 and its transpose A^* and its generalized inverse A^+

(1) 1. AA^*: $\mathbb{R}^2 \to \mathbb{R}^2$ is an invertible liner operator.
    2. A^*A: $\mathbb{R}^3 \to \text{Im}(A) \subseteq \mathbb{R}^3$ is an onto linear transformation with $\text{Im}(A)$ as an invariant subspace.
(2) $AA^* = 1_{\mathbb{R}^2}$: $\mathbb{R}^2 \to \mathbb{R}^2$ is the identity operator on $\mathbb{R}^2$, considered as the orthogonal projection of $\mathbb{R}^2$ onto itself along $\{\vec{0}\}$.

 $\Leftrightarrow$ A^*A: $\mathbb{R}^3 \to \mathbb{R}^3$ is the *orthogonal projection* of $\mathbb{R}^3$ onto $\text{Im}(A)$ along $\text{Ker}(A^*)$, i.e. A^*A is symmetric and $(A^*A)^2 = A^*A$.

In general, A^* does not have these properties except A^* is a right inverse of A.

(3) The generalized inverse

$$A^+ = A^*(AA^*)^{-1}$$

of A *orthogonalizes* A both on the right and on the left in the following sense:

 $AA^+ = 1_{\mathbb{R}^2}$ is the orthogonal projection of $\mathbb{R}^2$ onto itself along $\text{Ker}(A) = \{\vec{0}\}$.

 $\Leftrightarrow$ A^+A is the orthogonal projection of $\mathbb{R}^3$ onto $\text{Im}(A)$ along $\text{Ker}(A^*)$.

Therefore, A^+ can be defined as the peculiar inverse of the restricted linear isomorphism $A|_{\text{Im}(A^*)}$: $\text{Im}(A^*) = \mathbb{R}^2 \to \text{Im}(A)$, i.e.

$$A^+ = (A|_{\text{Im}(A^*)})^{-1}: \text{Im}(A) \to \text{Im}(A^*).$$

that makes A^+A an orthogonal projection. $\qquad\qquad$ (3.7.40)

These results are still valid for real $A_{m\times n}$ with rank equal to m. For a general setting, refer to Ex. 12 of Sec. B.8 and Secs. 4.5, 5.5.

Example 5 Let

$$A = \begin{bmatrix} 1 & 0 \\ 1 & 1 \\ 1 & -1 \end{bmatrix}.$$

Do the same problems as Example 4.

Solution Perform elementary row operations to

$$[A \mid \vec{b}^* \mid I_3] = \begin{bmatrix} 1 & 0 & \mid b_1 \mid & 1 & 0 & 0 \\ 1 & 1 & \mid b_2 \mid & 0 & 1 & 0 \\ 1 & -1 & \mid b_3 \mid & 0 & 0 & 1 \end{bmatrix}$$

$$\xrightarrow[\substack{E_{(2)-(1)} \\ E_{(3)-(1)}}]{} \begin{bmatrix} 1 & 0 & \mid b_1 & \mid & 1 & 0 & 0 \\ 0 & 1 & \mid b_2 - b_1 & \mid & -1 & 1 & 0 \\ 0 & -1 & \mid b_3 - b_1 & \mid & -1 & 0 & 1 \end{bmatrix}$$

$$\xrightarrow[E_{(3)+(2)}]{} \begin{bmatrix} 1 & 0 & \mid & b_1 & \mid & 1 & 0 & 0 \\ 0 & 1 & \mid & b_2 - b_1 & \mid & -1 & 1 & 0 \\ 0 & 0 & \mid & b_2 + b_3 - 2b_1 & \mid & -2 & 1 & 1 \end{bmatrix}. \qquad (*18)$$

From (*18)

$$A\vec{x}^* = \vec{b}^* \text{ has a solution } \vec{x} = (x_1, x_2) \in \mathbb{R}^2.$$
$$\Leftrightarrow b_2 + b_3 - 2b_1 = 0$$
$$\Rightarrow \text{ The solution is } x_1 = b_1, x_2 = b_2 - b_1.$$

The constrained condition $b_2 + b_3 - 2b_1 = 0$ can also be seen by eliminating x_1, x_2, x_3 from the set of equations $x_1 = b_1, x_1 + x_2 = b_2, x_1 - x_2 = b_3$.

(*18) also indicates that

$$BA = I_2, \quad \text{where } B = \begin{bmatrix} 1 & 0 & 0 \\ -1 & 1 & 0 \end{bmatrix}$$

i.e. B is a *left inverse* of A. In general, let $B = \begin{bmatrix} \vec{v}_1 \\ \vec{v}_2 \end{bmatrix}_{2 \times 3}$. Then

$$BA = \begin{bmatrix} \vec{v}_1 \\ \vec{v}_2 \end{bmatrix} A = \begin{bmatrix} \vec{v}_1 A \\ \vec{v}_2 A \end{bmatrix} = I_2$$
$$\Leftrightarrow \vec{v}_1 A = \vec{e}_1$$
$$\vec{v}_2 A = \vec{e}_2.$$

Suppose $\vec{v}_1 = (x_1, x_2, x_3)$. Then

$$\vec{v}_1 A = \vec{e}_1$$
$$\Leftrightarrow \begin{cases} x_1 + x_2 + x_3 = 1 \\ x_2 - x_3 = 0 \end{cases}$$
$$\Leftrightarrow \vec{v}_1 = (1, 0, 0) + t_1(-2, 1, 1), \quad t_1 \in \mathbb{R}.$$

Similarly, let $\vec{v}_2 = (x_1, x_2, x_3)$. Then

$$\vec{v}_2 A = \vec{e}_2$$

$$\Leftrightarrow \begin{cases} x_1 + x_2 + x_3 = 0 \\ x_2 - x_3 = 1 \end{cases}$$

$$\Leftrightarrow \vec{v}_2 = (-1, 1, 0) + t_2(-2, 1, 1), \quad t_2 \in \mathbb{R}.$$

Thus, the *left inverses* of A are

$$B = \begin{bmatrix} 1 - 2t_1 & t_1 & t_1 \\ -1 - 2t_2 & 1 + t_2 & t_2 \end{bmatrix}_{2\times 3} \quad \text{for } t_1, t_2 \in \mathbb{R}. \tag{*19}$$

On the other hand, (*18) says

$$E_{(3)+(2)} E_{(3)-(1)} E_{(2)-(1)} A = PA = \begin{bmatrix} 1 & 0 \\ 0 & 1 \\ 0 & 0 \end{bmatrix}$$

(row-reduced echelon matrix and

normal form of A),

$$P = E_{(3)+(2)} E_{(3)-(1)} E_{(2)-(1)} = \begin{bmatrix} 1 & 0 & 0 \\ -1 & 1 & 0 \\ -2 & 1 & 1 \end{bmatrix}$$

$$\Rightarrow A = E_{(2)-(1)}^{-1} E_{(3)-(1)}^{-1} E_{(3)+(2)}^{-1} \begin{bmatrix} 1 & 0 \\ 0 & 1 \\ 0 & 0 \end{bmatrix} = P^{-1} \begin{bmatrix} 1 & 0 \\ 0 & 1 \\ 0 & 0 \end{bmatrix}$$

$$= \begin{bmatrix} 1 & 0 & 0 \\ 1 & 1 & 0 \\ 1 & -1 & 1 \end{bmatrix} \begin{bmatrix} 1 & 0 \\ 0 & 1 \\ 0 & 0 \end{bmatrix} \quad \text{(LU-decomposition)}.$$

Refer to (1) in (2.7.70).

To investigate A^* and A^*A, consider $\vec{x} A = \vec{b}$ for $\vec{x} = (x_1, x_2, x_3) \in \mathbb{R}^3$ and $\vec{b} = (b_1, b_2) \in \mathbb{R}^2$.

By simple computation or by (*18),

$$\mathrm{Ker}(A) = \langle\langle(-2,1,1)\rangle\rangle = \mathrm{Im}(A^*)^\perp,$$
$$\mathrm{Im}(A) = \mathbb{R}^2 = \mathrm{Ker}(A^*)^\perp,$$
$$\mathrm{Ker}(A^*) = \{0\} = \mathrm{Im}(A)^\perp,$$
$$\mathrm{Im}(A^*) = \{(x_1,x_2,x_3) \in \mathbb{R}^3 \mid 2x_1 - x_2 - x_3 = 0\} = \langle\langle(1,2,0),(1,0,2)\rangle\rangle$$
$$= \langle\langle(1,1,1),(0,1,-1)\rangle\rangle = \mathrm{Ker}(A)^\perp,$$

and

$$A^*A = \begin{bmatrix} 3 & 0 \\ 0 & 2 \end{bmatrix},$$

$$(A^*A)^{-1} = \frac{1}{6}\begin{bmatrix} 2 & 0 \\ 0 & 3 \end{bmatrix}.$$

For any fixed $\vec{b} \in \mathbb{R}^2, \vec{x}A = \vec{b}$ always has a particular solution $\vec{b}(A^*A)^{-1}A^*$ and the solution set is

$$\vec{b}(A^*A)^{-1}A^* + \mathrm{Ker}(A), \qquad (*20)$$

which is a one-dimensional affine subspace of $\mathbb{R}^3$. Among so many solutions, it is $\vec{b}(A^*A)^{-1}A^*$ that has the shortest distance to the origin $\vec{0}$ (see Ex. <B> 7 of Sec. 3.7.3). For simplicity, let

$$A^+ = (A^*A)^{-1}A^*$$
$$= \frac{1}{6}\begin{bmatrix} 2 & 0 \\ 0 & 3 \end{bmatrix}\begin{bmatrix} 1 & 1 & 1 \\ 0 & -1 & -1 \end{bmatrix} = \frac{1}{6}\begin{bmatrix} 2 & 2 & 2 \\ 0 & 3 & -3 \end{bmatrix}_{2\times 3} \qquad (*21)$$

and can be considered as a linear transformation from $\mathbb{R}^2$ into $\mathbb{R}^3$ with the range space $\mathrm{Im}(A^*)$. Since

$$A^+A = I_2,$$

A^+ is a *left inverse* of A.

Therefore, it is reasonable to expect that one of the left inverses shown in (*19) should be A^+. Since the range of A^+ is $\text{Im}(A^*)$, then

$\quad$ B in (*19) is A^+.

$\Leftrightarrow$ The range space of $B = \text{Im}(A^*)$

$\Leftrightarrow (1 - 2t_1, t_1, t_1)$ and $(-1 - 2t_2, 1 + t_2, t_2)$ are in $\text{Im}(A^*)$.

$\Leftrightarrow \begin{cases} 2(1 - 2t_1) - t_1 - t_1 = 0 \\ 2(-1 - 2t_2) - (1 + t_2) - t_2 = 0 \end{cases}$

$\Rightarrow t_1 = \dfrac{1}{3} \quad \text{and} \quad t_2 = -\dfrac{1}{2}.$

In this case, B in (*19) is indeed equal to A^+. This A^+ is called the *generalized inverse* of A.

How about AA^*?

$$AA^* = \begin{bmatrix} 1 & 0 \\ 1 & 1 \\ 1 & -1 \end{bmatrix} \begin{bmatrix} 1 & 1 & 1 \\ 0 & 1 & -1 \end{bmatrix} = \begin{bmatrix} 1 & 1 & 1 \\ 1 & 2 & 0 \\ 1 & 0 & 2 \end{bmatrix} \qquad (*22)$$

is a linear operator on $\mathbb{R}^3$ with the range space $\text{Im}(A^*)$. Actual computation shows that $(AA^*)^2 \neq AA^*$. Therefore, AA^* is not a projection of $\mathbb{R}^3$ onto $\text{Im}(A^*)$ along $\text{Ker}(A)$. Also

eigenvalues of AA^*	eigenvectors
2	$\vec{v}_1 = \left(0, \dfrac{1}{\sqrt{2}}, -\dfrac{1}{\sqrt{2}}\right)$
3	$\vec{v}_2 = \left(\dfrac{1}{\sqrt{3}}, \dfrac{1}{\sqrt{3}}, \dfrac{1}{\sqrt{3}}\right)$
0	$\vec{v}_3 = \left(-\dfrac{2}{\sqrt{6}}, \dfrac{1}{\sqrt{6}}, \dfrac{1}{\sqrt{6}}\right)$

indicates that AA^* is not a projection (see (3.7.34)). Notice that

$$QAA^*Q^{-1} = \begin{bmatrix} 2 & 0 & 0 \\ 0 & 3 & 0 \\ 0 & 0 & 0 \end{bmatrix}, \quad \text{where } Q = \begin{bmatrix} \vec{v}_1 \\ \vec{v}_2 \\ \vec{v}_3 \end{bmatrix} \text{ is orthogonal.}$$

$$\Rightarrow QAA^*Q^* = \begin{bmatrix} \sqrt{2} & 0 & 0 \\ 0 & \sqrt{3} & 0 \\ 0 & 0 & 1 \end{bmatrix} \begin{bmatrix} 1 & 0 & 0 \\ 0 & 1 & 0 \\ 0 & 0 & 0 \end{bmatrix} \begin{bmatrix} \sqrt{2} & 0 & 0 \\ 0 & \sqrt{3} & 0 \\ 0 & 0 & 1 \end{bmatrix}$$

$$\Rightarrow R(AA^*)R^* = \begin{bmatrix} I_2 & 0 \\ 0 & 0 \end{bmatrix}_{3\times3},$$

where

$$R = \begin{bmatrix} \frac{1}{\sqrt{2}} & 0 & 0 \\ 0 & \frac{1}{\sqrt{3}} & 0 \\ 0 & 0 & 1 \end{bmatrix} \begin{bmatrix} 0 & \frac{1}{\sqrt{2}} & -\frac{1}{\sqrt{2}} \\ \frac{1}{\sqrt{3}} & \frac{1}{\sqrt{3}} & \frac{1}{\sqrt{3}} \\ -\frac{2}{\sqrt{6}} & \frac{1}{\sqrt{6}} & \frac{1}{\sqrt{6}} \end{bmatrix} = \begin{bmatrix} 0 & \frac{1}{2} & -\frac{1}{2} \\ \frac{1}{3} & \frac{1}{3} & \frac{1}{3} \\ -\frac{2}{\sqrt{6}} & \frac{1}{\sqrt{6}} & \frac{1}{\sqrt{6}} \end{bmatrix}.$$

Thus, the *index* of AA^* is 2 and the *signature* is equal to 2. By the way, we pose the question: What is the preimage of the unit circle (or disk) $y_1^2 + y_2^2 = 1$ (or ≤ 1) under A? Let $\vec{y} = (y_1, y_2) = \vec{x}A$. Then

$$y_1^2 + y_2^2 = \vec{y}\,\vec{y}^* = 1$$
$$\Leftrightarrow (\vec{x}A)(\vec{x}A)^* = \vec{x}AA^*\vec{x}^*$$
$$= x_1^2 + 2x_2^2 + 2x_3^2 + 2x_1x_2 + 2x_1x_3, \text{ in the natural basis}$$
$$\text{for } \mathbb{R}^3$$
$$= 2x_1'^2 + 3x_2'^2, \text{ in the basis } \{\vec{v}_1, \vec{v}_2, \vec{v}_3\} = \mathcal{B}$$
$$= x_1''^2 + x_2''^2, \text{ in the basis } \{R_{1*}, R_{2*}, R_{3*}\} = \mathcal{C}$$
$$= 1, \qquad\qquad (*23)$$

where $(x_1', x_2', x_3') = [\vec{x}]_\mathcal{B} = \vec{x}Q^{-1}$ and $(x_1'', x_2'', x_3'') = [\vec{x}]_\mathcal{C} = \vec{x}R^{-1}$. See Fig. 3.53.

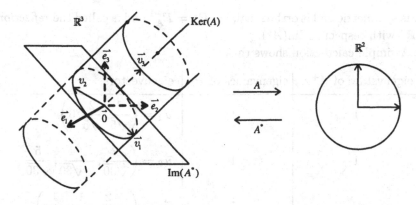

Fig. 3.53

Replace A^* in AA^* by A^+ and, in turn, consider

$$AA^+ = \begin{bmatrix} 1 & 0 \\ 1 & 1 \\ 1 & -1 \end{bmatrix} \cdot \frac{1}{6} \begin{bmatrix} 2 & 2 & 2 \\ 0 & 3 & -3 \end{bmatrix} = \frac{1}{6} \begin{bmatrix} 2 & 2 & 2 \\ 2 & 5 & -1 \\ 2 & -1 & 5 \end{bmatrix}.$$

AA^+ is symmetric and

$$(AA^+)^2 = AA^+ AA^+ = AI_2 A^+ = AA^+.$$

Therefore, $AA^+ \colon \mathbb{R}^3 \to \text{Im}(A^*) \subseteq \mathbb{R}^3$ is the *orthogonal projection* of $\mathbb{R}^3$ onto $\text{Im}(A^*)$ along $\text{Ker}(A)$. Note that

$$\text{Im}(AA^+) = \text{Im}(A^+) = \text{Ker}(A)^{\perp} = \text{Ker}(AA^+)^{\perp},$$
$$\text{Ker}(AA^+) = \text{Ker}(A) = \text{Im}(A^*)^{\perp} = \text{Im}(AA^+)^{\perp}.$$

What is the *reflection* or *symmetric point* of a point $\vec{x}$ in $\mathbb{R}^3$ with respect to the plane $\text{Im}(A^*)$? Notice that (see (*16))

$$\vec{x} \in \mathbb{R}^3$$
$$\to \vec{x} AA^+, \text{ the orthogonal projection of } \vec{x} \text{ on } \text{Im}(A^*)$$
$$\to \vec{x} AA^+ + (\vec{x} AA^+ - \vec{x}) = \vec{x}(2AA^+ - I_3), \text{ the reflection point.} \quad (*24)$$

Thus, denote the linear operator

$$P_A = 2AA^+ - I_3 = 2 \cdot \frac{1}{6} \begin{bmatrix} 2 & 2 & 2 \\ 2 & 5 & -1 \\ 2 & -1 & 5 \end{bmatrix} - \begin{bmatrix} 1 & 0 & 0 \\ 0 & 1 & 0 \\ 0 & 0 & 1 \end{bmatrix} = \begin{bmatrix} -\frac{1}{3} & \frac{2}{3} & \frac{2}{3} \\ \frac{2}{3} & \frac{2}{3} & -\frac{1}{3} \\ \frac{2}{3} & -\frac{1}{3} & \frac{2}{3} \end{bmatrix}.$$

P_A is symmetric and is orthogonal, i.e. $P_A^* = P_A^{-1}$ and is called the *reflection* of $\mathbb{R}^3$ with respect to $\text{Im}(A^*)$.

A simple calculation shows that

eigenvalues of A^+A	eigenvalues of P_A	eigenvectors
1	1	$\vec{u}_1 = \left(\dfrac{1}{\sqrt{5}}, \dfrac{2}{\sqrt{5}}, 0 \right)$
1	1	$\vec{u}_2 = \left(\dfrac{2}{\sqrt{30}}, -\dfrac{1}{\sqrt{30}}, \dfrac{5}{\sqrt{30}} \right)$
0	-1	$\vec{u}_3 = \left(-\dfrac{2}{\sqrt{6}}, \dfrac{1}{\sqrt{6}}, \dfrac{1}{\sqrt{6}} \right)$

$\mathcal{D} = \{\vec{u}_1, \vec{u}_2, \vec{u}_3\}$ is an orthonormal basis for $\mathbb{R}^3$. In $\mathcal{D}$,

$$[AA^+]_{\mathcal{D}} = S(AA^+)S^{-1} = \begin{bmatrix} 1 & 0 & 0 \\ 0 & 1 & 0 \\ 0 & 0 & 0 \end{bmatrix}, \quad \text{where } S = \begin{bmatrix} \vec{u}_1 \\ \vec{u}_2 \\ \vec{u}_3 \end{bmatrix} \text{ is orthogonal;}$$

$$[P_A]_{\mathcal{D}} = SP_A S^{-1} = \begin{bmatrix} 1 & 0 & 0 \\ 0 & 1 & 0 \\ 0 & 0 & -1 \end{bmatrix}.$$

Try to explain $[AA^+]_{\mathcal{D}}$ and $[P_A]_{\mathcal{D}}$ graphically.

As a counterpart of (3.7.40), we summarize in

A real matrix $A_{3\times 2}$ of rank 2 and its transpose A^* and its generalized inverse A^+

(1) 1. AA^*: $\mathbb{R}^3 \to \text{Im}(A^*) \subseteq \mathbb{R}^3$ is an onto linear transformation with $\text{Im}(A^*)$ as an invariant subspace.

2. A^*A: $\mathbb{R}^2 \to \mathbb{R}^2$ is an invertible linear operator.

(2) AA^*: $\mathbb{R}^3 \to \mathbb{R}^3$ is the orthogonal projection of $\mathbb{R}^3$ onto $\text{Im}(A^*)$ along $\text{Ker}(A)$, i.e. AA^* is symmetric and $(AA^*)^2 = AA^*$.

$\Leftrightarrow A^*A = 1_{\mathbb{R}^2}$: $\mathbb{R}^2 \to \mathbb{R}^2$ is the identity operator on $\mathbb{R}^2$, considered as the orthogonal projection of $\mathbb{R}^2$ onto itself along $\{\vec{0}\}$.

In general, A^* does not have these properties except A^* is a left inverse of A.

(3) The generalized inverse

$$A^+ = (A^*A)^{-1}A^*$$

of A *orthogonalizes* A both on the right and on the left in the following sense.

AA^+: $\mathbb{R}^3 \to \mathbb{R}^3$ is the orthogonal projection of $\mathbb{R}^3$ onto $\text{Im}(A^*)$ along $\text{Ker}(A)$.

$\Leftrightarrow A^+A = 1_{\mathbb{R}^2}$ is the orthogonal projection of $\mathbb{R}^2$ onto itself along $\text{Ker}(A^*) = \{\vec{0}\}$.

Therefore, A^+ can be defined as the inverse of the linear isomorphism $A|_{\text{Im}(A^*)}$: $\text{Im}(A^*) \subseteq \mathbb{R}^3 \to \mathbb{R}^2$, i.e.

$$A^+ = (A|_{\text{Im}(A^*)})^{-1}: \text{Im}(A) = \mathbb{R}^2 \to \text{Im}(A^*). \tag{3.7.41}$$

These results are still valid for real $A_{m\times n}$ with rank equal to n. (see Ex. 12 of Sec. B.8 and Sec. 5.5.)

Exercises

\<A\>

1. Prove (2.7.68) of Sec. 2.7.5 for $A_{3\times3}$.
2. Prove (2.7.69) of Sec. 2.7.5 for $A_{3\times3}$.
3. Prove (2.7.70) for $A_{2\times3}$ and $A_{3\times2}$.
4. Prove (2.7.71) for real symmetric matrix $A_{3\times3}$. For the invariance of the index and the signature of A, try the following methods.

   (1) A case-by-case examination. For example, try to prove that it is impossible for any invertible real matrix $P_{3\times3}$ so that

   $$P \begin{bmatrix} 1 & 0 & 0 \\ 0 & 0 & 0 \\ 0 & 0 & 0 \end{bmatrix} P^* = \begin{bmatrix} -1 & 0 & 0 \\ 0 & 0 & 0 \\ 0 & 0 & 0 \end{bmatrix}.$$

   (2) See Ex. \<B\> 3.

5. Prove Ex. \<A\> 7 of Sec. 2.7.5 for $A_{3\times3}$.
6. For each of the following matrices A:

   (1) Do problems as in Example 1.
   (2) Find the generalized inverse A^+ of A and explain it both algebraically and geometrically (one may refer to Exs. \<B\> 4 and 8 of Sec. 3.7.3, (3.7.40), (3.7.41) and Sec. B.8 if necessary).

   (a) $\begin{bmatrix} 3 & 1 & 1 \\ 2 & 4 & 2 \\ -1 & -1 & 1 \end{bmatrix}$. (b) $\begin{bmatrix} -1 & 0 & -3 \\ 0 & 1 & 2 \\ -1 & -1 & -5 \end{bmatrix}$. (c) $\begin{bmatrix} 1 & 2 & -1 \\ 2 & 4 & -2 \\ 3 & 6 & -3 \end{bmatrix}$.

7. For each of the following matrices, refer to Example 2 and do the same problems as in Ex. 6.

   (a) $\begin{bmatrix} 0 & -2 & 3 \\ 0 & 1 & -4 \\ 2 & 0 & 5 \end{bmatrix}$. (b) $\begin{bmatrix} 0 & 3 & 5 \\ -1 & 2 & 4 \\ -1 & -1 & -1 \end{bmatrix}$. (c) $\begin{bmatrix} 0 & -1 & -4 \\ 0 & 2 & 8 \\ 0 & 1 & 4 \end{bmatrix}$.

8. For each of the following matrices, refer to Example 3 and do the same problems as there.

   (a) $\begin{bmatrix} 0 & 1 & 1 \\ 1 & 0 & 1 \\ 1 & 1 & 0 \end{bmatrix}$. (b) $\begin{bmatrix} 2 & 1 & 1 \\ 1 & 2 & 1 \\ 1 & 1 & 2 \end{bmatrix}$. (c) $\begin{bmatrix} 2 & 3 & 0 \\ 3 & 5 & -1 \\ 0 & -1 & 2 \end{bmatrix}$.

   (d) $\begin{bmatrix} -2 & -2 & 0 \\ -2 & -1 & 1 \\ 0 & 1 & 1 \end{bmatrix}$.

9. Use (2.7.71) and Ex. 4 to determine the congruence among the following
   matrices:

$$\begin{bmatrix} 1 & 0 & 1 \\ 0 & 1 & 2 \\ 1 & 2 & 1 \end{bmatrix}, \begin{bmatrix} 0 & 1 & 2 \\ 1 & -1 & 3 \\ 2 & 3 & 4 \end{bmatrix} \text{ and } \begin{bmatrix} 1 & 2 & 3 \\ 2 & 4 & 5 \\ 3 & 5 & 6 \end{bmatrix}.$$

   If A and B are congruent, find an invertible matrix $P_{3\times3}$ so that
   $B = PAP^*$.

10. Let

$$A = \begin{bmatrix} 1 & 1 & 1 \\ 2 & 1 & 0 \end{bmatrix}.$$

   Do the same problems as in Example 4. One may also refer to Ex. <B>
   of Sec. 2.7.5.
11. Model after (*19) and (*21) and try to derive A^+ from (*11) in
    Example 4.
12. Let

$$A = \begin{bmatrix} 2 & 1 \\ 1 & 0 \\ 1 & 1 \end{bmatrix}.$$

   Do the same problems as in Example 5.

<B>

1. Prove (3.7.40).
2. Prove (3.7.41).
3. Let A be a nonzero real symmetric matrix of order n. Suppose there
   exist invertible matrices $P_{n\times n}$ and $Q_{n\times n}$ so that

$$PAP^* \text{ and } QAQ^*$$

   are diagonal matrices. Let p and q be the number of positive diagonal
   entries of PAP^* and QAQ^*, respectively. Suppose that $p < q$. For
   simplicity, let $\vec{x}_i = P_{i*}$, the ith row vector of P for $1 \le i \le n$ and
   $\vec{y}_j = Q_{j*}$ for $1 \le j \le n$. Let the rank $r(A) = r$. Note that $r \ge q > p$.
   (a) Define $f: \mathbb{R}^n \to \mathbb{R}^{r+p-q}$ by

$$f(\vec{x}) = (\vec{x}A\vec{x}_1^*, \ldots, \vec{x}A\vec{x}_p^*, \vec{x}A\vec{y}_{q+1}^*, \ldots, \vec{x}A\vec{y}_r^*).$$

   Show that f is linear and $r(f) \le r + p - q$ and hence

$$\dim \operatorname{Ker}(f) \ge n - (r + p - q) = n - r + (q - p) > n - r.$$

(b) There exists a nonzero $\vec{x}_0 \in \text{Ker}(f) - \langle\langle \vec{x}_{r+1}, \ldots, \vec{x}_n \rangle\rangle$. Hence, $f(\vec{x}_0) = 0$ implies that $\vec{x}_0 A \vec{x}_i^* = 0$ for $1 \le i \le p$ and $\vec{x}_0 A \vec{y}_j^* = 0$ for $q + 1 \le j \le r$.

(c) Let $\vec{x}_0 = \sum_{i=1}^n a_i \vec{x}_i = \sum_{j=1}^n b_j \vec{y}_j$. Use (b) to show that $a_i = 0$ for $1 \le i \le p$ and $b_j = 0$ for $q + 1 \le j \le r$.

(d) There exists some i_0 for $p + 1 \le i_0 \le r$ so that $a_{i_0} \ne 0$. Show that

$$\vec{x}_0 A \vec{x}_0^* = \sum_{i=p+1}^r a_i^2 \vec{x}_i A \vec{x}_i^* < 0 \quad \text{and}$$

$$\vec{x}_0 A \vec{x}_0^* = \sum_{j=1}^q b_j^2 \vec{y}_j A \vec{y}_j^* \ge 0,$$

a contradiction.

Hence, $p = q$ *should hold*. In fact, the above process can be further simplified as follows. Rewrite $\vec{x} = (x_1, \ldots, x_n) = \vec{y} P = \vec{z} Q$ where $\vec{y} = (y_1, \ldots, y_n)$ and $\vec{z} = (z_1, \ldots, z_n)$ so that

$$\vec{x} A \vec{x}^* = \vec{y}(PAP^*)\vec{y}^* = y_1^2 + \cdots + y_p^2 - y_{p+1}^2 - \cdots - y_r^2, \quad \text{and}$$

$$\vec{x} A \vec{x}^* = \vec{z}(QAQ^*)\vec{z}^* = z_1^2 + \cdots + z_q^2 - z_{q+1}^2 - \cdots - z_r^2.$$

Then

$$\vec{y} P = \vec{z} Q$$

$$\Leftrightarrow \vec{y} = \vec{z} Q P^{-1}$$

$$\Leftrightarrow y_j = \sum_{i=1}^n b_{ij} z_i \quad \text{for } 1 \le j \le n.$$

Now, consider the system of linear equations

$$y_j = \sum_{i=1}^n b_{ij} z_i = 0 \quad \text{for } 1 \le j \le p$$

$$z_{q+1} = 0$$

$$\vdots$$

$$z_n = 0$$

which has a nonzero solution $z_1^*, \ldots, z_q^*, z_{q+1}^* = 0, \ldots, z_n^* = 0$ since $q > p$. Corresponding to this set of solutions, the resulted $\vec{y}$ will induce that $\vec{x} A \vec{x}^* \le 0$ while the resulted $\vec{z}$ will induce that $\vec{x} A \vec{x}^* > 0$, a contradiction. Hence, $p = q$ should hold.

4. Let

$$A = \begin{bmatrix} 0 & 0 & 0 & 0 & 1 & 1 & 1 \\ 0 & 2 & 6 & 2 & 0 & 0 & 4 \\ 0 & 1 & 3 & 1 & 1 & 0 & 1 \\ 0 & 1 & 3 & 1 & 2 & 1 & 2 \end{bmatrix}_{4 \times 7} .$$

Find invertible matrices $P_{4 \times 4}$ and $Q_{7 \times 7}$ so that

$$PAQ = \begin{bmatrix} I_3 & 0 \\ 0 & 0 \end{bmatrix}_{4 \times 7} .$$

Express P and Q as products of elementary matrices.

5. Suppose a, b and c are positive numbers so that $aa' + bb' + cc' = 0$. Determine the row and column ranks of

$$A = \begin{bmatrix} 0 & c & -b & a' \\ -c & 0 & a & b' \\ b & -a & 0 & c' \\ -a' & -b' & -c' & 0 \end{bmatrix}$$

and find invertible matrices P and Q so that PAQ is the normal form of A.

6. For each of the following matrices $A_{m \times n}$, do the following problems.

(1) Find an invertible matrix P so that $PA = R$ is the row-reduced echelon matrix of A. Use this result to solve $A\vec{x}^* = \vec{b}^*$ where $\vec{x} \in \mathbb{R}^n$ and $\vec{b} \in \mathbb{R}^m$.

(2) Find the row and column ranks of A.

(3) Find an invertible matrix Q so that PAQ is the normal form of A. Also, express P and Q as products of elementary matrices.

(4) In case $m = n$, i.e. A is a square matrix, determine if A is invertible. If it is, find A^{-1} and express A and A^{-1} as products of elementary matrices and hence compute $\det A$ and $\det A^{-1}$.

(5) If A is a symmetric matrix, find invertible matrix S so that SAS^* is a diagonal matrix with diagonal entries equal to 1, -1 or 0. Determine the index and the signature of A.

(6) Find $\mathrm{Ker}(A), \mathrm{Im}(A), \mathrm{Ker}(A^*)$ and $\mathrm{Im}(A^*)$.

(7) Try to find the generalized inverse A^+ of A, if possible.

(a) $\begin{bmatrix} i & 1-i & 0 \\ 1 & -2 & 1 \\ 1 & 2i & -1 \end{bmatrix}$.

(b) $\begin{bmatrix} 2 & -2 & 1 & 1 & 0 & 0 & 3 \\ 2 & -1 & -1 & 0 & 1 & 0 & 5 \\ -1 & 2 & 2 & -1 & 0 & 1 & 6 \end{bmatrix}$.

(c) $\begin{bmatrix} 2 & 2 & -1 \\ -2 & 3 & 2 \\ -1 & 1 & -1 \\ 1 & 0 & 0 \\ 0 & 1 & 0 \\ 0 & 0 & 1 \end{bmatrix}$.

(d) $\begin{bmatrix} 0 & 0 & 0 & 0 & 1 & 1 & 1 \\ 0 & 2 & 6 & 2 & 0 & 0 & 4 \\ 0 & 1 & 3 & 1 & 1 & 0 & 1 \\ 0 & 1 & 3 & 1 & 2 & 1 & 2 \end{bmatrix}$.

(e) $\begin{bmatrix} 1 & 1 & 1 & -3 \\ 0 & 1 & 0 & 0 \\ 1 & 1 & 2 & -3 \\ 2 & 2 & 4 & -5 \end{bmatrix}$.

(f) $\begin{bmatrix} 1 & -3 & -1 & 5 & 6 \\ 2 & 0 & -1 & 2 & 3 \\ -1 & 0 & 2 & -1 & 3 \\ 3 & 1 & 3 & 2 & 1 \\ -2 & -1 & 0 & 1 & 2 \end{bmatrix}$.

(g) $\begin{bmatrix} 1 & \frac{1}{2} & \frac{1}{3} & \frac{1}{4} \\ \frac{1}{2} & \frac{1}{3} & \frac{1}{4} & \frac{1}{5} \\ \frac{1}{3} & \frac{1}{4} & \frac{1}{5} & \frac{1}{6} \\ \frac{1}{4} & \frac{1}{5} & \frac{1}{6} & \frac{1}{7} \end{bmatrix}$.

(h) $\begin{bmatrix} 0 & 1 & 1 & \cdots & 1 \\ 1 & 0 & 1 & \cdots & 1 \\ 1 & 1 & 0 & \cdots & 1 \\ \vdots & \vdots & \vdots & \ddots & \vdots \\ 1 & 1 & 1 & \cdots & 0 \end{bmatrix}_{n \times n}$.

7. Let

$$A = \begin{bmatrix} 1 & 0 & -1 & 2 & 1 \\ -1 & 1 & 3 & -1 & 0 \\ -2 & 1 & 4 & -1 & 3 \\ 3 & -1 & -5 & 1 & -6 \end{bmatrix}.$$

(a) Prove that the rank $r(A) = 3$ by showing that A_{*1}, A_{*2} and A_{*4} are linearly independent and $A_{*3} = -A_{*1} + 2A_{*2}, A_{*5} = -3A_{*1} - A_{*2} + 2A_{*4}$. Find $\alpha_1, \alpha_2, \alpha_3$ so that $A_{4*} = \alpha_1 A_{1*} + \alpha_2 A_{2*} + \alpha_3 A_{3*}$.

(b) Show that

$$\mathrm{Ker}(A) = \langle\langle (0, -1, 2, 1) \rangle\rangle,$$
$$\mathrm{Im}(A) = \langle\langle A_{1*}, A_{2*}, A_{3*} \rangle\rangle.$$

(c) A matrix $B_{4\times 4}$ has the property that $BA = O_{4\times 5}$ if and only if $\mathrm{Im}(B) \subseteq \mathrm{Ker}(A)$. Find all such matrices B. Is it possible to have a matrix $B_{4\times 4}$ of rank greater than one so that $BA = O$? Why?

(d) A matrix $C_{5\times5}$ has the property that $AC = O_{4\times5}$ if only if $\text{Im}(A) \subseteq \text{Ker}(C)$. Find all such matrices C. Why is $\text{r}(C) \leq 2$?

(e) Find matrices $B_{4\times4}$ so that BA have each possible rank. How about AC?

Also, try to do (a) and (b) by using elementary row operations on A. Can we solve (c) and (d) by this method? And (e)?

(*Note* Refer to Ex. <C> of Sec. 2.7.3, Ex. <C> 21 of Sec. 2.7.5 and Sec. B.5.)

8. Suppose

$$R = \begin{bmatrix} 1 & 4 & 0 & -1 & 0 & 2 & -5 \\ 0 & 0 & 1 & -3 & 0 & 5 & -2 \\ 0 & 0 & 0 & 0 & 1 & 0 & -1 \\ 0 & 0 & 0 & 0 & 0 & 0 & 0 \end{bmatrix}_{4\times7}$$

is the row-reduced echelon matrix of $A_{4\times7}$. Determine A if

$$A_{*1} = \begin{bmatrix} 3 \\ 2 \\ -9 \\ 5 \end{bmatrix}, \quad A_{*3} = \begin{bmatrix} 0 \\ 1 \\ -2 \\ 4 \end{bmatrix} \quad \text{and} \quad A_{*5} = \begin{bmatrix} -2 \\ -3 \\ 1 \\ -1 \end{bmatrix}$$

and find an invertible matrix $P_{4\times4}$ such that $PA = R$.

(*Note* Refer to Sec. B.5.)

9. Let A be any one of the following matrices

$$\begin{bmatrix} 1 & -1 & 1 & 1 \\ 1 & 0 & 2 & 1 \end{bmatrix}, \begin{bmatrix} 1 & -1 & 0 & 0 \\ 0 & 0 & 1 & -1 \\ 1 & -1 & 1 & -1 \end{bmatrix} \quad \text{and} \quad \begin{bmatrix} 1 & 0 & 0 & 0 \\ -1 & 1 & 0 & 0 \\ 0 & 2 & 1 & 1 \end{bmatrix}.$$

Do problems as in Example 4.

10. Let A be any one of the following matrices

$$\begin{bmatrix} 1 & 0 \\ -1 & 1 \\ 0 & 2 \\ 3 & 0 \end{bmatrix}, \begin{bmatrix} 1 & 1 & 0 \\ 1 & 1 & 0 \\ 0 & 1 & -1 \\ 1 & 0 & 1 \end{bmatrix} \quad \text{and} \quad \begin{bmatrix} 1 & 0 & 1 \\ 0 & -1 & 0 \\ 0 & 0 & -2 \\ 1 & 1 & 0 \end{bmatrix}.$$

Do problems as in Example 5.

<C> Abstraction and generalization

Read Ex. <C> of Sec. 2.7.5 and do problems listed there. Also, do the following problems.

1. Let

$$P_1(x) = 2 - 3x + 4x^2 - 5x^3 + 2x^4, \quad P_2(x) = -6 + 9x - 12x^2 + 15x^3 - 6x^4,$$
$$P_3(x) = 3 - 2x + 7x^2 - 9x^3 + x^4, \quad P_4(x) = 2 - 8x + 2x^2 - 2x^3 + 6x^4,$$
$$P_5(x) = 1 - x - 3x^2 + 2x^3 + x^4, \quad P_6(x) = -3 - 18x^2 + 12x^3 + 9x^4,$$
$$P_7(x) = -2 + 3x - 2x^2 + x^3, \quad P_8(x) = 2 - x - x^2 + 7x^3 - 9x^4$$

be polynomials in $P_4(\mathbb{R})$ which is isomorphic to $\mathbb{R}^5$. Find a subset of $\{P_1, P_2, \ldots, P_8\}$ which is a basis for $\langle\langle P_1, P_2, \ldots, P_8 \rangle\rangle$.

2. Let

$$S = \{(x_1, x_2, x_3, x_4, x_5) \in \mathbb{R}^5 \mid x_1 - x_2 + x_3 - x_4 + x_5 = 0\}$$

be a subspace of $\mathbb{R}^5$.

(a) Determine $\dim(S)$.
(b) Show that $(1, 1, 1, 1, 0) \in S$ and extend it to form a basis for S.

3. Let S be the set of solutions of the system of linear equations

$$3x_1 - x_2 + x_3 - x_4 + 2x_5 = 0$$
$$x_1 - x_2 - x_3 - 2x_4 - x_5 = 0.$$

(a) S is a subspace of $\mathbb{R}^5$. Determine $\dim(S)$.
(b) Show that $(1, -1, -2, 2, 0) \in S$ and extend it to form a basis for S.

4. The set

$$S = \left\{ \begin{bmatrix} 1 & 1 \\ 0 & 1 \end{bmatrix}, \begin{bmatrix} 0 & -1 \\ -1 & 1 \end{bmatrix}, \begin{bmatrix} 2 & 1 \\ 1 & 9 \end{bmatrix}, \begin{bmatrix} 1 & -2 \\ 4 & -2 \end{bmatrix}, \begin{bmatrix} 1 & 0 \\ -1 & 1 \end{bmatrix}, \begin{bmatrix} -1 & 2 \\ 2 & -1 \end{bmatrix} \right\}$$

generates a subspace V of $M(2; \mathbb{R})$ which is isomorphic to $\mathbb{R}^4$. Find a subset of S that is a basis for V.

3.7.6   *Diagonal canonical form*

We have encountered, up to now, many diagonalizable linear operators on $\mathbb{R}^3$ or real matrices $A_{3\times 3}$. Here in this subsection, we are going to prove (2.7.73) for square matrices of order 3 and hence realize partial results listed in (3.7.29).

Let $A = [a_{ij}]_{3\times 3}$ be a nonzero real matrix.

Suppose A is diagonalizable, i.e. there exists an invertible matrix $P_{3\times 3}$ so that

$$PAP^{-1} = \begin{bmatrix} \lambda_1 & & 0 \\ & \lambda_2 & \\ 0 & & \lambda_3 \end{bmatrix}, \quad \text{where } P = \begin{bmatrix} \vec{x}_1 \\ \vec{x}_2 \\ \vec{x}_3 \end{bmatrix}$$

$$\Leftrightarrow \vec{x}_i A = \lambda_i \vec{x}_i \quad \text{for } 1 \le i \le 3$$

$$\Leftrightarrow \vec{x}_i (A - \lambda_i I_3) = \vec{0} \quad \text{for } 1 \le i \le 3. \tag{3.7.42}$$

Remind that λ_1, λ_2 and λ_3 are *eigenvalues* of A and $\vec{x}_1, \vec{x}_2$ and $\vec{x}_3$ are associated *eigenvectors* of A, respectively. Note that

$$\mathcal{B} = \{\vec{x}_1, \vec{x}_2, \vec{x}_3\}$$

is a *basis* for $\mathbb{R}^3$, consisting entirely of eigenvectors.

For any vector $\vec{x} \in \mathbb{R}^3$, $\vec{x} = \alpha_1 \vec{x}_1 + \alpha_2 \vec{x}_2 + \alpha_3 \vec{x}_3$ for some unique scalars α_1, α_2 and α_3. Hence

$$\vec{x}(A - \lambda_1 I_3)(A - \lambda_2 I_3)(A - \lambda_3 I_3)$$
$$= \alpha_1 [\vec{x}_1 (A - \lambda_1 I_3)](A - \lambda_2 I_3)(A - \lambda_3 I_3)$$
$$+ \alpha_2 [\vec{x}_2 (A - \lambda_2 I_3)](A - \lambda_1 I_3)(A - \lambda_3 I_3)$$
$$+ \alpha_3 [\vec{x}_3 (A - \lambda_3 I_3)](A - \lambda_1 I_3)(A - \lambda_2 I_3)$$
$$= \alpha_1 \vec{0} + \alpha_2 \vec{0} + \alpha_3 \vec{0} = \vec{0} \quad \text{for all } \vec{x} \in \mathbb{R}^3$$
$$\Rightarrow (A - \lambda_1 I_3)(A - \lambda_2 I_3)(A - \lambda_3 I_3) = O_{3\times 3}. \tag{3.7.43}$$

A direct matrix computation as

$$(A - \lambda_1 I_3)(A - \lambda_2 I_3)(A - \lambda_3 I_3)$$

$$= P^{-1} \begin{bmatrix} 0 & & 0 \\ & \lambda_2 - \lambda_1 & \\ 0 & & \lambda_3 - \lambda_1 \end{bmatrix}$$

$$P \cdot P^{-1} \begin{bmatrix} \lambda_1 - \lambda_2 & & 0 \\ & 0 & \\ 0 & & \lambda_3 - \lambda_2 \end{bmatrix} P \cdot P^{-1} \begin{bmatrix} \lambda_1 - \lambda_3 & & 0 \\ & \lambda_2 - \lambda_3 & \\ 0 & & 0 \end{bmatrix} P$$

$$= P^{-1} O P = O$$

will work too. This result is a special case of *Cayley–Hamilton theorem* which states that A satisfies its characteristic polynomial

$$\det(A - tI_3) = -(t - \lambda_1)(t - \lambda_2)(t - \lambda_3).$$

Eigenvalues λ_1, λ_2 and λ_3 may be not distinct.

Case 1 $\lambda_1 \neq \lambda_2 \neq \lambda_3$

If $A_{3\times3}$ has three distinct eigenvalues λ_1, λ_2 and λ_3, then their respective eigenvectors $\vec{x}_1, \vec{x}_2$ and $\vec{x}_3$ should be linearly independent. For

$$\alpha_1 \vec{x}_1 + \alpha_2 \vec{x}_2 + \alpha_3 \vec{x}_3 = \vec{0}$$

$\Rightarrow$ (apply both sides by A) $\quad \alpha_1 \lambda_1 \vec{x}_1 + \alpha_2 \lambda_2 \vec{x}_2 + \alpha_3 \lambda_3 \vec{x}_3 = \vec{0}$

$\Rightarrow$ (eliminating, say $\vec{x}_1$, from the above two relations)

$$\alpha_2(\lambda_1 - \lambda_2)\vec{x}_2 + \alpha_3(\lambda_1 - \lambda_3)\vec{x}_3 = \vec{0}$$

$\Rightarrow$ (by inductive assumption) $\alpha_2(\lambda_1 - \lambda_2) = \alpha_3(\lambda_1 - \lambda_3) = 0$

$\Rightarrow$ (because $\lambda_1 \neq \lambda_2 \neq \lambda_3$) $\alpha_2 = \alpha_3 = 0$ and hence $\alpha_1 = 0$.

Now, $\{\vec{x}_1, \vec{x}_2, \vec{x}_3\}$ is a basis for $\mathbb{R}^3$ and thus (3.7.42) holds. In particular, A is diagonalizable and each *eigenspace*

$$E_{\lambda_i} = \{\vec{x} \in \mathbb{R}^3 \mid \vec{x}A = \lambda_i \vec{x}\} = \operatorname{Ker}(A - \lambda_i I_3)$$

is of dimension one.

Case 2 $\lambda_1 = \lambda_2 \neq \lambda_3$

(3.7.43) can be simplified as

$$(A - \lambda_1 I_3)(A - \lambda_3 I_3)$$

$$= P^{-1} \begin{bmatrix} 0 & & 0 \\ & 0 & \\ 0 & & \lambda_3 - \lambda_1 \end{bmatrix} P \cdot P^{-1} \begin{bmatrix} \lambda_1 - \lambda_3 & & 0 \\ & \lambda_1 - \lambda_3 & \\ 0 & & 0 \end{bmatrix} P$$

$$= P^{-1}OP = O. \tag{3.7.44}$$

In this case, $\vec{x}_1$ and $\vec{x}_2$ are eigenvectors associated to $\lambda_1 = \lambda_2$ and hence $\dim(E_{\lambda_1}) = 2$ while $\dim(E_{\lambda_3}) = 1$.

Conversely, if A has three eigenvalues λ_1, λ_2 and λ_3 with $\lambda_1 = \lambda_2 \neq \lambda_3$ such that $(A - \lambda_1 I_3)(A - \lambda_3 I_3) = O$ holds. We *claim* that

$$\dim(E_{\lambda_1}) = 2 \quad \text{and} \quad \dim(E_{\lambda_3}) = 1.$$

Suppose $\dim(E_{\lambda_3}) = 2$. Since $\dim(E_{\lambda_1}) \geq 1$ and $E_{\lambda_1} \cap E_{\lambda_3} = \{\vec{0}\}$, so $\dim(E_{\lambda_1}) = 1$ should hold. In a resulted basis $\mathcal{B} = \{\vec{x}_1, \vec{x}_2, \vec{x}_3\}$ consisting of eigenvectors $\vec{x}_1 \in E_{\lambda_1}$ and $\vec{x}_2, \vec{x}_3 \in E_{\lambda_3}$, A is diagonalizable with diagonal entries $\lambda_1, \lambda_3, \lambda_3$ which contradicts our original assumption. Therefore $\dim(E_{\lambda_3}) = 1$.

Now, $(A - \lambda_1 I_3)(A - \lambda_3 I_3) = O$ implies that

$$\text{Im}(A - \lambda_1 I_3) \subseteq \text{Ker}(A - \lambda_3 I_3)$$
$$\Rightarrow (\text{since } A - \lambda_1 I_3 \neq O_{3\times3} \text{ and } \dim(E_{\lambda_3}) = 1)$$
$$\dim \text{Im}(A - \lambda_1 I_3) = r(A - \lambda_1 I_3) = 1$$
$$\Rightarrow \dim \text{Ker}(A - \lambda_1 I_3) = \dim E_{\lambda_1} = 3 - 1 = 2.$$

Or, by using (3.7.31) (also, refer to Ex.<C> of Sec. 2.7.3),

$$r(A - \lambda_1 I_3) + r(A - \lambda_3 I_3) - 3 \leq r(O) = 0$$
$$\Rightarrow 1 \leq r(A - \lambda_1 I_3) \leq 3 - 2 = 1$$
$$\Rightarrow r(A - \lambda_1 I_3) = 1.$$

Hence $\dim(E_{\lambda_1}) = 2$.

Since $\mathbb{R}^3 = E_{\lambda_1} \oplus E_{\lambda_3}$, A is definitely diagonalizable.

Case 3 $\lambda_1 = \lambda_2 = \lambda_3$, say λ

Then (3.7.43) is simplified as

$$(A - \lambda I_3) = P^{-1} \begin{bmatrix} 0 & & 0 \\ & 0 & \\ 0 & & 0 \end{bmatrix} P = P^{-1}OP = O \qquad (3.7.45)$$

$$\Rightarrow A = \lambda I_3$$

i.e. A itself is a scalar matrix. $\qquad\qquad\square$

We summarize as (refer to (2.7.73))

The diagonalizability of a nonzero real matrix $A_{3\times3}$ and its canonical form

Suppose the characteristic polynomial of A is

$$\det(A - tI_3) = -(t - \lambda_1)(t - \lambda_2)(t - \lambda_3)$$

where λ_1, λ_2 and λ_3 are *real* numbers. Let

$$E_{\lambda_i} = \text{Ker}(A - \lambda_i I_3) = \{\vec{x} \in \mathbb{R}^3 \mid \vec{x}A = \lambda_i \vec{x}\}, \quad i = 1, 2, 3$$

be the *eigenspace* corresponding to λ_i. Then A is diagonalizable if and only if one of the following cases happens.

(1) $\lambda_1 \neq \lambda_2 \neq \lambda_3$.

    a. The minimal polynomial is $(t - \lambda_1)(t - \lambda_2)(t - \lambda_3)$.

b. Let $E_{\lambda_i} = \langle\langle \vec{x}_i \rangle\rangle$ for $1 \leq i \leq 3$. Then $\mathcal{B} = \{\vec{x}_1, \vec{x}_2, \vec{x}_3\}$ is a basis for $\mathbb{R}^3$ and

$$[A]_{\mathcal{B}} = PAP^{-1} = \begin{bmatrix} \lambda_1 & & 0 \\ & \lambda_2 & \\ 0 & & \lambda_3 \end{bmatrix}, \quad \text{where } P = \begin{bmatrix} \vec{x}_1 \\ \vec{x}_2 \\ \vec{x}_3 \end{bmatrix}.$$

Define the following matrices or linear operators:

$$A_1 = P^{-1} \begin{bmatrix} 1 & & 0 \\ & 0 & \\ 0 & & 0 \end{bmatrix} P, \quad A_2 = P^{-1} \begin{bmatrix} 0 & & 0 \\ & 1 & \\ 0 & & 0 \end{bmatrix} P,$$

$$A_3 = P^{-1} \begin{bmatrix} 0 & & 0 \\ & 0 & \\ 0 & & 1 \end{bmatrix} P.$$

Then,

1. $\mathbb{R}^3 = E_{\lambda_1} \oplus E_{\lambda_2} \oplus E_{\lambda_3}$.
2. Each $A_i: \mathbb{R}^3 \to \mathbb{R}^3$ is a projection of $\mathbb{R}^3$ onto E_{λ_i} along $E_{\lambda_j} \oplus E_{\lambda_k}$ for $1 \leq j < k \leq 3$ and $j, k \neq i$, i.e.

$$A_i^2 = A_i.$$

3. $A_i A_j = O_{3 \times 3}$ if $i \neq j, 1 \leq i, j \leq 3$.
4. $I_3 = A_1 + A_2 + A_3$.
5. $A = \lambda_1 A_1 + \lambda_2 A_2 + \lambda_3 A_3$.

See Fig. 3.31.

(2) $\lambda_1 = \lambda_2 \neq \lambda_3$ and their algebraic multiplicities are equal to their respective geometric dimensions, i.e. $\dim(E_{\lambda_1}) = 2$ and $\dim(E_{\lambda_3}) = 1$.

a. The minimal polynomial is $(t - \lambda_1)(t - \lambda_3)$.
b. Let $E_{\lambda_1} = \langle\langle \vec{x}_1, \vec{x}_2 \rangle\rangle$ and $E_{\lambda_3} = \langle\langle \vec{x}_3 \rangle\rangle$. Then $\mathcal{B} = \{\vec{x}_1, \vec{x}_2, \vec{x}_3\}$ is a basis for $\mathbb{R}^3$ and

$$[A]_{\mathcal{B}} = PAP^{-1} = \begin{bmatrix} \lambda_1 & & 0 \\ & \lambda_1 & \\ 0 & & \lambda_3 \end{bmatrix}, \quad \text{where } P = \begin{bmatrix} \vec{x}_1 \\ \vec{x}_2 \\ \vec{x}_3 \end{bmatrix}.$$

Define

$$A_1 = P^{-1} \begin{bmatrix} 1 & & 0 \\ & 1 & \\ 0 & & 0 \end{bmatrix} P \quad \text{and} \quad A_3 = P^{-1} \begin{bmatrix} 0 & & 0 \\ & 0 & \\ 0 & & 1 \end{bmatrix} P.$$

Then,

1. $\mathbb{R}^3 = E_{\lambda_1} \oplus E_{\lambda_3}$.
2. Each $A_i \colon \mathbb{R}^3 \to \mathbb{R}^3$ is a projection of $\mathbb{R}^3$ onto E_{λ_i} along E_{λ_j} for $j \neq i$, i.e. $A_i^2 = A_i$ for $i = 1, 3$.
3. $A_1 A_3 = A_3 A_1 = O_{3\times3}$.
4. $I_3 = A_1 + A_3$.
5. $A = \lambda_1 A_1 + \lambda_3 A_3$.

See Fig. 3.31 for $\lambda_1 = \lambda_2$.

(3) $\lambda_1 = \lambda_2 = \lambda_3$, say equal to λ and $\dim(E_\lambda) = 3$.

a. The minimal polynomial is $t - \lambda$.
b. For any basis $\mathcal{B} = \{\vec{x}_1, \vec{x}_2, \vec{x}_3\}$ for $\mathbb{R}^3$,

$$A = [A]_\mathcal{B} = PAP^{-1} = \lambda I_3, \quad \text{where } P = \begin{bmatrix} \vec{x}_1 \\ \vec{x}_2 \\ \vec{x}_3 \end{bmatrix}.$$

Namely, A is a scalar matrix. (3.7.46)

Notice that the following matrices

$$A = \begin{bmatrix} \lambda_1 & & 0 \\ 1 & \lambda_1 & \\ 0 & 0 & \lambda_2 \end{bmatrix}, \quad B = \begin{bmatrix} \lambda & & 0 \\ 1 & \lambda & \\ 0 & 0 & \lambda \end{bmatrix} \quad \text{and} \quad C = \begin{bmatrix} \lambda & & 0 \\ 1 & \lambda & \\ 0 & 1 & \lambda \end{bmatrix}$$

(3.7.47)

are not diagonalizable and their respective minimal polynomials are $(t - \lambda_1)^2 (t - \lambda_2), (t - \lambda)^2$ and $(t - \lambda)^3$. For details, see Sec. 3.7.7.

In what follows, we list four further examples to be concerned.

Note that similar matrices have the same characteristic polynomials and hence the same minimal polynomials, but not conversely for matrices of order $n \geq 4$. For example, the matrices

$$\begin{bmatrix} 1 & 0 & \vline & 0 \\ 1 & 1 & \vline & \\ \hline & & \vline & 1 & 0 \\ 0 & & \vline & 1 & 1 \end{bmatrix} \quad \text{and} \quad \begin{bmatrix} 1 & 0 & \vline & 0 \\ 1 & 1 & \vline & \\ \hline & & \vline & 1 & 0 \\ 0 & & \vline & 0 & 1 \end{bmatrix}$$

(3.7.48)

both have the same characteristic polynomial $(t-1)^4$ and the same minimal polynomial $(t-1)^2$, but they are not similar to each other (Why? One may prove this by contradiction or refer to Sec. 3.7.7).

Example 1 Test if

$$A = \begin{bmatrix} 0 & -1 & 0 \\ 3 & 3 & 1 \\ 1 & 1 & 1 \end{bmatrix} \quad \text{and} \quad B = \begin{bmatrix} -1 & 4 & 2 \\ -1 & 3 & 1 \\ -1 & 2 & 2 \end{bmatrix}$$

are similar.

Solution The characteristic polynomials are

$$\det(A - tI_3) = \begin{vmatrix} -t & -1 & 0 \\ 3 & 3-t & 1 \\ 1 & 1 & 1-t \end{vmatrix} = -t(3-t)(1-t) - 1 + 3(1-t) + t$$

$$= -(t-1)^2(t-2),$$

$$\det(B - tI_3) = -(t-1)^2(t-2).$$

So both are the same and A and B have common eigenvalues 1, 1 and 2.
  By computation,

$$(A - I_3)(A - 2I_3) = \begin{bmatrix} -1 & -1 & 0 \\ 3 & 2 & 1 \\ 1 & 1 & 0 \end{bmatrix} \begin{bmatrix} -2 & -1 & 0 \\ 3 & 1 & 1 \\ 1 & 1 & -1 \end{bmatrix}$$

$$= \begin{bmatrix} -1 & \cdots & \cdots \\ \cdots & \cdots & \cdots \\ \cdots & \cdots & \cdots \end{bmatrix} \neq O,$$

$$(B - I_3)(B - 2I_3) = \begin{bmatrix} -2 & 4 & 2 \\ -1 & 2 & 1 \\ -1 & 2 & 1 \end{bmatrix} \begin{bmatrix} -3 & 4 & 2 \\ -1 & 1 & 1 \\ -1 & 2 & 1 \end{bmatrix} = O.$$

So A has the minimal polynomial $(t-1)^2(t-2)$, while B has $(t-1)(t-2)$.
Therefore A and B are not similar and B is diagonalizable while A is not.
  Equivalently, $r(A-I_3) = 2$ implies that dim $E_1 = 3 - r(A-I_3) = 1$ which
is not equal to the algebraic multiplicity 2 of 1, meanwhile $r(B - I_3) = 1$
implies that dim $E_1 = 3 - r(B-I_3) = 2$. Thus A is not diagonalizable while
B is and they are not similar.   □

Example 2 Let $\vec{a}_1 = (-1, 1, 1)$, $\vec{a}_2 = (1, -1, 1)$ and $\vec{a}_3 = (1, 1, -1)$.

(1) Try to find linear operators mapping the tetrahedron $\triangle \vec{0}\, \vec{a}_1 \vec{a}_2 \vec{a}_3$ onto
   the tetrahedron $\triangle \vec{0}\,(-\vec{a}_1)(-\vec{a}_2)(-\vec{a}_3)$. See Fig. 3.54(a).

(2) Try to find a linear operator mapping the tetrahedron $\Delta \vec{0}\,\vec{a}_1\vec{a}_2\vec{a}_3$ onto the parallelogram $\square\,\vec{a}_1\vec{a}_2$. See Fig. 3.54(b).

Solution (1) There are six such possible linear operators. The simplest one, say f_1, among them is the one that satisfies

$$f_1(\vec{a}_i) = -\vec{a}_i \quad \text{for } 1 \le i \le 3.$$

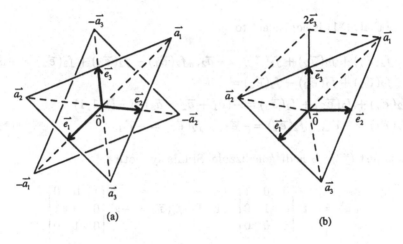

(a)   (b)

Fig. 3.54

In the natural basis $\mathcal{N} = \{\vec{e}_1, \vec{e}_2, \vec{e}_3\}$,

$$[f_1]_\mathcal{N} = \begin{bmatrix} \vec{a}_1 \\ \vec{a}_2 \\ \vec{a}_3 \end{bmatrix}^{-1} \begin{bmatrix} -1 & & 0 \\ & -1 & \\ 0 & & -1 \end{bmatrix} \begin{bmatrix} \vec{a}_1 \\ \vec{a}_2 \\ \vec{a}_3 \end{bmatrix} = -I_3$$

$$\Rightarrow f_1(\vec{x}) = -\vec{x} = -\vec{x}I_3.$$

It is possible that $\vec{a}_1$ and $\vec{a}_2$ are mapped into $-\vec{a}_2$ and $-\vec{a}_1$ respectively while $\vec{a}_3$ is to $-\vec{a}_3$. Denote by f_2 such a linear operator. Then

$$f_2(\vec{a}_1) = -\vec{a}_2,$$
$$f_2(\vec{a}_2) = -\vec{a}_1,$$
$$f_2(\vec{a}_3) = -\vec{a}_3. \tag{*1}$$

$$\Rightarrow [f_2]_{\mathcal{N}} = P^{-1} \begin{bmatrix} 0 & -1 & 0 \\ -1 & 0 & 0 \\ 0 & 0 & -1 \end{bmatrix} P, \quad \text{where } P = \begin{bmatrix} \vec{a}_1 \\ \vec{a}_2 \\ \vec{a}_3 \end{bmatrix} = \begin{bmatrix} -1 & 1 & 1 \\ 1 & -1 & 1 \\ 1 & 1 & -1 \end{bmatrix}$$

$$= \frac{1}{2} \begin{bmatrix} 0 & 1 & 1 \\ 1 & 0 & 1 \\ 1 & 1 & 0 \end{bmatrix} \begin{bmatrix} 0 & -1 & 0 \\ -1 & 0 & 0 \\ 0 & 0 & -1 \end{bmatrix} \begin{bmatrix} -1 & 1 & 1 \\ 1 & -1 & 1 \\ 1 & 1 & -1 \end{bmatrix} = \begin{bmatrix} 0 & -1 & 0 \\ -1 & 0 & 0 \\ 0 & 0 & -1 \end{bmatrix}$$

$$\Rightarrow f_2(\vec{x}) = \vec{x}[f_2]_{\mathcal{N}} = \vec{x} \begin{bmatrix} 0 & -1 & 0 \\ -1 & 0 & 0 \\ 0 & 0 & -1 \end{bmatrix} = -\vec{x} \begin{bmatrix} 0 & 1 & 0 \\ 1 & 0 & 0 \\ 0 & 0 & 1 \end{bmatrix}. \qquad (*2)$$

Notice that, (*1) is equivalent to

$$-f_2(\vec{e}_1) + f_2(\vec{e}_2) + f_2(\vec{e}_3) = -\vec{a}_2, \quad f_2(\vec{e}_1) - f_2(\vec{e}_2) + f_2(\vec{e}_3) = -\vec{a}_1,$$
$$f_2(\vec{e}_1) + f_2(\vec{e}_2) - f_2(\vec{e}_3) = -\vec{a}_3$$
$$\Rightarrow f_2(\vec{e}_1) + f_2(\vec{e}_2) + f_2(\vec{e}_3) = -(\vec{a}_1 + \vec{a}_2 + \vec{a}_3) = -(1,1,1)$$
$$\Rightarrow f_2(\vec{e}_1) = -\vec{e}_2, \quad f_2(\vec{e}_2) = -\vec{e}_1, \quad f_2(\vec{e}_3) = -\vec{e}_3. \qquad (*3)$$

This is just (*2). f_2 is diagonalizable. Similarly, both

$$f_3(\vec{x}) = -\vec{x} \begin{bmatrix} 0 & 0 & 1 \\ 0 & 1 & 0 \\ 1 & 0 & 0 \end{bmatrix} \quad \text{and} \quad f_4(\vec{x}) = -\vec{x} \begin{bmatrix} 1 & 0 & 0 \\ 0 & 0 & 1 \\ 0 & 1 & 0 \end{bmatrix}$$

are another two such linear operators.

The last two linear operators are

$$f_5(\vec{x}) = -\vec{x} \begin{bmatrix} 0 & 1 & 0 \\ 0 & 0 & 1 \\ 1 & 0 & 0 \end{bmatrix} \quad \text{and} \quad f_6(\vec{x}) = -\vec{x} \begin{bmatrix} 0 & 0 & 1 \\ 1 & 0 & 0 \\ 0 & 1 & 0 \end{bmatrix}.$$

Both are not diagonalizable. For details, see Sec. 3.7.8.

(2) The parallelogram $\square \vec{a}_1 \vec{a}_2$ has the vertices at $\vec{0}, \vec{a}_1, \vec{a}_1 + \vec{a}_2 = 2\vec{e}_3$ and $\vec{a}_2$.

Define a linear operator $g: \mathbb{R}^3 \to \mathbb{R}^3$ as

$$g(\vec{a}_1) = \vec{a}_1,$$
$$g(\vec{a}_2) = \vec{a}_2,$$
$$g(\vec{a}_3) = \vec{a}_1 + \vec{a}_2 = 2\vec{e}_3.$$

The process like (*2) or (*3) will lead to

$$g(\vec{x}) = \vec{x}[g]_{\mathcal{N}} = \vec{x} \begin{bmatrix} \frac{1}{2} & -\frac{1}{2} & \frac{3}{2} \\ -\frac{1}{2} & \frac{1}{2} & \frac{3}{2} \\ 0 & 0 & 1 \end{bmatrix} \quad \text{or}$$

$$[g]_{\mathcal{B}} = P[g]_{\mathcal{N}}P^{-1} = \begin{bmatrix} 1 & 0 & 0 \\ 0 & 1 & 0 \\ 1 & 1 & 0 \end{bmatrix},$$

where P is as above and $\mathcal{B} = \{\vec{a}_1, \vec{a}_2, \vec{a}_3\}$. g is diagonalizable and

$$Q[g]_{\mathcal{N}}Q^{-1} = \begin{bmatrix} 1 & 0 & 0 \\ 0 & 1 & 0 \\ 0 & 0 & 0 \end{bmatrix}, \quad \text{where } Q = \begin{bmatrix} -1 & 1 & 1 \\ 1 & -1 & 1 \\ 1 & 1 & -3 \end{bmatrix}.$$

g is a projection of $\mathbb{R}^3$ onto the subspace $\langle\langle \vec{a}_1, \vec{a}_2 \rangle\rangle$ along $\langle\langle (1, 1, -3) \rangle\rangle$ as can be visualized in Fig. 3.54(b).

The readers are urged to find more such linear operators. □

One of the main advantages of diagonalizable linear operators or matrices A is that it is easy to compute the power

$$A^n$$

for $n \geq 1$ and $n < 0$ if A is invertible. More precisely, suppose

$$A = P^{-1} \begin{bmatrix} \lambda_1 & & 0 \\ & \lambda_2 & \\ 0 & & \lambda_3 \end{bmatrix} P$$

$\Rightarrow$ 1. $\det(A) = \lambda_1\lambda_2\lambda_3$.

    2. A is invertible $\Leftrightarrow \lambda_1\lambda_2\lambda_3 \neq 0$. In this case,

$$A^{-1} = P^{-1} \begin{bmatrix} \lambda_1^{-1} & & 0 \\ & \lambda_2^{-1} & \\ 0 & & \lambda_3^{-1} \end{bmatrix} P.$$

    3. Hence

$$A^n = P^{-1} \begin{bmatrix} \lambda_1^n & & 0 \\ & \lambda_2^n & \\ 0 & & \lambda_3^n \end{bmatrix} P.$$

    4. $\operatorname{tr}(A) = \lambda_1 + \lambda_2 + \lambda_3$.

5. For any polynomial $g(t) \in P_n(\mathbb{R})$,

$$g(A) = P^{-1} \begin{bmatrix} g(\lambda_1) & & 0 \\ & g(\lambda_2) & \\ 0 & & g(\lambda_3) \end{bmatrix} P. \qquad (3.7.49)$$

These results still hold for any diagonalizable matrix of finite order.

Example 3 Use

$$A = \begin{bmatrix} 1 & -6 & 4 \\ -2 & -4 & 5 \\ -2 & -6 & 7 \end{bmatrix}$$

to justify (3.7.49).

Solution The characteristic polynomial is

$$\det(A - I_3) = -(t+1)(t-2)(t-3).$$

Hence A is diagonalizable.

For $\lambda_1 = -1$:

$$\vec{x}(A + I_3) = (x_1 \quad x_2 \quad x_3) \begin{bmatrix} 2 & -6 & 4 \\ -2 & -3 & 5 \\ -2 & -6 & 8 \end{bmatrix} = \vec{0}$$

$$\Rightarrow \vec{v}_1 = (1, 4, -3) \text{ is an associated eigenvector.}$$

For $\lambda_2 = 2$, $\vec{v}_2 = (0, 1, -1)$. For $\lambda_3 = 3$, $\vec{v}_3 = (1, 0, -1)$. Thus

$$A = P^{-1} \begin{bmatrix} -1 & & 0 \\ & 2 & \\ 0 & & 3 \end{bmatrix} P, \quad \text{where } P = \begin{bmatrix} \vec{v}_1 \\ \vec{v}_2 \\ \vec{v}_3 \end{bmatrix} = \begin{bmatrix} 1 & 4 & -3 \\ 0 & 1 & -1 \\ 1 & 0 & -1 \end{bmatrix}.$$

Now

$$P^{-1} = \frac{1}{2} \begin{bmatrix} 1 & -4 & 4 \\ 1 & -2 & -1 \\ 1 & -4 & -1 \end{bmatrix}$$

$$\Rightarrow A^n = P^{-1} \begin{bmatrix} (-1)^n & & 0 \\ & 2^n & \\ 0 & & 3^n \end{bmatrix} P$$

$$= \frac{1}{2} \begin{bmatrix} (-1)^n + 3^n & 4 \cdot (-1)^n - 4 \cdot 2^n & -3 \cdot (-1)^n + 4 \cdot 2^n - 3^n \\ (-1)^n - 3^n & 4 \cdot (-1)^n - 2 \cdot 2^n & -3 \cdot (-1)^n + 4 \cdot 2^n + 3^n \\ (-1)^n - 3^n & 4 \cdot (-1)^n - 4 \cdot 2^n & -3 \cdot (-1)^n + 4 \cdot 2^n + 3^n \end{bmatrix}$$

for $n = 0, \pm 1, \pm 2, \ldots$. □

In *Markov process* (see Applications <D₂>), one of the main themes is to compute

$$\lim_{n\to\infty} \vec{x}_0 A^n,$$

where A is a regular stochastic matrix and $\vec{x}_0$ is any initial probability vector. We give such an example.

Example 4 Let

$$A = \begin{bmatrix} \frac{2}{5} & \frac{1}{10} & \frac{1}{2} \\ \frac{1}{5} & \frac{7}{10} & \frac{1}{10} \\ \frac{1}{5} & \frac{1}{5} & \frac{3}{5} \end{bmatrix}.$$

(a) Compute $\lim_{n\to\infty} A^n$.

(b) For any *probability vector* $\vec{x}_0 = (\alpha_1, \alpha_2, \alpha_3)$, i.e. $\alpha_1 \geq 0, \alpha_2 \geq 0, \alpha_3 \geq 0$ and $\alpha_1 + \alpha_2 + \alpha_3 = 1$, compute $\lim_{n\to\infty} \vec{x}_0 A^n$.

Note that A is a *regular stochastic matrix*, i.e. each entry of it is positive and each *row* vector of it is a probability vector.

Solution The characteristic polynomial is

$$\det(A - tI_3) = \left(\frac{2}{5} - t\right)\left(\frac{7}{10} - t\right)\left(\frac{3}{5} - t\right) + \frac{1}{50} + \frac{1}{500}$$

$$- \frac{1}{10}\left(\frac{7}{10} - t\right) - \frac{1}{50}\left(\frac{3}{5} - t\right) - \frac{1}{50}\left(\frac{2}{5} - t\right)$$

$$= -\frac{1}{10}(10t^3 - 17t^2 + 8t - 1) = -\frac{1}{10}(t - 1)(2t - 1)(5t - 1).$$

Thus, A has eigenvalues 1, $\frac{1}{2}$ and $\frac{1}{5}$.

For $\lambda_1 = 1$:

$$\vec{x}(A - I_3) = (x_1 \quad x_2 \quad x_3)\begin{bmatrix} -\frac{3}{5} & \frac{1}{10} & \frac{1}{2} \\ \frac{1}{5} & -\frac{3}{10} & \frac{1}{10} \\ \frac{1}{5} & \frac{1}{5} & -\frac{2}{5} \end{bmatrix} = \vec{0}$$

$$\Rightarrow \vec{v}_1 = (5, 7, 8) \text{ is an associated eigenvector.}$$

For $\lambda_2 = \frac{1}{2}$, $\vec{v}_2 = (0, 1, -1)$ and for $\lambda_3 = \frac{1}{5}$, $\vec{v}_3 = (3, 1, -4)$. Therefore,

$$A = Q^{-1}\begin{bmatrix} 1 & & 0 \\ & \frac{1}{2} & \\ 0 & & \frac{1}{5} \end{bmatrix} Q, \quad \text{where } Q = \begin{bmatrix} 5 & 7 & 8 \\ 0 & 1 & -1 \\ 3 & 1 & -4 \end{bmatrix}.$$

By computation,

$$Q^{-1} = -\frac{1}{60} \begin{bmatrix} -3 & 36 & -15 \\ -3 & -44 & 5 \\ -3 & -16 & 5 \end{bmatrix}.$$

Then

$$\lim_{n \to \infty} A^n = \lim_{n \to \infty} Q^{-1} \begin{bmatrix} 1 & & 0 \\ & (\tfrac{1}{2})^n & \\ 0 & & (\tfrac{1}{5})^n \end{bmatrix} Q = Q^{-1} \begin{bmatrix} 1 & 0 & 0 \\ 0 & 0 & 0 \\ 0 & 0 & 0 \end{bmatrix} Q$$

$$= \begin{bmatrix} 0.25 & 0.35 & 0.40 \\ 0.25 & 0.35 & 0.40 \\ 0.25 & 0.35 & 0.40 \end{bmatrix}.$$

This limit matrix has three equal row vector

$$\vec{p} = (0.25, 0.35, 0.40) = \frac{1}{5+7+8} \vec{v}_1 = \frac{1}{20}(5, 7, 8),$$

which is the unique probability vector as an eigenvector associated to the eigenvalue 1 of A.

For any probability vector $\vec{x}_0 = (\alpha_1, \alpha_2, \alpha_3)$,

$$\lim_{n \to \infty} \vec{x}_0 A^n = \vec{x}_0 \left(\lim_{n \to \infty} A^n \right) = \vec{x}_0 \begin{bmatrix} \vec{p} \\ \vec{p} \\ \vec{p} \end{bmatrix}$$

$$= (0.25(\alpha_1 + \alpha_2 + \alpha_3), 0.35(\alpha_1 + \alpha_2 + \alpha_3),$$
$$0.40 \cdot (\alpha_1 + \alpha_2 + \alpha_3))$$
$$= (0.25, 0.35, 0.40) = \vec{p}.$$

This guarantees the uniqueness of $\vec{p}$.   □

Exercises

\<A\>

1. For each of the following matrices A, do the following problems.

   (1) Test if A is diagonalizable and justify your claim.
   (2) If A is diagonalizable, try to find an invertible matrix P so that PAP^{-1} is diagonal and justify (3.7.49).

(3) If A is not diagonalizable, try to decide its canonical form according to (3.7.29).

(a) $\begin{bmatrix} 10 & 11 & 3 \\ -3 & -4 & -3 \\ -8 & -8 & -1 \end{bmatrix}$. (b) $\begin{bmatrix} 3 & 3 & 3 \\ -3 & -3 & -3 \\ 3 & 3 & 3 \end{bmatrix}$. (c) $\begin{bmatrix} 0 & -2 & 1 \\ 1 & 3 & -1 \\ 0 & 0 & 1 \end{bmatrix}$.

(d) $\begin{bmatrix} -1 & 1 & 2 \\ 1 & 2 & 1 \\ 2 & 1 & -1 \end{bmatrix}$. (e) $\begin{bmatrix} 1 & -1 & 4 \\ 3 & 2 & -1 \\ 2 & 1 & -1 \end{bmatrix}$. (f) $\begin{bmatrix} 1 & 2 & 1 \\ 0 & 1 & 0 \\ 1 & 3 & 1 \end{bmatrix}$.

(g) $\begin{bmatrix} 3 & 2 & -2 \\ 2 & 2 & -1 \\ 2 & 1 & 0 \end{bmatrix}$. (h) $\begin{bmatrix} 4 & -3 & 3 \\ 0 & 1 & 4 \\ 2 & -2 & 1 \end{bmatrix}$. (i) $\begin{bmatrix} 0 & 0 & 1 \\ 0 & 1 & 2 \\ 0 & 0 & 1 \end{bmatrix}$.

2. Show that f_2, f_3 and f_4 in Example 2 are diagonalizable, while f_5 and f_6 are not diagonalizable. Guess what are their canonical forms.

3. Show that the two matrices in (3.7.48) are not similar.

<B>

1. Prove Ex. <B> 4 of Sec. 2.7.6 for 3×3 matrices.

2. Determine whether each of the following matrices A is diagonalizable. If it is, find invertible matrix $P_{4 \times 4}$ so that PAP^{-1} is diagonal. Also, justify (3.7.49).

(a) $\begin{bmatrix} 1 & 0 & 0 & 1 \\ 0 & 1 & 1 & 1 \\ 0 & 0 & 2 & 0 \\ 0 & 0 & 0 & 2 \end{bmatrix}$. (b) $\begin{bmatrix} 0 & -3 & 1 & 2 \\ -2 & 1 & -1 & 2 \\ -2 & 1 & -1 & 2 \\ -2 & -3 & 1 & 4 \end{bmatrix}$. (c) $\begin{bmatrix} 2 & 3 & 3 & 2 \\ 3 & 2 & 2 & 3 \\ 0 & 0 & 1 & 1 \\ 0 & 0 & 0 & 1 \end{bmatrix}$.

<C> Abstraction and generalization

Read Ex. <C> of Sec. 2.7.6 and do problems there. Also do the following problems.

1. Test the diagonalizability for each of the following linear operators f. If it is, find a basis $\mathcal{B}$ for which $[f]_{\mathcal{B}}$ is a diagonal matrix.

(a) $f: M(2; \mathbb{R}) \to M(2; \mathbb{R})$ defined by $f(A) = A^*$.

(b) $f: P_3(\mathbb{R}) \to P_3(\mathbb{R})$ defined by $f(p) = p'$, the first derivative of p.

(c) $f: P_3(\mathbb{R}) \to P_3(\mathbb{R})$ defined by $f(p)(x) = p(0) + p(1)(1 + x + x^2)$.

(d) $f: P_3(\mathbb{R}) \to P_3(\mathbb{R})$ defined by $f(p) = p' + p''$.

(e) Denote by $C^\infty(\mathbb{R})$ the vector space of all infinitely differentiable functions on $\mathbb{R}$. Let $V = \langle\langle e^x, xe^x, x^2 e^x \rangle\rangle$ be the subspace of $C^\infty(\mathbb{R})$

generated by e^x, xe^x and x^2e^x. $f: V \to V$ is defined by

$$f(p) = p'' - 2p' + p.$$

(f) $f: P_3(\mathbb{R}) \to P_3(\mathbb{R})$ defined by $f(p)(x) = \int_0^x p'(t)\, dt$.
(g) $f: P_3(\mathbb{R}) \to P_3(\mathbb{R})$ defined by $f(p)(x) = x^2 p''(x) - xp'(x)$.

2. To compute the eigenvalues of a matrix of larger order, one might adopt numerical methods such as Newton's one or the QR-decomposition of a matrix (see Secs. 4.4 and 5.4). For details, refer to Wilkinson [18] or Strang [15, 16]. In practice and in many occasions, bounds for eigenvalues are good enough to solve problems. Hirsch (1900) and Gersgörin (1931) gave some beginning results in this direction. For our purposes in Application $<D_3>$, we state *Gersgörin's disk theorem* as follows: Each eigenvalue of a complex matrix $A = [a_{ij}]_{n \times n}$ lies in some disk

$$|z - a_{jj}| \le \sum_{i=1}^{n} |a_{ij}| - |a_{jj}|, \quad j = 1, 2, \ldots, n$$

in the complex plane $\mathbb{C}$.

(a) To prove it, suppose λ is an eigenvalue of A and $\vec{x} = (x_1, \ldots, x_n)$ is an associated eigenvector. Let $|x_k| \ge |x_i|$ for $i = 1, 2, \ldots, n$. Then $x_k \ne 0$. Try to show that

$$\vec{x}A = \lambda\vec{x} \Leftrightarrow \sum_{i=1}^{n} a_{ij}x_i = \lambda x_j \quad j = 1, 2, \ldots, n$$

and hence

$$|\lambda x_k - a_{kk}x_k| = |x_k||\lambda - a_{kk}| \le |x_k| \left(\sum_{i=1}^{n} |a_{ik}| - |a_{kk}| \right).$$

(b) As a consequence of it, show that for any eigenvalue λ of A,

$$|\lambda| \le \min \left\{ \max_{1 \le j \le n} \sum_{i=1}^{n} |a_{ij}|, \ \max_{1 \le i \le n} \sum_{j=1}^{n} |a_{ij}| \right\}.$$

3. Let $A = [a_{ij}]_{n \times n} \in M(n; \mathbb{C})$. Define

$$\|A\|_r = \max_{1 \le i \le n} \sum_{j=1}^{n} |a_{ij}| \quad \text{and} \quad \|A\|_c = \max_{1 \le j \le n} \sum_{i=1}^{n} |a_{ij}|.$$

(a) Suppose A is a *positive* matrix, i.e. each entry $a_{ij} > 0$ for $1 \le i, j \le n$ and λ is an eigenvalue of A such that $|\lambda| = \|A\|_r$ or $\|A\|_c$, then

$$\lambda = \|A\|_r \quad \text{or} \quad \|A\|_c$$

respectively. Also, show that $E_\lambda = \{\vec{x} \in \mathbb{C}^n \mid \vec{x}A = \lambda\vec{x}\} = \langle\langle(1, 1, \ldots, 1)\rangle\rangle$.

(b) Furthermore, suppose A is a positive stochastic matrix (for definition, see Ex. <D$_3$>). Then any eigenvalue λ of A other than 1 satisfies $|\lambda| < 1$ and $\dim E_1 = 1$ where $E_1 = \{\vec{x} \in \mathbb{C}^n \mid \vec{x}A = \vec{x}\}$.

4. Let $A = [a_{ij}] \in M(n; \mathbb{F})$.

(a) Show that the *characteristic polynomial* of A

$$\det(A - tI_n)$$
$$= (-1)^n t^n + a_{n-1} t^{n-1} + \cdots + a_k t^k + \cdots + a_1 t + a_0$$
$$= (-1)^n t^n + \sum_{k=1}^{n} (-1)^{n-k} \left(\sum_{1 \le i_1 < \cdots < i_k \le n} \begin{vmatrix} a_{i_1 i_1} & \cdots & a_{i_1 i_k} \\ \vdots & & \vdots \\ a_{i_k i_1} & \cdots & a_{i_k i_k} \end{vmatrix} \right) t^{n-k}.$$

Namely, the coefficient a_{n-k} of $\det(A - tI_n)$ is equal to $(-1)^{n-k}$ times the sum of C_k^n principal subdeterminants (see Sec. B.6) of order k of A.

(b) If A has n eigenvalues $\lambda_1, \ldots, \lambda_n$ in $\mathbb{F}$ so that

$$\det(A - tI_n) = (-1)^n(t - \lambda_1) \ldots (t - \lambda_n),$$

then the *elementary symmetric function of order k of* $\lambda_1, \ldots, \lambda_n$ is

$$\sum_{1 \le i_1 < \cdots < i_k \le n} \lambda_{i_1}, \ldots, \lambda_{i_k} = (-1)^{n-k} a_{n-k}$$

$$= \text{sum of } C_k^n \text{ principal subdeterminants}$$
$$\text{of order } k \text{ of } A.$$

5. Let $A = [a_{ij}] \in M(n; \mathbb{C})$

(a) Suppose λ_0 is a simple root of $\det(A - tI_n) = 0$. Show that there exists some k so that

$$(A - \lambda_0 I_n)\begin{pmatrix} 1 & \cdots & k-1 & k+1 & \cdots & n \\ 1 & \cdots & k-1 & k+1 & \cdots & n \end{pmatrix} \ne 0$$

which is the principal subdeterminant of order $n-1$ of $\det(A - \lambda_0 I_n)$, obtained by deleting the kth row and kth column (see Sec. B.6).

(b) Let $\vec{x}_0$ be an eigenvector corresponding to the simple eigenvalue λ_0. Then $\vec{x}_0$ can be expressed as

$$(\Delta_1, \Delta_2, \ldots, \Delta_n)$$

if $\Delta_1 = (A - \lambda_0 I_n)\begin{pmatrix} 2 & \cdots & n \\ 2 & \cdots & n \end{pmatrix} \neq 0$ and $\Delta_i, 2 \leq i \leq n$, is obtained from Δ_1 by replacing the $(i-1)$st column by $-a_{12}, -a_{13}, \ldots, -a_{1n}$.

6. Let $\vec{x}_0 \in \mathbb{C}^n$ be a fixed vector. Show that

$$\det(\vec{x}_0^* \vec{x}_0 - \lambda I_n) = (-1)^n[\lambda^n - (\vec{x}_0 \overline{\vec{x}_0^*})\lambda^{n-1}],$$

where $\langle \vec{x}_0, \vec{x}_0 \rangle = \vec{x}_0 \overline{\vec{x}_0^*}$. Try to find an invertible matrix $P_{n \times n}$ so that $P(\vec{x}_0^* \vec{x}_0 - \lambda I_n)P^{-1} = \text{diag}[\langle \vec{x}_0, \vec{x}_0 \rangle, 0, \ldots, 0]$.

7. Let

$$A = \begin{bmatrix} 1 & 1 & 1 & 1 \\ 1 & i & -1 & -i \\ 1 & -1 & 1 & -1 \\ 1 & -i & -1 & i \end{bmatrix}.$$

(a) Find the characteristic polynomial, eigenvalues and eigenvectors of A. Diagonalize A.

(b) Compute A^k and their eigenvalues, for $2 \leq k \leq 4$.

8. Let

$$A = \begin{bmatrix} a_0 & a_1 & a_2 & a_3 \\ -a_1 & a_0 & -a_3 & a_2 \\ -a_2 & a_3 & a_0 & -a_1 \\ -a_3 & -a_2 & a_1 & a_0 \end{bmatrix}.$$

Find the characteristic polynomial of A and try to diagonalize A, if possible.

9. Find the characteristic polynomial of each of following matrices and its eigenvalues, if possible.

$$\begin{bmatrix} \alpha_1 & 1 & & & & & 0 \\ -1 & \alpha_2 & 1 & & & & \\ & -1 & \alpha_3 & 1 & & & \\ & & \ddots & \ddots & \ddots & & \\ & & & -1 & \alpha_{n-1} & 1 \\ 0 & & & & -1 & \alpha_n \end{bmatrix};$$

$$\begin{bmatrix} 0 & 1 & 1 & 0 & 0 & \cdots & 0 & 0 \\ 0 & 0 & 1 & 1 & 0 & \cdots & 0 & 0 \\ 0 & 0 & 0 & 1 & 1 & \cdots & 0 & 0 \\ \vdots & \vdots & \vdots & \vdots & \vdots & & \vdots & \vdots \\ 0 & 0 & 0 & 0 & 0 & \cdots & 1 & 1 \\ 1 & 0 & 0 & 0 & 0 & \cdots & 0 & 1 \\ 1 & 1 & 0 & 0 & 0 & \cdots & 0 & 0 \end{bmatrix}_{n\times n};$$

$$\begin{bmatrix} a_1 & 0 & \cdots & 0 & b_1 \\ 0 & a_2 & \cdots & 0 & b_2 \\ \vdots & \vdots & & \vdots & \vdots \\ 0 & 0 & \cdots & a_{n-1} & b_{n-1} \\ b_1 & b_2 & \cdots & b_{n-1} & a_n \end{bmatrix}.$$

10. Let

$$\varphi_0(t) = 1; \quad \varphi_k(t) = \prod_{j=1}^{k}(t - a_j), \quad 1 \le k \le n.$$

Show the characteristic polynomial of

$$\begin{bmatrix} a_1 & 1 & 0 & \cdots & 0 & 0 \\ 0 & a_2 & 1 & \cdots & 0 & 0 \\ \vdots & \vdots & \vdots & & \vdots & \vdots \\ 0 & 0 & 0 & \cdots & a_{n-1} & 1 \\ b_0 & b_1 & b_2 & \cdots & b_{n-2} & a_n \end{bmatrix}$$

is $(-1)^n[\varphi_n(t) - \sum_{k=0}^{n-2} b_k\varphi_k(t)]$.

11. Suppose $A = [a_{ij}] \in M(n; \mathbb{F})$ has the characteristic polynomial

$$\det(A - tI_n) = (-1)^n t^n + a_{n-1}t^{n-1} + \cdots + a_1 t + a_0$$
$$= (-1)^n(t - \lambda_1) \cdots (t - \lambda_n).$$

Let $s_k = \text{tr}A^k$ denote the *trace* of A^k, $k = 0, 1, 2, \ldots$. Show that if $2 \le k \le n$,

$$(-1)^{n-1}a_{n-1} = s_1,$$
$$k \cdot (-1)^{n-k}a_{n-k} = (-1)^{k-1}[s_k + (-1)^n a_{n-1}s_{k-1} + \cdots$$
$$+ (-1)^n a_{n-k+1}s_1];$$

and if $k > n$,

$$s_k - (-1)^{n-1}a_{n-1}s_{k-1} + (-1)^{n-2}a_{n-2}s_{k-2} + \cdots$$
$$+ (-1)^{n-1}(-1)a_1 s_{k-n+1} + (-1)^n a_0 s_{k-n}$$
$$= s_k + (-1)^n a_{n-1}s_{k-1} + (-1)^n a_{n-2}s_{k-2} + \cdots$$
$$+ (-1)^n a_1 s_{k-n+1} + (-1)^n a_0 s_{k-n} = 0.$$

These are called *Newton identities*.

12. (a) Let $A, B \in M(2; \mathbb{F})$. Show that

   (1) $\det(A - tI_2) = t^2 - (\text{tr } A)t + \det A$.

   (2) $AB + BA = A\text{tr } B + B\text{tr } A + I_2(\text{tr}(AB) - \text{tr } A \cdot \text{tr } B)$.

   (b) Let $A, B \in M(3; \mathbb{F})$. Show that

   (1) $\det(A - tI_3) = -t^3 + (\text{tr } A)t^2 - \text{tr}(\text{adj}A)t + \det A$.

   (2) $\det(A - tB) = \det A - \text{tr}((\text{adj}A)B)t + \text{tr}((\text{adj}B)A)t^2 - (\det B)t^3$.

   (3) $\det(A + B) = \det A + \text{tr}((\text{adj}A)B) + \text{tr}((\text{adj}B)A) + \det B$.

<D₁> Application (I): *The limit processes and the matrices*

Suppose $a_{ij}(t)$ is a real or complex valued function defined on a set S in the plane $\mathbb{R}^2$ or $\mathbb{C}$ for each $1 \le i \le m, 1 \le j \le n$. Then

$$A(t) = [a_{ij}(t)]_{m \times n} \in M(m, n; \mathbb{C}), \quad t \in S.$$

If t_0 is a limit point of S and for each $i, j, \lim_{t \to t_0} a_{ij}(t) = a_{ij}$ exists, then

$$A = [a_{ij}] = \lim_{t \to t_0} A(t)$$

is called the *limit matrix* of $A(t)$ as $t \to t_0$. Similarly, one defines

$$A'(t) = [a'_{ij}(t)]_{m \times n},$$

$$\int_a^b A(t)\, dt = \left[\int_a^b a_{ij}(t)\, dt \right]_{m \times n}.$$

Let $A^{(k)} = [a_{ij}^{(k)}]_{m \times n} \in M(m, n; \mathbb{C})$ be a sequence of matrices of the same order. Suppose for each $i, j, \lim_{k \to \infty} a_{ij}^{(k)} = a_{ij}$ exists, then

$$A = [a_{ij}]_{m \times n} = \lim_{k \to \infty} A^{(k)}$$

is called the *limit matrix* of the sequence matrices $A^{(k)}$ as $k \to \infty$.

Do the following problems.

1. Let $A = [a_{ij}]_{m \times n}$. Show that

$$\det A$$

   is a polynomial in n^2 variables $a_{ij}, 1 \le i, j \le n$. Hence, $\det A$ is a *continuous* function of its entries.
2. Try to use Ex. 1 to prove Ex. <C> 8(b) of Sec. 2.7.5.
3. Suppose A and B are $n \times n$ matrices.

   (a) If A or B is invertible, show that AB and BA have the same characteristic polynomial.
   (b) Use Ex. 1 to prove (a) when both A and B are non-invertible.

4. Prove

   (1) $\lim_{k \to \infty}(\alpha A^{(k)}) = \alpha \lim_{k \to \infty} A^{(k)}, \alpha \in \mathbb{C}$.
   (2) $\lim_{k \to \infty}(A^{(k)} + B^{(k)}) = \lim_{k \to \infty} A^{(k)} + \lim_{k \to \infty} B^{(k)}$.
   (3) $\lim_{k \to \infty} PA^{(k)} = P\left(\lim_{k \to \infty} A^{(k)} \right)$
      and $\lim_{k \to \infty} A^{(k)}Q = \left(\lim_{k \to \infty} A^{(k)} \right)Q$.
   (4) $\lim_{k \to \infty} PA^k P^{-1} = P\left(\lim_{k \to \infty} A^k \right)P^{-1}$.

5. Let $A = [a_{ij}] \in M(n; \mathbb{C})$. Then

   (i) $\lim_{k \to \infty} A^k$ exists.
   $\Leftrightarrow$ (ii) a. If λ is an eigenvalue of A, then either $|\lambda| < 1$ or $\lambda = 1$.

       b. If 1 is an eigenvalue of A, then
      the algebraic multiplicity of 1 = the geometric multiplicity of 1.

   (a) Try to prove (i) $\Rightarrow$ (ii) a.
   (b) Suppose A is diagonalizable and (ii) a holds. Try to show that $\lim_{k \to \infty} A^k$ exists.
      (For a complete proof for this result, one needs Jordan canonical forms of matrices (see Sec. 3.7.7 and Ex. <C> 13(e) there).)
   (c) Show that $\lim_{k \to \infty} \begin{bmatrix} 1 & 0 \\ 1 & 1 \end{bmatrix}^k$ does not exist.
   (d) Compute $\lim_{k \to \infty} A^k$, where A is either

$$\begin{bmatrix} 0 & -\frac{3}{2} & \frac{1}{2} \\ -\frac{1}{2} & -1 & \frac{1}{2} \\ -\frac{5}{2} & -\frac{15}{2} & 3 \end{bmatrix} \text{ or } \begin{bmatrix} -\frac{1}{2} & 0 & 0 \\ 0 & \frac{1}{3} & 1 \\ 0 & 0 & -\frac{1}{4} \end{bmatrix}.$$

6. Let $A \in M(n; \mathbb{C})$. Show that

$$\lim_{k \to \infty} A^{(k)} = O$$

if and only if

$$\rho(A) = \max_{1 \le j \le n} |\lambda_j| < 1$$

where $\lambda_1, \ldots, \lambda_n$ are eigenvalues of A. $\rho(A)$ is called the *spectral radius* of A.

<D₂> Application (II): *Matrix exponential, etc.*

It is well-known form analysis that the power series

$$e^z = \sum_{k=0}^{\infty} \frac{z^k}{k!} = 1 + z + \frac{1}{2!}z^2 + \cdots + \frac{1}{k!}z^k + \cdots$$

converses absolutely on the plane $\mathbb{C}$ and uniformly on every compact subsets of $\mathbb{C}$. For any fixed matrix $A = [a_{ij}] \in M(n; \mathbb{C})$, via *norms* for A such as $\|A\|_r$ and $\|A\|_c$ as defined in Ex. <C> 3 or others (see Ex. <C> 14 of Sec. 3.7.7), it is not hard to show that

$$\lim_{k \to \infty} \left(I_n + A + \frac{1}{2!}A^2 + \cdots + \frac{1}{k!}A^k \right) \quad \text{exists.}$$

Therefore, we define the *exponential* of A as this limit matrix and denote as

$$e^A = \sum_{k=0}^{\infty} \frac{1}{k!}A^k.$$

For example,

$$A = \begin{bmatrix} 1 & 0 & 1 \\ 2 & 1 & 3 \\ 1 & 0 & 1 \end{bmatrix}$$

$$\Rightarrow A = P^{-1} \begin{bmatrix} 0 & 0 & 0 \\ 0 & 1 & 0 \\ 0 & 0 & 2 \end{bmatrix} P, \quad \text{where } P = \begin{bmatrix} -1 & 0 & 1 \\ 3 & -1 & 2 \\ 1 & 0 & 1 \end{bmatrix}$$

$$\Rightarrow \sum_{l=0}^{k} \frac{1}{l!}A^l = P^{-1} \begin{bmatrix} 1 & & 0 \\ & \sum_{l=0}^{k} \frac{1}{l!} & \\ 0 & & \sum_{l=0}^{k} \frac{1}{l!} \cdot 2^l \end{bmatrix} P$$

$$\Rightarrow e^A = P^{-1} \begin{bmatrix} 1 & & 0 \\ & e & \\ 0 & & e^2 \end{bmatrix} P.$$

Do the following problems.

1. (a) Let $A = \begin{bmatrix} a & b \\ -b & a \end{bmatrix}$, $a, b \in \mathbb{R}$. Show that

$$e^{tA} = \begin{bmatrix} e^{at} \cos bt & e^{at} \sin bt \\ -e^{at} \sin bt & e^{at} \cos bt \end{bmatrix}, \quad t \in \mathbb{R}.$$

   (b) Let

$$A = \begin{bmatrix} 0 & a & -b \\ -a & 0 & c \\ b & -c & 0 \end{bmatrix}, \quad a, b, c \in \mathbb{R}.$$

   Show that there exists an invertible matrix P so that

$$e^A = P^{-1} \begin{bmatrix} e^{ui} & 0 & 0 \\ 0 & e^{-ui} & 0 \\ 0 & 0 & 1 \end{bmatrix} P,$$

   where $u = \sqrt{a^2 + b^2 + c^2}$.

2. Suppose $A_{n \times n}$ is diagonalizable and

$$PAP^{-1} = \begin{bmatrix} \lambda_1 & & 0 \\ & \ddots & \\ 0 & & \lambda_n \end{bmatrix}.$$

   Show that

$$e^A = P^{-1} \begin{bmatrix} e^{\lambda_1} & & 0 \\ & \ddots & \\ 0 & & e^{\lambda_n} \end{bmatrix} P.$$

   For e^A where A has Jordan canonical form, see Ex. <C> 13(a), (d) of Sec. 3.7.7.

3. Give examples to show that, in general,

   (1) $e^A \cdot e^B \neq e^B \cdot e^A$.
   (2) $e^{A+B} \neq e^A \cdot e^B$.

   Suppose $A, B \in M(n; \mathbb{C})$. Show that

$$e^{t(A+B)} = e^{tA} e^{tB}, \quad t \in \mathbb{C}$$

   if and only if $AB = BA$. Hence, in this case,

$$e^{A+B} = e^A e^B = e^B e^A.$$

4. Prove the following.

   (1) $\det e^A = e^{\operatorname{tr}(A)}$.

(2) For any A, e^A is invertible and

$$(e^A)^{-1} = e^{-A}.$$

(3) $e^{PAP^{-1}} = Pe^A P^{-1}$.

(4) If A is skew-symmetric, i.e. A is real and $A^* = -A$, then e^A is orthogonal, i.e. $(e^A)^* = e^{A^*} = e^{-A}$.

(5) $\frac{d}{dt}e^{tA} = Ae^{tA} = e^{tA} \cdot A$.

Suppose $A_m \in M(n; \mathbb{C})$ for $m = 1, 2, 3, \ldots$. If $\sum_{m=0}^{k} A_m = S_k$ converges to A as $k \to \infty$ (see Ex. <D$_1$>), then the series $\sum_{m=0}^{\infty} A_m$ is said to *converge* to the *sum* A and is denoted as

$$A = \sum_{m=0}^{\infty} A_m.$$

5. Let $A \in M(n; \mathbb{C})$ and its spectral radius $\rho(A) < 1$ (see Ex. <D$_1$> 6). Show that

(a) $I_n - A$ is invertible, and

(b) $(I_n - A)^{-1} = \sum_{m=0}^{\infty} A^m$ (Note that $A^0 = I_n$).

In case A is a nilpotent matrix of index k, then $(I_n - A)^{-1} = I_n + A + \cdots + A^{k-1}$.

6. (a) Let $\|A\|_1 = \sum_{i,j=1}^{n} |a_{ij}|$ where $A = [a_{ij}] \in M(n; \mathbb{C})$. Show that (see also Ex. <C> 14 of Sec. 3.7.7)

(1) $\|A\|_1 \geq 0$, and $= 0 \Leftrightarrow A = O$.

(2) $\|\alpha A\|_1 = |\alpha| \|A\|_1, \alpha \in \mathbb{C}$.

(3) $\|A + B\|_1 \leq \|A\|_1 + \|B\|_1$.

(4) $\|AB\|_1 \leq \|A\|_1 \|B\|_1$.

(5) $|\vec{x} A|_1 \leq |\vec{x}|_1 \|A\|_1$, where $\vec{x} = (x_1, \ldots, x_n) \in \mathbb{C}^n$ and $|\vec{x}|_1 = \sum_{k=1}^{n} |x_k|$.

Hence, $M(n; \mathbb{C})$ endowed with $\|\ \|_1$ is a Banach space and is also a Banach algebra.

(b) Suppose $A \in M(n; \mathbb{C})$ is invertible. For any $0 < \varepsilon < 1$, show that any matrix B in the set

$$\left\{ B \in M(n; \mathbb{C}) \,\big|\, \|B - A\|_1 < \frac{\varepsilon}{\|A^{-1}\|_1} \right\}$$

is also invertible. This means that the general linear group $GL(n; \mathbb{C})$ is an *open* set in $M(n; \mathbb{C})$.

7. (Weyr, 1887) Let $A \in M(n; \mathbb{C})$. Suppose the power series $\sum_{m=0}^{\infty} a_m z^m$ has positive *radius r of convergence*, where $r = \left(\overline{\lim}_{m \to \infty} \sqrt[m]{|a_m|}\right)^{-1}$. Show that

(1) if the spectral radius $\rho(A) < r$, then $\sum_{0}^{\infty} a_m A^m$ converges absolutely (namely, $\sum_{0}^{\infty} |a_m| \|A^m\|_1 < \infty$); and

(2) if $\rho(A) > r$, $\sum_{0}^{\infty} a_m A^m$ diverges (i.e. does not converge).

In particular, *in case $r = +\infty$*, then for *any* $A \in M(n; \mathbb{C})$, the power series

$$\sum_{m=0}^{\infty} a_m A^m$$

always converges absolutely. Suppose $\varphi(A) = \sum_{m=0}^{\infty} a_m A^m$, $\rho(A) < r$.

(a) Let $\lambda_1, \ldots, \lambda_n$ be eigenvalues of A. Then $\varphi(\lambda_1), \ldots, \varphi(\lambda_n)$ are eigenvalues of $\varphi(A)$.

(b) Suppose A is diagonalizable. Then A and $\varphi(A)$ are simultaneously diagonalizable.

(c) Let

$$A = \begin{bmatrix} \frac{1}{2} & 1 \\ 0 & \frac{1}{2} \end{bmatrix}.$$

Show that

$$(I_2 - A)^{-k} = \begin{bmatrix} 2^k & k \cdot 2^{k+1} \\ 0 & 2^k \end{bmatrix}, \quad k \geq 1.$$

(d) Let

$$A = \begin{bmatrix} 0 & a & -b \\ -a & 0 & c \\ b & -c & 0 \end{bmatrix}, \quad a, b, c \in \mathbb{R}.$$

Show that

$$\cos A = I_3 - \frac{\cosh u - 1}{u^2} A^2, \quad \sin A = \frac{\sinh u}{u} A$$

where $u^2 = a^2 + b^2 + c^2$.

8. Let $A \in M(n; \mathbb{C})$ be invertible. If there exists a matrix $X \in M(n; \mathbb{C})$ so that

$$e^X = A,$$

then X is called a *logarithmic matrix* of A and is denoted by

$$X = \log A.$$

$\log A$ is a *multiple-valued* function of A.

(a) If A is diagonalizable and

$$A = P^{-1} \begin{bmatrix} \lambda_1 & & 0 \\ & \ddots & \\ 0 & & \lambda_n \end{bmatrix} P,$$

then

$$\log A = P^{-1} \begin{bmatrix} \log \lambda_1 & & 0 \\ & \ddots & \\ 0 & & \log \lambda_n \end{bmatrix} P.$$

(b) Suppose

$$A = \begin{bmatrix} \lambda & & 0 \\ & \lambda & \\ & & \ddots & \\ & & & \lambda \end{bmatrix}, \quad \lambda \neq 0,$$

is a lower triangular matrix. Note that $(A - \lambda I_n)^k = O$ for $k \geq n$. Then

$$\log A = \log[\lambda I_n + (A - \lambda I_n)] = (\log \lambda)I_n + \sum_{m=1}^{\infty} \frac{(-1)^{m-1}}{m\lambda^m}(A - \lambda I_n)^m$$

$$= (\log \lambda)I_n + \sum_{m=1}^{n-1} \frac{(-1)^{m-1}}{m\lambda^m}(A - \lambda I_n)^m.$$

(c) Suppose

$$A = P^{-1} \begin{bmatrix} A_1 & & & 0 \\ & A_2 & & \\ & & \ddots & \\ 0 & & & A_k \end{bmatrix} P,$$

where each A_j is a lower triangular matrix of order r_j with diagonal entries all equal to λ_j for $1 \leq j \leq k$, and $r_1 + \cdots + r_k = n$. Then

$$\log A = P^{-1} \begin{bmatrix} \log A_1 & & & 0 \\ & \log A_2 & & \\ & & \ddots & \\ 0 & & & \log A_k \end{bmatrix} P.$$

(d) Try to determine $\log I_n$.

(e) Show that

$$\log(I_n + A) = \sum_{m=1}^{\infty} \frac{(-1)^{m-1}}{m} A^m, \quad \rho(A) < 1.$$

(f) Let

$$A = \begin{bmatrix} \lambda & & 0 \\ 1 & \lambda & \\ 0 & 1 & \lambda \end{bmatrix}, \quad \lambda \neq 0.$$

Show that

$$e^A = e^\lambda \begin{bmatrix} 1 & & 0 \\ 1 & 1 & \\ \frac{1}{2} & 1 & 1 \end{bmatrix} \quad \text{and} \quad \log A = \begin{bmatrix} \log \lambda & & 0 \\ \frac{1}{\lambda} & \log \lambda & \\ -\frac{1}{2\lambda^2} & \frac{1}{\lambda} & \log \lambda \end{bmatrix}.$$

<D3> Application (III): *Markov processes*

In a sequential experiment or data collection, each stage contains n states. Let

$a_{ij} =$ the probability that the ith state at time t causes the jth state at time $t + 1$,

for $1 \leq i, j \leq n$. Then $a_{ij} \geq 0$. The resulted matrix

$$M = \begin{bmatrix} a_{11} & \cdots & a_{1n} \\ \vdots & \ddots & \vdots \\ a_{n1} & \cdots & a_{nn} \end{bmatrix}, \quad \text{where } a_{ij} \geq 0, \ 1 \leq i, \ j \leq n \quad \text{and}$$

$$\sum_{j=1}^{n} a_{ij} = 1 \quad \text{for } 1 \leq i \leq n, \text{ i.e. each row is a } \textit{probability vector,}$$

is called the *transition matrix* or *stochastic matrix* or *Markov matrix* of this *Markov process*. M is called *positive* if each entry $a_{ij} > 0$ and is called *regular* if some non-negative power matrix M^r is positive.

For example, the matrix in Example 4 is a positive or regular stochastic matrix. The following matrices

$$\begin{bmatrix} 0.5 & 0.5 & 0 \\ 0 & 0 & 1 \\ 0 & 1 & 0 \end{bmatrix} \quad \text{and} \quad \begin{bmatrix} 0 & \frac{1}{2} & \frac{1}{4} & \frac{1}{4} \\ \frac{1}{2} & 0 & \frac{1}{4} & \frac{1}{4} \\ 0 & 0 & 1 & 0 \\ 0 & 0 & 0 & 1 \end{bmatrix}$$

are stochastic but not regular.

Do the following problems.

1. Let $M = [a_{ij}]_{n \times n}$ be a *non-negative* matrix, i.e. each $a_{ij} \geq 0$.

   (a) Show that

         (1) M is stochastic.

   $\Leftrightarrow$ (2) If $\vec{e}_0 = (1, 1, \ldots, 1) \in \mathbb{R}^n$, then $\vec{e}_0 M^* = \vec{e}_0$.

   $\Leftrightarrow$ (3) If $\vec{p}$ is a probability vector, so is $\vec{p}M$.

   (b) If M is stochastic, so are M^k for $k = 0, 1, 2, \ldots$.

   (c) If M_1 and M_2 are stochastic, then both

   $$(1 - t)M_1 + tM_2 \quad \text{for } 0 \leq t \leq 1 \text{ and } M_1 M_2$$

   are stochastic.

2. Let $M = [a_{ij}]_{n \times n}$ be a stochastic matrix.

   (a) M has eigenvalue 1.

   (b) If M is regular, then each eigenvalue λ of M is either $|\lambda| < 1$ or $\lambda = 1$. For the eigenvalue $\lambda = 1$, we have

   $$\dim\{\vec{x} \in \mathbb{R}^n \mid \vec{x}M = \vec{x}\} = 1$$

   i.e. the geometric multiplicity of 1 is equal to 1 (see Ex. 3(d) below).

3. Let $M = [a_{ij}]_{n \times n}$ be a regular stochastic matrix and $M^k = [a_{ij}^{(k)}]_{n \times n}$ for $k = 1, 2, 3, \ldots$. Then

   (a) M^k has limit matrix as $k \to \infty$. In fact,

   $$\lim_{k \to \infty} M^k = \begin{bmatrix} \vec{p}_0 \\ \vec{p}_0 \\ \vdots \\ \vec{p}_0 \end{bmatrix} = \begin{bmatrix} p_1 & p_2 & \cdots & p_n \\ p_1 & p_2 & \cdots & p_n \\ \vdots & \vdots & \ddots & \vdots \\ p_1 & p_2 & \cdots & p_n \end{bmatrix} = \vec{e}^* \vec{p}_0$$

   where $\vec{e}_0 = (1, 1, \ldots, 1)$ (see Ex. 1(a)) and $\vec{p}_0 = (p_1, \ldots, p_n)$ is a probability vector with

   $$p_j = \lim_{k \to \infty} a_{ij}^{(k)} \quad \text{for } 1 \leq i \leq n \text{ and } 1 \leq j \leq n.$$

(b) For any probability vector $\vec{p}$,

$$\lim_{k \to \infty} \vec{p} M^k = \vec{p}_0$$

always holds. This $\vec{p}_0$ is called the *limiting probability vector* of M.
(c) In particular, $\vec{p}_0 M^k = \vec{p}_0$ for $k = 1, 2, 3, \dots$. Hence, $\vec{p}_0$ is the *unique* probability vector that is also an eigenvector associated to the eigenvalue 1 of M. In this case, $\vec{p}_0$ is also called a *stability vector* or *steady-state vector* for M.
(d) Furthermore, as an eigenvalue of M,

the algebraic multiplicity of 1
= the geometric multiplicity of $1 = 1$.

(a) can be proved by using nested interval theorem in real analysis. (b) and (c) are then consequences of (a) (see also Ex. $<D_1>$ 5). While for (d), one needs Jordan canonical form, see Ex. $<C>$ 15 of Sec. 3.7.7 and Ex. 5 below.

4. Let

$$M = \begin{bmatrix} 0 & 1 \\ 1 & 0 \end{bmatrix}.$$

Show that $\lim_{k \to \infty} M^k$ does not exist but $\vec{p}_0 = \left(\frac{1}{2}, \frac{1}{2} \right)$ satisfies $\vec{p}_0 M^k = \vec{p}_0$ for $k = 1, 2, \dots$ and hence $\vec{p}_0$ is a stability vector for M. Note that M is not regular.

For $\vec{x} = (x_1, \dots, x_n) \in \mathbb{C}^n$, let

$$|\vec{x}|_1 = \sum_{i=1}^{n} |x_i|,$$

and for $A = [a_{ij}] \in M(n; \mathbb{C})$, let

$$\|A\|_1 = \sum_{i=1}^{n} |A_{i*}|_1 = \sum_{i,j=1}^{n} |a_{ij}|.$$

Then, just like Ex. $<D_2>$ 6(a), $M(n; \mathbb{C})$ is a Banach algebra with the norm $\|\ \|_1$. A vector $\vec{x} = (x_1, \dots, x_n)$ in $\mathbb{C}^n$ is called *positive* (or *non-negative*) if $x_i > 0$ (or $x_i \geq 0$) for $1 \leq i \leq n$ and is denoted as

$$\vec{x} > \vec{0} \quad (\text{or } \vec{x} \geq \vec{0}).$$

Hence, a probability vector is a nonzero non-negative vector. Define

$$\vec{x} > \vec{y} \ (\text{or } \vec{x} \geq \vec{y}) \Leftrightarrow \vec{x} - \vec{y} > \vec{0} \ (\text{or } \geq \vec{0}).$$

5. Let $A = [a_{ij}] \in M(n; \mathbb{C})$ be a positive matrix.

   (a) A has a unique positive eigenvalue $\lambda(A)$, greater than the absolute value of any other eigenvalue of A. Also

   $$\lambda(A) = \max\{\lambda \,|\, \lambda \geq 0 \text{ and there exists a nonzero } \vec{x} \geq \vec{0}$$
   $$\text{so that } \vec{x}A \geq \lambda\vec{x}\}$$
   $$= \min\{\lambda \,|\, \lambda > 0 \text{ and there exists a positive } \vec{x} > \vec{0}$$
   $$\text{so that } \vec{x}A \leq \lambda\vec{x}\}$$
   $$= \max_{\substack{\vec{x} \geq \vec{0} \\ \vec{x} \neq \vec{0}}} \min_{1 \leq j \leq n} \frac{\sum_{i=1}^{n} a_{ij}x_i}{x_j}$$
   $$= \min_{\substack{\vec{x} \geq \vec{0} \\ \vec{x} \neq \vec{0}}} \max_{1 \leq j \leq n} \frac{\sum_{i=1}^{n} a_{ij}x_i}{x_j}.$$

   (b) There exists a positive eigenvector of A corresponding to $\lambda(A)$. Also,

   the geometric multiplicity of $\lambda(A)$
   $$= \text{ the algebraic multiplicity of } \lambda(A) = 1.$$

   (c) $\min_{1 \leq j \leq n} \sum_{i=1}^{n} a_{ij} \leq \lambda(A) \leq \max_{1 \leq j \leq n} \sum_{i=1}^{n} a_{ij} = \|A\|_c$.

6. (*Positive matrix* and *positive stochastic matrix*) Let $A = [a_{ij}]_{n \times n}$ be a positive matrix.

   (a) $\vec{x}_0 = (d_1, \ldots, d_n)$ is a positive eigenvector of A corresponding to the largest positive eigenvalue $\lambda(A)$ if and only if there exists a positive Markov matrix M so that

   $$A = \lambda(A)D^{-1}M^*D, \quad \text{where } D = \text{diag}[d_1, \ldots, d_n].$$

   Note that $\vec{x}_0$ can be chosen as the unique positive probability vector.

   (b) Suppose $\vec{p}_0 = (p_1, \ldots, p_n)$ is the limiting probability vector of M in (a) corresponding to 1 and $\vec{e}_0 = (1, 1, \ldots, 1)$. Then

   $$\lim_{k \to \infty} \frac{A^k}{(\lambda(A))^k} = D^{-1}(\vec{p}_0^*\vec{e}_0)D = D^{-1}\vec{p}_0^*\vec{x}_0$$

   exists as a positive matrix. Note that $\vec{e}_0D = \vec{x}_0$, as in (a).

   (c) For any vector $\vec{x} \in \mathbb{R}^n$ (in particular, a non-negative vector), the limit vector

   $$\lim_{k \to \infty} \frac{\vec{x}A^k}{(\lambda(A))^k} = \vec{x}D^{-1}(\vec{p}_0^*\vec{e}_0)D = (\vec{x}D^{-1}\vec{p}_0^*)\vec{x}_0$$

   always exists and is an eigenvector of A corresponding to $\lambda(A)$.

(*Note*   For more information, refer to Chung [38, 39] and Doob [40] or simpler Kemeny and Snell [41].)

<D₄> Application (IV): *Differential equations*

Fundamental theorem of calculus says that the (simplest) differential equation

$$\frac{dx}{dt} = ax$$

has (all) the solutions

$$x(t) = \alpha e^{at}, \quad \alpha \text{ is any constant.}$$

The initial condition $x(0) = \alpha_0$ will restrict α to be α_0.

As a consequence, the homogeneous linear system of differential equations

$$\frac{dx_1}{dt} = \lambda_1 x_1(t),$$

$$\frac{dx_2}{dt} = \lambda_2 x_2(t),$$

with initial conditions $x_1(0) = \alpha_{10}, x_2(0) = \alpha_{20}$

has the unique solution

$$x_1(t) = \alpha_{10} e^{\lambda_1 t}, \quad x_2(t) = \alpha_{20} e^{\lambda_2 t}, \quad t \in \mathbb{R}.$$

In matrix notation, the system can be written as

$$\frac{d\vec{x}}{dt} = \vec{x}(t)A, \quad \text{where } \vec{x}(t) = (x_1(t), x_2(t)), \ A = \begin{bmatrix} \lambda_1 & 0 \\ 0 & \lambda_2 \end{bmatrix},$$

$$\vec{x}(0) = \vec{x}_0 = (\alpha_{10}, \alpha_{20})$$

and the solution is as

$$\vec{x}(t) = \vec{x}_0 e^{tA}.$$

Do the following problems.

1. *The homogeneous linear system of differential equations*

$$\frac{dx_j}{dt} = \sum_{i=1}^{n} a_{ij} x_i(t), \quad 1 \le j \le n,$$

with initial conditions $x_1(0) = \alpha_{10}, \ldots, x_n(0) = \alpha_{n0}$

can be written as, in matrix notation,

$$\frac{d\vec{x}}{dt} = \vec{x}(t)A, \quad \text{where } \vec{x}(t) = (x_1(t), \ldots, x_n(t)) \text{ and } A = [a_{ij}]_{n \times n},$$
$$\vec{x}(0) = \vec{x}_0 = (x_1(0), \ldots, x_n(0)).$$

Then, the *general solution* is

$$\vec{x}(t) = \vec{\alpha} e^{tA}, \quad \vec{\alpha} \in \mathbb{R}^n \text{ or } \mathbb{C}^n$$

and the initial value problem has the *unique* solution

$$\vec{x}(t) = \vec{x}_0 e^{tA}.$$

In particular, if A is *diagonalizable* and

$$A = P^{-1}DP, \quad \text{where } D = \begin{bmatrix} \lambda_1 & & & 0 \\ & \lambda_2 & & \\ & & \ddots & \\ 0 & & & \lambda_n \end{bmatrix} \quad \text{and } P = \begin{bmatrix} \vec{v}_1 \\ \vec{v}_2 \\ \vdots \\ \vec{v}_n \end{bmatrix},$$

then the solution is

$$\begin{aligned} \vec{x}(t) = \vec{\alpha} e^{tA} &= \vec{\alpha} P^{-1} e^{tD} P \\ &= (c_1, c_2, \ldots, c_n) e^{tD} P, \quad \text{where } \vec{\alpha} P^{-1} = (c_1, c_2, \ldots, c_n) \\ &= \sum_{i=1}^{n} c_i e^{\lambda_i t} \vec{v}_i \in \langle\langle \vec{v}_1, \vec{v}_2, \ldots, \vec{v}_n \rangle\rangle, \end{aligned}$$

where $\{\vec{v}_1, \vec{v}_2, \ldots, \vec{v}_n\}$ is the *fundamental system* of the solution space. In case A is not diagonalizable, the Jordan canonical form of A is needed (see Ex. <D> of Sec. 3.7.7).

2. For each of the following systems of differential equations, do the following problems.

(1) Find the general solutions, the fundamental system of solutions and the dimension of the solution space.

(2) Solve the initial value problem.

(a) $\dfrac{dx_1}{dt} = -x_2,$

$\dfrac{dx_2}{dt} = x_1,$ with $x_1(0) = 1, x_2(0) = -1.$

(b) $\dfrac{dx_1}{dt} = x_1 + 2x_2,$

$\dfrac{dx_2}{dt} = -2x_1 + x_2,$ with $x_1(0) = 2, x_2(0) = 0.$

(c) $\dfrac{dx_1}{dt} = x_2,$

$\dfrac{dx_2}{dt} = x_1,$ with $x_1(0) = -2, \; x_2(0) = 1.$

(d) $\dfrac{dx_1}{dt} = -x_2,$

$\dfrac{dx_2}{dt} = -x_1,$ with $x_1(0) = 2, x_2(0) = 0.$

(e) $\dfrac{d\vec{x}}{dt} = \vec{x}A,$ where $A = \begin{bmatrix} -1 & 1 & 2 \\ 1 & 2 & 1 \\ 2 & 1 & -1 \end{bmatrix}$ and $\vec{x}(0) = (2, 0, 4).$

(f) $\dfrac{d\vec{x}}{dt} = \vec{x}A, \vec{x}(0) = (-1, -2, 3),$ where $A = \begin{bmatrix} 2 & -1 & -1 \\ -1 & 2 & -1 \\ -1 & -1 & 2 \end{bmatrix}.$

(g) $\dfrac{d\vec{x}}{dt} = \vec{x}A, \vec{x}(0) = (0, 0, -2),$ where $A = \begin{bmatrix} 1 & -2 & 2 \\ -1 & 0 & -1 \\ 0 & 2 & -1 \end{bmatrix}.$

(h) $\dfrac{dx_1}{dt} = x_2 + x_3,$

$\dfrac{dx_2}{dt} = x_3 + x_1,$

$\dfrac{dx_3}{dt} = x_1 + x_2,$ with $x_1(0) = 1, x_2(0) = 0, x_3(0) = -1.$

3. Let

$$A = \begin{bmatrix} \lambda & 0 \\ 1 & \lambda \end{bmatrix} = \lambda I_2 + J, \quad \text{where } J = \begin{bmatrix} 0 & 0 \\ 1 & 0 \end{bmatrix}.$$

(a) Show that $J^2 = O$ and hence

$$(\lambda I_2 + J)^k = \lambda^k I_2 + k\lambda^{k-1} J \quad \text{for } k = 1, 2, 3, \ldots.$$

(b) Use definition and (a) to show that

$$e^{tA} = e^{\lambda t}I_2 + te^{\lambda t}J = \begin{bmatrix} e^{\lambda t} & 0 \\ te^{\lambda t} & e^{\lambda t} \end{bmatrix}.$$

(c) Solve

$$\frac{d\vec{x}}{dt} = \vec{x}A.$$

4. Solve

$$\frac{d\vec{x}}{dt} = \vec{x}A, \text{ where } A = \begin{bmatrix} \lambda & 0 & 0 \\ 1 & \lambda & 0 \\ 0 & 0 & \lambda \end{bmatrix} \text{ or } \begin{bmatrix} \lambda & 0 & 0 \\ 1 & \lambda & 0 \\ 0 & 1 & \lambda \end{bmatrix}.$$

5. To solve

$$\frac{d^3x}{dt^3} - 6\frac{d^2x}{dt^2} + 11\frac{dx}{dt} - 6x = 0,$$

let $x_1 = x, x_2 = \frac{dx}{dt}, x_3 = \frac{d^2x}{dt^2}$. Then, the original equation is equivalent to

$$\frac{dx_1}{dt} = x_2,$$

$$\frac{dx_2}{dt} = x_3,$$

$$\frac{dx_3}{dt} = \frac{d^3x}{dt^3} = 6x_1 - 11x_2 + 6x_3,$$

$$\Leftrightarrow \frac{d\vec{x}}{dt} = \vec{x}A, \text{ where } A = \begin{bmatrix} 0 & 0 & 6 \\ 1 & 0 & -11 \\ 0 & 1 & 6 \end{bmatrix} \text{ and } \vec{x} = (x_1, x_2, x_3).$$

The characteristic polynomial of A is $-(t-1)(t-2)(t-3)$. For $\lambda_1 = 1$, the associated eigenvector is $\vec{v}_1 = (1,1,1)$; for $\lambda_2 = 2, \vec{v}_2 = (1,2,4)$; for $\lambda_3 = 3, \vec{v}_3 = (1,3,9)$. Thus

$$A = P^{-1} \begin{bmatrix} 1 & 0 & 0 \\ 0 & 2 & 0 \\ 0 & 0 & 3 \end{bmatrix} P, \text{ where } P = \begin{bmatrix} 1 & 1 & 1 \\ 1 & 2 & 4 \\ 1 & 3 & 9 \end{bmatrix}.$$

Now, the general solution of the system is

$$\vec{x} = \vec{\alpha}e^{tA} = \vec{\alpha}P^{-1} \begin{bmatrix} e^t & 0 & 0 \\ 0 & e^{2t} & 0 \\ 0 & 0 & e^{3t} \end{bmatrix} P$$

and the solution of the original ordinary differential equation is $c_1 e^t + c_2 e^{2t} + c_3 e^{3t}$ for scalars c_1, c_2, c_3. If subject to the initial conditions

$$x(0) = 1, \quad x'(0) = 0, \quad x''(0) = 1,$$

then the *particular solution* is obtained by solving c_1, c_2, c_3 from

$$c_1 + c_2 + c_3 = 1, \quad c_1 + 2c_2 + 3c_3 = 0, \quad c_1 + 4c_2 + 9c_3 = 0.$$

The particular solution is $\frac{7}{2} e^t - 4e^{2t} + \frac{3}{2} e^{3t}$.

6. *The nth-order linear ordinary differential equation with constant coefficients*

$$\frac{d^n x}{dt^n} + a_{n-1} \frac{d^{n-1} x}{dt^{n-1}} + \cdots + a_1 \frac{dx}{dt} + a_0 x = 0$$

is equivalent to the homogeneous linear differential system

$$\frac{d\vec{x}}{dt} = \vec{x} A, \quad \text{where } A = \begin{bmatrix} 0 & \cdots & \cdots & 0 & -a_0 \\ 1 & 0 & \cdots & 0 & -a_1 \\ \vdots & \ddots & \ddots & \vdots & \vdots \\ 0 & \ddots & \ddots & 0 & -a_{n-2} \\ 0 & 0 & \cdots & 1 & -a_{n-1} \end{bmatrix}$$

and $\vec{x} = (x_1, \ldots, x_n)$ with $x_1 = x, x_2 = \frac{dx}{dt}, \ldots, x_n = \frac{d^{n-1} x}{dt^{n-1}}$.

(a) By Ex. 1, the system has the general solution

$$\vec{x}(t) = \vec{\alpha} e^{tA}$$

and the initial value problem $\vec{x}(0) = \vec{x}_0$ has the unique solution

$$\vec{x}(t) = \vec{x}_0 e^{tA}.$$

In fact, let $\vec{v}_i(t) = (b_{i1}(t), b_{i2}(t), \ldots, b_{in}(t))$ be the ith row vector of e^{tA}, then $\{\vec{v}_1(t), \ldots, \vec{v}_n(t)\}$ forms a fundamental system of solutions and

$$\vec{x}(t) = \sum_{i=1}^{n} \alpha_i \vec{v}_i(t), \quad \text{where } \vec{\alpha} = (\alpha_1, \ldots, \alpha_n).$$

(b) The original equation with the initial value conditions

$$x(0) = c_1, \frac{dx}{dt} = c_2, \ldots, \frac{d^{n-1}x}{dt^{n-1}}(0) = c_n$$

has the unique solution, for some scalars $\alpha_1, \ldots, \alpha_n$,

$$x(t) = \sum_{i=1}^{n} \alpha_i b_{i1}(t).$$

In particular, if the coefficient matrix A has n distinct eigenvalues $\lambda_1, \lambda_2, \ldots, \lambda_n$, then the general solution is

$$x(t) = \sum_{i=1}^{n} \alpha_i e^{\lambda_i t}.$$

These coefficients $\alpha_1, \ldots, \alpha_n$ can be expressed uniquely in terms of $c_1, c_2, \ldots, c_n$.

7. Model after Ex. 5 and solve each of the following equations.

(a) $\dfrac{d^2 x}{dt^2} - x = 0.$

(b) $\dfrac{d^2 x}{dt^2} + x = 0.$

(c) $\dfrac{d^3 x}{dt^3} + \dfrac{d^2 x}{dt^2} - \dfrac{dx}{dt} - x = 0,$ with $x(0) = 1, x'(0) = 0, x''(0) = -1.$

(d) $\dfrac{d^3 x}{dt^3} + 3\dfrac{d^2 x}{dt^2} + 3\dfrac{dx}{dt} + x = 0,$ with $x(0) = x'(0) = x''(0) = 1.$

(e) $\dfrac{d^3 x}{dt^3} - \dfrac{dx}{dt} = 0.$

8. Consider the 2nd-order system of ordinary differential equations

$$\frac{d^2 \vec{x}}{dt^2} = \vec{x} A, \quad \vec{x}(0) = \vec{x}_1, \quad \frac{d\vec{x}}{dt}(0) = \vec{x}_2, \quad \text{where } A = \begin{bmatrix} 1 & 5 \\ 2 & 4 \end{bmatrix}.$$

Since

$$PAP^{-1} = \begin{bmatrix} 6 & 0 \\ 0 & -1 \end{bmatrix}, \quad \text{where } P = \begin{bmatrix} 2 & 5 \\ 1 & -1 \end{bmatrix}$$

$$\Rightarrow \left(\frac{d^2 \vec{x}}{dt^2}\right) P^{-1} = \frac{d^2}{dt^2}(\vec{x} P^{-1}) = \vec{x} A P^{-1} = (\vec{x} P^{-1}) P A P^{-1}$$

$$\Rightarrow \frac{d^2 \vec{y}}{dt^2} = \vec{y} \begin{bmatrix} 6 & 0 \\ 0 & -1 \end{bmatrix}, \quad \text{where } \vec{y} = \vec{x} P^{-1},$$

whose general solution is

$$\begin{cases} y_1(t) = a_1 e^{\sqrt{6}t} + b_1 e^{-\sqrt{6}t}, \\ y_2(t) = a_2 \cos \sqrt{t} + b_2 \sin \sqrt{t}. \end{cases}$$

Then the original equation has the solution

$$\vec{x}(t) = \vec{y}(t)P = (2y_1(t) + y_2(t), 5y_1(t) - y_2(t)).$$

Finally, use $\vec{x}(0) = \vec{x}_1$ and $\vec{x}'(0) = \vec{x}_2$ to determine the coefficients a_1, b_1, a_2 and b_2.

9. Model after Ex. 8 and solve the following equations.

   (a) $\dfrac{d^2 \vec{x}}{dt^2} = \vec{x} \begin{bmatrix} 3 & -2 \\ -2 & 3 \end{bmatrix}$, $\vec{x}(0) = (0, 1)$, $\vec{x}'(0) = (-1, 2)$.

   (b) $\dfrac{d^2 \vec{x}}{dt^2} = \vec{x} \begin{bmatrix} 0 & 1 & 0 \\ 0 & 0 & 1 \\ 8 & -14 & 7 \end{bmatrix}$, $\vec{x}(0) = (1, 1, 1)$, $\vec{x}'(0) = (-1, -1, 0)$.

   (c) $\dfrac{d^2 \vec{x}}{dt^2} = \vec{x} \begin{bmatrix} 1 & 0 & 0 \\ 0 & 0 & 1 \\ 0 & 1 & 0 \end{bmatrix}$.

10. By introducing

$$\vec{y}(t) = (x_1(t), x_2(t), x_1'(t), x_2'(t)),$$

show that the 2nd-order system of differential equations $\frac{d^2 \vec{x}}{dt^2} = \vec{x} A$ in Ex. 8 can be expressed as a 1st-order system

$$\frac{d\vec{y}}{dt} = \vec{y}(t) \begin{bmatrix} 0 & 0 & 1 & 5 \\ 0 & 0 & 2 & 4 \\ 1 & 0 & 0 & 0 \\ 0 & 1 & 0 & 0 \end{bmatrix}.$$

Use method in Ex. 1 to the above equation to solve the original equation. Then, try to use this method to solve equations in Ex. 9.

(*Note* For further readings, see Boyce and Diprema [33] and Farlow [34] concerning dynamical systems.)

3.7.7 *Jordan canonical form*

For real matrix $A_{3\times 3}$ with three but coincident real eigenvalues, we are going to prove counterpart of (2.7.74) in this subsection, and hence realize partial results listed in (3.7.29).

Let $A = [a_{ij}]_{3\times3}$ be a nonzero real matrix.

Suppose A has three real eigenvalues λ_1, λ_2 and λ_3, but at least two of them are coincident and yet A is not diagonalizable (refer to (3.7.46)). Remind that

$$(A - \lambda_1 I_3)(A - \lambda_2 I_3)(A - \lambda_3 I_3) = O_{3\times3}$$

and the eigenspaces

$$E_{\lambda_i} = \{\vec{x} \in \mathbb{R}^3 \mid \vec{x}A = \lambda_i \vec{x}\} = \operatorname{Ker}(A - \lambda_i I_3), \quad i = 1, 2, 3$$

are invariant subspaces of at least dimension one.

Case 1 Suppose $\lambda_1 = \lambda_2 \neq \lambda_3$ and $(A - \lambda_1 I_3)(A - \lambda_3 I_3) \neq O$ but

$$(A - \lambda_1 I_3)^2(A - \lambda_3 I_3) = O. \tag{$*_1$}$$

Then $E_{\lambda_1} = E_{\lambda_2}$ and $\dim E_{\lambda_3} = 1$ (Why? One might refer to Case 2 in Sec. 3.7.6).

Just like the proof shown in Case 1 of Sec. 3.7.6, it follows easily that $E_{\lambda_1} \cap E_{\lambda_3} = \{\vec{0}\}$.

In case $\dim E_{\lambda_1} = 2$, then $\dim(E_{\lambda_1} + E_{\lambda_3}) = 3$ shows that $\mathbb{R}^3 = E_{\lambda_1} \oplus E_{\lambda_3}$. Via Case 2 in Sec 3.7.6, it follows that A is diagonalizable and thus, $(A - \lambda_1 I_3)(A - \lambda_3 I_3) = O$, contradicting to our assumption.

Hence $\dim E_{\lambda_1} = 1$. As a byproduct,

$$\mathbb{R}^3 \neq E_{\lambda_1} \oplus E_{\lambda_3}$$

so A is not diagonalizable.

Hence, we introduce the *generalized eigenspace*

$$G_{\lambda_1} = \{\vec{x} \in \mathbb{R}^3 \mid \vec{x}(A - \lambda_1 I_3)^2 = \vec{0}\} = \operatorname{Ker}((A - \lambda_1 I_3)^2)$$

corresponding to λ_1. G_{λ_1} is an invariant subspace of $\mathbb{R}^3$ and G_{λ_1} contains the eigenspace E_{λ_1} as its subspace. Thus, $\dim G_{\lambda_1} \geq 1$ holds.

We *claim* that dim $G_{\lambda_1} = 2$. To see this, notice that (*1) holds if and only if

$$\text{Im}(A - \lambda_1 I_3)^2 \subseteq \text{Ker}(A - \lambda_3 I_3)$$
$$\Rightarrow (\text{since } (A - \lambda_1 I_3)^2 \neq O_{3\times 3} \text{ and dim } E_{\lambda_3} = 1) \quad \text{r}((A - \lambda_1 I_3)^2) = 1$$
$$\Rightarrow \dim E_{\lambda_1} = 3 - \text{r}((A - \lambda_1 I_3)^2) = 3 - 1 = 2.$$

Or, by using (3.7.31) (also, refer to Ex. <C> of Sec. 2.7.3),

$$\text{r}((A - \lambda_1 I_3)^2) + \text{r}(A - \lambda_3 I_3) - 3 \leq \text{r}(O) = 0$$
$$\Rightarrow 1 \leq \text{r}((A - \lambda_1 I_3)^2) \leq 3 - 2 = 1$$
$$\Rightarrow \text{r}((A - \lambda_1 I_3)^2) = 1.$$

This means that dim $G_{\lambda_2} = 2$.

Furthermore, $G_{\lambda_1} \cap E_{\lambda_3} = \{\vec{0}\}$. In fact, take any $\vec{x}_1 \in G_{\lambda_1}$ and $\vec{x}_3 \in E_{\lambda_3}$ and suppose $\vec{x}_1 \neq \vec{0}$, $\vec{x}_3 \neq \vec{0}$ so that

$$\alpha_1 \vec{x}_1 + \alpha_3 \vec{x}_3 = 0$$

for some scalars α_1 and α_3. If $\vec{x}_1 \in E_{\lambda_1}$, since $\lambda_1 \neq \lambda_3$, it follows that $\alpha_1 = \alpha_3 = 0$. Thus, if $\vec{x}_1 \in G_{\lambda_1} - E_{\lambda_1}$, then $\vec{x}_1(A - \lambda_1 I_3) \neq \vec{0}$ and $\vec{x}_1(A - \lambda_1 I_3)^2 = \vec{0}$ which, in turn, implies that $\vec{x}_1(A - \lambda_1 I_3) \in E_{\lambda_1}$ which indicates that $\vec{x}_1(A - \lambda_1 I_3)$ is an eigenvector associated to λ_1. As a consequence,

$$\alpha_1 \vec{x}_1(A - \lambda_1 I_3) + \alpha_3 \vec{x}_3(A - \lambda_1 I_3) = \vec{0}$$
$$\Rightarrow \alpha_1 \vec{x}_1(A - \lambda_1 I_3)(A - \lambda_3 I_3) = \alpha_1(\lambda_1 - \lambda_3)(\vec{x}_1(A - \lambda_1 I_3))$$
$$= -\alpha_3 \vec{x}_3(A - \lambda_1 I_3)(A - \lambda_3 I_3)$$
$$= -\alpha_3 \vec{x}_3(A - \lambda_3 I_3)(A - \lambda_1 I_3) = -\alpha_3 \vec{0} = \vec{0}$$
$$\Rightarrow (\text{since } \lambda_1 \neq \lambda_3) \; \alpha_1 = 0$$
$$\Rightarrow \alpha_3 \vec{x}_3 = \vec{0} \text{ and hence } \alpha_3 = 0.$$

This proves that $G_{\lambda_1} \cap E_{\lambda_3} = \{\vec{0}\}$.

We conclude that

$$\mathbb{R}^3 = G_{\lambda_1} \oplus E_{\lambda_3}.$$

Take a vector $\vec{v}_2 \in G_{\lambda_1}$ so that $\vec{v}_2(A - \lambda_1 I_3) \neq \vec{0}$. Then $\vec{v}_1 = \vec{v}_2(A - \lambda_1 I_3)$ is an eigenvector of A associated to λ_1, i.e. $\vec{v}_2 \in E_{\lambda_1}$. Then $\{\vec{v}_1, \vec{v}_2\}$ is linearly independent and hence forms a basis for G_{λ_1}. Take any basis $\{\vec{v}_3\}$

for E_{λ_3}. Together, $\mathcal{B} = \{\vec{v}_1, \vec{v}_2, \vec{v}_3\}$ forms a basis for $\mathbb{R}^3$. Therefore,

$$\vec{v}_1 A = \lambda_1 \vec{v}_1,$$
$$\vec{v}_2 A = \vec{v}_2(A - \lambda_1 I_3) + \lambda_1 \vec{v}_2 = \vec{v}_1 + \lambda_1 \vec{v}_2,$$
$$\vec{v}_3 A = \lambda_3 \vec{v}_3$$

$$\Rightarrow \begin{bmatrix} \vec{v}_1 \\ \vec{v}_2 \\ \vec{v}_3 \end{bmatrix} A = \begin{bmatrix} \lambda_1 & 0 & 0 \\ 1 & \lambda_1 & 0 \\ 0 & 0 & \lambda_3 \end{bmatrix} \begin{bmatrix} \vec{v}_1 \\ \vec{v}_2 \\ \vec{v}_3 \end{bmatrix}$$

$$\Rightarrow PAP^{-1} = \begin{bmatrix} \lambda_1 & 0 & 0 \\ 1 & \lambda_1 & 0 \\ 0 & 0 & \lambda_3 \end{bmatrix}, \quad \text{where } P = \begin{bmatrix} \vec{v}_1 \\ \vec{v}_2 \\ \vec{v}_3 \end{bmatrix}.$$

Case 2   Suppose $\lambda_1 = \lambda_2 = \lambda_3$, say equal to λ and $A - \lambda I_3 \neq O$ but

$$(A - \lambda I_3)^2 = O. \tag{*2}$$

Let E_λ be the eigenspace and $G_\lambda = \{\vec{x} \in \mathbb{R}^3 \mid \vec{x}(A - \lambda I_3)^2 = \vec{0}\}$ the *generalized eigenspace* of A corresponding to λ. Then dim $E_\lambda \geq 1$ and $E_\lambda \subseteq G_\lambda$ holds. Also, (*2) says that dim $G_\lambda = 3$.

We *claim* that dim $E_\lambda = 2$. To see this, notice first that dim $E_\lambda \leq 2$ since $A - \lambda I_3 \neq O$. Since (*2) holds if and only if

$$\text{Im}(A - \lambda I_3) \subseteq \text{Ker}(A - \lambda I_3),$$

it is impossible that dim $E_\lambda = \dim \text{Ker}(A - \lambda I_3) = 1$ holds. Then, dim $E_\lambda = 2$. Or,

$$r(A - \lambda I_3) + r(A - \lambda I_3) - 3 \leq r(O) = 0$$
$$\Rightarrow 1 \leq r(A - \lambda I_3) \leq \frac{3}{2}$$
$$\Rightarrow r(A - \lambda I_3) = 1.$$

This is equivalent to say that dim $E_\lambda = 3 - r(A - \lambda I_3) = 3 - 1 = 2$.

Now, take a nonzero vector $\vec{v}_2 \in G_\lambda$ so that $\vec{v}_1 = \vec{v}_2(A - \lambda I_3) \neq \vec{0}$ is an eigenvector of A. There exists another nonzero vector $\vec{v}_3$ in E_λ which is linearly independent of $\vec{v}_1$. Then $\mathcal{B} = \{\vec{v}_1, \vec{v}_2, \vec{v}_3\}$ is a basis for $\mathbb{R}^3$ and

$$\vec{v}_1 A = \lambda \vec{v}_1,$$
$$\vec{v}_2 A = \vec{v}_1 + \lambda \vec{v}_2,$$
$$\vec{v}_3 A = \lambda \vec{v}_3$$

$$\Rightarrow \begin{bmatrix} \vec{v}_1 \\ \vec{v}_2 \\ \vec{v}_3 \end{bmatrix} A = \begin{bmatrix} \lambda & 0 & 0 \\ 1 & \lambda & 0 \\ 0 & 0 & \lambda \end{bmatrix} \begin{bmatrix} \vec{v}_1 \\ \vec{v}_2 \\ \vec{v}_3 \end{bmatrix}$$

$$\Rightarrow PAP^{-1} = \begin{bmatrix} \lambda & 0 & 0 \\ 1 & \lambda & 0 \\ 0 & 0 & \lambda \end{bmatrix}, \quad \text{where } P = \begin{bmatrix} \vec{v}_1 \\ \vec{v}_2 \\ \vec{v}_3 \end{bmatrix}.$$

Case 3 Suppose $\lambda_1 = \lambda_2 = \lambda_3 = \lambda$ and $A - \lambda I_3 \neq O$, $(A - \lambda I_3)^2 \neq O$ but

$$(A - \lambda I_3)^3 = O \tag{*3}$$

as it should be. The eigenspace is E_λ, while the *generalized eigenspace* is $G_\lambda = \{\vec{x} \in \mathbb{R}^3 \mid \vec{x}(A - \lambda I_3)^3 = \vec{0}\} = \mathbb{R}^3$. Note that $\dim E_\lambda \geq 1$.

We *claim* that $\dim E_\lambda = 1$. Obviously, it is not true that $\dim E_\lambda = 3$ since $A - \lambda I_3 \neq O$. Suppose that $\dim E_\lambda = 2$ does happen. Take a basis $\{\vec{u}_1, \vec{u}_2\}$ for E_λ and extend it to a basis $C = \{\vec{u}_1, \vec{u}_2, \vec{u}_3\}$ for $\mathbb{R}^3$. Then

$$[A]_C = QAQ^{-1} = \begin{bmatrix} \lambda & 0 & 0 \\ 0 & \lambda & 0 \\ c_{31} & c_{32} & c_{33} \end{bmatrix}, \quad \text{where } Q = \begin{bmatrix} \vec{u}_1 \\ \vec{u}_2 \\ \vec{u}_3 \end{bmatrix}$$

$\Rightarrow \det([A]_C - tI_3) = \det(A - tI_3) = (t - \lambda)^2(c_{33} - t) = -(t - \lambda)^3$

$\Rightarrow c_{33} = \lambda$

$\Rightarrow \vec{u}_3 A = c_{31}\vec{u}_1 + c_{32}\vec{u}_2 + \lambda\vec{u}_3$

$\Rightarrow \vec{u}_3(A - \lambda I_3) = c_{31}\vec{u}_1 + c_{32}\vec{u}_2 \in E_\lambda$

$\Rightarrow \vec{u}_3(A - \lambda I_3)^2 = \vec{0}$

$\Rightarrow$ (plus $\vec{u}_i(A - \lambda I_3)^2 = \vec{0}$ for $i = 1, 2)\vec{x}(A - \lambda I_3)^2 = \vec{0}$ for all $\vec{x} \in \mathbb{R}^3$.

$\Rightarrow (A - \lambda I_3)^2 = O$

which shows a contradiction to our original assumption. This proves the claim.

Since $G_\lambda = \mathbb{R}^3$, take any nonzero vector $\vec{v}_3 \in G_\lambda$ so that $\vec{v}_2 = \vec{v}_3(A - \lambda I_3) \neq \vec{0}$ and $\vec{v}_1 = \vec{v}_3(A - \lambda I_3)^2 \neq \vec{0}$. Note that $\vec{v}_1 \in E_\lambda$. Also, $\mathcal{B} = \{\vec{v}_1, \vec{v}_2, \vec{v}_3\}$ is linearly independent and hence is a basis for $\mathbb{R}^3$. Therefore,

$$\vec{v}_1 A = \lambda\vec{v}_1,$$
$$\vec{v}_2 A = \vec{v}_2(A - \lambda I_3) + \lambda\vec{v}_2 = \vec{v}_1 + \lambda\vec{v}_3,$$
$$\vec{v}_3 A = \vec{v}_3(A - \lambda I_3) + \lambda\vec{v}_3 = \vec{v}_2 + \lambda\vec{v}_3$$

$$\Rightarrow \begin{bmatrix} \vec{v}_1 \\ \vec{v}_2 \\ \vec{v}_3 \end{bmatrix} A = \begin{bmatrix} \lambda & 0 & 0 \\ 1 & \lambda & 0 \\ 0 & 1 & \lambda \end{bmatrix} \begin{bmatrix} \vec{v}_1 \\ \vec{v}_2 \\ \vec{v}_3 \end{bmatrix}$$

$$\Rightarrow A = P^{-1} \begin{bmatrix} \lambda & 0 & 0 \\ 1 & \lambda & 0 \\ 0 & 1 & \lambda \end{bmatrix} P, \quad \text{where } P = \begin{bmatrix} \vec{v}_1 \\ \vec{v}_2 \\ \vec{v}_3 \end{bmatrix}.$$

We summarize as (refer to (3.7.47))

The Jordan canonical form of a nonzero real matrix $A_{3\times 3}$
Suppose the characteristic polynomial of A is

$$\det(A - tI_3) = -(t - \lambda_1)(t - \lambda_2)(t - \lambda_3)$$

where λ_1, λ_2 and λ_3 are real numbers and at least two of them are coincident. Denote by $E_{\lambda_i} = \text{Ker}(A - \lambda_i I_3)$ the eigenspace for $i = 1, 2, 3$.

(1) $\lambda_1 = \lambda_2 \neq \lambda_3$. Let

$$G_{\lambda_1} = \text{Ker}(A - \lambda_1 I_3)^2$$

be the *generalized eigenspace* associated to λ_1. Then

    1. $(A - \lambda_1 I_3)(A - \lambda_3 I_3) \neq O$ but $(A - \lambda_1 I_3)^2(A - \lambda_3 I_3) = O$.

$\Leftrightarrow$ 2. $\dim G_{\lambda_1} = 2 = $ the algebraic multiplicity of λ_1,
      $\dim E_{\lambda_1} = \dim E_{\lambda_3} = 1$ and $G_{\lambda_1} \cap E_{\lambda_3} = \{\vec{0}\}$.

$\Leftrightarrow$ 3. $\mathbb{R}^3 = G_{\lambda_1} \oplus E_{\lambda_3}$

In this case, let $\vec{v}_2 \in G_{\lambda_1}$ be such that $\vec{v}_1 = \vec{v}_2(A - \lambda_1 I_3) \neq \vec{0}$ and $\vec{v}_3 \in E_{\lambda_3}$ a nonzero vector. Then $\mathcal{B} = \{\vec{v}_1, \vec{v}_2, \vec{v}_3\}$ is a basis for $\mathbb{R}^3$ and

$$[A]_{\mathcal{B}} = PAP^{-1} = \begin{bmatrix} \lambda_1 & 0 & 0 \\ 1 & \lambda_1 & 0 \\ 0 & 0 & \lambda_3 \end{bmatrix}, \quad \text{where } P = \begin{bmatrix} \vec{v}_1 \\ \vec{v}_2 \\ \vec{v}_3 \end{bmatrix}.$$

(2) $\lambda_1 = \lambda_2 = \lambda_3 = \lambda$. Then

    1. $A - \lambda I_3 \neq O$ but $(A - \lambda I_3)^2 = O$.

$\Leftrightarrow$ 2. $\dim G_\lambda = 3$, where $G_\lambda = \text{Ker}(A - \lambda I_3)^2$, and $\dim E_\lambda = 2$.

In this case, take $\vec{v}_2 \in G_\lambda$ so that $\vec{v}_1 = \vec{v}_2(A - \lambda I_3) \in E_\lambda$ is a nonzero vector and $\vec{v}_3 \in E_\lambda$ which is linearly independent of $\vec{v}_1$. Then $\mathcal{B} = \{\vec{v}_1, \vec{v}_2, \vec{v}_3\}$ is a basis for $\mathbb{R}^3$ and

$$[A]_{\mathcal{B}} = PAP^{-1} = \begin{bmatrix} \lambda & 0 & 0 \\ 1 & \lambda & 0 \\ 0 & 0 & \lambda \end{bmatrix}, \quad \text{where } P = \begin{bmatrix} \vec{v}_1 \\ \vec{v}_2 \\ \vec{v}_3 \end{bmatrix}.$$

(3) $\lambda_1 = \lambda_2 = \lambda_3 = \lambda$. Then

1. $A - \lambda I_3 \neq O, (A - \lambda I_3)^2 \neq O$ but $(A - \lambda I_3)^3 = O$.
$\Leftrightarrow$ 2. $\dim G_\lambda = 3$, where $G_\lambda = \text{Ker}(A - \lambda I_3)^3, \dim E_\lambda = 1$.

In this case, take $\vec{v}_3 \in G_\lambda$ so that $\vec{v}_2 = \vec{v}_3(A - \lambda I_3) \neq \vec{0}$ and $\vec{v}_1 = \vec{v}_3(A - \lambda I)^2 \neq \vec{0}$ which is in E_λ. Then $B = \{\vec{v}_1, \vec{v}_2, \vec{v}_3\}$ is a basis for $\mathbb{R}^3$ and

$$[A]_B = PAP^{-1} = \begin{bmatrix} \lambda & 0 & 0 \\ 1 & \lambda & 0 \\ 0 & 1 & \lambda \end{bmatrix}, \quad \text{where } P = \begin{bmatrix} \vec{v}_1 \\ \vec{v}_2 \\ \vec{v}_3 \end{bmatrix}. \quad (3.7.50)$$

For geometric mapping properties of these canonical forms, refer to examples in Sec. 3.7.2. For general setting concerned with Jordan canonical forms, refer to Sec. B.12. See also Exs. <C> 3, 16, 17 and 21.

Combining (3.7.46) and (3.7.50), we have

The criteria for similarity of two real matrices of order 3
Suppose both $A_{3\times3}$ and $B_{3\times3}$ have three *real* eigenvalues. Then

(1) A and B are similar.
$\Leftrightarrow$ (2) A and B have the same characteristic and minimal polynomials.
$\Leftrightarrow$ (3) A and B have the same Jordan canonical form (up to similarity, i.e. up to a permutation of their eigenvalues).

$$(3.7.51)$$

(3.7.48) indicates that (2) $\Rightarrow$ (1) is no more true for square matrices of order $n \geq 4$, but (1) $\Leftrightarrow$ (3) $\Rightarrow$ (2) still holds.

As an example, we continue Example 1 in Sec. 3.7.6 as

Example 1 Find the Jordan canonical forms of

$$A = \begin{bmatrix} 0 & -1 & 0 \\ 3 & 3 & 1 \\ 1 & 1 & 1 \end{bmatrix}, \quad B = \begin{bmatrix} -1 & 4 & 2 \\ -1 & 3 & 1 \\ -1 & 2 & 2 \end{bmatrix} \quad \text{and} \quad C = \begin{bmatrix} 2 & 0 & 0 \\ 0 & 1 & 1 \\ 0 & 0 & 1 \end{bmatrix}$$

and determine if they are similar.

Solution By direct computation, A, B and C have the same characteristic polynomial

$$\det(A - tI_3) = \det(B - tI_3) = \det(C - tI_3) = -(t-1)^2(t-2)$$

and the respective minimal polynomials

$$m_A(t) = m_C(t) = (t-1)^2(t-2), \quad \text{and}$$
$$m_B(t) = (t-1)(t-2).$$

Therefore, they have the respective Jordan canonical form

$$J_A = J_C = \begin{bmatrix} 1 & 0 & 0 \\ 1 & 1 & 0 \\ 0 & 0 & 2 \end{bmatrix}, \quad \text{and} \quad J_B = \begin{bmatrix} 1 & 0 & 0 \\ 0 & 1 & 0 \\ 0 & 0 & 2 \end{bmatrix}.$$

Thus, only A and C are similar.

To find an invertible $P_{3\times3}$ such that $PAP^{-1} = C$, let us find $R_{3\times3}$ and $Q_{3\times3}$ so that $RAR^{-1} = J_A$ and $QCQ^{-1} = J_C$ and hence

$$RAR^{-1} = QCQ^{-1}$$
$$\Rightarrow PAP^{-1} = C, \quad \text{where } P = Q^{-1}R.$$

Now,

$$A - I_3 = \begin{bmatrix} -1 & -1 & 0 \\ 3 & 2 & 1 \\ 1 & 1 & 0 \end{bmatrix}$$

$$\Rightarrow (A - I_3)^2 = \begin{bmatrix} -2 & -1 & -1 \\ 4 & 2 & 2 \\ 2 & 1 & 1 \end{bmatrix} \quad \text{with rank } r((A - I_3)^2) = 1.$$

Solve $\vec{x}(A-I_3)^2 = \vec{0}$ and we get $G_1 = \mathrm{Ker}((A-I_3)^2) = \langle\langle (1,0,1), (2,1,0) \rangle\rangle$. Take $\vec{v}_2 = (2,1,0)$, then

$$\vec{v}_1 = \vec{v}_2(A - I_3) = (2 \quad 1 \quad 0) \begin{bmatrix} -1 & -1 & 0 \\ 3 & 2 & 1 \\ 1 & 1 & 0 \end{bmatrix} = (1 \quad 0 \quad 1) \in E_1.$$

Take $\vec{v}_3 = (2,1,1) \in E_2 = \mathrm{Ker}(A - 2I_3)$. Then

$$RAR^{-1} = J_A, \quad \text{where } R = \begin{bmatrix} 1 & 0 & 1 \\ 2 & 1 & 0 \\ 2 & 1 & 1 \end{bmatrix}.$$

On the other hand,

$$QCQ^{-1} = J_C, \quad \text{where } Q = \begin{bmatrix} 0 & 0 & 1 \\ 0 & 1 & 0 \\ 1 & 0 & 0 \end{bmatrix} = Q^{-1}.$$

Hence, the required

$$P = Q^{-1}R = QR = \begin{bmatrix} 0 & 0 & 1 \\ 0 & 1 & 0 \\ 1 & 0 & 0 \end{bmatrix} \begin{bmatrix} 1 & 0 & 1 \\ 2 & 1 & 0 \\ 2 & 1 & 1 \end{bmatrix} = \begin{bmatrix} 2 & 1 & 1 \\ 2 & 1 & 0 \\ 1 & 0 & 1 \end{bmatrix}.$$

$\square$

Example 2 Let A be as in Example 1.

(a) Compute det A and A^{-1}.
(b) Compute A^n for $n = \pm 1, \pm 2, \ldots$.
(c) Compute e^A (see Ex. $<D_2>$ of Sec. 3.7.6).
(d) Solve the differential equation (refer to Ex. $<D_4>$ of Sec. 3.7.6).

$$\frac{d\vec{x}}{dt} = \vec{x} A.$$

Solution (a) By using $RAR^{-1} = J_A$ as shown in Example 1, det $A =$ det $J_A = 2$, and

$$A^{-1} = R^{-1}J_A^{-1}R = \begin{bmatrix} 1 & 1 & -1 \\ -2 & -1 & 2 \\ 0 & -1 & 1 \end{bmatrix} \begin{bmatrix} 1 & 0 & 0 \\ -1 & 1 & 0 \\ 0 & 0 & \frac{1}{2} \end{bmatrix} \begin{bmatrix} 1 & 0 & 1 \\ 2 & 1 & 0 \\ 2 & 1 & 1 \end{bmatrix}$$

$$= \begin{bmatrix} 1 & \frac{1}{2} & -\frac{1}{2} \\ -1 & 0 & 0 \\ 0 & -\frac{1}{2} & \frac{3}{2} \end{bmatrix}.$$

Notice that the above way to compute A^{-1} is not necessarily the best one. As a *summary*, one can try the following methods:

1. Direct computation by using the adjoint matrix adj$A$ (see (3.3.2) and Sec. B.6);
2. Elementary row operations (see Secs. 3.7.5 and B.5);
3. Caylay–Hamilton theorem (see, for example, (3.7.28)), to compute A^{-1}.

$(3.7.52)$

(b) Partition J_A into *Jordan blocks* as

$$J_A = \begin{bmatrix} D & \vdots & 0 \\ \cdots & \cdots & \cdots \\ 0 & \vdots & E \end{bmatrix}, \quad \text{where } D = \begin{bmatrix} 1 & 0 \\ 1 & 1 \end{bmatrix}_{2\times2} \text{ and } E = [2]_{1\times1}.$$

Then (refer to Ex. <C> of Sec. 2.7.5 if needed)

$$J_A^n = \begin{bmatrix} D^n & 0 \\ 0 & E^n \end{bmatrix} \quad \text{for } n = \pm1, \pm2, \dots.$$

Note that $E^n = [2^n]_{1\times1}$. To compute D^n, notice that

$$D = I_2 + N, \quad \text{where } N = \begin{bmatrix} 0 & 0 \\ 1 & 0 \end{bmatrix}$$

$$\Rightarrow (\text{since } N^k = 0 \text{ for } k \geq 2) D^2 = I_2 + 2N$$

$$\Rightarrow D^n = I_2 + nN = \begin{bmatrix} 1 & 0 \\ n & 1 \end{bmatrix} \quad \text{for } n \geq 1.$$

Since $D^{-1} = (I_2 + N)^{-1} = I_2 - N$, therefore

$$D^n = I_2 + nN = \begin{bmatrix} 1 & 0 \\ n & 1 \end{bmatrix} \quad \text{for } n \leq -1.$$

Hence,

$$J_A^n = \begin{bmatrix} 1 & 0 & 0 \\ n & 1 & 0 \\ 0 & 0 & 2^n \end{bmatrix} \quad \text{for } n = \pm1, \pm2, \dots.$$

$$\Rightarrow A^n = (R^{-1}J_A R)^n = R^{-1} J_A^n R$$

$$= \begin{bmatrix} 1 & 1 & -1 \\ -2 & -1 & 2 \\ 0 & -1 & 1 \end{bmatrix} \begin{bmatrix} 1 & 0 & 0 \\ n & 1 & 0 \\ 0 & 0 & 2^n \end{bmatrix} \begin{bmatrix} 1 & 0 & 1 \\ 2 & 1 & 0 \\ 2 & 1 & 1 \end{bmatrix}$$

$$= \begin{bmatrix} 3+n-2^{n+1} & 1-2^n & 1+n-2^n \\ -4-n+2^{n+2} & -1+2^{n+1} & -2-n+2^{n+1} \\ -2-n+2^{n+1} & -1+2^n & -n+2^n \end{bmatrix} \quad \text{for } n = \pm1, \pm2, \dots.$$

(c) By definition of e^{tA}, the partial sum

$$\sum_{k=0}^{n} \frac{t^k}{k!} A^k = R^{-1} \left(\sum_{k=0}^{n} \frac{t^k}{k!} J_A^k \right) R$$

$$= R^{-1} \begin{bmatrix} \sum_{k=0}^{n} \frac{t^k}{k!} & 0 & 0 \\ \sum_{k=1}^{n} \frac{t^k}{(k-1)!} & \sum_{k=0}^{n} \frac{t^k}{k!} & 0 \\ 0 & 0 & \sum_{k=0}^{n} \frac{1}{k!} (2t)^k \end{bmatrix} R$$

$$\rightarrow R^{-1} \begin{bmatrix} e^t & 0 & 0 \\ te^t & e^t & 0 \\ 0 & 0 & e^{2t} \end{bmatrix} R = R^{-1} e^{tJ_A} R \quad \text{as } n \rightarrow \infty.$$

$$\Rightarrow e^{tA} = R^{-1} \begin{bmatrix} e^t & 0 & 0 \\ te^t & e^t & 0 \\ 0 & 0 & e^{2t} \end{bmatrix} R \quad \text{for } t \in \mathbb{R}.$$

(d) Notice that

$$\frac{d\vec{x}}{dt} = \vec{x} A$$

$$\Leftrightarrow \frac{d\vec{x}}{dt} R^{-1} = \vec{x} A R^{-1} = (\vec{x} R^{-1})(R A R^{-1})$$

$$\Leftrightarrow \frac{d\vec{y}}{dt} = \vec{y} J_A, \quad \text{where } \vec{y} = \vec{x} R^{-1}.$$

Hence, according to Ex. <D_4> 1 of Sec. 3.7.6, the solution is

$$\vec{y}(t) = \vec{\alpha} e^{tJ_A}, \quad \text{or}$$
$$\vec{x}(t) = \vec{c} e^{tA} = \vec{\alpha} e^{tJ_A} R, \quad \text{where } \vec{c} = \vec{\alpha} R.$$

$$= (\alpha_1 \quad \alpha_2 \quad \alpha_3) \begin{bmatrix} e^t & 0 & 0 \\ te^t & e^t & 0 \\ 0 & 0 & e^{2t} \end{bmatrix} \begin{bmatrix} 1 & 0 & 1 \\ 2 & 1 & 0 \\ 2 & 1 & 1 \end{bmatrix}$$

$$= (\alpha_1 \quad \alpha_2 \quad \alpha_3) \begin{bmatrix} e^t & 0 & e^t \\ (2+t)e^t & e^t & te^t \\ 2e^{2t} & e^{2t} & e^{2t} \end{bmatrix}$$

$$\Rightarrow \vec{x}_1(t) = [\alpha_1 + (2+t)\alpha_2] e^t + 2\alpha_3 e^{2t},$$
$$\vec{x}_2(t) = \alpha_2 e^t + \alpha_3 e^{2t},$$
$$\vec{x}_3(t) = (\alpha_1 + \alpha_2 t) e^t + \alpha_3 e^{2t},$$

where α_1, α_2 and α_3 are arbitrary real constants. $\square$

Example 3 Let A be as in Example 1. Find a real matrix B so that

$$B^2 = A.$$

Such a B is called a *square root* of A. For the *nilpotent* matrices

$$N_1 = \begin{bmatrix} 0 & 0 & 0 \\ 1 & 0 & 0 \\ 0 & 1 & 0 \end{bmatrix} \quad \text{and} \quad N_2 = \begin{bmatrix} 0 & 0 & 0 \\ 0 & 0 & 0 \\ 1 & 0 & 0 \end{bmatrix},$$

show that $N_1^2 = N_2$, so N_1 is a *square root* of N_2 but N_1 does not have any square root.

Solution From Example 1, it is known that

$$A = R^{-1} J_A R, \quad \text{where } J_A = \begin{bmatrix} 1 & 0 & 0 \\ 1 & 1 & 0 \\ 0 & 0 & 2 \end{bmatrix}.$$

Suppose we can choose a matrix $\tilde{B}_{3\times 3}$ so that $\tilde{B}^2 = J_A$, then the matrix $B = R^{-1}\tilde{B}R$ is a required one since $B^2 = R^{-1}\tilde{B}^2 R = R^{-1}J_A R = A$ holds.

To find $\tilde{B}$ so that $\tilde{B}^2 = J_A$, consider firstly the Jordan block

$$D = \begin{bmatrix} 1 & 0 \\ 1 & 1 \end{bmatrix} = I_2 + N, \quad \text{where } N = \begin{bmatrix} 0 & 0 \\ 1 & 0 \end{bmatrix}$$

as in Example 2. Recall the power series expansion

$$\sqrt{1+x} = \sum_{k=0}^{\infty} C_k^{\frac{1}{2}} x^k, \quad C_k^{\frac{1}{2}} = \frac{\frac{1}{2}\left(\frac{1}{2}-1\right)\cdots\left(\frac{1}{2}-k+1\right)}{k!} \quad \text{for } k \geq 1 \text{ and } C_0^{\frac{1}{2}} = 1$$

which converges absolutely on $(-1, 1)$. Substituting N for x in this expansion leads to the polynomial in N (remember that $N^k = 0$ for $k \geq 2$)

$$I_2 + \frac{1}{2}N$$

which does satisfy

$$\left(I_2 + \frac{1}{2}N\right)^2 = I_2 + N + \frac{1}{4}N^2 = I_2 + N = D.$$

Therefore, we define the matrix

$$\tilde{B} = \begin{bmatrix} I_2 + \frac{1}{2}N & 0 \\ 0 & \sqrt{2} \end{bmatrix} = \begin{bmatrix} 1 & 0 & 0 \\ \frac{1}{2} & 1 & 0 \\ 0 & 0 & \sqrt{2} \end{bmatrix}.$$

Then $\tilde{B}^2 = J_A$.

Consequently, define

$$B = R^{-1}\tilde{B}R = \begin{bmatrix} 1 & 1 & -1 \\ -2 & -1 & 2 \\ 0 & -1 & 1 \end{bmatrix} \begin{bmatrix} 1 & 0 & 0 \\ \frac{1}{2} & 1 & 0 \\ 0 & 0 & \sqrt{2} \end{bmatrix} \begin{bmatrix} 1 & 0 & 1 \\ 2 & 1 & 0 \\ 2 & 1 & 1 \end{bmatrix}$$

$$= \begin{bmatrix} \frac{7}{2} - 2\sqrt{2} & 1 - \sqrt{2} & \frac{3}{2} - \sqrt{2} \\ -\frac{9}{2} + 4\sqrt{2} & -1 + 2\sqrt{2} & -\frac{5}{2} + 2\sqrt{2} \\ -\frac{5}{2} + 2\sqrt{2} & -1 + \sqrt{2} & -\frac{1}{2} + \sqrt{2} \end{bmatrix}.$$

Then $B^2 = A$ holds.

It is obvious that $N_1^2 = N_2$.

Suppose N_1 has a real square root S, i.e. $S^2 = N_1$. Then S, as a complex matrix, has three complex eigenvalues which are equal to zero. Therefore S, as a real matrix, is similar to

$$N_3 = \begin{bmatrix} 0 & 0 & 0 \\ 1 & 0 & 0 \\ 0 & 0 & 0 \end{bmatrix} \quad \text{or} \quad N_1 = \begin{bmatrix} 0 & 0 & 0 \\ 1 & 0 & 0 \\ 0 & 1 & 0 \end{bmatrix}.$$

Let P be an invertible matrix so that

$$PSP^{-1} = N_3$$
$$\Rightarrow PS^2P^{-1} = PN_1P^{-1} = N_3^2 = O.$$

which leads to $N_1 = O$, a contradiction. Similarly,

$$PSP^{-1} = N_1$$
$$\Rightarrow PS^2P^{-1} = PN_1P^{-1} = N_1^2 = N_2$$
$$\Rightarrow PN_1^2P^{-1} = PN_2P^{-1} = N_2^2 = O$$

which leads to $N_2 = O$, a contradiction. Hence, N_1 does not have any real square root.

Example 4 Let A be as in Example 1. Show that A and A^* are similar. Find an invertible matrix P, even a symmetric one, so that

$$A^* = PAP^{-1}.$$

Solution Since A and A^* have the same characteristic and minimal polynomials, according to (3.7.51), they are similar.

By Example 1,

$$A = R^{-1}J_AR, \quad \text{where } J_A = \begin{bmatrix} 1 & 0 & 0 \\ 1 & 1 & 0 \\ 0 & 0 & 2 \end{bmatrix}$$

$$\Rightarrow A^* = R^*J_A^*(R^*)^{-1}.$$

Now

$$J_A^* = \begin{bmatrix} 1 & 1 & 0 \\ 0 & 1 & 0 \\ 0 & 0 & 2 \end{bmatrix}$$

$$\Rightarrow SJ_A^*S^{-1} = J_A, \quad \text{where } S = \begin{bmatrix} 0 & 1 & 0 \\ 1 & 0 & 0 \\ 0 & 0 & 1 \end{bmatrix} = E_{(1)(2)} = S^{-1}.$$

Therefore,

$$\begin{aligned} A^* &= R^*S^{-1}(SJ_A^*S^{-1})S(R^*)^{-1} \\ &= R^*S^{-1}J_A S(R^*)^{-1} \\ &= R^*SRAR^{-1}S(R^*)^{-1} = PAP^{-1}, \end{aligned}$$

where

$$P = R^*SR = \begin{bmatrix} 1 & 2 & 2 \\ 0 & 1 & 1 \\ 1 & 0 & 1 \end{bmatrix}\begin{bmatrix} 0 & 1 & 0 \\ 1 & 0 & 0 \\ 0 & 0 & 1 \end{bmatrix}\begin{bmatrix} 1 & 0 & 1 \\ 2 & 1 & 0 \\ 2 & 1 & 1 \end{bmatrix} = \begin{bmatrix} 8 & 3 & 4 \\ 3 & 1 & 2 \\ 4 & 2 & 1 \end{bmatrix}$$

is an invertible symmetric matrix.   □

Example 5 Let A be as in Example 1. Show that there exist symmetric matrices A_1 and A_2 so that

$$A = A_1 A_2$$

and at least one of A_1 and A_2 is invertible.

Solution Using results in Example 4,

$$\begin{aligned} A &= P^{-1}A^*P \\ &= A_1 A_2, \quad \text{where} \end{aligned}$$

$$A_1 = P^{-1} = \begin{bmatrix} -3 & 5 & 2 \\ 5 & -8 & -4 \\ 2 & -4 & -1 \end{bmatrix}, \quad \text{and}$$

$$A_2 = A^*P = \begin{bmatrix} 0 & 3 & 1 \\ -1 & 3 & 1 \\ 0 & 1 & 1 \end{bmatrix}\begin{bmatrix} 8 & 3 & 4 \\ 3 & 1 & 2 \\ 4 & 2 & 1 \end{bmatrix} = \begin{bmatrix} 13 & 5 & 7 \\ 5 & 2 & 3 \\ 7 & 3 & 3 \end{bmatrix}.$$

Even by the definition, A_1 is symmetric and invertible; also, since $PA = A^*P$, $A_2^* = (A^*P)^* = P^*A = PA = A^*P = A_2$, so A_2 is symmetric and invertible in this case.

Let $A_1 = P^{-1}A^*$ and $A_2 = P$, then $A = A_1 A_2$ will work too. □

As a bonus of this subsection, in what follows we use examples to illustrate Jordan canonical forms for matrices of order larger than 3. For theoretical development in general, refer to Sec. B.12; for a concise argument, refer to Exs. <C> 3, 16, 17 and 21.

Example 6 Try to analyze the Jordan canonical form

$$
A = \begin{bmatrix}
2 & 0 & 0 & \vdots & & & & \\
1 & 2 & 0 & \vdots & & & & \\
0 & 1 & 2 & \vdots & & & & \\
\cdots & \cdots & \cdots & \vdots & \cdots & & & \\
 & & & \vdots & 2 & \vdots & & \\
 & & & \cdots & & \cdots & & \\
 & & & & \cdots & \cdots & & \\
 & & & & & \vdots & -1 & 0 & \vdots \\
 & & & & & \vdots & 1 & -1 & \vdots \\
 & & & & & \cdots & \cdots & \cdots \\
 & & & & & & & \vdots & 0 & 0 \\
 & & & & & & & \vdots & 1 & 0
\end{bmatrix}_{8\times8}
$$

$$
= \begin{bmatrix}
J_1 & & & O \\
 & J_2 & & \\
 & & J_3 & \\
O & & & J_4
\end{bmatrix}
$$

composed of 4 *Jordan blocks* of respective 3×3, 1×1, 2×2 and 2×2 orders.

Analysis The characteristic polynomial of A is

$$\det(A - tI_8) = (t-2)^4(t+1)^2 t^2.$$

Remember that we use $A_{i\cdot}$ to denote the ith row vector of A.

For J_1 and J_2:

$$\vec{e}_1 A = A_{1\cdot} = 2\vec{e}_1 \rightarrow \vec{e}_1(A - 2I_8) = \vec{e}_3(A - 2I_8)^3 = \vec{0},$$
$$\vec{e}_2 A = A_{2\cdot} = \vec{e}_1 + 2\vec{e}_2 \rightarrow \vec{e}_2(A - 2I_8) = \vec{e}_1 = \vec{e}_3(A - 2I_8)^2,$$
$$\vec{e}_3 A = A_{3\cdot} = \vec{e}_2 + 2\vec{e}_3 \rightarrow \vec{e}_3(A - 2I_8) = \vec{e}_2,$$
$$\vec{e}_4 A = A_{4\cdot} = 2\vec{e}_4 \rightarrow \vec{e}_4(A - 2I_8) = \vec{0}.$$

Also, by computation or inspection, the ranks (refer to (3.7.38))

$$r(A - 2I_8) = 6, \quad r((A - 2I_8)^2) = 5,$$
$$r((A - 2I_8)^k) = 4 \quad \text{for } k \geq 3.$$

These relations provide the following information:

1. The generalized eigenspace

$$G_2 = \{\vec{x} \in \mathbb{R}^8 \mid \vec{x}(A - 2I_8)^4 = \vec{0}\}$$
$$= \langle\langle \vec{e}_1, \vec{e}_2, \vec{e}_3, \vec{e}_4 \rangle\rangle = \langle\langle \vec{e}_1, \vec{e}_2, \vec{e}_3 \rangle\rangle \oplus \langle\langle \vec{e}_4 \rangle\rangle$$

is an invariant subspace of $\mathbb{R}^8$ of dimension 4, the *algebraic multiplicity* of the eigenvalue 2.
2. The eigenspace $E_2 = \text{Ker}(A - 2I_8) \subseteq G_2$ has the dimension

$$\dim E_2 = \dim \mathbb{R}^8 - r(A - 2I_8) = 8 - 6 = 2.$$

Thus, there is a basis for G_2 containing exactly two eigenvectors associated to 2.
3. Select a vector $\vec{v}_3$ satisfying $\vec{v}_3(A - 2I_8)^3 = \vec{0}$ but $\vec{v}_3(A - 2I_8)^2 \neq \vec{0}$. Denote $\vec{v}_2 = \vec{v}_3(A - 2I_8)$ and $\vec{v}_1 = \vec{v}_2(A - 2I_8) = \vec{v}_3(A - 2I_8)^2$, an eigenvector in E_2. Then

$$\{\vec{v}_1, \vec{v}_2, \vec{v}_3\}$$

is an *ordered* basis for $\text{Ker}[(A - 2I_8)^3 \mid \langle\langle \vec{e}_1, \vec{e}_2, \vec{e}_3 \rangle\rangle]$.
4. Take an eigenvector $\vec{v}_4 \in E_2$, linearly independent of $\vec{v}_1$. The $\{\vec{v}_4\}$ forms a basis for $\text{Ker}[(A - 2I_8) \mid \langle\langle \vec{e}_4 \rangle\rangle]$.
5. Combing 3 and 4,

$$\mathcal{B}_2 = \{\vec{v}_1, \vec{v}_2, \vec{v}_3, \vec{v}_4\}$$

is a basis for G_2 while $E_2 = \langle\langle \vec{v}_1, \vec{v}_4 \rangle\rangle$. Notice that $\text{Ker}(A - 2I_8) \subseteq \text{Ker}(A - 2I_8)^2 \subseteq \text{Ker}(A - 2I_8)^3$, and $\mathbb{R}^8 \supseteq \text{Im}(A - 2I_8) \supseteq \text{Im}(A - 2I_8)^2 \supseteq$

$\text{Im}(A - 2I_8)^3$, and then,

$$\vec{v}_1; \vec{v}_4 \leftarrow \cdots \text{ total number } 2 = 8 - r(A - 2I_8) = \dim \text{Ker}(A - 2I_8)$$

$$\vec{v}_2 \leftarrow \cdots \text{ total number } 1 = r(A - 2I_8) - r((A - 2I_8)^2)$$
$$= \dim \text{Ker}(A - 2I_8)^2 - \dim \text{Ker}(A - 2I_8)$$

$$\vec{v}_3 \leftarrow \cdots \text{ total number } 1 = r((A - 2I_8)^2) - r((A - 2I_8)^3)$$
$$= \dim \text{Ker}(A - 2I_8)^3 - \dim \text{Ker}(A - 2I_8)^2$$

Since

$$\vec{v}_1 A = \vec{v}_1(A - 2I_8) + 2\vec{v}_1 = 2\vec{v}_1,$$
$$\vec{v}_2 A = \vec{v}_2(A - 2I_8) + 2\vec{v}_2 = \vec{v}_1 + 2\vec{v}_2,$$
$$\vec{v}_3 A = \vec{v}_3(A - 2I_8) + 2\vec{v}_3 = \vec{v}_2 + 2\vec{v}_3,$$
$$\vec{v}_4 A = 2\vec{v}_4$$

$$\Rightarrow [A \mid G_2]_{B_2} = \begin{bmatrix} J_1 & \vdots & \\ \cdots & \vdots & \cdots \\ & \vdots & J_2 \end{bmatrix}.$$

For J_3:

$$\vec{e}_5 A = A_{5\cdot} = -\vec{e}_5 \rightarrow \vec{e}_5(A + I_8) = \vec{e}_6(A + I_8)^2 = \vec{0},$$
$$\vec{e}_6 A = A_{6\cdot} = \vec{e}_5 - \vec{e}_6 \rightarrow \vec{e}_6(A + I_8) = \vec{e}_5$$

and the ranks

$$r(A + I_8) = 7,$$
$$r((A + I_8)^k) = 6, \quad \text{for } k \geq 2.$$

These imply the following:

1. $G_{-1} = \text{Ker}((A + I_8)^2) = \langle\langle \vec{e}_5, \vec{e}_6 \rangle\rangle$ is an invariant subspace of dimension 2, the algebraic multiplicity of -1 as an eigenvalue of A.
2. $E_{-1} = \text{Ker}(A + I_8)$ has the dimension

$$\dim E_{-1} = \dim \mathbb{R}^8 - r(A + I_8) = 8 - 7 = 1.$$

Thus, since $E_{-1} \subseteq G_{-1}$, there exists a basis for G_{-1} containing exactly one eigenvector associated to -1.
3. Select a vector $\vec{v}_6$ satisfying $\vec{v}_6(A + I_8)^2 = \vec{0}$ but $\vec{v}_5 = \vec{v}_6(A + I_8) \neq \vec{0}$. Then $\vec{v}_5$ is an eigenvector in E_{-1}.

4. Now,

$$\mathcal{B}_{-1} = \{\vec{v}_5, \vec{v}_6\}$$

is an ordered basis for G_{-1} while $E_{-1} = \langle\langle \vec{v}_5 \rangle\rangle$. Notice that

$$\vec{v}_5 \leftarrow \cdots \text{ total number } 1 = 8 - r(A + I_8) = \dim \text{Ker}(A + I_8)$$
$$\vec{v}_6 \leftarrow \cdots \text{ total number } 1 = r(A + I_8) - r((A + I_8)^2)$$
$$= \dim \text{Ker}(A + I_8)^2 - \dim \text{Ker}(A + I_8).$$

Since

$$\vec{v}_5 A = \vec{v}_5 (A + I_8) - \vec{v}_5 = -\vec{v}_6,$$
$$\vec{v}_6 A = \vec{v}_6 (A + I_8) - \vec{v}_6 = \vec{v}_5 - \vec{v}_6$$
$$\Rightarrow [A \mid G_{-1}]_{\mathcal{B}_{-1}} = [J_3]$$

For J_4: similarly,

$$[A \mid G_0]_{\mathcal{B}_0} = [J_4]$$

where $\mathcal{B}_0 = \{\vec{v}_7, \vec{v}_8\}$ is an ordered basis for $G_0 = \text{Ker } A^2$, with $\vec{v}_8 \in G_0$ and $\vec{v}_7 = \vec{v}_8 A \neq \vec{0}$ is an eigenvector associated to 0.

Putting together,

$$\mathcal{B} = \mathcal{B}_2 \cup \mathcal{B}_{-1} \cup \mathcal{B}_0 = \{\vec{v}_1, \vec{v}_2, \vec{v}_3, \vec{v}_4, \vec{v}_5, \vec{v}_6, \vec{v}_7, \vec{v}_8\}$$

is an *ordered* basis for $\mathbb{R}^8$, called a *Jordan canonical basis* of A, and

$$[A]_{\mathcal{B}} = PAP^{-1} = \begin{bmatrix} [A \mid G_2]_{\mathcal{B}_2} & & O \\ & [A \mid G_{-1}]_{\mathcal{B}_{-1}} & \\ O & & [A \mid G_0]_{\mathcal{B}_0} \end{bmatrix}$$

$$= \begin{bmatrix} J_1 & & & \\ & J_2 & & \\ & & J_3 & \\ & & & J_4 \end{bmatrix},$$

where

$$P = \begin{bmatrix} \vec{v}_1 \\ \vdots \\ \vec{v}_8 \end{bmatrix}_{8 \times 8}.$$

□

Example 7 Find a Jordan canonical basis and the Jordan canonical form for the matrix

$$A = \begin{bmatrix} 7 & 1 & 2 & 2 \\ 1 & 4 & -1 & -1 \\ -2 & 1 & 5 & -1 \\ 1 & 1 & 2 & 8 \end{bmatrix}.$$

Solution We follow the process in Example 6. The characteristic polynomial is

$$\det(A - tI_4) = (t - 6)^4.$$

So A has only one eigenvalue 6, with algebraic multiplicity 4.

Compute the ranks:

$$A - 6I_4 = \begin{bmatrix} 1 & 1 & 2 & 2 \\ 1 & -2 & -1 & -1 \\ -2 & 1 & -1 & -1 \\ 1 & 1 & 2 & 2 \end{bmatrix} \Rightarrow r(A - 6I_4) = 2;$$

$$(A - 6I_4)^2 = \begin{bmatrix} 0 & 3 & 3 & 3 \\ 0 & 3 & 3 & 3 \\ 0 & -6 & -6 & -6 \\ 0 & 3 & 3 & 3 \end{bmatrix} \Rightarrow r((A - 6I_4)^2) = 1;$$

$r((A - 6I_4)^k) = 0$ for $k \geq 3$ (refer to (3.7.38)).

Therefore, there exists a basis $\mathcal{B} = \{\vec{v}_1, \vec{v}_2, \vec{v}_3, \vec{v}_4\}$ for $\mathbb{R}^4$ so that

$\vec{v}_1; \vec{v}_4 \leftarrow \cdots$ total number $= \dim E_6 = 4 - 2 = 2$,

$\vec{v}_2 \leftarrow \cdots$ total number $= r(A - 6I_4) - r((A - 6I_4)^2) = 1$,

$\vec{v}_3 \leftarrow \cdots$ total number $= r((A - 6I_4)^2) - r((A - 6I_4)^3) = 1$.

Thus, $\langle\langle \vec{v}_1, \vec{v}_2, \vec{v}_3, \vec{v}_4 \rangle\rangle = \text{Ker}((A - 6I_4)^3)$ and $\langle\langle \vec{v}_1, \vec{v}_4 \rangle\rangle = \text{Ker}(A - 6I_4)$. We want to choose such a basis $\mathcal{B}$ as a Jordan canonical basis of A.

To find a basis for $E_6 = \text{Ker}(A - 6I_8)$:

$$\vec{x}(A - 6I_4) = \vec{0}$$

$$\Rightarrow x_1 + x_2 - 2x_3 + x_4 = 0$$

$$x_1 - 2x_2 + 2x_3 + x_4 = 0$$

$$\Rightarrow x_2 = x_3, \quad x_1 = x_3 - x_4$$

$$\Rightarrow \vec{x} = (x_3 - x_4, x_3, x_3, x_4) = x_3(1, 1, 1, 0) + x_4(-1, 0, 0, 1).$$

We may take, $\{(1, 1, 1, 0), (-1, 0, 0, 1)\}$ as a basis for E_6.

To find a basis for $\text{Ker}((A - 6I_4)^3)$, solve

$$\vec{x}(A - 6I_4)^3 = \vec{0} \quad \text{and} \quad \vec{x}(A - 6I_4)^2 \neq \vec{0}$$
$$\Rightarrow \text{We may choose } \vec{e}_1 = (1, 0, 0, 0) \text{ as a solution.}$$

Let $\vec{v}_3 = \vec{e}_1$. Define

$$\vec{v}_2 = \vec{v}_3(A - 6I_4) = \vec{e}_1(A - 6I_4) = (1, 1, 2, 2),$$
$$\vec{v}_1 = \vec{v}_2(A - 6I_4) = (0, 3, 3, 3).$$

Then $\{\vec{v}_1, \vec{v}_2, \vec{v}_3\}$ is linearly independent. Therefore, we may select $\vec{v}_4 = (1, 1, 1, 0)$ or $(-1, 0, 0, 1)$, say the former. Then $G_6 = \text{Ker}(A - 6I_4)^4 = \text{Ker}(A - 6I_4)^3 = \mathbb{R}^4 = \langle\langle \vec{v}_1, \vec{v}_2, \vec{v}_3 \rangle\rangle \oplus \langle\langle \vec{v}_4 \rangle\rangle$.

Let $\mathcal{B} = \{\vec{v}_1, \vec{v}_2, \vec{v}_3, \vec{v}_4\}$. $\mathcal{B}$ is a Jordan canonical basis and

$$[A]_\mathcal{B} = PAP^{-1} = \begin{bmatrix} 6 & 0 & 0 & \vdots & \\ 1 & 6 & 0 & \vdots & \\ 0 & 1 & 6 & \vdots & \\ \cdots & \cdots & \cdots & \vdots & \cdots \\ & & & \vdots & 6 \end{bmatrix}, \quad \text{where } P = \begin{bmatrix} 0 & 3 & 3 & 3 \\ 1 & 1 & 2 & 2 \\ 1 & 0 & 0 & 0 \\ 1 & 1 & 1 & 0 \end{bmatrix}.$$

We can select a basis for $\text{Ker}((A - 6I_4)^3)$ in a reverse way. Take $\vec{v}_1 = (-1, 0, 0, 1) + (1, 1, 1, 0) = (0, 1, 1, 1)$. Solve

$$\vec{x}(A - 6I_4) = \vec{v}_1$$
$$\Rightarrow x_1 + x_2 - 2x_3 + x_4 = 0$$
$$x_1 - 2x_2 + x_3 + x_4 = 1$$
$$\Rightarrow \vec{v}_2 = \frac{1}{3}(1, 1, 2, 2).$$

Again, solve

$$\vec{x}(A - 6I_4) = \vec{v}_2$$
$$\Rightarrow \vec{v}_3 = \frac{1}{3}(1, 0, 0, 0).$$

Then $\{\vec{v}_1, \vec{v}_2, \vec{v}_3, \vec{v}_4\}$, where $\vec{v}_4 = (1, 1, 1, 0)$ or $(-1, 0, 0, 1)$, is also a Jordan canonical basis of A. $\square$

Example 8 Find a Jordan canonical basis and the Jordan canonical form for the matrix

$$A = \begin{bmatrix} 2 & 0 & 0 & 0 & 0 & 0 \\ 1 & 2 & 0 & 0 & 0 & 0 \\ -1 & 1 & 2 & 0 & 0 & 0 \\ 0 & 0 & 0 & 2 & 0 & 0 \\ 0 & 0 & 0 & 1 & 2 & 0 \\ 0 & 0 & 0 & 0 & 1 & 4 \end{bmatrix}_{6 \times 6}.$$

Compute e^{tA} and solve

$$\frac{d\vec{x}}{dt} = \vec{x}A.$$

Solution The characteristic polynomial is

$$\det(A - tI_6) = (t - 2)^5(t - 4).$$

So A has one eigenvalue 2 of multiplicity 5 and another eigenvalue 4 of multiplicity 1.

To compute the ranks:

$$r(A - 2I_6) = 4, \quad r((A - 2I_6)^2) = 2, \quad r((A - 2I_6)^k) = 1 \quad \text{for } k \geq 3;$$
$$r(A - 4I_6) = 7, \quad r((A - 4I_6)^k) = 1 \quad \text{for } k \geq 2.$$

These provide information about the number and the sizes of Jordan blocks in the Jordan canonical form:

1. For $\lambda = 2$, $\dim G_2 = 5$ and a basis $\{\vec{v}_1, \ldots, \vec{v}_5\}$ must be chosen so that

$$\vec{v}_1; \vec{v}_4 \leftarrow \cdots 2 = \dim \mathbb{R}^6 - r(A - 2I_6),$$
$$\vec{v}_2; \vec{v}_5 \leftarrow \cdots 2 = r(A - 4I_6) - r((A - 2I_6)^2),$$
$$\vec{v}_3 \leftarrow \cdots 1 = r((A - 4I_6)^2) - r((A - 2I_6)^3).$$

2. For $\lambda = 4$, $\dim E_4 = 1$ and a basis $\{\vec{v}_6\}$ is chosen for E_4.

Combining together, $\mathcal{B} = \{\vec{v}_1, \ldots, \vec{v}_5, \vec{v}_6\}$ is going to be a Jordan canonical basis of A and, in $\mathcal{B}$,

$$[A]_{\mathcal{B}} = PAP^{-1} = \begin{bmatrix} 2 & 0 & 0 & \vdots & & & \\ 1 & 2 & 0 & \vdots & & & \\ 0 & 1 & 2 & \vdots & & & \\ \cdots & \cdots & \cdots & \vdots & \cdots & \cdots & \\ & & & \vdots & 2 & 0 & \vdots \\ & & & \vdots & 1 & 2 & \vdots \\ & & & \vdots & \cdots & \cdots & \vdots & \cdots \\ & & & & & & \vdots & 4 \end{bmatrix}_{6\times6},$$

$$P = \begin{bmatrix} \vec{v}_1 \\ \vdots \\ \vec{v}_5 \\ \vec{v}_6 \end{bmatrix}.$$

To choose a Jordan canonical basis $\mathcal{B}$:
Solve

$$\vec{x}(A - 2I_6) = \vec{0}$$
$$\Rightarrow x_2 - x_3 = 0, \quad x_3 = 0, \quad x_5 = x_6 = 0$$
$$\Rightarrow \vec{x} = (x_1, 0, 0, x_4, 0, 0) = x_1\vec{e}_1 + x_4\vec{e}_4.$$

Therefore, $E_2 = \langle\langle \vec{e}_1, \vec{e}_4 \rangle\rangle$. Next, solve

$$\vec{x}(A - 2I_6)^3 = \vec{0} \quad \text{but} \quad \vec{x}(A - 2I_4)^2 \neq \vec{0}$$
$$\Rightarrow x_6 = 0 \quad \text{but at least one of } x_1, x_2, x_3, x_4 \text{ and } x_5 \text{ is not zero.}$$
$$\Rightarrow \text{Choose } \vec{v}_3 = \vec{e}_3 \quad \text{but not to be } \vec{e}_1 \text{ and } \vec{e}_4.$$
$$\Rightarrow \vec{v}_2 = \vec{v}_3(A - 2I_6) = (-1, 1, 0, 0, 0, 0) = -\vec{e}_1 + \vec{e}_2.$$
$$\Rightarrow \vec{v}_1 = \vec{v}_2(A - 2I_6) = (1, 0, 0, 0, 0, 0) = \vec{e}_1.$$

Really, it is $\vec{v}_1(A - 2I_6) = \vec{0}$. Next, solve

$$\vec{x}(A - 2I_6)^2 = \vec{0} \quad \text{but} \quad \vec{x}(A - 2I_6) \neq \vec{0}$$
$$\Rightarrow x_3 = 0, \quad x_6 = 0 \quad \text{but at least one of } x_1, x_2, x_4 \text{ and } x_5 \text{ is not zero.}$$
$$\Rightarrow \text{Choose } \vec{v}_5 = (0, 0, 0, 1, 0, 0) = \vec{e}_5 \quad (\text{why not } \vec{e}_2 \text{ and } \vec{e}_4?)$$
$$\Rightarrow \vec{v}_4 = \vec{v}_5(A - 2I_6) = (0, 0, 0, 1, 0, 0) = \vec{e}_4.$$

Therefore, $\{\vec{v}_1, \vec{v}_2, \vec{v}_3\}$ and $\{\vec{v}_4, \vec{v}_5\}$ are the required bases so that

$$\text{Ker}(A - 2I_6)^5 = \langle\langle \vec{v}_1, \vec{v}_2, \vec{v}_3 \rangle\rangle \oplus \langle\langle \vec{v}_4, \vec{v}_5 \rangle\rangle.$$

Finally, solve

$$\vec{x}(A - 4I_6) = \vec{0}$$
$$\Rightarrow -2x_1 + x_2 - x_3 = 0, \quad -2x_2 + x_3 = 0, \quad -2x_3 = 0,$$
$$-2x_4 + x_5 = 0, \quad -2x_5 + x_6 = 0$$
$$\Rightarrow x_1 = x_2 = x_3 = 0, \quad x_5 = 2x_4, \quad x_6 = 2x_5 = 4x_4$$
$$\Rightarrow \vec{x} = (0,0,0,x_4,2x_4,4x_4) = x_4(0,0,0,1,2,4)$$
$$\Rightarrow \vec{v}_6 = (0,0,0,1,2,4).$$

Combing together,

$$\mathcal{B} = \{\vec{v}_1, \vec{v}_2, \vec{v}_3, \vec{v}_4, \vec{v}_5, \vec{v}_6\}$$

is a Jordan canonical basis of A and the transition matrix is

$$P = \begin{bmatrix} 1 & 0 & 0 & 0 & 0 & 0 \\ -1 & 1 & 0 & 0 & 0 & 0 \\ 0 & 0 & 1 & 0 & 0 & 0 \\ 0 & 0 & 0 & 1 & 0 & 0 \\ 0 & 0 & 0 & 0 & 1 & 0 \\ 0 & 0 & 0 & 1 & 2 & 4 \end{bmatrix}_{6\times 6}.$$

To compute e^{tA}: let $J_A = [A]_{\mathcal{B}}$ for simplicity. Then

$$A = P^{-1}J_A P$$
$$\Rightarrow (\text{see Ex.}<\text{D}_2> 4 \text{ of Sec. } 3.7.6)$$
$$e^{tA} = P^{-1}e^{tJ_A}P.$$

The problem reduces to compute e^{tJ_A}. Thus

$$J_1 = \begin{bmatrix} 2 & 0 & 0 \\ 1 & 2 & 0 \\ 0 & 1 & 2 \end{bmatrix} = 2I_3 + N_1, \quad \text{where } N_1 = \begin{bmatrix} 0 & 0 & 0 \\ 1 & 0 & 0 \\ 0 & 1 & 0 \end{bmatrix} \quad \text{with } N_1^3 = O$$
$$\Rightarrow J_1^2 = 2^2 I_3 + 2^2 N_1 + N_1^2, \dots,$$

$$J_1^k = 2^k I_3 + C_1^k \cdot 2^{k-1} \cdot N_1 + C_2^k \cdot 2^{k-2} \cdot N_1^2$$

$$= \begin{bmatrix} 2^k & 0 & 0 \\ k \cdot 2^{k-1} & 2^k & 0 \\ \frac{k(k-1)}{2!} 2^{k-2} & k \cdot 2^{k-1} & 2^k \end{bmatrix} \quad \text{for } k \geq 2$$

$$\Rightarrow \sum_{k=0}^{n} \frac{1}{k!} J_1^k$$

$$= \begin{bmatrix} 1 + 2 + \sum_{k=2}^{n} \frac{1}{k!} 2^k & 0 & 0 \\ 1 + \sum_{k=2}^{n} \frac{1}{(k-1)!} 2^{k-1} & 1 + 2 + \sum_{k=2}^{n} \frac{1}{k!} 2^k & 0 \\ \sum_{k=2}^{n} \frac{1}{(k-2)!2!} 2^{k-2} & 1 + \sum_{k=2}^{n} \frac{1}{(k-1)!} 2^{k-1} & 1 + 2 + \sum_{k=2}^{n} \frac{1}{k!} 2^k \end{bmatrix}$$

$$\rightarrow \begin{bmatrix} e^2 & 0 & 0 \\ e^2 & e^2 & 0 \\ \frac{1}{2!} e^2 & e^2 & e^2 \end{bmatrix} = e^2 \begin{bmatrix} 1 & 0 & 0 \\ \frac{1}{1!} & 1 & 0 \\ \frac{1}{2!} & \frac{1}{1!} & 1 \end{bmatrix} \quad \text{as } n \rightarrow \infty$$

$$\Rightarrow e^{tJ_1} = e^{2t} \begin{bmatrix} 1 & 0 & 0 \\ \frac{1}{1!} & 1 & 0 \\ \frac{1}{2!} & \frac{1}{1!} & 1 \end{bmatrix}.$$

Similarly,

$$e^{tJ_2} = e^{2t} \begin{bmatrix} 1 & 0 \\ \frac{1}{1!} & 1 \end{bmatrix}, \quad \text{where } J_2 = \begin{bmatrix} 2 & 0 \\ 1 & 2 \end{bmatrix}, \quad \text{and}$$

$$e^{t[4]} = e^{4t} [1]_{1 \times 1}.$$

Therefore,

$$e^{tJ_A} = \begin{bmatrix} e^{tJ_1} & & \\ & e^{tJ_2} & \\ & & e^{t[4]} \end{bmatrix} = \begin{bmatrix} e^{2t} & 0 & 0 & \vdots & & & \\ e^{2t} & e^{2t} & 0 & \vdots & & & \\ \frac{1}{2}e^{2t} & e^{2t} & e^{2t} & \vdots & & & \\ \cdots & \cdots & \cdots & \vdots & \cdots & \cdots & \cdots \\ & & & \vdots & e^{2t} & 0 & \vdots \\ & & & \vdots & e^{2t} & e^{2t} & \vdots \\ \cdots & \cdots & \cdots & \vdots & \cdots \\ & & & & & \vdots & e^{4t} \end{bmatrix}_{6 \times 6}$$

The solution of $\frac{d\vec{x}}{dt} = \vec{x}A$ is

$$\vec{x}(t) = \vec{\alpha}e^{tA}, \quad \text{where } \vec{\alpha} \in \mathbb{R}^6 \text{ is any constanst vector.}$$

Remark

The method in Example 6 to determine the Jordan canonical form of a real matrix $A_{n \times n}$ for $n \geq 2$, once the characteristic polynomial splitting into linear factors as

$$(-1)^n(t - \lambda_1)^{r_1}(t - \lambda_2)^{r_2} \cdots (t - \lambda_k)^{r_k}, \quad r_1 + r_2 + \cdots + r_k = n,$$

is universally true for matrices over a field, in particular, the complex field $\mathbb{C}$.

For general setting in this direction, see Sec. B.12 and Ex. <C> 21.

Exercises

<A>

1. Try to use the method introduced in Example 6 to reprove (3.7.46) and (3.7.50) more systematically.

2. For each of the following matrices A, do the following problems.

   (1) Find a Jordan canonical basis of A and the Jordan canonical form of A. Try to use (3.7.32) and Ex. <A> 5 there to find a Jordan canonical basis and the Jordan canonical form of A^*.

   (2) Compute $\det A$.

   (3) In case A is invertible, compute A^{-1} by as many methods as possible and compare the advantage of one with the other (see (3.7.52)).

   (4) Compute A^n for $n \geq 1$ and e^{tA} and solve the equation

   $$\frac{d\vec{x}}{dt} = \vec{x}A \quad \text{(refer to Ex.<D_2> and <D_4> of Sec. 3.7.6).}$$

   (5) If A is invertible, find a square root of A (see Example 3).

   (6) Determine an invertible symmetric matrix R so that $A^* = RAR^{-1}$ (see Example 4).

   (7) Decompose A as a product of two symmetric matrices so that at least one of them is invertible (see Example 5).

$$\text{(a) } \begin{bmatrix} 1 & 2 & 3 \\ 2 & 4 & 6 \\ 3 & 6 & 9 \end{bmatrix}. \quad \text{(b) } \begin{bmatrix} 2 & 2 & -1 \\ 0 & 1 & 0 \\ -1 & -2 & 2 \end{bmatrix}. \quad \text{(c) } \begin{bmatrix} 0 & 0 & 2 \\ 1 & -2 & 1 \\ 2 & 0 & 0 \end{bmatrix}.$$

(d) $\begin{bmatrix} 0 & 4 & 2 \\ -3 & 8 & 3 \\ 4 & -8 & -2 \end{bmatrix}$.   (e) $\begin{bmatrix} 2 & 0 & 0 \\ 2 & 2 & 0 \\ -2 & 1 & 2 \end{bmatrix}$.   (f) $\begin{bmatrix} -1 & 1 & 0 \\ 0 & -1 & 1 \\ 0 & 0 & -1 \end{bmatrix}$.

(g) $\begin{bmatrix} -3 & 3 & -2 \\ -7 & 6 & -3 \\ 1 & -1 & 2 \end{bmatrix}$.   (h) $\begin{bmatrix} 1 & 0 & 1 \\ 1 & 0 & 2 \\ -1 & 0 & 3 \end{bmatrix}$.   (i) $\begin{bmatrix} 2 & -1 & -1 \\ 2 & -1 & -2 \\ -1 & 1 & 2 \end{bmatrix}$.

(j) $\begin{bmatrix} 5 & -3 & 2 \\ 6 & -4 & 4 \\ 4 & -4 & 5 \end{bmatrix}$.   (k) $\begin{bmatrix} 1 & -3 & 4 \\ 4 & -7 & 8 \\ 6 & -7 & 7 \end{bmatrix}$.   (l) $\begin{bmatrix} 4 & 6 & -15 \\ 3 & 4 & -12 \\ 2 & 3 & -8 \end{bmatrix}$.

<B>

1. For each of the following matrices A, do the same problems as indicated in Ex. <A> 2.

(a) $\begin{bmatrix} 3 & 0 & 0 & 0 \\ 1 & 3 & 0 & 0 \\ 0 & 1 & 3 & 0 \\ -1 & 1 & 0 & 3 \end{bmatrix}$.   (b) $\begin{bmatrix} 1 & 0 & 0 & 0 \\ 1 & 1 & 0 & 0 \\ 0 & 1 & 2 & 0 \\ -1 & 0 & 1 & 2 \end{bmatrix}$.

(c) $\begin{bmatrix} 0 & -2 & -2 & -2 \\ -3 & 1 & 1 & -3 \\ 1 & -1 & -1 & 1 \\ 2 & 2 & 2 & 4 \end{bmatrix}$.   (d) $\begin{bmatrix} 7 & 1 & -2 & 1 \\ 1 & 4 & 1 & 1 \\ 2 & -1 & 5 & 2 \\ 2 & -1 & -1 & 8 \end{bmatrix}$.

(e) $\begin{bmatrix} 2 & 0 & 0 & 0 \\ -1 & 3 & 1 & -1 \\ 0 & -1 & 1 & 0 \\ 1 & 0 & 0 & 3 \end{bmatrix}$.   (f) $\begin{bmatrix} 2 & -2 & -2 & -2 & 2 \\ -4 & 0 & -2 & -6 & 1 \\ 2 & 1 & 3 & 3 & 0 \\ 2 & 3 & 3 & 7 & 2 \\ 0 & 0 & 0 & 0 & 5 \end{bmatrix}_{5\times5}$.

(g) $\begin{bmatrix} 1 & 0 & 0 & 0 & 0 & 0 \\ 0 & 1 & 0 & 0 & 0 & 0 \\ 0 & 1 & 1 & 0 & 1 & 0 \\ 0 & 1 & 0 & 1 & 0 & 0 \\ 1 & 0 & 0 & 0 & 1 & 0 \\ 0 & -1 & 0 & 0 & -1 & 1 \end{bmatrix}_{6\times6}$.   (h) $\begin{bmatrix} 5 & 1 & 1 & 0 & 0 & 0 \\ 0 & 5 & 1 & 0 & 0 & 0 \\ 0 & 0 & 5 & 0 & 0 & 0 \\ 0 & 0 & 0 & 5 & 1 & -1 \\ 0 & 0 & 0 & 0 & 5 & 1 \\ 0 & 0 & 0 & 0 & 0 & 5 \end{bmatrix}_{6\times6}$.

$$\text{(i)} \begin{bmatrix} -1 & 0 & 2 & -2 & 0 & 0 & -1 \\ 0 & 1 & 1 & 0 & 0 & 0 & -1 \\ -1 & 0 & 2 & -1 & 0 & 0 & 0 \\ 1 & 0 & -1 & 2 & 0 & 0 & 1 \\ 1 & 0 & -1 & 1 & 1 & 0 & 2 \\ 3 & 0 & -6 & 3 & 0 & 1 & 4 \\ 0 & 0 & 0 & 0 & 0 & 0 & 1 \end{bmatrix}_{7 \times 7} .$$

2. A matrix $A \in M(n; \mathbb{R})$ (or $M(n; \mathbb{F})$) is called a *nilpotent matrix* of *index* k if $A^{k-1} \neq O$ but $A^k = O$ (see Ex. 7 in Sec. B.4). Verify that each of the following matrices is nilpotent and find its Jordan canonical form.

(a) $\begin{bmatrix} 0 & a & b \\ 0 & 0 & c \\ 0 & 0 & 0 \end{bmatrix}$. (b) $\begin{bmatrix} 1 & -1 & -1 \\ 2 & -2 & -2 \\ -1 & 1 & 1 \end{bmatrix}$. (c) $\begin{bmatrix} 0 & 0 & 0 & 0 & 1 & 0 \\ 0 & 0 & 1 & 1 & 0 & -1 \\ 0 & 0 & 0 & 0 & 0 & 0 \\ 0 & 0 & 0 & 0 & 0 & 0 \\ 0 & 0 & 1 & 0 & 0 & -1 \\ 0 & 0 & 0 & 0 & 0 & 0 \end{bmatrix}$.

<C> Abstraction and generalization

Read Sec. B.12 and try your best to do exercises there.
    Besides, do the following problems.

1. Find the Jordan canonical form for each of the following matrices.

(a) $\begin{bmatrix} 1 & -1 & & & O \\ & 1 & -1 & & \\ & & \ddots & \ddots & \\ & & & 1 & -1 \\ O & & & & 1 \end{bmatrix}_{n \times n}$. (b) $\begin{bmatrix} 0 & 1 & & & O \\ & 0 & 1 & & \\ & & \ddots & \ddots & \\ & & & 0 & 1 \\ O & & & & 0 \end{bmatrix}_{n \times n}$.

(c) The square power of the matrix in (b).

(d) $\begin{bmatrix} \alpha & 0 & 1 & 0 & & O \\ 0 & \alpha & 0 & 1 & & \\ & & \ddots & \ddots & \ddots & \\ & & \alpha & 0 & 1 \\ & & 0 & \alpha & 0 \\ O & & & 0 & 0 & \alpha \end{bmatrix}_{n \times n}$, $n \geq 3$.

(e) $\begin{bmatrix} 0 & 1 & 0 & \cdots & 0 & 0 \\ 0 & 0 & 1 & \cdots & 0 & 0 \\ \vdots & \vdots & \vdots & & \vdots & \vdots \\ & & & & & \\ 0 & 0 & 0 & \cdots & 0 & 1 \\ 1 & 0 & 0 & \cdots & 0 & 0 \end{bmatrix}_{n \times n}$.

(f) $\begin{bmatrix} 0 & 0 & \cdots & 0 & a_1 \\ 0 & 0 & \cdots & a_2 & 0 \\ \vdots & \vdots & & \vdots & \vdots \\ 0 & a_{n-1} & \cdots & 0 & 0 \\ a_n & 0 & \cdots & 0 & 0 \end{bmatrix}_{n \times n}$.

(g) $\begin{bmatrix} \alpha_0 & \alpha_1 & \alpha_2 & \cdots & \alpha_{n-2} & \alpha_{n-1} \\ \alpha_{n-1} & \alpha_0 & \alpha_1 & \cdots & \alpha_{n-3} & \alpha_{n-2} \\ \alpha_{n-2} & \alpha_{n-1} & \alpha_0 & \cdots & \alpha_{n-4} & \alpha_{n-3} \\ \vdots & \vdots & \vdots & & \vdots & \vdots \\ \alpha_1 & \alpha_2 & \alpha_3 & \cdots & \alpha_{n-1} & \alpha_0 \end{bmatrix}$, $\alpha_0, \alpha_1, \ldots, \alpha_{n-1} \in \mathbb{R}$ (or $\mathbb{C}$).

(h) $\begin{bmatrix} a_0 & a_1 & a_2 & \cdots & a_{n-1} \\ \mu a_{n-1} & a_0 & a_1 & \cdots & a_{n-2} \\ \mu a_{n-2} & \mu a_{n-1} & a_0 & \cdots & a_{n-3} \\ \vdots & \vdots & \vdots & & \vdots \\ \mu a_1 & \mu a_2 & \mu a_3 & \cdots & a_0 \end{bmatrix}$, $a_0, a_1, \ldots, a_{n-1}, \mu \in \mathbb{R}$ (or $\mathbb{C}$).

(i) $\begin{bmatrix} c_0 & c_1 & c_2 & c_3 & c_4 \\ c_1 & c_2 + c_1 a & c_3 + c_2 a & c_4 + c_3 a & c_0 + c_4 a \\ c_2 & c_3 + c_2 a & c_4 + c_3 a & c_0 + c_4 a & c_1 \\ c_3 & c_4 + c_3 a & c_0 + c_4 a & c_1 & c_2 \\ c_4 & c_0 + c_4 a & c_1 & c_2 & c_3 \end{bmatrix}$,

where $c_0, c_1, c_2, c_3, c_4, a \in \mathbb{R}$ (or $\mathbb{C}$). Try to extend to a matrix of order n.

(j) $P_{n \times n}$ is a permutation matrix of order n (see (2.7.67)), namely, for a permutation $\sigma: \{1, 2, \ldots, n\} \to \{1, 2, \ldots, n\}$,

$$P = \begin{bmatrix} \vec{e}_{\sigma(1)} \\ \vdots \\ \vec{e}_{\sigma(n)} \end{bmatrix}.$$

2. For each of the following linear operators f, do the following problems.

   (1) Compute the matrix representation $[f]_\mathcal{B}$, where $\mathcal{B}$ is the given basis.

   (2) Find a Jordan canonical basis and the Jordan canonical form of $[f]_\mathcal{B}$.

(3) Find the corresponding Jordan canonical basis for the original operator f.

(a) $f: P_2(\mathbb{R}) \rightarrow P_2(\mathbb{R})$ defined by $f(p)(x) = (x+1)p'(x)$, $\mathcal{B} = \{x^2 - x + 1,\ x + 1,\ x^2 + 1\}$.

(b) $f: P_2(\mathbb{R}) \rightarrow P_2(\mathbb{R})$ defined by $f(p) = p + p' + p''$, $\mathcal{B} = \{2x^2 - x + 1,\ x^2 + 3x - 2,\ -x^2 + 2x + 1\}$.

(c) $f: M(2;\mathbb{R}) \rightarrow M(2;\mathbb{R})$ defined by $f(A) = A^*$,

$$\mathcal{B} = \left\{ \begin{bmatrix} 1 & 1 \\ 1 & 0 \end{bmatrix},\ \begin{bmatrix} 1 & 1 \\ 0 & 1 \end{bmatrix},\ \begin{bmatrix} 1 & 0 \\ 1 & 1 \end{bmatrix},\ \begin{bmatrix} 0 & 1 \\ 1 & 1 \end{bmatrix} \right\}.$$

(d) $f: M(2;\mathbb{R}) \rightarrow M(2;\mathbb{R})$ defined by $f(A) = \begin{bmatrix} 1 & 2 \\ 2 & 1 \end{bmatrix} A$, where $\mathcal{B} = \{E_{11}, E_{12}, E_{21}, E_{22}\}$ is the natural basis for $M(2;\mathbb{R})$. What happens if $\begin{bmatrix} 1 & 2 \\ 2 & 1 \end{bmatrix}$ is replaced by $\begin{bmatrix} 1 & 2 \\ 0 & 1 \end{bmatrix}$ or $\begin{bmatrix} 1 & 2 \\ -2 & 1 \end{bmatrix}$?

(e) Let $V = \langle\langle e^x, xe^x, x^2e^x, e^{-x}, e^{-2x} \rangle\rangle$ be the vector subspace of $C[a,b]$, generated by $\mathcal{B} = \{x, xe^x, x^2e^x, e^{-x}, e^{-2x}\}$. $f: V \rightarrow V$ defined by $f(p) = p'$.

(f) Let V be the subspace, generated by $\mathcal{B} = \{1, x, y, x^2, y^2, xy\}$, of the vector space $P(x,y) = \{$polynomail functions over $\mathbb{R}$ in two variables x and $y\}$. $f: V \rightarrow V$ defined by $f(p) = \frac{\partial p}{\partial y}$.

3. Let f be a *nilpotent* operator on an n-dimensional vector space V over a field $\mathbb{F}$, i.e. there exists some positive integer k so that $f^k = 0$ and the smallest such k is called its *index* (refer to Ex. <B> 2). Note that the index $k \leq n$ (see (3.7.38)).

(a) Let $\mathcal{B}$ be any ordered basis for V. Prove that f is nilpotent of index k if and only if $[f]_\mathcal{B}$ is nilpotent of index k.

(b) Suppose $\mathbb{F} = \mathbb{C}$, the complex field. Prove that the following are equivalent.

(1) f is nilpotent of index k.

(2) There exist a sequence of bases $\mathcal{B}_1, \ldots, \mathcal{B}_k = \mathcal{B}$ such that $\mathcal{B}_i$ is a basis for $\text{Ker}(f^i)$ and $\mathcal{B}_{i+1}$ is an extension of $\mathcal{B}_i$ for $1 \leq i \leq k-1$. Hence $\mathcal{B}$ is a basis for $\text{Ker}(f^k) = V$ so that $[f]_\mathcal{B}$ is lower triangular with each diagonal entry equal to zero (see Ex. 5 of Sec. B.12).

(3) f has the characteristic polynomial

$$\det(f - t1_V) = (-1)^n t^n.$$

and the minimal polynomial t^k.

(4) For any ℓ, $1 \le \ell \le n$,

$$\mathrm{tr}(f^\ell) = 0.$$

(*Note* (b) is still valid even if $\mathbb{F}$ is an infinite field of characteristic zero. But (4) is no more true if $\mathbb{F}$ is not so. For example, in case $\mathbb{F} = I_2 = \{0, 1\}$, then

$$\begin{bmatrix} 1 & 0 \\ 1 & 1 \end{bmatrix}$$

satisfies $\mathrm{tr}[f^\ell] = 0$ for $\ell \ge 1$ but it is not nilpotent. See also Ex. <C> 11 of Sec. 2.7.6.)

Suppose f is nilpotent of index k. Let

$$\dim \mathrm{Ker}(f^i) = n_i, \quad 1 \le i \le k, \quad \text{with} \quad n_k = n, \ n_0 = 0 \quad \text{and}$$
$$r_i = n_i - n_{i-1} = \mathrm{r}(f^{i-1}) - \mathrm{r}(f^i), \quad 1 \le i \le k.$$

Note that $r_1 + r_2 + \cdots + r_k = n = \dim V$.

(c) For simplicity, let $V = \mathbb{F}^n$ and $f(\vec{x}) = \vec{x}A$, where $\vec{x} \in \mathbb{F}^n$ and $A \in M(n; \mathbb{F})$. Choose $\vec{x}_1, \ldots, \vec{x}_{r_k} \in \mathrm{Ker}(A^k) \setminus \mathrm{Ker}(A^{k-1})$, linearly independent, so that

$$\mathrm{Ker}(A^k) = V = \langle\langle \vec{x}_1, \ldots, \vec{x}_{r_k} \rangle\rangle \oplus \mathrm{Ker}(A^{k-1}).$$

Try to show that $\vec{x}_1 A, \ldots, \vec{x}_{r_k} A$ are linearly independent and

$$\langle\langle \vec{x}_1 A, \ldots, \vec{x}_{r_k} A \rangle\rangle \cap \mathrm{Ker}(A^{k-2}) = \{\vec{0}\}.$$

Hence, deduce that $r_k + n_{k-2} \le n_{k-1}$ and thus, $r_k \le r_{k-1}$. In case $r_k < r_{k-1}$, choose linearly independent vectors $\vec{x}_{r_k+1}, \ldots, \vec{x}_{r_{k-1}} \in \mathrm{Ker}(A^{k-1}) \setminus \mathrm{Ker}(A^{k-2})$ so that

$$\mathrm{Ker}(A^{k-1}) = \langle\langle \vec{x}_1 A, \ldots, \vec{x}_{r_k} A, \vec{x}_{r_k+1}, \ldots, \vec{x}_{r_{k-1}} \rangle\rangle \oplus \mathrm{Ker}(A^{k-2}).$$

Repeat the above process to show that $r_{k-1} \le r_{k-2}$. By induction, this process eventually leads to the following facts.

(1) $r_k \le r_{k-1} \le \cdots \le r_2 \le r_1, r_1 + r_2 + \cdots + r_k = n$. Call $\{r_k, \ldots, r_1\}$ the *invariant system* of f or A.

(2) $\text{Ker}(A^k) = V$ has a basis $\mathcal{B}$:

$$
\begin{array}{cccccc}
\vec{x}_1 & \cdots & \vec{x}_{r_k} & & & \\
\vec{x}_1 A & \cdots & \vec{x}_{r_k} A & \vec{x}_{r_k+1} & \cdots & \vec{x}_{r_{k-1}} \\
\vdots & & \vdots & \vdots & & \vdots \\
\vec{x}_1 A^{k-2} \cdots & \vec{x}_{r_k} A^{k-2} & \vec{x}_{r_k+1} A^{k-1} \cdots & \vec{x}_{r_{k-1}} A^{k-1} \cdots & \vec{x}_{r_3+1} \cdots & \vec{x}_{r_2} \\
\vec{x}_1 A^{k-1} \cdots & \vec{x}_{r_k} A^{k-1} & \vec{x}_{r_k+1} A^{k-2} \cdots & \vec{x}_{r_{k-1}} A^{k-2} \cdots & \vec{x}_{r_3+1} A \cdots & \vec{x}_{r_2} A \; \vec{x}_{r_2+1} \cdots \; \vec{x}_{r_1}
\end{array}
$$

(3) Denote

$$\mathcal{B}_j^{(i)} = \{\vec{x}_j A^{i-1}, \vec{x}_j A^{i-2}, \ldots, \vec{x}_j A, \vec{x}_j\},$$

$$1 \le i \le k, \; r_{i+1} + 1 \le j \le r_i \quad \text{with} \quad r_{k+1} = 0;$$

$$W_j^{(i)} = \text{the invariant subspace generated by } \mathcal{B}_j^{(i)}.$$

Then $\mathcal{B} = \mathcal{B}_1^{(k)} \cup \cdots \cup \mathcal{B}_{r_k}^{(k)} \cup \mathcal{B}_{r_k+1}^{(k-1)} \cup \cdots \cup \mathcal{B}_{r_2+1}^{(1)} \cup \cdots \cup \mathcal{B}_{r_1}^{(1)}$
and

$$[A]_\mathcal{B} = PAP^{-1}$$

$$
= \begin{bmatrix}
A_1^{(k)} & & & & & & & \\
& \ddots & & & & & 0 & \\
& & A_{r_k}^{(k)} & & & & & \\
& & & A_{r_k+1}^{(k-1)} & & & & \\
& & & & \ddots & & & \\
& & & & & A_{r_2}^{(2)} & & \\
& & & & & & A_{r_2+1}^{(1)} & \\
& 0 & & & & & & \ddots \\
& & & & & & & A_{r_1}^{(1)}
\end{bmatrix}_{n \times n}
$$

where

$$
A_j^{(i)} = \left[A \mid W_j^{(i)} \right]_{\mathcal{B}_j^{(i)}} = \begin{bmatrix}
0 & & & & & 0 \\
1 & 0 & & & & \\
& 1 & 0 & & & \\
& & & \ddots & & \\
& & & & 0 & \\
0 & & & & 1 & 0
\end{bmatrix}_{i \times i},
$$

$$2 \le i \le k, \; r_{i+1} + 1 \le j \le r_i;$$

$$A_j^{(1)} = [0]_{1 \times 1}, \quad r_2 + 1 \le j \le r_1.$$

Call $[A]_\mathcal{B}$ a (Jordan) *canonical form* of the nilpotent operator of f or A.

Compare with Ex. 5 of Sec. B.12.

(d) Show that, there exists a (nilpotent) operator g on V so that

$$g^2 = f$$

i.e. g is a *square root* of f, if and only if, there exist $2k$ non-negative integers $t_1, \ldots, t_{2k}$ satisfying

$$r_i = t_{2i-1} + t_{2i}, \quad 1 \le i \le k;$$
$$t_1 \ge t_2 \ge \cdots \ge t_{2k} \ge 0 \quad \text{and} \quad t_1 + t_2 + \cdots + t_{2k} = n.$$

In this case, if g is of index ℓ, then it is understood that $2k \ge \ell$ and $t_j = 0$ if $j > \ell$ so that $\{t_1, \ldots, t_\ell\}$ is the invariant system of g. (*Note*   Note that $r_i \ge 2t_{2i}$ and $r_{i+1} \le 2t_{2i+1}$ so that

$$\frac{r_i}{2} \ge t_{2i} \ge t_{2i+1} \ge \frac{r_{i+1}}{2}.$$

If both r_i and r_{i+1} are odd integers, then

$$\frac{r_i - 1}{2} \ge t_{2i} \ge t_{2i+1} \ge \frac{r_{i+1} + 1}{2} \Rightarrow r_i \ge r_{i+1} + 2.$$

Conversely, suppose these inequalities hold. If r_i is even, choose $t_{2i} = t_{2i+1} = \frac{r_i}{2}$; if r_i is odd, choose $t_{2i} = \frac{r_i+1}{2}$ and $t_{2i+1} = \frac{r_i-1}{2}$.)

(e) Let $A \in M(n; \mathbb{C})$ be nilpotent of index n. Show that A does not have square root (see Example 3). What happens to a nilpotent matrix $A_{n \times n}$ of index $n - 1$?

(f) Suppose $A_{n \times n}$ and $B_{n \times n}$ are nilpotent of the same index k, where $k = n$ or $n - 1$ and $n \ge 3$. Show that A and B are similar. What happens if $1 \le k \le n - 2$?

(g) Show that a nonzero nilpotent matrix is not diagonalizable.

(h) Find all canonical forms of a nilpotent matrix $A_{n \times n}$, where $n = 3, 4, 5$ and 6.

4. Let $A = [a_{ij}] \in M(n; \mathbb{C})$ be an invertible complex matrix.

(a) Show that A has a *square root*, i.e. there exists a matrix $B \in M(n; \mathbb{C})$ so that $B^2 = A$ (see Example 3).

(b) Let $p \ge 2$ be a positive integer, then there exists an invertible matrix $B \in M(n; \mathbb{C})$ so that $B^p = A$.

5. Any matrix $A \in M(n; \mathbb{C})$ is similar to its transpose A^* and there exists an invertible symmetric matrix P so that

$$A^* = PAP^{-1}$$

(see Example 4). Hence, A and A^* are diagonalizable at the same time (see Ex. <C> 9(f) of Sec. 2.7.6).

6. Any matrix $A \in M(n; \mathbb{C})$ can be decomposed as a product of two symmetric matrices where at least one of them can be designated as an invertible matrix (see Example 5).

7. Any matrix $A \in M(n; \mathbb{C})$ is similar to a lower triangular matrix whose nonzero nondiagonal entries are arbitrarily small positive number. Try to use the following steps:

   (1) Let

   $$PAP^{-1} = \begin{bmatrix} J_1 & & O \\ & \ddots & \\ O & & J_k \end{bmatrix}_{m \times n}, \quad \text{where } J_i = \begin{bmatrix} \lambda_i & & & O \\ 1 & \lambda_i & & \\ & \ddots & \ddots & \\ O & & 1 & \lambda_i \end{bmatrix}_{r_i \times r_i}$$

   for $1 \le i \le k$.

   (2) For any $\varepsilon > 0$, let $E_i = \text{diag}[\varepsilon, \varepsilon^2, \ldots, \varepsilon^{r_i}]$ be a diagonal matrix of order r_i. Then

   $$E_i J_i E_i^{-1} = \begin{bmatrix} \lambda_i & & & 0 \\ \varepsilon & \lambda_i & & \\ & \ddots & \ddots & \\ 0 & & \varepsilon & \lambda_i \end{bmatrix}_{r_i \times r_i} \quad \text{for } 1 \le i \le k.$$

   (3) Denote the pseudo-diagonal matrix $E = \text{diag}[E_1, E_2, \ldots, E_k]$. What is $(EP)A(EP)^{-1}$?

8. A matrix of the form

$$\begin{bmatrix} J_1 & & & O \\ & J_2 & & \\ & & \ddots & \\ O & & & J_k \end{bmatrix}_{n \times n}, \quad \text{where } J_i = \begin{bmatrix} \lambda & & & 0 \\ 1 & \lambda & & \\ & \ddots & \ddots & \\ 0 & & 1 & \lambda \end{bmatrix}_{r_i \times r_i} \quad \text{for } 1 \le i \le k$$

with *descending* sizes $r_1 \ge r_2 \ge \cdots \ge r_k$ has the characteristic polynomial $(-1)^n (t - \lambda)^n$ and the minimal polynomial $(t - \lambda)^{r_1}$. Prove this and use it to do the following problems.

(a) Find the minimal polynomial of a matrix A in its Jordan canonical form.

(b) A matrix is diagonalizable if and only if its minimal polynomial does not have repeated roots (refer to (2.7.73) and Ex. $<C>$ 9(e) in Sec. 2.7.6).

9. A complex matrix R similar to the matrix

$$\begin{bmatrix} 0 & 1 & \vdots & \\ 1 & 0 & \vdots & \\ \cdots & \cdots & \vdots & \cdots \\ & & \vdots & I_{n-2} \end{bmatrix}_{n \times 2}$$

is called a *reflection* of $\mathbb{C}^n$ with respect to a 2-dimensional subspace. Then, any *involutory* matrix $A_{n \times n}$, i.e. $A^2 = I_n$, which is not I_n or $-I_n$, can be decomposed as a product of finitely many reflections. Watch the following steps.

(1) For some invertible P and $1 \le r \le n-1$,

$$A = P^{-1} \begin{bmatrix} -I_r & 0 \\ 0 & I_{n-r} \end{bmatrix}_{n \times n} P$$

$$= P^{-1} \begin{bmatrix} -1 & 0 \\ 0 & I_{n-1} \end{bmatrix}_{n \times n} P \cdot P^{-1} \begin{bmatrix} 1 & & & \\ & -1 & & 0 \\ & & & \\ 0 & & & I_{n-2} \end{bmatrix}_{n \times n} P$$

$$\cdots P^{-1} \begin{bmatrix} I_{r-1} & & & \\ & -1 & & 0 \\ & & & \\ 0 & & & I_{n-r} \end{bmatrix}_{n \times n} P.$$

Let R_i denote the ith factor on the right, $1 \le i \le r$.

(2) Let

$$Q_i = \begin{bmatrix} I_{i-1} & & & O \\ & \begin{bmatrix} 0 & 1 & 0 \\ 1 & 0 & 1 \\ -1 & 0 & 1 \end{bmatrix} & \\ O & & & I_{n-i-2} \end{bmatrix}_{n \times n} \quad \text{for } i = 1, 2, \ldots, r.$$

Then

$$PR_iP^{-1} = Q_i^{-1} \begin{bmatrix} I_i & & O \\ & \begin{bmatrix} 0 & 1 \\ 1 & 0 \end{bmatrix} & \\ O & & I_{n-i-2} \end{bmatrix} Q_i \quad \text{for } 1 \le i \le r.$$

10. A matrix $A_{6\times 6}$ has the characteristic and the minimal polynomial

$$p(t) = (t-1)^4(t+2)^2,$$
$$m(t) = (t-1)^2(t+2)$$

respectively. Find the Jordan canonical form of A.

11. Suppose a matrix $A_{8\times 8}$ has the characteristic polynomial

$$(t+1)^2(t-1)^4(t-2)^2.$$

Find all possible Jordan canonical forms for such A and compute the minimal polynomial for each case.

12. Suppose a complex matrix $A_{n\times n}$ has all its eigenvalues equal to 1, then any power matrix $A^k(k \ge 1)$ is similar to A itself. What happens if $k \le -1$.

13. Let

$$J = \begin{bmatrix} \lambda & & & & \\ 1 & \lambda & & & \\ & 1 & \lambda & & \\ & & \ddots & \ddots & \\ & & & 1 & \lambda \end{bmatrix}_{n\times n}$$

and $p(x)$ be any polynomial in $P(\mathbb{R})$. Show that

$$p(J) = \begin{bmatrix} p(\lambda) & 0 & \cdots & \cdots & 0 & 0 \\ \frac{p'(\lambda)}{1!} & p(\lambda) & \cdots & \cdots & 0 & 0 \\ \frac{p''(\lambda)}{2!} & \frac{p'(\lambda)}{1!} & \cdots & \cdots & 0 & 0 \\ \vdots & \vdots & & & \vdots & \vdots \\ \frac{p^{(n-2)}(\lambda)}{(n-2)!} & \frac{p^{(n-3)}(\lambda)}{(n-3)!} & \cdots & \cdots & p(\lambda) & 0 \\ \frac{p^{(n-1)}(\lambda)}{(n-1)!} & \frac{p^{(n-2)}(\lambda)}{(n-2)!} & \cdots & \cdots & \frac{p'(\lambda)}{1!} & p(\lambda) \end{bmatrix}_{n\times n}.$$

In particular, compute J^k for $k \ge 1$. Show that

(a)

$$e^J = e^\lambda \begin{bmatrix} 1 & 0 & \cdots & \cdots & 0 & 0 \\ \frac{1}{1!} & 1 & \cdots & \cdots & 0 & 0 \\ \frac{1}{2!} & \frac{1}{1!} & \cdots & \cdots & 0 & 0 \\ \vdots & \vdots & & & & \\ \frac{1}{(n-2)!} & \frac{1}{(n-3)!} & \cdots & \cdots & 1 & 0 \\ \frac{1}{(n-1)!} & \frac{1}{(n-2)!} & \cdots & \cdots & \frac{1}{1!} & 1 \end{bmatrix}. \quad \text{What is } e^{tJ}?$$

(b) $\lim_{k\to\infty} J^k$ exists (see Ex. <D$_1$> of Sec. 3.7.6) if and only if one of the following holds:

   (1) $|\lambda| < 1$.
   (2) $\lambda = 1$ and $n = 1$.

(c) $\lim_{k\to\infty} J^k = O_{n\times n}$ if $|\lambda| < 1$ holds and $\lim_{k\to\infty} J^k = [1]_{1\times 1}$ if $\lambda = 1$ and $n = 1$.

(d) Let

$$PAP^{-1} = \begin{bmatrix} J_1 & & & 0 \\ & J_2 & & \\ & & \ddots & \\ 0 & & & J_k \end{bmatrix}$$

be the Jordan canonical form of a matrix $A_{n\times n}$ with k Jordan blocks $J_i, 1 \le i \le k$. Show that

$$e^{tA} = P^{-1} \begin{bmatrix} e^{tJ_1} & & & 0 \\ & e^{tJ_2} & & \\ & & \ddots & \\ 0 & & & e^{tJ_k} \end{bmatrix} P.$$

(e) Prove Ex. <D$_1$> 5 of Sec. 3.7.6.

14. For $A = [a_{ij}] \in M(n; \mathbb{C})$ and $p \ge 1$, define

$$\|A\|_p = \left[\sum_{i=1}^n \left(\sum_{j=1}^n |a_{ij}|^p \right) \right]^{\frac{1}{p}}, \quad \text{and}$$

$$\|A\|_\infty = \max_{1\le i,j\le n} |a_{ij}|.$$

Show that, for $1 \le p \le \infty$,

(1) $\|A\|_p \geq 0$ and equality holds if and only if $A = O_{n \times n}$.

(2) $\|\alpha A\|_p = |\alpha| \|A\|_p$, $\alpha \in \mathbb{C}$.

(3) $\|A + B\|_p \leq \|A\|_p + \|B\|_p$.

Hence, $\| \ \|_p$ is called a *p-norm* for M(n; $\mathbb{C}$). See also Exs. <C> 3 and <D$_2$> 6 of Sec. 3.7.6. Also,

(4) $\|AB\|_\infty \leq n \|A\|_\infty \cdot \|B\|_\infty$. Try to use (4) to show that

$$e^A = \lim_{k \to \infty} \sum_{\ell=0}^{k} \frac{1}{\ell!} A^\ell$$

exists for any $A \in$ M(n; $\mathbb{C}$). See Ex. <D$_2$> of Sec. 3.7.6.

15. Let $M = [a_{ij}] \in$ M(n; $\mathbb{R}$) be a stochastic matrix (see Ex. <D$_3$> of Sec. 3.7.6) and $PMP^{-1} = J$ be the Jordan canonical form of M. Let $\| \ \|_\infty$ be as in Ex. 14. Show that:

(1) $\|M^k\|_\infty \leq 1$ for all $k \geq 1$ and hence $\|J^k\|_\infty$ is bounded for all $k \geq 1$.

(2) Each Jordan block in J corresponding to the eigenvalue 1 is of the size 1×1, i.e. a matrix of order 1.

(3) $\lim_{k \to \infty} M^k$ exists if and only if, whenever λ is an eigenvalue of M with $|\lambda| = 1$, then $\lambda = 1$.

Use these results to prove Ex. <D$_3$> 3(d) of Sec. 3.7.6.

16. Let $A = [a_{ij}] \in$ M(n; $\mathbb{F}$). Show that the following are equivalent.

(a) A is *triangularizable*, i.e. there exists an invertible matrix P so that PAP^{-1} is a lower or upper triangular matrix.

(b) The characteristic polynomial of A can be factored as product of linear factors, i.e.

$$\det(A - tI_n) = (-1)^n (t - \lambda_1)^{r_1} \cdots (t - \lambda_k)^{r_k}$$

where $\lambda_1, \ldots, \lambda_k$ are distinct eigenvalues of A, $r_i \geq 1$, for $1 \leq i \leq k$ and $r_1 + \cdots + r_k = n$.

For (b) $\Rightarrow$ (a), see Ex. <C> 10(a) of Sec. 2.7.6. As a consequence, *any complex square matrix or real square matrix considered as a complex one is always triangularizable.* Yet, we still can give a more detailed account for (b) $\Rightarrow$ (a) as follows (refer to Ex. 3, and Exs. 2 and 3 of

Sec. B.12). Suppose the minimal polynomial of $A \in M(n; \mathbb{F})$ is

$$\psi I_A(t) = (t - \lambda_1)^{d_1} \cdots (t - \lambda_k)^{d_k}, \quad 1 \le d_i \le r_i \text{ for } 1 \le i \le k.$$

(See Ex. 8, and Ex. <C> 9 of Sec. 2.7.6.) Let

$$f_i(t) = \frac{\psi I_A(t)}{(t - \lambda_i)^{d_i}} = \prod_{\substack{\ell=1 \\ \ell \ne i}}^{k} (t - \lambda_\ell)^{d_\ell}, \quad 1 \le i \le k.$$

Then, $f_1(t), \ldots, f_k(t)$ are relatively prime. There exist polynomials $g_1(t), \ldots, g_k(t)$ so that

$$f_1(t)g_1(t) + \cdots + f_k(t)g_k(t) = 1.$$

(See Sec. A.5.) Hence

$$f_1(A)g_1(A) + \cdots + f_k(A)g_k(A) = I_n.$$

Let, for $1 \le i \le k$,

$$E_i = f_i(A)g_i(A), \quad \text{and}$$
$$W_i = \{ \vec{x} \in \mathbb{F}^n \mid \vec{x}(A - \lambda_i I_n)^{d_i} = \vec{0} \}$$
$$= \{ \vec{x} \in \mathbb{F}^n \mid \text{there exists some positive integer } \ell$$
$$\text{so that } \vec{x}(A - \lambda_i I_n)^\ell = \vec{0} \}.$$

Then

(1) *Each* $W_i, 1 \le i \le k$, *is an invariant subspace of* A *so that*

$$\mathbb{F}^n = W_1 \oplus \cdots \oplus W_k.$$

(2) $A \mid W_i = E_i \colon \mathbb{F}^n \to \mathbb{F}^n$ *is a projection (see (3.7.34)) onto* W_i *along* $W_1 \oplus \cdots \oplus W_{i-1} \oplus W_{i+1} \oplus \cdots \oplus W_k$, *namely,*

$$E_i^2 = E_i,$$
$$E_i E_j = O, \quad i \ne j, \, 1 \le i, \, j \le n.$$

Also, E_i *has*

the characteristic polynomial $= (t - \lambda_i)^{r_i}$ *or* $-(t - \lambda_i)^{r_i}$, *and*
the minimal polynomial $= (t - \lambda_i)^{d_i}$.

Note that $\dim W_i = r_i$, $1 \le i \le k$.

(3) *There exists a basis $\mathcal{B}_i$ for W_i so that*

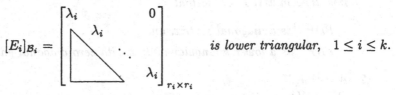

$$[E_i]_{\mathcal{B}_i} = \begin{bmatrix} \lambda_i & & & 0 \\ & \lambda_i & & \\ & & \ddots & \\ & & & \lambda_i \end{bmatrix}_{r_i \times r_i} \quad \text{is lower triangular,} \quad 1 \leq i \leq k.$$

(4) $\mathcal{B} = \mathcal{B}_1 \cup \cdots \cup \mathcal{B}_k$ *forms a basis* $\{\vec{x}_1, \ldots, \vec{x}_n\}$ *for* $\mathbb{F}^n$ *so that*

$$[A]_{\mathcal{B}} = PAP^{-1} = \begin{bmatrix} [E_1]_{\mathcal{B}_1} & & 0 \\ & \ddots & \\ 0 & & [E_k]_{\mathcal{B}_k} \end{bmatrix}_{n \times n}, \quad P = \begin{bmatrix} \vec{x}_1 \\ \vdots \\ \vec{x}_n \end{bmatrix}.$$

In this case,

$$E_i = P^{-1} \begin{bmatrix} 0 & & & & & & \\ & \ddots & & & & 0 & \\ & & 0 & & & & \\ & & & I_{r_i} & & & \\ & & & & 0 & & \\ & 0 & & & & \ddots & \\ & & & & & & 0 \end{bmatrix} P, \quad 1 \leq i \leq k.$$

17. (continued from Ex. 16) Define

$$D = \sum_{i=1}^{k} \lambda_i E_i = P^{-1} \begin{bmatrix} \lambda_1 I_{r_1} & & 0 \\ & \ddots & \\ 0 & & \lambda_k I_{r_k} \end{bmatrix}_{n \times n} P,$$

a diagonalizable matrix and

$$N = A - D = P^{-1} \begin{bmatrix} [E_1]_{\mathcal{B}_1} - \lambda_1 I_{r_1} & & 0 \\ & \ddots & \\ 0 & & [E_k]_{\mathcal{B}_k} - \lambda_k I_{r_k} \end{bmatrix}_{n \times n} P,$$

a nilpotent matrix (see Ex. 3).

If r is the least common multiple of $r_1, \ldots, r_k$, then $N^r = O$. Also, since both D and N are polynomials of A, so $DN = ND$ holds. We summarize as:

Let $A \in M(n; \mathbb{F})$ be a triangularizable matrix. Then there exist unique matrices D and N satisfying:

1. D is diagonalizable and N is nilpotent. Moreover, there exists an invertible matrix $P_{n \times n}$ so that

   PDP^{-1} is a diagonal matrix, and

   PNP^{-1} is a lower triangular matrix with zero diagonal entries.

2. $A = D + N$,

   $DN = ND$.

3. Both D and N can be expressed as polynomials in A. A and D have the same characteristic polynomial and hence, the set of eigenvalues.

If A is a real matrix having real eigenvalues $\lambda_1, \ldots, \lambda_k$, then P can be chosen to be a real matrix. Suppose

$$A = \begin{bmatrix} 3 & 2 & 2 \\ 1 & 2 & 2 \\ -1 & -1 & 0 \end{bmatrix}.$$

Show that the corresponding

$$D = \begin{bmatrix} 1 & 0 & -2 \\ 1 & 2 & 2 \\ 0 & 0 & 2 \end{bmatrix} \quad \text{and} \quad N = \begin{bmatrix} 2 & 2 & 4 \\ 0 & 0 & 0 \\ -1 & -1 & -2 \end{bmatrix}.$$

18. Suppose $A, B \in M(n; \mathbb{C})$ and B is nilpotent. In case $AB = BA$, show that

$$\det(A + B) = \det A.$$

In case $AB \neq BA$, give example to show that this identity may be not true.

19. (continued from Ex. 17) Suppose N is of index k. Then

$$A^m = (D + N)^m = \begin{cases} \sum_{i=0}^{m} C_{m-i}^m D^{m-i} N^i, & m < k \\ \sum_{i=0}^{k-1} C_{m-i}^m D^{m-i} N^i, & m \geq k \end{cases}$$

$$= \sum_{i=0}^{k-1} \frac{f^{(i)}(D)}{i!} N^i, \quad \text{if } f(z) = z^m \text{ and } m \geq k.$$

These suggest the following generalizations.

Suppose $A \in M(n; \mathbb{C})$ and $A = D + N$ as in Ex. 17, where $N^k = O$ but $N^{k-1} \neq O$.

(a) *For any polynomial $p(z)$,*

$$p(A) = p(D) + \sum_{i=1}^{k-1} \frac{p^{(i)}(D)}{i!} N^i,$$

where the former is diagonalizable and the latter is nilpotent, and both are commutative.

(b) *For any power series $f(z) = \sum_{n=0}^{\infty} a_n z^n$ with positive radius r of convergence, if the spectral radius (see Ex. <D$_1$> 6 of Sec. 3.7.6) $\rho(A) < r$, then*

$$f(A) = f(D) + \sum_{i=1}^{k-1} \frac{f^{(i)}(D)}{i!} N^i,$$

where the former is diagonalizable and the latter is nilpotent, and both are commutative.

For example, since $DN = ND$,

$$e^A = e^{D+N} = e^D e^N = e^D + e^D(e^N - I_n)$$

$$= e^D + e^D \sum_{i=1}^{k-1} \frac{1}{i!} N^i.$$

20. (continued from Ex. 19) Suppose $A \in M(n; \mathbb{C})$ is invertible and $\lambda_1, \ldots, \lambda_k$ are all distinct eigenvalues of A, with respective multiplicity $r_1, \ldots, r_k$ where $r_1 + \cdots + r_k = n$. To solve

$$e^X = A, \quad \text{where } X \in M(n; \mathbb{C})$$

is equivalent to solve

$$e^{D_1} = D,$$
$$e^{D_1}(e^{N_1} - I_n) = N \quad \text{or} \quad e^{N_1} = ND^{-1} + I_n,$$

where $A = D + N$ and $X = D_1 + N_1$ are decompositions as in Ex. 17. Choose invertible $P_{n \times n}$ so that

$$D = P^{-1} \begin{bmatrix} \lambda_1 I_{r_1} & & 0 \\ & \ddots & \\ 0 & & \lambda_k I_{r_k} \end{bmatrix} P.$$

Then

$$D_1 = P^{-1} \begin{bmatrix} (\log \lambda_1) I_{r_1} & & 0 \\ & \ddots & \\ 0 & & (\log \lambda_k) I_{r_k} \end{bmatrix} P.$$

Recall that each $\log \lambda_j$ is multiple-valued. On the other hand, $(ND^{-1})^k = O$ and $(ND^{-1})^{k-1} \neq O$ hold. Hence

$$N_1 = \log(ND^{-1} + I_n) = ND^{-1} - \frac{1}{2}(ND^{-1})^2 + \cdots + \frac{(-1)^{k-2}}{k}(ND^{-1})^{k-1}.$$

When D_1 and N_1 are given as above, the *matrix logarithm* of A is

$$\log A = D_1 + N_1,$$

which is multiple-valued. If $a \in \mathbb{C}$ and $a \neq 0$, the *matrix power* A^a is defined as

$$A^a = e^{a \log A}, \quad A \in \mathrm{GL}(n; \mathbb{C}).$$

In particular, if $a = \frac{1}{m}$ where $m \geq 1$ is a positive integer, then the mth root $A^{\frac{1}{m}}$ is defined as

$$A^{\frac{1}{m}} = e^{\frac{1}{m} \log A} = e^{\frac{1}{m}(D_1 + N_1)} = e^{\frac{1}{m} D_1} e^{\frac{1}{m} N_1}$$

$$= P^{-1} \begin{bmatrix} \lambda_1^{\frac{1}{m}} I_{r_1} & & 0 \\ & \ddots & \\ 0 & & \lambda_k^{\frac{1}{m}} I_{r_k} \end{bmatrix} P(I_n + ND^{-1})^{\frac{1}{m}}.$$

21. Try to use Exs. 16, 17 and 3 to prove that every triangularizable matrix $A \in M(n; \mathbb{F})$ has a *Jordan canonical basis* $\mathcal{B}$ so that $[A]_{\mathcal{B}}$ is the *Jordan canonical form* of A (refer to Sec. B.12).

<D> Application: *Differential equations*

For preliminary explanations concerned, please refer to Ex. <D4> of Sec. 3.7.6. Also, Exs. <C> 13 and 19 are helpful in the computation of e^{tA} in what follows.

We start from a concrete example. Solve

$$\frac{dx_1(t)}{dt} = 2x_1(t) - x_2(t) - x_3(t),$$

$$\frac{dx_2(t)}{dt} = 2x_1(t) - x_2(t) - 2x_3(t),$$

$$\frac{dx_3(t)}{dt} = -x_1(t) + x_2(t) + 2x_3(t).$$

Written in matrix form, this system is equivalent to

$$\frac{d\vec{x}}{dt} = \vec{x}(t)A, \quad \text{where } A = \begin{bmatrix} 2 & 2 & -1 \\ -1 & -1 & 1 \\ -1 & -2 & 2 \end{bmatrix} \quad \text{and}$$

$$\vec{x}(t) = (x_1(t), x_2(t), x_3(t)). \tag{*4}$$

A has characteristic polynomial $-(t-1)^3$. To compute the ranks:

$$A - I_3 = \begin{bmatrix} 1 & 2 & -1 \\ -1 & -2 & 1 \\ -1 & -2 & 1 \end{bmatrix} \Rightarrow r(A - I_3) = 1;$$

$$(A - I_3)^k = O_{3\times3} \quad \text{for } k \geq 2 \Rightarrow r((A - I_3)^k) = 0 \text{ for } k \geq 2.$$

Therefore,

$$A = P^{-1}JP, \quad \text{where } J = \begin{bmatrix} 1 & 0 & 0 \\ 1 & 1 & 0 \\ 0 & 0 & 1 \end{bmatrix} \quad \text{and } P = \begin{bmatrix} 1 & 2 & -1 \\ 1 & 0 & 0 \\ 1 & 1 & 0 \end{bmatrix}.$$

Compute e^{tJ} (see Ex. <C> 13):

$$e^{tJ} = \begin{bmatrix} e^t & 0 & 0 \\ te^t & e^t & 0 \\ 0 & 0 & e^t \end{bmatrix}.$$

Hence, the general solution is

$$\vec{x}(t) = \vec{\alpha}P^{-1}e^{tJ}P = \vec{c} \begin{bmatrix} e^t & 0 & 0 \\ te^t & e^t & 0 \\ 0 & 0 & e^t \end{bmatrix} \begin{bmatrix} 1 & 2 & -1 \\ 1 & 0 & 0 \\ 1 & 1 & 0 \end{bmatrix}$$

$$= ((c_1 + c_2 + c_3 + c_2t)e^t, \ (2c_1 + c_3 + 2c_2t)e^t, \ -(c_1 + c_2t)e^t).$$

If the initial condition $x_1(0) = 0$, $x_2(0) = 0$ and $x_3(0) = 1$ is imposed, then

$$c_1 + c_2 + c_3 = 0, \quad 2c_1 + c_3 = 0, \quad -c_1 = 1$$
$$\Rightarrow c_1 = -1, \quad c_3 = 2, \quad c_2 = -1.$$

The particular solution is

$$\vec{x}(t) = (-te^t, -2te^t, (1+t)e^t). \qquad \square$$

Consider the *non-homogeneous* equation

$$\frac{d\vec{x}}{dt} = \vec{x}A + \vec{f}(t), \quad \vec{x}(t_0) = \vec{c}_0, \qquad (*5)$$

where $A_{n \times n}$ is in $M(n;\ \mathbb{C})$. (*5) can be rewritten as

$$\frac{d\vec{x}}{dt} - \vec{x}A = \vec{f}(t)$$
$$\Rightarrow \left[\frac{d\vec{x}}{dt} - \vec{x}A\right]e^{-tA} = \vec{f}(t)e^{-tA}$$
$$\Rightarrow \frac{d}{dt}\left[\vec{x}(t)e^{-tA}\right] = \vec{f}(t)e^{-tA}$$
$$\Rightarrow \text{(integrate both sides entrywise form } t_0 \text{ to } t, \text{ refer to}$$
$$\text{Ex. } <D_1> \text{ of Sec. 3.7.6)}$$
$$\vec{x}(t)e^{-tA}|_{t=t_0}^t = \int_{t_0}^t \vec{f}(t)e^{-tA}\, dt$$
$$\Rightarrow \vec{x}(t) = \vec{c}_0 e^{(t-t_0)A} + \left(\int_{t_0}^t \vec{f}(t)e^{-tA}\, dt\right)e^{tA}. \qquad (*6)$$

This is the solution to (*5). Meanwhile, the solution to the *homogenous* equation (*5) with $\vec{f}(t) = \vec{0}$ is

$$\vec{x}(t) = \vec{c}_0 e^{(t-t_0)A}.$$

What are the general solution to (*5) without the initial condition?
Do the following problems.

1. For each of the following equations, do the following problems.
   (1) Solve the homogeneous equation $\frac{d\vec{x}}{dt} = \vec{x}A$, and then the solution to the initial value $\vec{x}(0) = \vec{x}_0$.
   (2) Solve the inhomogeneous equation $\frac{d\vec{x}}{dt} = \vec{x}A + \vec{f}(t)$, and then the solution to the initial value $\vec{x}(0) = \vec{x}_0$.

   (a) $\dfrac{d\vec{x}}{dt} = \vec{x}A + \vec{f}(t)$, where $A = \begin{bmatrix} 1 & 4 \\ 2 & 3 \end{bmatrix}$, $\vec{x}(0) = (1,2)$, $\vec{f}(t) = (1,-1)$.

(b) $\dfrac{d\vec{x}}{dt} = \vec{x}A + \vec{f}(t)$, where $A = \begin{bmatrix} 1 & -2 & -1 \\ -3 & -6 & -4 \\ 3 & 13 & 8 \end{bmatrix}$,

$$\vec{x}(0) = (1,0,0), \ \vec{f}(t) = (0,-1,1).$$

(c) $\dfrac{d\vec{x}}{dt} = \vec{x}A + \vec{f}(t)$, where $A = \begin{bmatrix} 3 & 4 & 0 & 0 \\ -4 & -5 & 0 & 0 \\ 0 & -2 & 3 & 2 \\ 2 & 4 & -2 & -1 \end{bmatrix}$,

$$\vec{x}(0) = (0,0,1,1), \ \vec{f}(t) = (1,1,1,1).$$

2. Consider the equation

$$\frac{d^2x}{dt^2} - 3\frac{dx}{dt} + 2x = e^{-3t}, \quad x(1) = 1, \ x'(1) = 0.$$

Rewrite it as

$$\frac{d\vec{x}}{dt} = \vec{x}A + \vec{f}(t), \quad A = \begin{bmatrix} 0 & -2 \\ 1 & 3 \end{bmatrix}, \ \vec{f}(t) = (0, e^{-3t}),$$

$$\vec{x}(t) = \left(x(t), \frac{dx}{dt} \right) \text{ and } \vec{x}(1) = (1,0).$$

Then (see Exs. <D$_4$> 5, 6 of Sec. 3.7.6 and (*6)),

$$e^{tA} = \begin{bmatrix} -e^{2t} + 2e^t & -2e^{2t} + 2e^t \\ e^{2t} - e^t & 2e^{2t} - e^t \end{bmatrix}$$

$$\Rightarrow \vec{f}(t)e^{-tA} = (e^{-5t} - e^{-4t}, 2e^{-5t} - e^{-4t})$$

$$\Rightarrow \int_1^t \vec{f}(t)e^{-tA}\,dt = \left(\int_1^t (e^{-5t} - e^{-4t})\,dt, \int_1^t (2e^{-5t} - e^{-4t})\,dt \right)$$

$$= \cdots$$

$$\Rightarrow \left(\int_1^t \vec{f}(t)e^{-tA}\,dt \right) e^{tA}$$

$$= \left(\frac{1}{20}e^{-3t} + \frac{1}{5}e^{(2t-5)} - \frac{1}{4}e^{t-4}, \ -\frac{2}{5}e^{-3t} + \frac{2}{5}e^{(2t-5)} - \frac{1}{4}e^{t-4} \right)$$

$$\Rightarrow \vec{x}(t) = \left(-e^{2(t-1)} + 2e^{t-1} + \frac{1}{20}e^{-3t} + \frac{1}{5}e^{(2t-5)} - \frac{1}{4}e^{t-4}, \ \cdots \right)$$

$$\Rightarrow x(t) = -e^{2(t-1)} + 2e^{t-1} + \frac{1}{20}e^{-3t} + \frac{1}{5}e^{(2t-5)} - \frac{1}{4}e^{t-4}.$$

Try to work out the details.

3. Model after Ex. 2 to solve each of the following equations.

(a) $\dfrac{d^2x}{dt^2} + 4x = \sin t$, $x(0) = 1$, $x'(0) = 0$.

(b) $\dfrac{d^3x}{dt^3} - \dfrac{d^2x}{dt^2} - \dfrac{dx}{dt} + x = 1$, $x(0) = 0$, $x'(0) = 1$, $x''(0) = -1$.

(c) $\dfrac{d^3x}{dt^3} - 6\dfrac{d^2x}{dt^2} - 7\dfrac{dx}{dt} - 6x = t$, $x(0) = 1$, $x'(0) = 0$, $x''(0) = 1$.

(d) $\dfrac{d^2x}{dt^2} = 2\dfrac{dx}{dt} + 5y + 3$,

$\dfrac{dy}{dt} = -\dfrac{dx}{dt} - 2y$, $x(0) = 0$, $x'(0) = 0$, $y(0) = 1$.

4. *Suppose each $a_{ij}(t)$ is a real or complex valued function and is continuous on $t \geq 0$, for $1 \leq i, j \leq n$. Let $A(t) = [a_{ij}(t)]_{n \times n}$. Then the differential equation*

$$\frac{d\vec{x}}{dt} = \vec{x}A(t), \quad \vec{x}(0) = \vec{a} \in \mathbb{R}^n \text{ (or } \mathbb{C}^n\text{)}$$

has a unique solution

$$\vec{x} = \vec{a}X(t), \quad t \geq 0,$$

where $X(t)_{n \times n}$ is the unique solution of the matrix differential equation

$$\frac{dX}{dt} = XA(t), \quad X(0) = I_n.$$

If the matrix equation has a solution $X(t)$, it must be of the form

$$X(t) = I_n + \int_0^t XA(s)\, ds$$

(see Ex.<D_1> of Sec. 3.7.6). This suggests the following proof, called the *method of successive approximation*. Define

$$X_0 = I_n$$

$$X_{m+1} = I_n + \int_0^t X_m A(s)\, ds, \quad m \geq 1.$$

Step 1 Fix $t_1 > 0$. Let $\alpha = \max_{0 \leq t \leq t_1} \|A(t)\|_1$ (for $\| \ \|_1$, see Ex. 14).
Then

$$\|X_{m+1} - X_m\|_1 \leq \cdots \leq \int_0^t \|X_m - X_{m-1}\|_1 \|A(s)\|_1 \, ds$$

$$\leq \alpha \int_0^t \|X_m - X_{m-1}\|_1 \, ds$$

$$\Rightarrow \|X_1 - X_0\|_1 \leq \alpha t,$$

$$\|X_2 - X_1\|_1 \leq \frac{\alpha^2}{2!} t^2,$$

$$\vdots$$

$$\|X_{m+1} - X_m\|_1 \leq \frac{(\alpha t)^{m+1}}{(m+1)!}, \quad m \geq 1$$

$$\Rightarrow \sum_{m=0}^{\infty} \|X_{m+1} - X_m\|_1 \leq \sum_{m=0}^{\infty} \frac{(\alpha t)^{m+1}}{(m+1)!} = e^{\alpha t} - 1 < \infty, \quad 0 \leq t \leq t_1$$

$$\Rightarrow \sum_{m=0}^{\infty} (X_{m+1} - X_m) = \lim_{m \to \infty} X_m = X \text{ exists on } [0, t_1] \text{ and hence,}$$
$$\text{on } t \geq 0.$$

Step 2 Suppose $Y_{n \times n}$ is another solution. Then

$$X - Y = \int_0^t (X - Y)A(s) \, ds.$$

Since both X and Y are differentiable, they are continuous on $[0, t_1]$.
Therefore $\alpha_1 = \max_{0 \leq t \leq t_1} \|X(t) - Y(t)\|_1 < \infty$. Now, for $0 \leq t \leq t_1$,

$$\|X(t) - Y(t)\|_1 \leq \int_0^t \|X(t) - Y(t)\|_1 \|A(s)\|_1 \, ds$$

$$\leq \alpha_1 \int_0^t \|A(s)\|_1 \, ds \leq \alpha_1 \alpha t.$$

$$\Rightarrow \|X(t) - Y(t)\|_1 \leq \int_0^t \alpha \alpha_1 s \|A(s)\|_1 \, ds \leq \alpha^2 \alpha_1 \int_0^t s \, ds = \frac{\alpha^2 \alpha_1}{2!} t^2$$

$$\vdots$$

$$\Rightarrow \|X(t) - Y(t)\|_1 \leq \alpha_1 \frac{(\alpha t)^m}{m!}, \quad 0 \leq t \leq t_1, \ m \geq 1$$

$$\Rightarrow \|X(t) - Y(t)\|_1 = 0, \quad 0 \leq t \leq t_1$$

$$\Rightarrow X(t) = Y(t) \quad \text{on} \quad 0 \leq t \leq t_1 \quad \text{and hence, on } t \geq 0.$$

Thus, such a solution X is *unique*.

Step 3 Let $\vec{x} = \vec{a}X(t)$. Then

$$\frac{d\vec{x}}{dt} = \frac{d}{dt}(\vec{a}X(t)) = \vec{a}\frac{d}{dt}(X(t)) = \vec{a}X(t)A(t) = \vec{x}A(t), \quad t \geq 0 \text{ and}$$
$$\vec{x}(0) = \vec{a}X(0) = \vec{a}I_n = \vec{a}.$$

Hence, this $\vec{x}$ is a solution. Just like Step 2, $\vec{x}$ can be shown to be unique.

5. (continued from Ex. 4) In case $A(t) = A$ is a *constant* matrix, then

$$x_0 = I_n,$$

$$x_0 = I_n + \int_0^t A\,ds = I_n + tA,$$

$$\vdots$$

$$x_m = I_n + tA + \frac{1}{2!}t^2A^2 + \cdots + \frac{1}{m!}t^mA^m, \quad t \geq 0.$$

These leads to the following important fact:

$$\frac{dX}{dt} = XA, \quad X(0) = I_n$$

has the unique solution

$$\sum_{m=0}^{\infty} \frac{1}{m!}t^mA^m \underset{(\text{def.})}{=} e^{tA}.$$

This definition coincides with that defined in Ex. <D$_2$> *of Sec. 3.7.6. Hence,*

$$\frac{d\vec{x}}{dt} = \vec{x}A, \quad \vec{x}(0) = \vec{a}$$

has the unique solution

$$\vec{x} = \vec{a}e^{tA}.$$

These results are useful in linear functional equation, Lie group and Lie algebra, quantum mechanics and probability, etc.

3.7.8 *Rational canonical form*

For real matrix $A_{3\times 3}$ having nonreal eigenvalues, this subsection is going to prove its counterpart of (2.7.75) and hence finishes the investigation of canonical forms listed in (3.7.29).

Let $A = [a_{ij}]_{3\times 3}$ be a nonzero real matrix.

A definitely has at least one real eigenvalue λ (refer to Sec. A.5). Hence, its characteristic polynomial is of the form

$$\det(A - tI_3) = -(t - \lambda)(t^2 + a_1 t + a_0),$$

where a_1 and a_0 are real constants. What we really care is the case that the quadratic polynomial $t^2 + a_1 t + a_0$ does not have real zeros, i.e. $a_1^2 - 4a_0 < 0$. But for completeness, we also discuss the cases $a_1^2 - 4a_0 \geq 0$.

As we have learned in Secs. 2.7.6–2.7.8, 3.7.6 and 3.7.7, a *canonical form* is a description of a matrix representation of a linear operator or square matrix, obtained by describing a certain kind of ordered basis for the space according to the features (e.g. eigenvalues, eigenvectors, etc.) of the given operator or matrix. In this section, we will establish the canonical forms for real 3×3 matrices *based on the irreducible monic factors of its characteristic polynomial instead of eigenvalues (see Sec. A.5).*

Four cases are considered as follows.

Case 1 The characteristic polynomial $\det(A - tI_3) = -(t - \lambda_1)(t - \lambda_2)(t - \lambda_3)$ where $\lambda_1 \neq \lambda_2 \neq \lambda_3$. The canonical form in this case is nothing new but the diagonal matrix studied in Sec. 3.7.6.

Case 2 $\det(A - tI_3) = -(t - \lambda_1)^2(t - \lambda_2)$ where $\lambda_1 \neq \lambda_2$ but $(A - \lambda_1 I_3) \cdot (A - \lambda_2 I_3) \neq O$. As in Secs. 3.7.6 and 3.7.7, let

$$G_{\lambda_1} = \{\vec{x} \in \mathbb{R}^3 \mid \vec{x}(A - \lambda_1 I_3)^2 = \vec{0}\} \quad \text{(the generalized eigenspace)},$$
$$E_{\lambda_2} = \{\vec{x} \in \mathbb{R}^3 \mid \vec{x}(A - \lambda_2 I_3) = \vec{0}\} \quad \text{(the eigenspace)}.$$

Case 1 in Sec. 3.7.7 showed that

1. $\dim G_{\lambda_1} = 2$, $\dim E_{\lambda_2} = 1$,
2. $G_{\lambda_1} \cap E_{\lambda_2} = \{\vec{0}\}$,
3. $\mathbb{R}^3 = G_{\lambda_1} \oplus E_{\lambda_2}$.

and a particular basis had been chosen for G_{λ_1} and hence a basis $\mathcal{B}$ for $\mathbb{R}^3$ so that $[A]_\mathcal{B}$ is the Jordan canonical form for A.

Here, we try to choose another basis for G_{λ_1} so that the induced basis $\mathcal{B}$ for $\mathbb{R}^3$ represents A in a canonical form. The central ideas behind this method are universally true for Cases 3 and 4 in the following.

Notice that

$$(A - \lambda_1 I_3)^2 = A^2 - 2\lambda_1 A + \lambda_1^2 I_3$$
$$\Rightarrow \vec{x}(A^2 - 2\lambda_1 A + \lambda_1^2 I_3) = \vec{0} \quad \text{for all } \vec{x} \in G_{\lambda_1}$$
$$\Rightarrow \vec{x}A^2 = -\lambda_1^2 \vec{x} + 2\lambda_1 \vec{x}A \quad \text{for all } \vec{x} \in G_{\lambda_1}.$$

This suggests that $\vec{x}A^2$ can be represented as a particular linear combination of $\vec{x}$ and $\vec{x}A$ for any $\vec{x} \in G_{\lambda_1}$. It is not necessary that $\vec{x}$ and $\vec{x}A$ should be linear independent for any $\vec{x} \in G_{\lambda_1}$, for example, that $\vec{x}$ is an eigenvector of A associated to λ_1 is the case. So, we try to find, if exists, a vector $\vec{x}$ so that $\{\vec{x}, \vec{x}A\}$ is linearly independent and hence forms a basis for G_{λ_1}. This is the basis we wanted.

Therefore, we pick up any vector $\vec{v}_1$ in G_{λ_1} but not in E_{λ_1}, then $\vec{v}_1 A \in G_{\lambda_1}$ but is linearly independent of $\vec{v}_1$. Thus

$$G_{\lambda_1} = \langle\langle \vec{v}_1, \vec{v}_1 A \rangle\rangle, \quad \text{and}$$

$$E_{\lambda_2} = \langle\langle \vec{v}_2 \rangle\rangle$$

$$\Rightarrow B = \{\vec{v}_1, \vec{v}_1 A, \vec{v}_2\} \quad \text{is a basis for } \mathbb{R}^3.$$

Since

$$\vec{v}_1 A = 0 \cdot \vec{v}_1 + \vec{v}_1 A,$$

$$(\vec{v}_1 A)A = \vec{v}_1 A^2 = -\lambda_1^2 \vec{v}_1 + 2\lambda_1 \vec{v}_1 A,$$

$$\vec{v}_2 A = \lambda_2 \vec{v}_2$$

$$\Rightarrow \begin{bmatrix} \vec{v}_1 \\ \vec{v}_1 A \\ \vec{v}_2 \end{bmatrix} A = \begin{bmatrix} 0 & 1 & 0 \\ -\lambda_1^2 & 2\lambda_1 & 0 \\ 0 & 0 & \lambda_2 \end{bmatrix} \begin{bmatrix} \vec{v}_1 \\ \vec{v}_1 A \\ \vec{v}_2 \end{bmatrix}$$

$$\Rightarrow [A]_B = PAP^{-1} = \begin{bmatrix} 0 & 1 & 0 \\ -\lambda_1^2 & 2\lambda_1 & 0 \\ 0 & 0 & \lambda_2 \end{bmatrix}, \quad \text{where } P = \begin{bmatrix} \vec{v}_1 \\ \vec{v}_1 A \\ \vec{v}_2 \end{bmatrix}.$$

This is the so-called *rational canonical form* of A.

Case 3 $\det(A - tI_3) = -(t - \lambda)^3$ but $A - \lambda I_3 \neq O$.

It might happen that $(A-\lambda I_3)^2 = O$. Then Case 2 in Sec. 3.7.7 indicated that

$$\dim E_\lambda = 2 \quad \text{and} \quad \dim G_\lambda = 3, \quad \text{where } G_\lambda = \text{Ker}(A - \lambda I_3)^2.$$

As in Case 2 above, take a vector $\vec{v}_1 \in G_\lambda = \mathbb{R}^3$ so that $\vec{v}_1 A$ is not in E_λ but is linearly independent of $\vec{v}_1$. Since $\dim E_\lambda = 2$, it is possible to choose another vector $\vec{v}_2 \in E_\lambda$ so that, as in Case 2 above,

$$\mathcal{B} = \{\vec{v}_1, \vec{v}_1 A, \vec{v}_2\}$$

is a basis for $\mathbb{R}^3$. In $\mathcal{B}$,

$$[A]_\mathcal{B} = PAP^{-1} = \begin{bmatrix} 0 & 1 & 0 \\ -\lambda^2 & 2\lambda & 0 \\ 0 & 0 & \lambda \end{bmatrix}, \quad \text{where } P = \begin{bmatrix} \vec{v}_1 \\ \vec{v}_1 A \\ \vec{v}_2 \end{bmatrix}.$$

In case $(A - \lambda I_3)^2 \neq O$ but $(A - \lambda I_3)^3 = O$. Case 3 in Sec. 3.7.7 showed that

$$\dim E_\lambda = 1 \quad \text{and} \quad \dim G_\lambda = 3, \quad \text{where } G_\lambda = \text{Ker}(A - \lambda I_3)^3 = \mathbb{R}^3.$$

Since

$$(A - \lambda I_3)^3 = O$$
$$\Rightarrow \vec{x}(A^3 - 3\lambda A^2 + 3\lambda^2 A - \lambda^3 I_3) = \vec{0} \quad \text{for all } \vec{x} \in G_\lambda$$
$$\Rightarrow \vec{x}A^3 = \lambda^3 \vec{x} - 3\lambda^2 \vec{x} A + 3\lambda \vec{x} A^2 \quad \text{for all } \vec{x} \in G_\lambda,$$

all we need to do is to choose a vector $\vec{v} \in G_\lambda$ so that $\{\vec{v}, \vec{v}A, \vec{v}A^2\}$ is linearly independent. Since $(A - \lambda I_3)^2 \neq O$, it is possible to choose a vector $\vec{v} \in G_\lambda = \mathbb{R}^3$ so that

$$\vec{v}(A - \lambda I_3)^2 \neq \vec{0} \quad \text{(which implicitly implies that } \vec{v}(A - \lambda I_3) \neq \vec{0})$$
$$\Rightarrow \vec{v}A^2 \neq -\lambda^2 \vec{v} + 2\lambda \vec{v} A \quad \text{and}$$
$$\vec{v}(A - \lambda I_3)^2 \quad \text{is an eigenvector of } A \text{ associated to } \lambda.$$

It follows that

$$\mathcal{B} = \{\vec{v}, \vec{v}A, \vec{v}A^2\}$$

is a basis for $\mathbb{R}^3$ (see Ex. <A> 1). In $\mathcal{B}$,

$$[A]_\mathcal{B} = PAP^{-1} = \begin{bmatrix} 0 & 1 & 0 \\ 0 & 0 & 1 \\ \lambda^3 & -3\lambda^2 & 3\lambda \end{bmatrix}, \quad \text{where } P = \begin{bmatrix} \vec{v} \\ \vec{v}A \\ \vec{v}A^2 \end{bmatrix}.$$

Case 4 $\det(A - tI_3) = -(t - \lambda)(t^2 + a_1 t + a_0)$ where $a_1^2 - 4a_0 < 0$. Still denote $E_\lambda = \mathrm{Ker}(A - \lambda I_3)$ and introduce

$$K_\lambda = \{\vec{x} \in \mathbb{R}^3 \mid \vec{x}(A^2 + a_1 A + a_0 I_3) = \vec{0}\}.$$

It is obvious that K_λ is an invariant subspace of $\mathbb{R}^3$, i.e. $\vec{x}A \in K_\lambda$ for each $\vec{x} \in K_\lambda$.

Since $a_1^2 - 4a_0 < 0$,   so $E_\lambda \cap K_\lambda = \{\vec{0}\}$.

We claim that $\dim K_\lambda = 2$. Take any nonzero vector $\vec{v} \in K_\lambda$, then $\{\vec{v}, \vec{v}A\}$ is linearly independent and hence, is a basis for K_λ. To see this, suppose on the contrary that $\vec{v}$ and $\vec{v}A$ are linearly dependent. Therefore, there exist scalars α and β, at least one of them is not equal to zero, so that

$$\alpha\vec{v} + \beta\vec{v}A = \vec{0}$$
$$\Rightarrow \alpha\vec{v}A + \beta\vec{v}A^2 = \vec{0}$$
$$\Rightarrow (\text{since } \vec{v}A^2 = -a_1\vec{v}A - a_0\vec{v}) - a_0\beta\vec{v} + (\alpha - a_1\beta)\vec{v}A = \vec{0}$$
$$\Rightarrow \alpha\colon (-a_0\beta) = \beta\colon (\alpha - a_1\beta) \quad \text{and hence } -a_0\beta^2 = \alpha^2 - a_1\alpha\beta$$
$$\Rightarrow \alpha^2 - a_1\alpha\beta + a_0\beta^2 = 0$$

which is impossible since $a_1^2 - 4a_0 < 0$.

Take any nonzero vector $\vec{u} \in E_\lambda$. Then

$$\mathcal{B} = \{\vec{v}, \vec{v}A, \vec{u}\}$$

is a basis for $\mathbb{R}^3$. In $\mathcal{B}$,

$$[A]_{\mathcal{B}} = PAP^{-1} = \begin{bmatrix} 0 & 1 & 0 \\ -a_0 & -a_1 & 0 \\ 0 & 0 & \lambda \end{bmatrix}, \quad \text{where } P = \begin{bmatrix} \vec{v} \\ \vec{v}A \\ \vec{u} \end{bmatrix}.$$

We summarize as (refer to (3.7.46) and compare with (3.7.50)

The rational canonical form of a nonzero real matrix $A_{3\times 3}$
Suppose the characteristic polynomial of A is

$$\det(A - tI_3) = -(t - \lambda)(t^2 + a_1 t + a_0),$$

where λ, a_1 and a_0 are real numbers.

(1) In case $a_1^2 - 4a_0 > 0$, then $\det(A - tI_3) = -(t - \lambda_1)(t - \lambda_2)(t - \lambda_3)$ where $\lambda = \lambda_1 \neq \lambda_2 \neq \lambda_3$ are real numbers. A is diagonalizable. See (1) in (3.7.46).

(2) In case $a_1^2 - 4a_0 = 0$, then $\det(A - tI_3) = -(t - \lambda_1)^2(t - \lambda_2)$ where λ_1 is a real number and $\lambda = \lambda_2$.

(a) $\lambda_1 \neq \lambda_2$ but $(A - \lambda_1 I_3)(A - \lambda_2 I_3) = O$. A is diagonalizable. See (2) in (3.7.46).

(b) $\lambda_1 \neq \lambda_2$ but $(A - \lambda_1 I_3)(A - \lambda_2 I_3) \neq O$. Let

$$G_{\lambda_1} = \text{Ker}(A - \lambda_1 I_3)^2 \quad \text{and} \quad E_{\lambda_2} = \text{Ker}(A - \lambda_2 I_3).$$

Then

1. $\dim G_{\lambda_1} = 2, \dim E_{\lambda_2} = 1$,
2. $G_{\lambda_1} \cap E_{\lambda_2} = \{\vec{0}\}$,
3. $\mathbb{R}^3 = G_{\lambda_1} \oplus E_{\lambda_2}$.

Take any vector $\vec{v}_1 \in G_{\lambda_1} \backslash E_{\lambda_1}$ where $E_{\lambda_1} = \text{Ker}(A - \lambda_1 I_3)$ and any nonzero vector $\vec{v}_2 \in E_{\lambda_2}$, then $\mathcal{B} = \{\vec{v}_1, \vec{v}_1 A, \vec{v}_2\}$ is a basis for $\mathbb{R}^3$ and

$$[A]_{\mathcal{B}} = PAP^{-1} = \begin{bmatrix} 0 & 1 & 0 \\ -\lambda_1^2 & 2\lambda_1 & 0 \\ 0 & 0 & \lambda_2 \end{bmatrix}, \quad \text{where } P = \begin{bmatrix} \vec{v}_1 \\ \vec{v}_1 A \\ \vec{v}_2 \end{bmatrix}.$$

(c) $\lambda_1 = \lambda_2 = \lambda$ but $A - \lambda I_3 = O$. $A = \lambda I_3$ is a scalar matrix. See (3) in (3.7.46).

(d) $\lambda_1 = \lambda_2 = \lambda$ but $(A - \lambda I_3) \neq O, (A - \lambda I_3)^2 = O$. Let

$$G_\lambda = \text{Ker}(A - \lambda I_3)^2 \quad \text{and} \quad E_\lambda = \text{Ker}(A - \lambda I_3).$$

Then $\dim E_\lambda = 2$ and $\dim G_\lambda = 3$. Take any vector $\vec{v}_1 \in G_\lambda \backslash E_\lambda$ so that $\vec{v}_1 A \notin E_\lambda$ and is linearly independent of $\vec{v}_1$. In the basis $\mathcal{B} = \{\vec{v}_1, \vec{v}_1 A, \vec{v}_2\}$ for $\mathbb{R}^3$, where $\vec{v}_2 \in E_\lambda$,

$$[A]_{\mathcal{B}} = PAP^{-1} = \begin{bmatrix} 0 & 1 & 0 \\ -\lambda^2 & 2\lambda & 0 \\ 0 & 0 & \lambda \end{bmatrix}, \quad \text{where } P = \begin{bmatrix} \vec{v}_1 \\ \vec{v}_1 A \\ \vec{v}_2 \end{bmatrix}.$$

(e) $\lambda_1 = \lambda_2 = \lambda$ but $(A - \lambda I_3)^2 \neq O, (A - \lambda I_3)^3 = O$. Let

$$G_\lambda = \text{Ker}(A - \lambda I_3)^3 \quad \text{and} \quad E_\lambda = \text{Ker}(A - \lambda I_3).$$

Then $\dim E_\lambda = 1$ and $\dim G_\lambda = 3$. Choose a vector $\vec{v} \in G_\lambda = \mathbb{R}^3$ so that $\vec{v}(A - \lambda I_3)^2 \neq \vec{0}$. Thus, in the basis $\mathcal{B} = \{\vec{v}, \vec{v} A, \vec{v} A^2\}$ for $\mathbb{R}^3$,

$$[A]_{\mathcal{B}} = PAP^{-1} = \begin{bmatrix} 0 & 1 & 0 \\ 0 & 0 & 1 \\ \lambda^3 & -3\lambda^2 & 3\lambda \end{bmatrix}, \quad \text{where } P = \begin{bmatrix} \vec{v} \\ \vec{v} A \\ \vec{v} A^2 \end{bmatrix}.$$

(3) In case $a_1^2 - 4a_0 < 0$. Let

$$K_\lambda = \text{Ker}(A^2 + a_1 A + a_0 I_3) \quad \text{and} \quad E_\lambda = \text{Ker}(A - \lambda I_3).$$

Thus

1. $\dim K_\lambda = 2, \dim E_\lambda = 1$,
2. $K_\lambda \cap E_\lambda = \{\vec{0}\}$,
3. $\mathbb{R}^3 = K_\lambda \oplus E_\lambda$.

Take any nonzero vector $\vec{v}_1 \in K_\lambda$ and any nonzero vector $\vec{v}_2 \in E_\lambda$. In the basis $\mathcal{B} = \{\vec{v}_1, \vec{v}_1 A, \vec{v}_2\}$,

$$[A]_{\mathcal{B}} = PAP^{-1} = \begin{bmatrix} 0 & 1 & 0 \\ -a_0 & -a_1 & 0 \\ 0 & 0 & \lambda \end{bmatrix}, \quad \text{where } P = \begin{bmatrix} \vec{v}_1 \\ \vec{v}_1 A \\ \vec{v}_2 \end{bmatrix}. \quad (3.7.53)$$

For geometric mapping properties for matrices in rational canonical forms, refer to Examples 4 and 5 in Sec. 3.7.2.

Example 1 Let

$$A = \begin{bmatrix} 0 & 0 & 1 \\ 1 & 0 & -1 \\ 0 & 1 & 1 \end{bmatrix}.$$

Find the rational canonical form of A. If A is considered as a complex matrix, what happens?

Solution The characteristic polynomial of A is

$$\det(A - tI_3) = t^2(1 - t) + 1 - t = -(t - 1)(t^2 + 1).$$

For $E = \text{Ker}(A - I_3)$: Solve

$$\vec{x}(A - \lambda I_3) = \vec{0}$$
$$\Rightarrow E = \langle\langle(1, 1, 1)\rangle\rangle.$$

For $K = \text{Ker}(A^2 + I_3)$: By computing

$$A^2 + I_3 = \begin{bmatrix} 0 & 1 & 1 \\ 0 & -1 & 0 \\ 1 & 1 & 0 \end{bmatrix} + \begin{bmatrix} 1 & 0 & 0 \\ 0 & 1 & 0 \\ 0 & 0 & 1 \end{bmatrix} = \begin{bmatrix} 1 & 1 & 1 \\ 0 & 0 & 0 \\ 1 & 1 & 1 \end{bmatrix},$$

then solve

$$\vec{x}(A^2 + I_3) = \vec{0}$$
$$\Rightarrow x_1 + x_3 = 0$$
$$\Rightarrow K = \{\vec{x} \in \mathbb{R}^3 \mid x_1 + x_3 = 0\} = \langle\langle(0,1,0),(1,0,-1)\rangle\rangle.$$

Choose $\vec{v}_1 = (0,1,0)$. Then $\vec{v}_1 A = (1,0,-1)$. Also, take $\vec{v}_2 = (1,1,1)$. Then $\mathcal{B} = \{\vec{v}_1, \vec{v}_1 A, \vec{v}_2\}$ is a basis for $\mathbb{R}^3$. Therefore,

$$\vec{v}_1 A = 0 \cdot \vec{v}_1 + 1 \cdot \vec{v}_1 A + 0 \cdot \vec{v}_2,$$
$$(\vec{v}_1 A)A = \vec{v}_1 A^2 = (0,-1,0) = -\vec{v}_1 + 0 \cdot \vec{v}_1 A + 0 \cdot \vec{v}_2,$$
$$\vec{v}_2 A = \vec{v}_2 = 0 \cdot \vec{v}_1 + 0 \cdot \vec{v}_1 A + 1 \cdot \vec{v}_2$$

$$\Rightarrow [A]_{\mathcal{B}} = PAP^{-1} = \begin{bmatrix} 0 & 1 & 0 \\ -1 & 0 & 0 \\ 0 & 0 & 1 \end{bmatrix}, \quad \text{where } P = \begin{bmatrix} 0 & 1 & 0 \\ 1 & 0 & -1 \\ 1 & 1 & 1 \end{bmatrix}.$$

This is the required canonical form.

On the other hand, if A is considered as a complex matrix,

$$\det(A - tI_3) = -(t-1)(t-i)(t+i)$$

and then A has three distinct complex eigenvalues $i, -i$ and 1.

For $\lambda_1 = i$: Solve

$$\vec{x}(A - iI_3) = \vec{0} \quad \text{where } \vec{x} = (x_1, x_2, x_3) \in \mathbb{C}^3.$$
$$\Rightarrow -ix_1 + x_2 = -ix_2 + x_3 = 0$$
$$\Rightarrow \vec{x} = (-x_3, -ix_3, x_3) = x_3(-1, -i, 1) \quad \text{for } x_3 \in \mathbb{C}$$
$$\Rightarrow E_i = \text{Ker}(A - iI_3) = \langle\langle(-1,-i,1)\rangle\rangle.$$

E_i is a one-dimensional subspace of $\mathbb{C}^3$. For $\lambda_1 = -i$: Solve

$$\vec{x}(A + iI_3) = \vec{0}$$
$$\Rightarrow ix_1 + x_2 = ix_2 + x_3 = 0$$
$$\Rightarrow \vec{x} = (-x_3, ix_3, x_3) = x_3(-1, i, 1) \quad \text{for } x_3 \in \mathbb{C}$$
$$\Rightarrow E_{-i} = \text{Ker}(A + iI_3) = \langle\langle(-1,i,1)\rangle\rangle$$

and $\dim E_{-i} = 1$. As before, we knew already that $E_1 = \text{Ker}(A - I_3) = \langle\langle(1,1,1)\rangle\rangle$. In the basis $\mathcal{C} = \{(-1,-i,1),(-1,i,1),(1,1,1)\}$ for $\mathbb{C}^3$,

$$[A]_{\mathcal{C}} = QAQ^{-1} = \begin{bmatrix} i & 0 & 0 \\ 0 & -i & 0 \\ 0 & 0 & 1 \end{bmatrix}, \quad \text{where } Q = \begin{bmatrix} -1 & -i & 1 \\ -1 & i & 1 \\ 1 & 1 & 1 \end{bmatrix}.$$

Thus, as a complex matrix, A is diagonalizable (refer to (3.7.46)). □

Example 2 Determine the rational canonical form of

$$A = \begin{bmatrix} 4 & 2 & 3 \\ 2 & 1 & 2 \\ -1 & -2 & 0 \end{bmatrix}.$$

What happens if A is considered as a complex matrix?

Solution The characteristic polynomial is

$$\det(A - tI_3) = -(t-1)^2(t-3).$$

Also

$$A - I_3 = \begin{bmatrix} 3 & 2 & 3 \\ 2 & 0 & 2 \\ -1 & -2 & -1 \end{bmatrix} \Rightarrow r(A - I_3) = 2$$

$$\Rightarrow \dim \mathrm{Ker}(A - I_3) = 1 < 2,$$

$$(A - I_3)^2 = \begin{bmatrix} 10 & 0 & 10 \\ 4 & 0 & 4 \\ -6 & 0 & -6 \end{bmatrix} \Rightarrow r(A - I_3)^2 = 1$$

$$\Rightarrow \dim \mathrm{Ker}(A - I_3)^2 = 2.$$

Hence A is not diagonalizable, even as a complex matrix.

Solve

$$\vec{x}(A - I_2)^2 = \vec{0}$$

$$\Rightarrow 5x_1 + 2x_2 - 3x_3 = 0$$

$$\Rightarrow \mathrm{Ker}(A - I_3)^2 = \langle\langle (3,0,5), (0,3,2) \rangle\rangle.$$

Take $\vec{v}_1 = (3,0,5)$ and compute $\vec{v}_1 A = (7,-4,9)$. Choose $\vec{v}_2 = (1,0,1) \in \mathrm{Ker}(A - 3I_3)$. Then

$$\vec{v}_1 A = 0 \cdot \vec{v}_1 + 1 \cdot \vec{v}_1 A + 0 \cdot \vec{v}_2$$

$$(\vec{v}_1 A)A = \vec{v}_1 A^2 = (11,-8,13) = 2(7,-4,9) - (3,0,5)$$

$$= -\vec{v}_1 + 2\vec{v}_1 A + 0 \cdot \vec{v}_2$$

$$\vec{v}_2 A = 3\vec{v}_2 = 0 \cdot \vec{v}_1 + 0 \cdot \vec{v}_1 A + 3 \cdot \vec{v}_2$$

$$\Rightarrow [A]_B = PAP^{-1} = \begin{bmatrix} 0 & 1 & 0 \\ -1 & 2 & 0 \\ 0 & 0 & 3 \end{bmatrix}, \quad \text{where } P = \begin{bmatrix} 3 & 0 & 5 \\ 7 & -4 & 9 \\ 1 & 0 & 1 \end{bmatrix}.$$

$\square$

Example 3 Determine the rational canonical forms of

$$A = \begin{bmatrix} 2 & 0 & 0 \\ 0 & 2 & 1 \\ 0 & 0 & 2 \end{bmatrix} \quad \text{and} \quad B = \begin{bmatrix} 2 & 1 & 0 \\ 0 & 2 & 1 \\ 0 & 0 & 2 \end{bmatrix}.$$

Solution Both A and B have the same characteristic polynomial

$$\det(A - tI_3) = \det(B - tI_3) = -(t - 2)^3.$$

For A:

$$A - 2I_3 = \begin{bmatrix} 0 & 0 & 0 \\ 0 & 0 & 1 \\ 0 & 0 & 0 \end{bmatrix} \Rightarrow r(A - 2I_3) = 1 \Rightarrow \dim \text{Ker}(A - 2I_3) = 2,$$

$$(A - 2I_3)^2 = O \Rightarrow \dim \text{Ker}(A - 2I_3)^2 = 3.$$

Solve

$$\vec{x}(A - 2I_3) = \vec{0}$$
$$\Rightarrow \text{Ker}(A - 2I_3) = \langle\langle (1,0,0), (0,0,1) \rangle\rangle.$$

Choose $\vec{v}_1 = (0,1,0)$ so that $\vec{v}_1 A = (0,2,1)$ is linearly independent of $\vec{v}_1$. Take $\vec{v}_2 = (1,0,0)$. In the basis $\mathcal{B} = \{\vec{v}_1, \vec{v}_1 A, \vec{v}_2\}$ for $\mathbb{R}^3$, since $(\vec{v}_1 A)A = \vec{v}_1 A^2 = (0,4,4) = -4(0,1,0) + 4(0,2,1) = -4\vec{v}_1 + 4\vec{v}_1 A$, so

$$[A]_\mathcal{B} = PAP^{-1} = \begin{bmatrix} 0 & 1 & 0 \\ -4 & 4 & 0 \\ 0 & 0 & 2 \end{bmatrix}, \quad \text{where } P = \begin{bmatrix} 0 & 1 & 0 \\ 0 & 2 & 1 \\ 1 & 0 & 0 \end{bmatrix}.$$

For B:

$$B - 2I_3 = \begin{bmatrix} 0 & 1 & 0 \\ 0 & 0 & 1 \\ 0 & 0 & 0 \end{bmatrix} \Rightarrow r(B - 2I_3) = 2 \Rightarrow \dim \text{Ker}(B - 2I_3) = 1,$$

$$(B - 2I_3)^2 = \begin{bmatrix} 0 & 0 & 1 \\ 0 & 0 & 0 \\ 0 & 0 & 0 \end{bmatrix} \Rightarrow r(B - 2I_3)^2 = 1,$$

$$(B - 2I_3)^3 = O_{3\times3} \Rightarrow \dim \text{Ker}(B - 2I_3)^3 = 3.$$

Take $\vec{v} = \vec{e}_1 = (1,0,0)$, then $\vec{v}(B - 2I_3)^2 = (0,0,1) \neq \vec{0}$ and so consider $\vec{v}B = (2,1,0)$ and $\vec{v}B^2 = (4,4,1)$. In the basis $\mathcal{C} = \{\vec{v}, \vec{v}B, \vec{v}B^2\}$ for $\mathbb{R}^3$,

$$\vec{v}B^3 = (8,12,6) = 6 \cdot (4,4,1) - 12 \cdot (2,1,0) + 8 \cdot (1,0,0)$$
$$= 8\vec{v} - 12\vec{v}B + 6\vec{v}B^2$$

$$\Rightarrow [B]_{\mathcal{C}} = QBQ^{-1} = \begin{bmatrix} 0 & 1 & 0 \\ 0 & 0 & 1 \\ 8 & -12 & 6 \end{bmatrix}, \quad \text{where } Q = \begin{bmatrix} 1 & 0 & 0 \\ 2 & 1 & 0 \\ 4 & 4 & 1 \end{bmatrix}.$$

$\square$

To the end, we use two concrete examples to illustrate the central ideas behind the determination of the rational canonical forms of matrices of order higher than 3. For details, read Sec. B.12.

Example 4 Analyze the rational canonical form

$$A = \begin{bmatrix} 0 & 1 & 0 & 0 & \vdots & & & \\ 0 & 0 & 1 & 0 & \vdots & & & \\ 0 & 0 & 0 & 1 & \vdots & & & \\ -1 & -2 & -3 & -2 & \vdots & & & \\ \cdots & \cdots & \cdots & \cdots & \vdots & \cdots & \cdots & \cdots \\ & & & & \vdots & 0 & 1 & \vdots \\ & & & & \vdots & -1 & -1 & \vdots \\ & & & & \vdots & \cdots & \cdots & \vdots \\ & & & & & & & 0 & 1 \\ & & & & & & & -1 & 1 \end{bmatrix}_{8 \times 8}$$

$$= \begin{bmatrix} R_1 & & 0 \\ & R_2 & \\ 0 & & R_3 \end{bmatrix}.$$

Analysis The characteristic polynomial is

$$\det(A - tI_8) = \det(R_1 - tI_4)\det(R_2 - tI_2)\det(R_3 - tI_2)$$
$$= (t^2 + t + 1)^2 \cdot (t^2 + t + 1) \cdot (t^2 - t + 1)$$
$$= (t^2 + t + 1)^3 (t^2 - t + 1).$$

For computations concerned with block matrices, refer to Ex. <C> of Sec. 2.7.5. Let $\mathcal{N} = \{\vec{e}_1, \vec{e}_2, \ldots, \vec{e}_8\}$ be the natural basis for $\mathbb{R}^8$.

For R_1 and R_2:

$$\vec{e}_1 A = \vec{e}_2,$$

$$\vec{e}_2 A = \vec{e}_3 = \vec{e}_1 A^2,$$

$$\vec{e}_3 A = \vec{e}_4 = \vec{e}_1 A^3,$$

$$\vec{e}_4 A = -\vec{e}_1 - 2\vec{e}_2 - 3\vec{e}_3 - 2\vec{e}_4$$

$$= -\vec{e}_1 - 2\vec{e}_1 A - 3\vec{e}_1 A^2 - 2\vec{e}_1 A^3 = \vec{e}_1 A^4$$

$$\Rightarrow \vec{e}_i(A^4 + 2A^3 + 3A^2 + 2A^2 + I_8) = \vec{e}_i(A^2 + A + I_8)^2 = \vec{0} \quad \text{for } 1 \leq i \leq 4;$$

$$\vec{e}_5 A = \vec{e}_6,$$

$$\vec{e}_6 A = -\vec{e}_5 A - \vec{e}_6 = -\vec{e}_5 - \vec{e}_5 A = \vec{e}_5(-I_8 - A) = \vec{e}_5 A^2$$

$$\Rightarrow \vec{e}_i(A^2 + A + I_8) = \vec{0} \quad \text{for } i = 5, 6.$$

Also, compute the ranks as follows:

$$A^2 = \begin{bmatrix} R_1^2 & & 0 \\ & R_2^2 & \\ 0 & & R_3^2 \end{bmatrix}$$

$$= \begin{bmatrix} 0 & 0 & 1 & 0 & \vdots & & & & \\ 0 & 0 & 0 & 1 & \vdots & & & & \\ -1 & -2 & -3 & -2 & \vdots & & & & \\ 2 & 3 & 4 & 1 & \vdots & & & & \\ \cdots & \cdots & \cdots & \cdots & \vdots & \cdots & \cdots & & \\ & & & & \vdots & -1 & -1 & \vdots & \\ & & & & \vdots & 1 & 0 & \vdots & \\ & & & & \vdots & \cdots & \cdots & \vdots & \cdots & \cdots \\ & & & & & & & \vdots & -1 & 1 \\ & & & & & & & \vdots & -1 & 0 \end{bmatrix}$$

$$\Rightarrow A^2 + A + I_8 = \begin{bmatrix} 1 & 1 & 1 & 0 & \vdots & & & \\ 0 & 1 & 1 & 1 & \vdots & & & \\ -1 & -2 & -2 & -1 & \vdots & & & \\ 1 & 1 & 1 & 0 & \vdots & & & \\ \cdots & \cdots & \cdots & \cdots & \vdots & \cdots & \cdots \\ & & & & \vdots & 0 & 0 & \vdots \\ & & & & \vdots & 0 & 0 & \vdots \\ & & & & \vdots & \cdots & \cdots & \vdots & \cdots & \cdots \\ & & & & & & & \vdots & 0 & 2 \\ & & & & & & & \vdots & -2 & 2 \end{bmatrix}$$

$$\Rightarrow r(A^2 + A + I_8) = 4;$$

$$\Rightarrow (A^2 + A + I_8)^2 = \begin{bmatrix} O_{4\times4} & & & & \\ & O_{2\times2} & & & \\ & & \vdots & \cdots & \cdots \\ & & \vdots & -4 & 4 \\ & & \vdots & -4 & 0 \end{bmatrix}_{8\times8}$$

$$\Rightarrow r(A^2 + A + I_8)^k = 2 \quad \text{for } k \geq 2.$$

These relations together provide the following information:

1. Let $p_1(t) = t^2 + t + 1$. The set

$$K_{p_1} = \{\vec{x} \in \mathbb{R}^8 \mid \vec{x}(A^2 + A + I_8)^3 = \vec{0}\} = \operatorname{Ker} p_1(A)^3$$
$$= \langle\langle \vec{e}_1, \ldots, \vec{e}_4, \vec{e}_5, \vec{e}_6 \rangle\rangle = \langle\langle \vec{e}_1, \vec{e}_2, \vec{e}_3, \vec{e}_4 \rangle\rangle \oplus \langle\langle \vec{e}_5, \vec{e}_6 \rangle\rangle$$

is an invariant subspace of $\mathbb{R}^8$ of dimension equal to

$3 \cdot 2 =$ (the algebraic multiplicity 3 of $p_1(t)$). (the degree of $p_1(t)$) $= 6$.

2. K_{p_1} contains an invariant subspace

$$\langle\langle \vec{e}_1, \vec{e}_2, \vec{e}_3, \vec{e}_4 \rangle\rangle = \langle\langle \vec{e}_1, \vec{e}_1 A, \vec{e}_1 A^2, \vec{e}_1 A^3 \rangle\rangle = C(\vec{e}_1)$$

which is an *A-cycle* of length $4 =$ degree of $p_1(t)^2$. Note that $p_1(A)^2$ is an *annihilator* of $C_A(\vec{e}_1)$, simply denoted by $C(\vec{e}_1)$, of least degree, i.e.

$$p_1(A)^2\big|_{C(\vec{e}_1)} = O_{4\times4}.$$

3. K_{p_1} contains another invariant subspace

$$\langle\langle \vec{e}_5, \vec{e}_6 \rangle\rangle = \langle\langle \vec{e}_5, \vec{e}_5 A \rangle\rangle = C(\vec{e}_5)$$

which is an *A-cycle* of length $2 = $ degree of $p_1(t)$. Note that $p_1(A)$ is an *annihilator* of $C_A(\vec{e}_5)$ of least degree, i.e.

$$p_1(A)|_{C(\vec{e}_5)} = O_{2\times 2}.$$

4. Solve

$$\vec{x}\, p_1(A) = \vec{0} \quad \text{for } \vec{x} \in \mathbb{R}^8$$
$$\Rightarrow \vec{x} = x_3(1,1,1,0,0,0,0,0) + x_4(-1,0,0,1,0,0,0,0)$$
$$+ (0,0,0,0,x_5,x_6,0,0)$$
$$\Rightarrow \text{Ker}\, p_1(A) = \langle\langle \vec{v}_1, \vec{v}_2, \vec{e}_5, \vec{e}_6 \rangle\rangle \quad \text{where}$$
$$\vec{v}_1 = (1,1,1,0,0,0,0,0) \text{ and } \vec{v}_2 = (-1,0,0,1,0,0,0,0).$$

5. Solve

$$\vec{x}\, p_1(A)^2 = \vec{0}$$
$$\Rightarrow \text{Ker}\, p_1(A)^2 = \langle\langle \vec{e}_1, \vec{e}_2, \vec{e}_3, \vec{e}_4, \vec{e}_5, \vec{e}_6 \rangle\rangle.$$

Hence

$$\text{Ker}\, p_1(A)^2 = \langle\langle \vec{e}_1, \vec{e}_2, \vec{e}_3, \vec{e}_4 \rangle\rangle \oplus \langle\langle \vec{e}_5, \vec{e}_6 \rangle\rangle$$
$$= \langle\langle \vec{v}_1, \vec{v}_2, \vec{e}_3, \vec{e}_4 \rangle\rangle \oplus \langle\langle \vec{e}_5, \vec{e}_6 \rangle\rangle.$$

6. Take any vector $\vec{v} \in \text{Ker}\, p_1(A)^2$ but not in $\text{Ker}\, p_1(A)$, say $\vec{v} = \vec{e}_3 + \vec{e}_4$. Then

$$\vec{v} A = (-1,-2,-3,-1,0,\dots,0),$$
$$\vec{v} A^2 = (1,1,1,-1,0,\dots,0),$$
$$\vec{v} A^3 = (1,3,4,3,0,\dots,0),$$
$$\vec{v} A^4 = (-3,-5,-6,-2,0,\dots,0) = -\vec{v} - 2\vec{v}A - 3\vec{v}A^2 - 2\vec{v}A^3.$$

Also, $\mathcal{B}_{\vec{v}} = \{\vec{v}, \vec{v}A, \vec{v}A^2, \vec{v}A^3\}$ is a basis for $\text{Ker}\, p_1(A)^2$. In $\mathcal{B}_{\vec{v}}$,

$$[A|_{\text{Ker}\, p_1(A)^2}]_{\mathcal{B}_{\vec{v}}} = R_1.$$

The matrix R_1 is called the *companion matrix* of $p_1(t)^2 = t^4 + 2t^3 + 3t^2 + 2t + 1$. Take any vector $\vec{u} \in \text{Ker}\, p_1(A)$, say $\vec{u} = \vec{v}_1$. Then

$$\vec{u} A = (0,1,1,1,0,0,0,0),$$
$$\vec{u} A^2 = (-1,-2,-2,-1,0,0,0,0) = -\vec{u} - \vec{u}A$$

and $\mathcal{B}_{\vec{u}} = \{\vec{u}, \vec{u}A\}$ is a basis for $\operatorname{Ker} p_1(A)$. In $\mathcal{B}_{\vec{u}}$,

$$[A|_{\operatorname{Ker} p_1(A)}]_{\mathcal{B}_{\vec{u}}} = R_2.$$

The matrix R_2 is called the *companion matrix* of $p_1(t) = t^2 + t + 1$.

7. It can be show that $\mathcal{B}_{\vec{v}} \cup \mathcal{B}_{\vec{u}}$ is linearly independent. Hence $\mathcal{B}_{\vec{v}} \cup \mathcal{B}_{\vec{u}}$ is a basis for $\operatorname{Ker} p_1(A)^3 = \operatorname{Ker} p_1(A)^2$. Notice that

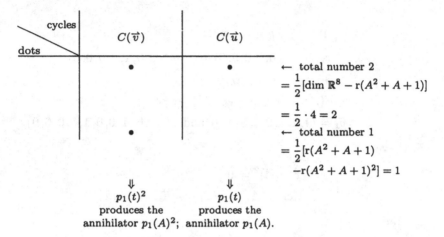

Where the 2 in $\frac{1}{2}$ is the degree of $p_1(t)$. Also,

$$\begin{aligned}
\dim \operatorname{Ker} p_1(A)^3 &= \deg p_1(t)^2 + \deg p_1(t) \\
&= 2 \cdot 2 + 2 \cdot 1 \\
&= 2 \cdot 3 \\
&= \text{(the degree of } p(t)) \cdot \text{(the number of dots)}
\end{aligned}$$

Combining together, we get

$$[A\,|_{\operatorname{Ker} p_1(A)^3}]_{\mathcal{B}_{\vec{v}} \cup \mathcal{B}_{\vec{u}}} = \begin{bmatrix} R_1 & 0 \\ 0 & R_2 \end{bmatrix}_{6 \times 6}. \qquad (*1)$$

For R_3:

$$\vec{e}_7 A = \vec{e}_8,$$
$$\vec{e}_8 A = (0, \ldots, 0, -1, 1) = -\vec{e}_7 + \vec{e}_8 = -\vec{e}_7 + \vec{e}_7 A = \vec{e}_7 A^2$$
$$\Rightarrow \vec{e}_i(A^2 - A + I_8) = \vec{0} \quad \text{for } i = 7, 8.$$

Also,

$$A^2 - A + I_8 = \begin{bmatrix} 0 & -1 & 1 & 0 & \vdots & & & \\ 0 & 0 & -1 & 0 & \vdots & & & \\ -1 & -2 & -3 & -3 & \vdots & & & \\ 3 & 5 & 7 & 3 & \vdots & & & \\ \cdots & \cdots & \cdots & \cdots & \vdots & \cdots & \cdots & \cdots \\ & & & & \vdots & -1 & 2 & \vdots \\ & & & & \vdots & 2 & 1 & \vdots \\ & & & & \cdots & \cdots & \vdots & \cdots \cdots \\ & & & & & & \vdots & 0 & 0 \\ & & & & & & \vdots & 0 & 0 \end{bmatrix}$$

$$\Rightarrow r(A^2 - A + I_8) = 6$$
$$\Rightarrow r(A^2 - A + I_8)^k = 2 \quad \text{for } k \geq 2.$$

These facts provide the following information:

1. Set $p_2(t) = t^2 - t + 1$. Then

$$\operatorname{Ker} p_2(A) = \{\vec{x} \in \mathbb{R}^8 \mid \vec{x} p_2(A) = \vec{0}\}$$

   is an invariant subspace of dimension 2, the degree of $p_2(t)$.

2. $\operatorname{Ker} p_2(A) = \langle\langle \vec{e}_7, \vec{e}_8 \rangle\rangle = \langle\langle \vec{e}_7, \vec{e}_7 A \rangle\rangle = C(\vec{e}_7)$ is an A-cycle of length 2 which is annihilated by $p_2(t)$, i.e.

$$p_2(A)|_{C(\vec{e}_7)} = O_{2\times2}.$$

3. Solve

$$\vec{x} p_2(A) = \vec{0} \quad \text{for } \vec{x} \in \mathbb{R}^8$$
$$\Rightarrow \operatorname{Ker} p_2(A) = \langle\langle \vec{e}_7, \vec{e}_8 \rangle\rangle \quad \text{as it should be by 2.}$$

4. Take any nonzero vector $\vec{w} \in \mathrm{Ker}\, p_2(A)$, say $\vec{w} = \alpha \vec{e_7} + \beta \vec{e_8}$. Then

$$\vec{w}A = \alpha \vec{e_7}A + \beta \vec{e_8}A = \alpha \vec{e_7}A + \beta \vec{e_7}A^2 = -\beta \vec{e_7} + (\alpha + \beta)\vec{e_8}$$
$$\Rightarrow \vec{w}A^2 = -\beta \vec{e_7}A + (\alpha + \beta)\vec{e_8}A = -\beta \vec{e_7}A + (\alpha + \beta)(-\vec{e_7} + \vec{e_7}A)$$
$$= -(\alpha + \beta)\vec{e_7} + \alpha \vec{e_7}A$$
$$= -\vec{w} + \vec{w}A.$$

Then $\mathcal{B}_{\vec{w}} = \{\vec{w}, \vec{w}A\}$ is a basis for $\mathrm{Ker}\, p_2(A)$. The matrix

$$\left[A \,\big|_{\mathrm{Ker}\, p_2(A)}\right]_{\mathcal{B}_{\vec{w}}} = R_3 \qquad (*2)$$

is called the *companion matrix* of $p_2(t) = t^2 - t + 1$.

5. Notice that

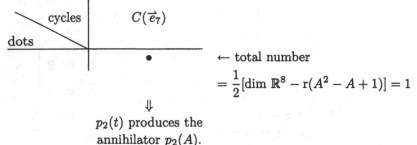

$$\leftarrow \text{total number}$$
$$= \frac{1}{2}[\dim \mathbb{R}^8 - r(A^2 - A + 1)] = 1$$

$\Downarrow$

$p_2(t)$ produces the
annihilator $p_2(A)$.

Putting (*1) and (*2) together, let

$$\mathcal{B} = \mathcal{B}_{\vec{v}} \cup \mathcal{B}_{\vec{u}} \cup \mathcal{B}_{\vec{w}} = \{\vec{v}, \vec{v}A, \vec{v}A^2, \vec{v}A^3, \vec{u}, \vec{u}A, \vec{w}, \vec{w}A\}.$$

$\mathcal{B}$ is basis for $\mathbb{R}^8$ and is called *a rational canonical basis* of A. In $\mathcal{B}$,

$$[A]_{\mathcal{B}} = PAP^{-1} = \begin{bmatrix} R_1 & & 0 \\ & R_2 & \\ 0 & & R_3 \end{bmatrix}, \quad \text{where } P = \begin{bmatrix} \vec{v} \\ \vec{v}A \\ \vdots \\ \vec{w}A \end{bmatrix}_{8\times 8}$$

is called *the rational canonical form* of A. $\qquad \square$

Example 5 Find a rational canonical basis and the rational canonical form
for the matrix

$$A = \begin{bmatrix} 0 & 2 & 0 & -6 & 2 \\ 1 & -2 & 0 & 0 & 2 \\ 1 & 0 & 1 & -3 & 2 \\ 1 & -2 & 1 & -1 & 2 \\ 1 & -4 & 3 & -3 & 4 \end{bmatrix}.$$

What happens if A is considered as a complex matrix?

Solution The characteristic polynomial of A is

$$\det(A - tI_5) = -(t^2 + 2)^2(t - 2).$$

Then $p_1(t) = t^2 + 2$ has algebraic multiplicity 2 and $p_2(t) = t - 2$ has algebraic multiplicity 1. Therefore

$$\dim \operatorname{Ker} p_1(A)^2 = 4 \quad \text{and} \quad \dim \operatorname{Ker} p_2(A) = 2.$$

To compute the ranks:

$$A^2 = \begin{bmatrix} -2 & 0 & 0 & 0 & 0 \\ 0 & -2 & 6 & -12 & 6 \\ 0 & 0 & 4 & -12 & 6 \\ 0 & 0 & 6 & -14 & 6 \\ 0 & 0 & 12 & -24 & 10 \end{bmatrix}$$

$$\Rightarrow A^2 + 2I_5 = \begin{bmatrix} 0 & 0 & 0 & 0 & 0 \\ 0 & 0 & 6 & -12 & 6 \\ 0 & 0 & 6 & -12 & 6 \\ 0 & 0 & 6 & -12 & 6 \\ 0 & 0 & 12 & -24 & 12 \end{bmatrix} \Rightarrow r(A^2 + 2I_5) = 1$$

$$\Rightarrow r(A^2 + 2I_5)^2 = 1.$$

Since $\frac{1}{2}(\dim \mathbb{R}^5 - r(A^2 + 2I_5)) = \frac{1}{2}(5 - 1) = 2$, so there exists a rational canonical basis $\mathcal{B} = \{\vec{v}_1, \vec{v}_1 A, \vec{v}_2, \vec{v}_2 A, \vec{v}_3\}$ of A so that

$$[A]_\mathcal{B} = QAQ^{-1} = \begin{bmatrix} 0 & 1 & \vdots & & \\ -2 & 0 & \vdots & & \\ \cdots & \cdots & \vdots & \cdots & \cdots \\ & & \vdots & 0 & 1 & \vdots \\ & & \vdots & -2 & 0 & \vdots \\ & & \vdots & \cdots & \cdots & \vdots & \cdots \\ & & & & & \vdots & 2 \end{bmatrix}, \quad \text{where } Q = \begin{bmatrix} \vec{v}_1 \\ \vec{v}_1 A \\ \vec{v}_2 \\ \vec{v}_2 A \\ \vec{v}_3 \end{bmatrix}. \quad (*3)$$

To find such a rational canonical basis $\mathcal{B}$: Solve

$$\vec{x}(A^2 + 2I_5) = \vec{0}$$

$$\Rightarrow x_2 + x_3 + x_4 + 2x_5 = 0$$

$$\Rightarrow \vec{x} = (x_1, -x_3 - x_4 - 2x_5, x_3, x_4, x_5)$$

$$= x_1(1, 0, 0, 0, 0) + x_3(0, -1, 1, 0, 0) + x_4(0, -1, 0, 1, 0)$$

$$+ x_5(0, -2, 0, 0, 1) \quad \text{where } x_1, x_3, x_4, x_5 \in \mathbb{R}.$$

Take $\vec{v}_1 = \vec{e}_1$, then

$$\vec{v}_1 A = (0, 2, 0, -6, 2)$$
$$\Rightarrow \vec{v}_1 A^2 = (-2, 0, 0, 0, 0) = -2\vec{e}_1 = -2\vec{v}_1.$$

Take $\vec{v}_2 = (0, -1, 1, 0, 0)$ which is linearly independent of $\vec{v}_1$ and $\vec{v}_1 A$. Then

$$\vec{v}_2 A = (0, 2, 1, -3, 0)$$
$$\Rightarrow \vec{v}_2 A^2 = (0, 2, -2, 0, 0) = -2\vec{v}_2.$$

Solve

$$\vec{x}(A - 2I_5) = \vec{0}$$
$$\Rightarrow \vec{x} = (0, 0, x_3, -2x_3, x_3) = x_3(0, 0, 1, -2, 1) \quad \text{for } x_3 \in \mathbb{R}.$$

Take the eigenvector $\vec{v}_3 = (0, 0, 1, -2, 1)$. Then $\mathcal{B} = \{\vec{v}_1, \vec{v}_1 A, \vec{v}_2, \vec{v}_2 A, \vec{v}_3\}$ is a rational canonical basis of A and a required Q is

$$Q = \begin{bmatrix} 1 & 0 & 0 & 0 & 0 \\ 0 & 2 & 0 & -6 & 2 \\ 0 & -1 & 1 & 0 & 0 \\ 0 & 2 & 1 & -3 & 0 \\ 0 & 0 & 1 & -2 & 1 \end{bmatrix}.$$

The transpose A^* is the matrix B mentioned in Ex. 15 of Sec. B.12. PA^*P^{-1} is the rational canonical form indicated in (*3). Hence, A and A^* have the same rational canonical form and thus, they are similar. In fact

$$A^* = RAR^{-1}, \quad \text{where } R = P^{-1}Q.$$

Suppose A is considered as a complex matrix. Then

$$\det(A - tI_5) = -(t - \sqrt{2}i)^2(t + \sqrt{2}i)^2(t - 2)$$

and A has the rational canonical form

$$\begin{bmatrix} 0 & 1 & \vdots & & & \\ 2 & 2\sqrt{2}i & \vdots & & & \\ \cdots & \cdots & \vdots & \cdots & \cdots & \cdots \\ & & \vdots & 0 & 1 & \vdots \\ & & \vdots & 2 & -2\sqrt{2}i & \vdots \\ & & \vdots & \cdots & \cdots & \vdots & \cdots \\ & & & & & \vdots & 2 \end{bmatrix}_{5 \times 5}$$

Readers are urged to find an invertible complex matrix $S_{5 \times 5}$ so that SAS^{-1} is the above-mentioned rational canonical form.

Exercises

<A>

1. In Case 3, try to show that $\mathcal{B} = \{\vec{v}, \vec{v}A, \vec{v}A^2\}$ is a basis for $\mathbb{R}^3$, where $\vec{v} \in G_\lambda$ but $\vec{v}(A - \lambda I_3)^2 \neq \vec{0}$.

2. For each of the following matrices A, do the following problems.

   (1) Find a rational canonical basis $\mathcal{B}$ and a matrix P so that PAP^{-1} is the rational canonical form of A.

   (2) Find a matrix Q so that QA^*Q^{-1} is the rational canonical form of A. Try to use (3.7.32) and Ex. <A> 5 there to derive the results from (1).

   (3) Show that A and A^* are similar and find an invertible R so that $A^* = RAR^{-1}$.

   (4) If A is considered as a complex matrix, find the rational canonical form of A. Is A diagonalizable? If it is, find a matrix S so that SAS^{-1} is diagonal.

$$\text{(a)} \begin{bmatrix} 0 & 3 & 1 \\ -1 & 3 & 1 \\ 0 & 1 & 1 \end{bmatrix}. \quad \text{(b)} \begin{bmatrix} 10 & 11 & 3 \\ -3 & -4 & -3 \\ -8 & -8 & -1 \end{bmatrix}. \quad \text{(c)} \begin{bmatrix} 2 & 0 & 0 \\ 3 & 2 & 0 \\ 0 & 0 & 5 \end{bmatrix}.$$

$$\text{(d)} \begin{bmatrix} 1 & 1 & 1 \\ 0 & 1 & 0 \\ 0 & 0 & 1 \end{bmatrix}. \quad \text{(e)} \begin{bmatrix} 7 & 3 & 3 \\ 0 & 1 & 0 \\ -3 & -3 & 1 \end{bmatrix}. \quad \text{(f)} \begin{bmatrix} 2 & 0 & 3 \\ 0 & 1 & 0 \\ 0 & 1 & 2 \end{bmatrix}.$$

$$\text{(g)} \begin{bmatrix} 3 & 0 & 0 \\ 0 & 3 & 1 \\ 0 & 0 & 3 \end{bmatrix}. \quad \text{(h)} \begin{bmatrix} 3 & 0 & 0 \\ 1 & 3 & 0 \\ 0 & 1 & 3 \end{bmatrix}.$$

<B>

1. For each of the following matrices, do the same problems as listed in Ex. <A> 2.

$$\text{(a)} \begin{bmatrix} 2 & 3 & 3 & 5 \\ 3 & 2 & 2 & 3 \\ 0 & 0 & 1 & 1 \\ 0 & 0 & 0 & 1 \end{bmatrix}. \quad \text{(b)} \begin{bmatrix} 2 & 0 & 0 & 0 \\ 0 & 2 & 0 & 0 \\ 0 & 0 & 3 & 1 \\ 0 & 0 & 0 & 3 \end{bmatrix}. \quad \text{(c)} \begin{bmatrix} 2 & 1 & 0 & 0 \\ 0 & 2 & 0 & 0 \\ 0 & 0 & 3 & 0 \\ 0 & 0 & 0 & 3 \end{bmatrix}.$$

2. In Example 5, find invertible complex matrix S so that SAS^{-1} is the rational canonical form over the complex field.

3. Let V be the vector space of all 2×2 real lower triangular matrices. Define $f\colon V \to V$ by

$$f\left(\begin{bmatrix} a_{11} & 0 \\ a_{21} & a_{22} \end{bmatrix}\right) = \begin{bmatrix} 7a_{11} + 3a_{21} + 3a_{22} & 0 \\ a_{21} & -3a_{11} - 3a_{21} + a_{22} \end{bmatrix}.$$

Show that f is linear and find its rational canonical form and a rational canonical basis.

<C> Abstraction and generalization

Read Sec. B.12 and try to do exercises there.

3.8   Affine Transformations

At this moment, one should turn to Sec. 2.8 for general description about affine space (2.8.1), affine subspace (2.8.2) and affine transformation or mapping (2.8.3).

$\mathbb{R}^3$ will play as a three-dimensional real vector space and, at the same time, as an affine space.

Results such as (2.8.4) and (2.8.8), including the procedures to obtain them, can be easily extended to $\mathbb{R}^3$.

For example, an affine transformation $T\colon \mathbb{R}^3 \to \mathbb{R}^3$ is *geometrically* characterized as the one-to-one mapping from $\mathbb{R}^3$ onto itself that *preserves ratios of signed lengths of line segments along the same or parallel lines in* $\mathbb{R}^3$.

While, the *affine group on the space* $\mathbb{R}^3$ is

$$G_a(3\colon \mathbb{R}) = \{\vec{x}_0 + f(\vec{x}) \mid \vec{x}_0 \in \mathbb{R}^3 \text{ and}$$
$$f\colon \mathbb{R}^3 \to \mathbb{R}^3 \text{ is an invertible linear operator}\} \qquad (3.8.1)$$

with the *identity element* $I\colon \mathbb{R}^3 \to \mathbb{R}^3$ defined as the identity operator $1_{\mathbb{R}^3}\colon \mathbb{R}^3 \to \mathbb{R}^3$, i.e.

$$I(\vec{x}) = 1_{\mathbb{R}^3}(\vec{x}) = \vec{x} \quad \text{for all } \vec{x} \in \mathbb{R}^3,$$

and the *inverse*
$$T^{-1}(\vec{x}) = -f^{-1}(\vec{x}_0) + f^{-1}(\vec{x})$$
of $T(\vec{x}) = \vec{x}_0 + f(\vec{x})$.

Elements in $G_a(3; \mathbb{R})$ are also called *affine motion* if their geometric aspects are going to be emphasized.

This section is divided into five subsections, each of them as a counterpart of the same numbered subsection of Sec. 2.8.

Section 3.8.1: Use examples to illustrate matrix representations of an affine transformation with respect to various affine bases for $\mathbb{R}^3$ and the relations among them.

Section 3.8.2: As an extension of Sec. 3.7.2 to affine transformations, here we introduce the following basic ones along with their geometric and algebraic characterizations:

Translations; Reflections;
(one-way, two-way) Stretches and Enlargement (Similarity);
Shearings; Rotations; Orthogonal reflections.

Section 3.8.3: Study the affine invariants under the affine group $G_a(3; \mathbb{R})$ and prove (3.7.15) and the geometric interpretation of the determinant of a linear operator on $\mathbb{R}^3$.

Section 3.8.4: Here we treat the affine geometry for $\mathbb{R}^3$ as a model for $\mathbb{R}^n$ or finite-dimensional affine space over a field or ordered field:

1. Affine independence and dependence; affine basis; affine coordinates and barycentric coordinates; affine subspaces and their operations including intersection theorem; Menelaus and Ceva theorems.
2. Half space, convex set, simplex and polyhedron.
3. Affine transformations.
4. Connections of $\mathbb{R}^3$ with the three-dimensional projective space $P^3(\mathbb{R})$.

Section 3.8.5: Introduce the classifications of quadrics in the affine (Euclidean) space $\mathbb{R}^3$ and in the projective space $P^3(\mathbb{R})$. *The algebraic characterization of each quadric and the study of such subjects as diametral plan, center, tangent plane, pole and polar, etc. will be postponed to Sec. 5.10.*

3.8.1 *Matrix representations*

Just like Sec. 3.7.3 to Sec. 2.7.2, the processes adopted, the definitions concerned and the results obtained in Sec. 2.8.1 can be easily generalized, almost verbatim, to affine transformations on $\mathbb{R}^3$.

For examples,

$\mathbb{R}^2$	$\mathbb{R}^3$
affine basis $\mathcal{B} = \{\vec{a}_0, \vec{a}_1, \vec{a}_2\}$ with base point $\vec{a}_0$.	affine basis $\mathcal{B} = \{\vec{a}_0, \vec{a}_1, \vec{a}_2, \vec{a}_3\}$ with base point $\vec{a}_0$ (see (3.6.1)).
$T \colon \mathbb{R}^2 \to \mathbb{R}^2$ is an affine transformation.	$T(\vec{x}) = \vec{x}_0 + f(\vec{x}) \colon \mathbb{R}^3 \to \mathbb{R}^3$ is an affine transformation.
(2.8.9)	$[T(\vec{x})]_\mathcal{C} = [T(\vec{a}_0)]_\mathcal{C} + [\vec{x}]_\mathcal{B}[f]_\mathcal{C}^\mathcal{B}$ is the *matrix representation* of T with aspect to $\mathcal{B}$ and $\mathcal{C}$. See *Explanation* below. (3.8.2)
(2.8.10)	$([T(\vec{x})]_\mathcal{C} 1) = ([\vec{x}]_\mathcal{B} 1)\begin{bmatrix} [f]_\mathcal{C}^\mathcal{B} & 0 \\ [T(\vec{a}_0)]_\mathcal{C} & 1 \end{bmatrix}.$ (3.8.3)
$G_a(2;\mathbb{R})$ in (2.8.11)	$G_a(3;\mathbb{R}) = \left\{ \begin{bmatrix} A & 0 \\ \vec{x}_0 & 1 \end{bmatrix} \middle| A \in GL(3;\mathbb{R}) \text{ and } \vec{x}_0 \in \mathbb{R}^3 \right\}$ with identity: $\begin{bmatrix} I_3 & 0 \\ 0 & 1 \end{bmatrix}$, and inverse: $\begin{bmatrix} A & 0 \\ \vec{x}_0 & 1 \end{bmatrix} = \begin{bmatrix} A^{-1} & 0 \\ -\vec{x}_0 A^{-1} & 1 \end{bmatrix}.$ (3.8.4)
(2.8.15)	$T = f_2 \circ f_1$ with $f_2(\vec{x}) = [T(\vec{x}_0) - \vec{x}_0] + \vec{x}$, a translation and $f_1(\vec{x}) = \vec{x}_0 + f(\vec{x} - \vec{x}_0)$, an affine transformation keeping $\vec{x}_0$ fixed. (3.8.5)
(2.8.16)	*The fundamental theorem*: For any two affine bases $\mathcal{B} = \{\vec{a}_0, \vec{a}_1, \vec{a}_2, \vec{a}_3\}$ and $\mathcal{C} = \{\vec{b}_0, \vec{b}_1, \vec{b}_2, \vec{b}_3\}$, there exists a unique $T \in G_a(3;\mathbb{R})$ such that $T(a_i) = b_i \quad \text{for } 0 \le i \le 3.$ (3.8.6)

Explanation about (3.8.2) and (3.8.3):

Let $\mathcal{B} = \{\vec{a}_0, \vec{a}_1, \vec{a}_2, \vec{a}_3\}$ and $\mathcal{C} = \{\vec{b}_0, \vec{b}_1, \vec{b}_2, \vec{b}_3\}$ be two affine bases for $\mathbb{R}^3$ and

$$T(\vec{x}) = \vec{x}_0 + f(\vec{x})$$

an affine transformation on $\mathbb{R}^3$.

Rewrite T as

$$T(\vec{x}) = T(\vec{a}_0) + f(\vec{x} - \vec{a}_0), \quad \text{where } T(\vec{a}_0) = \vec{x}_0 + f(\vec{a}_0).$$
$$\Rightarrow T(\vec{x}) - \vec{b}_0 = T(\vec{a}_0) - \vec{b}_0 + f(\vec{x} - \vec{a}_0).$$

Consider $\mathcal{B}$ and $\mathcal{C}$ as *bases* for $\mathbb{R}^3$, i.e.

$$\mathcal{B} = \{\vec{a}_1 - \vec{a}_0, \vec{a}_2 - \vec{a}_0, \vec{a}_3 - \vec{a}_0\}, \quad \text{and}$$
$$\mathcal{C} = \{\vec{b}_1 - \vec{b}_0, \vec{b}_2 - \vec{b}_0, \vec{b}_3 - \vec{b}_0\}. \tag{*1}$$

(see (3.6.1)). Then

$$[T(\vec{x}) - \vec{b}_0]_\mathcal{C} = [T(\vec{a}_0) - \vec{b}_0]_\mathcal{C} + [f(\vec{x} - \vec{a}_0)]_\mathcal{C}$$
$$= [T(\vec{a}_0) - \vec{b}_0]_\mathcal{C} + [\vec{x} - \vec{a}_0]_\mathcal{B}[f]_\mathcal{C}^\mathcal{B}$$

or, in short (see (3.6.5))

$$[T(\vec{x})]_\mathcal{C} = [T(\vec{a}_0)]_\mathcal{C} + [\vec{x}]_\mathcal{B}[f]_\mathcal{C}^\mathcal{B} \tag{3.8.2}$$

which is called the *matrix representation* of T with respect to the *affine bases* $\mathcal{B}$ and $\mathcal{C}$ in the sense (*1), i.e.

1. $[T(\vec{x})]_\mathcal{C} = (y_1, y_2, y_3) \Leftrightarrow T(\vec{x}) - \vec{b}_0 = \sum_{i=1}^3 y_i(\vec{b}_i - \vec{b}_0)$.
2. $[T(\vec{a}_0)]_\mathcal{C} = (p_1, p_2, p_3) \Leftrightarrow T(\vec{a}_0) - \vec{b}_0 = \sum_{i=1}^3 p_i(\vec{b}_i - \vec{b}_0)$.
3. $[\vec{x}]_\mathcal{B} = (x_1, x_2, x_3) \Leftrightarrow \vec{x} - \vec{a}_0 = \sum_{i=1}^3 x_i(\vec{a}_i - \vec{a}_0)$.
4.

$$[f]_\mathcal{C}^\mathcal{B} = \begin{bmatrix} [f(\vec{a}_1 - \vec{a}_0)]_\mathcal{C} \\ [f(\vec{a}_2 - \vec{a}_0)]_\mathcal{C} \\ [f(\vec{a}_3 - \vec{a}_0)]_\mathcal{C} \end{bmatrix}_{3 \times 3} \tag{*2}$$

where each $f(\vec{a}_i - \vec{a}_0)$ is a vector in $\mathbb{R}^3$ and

$$[f(\vec{a}_i - \vec{a}_0)]_\mathcal{C} = (\alpha_{i1}, \alpha_{i2}, \alpha_{i3})$$

$$\Leftrightarrow f(\vec{a}_i - \vec{a}_0) = \sum_{j=1}^3 \alpha_{ij}(\vec{b}_j - \vec{b}_0) \quad \text{for } 1 \leq i \leq 3.$$

(see (3.6.6)). While, (3.8.3) is just another representation of (3.8.2).

Example 1 In the *natural affine basis* $\mathcal{N} = \{\vec{0}, \vec{e}_1, \vec{e}_2, \vec{e}_3\}$, let affine transformation T on $\mathbb{R}^3$ be defined as

$$T(\vec{x}) = \vec{x}_0 + \vec{x}A, \quad \text{where } \vec{x}_0 = (1, -1, 0) \quad \text{and} \quad A = \begin{bmatrix} 4 & -3 & 3 \\ 0 & 1 & 4 \\ 2 & -2 & 1 \end{bmatrix}.$$

Let $\mathcal{B} = \{\vec{a}_0, \vec{a}_1, \vec{a}_2, \vec{a}_3\}$, where $\vec{a}_0 = (1, 0, 1)$, $\vec{a}_1 = (2, 2, 2)$, $\vec{a}_2 = (1, 1, 1)$, $\vec{a}_3 = (2, 3, 3)$ and $\mathcal{C} = \{\vec{b}_0, \vec{b}_1, \vec{b}_2, \vec{b}_3\}$, where $\vec{b}_0 = (0, -1, -1)$, $\vec{b}_1 = (2, -1, -1)$, $\vec{b}_2 = (1, 1, 0)$, $\vec{b}_3 = (0, 0, 1)$.

(a) Show that both $\mathcal{B}$ and $\mathcal{C}$ are affine bases for $\mathbb{R}^3$.
(b) Find the matrix representation of T with respect to $\mathcal{B}$ and $\mathcal{C}$.
(c) Show that A is diagonalizable. Then, try to find an affine basis $\mathcal{D}$ for $\mathbb{R}^3$ so that, in $\mathcal{D}$, T is as simple as possible.

Solution (a) Since

$$\det \begin{bmatrix} \vec{a}_1 - \vec{a}_0 \\ \vec{a}_2 - \vec{a}_0 \\ \vec{a}_3 - \vec{a}_0 \end{bmatrix} = \begin{vmatrix} 1 & 2 & 1 \\ 0 & 1 & 0 \\ 1 & 3 & 2 \end{vmatrix} = 1 \neq 0,$$

so $\{\vec{a}_0, \vec{a}_1, \vec{a}_2, \vec{a}_3\}$ is affinely independent (see Sec. 3.6) and hence $\mathcal{B}$ is an affine basis for $\mathbb{R}^3$. Similarly,

$$\det \begin{bmatrix} \vec{b}_1 - \vec{b}_0 \\ \vec{b}_2 - \vec{b}_0 \\ \vec{b}_3 - \vec{b}_0 \end{bmatrix} = \begin{vmatrix} 2 & 0 & 0 \\ 1 & 2 & 1 \\ 0 & 1 & 2 \end{vmatrix} = 6$$

says that $\mathcal{C}$ is an affine basis for $\mathbb{R}^3$.

(b) Here, we write $\mathcal{B} = \{\vec{a}_1 - \vec{a}_0, \vec{a}_2 - \vec{a}_0, \vec{a}_3 - \vec{a}_0\}$ and $\mathcal{C} = \{\vec{b}_1 - \vec{b}_0, \vec{b}_2 - \vec{b}_0, \vec{b}_3 - \vec{b}_0\}$ and consider $\mathcal{B}$ and $\mathcal{C}$ as bases for the vector space $\mathbb{R}^3$. In the formula (see (3.8.2))

$$[T(\vec{x})]_{\mathcal{C}} = [T(\vec{a}_0)]_{\mathcal{C}} + [\vec{x}]_{\mathcal{B}} A_{\mathcal{C}}^{\mathcal{B}}$$

we need to compute $[T(\vec{a}_0)]_{\mathcal{C}}$ and $A_{\mathcal{C}}^{\mathcal{B}}$. Now

$$T(\vec{a}_0) = \vec{x}_0 + \vec{a}_0 A$$
$$= (1, -1, 0) + (1, 0, 1)A = (1, -1, 0) + (6, -5, 4) = (7, -6, 4)$$
$$\Rightarrow T(\vec{a}_0) - \vec{b}_0 = (7, -6, 4) - (0, -1, -1) = (7, -5, 5)$$
$$\Rightarrow [T(\vec{a}_0)]_{\mathcal{C}} = (T(\vec{a}_0) - \vec{b}_0)[1_{\mathbb{R}^3}]_{\mathcal{C}}^{\mathcal{N}} \text{ (refer to (2.7.23) and (3.3.3)),}$$

where

$$[1_{\mathbb{R}^3}]_{\mathcal{N}}^{\mathcal{C}} = \begin{bmatrix} \vec{b}_1 - \vec{b}_0 \\ \vec{b}_2 - \vec{b}_0 \\ \vec{b}_3 - \vec{b}_0 \end{bmatrix} = \begin{bmatrix} 2 & 0 & 0 \\ 1 & 2 & 1 \\ 0 & 1 & 2 \end{bmatrix} \quad \text{and}$$

$$[1_{\mathbb{R}^3}]_{\mathcal{C}}^{\mathcal{N}} = \frac{1}{6} \begin{bmatrix} 3 & 0 & 0 \\ -2 & 4 & -2 \\ 1 & -2 & 4 \end{bmatrix} = ([1_{\mathbb{R}^3}]_{\mathcal{N}}^{\mathcal{C}})^{-1}.$$

Hence

$$[T(\vec{a}_0)]_C = (7 \quad -5 \quad 5) \cdot \frac{1}{6} \begin{bmatrix} 3 & 0 & 0 \\ -2 & 4 & -2 \\ 1 & -2 & 4 \end{bmatrix} = \frac{1}{6}(36, -30, 30) = (6, -5, 5).$$

On the other hand, $A_C^B = [1_{\mathbb{R}^3}]_N^B A_N^N [1_{\mathbb{R}^3}]_C^N$ means that (see (3.3.4))

$$A_C^B = \begin{bmatrix} [(\vec{a}_1 - \vec{a}_0)A]_C \\ [(\vec{a}_2 - \vec{a}_0)A]_C \\ [(\vec{a}_3 - \vec{a}_0)A]_C \end{bmatrix} = \begin{bmatrix} (\vec{a}_1 - \vec{a}_0)A[1_{\mathbb{R}^3}]_C^N \\ (\vec{a}_2 - \vec{a}_0)A[1_{\mathbb{R}^3}]_C^N \\ (\vec{a}_3 - \vec{a}_0)A[1_{\mathbb{R}^3}]_C^N \end{bmatrix} = \begin{bmatrix} \vec{a}_1 - \vec{a}_0 \\ \vec{a}_2 - \vec{a}_0 \\ \vec{a}_3 - \vec{a}_0 \end{bmatrix} A[1_{\mathbb{R}^3}]_C^N$$

$$= \begin{bmatrix} 1 & 2 & 1 \\ 0 & 1 & 0 \\ 1 & 3 & 2 \end{bmatrix} \begin{bmatrix} 4 & -3 & 3 \\ 0 & 1 & 4 \\ 2 & -2 & 1 \end{bmatrix} \cdot \frac{1}{6} \begin{bmatrix} 3 & 0 & 0 \\ -2 & 4 & -2 \\ 1 & -2 & 4 \end{bmatrix}$$

$$= \frac{1}{6} \begin{bmatrix} 36 & -36 & 54 \\ 2 & -4 & 14 \\ 49 & -50 & 76 \end{bmatrix}.$$

(c) The characteristic polynomial of A is $\det(A - tI_3) = -(t-1)(t-2)$ $(t-3)$. Hence A is diagonalizable and

$$PAP^{-1} = \begin{bmatrix} 1 & 0 & 0 \\ 0 & 2 & 0 \\ 0 & 0 & 3 \end{bmatrix}, \quad \text{where } P = \begin{bmatrix} 4 & -3 & -6 \\ 1 & -1 & -1 \\ -2 & 2 & 1 \end{bmatrix}.$$

Let $\vec{v}_1 = (4, -3, -6)$, $\vec{v}_2 = (1, -1, -1)$ and $\vec{v}_3 = (-2, 2, 1)$ be eigenvectors associated to 1, 2 and 3, respectively. Then $\mathcal{D} = \{\vec{x}_0, \vec{x}_0 + \vec{v}_1, \vec{x}_0 + \vec{v}_2, \vec{x}_0 + \vec{v}_3\}$ is an affine basis for $\mathbb{R}^3$. With $\vec{x}_0$ as a base point,

$$T(\vec{x}) = \vec{x}_0 + \vec{x}A = T(\vec{x}_0) + (\vec{x} - \vec{x}_0)A, \quad \text{where } T(\vec{x}_0) = \vec{x}_0 + \vec{x}_0 A$$
$$\Rightarrow T(\vec{x}) - \vec{x}_0 = T(\vec{x}_0) - \vec{x}_0 + (\vec{x} - \vec{x}_0)A$$
$$\Rightarrow [T(\vec{x})]_\mathcal{D} = [T(\vec{x}_0)]_\mathcal{D} + [(\vec{x} - \vec{x}_0)A]_\mathcal{D}, \quad \text{i.e.}$$
$$(T(\vec{x}) - \vec{x}_0)P^{-1} = (\vec{x}_0 A)P^{-1} + (\vec{x} - \vec{x}_0)P^{-1}(PAP^{-1}).$$

Let

$$[\vec{x}]_\mathcal{D} = (\vec{x} - \vec{x}_0)P^{-1} = (\alpha_1, \alpha_2, \alpha_3), \quad \text{and}$$
$$[T(\vec{x})]_\mathcal{D} = (T(\vec{x}) - \vec{x}_0)P^{-1} = (\beta_1, \beta_2, \beta_3).$$

By computing

$$P^{-1} = \begin{bmatrix} 1 & -9 & -3 \\ 1 & -8 & -2 \\ 0 & -2 & -1 \end{bmatrix}$$

$$\Rightarrow [T(\vec{x}_0)]_{\mathcal{D}} = \vec{x}_0 A P^{-1} = (1 \quad -1 \quad 0) \begin{bmatrix} 4 & -3 & 3 \\ 0 & 1 & 4 \\ 2 & -2 & 1 \end{bmatrix} \begin{bmatrix} 1 & -9 & -3 \\ 1 & -8 & -2 \\ 0 & -2 & -1 \end{bmatrix}$$

$$= (0 \quad 5 \quad -3).$$

Then, in terms of $\mathcal{D}$, T has the simplest form as

$$[T(\vec{x})]_{\mathcal{D}} = (0, 5, -3) + [\vec{x}]_{\mathcal{D}} \begin{bmatrix} 1 & 0 & 0 \\ 0 & 2 & 0 \\ 0 & 0 & 3 \end{bmatrix} \quad \text{or}$$

$$\beta_1 = \alpha_1,$$
$$\beta_2 = 5 + 2\alpha_2,$$
$$\beta_3 = -3 + 3\alpha_3.$$

See Fig. 3.55 and try to explain geometric mapping behavior of T in $\mathcal{D}$. For example, what is the image of the parallelepiped $\varXi(\vec{x}_0 + \vec{v}_1)(\vec{x}_0 + \vec{v}_2)(\vec{x}_0 + \vec{v}_3)$ with vertex at $\vec{x}_0$ under T?   □

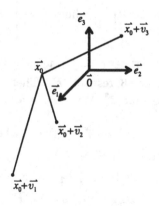

Fig. 3.55

Example 2 Let $\vec{a}_0 = (1, 1, 1)$, $\vec{a}_1 = (-1, 1, 1)$, $\vec{a}_2 = (1, -1, 1)$, $\vec{a}_3 = (1, 1, -1)$. How many affine transformations are there that map the tetrahedron $\Delta \vec{a}_0 \vec{a}_1 \vec{a}_2 \vec{a}_3$ onto the tetrahedron $\Delta(-\vec{a}_0)(-\vec{a}_1)(-\vec{a}_2)(-\vec{a}_3)$? Try to find out some of them and rewrite them in the simplest forms, if possible. See Fig. 3.56.

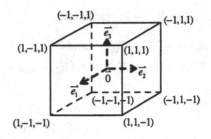

Fig. 3.56

Solution According to the fundamental theorem (3.8.6), there are $4! = 24$ different affine transformations that will meet the requirement.

Notice that, letting $\vec{b}_i = -\vec{a}_i$ for $0 \le i \le 3$,

$$\vec{a}_1 - \vec{a}_0 = (-2,0,0); \quad \vec{b}_1 - \vec{b}_0 = (2,0,0) = -(\vec{a}_1 - \vec{a}_0)$$
$$\vec{a}_2 - \vec{a}_0 = (0,-2,0); \quad \vec{b}_2 - \vec{b}_0 = (0,2,0) = -(\vec{a}_2 - \vec{a}_0)$$
$$\vec{a}_3 - \vec{a}_0 = (0,0,-2); \quad \vec{b}_3 - \vec{b}_0 = (0,0,2) = -(\vec{a}_3 - \vec{a}_0).$$

The simplest one among all is, even by geometric intuition, the one T_1 that satisfies

$$T_1(\vec{a}_i) = \vec{b}_i = -\vec{a}_i \quad \text{for } i = 0,1,2,3.$$

$\Leftrightarrow$ The unique invertible linear operator $f_1: \mathbb{R}^3 \to \mathbb{R}^3$, where $T_1(\vec{x}) = \vec{b}_0 + f_1(\vec{x} - \vec{x}_0)$, satisfies

$$f_1(\vec{a}_j - \vec{a}_0) = \vec{b}_j - \vec{b}_0 = -(\vec{a}_j - \vec{a}_0) \quad \text{for } j = 1,2,3.$$

$\Rightarrow T_1(\vec{x}) = \vec{b}_0 + (\vec{x} - \vec{a}_0)A_1$, where $A_1 = \begin{bmatrix} -1 & 0 & 0 \\ 0 & -1 & 0 \\ 0 & 0 & -1 \end{bmatrix}$.

$\Rightarrow$ (since $\vec{a}_0 A_1 = \vec{b}_0$)

$$T_1(\vec{x}) = f_1(\vec{x}) = \vec{x} A_1 = \vec{x} \begin{bmatrix} -1 & 0 & 0 \\ 0 & -1 & 0 \\ 0 & 0 & -1 \end{bmatrix} \quad \text{or}$$

$$\begin{cases} y_1 = -x_1 \\ y_2 = -x_2, \\ y_3 = -x_3 \end{cases} \quad \text{where } \vec{x} = (x_1, x_2, x_3) \text{ and } \vec{y} = T(\vec{x}) = (y_1, y_2, y_3).$$

Another one is T_2 that satisfies

$$T_2(\vec{a}_0) = -\vec{a}_0, \quad T_2(\vec{a}_1) = -\vec{a}_1 \quad \text{and} \quad T_2(\vec{a}_2) = -\vec{a}_3, \quad T_2(\vec{a}_3) = -\vec{a}_2.$$

$\Leftrightarrow$ The linear operator $f_2 \colon \mathbf{R}^3 \to \mathbf{R}^3$ satisfies

$$f_2(\vec{a}_1 - \vec{a}_0) = -(\vec{a}_1 - \vec{a}_0), \quad \text{and}$$
$$f_2(\vec{a}_2 - \vec{a}_0) = -(\vec{a}_3 - \vec{a}_0), \quad f_2(\vec{a}_3 - \vec{a}_0) = -(\vec{a}_2 - \vec{a}_0).$$

$\Rightarrow$ (since $\vec{a}_0 A_2 = \vec{b}_0$)

$$T_2(\vec{x}) = f_2(\vec{x}) = \vec{x} A_2 = \vec{x} \begin{bmatrix} -1 & 0 & 0 \\ 0 & 0 & -1 \\ 0 & -1 & 0 \end{bmatrix} \quad \text{or}$$

$$y_1 = -x_1, \quad y_2 = -x_3, \quad y_3 = -x_2.$$

If the vertex $\vec{a}_0$ is preassigned to $\vec{b}_0 = -\vec{a}_0$, then the total number of such mappings is $3! = 6$. Each one of them is represented by one of the following matrices A:

$$\begin{bmatrix} -1 & 0 & 0 \\ 0 & -1 & 0 \\ 0 & 0 & -1 \end{bmatrix}, \quad \begin{bmatrix} -1 & 0 & 0 \\ 0 & 0 & -1 \\ 0 & -1 & 0 \end{bmatrix}, \quad \begin{bmatrix} 0 & 0 & -1 \\ 0 & -1 & 0 \\ -1 & 0 & 0 \end{bmatrix},$$

$$\begin{bmatrix} 0 & -1 & 0 \\ -1 & 0 & 0 \\ 0 & 0 & -1 \end{bmatrix}, \quad \begin{bmatrix} 0 & 0 & -1 \\ -1 & 0 & 0 \\ 0 & -1 & 0 \end{bmatrix}, \quad \begin{bmatrix} 0 & -1 & 0 \\ 0 & 0 & -1 \\ -1 & 0 & 0 \end{bmatrix}.$$

The corresponding affine transformation is of the form $T(\vec{x}) = \vec{x} A$.

Suppose $\vec{a}_0$ is assigned to $\vec{b}_1 = -\vec{a}_1$. Construct a $T_3 \colon \mathbf{R}^3 \to \mathbf{R}^3$ satisfying

$$T_3(\vec{a}_0) = -\vec{a}_1, \quad T_3(\vec{a}_1) = -\vec{a}_0, \quad T_3(\vec{a}_2) = -\vec{a}_2 \quad \text{and} \quad T_3(\vec{a}_3) = -\vec{a}_3.$$

$\Leftrightarrow$ The unique invertible linear operator $f_3 \colon \mathbf{R}^3 \to \mathbf{R}^3$ satisfies

$$f_3(\vec{a}_1 - \vec{a}_0) = \vec{b}_0 - \vec{b}_1 = -(\vec{a}_0 - \vec{a}_1) = \vec{a}_1 - \vec{a}_0, \quad \text{and}$$
$$f_3(\vec{a}_j - \vec{a}_0) = \vec{b}_j - \vec{b}_1 = -(\vec{a}_j - \vec{a}_1)$$
$$= -(\vec{a}_j - \vec{a}_0) + (\vec{a}_1 - \vec{a}_0) \quad \text{for } j = 2, 3.$$

$\Rightarrow T_3(\vec{x}) = \vec{b}_1 + (\vec{x} - \vec{a}_0) A_3, \quad \text{where } A_3 = \begin{bmatrix} 1 & 0 & 0 \\ 1 & -1 & 0 \\ 1 & 0 & -1 \end{bmatrix}.$

$\Rightarrow$ (since $\vec{a}_0 A_3 = (3, -1, -1)$ and $\vec{b}_1 = -\vec{a}_1 = (1, -1, -1)$)

$$T_3(\vec{x}) = (-2, 0, 0) + \vec{x} \begin{bmatrix} 1 & 0 & 0 \\ 1 & -1 & 0 \\ 1 & 0 & -1 \end{bmatrix}.$$

Notice that A_3 is diagonalizable and hence is similar to

$$\begin{bmatrix} 1 & 0 & 0 \\ 0 & -1 & 0 \\ 0 & 0 & -1 \end{bmatrix} = PA_3P^{-1}, \quad \text{where } P = \begin{bmatrix} 1 & 0 & 0 \\ 1 & -2 & 0 \\ 1 & 0 & -2 \end{bmatrix}.$$

Try to find an affine basis $\mathcal{B}$ for $\mathbb{R}^3$ so that the linear part A_3 of T_3 is in this diagonal form.

Suppose T_4 satisfies

$$T_4(\vec{a}_0) = -\vec{a}_1, \quad T_4(\vec{a}_1) = -\vec{a}_2, \quad T_4(\vec{a}_2) = -\vec{a}_3 \quad \text{and} \quad T_4(\vec{a}_3) = -\vec{a}_0.$$

$\Leftrightarrow$ The unique invertible linear operator f_4 satisfies

$$\begin{aligned} f_4(\vec{a}_1 - \vec{a}_0) &= -\vec{a}_2 + \vec{a}_1 = -(\vec{a}_2 - \vec{a}_1) = -(\vec{a}_2 - \vec{a}_0) + (\vec{a}_1 - \vec{a}_0), \\ f_4(\vec{a}_2 - \vec{a}_0) &= -\vec{a}_3 + \vec{a}_1 = -(\vec{a}_3 - \vec{a}_0) + (\vec{a}_1 - \vec{a}_0), \\ f_4(\vec{a}_3 - \vec{a}_0) &= -\vec{a}_0 + \vec{a}_1 = \vec{a}_1 - \vec{a}_0 \end{aligned}$$

$$\Rightarrow T_4(\vec{x}) = -\vec{a}_1 + (\vec{x} - \vec{a}_0)A_4, \quad \text{where } A_4 = \begin{bmatrix} 1 & -1 & 0 \\ 1 & 0 & -1 \\ 1 & 0 & 0 \end{bmatrix}.$$

$\Rightarrow$ (since $-\vec{a}_1 = (1, -1, -1)$, $\vec{a}_0 A_4 = (3, -1, -1)$)

$$T_4(\vec{x}) = (-2, 0, 0) + \vec{x} \begin{bmatrix} 1 & -1 & 0 \\ 1 & 0 & -1 \\ 1 & 0 & 0 \end{bmatrix}.$$

There are four others of this type whose respective linear parts are

$$\begin{bmatrix} 1 & 0 & -1 \\ 1 & -1 & 0 \\ 1 & 0 & 0 \end{bmatrix}, \begin{bmatrix} 1 & -1 & 0 \\ 1 & 0 & 0 \\ 1 & 0 & -1 \end{bmatrix}, \begin{bmatrix} 1 & 0 & 0 \\ 1 & 0 & -1 \\ 1 & -1 & 0 \end{bmatrix}, \begin{bmatrix} 1 & 0 & -1 \\ 1 & 0 & 0 \\ 1 & -1 & 0 \end{bmatrix}.$$

Yet, there are another 12 such affine transformations which are left as Ex. <A> 2.   □

Exercises

<A>

1. Prove (3.8.4), (3.8.5) and (3.8.6) in details.
2. Find another 12 affine transformations as mentioned in Example 2.
3. Let $\mathcal{B} = \{\vec{a}_0, \vec{a}_1, \vec{a}_2, \vec{a}_3\}$, where $\vec{a}_0 = (2, 1, -1)$, $\vec{a}_1 = (3, 1, -2)$, $\vec{a}_2 = (2, 3, 2)$, $\vec{a}_3 = (3, 2, 0)$ and $C = \{\vec{b}_0, \vec{b}_1, \vec{b}_2, \vec{b}_3\}$, where $\vec{b}_0 = (1, 3, 1)$, $\vec{b}_1 = (0, 3, 2)$, $\vec{b}_2 = (2, 4, 2)$, $\vec{b}_3 = (2, 1, 2)$. Show that $\mathcal{B}$ and C are affine bases for $\mathbb{R}^3$. For each of the following affine transformations T in the natural affine basis $\mathcal{N} = \{\vec{0}, \vec{e}_1, \vec{e}_2, \vec{e}_3\}$, do the following problems:

(1) The matrix representation of T with respect to $\mathcal{B}$ and $\mathcal{B}$, denoted by $[T]_{\mathcal{B}}$, i.e.
$$[T]_{\mathcal{B}}(\vec{x}) = [T(\vec{x})]_{\mathcal{B}} = [T(\vec{a}_0)]_{\mathcal{B}} + [\vec{x}]_{\mathcal{B}} A_{\mathcal{B}}^{\mathcal{B}}.$$

(2) The matrix representation $[T]_C$ of T with respect to C and C.

(3) The matrix representation $[T]_C^{\mathcal{B}}$ of T with respect to $\mathcal{B}$ and C, i.e.
$$[T]_C^{\mathcal{B}}(\vec{x}) = [T(\vec{x})]_C = [T(\vec{a}_0)]_C + [\vec{x}]_{\mathcal{B}} A_C^{\mathcal{B}}.$$

(4) What is $[T]_{\mathcal{B}}^C$? Anything to do with $[T]_C^{\mathcal{B}}$?

(5) Find the (diagonal, Jordan or rational) canonical form of A. Thus, construct an affine basis $\mathcal{D}$ for $\mathbb{R}^3$ in which T is as simple as possible.

(a) $T(\vec{x}) = \vec{x}_0 + \vec{x} A$, where $\vec{x}_0 = (1, -1, 1)$ and $A = \begin{bmatrix} 0 & 1 & -1 \\ -4 & 4 & -2 \\ -2 & 1 & 1 \end{bmatrix}$.

(b) $T(\vec{x}) = \vec{x}_0 + \vec{x} A$, where $\vec{x}_0 = (-1, -1, 5)$ and $A = \begin{bmatrix} -3 & -3 & -2 \\ -7 & 6 & -3 \\ 1 & -1 & 2 \end{bmatrix}$.

(c) $T(\vec{x}) = \vec{x}_0 + \vec{x} A$, where $\vec{x}_0 = (1, 2, 3)$ and $A = \begin{bmatrix} -1 & 3 & 0 \\ 0 & -1 & 2 \\ 0 & 0 & -1 \end{bmatrix}$.

(d) $T(\vec{x}) = \vec{x}_0 + \vec{x} A$, where $\vec{x}_0 = (-2, 3, 2)$ and $A = \begin{bmatrix} 2 & 0 & 0 \\ 2 & 2 & 0 \\ -2 & 1 & 2 \end{bmatrix}$.

(e) $T(\vec{x}) = \vec{x}_0 + \vec{x} A$, where $\vec{x}_0 = (-2, 1, 0)$ and $A = \begin{bmatrix} 0 & 0 & 1 \\ 1 & 0 & -1 \\ 0 & 1 & 1 \end{bmatrix}$.

(f) $T(\vec{x}) = \vec{x}_0 + \vec{x}A$, where $\vec{x}_0 = (4, -1, -1)$ and $A = \begin{bmatrix} 4 & 2 & 3 \\ 2 & 1 & 2 \\ -1 & -2 & 0 \end{bmatrix}$.

(g) $T(\vec{x}) = \vec{x}_0 + \vec{x}A$, where $\vec{x}_0 = (2, 1, -1)$ and $A = \begin{bmatrix} 2 & 1 & 0 \\ 0 & 2 & 1 \\ 1 & 0 & 2 \end{bmatrix}$.

4. Let T and $\mathcal{B}$ and $\mathcal{D}$ be as in Ex. 3(a).

   (a) Compute T^{-1} and $[T^{-1}]_\mathcal{B}$. Is $[T^{-1}]_\mathcal{B} = [T]_\mathcal{B}^{-1}$ (refer to (3.3.3))?
   (b) Compute T^2 and $[T^2]_\mathcal{B}$, where $T^2 = T \cdot T$.
   (c) Try to compute T^3 in terms of $\mathcal{D}$.

5. Let T_1 be as in Ex. 3(b) and T_2 be as in Ex. 3(f). Also, $\mathcal{B}$ and $\mathcal{C}$ are as in Ex. 3.

   (a) Compute $[T_2 \circ T_1]_\mathcal{B}$. Is this equal to $[T_2]_\mathcal{B}[T_1]_\mathcal{B}$?
   (b) Compute $[T_1 \circ T_2^{-1}]_\mathcal{C}$.
   (c) Compute $[T_1 \circ T_2]_\mathcal{C}^\mathcal{B}$.

6. In Fig. 3.56, find all possible affine transformations mapping the tetrahedron $\Delta(-\vec{a}_1)\vec{a}_3\vec{a}_2(-\vec{a}_0)$ onto the tetrahedron $\Delta\vec{a}_0\vec{a}_1\vec{a}_2\vec{a}_3$, where $\vec{a}_1, \vec{a}_2, \ldots$, etc. are as in Example 2.

7. In Fig. 3.56, find all possible affine transformations mapping the tetrahedron $\Delta\vec{a}_0\vec{a}_1\vec{a}_2\vec{a}_3$ onto itself. Do they constitute a group under composite operation? Give precise reasons.

8. Try to extend Exs. <A> 2 through 8 of Sec. 2.8.1 to $\mathbb{R}^3$ and prove your statements.

<B>

1. We define the affine group $G_a(3; \mathbb{R})$ with respect to the natural affine basis $\mathcal{N}$ (see (3.8.4)). Show that the affine group with respect to an affine basis $\mathcal{B} = \{\vec{a}_0, \vec{a}_1, \vec{a}_2, \vec{a}_3\}$ is the conjugate group

$$\begin{bmatrix} A_0 & 0 \\ \vec{a}_0 & 1 \end{bmatrix}_{4 \times 4} G_a(3; \mathbb{R}) \begin{bmatrix} A_0 & 0 \\ \vec{a}_0 & 1 \end{bmatrix}_{4 \times 4}^{-1},$$

   where A_0 is the transition matrix from $\mathcal{B} = \{\vec{a}_1 - \vec{a}_0, \vec{a}_2 - \vec{a}_0, \vec{a}_3 - \vec{a}_0\}$ to $\mathcal{N} = \{\vec{e}_1, \vec{e}_2, \vec{e}_3\}$ as bases for the vector space $\mathbb{R}^3$.

2. Let $\mathcal{B} = \{\vec{a}_0, \vec{a}_1, \vec{a}_2, \vec{a}_3, \vec{a}_4\}$, where $\vec{a}_0 = (-1, -1, 1, -2)$, $\vec{a}_1 = (0, 0, 2, -1)$, $\vec{a}_2 = (-1, 0, 2, -1)$, $\vec{a}_3 = (-1, -1, 2, -1)$, $\vec{a}_4 = (-1, -1, 1, -1)$ and $\mathcal{C} = \{\vec{b}_0, \vec{b}_1, \vec{b}_2, \vec{b}_3\}$, where $\vec{b}_0 = (2, -3, 4, -5)$, $\vec{b}_1 = (3, -2, 5, -5)$, $\vec{b}_2 = (3, -3, 5, -4)$, $\vec{b}_3 = (3, -2, 4, -4)$, $\vec{b}_4 = $

$(2, -2, 5, -4)$. Show that both $\mathcal{B}$ and $\mathcal{C}$ are affine bases for $\mathbb{R}^4$. For each of the following affine transformations $T(\vec{x}) = \vec{x}_0 + \vec{x}A$, do the same problems as in Ex. <A> 3.

(a) $\vec{x}_0 = (2, 3, 5, 7), \quad A = \begin{bmatrix} 1 & 0 & 0 & 0 \\ 1 & 1 & 0 & 0 \\ 2 & 0 & 2 & 1 \\ -1 & 1 & -1 & 1 \end{bmatrix}$.

(b) $\vec{x}_0 = (0, -2, -2, -2), \quad A = \begin{bmatrix} 2 & -1 & 0 & 1 \\ 0 & 3 & -1 & 0 \\ 0 & 1 & 1 & 0 \\ 0 & -1 & 0 & 3 \end{bmatrix}$.

(c) $\vec{x}_0 = (-3, 1, 1, -3), \quad A = \begin{bmatrix} 2 & -4 & 2 & 2 \\ -2 & 0 & 1 & 3 \\ -2 & -2 & 3 & 3 \\ -2 & -6 & 3 & 7 \end{bmatrix}$.

(d) $\vec{x}_0 = (2, 0, -2, -2), \quad A = \begin{bmatrix} 1 & -2 & 0 & 0 \\ 2 & 1 & 0 & 0 \\ 1 & 0 & 1 & -2 \\ 0 & 1 & 2 & 1 \end{bmatrix}$.

3. Try to describe all possible affine transformations on $\mathbb{R}^4$ mapping the four-dimensional *tetrahedron* (or *simplex*).

$$\Delta\vec{a}_0\vec{a}_1\vec{a}_2\vec{a}_3\vec{a}_4 = \left\{ \sum_{i=0}^{4} \lambda_i \vec{a}_i \mid \lambda_i \geq 0 \text{ for } 0 \leq i \leq 4 \text{ and } \sum_{i=0}^{4} \lambda_i = 1 \right\}$$

onto another tetrahedron $\Delta\vec{b}_0\vec{b}_1\vec{b}_2\vec{b}_3\vec{b}_4$. Try to use matrices to write some of them explicitly.

<C> Abstraction and generalization

Read Ex. <C> of Sec. 2.8.1.

3.8.2 *Examples*

We are going to extend the contents of Sec. 2.8.2 to the affine space $\mathbb{R}^3$. The readers should review Sec. 2.8.2 for detailed introductions.

Remember that $\mathbb{R}^3$ also plays as a vector space.

We also need *spatial Euclidean concepts* such as angles, lengths, areas and volumes in some cases. One can refer to *Introduction* and *Natural Inner Product* in Part 2 and the beginning of Chap. 5 for preliminary knowledge.

Some terminologies are unified as follows.

Let $T(\vec{x}) = \vec{x}_0 + \vec{x}A$ be an affine transformation on $\mathbb{R}^3$. It is always understood that, letting $A = [a_{ij}]_{3\times 3}$,

$$\det A \neq 0$$

unless specified. If $\vec{x}_0 = (b_1, b_2, b_3)$, then the traditional algebraic equivalent of T is written as

$$y_j = \sum_{i=1}^{3} a_{ij} x_i + b_j \quad \text{for } 1 \leq j \leq 3, \tag{3.8.7}$$

where $\vec{x} = (x_1, x_2, x_3)$ and $\vec{y} = T(\vec{x}) = (y_1, y_2, y_3)$.

A point $\vec{x} \in \mathbb{R}^3$ is called a *fixed point* or an *invariant point* of T if

$$T(\vec{x}) = \vec{x}. \tag{3.8.8}$$

An affine subspace S of $\mathbb{R}^3$ is called an *invariant (affine) subspace* of T if

$$T(S) \subseteq S. \tag{3.8.9}$$

If, in addition, each point of S is an invariant point, then S is called an *(affine) subspace of invariant points*.

Case 1 Translation

Let $\vec{x}_0$ be a fixed vector in $\mathbb{R}^3$. The mapping

$$T(\vec{x}) = \vec{x}_0 + \vec{x}, \quad \vec{x} \in \mathbb{R}^3 \tag{3.8.10}$$

is called a *translation* of $\mathbb{R}^3$ along $\vec{x}_0$. Refer to Fig. 2.96. The set of all such translations form a subgroup of $G_a(3; \mathbb{R})$.

Translations preserve all geometric mapping properties listed in (3.7.15). A translation does not have fixed point unless $\vec{x}_0 = \vec{0}$, which in this case every point is a fixed point. Any line or plane parallel to $\vec{x}_0$ is invariant under translation.

Case 2 Reflection

Suppose a line OA and a plane Σ in space $\mathbb{R}^3$ intersect at the point O. For any point X in $\mathbb{R}^3$, draw a line XP, parallel to OA, intersecting Σ at the point P and extend it to a point X' so that $XP = PX'$. See Fig. 3.57. The mapping $T: \mathbb{R}^3 \to \mathbb{R}^3$ defined by

$$T(X) = X'$$

is called the *(skew) reflection* of space $\mathbb{R}^3$ along the direction $\overrightarrow{OA}$ with respect to the plane Σ. In case the line OA is perpendicular to the plane

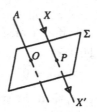

Fig. 3.57

Σ, T is called the *orthogonal reflection* or *symmetric motion* of the space $\mathbb{R}^3$ with respect to the plane Σ. For details, see (3.8.26).

Just like what we did in (2.8.22)–(2.8.25), we can summarize as

The reflection
Let $\vec{a}_0, \vec{a}_1, \vec{a}_2$ and $\vec{a}_3$ be non-coplanar points in the space $\mathbb{R}^3$. Denote by T the reflection of $\mathbb{R}^3$ along the direction $\vec{a}_3 - \vec{a}_0$ with respect to the plane $\Sigma = \vec{a}_0 + \langle\langle \vec{a}_1 - \vec{a}_0, \vec{a}_2 - \vec{a}_0 \rangle\rangle$. Then,

1. In the affine basis $\mathcal{B} = \{\vec{a}_0, \vec{a}_1, \vec{a}_2, \vec{a}_3\}$,

$$[T(\vec{x})]_\mathcal{B} = [\vec{x}]_\mathcal{B}[T]_\mathcal{B}, \quad \text{where } [T]_\mathcal{B} = \begin{bmatrix} 1 & 0 & 0 \\ 0 & 1 & 0 \\ 0 & 0 & -1 \end{bmatrix}.$$

2. In the natural affine basis $\mathcal{N} = \{\vec{0}, \vec{e}_1, \vec{e}_2, \vec{e}_3\}$,

$$T(\vec{x}) = \vec{a}_0 + (\vec{x} - \vec{a}_0)P^{-1}[T]_\mathcal{B}P, \quad \text{where } P = P_\mathcal{N}^\mathcal{B} = \begin{bmatrix} \vec{a}_1 - \vec{a}_0 \\ \vec{a}_2 - \vec{a}_0 \\ \vec{a}_3 - \vec{a}_0 \end{bmatrix}_{3 \times 3}.$$

Also,

(1) A reflection preserves all properties 1–7, c and d listed in (3.7.15).
(2) An orthogonal reflection, in addition to (1), preserves

    a. angle,
    b. length

but reverses the direction.   (3.8.11)

The details are left as Ex. $<$A$>$ 1.

T in (3.8.11) can be rewritten as

$$T(\vec{x}) = \vec{a}_0(I_3 - P^{-1}[T]_B P) + \vec{x}P^{-1}[T]_B P$$
$$= \vec{x}_0 + \vec{x}A, \quad \text{where } \vec{x}_0 = \vec{a}_0(I_3 - A) \text{ and } A = P^{-1}[T]_B P. \quad (3.8.12)$$

Notice that $\det A = \det[T]_B = -1$. (3.8.12) suggests how to test whether a given affine transformation $T(\vec{x}) = \vec{x}_0 + \vec{x}A$ is a reflection. Watch the following steps (compare with (2.8.27)):

1. $\det A = -1$ is a necessary condition.
2. If A has eigenvalues 1, 1 and -1, and A is diagonalizable, then T is a reflection if $\vec{x}(I_3 - A) = \vec{x}_0$ has a solution.
3. Compute linearly independent eigenvectors $\vec{v}_1$ and $\vec{v}_2$ corresponding to 1. Then the solution set of $\vec{x}(I_3 - A) = \vec{x}_0$, i.e.

$$\frac{1}{2}(\vec{0} + T(\vec{0})) + \langle\langle \vec{v}_1, \vec{v}_2 \rangle\rangle = \frac{1}{2}\vec{x}_0 + \langle\langle \vec{v}_1, \vec{v}_2 \rangle\rangle$$

   is the *plane of invariant points* of T.
4. Compute an eigenvector $\vec{v}_3$ corresponding to -1. Then $\vec{v}_3$ or $-\vec{v}_3$ is a *direction* of reflection T. In fact,

$$\vec{x}_0(A + I_3) = \vec{x}(I_3 - A)(I_3 + A) = \vec{0} \quad \text{(see (2)in (3.7.46))},$$

   so $\vec{x}_0$ is a direction if $\vec{x}_0 \neq \vec{0}$. Note that $\vec{v}_3$ and $\vec{x}_0$ are linearly dependent.

$$(3.8.13)$$

The details are left to the readers.

Example 1

(a) Find the reflection of $\mathbb{R}^3$ along the direction $\vec{v}_3 = (-1, 1, -1)$ with respect to the plane $(2, -2, 3) + \langle\langle(0, 1, 0), (0, -1, 1)\rangle\rangle$.
(b) Show that

$$T(\vec{x}) = \vec{x}_0 + \vec{x}A, \quad \text{where } \vec{x}_0 = (0, -2, -4) \text{ and } A = \begin{bmatrix} 1 & 0 & 0 \\ 0 & \frac{5}{3} & \frac{4}{3} \\ 0 & -\frac{4}{3} & -\frac{5}{3} \end{bmatrix}$$

   is a reflection. Determine its direction and plane of invariant points.

Solution (a) In the affine basis $\mathcal{B} = \{(2,-2,3),\ (2,-1,3),\ (2,-3,4),\ (1,-1,2)\}$, the required T has the representation

$$[T(\vec{x})]_\mathcal{B} = [\vec{x}]_\mathcal{B}[T]_\mathcal{B}, \text{ where } [T]_\mathcal{B} = \begin{bmatrix} 1 & 0 & 0 \\ 0 & 1 & 0 \\ 0 & 0 & -1 \end{bmatrix}.$$

While, in the natural affine basis $\mathcal{N}$,

$$T(\vec{x}) = (2,-2,3) + (\vec{x} - (2,-2,3))P^{-1}[T]_\mathcal{B}P$$

where

$$P = P_\mathcal{N}^\mathcal{B} = \begin{bmatrix} 0 & 1 & 0 \\ 0 & -1 & 1 \\ -1 & 1 & -1 \end{bmatrix} \Rightarrow P^{-1} = \begin{bmatrix} 0 & -1 & -1 \\ 1 & 0 & 0 \\ 1 & 1 & 0 \end{bmatrix}.$$

Therefore,

$$P^{-1}[T]_\mathcal{B}P = \begin{bmatrix} 0 & -1 & -1 \\ 1 & 0 & 0 \\ 1 & 1 & 0 \end{bmatrix} \begin{bmatrix} 1 & 0 & 0 \\ 0 & 1 & 0 \\ 0 & 0 & -1 \end{bmatrix} \begin{bmatrix} 0 & 1 & 0 \\ 0 & -1 & 1 \\ -1 & 1 & -1 \end{bmatrix}$$

$$= \begin{bmatrix} -1 & 2 & -2 \\ 0 & 1 & 0 \\ 0 & 0 & 1 \end{bmatrix}, \text{ and}$$

$$(2,-2,3) - (2,-2,3)P^{-1}[T]_\mathcal{B}P = (2,-2,3) - (-2,2,-1) = (4,-4,4)$$

$$\Rightarrow T(\vec{x}) = (4,-4,4) + \vec{x}\begin{bmatrix} -1 & 2 & -2 \\ 0 & 1 & 0 \\ 0 & 0 & 1 \end{bmatrix} \text{ for } \vec{x} \in \mathbb{R}^3, \quad \text{or}$$

$$y_1 = 4 - x_1, \quad y_2 = -4 + 2x_1 + x_2, \quad y_3 = 4 - 2x_1 + x_3.$$

(b) Since $\det A = -1$, it is possible that T is a reflection. To make certainty, compute the characteristic polynomial $\det(A - tI_3) = -(t-1)^2(t+1)$. So A has eigenvalues 1, 1 and -1. Moreover,

$$(A - I_3)(A + I_3) = \begin{bmatrix} 0 & 0 & 0 \\ 0 & \frac{2}{3} & \frac{4}{3} \\ 0 & -\frac{4}{3} & -\frac{8}{3} \end{bmatrix} \begin{bmatrix} 2 & 0 & 0 \\ 0 & \frac{8}{3} & \frac{4}{3} \\ 0 & -\frac{4}{3} & -\frac{2}{3} \end{bmatrix} = O_{3\times3}$$

indicates that A is diagonalizable and thus, the corresponding T is a reflection if $\vec{x}(I_3 - A) = \vec{x}_0$ has a solution. Now

$$\vec{x}(I_3 - A) = \vec{x}_0$$

$$\Rightarrow x_2 - 2x_3 - 3 = 0 \quad \text{(the plane of invariant points).}$$

So T is really a reflection.

Take eigenvectors $\vec{v}_1 = (2,0,0)$ and $\vec{v}_2 = (1,2,1)$ corresponding to 1 and $\vec{v}_3 = (0,1,2)$ corresponding to -1. Then

$$\frac{1}{2}\vec{x}_0 + \langle\langle\vec{v}_1, \vec{v}_2\rangle\rangle = (0,-1,-2) + \{(2\alpha_1 + \alpha_2, 2\alpha_2, \alpha_2)|\alpha_1, \alpha_2 \in \mathbb{R}\}$$

$$\Leftrightarrow \begin{cases} x_1 = 2\alpha_1 + \alpha_2 \\ x_2 = -1 + 2\alpha_2 \\ x_3 = -2 + \alpha_2, \quad \text{for } \alpha_1, \alpha_2 \in \mathbb{R} \end{cases}$$

$$\Leftrightarrow x_2 - 2x_3 - 3 = 0$$

indeed is the plane of invariant points. Also, $\vec{v}_3 = -\frac{1}{2}\vec{x}_0$ or just $\vec{x}_0$ itself is the direction of T.

In the affine basis $\mathcal{C} = \{\vec{v}_3, \vec{v}_3 + \vec{v}_1, \vec{v}_3 + \vec{v}_2, 2\vec{v}_3\}$,

$$[T(\vec{x})]_{\mathcal{C}} = [\vec{x}]_{\mathcal{C}} \begin{bmatrix} 1 & 0 & 0 \\ 0 & 1 & 0 \\ 0 & 0 & -1 \end{bmatrix}.$$

See Fig. 3.58.

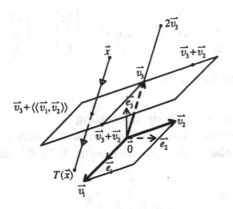

Fig. 3.58

Remark The linear operator

$$T(\vec{x}) = \vec{x}A, \quad \text{where } A = \begin{bmatrix} 0 & 1 & 0 \\ 1 & 0 & 0 \\ 0 & 0 & 1 \end{bmatrix} \tag{3.8.14}$$

is a standard *orthogonal reflection* of $\mathbb{R}^3$ in the direction $\vec{v}_3 = (\frac{1}{\sqrt{2}}, -\frac{1}{\sqrt{2}}, 0)$ with respect to the plane

$$\langle\langle \vec{v}_1, \vec{v}_2 \rangle\rangle = \{(x_1, x_2, x_3) \in \mathbb{R}^3 \mid x_1 = x_2\}, \quad \text{where}$$

$$\vec{v}_1 = \left(\frac{1}{\sqrt{2}}, \frac{1}{\sqrt{2}}, 0\right) \quad \text{and} \quad \vec{v}_2 = \vec{e}_3 = (0, 0, 1).$$

See Fig. 3.59 and compare with Fig. 3.32 with a = b = c = 1. Notice that

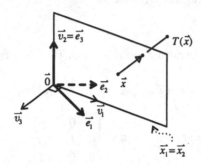

Fig. 3.59

$$A = P^{-1} \begin{bmatrix} 1 & 0 & 0 \\ 0 & 1 & 0 \\ 0 & 0 & -1 \end{bmatrix} P, \quad \text{where } P = \begin{bmatrix} \vec{v}_1 \\ \vec{v}_2 \\ \vec{v}_3 \end{bmatrix} = \begin{bmatrix} \frac{1}{\sqrt{2}} & \frac{1}{\sqrt{2}} & 0 \\ 0 & 0 & 1 \\ \frac{1}{\sqrt{2}} & -\frac{1}{\sqrt{2}} & 0 \end{bmatrix}.$$

For details, refer to Case 7 below.   □

Case 3   One-way stretch or stretching

Let the line OA intersect the plane Σ at the point O as indicated in Fig. 3.60. Take any fixed scalar $k \neq 0$. For any point X in $\mathbb{R}^3$, draw the line XP,

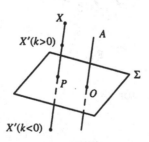

Fig. 3.60

parallel to OA and intersecting Σ at the point P. Pick up the point X' on XP so that $X'P = kXP$ in signed length. The mapping $T\colon \mathbb{R}^3 \to \mathbb{R}^3$ defined by

$$T(X) = X'$$

is an affine transformation and is called a *one-way stretch* with *scale factor* k of $\mathbb{R}^3$ in the *direction OA* with the plane Σ as the *plane of invariant points*. In case OA is perpendicular to Σ, T is called *orthogonal one-way stretch*. One-way stretch has the following features:

1. Each line parallel to the direction is an invariant line.
2. Each plane, parallel to the plane Σ of invariant points, moves to a new parallel plane with a distance proportional to $|k|$ from the original one.
3. Each line L, not parallel to the plane Σ, intersects with its image line $T(L)$ at a point lying on Σ.
4. Each plane Σ', not parallel to the plane Σ, intersects with its image plane $T(\Sigma')$ along a line lying on Σ.

See Fig. 3.61.

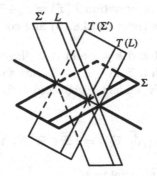

Fig. 3.61

As a counterpart of (2.8.28), we have

The one-way stretch

Let $\vec{a}_0, \vec{a}_1, \vec{a}_2$ and $\vec{a}_3$ be non-coplanar points in $\mathbb{R}^3$. Let T denote the one-way stretch, with scale factor k, of $\mathbb{R}^3$ along the direction $\vec{a}_3 - \vec{a}_0$ and the plane of invariant points $\Sigma\colon \vec{a}_0 + \langle\langle \vec{a}_1 - \vec{a}_0, \vec{a}_2 - \vec{a}_0 \rangle\rangle$. Then,

1. In the affine basis $\mathcal{B} = \{\vec{a}_0, \vec{a}_1, \vec{a}_2, \vec{a}_3\}$,

$$[T(\vec{x})]_{\mathcal{B}} = [\vec{x}]_{\mathcal{B}}[T]_{\mathcal{B}}, \quad \text{where } [T]_{\mathcal{B}} = \begin{bmatrix} 1 & 0 & 0 \\ 0 & 1 & 0 \\ 0 & 0 & k \end{bmatrix}.$$

2. In the natural affine basis $\mathcal{N} = \{\vec{0}, \vec{e}_1, \vec{e}_2, \vec{e}_3\}$,

$$T(\vec{x}) = \vec{a}_0 + (\vec{x} - \vec{a}_0)P^{-1}[T]_{\mathcal{B}}P, \quad \text{where } P = P_{\mathcal{N}}^{\mathcal{B}} = \begin{bmatrix} \vec{a}_1 - \vec{a}_0 \\ \vec{a}_2 - \vec{a}_0 \\ \vec{a}_3 - \vec{a}_0 \end{bmatrix}.$$

Also,

(1) A one-way stretch preserves all the properties 1–7 listed in (3.7.15),
(2) but enlarges the volume by the scale factor $|k|$ and preserves the orientation if $k > 0$ while reverses the orientation if $k < 0$.

$$(3.8.15)$$

In case $k = -1$, then (3.8.15) reduces to (3.8.13) for reflections.

To test if an affine transformation $T(\vec{x}) = \vec{x}_0 + \vec{x}A$ is a one-way stretch, follow the following steps (refer to (3.8.13) and compare with (2.8.29)):

1. A has eigenvalues 1, 1 and $k \neq 1$ and A is diagonalizable.
2. A is a one-way stretch if $\vec{x}(I_3 - A) = \vec{x}_0$ has a solution.
3. Compute linearly independent eigenvectors $\vec{v}_1$ and $\vec{v}_2$ corresponding to 1, then

$$\frac{1}{1-k}\vec{x}_0 + \langle\langle\vec{v}_1, \vec{v}_2\rangle\rangle = \{\vec{x} \in \mathbb{R}^3 \mid \vec{x}(A - I_3) = \vec{x}_0\}$$

is the plane of invariant points.
4. Compute an eigenvector $\vec{v}_3$ corresponding to k, then $\vec{v}_3$ is the direction of the stretch. In particular,

$$\frac{1}{1-k}\vec{x}_0 \text{ or } \vec{x}_0$$

is a direction (up to a nonzero scalar) if $\vec{x}_0 \neq \vec{0}$. $(3.8.16)$

The details are left to the readers.

Example 2

(a) (See Example 1(a).) Find the one-way stretch with scale factor $k \neq 1$ of $\mathbb{R}^3$ along the direction $\vec{v}_3 = (-1, 1, -1)$ with respect to the plane $(-1, 1, 1) + \langle\langle \vec{v}_1, \vec{v}_2 \rangle\rangle$, where $\vec{v}_1 = (0, 1, 0)$ and $\vec{v}_2 = (0, -1, 1)$.

(b) Show that

$$T(\vec{x}) = \vec{x}_0 + \vec{x}A, \quad \text{where } \vec{x}_0 = (-2, 4, 2) \text{ and } A = \begin{bmatrix} -1 & 4 & 2 \\ -1 & 3 & 1 \\ -1 & 2 & 2 \end{bmatrix}$$

represents a one-way stretch. Determine its direction and its plane of invariant points.

Solution   (a) In the affine basis $\mathcal{B} = \{(-1, 1, 1), (-1, 2, 1), (-1, 0, 2), (-2, 2, 0)\}$, the one-way stretch T is

$$[T(\vec{x})]_\mathcal{B} = [\vec{x}]_\mathcal{B}[T]_\mathcal{B}, \quad \text{where}[T]_\mathcal{B} = \begin{bmatrix} 1 & 0 & 0 \\ 0 & 1 & 0 \\ 0 & 0 & k \end{bmatrix}.$$

In the natural affine basis $\mathcal{N}$,

$$T(\vec{x}) = (-1, 1, 1) + (\vec{x} - (-1, 1, 1))P^{-1}[T]_\mathcal{B}P,$$

where

$$P = P_\mathcal{N}^B = \begin{bmatrix} 0 & 1 & 0 \\ 0 & -1 & 1 \\ -1 & 1 & -1 \end{bmatrix} \Rightarrow P^{-1} = \begin{bmatrix} 0 & -1 & -1 \\ 1 & 0 & 0 \\ 1 & 1 & 0 \end{bmatrix}.$$

Hence,

$$P^{-1}[T]_\mathcal{B}P = \begin{bmatrix} 0 & -1 & -1 \\ 1 & 0 & 0 \\ 1 & 1 & 0 \end{bmatrix} \begin{bmatrix} 1 & 0 & 0 \\ 0 & 1 & 0 \\ 0 & 0 & k \end{bmatrix} \begin{bmatrix} 0 & 1 & 0 \\ 0 & -1 & 1 \\ -1 & 1 & -1 \end{bmatrix}$$

$$= \begin{bmatrix} k & 1-k & -1+k \\ 0 & 1 & 0 \\ 0 & 0 & 1 \end{bmatrix},$$

$$(-1, 1, 1) - (-1, 1, 1)P^{-1}[T]_\mathcal{B}P = (-1, 1, 1) - (-k, k, 2-k)$$
$$= (-1+k, 1-k, -1+k)$$

$$\Rightarrow T(\vec{x}) = (-1+k, 1-k, -1+k)$$

$$+ \vec{x} \begin{bmatrix} k & 1-k & -1+k \\ 0 & 1 & 0 \\ 0 & 0 & 1 \end{bmatrix} \quad \text{for } \vec{x} \in \mathbb{R}^3, \quad \text{or}$$

$$\begin{cases} y_1 = -1+k+kx_1, \\ y_2 = 1-k+(1-k)x_1 + x_2, \\ y_3 = -1+k+(-1+k)x_1 + x_3, \end{cases}$$

where $\vec{x} = (x_1, x_2, x_3)$ and $\vec{y} = T(\vec{x}) = (y_1, y_2, y_3)$.

(b) The characteristic polynomial of A is $\det(A - tI_3) = -(t-1)^2(t-2)$. So A has eigenvalues 1, 1 and 2. Since $(A - I_3)(A - 2I_3) = O_{3\times 3}$ (see Example 1 of Sec. 3.7.6), A is diagonalizable and certainly, the corresponding T will represent a one-way stretch if $\vec{x}(I_3 - A) = \vec{x}_0$ has a solution. Solve

$$\vec{x}(I_3 - A) = \vec{x}_0, \quad \text{i.e.}$$

$$(x_1, x_2, x_3) \begin{bmatrix} 2 & -4 & -2 \\ 1 & -2 & -1 \\ 1 & -2 & -1 \end{bmatrix} = (-2, 4, 2)$$

$$\Rightarrow 2x_1 + x_2 + x_3 = -2 \quad \text{does have a (in fact, infinite) solution.}$$

Therefore, T is a one-way stretch with scale factor 2.

Note that $\vec{x}(A - I_3) = \vec{0} \Leftrightarrow 2x_1 + x_2 + x_3 = 0$. Take $\vec{v}_1 = (1, -2, 0)$ and $\vec{v}_2 = (1, 0, -2)$ as linearly independent eigenvectors corresponding to 1. On the other hand, $\vec{x}(A - 2I_3) = \vec{0} \Leftrightarrow 3x_1 + x_2 + x_3 = 4x_1 + x_2 + 2x_3 = 2x_1 + x_2 = 0$. Choose $\vec{v}_3 = (1, -2, -1)$ as an eigenvector corresponding to 2.

Since $\vec{x}_0 = -2\vec{v}_3$, both $\vec{x}_0$ and $\vec{v}_3$ can be chosen as a direction of the one-way stretch T. To find an invariant point of T, let such a point be denoted as $\alpha \vec{x}_0$. By the very definition of a one-way stretch along $\vec{x}_0$,

$$\alpha \vec{x}_0 - T(\vec{0}) = \alpha \vec{x}_0 - \vec{x}_0 = k(\alpha \vec{x}_0 - \vec{0}) \quad \text{with} \quad k = 2$$

$$\Rightarrow (\text{since } \vec{x}_0 \neq \vec{0})\alpha - 1 = k\alpha \quad \text{with} \quad k = 2$$

$$\Rightarrow \alpha = \frac{1}{1-k} = -1$$

$$\Rightarrow \text{An invariant point } \alpha \vec{x}_0 = -\vec{x}_0 = 2\vec{v}_3$$

$$\Rightarrow \text{The plane of invariant point is } 2\vec{v}_3 + \langle\langle \vec{v}_1, \vec{v}_2 \rangle\rangle.$$

Notice that

$$\vec{x} = (x_1, x_2, x_3) \in 2\vec{v}_3 + \langle\langle \vec{v}_1, \vec{v}_2 \rangle\rangle$$
$$\Rightarrow x_1 = 2 + \alpha_1 + \alpha_2, x_2 = -4 - 2\alpha_1, x_3 = -2 - 2\alpha_2 \quad \text{for } \alpha_1, \alpha_2 \in \mathbb{R}$$
$$\Rightarrow \text{(by eliminating the parameters } \alpha_1 \text{ and } \alpha_2) \ 2x_1 + x_2 + x_3 = -2$$
$$\Rightarrow \vec{x}(I_3 - A) = \vec{x}_0$$

as we claimed before. In the affine basis $\mathcal{C} = \{2\vec{v}_3, 2\vec{v}_3 + \vec{v}_1, 2\vec{v}_3 + \vec{v}_2, 2\vec{v}_3 + \vec{v}_3\}$,

$$[T(\vec{x})]_{\mathcal{C}} = [\vec{x}]_{\mathcal{C}} P A P^{-1} = [\vec{x}]_{\mathcal{C}} \begin{bmatrix} 1 & 0 & 0 \\ 0 & 1 & 0 \\ 0 & 0 & 2 \end{bmatrix}, \quad \text{where } P = \begin{bmatrix} 1 & -2 & 0 \\ 1 & 0 & -2 \\ 1 & -2 & -1 \end{bmatrix}.$$

See Fig. 3.62.

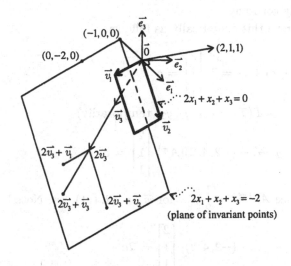

Fig. 3.62

We propose the following questions:

Q1 Where is the image of a plane, parallel to $2x_1 + x_2 + x_3 = -2$, under T?

Q2 Where is the image of a plane or a line, nonparallel to $2x_1 + x_2 + x_3 = -2$, under T? Where do these two planes or lines intersect?

Q3 What is the image of the tetrahedron $\Delta \vec{a}_0 \vec{a}_1 \vec{a}_2 \vec{a}_3$, where $\vec{a}_0 = \vec{0}$, $\vec{a}_1 = (1, 0, -1)$, $\vec{a}_2 = 2\vec{v}_3 = (2, -4, -2)$ and $\vec{a}_3 = (-2, 0, 0)$, under T? What are the volumes of these two tetrahedra?

To answer these questions, we compute firstly that

$$A^{-1} = \frac{1}{2} \begin{bmatrix} 4 & -4 & -2 \\ 1 & 0 & -1 \\ 1 & -2 & 1 \end{bmatrix}.$$

This implies the inverse affine transformation of T is

$$\vec{x} = (T(\vec{x}) - \vec{x}_0)A^{-1}.$$

For Q1 A plane, parallel to $2x_1 + x_2 + x_3 = -2$, is of the form $2x_1 + x_2 + x_3 = c$ which intercepts x_1-axis at the point $\left(\frac{c}{2}, 0, 0\right)$. Since $2x_1 + x_2 + x_3 = -2$ is the plane of invariant points of T, with scale factor 2, so the image plane (refer to Fig. 3.62) is

$$2x_1 + x_2 + x_3 = -2 + 2(c - (-2)) = 2(c+1),$$

where c is any constant.

We can prove this analytically as follows.

$$2x_1 + x_2 + x_3 = \vec{x} \begin{bmatrix} 2 \\ 1 \\ 1 \end{bmatrix} = c$$

$\Rightarrow$ (let $\vec{y} = T(\vec{x}) = (y_1, y_2, y_3)$ temporarily)

$$[(y_1, y_2, y_3) - (-2, 4, 2)]A^{-1} \begin{bmatrix} 2 \\ 1 \\ 1 \end{bmatrix} = c$$

$\Rightarrow$ $\left(\text{since } A^{-1}\vec{y}_0^* = \frac{1}{2}\vec{y}_0^* \text{ for } \vec{y}_0 = (2, 1, 1), \text{ see Note below}\right)$

$$[(y_1, y_2, y_3) - (-2, 4, 2)] \begin{bmatrix} 2 \\ 1 \\ 1 \end{bmatrix} = 2c$$

$\Rightarrow 2y_1 + y_2 + y_3 - (-4 + 4 + 2) = 2c$

$\Rightarrow$ (change y_1, y_2, y_3 back to x_1, x_2, x_3 respectively)

$$2x_1 + x_2 + x_3 = 2(c+1)$$

as claimed above. Therefore, such a plane has its image a plane parallel to itself and, of course, to $2x_1 + x_2 + x_3 = -2$.

Note It is well-known that $\mathrm{Ker}(A - I_3)^\perp = \mathrm{Im}(A^* - I_3)$ (see (3.7.32) and Ex. <A> 5 of Sec. 3.7.3). Since $\mathrm{Ker}(A - I_3)^\perp = \langle\langle (2, 1, 1) \rangle\rangle$, so $\mathrm{Im}(A^* - I_3) =$

$\langle\langle (2,1,1) \rangle\rangle$ holds. Since $(A^* - I_3)(A^* - 2I_3) = O_{3\times3}$, so $\vec{y}_0(A^* - 2I_3) = \vec{0}$ where $\vec{y}_0 = (2,1,1)$. Therefore

$$\vec{y}_0 A^* = 2\vec{y}_0$$
$$\Rightarrow A\vec{y}_0^* = 2\vec{y}_0^*$$
$$\Rightarrow A^{-1}\vec{y}_0^* = \frac{1}{2}\vec{y}^*$$

which can also be easily shown by direct computation.

For Q2 Take, for simplicity, the plane $x_1 - x_2 - x_3 = 5$. To find the image plane of this plane under T, observe that

$$x_1 - x_2 - x_3 = (x_1, x_2, x_3)\begin{bmatrix} 1 \\ -1 \\ -1 \end{bmatrix} = 5$$

$$\Rightarrow \text{(let } \vec{y} = T(\vec{x}) \text{ temporarily)} \ [(y_1, y_2, y_3) - (-2, 4, 2)]A^{-1}\begin{bmatrix} 1 \\ -1 \\ -1 \end{bmatrix} = 5$$

$$\Rightarrow (y_1 + 2, y_2 - 4, y_3 - 2)\begin{bmatrix} 5 \\ 1 \\ 1 \end{bmatrix} = 5(y_1 + 2) + y_2 - 4 + y_3 - 2 = 5$$

$$\Rightarrow \text{(replace } y_1, y_2, y_3 \text{ by } x_1, x_2, x) \quad 5x_1 + x_2 + x_3 = 1$$

which is the equation of the image plane.

The two planes $x_1 - x_2 - x_3 = 5$ and $5x_1 + x_2 + x_3 = 1$ do intersect along the line $x_1 = 1, x_2 + x_3 = -4$ which obviously lies on the plane $2x_1 + x_2 + x_3 = -2$. Refer to Fig. 3.61.

For Q3 By computation,

$T(\vec{a}_0) = T(\vec{0}) = \vec{x}_0 = (-2, 4, 2)$,
$T(\vec{a}_1) = (-2, 4, 2) + (1, 0, -1)A = (-2, 4, 2) + (0, 2, 0) = (-2, 6, 2)$,
$T(\vec{a}_2) = (-2, 4, 2) + (2, -4, -2)A = (-2, 4, 2) + (4, -8, -4) = (2, -4, -2)$,
$T(\vec{a}_3) = (-2, 4, 2) + (-2, 0, 0)A = (-2, 4, 2) + (2, -8, -4) = (0, -4, -2)$.

These four points form a tetrahedron $\Delta T(\vec{a}_0)T(\vec{a}_1)T(\vec{a}_2)T(\vec{a}_3)$ whose volume is equal to

$$\det\begin{bmatrix} T(\vec{a}_1) - T(\vec{a}_0) \\ T(\vec{a}_2) - T(\vec{a}_0) \\ T(\vec{a}_3) - T(\vec{a}_0) \end{bmatrix} = \begin{vmatrix} 0 & 2 & 0 \\ 4 & -8 & -4 \\ 2 & -8 & -4 \end{vmatrix} = 16.$$

While $\triangle \vec{a_0}\,\vec{a_1}\,\vec{a_2}\,\vec{a_3}$ has volume equal to

$$\det \begin{bmatrix} \vec{a_1} \\ \vec{a_2} \\ \vec{a_3} \end{bmatrix} = \begin{vmatrix} 1 & 0 & -1 \\ 2 & -4 & -2 \\ -2 & 0 & 0 \end{vmatrix} = 8.$$

Therefore,

$$\frac{\text{Volume of } \triangle T(\vec{a_0})\cdots T(\vec{a_3})}{\text{Volume of } \triangle \vec{a_0}\cdots \vec{a_3}} = \frac{16}{8} = 2 = \det A.$$

See Fig. 3.63.

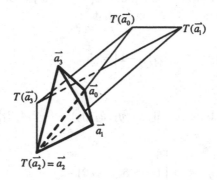

Fig. 3.63

Case 4   Two-way stretch

A two-way stretch is the composite of two one-way stretches whose planes of invariant points intersect along a line which is going to be the only line of invariant points if both scale factors are different from 1.

As a counterpart of (2.8.30), we have

The two-way stretch

Let $\vec{a_0}, \vec{a_1}, \vec{a_2}$ and $\vec{a_3}$ be non-coplanar points in $\mathbb{R}^3$. Denote by T the two-way stretch having scale factor k_1 along $\vec{a_0} + \langle\langle \vec{a_1} - \vec{a_0} \rangle\rangle$ and scale factor k_2 along $\vec{a_0} + \langle\langle \vec{a_2} - \vec{a_0} \rangle\rangle$ and $\vec{a_0} + \langle\langle \vec{a_3} - \vec{a_0} \rangle\rangle$ the only line of invariant points. Then

1. In the affine basis $\mathcal{B} = \{\vec{a_0}, \vec{a_1}, \vec{a_2}, \vec{a_3}\}$,

$$[T(\vec{x})]_{\mathcal{B}} = [\vec{x}]_{\mathcal{B}}[T]_{\mathcal{B}},$$

$$\text{where } [T]_{\mathcal{B}} = \begin{bmatrix} k_1 & 0 & 0 \\ 0 & k_2 & 0 \\ 0 & 0 & 1 \end{bmatrix} = \begin{bmatrix} k_1 & 0 & 0 \\ 0 & 1 & 0 \\ 0 & 0 & 1 \end{bmatrix} \begin{bmatrix} 1 & 0 & 0 \\ 0 & k_2 & 0 \\ 0 & 0 & 1 \end{bmatrix}.$$

2. In the natural affine basis $\mathcal{N} = \{\vec{0}, \vec{e}_1, \vec{e}_2, \vec{e}_3\}$,

$$T(\vec{x}) = \vec{a}_0 + (\vec{x} - \vec{a}_0)P^{-1}[T]_{\mathcal{B}}P, \quad \text{where } P = \begin{bmatrix} \vec{a}_1 - \vec{a}_0 \\ \vec{a}_2 - \vec{a}_0 \\ \vec{a}_3 - \vec{a}_0 \end{bmatrix}.$$

Also,

(1) A two-way stretch preserves all the properties 1–7 listed in (3.7.15),
(2) but enlarge the volume by the scale factor $|k_1 k_2|$ and preserves the orientation if $k_1 k_2 > 0$ while reverses the orientation if $k_1 k_2 < 0$.

$$(3.8.17)$$

See Fig. 3.64 (refer to Fig. 2.102).

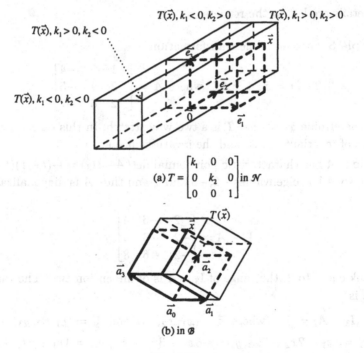

(a) $T = \begin{bmatrix} k_1 & 0 & 0 \\ 0 & k_2 & 0 \\ 0 & 0 & 1 \end{bmatrix}$ in $\mathcal{N}$

(b) in $\mathcal{B}$

Fig. 3.64

In case $k_1 = k_2 = k \neq 1$, then the plane $a_0 + \langle\langle \vec{a}_1 - \vec{a}_0, \vec{a}_2 - \vec{a}_0 \rangle\rangle$ is an invariant plane of T, on which T is an *enlargement* with scale factor k.

To see if $T(\vec{x}) = \vec{x}_0 + \vec{x}A$ is a two-way stretch, try the following steps:

1. A has eigenvalues 1, k_1 and k_2 where $k_1 \neq 1$ and $k_2 \neq 1$.
2. If A is diagonalizable (hence, the rank $r(I_3 - A) = 2$) and the equation $\vec{x}(I_3 - A) = \vec{x_0}$ has a nontrivial solution, then A is a two-way stretch.
3. Compute an eigenvector $\vec{v_3}$ associated to 1. Let $\vec{x_1}$ be a solution of $\vec{x}(I_3 - A) = \vec{x_0}$. Then

$$\vec{x_1} + \langle\langle \vec{v_3} \rangle\rangle = \{\vec{x} \in \mathbb{R}^3 \,|\, \vec{x}(I_3 - A) = \vec{x_0}\}$$

is the line of invariant points.
4. Compute an eigenvector $\vec{v_1}$ for k_1 and an eigenvector $\vec{v_2}$ for k_2 so that $\vec{v_1}$ and $\vec{v_2}$ are linearly independent in case $k_1 = k_2$. Then

$$\vec{x_1} + \langle\langle \vec{v_1}, \vec{v_2} \rangle\rangle$$

is an invariant plane.

$$(3.8.18)$$

The details are left to the readers.

Example 3 Give an affine transformation

$$T(\vec{x}) = \vec{x_0} + \vec{x}A, \quad \text{where } A = \begin{bmatrix} -1 & 6 & -4 \\ 2 & 4 & -5 \\ 2 & 6 & -7 \end{bmatrix}.$$

Try to determine $\vec{x_0}$ so that T is a two-way stretch. In this case, determine the line of invariant points and the invariant plane.

Solution A has characteristic polynomial $\det(A - tI_3) = -(t-1)(t+2) \times (t+3)$, so A has eigenvalues $-2, -3$ and 1 and thus A is diagonalizable.
Since

$$I_3 - A = \begin{bmatrix} 2 & -6 & 4 \\ -2 & -3 & 5 \\ -2 & -6 & 8 \end{bmatrix}$$

has rank equal to 2, the range of $I_3 - A$ is of dimension two. The range of $I_3 - A$ is

$\vec{x}(I_3 - A) = \vec{y}, \quad \text{where } \vec{x} = (x_1, x_2, x_3) \text{ and } \vec{y} = (y_1, y_2, y_3)$

$\Leftrightarrow y_1 = 2x_1 - 2x_2 - 2x_3, y_2 = -6x_1 - 3x_2 - 6x_3, y_3 = 4x_1 + 5x_2 + 8x_3$

$\Leftrightarrow y_1 + y_2 + y_3 = 0$

$\Leftrightarrow$ (replace y_1, y_2, y_3 by x_1, x_2, x_3 respectively)

$\text{Im}(I_3 - A) = \{\vec{x} = (\vec{x_1}, \vec{x_2}, \vec{x_3}) \,|\, x_1 + x_2 + x_3 = 0\}.$

So, any point $\vec{x_0}$ so that $\vec{x_0} \in \text{Im}(I_3 - A)$ will work.

Solving $\vec{x}(A - I_3) = \vec{0}$, get the corresponding eigenvector $\vec{v}_3 = (1, 4, -3)$. Solving $\vec{x}(A + 2I_3) = \vec{0}$, get $\vec{v}_1 = (0, 1, -1)$ and solving $\vec{x}(A + 3I_3) = \vec{0}$, get $\vec{v}_2 = (1, 0, -1)$.

For any $\vec{x}_1$ such that $\vec{x}_1(I_3 - A) = \vec{x}_0$ holds, T is a two-way stretch with

the line of invariant points: $\vec{x}_1 + \langle\langle \vec{v}_3 \rangle\rangle$, and
the invariant plane: $\vec{x}_1 + \langle\langle \vec{v}_1, \vec{v}_2 \rangle\rangle$.

For example, take $\vec{x}_0 = (2, -3, 1)$ and solve $\vec{x}(I_3 - A) = (2, -3, 1)$ so that

$$\vec{x} = (x_1, -3 + 4x_1, 2 - 3x_1) = (0, -3, 2) + x_1(1, 4, -3), \quad x_1 \in \mathbb{R}$$
$$\Rightarrow (0, -3, 2) + \langle\langle (1, 4, -3) \rangle\rangle, \quad \text{where } \vec{x}_1 = (0, -3, 2)$$

is the line L of invariant points. On the other hand, the plane

$$(0, -3, 2) + \langle\langle \vec{v}_1, \vec{v}_2 \rangle\rangle = \{(\alpha_2, -3 + \alpha_1, 2 - \alpha_1 - \alpha_2) \,|\, \alpha_1, \alpha_2 \in \mathbb{R}\}$$
$$= \{\vec{x} = (x_1, x_2, x_3) \,|\, x_1 + x_2 + x_3 = -1\}$$

is an invariant plane.

Note Careful readers might have observed in Example 2 that, since $(A - 2I_3)(A - I_3) = O, \text{Im}(A - 2I_3), \text{Ker}(A - I_3)$ and $\langle\langle \vec{v}_1, \vec{v}_2 \rangle\rangle$ are the same plane $2x_1 + x_2 + x_3 = 0$ and hence, the plane of invariant points is just a translation of it, namely $2x_1 + x_2 + x_3 = -2$. Refer to Fig. 3.62.

This is not accidental. This does happen in Example 3 too. Since

$$(A - I_3)(A + 2I_3)(A + 3I_3) = O_{3\times3}$$
$$\Rightarrow (\text{since } r(A - I_3) = 2) \, \text{Im}(A - I_3) = \text{Ker}(A + 2I_3)(A + 3I_3)$$
$$\Rightarrow (\text{since } \text{Ker}(A + 2I_3) \oplus \text{Ker}(A + 3I_3) \subseteq \text{Ker}(A + 2I_3)(A + 3I_3))$$
$$\text{Im}(A - I_3) = \text{Ker}(A + 2I_3) \oplus (A + 3I_3) = \langle\langle \vec{v}_1, \vec{v}_2 \rangle\rangle.$$

This justifies that both $\text{Im}(A - I_3)$ and $\langle\langle \vec{v}_1, \vec{v}_2 \rangle\rangle$ are the same plane $x_1 + x_2 + x_3 = 0$ as indicated above. As a consequence, all the planes parallel to it are invariant planes. See Fig. 3.65.

In the affine basis $\mathcal{C} = \{\vec{x}_1, \vec{x}_1 + \vec{v}_1, \vec{x}_1 + \vec{v}_2, \vec{x}_1 + \vec{v}_3\}$,

$$[T(\vec{x})]_\mathcal{C} = [\vec{x}]_\mathcal{C}[T]_\mathcal{C}, \quad \text{where } [T]_\mathcal{C} = \begin{bmatrix} -2 & 0 & 0 \\ 0 & -3 & 0 \\ 0 & 0 & 1 \end{bmatrix}.$$

Try to answer the following questions:

Q1 What is the image of a plane parallel to the line of invariant points?
Q2 Is each plane containing the line of invariant points an invariant plane?

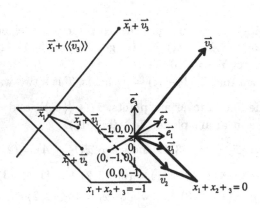

Fig. 3.65

Q3 Where is the intersecting line of a plane, nonparallel to $\vec{x}_1 + \langle\langle\vec{v}_1, \vec{v}_2\rangle\rangle$, with its image?

Q4 Let $\vec{a}_0 = \vec{x}_1 = (0, -3, 2)$, $\vec{a}_1 = (-1, -1, -1)$, $\vec{a}_2 = (1, -2, 1)$ and $\vec{a}_3 = \vec{x}_1 + \vec{v}_3 = (1, 1, -1)$. Find the image of the tetrahedron $\triangle \vec{a}_0 \vec{a}_1 \vec{a}_2 \vec{a}_3$ and compute its volume.

For these, note that the inverse transformation of T is

$$\vec{x} = (\vec{y} - (2, -3, 1))A^{-1}, \quad \text{where } \vec{y} = T(\vec{x}) \text{ and } A^{-1} = \frac{1}{6}\begin{bmatrix} 2 & 18 & -14 \\ 4 & 15 & -13 \\ 4 & 18 & -16 \end{bmatrix}.$$

For Q1 Geometric intuition (see Fig. 3.65) suggests that the image of any such plane would be parallel to the plane. This can be proved analytically as follows.

A plane Σ parallel to the line L of invariant points has equation like

$$(-4a_2 + 3a_3)x_1 + a_2 x_2 + a_3 x_3 = b, \quad \text{where } b \neq -3a_2 + 2a_3, a_2, a_3 \in \mathbb{R}.$$

The condition $b \neq -3a_2 + 2a_3$ means that L is not coincident on Σ. Then

$$\Sigma \parallel L$$

$$\Leftrightarrow \vec{x} \begin{bmatrix} -4a_2 + 3a_3 \\ a_2 \\ a_3 \end{bmatrix} = b$$

$$\Leftrightarrow [\vec{y} - (2, -3, 1)]A^{-1} \begin{bmatrix} -4a_2 + 3a_3 \\ a_2 \\ a_3 \end{bmatrix} = \frac{1}{6}[\vec{y} - (2, -3, 1)] \begin{bmatrix} 10a_2 - 8a_3 \\ -a_2 - a_3 \\ 2a_2 - 4a_3 \end{bmatrix} = b$$

$$\Leftrightarrow (10a_2 - 8a_3)y_1 - (a_2 + a_3)y_2 + (2a_2 - 4a_3)y_3 = 6b + 25a_2 - 17a_3$$
$$\Rightarrow (\text{let } \alpha_2 = -a_2 - a_3, \alpha_3 = 2a_2 - 4a_3)$$
$$(-4\alpha_2 + 3\alpha_3)x_1 + \alpha_2 x_2 + \alpha_3 x_3 = 6b + 25a_2 - 17a_3,$$

where $6b + 25a_2 - 17a_3 \neq -18a_2 + 12a_3 + 25a_2 - 17a_3 = -3\alpha_2 + 2\alpha_3$. Hence, the image plane $T(\Sigma)\|\Sigma\|L$. See Note below.

Note (refer to Note in Example 2) It is easy to see that

$$\Sigma \parallel \langle\langle \vec{v}_3, \vec{v}_1 \rangle\rangle \Leftrightarrow a_2 = a_3.$$

Hence, Σ has equation $-ax_1 + ax_2 + ax_3 = b$, where $a \neq 0$, or equivalently, $x_1 - x_2 - x_3 = b$ with $-\frac{b}{a}$ replacing by $b \neq 1$. Let $\vec{n} = (1, -1, -1)$ be the normal to these parallel planes. Then

$$\vec{v}_3 \in \text{Ker}(A - I_3) = \text{Im}(A^* - I_3)^{\perp} \quad \text{and}$$
$$\vec{v}_1 \in \text{Ker}(A + 2I_3) = \text{Im}(A^* + 2I_3)^{\perp}$$

$$\Rightarrow \vec{n} \in \text{Im}(A^* - I_3) \cap \text{Im}(A^* + 2I_3) = \text{Ker}(A^* + 3I_3)$$
$$\Rightarrow \vec{n}A^* = -3\vec{n}$$
$$\Rightarrow A\vec{n}^* = -3\vec{n}^*$$
$$\Rightarrow A^{-1}\vec{n}^* = -\frac{1}{3}\vec{n}^*.$$

Similarly,

$$\Sigma \parallel \langle\langle \vec{v}_3, \vec{v}_2 \rangle\rangle \Leftrightarrow 2a_2 - a_3 = 0.$$
$$\Rightarrow \Sigma \text{ has equation } 2x_1 + x_2 + 2x_3 = b, \quad b \neq 1.$$
$$\Rightarrow A^{-1}\vec{u}^* = -\frac{1}{2}\vec{u}^*, \text{ where } \vec{u} = (2, 1, 2), \text{ the normal vector.}$$

For Q2 The answer is negative in general except the planes $\vec{x}_1 + \langle\langle \vec{v}_3, \vec{v}_1 \rangle\rangle$ and $\vec{x}_1 + \langle\langle \vec{v}_3, \vec{v}_2 \rangle\rangle$ which are invariant planes. But each such plane and its image plane intersect along the line L of invariant points. Q1 answers all these claims.

For Q3 For simplicity, take $x_3 = 0$ as a sample plane. The plane $x_3 = 0$ intercepts the line L at the point $(\frac{2}{3}, -\frac{1}{3}, 0)$ and intersects the plane $\vec{x}_1 + \langle\langle \vec{v}_1, \vec{v}_2 \rangle\rangle$ along the line $x_3 = 0, x_1 + x_2 + x_3 = -1$. To find its image under T,

$$x_3 = \vec{x} \begin{bmatrix} 0 \\ 0 \\ 1 \end{bmatrix} = 0$$

$$\Leftrightarrow [\vec{y} - (2, -3, 1)]A^{-1} \begin{bmatrix} 0 \\ 0 \\ 1 \end{bmatrix} = \frac{1}{6}[\vec{y} - (2, -3, 1)] \begin{bmatrix} -14 \\ -13 \\ -16 \end{bmatrix} = 0$$

$$\Rightarrow \text{(replace } \vec{y} \text{ by } \vec{x}) \quad 14x_1 + 13x_2 + 16x_3 = 5.$$

This image plane does intercept the line L at the invariant point $\left(\frac{2}{3}, -\frac{1}{3}, 0\right)$ and intersects the original plane $x_3 = 0$ along the line $x_3 = 0, 14x_1 + 13x_2 + 16x_3 = 5$.

For Q4 By computation,

$$T(\vec{a}_0) = \vec{x}_1 = (0, -3, 2),$$
$$T(\vec{a}_1) = (0, -3, 2) + (-1, -1, -1)A = (-3, -19, 18),$$
$$T(\vec{a}_2) = (0, -3, 2) + (1, -2, 1)A = (-3, 1, 1),$$
$$T(\vec{a}_3) = \vec{x}_1 + \vec{v}_3 = (1, 1, -1).$$

Then,

The signed volume of $\triangle \vec{a}_0 \vec{a}_1 \vec{a}_2 \vec{a}_3$

$$= \begin{vmatrix} \vec{a}_1 - \vec{a}_0 \\ \vec{a}_2 - \vec{a}_0 \\ \vec{a}_3 - \vec{a}_0 \end{vmatrix} = \begin{vmatrix} -1 & 2 & -3 \\ 1 & 1 & -1 \\ 1 & 4 & -3 \end{vmatrix} = -6, \quad \text{and}$$

the signed volume of $\triangle T(\vec{a}_0) T(\vec{a}_1) T(\vec{a}_2) T(\vec{a}_3)$

$$= \begin{vmatrix} -3 & -16 & 16 \\ -3 & 4 & -1 \\ 1 & 4 & -3 \end{vmatrix} = -36.$$

$$\Rightarrow \frac{\text{The signed volume of } \triangle T(\vec{a}_0) \cdots T(\vec{a}_3)}{\text{The signed volume of } \triangle \vec{a}_0 \cdots \vec{a}_3} = \frac{-36}{-6} = 6 = \det A. \qquad \Box$$

Case 5   (Three-way) stretch

The composite of three one-way stretches with scale factors all different from 1 is called a *three-way stretch* or simply a *stretch*.

As an extension of (3.8.17), we have

The stretch

Let $\vec{a}_0, \vec{a}_1, \vec{a}_2$ and $\vec{a}_3$ be non-coplanar points in $\mathbb{R}^3$. Denote by T the stretch with scale factor k_i along the line $\vec{a}_0 + \langle\langle \vec{a}_i - \vec{a}_0 \rangle\rangle$ for $i = 1, 2, 3$, where k_1, k_2 and k_3 are all nonzero and different from 1. Then

1. In the affine basis $\mathcal{B} = \{\vec{a}_0, \vec{a}_1, \vec{a}_2, \vec{a}_3\}$,

$$[T(\vec{x})]_\mathcal{B} = [\vec{x}]_\mathcal{B}[T]_\mathcal{B}, \quad \text{where}$$

$$[T]_\mathcal{B} = \begin{bmatrix} k_1 & 0 & 0 \\ 0 & k_2 & 0 \\ 0 & 0 & k_3 \end{bmatrix} = \begin{bmatrix} k_1 & 0 & 0 \\ 0 & 1 & 0 \\ 0 & 0 & 1 \end{bmatrix} \begin{bmatrix} 1 & 0 & 0 \\ 0 & k_2 & 0 \\ 0 & 0 & 1 \end{bmatrix} \begin{bmatrix} 1 & 0 & 0 \\ 0 & 1 & 0 \\ 0 & 0 & k_3 \end{bmatrix}.$$

2. In the natural affine basis $\mathcal{N} = \{\vec{0}, \vec{e}_1, \vec{e}_2, \vec{e}_3\}$,

$$T(\vec{x}) = \vec{a}_0 + (\vec{x} - \vec{a}_0)P^{-1}[T]_\mathcal{B}P, \quad \text{where } P = \begin{bmatrix} \vec{a}_1 - \vec{a}_0 \\ \vec{a}_2 - \vec{a}_0 \\ \vec{a}_3 - \vec{a}_0 \end{bmatrix}.$$

If $k_1 \neq k_2 \neq k_3$, then $\vec{a}_0$ is the only invariant point and there is no invariant plane. If $k_1 = k_2 \neq k_3$, then $\vec{a}_0 + \langle\langle \vec{a}_1 - \vec{a}_0, \vec{a}_2 - \vec{a}_0 \rangle\rangle$ is an invariant plane. If $k_1 = k_2 = k_3 = k$, then T is called an *enlargement* with scale factor k and any plane parallel to each of the three coordinate planes (see Sec. 3.6) is an invariant plane. In case $\vec{a}_1 - \vec{a}_0, \vec{a}_2 - \vec{a}_0$ and $\vec{a}_3 - \vec{a}_0$ are perpendicular to each other, T is called an *orthogonal stretch*; if, in addition, the lengths $|\vec{a}_1 - \vec{a}_0| = |\vec{a}_2 - \vec{a}_0| = |\vec{a}_3 - \vec{a}_0|$ and $k_1 = k_2 = k_3 = k$, then T is called a *similarity* with scale factor k. In general,

(a) A stretch preserves all properties 1–7 listed is (3.7.15),
(b) but enlarge the volumes by the scalar factor $|k_1 k_2 k_3|$ and preserves the orientation if $k_1 k_2 k > 0$ while reverses the orientation if $k_1 k_2 k < 0$.

$$(3.8.19)$$

See Fig. 3.66.

To test if an affine transformation $T(\vec{x}) = \vec{x}_0 + \vec{x}A$ is a stretch, all one needs to do is to compute the eigenvalues of A and to see if it is diagonalizable (see (3.7.46)).

Case 6 Shearing (Euclidean notions are needed)

Let Σ be a plane in space and $\vec{v} \neq \vec{0}$ be a space vector which is *parallel* to Σ. Take any fixed scalar $k \neq 0$. Each point X in space moves in the

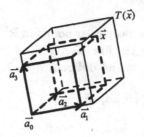

Fig. 3.66

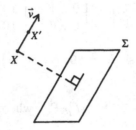

Fig. 3.67

direction $\vec{v}$ to a new point X' so that

$$\frac{\text{the signed distance } XX' \text{ from } X \text{ to } X'}{\text{the (perpendicular) distance from } X \text{ to } \Sigma} = k.$$

It is understood that, if $X \in \Sigma$, then $X' = X$. Then, the mapping $T: \mathbb{R}^3 \to \mathbb{R}^3$ defined by

$$T(X) = X'$$

is an affine transformation and is called a *shearing* with *coefficient* k in the *direction* $\vec{v}$ with respect to the *plane* Σ *of invariant points*. Obviously,

1. Any plane parallel to Σ or the direction is an invariant plane.
2. Points on opposite sides of Σ move in opposite directions $\vec{v}$ and $-\vec{v}$ (refer to Fig. 2.105).
3. Each plane which intersects with Σ along a line L will intersect its image plane, under T, along the same line L.
4. T preserves the volume of a parallelepiped, and hence volumes of a tetrahedron and any solid domain in space (refer to Case 5 in Sec. 2.8.2 and the right figure in Fig. 3.34).

$$(3.8.20)$$

The details are left as Ex. <A> 12.

As a counterpart of (2.8.32) in $\mathbb{R}^3$, we have

The shearing
Let $\vec{a}_0, \vec{a}_1, \vec{a}_2$ and $\vec{a}_3$ be non-coplanar points in $\mathbb{R}^3$ so that

1. (in lengths) $|\vec{a}_1 - \vec{a}_0| = |\vec{a}_2 - \vec{a}_0| = |\vec{a}_3 - \vec{a}_0| = 1$, and
2. (perpendicularity) $(\vec{a}_1 - \vec{a}_0) \perp (\vec{a}_2 - \vec{a}_0) \perp (\vec{a}_3 - \vec{a}_0)$,

i.e. $\mathcal{B} = \{\vec{a}_0, \vec{a}_1, \vec{a}_2, \vec{a}_3\}$ is an *orthonormal affine basis* for $\mathbb{R}^3$. Denote by T the shearing with coefficient k in the direction $\vec{a}_1 - \vec{a}_0$ with respect to the plane $\vec{a}_0 + \langle\langle \vec{a}_1 - \vec{a}_0, \vec{a}_3 - \vec{a}_0 \rangle\rangle$ as plane of invariant points. Then

a. In $\mathcal{B}$,

$$[T(\vec{x})]_\mathcal{B} = [\vec{x}]_\mathcal{B}[T]_\mathcal{B}, \quad \text{where } [T]_\mathcal{B} = \begin{bmatrix} 1 & 0 & 0 \\ k & 1 & 0 \\ 0 & 0 & 1 \end{bmatrix}.$$

b. In $\mathcal{N} = \{\vec{0}, \vec{e}_1, \vec{e}_2, \vec{e}_3\}$,

$$T(\vec{x}) = \vec{a}_0 + (\vec{x} - \vec{a}_0)P^{-1}[T]_\mathcal{B}P, \quad \text{where } P = \begin{bmatrix} \vec{a}_1 - \vec{a}_0 \\ \vec{a}_2 - \vec{a}_0 \\ \vec{a}_3 - \vec{a}_0 \end{bmatrix}.$$

Here P is an orthogonal matrix, i.e. $P^* = P^{-1}$.

Also,

(1) A shearing preserves all the properties 1–7 listed in (3.7.15), and
(2) preserves the volumes and the orientations. $\hspace{3cm}$ (3.8.21)

For skew shearing, refer to Ex. <A> 17 of Sec. 2.8.2 and Ex. <B> 1.

To see if an affine transformation $T(\vec{x}) = \vec{x}_0 + \vec{x}A$ is a shearing, try the following steps (refer to (2.8.33)):

1. Compute the eigenvalues of A. In case A has only one eigenvalue 1 with multiplicity 3 and $(A - I_3)^2 = O, A$ is not diagonalizable (see (2) in (3.7.50)) and then, the associated T is a shearing if $\vec{x}(I_3 - A) = \vec{x}_0$ has a solution.
2. The eigenspace

$$E = \text{Ker}(A - I_3)$$

has dimension equal to 2. Determine eigenvectors $\vec{v}_1$ and $\vec{v}_3$ so that $|\vec{v}_1| = |\vec{v}_3| = 1$ and $\vec{v}_1 \perp \vec{v}_3$ in order to form an *orthonormal basis* for E.

3. Take a vector $\vec{v}_2$, of unit length and perpendicular to E, then $\vec{v}_2 A - \vec{v}_2 = k\vec{v}_1$ holds and k is the coefficient.

4. Take $\vec{a}_0$ as a solution of $\vec{x}(I_3 - A) = \vec{x}_0$, then $\vec{v}_1$ is the direction and $\vec{a}_0 + \langle\langle \vec{v}_1, \vec{v}_3 \rangle\rangle$ or $\vec{x}(I_3 - A) = \vec{x}_0$ is the plane of invariant points.

$$(3.8.22)$$

The details are left as Ex. <A> 12.

Example 4 Let $\vec{a}_0 = (1,1,0)$, $\vec{a}_1 = (2,0,-1)$ and $\vec{a}_2 = (0,-1,1)$. Try to construct a shearing with coefficient $k \neq 0$ in the direction $\vec{a}_1 - \vec{a}_0$ with $\vec{a}_0 + \langle\langle (\vec{a}_2 - \vec{a}_0) \rangle\rangle^{\perp}$ as the plane of invariant points. Note that $(\vec{a}_1 - \vec{a}_0) \perp (\vec{a}_2 - \vec{a}_0)$.

Solution Since $\vec{a}_1 - \vec{a}_0 = (1,-1,-1)$ has length $\sqrt{3}$, take the unit vector $\vec{v}_1 = \frac{1}{\sqrt{3}}(1,-1,-1)$. Since $\vec{a}_2 - \vec{a}_0 = (-1,-2,1)$ happens to be perpendicular to $\vec{a}_1 - \vec{a}_0$, i.e.

$$(\vec{a}_2 - \vec{a}_0)(\vec{a}_1 - \vec{a}_0)^* = (-1 \quad -2 \quad 1)\begin{bmatrix} 1 \\ -1 \\ -1 \end{bmatrix} = -1 + 2 - 1 = 0,$$

so we can choose $\vec{v}_2$ to be equal to $\vec{a}_2 - \vec{a}_0$ dividing by its length $\sqrt{6}$, i.e. $\vec{v}_2 = \frac{1}{\sqrt{6}}(-1,-2,1)$. Then, choose a vector $\vec{v}_3 = (\alpha_1, \alpha_2, \alpha_3)$ of unit length so that

$$\vec{v}_3 \perp \vec{v}_1 \quad \text{and} \quad \vec{v}_3 \perp \vec{v}_2$$
$$\Rightarrow \alpha_1 - \alpha_2 - \alpha_3 = 0 \quad \text{and} \quad -\alpha_1 - 2\alpha_2 + \alpha_3 = 0$$
$$\Rightarrow \alpha_1 = \alpha_3 \text{ and } \alpha_2 = 0$$
$$\Rightarrow \vec{v}_3 = \frac{1}{\sqrt{2}}(1,0,1).$$

In the orthonormal affine basis $\mathcal{B} = \{\vec{a}_0, \vec{a}_0 + \vec{v}_1, \vec{a}_0 + \vec{v}_2, \vec{a}_0 + \vec{v}_3\}$, the required shearing T has the representation

$$[T(\vec{x})]_{\mathcal{B}} = [\vec{x}]_{\mathcal{B}}[T]_{\mathcal{B}}, \quad \text{where } [T]_{\mathcal{B}} = \begin{bmatrix} 1 & 0 & 0 \\ k & 1 & 0 \\ 0 & 0 & 1 \end{bmatrix}, \text{ simply denoted as } A.$$

While in $\mathcal{N} = \{\vec{0}, \vec{e}_1, \vec{e}_2, \vec{e}_3\}$,

$$T(\vec{x}) = \vec{a}_0 + (\vec{x} - \vec{a}_0)P^{-1}AP,$$

where

$$P = \begin{bmatrix} \vec{v}_1 \\ \vec{v}_2 \\ \vec{v}_3 \end{bmatrix} = \begin{bmatrix} \frac{1}{\sqrt{3}} & -\frac{1}{\sqrt{3}} & -\frac{1}{\sqrt{3}} \\ -\frac{1}{\sqrt{6}} & -\frac{2}{\sqrt{6}} & \frac{1}{\sqrt{6}} \\ \frac{1}{\sqrt{2}} & 0 & \frac{1}{\sqrt{2}} \end{bmatrix} \quad \text{with} \quad P^{-1} = P^*.$$

By computation,

$$P^{-1}AP = \begin{bmatrix} 1 - \frac{k}{3\sqrt{2}} & \frac{k}{3\sqrt{2}} & \frac{k}{3\sqrt{2}} \\ \frac{-2k}{3\sqrt{2}} & 1 + \frac{2k}{3\sqrt{2}} & \frac{2k}{3\sqrt{2}} \\ \frac{k}{3\sqrt{2}} & \frac{-k}{3\sqrt{2}} & 1 - \frac{k}{3\sqrt{2}} \end{bmatrix}, \quad \text{and}$$

$$\vec{x}_0 = \vec{a}_0 - \vec{a}_0 P^{-1}AP = \vec{a}_0 P^{-1}(I_3 - A)P = \frac{k}{\sqrt{2}}(1, -1, -1)$$

$$\Rightarrow T(\vec{x}) = \vec{x}_0 + \vec{x}P^{-1}AP.$$

See Fig. 3.68.

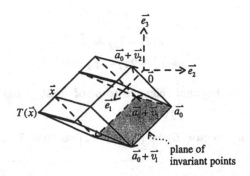

Fig. 3.68

For simplicity, take $k = \sqrt{2}$ and consider the converse problem. Let

$$T(\vec{x}) = \vec{x}_0 + \vec{x}B, \quad \text{where} \quad \vec{x}_0 = (1, -1, -1) \quad \text{and}$$

$$B = P^{-1}AP = \begin{bmatrix} \frac{2}{3} & \frac{1}{3} & \frac{1}{3} \\ -\frac{2}{3} & \frac{5}{3} & \frac{2}{3} \\ \frac{1}{3} & -\frac{1}{3} & \frac{2}{3} \end{bmatrix}.$$

We want to check if this T is a shearing. Follow the steps in (3.8.22).

1. B has characteristic polynomial

$$\det(B - tI_3) = \frac{1}{27}(-27t^3 + 81t^2 - 81t + 27) = -(t-1)^3.$$

So B has eigenvalue 1 of algebraic multiplicity 3. Furthermore, $B - I_3 \neq O$ but

$$(B - I_3)^2 = \begin{bmatrix} -\frac{1}{3} & \frac{1}{3} & \frac{1}{3} \\ -\frac{2}{3} & \frac{2}{3} & \frac{2}{3} \\ \frac{1}{3} & -\frac{1}{3} & -\frac{1}{3} \end{bmatrix}^2 = O_{3\times3}.$$

Hence B is not diagonalizable. Solve

$$\vec{x}(I_3 - B) = \vec{x}_0 = (1, -1, -1), \quad \text{where } \vec{x} = (x_1, x_2, x_3)$$
$$\Rightarrow x_1 + 2x_2 - x_3 = 3 \quad \text{has infinitely many solutions.}$$

Pick any such a solution, say $\vec{a}_0 = (1, 1, 0)$. These information guarantee that T is a shearing.

2. Solve

$$\vec{x}(B - I_3) = \vec{0}$$
$$\Rightarrow x_1 + 2x_2 - x_3 = 0.$$

Pick up two orthogonal eigenvectors of unit length, say $\vec{v}_1 = \frac{1}{\sqrt{3}}(1, -1, -1)$ and $\vec{v}_3 = \frac{1}{\sqrt{2}}(1, 0, 1)$.

3. To choose a unit vector $\vec{v}_2 = (\alpha_1, \alpha_2, \alpha_3)$ so that $\vec{v}_2 \perp \vec{v}_1$ and $\vec{v}_2 \perp \vec{v}_3$ hold, solve

$$\alpha_1 - \alpha_2 - \alpha_3 = \alpha_1 + \alpha_3 = 0 \Rightarrow (\alpha_1, \alpha_2, \alpha_3) = \alpha_1(1, 2, -1).$$

Take $\vec{v}_2 = \frac{1}{\sqrt{6}}(1, 2, -1)$. Notice that this $\vec{v}_2$ is not an eigenvector associated to 1 since $\dim \text{Ker}(B - I_3) = 2$. By computation,

$$\vec{v}_2 B = \frac{1}{\sqrt{6}}(1, 2, -1) \begin{bmatrix} \frac{2}{3} & \frac{1}{3} & \frac{1}{3} \\ -\frac{2}{3} & \frac{5}{3} & \frac{2}{3} \\ \frac{1}{3} & -\frac{1}{3} & \frac{2}{3} \end{bmatrix} = \frac{1}{\sqrt{6}}(-1, 4, 1)$$

$$\Rightarrow \vec{v}_2 B - \vec{v}_2 = \frac{2}{\sqrt{6}}(-1, 1, 1) = -\sqrt{2}\vec{v}_1.$$

Therefore, $-\sqrt{2}$ is the coefficient.

Hence, in $C = \{\vec{a}_0, \vec{a}_0 + \vec{v}_1, \vec{a}_0 + \vec{v}_2, \vec{a}_0 + \vec{v}_3\}$,

$$[T(\vec{x})]_C = [\vec{x}]_C [T]_C, \quad \text{where } [T]_C = \begin{bmatrix} 1 & 0 & 0 \\ -\sqrt{2} & 1 & 0 \\ 0 & 0 & 1 \end{bmatrix}.$$

Notice that the $\vec{v}_2$ in C is equal to $-\vec{v}_2$ in B metioned above. This is the reason why we have $-\sqrt{2}$ here instead of the original $\sqrt{2}$ (see 2 in (3.8.20)). Two questions are raised as follows.

Q1 What is the image plane of the plane $x_1 + x_2 = 0$ under T? Where do they intersect?

Q2 Where is the image of the tetrahedron $\triangle \vec{b}_0 \vec{b}_1 \vec{b}_2 \vec{b}_3$, where $\vec{b}_0 = (1,1,1)$, $\vec{b}_1 = (-1,1,1)$, $\vec{b}_2 = (1,-1,1)$ and $\vec{b}_3 = (1,1,-1)$? How are their volumes related?

We need to use the inverse of $T(\vec{x}) = \vec{x}_0 + \vec{x}B$, namely,

$$\vec{x} = (\vec{y} - \vec{x}_0)B^{-1}, \quad \text{where } \vec{x}_0 = (1,-1,-1) \text{ and } \vec{y} = T(\vec{x}) \text{ and}$$

$$B^{-1} = P^{-1}A^{-1}P = \frac{1}{3} \begin{bmatrix} 4 & -1 & -1 \\ 2 & 1 & -2 \\ -1 & 1 & 4 \end{bmatrix}.$$

For Q1

$$x_1 + x_2 = \vec{x} \begin{bmatrix} 1 \\ 1 \\ 0 \end{bmatrix} = 0 \quad \text{if } \vec{x} = (x_1, x_2, x_3)$$

$$\Rightarrow [(y_1, y_2, y_3) - (1,-1,-1)] \cdot \frac{1}{3} \begin{bmatrix} 4 & -1 & -1 \\ 2 & 1 & -2 \\ -1 & 1 & 4 \end{bmatrix} \begin{bmatrix} 1 \\ 1 \\ 0 \end{bmatrix} = 0$$

$$\Rightarrow (\text{replace } y_1, y_2, y_3 \text{ by } x_1, x_2, x_3) \; x_1 + x_2 = 0.$$

Hence, $x_1 + x_2 = 0$ and its image plane are coincident. This is within our expectation because the normal vector $(1,1,0)$ of $x_1 + x_2 = 0$ is perpendicular to $\vec{v}_1 = \frac{1}{\sqrt{3}}(1,-1,-1)$, the direction of the shearing T (see 1 in (3.8.20)).

Any plane parallel to $x_1 + 2x_2 - x_3 = 3$ has equation $x_1 + 2x_2 - x_3 = c$ where c is a constant. Then

$$x_1 + 2x_2 - x_3 = \vec{x} \begin{bmatrix} 1 \\ 2 \\ -1 \end{bmatrix} = c$$

$$\Rightarrow [(y_1, y_2, y_3) - (1, -1, -1)] \cdot \frac{1}{3} \begin{bmatrix} 4 & -1 & -1 \\ 2 & 1 & -2 \\ -1 & 1 & 4 \end{bmatrix} \begin{bmatrix} 1 \\ 2 \\ -1 \end{bmatrix} = c$$

$$\Rightarrow y_1 + 2y_2 - y_3 = c,$$

which means that it is an invariant plane.

For Q2 By computation,

$$T(\vec{b_0}) = (1, -1, -1) + (1, 1, 1)B = (1, -1, -1) + \left(\frac{1}{3}, \frac{5}{3}, \frac{5}{3}\right) = \left(\frac{4}{3}, \frac{2}{3}, \frac{2}{3}\right),$$

$$T(b_1) = (1, -1, -1) + (-1, 1, 1)B = (1, -1, -1) + (-1, 1, 1) = (0, 0, 0),$$

$$T(b_2) = (1, -1, -1) + (1, -1, 1)B$$
$$= (1, -1, -1) + \left(\frac{5}{3}, -\frac{5}{3}, \frac{1}{3}\right) = \left(\frac{8}{3}, -\frac{8}{3}, -\frac{2}{3}\right),$$

$$T(b_3) = (1, -1, -1) + (1, 1, -1)B$$
$$= (1, -1, -1) + \left(-\frac{1}{3}, \frac{7}{3}, \frac{1}{3}\right) = \left(\frac{2}{3}, \frac{4}{3}, -\frac{2}{3}\right).$$

These four image points form a tetrahedron $\Delta T(\vec{b_0}) \cdots T(\vec{b_3})$. Thus

$$\text{the signed volume of } \Delta \vec{b_0}\vec{b_1}\vec{b_2}\vec{b_3} = \frac{1}{6} \begin{vmatrix} \vec{b_1} - \vec{b_0} \\ \vec{b_2} - \vec{b_0} \\ \vec{b_3} - \vec{b_0} \end{vmatrix}$$

$$= \frac{1}{6} \begin{vmatrix} -2 & 0 & 0 \\ 0 & -2 & 0 \\ 0 & 0 & -2 \end{vmatrix} = -\frac{4}{3},$$

the signed volume of $\Delta T(\vec{b}_0) \cdots T(\vec{b}_3) = \dfrac{1}{6} \begin{vmatrix} T(\vec{b}_1) - T(\vec{b}_0) \\ T(\vec{b}_2) - T(\vec{b}_0) \\ T(\vec{b}_3) - T(\vec{b}_0) \end{vmatrix}$

$$= \frac{1}{6} \begin{vmatrix} -\frac{4}{3} & -\frac{2}{3} & -\frac{2}{3} \\ \frac{4}{3} & -\frac{10}{3} & -\frac{4}{3} \\ -\frac{2}{3} & \frac{2}{3} & -\frac{4}{3} \end{vmatrix}$$

$$= \frac{1}{6} \cdot \left(-\frac{8}{27} \right) \cdot 27 = -\frac{4}{3}.$$

So both have the same volumes. □

Case 7 Rotation (Euclidean notions are needed)

Let the line OA be perpendicular to a plane Σ in space $\mathbb{R}^3$, as indicated in Fig. 3.69. With OA as the line of invariant points, rotate the whole space through an angle θ so that a point X is moved to its new position X'. Define a mapping $T: \mathbb{R}^3 \to \mathbb{R}^3$ by

$$T(X) = X'.$$

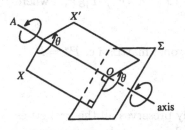

Fig. 3.69

Then T is an affine transformation and is called the *rotation* of the space with the line OA as *axis* and through the *angle* θ. Then,

1. The axis is the only line of invariant points if $\theta \neq 0$.
2. The rotation in $\mathbb{R}^3$ results in a rotation on any plane, perpendicular to the axis at O, with O as center and through the angle θ (see Fig. 2.108).
3. Any plane containing the axis is rotated through the angle θ just like the action of opening a door.

(3.8.23)

The details are left to the readers.

As a counterpart of (2.8.34) in space, we have

The rotation

Let $\vec{a}_0, \vec{a}_1, \vec{a}_2$ and $\vec{a}_3$ be non-coplanar points in $\mathbb{R}^3$ so that

1. $|\vec{a}_1 - \vec{a}_0| = |\vec{a}_2 - \vec{a}_0| = |\vec{a}_3 - \vec{a}_0| = 1$, and
2. $(\vec{a}_1 - \vec{a}_0) \perp (\vec{a}_2 - \vec{a}_0) \perp (\vec{a}_3 - \vec{a}_0)$.

Let T denote the rotation of the space with $\vec{a}_0 + \langle\langle \vec{a}_1 - \vec{a}_0 \rangle\rangle$ as axis and through the angle θ. Then

1. In the orthonormal affine basis $\mathcal{B} = \{\vec{a}_0, \vec{a}_1, \vec{a}_2, \vec{a}_3\}$,

$$[T(\vec{x})]_\mathcal{B} = [\vec{x}]_\mathcal{B}[T]_\mathcal{B}, \quad \text{where } [T]_\mathcal{B} = \begin{bmatrix} 1 & 0 & 0 \\ 0 & \cos\theta & \sin\theta \\ 0 & -\sin\theta & \cos\theta \end{bmatrix}.$$

2. In $\mathcal{N} = \{\vec{0}, \vec{e}_1, \vec{e}_2, \vec{e}_3\}$,

$$T(\vec{x}) = \vec{a}_0 + (\vec{x} - \vec{a}_0)P^{-1}[T]_\mathcal{B}P, \quad \text{where } P = \begin{bmatrix} \vec{a}_1 - \vec{a}_0 \\ \vec{a}_2 - \vec{a}_0 \\ \vec{a}_3 - \vec{a}_0 \end{bmatrix}.$$

Here P is an orthogonal matrix, i.e. $P^* = P^{-1}$.

Also,

(a) a rotation not only preserves all the properties 1–7 listed in (3.7.15),
(b) but also preserves length, angle, area, volume and orientation.

$$(3.8.24)$$

The details are left as Ex. <A> 17.

To test if an affine transformation $T(\vec{x}) = \vec{x}_0 + \vec{x}A$ is a rotation in $\mathbb{R}^3$, try the following steps (refer to (2.8.35)):

1. Justify if $A^* = A^{-1}$, i.e. A is an orthogonal matrix, and $\det A = 1$. Then, the associated T is a rotation if $\vec{x}(I_3 - A) = \vec{x}_0$ has a solution.
2. Solve $\det(A - I_3) = 0$ and see what is the algebraic multiplicity m of the eigenvalue 1 which could only be 1 and 3 (see (3) in (3.7.46) and (2), (3) in (3.7.50)). In case $m = 3$, then $A = I_3$ holds. If $m = 1$, determine an eigenvector $\vec{v}_1$ of *unit length* associated to 1.

3. Extend $\vec{v}_1$ to an *orthonormal* basis $\mathcal{B} = \{\vec{v}_1, \vec{v}_2, \vec{v}_3\}$ for $\mathbb{R}^3$. Then

$$[A]_{\mathcal{B}} = \begin{bmatrix} 1 & 0 \\ 0 & B \end{bmatrix},$$

where $B_{2\times2}$ is an orthogonal matrix with $\det B = 1$ and hence (see (2.8.35))

$$B = \begin{bmatrix} \cos\theta & \sin\theta \\ -\sin\theta & \cos\theta \end{bmatrix} \quad \text{for some } \theta \in \mathbb{R}.$$

4. In case $\theta \neq 0$, i.e. $A \neq I_3$, take any solution $\vec{a}_0$ of the equation $\vec{x}(I_3 - A) = \vec{x}_0$. Then,

   (a) The axis of rotation: $\vec{a}_0 + \langle\langle \vec{v}_1 \rangle\rangle$.
   (b) The (perpendicular) *rotational plane*:

   $$\vec{a}_0 + \langle\langle \vec{v}_2, \vec{v}_3 \rangle\rangle = \vec{a}_0 + \text{Im}(A - I_3).$$

   (c) The angle of rotation: θ.

$$(3.8.25)$$

The details are left as Ex. <A> 17.

Example 5 Give an affine transformation

$$T(\vec{x}) = \vec{x}_0 + \vec{x}A, \quad \text{where } A = \begin{bmatrix} -\frac{1}{3} & \frac{2}{3} & -\frac{2}{3} \\ \frac{2}{3} & \frac{2}{3} & \frac{1}{3} \\ \frac{2}{3} & -\frac{1}{3} & -\frac{2}{3} \end{bmatrix}.$$

Try to determine these $\vec{x}_0$ so that T is a rotation. In this case, determine the axis and the angle of the rotation, also the rotational plane.

Solution The three row (or column) vectors of A are of unit length and are perpendicular to each other. So $A^* = A^{-1}$ and A is orthogonal. Direct computation shows that

$$\det A = \frac{1}{27}(4 + 4 + 4 + 8 + 8 - 1) = \frac{1}{27} \cdot 27 = 1.$$

Therefore the associated T is a rotation of $\mathbb{R}^3$ if $\vec{x}_0$ is chosen from $\text{Im}(I_3 - A)$.

The characteristic polynomial of A is

$$\det(A - tI_3) = -\frac{1}{3}(t - 1)(3t^2 + 4t + 3)$$

and thus A has a real eigenvalue 1 of multiplicity one. Solve

$$\vec{x}(A - I_3) = (x_1, x_2, x_3)\begin{bmatrix} -\frac{4}{3} & \frac{2}{3} & -\frac{2}{3} \\ \frac{2}{3} & -\frac{1}{3} & \frac{1}{3} \\ \frac{2}{3} & -\frac{1}{3} & -\frac{5}{3} \end{bmatrix} = \vec{0}$$

$$\Rightarrow 2x_1 - x_2 - x_3 = 0, \quad 2x_1 - x_2 + 5x_3 = 0$$

$$\Rightarrow \vec{x} = x_1(1, 2, 0) \quad \text{for } x_1 \in \mathbb{R}.$$

Hence, take an eigenvector $\vec{v}_1 = \frac{1}{\sqrt{5}}(1, 2, 0)$.
   Let $\vec{v} = (\alpha_1, \alpha_2, \alpha_3)$. Then

$$\vec{v} \perp \vec{v}_1$$

$$\Leftrightarrow \alpha_1 + 2\alpha_2 = 0.$$

Therefore, we may choose $\vec{v}_2 = \frac{1}{\sqrt{5}}(-2, 1, 0)$. Again, solve $\alpha_1 + 2\alpha_2 = -2\alpha_1 + \alpha_2 = 0$, we may choose $\vec{v}_3 = (0, 0, 1) = \vec{e}_3$. Now, $\mathcal{B} = \{\vec{v}_1, \vec{v}_2, \vec{v}_3\}$ forms an orthonormal basis for $\mathbb{R}^3$.
   By computation,

$$\vec{v}_2 A = \frac{1}{3\sqrt{5}}(4, -2, 5) = \frac{2}{3}\vec{v}_2 + \frac{\sqrt{5}}{3}\vec{v}_3,$$

$$\vec{v}_3 A = \frac{1}{3}(2, -1, -2) = -\frac{\sqrt{5}}{3}\vec{v}_2 + \frac{2}{3}\vec{v}_3$$

$$\Rightarrow [A]_{\mathcal{B}} = PAP^{-1} = \begin{bmatrix} 1 & 0 & 0 \\ 0 & \frac{2}{3} & \frac{\sqrt{5}}{3} \\ 0 & -\frac{\sqrt{5}}{3} & \frac{2}{3} \end{bmatrix}, \quad \text{where } P = \begin{bmatrix} \frac{1}{\sqrt{5}} & \frac{2}{\sqrt{5}} & 0 \\ -\frac{2}{\sqrt{5}} & \frac{1}{\sqrt{5}} & 0 \\ 0 & 0 & 1 \end{bmatrix}.$$

To find $\text{Im}(I_3 - A)$, let $\vec{y} = \vec{x}(I_3 - A)$. Then

$$\vec{y} = \vec{x}(I_3 - A)$$

$$\Leftrightarrow y_1 = \frac{2}{3}(2x_1 - x_2 - x_3), \quad y_2 = \frac{1}{3}(-2x_1 + x_2 + x_3) \quad \text{and}$$

$$y_3 = \frac{1}{3}(2x_1 - x_2 + 5x_3)$$

$$\Leftrightarrow y_1 + 2y_2 = 0.$$

So the image subspace is $x_1 + 2x_2 = 0$ with a unit normal vector $\vec{v}_1$. Note that, since $A^* = A^{-1}, \mathrm{Im}(A - I_3)^\perp = \mathrm{Ker}(A^* - I_3) = \mathrm{Ker}(A^{-1} - I_3) = \mathrm{Ker}(A - I_3)$. This is the theoretical reason why $\vec{v}_1 \perp \mathrm{Im}(A - I_3) = \mathrm{Ker}(A - I_3)^\perp = \langle\langle \vec{v}_2, \vec{v}_3 \rangle\rangle$.

Take any point $\vec{x}_0$ on $\mathrm{Im}(A - I_3)$, say $\vec{x}_0 = \vec{0}$ for simplicity. Then, the axis of rotation is $\langle\langle \vec{v}_1 \rangle\rangle$, the rotational plane is $\langle\langle \vec{v}_2, \vec{v}_3 \rangle\rangle = \mathrm{Im}(A - I_3)$ and the angle of rotation is $\theta = \tan^{-1} \frac{\sqrt{5}}{2}$. See Fig. 3.70.

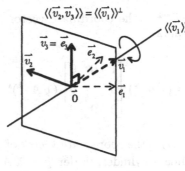

Fig. 3.70

We have two questions:

Q1 What is the image of the unit sphere $x_1^2 + x_2^2 + x_3^2 = 1$ under $\vec{y} = \vec{x}A$?
Q2 What is the image of the cylinder $x_1^2 + x_2^2 = 1$ under $\vec{y} = \vec{x}A$?

For Q1 It is the unit sphere itself both by geometric intuition and by analytic proof.

For Q2 Geometric intuition tells as that it is still a cylinder with central axis along the vector $\vec{e}_3 A = \left(\frac{2}{3}, -\frac{1}{3}, -\frac{2}{3}\right)$ and the base circle lying on the plane $2x_1 - x_2 - 2x_3 = 0$. And computation shows that the image has the equation, a complicated one,

$$5x_1^2 + 8x_2^2 + 5x_3^2 + 4x_1x_2 + 8x_1x_3 - 4x_2x_3 = 9 \tag{*}$$

in the eyes of $\mathcal{N} = \{\vec{0}, \vec{e}_1, \vec{e}_2, \vec{e}_3\}$.

Hence, we raise a third question concerning Q2 and (*).

Q3 What is the equation of the image cylinder (*) in the orthonormal affine basis $\mathcal{C} = \{\vec{0}, A_{1*}, A_{2*}, A_{3*}\}$ where A_{i*} denotes the ith row vector of A for $i = 1, 2, 3$?

In $\mathcal{B}$, one needs to compute

$$\vec{x} \begin{bmatrix} 1 & 0 & 0 \\ 0 & 1 & 0 \\ 0 & 0 & 0 \end{bmatrix} \vec{x}^*$$

$$= (\vec{y} P^{-1})(PAP^{-1})P \begin{bmatrix} 1 & 0 & 0 \\ 0 & 1 & 0 \\ 0 & 0 & 0 \end{bmatrix} P^{-1}(PAP^{-1})^{-1}(\vec{y} P^{-1})^* = 1,$$

where $\vec{y} = \vec{x} A$ and $\vec{y} P^{-1}$ is the coordinate vector of $\vec{y}$ in the basis $\mathcal{B}$. In $\mathcal{C}$, one needs to compute

$$\vec{x} \begin{bmatrix} 1 & 0 & 0 \\ 0 & 1 & 0 \\ 0 & 0 & 0 \end{bmatrix} \vec{x}^* = (\vec{y} A^{-1}) \begin{bmatrix} 1 & 0 & 0 \\ 0 & 1 & 0 \\ 0 & 0 & 0 \end{bmatrix} (\vec{y} A^{-1})^* = \alpha_1^2 + \alpha_2^2 = 1,$$

where $\vec{y} A^{-1} = (\alpha_1, \alpha_2, \alpha_3)$ is the coordinate vector of $\vec{y} = \vec{x} A$ in the basis $\mathcal{C}$. Hence, to view the image cylinder, under $\vec{y} = \vec{x} A$, in $\mathcal{C}$ is the same as to view the original cylinder in $\mathcal{N}$.

Case 8   Orthogonal reflection (Euclidean notions are needed)

Let us give a detailed account about the orthogonal reflection introduced in Case 2.

The orthogonal projection and reflection

Let $\vec{a}_0, \vec{a}_1$ and $\vec{a}_2$ be three non-collinear points in the space $\mathbb{R}^3$.

(1) The *orthogonal projection* of $\mathbb{R}^3$ onto the affine subspace S: $\vec{a}_0 + \langle\langle \vec{a}_1 - \vec{a}_0, \vec{a}_2 - \vec{a}_0 \rangle\rangle$ is the mapping P_{proj} defined by

$$\vec{x} \to \vec{a}_0 + (\vec{x} - \vec{a}_0)A_{\text{proj}} = P_{\text{proj}}(\vec{x}).$$

where

$$A_{\text{proj}} = A^*(AA^*)^{-1}A \quad \text{and} \quad A = \begin{bmatrix} \vec{a}_1 - \vec{a}_0 \\ \vec{a}_2 - \vec{a}_0 \end{bmatrix}_{2 \times 3}$$

(see (*14) in Sec. 3.7.5). Note that $AA^* = I_2$ if $\{\vec{a}_1 - \vec{a}_0, \vec{a}_2 - \vec{a}_0\}$ is an orthonormal affine basis for the subspace S. See Fig. 3.71.

(2) The *orthogonal reflection* of $\mathbb{R}^3$ with respect to the affine subspace S is the affine transformation T defined by

$$\vec{x} \rightarrow \vec{a_0} + (\vec{x} - \vec{a_0})R = T(\vec{x}),$$

where

$$R = 2A^*(AA^*)^{-1}A - I_3 = 2A_{\text{proj}} - I_3,$$

satisfies

1. R is symmetric, i.e. $R^* = R$, and
2. R is orthogonal, i.e. $R^* = R^{-1}$.

and thus $R^2 = I_3$ (see (*17) and (*24) in Sec. 3.7.5). See Fig. 3.71.

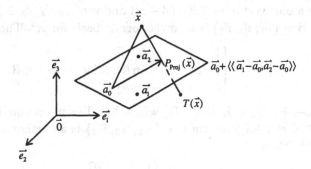

Fig. 3.71

(3) Let $\mathcal{B} = \{\vec{a_0}, \vec{a_1}, \vec{a_2}, \vec{a_3}\}$ be an orthonormal affine basis for $\mathbb{R}^3$ so that S is $\vec{a_0} + \langle\langle \vec{a_1} - \vec{a_0}, \vec{a_2} - \vec{a_0} \rangle\rangle$. Let T be the orthogonal reflection defined as in (2). Then,

1. In $\mathcal{B}$,

$$[T(\vec{x})]_{\mathcal{B}} = [\vec{x}]_{\mathcal{B}}[T]_{\mathcal{B}}, \quad \text{where } [T]_{\mathcal{B}} = \begin{bmatrix} 1 & 0 & 0 \\ 0 & 1 & 0 \\ 0 & 0 & -1 \end{bmatrix}.$$

2. In $\mathcal{N} = \{\vec{0}, \vec{e_1}, \vec{e_2}, \vec{e_3}\}$,

$$T(\vec{x}) = \vec{a_0} + (\vec{x} - \vec{a_0})P^{-1}[T]_{\mathcal{B}}P, \quad \text{where } P = \begin{bmatrix} \vec{a_1} - \vec{a_0} \\ \vec{a_2} - \vec{a_0} \\ \vec{a_3} - \vec{a_0} \end{bmatrix}.$$

Here P is orthogonal, i.e. $P^* = P^{-1}$. Also, $R = P^{-1}[T]_{\mathcal{B}}P$.

Also, an orthogonal reflection

(a) preserves all the properties 1–7 listed in (3.7.15), and
(b) preserves the length, angle, area and volume but reverses the orientations.   (3.8.26)

The details are left as Ex. <A> 21.

To test if an affine transformation $T(\vec{x}) = \vec{x}_0 + \vec{x}A$ is an orthogonal reflection, try the following steps:

1. Justify if A is orthogonal and symmetric and $\det A = -1$. Then the associated T is an orthogonal reflection if $\vec{x}(I_3 - A) = \vec{x}_0$ has a solution.
2. A has eigenvalues 1, 1, -1. Two kinds of normalized A are possible.

   (a) Take a unit vector $\vec{v}_1 \in \text{Ker}(A - I_3)$ and vectors $\vec{v}_2, \vec{v}_3 \in \langle\langle\vec{v}_1\rangle\rangle^{\perp}$ so that $\mathcal{B} = \{\vec{v}_1, \vec{v}_2, \vec{v}_3\}$ is an orthonormal basis for $\mathbb{R}^3$. Then,

   $$[A]_{\mathcal{B}} = \begin{bmatrix} 1 & 0 & 0 \\ 0 & \cos\theta & \sin\theta \\ 0 & \sin\theta & -\cos\theta \end{bmatrix} \quad \text{for some } \theta \in \mathbb{R}.$$

   (b) Choose $\vec{u}_1, \vec{u}_2 \in \text{Ker}(A - I_3)$ which has dimension equal to 2 and $\vec{u}_3 \in \text{Ker}(A + I_3)$ so that $\mathcal{C} = \{\vec{u}_1, \vec{u}_2, \vec{u}_3\}$ is an orthonormal basis for $\mathbb{R}^3$. Then

   $$[A]_{\mathcal{C}} = \begin{bmatrix} 1 & 0 & 0 \\ 0 & 1 & 0 \\ 0 & 0 & -1 \end{bmatrix}.$$

3. Take any solution $\vec{a}_0$ of the equation $\vec{x}(I_3 - A) = \vec{x}_0$. Then,
   (a) The direction of orthogonal reflection: $\vec{a}_0 + \langle\langle\vec{v}_3\rangle\rangle$.
   (b) The plane of invariant points: $\vec{a}_0 + \langle\langle\vec{v}_2, \vec{v}_3\rangle\rangle = \vec{a}_0 + \text{Im}(A - I_3)$.

                                                              (3.8.27)

The details are left as Ex. <A> 21.

Example 6 Give an affine transformation

$$T(\vec{x}) = \vec{x}_0 + \vec{x}A, \quad \text{where } A = \begin{bmatrix} \frac{1}{3} & \frac{2}{3} & \frac{2}{3} \\ \frac{2}{3} & \frac{1}{3} & -\frac{2}{3} \\ \frac{2}{3} & -\frac{2}{3} & \frac{1}{3} \end{bmatrix}.$$

Determine these $\vec{x}_0$ so that each such T is an orthogonal reflection, and the direction and the plane of invariant points.

Solution It is obvious that A is orthogonal and $\det A = -1$. If we pick $\vec{x}_0 \in \mathrm{Im}(A - I_3)$, then the associated T is an orthogonal reflection.

The characteristic polynomial of A is

$$\det(A - tI_3) = \frac{1}{27}(-27t^3 + 27t^2 + 27t - 27) = -(t-1)^2(t+1)$$

and hence A has eigenvalue 1 of multiplicity 2 and another eigenvalue -1. Solve

$$\vec{x}(A - I_3) = \vec{0}$$

$$\Rightarrow x_1 - x_2 - x_3 = 0.$$

Take unit eigenvectors $\vec{u}_1 = \frac{1}{\sqrt{2}}(1,1,0)$ and $\vec{u}_2 = \frac{1}{\sqrt{6}}(1,-1,2)$ so that $\vec{u}_1 \perp \vec{u}_2$. Solve

$$\vec{x}(A + I_3) = \vec{0}$$

$$\Rightarrow 2x_1 + x_2 + x_3 = 0, \quad x_1 + 2x_2 - x_3 = 0.$$

Take a unit vector $\vec{u}_3 = \frac{1}{\sqrt{3}}(1,-1,-1)$. Then, $\mathcal{C} = \{\vec{u}_1, \vec{u}_2, \vec{u}_3\}$ is an orthonormal basis for $\mathbb{R}^3$. In $\mathcal{C}$,

$$[A]_{\mathcal{C}} = QAQ^{-1} = \begin{bmatrix} 1 & 0 & 0 \\ 0 & 1 & 0 \\ 0 & 0 & -1 \end{bmatrix}, \quad \text{where } Q = \begin{bmatrix} \frac{1}{\sqrt{2}} & \frac{1}{\sqrt{2}} & 0 \\ \frac{1}{\sqrt{6}} & -\frac{1}{\sqrt{6}} & \frac{2}{\sqrt{6}} \\ \frac{1}{\sqrt{3}} & -\frac{1}{\sqrt{3}} & -\frac{1}{\sqrt{3}} \end{bmatrix}.$$

Note that $\mathrm{Im}(A - I_3) = \mathrm{Ker}(A + I_3) = \mathrm{Ker}(A - I_3)^{\perp}$. So there are three different ways to compute $\mathrm{Im}(A - I_3) = \langle\langle(1,-1,-1)\rangle\rangle$. Pick up any point $\vec{x}_0$ on $\mathrm{Im}(A - I_3)$, say $\vec{x}_0 = \vec{0}$ for simplicity, then the *direction* of the reflection $T = \vec{x}A$ is $\vec{u}_3$ and the *plane of invariant points* is $\langle\langle \vec{u}_1, \vec{u}_2 \rangle\rangle = \{\vec{x} \in \mathbb{R}^3 \mid x_1 - x_2 - x_3 = 0\} = \langle\langle \vec{u}_3 \rangle\rangle^{\perp}$. See Fig. 3.72(a).

On the other hand, take $\vec{v}_1 = \vec{u}_1 = \frac{1}{\sqrt{2}}(1,1,0)$. Then $\langle\langle \vec{v}_1 \rangle\rangle^{\perp} = \{\vec{x} \mid x_1 + x_2 = 0\}$. Choose $\vec{v}_2 = \frac{1}{\sqrt{2}}(1,-1,0)$, $\vec{v}_3 = \vec{e}_3 = (0,0,1)$ so that $\mathcal{B} = \{\vec{v}_1, \vec{v}_2, \vec{v}_3\}$ forms an orthonormal basis for $\mathbb{R}^3$. Since

$$\vec{v}_1 A = \vec{v}_1,$$

$$\vec{v}_2 A = \frac{1}{\sqrt{2}}\left(-\frac{1}{3}, \frac{1}{3}, \frac{4}{3}\right) = -\frac{1}{3}\vec{v}_2 + \frac{4}{3\sqrt{2}}\vec{v}_3,$$

$$\vec{v}_3 A = \left(\frac{2}{3}, -\frac{2}{3}, \frac{1}{3}\right) = \frac{2\sqrt{2}}{3}\vec{v}_2 + \frac{1}{3}\vec{v}_3$$

$$\Rightarrow [A]_B = PAP^{-1} = \begin{bmatrix} 0 & 0 \\ 0 & -\frac{1}{3} & \frac{2\sqrt{2}}{3} \\ 0 & \frac{2\sqrt{2}}{3} & \frac{1}{3} \end{bmatrix}, \quad \text{where } P = \begin{bmatrix} \frac{1}{\sqrt{2}} & \frac{1}{\sqrt{2}} & 0 \\ \frac{1}{\sqrt{2}} & -\frac{1}{\sqrt{2}} & 0 \\ 0 & 0 & 1 \end{bmatrix}.$$

Also, the submatrix

$$\begin{bmatrix} -\frac{1}{3} & \frac{2\sqrt{2}}{3} \\ \frac{2\sqrt{2}}{3} & \frac{1}{3} \end{bmatrix} = \begin{bmatrix} -1 & 0 \\ 0 & 1 \end{bmatrix} \begin{bmatrix} \cos\theta & \sin\theta \\ -\sin\theta & \cos\theta \end{bmatrix}$$

where $\cos\theta = \frac{1}{3}, \sin\theta = -\frac{2\sqrt{2}}{3}$. To interpret $T(\vec{x}) = \vec{x}A$ in B, take any point $\vec{x} \in \mathbb{R}^3$. Then, letting $\vec{y} = \vec{x}A$,

$$\vec{y} = \vec{x}P^{-1}(PAP^{-1})P$$
$$\Rightarrow [\vec{y}]_B = [\vec{x}]_B(PAP^{-1}), \quad \text{i.e. letting } [\vec{x}]_B = (\alpha_1, \alpha_2, \alpha_3) \text{ and}$$
$$[\vec{y}]_B = (\beta_1, \beta_2, \beta_3),$$
$$\beta_1 = \alpha_1,$$
$$(\beta_2, \beta_3) = (\alpha_2, \alpha_3) \begin{bmatrix} -1 & 0 \\ 0 & 1 \end{bmatrix} \begin{bmatrix} \cos\theta & \sin\theta \\ -\sin\theta & \cos\theta \end{bmatrix}.$$

This means that, when the height α_1 of the point $\vec{x}$ to the plane $\langle\langle \vec{v}_2, \vec{v}_3 \rangle\rangle$ being fixed, the orthogonal projection (α_2, α_3) of $\vec{x}$ on $\langle\langle \vec{v}_2, \vec{v}_3 \rangle\rangle$ is subject to a reflection with respect to the axis $\langle\langle \vec{v}_3 \rangle\rangle$ to the point $(-\alpha_2, \alpha_3)$ and then is rotated through the angle θ in the plane $\langle\langle \vec{v}_2, \vec{v}_3 \rangle\rangle$ with $\vec{0}$ as center to the point (β_2, β_3). Therefore, the point $(\beta_1, \beta_2, \beta_3)$, where $\beta_1 = \alpha_1$, is the coordinate vector of $\vec{x}A = T(\vec{x})$ in B. See Fig. 3.72(b). Compare with Figs. 2.109 and 2.110. □

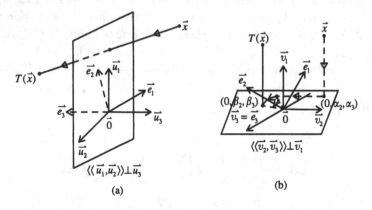

(a)

(b)

Fig. 3.72

Just like (2.8.38) and (2.8.39), an affine transformation $T(\vec{x}) = \vec{x}_0 + \vec{x}A$ on $\mathbb{R}^3$ can be expressed as a composite of a finite number of reflections, stretches and shearings and then followed by a translation.

For example, let $T(\vec{x}) = \vec{x}_0 + \vec{x}A$ where A is the one in Example 1 of Sec. 3.7.5. Then, A is the composite of the following affine transformations in $\mathcal{N} = \{\vec{0}, \vec{e}_1, \vec{e}_2, \vec{e}_3\}$,

1. $\begin{bmatrix} 1 & 0 & 0 \\ 4 & 1 & 0 \\ 0 & 0 & 1 \end{bmatrix}$: a shearing in the direction $\vec{e}_1$ with $\langle\langle \vec{e}_1, \vec{e}_3 \rangle\rangle$ as the plane of invariant points and coefficient 4.

2. $\begin{bmatrix} 1 & 0 & 0 \\ 0 & 1 & 0 \\ 1 & 0 & 1 \end{bmatrix}$: a shearing in the direction $\vec{e}_1$ with $\langle\langle \vec{e}_1, \vec{e}_2 \rangle\rangle$ as the plane of invariant points and coefficient 1.

3. $\begin{bmatrix} 1 & 0 & 0 \\ 0 & -1 & 0 \\ 0 & 0 & 1 \end{bmatrix} \begin{bmatrix} 1 & 0 & 0 \\ 0 & 1 & 0 \\ 0 & 0 & 5 \end{bmatrix} = \begin{bmatrix} 1 & 0 & 0 \\ 0 & -1 & 0 \\ 0 & 0 & 5 \end{bmatrix}$: a two-way stretch having scale factor -1
along $\langle\langle \vec{e}_2 \rangle\rangle$ and 5 along $\langle\langle \vec{e}_3 \rangle\rangle$ and $\langle\langle \vec{e}_1 \rangle\rangle$ the line of invariant points. Note that the former factor represents an orthogonal reflection in the direction $\vec{e}_2$.

4. $\begin{bmatrix} 1 & 1 & 0 \\ 0 & 1 & 0 \\ 0 & 0 & 1 \end{bmatrix}$: a shearing in the direction $\vec{e}_2$ with $\langle\langle \vec{e}_2, \vec{e}_3 \rangle\rangle$ as the plane of invariant points and coefficient 1.

5. $\begin{bmatrix} 1 & 0 & 0 \\ 0 & 1 & 5 \\ 0 & 0 & 1 \end{bmatrix}$: a shearing in the direction $\vec{e}_3$ with $\langle\langle \vec{e}_1, \vec{e}_3 \rangle\rangle$ as the plane of invariant points and coefficient 5.

6. $\begin{bmatrix} 1 & 0 & -5 \\ 0 & 1 & 0 \\ 0 & 0 & 1 \end{bmatrix}$: a shearing in the direction $\vec{e}_3$ with $\langle\langle \vec{e}_2, \vec{e}_3 \rangle\rangle$ as the plane of invariant points and coefficient -5.

7. $\vec{x} \to \vec{x}_0 + \vec{x}$: a translation along $\vec{x}_0$.

Readers shall practice more examples by themselves.

Exercises

<A>

1. Prove (3.8.11) and (3.8.13).
2. Let $\vec{a}_0 = (1,1,1)$ and the vectors $\vec{x}_1 = (1,2,3)$, $\vec{x}_2 = (2,2,1)$ and $\vec{x}_3 = (3,4,3)$.

(a) Determine the reflection T_i of $\mathbb{R}^3$ in the direction $\vec{x}_i$ with respect to the plane $\vec{a}_0 + \langle\langle \vec{x}_j, \vec{x}_k \rangle\rangle$, where $j \neq k \neq i$ for $i = 1, 2, 3$.

(b) Describe $T_1 \circ T_2$, $T_2 \circ T_3$ and $T_3 \circ T_1$ respectively their mapping properties.

(c) Compute $T_1 \circ T_2 \circ T_3$, $T_2 \circ T_3 \circ T_1$ and $T_3 \circ T_1 \circ T_2$. Are they coincident? Any geometric interpretation?

3. Let

$$T(\vec{x}) = \vec{x}_0 + \vec{x}A, \quad \text{where } A = \begin{bmatrix} 2 & 3 & 0 \\ -1 & -2 & 0 \\ 0 & 0 & 1 \end{bmatrix}.$$

Use this T to justify (3.8.13). For a particular choice of $\vec{x}_0$, do the following questions.

(1) Find the image of the plane $x_1 - x_2 = c$ under T.
(2) Find the image of the plane $a_1 x_1 + a_2 x_2 + a_3 x_3 = b$, where $a_1 + a_2 \neq 0$, under T. Where do they intersect?
(3) Find the image of the tetrahedron $\triangle \vec{a}_0 \vec{a}_1 \vec{a}_2 \vec{a}_3$ and compute their volumes.

4. Prove (3.8.15) and (3.8.16).

5. Let $\vec{a}_0 = (-3, -6, 1)$ and the vectors $\vec{x}_1 = (1, 2, 2)$, $\vec{x}_2 = (1, -1, 1)$ and $\vec{x}_3 = (4, -12, 1)$.

(a) Determine the one-way stretch T_i of $\mathbb{R}^3$ with scale factor k_i in the direction $\vec{x}_i$ with respect to the plane $\vec{a}_0 + \langle\langle \vec{x}_j, \vec{x}_k \rangle\rangle$, where $j \neq k \neq i$ for $i = 1, 2, 3$.

(b) Determine the two-way stretches $T_1 \circ T_2$, $T_2 \circ T_3$ and $T_3 \circ T_1$ respectively and their lines of invariant points.

(c) Give geometric interpretation and analytic proof to see if $T_1 \circ T_2$ is equal to $T_2 \circ T_1$.

(d) Compute $T_1 \circ T_2 \circ T_3$, $T_2 \circ T_3 \circ T_1$ and $T_3 \circ T_1 \circ T_2$. Are they equal?

6. Use

$$T(\vec{x}) = \vec{x}_0 + \vec{x}A, \quad \text{where } A = \begin{bmatrix} 4 & -3 & -3 \\ 6 & -5 & -6 \\ 0 & 0 & 1 \end{bmatrix}$$

to justify (3.8.16). For a particular choice of $\vec{x}_0$, do the following questions.

(1) Find the image of the plane $x_1 + 2x_2 = c$ under T.

(2) Find the image of the plane $a_1x_1 + a_2x_2 + a_3x_3 = b$ under T, where $a_1 - a_2 - a_3 = 0$.

(3) Find the image of the plane $a_1x_1 + a_2x_2 + a_3x_3 = b$ under T, where the vector (a_1, a_2, a_3) is linearly independent of $(1, 2, 0)$.

(4) Find the image of the tetrahedron $\Delta \vec{a}_0 \vec{a}_1 \vec{a}_2 \vec{a}_3$ and the ratio of their volumes.

Do the same problem to $\vec{x} \to \vec{x}_0 + \vec{x}A^*$.

7. Prove (3.8.17) and (3.8.18).

8. Let $\vec{a}_0 = (4, -3, -3)$ and the vectors $\vec{x}_1 = (1, 1, 4)$, $\vec{x}_2 = (2, -1, -12)$ and $\vec{x}_3 = (2, 1, 1)$.

   (a) Find the two-way stretch T_i of $\mathbb{R}^3$ with scale factor k_j along $\vec{a}_0 + \langle\langle \vec{x}_j \rangle\rangle$ and scale factor k_l along $\vec{a}_0 + \langle\langle \vec{x}_l \rangle\rangle$ and $\vec{a}_0 + \langle\langle \vec{x}_i \rangle\rangle$ the line of invariant points, where $j \neq l \neq i$ for $i = 1, 2, 3$.

   (b) Do problems (a), (b) and (d) as in Ex. 5.

9. Use

$$T(\vec{x}) = \vec{x}_0 + \vec{x}A, \quad \text{where } A = \begin{bmatrix} 5 & 4 & 4 \\ -7 & -3 & -1 \\ 7 & 4 & 2 \end{bmatrix}$$

to justify (3.8.18). For a particular choice of $\vec{x}_0$, do the following questions.

(1) Do the same questions Q1, Q2 and Q3 as in Example 3.

(2) Find the image of the tetrahedron $\Delta \vec{a}_0 \vec{a}_1 \vec{a}_2 \vec{a}_3$ under T and compute there volumes.

Do the same problem to $\vec{x} \to \vec{x}_0 + \vec{x}A^*$.

10. Prove (3.8.19).

11. Use both

$$T(\vec{x}) = \vec{x}_0 + \vec{x}A, \quad \text{where } A = \begin{bmatrix} 3 & 1 & -1 \\ 1 & 3 & -1 \\ -1 & -1 & 5 \end{bmatrix}$$

and

$$T(\vec{x}) = \vec{x}_0 + \vec{x}A, \quad \text{where } A = \begin{bmatrix} 3 & 0 & 1 \\ 2 & 2 & 2 \\ 1 & 0 & 3 \end{bmatrix}$$

to justify (3.8.19).

12. Prove (3.8.20)–(3.8.22).

13. Let $\vec{a}_0 = (4, 2, 1)$ and the vectors $\vec{x}_1 = (0, \frac{3}{5}, \frac{4}{5})$, $\vec{x}_2 = (\frac{3}{5}, \frac{16}{25}, -\frac{12}{25})$ and $\vec{x}_3 = (\frac{4}{5}, -\frac{12}{25}, \frac{9}{25})$.

   (a) Show that $|\vec{x}_1| = |\vec{x}_2| = |\vec{x}_3| = 1$ and $\vec{x}_1 \perp \vec{x}_2 \perp \vec{x}_3$.
   (b) Find the shearing T_i of $\mathbb{R}^3$ with coefficient $k_i \neq 0$ in the direction $\vec{x}_i$ with $\vec{a}_0 + \langle\langle \vec{x}_j, \vec{x}_k \rangle\rangle$ as the plane of invariant points, where $j \neq k \neq i$ for $i = 1, 2, 3$.
   (c) Do problems (b), (c) and (d) as in Ex. 5.

14. Let $\vec{a}_0 = (2, 2, 1)$ and the vectors $\vec{y}_1 = (2, 1, 0)$, $\vec{y}_2 = (0, 1, 1)$ and $\vec{y}_3 = (2, 0, 2)$.

   (a) Try to construct vectors $\vec{x}_1$, $\vec{x}_2$ and $\vec{x}_3$ form $\vec{y}_1, \vec{y}_2, \vec{y}_3$ so that $|\vec{x}_1| = |\vec{x}_2| = |\vec{x}_3| = 1$ and $\vec{x}_1 \perp \vec{x}_2 \perp \vec{x}_3$.
   (b) Do the same problems as in Ex. 13.

15. For each of the following, do the same problems as in Exs. 13 and 14.

   (a) $\vec{a}_0 = (2, 5, 5)$ and the vectors $\vec{y}_1 = (1, 1, 0)$, $\vec{y}_2 = (0, 1, 1)$, $\vec{y}_3 = (1, 0, 1)$.
   (b) $\vec{a}_0 = (-1, -2, 1)$ and the vectors $\vec{y}_1 = (1, 2, 1)$, $\vec{y}_2 = (1, 0, 1)$, $\vec{y}_3 = (1, 0, 2)$.
   (c) $\vec{a}_0 = (2, 5, 5)$ and the vectors $\vec{y}_1 = (1, 1, 0)$, $\vec{y}_2 = (2, 0, 1)$, $\vec{y}_3 = (2, 2, 1)$.

16. Use

$$T(\vec{x}) = \vec{x}_0 + \vec{x}A, \quad \text{where } A = \begin{bmatrix} 0 & 1 & 1 \\ -2 & 3 & 2 \\ 1 & -1 & 0 \end{bmatrix}$$

to justify (3.8.22). For a particular choice of $\vec{x}_0$, do the following questions.

   (1) Find the image plane of any plane parallel to the direction of the shearing.
   (2) Find the image plane of any plane parallel to the plane of invariant points.
   (3) Find the image of the tetrahedron $\Delta \vec{a}_0 \vec{a}_1 \vec{a}_2 \vec{a}_3$ under the shearing and compute its volume.

17. Prove (3.8.23)–(3.8.25).

18. Use

$$T(\vec{x}) = \vec{x}_0 + \vec{x}A, \quad \text{where } A = \begin{bmatrix} \frac{1}{\sqrt{2}} & \frac{1}{\sqrt{2}} & 0 \\ -\frac{1}{\sqrt{6}} & \frac{1}{\sqrt{6}} & \frac{2}{\sqrt{6}} \\ \frac{1}{\sqrt{3}} & -\frac{1}{\sqrt{3}} & \frac{1}{\sqrt{3}} \end{bmatrix}$$

to justify (3.8.25). For a particular choice of $\vec{x}_0$, do the following questions.

(1) Find the image of a line, parallel to or intersecting with the axis of rotation, under T.

(2) Find the image of a plane, containing or parallel to the axis of rotation.

(3) Find the image of a plane, perpendicular to the axis of rotation.

(4) Find the image of a plane, intersecting the axis of rotation at a point.

(5) Find the image of a tetrahedron $\triangle \vec{a}_0 \vec{a}_1 \vec{a}_2 \vec{a}_3$ and compute its volume.

(6) Find the image of the unit sphere $x_1^2 + x_2^2 + x_3^2 = 1$. How about the ellipsoid $\frac{x_1^2}{a_1^2} + \frac{x_2^2}{a_2^2} + \frac{x_3^2}{a_3^2} = 1$? How about $|x_1| + |x_2| + |x_3| = 1$?

19. Let

$$T(\vec{x}) = \vec{x}_0 + \vec{x}A, \quad \text{where } A = \begin{bmatrix} \frac{1}{3} & \frac{2}{3} & \frac{2}{3} \\ -\frac{2}{3} & \frac{2}{3} & -\frac{1}{3} \\ -\frac{2}{3} & -\frac{1}{3} & \frac{2}{3} \end{bmatrix}.$$

Do the same problems as Ex. 18.

20. Let

$$T(\vec{x}) = \vec{x}_0 + \vec{x}A, \quad \text{where } A = \begin{bmatrix} -\frac{7}{9} & \frac{4}{9} & \frac{4}{9} \\ \frac{4}{9} & -\frac{1}{9} & \frac{8}{9} \\ \frac{4}{9} & \frac{8}{9} & -\frac{1}{9} \end{bmatrix}.$$

Do the same problems as Ex. 18.

21. Prove (3.8.26) and (3.8.27).

22. Use

$$T(\vec{x}) = \vec{x}_0 + \vec{x}A, \quad \text{where } A = \frac{2}{3}\begin{bmatrix} \frac{1}{2} & -1 & -1 \\ -1 & \frac{1}{2} & -1 \\ -1 & -1 & \frac{1}{2} \end{bmatrix}$$

to justify (3.8.27). For a particular choice of $\vec{x}_0$, do the following questions.

(1) Find the image of a line, parallel or skew to the direction of the orthogonal reflection, under T.

(2) Find the image of a plane, containing or parallel to the direction of reflection.

(3) Find the image of a plane, perpendicular to the direction of reflection.

(4) Find the image of a plane, intersecting with the plane of invariant points along a line.

(5) Find the image of a tetrahedron $\Delta \vec{a}_0 \vec{a}_1 \vec{a}_2 \vec{a}_3$ and compute its volume.

(6) Find the images of $|x_1| + |x_2| + |x_3| = 1$ and $x_1^2 + x_2^2 + x_3^2 = 1$.

23. Let

$$T(\vec{x}) = \vec{x}_0 + \vec{x}A, \quad \text{where } A = \frac{1}{3}\begin{bmatrix} 2 & -2 & -1 \\ -2 & -1 & -2 \\ -1 & -2 & 2 \end{bmatrix}.$$

Do the same problems as Ex. 22.

24. Let

$$A = \begin{bmatrix} 0 & \frac{1}{3} & 0 \\ 0 & 0 & \frac{1}{3} \\ 9 & -9 & 3 \end{bmatrix}.$$

(a) Express A as a product of a finite number of elementary matrices, each representing either a reflection, a stretch or a shearing in $\mathcal{N} = \{\vec{0}, \vec{e}_1, \vec{e}_2, \vec{e}_3\}$.

(b) Show that A has only one eigenvalue 1 of multiplicity 3 and $(A - I_3)^2 \neq O$ but $(A - I_3)^3 = O$.

(1) Find a basis $\mathcal{B} = \{\vec{v}_1, \vec{v}_2, \vec{v}_3\}$ so that

$$[A]_{\mathcal{B}} = PAP^{-1} = \begin{bmatrix} 1 & 0 & 0 \\ 1 & 1 & 0 \\ 0 & 1 & 1 \end{bmatrix}$$

is the Jordan canonical form (see (3) in (3.7.50)).

(2) In $\mathcal{B}$, A can be expressed as a product of shearing matrices, i.e.

$$[A]_{\mathcal{B}} = \begin{bmatrix} 1 & 0 & 0 \\ 1 & 1 & 0 \\ 0 & 0 & 1 \end{bmatrix}\begin{bmatrix} 1 & 0 & 0 \\ 0 & 1 & 0 \\ 0 & 1 & 1 \end{bmatrix}.$$

Note that shearing here means skew shearing (refer to Ex. <A> 17 of Sec. 2.8.2 and Ex. <B> 1).

(c) Try to find the image of the cube with vertices at $(\pm 1, \pm 1, \pm 1)$ under $\vec{x} \to \vec{x}A$ by the following methods.

(1) Direct computation.
(2) Factorizations in (b) and (a), respectively.

25. Let

$$A = \begin{bmatrix} -1 & 1 & 0 \\ -4 & 3 & 0 \\ 1 & 0 & 2 \end{bmatrix}.$$

(a) Do as (a) of Ex. 24.
(b) Show that A has eigenvalues 2 and 1, 1 but A is not diagonalizable.

(1) Find a basis $\mathcal{B} = \{\vec{v}_1, \vec{v}_2, \vec{v}_3\}$ so that

$$[A]_{\mathcal{B}} = P^{-1}AP = \begin{bmatrix} 2 & 0 & 0 \\ 0 & 1 & 0 \\ 0 & 1 & 1 \end{bmatrix}$$

is the Jordan canonical form.

(2) In $\mathcal{B}$, show that A can be expressed as a product of a one-way stretch and a shearing, i.e.

$$[A]_{\mathcal{B}} = \begin{bmatrix} 2 & 0 & 0 \\ 0 & 1 & 0 \\ 0 & 0 & 1 \end{bmatrix} \begin{bmatrix} 1 & 0 & 0 \\ 0 & 1 & 0 \\ 0 & 1 & 1 \end{bmatrix}.$$

(c) Do as (c) of Ex. 24.

<B>

1. Try to define *skew* shearing in $\mathbb{R}^3$ (refer to Ex. <A> 17 of Sec. 2.8.2).
2. Model after (1) and (2) in (3.8.26) and try to define the *orthogonal reflection* of the space $\mathbb{R}^3$ with respect to a straight line. Then, show that the orthogonal reflection of $\mathbb{R}^3$ with respect to the line

$$\frac{x_1}{a_{11}} = \frac{x_2}{a_{12}} = \frac{x_3}{a_{13}}$$

is given by the matrix

$$R = \frac{2}{a_{11}^2 + a_{12}^2 + a_{13}^2} \begin{bmatrix} \frac{a_{11}^2 - a_{12}^2 - a_{13}^2}{2} & a_{11}a_{12} & a_{11}a_{13} \\ a_{11}a_{12} & \frac{-a_{11}^2 + a_{12}^2 - a_{13}^2}{2} & a_{12}a_{13} \\ a_{11}a_{13} & a_{12}a_{13} & \frac{-a_{11}^2 - a_{12}^2 + a_{13}^2}{2} \end{bmatrix}.$$

3. Show that the orthogonal reflection of $\mathbb{R}^3$ with respect to the plane

$$a_{11}x_1 + a_{12}x_2 + a_{13}x_3 = 0$$

is given by $-R$, where R is as in Ex 2. Explain this geometrically.

4. Try to define

   (1) translation,
   (2) reflection,
   (3) (one-way, two-way, three-way) stretch,
   (4) shearing,
   (5) rotation, and
   (6) orthogonal reflection

on $\mathbb{R}^4$ and then try to explore their respective features.

<C> Abstraction and generalization

Read Ex.<C> of Sec. 2.8.2.

   Try to do Ex. <B> 4 on $\mathbb{F}^n$ over a field $\mathbb{F}$, in particular, on $\mathbb{R}^n$ endowed with the natural inner product.

3.8.3 *Affine invariants*

We come back to the proof of (3.7.15) in the content of an affine transformation $T(\vec{x}) = \vec{x}_0 + \vec{x}A$ on $\mathbb{R}^3$. Remind that $A = [a_{ij}]_{3 \times 3}$ is an invertible real matrix.

   Only these different from the proofs of (2.7.9) (see Sec. 2.8.3) are needed to be touched here.

2. *T preserves the planes.*

Give a fixed plane (see (1) in (3.5.3))

$$\Sigma: \vec{x} = \vec{a} + t_1 \vec{b}_1 + t_2 \vec{b}_2, \quad t_1, t_2 \in \mathbb{R},$$

where $\vec{b}_1$ and $\vec{b}_2$ are linearly independent vectors in $\mathbb{R}^3$. To find its image $T(\Sigma)$ under T, notice that

$$\Sigma: \vec{x} = \vec{a} + (t_1, t_2) \begin{bmatrix} \vec{b}_1 \\ \vec{b}_2 \end{bmatrix}$$

$$\Rightarrow T(\vec{x}) = \vec{x}_0 + \left(\vec{a} + (t_1, t_2) \begin{bmatrix} \vec{b}_1 \\ \vec{b}_2 \end{bmatrix} \right) A$$

$$= \vec{x}_0 + \vec{a}A + (t_1, t_2) \begin{bmatrix} \vec{b}_1 \\ \vec{b}_2 \end{bmatrix} A$$

$$= \vec{y}_0 + t_1 \vec{c}_1 + t_2 \vec{c}_2, \qquad (*)$$

where $\vec{c}_1 = \vec{b}_1 A$, $\vec{c}_2 = \vec{b}_2 A$ and $\vec{y}_0 = T(\vec{a}) = \vec{x}_0 + \vec{a}A$ and $\vec{c}_1$ and $\vec{c}_2$ are linearly independent vectors in $\mathbb{R}^3$. The latter is easily seen by showing

$$(t_1, t_2) \begin{bmatrix} \vec{c}_1 \\ \vec{c}_2 \end{bmatrix} = (t_1, t_2) \begin{bmatrix} \vec{b}_1 \\ \vec{b}_2 \end{bmatrix} A = \vec{0} \text{ for same scalar } t_1, t_2$$

$$\Leftrightarrow \text{(since } A \text{ is invartible)} \quad (t_1, t_2) \begin{bmatrix} \vec{b}_1 \\ \vec{b}_2 \end{bmatrix} = \vec{0}$$

$$\Leftrightarrow \text{(since } \vec{b}_1 \text{ and } \vec{b}_2 \text{ are linearly independent)} \quad t_1 = t_2 = 0.$$

4. *T preserves relative positions of lines and planes.*
Give two lines

$$L_1: \vec{x} = \vec{a}_1 + t\vec{b}_1,$$

$$L_2: \vec{x} = \vec{a}_2 + t\vec{b}_2, \quad t \in \mathbb{R}.$$

They might be coincident, parallel, intersecting or skew to each other (see (3.4.5)). We only prove the *skew* case as follows.

L_1 and L_2 are skew to each other, i.e. $\vec{b}_1$, $\vec{b}_2$ and $\vec{a}_2 - \vec{a}_1$ are linearly independent.
$\Leftrightarrow$ (since A is invertible) $\vec{b}_1 A$, $\vec{b}_2 A$ and $(\vec{a}_2 - \vec{a}_1)A = (\vec{x}_0 + \vec{a}_2 A) - (\vec{x}_0 + \vec{a}_1 A)$ are linear independent.
$\Leftrightarrow$ (see (*)) The image lines

$$T(L_1): \vec{x} = (\vec{x}_0 + \vec{a}_1 A) + t\vec{b}_1 A,$$

$$T(L_2): \vec{x} = (\vec{x}_0 + \vec{a}_2 A) + t\vec{b}_2 A, \quad t \in \mathbb{R}$$

are skew to each other.

This finishes the proof.

Give a line L and a plane Σ. They might be coincident, parallel or intersecting at a point (see (3.5.5)). Obviously T will preserve these relative positions. Details are left to the readers. The same is true for the relative positions of two planes (see (3.5.6)).

3. *T preserves tetrahedron and parallelepiped.*
This is an easy consequence of 4 and 2.

7. *T preserves the ratio of solid volumes.*

Let $\vec{a}_0, \vec{a}_1, \vec{a}_2$ and $\vec{a}_3$ be four different points in space $\mathbb{R}^3$. The set

$$\square\,\vec{a}_0\,\vec{a}_1\,\vec{a}_2\,\vec{a}_3 = \left\{ \sum_{i=0}^{3} \lambda_i \vec{a}_i \mid 0 \le \lambda_i \le 1 \text{ for } i = 0, 1, 2, 3 \right\} \tag{3.8.28}$$

is called a *parallelepiped* with *vertex* at $\vec{a}_0$ and *side vectors* $\vec{a}_1 - \vec{a}_0$, $\vec{a}_2 - \vec{a}_0$ and $\vec{a}_3 - \vec{a}_0$ and is called a *degenerated* one if the side vectors are linearly dependent. See Fig. 3.73 (see also Fig. 3.2). While the tetrahedron (see Secs. 3.5 and 3.6)

$$\Delta\,\vec{a}_0\,\vec{a}_1\,\vec{a}_2\,\vec{a}_3 = \left\{ \sum_{i=0}^{3} \lambda_i \vec{a}_i \mid 0 \le \lambda_i \le 1 \text{ for } 0 \le i \le 3 \text{ and} \right.$$

$$\left. \lambda_0 + \lambda_1 + \lambda_2 + \lambda_3 = 1 \right\} \tag{3.8.29}$$

is contained in the parallelepiped $\square\,\vec{a}_0\,\vec{a}_1\,\vec{a}_2\,\vec{a}_3$ and has its volume one-sixth of the latter. See Fig. 3.73 (also, see Fig. 3.17).

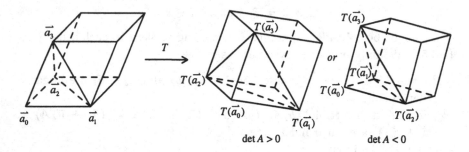

$$\det A > 0 \qquad\qquad \det A < 0$$

Fig. 3.73

Thus (see Sec. 5.3),

the signed volume of $\square\,T(\vec{a}_0)T(\vec{a}_1)T(\vec{a}_2)T(\vec{a}_3)$

$$= \det \begin{bmatrix} T(\vec{a}_1) - T(\vec{a}_0) \\ T(\vec{a}_2) - T(\vec{a}_0) \\ T(\vec{a}_3) - T(\vec{a}_0) \end{bmatrix} = \det \begin{bmatrix} (\vec{a}_1 - \vec{a}_0)A \\ (\vec{a}_2 - \vec{a}_0)A \\ (\vec{a}_3 - \vec{a}_0)A \end{bmatrix} = \det \begin{bmatrix} \vec{a}_1 - \vec{a}_0 \\ \vec{a}_2 - \vec{a}_0 \\ \vec{a}_3 - \vec{a}_0 \end{bmatrix} \det A$$

$$= \text{(the signed volume of } \square\,\vec{a}_0\,\vec{a}_1\,\vec{a}_2\,\vec{a}_3)\det A. \tag{3.8.30}$$

From this it follows immediately (refer to (2.8.44))

The geometric interpretation of the determinant of a linear operator or a square matrix

Let $f(\vec{x}) = \vec{x}A\colon \mathbb{R}^3 \to \mathbb{R}^3$ be a linear operator and $T(\vec{x}) = \vec{x}_0 + \vec{x}A$ the associated affine transformation (A is allowed to be non-invertible here, see (2.8.20)) in $\mathcal{N} = \{\vec{e}_1, \vec{e}_2, \vec{e}_3\}$, where $A = [a_{ij}]_{3\times3}$ is a real matrix. Then the determinant

$$\det f = \det A = \begin{vmatrix} a_{11} & a_{12} & a_{13} \\ a_{21} & a_{22} & a_{23} \\ a_{31} & a_{32} & a_{33} \end{vmatrix}$$

= the signed volume of the parallelepiped $\square \vec{a}_0 \vec{a}_1 \vec{a}_2 \vec{a}_3$ where

$\vec{a}_i = f(\vec{e}_i) = (a_{i1}, a_{i2}, a_{i3})$, the ith row of A, $i = 1, 2, 3$.

$$= \frac{\text{the signed volume of } \square \vec{a}_0 \vec{a}_1 \vec{a}_2 \vec{a}_3}{\text{the volume of the unit cube } \square \vec{0} \vec{e}_1 \vec{e}_2 \vec{e}_3}.$$

Therefore, for any such affine transformation $T(\vec{x}) = \vec{x}_0 + \vec{x}A$,

1. $\dfrac{\text{the signed volume of the image domain } T(\Omega)}{\text{the signed volume of measurable space domain } \Omega} = \det A$, in particular, if Ω is a tetrahedron or a parallelepiped.
2. T preserves orientation $\Leftrightarrow \det A > 0$; and reverses the orientation $\Leftrightarrow \det A < 0$.
3. The image tetrahedron or parallelepiped is degenerated $\Leftrightarrow \det A = 0$.

$$(3.8.31)$$

We have encountered lots of examples concerned with (3.8.31) in Secs. 3.8.1 and 3.8.2. No further examples will be presented here.

Exercises

<A>

1. Prove (3.7.15) in detail by using the following two methods.

   (1) Complete the proof shown in the content.

   (2) Use matrix factorizations of A, see Secs. 3.7.5 and 3.8.2.

   Then, give concrete numerical examples to show that a–d are not necessarily true in general.

<B>

Try to formulate a result for $\mathbb{R}^4$ similar to (3.7.15) and prove it.

<C> Abstraction and generalization

Read Ex. <C> of Sec. 2.8.3.

3.8.4 *Affine geometry*

Readers should go back to Sec. 2.8.4 for fundamental concepts about (planar) affine geometry.

According to F. Klein, the objects we study in spatial affine geometry are the properties that are invariant under the affine transformations, i.e. under the group $G_a(3;\mathbb{R})$ of affine transformations (see Sec. 3.8.1). Section 3.8.3 indicates that affine geometry deals with the following topics:

1. Barycenters (see Sec. 3.6) such as centroid or median points.
2. Parallelism.
3. Ratios of each pair of parallel segments (see Sec. 2.8).
4. Collinearity of affine subspaces (i.e. Menelaus theorem).
5. Concurrence of affine subspaces (i.e. Ceva theorem).

But it does not deal with such objects as lengths, angles, areas and volumes which are in the realm of spatial Euclidean geometry (refer to Part 2).

In what follows we will introduce a sketch of affine geometry for $\mathbb{R}^3$ to the content and by the method that are universally true for more general affine space over a field or ordered field.

3.8.4.1 *Affine independence and dependence (Sec. 3.6 revisited)*

A set of points $\{\vec{a}_0, \vec{a}_1, \ldots, \vec{a}_k\}$ in $\mathbb{R}^3$ (or any *affine space* V) is said to be *affinely independent* if the vectors

$$\vec{a}_1 - \vec{a}_0, \vec{a}_2 - \vec{a}_0, \ldots, \vec{a}_k - \vec{a}_0 \qquad (3.8.32)$$

are linearly independent in the vector space $\mathbb{R}^3$ (or V) and *affinely dependent* otherwise. See Ex. <A> 1.

For $\mathbb{R}^3$, an affinely independent set $\mathcal{B} = \{\vec{a}_0, \vec{a}_1, \vec{a}_2, \vec{a}_3\}$ is called an *affine basis* or *affine frame* with the point $\vec{a}_0$ as the *base point* or *origin* and $\vec{a}_i - \vec{a}_0$ the *i*th *unit vector* or *coordinate vector*.

3.8.4.2 *Affine subspaces*

Let S be any vector subspace of $\mathbb{R}^3$ (or V) of dimension k for $k = 0, 1, 2, 3$ and $\vec{x}_0$ any point in $\mathbb{R}^3$. The set

$$S^k = \vec{x}_0 + S = \{\vec{x}_0 + \vec{v} \mid \vec{v} \in S\} \qquad (3.8.33)$$

is called a *k-dimension affine subspace* of $\mathbb{R}^3$. In particular,

a *point*: $S^0 = \{\vec{x}_0\}$,
a *line*: S^1,
a *plane* or *hyperplane*: S^2,
the *space*: $S^3 = \mathbb{R}^3$.

For any basis $\{\vec{v}_1, \ldots, \vec{v}_k\}$ for S, the set $\{\vec{x}_0, \vec{x}_0 + \vec{v}_1, \ldots, \vec{x}_0 + \vec{v}_k\}$ is an affine basis for S^k and can be extended to an affine basis for $\mathbb{R}^3$.

3.8.4.3 Affine coordinates and barycentric coordinates (Sec. 3.6 revisited)

Let $\mathcal{B} = \{\vec{a}_0, \vec{a}_1, \ldots, \vec{a}_k\}$ be an affine basis for S^k. Then

$$\vec{x} \in S^k$$

$\Leftrightarrow$ The exists unique scalars $x_1, x_2, \ldots, x_k \in \mathbb{R}$ so that

$$\vec{x} = \vec{a}_0 + \sum_{i=1}^{k} x_i(\vec{a}_i - \vec{a}_0). \tag{3.8.34}$$

The associated terminologies are as follows:

1. The *affine coordinate (vector)* of $\vec{x}$ with respect to $\mathcal{B}$:

$$[\vec{x}]_\mathcal{B} = (x_1, x_2, \ldots, x_k).$$

2. The *i*th *coordinate* of $\vec{x}$ in $\mathcal{B}$: x_i for $1 \leq i \leq k$.
3. The *i*th *unit* or *coordinate point*: $\vec{a}_i$ for $1 \leq i \leq k$.
4. The *i*th *coordinate axis*: $\vec{a}_0 + \langle\langle \vec{a}_i - \vec{a}_0 \rangle\rangle$ for $1 \leq i \leq k$.
5. The *i*th *coordinate hyperplane*: $\vec{a}_0 + \langle\langle \vec{a}_1 - \vec{a}_0, \ldots, \vec{a}_{i-1} - \vec{a}_0, \vec{a}_{i+1} - \vec{a}_0, \ldots, \vec{a}_k - \vec{a}_0 \rangle\rangle$ for $1 \leq i \leq k$.
6. For any $(k+1)$ points $\vec{p}_i$ in S^k, $0 \leq i \leq k$, let $[\vec{p}_i]_\mathcal{B} = (\alpha_{i1}, \alpha_{i2}, \ldots, \alpha_{ik})$ be the coordinate of $\vec{p}_i$ in $\mathcal{B}$. Then the quantity

$$V(\vec{p}_0, \vec{p}_1, \ldots, \vec{p}_k) = \frac{1}{k!} \det \begin{bmatrix} [\vec{p}_1 - \vec{p}_0]_\mathcal{B} \\ \vdots \\ [\vec{p}_k - \vec{p}_0]_\mathcal{B} \end{bmatrix}_{k \times k}$$

$$= \frac{1}{k!} \begin{vmatrix} \alpha_{01} & \alpha_{02} & \cdots & \alpha_{0k} & 1 \\ \alpha_{11} & \alpha_{12} & \cdots & \alpha_{1k} & 1 \\ \vdots & \vdots & & \vdots & \vdots \\ \alpha_{k1} & \alpha_{k2} & \cdots & \alpha_{kk} & 1 \end{vmatrix} \tag{3.8.35}$$

is called the signed *volume* with respect to $\mathcal{B}$ of the point set $\{\vec{p}_0, \vec{p}_1, \ldots, \vec{p}_k\}$ or the *k-simplex* $\Delta \vec{p}_0 \vec{p}_1 \cdots \vec{p}_k$ if $\vec{p}_0, \vec{p}_1, \ldots, \vec{p}_k$ are affinely independent (see (3.8.38) and (3.8.30)).

Refer to Figs. 3.73 and Fig. 3.22.

Suppose $\vec{a}_0, \vec{a}_1, \ldots, \vec{a}_k$ are $k+1$ points, not necessarily affine independent, in $\mathbb{R}^3$. Then a point $\vec{x} \in \mathbb{R}^3$ can be expressed as

$$\vec{x} = \sum_{i=0}^{k} \lambda_i \vec{a}_i, \quad \text{where } \lambda_0 + \lambda_1 + \cdots + \lambda_k = 1.$$

$\Leftrightarrow$ for any fixed point $\vec{p}_0$ in $\mathbb{R}^3$, the vector $\vec{x} - \vec{p}_0$

can be expressed as $\vec{x} - \vec{p}_0 = \sum_{i=0}^{k} \lambda_i (\vec{a}_i - \vec{p}_0).$ (3.8.36)

Then the ordered scalars $\lambda_0, \lambda_1, \ldots, \lambda_k$ with $\sum_{i=1}^{k} \lambda_i = 1$ is called a *barycentric coordinate* of $\vec{x}$, in the affine subspace $\vec{a}_0 + \langle\langle \vec{a}_1 - \vec{a}_0, \ldots, \vec{a}_k - \vec{a}_0 \rangle\rangle$,
with respect to $\{\vec{a}_0, \vec{a}_1, \ldots, \vec{a}_k\}$. In particular, if $\mathcal{B} = \{\vec{a}_0, \vec{a}_1, \ldots, \vec{a}_k\}$ is
affinely independent, such $\lambda_0, \lambda_1, \ldots, \lambda_k$ are *uniquely* determined as long as
$\vec{x}$ lies in the subspace and hence,

$$(\vec{x})_{\mathcal{B}} = (\lambda_0, \lambda_1, \ldots, \lambda_k) \quad \text{with} \quad \sum_{i=0}^{k} \lambda_i = 1 \quad\quad (3.8.37)$$

is called *the barycentric coordinate* of $\vec{x}$ with respect to $\mathcal{B}$ or a *barycenter*
of $\mathcal{B}$ with *weight* $\lambda_0, \lambda_1, \ldots, \lambda_k$.

In case $\{\vec{a}_0, \vec{a}_1, \ldots, \vec{a}_k\}$ is affinely independent in $\mathbb{R}^3$. The set

$$\Delta \vec{a}_0 \vec{a}_1 \cdots \vec{a}_k = \left\{ \sum_{i=0}^{k} \lambda_i \vec{a}_i \mid \lambda_i \geq 0 \text{ for } 0 \leq i \leq k \text{ and } \sum_{i=0}^{k} \lambda_i = 1 \right\}$$

$$(3.8.38)$$

is called the *k-simplex* (see (3.8.53) below) with

vertices: $\vec{a}_i$ for $0 \leq i \leq k$,
edges: $\vec{a}_i \vec{a}_j$ for $i \neq j$ and $0 \leq i, j \leq k$,
2-dimensional *faces*: $\Delta \vec{a}_0 \vec{a}_1 \vec{a}_2$, $\Delta \vec{a}_0 \vec{a}_1 \vec{a}_3$, $\Delta \vec{a}_0 \vec{a}_2 \vec{a}_3$ and $\Delta \vec{a}_1 \vec{a}_2 \vec{a}_3$
    if $k = 3$.

In particular,

 $\Delta \vec{a}_0$: the point $\{\vec{a}_0\}$,

 $\Delta \vec{a}_0 \vec{a}_1$: the line segment $\vec{a}_0 \vec{a}_1$,

 $\Delta \vec{a}_0 \vec{a}_1 \vec{a}_2$: the triangle (see Fig. 2.26),

 $\Delta \vec{a}_0 \vec{a}_1 \vec{a}_2 \vec{a}_3$: the tetrahedron (see Figs. 3.17 and 3.74).

Note that a tetrahedron has 4 vertices, 6 edges and 4 faces which satisfy
the Euler formula $V - E + F = 2$ for polyhedron (see Ex. <B> 8), and the
point $\left(\frac{1}{3}, \frac{1}{3}, \frac{1}{3}\right)$ is the barycenter.

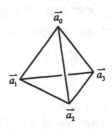

Fig. 3.74

Let $\mathcal{B} = \{\vec{a}_0, \vec{a}_1, \ldots, \vec{a}_k\}$ be affinely independent in $\mathbb{R}^3$ and $\vec{x}, \vec{b}_0,$ $\vec{b}_1, \ldots, \vec{b}_r$ be points in the subspace $\vec{a}_0 + \langle\langle \vec{a}_1 - \vec{a}_0, \ldots, \vec{a}_k - \vec{a}_0 \rangle\rangle$ so that $\mathcal{C} = \{\vec{b}_0, \vec{b}_1, \ldots, \vec{b}_r\}$ is affinely independent. Denote

$$[\vec{x}]_\mathcal{B} = (x_1, \ldots, x_k), \quad \text{and} \quad (\vec{x})_\mathcal{C} = (\lambda_0, \lambda_1, \ldots, \lambda_r);$$
$$[\vec{b}_i]_\mathcal{B} = (\alpha_{i1}, \ldots, \alpha_{ik}) \quad \text{for } 0 \le i \le r.$$

Then,

$$\vec{x} \in \vec{b}_0 + \langle\langle \vec{b}_1 - \vec{b}_0, \ldots, \vec{b}_r - \vec{b}_0 \rangle\rangle$$

$$\Leftrightarrow (x_1 \cdots x_k) = (\lambda_0 \lambda_1 \cdots \lambda_r) \begin{bmatrix} \alpha_{01} & \alpha_{02} & \cdots & \alpha_{0k} \\ \alpha_{11} & \alpha_{12} & \cdots & \alpha_{1k} \\ \vdots & \vdots & & \vdots \\ \alpha_{r1} & \alpha_{r2} & \cdots & \alpha_{rk} \end{bmatrix}_{(r+1)\times k} \quad \text{or, in short,}$$

$$[\vec{x}]_\mathcal{B} = (\vec{x})_\mathcal{C} \begin{bmatrix} [\vec{b}_0]_\mathcal{B} \\ [\vec{b}_1]_\mathcal{B} \\ \vdots \\ [\vec{b}_k]_\mathcal{B} \end{bmatrix}_{(r+1)\times k}. \tag{3.8.39}$$

3.8.4.4 Operations of affine subspaces

Let S^r and S^k be two affine subspaces of $\mathbb{R}^3$ (or V), say

$$S^r = \vec{y}_0 + S_1 \quad \text{and} \quad S^k = \vec{x}_0 + S_2$$

for some vector subspaces S_1 and S_2 of $\mathbb{R}^3$ (or V) of respective dimension r and k.

The *intersection* subspace of S^r and S^k is defined to be the intersection of S^r and S^k as subsets of $\mathbb{R}^3$ and is denoted by

$$S^r \cap S^k. \tag{3.8.40}$$

It can be shown easily that

1. $S^r \cap S^k \neq \emptyset \Leftrightarrow \vec{x}_0 - \vec{y}_0 \in S_1 + S_2$;
2. in case $S^r \cap S^k \neq \emptyset$, one may choose $\vec{x}_0 = \vec{y}_0 \in S^r \cap S^k$ and

$$S^r \cap S^k = \vec{x}_0 + (S_1 \cap S_2). \tag{3.8.41}$$

The *sum* space of S^r and S^k is defined as and denoted by

$$S^r + S^k = \cap\{W \text{ is affine subspace of } \mathbb{R}^3 \text{ containing both } S^r \text{ and } S^k\}. \tag{3.8.42}$$

It can be shown that

$$S^r + S^k = \begin{cases} \vec{x}_0 + (S_1 + S_2), & \text{if } S^r \cap S^k \neq \emptyset \text{ and } \vec{x}_0 \in S^r \cap S^k \\ \vec{x}_0 + \langle\langle \vec{y}_0 - \vec{x}_0 \rangle\rangle \oplus (S_1 + S_2), & \text{if } S^r \cap S^k = \emptyset \end{cases}. \tag{3.8.43}$$

The details are left as Ex. <A> 4.

3.8.4.5 *Dimension (or intersection) theorem*

Let $S^r = \vec{y}_0 + S_1$ and $S^k = \vec{x}_0 + S_2$ be affine subspaces of $\mathbb{R}^3$ (or any finite dimensional affine space V).

(a) If $S^r \cap S^k \neq \emptyset$, then

$$\dim S^r + \dim S^k = r + k$$
$$= \dim S^r \cap S^k + \dim(S^r + S^k).$$

(b) If $S^r \cap S^k = \emptyset$, define $\dim \emptyset = -1$, then

$$\dim S^r \cap S^k + \dim(S^r + S^k) = \dim S^r + \dim S^k - \dim(S_1 \cap S_2).$$

In particular,

1. in case $S_1 \cap S_2 = \{\vec{0}\}$,

$$\dim(S^r + S^k) = \dim S^r + \dim S^k + 1;$$

2. in case $S_1 \cap S_2 \not\supseteq \{\vec{0}\}$,

$$\dim(S^r + S^k) < \dim S^r + \dim S^k + 1. \tag{3.8.44}$$

The details are left as Ex. <A> 5.

3.8.4.6   *Relative positions of affine subspaces*

Suppose $S^r = \vec{y}_0 + S_1$ and $S^k = \vec{x}_0 + S_2$ are two affine subspaces of $\mathbb{R}^3$ (or V). For their relative positions, we consider the following four cases.

Case 1 If $S^r \subseteq S^k$ (or $S^r \supseteq S^k$) happens, S^r is said to be *coincident* with a subspace, namely S^r itself, of S^k.

Case 2 If $S^r \cap S^k = \emptyset$ and $S_1 \cap S_2 = \{\vec{0}\}$ happen, S^r and S^k are said to be *skew* to each other. Then, by (3.8.41) and (3.8.44),

$\quad S^r$ and S^k are skew to each other

$\Leftrightarrow \vec{y}_0 - \vec{x}_0 \notin S_1 \oplus S_2$ (and hence, $S^r + S^k = \vec{x}_0 + \langle\langle \vec{y}_0 - \vec{x}_0 \rangle\rangle \oplus (S_1 \oplus S_2)$).

$\Leftrightarrow \dim(S^r + S^k) = \dim S^r + \dim S^k + 1.$ \hfill (3.8.45)

The details are left as Ex. $<\!A\!>$ 6.

Case 3 If $S^r \cap S^k = \emptyset$ but either $S_1 \supseteq S_2$ or $S_1 \subseteq S_2$ happens, then S^r and S^k are said to be *parallel* to each other and is denoted as

$$S^r \parallel S^k. \hspace{3cm} (3.8.46)$$

In case $S_1 \subseteq S_2$, the translation $\vec{x} \rightarrow \vec{x}_0 - \vec{y}_0 + \vec{x}$ will transform $S^r = \vec{y}_0 + S_1$ to an affine subspace $\vec{x}_0 + S_1$ of $S^k = \vec{x}_0 + S_2$.

Case 4 If $S^r \cap S^k = \emptyset$, $S_1 \cap S_2 \not\supseteq \{\vec{0}\}$ and $S_1 \not\subseteq S_2$ and $S_2 \not\subseteq S_1$ happen, then S^r and S^k are *neither* coincident, skew *nor* parallel. In this case, S^r contains a subspace $S_1^p = \vec{y}_0 + S_1 \cap S_2$, where $p = \dim S_1 \cap S_2$ and S^k contains a subspace $S_2^p = \vec{x}_0 + S_1 \cap S_2$ so that

$$S_1^p \parallel S_2^p.$$

Does this happen in $\mathbb{R}^3$?

$\quad$ We can use concepts introduced here to recapture (2.5.9), (3.4.5), (3.5.5) and (3.5.6).

$\quad$ Take (3.4.5) and (3.5.6) as examples.

Example 1 Determine the relative positions of two lines $S_1^1 = \vec{x}_0 + S_1$ and $S_2^1 = \vec{y}_0 + S_2$ in $\mathbb{R}^3$ (see Fig. 3.13).

Solution Remind that both S_1 and S_2 are one-dimensional vector subspaces of $\mathbb{R}^3$ and $\vec{y}_0 - \vec{x}_0$ is a vector in $\mathbb{R}^3$.

$\quad$ In case $S_1 = S_2$: then $\vec{x}_0 = \vec{y}_0$ will induce that S_1^1 is coincident with S_2^1, while $\vec{x}_0 \neq \vec{y}_0$ will result in the parallelism of S_1^1 with S_2^1.

In case $S_1 \cap S_2 = \{\vec{0}\}$: then the condition $\vec{x}_0 - \vec{y}_0 \in S_1 \oplus S_2$ implies that S_1^1 intersects S_2^1 at a point, while $\vec{x}_0 - \vec{y}_0 \notin S_1 \oplus S_2$ implies that S_1^1 is skew to S_2^1. □

Example 2 Determine the relative positions of two planes $S_1^2 = \vec{x}_0 + S_1$ and $S_2^2 = \vec{y}_0 + S_2$ in $\mathbb{R}^3$ (see Fig. 3.16).

Solution Both S_1 and S_2 are two-dimensional vector subspaces of $\mathbb{R}^3$.

In case $S_1 = S_2$: if $\vec{x}_0 - \vec{y}_0 \in S_1$, then $S_1^2 = S_2^2$ is coincident; if $\vec{x}_0 - \vec{y}_0 \notin S_1$, then $S_1^2 \parallel S_2^2$.

In case $S_1 \cap S_2$ is one-dimensional: no matter $\vec{x}_0 - \vec{y}_0 \in S_1 \cap S_2$ or not, S_1^2 intersects with S_2^2 along a line, namely $\vec{x}_0 + S_1 \cap S_2$ if $\vec{y}_0$ is chosen to be equal to $\vec{x}_0$.

By dimension theorem for vector spaces: $\dim S_1 + \dim S_2 = \dim S_1 \cap S_2 + \dim(S_1 + S_2)$ (see Ex. <A> 11 of Sec. 3.2 or Sec. B.2), since $\dim(S_1 + S_2) \le 3$, therefore $\dim S_1 \cap S_2 \ge 1$. Hence $S_1 \cap S_2 = \{\vec{0}\}$ never happens in $\mathbb{R}^3$ and S_1 and S_2 can not be skew to each other. □

As a consequence of Examples 1 and 2, in a tetrahedron $\triangle \vec{a}_0 \vec{a}_1 \vec{a}_2 \vec{a}_3$ (see Fig. 3.74), we have the following information:

1. Vertices are skew to each other.
2. Vertex $\vec{a}_3$ is skew to the face $\triangle \vec{a}_0 \vec{a}_1 \vec{a}_2$, etc.
3. The edge $\vec{a}_0 \vec{a}_1$ is skew to the edge $\vec{a}_2 \vec{a}_3$ but intersects with the face $\vec{a}_1 \vec{a}_2 \vec{a}_3$, etc.
4. Any two faces intersect along their common edge.

Do you know what happens to $\triangle \vec{a}_0 \vec{a}_1 \vec{a}_2 \vec{a}_3 \vec{a}_4$ in $\mathbb{R}^4$?

To cultivate our geometric intuition, we give one more example.

Example 3 Determine the relative positions of

(a) two straight lines in $\mathbb{R}^4$,
(b) two (two-dimensional) planes in $\mathbb{R}^4$, and
(c) one (two-dimensional) plane and one (three-dimensional) hyperplane in $\mathbb{R}^4$.

Solution

(a) The answer is like Example 1. Why? Prove it.

(b) Let $S_1^2 = \vec{x}_0 + S_1$ and $S_2^2 = \vec{y}_0 + S_2$ be two planes in $\mathbb{R}^4$.

Besides the known three relative positions in Example 2, namely, coincident, parallel and intersecting along a line, there does exist another two possibilities in $\mathbb{R}^4$: intersecting at a point, and Case 4 but never skew to each other.

In case $\dim S_1 \cap S_2 = 1$: then

$$\dim S_1 = \dim S_2 = 2$$
$$\Rightarrow \dim(S_1 + S_2) = 2 + 2 - 1 = 3.$$

Hence, it is possible to choose $\vec{x}_0, \vec{y}_0 \in \mathbb{R}^4$ so that $\vec{x}_0 - \vec{y}_0 \notin S_1 + S_2$. According to (3.8.41), $S_1^2 \cap S_2^2 = \emptyset$ holds. So Case 4 does happen.

In case $S_1 \cap S_2 = \{\vec{0}\}$: then

$$\dim S_1 = \dim S_2 = 2$$
$$\Rightarrow \dim(S_1 + S_2) = \dim(S_1 \oplus S_2) = 2 + 2 - 0 = 4,$$

namely, $S_1 \oplus S_2 = \mathbb{R}^4$ holds. No matter how $\vec{x}_0$ and $\vec{y}_0$ are chosen, $\vec{x}_0 - \vec{y}_0 \in \mathbb{R}^4$ is always true. Therefore, S_1^2 and S_2^2 intersect at a point. Since $\vec{x}_0 - \vec{y}_0 \notin \mathbb{R}^4$ does not happen in this case, so S_1^2 and S_2^2 are never skew to each other.

(c) Let $S^2 = \vec{x}_0 + S_1$ and $S^3 = \vec{y}_0 + S_2$ where $\dim S_1 = 2, \dim S_2 = 3$. Since

$$\dim S_1 \cap S_2 = \dim S_1 + \dim S_2 - \dim(S_1 + S_2)$$
$$= 2 + 3 - \dim(S_1 + S_2) = 5 - \dim(S_1 + S_2)$$

and $3 \le \dim(S_1 + S_2) \le 4$, therefore $\dim S_1 \cap S_2$ could be 1 or 2 only.

In case $\dim S_1 \cap S_2 = 1$: then $\dim(S_1 + S_2) = 4$. For any two points $\vec{x}_0, \vec{y}_0$ in $\mathbb{R}^4$, $\vec{x}_0 - \vec{y}_0 \in S_1 + S_2 = \mathbb{R}^4$ always hold. Then S^2 and S^3 will intersect along the line $\vec{x}_0 + S_1 \cap S_2$ if $\vec{y}_0 = \vec{x}_0$.

In case $\dim S_1 \cap S_2 = 2$: then $S_1 \subseteq S_2$ and $S_1 + S_2 = S_2$ holds. For any two points $\vec{x}_0, \vec{y}_0 \in \mathbb{R}^4$, either

$$\vec{x}_0 - \vec{y}_0 \in S_1 + S_2 \Rightarrow S^2 \text{ and } S^3 \text{ are coincident because } S^2 \subseteq S^3,$$

or

$$\vec{x}_0 - \vec{y}_0 \notin S_1 + S_2 = S_2 \Rightarrow S^2 \,\|\, S^3, \text{ parallel to each other.}$$

Remark In $\mathbb{R}^4$.

S_1^1 and S_2^2 can be only coincident, parallel or intersecting at a point.

How about $S^1 = \vec{x}_0 + S_1$ and $S^3 = \vec{y}_0 + S_2$? Since

$$\dim S_1 \cap S_2 = 1 + 3 - \dim(S_1 + S_2) = 4 - \dim(S_1 + S_2),$$

and $3 \leq \dim(S_1 + S_2) = 4, \dim S_1 \cap S_2$ can be only 1 or 0.

1. $\dim S_1 \cap S_2 = 1$: then $S_1 \subseteq S_2$ and $S_1 + S_2 = S_2$ hold. For $\vec{x}_0, \vec{y}_0$ in $\mathbb{R}^4$, either

$$\vec{x}_0 - \vec{y}_0 \in S_1 + S_2 \Rightarrow S^1 \text{ and } S^3 \text{ are coincident with } S^1 \subseteq S^3,$$

or

$$\vec{x}_0 - \vec{y}_0 \notin S_1 + S_2 \Rightarrow S^1 \parallel S^3.$$

2. $\dim S_1 \cap S_2 = 0$: then $S_1 + S_2 = S_1 \oplus S_2 = \mathbb{R}^4$. No matter how $\vec{x}_0$ and $\vec{y}_0$ are chosen in $\mathbb{R}^4, \vec{x}_0 - \vec{y}_0 \in S_1 + S_2$ always holds. Then S^1 and S^3 will intersect at one point.

See Exs. <A> 7 and 8 for more practice.

3.8.4.7   Menelaus and Ceva theorems

As a generalization of (2.8.45), we have

Menelaus Theorem *Suppose* $\{\vec{a}_0, \vec{a}_1, \ldots, \vec{a}_k\}$ *are affinely independent in* $\mathbb{R}^3$ *(or* V *with* $\dim V < \infty$*), and put* $\vec{a}_{k+1} = \vec{a}_0$*. Let* $\vec{b}_i$ *be an arbitrary point on the line segment* $\vec{a}_i \vec{a}_{i+1}$ *other than the end points* $\vec{a}_i$ *and* $\vec{a}_{i+1}$ *for* $0 \leq i \leq k$*. Let* $\alpha_i \in \mathbb{R}$ *(or a field* $\mathbb{F}$*) so that, for the vectors* $\vec{a}_i \vec{b}_i = \vec{b}_i - \vec{a}_i$ *and* $\vec{b}_i \vec{a}_{i+1} = \vec{a}_{i+1} - \vec{b}_i$*,*

$$\alpha_i \vec{a}_i \vec{b}_i = \vec{b}_i \vec{a}_{i+1} \quad \text{for } 0 \leq i \leq k.$$

Then,

$$\vec{b}_0, \vec{b}_1, \vec{b}_2, \ldots, \vec{b}_k \text{ are affinely dependent.}$$
$$\Leftrightarrow \alpha_0 \alpha_1 \alpha_2 \cdots \alpha_k = (-1)^{k+1}. \tag{3.8.47}$$

Note that, for $k = 2$, this is the Menelaus theorem in (2.8.45). For $k = 3$ in $\mathbb{R}^3$, see Fig. 3.75.

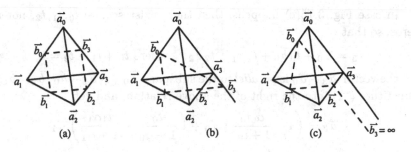

Fig. 3.75

Sketch of proof Suppose $k = 3$.

By assumptions, $\alpha_0\alpha_1\alpha_2\alpha_3 \neq 0$ and

$$
\begin{aligned}
\vec{b}_0 &= \frac{\alpha_0}{1+\alpha_0}\vec{a}_0 + \frac{1}{1+\alpha_0}\vec{a}_1, \\
\vec{b}_1 &= \frac{\alpha_1}{1+\alpha_1}\vec{a}_1 + \frac{1}{1+\alpha_1}\vec{a}_2, \\
\vec{b}_2 &= \frac{\alpha_2}{1+\alpha_2}\vec{a}_2 + \frac{1}{1+\alpha_2}\vec{a}_3, \\
\vec{b}_3 &= \frac{\alpha_3}{1+\alpha_3}\vec{a}_3 + \frac{1}{1+\alpha_3}\vec{a}_0.
\end{aligned}
\tag{*1}
$$

The necessity In case Fig. 3.75(a) or (b) happens, by (3.8.36) it follows that there exist scalars $\lambda_0, \lambda_1, \lambda_2$ so that

$$
\vec{b}_3 = \lambda_0\vec{b}_0 + \lambda_1\vec{b}_1 + \lambda_2\vec{b}_2, \quad \text{where } \lambda_0 + \lambda_1 + \lambda_2 = 1.
$$

By (*1), this implies that

$$
\frac{\lambda_0\alpha_0}{1+\alpha_0}\vec{a}_0 + \left(\frac{\lambda_0}{1+\alpha_0} + \frac{\lambda_1\alpha_1}{1+\alpha_1}\right)\vec{a}_1 + \left(\frac{\lambda_1}{1+\alpha_1} + \frac{\lambda_2\alpha_2}{1+\alpha_2}\right)\vec{a}_2
$$

$$
+ \frac{\lambda_2}{1+\alpha_2}\vec{a}_3 = \frac{\alpha_3}{1+\alpha_3}\vec{a}_3 + \frac{1}{1+\alpha_3}\vec{a}_0
$$

$$
\Rightarrow \frac{\lambda_0\alpha_0}{1+\alpha_0} = \frac{1}{1+\alpha_3}, \quad \frac{\lambda_2}{1+\alpha_2} = \frac{\alpha_3}{1+\alpha_3},
$$

$$
\frac{\lambda_0}{1+\alpha_0} + \frac{\lambda_1\alpha_1}{1+\alpha_1} = 0 \quad \text{and} \quad \frac{\lambda_1}{1+\alpha_1} + \frac{\lambda_2\alpha_2}{1+\alpha_2} = 0
$$

$$
\Rightarrow \lambda_0 = \frac{1+\alpha_0}{\alpha_0(1+\alpha_3)}, \lambda_2 = \frac{\alpha_3(1+\alpha_2)}{1+\alpha_3} \quad \text{and} \quad \lambda_1 = -\frac{1+\alpha_1}{\alpha_0\alpha_1(1+\alpha_3)}
$$

$$
\Rightarrow (\text{since } \lambda_0 + \lambda_1 + \lambda_2 = 1) \ \alpha_0\alpha_1\alpha_2\alpha_3 = 1.
$$

In case Fig. 3.75(c) happens, then there exist scalars t_0, t_1, t_2, not all zeros, so that

$$\vec{a}_3 = \vec{a}_0 + t_0\vec{b}_0 + t_1\vec{b}_1 + t_2\vec{b}_2, \quad \text{where } t_0 + t_1 + t_2 = 0$$

i.e. the vector $\vec{a}_3 - \vec{a}_0$ is *parallel* to the subspace $\vec{b}_0 + \langle\langle \vec{b}_1 - \vec{b}_0, \vec{b}_2 - \vec{b}_0 \rangle\rangle$. Substitute (*1) into the right of the above relation, and get

$$\vec{a}_3 = \left(1 + \frac{\alpha_0 t_0}{1 + \alpha_0}\right)\vec{a}_0 + \left(\frac{t_0}{1 + \alpha_0} + \frac{t_1\alpha_1}{1 + \alpha_1}\right)\vec{a}_1$$

$$+ \left(\frac{t_1}{1 + \alpha_1} + \frac{t_2\alpha_2}{1 + \alpha_2}\right)\vec{a}_2 + \frac{t_2}{1 + \alpha_2}\vec{a}_3$$

$$\Rightarrow 1 + \frac{\alpha_0 t_0}{1 + \alpha_0} = 0, \quad \frac{t_0}{1 + \alpha_0} + \frac{t_1\alpha_1}{1 + \alpha_1} = 0,$$

$$\frac{t_1}{1 + \alpha_1} + \frac{t_2\alpha_2}{1 + \alpha_2} = 0 \quad \text{and} \quad \frac{t_2}{1 + \alpha_2} = 1$$

$$\Rightarrow (\text{since } t_0 + t_1 + t_2 = 0) \; \alpha_0\alpha_1\alpha_2 = -1.$$

By imagination, the extended line $\vec{a}_0\vec{a}_3$ will intersect the subspace $\vec{b}_0 + \langle\langle \vec{b}_1 - \vec{b}_0, \vec{b}_2 - \vec{b}_0 \rangle\rangle$ at a point $\vec{b}_3 = \infty$ *at infinity* (see Sec. 3.8.4.10 below) and in this situation, α_3 should be considered as -1. Therefore,

$$\alpha_0\alpha_1\alpha_2\alpha_3 = 1$$

still holds.

The sufficiency We are going to prove that

$\vec{b}_0, \vec{b}_1, \vec{b}_2$ and $\vec{b}_3$ are affinely dependent.

$\Leftrightarrow$ There exist scalars t_0, t_1, t_2, t_3, not all zeros, so that

$$t_0\vec{b}_0 + t_1\vec{b}_1 + t_2\vec{b}_2 + t_3\vec{b}_3 = \vec{0}, \quad \text{where } t_0 + t_1 + t_2 + t_3 = 0. \quad (*2)$$

Now, (*1) and (*2) imply that

$$\left(\frac{\alpha_0 t_0}{1 + \alpha_0} + \frac{t_3}{1 + \alpha_3}\right)\vec{a}_0 + \left(\frac{t_0}{1 + \alpha_0} + \frac{t_1\alpha_1}{1 + \alpha_1}\right)\vec{a}_1$$

$$+ \left(\frac{t_1}{1 + \alpha_1} + \frac{t_2\alpha_2}{1 + \alpha_2}\right)\vec{a}_2 + \left(\frac{t_2}{1 + \alpha_2} + \frac{t_3\alpha_3}{1 + \alpha_3}\right)\vec{a}_3 = \vec{0}.$$

Since $\alpha_0\alpha_1\alpha_2\alpha_3 = 1$, so $\vec{b}_0, \vec{b}_1, \vec{b}_2$ and $\vec{b}_3$ are all distinct and t_0, t_1, t_2 and t_3 can be so chosen, all not equal to zeros, so that

$$\frac{t_3}{t_0} = -\frac{(1 + \alpha_3)\alpha_0}{1 + \alpha_0}, \quad \frac{t_0}{t_1} = -\frac{(1 + \alpha_0)\alpha_1}{1 + \alpha_1},$$

$$\frac{t_1}{t_2} = -\frac{(1 + \alpha_1)\alpha_2}{1 + \alpha_2} \quad \text{and} \quad \frac{t_2}{t_3} = -\frac{(1 + \alpha_2)\alpha_3}{1 + \alpha_3}.$$

In this case,

$$t_0 \vec{b}_0 + t_1 \vec{b}_1 + t_2 \vec{b}_2 + t_3 \vec{b}_3 = \vec{0}$$

does happen and

$$t_0 + t_1 + t_2 + t_3 = \sum_{i=0}^{3} \left(\frac{1}{1+\alpha_i} + \frac{\alpha_i}{1+\alpha_i} \right) t_i, \quad \text{where } \alpha_4 = \alpha_0$$

$$= \left(\frac{\alpha_0 t_0}{1+\alpha_0} + \frac{t_3}{1+\alpha_3} \right) + \left(\frac{t_0}{1+\alpha_0} + \frac{t_1 \alpha_1}{1+\alpha_1} \right) + \left(\frac{t_1}{1+\alpha_1} + \frac{t_2 \alpha_2}{1+\alpha_2} \right)$$

$$+ \left(\frac{t_2}{1+\alpha_2} + \frac{t_3 \alpha_3}{1+\alpha_3} \right)$$

$$= 0.$$

In case $\vec{b}_3 = \infty$, then $t_3 = 0$ if and only if $\alpha_3 = -1$. Then, the condition $\alpha_0 \alpha_1 \alpha_2 = -1$ would imply that $\vec{b}_0, \vec{b}_1$ and $\vec{b}_2$ are coplanar. $\quad \square$

As a generalization of (2.8.46), we have

Ceva Theorem *Let* $\vec{a}_0, \vec{a}_1, \ldots, \vec{a}_k;\ \vec{b}_0, \vec{b}_1, \ldots, \vec{b}_k$ *and* $\alpha_0, \alpha_1, \ldots, \alpha_k$ *be as in the Menelaus theorem (3.8.47). Suppose* $k \geq 2$. *Let the affine subspace*

$$\pi_i \colon \vec{b}_i + \langle\langle \vec{a}_{i+2} - \vec{b}_i, \ldots, \vec{a}_{i-1} - \vec{b}_i \rangle\rangle \quad \text{for } 0 \leq i \leq k,$$

where $\vec{a}_{-1} = \vec{a}_k, \vec{a}_{k+1} = \vec{a}_1$ *and* $\vec{a}_{k+2} = \vec{a}_2$. *Then*

$$\pi_0, \pi_1, \ldots, \pi_\kappa \text{ are concurrent at a point.}$$
$$\Leftrightarrow \alpha_0 \alpha_1 \cdots \alpha_k = 1. \tag{3.8.48}$$

Note that the case $k = 2$ is the Ceva theorem in (2.8.46). See Fig. 3.76.

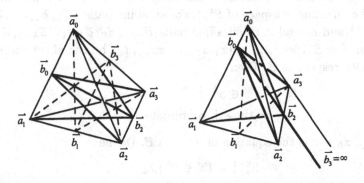

Fig. 3.76

Sketch of proof Suppose $k = 3$. Adopt (*1) in what follows.

π_0 and π_2 intersect along the line $\overrightarrow{b}_0\overrightarrow{b}_2$, while π_1 and π_3 intersect along the line $\overrightarrow{b}_1\overrightarrow{b}_3$. Then

π_0, π_1, π_2 and π_3 are concurrent at a point.

$\Leftrightarrow$ The lines $\overrightarrow{b}_0\overrightarrow{b}_2$ and $\overrightarrow{b}_1\overrightarrow{b}_3$ are concurrent at a point.

$\Leftrightarrow$ There exist scalars t_0 and t_1 such that

$$(1 - t_0)\overrightarrow{b}_0 + t_0\overrightarrow{b}_2 = (1 - t_1)\overrightarrow{b}_1 + t_1\overrightarrow{b}_3.$$

$$\Leftrightarrow \frac{(1 - t_0)\alpha_0}{1 + \alpha_0}\overrightarrow{a}_0 + \frac{1 - t_0}{1 + \alpha_0}\overrightarrow{a}_1 + \frac{t_0\alpha_2}{1 + \alpha_2}\overrightarrow{a}_2 + \frac{t_0}{1 + \alpha_2}\overrightarrow{a}_3$$
$$= \frac{t_1}{1 + \alpha_3}\overrightarrow{a}_0 + \frac{(1 - t_1)\alpha_1}{1 + \alpha_1}\overrightarrow{a}_1 + \frac{1 - t_1}{1 + \alpha_1}\overrightarrow{a}_2 + \frac{t_1\alpha_3}{1 + \alpha_3}\overrightarrow{a}_3.$$

$$\Leftrightarrow \frac{(1 - t_0)\alpha_0}{1 + \alpha_0} = \frac{t_1}{1 + \alpha_3}, \quad \frac{1 - t_0}{1 + \alpha_0} = \frac{(1 - t_1)\alpha_1}{1 + \alpha_1},$$
$$\frac{t_0\alpha_2}{1 + \alpha_2} = \frac{1 - t_1}{1 + \alpha_1} \quad \text{and} \quad \frac{t_0}{1 + \alpha_2} = \frac{t_1\alpha_3}{1 + \alpha_3}.$$

$$\Leftrightarrow \frac{1 - t_1}{t_1} = \frac{\alpha_2\alpha_3(1 + \alpha_1)}{1 + \alpha_3} \quad \text{and} \quad \frac{t_1}{1 - t_1} = \frac{\alpha_0\alpha_1(1 + \alpha_3)}{1 + \alpha_1}.$$

$$\Leftrightarrow \alpha_0\alpha_1\alpha_2\alpha_3 = 1. \qquad \square$$

This finishes the proof.

3.8.4.8 *Half-space, convex set, simplex and polyhedron*

Concepts introduced here are still valid for finite-dimensional affine space V over an *ordered* field $\mathbb{F}$.

Let S^k be a k-dimensional affine subspace of $\mathbb{R}^3$ (or V) and, in turn, S^{k-1} be an affine subspace of S^k. Take an affine basis $\{\overrightarrow{a}_0, \overrightarrow{a}_1, \ldots, \overrightarrow{a}_{k-1}\}$ for S^{k-1} and extend it to an affine basis $\mathcal{B} = \{\overrightarrow{a}_0, \overrightarrow{a}_1, \ldots, \overrightarrow{a}_{k-1}, \overrightarrow{a}_k\}$ for S^k. For $\overrightarrow{x} \in S^k$, let $[\overrightarrow{x}]_{\mathcal{B}} = (x_1, x_2, \ldots, x_{k-1}, x_k)$ be the affine coordinate of $\overrightarrow{x}$ with respect to $\mathcal{B}$. Then

$$\overrightarrow{x} \in S^{k-1}$$

$$\Leftrightarrow \text{the } k\text{th coordinate } x_k = 0,$$

namely, $x_k = 0$ is the equation of S^{k-1} in $\mathcal{B}$. Define

$$S_+^{k-1} = \{\overrightarrow{x} \in S^k \mid x_k > 0\},$$
$$S_-^{k-1} = \{\overrightarrow{x} \in S^k \mid x_k < 0\}$$

(3.8.49)

and are called the *open half-spaces of* S^k *divided by* S^{k-1}. Both $S^{k-1}_+ \cup S^{k-1}$
and $S^{k-1}_- \cup S^{k-1}$ are called the *closed half-spaces*. For a point $\vec{x}$ of S^k that
does not lie on S^{k-1}, the half-space containing $\vec{x}$ is called the *side of* $\vec{x}$
with respect to S^{k-1}. Let $\vec{p}$ and $\vec{q}$ be two points on a line S^1. The closed
side of $\vec{q}$ with respect to $\vec{p}$ is called the (*closed*) *half line* or *ray* from $\vec{p}$ to
$\vec{q}$ and is denoted by $\overrightarrow{\vec{p}\,\vec{q}}$. The set

$$\overline{\vec{p}\,\vec{q}} = \overrightarrow{\vec{p}\,\vec{q}} \cap \overrightarrow{\vec{q}\,\vec{p}} = \{\vec{x} = (1-t)\vec{p} + t\vec{q} \mid 0 \le t \le 1\} \qquad (3.8.50)$$

is called a *segment* joining $\vec{p}$ and $\vec{q}$. Note that $\overline{\vec{p}\,\vec{q}} = \overline{\vec{q}\,\vec{p}}$. See Fig. 3.77.

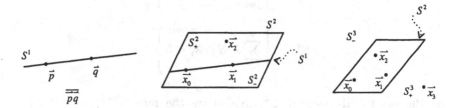

Fig. 3.77

A subset C of $\mathbb{R}^3$ (or V) is called *convex* if the segment joining any two
points of C is contained in C. Obviously,

1. an open or closed half space of each dimension is convex, and
2. the intersection $C = \cap_\alpha C_\alpha$ is convex if each C_α is convex.

Therefore, for any nonempty subset D of $\mathbb{R}^3$ (or V),

$$\mathrm{Con}(D) = \cap\{C \text{ is convex in } \mathbb{R}^3 \text{ and } C \supseteq D\} \qquad (3.8.51)$$

is a convex set and is called the *convex hull* or *closure* of D. Note that

$$\mathrm{Con}(D) = \left\{ \sum_{i=0}^{k} \lambda_i \vec{x}_i \mid \vec{x}_i \in D, \lambda_i \ge 0 \text{ for } 0 \le i \le k \text{ and } \sum_{i=0}^{k} \lambda_i = 1, \right.$$

$$\left. \text{where } k \ge 1 \text{ is arbitrary} \right\}.$$

In particular, in case $D = \{\vec{a}_0, \vec{a}_1, \dots, \vec{a}_k\}$ is a finite subset of $\mathbb{R}^3$ (or V),
then

$$\mathrm{Con}(\vec{a}_0, \vec{a}_1, \dots, \vec{a}_k) \qquad (3.8.52)$$

is called *convex cell* with its *dimension* the dimension of the affine subspace
$\vec{a}_0 + \langle\langle \vec{a}_1 - \vec{a}_0, \dots, \vec{a}_k - \vec{a}_0 \rangle\rangle$.

In case $\{\vec{a}_0, \vec{a}_1, \ldots, \vec{a}_k\}$ is affinely independent,

$$\mathrm{Con}(\vec{a}_0, \vec{a}_1, \ldots, \vec{a}_k) = \Delta\vec{a}_0\vec{a}_1 \cdots \vec{a}_k \tag{3.8.53}$$

is called *k-simplex* as mentioned in (3.8.38). This k-simplex is contained in the k-dimensional affine subspace S^k: $\vec{a}_0 + \langle\langle \vec{a}_1 - \vec{a}_0, \ldots, \vec{a}_k - \vec{a}_0 \rangle\rangle$. For each i, $0 \le i \le k$, let

$$S_{i+}^k = \left\{ \sum_{j=0}^k \lambda_j \vec{a}_j \mid \lambda_i > 0 \right\},$$

$$S_{i-}^k = \left\{ \sum_{j=0}^k \lambda_j \vec{a}_j \mid \lambda_i < 0 \right\}$$

be the open half-spaces of S^k divided by the face S_i^{k-1}: $\vec{a}_i + \langle\langle \vec{a}_1 - \vec{a}_i, \ldots, \vec{a}_{i-1} - \vec{a}_i, \vec{a}_{i+1} - \vec{a}_i, \ldots, \vec{a}_k - \vec{a}_i \rangle\rangle$. Then the corresponding closed half-spaces are

$$\overline{S_{i+}^k} = S_{i+}^k \cup S_i^{k-1} \quad \text{and} \quad \overline{S_{i-}^k} = S_{i-}^k \cup S_i^{k-1}.$$

Hence

$$\Delta\vec{a}_0\vec{a}_1 \cdots \vec{a}_k = \bigcap_{i=0}^k \overline{S_{i+}^k}, \quad \text{and}$$

$$\mathrm{Int}\,\Delta\vec{a}_0\vec{a}_1 \cdots \vec{a}_k = \bigcap_{i=0}^k S_{i+}^k \tag{3.8.54}$$

while the latter is called an *open* k-simplex. As a consequence, $\mathbb{R}^3$ (or V) can be endowed with a suitable topological structure so that $\mathbb{R}^3$ (or V) becomes a *Hausdorff topological space*.

A subset of $\mathbb{R}^3$ (or V) is called *bounded* if it is contained in some simplex. A *polyhedron* in $\mathbb{R}^3$ (or V) is a bounded subset obtained via a finite process of constructing intersections and unions from a finite number of closed half-spaces of various dimensions. In *algebraic topology*, it can be shown that any polyhedron admits simplicial decomposition. Hence, a polyhedron can also be defined as the set-theoretic union of a finite number of simplexes. A *convex polyhedron* can be characterized algebraically by several linear inequalities satisfied by coordinates of points contained in it. For example,

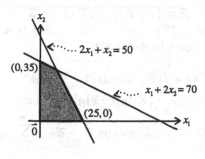

Fig. 3.78

in $\mathbb{R}^2$,

$$2x_1 + x_2 \leq 50,$$
$$x_1 + 2x_2 \leq 70,$$
$$x_1 \geq 0,$$
$$x_2 \geq 0$$

all together represent the shaded point set (a polyhedron) shown in Fig. 3.78. A general form of *linear programming* problems is to find values of $x_0, x_1, \ldots, x_k$ that will maximize or minimize the function

$$f(x_0, x_1, \ldots, x_k) = \sum_{i=0}^{k} \alpha_i x_i$$

subject to the constrained conditions

$$\sum_{j=0}^{k} a_{ij} x_j \leq \text{ or } \geq \text{ or } = b_i \quad \text{for } 1 \leq i \leq m.$$

3.8.4.9 *Affine mappings and affine transformations*

A mapping $T \colon \mathbb{R}^m$ (or V with $\dim V < \infty$) $\to \mathbb{R}^n$ (or W with $\dim W < \infty$) is called an *affine mapping* if

$$T(\vec{x}) = \vec{x}_0 + f(\vec{x}), \tag{3.8.55}$$

where $f \colon \mathbb{R}^m$ (or V) $\to \mathbb{R}^n$ (or W) is a linear transformation. In case $\mathbb{R}^m = \mathbb{R}^n$ (or $V = W$), it is called *affine transformation* or a *regular* or *proper* one if its linear part f is an invertible linear operator (refer to the convention mentioned in (2.8.20)).

Fix an affine basis $\mathcal{B} = \{\vec{a}_0 \ \vec{a}_1, \ldots, \vec{a}_n\}$ for $\mathbb{R}^n$ ($n = 1, 2, 3 \ldots$) so that the same notation $\mathcal{B} = \{\vec{a}_1 - \vec{a}_0, \ldots, \vec{a}_n - \vec{a}_0\}$ is a basis for the vector space $\mathbb{R}^n$. Then, as shown in Sec. 3.8.1,

$$T(\vec{x}) = \vec{x}_0 + f(\vec{x}) \text{ (regular)}, \quad \vec{x} \in \mathbb{R}^n$$
$$\Leftrightarrow (\text{in } \mathcal{B})[T(\vec{x})]_\mathcal{B} = [\vec{x}_0]_\mathcal{B} + [\vec{x}]_\mathcal{B}[f]_\mathcal{B} \quad \text{i.e. let}$$

$$[\vec{x}]_\mathcal{B} = (x_1, \ldots, x_n) \text{ means } \vec{x} - \vec{a}_0 = \sum_{i=1}^{n} x_i(\vec{a}_i - \vec{a}_0),$$

$$[\vec{x}_0]_\mathcal{B} = (b_1, \ldots, b_n),$$

$$[f]_\mathcal{B} = [a_{ij}]_{n \times n} = \begin{bmatrix} [f(\vec{a}_1 - \vec{a}_0)]_\mathcal{B} \\ \vdots \\ [f(\vec{a}_n - \vec{a}_0)]_\mathcal{B} \end{bmatrix}, \quad \text{and}$$

$$[T(\vec{x})]_\mathcal{B} = (y_1, \ldots, y_n),$$

then

$$y_j = b_j + \sum_{i=1}^{n} a_{ij} x_i \quad \text{for } 1 \leq j \leq n. \tag{3.8.56}$$

Refer to (3.8.2) and (3.8.7). In particular, an (regular) affine transformation represented by

$$y_j = a x_j \quad \text{for } 1 \leq j \leq n \text{ and } a \neq 0,$$

is called a *similarity* or *homothety* or *enlargement* (in particular, in $\mathbb{R}^n$, see Sec. 3.8.3) with the base point $\vec{a}_0$ as center.

Let (see(3.8.4))

$$\mathrm{G_a}(n; \mathbb{R}) = \{\text{regular affine transformations on } \mathbb{R}^n\}, \text{ and}$$
$$\mathrm{T_a}(n; \mathbb{R}) = \{\text{translations on } \mathbb{R}^n\}. \tag{3.8.57}$$

Then $\mathrm{T_a}(n; \mathbb{R})$ is a group, called the *group of translations* of order n, and is a normal subgroup of $\mathrm{G_a}(n; \mathbb{R})$, the *group of affine transformations* of order n. Also,

1. $\mathrm{T_a}(n; \mathbb{R})$, as a *vector group* (i.e. an additive group of vector space), acts transitively on $\mathbb{R}^n$.
2. $\mathrm{T_a}(n; \mathbb{R})$, as an additive group, is isomorphic to $\mathbb{R}^n$.

Hence, the quotient group

$$G_a(n; \mathbb{R})/T_a(n; R) \text{ is isomorphic to } GL(n;\mathbb{R}),$$

the *general linear group* over $\mathbb{R}$ of order n (see Sec. 3.7.3). Furthermore, for a fixed point $\vec{x}_0$ in $\mathbb{R}^n$ (or V), the *isotropy group* at $\vec{x}_0$

$$I_a(\vec{x}_0) = \{T \in G_a(n; \mathbb{R}) \mid T(\vec{x}_0) = \vec{x}_0\}$$

is a normal subgroup of $G_a(n; \mathbb{R})$ and is isomorphic to $GL(n; \mathbb{R})$. See Ex. <A> 5 of Sec. 2.8.1.

3.8.4.10 *Projectionalization of an affine space*

Exs. <B> of Secs. 2.6 and 3.6 might help understanding the material in this subsection.

Parallelism between affine subspaces of $\mathbb{R}^3$ (or $V, \dim V < \infty$) is an equivalent relation (see Sec. A.1).

The equivalent class of a one-dimensional subspace S^1 is called a *point at infinity* and is denoted as S^0_∞. See Fig. 3.79.

For a fixed subspace S^k, let

$$S^{k-1}_\infty = \{S^0_\infty \mid S^1 \text{ is a line contained in } S^k\}$$
$$= \text{the equivalent class of } S^k \text{ under parallelism} \qquad (3.8.58)$$

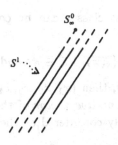

Fig. 3.79

and called a $(k-1)$-*dimensional subspace at infinity*. Note that, for two subspaces S^k and A^k, $S^k \parallel A^k \Leftrightarrow S^{k-1}_\infty = A^{k-1}_\infty$. In particular, S^2_∞ is called the *hyperplane at infinity*. See Fig. 3.80.

The set-theoretic union

$$P^3(\mathbb{R}) = \mathbb{R}^3 \cup S^2_\infty \qquad (3.8.59)$$

is supplied with the structure of a *projective space*:

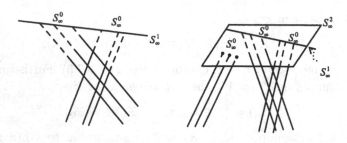

Fig. 3.80

1. points: elements of $P^3(\mathbb{R})$;
2. lines: $S^1 \cup S^0_\infty$ where $S^1 \subseteq \mathbb{R}^3$, and S^1_∞; and
3. planes: $S^2 \cup S^1_\infty$ where $S^2 \subseteq \mathbb{R}^3$, and S^2_∞.

$P^3(\mathbb{R})$ can be endowed with a *linear structure* introduced as follows.

Let $\mathbb{R}^4$ be the four-dimensional vector space over the real field $\mathbb{R}$ and fix the natural basis $\mathcal{N} = \{\vec{e}_1, \vec{e}_2, \vec{e}_3, \vec{e}_4\}$ for simplicity. For nonzero vectors $\vec{x}, \vec{y}$ in $\mathbb{R}^4$, we say that $\vec{x}$ and $\vec{y}$ are *equivalent* if there exists a nonzero scalar λ so that

$$\vec{y} = \lambda \vec{x}$$

$$\Leftrightarrow \langle\langle \vec{y} \rangle\rangle = \langle\langle \vec{x} \rangle\rangle \quad \text{as a one-dimensional subspace of } \mathbb{R}^4.$$

Then, the set of equivalent classes can be considered as the projective space, i.e.

$$P^3(\mathbb{R}) = \{\langle\langle \vec{x} \rangle\rangle \mid \vec{x} \in \mathbb{R}^4 \text{ and } \vec{x} \neq \vec{0}\}. \tag{3.8.60}$$

In case $\vec{x} = (x_1, x_2, x_3, x_4)$, then (x_1, x_2, x_3, x_4) or $\vec{x}$ is called a *homogeneous coordinate* or *representative vector* of the point $\langle\langle \vec{x} \rangle\rangle$ with respect to the *basis* $\mathcal{N}$ and is simply considered as the point $\langle\langle \vec{x} \rangle\rangle$ itself in many occasions. Define

an *ideal point* or *point at infinity:* $(x_1, x_2, x_3, 0)$,
an *ordinary* or *affine point:* (x_1, x_2, x_3, x_4) with $x_4 \neq 0$, and
the *affine* or *inhomogeneous coordinate* of an affine point:

$$\left(\frac{x_1}{x_4}, \frac{x_2}{x_4}, \frac{x_3}{x_4}, 1 \right) \quad \text{or} \quad \left(\frac{x_1}{x_4}, \frac{x_2}{x_4}, \frac{x_3}{x_4} \right).$$

Therefore, in (3.8.59),
the affine subspace $\mathbb{R}^3 = \{(x_1, x_2, x_3, x_4) \mid x_4 \neq 0\}$,
the hyperplane at infinity S^2_∞ or π_∞: $x_4 = 0$, and

the line: $\langle\langle \vec{x}_1, \vec{x}_2 \rangle\rangle = \langle\langle \vec{x}_1 \rangle\rangle \oplus \langle\langle \vec{x}_2 \rangle\rangle$,

where $\vec{x}_1$ and $\vec{x}_2$ are linearly independent vectors in $\mathbb{R}^4$.

Call $\tilde{N} = \{\vec{e}_1, \vec{e}_2, \vec{e}_3, \vec{e}_4; \vec{e}_1 + \vec{e}_2 + \vec{e}_3 + \vec{e}_4\}$ the *natural projective basis* for $P^3(\mathbb{R})$ in (3.8.60) with $\vec{e}_1, \ldots, \vec{e}_4$ as *vertices* and $\vec{e}_1 + \cdots + \vec{e}_4$ as *unit*. Then the *projective coordinate* of $\langle\langle \vec{x} \rangle\rangle$ with respect to $\tilde{N}$ is defined as

$$[\vec{x}]_{\tilde{N}} = [\vec{x}]_N = (x_1, x_2, x_3, x_4) \quad \text{if } \vec{x} = \sum_{i=1}^{4} x_i \vec{e}_i \qquad (3.8.61)$$

and is equal to a homogeneous coordinate of $\langle\langle \vec{x} \rangle\rangle$.

A one-to-one mapping F from $P^3(\mathbb{R})$ onto itself is called a *projective transformation* if, for any $\sigma \neq 0$,

$$[F(\vec{x})]_{\tilde{N}} = \sigma[\vec{x}]_{\tilde{N}}[F]_N, \quad \text{where } [F]_N = \begin{bmatrix} [F(\vec{e}_1)]_N \\ [F(\vec{e}_2)]_N \\ [F(\vec{e}_3)]_N \\ [F(\vec{e}_4)]_N \end{bmatrix}_{4 \times 4}$$

$$= [a_{ij}]_{4 \times 4} \text{ is invertible.} \qquad (3.8.62)$$

$[F]_N$ or $\sigma[F]_N$ for any $\sigma \neq 0$ is called a *matrix representation* of F with respect to $\tilde{N}$. (3.8.62) can be expressed as

$$y_j = \sigma \sum_{i=1}^{4} a_{ij}x_i, \quad \sigma \neq 0, \ 1 \leq j \leq 4,$$

where $[F(\vec{x})]_{\tilde{N}} = (y_1, y_2, y_3, y_4)$.

Suppose a projective transformation F on $P^3(\mathbb{R})$ preserves the hyperplane at infinity π_∞ invariant. Then

$$x_4 = 0 \quad \text{if and only if } y_4 = 0$$

$$\Leftrightarrow a_{14}x_1 + a_{24}x_2 + a_{34}x_3 = 0 \quad \text{for all } x_1, x_2, x_3 \in \mathbb{R}$$

$$\Leftrightarrow a_{14} = a_{24} = a_{34} = 0.$$

Taking $\sigma = \frac{1}{a_{44}}$ ($a_{44} \neq 0$, why?), the transformation reduces to

$$y_j = \frac{1}{a_{44}} \sum_{i=1}^{4} a_{ij}x_i \quad \text{for } 1 \leq j \leq 3,$$

$$y_4 = x_4.$$

which, in terms of inhomogeneous coordinates (i.e. replacing x_i by $\frac{x_i}{x_4}$, etc.), can be rewritten as

$$y_j = \frac{a_{4j}}{a_{44}} + \sum_{i=1}^{3} \frac{a_{ij}}{a_{44}} x_i \quad \text{for } 1 \le j \le 3.$$

This represents an affine transformation on the affine space $\mathbb{R}^3$. The reverse process tells us that an affine transformation on $\mathbb{R}^3$ induces a projective transformation on $P^3(\mathbb{R})$ leaving π_∞ invariant.

If, in addition, F preserves each point at infinity invariant, then $a_{ij} = 0$, for $1 \le i,j \le 3$ and $i \ne j$ while $a_{ii} = 1, 1 \le i \le 3$. In this case, the corresponding affine transformation is a translation.

We summarize as

The Group $G_p(3;\mathbb{R})$ of projective transformations of order 3 over $\mathbb{R}$

The set

$$G_p(3;\mathbb{R})$$

of all projective transformations on $P^3(\mathbb{R})$ constitutes a group.

(1) The set of all projective transformations on $P^3(\mathbb{R})$ that leave π_∞ invariant forms a *subgroup* of $G_p(3;\mathbb{R})$ and is isomorphic to $G_a(3;\mathbb{R})$, the group of affine transformations on $\mathbb{R}^3$.

(2) The set of all projective transformations on $P^3(\mathbb{R})$ that leave each point at infinity invariant forms a *subgroup* of $G_p(3;\mathbb{R})$ and is isomorphic to

$$T_a(3;\mathbb{R}), \text{ the group of translations on } \mathbb{R}^3 \text{ (see(3.8.57)).} \qquad (3.8.63)$$

For detailed account about projective line, plane and space, refer to [6, pp. 1–218].

Exercises

<A>

1. Suppose $\{\vec{a}_0, \vec{a}_1, \ldots, \vec{a}_k\}$ is affinely independent in $\mathbb{R}^3$ as defined in (3.8.32). Let $\pi\colon \{0,1,\ldots,k\} \to \{0,1,\ldots,k\}$ be any permutation. Show that $\{\vec{a}_{\pi(0)}, \vec{a}_{\pi(1)}, \ldots, \vec{a}_{\pi(k)}\}$ is affinely independent.
2. Prove (3.8.36).
3. Prove (3.8.39).
4. Prove (3.8.41) and (3.8.43).

5. Prove (3.8.44).

6. Prove (3.8.45).

7. In $\mathbb{R}^3$, determine the relative positions of

   (1) three lines S_1^1, S_2^1 and S_3^1,

   (2) three planes S_1^2, S_2^2 and S_3^2,

   (3) two lines S_1^1, S_2^1 and one plane S^2, and

   (4) one line S^1 and two planes S_1^2, S_2^2.

8. In $\mathbb{R}^5$, determine the relative positions of

   (1) two lines S_1^1, S_2^1,

   (2) two planes S_1^2, S_2^2,

   (3) one line S^1 and one plane S^2,

   (4) S^1 and S^3, S^1 and S^4,

   (5) S^2 and S^3, S^2 and S^4,

   (6) S_1^3 and S_2^3, S^3 and S^4, and

   (7) S_1^4 and S_2^4.

9. Prove (3.8.47) in $\mathbb{R}^5$ for $k = 4$.

10. Prove (3.8.48) in $\mathbb{R}^5$ for $k = 4$.

11. Fix two scalars $a_i, b_i \in \mathbb{R}$ (or an ordered field) with $a_i < b_i$ for $1 \leq i \leq 3$. The set

$$\{(x_1, x_2, x_3) \in \mathbb{R}^3 \mid a_i \leq x_i \leq b_i \text{ for } 1 \leq i \leq 3\}$$

is generally called a *parallelotope* (especially in $\mathbb{R}^n$). Show that a parallelotope is a polyhedron. Is it convex?

12. Show that any convex cell is a polyhedron.

13. Try to model after Sec. 3.8.4.10 to define the projective line $P^1(\mathbb{R})$ and the projective plane $P^2(\mathbb{R})$.

14. Let $\tilde{\mathcal{B}} = \{\vec{a}_1, \vec{a}_2, \vec{a}_3, \vec{a}_4, \vec{a}_5\}$ be a set of vectors in $\mathbb{R}^4$ so that any four of them are linearly independent. Try to use this $\tilde{\mathcal{B}}$ to replace $\tilde{\mathcal{N}}$ in (3.8.61) to reprove (3.8.62) and (3.8.63).

15. State and prove similar results for $P^1(\mathbb{R})$ and $P^2(\mathbb{R})$ as (3.8.63).

16. (*cross ratio*) Let $A_1 = \langle\langle \vec{x}_1 \rangle\rangle$, $A_2 = \langle\langle \vec{x}_2 \rangle\rangle$, $A_3 = \langle\langle \varepsilon_1 \vec{x}_1 + \varepsilon_2 \vec{x}_2 \rangle\rangle$ and $A_4 = \langle\langle \alpha_1 \vec{x}_1 + \alpha_2 \vec{x}_2 \rangle\rangle$ be four distinct points in $P^3(\mathbb{R})$ (or $P^1(\mathbb{R})$ or $P^2(\mathbb{R})$). Note that they are collinear. The *cross ratio* of A_1, A_2, A_3 and A_4, in this ordering, is defined and denoted by

$$(A_1, A_2; A_3, A_4) = \frac{\alpha_1}{\varepsilon_1} : \frac{\alpha_2}{\varepsilon_2} = \frac{\varepsilon_2 \alpha_1}{\varepsilon_1 \alpha_2}.$$

Let $F: P^3(\mathbb{R}) \to P^3(\mathbb{R})$ be a projective transformation. Denote $A'_i = F(A_i)$ for $1 \leq i \leq 4$. Show that

$$(A'_1, A'_2; A'_3, A'_4) = (A_1, A_2; A_3, A_4)$$

holds. This means that cross ratio is a *projective invariant*. Hence, cross ratio plays one of the essential roles in projective geometry. Refer to Ex. <A> 4 of Sec. 2.8.4 and Fig. 2.125.

<B>

The following problems are all true for $\mathbb{R}^n, n = 1, 2, 3, \ldots$, even for n-dimensional affine space V over the real field (or an ordered field). The readers are encouraged to prove the case $n = 3$, at least.

1. For a k-simplex $\triangle \vec{a}_0 \vec{a}_1 \cdots \vec{a}_k$ in $\mathbb{R}^n$, define its

$$boundary \ \partial\Delta = \left\{ \sum_{j=0}^{k} \lambda_j \vec{a}_j \mid \lambda_j \geq 0 \text{ for } 0 \leq j \leq k \text{ but at least one of} \right.$$

$$\left. \lambda_j \text{ is equal to zero and } \sum_{j=0}^{k} \lambda_j = 1 \right\},$$

$$interior \ \text{Int}\,\Delta = \left\{ \sum_{j=0}^{k} \lambda_j \vec{a}_j \mid \lambda_j > 0 \text{ for } 0 \leq j \leq k \text{ and } \sum_{j=0}^{k} \lambda_j = 1 \right\},$$

and

$$exterior \ \text{Ext}\,\Delta = \left\{ \sum_{j=0}^{k} \lambda_j \vec{a}_j \mid \text{ at least one of } \lambda_0, \lambda_1, \ldots, \lambda_k \text{ is less than} \right.$$

$$\left. \text{zero and } \sum_{j=0}^{k} \lambda_j = 1 \right\}.$$

(a) In case $k = n$, show that $\partial\Delta$, Int Δ and Ext Δ are pairwise disjoint and

$$\mathbb{R}^n = \text{Int}\,\Delta \cup \partial\Delta \cup \text{Ext}\,\Delta.$$

(b) In case $k = n$, show that

    (1) Int Δ is a convex set,

    (2) any two points in Ext Δ can be connected by a polygonal curve, composed of line segments, which is contained entirely in Ext Δ, and

(3) any polygonal curve connecting a point in Int Δ to another point in Ext Δ will intersect the boundary $\partial\Delta$ at at least one point.

See Fig. 3.81. Therefore, $\mathbb{R}^n$ is said to be *separated* by any n-simplex into three parts: Int $\Delta, \partial\Delta$ and Ext Δ.

(c) In case $0 \le k \le n-1$, a k-simplex cannot separate $\mathbb{R}^n$ in the following sense: for any point $\vec{x}_0 \in \Delta\vec{a}_0\vec{a}_1 \cdots \vec{a}_k$ and any point $\vec{y}_0$ outside of it but in $\mathbb{R}^n$, there exists a polygonal curve connecting $\vec{x}_0$ to $\vec{y}_0$ which does not contain points in $\Delta\vec{a}_0\vec{a}_1 \cdots \vec{a}_k$ except $\vec{x}_0$. See Fig. 3.82.

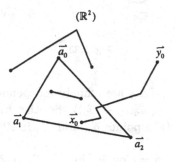

$(\mathbb{R}^2)$

Fig. 3.81

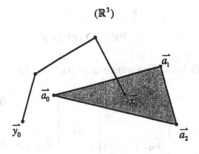

$(\mathbb{R}^3)$

Fig. 3.82

2. Let K be a convex set on a k-dimensional subspace S^k of $\mathbb{R}^n$, where $0 \le k \le n-1$. Take any point $\vec{x}_0 \notin K$. The set

$$\mathrm{Con}(\vec{x}_0; K) = \{(1-t)\vec{x}_0 + t\vec{x} \mid 0 \le t \le 1 \text{ and } \vec{x} \in K\}$$

is called a *cone* with K as *base* and $\vec{x}_0$ as the *vertex*. See Fig. 3.83. Show that $\mathrm{Con}(\vec{x}_0; K)$ is a convex set.

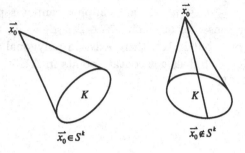

Fig. 3.83

3. Suppose $\vec{a}_0, \vec{a}_1, \ldots, \vec{a}_{n-1}$ are affinely independent in $\mathbb{R}^n$. Let $S^{n-1} = \vec{a}_0 + \langle\langle \vec{a}_1 - \vec{a}_0, \ldots, \vec{a}_{n-1} - \vec{a}_0 \rangle\rangle$ be the hyperplane spanned by $\vec{a}_0, \vec{a}_1, \ldots, \vec{a}_{n-1}$. Take points $\vec{a}_n$ and $\vec{b}_n$ in $\mathbb{R}^n$ and denote

$$A = \mathrm{Con}(\vec{a}_0, \vec{a}_1, \ldots, \vec{a}_{n-1}, \vec{a}_n),$$
$$B = \mathrm{Con}(\vec{a}_0, \vec{a}_1, \ldots, \vec{a}_{n-1}, \vec{b}_n).$$

(a) Suppose $\vec{a}_n$ and $\vec{b}_n$ lie on the same side of S^{n-1} (see (3.8.49)). Show that

$$\mathrm{Int}\, A \cap \mathrm{Int}\, B \ne \phi,$$

and A and B have the same orientation, i.e.

$$\det \begin{bmatrix} \vec{a}_1 - \vec{a}_0 \\ \vdots \\ \vec{a}_{n-1} - \vec{a}_0 \\ \vec{a}_n - \vec{a}_0 \end{bmatrix} \cdot \det \begin{bmatrix} \vec{a}_1 - \vec{a}_0 \\ \vdots \\ \vec{a}_{n-1} - \vec{a}_0 \\ \vec{b}_n - \vec{a}_0 \end{bmatrix} > 0.$$

See Fig. 3.84(a).

(b) Suppose $\vec{a}_n$ and $\vec{b}_n$ lie on the opposite sides of S^{n-1}. Show that

$$A \cap B = \text{Con}(\vec{a}_0, \vec{a}_1, \ldots, \vec{a}_{n-1}),$$

and A and B have the opposite orientations, i.e.

$$\det \begin{bmatrix} \vec{a}_1 - \vec{a}_0 \\ \vdots \\ \vec{a}_{n-1} - \vec{a}_0 \\ \vec{a}_n - \vec{a}_0 \end{bmatrix} \cdot \det \begin{bmatrix} \vec{a}_1 - \vec{a}_0 \\ \vdots \\ \vec{a}_{n-1} - \vec{a}_0 \\ \vec{b}_n - \vec{a}_0 \end{bmatrix} < 0.$$

See Fig. 3.84(b).

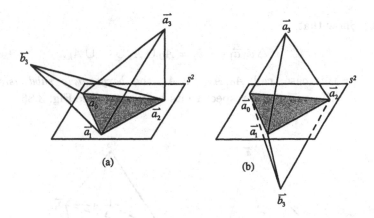

(a)   (b)

Fig. 3.84

4. Let $\triangle \vec{a}_0 \vec{a}_1 \cdots \vec{a}_k$ be a k-simplex in $\mathbb{R}^n$ where $k \geq 2$ and $\vec{x}$ be its barycenter. Suppose $\vec{x}_i$ is the barycenter of its ith *face* $\triangle \vec{a}_0 \vec{a}_1 \cdots \vec{a}_{i-1} \vec{a}_{i+1} \cdots \vec{a}_k$ for $0 \leq i \leq k$.

(a) Show that the line segments $\overline{\vec{a}_i \vec{x}_i}, 0 \leq i \leq k$, meet at $\vec{x}$.

(b) Show that $\overline{\vec{x}_i \vec{x}} = \frac{1}{k+1} \overline{\vec{x} \vec{a}_i}$ (in signed length).

See Fig. 3.85.

5. Let $\vec{x}$ be the barycenter of the k-simplex $\triangle \vec{a}_0 \vec{a}_1 \cdots \vec{a}_k$ in $\mathbb{R}^n$ where $k \geq 2$. Construct k-simplexes as follows:

$$A_0 = \triangle \vec{x} \vec{a}_1 \cdots \vec{a}_k,$$
$$A_i = \triangle \vec{a}_0 \vec{a}_1 \cdots \vec{a}_{i-1} \vec{x} \vec{a}_{i+1} \cdots \vec{a}_k \quad \text{for } 1 \leq i \leq k.$$

(a) If $i \neq j$, show that A_i and A_j have at most one face in common but do not have common interior points.

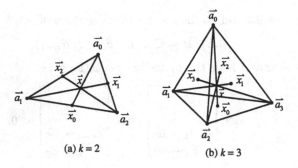

(a) $k = 2$ (b) $k = 3$

Fig. 3.85

(b) Show that

$$\triangle \vec{a}_0 \vec{a}_1 \cdots \vec{a}_k = A_0 \cup A_1 \cup \cdots \cup A_k.$$

In this case, call $A_0, A_1, \ldots, A_k$ the *barycentric subdivision* of $\triangle \vec{a}_0 \vec{a}_1 \cdots \vec{a}_k$ with respect to its barycenter. See Fig. 3.86.

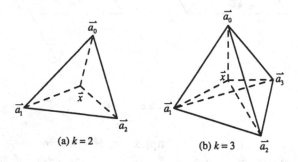

(a) $k = 2$ (b) $k = 3$

Fig. 3.86

6. Let $\vec{x}$ be the barycenter of the n-simplex $\triangle \vec{a}_0 \vec{a}_1 \cdots \vec{a}_n$ in $\mathbb{R}^n$. The remaining vertices of $\vec{a}_0, \vec{a}_1, \ldots, \vec{a}_n$, after eliminating from them the vertices $\vec{a}_{i_1}, \ldots, \vec{a}_{i_k}$ with $0 \leq i_1 < i_2 < \cdots < i_k \leq n + 1$, span an $(n - k)$-simplex. Use $\vec{x}_{i_1 \cdots i_k}$ to denote the barycenter of this $(n-k)$-simplex. Take n distinct numbers $i_1, i_2, \ldots, i_n$ from $0, 1, 2, \ldots, n$. Define an affine transformation $T \colon \mathbb{R}^n \to \mathbb{R}^n$ as

$$T(\vec{x}) = \vec{a}_0,$$
$$T(\vec{x}_{i_1 \cdots i_k}) = \vec{a}_k \quad \text{for } 1 \leq k \leq n.$$

Let $\det f$ denote the determinant of the linear part f of T^{-1}. Show that

$$\det f = \frac{1}{(n+1)!}(-1)^{\sigma(i_0, i_1, \dots, i_n)},$$

where $(i_0, i_1, \dots, i_n)$ is a permutation of $0, 1, 2, \dots, n$, defined as $k \rightarrow i_k$ for $1 \le k \le n$ and $0 \rightarrow i_0$ (i_0 is different from $i_1, \dots, i_n$); $(-1)^{\sigma(i_0, i_1, \dots, i_n)} = 1$ or -1 according $(i_0, i_1, \dots, i_n)$ is an even permutation or odd.

7. Let $\vec{a}_1, \dots, \vec{a}_k$ be linearly independent in $\mathbb{R}^n$. Fix a point $\vec{a}_0$ in $\mathbb{R}^n$. Define, for given positive scalars $c_1, \dots, c_k$,

$$P = \left\{ \vec{a}_0 + \sum_{i=1}^{k} \lambda_i \vec{a}_i \mid |\lambda_i| \le c_i, \text{ for } 1 \le i \le k \right\}.$$

This parallelotope (see Ex. <A> 11) is sometime called a *k-parallelogram* with *center* at $\vec{a}_0$ and *side vectors* $\vec{a}_1, \dots, \vec{a}_k$. Concerned terminologies are:

Vertex: $\vec{x} = \vec{a}_0 + \varepsilon_1 \vec{a}_1 + \cdots + \varepsilon_k \vec{a}_k$, where $\varepsilon_1 = \pm c_1, \dots, \varepsilon_k = \pm c_k$;

Face: $\{\vec{x} \in P \mid \lambda_i = c_i \text{ or } -c_i \text{ and } |\lambda_j| \le c_j \text{ for } j \ne i, 1 \le j \le k\}$.

There are 2^k vertices and $2k$ faces. The set of all its faces constitutes its *boundary* ∂P. Furthermore, the set

$$\text{Int } P = \{\vec{x} \in P \mid |\lambda_j| < c_j \text{ for } 1 \le j \le k\}$$

is called the *interior* of P and the set $\text{Ext } P = \mathbb{R}^n - P$ is called the *exterior* of P. See Fig. 3.87.

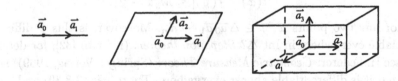

Fig. 3.87

Use P to replace $\triangle \vec{a}_0 \vec{a}_1 \cdots \vec{a}_n$ in Ex. 1 to prove (a)–(c) there.

8. *Generalized Euler formula:* $V - F + E = 2$.

   (a) Let $\triangle \vec{a}_0 \vec{a}_1 \cdots \vec{a}_k$ be a k-simplex in $\mathbb{R}^n$. Any $s + 1$ distinct points out of $\vec{a}_0, \vec{a}_1, \dots, \vec{a}_k$ can be used to construct a s-simplex. The total of them is $\alpha_s = C_{s+1}^{k+1}$ for $0 \le s \le k$. Show that

   $$\alpha_0 - \alpha_1 + \alpha_2 - \cdots + (-1)^{k-1}\alpha_{k-1} + (-1)^k \alpha_k = 1.$$

(b) Let P be as in Ex. 7. Let, for $0 \leq s \leq k$,

$$\alpha_s = C_s^k \cdot 2^s$$

$$= \text{the number of } s\text{-parallelograms contained}$$
$$\text{in any one of its } 2k \text{ faces.}$$

Show that

$$\alpha_0 - \alpha_1 + \alpha_2 - \cdots + (-1)^{k-1}\alpha_{k-1} + (-1)^k\alpha_k = 1.$$

<C> Abstraction and generalization

1. Prove Menelaus theorem (3.8.47) and Ceva theorem (3.8.48) in $\mathbb{R}^n$ or any finite-dimensional affine space V over a field.
2. Try to model after Sec. 3.8.4.10 to construct the n-dimensional projective space $P^n(\mathbb{R})$ and develop its basic properties.

<D> Application

1. Let $\triangle \vec{a}_0 \vec{a}_1 \cdots \vec{a}_n$ be an n-simplex in $\mathbb{R}^n$ and $\alpha_0, \alpha_1, \ldots, \alpha_n$ be fixed real numbers. Define a function $f \colon \triangle \vec{a}_0 \vec{a}_1 \cdots \vec{a}_n \to \mathbb{R}$ by

$$f\left(\sum_{i=0}^{n} \lambda_i \vec{a}_i\right) = \sum_{i=0}^{n} \lambda_i \alpha_i,$$

where $\lambda_i \geq 0$ for $0 \leq i \leq n$ and $\lambda_0 + \lambda_1 + \cdots + \lambda_n = 1$. Show that f is *uniformly Lipschitzen* on $\triangle \vec{a}_0 \vec{a}_1 \cdots \vec{a}_n$, i.e. there exists a constant $M > 0$ so that

$$|f(\vec{x}) - f(\vec{y})| \leq M|\vec{x} - \vec{y}|$$

for any two points $\vec{x}, \vec{y} \in \triangle \vec{a}_0 \vec{a}_1 \cdots \vec{a}_n$. Moreover, is this f differentiable everywhere in Int $\triangle$? *Stepanov theorem* (refer to [32]; for details, see H. Federer: *Geometric Measure Theory* (Springer-Verlag, 1969)) says that it is differentiable almost everywhere. Try to use (3.8.39) and compute its total differential at point where it is differentiable.

3.8.5 *Quadrics*

Here in this subsection, $\mathbb{R}^3$ is considered as an affine space in general, and as a vector space in particular. In some cases, $\mathbb{R}^3$ as an Euclidean space (see Part 2 and Chap. 5) is implicitly understood.

$\mathcal{N} = \{\vec{0}, \vec{e}_1, \vec{e}_2, \vec{e}_3\}$ always represents the natural affine basis for $\mathbb{R}^3$, and $\vec{x} = [\vec{x}]_{\mathcal{N}} = (x_1, x_2, x_3)$ is used.

The set of points $\vec{x}$ in $\mathbb{R}^3$ that satisfies the equation of the second degree in three real variables x_1, x_2 and x_3 with real coefficients

$$b_{11}x_1^2 + b_{22}x_2^2 + b_{33}x_3^2 + 2b_{12}x_1x_2 + 2b_{13}x_1x_3$$
$$+ 2b_{23}x_3x_3 + 2b_1x_1 + 2b_2x_2 + 2b_3x_3 + b = 0 \qquad (3.8.64)$$

is called a *quadric surface* or *surface of the second order* or simply *quadric*, where the coefficients b_{11}, b_{22} and b_{33} are not all zeros. Adopt the natural inner product $\langle , \rangle$ notation

$$\langle \vec{x}, \vec{y} \rangle = \vec{x}\,\vec{y}^*.$$

Then (3.8.64) can be expressed as

$$\langle \vec{x}, \vec{x}B \rangle + 2\langle \vec{x}, \vec{b} \rangle + b = 0,$$

where

$$B = \begin{bmatrix} b_{11} & b_{12} & b_{13} \\ b_{21} & b_{22} & b_{23} \\ b_{31} & b_{32} & b_{33} \end{bmatrix} \text{ is symmetric, i.e. } b_{ij} = b_{ji} \text{ for } 1 \le i, j \le 3,$$

$$\vec{b} = (b_1, b_2, b_3). \qquad (3.8.65)$$

According to (1) in (2.7.71) and better referring to Example 3 in Sec. 3.7.5, there exists an invertible matrix P so that

$$PBP^* = \begin{bmatrix} \lambda_1 & & 0 \\ & \lambda_2 & \\ 0 & & \lambda_3 \end{bmatrix}_{3\times 3}. \qquad (*1)$$

Thus, (3.8.65) can be rewritten as

$$\vec{x}P^{-1}(PBP^*)(\vec{x}P^{-1})^* + (\vec{x}P^{-1})(\vec{b}P^*)^* + b = 0. \qquad (*2)$$

Since $\mathcal{B} = \{P_{1*}, P_{2*}, P_{3*}\}$, composed of row vectors of P, is a basis for $\mathbb{R}^3$, then

$$\vec{x}P^{-1} = (y_1, y_2, y_3) = [\vec{x}]_{\mathcal{B}}.$$

(*2) is reduced to, in $\mathcal{B}$,

$$(y_1 \quad y_2 \quad y_3) \begin{bmatrix} \lambda_1 & & 0 \\ & \lambda_2 & \\ 0 & & \lambda_3 \end{bmatrix} \begin{bmatrix} y_1 \\ y_2 \\ y_3 \end{bmatrix} + (y_1 \quad y_2 \quad y_3) \begin{bmatrix} c_1 \\ c_2 \\ c_3 \end{bmatrix} + b = 0, \quad \text{where}$$

$$\vec{b}P^* = \vec{c} = (c_1, c_2, c_3)$$

$$\Rightarrow \lambda_1 y_1^2 + \lambda_2 y_2^2 + \lambda_3 y_3^2 + c_1 y_1 + c_2 y_2 + c_3 y_3 + b = 0. \qquad (*3)$$

Therefore, the $x_i x_j$ terms where $i \ne j$ are eliminated from (3.8.65).

Suppose $\lambda_1 c_1 \neq 0$ in (*3). The complete square of $\lambda_1 y_1^2 + c_1 y_1$ as

$$\lambda_1 y_1^2 + c_1 y_1 = \lambda_1 \left(y_1^2 + \frac{c_1}{\lambda_1} y_1 \right) = \lambda_1 \left(y_1 + \frac{c_1}{2\lambda_1} \right)^2 - \frac{c_1^2}{4\lambda_1}$$

suggests that the affine transformation

$$\vec{x} \to \left(\frac{c_1}{2\lambda_1}, 0, 0 \right) + \vec{x} P^{-1} = \vec{z} = (z_1, z_2, z_3)$$

will reduce (*3) to

$$\lambda_1 z_1^2 + \lambda_2 z_2^2 + \lambda_3 z_3^2 + c_2 z_2 + c_3 z_3 + b' = 0. \tag{*4}$$

In case $\lambda_2 c_2 \neq 0$ or (and) $\lambda_3 c_3 \neq 0$, the same method can be used to eliminate the first order terms concerning x_2, x_3.

Since $B \neq O_{3\times3}$, so λ_1, λ_2 and λ_3 in (*3) cannot be all equal to zero. This means that there always has at least one term of second order left no matter what affine transformation is applied.

Hence, after a suitable affine transformation, (3.8.65) can be reduced to one of the following types.

The standard forms of quadrics
In the *Cartesian coordinate system* $\mathcal{N} = \{\vec{e}_1, \vec{e}_2, \vec{e}_3\}$, the quadrics (3.8.64) are classified into the following 17 standard forms, where $a_1 > 0, a_2 > 0$, $a_3 > 0$ and $a \neq 0$.

1. Ellipsoid (Fig. 3.88) $\dfrac{x_1^2}{a_1^2} + \dfrac{x_2^2}{a_2^2} + \dfrac{x_3^2}{a_3^2} = 1$.

2. Imaginary ellipsoid $\dfrac{x_1^2}{a_1^2} + \dfrac{x_2^2}{a_2^2} + \dfrac{x_3^2}{a_3^2} = -1$.

3. Point ellipsoid or imaginary elliptic cone $\dfrac{x_1^2}{a_1^2} + \dfrac{x_2^2}{a_2^2} + \dfrac{x_3^2}{a_3^2} = 0$.

4. Hyperboloid of two-sheets (Fig. 3.89) $\dfrac{x_1^2}{a_1^2} + \dfrac{x_2^2}{a_2^2} - \dfrac{x_3^2}{a_3^2} = -1$.

5. Hyperboloid of one-sheet (Fig. 3.90) $\dfrac{x_1^2}{a_1^2} + \dfrac{x_2^2}{a_2^2} - \dfrac{x_3^2}{a_3^2} = 1$.

6. Elliptic cone (Fig. 3.91) $\dfrac{x_1^2}{a_1^2} + \dfrac{x_2^2}{a_2^2} - \dfrac{x_3^2}{a_3^2} = 0$.

7. Elliptic paraboloid (Fig. 3.92) $\dfrac{x_1^2}{a_1^2} + \dfrac{x_2^2}{a_2^2} + 2ax_3 = 0$.

8. Hyperbolic paraboloid (Fig. 3.93) $\dfrac{x_1^2}{a_1^2} - \dfrac{x_2^2}{a_2^2} + 2ax_3 = 0.$

9. Elliptic cylinder (Fig. 3.94) $\dfrac{x_1^2}{a_1^2} + \dfrac{x_2^2}{a_2^2} - 1 = 0.$

10. Imaginary elliptic cylinder $\dfrac{x_1^2}{a_1^2} + \dfrac{x_2^2}{a_2^2} + 1 = 0.$

11. Imaginary intersecting planes $\dfrac{x_1^2}{a_1^2} + \dfrac{x_2^2}{a_2^2} = 0.$

12. Hyperbolic cylinder (Fig. 3.95) $\dfrac{x_1^2}{a_1^2} - \dfrac{x_2^2}{a_2^2} = 1.$

13. Intersection planes (Fig. 3.96) $\dfrac{x_1^2}{a_1^2} - \dfrac{x_2^2}{a_2^2} = 0.$

14. Parabolic cylinder (Fig. 3.97) $x_1^2 + 2ax_2 = 0.$
15. Parallel planes (Fig. 3.98) $x_1^2 - a^2 = 0.$
16. Imaginary parallel planes $x_1^2 + a^2 = 0.$
17. Coincident planes (Fig. 3.99) $x_1^2 = 0.$

In the *Euclidean space* $\mathbb{R}^3$, the coefficients of terms of second order can be chosen as eigenvalues of A (see Remark 1 below), while in the *affine space* $\mathbb{R}^3$, they are not necessarily so (see Remark 2 below).

$$(3.8.66)$$

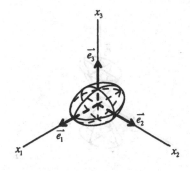

Fig. 3.88

Several remarks are provided.

Remark 1 *The standard forms in the Euclidean space* $\mathbb{R}^3$.

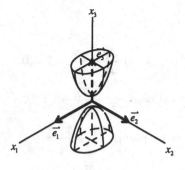

Fig. 3.89

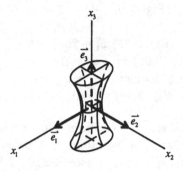

Fig. 3.90

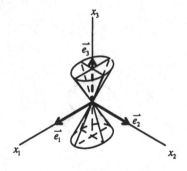

Fig. 3.91

In (*1), P can be chosen as an orthogonal matrix and hence, λ_1, λ_2 and λ_3 are *eigenvalues* of B (see (5.7.3)). In this case, λ_1, λ_2 and λ_3 are kept all the way down to (*3) and (*4).

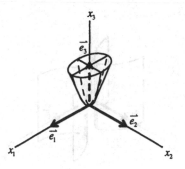

Fig. 3.92

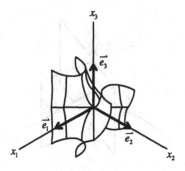

Fig. 3.93

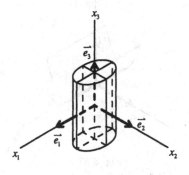

Fig. 3.94

Since λ_1, λ_2 and λ_3 in (*3) cannot be all zeros, in case (*3) is of the form

$$\lambda_1 y_1^2 + c_2 y_2 + b = 0, \quad \text{where } \lambda_1 \neq 0, \ c_2 \neq 0,$$

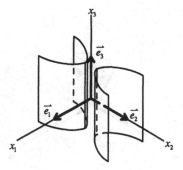

Fig. 3.95

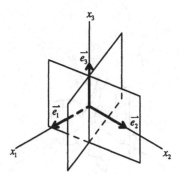

Fig. 3.96

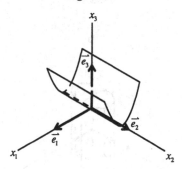

Fig. 3.97

a translation will reduce it to the form

$$\lambda_1 z_1^2 + c_2 z_2 = 0$$

$\Rightarrow$ (replace z_1 and z_2 by x_1 and x_2, respectively)

$$x_1^2 + 2ax_2 = 0, \quad \text{where } 2a = \frac{c_2}{\lambda_1}.$$

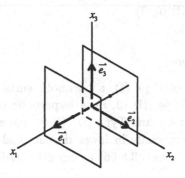

Fig. 3.98

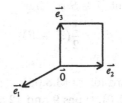

Fig. 3.99

This is the standard form 14. The readers should practice all other cases.

Instead of affine transformations, in Euclidean space $\mathbb{R}^3$ we use the *rigid motions* to reduce the quadrics (3.8.64) to their standard forms listed in (3.8.66). For details, refer to Sec. 5.10.

Remark 2 *The standard forms in the affine space* $\mathbb{R}^3$.
If we adopt the Sylvester's law of inertia (see (2) and (3) in (2.7.71)), let k be the *index* of B and $r = r(B)$ the *rank* of B. Then the quadrics (3.8.64) have the following *canonical forms*:

$$\text{Type (I) } (k, r-k): \sum_{i=1}^{k} x_i^2 - \sum_{i=k+1}^{r} x_i^2 = 0,$$

$$\text{Type(II) } (k, r-k): \sum_{i=1}^{k} x_i^2 - \sum_{i=k+1}^{r} x_i^2 + 1 = 0, \qquad (3.8.67)$$

$$\text{Type(III) } (k, r-k): \sum_{i=1}^{k} x_i^2 - \sum_{i=k+1}^{r} x_i^2 + 2x_{r+1} = 0,$$

in the affine space $\mathbb{R}^3$ in a strict sense. For example,

1 (Ellipsoid): Type (II) (0, 3),

2 : Type (II) (3, 0).

3 : Type (I) (3, 0).

4 : Type (II) (2, 1), etc.

In general, a cone is of Type (I), a parabolic surface is of Type (II), an elliptic surface is of Type (II) $(0, r)$, a hyperbolic surface is of Type (II) $(k, r - k)$ where $0 < k < r$ and Type (II) $(k, 0)$ represents empty set.

But, in practice, especially in lower dimensional affine spaces such as $\mathbb{R}^2$ and $\mathbb{R}^3$, we prefer to use (3.8.66) as our standard forms for quadrics.

Remark 3 *Centers. Nondegenerated quadrics.*
A point $\vec{c}$ in the space $\mathbb{R}^3$ is called a *center* of a quadric S if for each point $\vec{x} \in S$, there is another point $\vec{y} \in S$ so that

$$\vec{c} = \frac{1}{2}(\vec{x} + \vec{y}). \tag{3.8.68}$$

(refer (2.8.53)).

According to the standard forms (3.8.66), types 1, 3, 4, 5 and 6 all have a unique center at $\vec{0} = (0, 0, 0)$, types 9 and 12 all have the x_3-axis as their sets of centers, type 15 has the x_2x_3-coordinate plane as its set of centers, while type 17 has every point of it as a center. Types 7, 8 and 14 do not have any center.

Among these, types 1, 4, 5, 7 and 8 are called *nondegenerated* or *proper* quadrics, and the others *degenerated quadrics*.

Remark 4 *Regular and singular points of a surface (calculus are required).*
Let $F(x_1, x_2, x_3)$ be a real-valued $\mathbb{C}^\infty$ (or $\mathbb{C}^3$) function defined on an open subset of the Euclidean space $\mathbb{R}^3$. Roughly speaking, the set of points satisfying

$$F(x_1, x_2, x_3) = 0, \quad \vec{x} = (x_1, x_2, x_3)$$

is called a *surface* in $\mathbb{R}^3$. For simplicity, usually call $F(x_1, x_2, x_3) = 0$ is (the equation of) a surface S.

Suppose a certain neighborhood of a point $\vec{x}_0$ on S is given by a $\mathbb{C}^3$ vector-valued mapping

$$(u, v) \to \vec{\gamma}(u, v) = (x_1(u, v), x_2(u, v), x_3(u, v))$$

so that the tangent vectors at $\vec{\gamma}(u_0, v_0) = \vec{x}_0$,

$$\frac{\partial \vec{\gamma}}{\partial u}(u_0, v_0) \text{ and } \frac{\partial \vec{\gamma}}{\partial v}(u_0, v_0) \text{ are linearly independent,}$$

then $\vec{x}_0$ is called a *regular point* of S. A point on S that is not a regular point is called a *singular point*.

It can be shown that, in case $\vec{x}_0 = \vec{\gamma}(u_0, v_0)$ is a regular point, there exist an open neighborhood O of (u_0, v_0) and a neighborhood, say $\vec{\gamma}(O)$, of $\vec{x}_0$ so that

$$(u, v) \in O \to (u, v) \in \vec{\gamma}(O)$$

is a diffeomorphism (i.e. one-to-one, onto, $\vec{\gamma}$ and $\vec{\gamma}^{-1}$ are $\mathbb{C}^3$). See Fig. 3.100.

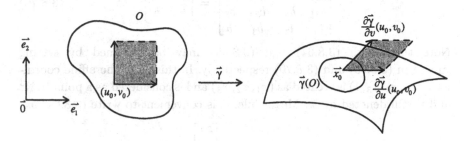

Fig. 3.100

Geometrically, this means that the part $\vec{\gamma}(O)$ of S is smooth, single-layered and contains no cusp or self-intersecting subset of S.

By a *double point* $\vec{x}_0$ of a quadric $F(x_1, x_2, x_3) = F(\vec{x}) = 0$, we mean the point $\vec{x}_0$ with the property that $F(\vec{x} + \vec{x}_0)$ does not contain first order terms in x_1, x_2 and x_3.

A quadric without singular points is called a *central quadric* if it has a center or centers. According to (3.8.66), they are types 1, 4, 5, 9, 12 and 15.

Noncentral quadrics without singular points are types 7, 8 and 14.

A singular point of a quadric is a double point. The set of singular points of a quadric is either

1. a point, such as type 6 with $\vec{0}$ as the singular point,
2. a line, such as type 13 with $\vec{x}_3$-axis as the singular line, or
3. a plane, such as type 17 with itself as the singular plane.

See Figs. 3.91, 3.96 and 3.99, respectively.

In what follows, we adopt homogeneous coordinates and affine coordinates for $\mathbb{R}^3$, introduced in Sec. 3.8.4.10 to determine affine invariants about quadrics (refer to (2.8.59)).

In homogeneous coordinates (x_1, x_2, x_3, x_4), the quadrics (3.8.64) is expressed as

$$\sum_{i,j=1}^{4} b_{ij} x_i x_j = 0 \quad \text{with } b_{4j} = b_{j4} = b_j \text{ for } 1 \leq j \leq 3 \text{ and } b_{44} = b, \quad (3.8.69)$$

or (3.8.65) as

$$\langle \vec{x}, \vec{x}\tilde{B} \rangle = 0, \quad \text{with} \quad \vec{x} = (x_1, x_2, x_3, x_4) \quad \text{and}$$

$$\tilde{B} = \begin{bmatrix} b_{11} & b_{12} & b_{13} & b_1 \\ b_{21} & b_{22} & b_{23} & b_2 \\ b_{31} & b_{32} & b_{33} & b_3 \\ b_1 & b_2 & b_3 & b \end{bmatrix} = \begin{bmatrix} B & \vec{b}^* \\ \vec{b} & b \end{bmatrix}_{4\times4}. \quad (3.8.70)$$

Note that, both (3.8.64) and (3.8.65) may be obtained by setting $x_4 = 1$ in (3.8.69) and (3.8.70) respectively. In this case, the affine coordinate $(x_1, x_2, x_3, 1)$ is treated as (x_1, x_2, x_3) and is considered as a point in $\mathbb{R}^3$ and is still denoted by $\vec{x}$. Meanwhile, it is convenient to write (3.8.64) as

$$(x_1, x_2, x_3, 1)\tilde{B} \begin{bmatrix} x_1 \\ x_2 \\ x_3 \\ 1 \end{bmatrix} = 0. \quad (3.8.71)$$

According to (1) in (3.8.63), an affine transformation $T(\vec{y}) = \vec{y}_0 + \vec{y}A$ on $\mathbb{R}^3$ can be written, in homogeneous coordinates, as

$$\vec{x} = \vec{y}\tilde{A}, \quad \text{with}$$
$$\vec{y} = (y_1, y_2, y_3, y_4),$$
$$\vec{x} = T(\vec{y}) = (x_1, x_2, x_3, x_4),$$

$$\tilde{A} = \begin{bmatrix} A & 0 \\ \vec{y}_0 & 1 \end{bmatrix} = \begin{bmatrix} a_{11} & a_{12} & a_{13} & 0 \\ a_{21} & a_{22} & a_{23} & 0 \\ a_{31} & a_{32} & a_{33} & 0 \\ \alpha_1 & \alpha_2 & \alpha_3 & 1 \end{bmatrix}_{4\times4}, \quad (3.8.72)$$

where $\vec{y}_0 = (\alpha_1, \alpha_2, \alpha_3)$ and $A = [a_{ij}]_{3\times3}$ is invertible. Or, in the corresponding affine coordinates,

$$(x_1, x_2, x_3, 1) = (y_1, y_2, y_3, 1)\tilde{A},$$

where

$$\vec{y} = (y_1, y_2, y_3) \quad \text{and} \quad \vec{x} = T(\vec{y}) = (x_1, x_2, x_3). \quad (3.8.73)$$

Note that $\tilde{A}$ is invertible.

By the same process leading to (*14) and beyond in Sec. 2.8.5, we have the following counterpart of (2.8.59).

The affine invariants of quadrics
For a quadric

$$\langle \vec{x}, \vec{x}B \rangle + 2\langle \vec{b}, \vec{x} \rangle + b = 0,$$

the signs or zeros of

1. $\det B$, and

2. $\det \begin{bmatrix} B & \vec{b}^* \\ \vec{b} & b \end{bmatrix}$

are affine invariants. In case these two quantities are positive, the positiveness of $\operatorname{tr} B$ is also an affine invariant.

$$(3.8.74)$$

These three quantities are Euclidean invariants under the rigid motions on $\mathbb{R}^3$ (see Sec. 5.7).

We postpone the characterizations of quadrics by means of Euclidean concepts to Sec. 5.10 (refer to (2.8.52) for quadratic curves), including there many computational examples.

Remark 5 *The standard (or canonical) forms of quadrics in the projective space $P^3(\mathbb{R})$, introduced in (3.8.59) and (3.8.60).*
By applying the Sylvester's law of inertia (see (2.7.71)) to the symmetric matrix $\tilde{B}$ in (3.8.70), we obtain the canonical forms of quadrics in $P^3(\mathbb{R})$ as

$$\sum_{i=1}^{k} x_i^2 - \sum_{i=k+1}^{r} x_i^2 = 0, \qquad (3.8.75)$$

where k is the signature of $\tilde{B}$ and $r = r(\tilde{B})$, the rank of $\tilde{B}$ (compare with (3.8.67)).

Also, a regular quadric is central or noncentral (see Remark 4) if and only its center belongs to the affine space $\mathbb{R}^3$ or is a point at infinity (refer to Ex. <B> 8 of Sec. 2.8.5).

Exercises

\<A\>

1. Prove (3.8.66) in detail.
2. Prove (3.8.67) in detail.
3. Prove (3.8.74) in detail.
4. Prove (3.8.75) and the statement in the last paragraph in detail.

APPENDIX A

Some Prerequisites

A.1 Sets

A *set* is a collection of objects, called *members* or *elements* of the set. When a set is to be referred to more than once, it is convenient to label it, usually, by a capital letter such as $A, B, \ldots$.

There are two distinct ways to describe a set:

1. By listing the elements of the set between curly brackets { }.
2. By giving a rule or characteristic property, which the elements of the set must satisfy.

For example, the set of all even positive integers less than 10 is written as $\{2, 4, 6, 8\} = \{6, 2, 8, 4\}$ or

$$\{x \mid x \text{ is an even positive integer less than } 10\}.$$

Note that each element of a set is not repeated within the set itself and the order in which the elements of a set are listed is immaterial.

Some definitions and notations are listed as follows:

1. $A \subseteq B$ or $B \supseteq A$ (A is a *subset* of B): every element of A is an element of B. $A \subsetneq B$ (A is a *proper* subset of B).
2. $A = B$ (A is equal to B): if and only if $A \subseteq B$ and $B \subseteq A$.
3. $\emptyset$ (*empty* set): the set that contains no elements.
4. $x \in A$: x is an element of the set A.
5. $x \notin A$: x is not an element of the set A.
6. $A \cup B$ (the *union* of A and B): $A \cup B = \{x \mid x \in A \text{ or } x \in B\}$.
7. $A \cap B$ (the *intersection* of A and B): $A \cap B = \{x \mid x \in A \text{ and } x \in B\}$.
8. A and B are *disjoint* if $A \cap B = \emptyset$.
9. $A - B$ (the *difference* of A by B): $\{x \mid x \in A \text{ and } x \notin B\}$.
10. $A \times B$ (the *Cartesian product* of A and B): $\{(x, y) \mid x \in A \text{ and } y \in B\}$.

Let $\wedge$ be an index set and $\{A_\lambda \mid \lambda \in \wedge\}$ be a collection of sets, the union and intersection of these sets are defined respectively by

$$\bigcup_{\lambda \in \wedge} A_\lambda = \{x \mid x \in A_\lambda \text{ for some } \lambda \in \wedge\}, \quad \text{and}$$

$$\bigcap_{\lambda \in \wedge} A_\lambda = \{x \mid x \in A_\lambda \text{ for all } \lambda \in \wedge\}.$$

A rule for determining if, for each ordered pair (x, y) of elements of a set A, x stands in a given relationship to y, is said to define a *relation* R on A. A relation R on a set A is called an *equivalent relation* if the following three conditions hold.

1. (Reflexivity) xRx (x is in relation R to itself).
2. (Symmetry) $xRy \Rightarrow yRx$.
3. (Transitivity) xRy and $yRz \Rightarrow xRz$.

Let R be an equivalent relation, $x \sim y$ is usually written in place of xRy. That $x - y$ is divisible by a fixed integer is an equivalent relation on the set of integers.

A.2   Functions

A *function* f from a set A into a set B, denoted by

$$f: A \to B,$$

is a rule that associates each element $x \in A$ to a unique element, denoted by $f(x)$, in B. Equivalently, a *function* is a set of ordered pairs (as a subset of $A \times B$) with the property that no two ordered pairs have the same first element.

Some terminologies are at hand.

1. $f(x)$: the *image* of x under f.
2. x: a *preimage* of $f(x)$ under f.
3. A: the *domain* of f.
4. $f(A) = \{f(x) \mid x \in A\}$: the *range* of f, a subset of B.
5. $f^{-1}(S) = \{x \in A \mid f(x) \in S\}$: the *preimage* of $S \subseteq B$.
6. $f = g$ (f is *equal to* g): $f(x) = g(x)$ for all $x \in A$ if $f: A \to B$ and $g: A \to B$.
7. f is *one-to-one*: If $f(x) = g(y)$ implies $x = y$, or, equivalently, if $x \neq y$ implies $f(x) \neq f(y)$.
8. f is *onto*: $f(A) = B$ if $f: A \to B$, and said that f *is onto* B.

9. $f|_S$ or $f\,|\,S$ (*restriction* of f on a subset $S \subseteq A$): $f|_S(x) = f(x)$ for each $x \in S$ if $f\colon A \to B$.

Let A, B and C be sets and $f\colon A \to B$ and $g\colon B \to C$ be functions. Note that the range $f(A)$ of f lies in the domain B of the definition of g as a subset. See Fig. A.1. Then, the *composite* of g and f is the function $g \circ f\colon A \to C$ defined by

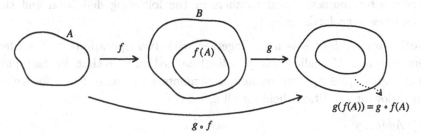

Fig. A.1

$$(g \circ f)(x) = g(f(x)), \quad x \in A.$$

Usually, $g \circ f \neq f \circ g$, even if both are defined. But associative law $h \circ (g \circ f) = (h \circ g) \circ f$ is true.

The *identity* function $1_A\colon A \to A$ is the function

$$1_A(x) = x, \quad x \in A.$$

It keeps every element of A fixed.

Suppose that $f\colon A \to B$ is a function such that there exists another function $g\colon B \to A$ satisfying

$$g \circ f = 1_A, \quad \text{i.e. } g(f(x)) = x, \quad x \in A.$$

Then f is one-to-one, and g is onto and is called a *left inverse* function of f. In case

$$f \circ g = 1_B, \quad \text{i.e. } f(g(y)) = y, \quad y \in B$$

then f is onto, and g is one-to-one and is called a *right inverse* function of f.

The following are equivalent:

1. $f\colon A \to B$ has a function $g\colon B \to A$ as both left and right inverse functions, i.e.

$$g(f(x)) = x, \ x \in A \quad \text{and} \quad f(g(y)) = y, \ y \in B.$$

2. $f\colon A \to B$ is one-to-one and onto.

Such a function f is called *invertible*. Being unique if f is invertible, $g: B \to A$ is called the *inverse* function of f and is denoted by

$$f^{-1}: B \to A.$$

A.3 Fields

Model after the set of real numbers in the following definition and the properties henceforth derived.

Definition A *field* $\mathbb{F}$ is a set together with two operations "+" (called *addition*) and "·" (called *multiplication*) defined on it so that, for each pair of elements $a, b \in \mathbb{F}$, there are unique elements $a + b$ and $a \cdot b$ in $\mathbb{F}$ for which the following conditions hold for all $a, b, c \in \mathbb{F}$:

(1) *Addition*

    (a) (commutative) $a + b = b + a$.

    (b) (associative) $(a + b) + c = a + (b + c)$.

    (c) (identity element) There exists an element $0 \in \mathbb{F}$, called *zero*, such that

$$a + 0 = a.$$

    (d) (inverse element) For each $a \in \mathbb{F}$, there exists an element, denoted by $-a$, in $\mathbb{F}$ such that

$$a + (-a) \underset{\text{(def.)}}{=} a - a = 0.$$

(2) *Multiplication*

    (a) (commutative) $a \cdot b = b \cdot a$.

    (b) (associative) $(a \cdot b) \cdot c = a \cdot (b \cdot c)$.

    (c) (identity element) There exists an element $1 \in \mathbb{F}$, called *unity*, such that $1 \cdot a = a$.

    (d) (inverse element) For each *nonzero* element $a \in \mathbb{F}$, there exists an element, denoted by a^{-1}, in $\mathbb{F}$ such that

$$a^{-1} \cdot a = 1.$$

(3) *Addition and multiplication*

    (distributive) $a \cdot (b + c) = a \cdot b + a \cdot c$.

The elements $a + b$ and $a \cdot b$ (also denoted as ab) are called the *sum* and *product*, respectively, of a and b. If $b \neq 0$, then $a \cdot b^{-1}$ also denote as a/b or $\frac{a}{b}$ and is called *division* by b.

In a $\mathbb{F}$, the following operational properties hold:

1. *Cancellation laws*

$$a + b = c + b \Rightarrow a = c.$$
$$ab = cb \quad \text{and} \quad b \neq 0 \Rightarrow a = c.$$

2. Hence, 0 and 1 are unique.
3. $a \cdot 0 = 0$ for any $\in \mathbb{F}$.
4. $a \cdot (-b) = (-a) \cdot b = -a \cdot b$.
5. $(-a) \cdot (-b) = a \cdot b$.

Proofs are left to the readers.

Standard fields and their fixed symbols

The Real Field $\mathbb{R}$:

$\mathbb{R}$ is an *ordered* field too and has Archimedean property. Readers of this book are supposed to be familiar with all the operational properties of $\mathbb{R}$.

The Complex Field $\mathbb{C}$:

The set

$$\mathbb{C} = \{a + ib \mid a, b \in \mathbb{R}\} \quad \text{where } i = \sqrt{-1}$$

with the

addition: $(a_1 + ib_1) + (a_2 + ib_2) = (a_1 + a_2) + i(b_1 + b_2)$, and

multiplication: $(a_1 + ib_1)(a_2 + ib_2) = (a_1 a_2 - b_1 b_2) + i(a_1 b_2 + a_2 b_1)$,

is a field where

$$0 = 0 + i0, \quad 1 = 1 + i0$$

are the respective zero and identity elements and, if $z = a + bi \neq 0$ (means at least one of a and b is not zero), the multiplicative inverse is

$$\frac{1}{z} = z^{-1} = \frac{a - bi}{a^2 + b^2}.$$

$\mathbb{C}$ is called the *complex Field* and its elements *complex numbers*. $\mathbb{C}$ cannot be ordered. A complex number is usually denoted by $z = a + ib$ with $a = \text{Re}\, z$ the *real part* and $b = \text{Im}\, z$ the *imaginary part* of it. The following are common sense in $\mathbb{C}$.

1. $\bar{z} = a - ib$ is called the *conjugate* of $z = a + bi = a + ib$.
2. $z\bar{z} = a^2 + b^2 \geq 0$.
3. Define the *absolute value* of z by

$$|z| = \sqrt{z\bar{z}} = \sqrt{a^2 + b^2}.$$

Then,

(a) $|z| \geq 0$, and $= 0 \Leftrightarrow z = 0$.

(b) $|z| = |\bar{z}|; |z_1 z_2| = |z_2 z_1| = |z_1||z_2|$ and $\left|\frac{z_1}{z_2}\right| = \frac{|z_1|}{|z_2|}$.

(c) $|z_1 + z_2| \leq |z_1| + |z_2|$ with equality if and only if there exists $\alpha \geq 0$ such that $z_1 = \alpha z_2$ or $z_2 = \alpha z_1$.

Geometrically, a complex number $z = a + ib$ is better interpreted as the *point* with coordinate (a, b) in the Cartesian coordinate plane or is considered as a *plane vector* pointing from $(0, 0)$ to the point (a, b).

The Finite Field $\mathbb{I}_p$ (p is a prime):

Let p be a fixed prime number. The residue classes of remainders under division by p

$$\mathbb{I}_p = \{0, 1, 2, \ldots, p - 1\}$$

together with usual addition and multiplication of integers modulus p is a field with finitely many elements.

$\mathbb{I}_p$ has a finite *characteristic p*. This means that $1 + 1 + \cdots + 1 = 0$ if and only if, the exact time of the appearance of 1 is a multiple of p in the summand.

A field having no finite characteristic is said to be of characteristic 0, such as $\mathbb{R}, \mathbb{C}$, the rational field Q and the radical field $Q(\sqrt{2}) = \{a + b\sqrt{2} \,|\, a, b \in Q\}$ with addition and multiplication as in $\mathbb{R}$.

A.4   Groups

Model partially after the addition properties of the set of integers or the multiplication properties of the set of nonzero real numbers.

Definition A *group* G is a set on which an operator $\circ$ is defined so that associates to each pair of elements a, b in G, a unique element $a \circ b$ in G for which the following properties hold for all a, b, c in G:

1. (associative) $(a \circ b) \circ c = a \circ (b \circ c)$.
2. (identity element) There exists an element e in G so that

$$a \circ e = e \circ a = a.$$

3. (inverse element) For each a in G, there exists an element a^{-1} in G so that

$$a \circ a^{-1} = a^{-1} \circ a = e.$$

If, in addition, $a \circ b = b \circ a$ holds for all a, b in G, then the group G is called *commutative* or *abelian*.

A nonempty subset S of a group G is called a *subgroup* if $a, b \in S$, then $a \circ b \in S$ and if $a \in S$, then $a^{-1} \in S$.

For example, the set of nonzero elements in any field is a group under the operation of field multiplication, while the field itself is a group by using addition operation.

The set of invertible real $n \times n$ matrices forms a nonabelian group $GL(n; \mathbb{R})$ under matrix multiplication. Various subgroups of $GL(n; R)$, especially when $n = 2, 3$, will be introduced in this book.

A.5 Polynomials

A polynomial in indeterminate t with coefficients from a field $\mathbb{F}$ is an expression of the form

$$p(t) = a_n t^n + a_{n-1} t^{n-1} + \cdots + a_1 t + a_0,$$

where n is a non-negative integer and $a_n, a_{n-1}, \ldots, a_1, a_0$ are elements of $\mathbb{F}$. If $\mathbb{F} = \mathbb{R}$, $p(t)$ is called a polynomial with *real* coefficients; if $\mathbb{F} = \mathbb{C}$, a polynomial with *complex* coefficients.

Here are some conventions:

1. If $a_n = \cdots = a_0 = 0$, then $p(t) = 0$ is called the *zero* polynomial and of *degree* -1.
2. The *degree* of a nonzero polynomial is defined to be the largest exponent of t in the expression of $p(t)$ with a nonzero coefficient.
3. Two polynomials $p(t)$ and $g(t)$ are *equal*

$$p(t) = g(t)$$

if both have the same degree and the coefficients of like powers of t are equal.
4. If $\mathbb{F}$ is an *infinite field* (i.e. a field containing infinite elements), a polynomial $p(t)$ with coefficients in $\mathbb{F}$ are often regarded as a function $p \colon \mathbb{F} \to \mathbb{F}$ and p or $p(t), t \in \mathbb{F}$ is called a *polynomial function*.

Furthermore, for a pair of polynomials $p(t)$ as above and $q(t) = b_m t^m + \cdots + b_1 t + b_0$ with $n \geq m$, the *sum* $p + q$ of p and q is defined as the addition

$$(p + q)(t) = p(t) + q(t)$$
$$= a_n t^n + \cdots + a_{m+1} t^{m+1} + (a_m + b_m) t^m$$
$$+ \cdots + (a_1 + b_1) t + a_0 + b_0.$$

The *scalar product* or *multiplication* αp of a scalar $\alpha \in \mathbb{F}$ and p is defined as

$$(\alpha p)(t) = \alpha p(t)$$
$$= \alpha a_n t^n + \cdots + \alpha a_1 t + \alpha a_0.$$

Therefore, the set of all polynomials with coefficients from a field $\mathbb{F}$

$$P(\mathbb{F})$$

is a vector space over $\mathbb{F}$ (see Sec. B.1), while the set of such polynomials of degrees no more than n, a nonnegative integer,

$$P_n(\mathbb{F})$$

forms a vector subspace of dimension $n + 1$.

Let $f(t)$ be a polynomial and $g(t)$ a polynomial of non-negative degree. Then, there exist unique polynomials $q(t)$ and $r(t)$ such that

1. the degree of $r(t)$ is less than that of $f(t)$, and
2. $f(t) = q(t)g(t) + r(t)$.

This is the so-called *Division Algorithm for Polynomials*. It follows that, $t - a$ divides $f(t)$ if and only if $f(a) = 0$, and such a is called a *zero* of $f(t)$. Any polynomial of degree $n \geq 1$ has at most n distinct zeros.

A polynomial $p(t)$ of positive degree is called *irreducible* if it cannot be factored as a product of polynomials with coefficients from the *same* field $\mathbb{F}$, each having positive degree. If $f(t)$ is irreducible and $f(x)$ does not divide another polynomial $g(t)$ with coefficients from the same field, then $f(t)$ and $g(t)$ are *relatively prime*. This means that no polynomial of positive degree can divide both of them. In this case, there exist polynomials $f_1(t)$ and $g_1(t)$ such that

$$f(t)f_1(t) + g(t)g_1(t) = 1 \quad \text{(constant polynomial 1)}.$$

If an irreducible polynomial $p(t)$ divides the product $g(t)h(t)$ of two polynomials $g(t)$ and $h(t)$ over the same field $\mathbb{F}$, in symbol

$$f(t) \,|\, g(t)h(t),$$

then $f(t) \,|\, g(t)$ or $f(t) \,|\, h(t)$.

Unique Factorization Theorem for Polynomials

For any polynomial $p(t) \in P(\mathbb{F})$ of positive degree, there exist

1. a unique constant c,
2. unique irreducible polynomials $p_1(t), \ldots, p_n(t)$ whose leading coefficients are equal to 1 (the so-called *monic polynomial*), and
3. unique positive integers $r_1, \ldots, r_n$,

such that

$$p(t) = cp_1(t)^{r_1} \cdots p_n(t)^{r_n}.$$

For an infinite field $\mathbb{F}$ such as $Q, Q(\sqrt{2}), \mathbb{R}$ and $\mathbb{C}$, two polynomials $f(t)$ and $g(t)$ are equal if and only if there are more than n distinct scalars $a_1, \ldots, a_m$ such that

$$f(a_i) = g(a_i), \quad 1 \le i \le m$$

where n is the larger of the degrees of $f(t)$ and $g(t)$.

In this book, polynomials concerned in most cases are polynomials with real coefficients and of degree 1, 2 or 3.

For quadratic polynomial $p(x) = ax^2 + bx + c$ with a, b, c in $\mathbb{R}$ and x as a real variable,

$$p(x) = a\left(x - \frac{-b + \sqrt{b^2 - 4ac}}{2a}\right)\left(x - \frac{-b - \sqrt{b^2 - 4ac}}{2a}\right).$$

In case

1. $b^2 - 4ac > 0$, $p(x)$ has two distinct real zeros;
2. $b^2 - 4ac = 0$, $p(x)$ has two equal real zeros; and
3. $b^2 - 4ac < 0$, $p(x)$ has two conjugate complex zeros.

As for cubic polynomial $p(x) = ax^3 + bx^2 + cx + d$ with a, b, c, d in $\mathbb{R}$ and x as a real variable,

1. it always has at least one real zero, and
2. if it has only one real zero, the other two zeros are conjugate complex numbers.

APPENDIX B

Fundamentals of Algebraic Linear Algebra

This appendix is divided into twelve sections. Among them, Secs. B.1–B.6 are devoted to static structures of vector spaces themselves, while Secs. B.7–B.12 are mainly concerned with dynamic relations between vector spaces, namely the study of linear transformations. Most topics are stated in the realm of finite dimensional vector spaces. From Sec. B.4 on, some exercise problems are attached as parts of the contents. Few geometric interpretations of abstract results are touched and the methods adopted are almost purely algebraic. Essentially no proofs are given. In short, the manner presented is mostly adopted in the nowadays linear algebra books.

A vector space V over a field $\mathbb{F}$ is defined axiomatically in Sec. B.1, with $\mathbb{F}^n, I_p^n$ and $M(m, n; \mathbb{F})$ as concrete examples along with subspace operations. Via the techniques of linear combination, dependence and independence introduced in Sec. B.2, Sec. B.3 introduces the basis and dimension for a vector space. The concept of matrices over a field and their algebraic operations are in Sec. B.4, and the elementary row or column operations on a matrix are in Sec. B.5. The determinant function on square matrices is sketched in Sec. B.6.

Section B.7 is devoted to linear transformation (functional, operator, or isomorphism) and its matrix representation with respect to bases. Section B.8 investigates a matrix and its transpose, mostly from the viewpoint of linear transformations. Inner product spaces with specified linear operators on them such as orthogonal, normal, etc. are in Sec. B.9. Eigenvalues and eigenvectors are in Sec. B.10, while Sec. B.11 investigates the diagonalizability of a matrix. For nondiagonalizable matrices, their Jordan and rational canonical forms are sketched in Sec. B.12. That is all!

B.1 Vector (or Linear) Spaces

Definition A *vector* or *linear space* V over a field $\mathbb{F}$ consists of a set (usually, still denoted by V) on which two operations (called *addition* and *scalar multiplication*, respectively) are defined so that for each pair of elements $\vec{x}, \vec{y}$ in V, there is a unique element

$$\vec{x} + \vec{y} \quad \text{(called the } sum \text{ of } \vec{x} \text{ and } \vec{y})$$

in V, and for each element α in $\mathbb{F}$ and each element $\vec{x}$ in V, there is a unique element

$$\alpha\vec{x} \quad \text{(called the scalar } product \text{ of } \vec{x} \text{ by } \alpha)$$

in V, such that the following conditions hold for all $\vec{x}, \vec{y}, \vec{z} \in V$ and $\alpha, \beta \in \mathbb{F}$:

(1) *Addition*

    (a) (commutative) $\vec{x} + \vec{y} = \vec{y} + \vec{x}$.
    (b) (associative) $(\vec{x} + \vec{y}) + \vec{z} = \vec{x} + (\vec{y} + \vec{z})$.
    (c) (zero vector) There is an element, denoted by $\vec{0}$, in V such that

$$\vec{x} + \vec{0} = \vec{x}.$$

    (d) (negative or inverse vector of a vector) For each $\vec{x} \in V$, there exists another element, denoted by $-\vec{x}$, in V such that

$$\vec{x} + (-\vec{x}) = \vec{0}.$$

(2) *Scalar multiplication*

    (a) $1\vec{x} = \vec{x}$.
    (b) $\alpha(\beta\vec{x}) = (\alpha\beta)\vec{x}$.

(3) The addition and scalar multiplication satisfy the distributive laws:

$$(\alpha + \beta)\vec{x} = \alpha\vec{x} + \beta\vec{x},$$
$$\alpha(\vec{x} + \vec{y}) = \alpha\vec{x} + \alpha\vec{y}.$$

The elements of the field $\mathbb{F}$ are called *scalars* and the elements of the vector space V are called *vectors*. The word "vector", without any practical meaning such as displacement or acting force, is now being used to describe any element of a vector space.

If the underlying field $\mathbb{F}$ is the real field $\mathbb{R}$ or the complex field $\mathbb{C}$, then the corresponding vector space is called specifically a *real* or a *complex* vector space, respectively.

A vector space will frequently be discussed without explicitly mentioning its field of scalars.

Some elementary consequences of the definition of a vector space are listed as follows:

1. (cancellation law) If $\vec{x}, \vec{y}, \vec{z} \in V$ and $\vec{x} + \vec{z} = \vec{y} + \vec{z}$, then $\vec{x} = \vec{y}$.
2. Then, zero vector $\vec{0}$ is unique.
3. Negative $-\vec{x}$ of a vector $\vec{x}$ is unique.
4. $\alpha\vec{x} = \vec{0} \Leftrightarrow \alpha = 0$ or $\vec{x} = 0$.
5. $(-\alpha)\vec{x} = -(\alpha\vec{x}) = \alpha(-\vec{x})$.

Examples

For a given field $\mathbb{F}$ and a positive integer n, define the set of all n-tuples with entries from $\mathbb{F}$ as the set

$$\mathbb{F}^n = \{(x_1, \ldots, x_n) \mid x_i \in \mathbb{F} \text{ for } 1 \le i \le n\}.$$

If $\vec{x} = (x_1, \ldots, x_n)$ and $\vec{y} = (y_1, \ldots, y_n)$ are in $\mathbb{F}^n$, define the operations of componentwise addition and multiplication as

$$\vec{x} + \vec{y} = (x_1 + y_1, \ldots, x_n + y_n),$$
$$\alpha\vec{x} = (\alpha x_1, \ldots, \alpha x_n), \quad \alpha \in \mathbb{F}.$$

$\mathbb{F}^n$ is then a vector space over $\mathbb{F}$ with

$$\vec{0} = (0, \ldots, 0),$$
$$-\vec{x} = (-x_1, \ldots, -x_n).$$

The following specified vectors and notations for them will be used throughout the whole book:

$$\vec{e}_i = (0, \ldots, 0, 1, 0, \ldots, 0), \quad 1 \le i \le n.$$
$$\uparrow \text{\scriptsize ith component}$$

In particular, we have vector spaces

$$\mathbb{R}^n, \quad \mathbb{C}^n \text{ and } \mathbb{I}_p^n \quad (p \text{ a prime}), \quad n \ge 1.$$

By Sec. A.5, the set $P(\mathbb{F})$ of all polynomials with coefficients from a field $\mathbb{F}$ is a vector space over $\mathbb{F}$.

Let X be a nonempty set and $\mathbb{F}$ a field. The set of all functions from X into $\mathbb{F}$

$$\mathcal{F}(X, \mathbb{F}) = \{\text{function } f \colon X \to \mathbb{F}\}$$

is a vector space over $\mathbb{F}$ under the operations $f + g$ and αf defined by

$$(f + g)(x) = f(x) + g(x),$$
$$(\alpha f)(x) = \alpha f(x), \quad \alpha \in \mathbb{F}, \ x \in X.$$

In real analysis, the set $C[a, b]$ of continuous functions on an interval $[a, b]$ forms a vector space. This is an important vector space.

The set $M(m, n; \mathbb{F})$ of $m \times n$ matrices with entries from a field $\mathbb{F}$ forms a vector space over $\mathbb{F}$. See Sec. B.4.

A nonempty subset S of a vector space V over a field $\mathbb{F}$ is called a *vector or linear subspace* of V if S is a vector space over $\mathbb{F}$ under the original addition and scalar multiplication defined on V. This is equivalent to say that, if

1. $\vec{x}$ and $\vec{y}$ are in S, then $\vec{x} + \vec{y}$ is in S, and
2. for each $\alpha \in \mathbb{F}$ and $\vec{x} \in S$, then $\alpha \vec{x} \in S$.

Then S is simply called a *subspace* of V.

Any vector space V has the *zero subspace* $\{\vec{0}\}$ and itself as trivial subspaces. Subspaces carry the following three operations with themselves:

1. *Intersection subspace* If S_1 and S_2 are subspaces of V, then

$$S_1 \cap S_2 = \{\vec{x} \in V \mid \vec{x} \in S_1 \text{ and } \vec{x} \in S_2\}$$

is a subspace of V. Actually, for an arbitrary family $\{S_\lambda\}_{\lambda \in \Lambda}$ of subspaces of V, the intersection

$$\bigcap_{\lambda \in \Lambda} S_\lambda$$

is always a subspace of V.

2. *Sum subspace* If S_1 and S_2 are subspaces of V, then

$$S_1 + S_2 = \{\vec{x}_1 + \vec{x}_2 \mid \vec{x}_1 \in S_1 \text{ and } \vec{x}_2 \in S_2\}$$

is a subspace which has S_1 and S_2 as its subspaces. In case $S_1 \cap S_2 = \{\vec{0}\}$, denote $S_1 + S_2$ by

$$S_1 \oplus S_2$$

and is called the *direct sum* of S_1 and S_2.

3. *Quotient space of V modulus a subspace S* For any $\vec{x} \in V$, the *coset* of S containing $\vec{x}$ is the set

$$\vec{x} + S = \{\vec{x} + \vec{v} \mid \vec{v} \in S\}.$$

Then $\vec{x} + S = \vec{y} + S$ if and only if $\vec{x} - \vec{y} \in S$. The quotient set

$$V/S = \{\vec{x} + S \mid x \in V\}$$

forms a vector space over $\mathbb{F}$ under the well-defined operations:

$$(\vec{x}_1 + S) + (\vec{x}_2 + S) = (\vec{x}_1 + \vec{x}_2) + S, \quad \text{and}$$

$$\alpha(\vec{x} + S) = \alpha\vec{x} + S,$$

with $\vec{0} + S = S$ acts as zero vector in it.

For example, let m and n be integers with $m < n$ and

$$\tilde{\mathbb{F}}^m = \{(x_1, \ldots, x_m, 0, \ldots, 0) \mid x_i \in \mathbb{F} \text{ for } 1 \leq i \leq m\}.$$

Then $\tilde{\mathbb{F}}^m$ is a subspace of $\mathbb{F}^n$. $\tilde{\mathbb{F}}^m$ can be identified with $\mathbb{F}^m$ in isomorphic sense (see Sec. B.7).

In $\mathbb{R}^2$, let S be the subspace defined by the line $ax + by = 0$. Then, the quotient space $\mathbb{R}^2/S$ is the set of all lines (including S itself) parallel to S, which, in turn, can be identified with the line $bx - ay = 0$ in isomorphic sense.

B.2 Main Techniques: Linear Combination, Dependence and Independence

How to construct subspaces of a vector space? What are the forms they exist in the simplest manner?

For any *finite* number of vectors $\vec{x}_1, \ldots, \vec{x}_k$ in a vector space V over a field $\mathbb{F}$ and any scalars $\alpha_1, \ldots, \alpha_k \in \mathbb{F}$, the *finite* sum

$$\alpha_1\vec{x}_1 + \cdots + \alpha_k\vec{x}_k \quad \text{or in short} \quad \sum_{i=1}^{k} \alpha_i\vec{x}_i$$

is called a *linear combination* of the vectors $\vec{x}_1, \ldots, \vec{x}_k$ with *coefficients* $\alpha_1, \ldots, \alpha_k$.

Let S be any nonempty subset of V. The set of linear combinations of (finitely many) vectors in S

$$\langle\langle S \rangle\rangle = \{\text{finite linear combinations of vectors in } S\}$$

is a subspace of V and is called the *subspace generated* or *spanned* by S. In case $S = \{\vec{x}_1, \ldots, \vec{x}_k\}$ is a finite set, we would write

$$\langle\langle \vec{x}_1, \ldots, \vec{x}_k \rangle\rangle$$

instead of $\langle\langle \{\vec{x}_1, \ldots, \vec{x}_k\} \rangle\rangle$.

It could happen that most of vectors in $\langle\langle S \rangle\rangle$ can be written as linear combinations of elements from a certain special subset of S. Suppose $\vec{x} \in S$ is of the form

$$\vec{x} = \alpha_1 \vec{x}_1 + \cdots + \alpha_k \vec{x}_k.$$

Then, in a certain problem concerning the set of vectors $\{\vec{x}, \vec{x}_1, \ldots, \vec{x}_n\}$, the role of $\vec{x}$ becomes inessential because it can be determined by the remaining $\vec{x}_1, \ldots, \vec{x}_k$ through the process of linear combination. Note also that, the formula can be equivalently written as

$$\alpha \vec{x} + \alpha_1 \vec{x}_1 + \cdots + \alpha_k \vec{x}_k = \vec{0},$$

where $\alpha = -1, \alpha_2, \ldots, \alpha_k$ are not all equal to zero. This latter expression indicates implicitly that one of $\vec{x}, \vec{x}_1, \ldots, \vec{x}_k$, say $\vec{x}$ in this case, depends linearly on the others. Hence, here comes naturally the

Definition A nonempty subset S of a vector space V is said to be *linearly dependent* if there exist a *finite* number of *distinct* vectors $\vec{x}_1, \ldots, \vec{x}_k$ in S and scalars $\alpha_1, \ldots, \alpha_k$, *not all zero*, such that

$$\alpha_1 \vec{x}_1 + \cdots + \alpha_k \vec{x}_k = \vec{0}.$$

Vectors in S are also said to be *linearly dependent*.

If S is linearly dependent, then at least one of the vectors in S can be written as a linear combination of other vectors in S. This vector is said to be *linearly dependent* on these others. Thus, those vectors in S that are linearly dependent on the others in S play no role in spanning the subspace $\langle\langle S \rangle\rangle$ and hence, can be *eliminated* from S without affecting the generation of $\langle\langle S \rangle\rangle$. This process can be proceeded until what remain in S are not linearly dependent any more. These remaining vectors, if any, are linearly independent and are good enough to span the same $\langle\langle S \rangle\rangle$.

Definition A nonempty subset S of a vector space V is said to be *linearly independent* if for *any finite* number of distinct vectors $\vec{x}_1, \ldots, \vec{x}_k$ in S, the relation

$$\alpha_1 \vec{x}_1 + \cdots + \alpha_k \vec{x}_k = \vec{0}$$

holds if and only if the scalars $\alpha_1, \ldots, \alpha_k$ are *all zero*. In this case, vectors in S are said to be *linearly independent*.

Note the following trivial facts:

1. $\{\vec{0}\}$ is linearly dependent.
2. Any *nonzero* vector alone is linearly independent.

3. Let $S_1 \subseteq S_2 \subseteq V$. If S_1 is linearly dependent, so is S_2. Thus, if S_2 is linearly independent, so is S_1.

B.3 Basis and Dimension

We start from the vector space $\mathbb{F}^n$ stated in Sec. B.1.

Each vector $\vec{x} = \{x_1, \ldots, x_n\} \in \mathbb{F}^n$ can be expressed as a linear combination of $\vec{e}_1, \ldots, \vec{e}_n$ as

$$\vec{x} = x_1 \vec{e}_1 + \cdots + x_n \vec{e}_n = \sum_{i=1}^{n} x_i \vec{e}_i.$$

Also, $\{\vec{e}_1, \ldots, \vec{e}_n\}$ is linearly independent, i.e.

$$\sum_{i=1}^{n} x_i \vec{e}_i = \vec{0}$$

would imply that $x_1 = \cdots = x_n = 0$. Such a linearly independent set $\{\vec{e}_1, \ldots, \vec{e}_n\}$, that generates the whole space $\mathbb{F}^n$ is called a *basis* for $\mathbb{F}^n$, and the number n of vectors in a basis is called the *dimension* of $\mathbb{F}^n$.

Vector spaces are divided into *finite-dimensional* and *infinite-dimensional* according to the fact that if they can be generated by a finite subset of itself or not.

A *basis* $\mathcal{B}$ for a vector space V is a linearly independent subset that generates V. Vectors in $\mathcal{B}$ are called *basis vectors* of V. Note that every vector in V can be expressed *uniquely* as a linear combination of vectors of $\mathcal{B}$.

Every vector space would have at least one basis.

The most powerful tool to determine if a basis exists is the following

Steinitz's Replacement Theorem *Let $\{\vec{x}_1, \ldots, \vec{x}_n\}$ be a finite set of vectors in a vector space V. Suppose $\vec{y}_1, \ldots, \vec{y}_m \in \langle\langle \vec{x}_1, \ldots, \vec{x}_n \rangle\rangle$ are linearly independent. Then*

1. *$m \leq n$, and*
2. *m vectors among $\vec{x}_1, \ldots, \vec{x}_n$, say $\vec{x}_1, \ldots, \vec{x}_m$, can be replaced by $\vec{y}_1, \ldots, \vec{y}_m$ so that $\{\vec{y}_1, \ldots, \vec{y}_m, \vec{x}_{m+1}, \ldots, \vec{x}_n\}$ still generates the same subspace $\langle\langle \vec{x}_1, \ldots, \vec{x}_n \rangle\rangle$, i.e.*

$$\langle\langle \vec{y}_1, \ldots, \vec{y}_m, \vec{x}_{m+1}, \ldots, \vec{x}_n \rangle\rangle = \langle\langle \vec{x}_1, \ldots, \vec{x}_m, \vec{x}_{m+1}, \ldots, \vec{x}_n \rangle\rangle.$$

Now, suppose that V is finite-dimensional and V has at least one nonzero vector.

Take any nonzero vector $\vec{x}_1 \in V$. Construct subspace $\langle\langle \vec{x}_1 \rangle\rangle$. In case $V = \langle\langle \vec{x}_1 \rangle\rangle$ and $\mathcal{B} = \{\vec{x}_1\}$ is a basis for V. If $\langle\langle \vec{x}_1 \rangle\rangle \subsetneq V$ (this means

$\langle\langle\vec{x}_1\rangle\rangle \subseteq V$ but $\langle\langle\vec{x}_1\rangle\rangle \neq V$), then there exists a vector $\vec{x}_2 \in V - \langle\langle\vec{x}_1\rangle\rangle$ and, $\vec{x}_1$ and $\vec{x}_2$ are linearly independent. Construct subspace $\langle\langle\vec{x}_1, \vec{x}_2\rangle\rangle$. Then, either $V = \langle\langle\vec{x}_1, \vec{x}_2\rangle\rangle$ and $\mathcal{B} = \{\vec{x}_1, \vec{x}_2\}$ is a basis for V, or there exists $\vec{x}_3 \in V - \langle\langle\vec{x}_1, \vec{x}_2\rangle\rangle$ so that $\vec{x}_1, \vec{x}_2, \vec{x}_3$ are linearly independent. Continue this procedure as it will stop after a finite number of steps, say n. Therefore,

$$V = \langle\langle\vec{x}_1, \ldots, \vec{x}_n\rangle\rangle,$$

where $\vec{x}_1, \ldots, \vec{x}_n$ are linearly independent. Then $\mathcal{B} = \{\vec{x}_1, \ldots, \vec{x}_n\}$ is a *basis* for V, and is called an *ordered basis* if the ordering $\vec{x}_1, \ldots, \vec{x}_n$ is emphasized.

The number of basis vectors in any basis for V is the same positive integer for any two bases for V, say n, and is called the *dimension* of V and denoted by

$$\dim V = n.$$

In this case, V is said to be an *n-dimensional vector space* over a field $\mathbb{F}$.

If $V = \{\vec{0}\}$, define $\dim V = 0$ and V is called 0-*dimensional*.

If W is subspace of V, then

$$\dim W \leq \dim V$$

and equality holds if and only if $W = V$. Furthermore, any basis for W can be extended to a basis $\mathcal{B}$ for V. In other words, there exists another vector subspace U of V such that

$$V = W \oplus U.$$

For any two subspaces W_1 and W_2 of V, where $\dim V$ is finite

$$\dim(W_1 \cap W_2) + \dim(W_1 + W_2) = \dim W_1 + \dim W_2$$

holds.

If V is infinite-dimensional, it can be proved, by Zorn's lemma from set theory, that V has a basis which is a *maximal* linearly independent subset of V. Every basis has the same cardinality.

In such an elementary book like this one, *we will study finite-dimensional vector spaces only unless specified.*

Example

Let $P_n(\mathbb{F}), n \geq 0$, be as in Sec. A.4 where $\mathbb{F}$ is an infinite field.

The set $\{1, t, \ldots, t^n\}$ forms a basis for $P_n(\mathbb{F})$.

Take any fixed but distinct scalars $a_0, a_1, a_2, \ldots, a_n$ in $\mathbb{F}$. The polynomial of degree n

$$p_i(t) = \frac{(t - a_0) \cdots (t - a_{i-1})(t - a_{i+1}) \cdots (t - a_n)}{(a_i - a_0) \cdots (a_i - a_{i-1})(a_i - a_{i+1}) \cdots (a_i - a_n)}$$

$$= \prod_{\substack{j=0 \\ j \neq i}}^{n} \frac{t - a_j}{a_i - a_j}, \quad 0 \leq i \leq n$$

is uniquely characterized as the polynomial function $p_i \colon \mathbb{F} \to \mathbb{F}$ satisfying

$$p_i(a_j) = \begin{cases} 1, & i = j \\ 0, & i \neq j \end{cases}.$$

The polynomials $p_0, p_1, \ldots, p_n$ are called the *Lagrange polynomials* associated with $a_0, a_1, \ldots, a_n$. It is easily seen that $\{p_0, p_1, \ldots, p_n\}$ is a basis for the $(n+1)$-dimensional vector space $\mathrm{P}_n(\mathbb{F})$. Every polynomial $p \in \mathrm{P}_n(\mathbb{F})$ is uniquely expressed as

$$p = \sum_{i=0}^{n} p(a_i) p_i,$$

which is called the *Lagrange interpolation formula* in the sense that, if $\alpha_0, \alpha_1, \ldots, \alpha_n$ are arbitrarily given $(n+1)$ scalars in $\mathbb{F}$, then this p is the unique polynomial having the property that

$$p(a_i) = \alpha_i, \quad 0 \leq i \leq n.$$

B.4   Matrices

Let m and n be positive integers.

An $m \times n$ *matrix* with entries from a field $\mathbb{F}$ is an ordered rectangular array of the form

$$\begin{bmatrix} a_{11} & a_{12} & \cdots & a_{1n} \\ a_{21} & a_{22} & \cdots & a_{2n} \\ \vdots & \vdots & & \vdots \\ a_{m1} & a_{m2} & \cdots & a_{mn} \end{bmatrix} \quad \text{or} \quad \begin{pmatrix} a_{11} & a_{12} & \cdots & a_{1n} \\ a_{21} & a_{22} & \cdots & a_{2n} \\ \vdots & \vdots & & \vdots \\ a_{m1} & a_{m2} & \cdots & a_{mn} \end{pmatrix},$$

where each entry $a_{ij}, 1 \leq i \leq m, 1 \leq j \leq n$, is an element of $\mathbb{F}$. Capital letters such as A, B and C, etc. are used to denote matrices. The entries $a_{i1}, a_{i2}, \ldots, a_{in}$ of the matrix A above compose the *i*th *row* of A and is denoted by

$$A_{i*} = (a_{i1}a_{i2}\cdots a_{in}), \quad 1 \le i \le m$$

and is called a *row matrix*. Similarly, the entries $a_{1j}, a_{2j}, \ldots, a_{mj}$ of A compose the jth *column* of A and is denoted by

$$A_{*j} = \begin{bmatrix} a_{1j} \\ a_{2j} \\ \vdots \\ a_{mj} \end{bmatrix} \text{ or } \begin{pmatrix} a_{1j} \\ a_{2j} \\ \vdots \\ a_{mj} \end{pmatrix}, \quad 1 \le j \le n$$

and is called a *column matrix*. The entry a_{ij} which lies in the ith row and jth column is called the (i, j) entry of A. Then, the matrix A is often written in shorthand as

$$A_{m \times n} \text{ or } A = [a_{ij}]_{m \times n} \quad \text{or} \quad (a_{ij})_{m \times n} \text{ or } A = [a_{ij}].$$

Matrices are used to describe route maps in topology and networks, and to store a large quantity of numerical data on many occasions. They appear seemingly naturally in the treatment of some geometrical problems. Actually, if we endow this notation with suitable operations, the static and dynamic properties of matrices will play the core of study about finite-dimensional vector spaces in linear algebra.

Matrices with entries in $\mathbb{R}$ or $\mathbb{C}$ are called *real* or *complex* matrices, respectively.

Two $m \times n$ matrices $A = [a_{ij}]$ and $B = [b_{ij}]$ are defined to be *equal* if and only if $a_{ij} = b_{ij}$ for $1 \le i \le m, 1 \le j \le n$ and is denoted as

$$A = B.$$

The $m \times n$ matrix having each entry a_{ij} equal to zero is called the *zero matrix* and is denoted by

$$O.$$

The $n \times m$ matrix obtained by interchange m rows into m columns and n columns into n rows of a $m \times n$ matrix $A = [a_{ij}]$ is called the *transpose* of A and is denoted by

$$A^* = [b_{ji}] \quad \text{where } b_{ji} = a_{ij}, \ 1 \le i \le m, \ 1 \le j \le n.$$

An $n \times n$ matrix $A = [a_{ij}]_{n \times n}$ is called a *square matrix of order n* with $a_{ii}, 1 \le i \le n$, as its *(main) diagonal* entries.

The following are some special square matrices.

1. *Diagonal matrix $a_{ij} = 0$ for $1 \leq i, j \leq n$ but $i \neq j$.* It is of the form

$$\begin{bmatrix} a_{11} & & & 0 \\ & a_{22} & & \\ & & \ddots & \\ 0 & & & a_{nn} \end{bmatrix}$$

and is denoted by $\text{diag}[a_{11}, \ldots, a_{nn}]$.

2. *Identity matrix I_n of order n* This is the diagonal matrix with diagonal entries $a_{ii} = 1, 1 \leq i \leq n$ and is denoted by

$$I_n = \begin{bmatrix} 1 & & & 0 \\ & 1 & & \\ & & \ddots & \\ 0 & & & 1 \end{bmatrix}.$$

3. *Scalar matrix αI_n with $\alpha \in \mathbb{F}$* This is the diagonal matrix with $a_{11} = a_{22} = \cdots = a_{nn}$, say equal to α, and is denoted by

$$\alpha I_n = \begin{bmatrix} \alpha & & & 0 \\ & \alpha & & \\ & & \ddots & \\ 0 & & & \alpha \end{bmatrix}.$$

4. *Upper triangular matrix $a_{ij} = 0$ for $1 \leq j < i \leq n$.* It is of the form

$$\begin{bmatrix} a_{11} & a_{12} & a_{13} & \cdots & a_{1n} \\ & a_{22} & a_{23} & \cdots & a_{2n} \\ & & a_{33} & \cdots & a_{3n} \\ & 0 & & \ddots & \vdots \\ & & & & a_{nn} \end{bmatrix}.$$

5. *Lower triangular matrix $a_{ij} = 0$ for $1 \leq i < j \leq n$.*

6. *Symmetric matrix $a_{ij} = a_{ji}, 1 \leq i, j \leq n$,* i.e. $A^* = A$.

7. *Skew-symmetric matrix* $a_{ij} = -a_{ji}, 1 \leq i, j \leq n$, i.e. $A^* = -A$. For infinite field $\mathbb{F}$ or a finite field with characteristic $p \neq 2$, the main diagonal entries a_{ii} of a skew-symmetric matrix are all zero.

Let $A = [a_{ij}]_{m \times n}$ be a complex matrix. Then $\bar{A} = [\bar{a}_{ij}]_{m \times n}$ is called the *conjugate matrix* of A and the $n \times m$ matrix

$$\bar{A}^* = [\bar{a}_{ji}]_{n \times m}$$

the *conjugate transpose* of A.

8. *Hermitian matrix* A complex square matrix A satisfying

$$\bar{A}^* = A$$

is called Hermitian.

9. *Skew-Hermitian matrix* This means a complex square matrix A having the property

$$\bar{A}^* = -A.$$

In this case, the main diagonal entries of A are all pure imaginaries.

Hermitian or skew-Hermitian matrices with real entries are just real symmetric or real skew-symmetric matrices, respectively.

Let

$$M(m, n; \mathbb{F}) = \{m \times n \text{ matrices with entries in the field } \mathbb{F}\}.$$

On which two operations are defined as follows.

1. *Addition* For each pair of matrices $A = [a_{ij}], B = [b_{ij}] \in M(m, n; \mathbb{F})$, the *sum* $A + B$ of A and B is the $m \times n$ matrix

$$A + B = [a_{ij} + b_{ij}]_{m \times n}$$

obtained by entry-wise addition of corresponding entries of A and B.

2. *Scalar multiplication* For each $\alpha \in \mathbb{F}$ and $A = [a_{ij}] \in M(m, n; \mathbb{F})$, the *scalar product* of A by α is the $m \times n$ matrix

$$\alpha A = [\alpha a_{ij}]_{m \times n}.$$

They enjoy the following properties:

$$A + B = B + A;$$
$$(A + B) + C = A + (B + C);$$
$$A + O = A \quad (O \text{ is zero } m \times n \text{ matrix});$$
$$A + (-A) = O \quad (-A \text{ means } (-1)A);$$
$$1A = A;$$
$$\alpha(\beta A) = (\alpha\beta)A;$$
$$\alpha(A + B) = \alpha A + \alpha B;$$
$$(\alpha + \beta)A = \alpha A + \beta A.$$

Therefore, $\mathrm{M}(m, n; \mathbb{F})$ is an mn-dimensional vector space over $\mathbb{F}$ with

$$
E_{ij} = \begin{bmatrix}
 & & 0 & & \\
 & 0 & \vdots & 0 & \\
 & & 0 & & \\
0 & \cdots & 0 \; 1 \; 0 & \cdots & 0 \\
 & & 0 & & \\
 & 0 & \vdots & 0 & \\
 & & 0 & &
\end{bmatrix} \leftarrow i\text{th row}, \quad 1 \le i \le m, \; 1 \le j \le n
$$

$$\uparrow$$
$$j\text{th column}$$

as basis vectors.

Owing to the same operational properties, a $1 \times n$ row matrix is also regarded as a *row vector* in $\mathbb{F}^n$ and vice versa, while a $m \times 1$ column matrix as a *column vector* in $\mathbb{F}^m$ and vice versa.

In order to define matrix multiplication in a reasonable way, we start from a simple example.

Suppose a resort island has two towns A and B. The island bus company runs a single bus which operates on two routes:

1. from A to B in either direction;
2. a circular route from B to B along the coast.

See the network on the Fig. B.1. By a single stage journey we mean either $A \rightarrow B$ or $B \rightarrow A$, or $B \rightarrow B$ and can be expressed in matrix form

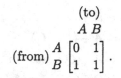

$$\text{(from)} \begin{array}{c} \\ A \\ B \end{array} \overset{\overset{\text{(to)}}{A \ \ B}}{\begin{bmatrix} 0 & 1 \\ 1 & 1 \end{bmatrix}}.$$

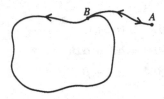

Fig. B.1

How many routes can a tourist choose for a second-stage journey? Suppose the bus starts at A, then the possible routes are

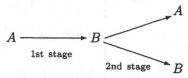

If the bus starts at B, then the possible routes are

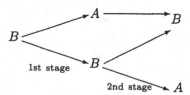

The situation can be described simply by using matrix notation such as

$$\begin{array}{c} A \\ B \end{array} \overset{A \ \ B}{\begin{bmatrix} 0 & 1 \\ 1 & 1 \end{bmatrix}} \begin{array}{c} A \\ B \end{array} \overset{A \ \ B}{\begin{bmatrix} 0 & 1 \\ 1 & 1 \end{bmatrix}} = \begin{array}{c} A \\ B \end{array} \overset{A \ \ B}{\begin{bmatrix} 1 & 1 \\ 1 & 2 \end{bmatrix}}$$

or just

$$\begin{bmatrix} 0 & 1 \\ 1 & 1 \end{bmatrix} \begin{bmatrix} 0 & 1 \\ 1 & 1 \end{bmatrix} = \begin{bmatrix} 1 & 1 \\ 1 & 2 \end{bmatrix}.$$

Do you see how entries $1, 1, 1, 2$ of the right matrix came out? Readers are urged to try more complicated examples from your daily livings.

For an $m \times n$ matrix $A = [a_{ij}]_{m \times n}$ and an $n \times p$ matrix $B = [b_{jk}]_{n \times p}$ the *product* AB of A following by B is an $m \times p$ matrix whose (i, k) entry is equal to the sum of the componentwise product of the corresponding entries of ith row of A and kth column of B, i.e.

$$AB = [c_{ik}]_{m \times p}, \quad \text{where } c_{ik} = \sum_{j=1}^{n} a_{ij} b_{jk} \underset{(\text{def.})}{=} A_{i*} B_{*k}.$$

Note that this *multiplication* operation is defined only if the number of columns of the left-handed matrix A is equal to the number of rows of the right-handed matrix B. BA is not defined except $p = m$. Multiplication has the following properties:

1. $(AB)C = A(BC)$;
2. $A(B + C) = AB + AC, (A + B)C = AC + BC$;
3. (Identity matrix) If $A \in \text{M}(m, n; \mathbb{F})$, then

$$I_m A = A I_n = A.$$

In case $m = n$, the vector space $\text{M}(n; \mathbb{F})$ with this operation of multiplication between pair of its elements is said to be an *associative algebra with identity* I_n.

The problems in the following exercises are designed for the reader to increase their abilities in the manipulation of matrices. Some of them will be discussed in the text.

Exercises

1. *Unusual properties of matrix multiplication* In $\text{M}(2; \mathbb{R})$ or even $\text{M}(m, n; \mathbb{F}), m, n \geq 2$, the followings happen (refer to (1) in (1.2.4) and Sec. A.3).

   (a) $AB \neq BA$. For example,

   $$A = \begin{bmatrix} 1 & 1 \\ 0 & 0 \end{bmatrix} \text{ and } B = \begin{bmatrix} 1 & 0 \\ 1 & 0 \end{bmatrix}, \quad \text{or}$$

   $$A = \begin{bmatrix} 1 & 1 \\ 0 & 0 \\ 1 & 0 \end{bmatrix} \text{ and } B = \begin{bmatrix} 1 & 0 & 0 \\ 1 & 0 & 1 \end{bmatrix}.$$

   (b) There exist $A \neq O$ and $B \neq O$ for which $AB = O$. For example,

   $$A = \begin{bmatrix} 1 & 0 \\ 0 & 0 \end{bmatrix} \text{ and } B = \begin{bmatrix} 0 & 0 \\ 0 & 1 \end{bmatrix}.$$

   In this case, $BA = O$ holds too.

(c) There exist $A \neq O$ and $B \neq O$ so that $AB = O$ but $BA \neq O$. For example,

$$A = \begin{bmatrix} 1 & 0 \\ 0 & 0 \end{bmatrix} \quad \text{and} \quad B = \begin{bmatrix} 0 & 0 \\ 1 & 1 \end{bmatrix}.$$

(d) There exist $A \neq O, B$ and C such that $AB = AC$ but $B \neq C$. For example,

$$A = \begin{bmatrix} 1 & 0 \\ 0 & 0 \end{bmatrix}, \quad B = \begin{bmatrix} 0 & 0 \\ 1 & 1 \end{bmatrix} \quad \text{and} \quad C = \begin{bmatrix} 0 & 0 \\ 0 & 1 \end{bmatrix}.$$

(e) $A^2 = O$ but $A \neq O$ where $A^2 = AA$. For example,

$$A = \begin{bmatrix} 0 & 1 \\ 0 & 0 \end{bmatrix}.$$

2. (continued) Do the following problems.

   (a) Construct 2×2 matrices $A \neq O, X$ and Y such that $AX = XA = AY = YA$ but $X \neq Y$.

   (b) Construct 2×2 matrices A and B with nonzero entries such that $AB = O$.

   (c) Construct 2×2 matrices $A \neq O$ and $B \neq O$ for which $A^2 + B^2 = O$.

   (d) Construct a 2×2 matrix A with nonzero entries such that $AB \neq I_2$ and $BA \neq I_2$ for any 2×2 matrix B.

3. Diagonal matrices.

   (a) Show that the following are equivalent: for a square matrix A of order n,

     (1) A *commutes* with all $n \times n$ matrices B, i.e. $AB = BA$.
     (2) A commutes with all $E_{ij}, 1 \leq i, j \leq n$.
     (3) A is a scalar matrix.

   (b) Any two diagonal matrices (of the same order) commute.

   (c) If a matrix A commutes with a diagonal matrix $\text{diag}[a_1, \ldots, a_n]$ where $a_i \neq a_j, i \neq j$, then A itself is diagonal.

   (d) Show that the set V of all diagonal matrices of order n forms a vector subspace of $M(n; \mathbb{F})$ with $E_{ii}, 1 \leq i \leq n$, as a basis. Try to find another subspace W so that

$$M(n; \mathbb{F}) = V \oplus W.$$

4. Fix a matrix $A \in M(n; \mathbb{F})$. The set of matrices that commute with A

$$V = \{Z \in M(n; \mathbb{F}) \mid AZ = ZA\}$$

is a subspace.

(a) Let $A = \begin{bmatrix} 1 & 0 \\ -1 & 0 \end{bmatrix}$. Show that

$$V = \left\{ \begin{bmatrix} a_{22} - a_{21} & 0 \\ a_{21} & a_{22} \end{bmatrix} \Big| \, a_{21}, a_{22} \in \mathbb{F} \right\}.$$

Find a basis for V. What is $\dim V$?

(b) Let

$$A = \begin{bmatrix} 0 & 1 & 0 \\ 0 & 0 & 1 \\ 0 & 0 & 0 \end{bmatrix}.$$

Determine the corresponding V and find a basis for V.

5. Let $A \in \mathrm{M}(n; \mathbb{F})$ and $p \geq 1$ be an integer. The pth *power* of A is

$$A^p = AA \cdots A \quad (p \text{ times})$$

and A^0 is defined to be the identity matrix I_n. Obviously,

$$A^p A^q = A^q A^p = A^{p+q}, \quad \text{and}$$
$$(A^p)^q = A^{pq}$$

hold for any nonnegative integers p, q. In the following, $A = [a_{ij}] \in \mathrm{M}(2; \mathbb{F})$.

(a) Find all such A so that $A^2 = O$.

(b) All A so that $A^3 = O$.

(c) All A so that $A^p = O$ where $p \geq 3$ is an integer.

6. A matrix $A \in \mathrm{M}(n; \mathbb{F})$ is said to be *idempotent* if

$$A^2 = A.$$

(a) Show that the matrices

$$\begin{bmatrix} 1 & 0 \\ 0 & 0 \end{bmatrix} \quad \text{and} \quad \begin{bmatrix} 1 & 2 & 2 \\ 0 & 0 & -1 \\ 0 & 0 & 1 \end{bmatrix}$$

are idempotent.

(b) Show that

$$A = \begin{bmatrix} 2 & -2 & 4 \\ -1 & 3 & 4 \\ 1 & -2 & -3 \end{bmatrix}, \quad B = \begin{bmatrix} -1 & 2 & 4 \\ 1 & -2 & -4 \\ 1 & 2 & -4 \end{bmatrix}$$

are idempotent and $AB = O$. What is BA?

(c) Determine all diagonal matrices in $M(n; \mathbb{R})$ that are idempotent.

(d) If A is idempotent and p is a positive integer, then $A^p = A$.

(e) Suppose that $AB = A$ and $BA = B$. Show that A and B are idempotent.

(f) Suppose A and B are idempotent. Show that $A + B$ is idempotent if and only if $AB + BA = O$.

(g) Determine all real 2×2 matrices which are idempotent.

7. A matrix $A \in M(n; \mathbb{F})$ is called *nilpotent* if there exists a positive integer p such that

$$A^p = O.$$

Then, for any integer $q \geq p$, $A^q = O$ also holds. Hence, the least positive integer p for which $A^p = O$ is referred to as the *degree or index of nilpotency* of A.

(a) Show that

$$\begin{bmatrix} 1 & 2 & 3 \\ 1 & 2 & 3 \\ -1 & -2 & -3 \end{bmatrix} ; \quad \begin{bmatrix} -4 & 4 & -4 \\ 1 & -1 & 1 \\ 5 & -5 & 5 \end{bmatrix} ; \quad \begin{bmatrix} 0 & 1 & 0 & 0 \\ 0 & 0 & 1 & 0 \\ 0 & 0 & 0 & 1 \\ 0 & 0 & 0 & 0 \end{bmatrix}$$

are nilpotent of respective degree $2, 3, 4$.

(b) A square matrix $A = [a_{ij}]_{n \times n}$, $a_{ij} = 0$ for $1 \leq j \leq i \leq n$, i.e. an upper triangular matrix having zeros on the main diagonal, is nilpotent.

(c) Suppose A is nilpotent of degree 2. Show that for any positive integer p, $A(I_n + A)^p = A$ holds.

(d) If A is a nilpotent $n \times n$ matrix and $A \neq O$, show that there exists a nonzero square matrix B so that $AB = O$. Try to prove the case that $n = 2$.

(e) Determine all real nilpotent 2×2 matrices of degree 2. Find some complex nilpotent 2×2 matrices.

8. Let $A = [a_{ij}] \in M(2; \mathbb{R})$. Solve the matrix equation

$$XA = O,$$

where $X \in M(2; \mathbb{R})$.

9. A matrix A is called *involutory* if

$$A^2 = I_n,$$

which is equivalent to $(I_n - A)(I_n + A) = O$.

(a) Show that

$$\begin{bmatrix} \cos\theta & \sin\theta \\ -\sin\theta & \cos\theta \end{bmatrix}, \quad \theta \in \mathbb{R}$$

is involutory.

(b) Show that

$$\begin{bmatrix} 3-4a & 2-4a & 2-4a \\ -1+2a & 2a & -1+2a \\ -3+2a & -3+2a & -2+2a \end{bmatrix}, \quad a \in \mathbb{C}$$

is involutory.

(c) Show that $\frac{1}{2}(I_n + A)$ is idempotent if A is involutory.

(d) Determine all real 2×2 involutory matrices and find some complex ones.

10. Find all real 2×2 matrices A for which $A^3 = I_2$. How complicated it might be if A is a complex matrix?

11. Find all real 2×2 matrices A such that $A^4 = I_2$.

12. *Upper and lower triangular matrices*

(a) A product of two upper (or lower) triangular matrices is upper (or lower) triangular.

(b) Show that

$$\begin{bmatrix} a_{11} & a_{12} \\ a_{21} & a_{22} \end{bmatrix} = \begin{bmatrix} 1 & 0 \\ b_{21} & 1 \end{bmatrix}\begin{bmatrix} c_{11} & c_{12} \\ 0 & c_{22} \end{bmatrix}.$$

Find out the relations among a_{ij}, b_{21} and c_{ij}.

(c) Show that the set of upper triangular matrices

$$V = \{A \in M(n; \mathbb{F}) \mid A \text{ is upper triangular}\}$$

is a vector subspace of $M(n; \mathbb{F})$. Find a basis for V and another subspace W of $M(n; \mathbb{F})$ such that

$$M(n; \mathbb{F}) = V \oplus W.$$

(d) Do (c) for lower triangular matrices.

13. *Transpose*

(a) Show that the operator $*: M(m, n; \mathbb{F}) \to M(n, m; \mathbb{F})$ defined by

$$*(A) = A^*$$

is a linear isomorphism (see Sec. B.7), i.e. $*$ is 1-1, onto and

$$(A + B)^* = A^* + B^*,$$
$$(\alpha A)^* = \alpha A^* \quad \text{for } \alpha \in \mathbb{F}.$$

(b) Show that

$$(A^*)^* = A,$$
$$(AB)^* = B^*A^*.$$

14. *Symmetric and skew-symmetric matrices*

(a) Let $A = [a_{ij}] \in M(n; \mathbb{R})$. Thus,

$$A = \frac{1}{2}(A + A^*) + \frac{1}{2}(A - A^*),$$

where $\frac{1}{2}(A + A^*)$ is symmetric while $\frac{1}{2}(A - A^*)$ is skew-symmetric.

(b) A is symmetric and skew-symmetric simultaneously if and only if $A = O$.

(c) Find symmetric matrices A and B for which AB is not symmetric. If A and B are symmetric such that $AB = BA$, then AB is symmetric. Is $AB = BA$ a necessary condition? Are there similar results for skew-symmetric matrices?

(d) Suppose A and B are symmetric matrices.

   (1) $A + B, AB + BA$ and ABA are symmetric but $AB - BA$ is skew-symmetric.
   (2) αA is symmetric for $\alpha \in \mathbb{R}$.
   (3) A^p is symmetric for any positive integer p.
   (4) PAP^* is symmetric for any $n \times n$ matrix P.
   (5) PAP^{-1} is symmetric for any orthogonal $n \times n$ matrix P (see Ex. 18 below).

(e) Show that $(1), (2), (4)$ and (5) in (d) still hold if "symmetric" is replaced by "skew-symmetric". How about ABA and (3)?

(f) Let

$$V_1 = \{A \in M(n; \mathbb{R}) \mid A = A^*\}, \quad \text{and}$$
$$V_2 = \{A \in M(n; \mathbb{R}) \mid A = -A^*\}.$$

Show that V_1 is a $\frac{1}{2}n(n+1)$-dimensional subspace of $M(n; \mathbb{R})$, while V_2 is $\frac{1}{2}n(n-1)$-dimensional. Find a basis for V_1 and a basis for V_2. Then, prove that

$$M(n; \mathbb{R}) = V_1 \oplus V_2.$$

(g) Let

$$V_3 = \{A = [a_{ij}] \in M(n; \mathbb{R}) \,|\, a_{ij} = 0 \text{ for } 1 \leq i \leq j \leq n\}$$

and V_1 be as in (f). Show that V_3 is a $\frac{1}{2}n(n-1)$-dimensional subspace of $M(n; \mathbb{R})$ and

$$M(n; \mathbb{R}) = V_1 \oplus V_3.$$

15. *Rank of a matrix* (see Sec. B.5 and Ex. 2 of Sec. B.7)

Let $A = [a_{ij}] \in M(m, n; \mathbb{F})$ and $A \neq O$. The maximal number of linear independent row (or column) vectors of A is defined as the *row* (or *column*) *rank* of A. It can be show that, for any matrix $A \neq O$,

row rank of $A = $ column rank of $A = \mathrm{r}(A)$.

This common number $\mathrm{r}(A)$ is called the *rank* of A. Define the rank of O to be zero. Therefore $0 \leq \mathrm{r}(A) \leq m, n$. Try to prove this result in case A is a 2×3 or 3×3 real matrix.

16. *Invertible matrix and its inverse matrix*

A matrix $A = [a_{ij}] \in M(n; \mathbb{F})$ is said to be *invertible* (refer to (2.4.2)) if there exists another $n \times n$ matrix B such that

$$AB = BA = I_n.$$

In this case, B is called the *inverse matrix* of A and is denoted by

$$A^{-1}.$$

(a) A^{-1} is unique if A is invertible, and $(A^{-1})^{-1} = A$.

(b) A is invertible if and only if A^* is invertible, and $(A^*)^{-1} = (A^{-1})^*$.

(c) If A and B are invertible, then AB is invertible and

$$(AB)^{-1} = B^{-1}A^{-1}.$$

(d) The following are equivalent: $A = [a_{ij}]_{n \times n}$

(1) A is invertible.

(2) The row rank of A is n.

(3) The column rank of A is n.

(4) The linear transformation $f \colon \mathbb{F}^n \to \mathbb{F}^n$ defined by

$$f(\vec{x}) = \vec{x} A$$

is a linear isomorphism (see Sec. B.7).

(5) The linear transformation $g \colon \mathbb{F}^n \to \mathbb{F}^n$ defined by

$$g(\vec{y}) = A\vec{y} \quad (\vec{y} \text{ is considered as a column vector})$$

is a linear isomorphism (see Sec. B.7).

(6) The *homogenous equation* $\vec{x}A = \vec{0}$ ($\vec{x} \in \mathbb{F}^n$) has zero solution $\vec{0}$ only. So is $A\vec{y} = \vec{0}$.

(7) There exists an $n \times n$ matrix B such that $AB = I_n$.

(8) There exists an $n \times n$ matrix B such that $BA = I_n$.

(9) The *matrix equation* $XA = O$ has zero solution $X = O$ only. So is $AY = O$.

(10) The matrix equation $XA = B$ always has a unique solution. So is $AY = B$.

(11) The determinant of A satisfies

$$\det A \neq 0$$

and thus $\det A^{-1} = (\det A)^{-1}$.

These results extend those stated in Exs. <A> 2 and <B> 4 of Sec. 2.4. Try to prove these in case A is a 2×2 or 3×3 real matrix.

(e) State equivalent conditions for an $n \times n$ matrix A to be not invertible which is called *singular*.

(f) Suppose A is invertible. For positive integer p, A^p is invertible and

$$(A^p)^{-1} = (A^{-1})^p.$$

Therefore, extend the power of an invertible matrix A to negative exponent:

$$A^{-p} = (A^{-1})^p, \quad p > 0.$$

(g) Suppose $A, B \in M(n; \mathbb{F})$ such that AB is invertible. Then A and B are invertible.

17. Suppose A and B are invertible. Find such A and B so that $A + B$ is invertible. If, in addition, $A + B$ is invertible, then show that $A^{-1} + B^{-1}$ is invertible and

$$(A^{-1} + B^{-1})^{-1} = A(A + B)^{-1}B = B(A + B)^{-1}A.$$

18. Let $A \in M(n; \mathbb{R})$ be skew-symmetric.

(a) Then $I_n - A$ is invertible. Try to prove this directly if $n = 2$. Is $aI_n + A$ invertible for all $a \neq 1$ in $\mathbb{R}$?

(b) The matrix

$$B = (I_n + A)(I_n - A)^{-1}$$

satisfies $BB^* = B^*B = I_n$, i.e. $B^* = B^{-1}$. A real matrix such as B is called an *orthogonal* matrix.

19. Let $A \in M(m, n; \mathbb{F})$ and $B \in M(n, m; \mathbb{F})$.

    (a) In case $m = 2$ and $n = 1$, show directly that AB is singular.
    (b) Suppose $m > n$. Then AB is singular.
    (c) If the rank $r(A) = m < n$, then AA^* is invertible while A^*A is singular.
    (d) If the rank $r(A) = n < m$, then AA^* is singular while A^*A is invertible.
    (e) $AA^* - aI_n$ is invertible for $a < 0$ in $\mathbb{R}$ if A is a real matrix.
    (f) $r(AA^*) = r(A^*A) = r(A)$.

20. Suppose $A \in M(n; \mathbb{F})$ and there exists a positive integer k such that $A^k = O$. Prove that $I_n - A$ is invertible and

$$(I_n - A)^{-1} = I_n + A + A^2 + \cdots + A^{k-1}.$$

21. (a) If symmetric matrix A is invertible, then A^{-1} is symmetric.
    (b) If skew-symmetric matrix A is invertible, then so is A^{-1}.

22. If $A = \text{diag}[a_1, a_2, \ldots, a_n]$ is a diagonal matrix, show that A is invertible if and only if its main diagonal entries $a_i \neq 0, 1 \leq i \leq n$. In this case,

$$A^{-1} = \text{diag}[a_1^{-1}, a_2^{-1}, \ldots, a_n^{-1}].$$

23. An upper (or lower) triangular matrix is invertible if and only if its main diagonal entries are nonzero.

24. Let $A = [a_{ij}]$, $B = [b_{ij}] \in M(2; \mathbb{R})$ or $M(2; \mathbb{C})$.

    (a) Compute AB and BA.
    (b) Show that $AB - BA = \begin{bmatrix} a & b \\ c & -a \end{bmatrix}$ for some $a, b, c \in \mathbb{R}$.
    (c) Prove that there does not exist any $\alpha \neq 0$ in $\mathbb{R}$ such that

$$AB - BA = \alpha I_2.$$

    (d) Show that $(AB - BA)^2$ is a scalar matrix by direct computation.
    (e) Calculate $(AB - BA)^n$ for all positive integers $n \geq 3$.

25. *Trace of a square matrix*
    The mapping $\text{tr}: M(n; \mathbb{F}) \to \mathbb{F}$ defined by, for $A = [a_{ij}]_{n \times n}$,

$$\text{tr} A = \sum_{i=1}^{n} a_{ii}$$

    i.e. the sum of entries on A's main diagonal, is called the *trace* of A.

(a) tr is a linear transformation (functional), i.e.

$$\text{tr}(A + B) = \text{tr}\,A + \text{tr}\,B;$$
$$\text{tr}(\alpha A) = \alpha \text{tr}\,A$$

for $A, B \in M(n; \mathbb{F})$ and $\alpha \in \mathbb{F}$.

(b) $\text{tr}(AB) = \text{tr}(BA)$. In fact, if $A = [a_{ij}]_{n \times n}$ and $B = [b_{ij}]_{n \times n}$, then

$$\text{tr}(AB) = \sum_{i,j=1}^{n} a_{ij} b_{ji}.$$

(c) If P is invertible, then $\text{tr}(PAP^{-1}) = \text{tr}\,A$.

(d) $\text{tr}\,A = \text{tr}\,A^*$.

26. In $M(n; \mathbb{R})$, define

$$\langle A, B \rangle = \text{tr}\,AB^*.$$

Then $\langle\,,\,\rangle$ has the following properties:

(1) $\langle A, A \rangle \geq 0$ with equality if and only if $A = O$.

(2) $\langle A, B \rangle = \langle B, A \rangle$.

(3) $\langle \alpha_1 A_1 + \alpha_2 A_2, B \rangle = \alpha_1 \langle A_1, B \rangle + \alpha_2 \langle A_2, B \rangle$ for $\alpha_1, \alpha_2 \in \mathbb{R}$.

It is said that $\langle\,,\,\rangle$ defines an *inner product* on the vector space $M(n; \mathbb{R})$ (see Sec. B.9).

(*Note* In the *complex* vector space $M(m, n; \mathbb{C})$, define

$$\langle A, B \rangle = \text{tr}\,A\bar{B}^*$$

where $\bar{B}^* = [\bar{b}_{ji}]$ is the *conjugate transpose* of $B = [b_{ij}]$. Then,

(1) $\langle A, A \rangle \geq 0$ and $= 0 \Leftrightarrow A = O$.

(2) $\langle A, B \rangle = \overline{\langle B, A \rangle}$.

(3) $\langle \alpha_1 A_1 + \alpha_2 A_2, B \rangle = \alpha_1 \langle A_1, B \rangle + \alpha_2 \langle A_2, B \rangle$.

And $\langle\,,\,\rangle$ is said to define an *inner product* on $M(m, n; \mathbb{C})$.)

27. *Miscellanea about trace* Do the following problems, at least for $n = 2$ or $n = 3$.

(a) If $A, B \in M(n; \mathbb{F})$ and A is idempotent, then $\text{tr}(AB) = \text{tr}(ABA)$.

(b) If $\text{tr}(AB) = 0$ for all matrices $B \in M(n; \mathbb{F})$, then $A = O$.

(c) Suppose both $\text{tr}\,A = 0$ and $\text{tr}\,A^2 = 0$ for a 2×2 matrix A. Then, $A^2 = O$ holds.

(d) If $A^2 = O$ for $A \in M(2; \mathbb{F})$, then $\text{tr}\,A = 0$.

(e) If $A \in M(2; \mathbb{R})$ is idempotent, then $\text{tr}\,A$ is an integer and $0, 1, 2$ are the only possibilities. How about idempotent matrix $A \in M(n; \mathbb{R})$?

(f) If $A \in M(2; \mathbb{R})$ is symmetric and nilpotent, then $A = O$.

(g) If $A_1, A_2, \ldots, A_k$ are real $n \times n$ symmetric matrices and $\text{tr}\left(\sum_{i=1}^{k} A_i^2\right) = 0$, then, $A_1 = A_2 = \cdots = A_k = O$.

(h) If $\text{tr}(ABC) = \text{tr}(CBA)$ for all C in $M(n; \mathbb{F})$, then $AB = BA$.

28. Suppose $A \in M(2; \mathbb{R})$ and $\text{tr}\, A = 0$. Show that there exists an invertible matrix P such that

$$PAP^{-1} = \begin{bmatrix} 0 & \alpha \\ \beta & 0 \end{bmatrix}$$

for some $\alpha, \beta \in \mathbb{R}$. Are α, β and P unique? Justify this result if

$$A = \begin{bmatrix} 1 & 2 \\ 3 & -1 \end{bmatrix}$$

and try to explain it geometrically if possible.

29. Let

$$V = \{A \in M(n; \mathbb{F}) \mid \text{tr}\, A = 0\}.$$

Then V is an $(n^2 - 1)$-dimensional vector subspace of $M(n; \mathbb{F})$ and $\{-E_{11} + E_{ii} \mid 2 \leq i \leq n\} \cup \{E_{ij} \mid 1 \leq i, j \leq n, \text{ but } i \neq j\}$ forms a basis for V. Also

$$M(n; \mathbb{F}) = V \oplus \langle\langle E_{11} \rangle\rangle.$$

30. In $M(2; \mathbb{R})$, let

$$W = \{AB - BA \mid A, B \in M(2; \mathbb{R})\}.$$

(a) Show that W is a subspace of V mentioned in Ex. 29.

(b) Let $B = [b_{ij}]_{2 \times 2}$. Shows that

$$E_{11}B - BE_{11} = b_{12}E_{12} - b_{21}E_{21},$$
$$E_{12}B - BE_{12} = b_{21}(E_{11} - E_{22}) + (b_{22} - b_{11})E_{12},$$
$$E_{21}B - BE_{21} = b_{12}(-E_{11} + E_{22}) + (b_{11} - b_{22})E_{21},$$
$$E_{22}B - BE_{22} = -b_{12}E_{12} + b_{21}E_{21}$$

and hence shows that

$$\{E_{11} - E_{22}, E_{12}, E_{21}\}$$

forms a basis for W.

(c) Therefore, $W = V$ holds. This means that, for any $C \in M(2; \mathbb{R})$ with $\text{tr}\, C = 0$, there exists $A, B \in M(2; \mathbb{R})$ such that

$$C = AB - BA.$$

$\Bigg($ *Note* For any field $\mathbb{F}$ and $n \geq 2$,

$$\{A \in M(n; \mathbb{F}) \mid \operatorname{tr} A = 0\}$$
$$= \{AB - BA \mid A, B \in M(n; \mathbb{F})\}$$

still holds as subspaces of $M(n; \mathbb{F})$. In case $\mathbb{F}$ is a field of characteristic 0 (i.e. $1 + 1 + \cdots + 1 \neq 0$ for any finite number of 1) such as $\mathbb{R}$ and $\mathbb{C}$, it is not possible to find matrices $A, B \in M(n; \mathbb{F})$ such that

$$AB - BA = I_n,$$

(refer to Ex. 25). For field $\mathbb{F}$ of characteristic p (i.e. $1 + 1 + \cdots + 1 = 0$ for p's 1) such as $I_p = \{1, 2, \ldots, p - 1\}$ where p is a prime, this does happen. For example, in $I_3 = \{0, 1, 2\}$, let

$$A = \begin{bmatrix} 0 & 1 & 0 \\ 0 & 0 & 1 \\ 0 & 0 & 0 \end{bmatrix} \quad \text{and} \quad B = \begin{bmatrix} 0 & 0 & 0 \\ 1 & 0 & 0 \\ 0 & 2 & 0 \end{bmatrix}.$$

Then

$$AB - BA = \begin{bmatrix} 1 & 0 & 0 \\ 0 & 2 & 0 \\ 0 & 0 & 0 \end{bmatrix} - \begin{bmatrix} 0 & 0 & 0 \\ 0 & 1 & 0 \\ 0 & 0 & 2 \end{bmatrix} = \begin{bmatrix} 1 & 0 & 0 \\ 0 & 1 & 0 \\ 0 & 0 & -2 \end{bmatrix} = I_3. \Bigg)$$

31. *Similarity*

Two square matrices $A, B \in M(n; \mathbb{F})$ are said to be *similar* if there exists an invertible matrix $P \in M(n; \mathbb{F})$ such that

$$B = PAP^{-1} \quad (\text{or } A = P^{-1}BP)$$

(refer to (2.7.25)). Use $A \sim B$ to denote that A is similar to B.

(a) Similarity is an *equivalent relation* (see Sec. A.1) among matrices of the same order. That is,

 (1) $A \sim A$.
 (2) $A \sim B \Rightarrow B \sim A$.
 (3) $A \sim B$ and $B \sim C \Rightarrow A \sim C$.

(b) Similarity provides a useful tool to study the *geometric* behavior of linear or affine transformations by suitable choices of bases (see Secs. 2.7.6, 3.7.6, etc.). *Algebraically*, similarity has many advantages in computation. For example,

(1) Orthogonal similarity keeps symmetric and skew-symmetric properties of matrices. (See Ex. 6(c) (4)–(6) of Sec. B.9.)

(2) Similarity preserves the determinants of matrices, i.e.

$$\det A = \det B.$$

(3) Similarity raises the power of a matrix easier, i.e.

$$B^n = PA^n P^{-1},$$

where one of A^n and B^n such as diagonal matrices can be easily computed.

(4) If $B = PAP^{-1}$ is a scalar matrix, then so is A and $AB = BA$ holds.

(c) *Fibonacci sequence* $a_0, a_1, a_2, \ldots, a_n, \ldots$ is defined by a

(1) recursive equation: $a_n = a_{n-2} + a_{n-1}, n \geq 2$ with a

(2) boundary condition: $a_0 = a_1 = 1$.

How to determine explicitly the general term a_n which is expected to be a positive integer? With matrix form,

$$(a_{n-1} \quad a_n) = (a_{n-2} \quad a_{n-1}) \begin{bmatrix} 0 & 1 \\ 1 & 1 \end{bmatrix} = (a_{n-3} \quad a_{n-2}) \begin{bmatrix} 0 & 1 \\ 1 & 1 \end{bmatrix}^2$$

$$= \cdots = (a_0 \quad a_1) \begin{bmatrix} 0 & 1 \\ 1 & 1 \end{bmatrix}^{n-1}, \quad n \geq 2.$$

The problem reduces to compute A^n where

$$A = \begin{bmatrix} 0 & 1 \\ 1 & 1 \end{bmatrix}.$$

Does A^n seem to be easily computed? If yes, try it. Fortunately, it can be shown (see Ex. 8 of Sec. B.9 and Sec. B.11) as follows:

$$A = \begin{bmatrix} 1 & \frac{1+\sqrt{5}}{2} \\ 1 & \frac{1-\sqrt{5}}{2} \end{bmatrix}^{-1} \begin{bmatrix} \frac{1+\sqrt{5}}{2} & 0 \\ 0 & \frac{1-\sqrt{5}}{2} \end{bmatrix} \begin{bmatrix} 1 & \frac{1+\sqrt{5}}{2} \\ 1 & \frac{1-\sqrt{5}}{2} \end{bmatrix}$$

$$\Rightarrow A^n = \begin{bmatrix} 1 & \frac{1+\sqrt{5}}{2} \\ 1 & \frac{1-\sqrt{5}}{2} \end{bmatrix}^{-1} \begin{bmatrix} \left(\frac{1+\sqrt{5}}{2}\right)^n & 0 \\ 0 & \left(\frac{1-\sqrt{5}}{2}\right)^n \end{bmatrix} \begin{bmatrix} 1 & \frac{1+\sqrt{5}}{2} \\ 1 & \frac{1-\sqrt{5}}{2} \end{bmatrix}$$

By simple computation, the general term is

$$a_n = \frac{1}{\sqrt{5}} \left\{ \left(\frac{1+\sqrt{5}}{2}\right)^{n+1} - \left(\frac{1-\sqrt{5}}{2}\right)^{n+1} \right\}, \quad n \geq 0.$$

Is this really a positive integer?

(d) Not every square matrix can be similar to a diagonal matrix. For example, there does not exist an invertible matrix P such that

$$P \begin{bmatrix} 1 & 0 \\ 1 & 1 \end{bmatrix} P^{-1}$$

is a diagonal matrix. Why? Is

$$\begin{bmatrix} 1 & 0 \\ 1 & 1 \end{bmatrix}^n = \begin{bmatrix} 1 & 0 \\ n & 1 \end{bmatrix}, \quad n \geq 1$$

correct?

32. *Necessary condition for a matrix to be similar to a diagonal matrix*
Let $A = [a_{ij}] \in M(n; \mathbb{F})$ and P be an invertible matrix so that

$$PAP^{-1} = \begin{bmatrix} \lambda_1 & & 0 \\ & \ddots & \\ 0 & & \lambda_n \end{bmatrix}.$$

Fix $i, 1 \leq i \leq n$, then

$$PAP^{-1} - \lambda_i I_n = PAP^{-1} - P(\lambda_i I_n)P^{-1} = P(A - \lambda_i I_n)P^{-1}$$

$$= \begin{bmatrix} \lambda_1 - \lambda_i & & & & & & 0 \\ & \ddots & & & & & \\ & & \lambda_{i-1} - \lambda_i & & & & \\ & & & 0 & & & \\ & & & & \lambda_{i+1} - \lambda_i & & \\ & & & & & \ddots & \\ 0 & & & & & & \lambda_n - \lambda_i \end{bmatrix}$$

$\Rightarrow \det(A - \lambda_i I_n) = 0, 1 \leq i \leq n.$

This means that the entries $\lambda_i, 1 \leq i \leq n$, of the diagonal matrix $\operatorname{diag}[\lambda_1, \ldots, \lambda_n]$ are zeros of the polynomial

$$\det(A - tI_n) = \begin{vmatrix} a_{11} - t & a_{12} & \cdots & a_{1n} \\ a_{21} & a_{22} - t & \cdots & a_{2n} \\ \vdots & \vdots & & \vdots \\ a_{n1} & a_{n2} & \cdots & a_{nn} - t \end{vmatrix}$$

$$= (-1)^n t^n + \alpha_{n-1} t^{n-1} + \cdots + \alpha_1 t + \alpha_0$$

which is called the *characteristic polynomial* of A. Note that, if $\det(A - \lambda I_n) = 0$, then the homogeneous equation $\vec{x}(A - \lambda I_n) = 0$, i.e.

$$\vec{x}A = \lambda\vec{x},$$

has *nonzero* solutions (see Ex. 16(d) (11)). Such a nonzero $\vec{x}$ is called an *eigenvector* corresponding to the *eigenvalue* λ. In particular, if $\vec{x}_i$ is the ith *row vector* of P, then $\vec{x}_i \neq \vec{0}$ and

$$\vec{x}_i A = \lambda_i \vec{x}_i, \quad 1 \le i \le n.$$

(a) Justify the above by the example

$$\begin{bmatrix} 3 & 2 \\ 2 & -3 \end{bmatrix} \begin{bmatrix} 5 & 12 \\ 12 & -5 \end{bmatrix} \begin{bmatrix} 3 & 2 \\ 2 & -3 \end{bmatrix}^{-1} = \begin{bmatrix} 13 & 0 \\ 0 & -13 \end{bmatrix}.$$

(b) Suppose that there exists an invertible matrix $P = [p_{ij}]$ such that

$$P \begin{bmatrix} 1 & 2 \\ 3 & 4 \end{bmatrix} P^{-1} = \begin{bmatrix} \frac{5+\sqrt{33}}{2} & 0 \\ 0 & \frac{5-\sqrt{33}}{2} \end{bmatrix}.$$

Determine $p_{ij}, 1 \le i, j \le 2$. Is such a P unique? If yes, why? If no, how many of them are there?

(c) Suppose

$$\begin{bmatrix} 2 & 7 \\ 0 & 1 \end{bmatrix} \begin{bmatrix} a_{11} & a_{12} \\ a_{21} & a_{22} \end{bmatrix} \begin{bmatrix} 2 & 7 \\ 0 & 1 \end{bmatrix}^{-1} = \begin{bmatrix} 1 & 0 \\ 0 & -1 \end{bmatrix}.$$

Determine $a_{ij}, 1 \le i, j \le 2$.

(d) Can you say more precisely why $\begin{bmatrix} 1 & 0 \\ 1 & 1 \end{bmatrix}$ is not similar to a diagonal matrix than you did in Ex. 31(d)?

B.5   Elementary Matrix Operations and Row-Reduced Echelon Matrices

Based on the essence of *elimination method* of variables in solving systems of linear equations, we define three types of *elementary row* (or *column*) operations on matrices as follows:

Type 1: Interchanging any two rows (or columns) of a matrix.

Type 2: Multiplying any one row (or column) of a matrix by a *nonzero* constant.

Type 3: Adding any constant multiple of a row (or column) of a matrix to another row (or column).

An $n \times n$ *elementary matrix* of type 1, 2 or 3 is a matrix obtained by performing an elementary operation on I_n, of type 1, 2 or 3 respectively. We adopt the following notations for elementary matrices:

Type 1: $E_{(i)(j)}$ (interchanging ith row and jth row, $i \neq j$);
Type 2: $E_{\alpha(i)}$ (multiplying ith row by $\alpha \neq 0$);
Type 3: $E_{(j)+\alpha(i)}$ (adding α multiple of ith row to jth row);

and $F_{(i)(j)}, F_{\alpha(i)}, F_{(j)+\alpha(i)}$ for corresponding elementary matrices obtained by perform elementary column operations. For example, 2×2 elementary matrices are:

$$E_{(i)(j)} = \begin{bmatrix} 0 & 1 \\ 1 & 0 \end{bmatrix}; \quad E_{\alpha(1)} = \begin{bmatrix} \alpha & 0 \\ 0 & 1 \end{bmatrix}; \quad E_{\alpha(2)} = \begin{bmatrix} 1 & 0 \\ 0 & \alpha \end{bmatrix};$$

$$E_{(2)+\alpha(1)} = \begin{bmatrix} 1 & 0 \\ \alpha & 1 \end{bmatrix}; \quad E_{(1)+\alpha(2)} = \begin{bmatrix} 1 & \alpha \\ 0 & 1 \end{bmatrix}.$$

Some 3×3 elementary matrices are:

$$E_{(1)(3)} = \begin{bmatrix} 0 & 0 & 1 \\ 0 & 1 & 0 \\ 1 & 0 & 0 \end{bmatrix}; \quad E_{2(\alpha)} = \begin{bmatrix} 1 & 0 & 0 \\ 0 & \alpha & 0 \\ 0 & 0 & 1 \end{bmatrix};$$

$$E_{(3)+\alpha(1)} = \begin{bmatrix} 1 & 0 & 0 \\ 0 & 1 & 0 \\ \alpha & 0 & 1 \end{bmatrix} = F_{(1)+\alpha(3)}; \quad E_{(1)+\alpha(2)} = \begin{bmatrix} 1 & \alpha & 0 \\ 0 & 1 & 0 \\ 0 & 0 & 1 \end{bmatrix} = F_{(2)+\alpha(1)}.$$

The following are some basic properties of elementary matrices:

1. $E_{(i)(j)} = F_{(i)(j)}; E_{\alpha(i)} = F_{\alpha(i)}; E_{(j)+\alpha(i)} = F_{(i)+\alpha(j)}.$
2. The determinants (see Sec. B.6) are $\det E_{(i)(j)} = -1, \det E_{\alpha(i)} = \alpha$; $\det E_{(j)+\alpha(i)} = 1.$
3. Elementary matrices are invertible. In particular,

$$E_{(i)(j)}^{-1} = E_{(j)(i)}; \quad E_{\alpha(i)}^{-1} = E_{\frac{1}{\alpha}(i)}; \quad E_{(j)+\alpha(i)}^{-1} = E_{(j)-\alpha(i)}.$$

4. The matrix obtained by performing an elementary row operation on a given matrix A, of the respective type $E_{(i)(j)}, E_{\alpha(i)}$ or $E_{(j)+\alpha(i)}$, is

$$E_{(i)(j)}A, \quad E_{\alpha(i)}A \quad \text{or} \quad E_{(j)+\alpha(i)}A,$$

respectively.

Echelon and row-reduced echelon matrices
Give a matrix $A = [a_{ij}] \in M(m, n; \mathbb{F})$.

The *first nonzero* entry of a row is called its *leading entry*.
A matrix is called an *echelon matrix* if

1. the leading entries move to the right in successive rows,
2. the entries of the column passing a leading entry are all zero below that leading entry, and
3. all zero rows, if any, are at the bottom of the matrix.

For example, the following are echelon matrices:

$$
\begin{bmatrix} 0 & -1 & 3 & 1 \\ 0 & 0 & 2 & 1 \\ 0 & 0 & 0 & 1 \end{bmatrix}, \quad
\begin{bmatrix} 2 & 0 & 1 & 5 & 0 \\ 0 & -1 & 3 & 4 & 0 \\ 0 & 0 & 0 & -2 & 1 \\ 0 & 0 & 0 & 0 & 4 \\ 0 & 0 & 0 & 0 & 0 \end{bmatrix}.
$$

An echelon matrix is called *row-reduced* if

1. Every leading entry is 1, and
2. the entries of the column passing a leading entry are all zero above that leading entry.

For examples,

$$
\begin{bmatrix} 0 & 1 & -3 & 0 & \frac{1}{2} \\ 0 & 0 & 0 & 1 & 2 \\ 0 & 0 & 0 & 0 & 0 \end{bmatrix}, \quad
\begin{bmatrix} 0 & 0 & 1 & -11 \\ 1 & 0 & 0 & 17 \\ 0 & 1 & 0 & -5 \end{bmatrix}
$$

the former is a row-reduced echelon matrix, while the latter is not.

The leading entry of a certain row of a matrix is called a *pivot* if there is no leading entry above it in the same column. A column of an echelon matrix A in which a pivot appears is called a *pivot column* and the corresponding variable in the equation $A\vec{x}^* = \vec{0}^*$, where $\vec{x} \in \mathbb{F}^n$, a *pivot variable*, while all remaining variables are called *free variables*.

Give a nonzero matrix $A \in M(m, n; \mathbb{F})$. After performing a finite number of elementary row operations on A, A can be reduced to a matrix of the following type: there exists a unique positive integer r for $1 \le r \le m, n$ and a sequence of positive integers $1 \le k_1 < k_2 < \cdots < k_r \le n$ such that

$$
R = \begin{bmatrix}
1 * \cdots * & 0 * \cdots * & 0 * \cdots \cdots * & 0 * \cdots \cdots * \\
0 \cdots \cdots 0 & 1 * \cdots * & 0 * \cdots \cdots * & 0 * \cdots \cdots * \\
\vdots & 0 \cdots \cdots 0 & 1 * \cdots \cdots * & 0 * \cdots \cdots * \\
& & 0 * \cdots \cdots * & \vdots \\
0 \; \vdots & \vdots & \vdots & 0 * \cdots \cdots * \\
\vdots & \vdots & \vdots 0 \cdots \cdots 0 & 1 * \cdots \cdots * \\
\vdots & \vdots & \vdots \vdots & 0 0 \cdots \cdots 0 \\
0 \cdots \cdots \cdots 0 & \cdots \cdots 0 0 & \cdots \cdots 0 0 & \cdots 0 0 \cdots \cdots 0
\end{bmatrix}_{m \times n}
\quad \leftarrow r\text{th row}
$$

$$(\ast)$$

where $\ast$ could be any scalars and zeros elsewhere. Such a R is unique once A is given, and is called the *row-reduced echelon matrix* of A.

The following are some basic results about row-reduced echelon matrices:

1. There exists an invertible $m \times m$ matrix P which is a product of finitely many elementary matrices such that

$$PA = R$$

is the row-reduced echelon matrix.

2. The first r rows $R_{1*}, \ldots, R_{r*}$ of R are linearly independent. The k_1th column $R_{*k_1}, \ldots, k_r$th column R_{*k_r} are linearly independent. Therefore, owing to the invertibility of P, the k_1th column vector $A_{*k_1}, \ldots, k_r$th column vector A_{*k_r} are linearly independent. Since R has rank r, so

$$r(A) = r(R) = r.$$

3. Suppose $k \ne k_1, \ldots, k_r$ and $1 \le k \le n$. If the kth column vector R_{*k} of R is the linear combination $R_{*k} = \alpha_1 R_{*k_1} + \cdots + \alpha_r R_{*k_r}$, then the kth column vector A_{*k} has the linear combination

$$A_{*k} = \alpha_1 A_{*k_1} + \cdots + \alpha_r A_{*k_r}.$$

This is true because $PA_{*j} = R_{*j}$ for $1 \le j \le n$. Let $P = [p_{ij}]_{m \times m}$. Then, for $r+1 \le i \le m, P_{i*}A = \sum_{j=1}^{m} p_{ij} A_{j*} = R_{i*} = \vec{0}$ holds.

4. *Normal form of a matrix* Suppose $A_{m \times n} \neq O$. Then, the exist invertible matrix $P_{m \times m}$ and invertible matrix $Q_{n \times n}$ (which is a product of finitely many elementary $n \times n$ matrices) such that

$$PAQ = \begin{bmatrix} I_r & 0 \\ 0 & 0 \end{bmatrix},$$

where $r = \mathrm{r}(A)$.

Readers should try your best to give proofs of the above-mentioned results, at least for 2×2 or 3×3 or 2×3 matrices.

Applications of row-reduced echelon matrix (a few supplement)

Application 1 *To solve system of homogenous linear equations*
For $A \in M(m, n; \mathbb{F})$ and $\vec{x} \in \mathbb{F}^n$, considered as a *column* vector (here and Application 2 only), the matrix equation

$$A\vec{x} = \vec{0}, \quad \text{where } A = [a_{ij}]_{m \times n} \text{ and } \vec{x} = \begin{bmatrix} x_1 \\ \vdots \\ x_n \end{bmatrix},$$

represents a system of homogenous m linear equations in n unknowns $x_1, \ldots, x_n$. Let $PA = R$ be row-reduced. Since P is invertible, $A\vec{x} = \vec{0}$ is *consistent* with $R\vec{x} = \vec{0}$, i.e. both have the same set of solutions. Let $r = \mathrm{r}(A) \geq 1$.

1. Let $y_1, \ldots, y_{n-r}$, be these unknowns among $x_1, \ldots, x_n$ but other than $x_{k_1}, \ldots, x_{k_r}$. Note that $y_1, \ldots, y_{n-r}$ are free variables.
2. Write out $R\vec{x} = \vec{0}$ as

$$x_{k_i} = \sum_{j=1}^{n-r} b_{ij} y_j, \quad 1 \leq i \leq r.$$

3. For each $j, 1 \leq j \leq n - r$, let $y_j = 1$ but $y_k = 0$ if $k \neq j, 1 \leq k \leq n - r$. The resulted *fundamental solution* is

$$\vec{v}_j = (0, \ldots, 0, b_{1j}, 0, \ldots, 0, b_{2j}, 0, \ldots, 0, b_{rj}, 0, \ldots, 0).$$
$$\uparrow \qquad\qquad \uparrow \qquad\qquad \uparrow$$
$$k_1 \text{th} \qquad\quad k_2 \text{th} \qquad\quad k_r \text{th}$$

4. The vectors $\vec{v}_1, \ldots, \vec{v}_{n-r}$ are linearly independent. The *general solution* is a linear combination of $\vec{v}_1, \ldots, \vec{v}_{n-r}$ such as

$$\vec{v} = \sum_{j=1}^{n-r} y_j \vec{v}_j, y_1, \ldots, y_{n-r} \in \mathbb{F}.$$

5. Therefore, the *solution space*

$$V = \{\vec{x} \in \mathbb{F}^n \mid A\vec{x} = \vec{0}\}$$

is an $(n - r(A))$-dimensional subspace of $\mathbb{F}^n$.

Application 2 *To solve system of non-homogenous linear equations*
For a given *column* vector $\vec{b} \in \mathbb{F}^m$, the matrix equation

$$A\vec{x} = \vec{b}$$

is a system of non-homogenous m linear equations in n unknowns $x_1, \ldots, x_n$. $A\vec{x} = \vec{b}$ is consistent with $R\vec{x} = P\vec{b}$. Suppose $r(A) = r$.

1. Perform P on the augmented $m \times (n+1)$ matrix $[A \mid \vec{b}]$ to obtain

$$P[A \mid \vec{b}] = [PA \mid P\vec{b}] = [R \mid P\vec{b}].$$

2. Thus,

$A\vec{x} = \vec{b}$ has a solution.

$\Leftrightarrow r(A) = r([A \mid \vec{b}])$

$\Leftrightarrow$ The last $(m - r)$ components $P_{(r+1)*}\vec{b}, \ldots, P_{m*}\vec{b}$ of the column
vector $P\vec{b}$ are all equal to zero.

3. In case having a solution, write $R\vec{x} = P\vec{b}$ as

$$x_{k_i} = P_{i*}\vec{b} + \sum_{j=1}^{n-r} b_{ij}y_j, \quad 1 \leq i \leq r$$

4. Letting $y_1 = \cdots = y_{n-r} = 0$, a *particular solution* is

$$\vec{v}_0 = (0, \ldots, 0, P_{1*}\vec{b}, 0, \ldots, 0, P_{2*}\vec{b}, 0, \ldots, 0, P_{r*}\vec{b}, 0, \ldots, 0)$$
$$\quad\quad\quad\quad \uparrow \quad\quad\quad\quad\quad \uparrow \quad\quad\quad\quad \uparrow$$
$$\quad\quad\quad k_1\text{th} \quad\quad\quad\quad k_2\text{th} \quad\quad\quad k_r\text{th}$$

5. The *general solution* is

$$\vec{v} = \vec{v}_0 + \sum_{j=1}^{n-r} y_j\vec{v}_j, \quad y_1, \ldots, y_{n-r} \in \mathbb{F}.$$

6. Therefore, the set of solutions

$$\{\vec{x} \in \mathbb{F}^m \mid \vec{x}A = \vec{b}\}$$

is the $(n - r(A))$-dimensional *affine subspace* of $\mathbb{F}^n$

$$\vec{v}_0 + \{\vec{x} \in \mathbb{F}^n \mid A\vec{x} = \vec{0}\} = \vec{v}_0 + V.$$

See Fig. B.2.

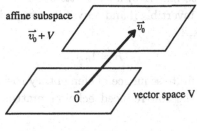

Fig. B.2

For example, let

$$A = \begin{bmatrix} 2 & 3 & 1 & 4 & -9 \\ 1 & 1 & 1 & 1 & -3 \\ 1 & 1 & 1 & 2 & -5 \\ 2 & 2 & 2 & 3 & -8 \end{bmatrix}.$$

Then A has row-reduced echelon matrix

$$R = \begin{bmatrix} 1 & 0 & 2 & 0 & -2 \\ 0 & 1 & -1 & 0 & 1 \\ 0 & 0 & 0 & 1 & -2 \\ 0 & 0 & 0 & 0 & 0 \end{bmatrix}.$$

Do the following problems:

1. Find invertible matrix P, expressed as a product of finitely many 4×4 elementary matrices, so that $PA = R$.
2. Find invertible matrix Q, expressed as a product of finitely many 5×5 elementary matrices, so that

$$PAQ = \begin{bmatrix} I_3 & 0 \\ 0 & 0 \end{bmatrix}.$$

3. Solve $A\vec{x} = \vec{0}$.
4. Find necessary and sufficient condition for $A\vec{x} = \vec{b}$ to have a solution.
5. In case $\vec{b} = \begin{bmatrix} 17 \\ 6 \\ 8 \\ 14 \end{bmatrix}$, show that $A\vec{x} = \vec{b}$ has a solution. Then, find a particular solution of it and write out the solution set as an affine subspace of $\mathbb{R}^5$.

Application 3 *To determine the rank of a nonzero matrix*

Application 4 *To determine the invertibility of a square matrix and how to compute the inverse matrix of an invertible matrix*
A square matrix A is invertible if and only if its row-reduced echelon matrix is the identity matrix, i.e.

$$R = I_n.$$

Thus, just perform a finite sequence of elementary row operations to $n \times 2n$ matrix $[A \,|\, I_n]$ until the row-reduced echelon matrix R of A comes out, such as

$$[A \,|\, I_n] \xrightarrow{E_1} E_1[A \,|\, I_n]$$
$$= [E_1 A_n \,|\, E_1] \xrightarrow{E_2} E_2[E_1 A \,|\, E_1]$$
$$= [E_2 E_1 A \,|\, E_2 E_1] \xrightarrow{E_3} \cdots \xrightarrow{E_k} E_k[E_{k-1} \cdots E_2 E_1 A \,|\, E_{k-1} \cdots E_2 E_1]$$
$$= [E_k E_{k-1} \cdots E_2 E_1 A \,|\, E_k E_{k-1} \cdots E_2 E_1] = [PA \,|\, P] = [R \,|\, P],$$

where $P = E_k E_{k-1} \cdots E_2 E_1$.

If $R \neq I_n$, then A is definitely not invertible. In case $R = I_n$, then A is invertible and, at the same time,

$$A^{-1} = P = E_k E_{k-1} \cdots E_2 E_1.$$

Application 5 *To express an invertible matrix as a product of finitely many elementary matrices*
See Application 4, then

$$A = E_1^{-1} E_2^{-1} \cdots E_{k-1}^{-1} E_k^{-1}.$$

Application 6 *To compute the determinant of an invertible matrix*
See Application 5, then

$$\det A = \det E_1^{-1} \cdot \det E_2^{-1} \cdots \det E_{k-1}^{-1} \cdot \det E_k^{-1}.$$

For example,

$$A = \begin{bmatrix} 1 & 1 & 1 & -3 \\ 0 & 1 & 0 & 0 \\ 1 & 1 & 2 & -3 \\ 2 & 2 & 4 & -5 \end{bmatrix} = \begin{bmatrix} 1 & 0 & 0 & 0 \\ 0 & 1 & 0 & 0 \\ 0 & 0 & 1 & 0 \\ 0 & 0 & 2 & 1 \end{bmatrix} \cdot \begin{bmatrix} 1 & 0 & 0 & 0 \\ 0 & 1 & 0 & 0 \\ 1 & 0 & 1 & 0 \\ 0 & 0 & 0 & 1 \end{bmatrix} \cdot$$

$$\begin{bmatrix} 1 & 1 & 0 & 0 \\ 0 & 1 & 0 & 0 \\ 0 & 0 & 1 & 0 \\ 0 & 0 & 0 & 1 \end{bmatrix} \cdot \begin{bmatrix} 1 & 0 & 1 & 0 \\ 0 & 1 & 0 & 0 \\ 0 & 0 & 1 & 0 \\ 0 & 0 & 0 & 1 \end{bmatrix} \cdot \begin{bmatrix} 1 & 0 & 0 & -3 \\ 0 & 1 & 0 & 0 \\ 0 & 0 & 1 & 0 \\ 0 & 0 & 0 & 1 \end{bmatrix}.$$

and $\det A = 1$.

Application 7 *See Sec. 5.9.2 and Ex. 2(f) of Sec. B.7.*

Application 8 *To compute the row space $R(A)$, the left kernel $N(A)$, the column space $R(A^*)$ and the right kernel $N(A^*)$ to be introduced in Sec. B.8 Adopt notation in (*). Then*

$$R(A) = \langle\langle R_{1*}, \ldots, R_{r*} \rangle\rangle;$$
$$R(A^*) = \langle\langle A_{*k_1}, \ldots, A_{*k_r} \rangle\rangle;$$
$$N(A) = \langle\langle P_{r+1,*}, \ldots, P_{m*} \rangle\rangle, \text{ where } PA = R;$$
$$N(A^*) = \langle\langle \vec{v}_1, \ldots, \vec{v}_{n-r} \rangle\rangle, \text{ where } \vec{v}_1, \ldots, \vec{v}_{n-r} \text{ are fundamental}$$

solution of $A\vec{x}^* = \vec{0}^*$ in Application 1 above.

In fact, we are able to determine these four subspaces once we reduce the original matrix A to its echelon form S, i.e. $QA = S$ for some invertible matrix $Q_{m \times m}$.

B.6 Determinants

Let $A = [a_{ij}]$ be an $n \times n$ matrix with entries from a field $\mathbb{F}$. The *determinant* of A is an element of $\mathbb{F}$, denote by

$$\det A,$$

which can be defined inductively on n as follows:

1. $n = 1$. The determinant $\det A$ of order 1 is

$$\det A = a_{11}.$$

2. $n = 2$. The determinant $\det A$ of order 2 is

$$\det A \underset{\text{(def.)}}{=} \begin{vmatrix} a_{11} & a_{12} \\ a_{21} & a_{22} \end{vmatrix} \underset{\text{(def.)}}{=} a_{11}a_{22} - a_{12}a_{21}.$$

Suppose the determinant of order $n-1$ has been defined and $A = [a_{ij}]_{n \times n}$ for $n \geq 3$ is given. Let

$$A_{ij} = \begin{bmatrix} a_{11} & a_{12} & \cdots & a_{1j} & \cdots & a_{1n} \\ \vdots & \vdots & & \vdots & & \vdots \\ a_{i1} & a_{i2} & \cdots & a_{ij} & \cdots & a_{in} \\ \vdots & \vdots & & \vdots & & \vdots \\ a_{n1} & a_{n2} & \cdots & a_{nj} & \cdots & a_{nn} \end{bmatrix} \leftarrow i\text{th row deleted}$$

$$\underset{j\text{th column deleted}}{\uparrow}$$

be the $(n-1) \times (n-1)$ matrix obtained from A by deleting the ith row and the jth column. The defined determinant det A_{ij} of order $n-1$ is called the *minor* of the (i,j)-entry a_{ij} in the matrix A and

$$(-1)^{i+j} \det A_{ij}, \quad 1 \le i, \; j \le n$$

is called the *cofactor* of a_{ij} in A.

3. *The determinant* det A *of order* n is defined as

$$\det A = \sum_{j=1}^{n} (-1)^{i+j} a_{ij} \det A_{ij}.$$

and is called the *expansion of the determinant* det A *along the ith row*.

This inductive definition for determinants provide, at the same time, a technique needed to compute a determinant.

[3] gives five different definitions for determinants. One of these definitions is the following

Characteristic Properties of Determinants
There exists a unique function det: $M(n; \mathbb{F}) \to \mathbb{F}$ satisfying the following properties:

1. (multiplinear) det is a linear function of each row of an $n \times n$ matrix when the remaining $n-1$ rows are held fixed, i.e.

$$\det \begin{bmatrix} A_{1*} \\ \vdots \\ \alpha A_{i*} + B_{i*} \\ \vdots \\ A_{n*} \end{bmatrix} = \alpha \det \begin{bmatrix} A_{1*} \\ \vdots \\ A_{i*} \\ \vdots \\ A_{n*} \end{bmatrix} + \det \begin{bmatrix} A_{1*} \\ \vdots \\ B_{i*} \\ \vdots \\ A_{n*} \end{bmatrix}$$

for $1 \le i \le n$ and $\alpha \in \mathbb{F}$.

2. (alternating) If $B_{n \times n}$ is obtained by interchanging any two rows of an $n \times n$ matrix A, then

$$\det B = -\det A.$$

3. (unit) For the $n \times n$ identity matrix I_n,

$$\det I_n = 1.$$

Try to prove this result for $n = 2$ or $n = 3$.

From the very definition for determinants, the following basic properties can be deduced.

1. The determinants of elementary matrices (see Sec. B.5) are

$$\det E_{(i)(j)} = -1;$$
$$\det E_{\alpha(i)} = \alpha;$$
$$\det E_{(j)+\alpha(i)} = 1.$$

2. $A_{n \times n}$ is not invertible (i.e. singular) if and only if

$$\det A = 0$$

and hence is equivalent to the rank $r(A) < n$.

3. The $\det A$ can be expanded along any row, i.e.

$$\det A = \sum_{j=1}^{n} (-1)^{i+j} a_{ij} \det A_{ij}, \quad 1 \le i \le n.$$

4. The interchange of rows and columns does not change the value of $\det A$, i.e.

$$\det A^* = \det A.$$

Thus, $\det A$ can be expanded along any column.

5. For $A = \big[a_{ij}^{(1)}\big]_{n \times n}, B = \big[a_{ij}^{(2)}\big]_{n \times n}$ and $\alpha \in \mathbb{F}$,

$$\det(\alpha A) = \alpha^n \det A;$$
$$\det(A + B) = \det \big[a_{ij}^{(1)} + a_{ij}^{(2)}\big]$$
$$= \sum_{k_1, \ldots, k_n = 1}^{2} \begin{vmatrix} a_{11}^{(k_1)} & a_{12}^{(k_2)} & \cdots & a_{1n}^{(k_n)} \\ \vdots & \vdots & & \vdots \\ a_{n1}^{(k_1)} & a_{n2}^{(k_2)} & \cdots & a_{nn}^{(k_n)} \end{vmatrix}.$$

(a sum of 2^n terms)

6. Let $A = [a_{ij}]_{m \times n}$ and $B = [b_{ij}]_{n \times m}$.
   If $m = n$: $\det AB = \det A \cdot \det B$.
   If $m > n$: $\det AB = 0$.

If $m < n$:

$$\det AB$$

$$= \sum_{1 \le j_1 < j_2 < \cdots < j_m \le n} \begin{vmatrix} a_{1j_1} & \cdots & a_{1j_m} \\ \vdots & & \vdots \\ a_{mj_1} & \cdots & a_{mj_m} \end{vmatrix} \begin{vmatrix} b_{j_11} & \cdots & b_{j_1m} \\ \vdots & & \vdots \\ b_{j_m1} & \cdots & b_{j_mm} \end{vmatrix}.$$

(a sum of C_m^n terms)

This is called *Canchy–Binnet formula*.

To extend the expansion of a determinant along a row or a column to that along some rows or columns, say along i_1th, $\ldots$, i_kth rows, where $1 \le i_1 < \cdots < i_k \le n$, let

$$A \begin{pmatrix} i_1 & \cdots & i_k \\ j_1 & \cdots & j_k \end{pmatrix} = \det \begin{bmatrix} a_{i_1j_1} & \cdots & a_{i_1j_k} \\ \vdots & & \vdots \\ a_{i_kj_1} & \cdots & a_{i_kj_k} \end{bmatrix}, \quad 1 \le j_1 < \cdots j_k \le n$$

denote a *subdeterminant* of order k of $\det A$ and is called *principal* if $i_l = j_l$ for $1 \le l \le k$. Use $i_{k+1}, \ldots, i_n$, where $1 \le i_{k+1} < \cdots < i_n \le n$ to denote those integers among $1, 2, \ldots, n$ that are different from $i_1, \ldots, i_k$. Similarly, $j_{k+1}, \ldots, j_n$ are these integers among $1, 2, \ldots, n$ but are different from $j_1, \ldots, j_k$. Then, the subdeterminant of $n - k$

$$\tilde{A} \begin{pmatrix} i_1 & \cdots & i_k \\ j_1 & \cdots & j_k \end{pmatrix} = (-1)^{i_1 + \cdots + i_k + j_1 + \cdots + j_k} A \begin{pmatrix} i_{k+1} & \cdots & i_n \\ j_{k+1} & \cdots & j_n \end{pmatrix}$$

is called the *cofactor* of $A \begin{pmatrix} i_1 & \cdots & i_k \\ j_1 & \cdots & j_k \end{pmatrix}$ in $\det A$. By using 5, we have

7. *Laplace expansion* Give an $n \times n$ matrix $A = [a_{ij}]$. For any positive integer $1 \le k \le n$ and $1 \le i_1 < \cdots < i_k \le n$,

$$\det A = \sum_{1 \le j_1 < \cdots < j_k \le n} A \begin{pmatrix} i_1 & \cdots & i_k \\ j_1 & \cdots & j_k \end{pmatrix} \tilde{A} \begin{pmatrix} i_1 & \cdots & i_k \\ j_1 & \cdots & j_k \end{pmatrix}.$$

In most cases, it is not easy to calculate the value of a determinant of order $n \ge 4$ by hand-computing. A compensative method to do so is to use elementary row operations, the so-called *Gaussian elimination method*, to reduce an $n \times n$ square matrix to an upper triangular matrix whose determinant is the product of its main diagonal entries. Also, see Application 6 to Sec. B.5.

Laplace expansion formula shows that

$$\det \begin{bmatrix} A_{11} & A_{12} \\ 0 & A_{22} \end{bmatrix} = \det A_{11} \cdot \det A_{22},$$

where A_{11} and A_{22} are square matrices. Furthermore, if A_{11} is invertible, then by use of the identity

$$\begin{bmatrix} I_{n_1} & 0 \\ -A_{21}A_{11}^{-1} & I_{n_2} \end{bmatrix} \begin{bmatrix} A_{11} & A_{12} \\ A_{21} & A_{22} \end{bmatrix} = \begin{bmatrix} A_{11} & A_{12} \\ 0 & -A_{21}A_{11}^{-1}A_{12} + A_{22} \end{bmatrix},$$

it follows that

$$\det \begin{bmatrix} A_{11} & A_{12} \\ A_{21} & A_{22} \end{bmatrix} = \det A_{11} \cdot \det(-A_{21}A_{11}^{-1}A_{12} + A_{22}).$$

Therefore, if $\det A_{11} \neq 0$ and $A_{11}A_{21} = A_{21}A_{11}$ holds, then

$$\det \begin{bmatrix} A_{11} & A_{12} \\ A_{21} & A_{22} \end{bmatrix} = \det(A_{11}A_{22} - A_{21}A_{12}).$$

Let $A = [a_{ij}]_{n \times n}$. By basic properties 2 and 3 above, the following identities

$$\sum_{j=1}^{n} (-1)^{i+j} a_{kj} \cdot \det A_{ij} = \delta_{ik} \det A, \quad 1 \leq i, \ k \leq n \qquad (*)$$

hold. To put these identities in a compact form, we define the *adjoint matrix* of A as

$$\text{adj } A = [b_{ij}]_{n \times n},$$
$$b_{ij} = (-1)^{j+i} \det A_{ji}, \quad 1 \leq i, \ j \leq n.$$

Thus, $(*)$ can be written as a single identity

$$A \cdot \text{adj } A = \text{adj } A \cdot A = (\det A) I_n.$$

We conclude that *a square matrix A is invertible if and only if* $\det A \neq 0$. In this case, the *inverse matrix* is

$$A^{-1} = \frac{1}{\det A} \text{adj } A.$$

Let $A = [a_{ij}]_{n \times n}$. The system of linear equations

$$A\vec{x} = \vec{b}$$

in n unknowns $x_1, \ldots, x_n$, where $\vec{x}$ is the $n \times 1$ column vector $\begin{bmatrix} x_1 \\ \vdots \\ x_n \end{bmatrix} \in \mathbb{F}^n$

and $\vec{b} = \begin{bmatrix} b_1 \\ \vdots \\ b_n \end{bmatrix} \in \mathbb{F}^n$, *has a unique solution if and only if* $\det A \neq 0$. Let X_k

be the matrix obtained from I_n by replacing its kth column by $\vec{x}$. Then

$$AX_k = A[\vec{e}_1 \cdots \vec{e}_{k-1} \vec{x} \vec{e}_{k+1} \cdots \vec{e}_n]$$
$$= [A\vec{e}_1 \cdots A\vec{e}_{k-1} A\vec{x} A\vec{e}_{k+1} \cdots A\vec{e}_n]$$
$$= [A_{*1} \cdots A_{*,k-1} \vec{b} A_{*,k+1} \cdots A_{*n}]$$
$$\Rightarrow \det AX_k = \det A \cdot \det X_k = x_k \det A$$
$$= \det[A_{*1} \cdots A_{*,k-1} \vec{b} A_{*,k+1} \cdots A_{*n}]$$
$$\Rightarrow x_k = \frac{1}{\det A} \det[A_{*1} \cdots A_{*,k-1} \vec{b} A_{*,k+1} \cdots A_{*n}], \quad 1 \le k \le n,$$

where $[A_{*1} \cdots A_{*,k-1} \vec{b} A_{*,k+1} \cdots A_{*n}]$ is the matrix obtained from A by replacing its kth column A_{*k} by $\vec{b}$. This is the *Cramer's Rule* for solutions of the equation $A\vec{x} = \vec{b}$. Note that, in this case, $\vec{x} = A^{-1}\vec{b}$.

B.7 Linear Transformations and Their Matrix Representations

Definition Suppose V and W are vector spaces over the same filed $\mathbb{F}$. A function (see Sec. A.2)

$$f\colon V \to W$$

is called a *linear transformation* or *mapping* from V into W if it preserves linear structures of vector spaces, i.e. for any $\vec{x}, \vec{y} \in V$ and $\alpha \in \mathbb{F}$, the following properties hold:

1. $f(\alpha\vec{x}) = \alpha f(\vec{x})$.
2. $f(\vec{x} + \vec{y}) = f(\vec{x}) + f(\vec{y})$.

If, in addition, f is both one-to-one and onto, then f is called a *linear isomorphism* from V onto W, and V and W are called *isomorphic*. In case $W = V$, a linear transformation $f\colon V \to V$ is specially called a *linear operator*; while if $W = \mathbb{F}$, $f\colon V \to \mathbb{F}$ is called a *linear functional*.

Remark
In general situations, both conditions 1 and 2 in the definition are independent of each other and hence are needed simultaneously in the definition of a linear transformation.

For example, define a mapping $f\colon \mathbb{R}^2$ (vector space) $\to \mathbb{R}^2$ by

$$f(\vec{x}) = \begin{cases} \vec{x}, & \text{if } x_1 x_2 \ge 0 \\ -\vec{x}, & \text{if } x_1 x_2 < 0, \end{cases}$$

where $\vec{x} = (x_1, x_2) \in \mathbb{R}^2$. Then f obviously satisfies condition 1 but not 2 and hence is not linear.

When $\mathbb{R}$ is considered as a vector space over the rational filed $\mathbb{Q}$, $\mathbb{R}$ is an infinite-dimensional vector space and has a basis $\mathcal{B}$ (see Sec. B.3). Let x_1 and x_2 be distinct elements of $\mathcal{B}$, and define $\tau: \mathcal{B} \to \mathbb{R}$ by

$$\tau(x) = \begin{cases} x_1, & \text{if } x = x_2, \\ x_2, & \text{if } x = x_1, \\ x, & \text{otherwise.} \end{cases}$$

Then, there exists a linear transformation $f: \mathbb{R}(\text{over } \mathbb{Q}) \to \mathbb{R}$ such that $f(x) = \tau(x)$ for all $x \in \mathcal{B}$. Of course, f is *additive*, i.e. condition 2 in definition holds but for $\alpha = \frac{x_2}{x_1}$ which is an irrational number, $f(\alpha x_1) = f(x_2) = \tau(x_2) = x_1 \neq \frac{x_2}{x_1} \cdot x_2 = \alpha f(x_1)$.

It is worth mentioning that a continuous or even bounded additive mapping $f: \mathbb{R}(\text{over } \mathbb{R} \text{ itself}) \to \mathbb{R}$ should be linear.

The existence of linear transformations

The function $0: V \to W$ mapping every vector in V into zero vector $\vec{0}$ in W, i.e.

$$0(\vec{x}) = \vec{0}, \quad \vec{x} \in V$$

is a linear transformation and is called the *zero transformation*.

For simplicity, suppose that V is finite-dimensional and $\dim V = m$. Choose any fixed basis $\mathcal{B} = \{\vec{x}_1, \ldots, \vec{x}_m\}$ for V and *any* m vectors $\vec{y}_1, \ldots, \vec{y}_m$ in W. There exists exactly one linear transformation $f: V \to W$ such that

$$f(\vec{x}_i) = \vec{y}, \quad 1 \leq i \leq m.$$

All we need to do is to define a function $f: V \to W$ by assigning $f(\vec{x}_i) = \vec{y}_i$, $1 \leq i \leq m$ and then extending it *linearly* to all vectors $\sum_{i=1}^{m} \alpha_i \vec{x}_i$ in V by

$$f\left(\sum_{i=1}^{m} \alpha_i \vec{x}_i\right) = \sum_{i=1}^{m} \alpha_i \vec{y}_i.$$

This result still holds for infinite-dimensional space V.

The set of linear transformations from V into W

$$L(V, W) \quad \text{or} \quad \text{Hom}(V, W)$$

forms a *vector space* over $\mathbb{F}$, where $(f+g)(\vec{x}) = f(\vec{x}) + g(\vec{x})$ and $(\alpha f)(\vec{x}) = \alpha f(\vec{x})$, $\vec{x} \in V$. $\dim L(V, W) = mn$ if $\dim V = m$ and $\dim W = n$ (see Ex. 6).

Suppose $f \in L(V, W)$ and $g \in L(W, U)$, then the composite $g \circ f \in L(V, U)$.

Kernel (space) and range (space) of a linear transformation

Suppose $f: V \to W$ is a linear transformation. Then

$$\textit{Kernel: } \mathrm{Ker}(f) = \{\vec{x} \in V \mid f(\vec{x}) = \vec{0}\}, \text{ or denoted by } \mathrm{N}(f),$$
$$\textit{Range: } \mathrm{Im}(f) = \{f(\vec{x}) \in W \mid \vec{x} \in V\}, \text{ or denoted by } \mathrm{R}(f)$$

are subspaces of V and W respectively. In case $\dim V < \infty$, then

$$\dim \mathrm{Ker}(f) + \dim \mathrm{Im}(f) = \dim V$$

holds with $\dim \mathrm{Ker}(f)$ called the *nullity* of f and $\dim \mathrm{Im}(f)$ the *rank* of f.

As a consequence, f is one-to-one if and only if $\mathrm{Ker}(f) = \{\vec{0}\}$.

Suppose $\dim V = m < \infty$ and $\dim W = n < \infty$. Then,

1. If $m < n$, f can only be one-to-one but never onto.
2. If $m > n$, f can only be onto but never one-to-one.
3. If $m = n$, f is one-to-one if and only if f is onto.

In Case 3, such a $f: V \to W$ is a *linear isomorphism*.

Matrix representations of linear transformations between finite-dimensional spaces

Let $\dim V = m$ and $\dim W = n$. Fix a basis $\mathcal{B} = \{\vec{a}_1, \ldots, \vec{a}_m\}$ for V and a basis $\mathcal{C} = \{\vec{b}_1, \ldots, \vec{b}_n\}$ for W.

For $\vec{x} \in V$, there exists unique scalars $x_1, \ldots, x_n \in \mathbb{F}$ such that

$$\vec{x} = \sum_{i=1}^{m} x_i \vec{a}_i.$$

The coefficients $x_1, \ldots, x_m$ forms a vector

$$[\vec{x}]_{\mathcal{B}} = (x_1, \ldots, x_m) \in \mathbb{F}^m$$

and is called the *coordinate vector* of $\vec{x}$ relative to the basis $\mathcal{B}$. Similarly, for $\vec{y} \in W$, it has coordinate vector relative to $\mathcal{C}$

$$[\vec{y}]_{\mathcal{C}} = (y_1, \ldots, y_n) \in \mathbb{F}^n$$

if and only if $\vec{y} = \sum_{j=1}^{n} y_j \vec{b}_j$.

Let $f: V \to W$ be linear. Then $f(\vec{a}_i) \in W$ and

$$f(\vec{a}_i) = \sum_{j=1}^{n} a_{ij} \vec{b}_j, \quad 1 \le i \le m$$

$$\Rightarrow [f(\vec{a}_i)]_C = (a_{i1}, a_{i2}, \dots, a_{in}), \quad 1 \le i \le m.$$

For any $\vec{x} = \sum_{i=1}^{m} x_i \vec{a}_i$,

$$f(\vec{x}) = \sum_{i=1}^{m} x_i f(\vec{a}_i) = \sum_{i=1}^{m} x_i \sum_{j=1}^{n} a_{ij} \vec{b}_j = \sum_{j=1}^{n} \left(\sum_{i=1}^{m} x_i a_{ij} \right) \vec{b}_j$$

$$\Rightarrow [f(\vec{x})]_C = \left(\sum_{i=1}^{m} x_i a_{i1}, \sum_{i=1}^{m} x_i a_{i2}, \dots, \sum_{i=1}^{m} x_i a_{im} \right)$$

$$= (x_1 \cdots x_m) \begin{bmatrix} a_{11} & a_{12} & \cdots & a_{1n} \\ a_{21} & a_{22} & \cdots & a_{2n} \\ \vdots & \vdots & & \vdots \\ a_{m1} & a_{m2} & \cdots & a_{mn} \end{bmatrix}$$

$$= [\vec{x}]_B [f]_C^B,$$

where the $m \times n$ matrix

$$[f]_C^B = \begin{bmatrix} [f(\vec{a}_1)]_C \\ \vdots \\ [f(\vec{a}_m)]_C \end{bmatrix}$$

is called the *matrix representation* of f relative to *ordered* bases B and C. In case $W = V$ and $C = B$, simply denote it by $[f]_B$.

Let $B' = \{\vec{a}'_1, \dots, \vec{a}'_m\}$ be another basis for V. The identity linear operator $1_V: V \to V$ can be written in the matrix form as

$$[\vec{x}]_{B'} = [\vec{x}]_B [1_V]_{B'}^B,$$

where

$$[1_V]_{B'}^B = \begin{bmatrix} [\vec{a}_1]_{B'} \\ \vdots \\ [\vec{a}_m]_{B'} \end{bmatrix}$$

is the matrix representation of 1_V relative to B and B' and is called the *change of coordinate matrix* or *transition matrix changing B into B'*. Similarly, for another basis C' for W and $\vec{y} \in W$, we have

$$[\vec{y}]_{C'} = [\vec{y}]_C [1_W]_{C'}^C.$$

Both $[1_V]_{B'}^B$ and $[1_W]_{C'}^C$ are invertible.

What are the possible relations among $[f]_C^B, [f]_{C'}^{B'}, [1_V]_{B'}^B$ and $[1_W]_{C'}^C$? Since $[f(\vec{x})]_{C'} = [\vec{x}]_{B'}[f]_{C'}^{B'}$, therefore

$$
\begin{aligned}
[f(\vec{x})]_{C'} &= [f(\vec{x})]_C [1_W]_{C'}^C \\
&= [\vec{x}]_B [f]_C^B [1_W]_{C'}^C \\
&= [\vec{x}]_B [1_V]_{B'}^B [f]_{C'}^{B'}, \quad \vec{x} \in V \\
\Rightarrow [f]_C^B [1_W]_{C'}^C &= [1_V]_{B'}^B [f]_{C'}^{B'} \\
\Rightarrow [f]_{C'}^{B'} &= ([1_V]_{B'}^B)^{-1} [f]_C^B [1_W]_{C'}^C = [1_V]_B^{B'} [f]_C^B [1_W]_{C'}^C.
\end{aligned}
$$

This means the following diagram is commutative.

$$
\begin{array}{ccc}
V & \xrightarrow{\ [f]_C^B\ } & W \\
(B) & f & (C) \\
1_V \uparrow [1_V]_B^{B'} & [1_W]_{C'}^C & \downarrow 1_W \\
(B') & \xrightarrow[\ f\ ]{[f]_{C'}^{B'}} & (C') \\
V & & W
\end{array}
$$

Summarize above results as partial of

The Relations between $L(V, W)$ and $M(m, n; \mathbb{F})$
Let $\dim V = m$, $\dim W = n$ and $\dim U < \infty$.

1. For each basis B for V,

$$
\vec{x} \in V \to [\vec{x}]_B \in \mathbb{F}^m
$$

is a linear isomorphism.
2. For each basis B for V and basis C for W, each $f \in L(V, W)$ has a unique matrix representation $[f]_C^B$ relative to B and C:

$$
[f(\vec{x})]_C = [\vec{x}]_B [f]_C^B.
$$

This is equivalent to say that the following diagram is commutative.

$$
\begin{array}{ccc}
V & \xrightarrow{f} & W \\
(B) & & (C) \\
\text{iso} \updownarrow & & \updownarrow \text{iso} \\
\mathbb{F}^m & \xrightarrow{[f]_C^B} & \mathbb{F}^n
\end{array}
$$

3. For another basis $\mathcal{B}'$ for V and $\mathcal{C}'$ for W, $[f]_{\mathcal{C}'}^{\mathcal{B}'}$ and $[f]_{\mathcal{C}}^{\mathcal{B}}$ are related to each other, subject to changes of coordinate matrices $[1_V]_{\mathcal{B}}^{\mathcal{B}'}$ and $[1_W]_{\mathcal{C}}^{\mathcal{C}'}$, as

$$[f]_{\mathcal{C}'}^{\mathcal{B}'} = [1_V]_{\mathcal{B}}^{\mathcal{B}'} [f]_{\mathcal{C}}^{\mathcal{B}} [1_W]_{\mathcal{C}'}^{\mathcal{C}}.$$

In case $W = V$, let $\mathcal{B} = \mathcal{C}$ and $\mathcal{B}' = \mathcal{C}'$, then $[f]_{\mathcal{B}}$ and $[f]_{\mathcal{B}'}$ are similar, i.e.

$$[f]_{\mathcal{B}'} = [1_V]_{\mathcal{B}}^{\mathcal{B}'} [f]_{\mathcal{B}} \left([1_V]_{\mathcal{B}}^{\mathcal{B}'}\right)^{-1}.$$

4. The mapping (refer to Ex. 6)

$$f \in L(V, W) \to [f]_{\mathcal{C}}^{\mathcal{B}} \in M(m, n; \mathbb{F})$$

is a linear isomorphism, i.e.

$$[f + g]_{\mathcal{C}}^{\mathcal{B}} = [f]_{\mathcal{C}}^{\mathcal{B}} + [g]_{\mathcal{C}}^{\mathcal{B}} \quad \text{and} \quad [\alpha f]_{\mathcal{C}}^{\mathcal{B}} = \alpha [f]_{\mathcal{C}}^{\mathcal{B}} \text{ for } \alpha \in \mathbb{F}.$$

5. Suppose $f \in L(V, W)$ and $g \in L(W, U)$ and $\mathcal{D}$ is a basis for U. Then,

$$[g \circ f]_{\mathcal{D}}^{\mathcal{B}} = [f]_{\mathcal{C}}^{\mathcal{B}} [g]_{\mathcal{D}}^{\mathcal{C}}.$$

6. Suppose $W = V$ and $\mathcal{B} = \mathcal{C}$. Let

$$GL(V, V) = \{f \in L(V, V) \mid f \text{ is invertible}\}, \quad \text{and}$$
$$GL(n; \mathbb{F}) = \{A \in M(n; \mathbb{F}) \mid A \text{ is invertible}\}.$$

The former is a *group* under the operation of the composite of functions, while the latter is a *group* under the multiplication of matrices. Both are called the *general linear group* of order n. The mapping, for each basis $\mathcal{B}$ for V,

$$f \in GL(V, V) \to [f]_{\mathcal{B}} \in GL(n; \mathbb{F})$$

is a group isomorphism, i.e.

$$[g \circ f]_{\mathcal{B}} = [f]_{\mathcal{B}} [g]_{\mathcal{B}} \quad \text{and} \quad [f^{-1}]_{\mathcal{B}} = [f]_{\mathcal{B}}^{-1}.$$

These results enable us to reduce the study of linear transformations between finite-dimensional vector spaces to the study of matrices, subject to "similarity" (see Ex. 31 in Sec. B.4). More precisely, for a matrix $A \in M(m, n; \mathbb{F})$, we would consider A as the *linear transformation* defined by

$$\vec{x} \in \mathbb{F}^m \to \vec{x}A \in \mathbb{F}^n$$

whose matrix representation relative to the natural basis $\mathcal{N} = \{\vec{e}_1, \ldots, \vec{e}_m\}$ for $\mathbb{F}^m$ and the natural basis $\mathcal{N}'$ for $\mathbb{F}^n$ is A itself, while relative to another basis $\mathcal{B}$ for $\mathbb{F}^m$ and basis $\mathcal{B}'$ for $\mathbb{F}^n$ is

$$[1_{\mathbb{F}^m}]_{\mathcal{N}}^{\mathcal{B}} A \left([1_{\mathbb{F}^n}]_{\mathcal{N}'}^{\mathcal{B}'}\right)^{-1}.$$

Exercises

Let V and W be vector spaces over the same field $\mathbb{F}$ throughout the following problems.

1. Suppose $\dim V < \infty$ and $f \in L(V, W)$.

   (a) If S is a subspace of V, then

   $$\dim(f^{-1}(\vec{0}) \cap S) + \dim f(S) = \dim S.$$

   Hence $\dim f(S) \leq \dim S$ with equality if and only if $f^{-1}(\vec{0}) \cap S = \{\vec{0}\}$.

   (b) If T is a subspace of W, then

   $$\dim f^{-1}(\vec{0}) + \dim(f(V) \cap T) = \dim f^{-1}(T).$$

   In particular, $\dim(f(V) \cap T) \leq \dim f^{-1}(T)$ with equality if and only if $f^{-1}(\vec{0}) = \{\vec{0}\}$, i.e. f is one-to-one.

2. Let $A \in M(m, n; \mathbb{F})$ be a nonzero matrix. Consider $\vec{x} \in \mathbb{F}^m \to \vec{x}A \in \mathbb{F}^n$ as a linear transformation and $\vec{y} \in \mathbb{F}^n \to \vec{y}A^* \in \mathbb{F}^m$ as a linear transformation.

   (a) Then (refer to Ex. 15 of Sec. B.4)

   the maximal number of linearly independent row vectors of A

   $=$ the dimension of the range space $\{\vec{x}A \mid \vec{x} \in \mathbb{F}^m\}$.

   This common number is called the *row rank* of A.

   (b) Similarly,

   the maximal number of linearly independent column vectors of A

   $=$ the dimension of the range space $\{\vec{y}A^* \mid \vec{y} \in \mathbb{F}^n\}$.

   This common number is called the *column rank* of A.

   (c) Therefore,

   $$\dim \operatorname{Ker} A^* = n - \text{the column rank of } A.$$

   (d) On the other hand, $\operatorname{Ker} A^* = \{\vec{y} \in \mathbb{F}^n \mid \vec{y}A^* = \vec{0}\}$ is the solution space of the system of homogenous linear equations $A\vec{y}^* = \vec{0}$ in n

unknowns. From Application 1 in Sec. B.5, we have already known that

$$\dim \operatorname{Ker} A^* = n - \text{the row rank of } A.$$

(*Note* The invertible matrix P such that $PA = R$ is row-reduced echelon matrix preserves row rank of A, because the range space $\{\vec{x}A \mid \vec{x} \in \mathbb{F}^m\} = \{\vec{x}(PA) \mid \vec{x} \in \mathbb{F}^m\} = \{\vec{x}R \mid \vec{x} \in \mathbb{F}^m\}$.)

(e) Combining (c) and (d), the following holds:

$$\text{row rank of } A = \text{column rank of } A.$$

(f) Let $\vec{a}_i = (a_{i1}, a_{i2}, \ldots, a_{im}) \in \mathbb{F}^m, 1 \le i \le k \le m$. Show that $\vec{a}_1, \vec{a}_2, \ldots, \vec{a}_k$ are linearly independent if and only if there exist integers $1 \le j_1 < j_2 < \cdots < j_k \le m$ such the $k \times k$ submatrix

$$\begin{bmatrix} a_{1j_1} & \cdots & a_{1j_k} \\ a_{2j_1} & \cdots & a_{2j_k} \\ \vdots & & \vdots \\ a_{kj_1} & \cdots & a_{kj_k} \end{bmatrix}$$

is invertible.

(g) Let $r(1 \le r \le m, n)$ denote the largest integer such that some $r \times r$ submatrix of A are invertible, or equivalently, has a nonzero determinant. This r is defined to be the *rank* of A.

(h) Show that

$$\text{row rank of } A = \text{column rank of } A = \text{rank of } A.$$

Usually, this common number is denoted by $r(A)$. Note $r(O) = 0$.

3. (a) Let $P \in M(k, m; \mathbb{F})$ and $A \in M(m, n; \mathbb{F})$. Show that

$$r(P) + r(A) - m \le r(PA) \le \min\{r(P), r(A)\}$$

with

(1) $r(PA) = r(A) \Leftrightarrow \mathbb{F}^m = \operatorname{Im}(P) + \operatorname{Ker}(A)$,
(2) $r(PA) = r(P) \Leftrightarrow \operatorname{Im}(P) \cap \operatorname{Ker}(A) = \{\vec{0}\}$,
(3) $r(PA) = r(P) + r(A) - m \Leftrightarrow \operatorname{Ker}(A) \subseteq \operatorname{Im}(P)$.

(b) Suppose $P \in M(m; \mathbb{F})$ and $Q \in M(n; \mathbb{F})$ are invertible, and $A \in M(m, n; \mathbb{F})$. Then

$$r(PAQ) = r(PA) = r(AQ) = r(A).$$

(c) Suppose $A, B \in \mathrm{M}(m, n; \mathbb{F})$. Then

$$r(A + B) \leq r(A) + r(B)$$

with equality if and only if $\mathrm{Im}(A + B) = \mathrm{Im}(A) \oplus \mathrm{Im}(B)$.

4. Suppose $A \in \mathrm{M}(m, n; \mathbb{F})$, where $\mathbb{F}$ is a subfield of $\mathbb{C}$. Show that

$$r(A) = r(A^*) = r(AA^*) = r(A^*A).$$

5. Let $f \in \mathrm{L}(V, W)$.

(a) f is onto $\Leftrightarrow$ f is right-invertible, i.e. there exists $g \in \mathrm{L}(W, V)$ such that

$$f \circ g = 1_W.$$

(b) f is one - to - one $\Leftrightarrow$ f is left-invertible, i.e. there exists $h \in \mathrm{L}(W, V)$ such that

$$h \circ f = 1_V.$$

In case $\dim V = m$ and $\dim W = n$, these results are equivalent to say that, for $A \in \mathrm{M}(m, n; \mathbb{F})$ considered as the mapping $\vec{x} \in \mathbb{F}^m \to \vec{x}A \in \mathbb{F}^n$,

(c) A has rank n $\Leftrightarrow$ there exists a matrix $B \in \mathrm{M}(n, m; \mathbb{F})$ such that

$$BA = \mathrm{I}_n.$$

(d) A has rank m $\Leftrightarrow$ there exists a matrix $C \in \mathrm{M}(n, m; \mathbb{F})$ such that

$$AC = \mathrm{I}_m.$$

Try to prove (c) and (d) by the concept of row-reduced echelon matrix or some other methods without recourse to (a) and (b).

6. Suppose $\dim V = m, \dim W = n$ and $\mathcal{B} = \{\vec{x}_1, \ldots, \vec{x}_m\}$ is a basis for V and $\mathcal{C} = \{\vec{y}_1, \ldots, \vec{y}_n\}$ a basis for W. Define $f_{ij} \in \mathrm{L}(V, W)$ by

$$f_{ij}(\vec{x}_k) = \delta_{ki}\vec{y}_j, \quad 1 \leq i, \ k \leq m, \ 1 \leq j \leq n.$$

Then $\mathcal{N} = \{f_{ij} \mid 1 \leq i \leq m, 1 \leq j \leq n\}$ forms a basis for $\mathrm{L}(V, W)$. In particular, if $f \in \mathrm{L}(V, W)$ and $f = \sum_{i,j=1}^{m,n} a_{ij} f_{ij}$, then

$$[f]_{\mathcal{C}}^{\mathcal{B}} = [a_{ij}]_{m \times n}.$$

Therefore, $\dim \mathrm{L}(V, W) = \dim V \cdot \dim W$.

7. Suppose $\dim V = n$. Let $f \in \mathrm{L}(V, V)$ be an *idempotent* linear operator (refer to Ex. 6 of Sec. B.4), i.e.

$$f^2 = f \circ f = f \quad \text{or} \quad f \circ (f - 1_V) = (f - 1_V) \circ f = 0.$$

(a) Show that

$$R(f) = \{\vec{x} \in V \mid f(\vec{x}) = \vec{x}\} = \text{Ker}(f - 1_V),$$
$$R(f - 1_V) = \{\vec{x} \mid f(\vec{x}) = \vec{0}\} = \text{Ker}(f)$$

and hence

$$V = \text{Ker}(f - 1_V) \oplus \text{Ker}(f) = R(f - 1_V) \oplus R(f).$$

(b) Let $\{\vec{x}_1, \ldots, \vec{x}_r\}$ be a basis for $R(f) = \text{Ker}(f - 1_V)$ and $\{\vec{x}_{r+1}, \ldots, \vec{x}_n\}$ be a basis for $\text{Ker}(f) = R(f - 1_V)$ so that $\mathcal{B} = \{\vec{x}_1, \ldots, \vec{x}_r, \vec{x}_{r+1}, \ldots, \vec{x}_n\}$ is a basis for V. Then

$$[f]_{\mathcal{B}} = \begin{bmatrix} I_r & 0 \\ 0 & 0 \end{bmatrix}_{n \times n}.$$

8. Suppose $\mathbb{F}$ is a field with $1 + 1 \neq 0$ and $\dim V = n$. Let $f \in L(V, V)$ be an *involutory* linear operator (see Ex. 9 of Sec. B.4), i.e.

$$f^2 = 1_V \text{ or } (f - 1_V) \circ (f + 1_V) = (f + 1_V) \circ (f - 1_V) = 0.$$

(a) Then

$$\text{Ker}(f - 1_V) = \{\vec{x} \in V \mid f(\vec{x}) = \vec{x}\},$$
$$\text{Ker}(f + 1_V) = \{\vec{x} \in V \mid f(\vec{x}) = -\vec{x}\}$$

are subspaces of V. Let $\{\vec{x}_1, \ldots, \vec{x}_r\}$ be a basis for $\text{Ker}(f - 1_V)$ and extend it to a basis $\{\vec{x}_1, \ldots, \vec{x}_r, \vec{x}_{r+1}, \ldots, \vec{x}_n\}$ for V. Thus, $\{f(\vec{x}_{r+1}) - \vec{x}_{r+1}, \ldots, f(\vec{x}_n) - \vec{x}_n\}$ is linearly independent in V.

(b) It is clear that $\text{Ker}(f - 1_V) \cap \text{Ker}(f + 1_V) = \{\vec{0}\}$ and, for $\vec{x} \in V$,

$$\vec{x} = \frac{1}{2}(\vec{x} + f(\vec{x})) + \frac{1}{2}(\vec{x} - f(\vec{x})).$$

It follows that

$$V = \text{Ker}(f - 1_V) \oplus \text{Ker}(f + 1_V).$$

Hence $\{f(\vec{x}_{r+1}) - \vec{x}_{r+1}, \ldots, f(\vec{x}_n) - \vec{x}_n\}$ forms a basis for

$$R(f - 1_V) = \text{Ker}(f + 1_V).$$

Similarly, $R(f + 1_V) = \text{Ker}(f - 1_V)$.

(c) Take a basis $\{\vec{x}_1, \ldots, \vec{x}_r\}$ for $\text{Ker}(f - 1_V)$ and a basis $\{\vec{x}_{r+1}, \ldots, \vec{x}_n\}$ for $\text{Ker}(f + 1_V)$ so that $\mathcal{B} = \{\vec{x}_1, \ldots, \vec{x}_r, \vec{x}_{r+1}, \ldots, \vec{x}_n\}$ is a basis for V. then

$$[f]_{\mathcal{B}} = \begin{bmatrix} I_r & 0 \\ 0 & -I_{n-r} \end{bmatrix}_{n \times m},$$

where r is the rank of $f + 1_V$ which is equal to $\dim \text{Ker}(f - 1_V)$.

9. Let V be an n-dimensional real vector space and $f \in L(V, V)$ satisfy

$$f^2 = f \circ f = -1_V.$$

(a) *The complexification of V* Define two operations on V as follows:

1. *Addition*: as the one originally defined on V.
2. *Scalar multiplication*: for $\vec{x} \in V$ and $a, b \in \mathbb{R}$,

$$(a + bi)\vec{x} = a\vec{x} + bf(\vec{x}).$$

Then V is an n-dimensional *complex* vector space.

(b) Now, consider V as a complex vector space. Then f can be regarded as a complex linear operator in the following sense:

$$f((a + ib)\vec{x}) = f(a\vec{x} + bf(\vec{x})) = af(\vec{x}) + bf^2(\vec{x}) = af(\vec{x}) - b\vec{x}$$
$$= (a + bi)f(\vec{x}), \quad a, b \in \mathbb{R}.$$

Therefore $f^2 = -1_V$ is equivalent to $(f - i1_V)(f + i1_V) = 0$.

(c) Note that

$$\mathrm{Ker}(f - i1_V) = \{\vec{x} \in V \mid f(\vec{x}) = i\vec{x}\},$$
$$\mathrm{Ker}(f + i1_V) = \{\vec{x} \in V \mid f(\vec{x}) = -i\vec{x}\}$$

and $\vec{x} = \frac{1}{2}(\vec{x} + if(\vec{x})) + \frac{1}{2}(\vec{x} - if(\vec{x}))$ for each $\vec{x} \in V$. Thus

$$V = \mathrm{Ker}(f - i1_V) \oplus \mathrm{Ker}(f + i1_V).$$

(d) Suppose $r = \dim \mathrm{Ker}(f - i1_V)$. There exist vectors $\vec{x}_{r+1}, \ldots,$ $\vec{x}_n \in V$ such that $(f - i1_V)(\vec{x}) = f(\vec{x}_j) - i\vec{x}_j, r + 1 \leq j \leq n$, from a basis for the range space $\mathrm{R}(f - i1_V)$. At the same time, for $r + 1 \leq j \leq n, (f - i1_V)(\vec{x}) \in \mathrm{Ker}(f + i1_V)$ which has dimension $n - r$. Therefore,

$$\mathrm{R}(f - i1_V) = \mathrm{Ker}(f + i1_V).$$

Similarly,

$$\mathrm{R}(f + i1_V) = \mathrm{Ker}(f - i1_V).$$

The vectors $(f + i1_V)(\vec{x}_j), r + 1 \leq j \leq n$, are linearly independent in $\mathrm{R}(f + i1_V)$. Then

$$n - r \leq r$$

holds. Similarly, $r \leq n - r$. Hence

$$n = 2r.$$

In particular, n is even and $r = \frac{n}{2}$.

(e) By (d), there exist linearly independent vectors $\vec{x}_1, \ldots, \vec{x}_r$ such that $(f - i1_V)(\vec{x}_j), 1 \leq j \leq r$, forms a basis for $\mathrm{R}(f - i1_V)$ and $(f + i1_V)(\vec{x}_j), 1 \leq j \leq r$, forms a basis for $\mathrm{R}(f + i1_V)$. They together form a basis $\mathcal{C}$ for V. Note that

$$x_j = \frac{1}{2i}[(f(\vec{x}_j) + i\vec{x}_j) - (f(\vec{x}_j) - i\vec{x}_j)],$$

$$f(\vec{x}_j) = \frac{1}{2}[(f(\vec{x}_j) + i\vec{x}_j) + (f(\vec{x}_j) - i\vec{x}_j)], \quad 1 \leq j \leq r.$$

Hence,

$$\mathcal{B} = \{\vec{x}_1, \ldots, \vec{x}_r, f(\vec{x}_1), \ldots, f(\vec{x}_r)\}$$

forms a basis for the *real* space V.

(f) Let $\mathcal{C}$ and $\mathcal{B}$ be as in (e). When consider V as an n-dimensional complex vector space, then

$$[f]_{\mathcal{C}} = \begin{bmatrix} i\mathrm{I}_n & O \\ O & -i\mathrm{I}_n \end{bmatrix}_{n \times n}, \quad r = \frac{n}{2} = \dim \mathrm{Ker}(f - i1_V);$$

while as a real vector space,

$$[f]_{\mathcal{B}} = \begin{bmatrix} O & \mathrm{I}_n \\ -\mathrm{I}_n & O \end{bmatrix}_{n \times n}.$$

10. Let $A = [a_{ij}]_{n \times n} \in \mathrm{M}(n; \mathbb{F})$. Define $g, h \colon \mathrm{M}(n; \mathbb{F}) \to \mathrm{M}(n; \mathbb{F})$ as

$$g(X) = XA,$$
$$h(X) = AX$$

for $X \in \mathrm{M}(n; \mathbb{F})$. Let $\mathcal{N} = \{E_{11}, E_{12}, \ldots, E_{1n}, \ldots, E_{n1}, \ldots, E_{nn}\}$ be the natural basis for $\mathrm{M}(n; \mathbb{F})$.

(a) Show that

$$[g]_{\mathcal{N}} = \begin{bmatrix} A & & & \\ & A & & 0 \\ & & \ddots & \\ 0 & & & A \end{bmatrix}_{n^2 \times n^2}$$

and $g(X) = [X]_{\mathcal{N}}[g]_{\mathcal{N}}$. $[g]_{\mathcal{N}}$ or g has rank

$$\mathrm{r}([g]_{\mathcal{N}}) = n\,\mathrm{r}(A)$$

and hence, g is invertible if and only if $\mathrm{r}(A) = n$, i.e. $\det A \neq 0$. The kernel space $\mathrm{Ker}(g)$ has dimension $n^2 - n\,\mathrm{r}(A) = n(n - \mathrm{r}(A))$.

(b) Show that

$$[h]_{\mathcal{N}} = \begin{bmatrix} a_{11}I_n & a_{21}I_n & \cdots & a_{n1}I_n \\ a_{12}I_n & a_{22}I_n & \cdots & a_{n2}I_n \\ \vdots & \vdots & & \vdots \\ a_{1n}I_n & a_{2n}I_n & \cdots & a_{nn}I_n \end{bmatrix}_{n^2 \times n^2}$$

and

$$r([h]_{\mathcal{N}}) = n\, r(A),$$
$$\det[h]_{\mathcal{N}} = (\det A)^n.$$

(c) Let $\mathcal{B} = \{E_{11}, E_{21}, \ldots, E_{n1}, \ldots, E_{1n}, E_{2n}, \ldots, E_{nn}\}$ be another ordered basis for $M(n; \mathbb{F})$. Show that

$$[g]_{\mathcal{B}} = [h]_{\mathcal{N}}^*.$$

11. Suppose $T: M(n; \mathbb{F}) \to M(n; \mathbb{F})$ is an *algebra homomorphism*, i.e.

1. T is a linear operator, and
2. T preserves matrix multiplication

$$T(XY) = T(X)T(Y), \quad X, Y \in M(n; \mathbb{F}).$$

Then, either $T = 0$ (zero linear operator) or there exists an invertible matrix $P \in M(n; \mathbb{F})$ such that

$$T(X) = PXP^{-1}, \quad X \in M(n; \mathbb{F}).$$

We may suppose that $T \neq 0$. Prove this result by the following steps.

(a) Show that

$$T(I_n) = I_n.$$
$$T(X^{-1}) = T(X)^{-1} \quad \text{if } X \text{ is invertible,}$$
$$T(QXQ^{-1}) = T(Q)T(X)T(Q)^{-1}.$$

(b) Now $I_n = \sum_{i=1}^{n} T(E_{ii})$. So at least one of $T(E_{ii}), 1 \leq i \leq n$, is not a zero matrix, say $T(E_{11}) \neq O$. For some suitable elementary matrices Q_1, Q_2 of type 1, $E_{ij} = Q_1 E_{11} Q_2$ holds for $1 \leq i, j \leq n$. Hence $T(E_{ij}) \neq O, 1 \leq i, j \leq n$.

(c) The rank $r(T(E_{ii})) = 1, 1 \leq i \leq n$. Let $\vec{x}_1 \in \mathbb{F}^n$ be such that $\vec{x}_1 \neq \vec{0}$ and $\vec{e}_1 T(E_{11}) = \vec{x}_1$. Let $\vec{x}_i = \vec{x}_1 T(E_{1i}), 2 \leq i \leq n$.

By using $E_{1k}E_{ij} = \delta_{ik}E_{1j}$,

$$\vec{x}_k T(E_{ij}) = \vec{x}_1 T(E_{1k})T(E_{ij}) = \vec{x}_1 T(E_{1k}E_{ij}) = \vec{x}_1 T(\delta_{ik}E_{1j})$$
$$= \delta_{ik}\vec{x}_1 T(E_{1j}) = \delta_{ik}\vec{x}_j, \quad 1 \le i, \ j \le n, 1 \le k \le n.$$

In particular, $\vec{x}_i T(E_{i1}) = \vec{x}_1 \ne \vec{0}$ implies that $\vec{x}_i \ne \vec{0}, 2 \le k \le n$. Also $\vec{x}_i T(E_{ii}) = \vec{x}_i, 2 \le i \le n$. Therefore, $\{\vec{x}_1, \vec{x}_2, \ldots, \vec{x}_n\}$ forms a basis for $\mathbb{F}^n$.

(d) Let

$$P = \begin{bmatrix} \vec{x}_1 \\ \vec{x}_2 \\ \vdots \\ \vec{x}_n \end{bmatrix}.$$

Then P is invertible and

$$T(E_{ij}) = PE_{ij}P^{-1}, \quad 1 \le i, j \le n$$

and hence $T(X) = PXP^{-1}, X \in \mathrm{M}(n; \mathbb{F})$.

12. Prove that the following are equivalent.

(a) V is a k-dimensional subspace of $\mathbb{F}^m$.
(b) There exist linearly independent vectors $\vec{x}_1, \vec{x}_2, \ldots, \vec{x}_k$ in V such that

$$V = \langle\langle \vec{x}_1, \vec{x}_2, \ldots, \vec{x}_k \rangle\rangle.$$

(c) There exists an $m \times n$ matrix A of rank $m - k$ such that

$$V = \{\vec{x} \in \mathbb{F}^m \mid \vec{x}A = \vec{0}\}.$$

(d) There exists an $n \times m$ matrix B of rank k such that

$$V = \{\vec{x}B \mid \vec{x} \in \mathbb{F}^n\}$$

Is it possible that $n < k$ or $n = k$?

13. Suppose V_1 and V_2 are subspaces of V and $V = V_1 \oplus V_2$. For each $\vec{x} \in V$, there exist unique $\vec{x}_1 \in V_1$ and $\vec{x}_2 \in V_2$ such that $\vec{x} = \vec{x}_1 + \vec{x}_2$. Define $p \colon V \to V_1$ by

$$p(\vec{x}) = \vec{x}_1.$$

Then $p \in \mathrm{L}(V, V)$. Such a p has the following geometric properties:

(1) p keeps vectors in V_1 fixed, i.e.

$$p(\vec{x}) = \vec{x}, \quad \vec{x} \in V_1.$$

(2) p maps each vector in V_2 into zero vector $\vec{0}$, i.e. $p(x) = \vec{0}$ for $\vec{x} \in V_2$.

(3) p projects each vector $\vec{x} = \vec{x}_1 + \vec{x}_2$ in V into $\vec{x}_1$ along the direction parallel to the complementary subspace V_2 of V_1 in V.

See Fig. B.3. Hence, p is called the *projection* (operator) of V onto V_1 along V_2. Note that $p(p(\vec{x})) = p(\vec{x})$ for each $\vec{x} \in V$, i.e. in short,

$$p^2 = p$$

(Refer to Ex. 7). Conversely, if $p \in L(V, V)$ is such that $p^2 = p$, then

(1) $V = \text{Ker}(p) \oplus \text{R}(p)$ where $\text{R}(p) = \{\vec{x} \in V \mid p(\vec{x}) = \vec{x}\}$, and

(2) p is the projection of V onto $\text{R}(p)$ along $\text{Ker}(p)$.

Try to explain why $1_V - p$ is the projection of V onto $\text{Ker}(p)$ along $\text{R}(p)$.

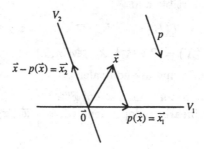

Fig. B.3

14. Prove that the following are equivalent. In cases (d), (e) and (f), V is supposed to be finite-dimensional. Let $V_1, \ldots, V_k$ be subspaces of V.

(a) $V = V_1 \oplus \cdots \oplus V_k$ and is called the *direct sum* of $V_1, \ldots, V_k$ if

(1) $V = \sum_{i=1}^{k} V_i$,

(2) $V_i \cap \left(\sum_{j \neq i} V_j \right) = \{\vec{0}\}$ for each $i, 1 \leq i \leq k$.

(b) $V = \sum_{i=1}^{k} V_i$, and, for any vector $\vec{x}_i \in V_i, 1 \leq i \leq k$ such that, if $\vec{x}_1 + \cdots + \vec{x}_k = \vec{0}$, then $\vec{x}_1 = \cdots = \vec{x}_k = \vec{0}$.

(c) Each vector $\vec{x} \in V$ can be uniquely expressed as $\vec{x} = \vec{x}_1 + \cdots + \vec{x}_k$, where $\vec{x}_i \in V_i, 1 \leq i \leq k$.

(d) If $\mathcal{B}_i$ is *any* ordered basis for $V_i, 1 \leq i \leq k$, then $\mathcal{B}_1 \cup \cdots \cup \mathcal{B}_k$ is an ordered basis for V.

(e) There exists *an* ordered basis $\mathcal{B}_i$ for $V_i, 1 \leq i \leq k$, such that $\mathcal{B}_1 \cup \cdots \cup \mathcal{B}_k$ is an ordered basis for V.

(f) $V = \sum_{i=1}^{k} V_i$ and $\dim V = \dim V_1 + \cdots + \dim V_k$.

(g) There exist linear operators $p_1, \ldots, p_k \in L(V, V)$ with $R(p_i) = V_i$ for $1 \le i \le k$ and satisfy:

(1) p_i is a projection, i.e. $p_i^2 = p_i, 1 \le i \le k$.

(2) $p_i \circ p_j = p_j \circ p_i = 0$ for $i \ne j, 1 \le i, j \le k$, i.e.

$$\mathrm{Ker}(p_i) = \bigoplus_{j \ne i} V_j.$$

(3) $1_V = p_1 + \cdots + p_k$, i.e.

$$\vec{x} = p_1(\vec{x}) + \cdots + p_k(\vec{x}), \quad \vec{x} \in V.$$

15. Suppose $\dim V < \infty$ and $f \in L(V, V)$. Prove the following:

(a) $\mathrm{Ker}(f) \subseteq \mathrm{Ker}(f^2) \subseteq \cdots \subseteq \mathrm{Ker}(f^k) \subseteq \mathrm{Ker}(f^{k+1}) \subseteq \cdots$ and $R(f) \supseteq R(f^2) \supseteq \cdots \supseteq R(f^k) \supseteq R(f^{k+1}) \supseteq \cdots$.

(b) There exists a positive integer k such that $R(f^m) = R(f^k)$ and $\mathrm{Ker}(f^m) = \mathrm{Ker}(f^k)$ for any positive integer $m \ge k$.

(c) There exists a positive integer n such that

$$\mathrm{Ker}(f^n) \cap R(f^n) = \{\vec{0}\}.$$

16. Let S be a vector subspace of V and V/S be the quotient space (refer to Sec. B.1).

(a) Define $\pi \colon V \to V/S$ by

$$\pi(\vec{x}) = \vec{x} + S \quad \text{for } \vec{x} \in V.$$

Then π is well-defined and is a linear transformation which is called the *natural projection* of V onto V/S.

(b) $V = S \oplus U$ if and only if the restriction $\pi|_U \colon U \to V/S$ is a linear isomorphism.

(c) In case $\dim V < \infty$, then

$$\dim V/S = \dim V - \dim S.$$

Once $\dim V/S < \infty$, it is called the *co-dimension* of S in V and is denoted as co-dim S.

17. Suppose $f \in L(V, W)$ and S, T are respective subspace of V, W such that $f(S) \subseteq T$ holds.

(a) Then there exists a unique linear transformation $\tilde{f}: V/S \to W/T$ such that the following diagram is commutative.

$$
\begin{array}{ccc}
V & \xrightarrow{\ f\ } & W \\
{\scriptstyle \pi_V}\downarrow & & \downarrow{\scriptstyle \pi_W} \\
V/S & \xrightarrow[\tilde{f}]{} & W/T
\end{array}
$$

Where π_V, π_W are natural projections (see Ex. 16).

(b) In case $W = V, T = S$, and S is an *f-invariant subspace*, i.e. $f(S) \subseteq S$, then there exists a unique linear transformation $\tilde{f}: V/S \to V/S$ such that $\tilde{f} \circ \pi = \pi \circ f$ where $\pi = \pi_V$ is the projection.

(c) There is a unique linear isomorphism $\tilde{f}: V/\mathrm{Ker}(f) \to R(f)$ defined by

$$\tilde{f}(\vec{x} + \mathrm{Ker}(f)) = f(\vec{x})$$

so that $f = \tilde{f} \circ \pi$ where $\pi: V \to V/\mathrm{Ker}(f)$ is the projection. This means that *every linear transformation can be expressed as the composite of an onto linear transformation and a linear isomorphism.*

18. Suppose $f \in L(V,W)$ and $g \in L(V,U)$ such that $\mathrm{Ker}(f) \subseteq \mathrm{Ker}(g)$ holds. Try to find $h \in L(W,U)$ such that $g = h \circ f$. The following diagram will guarantee the existence of such a linear transformation h

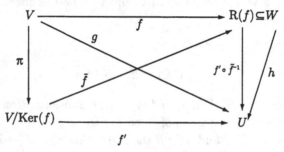

where

$\pi: V \to V/\mathrm{Ker}(f)$ is the natural projection;

$\tilde{f}: V/\mathrm{Ker}(f) \to R(f)$ is the linear isomorphism as in Ex. 17(c);

$f': V/\mathrm{Ker}(f) \to U$ is the linear transformation such that $g = f' \circ \pi$.
     This is possible because $\mathrm{Ker}(f) \subseteq \mathrm{Ker}(g)$ and f' is defined by
         $f'(\vec{x} + \mathrm{Ker}(f)) = g(\vec{x})$ for $\vec{x} \in V$;

$h' = f' \circ \tilde{f}^{-1} : R(f) \to U$ is linear; and

$h: W \to U$ is a linear extension of h' from $R(f)$ to the whole space W.

Then this h is a required one.

(a) $\text{Ker}(f) \subseteq \text{Ker}(g)$ if and only if there exists a linear transformation $h: W \to U$ such that

$$g = h \circ f.$$

In this case, g is said to be *decomposable* through f.

An element in $L(V, \mathbb{F})$ is specifically called a *linear functional*.

(b) Suppose $f_i \in L(V; \mathbb{F})$ for $1 \le i \le n$. Consider $\varphi = (f_1, \ldots, f_n) \in L(V; \mathbb{F}^n)$ and use Ex. 17(c), then

$$\text{co-dim} \bigcap_{i=1}^{n} \text{Ker}(f_i) \le n < +\infty.$$

(c) Suppose $f, f_i \in L(V, \mathbb{F})$ for $1 \le i \le n$. Then, by (a),

$$\bigcap_{i=1}^{n} \text{Ker}(f_i) \subseteq \text{Ker}(f)$$

if and only if f is a linear combination of $f_1, \ldots, f_n$, i.e. there exist scalars $a_1, \ldots, a_n \in \mathbb{F}$ such that

$$f = a_1 f_1 + \cdots + a_n f_n.$$

(*Note* This result lays the linearly algebraic foundation for the *Lagrange multiplier method* in solving constrained extremum problems (refer to Ex. <D> 8 of Sec. 5.6).)

19. The vector space $L(V, \mathbb{F})$ is usually denoted by

$$V^*$$

and is called the (first) *dual space* of V. For $\vec{x} \in V$ and $f \in V^*$, the scalar $f(\vec{x})$ is also denoted as $\langle \vec{x}, f \rangle$.

(a) For any $\vec{x} \in V$ and $\vec{x} \ne \vec{0}$, there exists an $f \in V^*$ such that $\langle \vec{x}, f \rangle \ne 0$ but $\langle \vec{y}, f \rangle = 0$, for any other $\vec{y} \in V$.

(b) Suppose $\dim V = n$ and $\mathcal{B} = \{\vec{x}_1, \ldots, \vec{x}_n\}$ is a basis for V. Then, there exists a unique basis $\mathcal{B}^* = \{f_1, \ldots, f_n\}$ for V^* such that

$$\langle \vec{x}_i, f_j \rangle = \delta_{ij}, \quad 1 \le i, \ j \le n.$$

$\mathcal{B}^*$ is called the *dual basis* of $\mathcal{B}$ in V^*. Therefore,

$$\vec{x} = \sum_{j=1}^{n} \langle \vec{x}, f_j \rangle \vec{x}_j, \quad \vec{x} \in V;$$

$$f = \sum_{i=1}^{n} \langle \vec{x}_i, f \rangle f_i, \quad f \in V^*.$$

In particular, $\dim V^* = \dim V = n$ and hence V^* is isomorphic to V.

(c) Let $P_n(\mathbb{F}), \mathcal{N} = \{1, t, t^2, \ldots, t^n\}$ and $\mathcal{B} = \{p_0, p_1, p_2, \ldots, p_n\}$ be as in Sec. B.3, where $p_i, 1 \leq p \leq n$, are Lagrange polynomials associated with the scalars $a_0, a_1, a_2, \ldots, a_n \in \mathbb{F}$. Let $\mathcal{B}^* = \{f_0, f_1, f_2, \ldots, f_n\}$ be the dual basis of $\mathcal{B}$ in $P_n(\mathbb{F})^*$. Note that $f_i(p) = p(a_i), 0 \leq i \leq n$, for $p \in P_n(\mathbb{F})$. Therefore, by (b), for any $p \in P_n(\mathbb{F})$,

$$p = \sum_{j=0}^{n} \langle p, f_j \rangle p_j = \sum_{j=0}^{n} f_j(p) p_j = \sum_{j=0}^{n} p(a_j) p_j$$

$$\Rightarrow p(t) = \sum_{j=0}^{n} p(a_j) p_j(t).$$

In particular, take $p = t^i$, then

$$t^i = \sum_{j=0}^{n} a_j^i p_j(t), \quad 0 \leq i \leq n.$$

The coefficient matrix

$$[a_j^i] = \begin{bmatrix} 1 & 1 & 1 & \cdots & 1 \\ a_0 & a_1 & a_2 & \cdots & a_n \\ \vdots & \vdots & \vdots & & \vdots \\ a_0^n & a_1^n & a_2^n & \cdots & a_n^n \end{bmatrix}$$

is the transition matrix from the basis $\mathcal{N}$ to the basis $\mathcal{B}$ and hence is invertible. This matrix is called *Vandermonde matrix* of order $n + 1$ and has the determinant

$$\det[a_j^i] = \prod_{0 \leq i < j \leq n} (a_j - a_i).$$

In general, $p(t) = \sum_{j=0}^{n} p(a_j) p_j(t)$ is nothing but the Lagrange interpolation formula mentioned in Sec. B.3.

20. Suppose S is a proper subspace of V such that, for any subspace U satisfying $S \subseteq U \subseteq V$, then it is necessarily that $U = S$ or $U = V$. Such a subspace S is called a *hypersubspace* of V.

(a) Let $f \in V^*$ and $f \neq 0$. Then, for any $\vec{x}_0 \in V \setminus \mathrm{Ker}(f)$,

$$V = \mathrm{Ker}(f) \oplus \langle\langle \vec{x}_0 \rangle\rangle$$

holds and $f(\alpha\vec{x}_0 + \mathrm{Ker}(f)) = \alpha f(\vec{x}_0)$, for any $\alpha \in \mathbb{F}$.

(b) S is a hypersubspace of V if and only if there exists a nonzero $f \in V^*$ such that

$$S = \mathrm{Ker}(f).$$

(c) Suppose $f, g \in V^*$. Then, $\mathrm{Ker}(f) \subseteq \mathrm{Ker}(g)$ if and only if there exists a scalar $\alpha \in \mathbb{F}$ such that

$$g = \alpha f.$$

In case $g = 0$, take $\alpha = 0$; if $g \neq 0$, then $\mathrm{Ker}(f) = \mathrm{Ker}(g)$ and $\alpha \neq 0$ can be taken as $g(\vec{x}_0)/f(\vec{x}_0)$ for any $\vec{x}_0 \in V \setminus \mathrm{Ker}(f)$.

(*Note* For $f \in V^*$ and $f \neq 0$, the mapping $\mathrm{Ker}(f) \leftrightarrow \langle\langle f \rangle\rangle$ sets up a one-to-one correspondence between the set of all hypersubspaces of V and the set of all one-dimensional subspaces of V^*.)

(d) Suppose $\dim V = n$. U is a k-dimensional subspace of V if and only if U is the intersection of $(n - k)$ hypersubspaces of V. That is to say, there exist linearly independent $f_1, \ldots, f_{n-k} \in V^*$ such that

$$U = \bigcap_{i=1}^{n-k} \mathrm{Ker}(f).$$

(e) Suppose $\dim V = n$. Let U and $f_1, \ldots, f_{n-k}$ be as in (d). For $f \in V^*$, then

$$U = \bigcap_{i=1}^{n-k} \mathrm{Ker}(f_i) \subseteq \mathrm{Ker}(f)$$

if and only if there exist scalars $\alpha_1, \ldots, \alpha_{n-k} \in \mathbb{F}$ such that $f = \alpha_1 f_1 + \cdots + \alpha_{n-k} f_{n-k}$, i.e. $f \in \langle\langle f_1, \ldots, f_{n-k} \rangle\rangle$.

21. For a vector space V, the dual space

$$(V^*)^* = V^{**}$$

of its first dual space V^* is called the *second dual space* of V.

(a) For each vector $\vec{x} \in V$, define $\vec{x}^{**}: V^* \to \mathbb{F}$ by

$$\vec{x}^{**}(f) = f(x) = \langle f, \vec{x} \rangle.$$

Then, $\vec{x}^{**} \in V^{**}$.

(b) Define $\Phi: V \to V^{**}$ by

$$\Phi(\vec{x}) = \vec{x}^{**}.$$

Then, Φ is a one-to-one linear transformation from V *into* V^{**}.

(c) Suppose $\dim V = n < \infty$. Then $\Phi: V \to V^{**}$ is a *natural isomorphism*. Therefore, every basis for V^* is a dual basis of some basis for V in the sense that

$$\langle \vec{x}, f \rangle = \langle f, \vec{x} \rangle, \quad \vec{x} \in V \text{ and } f \in V^*.$$

We call V and V^* are *dual to each other* relative to the symmetric bilinear functional $\langle, \rangle: V \times V^* \to \mathbb{F}$ defined by $\langle \vec{x}, f \rangle = f(\vec{x})$.

(d) Suppose $\dim V = n$. Let $\mathcal{B}$ and $\mathcal{B}^*, \mathcal{C}$ and $\mathcal{C}^*$ be two pairs of dual bases for V and V^* respectively. Then

$$Q = P^{*-1},$$

where $P = [1_V]_{\mathcal{B}}^{\mathcal{C}}$ and $Q = [1_{V^*}]_{\mathcal{B}^*}^{\mathcal{C}^*}$.

(e) Let $\mathcal{B} = \{\vec{x}_1, \ldots, \vec{x}_n\}$ and $\mathcal{B}^* = \{f_1, \ldots, f_n\}$, $\mathcal{C} = \{\vec{y}_1, \ldots, \vec{y}_n\}$ and $\mathcal{C}^* = \{g_1, \ldots, g_n\}$. Let $\phi, \psi: V \to V^*$ be the unique linear isomorphisms such that

$$\varphi(\vec{x}_i) = f_i \text{ and } \psi(\vec{y}_i) = g_i, \quad \text{for } 0 \le i \le n.$$

Then, $\varphi = \psi$ if and only if $P = Q$, i.e.

$$P = P^{*-1}.$$

22. Let S be a nonempty subset of a vector space V. The set

$$S^0 = \{f \in V^* \mid f(\vec{x}) = \vec{0} \text{ for all } \vec{x} \in S\}$$

is a vector subspace of V^* and is called the *annihilator* of S in V^*. In particular,

$$\{\vec{0}\}^0 = V^*,$$
$$V^0 = \{0\}.$$

(a) For a general vector space V, the following hold.

    (1) If S is a subspace of V and $\vec{x} \notin S$, then there exists $f \in S^0$ such that $f(\vec{x}) \ne 0$.

(2) If $S \subseteq V$ is a subspace, then V is isomorphic to the external direct sum space $S \oplus S^0 = \{(\vec{x}, f) \mid \vec{x} \in S \text{ and } f \in S^0\}$.

(3) Suppose S_1 and S_2 are subspaces of V. Then $S_1 = S_2$ if and only if $S_1^0 = S_2^0$.

(4) The subspace $\langle\langle S \rangle\rangle$ generated by S satisfies

$$\langle\langle S \rangle\rangle \subseteq (S^0)^0 = S^{00}.$$

(b) Suppose $\dim V = n < \infty$. The following hold.

1. Let S be a subspace of V. Then,

$$\dim S + \dim S^0 = \dim V.$$

In this case, for any basis $\{f_1, \ldots, f_k\}$ for S^0, $S = \bigcap_{i=1}^{k} \mathrm{Ker}(f_i)$.

2. $\langle\langle S \rangle\rangle = S^{00}$ for any nonempty subset S of V.

3. Suppose S_1 and S_2 are subspaces of V. Then,

$$(S_1 + S_2)^0 = S_1^0 \cap S_2^0;$$
$$(S_1 \cap S_2)^0 = S_1^0 + S_2^0.$$

(*Note*   For $f \in V^*$ and $f \neq 0$,

$$\mathrm{Ker}(f)^0 = \langle\langle f \rangle\rangle$$

holds. Hence (refer to Note in Ex. 20(c)), $S \to S^0$ sets up a one-to-one correspondence between the family of subspaces of V and the family of subspaces of V^*, but reverses the inclusion relation (i.e. $S_1 \subseteq S_2 \Rightarrow S_1^0 \supseteq S_2^0$). This reflects geometrically the *duality principal* between V and V^*.)

23. Suppose $\dim V = n < \infty$.

(a) In case, $V = S_1 \oplus S_2$, then S_1^0 is isomorphic to $S_2^* = \{f|_{s_2} \mid f \in V^*\}$ and S_2^0 is isomorphic to S_1^* and $V^* = S_1^0 \oplus S_2^0 = S_1^* \oplus S_2^*$.

(b) Conversely, if S is a subspace of V, then $V = S \oplus (S^*)^0$; $V^* = S^* \oplus S^0$.

(c) Suppose S is a subspace of V. Then

(1) S^0 is naturally isomorphic to $(V/S)^*$.

(2) S^* is naturally isomorphic to V^*/S^0.

24. Suppose $\varphi \in L(V, W)$. Then, there exists a unique $\varphi^* \in L(W^*, V^*)$ such that

$$\varphi^*(g) = g \circ \varphi, \quad g \in W^*$$

or

$$\langle \vec{x}, \varphi^*(g) \rangle = \langle \varphi(\vec{x}), g \rangle, \quad \vec{x} \in V, \ g \in W^*.$$

See the following diagram.

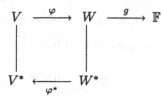

Such a φ^* is called the *adjoint* or the *dual* of φ. For matrix counterpart, see Sec. B.8. The mapping $\varphi \to \varphi^*$ has the following properties:

$$(\varphi + \psi)^* = \varphi^* + \psi^*;$$
$$(\alpha\varphi)^* = \alpha\varphi^*, \quad \text{for } \alpha \in \mathbb{F};$$
$$(\varphi \circ \psi)^* = \psi^* \circ \varphi^*, \quad \text{if } \psi \in L(W, U);$$
$$\varphi^{**} = (\varphi^*)^* = \varphi, \quad \text{if } \dim V < \infty \text{ and } \dim W < \infty.$$

(a) Suppose $\dim V = m < \infty$ and $\dim W = n < \infty$. Let $\mathcal{B}$ be a basis for V and $\mathcal{B}^*$ its dual basis in V^*, and $\mathcal{C}$ a basis for W and $\mathcal{C}^*$ its dual basis in W^*. Then, for $\varphi \in L(V, W)$,

$$[\varphi^*]_{\mathcal{B}^*}^{\mathcal{C}^*} = ([\varphi]_{\mathcal{C}}^{\mathcal{B}})^*,$$

that is, $n \times m$ matrix $[\varphi^*]_{\mathcal{B}^*}^{\mathcal{C}^*}$ is the transpose of $m \times n$ matrix $[\varphi]_{\mathcal{C}}^{\mathcal{B}}$.

(b) Suppose $\dim V = m < \infty$ and $\dim W = n < \infty$. Then, for any $\varphi \in L(V, W)$,

$$\mathrm{Ker}(\varphi) = R(\varphi^*)^0, \quad R(\varphi) = \mathrm{Ker}(\varphi^*)^0;$$
$$\mathrm{Ker}(\varphi^*) = R(\varphi)^0, \quad R(\varphi^*) = \mathrm{Ker}(\varphi)^0;$$

by using the dimension theorems, $\dim \mathrm{Ker}(\varphi) + \dim R(\varphi) = \dim V = m$ and $\mathrm{Ker}(\varphi) + \dim \mathrm{Ker}(\varphi)^0 = \dim V = m$. See Fig. B.4.

(c) Suppose $\dim V < \infty$ and $\dim W < \infty$. Then, for $\varphi \in L(V, W)$,

$$\mathrm{r}(\varphi) = \mathrm{r}(\varphi^*),$$

i.e. φ and its adjoint φ^* have the same rank. In matrix terminology, this means that the row rank of a matrix is equal to its column rank (refer to Sec. B.5 and Ex. 2).

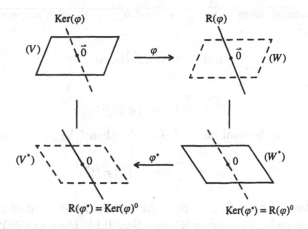

$$R(\varphi^*) = \text{Ker}(\varphi)^0 \qquad\qquad \text{Ker}(\varphi^*) = R(\varphi)^0$$

Fig. B.4

(d) Let $f_1, f_2, f_3 \in (\mathbb{C}^4)^*$ be defined as

$$f_1(\vec{x}) = ix_1 + x_2 - 3ix_3 + 4x_4,$$
$$f_2(\vec{x}) = -x_1 + x_3 - ix_4,$$
$$f_3(\vec{x}) = -2x_2 + (1+i)x_3 + ix_4.$$

Find the unique subspace S of $\mathbb{C}^4$ such that $S^0 = \langle\langle f_1, f_2, f_3 \rangle\rangle$.
Define $f \colon \mathbb{C}^4 \to \mathbb{C}^3$ by $f(\vec{x}) = (f_1(\vec{x}), f_2(\vec{x}), f_3(\vec{x}))$ and use it to
justify results stated in (b).

(e) Let S be the subspace of $\mathbb{R}^5$ generated by

$$\vec{x}_1 = (1, -1, 2, 3, 0), \quad \vec{x}_2 = (-1, 1, 2, 5, 2),$$
$$\vec{x}_3 = (0, 0, -1, -2, 3), \quad \vec{x}_4 = (2, -2, 3, 4, -1)$$

Find S^0.

It is worth saying something more about $\langle \vec{x}, f \rangle$ where $\vec{x} \in V$ and $f \in V^*$
in case $\dim V = n < \infty$.

Suppose $\mathcal{B} = \{\vec{x}_1, \dots, \vec{x}_n\}$ is a basis for V, and $\mathcal{D} = \{f_1, \dots, f_n\}$ is
a basis for V^* which is not necessarily the dual basis of $\mathcal{B}$ in V^*. For
$\vec{x} = \sum_{i=1}^n a_i \vec{x}_i$ and $f = \sum_{j=1}^n b_j f_j$, by bilinearity of $\langle , \rangle$,

$$\langle \vec{x}, f \rangle = f(\vec{x}) = \sum_{i,j=1}^n a_i b_j \langle \vec{x}_i, f_j \rangle,$$

which is a *bilinear form* in $a_1, \ldots, a_n$ and $b_1, \ldots, b_n$ with *symmetric* coefficient matrix

$$\langle,\rangle_{\mathcal{B},\mathcal{D}} = [\langle \vec{x}_i, f_j \rangle]_{n \times n}.$$

Thus,

$$\langle \vec{x}, f \rangle = [\vec{x}]_{\mathcal{B}} \langle,\rangle_{\mathcal{B},\mathcal{D}} [f]_{\mathcal{D}}^*.$$

In case $\mathcal{D} = \mathcal{B}^*$ is the dual basis of $\mathcal{B}$ in V^*, then $\langle,\rangle_{\mathcal{B},\mathcal{B}^*} = I_n$ and

$$\langle \vec{x}, f \rangle = \sum_{i=1}^{n} a_i b_i = [\vec{x}]_{\mathcal{B}} [f]_{\mathcal{B}^*}^*.$$

Just a stone's throw from here is the *natural inner product* on pairs of vectors in the real vector space $\mathbb{R}^n$ (see Sec. B.9). Identify $(\mathbb{R}^n)^*$ with $\mathbb{R}^n$ in the sense that the dual basis $\mathcal{N}^* = \{f_1, \ldots, f_n\}$ of the natural basis $\mathcal{N} = \{\vec{e}_1, \ldots, \vec{e}_n\}$ is so that $f_i = \vec{e}_i$ for $1 \leq i \leq n$ and $f = \sum_{i=1}^{n} y_i f_i \in (\mathbb{R}^n)^*$ is equal to $\sum_{i=1}^{n} y_i \vec{e}_i = \vec{y} \in \mathbb{R}^n$. Then,

$$\langle \vec{x}, \vec{y} \rangle = \sum_{i=1}^{n} x_i y_i$$

is the required inner product in $\mathbb{R}^n$. $\vec{x}$ is said to be *perpendicular* to $\vec{y}$ if $\langle \vec{x}, \vec{y} \rangle = 0$. Isomorphism between $\mathbb{R}^n$ and $(\mathbb{R}^n)^*$ will guarantee the justification of this viewpoint. See Ex. 5 of Sec. B.9.

Bilinear functionals or the resulted quadratic forms by themselves form an important topic in linear algebra, especially related to algebraic geometry. These materials are beyond the scope of an elementary book such as this one.

B.8 A Matrix and its Transpose

Theory concerning the transpose of a linear transformation as present in Ex. 24 of Sec. B.7 seems simpler and neater in the eyes of matrices. The pictures here aim to illustrate this by centering around the solution of the system of linear equations.

Let $A = [a_{ij}] \in M(m, n; \mathbb{F})$. Then its transpose $A^* \in M(n, m; \mathbb{F})$.

For simplicity, vectors $\vec{x}$ in $\mathbb{F}^m$ would view as a $1 \times m$ matrix so that $\vec{x}A$ is meaningful both as a matrix product and as the action of the linear transformation A acting on the vector $\vec{x}$. Instead of emphasizing $\vec{x}$ as a column vector whenever needed, we will use $\vec{x}^*$ to represent both as a vector and as an $m \times 1$ matrix, the transpose of $1 \times m$ matrix $\vec{x}$.

Associated with A are the following two linear transformations:

A: $F^m \to F^n$ defined as $\vec{x} \to \vec{x}A$,

A^*: $F^n \to F^m$ defined as $\vec{y} \to \vec{y}A^*$ which is equivalently to say that $\vec{y}^* \to A\vec{y}^*$.

and four vector subspaces (refer to Application 8 in Sec. B.5 for their computations):

$R(A) = \{\vec{x}A \mid \vec{x} \in F^n\}$: the row space of A, i.e. $\mathrm{Im}(A)$,

$N(A) = \{\vec{x} \in F^m \mid \vec{x}A = \vec{0}\}$: the *left kernel space* of A, i.e. $\mathrm{Ker}(A)$,

$R(A^*) = \{\vec{y}A^* \mid \vec{y} \in F^n\}$: the column space of A, i.e. $\mathrm{Im}(A^*)$,

$N(A^*) = \{\vec{y} \in F^n \mid \vec{y}A^* = \vec{0}\}$: the *right kernel space* of A, i.e. $\mathrm{Ker}(A^*)$.

In order to simulate the concept of perpendicularity of vectors in $\mathbb{R}^n$, we would use the notation

$$S^\perp$$

to denote the *annihilator* of a nonempty subset S of $\mathbb{F}^m$ instead of S^0 as we did in Sec. B.7. Also, dual space $(\mathbb{F}^m)^*$ is always regarded as $\mathbb{F}^m$ itself under isomorphism.

For any $\vec{x} \in \mathbb{F}^m$ and $\vec{y} \in N(A^*) \subseteq \mathbb{F}^n$,

$$\langle \vec{x}A, \vec{y} \rangle = (\vec{x}A)\vec{y}^* = \vec{x}(A\vec{y}^*) = \vec{x}\vec{0}^* = 0$$

always holds. It comes immediately the following facts:

1. $R(A)^\perp = N(A^*)$ and $R(A) = N(A^*)^\perp$, and $\mathbb{F}^n = N(A^*) \oplus R(A)$ with $\dim N(A^*) = n - r, r = \dim R(A) = \mathrm{r}(A)$, the rank of A.
2. $R(A^*)^\perp = N(A)$ and $R(A^*) = N(A)^\perp$, and $\mathbb{F}^m = N(A) \oplus R(A^*)$ with $\dim N(A) = m - r, r = \dim R(A^*) = \mathrm{r}(A)$.

For geometric intuition, remember that every vector in the left kernel space is "perpendicular" to every vector in the column space.

For every $\vec{x} \in \mathbb{F}^m$, $\vec{x}A \in R(A) \subseteq \mathbb{F}^n$. On the other hand, there exists $\vec{y} \in \mathbb{F}^n$ such that the expression

$$\vec{x} = \vec{x} - \vec{y}A^* + \vec{y}A^*, \quad \vec{x} - \vec{y}A^* \in N(A) \text{ and } \vec{y}A^* \in R(A^*)$$

is unique. Note that there are many such $\vec{y}$. Therefore

$$\vec{x}A = (\vec{y}A^*)A = \vec{y}A^*A$$

holds. This implies that

$$A|_{R(A^*)}\colon R(A^*) \to R(A)$$

is a linear transformation and is an isomorphism too. This is because that, if $\vec{x}A = \vec{0}$, then $\vec{x} \in N(A)$ which induces that $\vec{y}A^* \in N(A)$ and in turn $\vec{y}A^* = \vec{0}$.

Dual results also hold for A^*.

We summarize these in the following

Theorem *Let $A \in M(m, n; \mathbb{F})$. Consider $A: \mathbb{F}^m \to \mathbb{F}^n$ as the linear transformation defined as $\vec{x} \to \vec{x}A$ and $A^*: \mathbb{F}^n \to \mathbb{F}^m$ as $\vec{y} \to \vec{y}A^*$.*

1. $A|_{R(A^*)}: R(A^*) \to R(A)$ *defined by*

$$\vec{y}A^* \to (\vec{y}A^*)A = \vec{y}A^*A = \vec{x}A, \quad \vec{y} \in \mathbb{F}^n$$

is a linear isomorphism where $\vec{x} = \vec{x} - \vec{y}A^* + \vec{y}A^*, \vec{x} - \vec{y}A^* \in N(A)$ and $\vec{y}A^* \in R(A^*)$ for some $\vec{y} \in \mathbb{F}^n$ if $\vec{x}$ is given.

2. $A^*|_{R(A)}: R(A) \to R(A^*)$ *defined by*

$$\vec{x}A \to (\vec{x}A)A^* = \vec{x}AA^* = \vec{y}A^*, \quad \vec{x} \in \mathbb{F}^m$$

is a linear isomorphism where $\vec{y} = \vec{y} - \vec{x}A + \vec{x}A, \vec{y} - \vec{x}A \in N(A^*)$ and $\vec{x}A \in R(A)$ for some $\vec{x} \in \mathbb{F}^m$ if $\vec{y}$ is given.

See Fig. B.5.

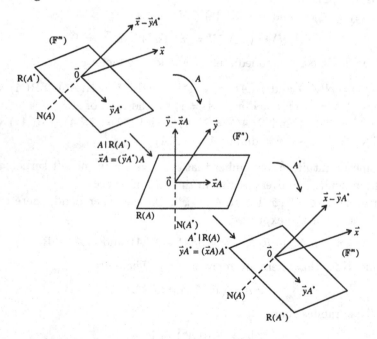

Fig. B.5

The materials here and the problems in the forthcoming exercises are briefly taken from [4]. And readers are highly recommended to catch a glimpse of Gilbert Strang's article [17].

Exercises

$A = [a_{ij}] \in M(m, n; \mathbb{F})$ throughout the following problems unless specified noted .

1. Let

$$A = \begin{bmatrix} 1 & -1 & 2 & 3 & 0 \\ -1 & 1 & 2 & 5 & 2 \\ 0 & 0 & -1 & -2 & 3 \\ 2 & -2 & 3 & 4 & -1 \end{bmatrix}.$$

(a) Find $R(A)$ and $N(A^*)$ so that $\mathbb{R}^5 = R(A) \oplus N(A^*)$.

(b) Find $R(A^*)$ and $N(A)$ so that $\mathbb{R}^4 = R(A^*) \oplus N(A)$.

(c) Give $\vec{x} = (x_1, x_2, x_3, x_4) \in \mathbb{R}^4$ and determine all possible $\vec{y} = (y_1, y_2, y_3, y_4, y_5) \in \mathbb{R}^5$ such that $\vec{x} - \vec{y}A^* \in N(A)$ and hence, $\vec{y}A^*A = \vec{x}A$ holds.

2. Consider the system of linear equations: $A\vec{y}^* = \vec{b}^*$, where $\vec{y} \in \mathbb{F}^n$ and $\vec{b} \in \mathbb{F}^m$, which is equivalent to $\vec{y}A^* = \vec{b}$.

(a) *Existence of solution* The following are equivalent.

   (1) $A\vec{y}^* = \vec{b}^*$ has at least one solution for any $\vec{b} \in \mathbb{F}^m$.

   (2) The column space $R(A^*) = \mathbb{F}^m$, i.e. $r(A) = m$.

   (3) A is right-invertible, i.e. there exists $B \in M(n, m; \mathbb{F})$ such that $AB = I_m$.

(b) *Uniqueness of solution* The following are equivalent.

   (1) $A\vec{y}^* = \vec{b}^*$ has at most one solution for any $\vec{b} \in \mathbb{F}^m$.

   (2) The n column vectors of A form a basis for the column space $R(A^*)$, i.e. $r(A) = n \leq m$.

   (3) A is left-invertible, i.e. there exists $C \in M(n, m; \mathbb{F})$ such that $CA = I_n$.

(c) *Existence and uniqueness of the solution* The following are equivalent.

(1) $A\vec{y}^* = \vec{b}^*$ has a unique solution for any $\vec{b} \in \mathbb{F}^m$.
(2) $r(A) = n = m$.
(3) A is invertible.

In this case, the unique solution is $\vec{y}^* = A^{-1}\vec{b}^*$ or $\vec{y} = \vec{b}\,A^{*-1}$.

(*Note* Change $A\vec{y}^* = \vec{b}^*$ back to $\vec{y}A^* = \vec{b}$. We use part of Fig. B.5 to interpret results obtained here. See Fig. B.6. For any given $\vec{b} \in R(A^*) \subseteq \mathbb{F}^m$, the equation $\vec{y}A^* = \vec{b}$ certainly has a solution $\vec{x}_0 A \in R(A)$ and its solution set is an $(n - r(A))$-dimensional affine subspace $\vec{x}_0 A + N(A^*)$ of $\mathbb{F}^n$.)

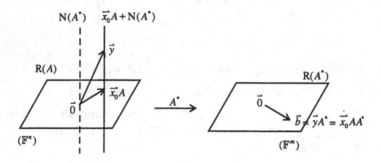

Fig. B.6

3. *Right and left invertible matrices of A*

(a) Suppose $r(A) = m < n$. Then the set of all right invertible matrices of A is

$$\mathcal{R} = \{(\vec{y}_1^*, \ldots, \vec{y}_m^*) \in M(n, m; \mathbb{F}) \mid \text{each } \vec{y}_i \in \mathbb{F}^n \text{ is a solution of}$$
$$A\vec{y}^* = \vec{e}_i^* \in \mathbb{F}^m \text{ for } 1 \leq i \leq m\}.$$

For each $i, 1 \leq i \leq m$, the solution set of $A\vec{y}^* = \vec{e}_i^* \in \mathbb{F}^m$ is an $(n - m)$-dimensional affine subspace of $\mathbb{F}^n$. Therefore, A has infinitely many right invertible matrices if $\mathbb{F}$ is an infinite field. In particular, a right invertible matrix is

$$A^*(AA^*)^{-1}.$$

(b) Suppose $r(A) = n < m$. Then the set of all left invertible matrices of A is

$$L = \left\{ \begin{bmatrix} \vec{x}_1 \\ \vec{x}_2 \\ \vdots \\ \vec{x}_n \end{bmatrix} \in M(n, m; \mathbb{F}) \middle|\ \text{each } \vec{x}_i \in \mathbb{F}^m \text{ is a solution of} \right.$$

$$\left. \vec{x} A = \vec{e}_i \in \mathbb{F}^n \text{ for } 1 \leq i \leq n \right\}.$$

A particular left invertible matrix is

$$(A^*A)^{-1}A^*.$$

4. *Solution set of* $A\vec{y}^* = \vec{b}^*$ *where* $\vec{y} \in \mathbb{F}^n$ *and* $\vec{b} \in \mathbb{F}^m, \vec{b} \neq \vec{0}$ Suppose $r(A) = r$ and

$$A = \begin{cases} (A_{11} \quad A_{12}), & r(A) = r(A_{11}) = m, \\ \begin{bmatrix} A_{11} & A_{12} \\ A_{21} & A_{22} \end{bmatrix}, & r(A_{11}) = r < m, \end{cases}$$

$$\vec{b}^* = \begin{cases} \vec{b}^*, & r(A) = m, \\ \begin{bmatrix} \vec{b}_1^* \\ \vec{b}_2^* \end{bmatrix}, & r(A) = r < m, \quad \vec{b}_1 \in \mathbb{F}^r, \ \vec{b}_2 \in \mathbb{F}^{m-r}. \end{cases}$$

(a) $A\vec{y}^* = \vec{b}^*$ has a solution if and only if

  (1) in case $r(A) = m$, or
  (2) in case $r(A) = r < m$, then $\vec{b}_2^* = A_{21}A_{11}^{-1}\vec{b}_1^*$ holds.

(b) The solution set is

$$\left\{ \vec{y} \in \mathbb{F}^n \middle|\ \vec{y}^* = \begin{bmatrix} A_{11}^{-1}\vec{b}_1^* - A_{11}^{-1}A_{12}\vec{y}_1^* \\ \vec{y}_1^* \end{bmatrix}, \vec{y}_1 \in \mathbb{F}^{n-r} \right\}$$

$$= (\vec{b}_1(A_{11}^*)^{-1}, \vec{0})_{1 \times n}$$
$$+ \{((-\vec{y}_1 A_{12}^*(A_{11}^*)^{-1}, \vec{y}_1)_{1 \times n} \in \mathbb{F}^n \mid \vec{y}_1 \in \mathbb{F}^{n-r}\}$$

which is an $(n-r)$-dimensional affine subspace of $\mathbb{F}^n$. Note that, if $r(A) = n = r, \mathbb{F}^{n-r} = \{\vec{0}\}$.

5. *Solution set of* $A\vec{y}^* = \vec{b}^*$ *where* $\vec{y} \in \mathbb{F}^n$ *and* $\vec{b} \in \mathbb{F}^m, \vec{b} \neq \vec{0}$

(a) Suppose $r(A) = m < n$. The solution set

$$\{\vec{y} \in \mathbb{F}^n \mid \vec{y}^* = B\vec{b}^* \text{ where } B \text{ is any right invertible matrix of } A\}$$

is an $(n - m)$-dimensional affine subspace of $\mathbb{F}^n$ with a particular solution

$$(A^*(AA^*)^{-1}\vec{b}^{\,*})^* = \vec{b}(AA^*)^{-1}A.$$

(b) Suppose $r(A) = n < m$. If $A\vec{y}^{\,*} = \vec{b}^{\,*}$ has a solution (i.e. $r[A \mid \vec{b}^{\,*}] = r(A)$), then the unique solution is the solution of $A^*A\vec{y}^{\,*} = A^*\vec{b}^{\,*}$, which is

$$((A^*A)^{-1}A^*\vec{b}^{\,*})^* = \vec{b}A(A^*A)^{-1}$$

(refer to Ex. 2(b) and Ex. 3(b)).

6. Examples for Exs. 4 and 5.

(a) Let

$$A = [A_{11} \ \ A_{12}]_{2\times 4}, \quad \text{where } A_{11} = \begin{bmatrix} 1 & 1 \\ 1 & -2 \end{bmatrix} \text{ and } A_{12} = \begin{bmatrix} -2 & 3 \\ 1 & -1 \end{bmatrix}.$$

Then $r(A) = 2 < 4$. The right invertible matrices of A are of the form

$$B = \begin{bmatrix} \frac{2}{3} + z_1 - \frac{5}{3}z_2 & \frac{1}{3} + z_3 - \frac{5}{3}z_4 \\ \frac{1}{3} + z_1 - \frac{4}{3}z_2 & -\frac{1}{3} + z_3 - \frac{4}{3}z_4 \\ z_1 & z_3 \\ z_2 & z_4 \end{bmatrix}_{4\times 2}, \quad z_1, z_2, z_3, z_4 \in \mathbb{F}.$$

For $\vec{b} = (b_1, b_2) \in \mathbb{F}^2$, $A\vec{y}^{\,*} = \vec{b}^{\,*}$ has solutions

$$\vec{y} = \left(\frac{2}{3}b_1 + \frac{1}{3}b_2, \frac{1}{3}b_1 - \frac{1}{3}b_2, 0, 0\right)$$

$$+ \alpha(1, 1, 1, 0) + \beta\left(-\frac{5}{3}, -\frac{4}{3}, 0, 1\right),$$

where $\alpha, \beta \in \mathbb{F}$.

(b) Let

$$A = \begin{bmatrix} 0 & 1 & 1 \\ 1 & 0 & 1 \\ 1 & 1 & 0 \\ 4 & 1 & -1 \end{bmatrix}_{4\times 3}.$$

Then $r(A) = 3 < 4$. A has left invertible matrices of the form

$$B = \begin{bmatrix} -\frac{1}{2} + 2z_1 & \frac{1}{2} - z_1 & \frac{1}{2} - 3z_1 & z_1 \\ \frac{1}{2} + 2z_2 & -\frac{1}{2} - z_2 & \frac{1}{2} - 3z_2 & z_2 \\ \frac{1}{2} + 2z_3 & \frac{1}{2} - z_3 & -\frac{1}{2} - 3z_3 & z_3 \end{bmatrix}, \quad z_1, z_2, z_3 \in \mathbb{F}.$$

For $\vec{b} = (b_1, b_2, b_3, b_4) \in \mathbb{F}^4$, the possible solution of $A\vec{y}^* = \vec{b}^*$ is

$$\vec{y}^* = B\vec{b}^* = \begin{bmatrix} -\frac{1}{2}b_1 + \frac{1}{2}b_2 + \frac{1}{2}b_3 + z_1(2b_1 - b_2 - 3b_3 + b_4) \\ \frac{1}{2}b_1 - \frac{1}{2}b_2 + \frac{1}{2}b_3 + z_2(2b_1 - b_2 - 3b_3 + b_4) \\ \frac{1}{2}b_1 + \frac{1}{2}b_2 - \frac{1}{2}b_3 + z_3(2b_1 - b_2 - 3b_3 + b_4) \end{bmatrix}.$$

Therefore, $A\vec{y}^* = \vec{b}^*$ has a solution if and only if $r(A) = r[A \mid \vec{b}^*]$ which induces that $2b_1 - b_2 - 3b_3 + b_4 = 0$. In this case, the unique solution is

$$\vec{y} = \vec{b}\,A(A^*A)^{-1}$$

$$= (b_1 \quad b_2 \quad b_3 \quad b_4) \begin{bmatrix} 0 & 1 & 1 \\ 1 & 0 & 1 \\ 1 & 1 & 0 \\ 4 & 1 & -1 \end{bmatrix} \cdot \frac{1}{60} \begin{bmatrix} 9 & -15 & 9 \\ -15 & 45 & -15 \\ 9 & -15 & 29 \end{bmatrix}$$

$$= \left(-\frac{1}{2}b_1 + \frac{1}{2}b_2 + \frac{1}{2}b_3, \frac{1}{2}b_1 - \frac{1}{2}b_2 + \frac{1}{2}b_3, \frac{1}{2}b_1 + \frac{1}{2}b_2 - \frac{1}{2}b_3 \right),$$

which coincides with $\vec{y}^* = B\vec{b}^*$ or $\vec{y} = \vec{b}B^*$ above.

Exercises 7–14 are concerned with real or complex $m \times n$ matrix $A = [a_{ij}]_{m \times n}$. By the way, the readers are required to possess basic knowledge about inner product $\langle , \rangle$ in $\mathbb{R}^n$ or $\mathbb{C}^n$ (see Sec. B.9, if necessary).

7. *The geometric interpretation of the solution* $\vec{b}\,(AA^*)^{-1}A$ *of* $A\vec{y}^* = \vec{b}^*$ *with* $\vec{b} \neq \vec{0}$ Suppose $r(A) = m < n$. Thus, the solution $\vec{b}\,(AA^*)^{-1}A$ is the one that makes $|\vec{y}|$ minimum among so many solutions of $A\vec{y}^* = \vec{b}^*$ (see Exs. 4 and 5). That is, the *distance* from $\vec{0}$ to the solution set of $A\vec{y}^* = \vec{b}^*$

$$\min_{A\vec{y}^* = \vec{b}^*} |\vec{y}|$$

is obtained at $\vec{b}\,(AA^*)^{-1}A$ with

   perpendicular vector: $\vec{b}\,(AA^*)^{-1}A,$
   distance: $|\vec{b}\,(AA^*)^{-1}A|.$

See Fig. B.7.

8. *The geometric interpretation of* $\vec{b}\,A(A^*A)^{-1}$ *in case* $r(A) = n < m$ *and* $\vec{b} \neq \vec{0}$ For a given $\vec{b} \in \mathbb{R}^m$ (or $\mathbb{C}^m$), as an *approximate solution* $\vec{y} \in \mathbb{R}^n$ to $A\vec{y}^* = \vec{b}^*$, the *error function* $|\vec{b}^* - A\vec{y}^*| = |\vec{b} - \vec{y}A^*|$ attains its minimum at $\vec{b}\,A(A^*A)^{-1}$. In particular, if $A\vec{y}^* = \vec{b}^*$ or

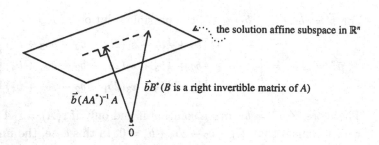

the solution affine subspace in $\mathbb{R}^n$

$\vec{b}B^*$ (B is a right invertible matrix of A)

$\vec{b}(AA^*)^{-1}A$

$\vec{0}$

Fig. B.7

$\vec{y}A^* = \vec{b}$ has a solution, then $|\vec{b} - \vec{y}A^*|$ attains its minimum 0 at the unique solution $\vec{b}A(AA^*)^{-1}$. That is,

$$\min_{\vec{y} \in \mathbb{R}^n} |\vec{b} - \vec{y}A^*|$$

is obtained at $\vec{b}A(A^*A)^{-1}$. The linear operator $A(A^*A)^{-1}A^*$: $\mathbb{R}^m \to R(A^*) \subseteq \mathbb{R}^m$ is the *orthogonal projection* of $\mathbb{R}^m$ onto the column space $R(A^*)$ of A with

$$\text{projection vector: } \vec{b}A(AA^*)^{-1}A^* \in R(A^*),$$
$$\text{distance vector: } \vec{b} - \vec{b}A(AA^*)^{-1}A^* \in N(A),$$
$$\text{the left kernel of } A, \text{ and}$$
$$\text{distance from } \vec{b} \text{ to } R(A^*): |\vec{b} - \vec{b}A(AA^*)^{-1}A^*|.$$

See Fig. B.8.

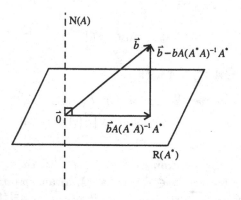

$N(A)$

$\vec{b}$

$\vec{b} - bA(A^*A)^{-1}A^*$

$\vec{0}$

$\vec{b}A(A^*A)^{-1}A^*$

$R(A^*)$

Fig. B.8

In short, $\vec{y}A^* = \vec{b}$ is not possible for general $\vec{b} \in \mathbb{R}^m$ but $\vec{b}A(A^*A)^{-1}$ is the optimal solution of $\vec{y}A^* = \vec{b}$ in the sense that it minimizes the quantities $|\vec{b} - \vec{y}A^*|$ as $\vec{y}$ varies in $\mathbb{R}^n$ by solving $\vec{y}A^*A = \vec{b}A$. This is the so-called *least square problems*. Take Ex. 6(b) as our example here. $|\vec{b} - \vec{y}A^*|$ has its minimum at

$$\vec{b}A(A^*A)^{-1} = \frac{1}{60}(b_1 \quad b_2 \quad b_3 \quad b_4) \begin{bmatrix} -6 & 30 & 14 \\ 18 & -30 & 38 \\ -6 & 30 & -6 \\ 12 & 0 & -8 \end{bmatrix}$$

$$= \frac{1}{30}(-3b_1 + 9b_2 - 3b_3 + 6b_4,$$

$$15b_1 - 15b_2 + 15b_3, 7b_1 + 19b_2 - 3b_3 - 4b_4).$$

As a summary, it is worth reviewing some of main results in order to picture what will be going on in the sequel. For $A_{m \times n}$,

(1) If $r(A) = m = n$, then A has invertible matrix A^{-1}.
(2) If $r(A) = m < n$, then A has right invertible matrix $A^* (AA^*)^{-1}$.
(3) If $r(A) = n < m$, then A has left invertible matrix $(A^*A)^{-1}A^*$.

What happens if $r(A) = r < \min(m, n)$? To answer this question, Exercises 9–14 investigate the *generalized inverse matrix* or *pseudoinverse matrix* introduced by E. H. Moore (1935) and R. Penrose (1953).

9. *Geometric mapping behavior of AA^**: $\mathbb{R}^m \to R(A^*) \subseteq \mathbb{R}^m$ The following are equivalent (refer to Fig. B.5).

   (a) $\vec{x}AA^* = \vec{y}A^*, \vec{x} \in \mathbb{R}^m$ and $\vec{y} \in \mathbb{R}^n$.
   (b) $\vec{y} - \vec{x}A \in N(A^*) = R(A)^\perp$. Therefore, $\vec{y} = \vec{y} - \vec{x}A + \vec{x}A$, $\vec{x}A \in R(A)$.
   (c) $\vec{x}A$ is the *orthogonal projection* (see Sec. B.9) of $\vec{y}$ onto the row space $R(A)$ of A.
   (d) $|\vec{y} - \vec{x}A| = \min_{\vec{z} \in \mathbb{R}^m} |\vec{y} - \vec{z}A|$, i.e. $|\vec{y} - \vec{x}A|$ is the distance from $\vec{y}$ to the row space $R(A)$.

   Try to figure out corresponding results for the mapping A^*A: $\mathbb{R}^n \to R(A) \subseteq \mathbb{R}^n$.

10. *Characterization of AA^**: $\mathbb{R}^m \to R(A^*)$ *as an orthogonal projection* (see Sec. B.9) The following are equivalent (refer to Fig. B.5).

   (a) (geometric) AA^*: $\mathbb{R}^m \to R(A^*) \subseteq \mathbb{R}^m$ is an orthogonal projection onto the column space $R(A^*)$ of A (i.e. AA^* is symmetric and $(AA^*)^2 = AA^*$).

(b) (algebraic) $A^*AA^* = A^*$ or $AA^* \mid R(A^*)$ is an identity mapping.

(c) (geometric) $\vec{y} - \vec{x}A \in N(A^*) = R(A)^\perp$, $\vec{x} - \vec{y}A^* \in N(A) = R(A^*)^\perp$ for $\vec{x} \in \mathbb{R}^m$ and $\vec{y} \in \mathbb{R}^n$.

(d) (geometric) $\min_{\vec{x}AA^* = \vec{y}A^*} |\vec{x}| = |\vec{y}A^*|$ for a given $\vec{y} \in \mathbb{R}^n$ and all such $\vec{x} \in \mathbb{R}^m$.

(e) (geometric) For $\vec{x} \in \mathbb{R}^m$ and $\vec{y} \in \mathbb{R}^n$ such that $\vec{x}AA^* = \vec{y}A^*$, $\vec{x}A$ is the orthogonal projection of $\vec{y}$ on the row space $R(A)$ of A and

$$\vec{x}A = \vec{y}A^*A.$$

Therefore,

$$|\vec{y} - \vec{x}A| = \min_{\vec{z} \in \mathbb{R}^m} |\vec{y} - \vec{z}A|.$$

(f) (algebraic) $AA^*A = A$ or $A^*A \mid R(A)$ is an identity mapping.

(g) (geometric) $A^*A: \mathbb{R}^n \to R(A) \subseteq \mathbb{R}^n$ is an orthogonal projection from $\mathbb{R}^n$ onto the row space $R(A)$ of A (i.e. A^*A is symmetric and $(A^*A)^2 = A^*A$).

In case $AA^*: \mathbb{R}^m \to \mathbb{R}^m$ is not an orthogonal projection, equivalent statements in Ex. 10 are no more true and we shall withdraw our knowledge from here to what stated in Ex. 9. For a given $A_{m \times n}$, it is the deviation of the geometric mapping properties of A^* that makes AA^* not an orthogonal one. How to compensate this shortcoming of A^* is to replace A^* by an $n \times m$ matrix A^+ so that AA^+ will recover many of the equivalent properties stated in Ex. 10. This A^+ is the Moore–Penrose *generalized inverse matrix* of A.

11. *Equivalent definitions of the generalized inverse matrix* Suppose $A_{m \times n}$ is a real or complex matrix. For any real or complex $n \times m$ matrix A^+, the following possible properties of A and A^+ are equivalent.

(a) (algebraic)

    (1) $AA^+A = A$.

    (2) $A^+AA^+ = A^+$.

    (3) AA^+ and A^+A are symmetric matrices (or Hermitian in case complex matrices).

(b) (geometric) AA^+ and A^+A are symmetric, and for $\vec{x} \in \mathbb{R}^m$ and $\vec{y} \in \mathbb{R}^n$, then

$$\vec{x}A - \vec{y} \in N(A^+), \text{ the left kernel of } A^+$$
$$\Leftrightarrow \vec{x} - \vec{y}A^+ \in N(A).$$

(c) (geometric and algebraic) For given $\vec{b} \in \mathbb{R}^n$, then among all solutions or approximate solutions $\vec{x}$ of $\vec{x}A = \vec{b}$, it is the vector

$$\vec{x}_0 = bA^+$$

that satisfies the following restricted conditions simultaneously,

(1) $\vec{x}A$ is the orthogonal projection of $\vec{b}$ on the row space R(A),
(2) $\vec{x} \in R(A^*)$, the column space of A.

(d) (geometric and algebraic) For given $\vec{b} \in \mathbb{R}^n$, then among all solutions or approximate solutions $\vec{x}$ of $\vec{x}A = \vec{b}$, it is the vector

$$\vec{x}_0 = bA^+$$

that satisfies

(1) $|\vec{b} - \vec{x}A| = \min_{\vec{z} \in \mathbb{R}^m} |\vec{b} - \vec{z}A|$,
(2) $|\vec{x}|$ is the minimum.

$\vec{x}_0$ is usually called the *optimal solution* of the equations $\vec{x}A = \vec{b}$ under the constrained conditions 1 and 2.

Such an A^+ exists uniquely once A is given (see Ex. 12), and is called the *generalized inverse (matrix)* or *pseudoinverse* of A.

12. *Existence   and   uniqueness   of A^+*   For   any   permutation $\sigma: \{1, 2, \ldots, m\} \to \{1, 2, \ldots, m\}$, the $m \times m$ matrix

$$\begin{bmatrix} \vec{e}_{\sigma(1)} \\ \vdots \\ \vec{e}_{\sigma(m)} \end{bmatrix}_{m \times m}$$

obtained by performing σ on rows of the identity matrix I_m is called a *permutation matrix* of order m. For an $m \times n$ matrix A of rank r, there exist permutation matrices $P_{m \times m}$ and $Q_{n \times n}$ such that

$$PAQ = \begin{bmatrix} A_{11} & A_{12} \\ A_{21} & A_{22} \end{bmatrix}_{m \times n},$$

where A_{11} is an invertible $r \times r$ matrix so that $r(A_{11}) = r(A) = r$. Then,

$$A = BC, \quad \text{where}$$

$$B = P^{-1} \begin{bmatrix} I_r \\ A_{21}A_{11}^{-1} \end{bmatrix} \quad \text{or} \quad P^{-1} \begin{bmatrix} A_{11} \\ A_{21} \end{bmatrix},$$

$$C = (A_{11} \quad A_{12})Q^{-1} \quad \text{or} \quad (I_r \quad A_{11}^{-1}A_{21})Q^{-1}.$$

Note that $B_{m \times r}$ and $C_{r \times n}$ are such that $r(B) = r(C) = r$. Therefore, the generalized inverse is

$$A^+ = C^*(CC^*)^{-1}(B^*B)^{-1}B^*.$$

In particular,

$$A^+ = \begin{cases} A^*(AA^*)^{-1}, & \text{if } r(A) = m < n \\ (A^*A)^{-1}A, & \text{if } r(A) = n < m . \\ A^{-1}, & \text{if } r(A) = m = n \end{cases}$$

Consider $A^+: \mathbb{R}^n \to \mathbb{R}^m$ as the linear transformation defined by $\vec{y} \to \vec{y}A^+$. Then

$$R(A^+) = R(A^*),$$
$$R((A^+)^*) = R(A),$$

and, of course, $r(A^+) = r(A)$. Hence,

(1) $AA^+: \mathbb{R}^m \to \mathbb{R}^m$ is the orthogonal projection of $\mathbb{R}^m$ onto the column space $R(A^*)$ of A, i.e.

$$AA^+ \text{ is symmetric and } (AA^+)^2 = AA^+.$$

(2) $A^+A: \mathbb{R}^n \to \mathbb{R}^n$ is the orthogonal projection of $\mathbb{R}^n$ onto the row space $R(A)$ of A, i.e.

$$A^+A \text{ is symmetric and } (A^+A)^2 = A^+A.$$

Remember that $\mathbb{R}^n = N(A^*) \oplus R(A)$ and $R(A)^\perp = N(A^*)$. (1) suggests that $A^+: \mathbb{R}^n \to \mathbb{R}^m$ takes the row space $R(A)$ back to the column space $R(A^*)$ and its restriction to the right kernel $N(A^*)$ is zero. Because $r(A) = r(A^*)$, its restriction $A \mid R(A^*): R(A^*) \to R(A)$ is invertible. Therefore,

$$A^+ = (A \mid R(A^*))^{-1}: R(A) \to R(A^*)$$

inverts $A \mid R(A^*)$. I agree with Gilbert Strang's comment: *that this is the one natural best definition of an inverse* (see [17, p. 853]). See Fig. B.9 (compare with Fig. B.5).

Notice that for $\vec{x} \in \mathbb{R}^m$, $\vec{x}AA^*$ is not necessarily the orthogonal projection of $\vec{x}$ onto $R(A^*)$ and hence $\vec{x}AA^*$ may be not equal to $\vec{x}AA^+$. If it is for each $\vec{x} \in \mathbb{R}^m$, then $AA^* = AA^+$ holds. This is equivalent to say that, for $\vec{y} \in \mathbb{R}^n$ such that $\vec{y} = \vec{y} - \vec{x}A + \vec{x}A$, $\vec{y}A^* = \vec{x}AA^*$ which is equivalent to $\vec{y}A^+ = \vec{x}AA^+$ (see Exs. 10 and 11).

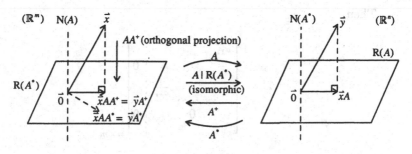

Fig. B.9

13. *Some basic properties of generalized inverses*

   (1) $(\alpha A)^+ = \frac{1}{\alpha} A^+$ where $\alpha \neq 0$.
   (2) $(A^+)^* = (A^*)^+$. Hence A^+ is symmetric if A is.
   (3) $(A^+)^+ = A$.
   (4) Suppose $A_{m \times r}$ and $B_{r \times n}$ and $\mathrm{r}(A) = \mathrm{r}(B) = r$, then

$$(AB)^+ = B^+ A^+.$$

   But, in general, $(AB)^+ \neq B^+ A^+$ (see (b) below).
   (5) Suppose $A_{n \times n}$ and $A^* = A$ and $A^2 = A$. Then

$$A^+ = A.$$

   Hence, $O_{n \times n}^+ = O$.
   (6) $O_{m \times n}^+ = O_{n \times m}$.
   (7) $R_{m \times m}$ and $Q_{n \times n}$ are invertible. Then

$$(PAQ)^+ = Q^{-1} A^+ P^{-1}.$$

(8) Suppose

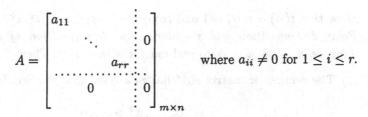

where $a_{ii} \neq 0$ for $1 \leq i \leq r$.

Then

$$A^+ = \begin{bmatrix} \frac{1}{a_{11}} & & & \vdots & \\ & \ddots & & \vdots & 0 \\ & & \frac{1}{a_{rr}} & \vdots & \\ \cdots\cdots\cdots\cdots & \cdots & \\ 0 & & & \vdots & 0 \\ & & & \vdots & \end{bmatrix}_{n \times m}.$$

Do the following two problems.

(a) Let

$$A = \begin{bmatrix} -1 & 0 & 1 & 2 \\ 0 & 1 & -1 & -1 \\ 1 & 1 & -2 & -3 \end{bmatrix}.$$

Show that

$$A^+ = \frac{1}{27} \begin{bmatrix} -9 & -9 & 0 \\ 0 & -9 & 9 \\ -3 & -6 & -3 \\ 6 & 3 & -3 \end{bmatrix}$$

and, for $\vec{b} \in \mathbb{R}^3$, find

$$\min_{\substack{\vec{x} \in \mathbb{R}^3, \\ \vec{x}AA^* = \vec{b}A^*}} |\vec{x}|.$$

(b) Let

$$A = \begin{bmatrix} 1 & 1 & 1 \\ -1 & -1 & -1 \end{bmatrix} \quad \text{and} \quad B = \begin{bmatrix} -1 & 1 \\ 1 & -1 \\ 1 & -1 \end{bmatrix}.$$

Show that $r(A) = r(B) = 1$ and compute $(AB)^+$ and $B^+ A^+$.

14. *Polar decomposition and singular value decomposition of a matrix*
Suppose $A_{m \times n}$ is a nonzero real matrix with rank r. Then

(1) The symmetric matrix AA^* has m eigenvalues (see Sec. B.10)

$$\lambda_1^2, \lambda_2^2, \ldots, \lambda_r^2, 0, \ldots, 0$$

with corresponding eigenvectors $\vec{x}_1, \vec{x}_2, \ldots, \vec{x}_r, \vec{x}_{r+1}, \ldots, \vec{x}_m \in \mathbb{R}^m$ such that $\vec{x} AA^* = \lambda_i^2 \vec{x}_i, 1 \leq i \leq r$ and $\vec{x} AA^* = \vec{0}, r + 1 \leq i \leq m,$ where

$$\lambda_i = |\vec{x}_i A| > 0 \quad \text{for } 1 \leq i \leq r.$$

(2) The eigenvectors $\{\vec{x}_1, \ldots, \vec{x}_r, \vec{x}_{r+1}, \ldots, \vec{x}_m\}$ from an *orthonormal basis* for $\mathbb{R}^m = R(A^*) \oplus N(A)$ such that $\{\vec{x}_1, \ldots, \vec{x}_r\}$ is a basis for $R(A^*) = N(A)^\perp$ and $\{\vec{x}_{r+1}, \ldots, \vec{x}_m\}$ is a basis for $N(A)$. Thus,

$$P = \begin{bmatrix} \vec{x}_1 \\ \vdots \\ \vec{x}_r \\ \vec{x}_{r+1} \\ \vdots \\ \vec{x}_m \end{bmatrix}_{m \times m}$$

is an *orthogonal matrix*, i.e. $P^* = P^{-1}$.

(3) Take an arbitrary orthonormal basis $\{\vec{y}_{r+1}, \ldots, \vec{y}_n\}$ for $N(A^*)$. Then, $\{\frac{\vec{x}_1 A}{\lambda_1}, \ldots, \frac{\vec{x}_r A}{\lambda_r}, \vec{y}_{r+1}, \ldots, \vec{y}_n\}$ is an orthonormal basis for $\mathbb{R}^n = R(A) \oplus N(A^*)$ and the basis vectors in that order are eigenvectors of A^*A corresponding to eigenvalues $\lambda_1^2, \ldots, \lambda_r^2, 0, \ldots, 0$. Then

$$Q = \begin{bmatrix} \frac{\vec{x}_1 A}{\lambda_1} \\ \vdots \\ \frac{\vec{x}_r A}{\lambda_r} \\ \vec{y}_{r+1} \\ \vdots \\ \vec{y}_n \end{bmatrix}_{n \times n}$$

is an orthogonal matrix.

Therefore, A can be written as the following *singular value decomposition*

$$A = P^{-1}DQ, \quad \text{where } D = \begin{bmatrix} \lambda_1 & & & & & \\ & \ddots & & & 0 & \\ & & \lambda_r & & & \\ & & & 0 & & \\ & 0 & & & \ddots & \\ & & & & & 0 \end{bmatrix}.$$

See Fig. B.10.

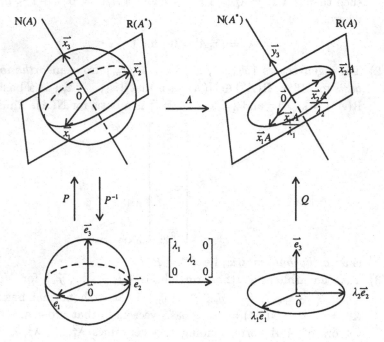

Fig. B.10

While the *polar decomposition* is $A = NP$ where $N = R^{-1}DR$ is positive semidefinite and $P = R^{-1}Q$ is orthogonal.

(*Note*   This result is usually proved by using general theory of diagonalizability about symmetric matrix (Ex. 8 of Sec. B.9). [7] provides a direct geometric proof without recourse to that general theory . We will encounter this for $n = 2$ in Sec. 4.5 and $n = 3$ in Sec. 5.5.)

Eventually, the generalized inverse of A is

$$A^+ = Q^{-1} \begin{bmatrix} \frac{1}{\lambda_1} & & & & \\ & \ddots & & & 0 \\ & & \frac{1}{\lambda_r} & & \\ & & & 0 & \\ 0 & & & & \ddots \\ & & & & & 0 \end{bmatrix} P.$$

B.9   Inner Product Spaces

Here in this section, the field $\mathbb{F}$ will always denote the real field $\mathbb{R}$ or the complex field $\mathbb{C}$.

Definition An *inner product* $\langle , \rangle$ on a vector space V over the field $\mathbb{F}$ is a function that assigns to each ordered pair of vectors $\vec{x}$ and $\vec{y}$ in V a scalar $\langle \vec{x}, \vec{y} \rangle$ in $\mathbb{F}$, such that for all $\vec{x}, \vec{y}, \vec{z}$ in V and all scalars α in $\mathbb{F}$, the following properties hold.

1. (positive) $\langle \vec{x}, \vec{x} \rangle \geq 0$ with equality holds if and only if $\vec{x} = \vec{0}$.
2. (conjugate symmetric) $\langle \vec{y}, \vec{x} \rangle = \overline{\langle \vec{x}, \vec{y} \rangle}$ (the complex conjugate of $\langle \vec{x}, \vec{y} \rangle$); in case $\mathbb{F} = \mathbb{R}$, (symmetric) $\langle \vec{y}, \vec{x} \rangle = \langle \vec{x}, \vec{y} \rangle$.
3. (linear) $\langle \alpha \vec{x} + \vec{y}, \vec{z} \rangle = \alpha \langle \vec{x}, \vec{z} \rangle + \langle \vec{y}, \vec{z} \rangle$.

The scalar $\langle \vec{x}, \vec{y} \rangle$ is called the inner product of $\vec{x}$ and $\vec{y}$.

A vector space V over the field $\mathbb{F}$ endowed with a specific inner product $\langle , \rangle$ is called an *inner product space* and is denoted as $(V, \langle , \rangle)$ or simply as V. If $\mathbb{F} = \mathbb{R}, V$ is called a *real inner product space*, whereas if $\mathbb{F} = \mathbb{C}$, a *complex inner product space* or *unitary space*.

Examples For $\vec{x} = (x_1, \ldots, x_n)$ and $\vec{y} = (y_1, \ldots, y_n)$ in $\mathbb{R}^n$ (see Sec. B.1)

$$\langle \vec{x}, \vec{y} \rangle = \vec{x} \, \vec{y}^* = \sum_{k=1}^{n} x_k y_k$$

is an inner product in $\mathbb{R}^n$, and is called the *natural* or *standard* inner product on the *real* vector space $\mathbb{R}^n$.

For $\vec{x}$ and $\vec{y} \in \mathbb{C}^n$ (see Sec. B.1), the inner product

$$\langle \vec{x}, \vec{y} \rangle = \vec{x} \, \vec{\bar{y}}^* = \sum_{k=1}^{n} x_k \bar{y}_k$$

is called the *natural* or *standard* inner product on the *complex* vector space $\mathbb{C}^n$.

$\mathbb{R}^n$ and $\mathbb{C}^n$ are always endowed with natural inner products unless otherwise specified.

Another example is shown in Ex. 26 of Sec. B.4.

Let $C[a, b]$ be the complex vector space of complex-valued continuous functions on $[a, b]$. For $f, g \in C[a, b]$,

$$\langle f, g \rangle = \int_a^b f(t) \overline{g(t)} \, dt$$

is an inner product.

ℓ^2 (square convergent series) and $L^2[a, b]$ (square Lebesgue integrable functions) are important inner product (or Hilbert) spaces in Real Analysis.

For a *complex* inner product space V, notice that

$$\langle \vec{x}, \beta \vec{y} \rangle = \bar{\beta} \langle \vec{x}, \vec{y} \rangle \quad \beta \in \mathbb{C}$$

and hence

$$\left\langle \sum_{j=1}^{m} \alpha_j \vec{x}_j, \sum_{k=1}^{n} \beta_k \vec{y}_k \right\rangle = \sum_{j=1}^{m} \sum_{k=1}^{n} \alpha_j \bar{\beta}_k \langle \vec{x}_j, \vec{y}_k \rangle.$$

In case V is real, $\bar{\beta}$ and $\bar{\beta}_k$ are replaced by β and β_k and $\langle , \rangle$ becomes bilinear.

Define the *norm* or *length* of $\vec{x} \in V$ by

$$|\vec{x}| = \sqrt{\langle \vec{x}, \vec{x} \rangle}.$$

Call $\vec{x}, \vec{y} \in V$ *orthogonal* or *perpendicular* to each other if $\langle \vec{x}, \vec{y} \rangle = 0$ and is denoted as

$$\vec{x} \perp \vec{y}.$$

Suppose $\vec{y} \neq \vec{0}$. Then for $\beta \in \mathbb{F}$ (remember $\mathbb{F} = \mathbb{C}$ or $\mathbb{R}$),

$$|\vec{x} - \beta \vec{y}|^2 = |\vec{x}|^2 - \bar{\beta}\langle \vec{x}, \vec{y} \rangle - \beta \langle \vec{y}, \vec{x} \rangle + |\beta|^2 |\vec{y}|^2 \geq 0.$$

By choosing $\beta = \frac{\langle \vec{x}, \vec{y} \rangle}{|\vec{y}|^2}$, then the above inequality becomes

$$\left| \vec{x} - \frac{\langle \vec{x}, \vec{y} \rangle}{|\vec{y}|^2} \vec{y} \right|^2 = |\vec{x}|^2 - \frac{|\langle \vec{x}, \vec{y} \rangle|^2}{|\vec{y}|^2} \geq 0$$

$$\Rightarrow |\langle \vec{x}, \vec{y} \rangle| \leq |\vec{x}||\vec{y}|,$$

where equality holds if and only if $\vec{y} = \beta \vec{x}$ or $\vec{x} = \beta \vec{y}$ for $\beta \in \mathbb{F}$. This is called the *Cauchy–Schwarz inequality*. Thus, it follows the *triangle inequality*

$$|\vec{x} + \vec{y}| \leq |\vec{x}| + |\vec{y}|.$$

In case V is a *real* inner product space, then for $\vec{x} \neq \vec{0}$ and $\vec{y} \neq \vec{0}$,

$$-1 \leq \frac{\langle \vec{x}, \vec{y} \rangle}{|\vec{x}||\vec{y}|} \leq +1.$$

Therefore, it is reasonable to define the cosine of the *angle* θ between $\vec{x}$ and $\vec{y}$ by

$$\cos \theta = \frac{\langle \vec{x}, \vec{y} \rangle}{|\vec{x}||\vec{y}|}.$$

We now can reinterpret $\langle \vec{x}, \vec{y} \rangle$ as the *signed orthogonal projection* $|\vec{y}| \cos \theta$ of $\vec{y}$ along $\vec{x}$ multiplied by the length $|\vec{x}|$ of $\vec{x}$ itself. Of course, if at least one of $\vec{x}$ and $\vec{y}$ is a zero vector, $\langle \vec{x}, \vec{y} \rangle = |\vec{x}||\vec{y}| \cos \theta$ still holds.

For inner product space V and $\vec{x}, \vec{y} \in V$ with $\vec{x} \neq 0$, call

1. $\left\langle \vec{y}, \frac{\vec{x}}{|\vec{x}|} \right\rangle \frac{\vec{x}}{|\vec{x}|} = \frac{\langle \vec{y}, \vec{x} \rangle}{|\vec{x}|^2} \vec{x}$ the *orthogonal projection vector* of $\vec{y}$ along $\vec{x}$, and

2. $\vec{y} - \frac{\langle \vec{y}, \vec{x} \rangle}{|\vec{x}|^2} \vec{x}$ the *perpendicular* or *orthogonal* vector from $\vec{y}$ to $\vec{x}$.

See Fig. B.11. Try to use these concepts to catch the idea used in the proof of Cauchy–Schwarz inequality above.

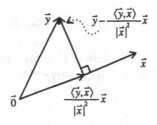

Fig. B.11

Orthogonality is the main feature in almost all topics concerned with inner products.

By the very definition, zero vector $\vec{0}$ is the only vector orthogonal to all vectors in the space V.

A nonempty subset S of V consisting of nonzero vectors is said to be *orthogonal* if any two distinct elements of S are orthogonal. Such a set is easily seem to be linearly independent.

An orthogonal set S consisting entirely of vectors of unit length is called *orthonormal*.

For example, in $\mathbb{R}^n$ or $\mathbb{C}^n, \mathcal{N} = \{\vec{e}_1, \ldots, \vec{e}_n\}$ satisfies

$$\langle \vec{e}_i, \vec{e}_j \rangle = \delta_{ij}, \quad 1 \leq i, \ j \leq n.$$

Therefore, $\mathcal{N}$ is an *orthonormal basis*, i.e. $\mathcal{N}$ is a basis that is orthonormal.

Exercises

In what follows, V will always denote an inner product space with inner product $\langle , \rangle$, usually not particularly mentioned.

1. Suppose $\dim V = n$ and $\mathcal{B} = \{\vec{x}_1, \ldots, \vec{x}_n\}$ is a basis for V. For $\vec{x} = \sum_{i=1}^{n} \alpha_i \vec{x}_i$ and $\vec{y} = \sum_{i=1}^{n} \beta_i \vec{x}_i$, then

$$\langle \vec{x}, \vec{y} \rangle = \sum_{i,j=1}^{n} \alpha_i \bar{\beta}_j \langle \vec{x}_i, \vec{x}_j \rangle = [\vec{x}]_{\mathcal{B}} A_{\mathcal{B}} [\vec{y}]_{\mathcal{B}}^*,$$

where

$$A_{\mathcal{B}} = [\langle \vec{x}_i, \vec{x}_j \rangle]_{n \times n} = \begin{bmatrix} \langle \vec{x}_1, \vec{x}_1 \rangle & \langle \vec{x}_1, \vec{x}_2 \rangle & \cdots & \langle \vec{x}_1, \vec{x}_n \rangle \\ \langle \vec{x}_2, \vec{x}_1 \rangle & \langle \vec{x}_2, \vec{x}_2 \rangle & \cdots & \langle \vec{x}_2, \vec{x}_n \rangle \\ \vdots & \vdots & & \vdots \\ \langle \vec{x}_n, \vec{x}_1 \rangle & \langle \vec{x}_n, \vec{x}_2 \rangle & \cdots & \langle \vec{x}_n, \vec{x}_n \rangle \end{bmatrix}$$

is called the *matrix representation* of the inner product $\langle , \rangle$ related to the basis $\mathcal{B}$. $A_{\mathcal{B}}$ is *Hermitian* (or *symmetric* if V is real), i.e. its conjugate transpose

$$\bar{A}_{\mathcal{B}}^* = A_{\mathcal{B}} \quad \text{or} \quad (A_{\mathcal{B}}^* = A_{\mathcal{B}})$$

and is *positive-definite*, i.e.

$$\vec{v} A_{\mathcal{B}} \vec{v}^* \geq 0 \quad \text{for any } \vec{v} \in \mathbb{F}^n$$

with equality only if $\vec{v} = \vec{0}$. Conversely, for any positive-definite Hermitian matrix $A \in M(n; \mathbb{C})$ and a basis $\mathcal{B}$ for V,

$$\langle \vec{x}, \vec{y} \rangle = [\vec{x}]_{\mathcal{B}} A [\vec{y}]_{\mathcal{B}}^*$$

defines an inner product $\langle , \rangle$ on V whose matrix representation is A itself.

Two matrix representations $A_{\mathcal{B}}$ and $A_{\mathcal{B}'}$ of an inner product $\langle , \rangle$ related to bases $\mathcal{B}$ and $\mathcal{B}'$ for V are *congruent*, i.e.

$$A_{\mathcal{B}'} = P A_{\mathcal{B}} \bar{P}^*,$$

where $P = [1_V]_{\mathcal{B}}^{\mathcal{B}'}$ is the transition matrix from $\mathcal{B}'$ to $\mathcal{B}$ (see Sec. B.7).

2. *The Gram–Schmidt orthogonalization process* Suppose $\{\vec{y}_1, \vec{y}_2, \ldots, \vec{y}_k, \ldots\}$ is linearly independent in V.

(a) Let (see Fig. B.11)

$$\vec{x}_1 = \vec{y}_1,$$
$$\vec{x}_2 = \vec{y}_2 - \frac{\langle \vec{y}_2, \vec{x}_1 \rangle}{|\vec{x}_1|^2} \vec{x}_1,$$
$$\vdots$$
$$\vec{x}_k = \vec{y}_k - \sum_{j=1}^{k-1} \frac{\langle \vec{y}_k, \vec{x}_j \rangle}{|\vec{x}_j|^2} \vec{x}_j,$$
$$\vdots$$

Then,

1. $\langle\langle \vec{x}_1, \ldots, \vec{x}_k \rangle\rangle = \langle\langle \vec{y}_1, \ldots, \vec{y}_k \rangle\rangle,\ k \geq 1$.
2. $\vec{x}_{k+1}$ is orthogonal to every vector in $\langle\langle \vec{y}_1, \ldots, \vec{y}_k \rangle\rangle$; in symbol, $\vec{x}_{k+1} \perp \langle\langle \vec{y}_1, \ldots, \vec{y}_k \rangle\rangle,\ k \geq 1$; moreover $\vec{x}_{k+1}$ is the orthogonal vector from $\vec{y}_{k+1}$ to the subspace $\langle\langle \vec{y}_1, \ldots, \vec{y}_k \rangle\rangle$.

Hence $\{\vec{x}_1, \vec{x}_2, \ldots, \vec{x}_k, \ldots\}$ is an orthogonal set and $\{\frac{\vec{x}_1}{|\vec{x}_1|}, \frac{\vec{x}_2}{|\vec{x}_2|}, \ldots, \frac{\vec{x}_k}{|\vec{x}_k|}, \ldots\}$ is an orthonormal set.

(b) The $(n+1)$-dimensional real vector space $P_n(\mathbb{R})$ (see Sec. B.3) has a basis $\mathcal{B} = \{1, x, x^2, \ldots, x^n\}$. Use this to show that $(n+1) \times (n+1)$ matrix

$$\begin{bmatrix} 1 & \frac{1}{2} & \frac{1}{3} & \cdots & \frac{1}{n} & \frac{1}{n+1} \\ \frac{1}{2} & \frac{1}{3} & \frac{1}{4} & \cdots & \frac{1}{n+1} & \frac{1}{n+2} \\ \vdots & \vdots & \vdots & & \vdots & \vdots \\ \frac{1}{n} & \frac{1}{n+1} & \frac{1}{n+2} & \cdots & \frac{1}{2n-1} & \frac{1}{2n} \\ \frac{1}{n+1} & \frac{1}{n+2} & \frac{1}{n+3} & \cdots & \frac{1}{2n} & \frac{1}{2n+1} \end{bmatrix}$$

is positive-definite. Also, $\{\vec{y}_0, \vec{y}_1, \ldots, \vec{y}_n\}$ where

$$y_0(t) = 1,$$
$$y_1(t) = t,$$
$$y_2(t) = t^2 - \frac{1}{3},$$
$$\vdots$$
$$y_k(t) = \frac{d^k}{dt^k}(t^2 - 1)^k, \quad 1 \leq k \leq n$$

is an orthogonal basis. How to get an orthonormal basis?

(c) V has orthogonal (or orthonormal) sets of vectors.

(d) Suppose $\dim V = n < \infty$. Then V has an orthonormal basis $\mathcal{B} = \{\vec{x}_1, \vec{x}_2, \ldots, \vec{x}_n\}$. For any $\vec{x} \in V$, then

$$\vec{x} = \sum_{i=1}^{n} \langle \vec{x}, \vec{x}_i \rangle \vec{x}_i$$

and for any linear operator $f\colon V \to V$, the matrix representation of f related to $\mathcal{B}$ is

$$[f]_\mathcal{B} = [\langle f(\vec{x}_i), \vec{x}_j \rangle]_{n \times n}.$$

If S_1 and S_2 are nonempty subsets of V such that each vector in S_1 is orthogonal to each vector in S_2, then S_1 and S_2 are said to be *orthogonal* or *perpendicular* to each other and is denoted as

$$S_1 \perp S_2.$$

In particular, if $S_1 = \{\vec{x}\}$, this is briefly as $\vec{x} \perp S_2$. Let S be a nonempty subset of V. The subspace

$$S^\perp = \{\vec{x} \in V \mid \vec{x} \perp S\}$$

is called the *orthogonal complement* of S in V and has the property that $(S^\perp)^\perp \supseteq \langle\langle S \rangle\rangle$ which is equal to S if S is already a subspace.

3. *The orthogonal decomposition* of an inner product space.

(a) Let S be a subspace of V and $\vec{x}_0 \in V$. Then there exist a $\vec{y}_0 \in S$ such that

$$|\vec{x}_0 - \vec{y}_0| = \min_{\vec{y} \in S} |\vec{x}_0 - \vec{y}|$$

if and only if $(\vec{x}_0 - \vec{y}_0) \perp S$. In this case, $\vec{y}_0$ is unique and is called the *orthogonal projection* of $\vec{x}_0$ on S and $\vec{x}_0 - \vec{y}_0$ the *orthogonal vector* from $\vec{x}_0$ to S. Moreover, the *Pythagorean theorem*

$$|\vec{x}_0|^2 = |\vec{y}_0|^2 + |\vec{x}_0 - \vec{y}_0|^2$$

holds.

(b) Suppose $\{\vec{x}_1, \vec{x}_2, \ldots, \vec{x}_n, \ldots\}$ is an orthonormal system in V. Then, for any $\vec{x} \in V$, the *optimal approximation inequality* is

$$\left| \vec{x} - \sum_{i=1}^{n} \alpha_i \vec{x}_i \right| \geq \left| \vec{x} - \sum_{i=1}^{n} \langle \vec{x}, \vec{x}_i \rangle \vec{x}_i \right|$$

$$= \left\{ |\vec{x}|^2 - \sum_{i=1}^{n} |\langle \vec{x}, \vec{x}_i \rangle|^2 \right\}^{1/2},$$

$$\alpha_1, \ldots, \alpha_n \in \mathbb{F} \text{ and } n \geq 1$$

with equality if and only if $\alpha_i = \langle \vec{x}, \vec{x}_i \rangle$ for $1 \leq i \leq n$. That is, the minimum is obtained at the orthogonal projection vector $\sum_{i=1}^{n} \langle \vec{x}, \vec{x}_i \rangle \vec{x}_i$ of $\vec{x}$. By the way,

$$\sum_{i=1}^{n} |\langle \vec{x}, \vec{x}_i \rangle|^2 \leq |\vec{x}|^2, \quad n \geq 1$$

is called the *Bessel inequality*. Try to show that

$$\min_{a,b,c} \int_{-1}^{1} |t^3 - a - bt - ct^2|^2 \, dt = \frac{38}{525}.$$

(c) Suppose S is a finite-dimensional subspace of V. Then

$$V = S \oplus S^{\perp}.$$

4. *The orthogonal projection (operator) of V onto a finite-dimensional subspace.* Suppose S is a finite-dimensional subspace of V.

(a) There exists a unique linear operator $p \colon V \to V$ such that for $\vec{x} \in V$,

$$\vec{x} = p(\vec{x}) + \vec{x} - p(\vec{x}), \ p(\vec{x}) \in S \quad \text{and}$$
$$(1_V - p)(\vec{x}) = \vec{x} - p(\vec{x}) \in S^{\perp}.$$

Such an operator p is called the *orthogonal projection* of V onto S; meanwhile, $1_V - p$ is the orthogonal projection of V onto $S^{\perp}$. Moreover,

$$p(\vec{x}) = \vec{x} \Leftrightarrow \vec{x} \in S,$$
$$p(\vec{x}) = \vec{0} \Leftrightarrow \vec{x} \in S^{\perp}, \quad \text{and}$$
$$|\vec{x}|^2 = |p(\vec{x})|^2 + |\vec{x} - p(\vec{x})|^2.$$

(b) Suppose $\dim V < \infty$ and $p \colon V \to V$ is a linear operator. Then the following are equivalent:

(1) p is the orthogonal projection of V onto its range space $S = p(V)$ along $S^{\perp}$.

(2) (algebraic) p is self-adjoint and idempotent, i.e.

$$\bar{p}^* = p,$$
$$p^2 = p.$$

(3) (geometric) p is a projection, i.e. $p^2 = p$ and $|p(\vec{x})| \le |\vec{x}|$ for $\vec{x} \in V$.

(4) (geometric) $p(\vec{x}) \perp (\vec{x} - p(\vec{x}))$ or $|p(\vec{x})|^2 = \langle p(\vec{x}), \vec{x} \rangle$ for $\vec{x} \in V$.

(5) (algebraic and geometric) p is a projection and

$$\mathrm{Im}(p) = \mathrm{Ker}(p)^\perp,$$
$$\mathrm{Ker}(p) = \mathrm{Im}(p)^\perp.$$

(c) Suppose $\dim S = k$ and $\{\vec{y}_1, \ldots, \vec{y}_k\}$ is a basis for S. Suppose $\dim V = n$ and $\mathcal{B} = \{\vec{x}_1, \ldots, \vec{x}_n\}$ is an orthonormal basis for V. Let

$$A = \begin{bmatrix} [\vec{y}_1]_{\mathcal{B}} \\ \vdots \\ [\vec{y}_k]_{\mathcal{B}} \end{bmatrix}_{k \times n}.$$

Then, the orthogonal projection of V onto S is

$$\bar{A}^* (A\bar{A}^*)^{-1} A \colon V \to S \subseteq V$$

defined by $[\vec{x}]_{\mathcal{B}} \to [\vec{x}]_{\mathcal{B}} \bar{A}^* (A\bar{A}^*)^{-1} A$. Try to explain what linear operator $2\bar{A}^* (A\bar{A}^*)^{-1} A - 1_n$ means?

5. Suppose $\dim V = n$.

(a) *Riesz representation theorem* For each $f \in V^*$ (the dual space of V, see Ex. 19 of Sec. B.7), there exists a unique $\vec{y} \in V$ such that

$$f(\vec{x}) = \langle \vec{x}, \vec{y} \rangle \quad \text{for } \vec{x} \in V.$$

Denote this f temporarily by $f_{\vec{y}}$. Conversely, $f_{\vec{y}}(\ ) = \langle, \vec{y} \rangle$ is in V^* for each $\vec{y} \in V$.

(b) Define $\langle, \rangle$ on V^* by $\langle f_{\vec{x}}, f_{\vec{y}} \rangle = \langle \vec{x}, \vec{y} \rangle$. Then V^* is an n-dimensional inner product space. Let $\varphi \colon V \to V^*$ be defined by

$$\varphi(\vec{x}) = f_{\vec{x}}.$$

Then,

(1) For each orthonormal basis $\mathcal{B} = \{\vec{x}_1, \ldots, \vec{x}_n\}$ for V, its dual basis $\mathcal{B}^* = \{f_1, \ldots, f_n\}$ is an orthonormal basis for V^*. Also, $f_j = f_{x_j}$ for $1 \le j \le n$.

(2) φ is *conjugate linear*, i.e. for $\vec{x}, \vec{y} \in V$ and $\alpha \in \mathbb{F}$,

$$\varphi(\alpha\vec{x} + \vec{y}) = \bar{\alpha}f_{\vec{x}} + f_{\vec{y}} = \bar{\alpha}\varphi(\vec{x}) + \varphi(\vec{y}).$$

(3) φ preserves inner products, i.e.

$$\langle \varphi(\vec{x}), \varphi(\vec{y}) \rangle = \langle \vec{x}, \vec{y} \rangle, \quad \vec{x}, \vec{y} \in V.$$

Thus, we identify $\vec{x}$ and $\varphi(\vec{x}) = f_{\vec{x}}$, and V and V^* and call V a *self-dual* inner product space. We will *adopt this convention when dealing with inner product spaces.*

(c) For each linear operator $f \colon V \to V$, there exists a unique linear operator $f^* \colon V \to V$, called the *adjoint* of f, such that

$$\langle f(\vec{x}), \vec{y} \rangle = \langle \vec{x}, f^*(\vec{y}) \rangle, \quad \vec{x}, \vec{y} \in V.$$

Moreover, for each orthonormal basis $\mathcal{B} = \{\vec{x}_1, \ldots, \vec{x}_n\}$,

$$[f^*]_{\mathcal{B}} = \overline{[f]_{\mathcal{B}}}^{\,*}.$$

In case V is real, then $[f^*]_{\mathcal{B}} = [f]_{\mathcal{B}}^*$. *We already treat its counterpart in matrices in Exs. 1–14 of Sec. B.8.*

(*Note* The following diagram shows the difference between the dual mapping $g \colon V^* \to V^*$ as indicated in Ex. 24 of Sec. B.7 and the adjoint $f^* \colon V \to V$ discussed here, where $\varphi \colon V \to V^*$ is defined as in (b).

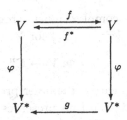

Notice that $f^* = \varphi^{-1} \circ g \circ \varphi$.)

(d) Suppose $f, g \colon V \to V$ are linear operators. Then

(1) $(f + g)^* = f^* + g^*$.

(2) $(\alpha f)^* = \bar{\alpha}f^*$ for $\alpha \in \mathbb{F}$.

(3) $(g \circ f)^* = f^* \circ g^*$.

(4) $(f^*)^* = f^{**} = f$.

(5) $1_V^* = 1_V$.

(e) Some special linear operators. Suppose $\mathcal{B}$ is an orthonormal basis for V and $f \colon V \to V$ is a linear operator.

(1) $\langle f(\vec{x}), f(\vec{y}) \rangle = \langle f^*(\vec{x}), f^*(\vec{y}) \rangle$ for $\vec{x}, \vec{y} \in V$.

$\Leftrightarrow f \circ f^* = f^* \circ f$.

$\Leftrightarrow [f]_B [\bar{f}]_B^* = [\bar{f}]_B^* [f]_B$.

Such an f is called a *normal operator* and $[f]_B$ a *normal matrix*.

(2) $\langle f(\vec{x}), f(\vec{y}) \rangle = \langle \vec{x}, \vec{y} \rangle$ for $\vec{x}, \vec{y} \in V$.

$\Leftrightarrow f \circ f^* = f^* \circ f = 1_V$.

$\Leftrightarrow [f]_B [\bar{f}]_B^* = I_n$ or $[\bar{f}]_B^* = [f]_B^{-1}$.

Such an f is called a *unitary operator* and $[f]_B$ a *unitary matrix*.

(3) $\langle f(\vec{x}), \vec{y} \rangle = \langle \vec{x}, f(\vec{y}) \rangle$ for $\vec{x}, \vec{y} \in V$.

$\Leftrightarrow f^* = f$.

$\Leftrightarrow [\bar{f}]_B^* = [f]_B$.

f is called a *Hermitian operator* or *self-adjoint operator* and $[f]_B$ a *Hermitian matrix*.

(4) $\langle f(\vec{x}), \vec{y} \rangle = -\langle \vec{x}, f(\vec{y}) \rangle$ for $\vec{x}, \vec{y} \in V$.

$\Leftrightarrow f^* = -f$.

$\Leftrightarrow [\bar{f}]_B^* = -[f]_B$.

f is called a *skew-Hermitian operator* and $[f]_B$ a *skew-Hermitian matrix*.

In case V is a real vector space, a unitary operator (matrix) is usually called an *orthogonal operator* (*matrix*), Hermitian operator (matrix) called *symmetric operator* (*matrix*) and skew-Hermitian operator (matrix) called *skew-symmetric operator* (*matrix*).

For an n-dimensional complex inner product space V and any fixed orthonormal basis $B = \{\vec{x}_1, \ldots, \vec{x}_n\}$ for V, the linear isomorphism

$$\Phi \colon V \to \mathbb{C}^n$$

defined by $\Phi(\vec{x}) = [\vec{x}]_B$ carries the inner product $\langle , \rangle$ on V into the natural inner product $\langle , \rangle$ on $\mathbb{C}^n$, and thus preserves inner products and any linear operator $f \colon V \to V$ is transformed into a matrix $[f]_B \in M(n; \mathbb{C})$ such that

$$[f(\vec{x})]_B = [\vec{x}]_B [f]_B, \quad \vec{x} \in V.$$

Conversely, any result concerning $\mathbb{C}^n$ and $M(n; \mathbb{C})$ can be reinterpreted as a corresponding result, uniquely up to similarity, in V with a fixed orthonormal basis. Henceforth, we focus our study on $\mathbb{C}^n$ endowed with the natural inner product and a matrix $A \in M(n; \mathbb{C})$ is considered as a linear operator on $\mathbb{C}^n$ defined by $\vec{x} \to \vec{x}A$. Since $\mathbb{R}$ is a subfield of $\mathbb{C}$, a real matrix $A \in M(n; \mathbb{R})$ may be considered as a complex matrix in many occasions and hence will inherit directly many valuable results from the complex one. One shall refer to Ex. 32 of Sec. B.4 and Sec. B.10 for concepts of eigenvalues and eigenvectors.

6. *Unitary matrices and orthogonal matrices*

 (a) For a matrix $U \in M(n; \mathbb{C})$, the following are equivalent:

  (1) U is unitary, i.e. $U\bar{U}^* = \bar{U}^*U = I_n$.
  (2) The n row vectors of U form an orthonormal basis for $\mathbb{C}^n$.
  (3) The n column vectors of U form an orthonormal basis for $\mathbb{C}^n$.
  (4) U(as a linear operator) transforms any orthonormal basis for $\mathbb{C}^n$ into an orthonormal basis for $\mathbb{C}^n$.
  (5) U transforms an orthonormal basis for $\mathbb{C}^n$ into another one.
  (6) U preserves inner products (and hence, orthogonality), i.e.

$$\langle \vec{x}U, \vec{y}U \rangle = \langle \vec{x}, \vec{y} \rangle, \quad \vec{x}, \vec{y} \in \mathbb{C}^n.$$

  (7) U preserves lengths, i.e.

$$|\vec{x}U| = |\vec{x}|, \quad \vec{x} \in \mathbb{C}^n.$$

 Note that these results are still valid for orthogonal matrix and $\mathbb{R}^n$.

 (b) Suppose U and V are unitary matrices. Then

$$\bar{U}, \ U^*, \ U^{-1} \text{ and } UV$$

 are unitary too. Also, $|\det U| = 1$ and in particular, $\det P = \pm 1$ if P is orthogonal.

 (c) The eigenvalues λ (see Sec. B.10) of a unitary matrix all have absolute value $|\lambda| = 1$, i.e. $\lambda = \frac{1}{\bar{\lambda}}$.
 (*Note* Let $P_{n \times n}$ be an orthogonal matrix. Then

  (1) If $\det P = 1$ and n is odd, or $\det P = -1$ and n is even, P has eigenvalue 1; whereas if $\det P = -1$, P always has eigenvalue -1. P has no other *real* eigenvalues except ± 1.
  (2) If λ is an eigenvalue of P, so is λ^{-1}. Hence, *complex* eigenvalues of P are in conjugate pairs

$$e^{i\theta}, e^{-i\theta} \quad (\theta \in \mathbb{R} \text{ and } 0 < \theta < \pi).$$

  (3) The set of all orthogonal matrices

$$O(n; \mathbb{R}) = \{P \in M(n; \mathbb{R}) \mid P^* = P^{-1}\}$$

  forms a *subgroup* of $GL(n; \mathbb{R})$, and has $\{P \in O(n; \mathbb{R}) \mid \det P = 1\}$ as its subgroup. These two groups handle the rigid motions in space $\mathbb{R}^n$.

(4) Symmetric and skew-symmetric matrices are invariant under orthogonal similarity (see Ex. 31(b) of Sec. B.4)

(5) (Cayley, 1846) Let $A_{n \times n}$ be a real skew-symmetric matrix. Then $I_n + A$ and $I_n - A$ are invertible and

$$P = (I_n - A)(I_n + A)^{-1}$$

is orthogonal. P does not have eigenvalue -1 and $\det P = 1$.

(6) Conversely, if orthogonal matrix P does not have -1 as its eigenvalue, then there exists a skew-symmetric matrix

$$A = (I_n + P)^{-1}(I_n - P)$$

such that $P = (I_n - A)(I_n + A)^{-1}$.)

(d) A transformation $f \colon \mathbb{C}^n \to \mathbb{C}^n$ is called a *rigid motion* or an *isometry* if

$$|f(\vec{x}) - f(\vec{y})| = |\vec{x} - \vec{y}|, \quad \vec{x}, \vec{y} \in \mathbb{C}^n.$$

For any fixed orthonormal basis $\mathcal{B}$ for $\mathbb{C}^n$, there exsits a unique unitary matrix U and a point $\vec{x}_0 \in \mathbb{C}^n$ such that

$$[f(\vec{x})]_\mathcal{B} = [\vec{x}_0]_\mathcal{B} + [\vec{x}]_\mathcal{B} U, \quad \vec{x} \in \mathbb{C}^n.$$

(*Note* A rigid motion $f \colon \mathbb{R}^n \to \mathbb{R}^n$ is of the form

$$f(\vec{x}) = \vec{x}_0 + \vec{x} P, \quad \vec{x} \in \mathbb{R}^n,$$

where $P_{n \times n}$ is orthogonal. Whereas

$$\vec{x} \to \vec{x}_0 + \vec{x}(rP), \quad r > 0$$

is a *similar* (i.e. angle-preserving) transformation.)

(e) Any unitary matrix $U_{n \times n}$ has n (complex) eigenvalues $\lambda_1, \ldots, \lambda_n$ with eigenvectors $\vec{x}_1, \ldots, \vec{x}_n \in \mathbb{C}^n$ such that

$$\vec{x}_j U = \lambda_j x_j, \quad |\lambda_j| = 1 \text{ for } 1 \le j \le n$$

and $\mathcal{B} = \{\vec{x}_1, \ldots, \vec{x}_n\}$ is an orthonormal basis for $\mathbb{C}^n$. Then

$$[U]_\mathcal{B} = Q U Q^{-1} = \begin{bmatrix} \lambda_1 & & 0 \\ & \ddots & \\ 0 & & \lambda_n \end{bmatrix} \quad \text{with} \quad Q = \begin{bmatrix} \vec{x}_1 \\ \vec{x}_2 \\ \vdots \\ \vec{x}_n \end{bmatrix}.$$

(f) Suppose $A \in M(n; \mathbb{C})$ has rank $r = r(A) \geq 1$ and its first r rows are linearly independent. Then there exists a unitary matrix U such that

$$A = BU = U^{-1}C,$$

where B is a lower-triangle matrix with first r diagonal entries positive and the remaining elements all zeros, whereas C is a upper-triangle matrix having the same diagonal entries as B. Such B and C are unique. In case A is invertible, U is unique too.

(g) (Schur, 1909) Every complex square matrix is *unitarily similar* (or *equivalent*) to a triangular matrix whose main diagonal entries are (complex) eigenvalues. That is to say, for $A \in M(n; \mathbb{C})$, there exists a unitary matrix U such that

$$UAU^{-1} = \begin{bmatrix} \lambda_1 & & & & \\ b_{21} & \lambda_2 & & 0 & \\ b_{31} & b_{32} & \lambda_3 & & \\ \vdots & \vdots & \vdots & \ddots & \\ b_{n1} & b_{n2} & b_{n3} & \cdots & \lambda_n \end{bmatrix}_{n \times n} \quad \text{with} \quad U = \begin{bmatrix} \vec{x}_1 \\ \vec{x}_2 \\ \vec{x}_3 \\ \vdots \\ \vec{x}_n \end{bmatrix}.$$

Note that the first row vector $\vec{x}_1$ of U is an eigenvector of A corresponding to the eigenvalue λ_1. Refer to Ex. <C> 10(a) of Sec. 2.7.6.

7. *Normal matrix and its spectral decomposition* Let $N \in M(n; \mathbb{C})$.

   (a) Suppose that N is normal. Then

      (1) N has n complex eigenvalues.
      (2) Eigenvectors of N corresponding to *distinct* eigenvalues are orthogonal.
      (3) If $\vec{x}N = \lambda\vec{x}$ for nonzero $\vec{x}$, then $\vec{x}\bar{N}^* = \bar{\lambda}\vec{x}$.

   (b) (Schur and Toeplitz, 1910) N is normal if and only if N is unitarily similar to a diagonal matrix. In fact, let $\lambda_1, \ldots, \lambda_n$ be eigenvalues of N with corresponding eigenvectors $\vec{x}_1, \ldots, \vec{x}_n$, i.e.

$$\vec{x}_j N = \lambda_j \vec{x}_j \quad \text{for } 1 \leq j \leq n$$

such that $\mathcal{B} = \{\vec{x}_1, \ldots, \vec{x}_n\}$ is an orthonormal basis for $\mathbb{C}^n$. Then

$$UNU^{-1} = \begin{bmatrix} \lambda_1 & & 0 \\ & \ddots & \\ 0 & & \lambda_n \end{bmatrix}, U = \begin{bmatrix} \vec{x}_1 \\ \vec{x}_2 \\ \vdots \\ \vec{x}_n \end{bmatrix}$$

where U is unitary.

(c) In particular, in case N is normal, then

   (1) N is unitary $\Leftrightarrow N$ is unitarily similar to a diagonal matrix with main diagonal entries of absolute value 1 (see Ex. 6(e)).

                 $\Leftrightarrow$ all eigenvalues of N are of absolute value 1.

   (2) N is Hermitian $\Leftrightarrow n$ eigenvalues of N are all *real*. Thus, N is unitarily similar to a real diagonal matrix.

   (3) N is *positive-definite* Hermitian, i.e. N is Hermitian and $\vec{x} N \vec{x}^* > 0$ for any $\vec{x} \neq \vec{0}$ in $\mathbb{C}^n \Leftrightarrow n$ eigenvalues of N are all *positive*.

   (4) N is skew-Hermitian $\Leftrightarrow n$ eigenvalues of N are all *pure imaginary*.

Moreover,

   (5) N is orthogonal $\Leftrightarrow N$ is a *real* matrix and its eigenvalues are of the forms $e^{i\theta}$, $e^{-i\theta}$, $0 \leq \theta \leq \pi$.

   (6) N is symmetric $\Leftrightarrow N$ is a *real* matrix and all its eigenvalues are real.

                 $\Leftrightarrow N$ is a real matrix and is *orthogonally similar* to a real diagonal matrix.

   (7) N is *positive semidefinite* symmetric, i.e. N is symmetric and $\vec{x} N \vec{x}^* \geq 0$ for any $\vec{x} \in \mathbb{R}^n \Leftrightarrow n$ eigenvalues are all nonnegative.

(d) *The spectral decomposition theorem* for normal matrix. Let $\lambda_1, \ldots, \lambda_r$ be *distinct* eigenvalues of a normal matrix N with respective algebraic multiplicity $k_1, \ldots, k_r$ where $k_1 + \cdots + k_r = n$. Let

$$W_j = \{\vec{x} \in \mathbb{C}^n \mid \vec{x} N = \lambda_j \vec{x}\}, \quad 1 \leq j \leq r$$

be the eigenspace corresponding to λ_i. Then

(1)

$$\mathbb{C}^n = W_1 \oplus W_2 \oplus \cdots \oplus W_r$$
$$= W_j \oplus W_j^\perp, \quad 1 \leq j \leq r,$$

where

$$W_j^\perp = \bigoplus_{\substack{l=1 \\ l \neq j}}^{r} W_l.$$

(2) The mapping

$$N_j = \prod_{\substack{l=1 \\ l \neq j}}^{r} \left(\frac{N - \lambda_l I_n}{\lambda_j - \lambda_l} \right) : \mathbb{C}^n \to \mathbb{C}^n$$

is the orthogonal projection of $\mathbb{C}^n$ onto W_j for $1 \leq j \leq r$ which is *positive semi-definite Hermitian.*

(3) $N_i N_j = \delta_{ij} N_i$ for $1 \leq i, j \leq r$.

(4) $N_j's$ decompose I_n as

$$I_n = N_1 + \cdots + N_r.$$

(5) $N_j's$ decompose N as

$$N = \lambda_1 N_1 + \cdots + \lambda_r N_r.$$

See Fig. B.12.

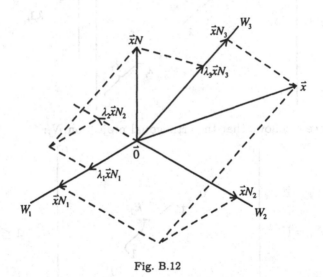

Fig. B.12

(*Note* It can be shown that the *algebraic multiplicity* k_j of λ_j is equal to its *geometric multiplicity* $\dim W_j$ (see Sec. B.10). For each $j, 1 \leq j \leq r$, take an orthonormal basis $\mathcal{B}_j = \{x_{j1}, \dots, x_{jk_j}\}$ for W_j so that $\mathcal{B} = \mathcal{B}_1 \cup \mathcal{B}_2 \cup \cdots \cup \mathcal{B}_r$ forms an orthonormal basis for $\mathbb{C}^n$.

Let

$$U = \begin{bmatrix} \vec{x}_{11} \\ \vdots \\ \vec{x}_{1k_1} \\ \vec{x}_{21} \\ \vdots \\ \vec{x}_{2k_2} \\ \vdots \\ \vec{x}_{r1} \\ \vdots \\ \vec{x}_{rk_r} \end{bmatrix}.$$

Then U is unitary and

$$UNU^{-1} = \begin{bmatrix} \lambda_1 & & & & & & \\ & \ddots & \overbrace{}^{k_1} & & & \mathbf{0} & \\ & & \lambda_1 & & & & \\ & & & \lambda_2 & \overbrace{}^{k_2} & & \\ & & & & \ddots & & \\ & & & & & \lambda_2 & \\ & & & & & & \ddots \\ & \mathbf{0} & & & & & \lambda_r \overbrace{}^{k_r} \\ & & & & & & & \ddots \\ & & & & & & & & \lambda_r \end{bmatrix} = \begin{bmatrix} \lambda_1 I_{k_1} & & & \mathbf{0} \\ & \lambda_2 I_{k_2} & & \\ & & \ddots & \\ \mathbf{0} & & & \lambda_r I_{k_r} \end{bmatrix}_{n \times n}.$$

Hence it follows that the orthogonal projection $N_j\colon \mathbb{C}^n \to W_j$ is

$$N_j = U^{-1} \begin{bmatrix} 0 & & & & & & & 0 \\ & \ddots & & & & & & \\ & & 0 & & & & & \\ & & & 1 \overbrace{}^{k_j} & & & & \\ & & & & \ddots & & & \\ & & & & & 1 & & \\ & & & & & & 0 & \\ & & & & & & & \ddots \\ 0 & & & & & & & 0 \end{bmatrix} U, \quad 1 \le j \le r$$

which is equal to $P_j(N)$ where $P_j(t)$, $1 \le j \le k$, are the Lagrange polynomials associated with $\lambda_1, \lambda_2, \ldots, \lambda_k$ (see Sec. B.3).)

(e) Prove the following:

(1) If N is normal and $N = \lambda_1 N_1 + \cdots + \lambda_r N_r$ is the spectral decomposition as in (d), then for any polynomial $f(t)$,

$$f(N) = \sum_{j=1}^{r} f(\lambda_j) N_j$$

holds. Therefore, deduce that if $N^l = 0$, then $N = 0$.

(2) If N is normal, then a matrix M commutes with N if and only if M commutes with each N_j.

(3) If N is normal, there exists a normal matrix M such that $M^2 = N$.

(4) N is normal if and only if $\bar{N}^* = f(N)$ for some polynomial f. For the necessity, take $f(t)$ to be the Lagrange interpolation polynomial such that $f(\lambda_j) = \bar{\lambda}_j$ for $1 \le j < r$.

(5) If N is normal, then a matrix M commutes with N if and only if M commutes with $\bar{N}^*$.

(f) Show that

$$N = \begin{bmatrix} 0 & 1+i & 0 \\ 1+i & 0 & 0 \\ 0 & 0 & i \end{bmatrix}$$

is a normal matrix and use it to justify (d).

8. Let $A = [a_{ij}]_{n \times m}$ be a symmetric real matrix. Then A is

(1) *Positive definite* $\Leftrightarrow \vec{x}A\vec{x}^* > 0$ for any $\vec{x} \in \mathbb{R}^n$ and $\vec{x} \ne 0$.

(2) *Negative definite* $\Leftrightarrow \vec{x}A\vec{x}^* < 0$ for any $\vec{x} \in \mathbb{R}^n$ and $\vec{x} \ne 0$.

(3) *Positive semidefinite* $\Leftrightarrow \vec{x}A\vec{x}^* \ge 0$ for any $\vec{x} \in \mathbb{R}^n$.

(4) *Negative semidefinite* $\Leftrightarrow \vec{x}A\vec{x}^* \le 0$ for any $\vec{x} \in \mathbb{R}^n$.

(5) *Indefinite* $\Leftrightarrow$ there exist $\vec{x}_1, \vec{x}_2 \in \mathbb{R}^n$ such that

$$\vec{x}_1 A \vec{x}_1^* > 0 \quad \text{and} \quad \vec{x}_2 A \vec{x}_2^* < 0.$$

There are corresponding definitions for Hermitian matrix with $\vec{x} \in \mathbb{C}^n$.

(a) Suppose $A_{n \times n} = [a_{ij}]_{n \times n}$ is symmetric. The following are equivalent.

(1) A is positive definite.

(2) There exists an invertible real matrix $P_{n \times n}$ such that

$$A = PP^*.$$

Moreover, P may be taken as a lower triangular matrix.

(3) All n eigenvalues of A are positive.

(4) (Frobenius, 1894) The leading principal submatrices A_k of A have positive determinants, i.e.

$$\det A_k = \det \begin{bmatrix} a_{11} & \cdots & a_{1k} \\ \vdots & & \vdots \\ a_{k1} & \cdots & a_{kk} \end{bmatrix} > 0, \quad \leq k \leq n.$$

(5) There exists a unique real lower-triangular matrix Δ with main diagonal entries all equal to 1 and a diagonal matrix $\mathrm{diag}[d_1, \ldots, d_n]$ such that

$$A = \Delta \begin{bmatrix} d_1 & & 0 \\ & \ddots & \\ 0 & & d_n \end{bmatrix} \Delta^*$$

where ($\det A_k$ as in (4))

$$d_k = \frac{\det A_k}{\det A_{k-1}} > 0 \quad \text{for } \leq k \leq n; \quad \det A_0 = 1.$$

Try to figure out criteria for negative definite and semidefinite matrices.

(b) Probably, the simplest criteria for a symmetric matrix to be indefinite is that it possesses at least one positive eigenvalue and one negative eigenvalue. Can you figure out others like (a)?

(c) Use

$$A = \begin{bmatrix} 2 & -1 & 0 \\ -1 & 2 & -1 \\ 0 & -1 & 2 \end{bmatrix}$$

to justify (a).

B.10   Eigenvalues and Eigenvectors

Let $A = [a_{ij}] \in \mathrm{M}(n; \mathbb{F})$ and $\lambda \in \mathbb{F}$.

If there exists a *nonzero* vector $\vec{x} \in \mathbb{F}^n$ such that

$$\vec{x} A = \lambda \vec{x},$$

then $\vec{x}$ is called an *eigenvector* of A corresponding to the *eigenvalue* λ. An eigenvector is also called a *characteristic vector* and an eigenvalue called a *characteristic value*. The eigenvalue λ of a linear transformation $f \colon \mathbb{F}^n \to \mathbb{F}^n$

is defined to be that of the matrix $[f]_B$ with respect to any fixed basis B for $\mathbb{F}^n$ (see Sec. B.7), while the corresponding eigenvector $\vec{x}$ is the one that satisfies $[\vec{x}]_B[f]_B = \lambda[\vec{x}]_B$ which is equivalent to $f(\vec{x}) = \lambda\vec{x}$. Similar definition is still valid for linear transformation $f\colon V \to V$ where $\dim V < \infty$.

Exercises

1. The following are equivalent:

   (1) λ is an eigenvalue of A.

   (2) λ is a zero of the characteristic polynomial $\det(A - tI_n)$ of A (see Ex. 32 of Sec. B.4), i.e. $\det(A-\lambda I_n) = 0$ or $A-\lambda I_n$ is not invertible.

   (3) The kernel $\operatorname{Ker}(A - \lambda I_n)$ of the linear transformation $A - \lambda I_n\colon$ $\mathbb{F}^n \to \mathbb{F}^n$ defined by $\vec{x} \to \vec{x}(A - \lambda I_n)$ has dimension ≥ 1.

2. *Some basic facts* Let λ be an eigenvalue of A with corresponding eigenvector $\vec{x}$.

   (1) λ^k is an eigenvalue of A^k for any positive integer k, with the same eigenvector $\vec{x}$.

   (2) λ^{-1} is an eigenvalue of A^{-1} if A is invertible, with the same eigenvector $\vec{x}$.

   (3) The eigenspace corresponding to λ

   $$E_\lambda = \{\vec{x} \in \mathbb{F}^n \mid \vec{x}A = \lambda\vec{x}\} \quad (\text{including } \vec{x} = \vec{0}) = \operatorname{Ker}(A - \lambda I_n)$$

   is a subspace of $\mathbb{F}^n$, of positive dimension.

   Also

   (4) Similar matrices have the same characteristic polynomial and hence the same eigenvalues.

   (5) A matrix A and its transpose A^* have the same characteristic polynomial and eigenvalues.

   (6) Let $\det(A - tI_n) = (-1)^n t^n + \alpha_{n-1}t^{n-1} + \cdots + \alpha_1 t + \alpha_0$ be the *characteristic polynomial* of $A = [a_{ij}]_{n\times n}$. Then (refer to Ex. <C> 4 of Sec. 3.7.6)

   $$\alpha_{n-1} = (-1)^{n-1}\operatorname{tr} A,$$
   $$\alpha_0 = \det A.$$

   Hence (added to Ex. 16(d) of Sec. B.4)

   $$A \text{ is invertible.}$$

   $$\Leftrightarrow \text{Zero is not an eigenvalue of } A.$$

   $$\Leftrightarrow \text{The constant term } \alpha_0 \neq 0.$$

3. *Some basic facts*

    (a) The eigenvalues of an upper (or lower) triangular matrix are the diagonal entries.

    (b) Eigenvectors associated with distinct eigenvalues are linearly independent.

    (c) For any $A, B \in M(n; \mathbb{C})$ such that B is invertible, there exists a scalar $\alpha \in \mathbb{C}$ such that $A + \alpha B$ is not invertible. There exist at most n distinct such scalars α.

    (d) For any $A, B \in M(n; \mathbb{C})$, the characteristic polynomials for the products AB and BA are equal.

4. *Cayley–Hamilton Theorem* Let $A = [a_{ij}] \in M(n; \mathbb{F})$ with characteristic polynomial $\det(A - tI_n) = (-1)^n t^n + \alpha_{n-1} t^{n-1} + \cdots + \alpha_1 t + \alpha_0$, then

$$(-1)^n A^n + \alpha_{n-1} A^{n-1} + \cdots + \alpha_1 A + \alpha_0 I_n = O.$$

In short, a matrix satisfies its characteristic polynomial. For example, let

$$A = \begin{bmatrix} 1 & -1 \\ 2 & 4 \end{bmatrix}.$$

Then, the characteristic polynomial is

$$\det(A - tI_2) = \begin{vmatrix} 1-t & -1 \\ 2 & 4-t \end{vmatrix} = (1-t)(4-t) + 2 = t^2 - 5t + 6.$$

Actual computation shows that

$$A^2 = \begin{bmatrix} 1 & -1 \\ 2 & 4 \end{bmatrix} \begin{bmatrix} 1 & -1 \\ 2 & 4 \end{bmatrix} = \begin{bmatrix} -1 & -5 \\ 10 & 14 \end{bmatrix}$$

$$\Rightarrow A^2 - 5A + 6I_2 = \begin{bmatrix} -1 & -5 \\ 10 & 14 \end{bmatrix} - 5 \begin{bmatrix} 1 & -1 \\ 2 & 4 \end{bmatrix} + 6 \begin{bmatrix} 1 & 0 \\ 0 & 1 \end{bmatrix} = \begin{bmatrix} 0 & 0 \\ 0 & 0 \end{bmatrix} = O.$$

    (a) Prove this theorem in cases $n = 2, 3$.

    (b) Justify this theorem by using

$$A = \begin{bmatrix} 3 & -1 & -1 \\ 1 & 0 & -1 \\ -2 & 5 & 4 \end{bmatrix}.$$

    (c) Prove this theorem in case A is diagonalizable (see Sec. B.11) and for general matrix A.

(d) If $\alpha_0 = \det A \neq 0$, then A is invertible (see Ex. 2(6)). Thus

$$[(-1)^n A^{n-1} + \alpha_{n-1} A^{n-2} + \cdots + \alpha_2 A + \alpha_1 I_n] A = -\alpha_0 I_n$$

shows that the inverse matrix is

$$A^{-1} = -\frac{1}{\alpha_0} [(-1)^n A^{n-1} + \alpha_{n-1} A^{n-2} + \cdots + \alpha_2 A + \alpha_1 I_n].$$

(e) Show that $A = \begin{bmatrix} 1 & 7 \\ -6 & -43 \end{bmatrix}$ satisfies $A^2 + 42A - I_2 = O$. Compute A^{-1}.

(f) If $A \in M(3; \mathbb{R})$ satisfies $A^5 - 4A^4 + 7A^3 - 6A^2 - 14I_2 = O$, show that A is invertible, and express A^{-1} in terms of A.

(g) If $A = [a_{ij}] \in M(2; \mathbb{F})$ is invertible, show that

$$A^{-1} = \alpha A + \beta I_2$$

for some $\alpha, \beta \in \mathbb{F}$. Determine α and β, and use a_{ij}'s to express entries of A^{-1}.

(h) If $A \in M(2; \mathbb{F})$ satisfies $A^2 + A + I_2 = O$, find A^{3n} explicitly in terms of A and I_2 for every integer n, positive or negative.

5. Compute *real* or *complex* eigenvalues for the following matrices and find out associated eigenvectors of the eigenvalues. Check Cayley–Hamilton theorem for each matrix.

$$\begin{bmatrix} 1 & 0 \\ 1 & 1 \end{bmatrix}; \begin{bmatrix} 4 & 1 \\ 2 & 3 \end{bmatrix}; \begin{bmatrix} 4 & -1 \\ 2 & 3 \end{bmatrix}; \begin{bmatrix} \frac{1}{2} & \frac{1}{5} \\ \frac{1}{5} & \frac{1}{3} \end{bmatrix};$$

$$\begin{bmatrix} 1 & 0 & 0 \\ 1 & 1 & 0 \\ 1 & 1 & 1 \end{bmatrix}; \begin{bmatrix} 0 & 0 & 1 \\ 0 & 1 & 0 \\ 1 & 0 & 0 \end{bmatrix}; \begin{bmatrix} 3 & 1 & 1 \\ 2 & 4 & 2 \\ -1 & -1 & 1 \end{bmatrix}; \begin{bmatrix} 6 & -3 & -2 \\ 4 & -1 & -2 \\ 10 & -5 & -3 \end{bmatrix};$$

AA^* and A^*A where $A = \begin{bmatrix} -1 & 2 & 0 \\ 3 & 1 & 4 \end{bmatrix}$.

B.11 Diagonalizability of a Square Matrix or a Linear Operator

We have touched this concept and some of its related results in quite a few places before, such as in Exs. 31 and 32 of Sec. B.4, Exs. 7–9 of Sec. B.7, Ex. 14 of Sec. B.8, Exs. 6–8 of Sec. B.9 and Sec. B.10. Now, we formally introduce the diagonalizability of a square matrix or a linear operator on a finite-dimensional vector space.

A square matrix is said to be *diagonalizable* if it is similar to a diagonal matrix (see Ex. 31 of Sec. B.4). That is to say, for $A_{n \times n}$, if there exists an invertible matrix P such that

$$PAP^{-1}$$

is a diagonal matrix, then A is called diagonalizable. If $\dim V < \infty$, a linear operator $f: V \to V$ is *diagonalizable* if there exists a basis $\mathcal{B}$ for V such that $[f]_\mathcal{B}$ is diagonalizable.

Suppose λ is an eigenvalue of a matrix $A_{n \times n}$. In addition to the eigenspace E_λ introduced in Ex. 2 of Sec. B.10, we introduce the *generalized eigenspace* of A corresponding to λ as the subspace

$$G_\lambda = \{\vec{x} \in \mathbb{F}^n \mid \vec{x}(A - \lambda I_n)^p = \vec{0} \text{ for some positive integer } p\}.$$

Nonzero vectors in G_λ are called *generalized eigenvectors* of A corresponding to λ.

Note that $E_\lambda \subseteq G_\lambda$ holds. The role played by G_λ will become much more clear in the next section B.12.

Using division algorithm for polynomials (see Sec. A.5), Cayley–Hamilton theorem (see Ex. 4 of Sec. B.10) allows us to find a polynomial $p(t)$ with the following properties:

1. The leading coefficient of $p(t)$ is 1.
2. If $g(t)$ is any polynomial such that $g(A) = O$, then $p(t)$ divides $g(t)$.

Such a $p(t)$ does exist and has the smallest degree among all such polynomials $g(t)$ and is called the *minimal polynomial* of the given square matrix A. Note that the minimal polynomial divides the characteristic polynomial.

For example, the matrix

$$A = \begin{bmatrix} -9 & 4 & 4 \\ -8 & 3 & 4 \\ -16 & 8 & 7 \end{bmatrix}$$

has characteristic polynomial $\det(A - tI_3) = -(t + 1)^2(t - 3)$ and minimal polynomial $(t + 1)(t - 3) = t^2 - 2t - 3$. What are generalized eigenspaces G_λ for $\lambda = -1$ and $\lambda = 3$?

While the matrix

$$B = \begin{bmatrix} 6 & -3 & -2 \\ 4 & -1 & -2 \\ 10 & -5 & -3 \end{bmatrix}$$

has characteristic polynomial $-(t - 2)(t^2 + 1)$ and minimal polynomial $(t - 2)(t^2 + 1)$. Readers are strongly urged to decide what are the generalized eigenspaces G_λ for $\lambda = 2$ and $\lambda = i$ and compare all possible differences between these and those for A above.

Exercises

1. Prove the following.

   (a) A matrix is similar to a scalar matrix αI_n if and only if it is αI_n itself.

   (b) A diagonalizable matrix having only one eigenvalue is a scalar matrix. Therefore,

$$\begin{bmatrix} 1 & 0 \\ 1 & 1 \end{bmatrix}, \quad \begin{bmatrix} 1 & 0 & 0 \\ 1 & 1 & 0 \\ 0 & 1 & 1 \end{bmatrix}, \text{ etc.}$$

   are not diagonalizable.

2. Let

$$A = \begin{bmatrix} a_{11} & a_{12} \\ a_{21} & a_{22} \end{bmatrix}$$

   be a real matrix. Then, there exist at most two distinct real numbers λ such that $A - \lambda I_2$ is not invertible. If such a number exists, it should be an eigenvalue of A.

   (a) Suppose $A \neq I_2$. Then A has two distinct real eigenvalues λ_1 and λ_2, if and only if there exists an invertible matrix P such that

$$PAP^{-1} = \begin{bmatrix} \lambda_1 & 0 \\ 0 & \lambda_2 \end{bmatrix}.$$

   (b) If $\det A < 0$, then A has two distinct real eigenvalues.

   (c) If $a_{11} = a_{22}$ with $a_{12} = a_{21} \neq 0$, then the *symmetric matrix* A has two real eigenvalues and is diagonalizable.

   (d) If $a_{11} = a_{22}$ with $a_{12} = -a_{21} \neq 0$, then A has no real eigenvalue and hence, $A - \lambda I_2$ is invertible for any real number λ.

   (e) In case $A^2 = A$, then

$$\mathbb{R}^2 = \{\vec{x} \in \mathbb{R}^2 | \vec{x}A = \vec{x}\} \oplus \{\vec{x} \in \mathbb{R}^2 | \vec{x}A = \vec{0}\}.$$

3. *Characteristic and minimal polynomials* Let $p(t)$ be the minimal polynomial of a given matrix $A_{n \times n}$.

   (a) A scalar λ is an eigenvalue of A if and only if $p(\lambda) = 0$. Hence the characteristic polynomial and the minimal polynomial for A has the same zeros.

(b) In case the characteristic polynomial is

$$\det(A - tI_n) = (-1)^n(t - \lambda_1)^{r_1} \cdots (t - \lambda_k)^{r_k},$$

where $\lambda_1, \ldots, \lambda_k$ are distinct eigenvalues of A and $r_1 \geq 1, \ldots, r_k \geq 1$ with $r_1 + \cdots + r_k = n$, then there exist integers $n_1, n_2, \ldots, n_k$ such that $1 \leq n_i \leq r_i$ for $1 \leq i \leq k$ and the minimal polynomial is

$$p(t) = (t - \lambda_1)^{n_1} \cdots (t - \lambda_k)^{n_k}.$$

(c) Moreover, if $\varphi(t)$ is an irreducible factor polynomial of $\det(A - tI_n)$, then $\varphi(t) \,|\, p(t)$ holds.

(d) For any polynomial

$$p(t) = t^n + a_{n-1}t^{n-1} + \cdots + a_1 t + a_0,$$

the so-called *companion matrix* of $p(t)$

$$A = \begin{bmatrix} 0 & 1 & 0 & \cdots & 0 \\ 0 & 0 & 1 & \cdots & 0 \\ \vdots & \vdots & \vdots & \cdots & \vdots \\ 0 & 0 & 0 & \cdots & 1 \\ -a_0 & -a_1 & -a_2 & \cdots & -a_{n-1} \end{bmatrix}$$

has the characteristic polynomial $(-1)^n p(t)$ and the minimal polynomial $p(t)$ itself.

4. *Criteria for diagonalizability* For $A_{n \times n}$, let

$$\det(A - tI_n) = (-1)^n(t - \lambda_1)^{r_1} \cdots (t - \lambda_k)^{r_k},$$

where $\lambda_1, \ldots, \lambda_k$ are distinct eigenvalues of A and $r_1 \geq 1, \ldots, r_k \geq 1$ with $r_1 + \cdots + r_k = n$. Then the following are equivalent.

(a) A is diagonalizable.

(b) There exists a basis $\mathcal{B} = \{\vec{x}_1, \ldots, \vec{x}_n\}$ for $\mathbb{F}^n$ consisting of eigenvectors of A such that

$$PAP^{-1} = \begin{bmatrix} \lambda_1 & & & & & & \\ & \ddots & {\scriptstyle r_1} & & & 0 & \\ & & \lambda_1 & & & & \\ & & & \ddots & & & \\ & & & & \lambda_k & {\scriptstyle r_k} & \\ & 0 & & & & \ddots & \\ & & & & & & \lambda_k \end{bmatrix}_{n\times n}$$

where

$$P = \begin{bmatrix} \vec{x}_1 \\ \vdots \\ \vec{x}_{r_1} \\ \vdots \\ \vec{x}_{r_1+\cdots+r_{k-1}+1} \\ \vdots \\ \vec{x}_{r_1+\cdots+r_{k-1}+r_k} \end{bmatrix}_{n\times n}$$

with $\vec{x}_l A = \lambda_i \vec{x}_l$ for $r_1 + \cdots + r_{i-1} + 1 \le l \le r_1 + \cdots + r_{i-1} + r_i$ and $1 \le i \le k$.

(c) $\mathbb{F}^n$ is the direct sum of eigenspaces E_{λ_i}, $1 \le i \le k$, of A, i.e.

$$\mathbb{F}^n = E_{\lambda_1} \oplus \cdots \oplus E_{\lambda_k}.$$

(d) Each eigenspace E_{λ_i} has dimension

$$\dim E_{\lambda_i} = r_i = n - \mathrm{r}(A - \lambda_i I_n),$$

i.e. the *algebraic multiplicity* r_i of the eigenvalue λ_i is equal to its *geometric multiplicity* $\dim E_{\lambda_i}$ for $1 \le i \le k$.

(e) Each eigenspace E_{λ_i} of A is equal to the corresponding generalized eigenspace G_{λ_i} for each i, $1 \le i \le k$, i.e.

$$E_{\lambda_i} = G_{\lambda_i}.$$

(f) The ranks $r(A - \lambda_i I_n) = r((A - \lambda_i I_n)^2)$ for $1 \leq i \leq k$ (this would imply that $G_{\lambda_i} = \text{Ker}(A - \lambda_i I_n) = E_{\lambda_i}$).

(g) The minimal polynomial $p(t)$ is a product of distinct linear factors, i.e.

$$p(t) = (t - \lambda_1) \cdots (t - \lambda_k).$$

In particular, an $n \times n$ matrix having n distinct eigenvalues is diagonalizable.

5. *The process for diagonalization* Suppose $A \in M(n; \mathbb{F})$. Then,

(1) Compute the characteristic polynomial $\det(A - tI_n)$ and try to factor it into product of linear factors such as $(-1)^n (t - \lambda_1)^{r_1} \cdots (t - \lambda_k)^{r_k}$ with distinct $\lambda_1, \ldots, \lambda_k$ and $r_1 + \cdots + r_k = n$.

(2) Use Ex. 4 to test for diagonalizability.

(3) If A is diagonalizable, choose a basis $\mathcal{B}_i = \{\vec{x}_{i1}, \ldots, \vec{x}_{ir_i}\}$ for the eigenspace E_{λ_i} with dimension r_i, $1 \leq i \leq k$. By Ex. 4(c), $\mathcal{B}_1 \cup \cdots \cup \mathcal{B}_k$ forms a basis for $\mathbb{F}^n$.

(4) Form an $n \times n$ matrix P with basis vectors in the *ordered* basis $\mathcal{B}_1 \cup \cdots \cup \mathcal{B}_k$ as row vectors, i.e.

$$P = \begin{bmatrix} \vec{x}_{11} \\ \vdots \\ \vec{x}_{1r_1} \\ \vdots \\ \vec{x}_{k1} \\ \vdots \\ \vec{x}_{kr_k} \end{bmatrix}.$$

This P is invertible.

(5) Then one can check that

$$PAP^{-1} = \begin{bmatrix} \lambda_1 & & r_1 & & & & 0 \\ & \ddots & & & & & \\ & & \lambda_1 & & & & \\ & & & \ddots & & & \\ & & & & \lambda_k & r_k & \\ & & & & & \ddots & \\ 0 & & & & & & \lambda_k \end{bmatrix}_{n \times n}.$$

For a linear operator $f: V \to V$, where $\dim V = n$, choose a suitable basis $\mathcal{B}$ for V and start from the matrix representation $[f]_{\mathcal{B}}$. Use each of the following matrices to justify the process stated above:

$$\begin{bmatrix} 1 & 5 \\ 3 & 4 \end{bmatrix}; \quad \begin{bmatrix} 0 & 1 & 0 \\ 0 & 0 & 1 \\ 1 & -1 & 1 \end{bmatrix}; \quad \begin{bmatrix} 1 & 3 & 1 \\ 2 & 2 & 4 \\ 1 & 1 & -1 \end{bmatrix}; \quad \begin{bmatrix} 2 & 1 & 1 \\ 1 & 2 & 1 \\ 1 & 1 & 2 \end{bmatrix};$$

$$\begin{bmatrix} 1 & 0 & -4 \\ 0 & 5 & 4 \\ -4 & 4 & 3 \end{bmatrix}; \quad \begin{bmatrix} 11 & 21 & 3 \\ -4 & -8 & -1 \\ -5 & -11 & 0 \end{bmatrix}; \quad \begin{bmatrix} 0 & 1 & 0 & 1 \\ 1 & 0 & 1 & 0 \\ 0 & 1 & 0 & 1 \\ 1 & 0 & 1 & 0 \end{bmatrix}.$$

B.12 Canonical Forms for Matrices: Jordan Form and Rational Form

As noted before, not every square matrix is diagonalizable.

For a given $A \in M(n; \mathbb{F})$, by unique factorization theorem for polynomials (see Sec. A.5), the characteristic polynomial is factored as

$$\det(A - tI_n) = (-1)^n p_1(t)^{r_1} \cdots p_k(t)^{r_k},$$

where $p_1(t), \ldots, p_k(t)$ are irreducible monic polynomials and $r_1 \geq 1, \ldots, r_k \geq 1$ are positive integers with $r_1 \deg p_1(t) + \cdots + r_k \deg p_k(t) = n$. In case each $p_i(t) = t - \lambda_i$ is a linear polynomial, a basis can be chosen from $\mathbb{F}^n$ according to which A can be expressed as the so-called Jordan canonical form; otherwise the rational canonical form.

This section is divided into two subsections.

B.12.1 *Jordan canonical form*

The main the result is the following theorem.

Theorem *Suppose $A \in M(n; \mathbb{F})$ is a nonzero matrix having characteristic polynomial factored over the field $\mathbb{F}$ as*

$$\det(A - tI_n) = (-1)^n (t - \lambda_1) \cdots (t - \lambda_n)$$

where $\lambda_1, \ldots, \lambda_n$ are eigenvalues of A which are not necessarily distinct. Then there exists a basis $\mathcal{B} = \{\vec{x}_1, \ldots, \vec{x}_n\}$ for $\mathbb{F}^n$ such that

$$PAP^{-1} = \begin{bmatrix} J_1 & & & \\ & J_2 & & \\ & & \ddots & \\ & & & J_k \end{bmatrix}_{n \times n} \qquad \text{(called the Jordan canonical form of A)}$$

where each J_i is a square matrix of the form $[\lambda_j]_{1\times 1}$ or the form

$$\begin{bmatrix} \lambda_j & 0 & 0 & \cdots & 0 & 0 \\ 1 & \lambda_j & 0 & \cdots & 0 & 0 \\ 0 & 1 & \lambda_j & \cdots & 0 & 0 \\ \vdots & \vdots & \vdots & \cdots & \vdots & \vdots \\ 0 & 0 & 0 & \cdots & \lambda_j & 0 \\ 0 & 0 & 0 & \cdots & 1 & \lambda_j \end{bmatrix} \quad (\text{called a Jordan block corresponding to } \lambda_j)$$

for some eigenvalue λ_j of A and

$$P = \begin{bmatrix} \vec{x}_1 \\ \vdots \\ \vec{x}_n \end{bmatrix}_{n\times n}.$$

If $\det(A - tI_n) = (-1)^n (t - \lambda_1)^{r_1} \cdots (t - \lambda_k)^{r_k}$ where $\lambda_1, \ldots, \lambda_k$ are distinct and $r_1 + \cdots + r_k = n$, the Jordan canonical form can be put in a neater way (see Ex. 7 below). We will use this factorization of $\det(A - tI_n)$ throughout the exercises.

Exercises

1. Suppose $\mathbb{R}^8$ has a basis $\mathcal{B} = \{\vec{x}_1, \ldots, \vec{x}_8\}$ so that, for a matrix $A_{8\times 8}$,

$$PAP^{-1} = \begin{bmatrix} 2 & 0 & 0 & & & & & \\ 1 & 2 & 0 & & & & & \\ 0 & 1 & 2 & & & & & \\ & & & 2 & & & & \\ & & & & 3 & 0 & & \\ & & & & 1 & 3 & & \\ & & & & & & 0 & 0 \\ & & & & & & 1 & 0 \end{bmatrix}_{8\times 8}, \quad P = \begin{bmatrix} \vec{x}_1 \\ \vdots \\ \vec{x}_8 \end{bmatrix}.$$

(a) Show that $\det(A - tI_8) = (t - 2)^4 (t - 3)^2 t^2$. Note that, among the basis vectors $\vec{x}_1, \ldots, \vec{x}_8$, only $\vec{x}_1, \vec{x}_4, \vec{x}_5$ and $\vec{x}_7$ are eigenvectors of A corresponding to eigenvalues $\lambda_1 = 2, \lambda_2 = 3$ and $\lambda_3 = 0$ with respective multiplicity 4, 2 and 2 which are the number of times that eigenvalues appear on the diagonal of PAP^{-1}.

(b) Determine the eigenspace E_{λ_i} and the generalized eigenspace G_{λ_i} (see Sec. B.11) for $1 \leq i \leq 3$ and see, if any, $E_{\lambda_i} = G_{\lambda_i}$ or not.

(c) For each λ_i, find the smallest positive integer p_i for which

$$G_{\lambda_i} = \text{Ker}((A - \lambda_i I_8)^{p_i}), \quad 1 \le i \le 3.$$

(d) Show that $\vec{x}_3, \vec{x}_3(A - \lambda_1 I_8) = \vec{x}_2, \vec{x}_3(A - \lambda_1 I_8)^2 = \vec{x}_1$, and $\vec{x}_4$ are linearly independent and hence,

$$G_{\lambda_1} = \langle\langle \vec{x}_1, \vec{x}_2, \vec{x}_3, \vec{x}_4 \rangle\rangle.$$

Similarly, $G_{\lambda_2} = \langle\langle \vec{x}_5, \vec{x}_6 \rangle\rangle$ with $\vec{x}_5 = \vec{x}_6(A - \lambda_2 I_8)$ and $G_{\lambda_3} = \langle\langle \vec{x}_7, \vec{x}_8 \rangle\rangle$ with $\vec{x}_7 = \vec{x}_8(A - \lambda_3 I_8)$.

(e) Let $A_i = A|G_{\lambda_i}$ denote the restriction of the linear operator $A: \mathbb{R}^8 \to \mathbb{R}^8$ to its invariant subspace G_{λ_i} for $1 \le i \le 3$. Compute the following for each i:

  1. $\text{r}(A_i), \text{r}(A_i^2), \text{r}(A_i^3)$ and $\text{r}(A_i^4)$.
  2. $\dim \text{Ker}(A_i^l)$ for $l = 1, 2, 3, 4$.

2. *Generalized eigenspace* (or *root space*) Suppose λ is an eigenvalue of a matrix $A_{n \times n}$.

(a) Since $\text{Ker}(A - \lambda I_n) \subseteq \text{Ker}(A - \lambda I_n)^2 \subseteq \cdots \subseteq \mathbb{F}^n$, there exists a smallest positive integer q such that $\text{Ker}(A - \lambda I_n)^l = \text{Ker}(A - \lambda I_n)^q$ for all $l \ge q$. Hence,

$$G_\lambda = \text{Ker}(A - \lambda I_n)^q$$

is an A-invariant subspace of $\mathbb{F}^n$ containing the eigenspace E_λ (see Ex. 15 of Sec. B.7).

(b) For each generalized eigenvector $\vec{x} \in G_\lambda$, there exists a smallest positive integer p such that $1 \le p \le q$ and $\vec{x}(A - \lambda I_n)^p = \vec{0}$. p is called the *order* of $\vec{x}$. The set

$$\{\vec{x}, \vec{x}(A - \lambda I_n), \vec{x}(A - \lambda I_n)^2, \ldots, \vec{x}(A - \lambda I_n)^{p-1}\}$$

is linearly independent and is called a *cycle* of generalized vectors of A corresponding to λ. Note that $\vec{x}(A - \lambda I_n)^{p-1} \in E_\lambda$.

(c) The restriction $A|G_\lambda$ of A to the invariant subspace G_λ has the minimal polynomial

$$(t - \lambda)^q.$$

Also, $q \mid \dim G_\lambda$ (see Ex. 3).

(d) Suppose λ_1 and λ_2 are distinct eigenvalues of A. Then

$$G_{\lambda_1} \cap G_{\lambda_2} = \{\vec{0}\}$$

and $G_{\lambda_1} \cup G_{\lambda_2}$ is linearly independent.

3. *First decomposition theorem for the space* Suppose $\lambda_1, \ldots, \lambda_k$ are distinct eigenvalues of $A_{n \times n}$, i.e.

$$\det(A - tI_n) = (-1)^n (t - \lambda_1)^{r_1} \cdots (t - \lambda_k)^{r_k}$$

where $r_1 + \cdots + r_k = n$. Then

$$\mathbb{F}^n = G_{\lambda_1} \oplus \cdots \oplus G_{\lambda_k}$$

and the restriction $A \,|\, G_{\lambda_i}$ has the characteristic polynomial $(t - \lambda_i)^{r_i}$ with

$$r_i = \dim G_{\lambda_i}$$

for $1 \leq i \leq k$.

4. *Cyclic invariant subspace* Suppose $A \in \mathrm{M}(n; \mathbb{F})$. For $\vec{x} \in \mathbb{F}^n$ and $\vec{x} \neq \vec{0}$, let

$$C(\vec{x}) = \cap \{W \,|\, W \text{ is } A\text{-invariant subspace of } \mathbb{F}^n \text{ containing } \vec{x}\}$$
$$= \langle\langle \vec{x}, \vec{x}A, \vec{x}A^2, \ldots, \vec{x}A^k, \ldots \rangle\rangle.$$

Then $C(\vec{x})$ is an A-invariant subspace and is the smallest such one. $C(\vec{x})$ is called the *cycle generated by $\vec{x}$ related to A* or *A-cycle generated by $\vec{x}$*. A polynomial $g(t) \in P(\mathbb{F})$ (see Sec. A.5) for which $\vec{x}g(A) = \vec{0}$ is called an *annihilator of $\vec{x}$ related to A* or just an *A-annihilator of $\vec{x}$*. There exists a unique annihilator of $\vec{x}$ with minimal degree and leading coefficient 1, and it is called the *minimal polynomial*, denoted $d_{\vec{x}}(t)$, of $\vec{x}$ related to A.

(a) Suppose the degree $\deg d_{\vec{x}}(t) = k$. Then

$$\dim C(\vec{x}) = k$$

and $\mathcal{B}_{\vec{x}} = \{\vec{x}, \vec{x}A, \ldots, \vec{x}A^{k-1}\}$ is a basis for $C(\vec{x})$ for which

$$[A \,|\, C(\vec{x})]_{\mathcal{B}_{\vec{x}}} = \begin{bmatrix} 0 & 1 & \cdots & 0 & 0 \\ 0 & 0 & \cdots & 0 & 0 \\ \vdots & \vdots & & \vdots & \vdots \\ 0 & 0 & \cdots & 1 & 0 \\ 0 & 0 & \cdots & 0 & 1 \\ -a_0 & -a_1 & \cdots & -a_{k-2} & -a_{k-1} \end{bmatrix}_{k \times k},$$

where $d_{\vec{x}}(t) = t^k + a_{k-1}t^{k-1} + a_{k-2}t^{k-2} + \cdots + a_1 t + a_0$.

(b) $(-1)^k d_{\vec{x}}(t)$ is the characteristic polynomial and $d_{\vec{x}}(t)$ itself the minimal polynomial for the restriction $A \,|\, C(\vec{x})$.

5. *Nilpotent matrix* A matrix $B_{n \times n}$ is said to be a *nilpotent matrix of power or degree or index* m if $B^m = O$ but $B^{m-1} \neq O$ for positive integer m. According to Ex. 2, the restriction of $A - \lambda I_n$ to G_λ is nilpotent of power q.

(a) There exists a nonzero vector $\vec{x}_1 \in \mathbb{F}^n$ such that the cycle $C(\vec{x}_1)$ satisfies

(1) $\dim C(\vec{x}_1) = m$, and

(2) the restriction $B \,|\, C(\vec{x}_1)$ is nilpotent of power $m_1 = m$.

(b) There exists a B-invariant subspace V of $\mathbb{F}^n$ such that

$$\mathbb{F}^n = C(\vec{x}_1) \oplus V$$

and $B \,|\, V$ is nilpotent of power $m_2 \leq m_1$.

(c) There exist nonzero vectors $\vec{x}_1, \vec{x}_2, \ldots, \vec{x}_k \in \mathbb{F}^n$ such that

$$\mathbb{F}^n = C(\vec{x}_1) \oplus C(\vec{x}_2) \oplus \cdots \oplus C(\vec{x}_k)$$

where,

(1) the restriction $B \,|\, C(\vec{x}_i)$ is nilpotent of power m_i for $1 \leq i \leq k$ with

$$m = m_1 \geq m_2 \geq \cdots \geq m_k, \quad \text{and} \quad \dim C(\vec{x}_i) = m_i;$$

(2) $B \,|\, C(\vec{x}_i)$ has minimal polynomial

$$d_i(t) = t^{m_i}, \quad 1 \leq i \leq k$$

and has the basis $\mathcal{B}_i = \{\vec{x}_i B^{m_i - 1}, \vec{x}_i B^{m_i - 2}, \ldots, \vec{x}_i B, \vec{x}_i\}$.

Therefore, $\mathcal{B} = \mathcal{B}_1 \cup \cdots \cup \mathcal{B}_k$ is a basis for $\mathbb{F}^n$ and the matrix representation of B related to $\mathcal{B}$ is

$$[B]_{\mathcal{B}} = \begin{bmatrix} N_1 & & & 0 \\ & N_2 & & \\ & & \ddots & \\ 0 & & & N_k \end{bmatrix}_{n \times n},$$

where

$$N_i = \begin{bmatrix} 0 & 0 & \cdots & 0 & 0 & 0 \\ 1 & 0 & \cdots & 0 & 0 & 0 \\ \vdots & \vdots & & \vdots & \vdots & \vdots \\ 0 & 0 & \cdots & 1 & 0 & 0 \\ 0 & 0 & \cdots & 0 & 1 & 0 \end{bmatrix}_{m_i \times m_i}, \quad 1 \leq i \leq k.$$

Such a direct sum decomposition of $\mathbb{F}^n$ is unique up to the ordering of $C(\vec{x}_i)$. Refer to Ex. <C> 3 of Sec. 3.7.7 for a different treatment.

6. *Second decomposition theorem for the space* Suppose λ is an eigenvalue of algebraic multiplicity r of a matrix $A_{n \times n}$. Then, there exist linear independent vectors $\vec{x}_1, \ldots, \vec{x}_k \in G_\lambda$, the generalized eigenspace of A corresponding to λ, such that

$$G_\lambda = C(\vec{x}_1) \oplus \cdots \oplus C(\vec{x}_k) = \mathrm{Ker}(A - \lambda I_n)^q,$$

where each $C(\vec{x}_i)$ is the cycle generated by $\vec{x}_i$ related to $(A - \lambda I_n) \,|\, G_\lambda$. Moreover, let $m_i = \dim C(\vec{x}_i)$ for $1 \le i \le k$ and $q = m_1 \ge m_2 \ge \cdots \ge m_k$. Note that $m_1 + \cdots + m_k = r = \dim G_\lambda$. Then

cycles bases	$C(\vec{x}_1)$	$C(\vec{x}_2)$	$\cdots$	$C(\vec{x}_k)$	
	$\vec{x}_1(A-\lambda I_n)^{m_1-1}$	$\vec{x}_2(A-\lambda I_n)^{m_2-1}$	$\cdots$	$\vec{x}_k(A-\lambda I_n)^{m_k-1}$	$\leftarrow$ 1st row
	$\vec{x}_1(A-\lambda I_n)^{m_1-2}$	$\vec{x}_2(A-\lambda I_n)^{m_2-2}$	$\cdots$	$\vec{x}_k(A-\lambda I_n)^{m_k-2}$	$\leftarrow$ 2nd row
	$\vdots$	$\vdots$		$\vdots$	
	$\vec{x}_1(A-\lambda I_n)$	$\vec{x}_2(A-\lambda I_n)$	$\cdots$	$\vec{x}_k(A-\lambda I_n)$	$\leftarrow (m_1-1)$th
	$\vec{x}_1$	$\vec{x}_2$	$\cdots$	$\vec{x}_k$	$\leftarrow m_1$th row
	$\uparrow$	$\uparrow$		$\uparrow$	
	$\mathcal{B}_1$	$\mathcal{B}_2$		$\mathcal{B}_k$	

where $k = \dim E_\lambda$, the dimension of the eigenspace E_λ of A corresponding to λ and $E_\lambda = \langle \langle \vec{x}_1(A - \lambda I_n)^{m_1-1}, \ldots, \vec{x}_k(A - \lambda I_n)^{m_k-1} \rangle \rangle$. Therefore, $\mathcal{B} = \mathcal{B}_1 \cup \cdots \cup \mathcal{B}_k$ forms a *Jordan canonical basis* for $A \,|\, G_\lambda$ and $(t - \lambda)^{m_i}$, $1 \le i \le k$, are called *elementary divisors* of $A \,|\, G_\lambda$ (or, of A). The matrix representation of $A \,|\, G_\lambda$ with respect to $\mathcal{B}$ is

$$[A \,|\, G_\lambda]_{\mathcal{B}} = \begin{bmatrix} J_1 & & & 0 \\ & J_2 & & \\ & & \ddots & \\ 0 & & & J_k \end{bmatrix}_{r \times r} ;$$

$$J_i = \begin{bmatrix} \lambda & & & & 0 \\ 1 & \lambda & & & \\ 0 & 1 & \lambda & & \\ \vdots & \vdots & \vdots & \ddots & \\ 0 & 0 & 0 & \cdots 1 & \lambda \end{bmatrix}_{m_i \times m_i}, \quad 1 \le i \le k.$$

Note that $[A\,|\,G_\lambda]_\mathcal{B}$ is a direct sum of k Jordan canonical blocks of decreasing sizes and $m_1 + \cdots + m_k = r = \dim G_\lambda$, the algebraic multiplicity of λ.

Remark In the Table above, let

$$l_j = \text{ the number of vectors in the } j\text{th row,} \quad 1 \le j \le m_1.$$

Then,

$$l_1 = \dim \mathbb{F}^n - r(A - \lambda I_n) = \dim \text{Ker}(A - \lambda I_n) = \dim E_\lambda;$$
$$l_j = r(A - \lambda I_n)^{j-1} - r(A - \lambda I_n)^j \quad \text{for } j > 1.$$

This means the size of the table is completely determined by $A\,|\,G_\lambda$ or by A itself.

7. *Jordan canonical form for matrix* Combined Ex. 3 and Ex. 6, here comes the main result. Suppose that $\lambda_1, \ldots, \lambda_k$ are distinct eigenvalues of $A_{n \times n}$ such that

$$\det(A - tI_n) = (-1)^n (t - \lambda_1)^{r_1} \cdots (t - \lambda_k)^{r_k} \quad \text{with}$$
$$r_1 + \cdots + r_k = n \text{ and } r_1 \ge r_2 \ge \cdots \ge r_k.$$

Then

(1) Each generalized eigenspace

$$G_{\lambda_i} = C(\vec{x}_{i1}) \oplus \cdots \oplus C(\vec{x}_{ik_i}), \quad 1 \le i \le k$$

such that $\dim C(\vec{x}_{ij}) = m_{ij}$ for $1 \le j \le k_i$ and $m_{i1} \ge m_{i2} \ge \cdots \ge m_{ik_i}$. Choose a basis $\mathcal{B}_{ij}$ for each $C(\vec{x}_{ij})$ for $1 \le j \le k_i$ so that $\mathcal{B}_i = \mathcal{B}_{i1} \cup \cdots \cup \mathcal{B}_{ik_i}$ is a Jordan canonical basis for G_{λ_i}. Thus

$$\cdot\, [A\,|\,G_{\lambda_i}]_{\mathcal{B}_i}$$

is as in Ex. 6.

(2) Moreover

$$\mathbb{F}^n = G_{\lambda_1} \oplus \cdots \oplus G_{\lambda_k}$$

and $\mathcal{B} = \mathcal{B}_1 \cup \cdots \cup \mathcal{B}_k$ is a Jordan canonical basis for $\mathbb{F}^n$. Thus, A has the *Jordan canonical form*

$$[A]_{\mathcal{B}} = PAP^{-1} = \begin{bmatrix} [A\,|\,G_{\lambda_1}]_{\mathcal{B}_1} & & & 0 \\ & [A\,|\,G_{\lambda_2}]_{\mathcal{B}_2} & & \\ & & \ddots & \\ 0 & & & [A\,|\,G_{\lambda_k}]_{\mathcal{B}_k} \end{bmatrix}_{n \times n}$$

where P is the invertible matrix where row vectors are basis vectors of $\mathcal{B}$ but arranged in a definite ordering.

By Ex. 6,

the characteristic polynomial $\det(A - tI_n)$

$= (-1)^n$ times the product of all elementary divisors of $A\,|\,G_{\lambda_i}$, $1 \le i \le k$; the minimal polynomial $= (t - \lambda_1)^{m_{11}}(t - \lambda_2)^{m_{21}} \cdots (t - \lambda_k)^{m_{k1}}$.

8. *The process of finding Jordan canonical form of a matrix* Given $A_{n \times n}$, the following process is suggested.

(1) Compute $\det(A - \lambda I_n)$ to see if it splits as $(-1)^n(t - \lambda_1)^{r_1} \cdots (t - \lambda_k)^{r_k}$ where $\lambda_1, \ldots, \lambda_k$ are distinct and $r_1 + \cdots + r_k = n$.

(2) Note $\dim G_{\lambda_i} = r_i$ so that the resulted Table for $A\,|\,G_{\lambda_i}$ as shown in Ex. 6 contains r_i terms. Then, decide the corresponding numbers l_{ij} of jth row in the Table as indicated inside the Remark beneath Ex. 6.

(3) Write out the Jordan block $[A\,|\,G_{\lambda_i}]_{\mathcal{B}_i}$ where $\mathcal{B}_i$ is the Jordan canonical basis for G_{λ_i} consisting of vectors from the Table. Then, construct the Jordan canonical form

$$[A]_{\mathcal{B}} = PAP^{-1} = \begin{bmatrix} [A\,|\,G_{\lambda_1}]_{\mathcal{B}_1} & & 0 \\ & \ddots & \\ 0 & & [A\,|\,G_{\lambda_k}]_{\mathcal{B}_k} \end{bmatrix}.$$

(4) Step 2 indicates how to determine the basis $\mathcal{B}_i$ and finally $\mathcal{B} = \mathcal{B}_1 \cup \cdots \cup \mathcal{B}_k$ when determining l_{ij} for $1 \le i \le k$ and $1 \le j$.

For example, let

$$A = \begin{bmatrix} 2 & 0 & 0 & 0 \\ -1 & 3 & 1 & -1 \\ 0 & -1 & 1 & 0 \\ 1 & 0 & 0 & 3 \end{bmatrix}.$$

The characteristic polynomial is

$$\det(A - tI_4) = (2 - t) \det \begin{vmatrix} 3-t & 1 & -1 \\ -1 & 1-t & 0 \\ 0 & 0 & 3-t \end{vmatrix} = (t-2)^3(t-3).$$

Then A has distinct eigenvalues $\lambda_1 = 2$ and $\lambda_2 = 3$ with respective multiplicity 3 and 2.

For $\lambda_1 = 2$ Note $\dim G_{\lambda_1} = 3$ and

$$A - \lambda_1 I_4 = \begin{bmatrix} 0 & 0 & 0 & 0 \\ -1 & 1 & 1 & -1 \\ 0 & -1 & -1 & 0 \\ 1 & 0 & 0 & 1 \end{bmatrix} \Rightarrow r(A - \lambda_1 I_4) = 2;$$

$$(A - \lambda_1 I_4)^2 = \begin{bmatrix} 0 & 0 & 0 & 0 \\ -2 & 0 & 0 & -2 \\ 1 & 0 & 0 & 1 \\ 1 & 0 & 0 & 1 \end{bmatrix} \Rightarrow r(A - \lambda_1 I_4)^2 = 1.$$

Hence $\text{Ker}(A - \lambda_1 I_4)^2 = \{\vec{x} = (x_1, x_2, x_3, x_4)| -2x_2 + x_3 + x_4 = 0\} = \langle\langle(1,0,0,0),(0,1,2,0),(0,1,0,2)\rangle\rangle$ has dimension 3. Take $\vec{x}_1 = (0,1,0,2)$. Hence $\vec{x}_1 \notin \text{Ker}(A - \lambda_1 I_4)$ and $\{\vec{x}_1(A - \lambda_1 I_4), \vec{x}_1\}$ is linearly independent. Choose $\vec{x}_2 = (1,0,0,0) \in \text{Ker}(A - \lambda I_4)$ which is linearly independent of $\vec{x}_1(A - \lambda_1 I_4) = (1,1,1,1)$.
  Then

$$G_{\lambda_1} = \langle\langle\vec{x}_1(A - \lambda_1 I_4), \vec{x}_1\rangle\rangle \oplus \langle\langle\vec{x}_2\rangle\rangle$$

and $\mathcal{B}_1 = \{\vec{x}_1(A - \lambda_1 I_4), \vec{x}_1, \vec{x}_2\}$ is a basis for G_{λ_1}. Therefore

$$[A \,|\, G_{\lambda_1}]_{\mathcal{B}_1} = \begin{bmatrix} 2 & 0 & 0 \\ 1 & 2 & 0 \\ 0 & 0 & 2 \end{bmatrix}.$$

For $\lambda_2 = 3$ Note that $\dim G_{\lambda_2} = 1 = \dim E_{\lambda_2}$. Take a corresponding eigenvector $\vec{x}_3 = (1, 0, 0, 1)$ so that $G_{\lambda_2} = \langle\langle \vec{x}_3 \rangle\rangle$ with basis $\mathcal{B}_2 = \{\vec{x}_3\}$. Therefore,

$$[A \mid G_{\lambda_2}]_{\mathcal{B}_2} = [3].$$

Combining these results together, we have the Jordan canonical form

$$PAP^{-1} = \begin{bmatrix} 2 & 0 & & \\ 1 & 2 & & \\ & & 2 & \\ & & & 3 \end{bmatrix}_{4 \times 4},$$

where

$$P = \begin{bmatrix} \vec{x}_1(A - \lambda_1 I_4) \\ \vec{x}_1 \\ \vec{x}_2 \\ \vec{x}_3 \end{bmatrix} = \begin{bmatrix} 1 & 1 & 1 & 1 \\ 0 & 1 & 0 & 2 \\ 1 & 0 & 0 & 0 \\ 1 & 0 & 0 & 1 \end{bmatrix}.$$

For each of the following matrices A, find a Jordan canonical basis $\mathcal{B}$ so that $[A]_{\mathcal{B}}$ is the Jordan canonical form of A.

(a) $\begin{bmatrix} 13 & -5 & -6 \\ 16 & -7 & -8 \\ 16 & -6 & -7 \end{bmatrix}$.   (b) $\begin{bmatrix} 3 & 3 & -2 \\ 0 & -1 & 0 \\ 8 & 6 & 5 \end{bmatrix}$.

(c) $\begin{bmatrix} -1 & -5 & 6 \\ 1 & 21 & -26 \\ 1 & 17 & -21 \end{bmatrix}$.   (d) $\begin{bmatrix} 4 & -2 & -1 \\ 5 & -2 & -1 \\ -2 & 1 & 1 \end{bmatrix}$.

(e) $\begin{bmatrix} 3 & -2 & -4 \\ 7 & -5 & -10 \\ -3 & 2 & 3 \end{bmatrix}$.   (f) $\begin{bmatrix} 3 & -4 & 7 & -17 \\ 1 & -1 & 1 & -6 \\ 0 & 0 & 2 & -1 \\ 0 & 0 & 1 & 0 \end{bmatrix}$.

(g) $\begin{bmatrix} 1 & 0 & 0 & 0 \\ 2 & 1 & 0 & 0 \\ 3 & 2 & 1 & 0 \\ 4 & 3 & 2 & 1 \end{bmatrix}$.   (h) $\begin{bmatrix} 0 & -2 & -2 & -2 \\ -3 & 1 & 1 & -3 \\ 1 & -1 & -1 & 1 \\ 2 & 2 & 2 & 4 \end{bmatrix}$.

9. *Similarity of two matrices* Two matrices $A_{n \times n}$ and $B_{n \times n}$, each having its Jordan canonical form, are similar if and only if they have the same Jordan canonical form, up to the ordering of their eigenvalues. Use this

result to determine which of the following matrices

$$\begin{bmatrix} 0 & -3 & 7 \\ -1 & -1 & 5 \\ -1 & -2 & 6 \end{bmatrix} ; \quad \begin{bmatrix} -3 & -7 & 1 \\ 3 & 6 & -1 \\ -2 & -3 & 2 \end{bmatrix} ; \quad \begin{bmatrix} 0 & -4 & -2 \\ 1 & 4 & 1 \\ -1 & -2 & 1 \end{bmatrix}$$

are similar.

10. Write out all Jordan canonical matrices (up to the orderings of Jordan blocks and eigenvalues) whose characteristic polynomials are the same polynomial

$$(t-2)^4(t-3)^2t^2(t+1).$$

B.12.2 *Rational canonical form*

The main result is the following theorem.

Theorem *Suppose $A \in M(n; \mathbb{F})$ is a nonzero matrix having its characteristic polynomials factored as*

$$\det(A - tI_n) = (-1)^n p_1(t)^{r_1} \dots p_k(t)^{r_k}$$

where $p_1(t), \dots, p_k(t)$ are distinct irreducible monic polynomials and $r_1 \geq 1, \dots, r_k \geq 1$ are positive integers. Then there exists a basis $\mathcal{B} = \{\vec{x}_1, \dots, \vec{x}_n\}$, called rational canonical basis *of A, such that*

$$[A]_{\mathcal{B}} = PAP^{-1} = \begin{bmatrix} R_1 & & & \\ & R_2 & & \\ & & \ddots & \\ & & & R_l \end{bmatrix}_{n \times n}$$

(*called the* rational canonical form *of A*)

where each R_i is the companion matrix (refer to Ex. 3(d) of Sec. B.11) of some polynomial $p(t)^m$, where $p(t)$ is a monic divisor of the characteristic polynomial $\det(A - tI_n)$ of A and m is a positive integer, or R_i is a 1×1 matrix $[\lambda]$, where λ is an eigenvalue of A, and

$$P = \begin{bmatrix} \vec{x}_1 \\ \vec{x}_2 \\ \vdots \\ \vec{x}_n \end{bmatrix}.$$

Exercises (continued)

11. For a matrix $A_{9\times9}$, suppose $\mathbb{R}^9$ has a basis $\mathcal{B} = \{\vec{x}, \ldots, \vec{x}_9\}$ so that

$$PAP^{-1} = \begin{bmatrix} 0 & 1 & 0 & 0 & & & & & \\ 0 & 0 & 1 & 0 & & & & & \\ 0 & 0 & 0 & 1 & & & & & \\ -1 & -2 & -3 & -2 & & & & & \\ & & & & 0 & -1 & & & \\ & & & & -1 & -1 & & & \\ & & & & & & 0 & 1 & \\ & & & & & & -1 & 0 & \\ & & & & & & & & 3 \end{bmatrix}_{9\times9}, \quad P = \begin{bmatrix} \vec{x}_1 \\ \vec{x}_2 \\ \vdots \\ \vec{x}_8 \\ \vec{x}_9 \end{bmatrix}.$$

PAP^{-1} is the rational canonical form for A with $\mathcal{B}$ the corresponding rational canonical basis.

(a) Show that the characteristic polynomial $\det(A - tI_9) = -p_1(t)^3 p_2(t)p_3(t)$ where $p_1(t) = t^2 + t + 1, p_2(t) = t^2 + 1$ and $p_3(t) = t - 3$ with the consecutive submatrices R_1, R_2, R_3 and R_4 as the respective companion matrix of $p_1(t)^2, p_1(t), p_2(t)$ and $p_3(t)$. Among the diagonal entries of PAP^{-1}, only 3 is an eigenvalue of A with $\vec{x}_9$ the corresponding eigenvector.

(b) Determine A-invariant subspaces

$$E_{p_i} = \{\vec{x} \in \mathbb{R}^9 \mid \vec{x}p_i(A)^m = \vec{0} \text{ for some positive integer } m\},$$

for $i = 1, 2, 3$. Try to find a smallest positive integer m_i such that $E_{p_i} = \operatorname{Ker} p_i(A)^{m_i}$ for $i = 1, 2, 3$.

(c) Show that $\vec{x}_1, \ldots, \vec{x}_6 \in E_{p_1}$ and $\vec{x}_2 = \vec{x}_1 A, \vec{x}_3 = \vec{x}_2 A = \vec{x}_1 A^2, \vec{x}_4 = \vec{x}_1 A^3$ and $\vec{x}_6 = \vec{x}_5 A$. Also, $\vec{x}_7, \vec{x}_8 \in E_{p_2}$ and $\vec{x}_8 = \vec{x}_7 A$. Therefore,

$$E_{p_1} = \langle\langle \vec{x}_1, \vec{x}_1 A, \vec{x}_1 A^2, \vec{x}_1 A^3 \rangle\rangle \oplus \langle\langle \vec{x}_5, \vec{x}_5 A \rangle\rangle,$$
$$E_{p_2} = \langle\langle \vec{x}_7, \vec{x}_7 A \rangle\rangle,$$
$$E_{p_3} = \langle\langle \vec{x}_9 \rangle\rangle.$$

(d) Let $A_i = A\,|\,E_{p_i}$ denote the restriction of the linear mapping $A\colon \mathbb{R}^9 \to \mathbb{R}^9$ to its invariant subspace E_{p_i} for $1 \le i \le 3$. Compute the following for each i

(1) $\operatorname{r}(A_i^l)$ for $l = 1, 2, 3, 4$,
(2) $\dim \operatorname{Ker}(A_i^l)$ for $l = 1, 2, 3, 4$.

12. (compare with Ex. 2) Let $p(t)$ be an irreducible monic factor of $\det(A - tI_n)$. Define

$$E_p = \{\vec{x} \in \mathbb{F}^n \mid \vec{x} p(A)^r = \vec{0} \text{ for some positive integer } r\}.$$

(a) E_p is an A-invariant subspace of $\mathbb{F}^n$. Also, there exists a smallest positive integer m for which

$$E_p = \operatorname{Ker} p(A)^m.$$

(b) The restriction $A \mid E_p$ has the minimal polynomial

$$(p(t))^m.$$

Also, $md \mid \dim E_p$, where $d = \deg p(t)$ (see Ex. 3(b) of Sec. B.11).

(c) Suppose $p_1(t)$ and $p_2(t)$ are distinct irreducible monic factors of $\det(A - tI_n)$, then

(1) $E_{p_1} \cap E_{p_2} = \{\vec{0}\}$ and $E_{p_1} \cup E_{p_2}$ is linearly independent.
(2) E_{p_1} is invariant under $p_2(A)$ and the restriction of $p_2(A)$ to E_{p_1} is one-to-one and onto.

Recall the A-cycle invariant subspace $C(\vec{x})$ generated by a nonzero vector $\vec{x} \in \mathbb{F}^n$ (related to A) and its basis $\mathcal{B}_{\vec{x}} = \{\vec{x}, \vec{x}A, \ldots, \vec{x}A^{k-1}\}$ if $k = \dim C(\vec{x})$ (see Ex. 4).

13. (compare with Ex. 6) Suppose $p(t)^r$ is a divisor of the characteristic polynomial of $A_{n \times n}$ where $p(t)$ is an irreducible monic polynomial such that $p(t)^{r+1}$ is no more a divisor. Let $d = \deg p(t)$.

(a) Let V be any A-invariant subspace of E_p and $\mathcal{B}$ be a basis for V. Then, for any $\vec{x} \in \operatorname{Ker} p(A)$ but $\vec{x} \notin V, \mathcal{B} \cup \mathcal{B}_{\vec{x}}$ is linearly independent.

(b) There exist linearly independent vectors $\vec{x}_1, \ldots, \vec{x}_k$ in $\operatorname{Ker} p(A) \subseteq E_p = \operatorname{Ker} p(A)^m$ such that $\mathcal{B} = \mathcal{B}_{\vec{x}_1} \cup \cdots \cup \mathcal{B}_{\vec{x}_1}$ forms a basis for E_p and

$$E_p = C(\vec{x}_1) \oplus \cdots \oplus C(\vec{x}_k).$$

(c) Let $p(t)^{m_i}$ be the A-annihilator of $\vec{x}_i$ with $m_1 \geq m_2 \geq \cdots \geq m_k$. Construct the following table.

cycles bases	$C(\vec{x}_1)$	$C(\vec{x}_2)$	$\ldots$	$C(\vec{x}_k)$	
	$\bullet$	$\bullet$		$\bullet$	$\leftarrow$ 1st row
	$\bullet$	$\bullet$	$\cdots$	$\bullet$	$\leftarrow$ 2nd row
	$\bullet$	$\bullet$		$\bullet$	
	$\vdots$	$\vdots$	$\cdots$	$\vdots$	
	$\bullet$	$\bullet$	$\cdots$	$\bullet$	$\leftarrow d\,m_1$th row

$$\mathcal{B}_{\vec{x}_1}\text{-with} \quad \mathcal{B}_{\vec{x}_2}\text{-with} \quad \mathcal{B}_{\vec{x}_k}\text{-with}$$

$$\vec{x}_1 p(A)^{m_1} = \vec{0}, \quad \vec{x}_2 p(A)^{m_2} = \vec{0}, \quad \vec{x}_k p(A)^{m_k} = \vec{0}.$$

Notice that $\dim E_p = r \cdot d = m_1 d + m_2 d + \cdots + m_k d$ and hence $r = m_1 + m_2 + \cdots + m_k$.

Let l_j be the number of dots in the jth row for $1 \le j \le d\,m_1$. Then

$$l_1 = \frac{1}{d}[\dim \mathbb{F}^n - \mathrm{r}[p(A)]] = \frac{1}{d}\dim \mathrm{Ker}\, p(A),$$

$$l_j = \frac{1}{d}[\mathrm{r}[p(A)^{j-1}] - \mathrm{r}[p(A)^j]] \quad \text{for } j > 1,$$

where d is the degree of $p(t)$. Therefore, the size of the Table is completely determined by $A \mid E_p$ or by A itself. Each dot contributes a companion matrix.

(d) The matrix representation of $A \mid E_p$ with respect to $\mathcal{B}$ is

$$[A \mid E_p]_{\mathcal{B}} = \begin{bmatrix} R_1 & & & 0 \\ & R_2 & & \\ & & \ddots & \\ 0 & & & R_k \end{bmatrix}$$

where each $R_j, 1 \le j \le k$, is the companion matrix of $p(t)^{m_j}$ and $dr = \dim E_p = d(m_1 + \cdots + m_k)$ if $E_p = \mathrm{Ker}(p(A))^r = \mathrm{Ker}(p(A))^m$ as in Ex. 12.

14. (compare with Ex. 7) *Fundamental decomposition theorem for square matrix or linear operator* Let the characteristic polynomial

of a nonzero matrix $A_{n \times n}$ be

$$\det(A - tI_n) = (-1)^n p_1(t)^{r_1} \cdots p_k(t)^{r_k}$$

where $p_1(t), \ldots, p_k(t)$ are distinct irreducible monic polynomials with respective degree $d_1, \ldots, d_k$ and $d_1 \geq d_2 \geq \cdots \geq d_k$.

(a) For $1 \leq i \leq k$,

$$E_{p_i} = C(\vec{x}_{i1}) \oplus \cdots \oplus C(\vec{x}_{ik_i})$$

such that $\dim C(\vec{x}_{ij}) = d_i m_{ij}$ for $1 \leq j \leq k_i$ and $m_{i1} \geq m_{i2} \geq \cdots \geq m_{ik_i}$ with $d_i r_i = \dim E_{p_i} = (m_{i1} + \cdots + m_{ik_i})d_i$. Take basis $\mathcal{B}_{ij} = \{\vec{x}_{ij}, \vec{x}_{ij}A, \ldots, \vec{x}_{ij}A^{d_i m_{ij}-1}\}$ for $C(\vec{x}_{ij})$, $1 \leq j \leq k_i$. Then $\mathcal{B}_i = \mathcal{B}_{i1} \cup \cdots \cup \mathcal{B}_{ik_i}$ is a rational canonical basis for E_{p_i} and

$$[A \,|\, E_{p_i}]_{\mathcal{B}_i}$$

is the rational canonical form of $A \,|\, E_{p_i}$ as in Ex. 13.

(b) Moreover,

$$\mathbb{F}^n = E_{p_1} \oplus \cdots \oplus E_{p_k}$$

and $\mathcal{B} = \mathcal{B}_1 \cup \cdots \cup \mathcal{B}_k$ is a rational canonical basis for $\mathbb{F}^n$ so that the matrix representation of A with respect to $\mathcal{B}$ is the rational canonical form

$$[A]_{\mathcal{B}} = PAP^{-1} = \begin{bmatrix} [A \,|\, E_{p_1}]_{\mathcal{B}_1} & & & 0 \\ & [A \,|\, E_{p_2}]_{\mathcal{B}_2} & & \\ & & \ddots & \\ 0 & & & [A \,|\, E_{p_k}]_{\mathcal{B}_k} \end{bmatrix}_{n \times n}$$

where P is the invertible $n \times n$ matrix where row vectors are basis vectors of $\mathcal{B}$ arranged in a definite order.

(c) For each $i, 1 \leq i \leq k$, the restriction $A \,|\, E_{p_i}$ has characteristic polynomial $p_i(t)^{r_i}$ and minimal polynomial $p_i(t)^{m_{i1}}$. Therefore, the minimal polynomial of A is

$$p_1(t)^{m_{11}} p_2(t)^{m_{21}} \cdots p_k(t)^{m_{k1}}.$$

15. (compare with Ex. 8) Ex. 13(d) indicates how to compute the rational canonical form of a matrix. For example, let

$$A = \begin{bmatrix} 0 & 1 & 0 & 0 \\ -4 & -1 & -1 & -1 \\ 12 & 3 & 6 & 8 \\ -7 & -3 & -4 & -5 \end{bmatrix} \in M(4; \mathbb{R}).$$

Step 1 Compute the characteristic polynomial.

By actual computation,

$$\det(A - tI_4) = t^4 + 5t^2 + 6 = (t^2 + 2)(t^2 + 3).$$

Let $p_1(t) = t^2 + 2$ and $p_2(t) = t^2 + 3$.

Step 2 Determine $[A \,|\, E_{p_i}]_{B_i}$.

Now

$$A^2 = \begin{bmatrix} -4 & -1 & -1 & -1 \\ -1 & -3 & -1 & -2 \\ 4 & 3 & 1 & 5 \\ -1 & -1 & -1 & -4 \end{bmatrix}$$

$$\Rightarrow p_1(A) = A^2 + 2I_4 = \begin{bmatrix} -2 & -1 & -1 & -1 \\ -1 & -1 & -1 & -2 \\ 4 & 3 & 3 & 5 \\ -1 & -1 & -1 & -2 \end{bmatrix} \quad \text{with } r(A^2 + 2I_4) = 2;$$

$$p_2(A) = A^2 + 3I_4 = \begin{bmatrix} -1 & -1 & -1 & -1 \\ -1 & 0 & -1 & -2 \\ 4 & 3 & 4 & 5 \\ -1 & -1 & -1 & -1 \end{bmatrix} \quad \text{with } r(A^2 + 3I_4) = 2.$$

Hence, there exist $\vec{x}_1 \in \mathrm{Ker}(A^2 + 2I_4)$ and $\vec{x}_2 \in \mathrm{Ker}(A^2 + 3I_4)$ so that $B_1 = \{\vec{x}_1, \vec{x}_1 A\}$ is a basis for E_{p_1} and $B_2 = \{\vec{x}_2, \vec{x}_2 A\}$ is a basis for E_{p_2}. Therefore,

$$[A \,|\, E_{p_1}]_{B_1} = \begin{bmatrix} 0 & 1 \\ -2 & 0 \end{bmatrix} \quad \text{and} \quad [A \,|\, E_{p_2}]_{B_2} = \begin{bmatrix} 0 & 1 \\ -3 & 0 \end{bmatrix}$$

and the rational canonical form of A is

$$[A]_B = PAP^{-1} = \begin{bmatrix} 0 & 1 & & \\ -2 & 0 & & \\ \hline & & 0 & 1 \\ & & -3 & 0 \end{bmatrix}, \quad P = \begin{bmatrix} \vec{x}_1 \\ \vec{x}_1 A \\ \vec{x}_2 \\ \vec{x}_2 A \end{bmatrix}$$

where $B = B_1 \cup B_2$ is a rational canonical basis.

Step 3 Determine $\mathcal{B}$ and hence P.

Let $\vec{x} = (x_1, x_2, x_3, x_4)$. Solving $\vec{x}p_1(A) = \vec{0}$, we get

$$\operatorname{Ker} p_1(A) = \langle\langle (1, 2, 1, 0), (0, -1, 0, 1)\rangle\rangle.$$

Solving $\vec{x}p_2(A) = \vec{0}$, we get

$$\operatorname{Ker} p_2(A) = \langle\langle (3, 1, 1, 0), (-1, 0, 0, 1)\rangle\rangle.$$

Choose $\vec{x}_1 = (0, -1, 0, 1)$ and $\vec{x}_2 = (-1, 0, 0, 1)$. Then $\vec{x}_1 A = (-3, -2, -3, -4)$ and $\vec{x}_2 A = (-7, -4, -4, -5)$. Thus, the rational canonical basis $\mathcal{B} = \{(0, -1, 0, 1), (-3, -2, -3, -4), (-1, 0, 0, 1), (-7, -4, -4, -5)\}$ and

$$P = \begin{bmatrix} 0 & -1 & 0 & -1 \\ -3 & -2 & -3 & -4 \\ -1 & 0 & 0 & 1 \\ -7 & -4 & -4 & -5 \end{bmatrix}.$$

Notice that, if A is considered as a complex matrix, then A is diagonalizable. Try to find an invertible complex matrix $Q_{4\times 4}$ so that

$$QAQ^{-1} = \begin{bmatrix} \sqrt{2i} & & & 0 \\ & -\sqrt{2i} & & \\ & & \sqrt{3i} & \\ 0 & & & -\sqrt{3i} \end{bmatrix}.$$

For another example, let

$$B = \begin{bmatrix} 0 & 1 & 1 & 1 & 1 \\ 2 & -2 & 0 & -2 & -4 \\ 0 & 0 & 1 & 1 & 3 \\ -6 & 0 & -3 & -1 & -3 \\ 2 & 2 & 2 & 2 & 4 \end{bmatrix} \in M(5; \mathbb{R}).$$

Then $\det(A - tI_5) = -(t^2 + 2)^2(t - 2)$. Let $p_1(t) = t^2 + 2$ with multiplicity $r_1 = 2$ and $p_2(t) = t - 2$ with $r_2 = 1$. Hence $\dim E_{p_i} = 2.2 = 4$, and $r_1 = 2 = m_1 + \cdots + m_k$ implies that $m_1 = m_2 = 1$ or

$m_1 = 2$. Now

$$B^2 = \begin{bmatrix} -2 & 0 & 0 & 0 & 0 \\ 0 & -2 & 0 & 0 & 0 \\ 0 & 6 & 4 & 6 & 12 \\ 0 & -12 & -12 & -14 & -24 \\ 0 & 6 & 6 & 6 & 10 \end{bmatrix}$$

$$\Rightarrow B^2 + 2I_5 = \begin{bmatrix} 0 & 0 & 0 & 0 & 0 \\ 0 & 0 & 0 & 0 & 0 \\ 0 & 6 & 6 & 6 & 12 \\ 0 & -12 & -12 & -12 & -24 \\ 0 & 6 & 6 & 6 & 12 \end{bmatrix}$$

with $\quad \mathrm{r}(p_1(B)) = \mathrm{r}(B^2 + 2I_5) = 1.$

Thus, the first row in the Table for E_p has $\frac{1}{2}(5 - \mathrm{r}(p_1(B))) = 2$ dots.

This implies that the only possibility is $m_1 = m_2 = 1$ and each dot contributes the companion matrix

$$\begin{bmatrix} 0 & 1 \\ -2 & 0 \end{bmatrix}$$

of $p_1(t) = t^2 + 2$ to the rational canonical form. On the other hand, $\dim E_{p_2} = 1$ and $p_2(t)$ results in the 1×1 matrix $[2]$ to the canonical form. Combining together, the rational canonical form of B is

$$PBP^{-1} = \begin{bmatrix} 0 & 1 & & \\ -2 & 0 & & \\ \hline & & 0 & 1 \\ & & -2 & 0 \\ \hline & & & & 2 \end{bmatrix}, \quad P = \begin{bmatrix} \vec{x}_1 \\ \vec{x}_1 A \\ \vec{x}_2 \\ \vec{x}_2 A \\ \vec{x}_3 \end{bmatrix}$$

where $\vec{x}_1 = \vec{e}_1 = (1, 0, 0, 0, 0)$ and $\vec{x}_2 = \vec{e}_2 = (0, 1, 0, 0, 0)$ are in $\mathrm{Ker}\, p_1(B)$, and $\vec{x}_1 A = (0, 1, 1, 1, 1)$ and $\vec{x}_2 A = (2, -2, 0, -2, 4)$, while $\vec{x}_3 = (0, 1, 1, 1, 2) \in \mathrm{Ker}\, p_2(B)$.

For each of the following real matrices A, find a rational canonical basis B so that $[A]_B$ is the rational canonical form of A.

(a) $\begin{bmatrix} 6 & -3 & -2 \\ 4 & -1 & -2 \\ 10 & -5 & -3 \end{bmatrix}$;   (b) $\begin{bmatrix} 1 & 0 & 3 \\ 2 & 1 & 2 \\ 0 & 0 & 2 \end{bmatrix}$;   (c) $\begin{bmatrix} 1 & -2 & 0 & 0 \\ 2 & 1 & 0 & 0 \\ 1 & 0 & 1 & -2 \\ 0 & 1 & 2 & 1 \end{bmatrix}$;

(d) $\begin{bmatrix} 2 & 0 & 0 & 0 \\ 1 & 2 & 0 & 0 \\ 0 & 1 & 2 & 0 \\ 0 & 0 & 0 & 2 \end{bmatrix}$.

References

On Linear Algebra

[1] 左銓如・季素月: 初等幾何研究 (Elementary Geometry), 九章出版社, 台北, 1998.

[2] 李炯生・查建國: 線性代數 (Linear Algebra), 中國科學技術大學出版社, 合肥, 1989.

[3] 林義雄: 初等線性代數(1): 行列式 (Determinants), 台北, 1982, 1987.

[4] 林義雄: 初等線性代數(2): 矩陣 (Matrices), 台北, 1982, 1987.

[5] 林義雄: 初等線性代數(3): 向量, 仿射空間 (Vector and Affine Spaces), 台北, 1983.

[6] 林義雄: 初等線性代數(4): 射影空間 (Projective Spaces), 台北, 1984.

[7] 林義雄: 初等線性代數(5): 內積空間 (Inner Product Spaces), 台北, 1984.

[8] 許以超: 線性代數與矩陣論 (Linear Algebra and Theory of Matrices), 高等教育出版社, 北京, 1992.

[9] A. R. Amir-Moe'z and A. L. Fass: Elements of Linear Algebra, Pergamon Press, The MacMillan Co., New York, 1962.

[10] N. V. Efimov and E. R. Rozendorn: Linear Algebra and Multidimensional Geometry (English translation), Mir Publishers, Moscow, 1975.

[11] A. E. Fekete: Real Linear Algebra, Marcel Dekkel, Inc., New York, Basel, 1985.

[12] H. Gupta: Matrices in n-dimensional Geometry, South Asian Publishers, New Delhi, 1985.

[13] M. Koecher: Lineare Algebra und Analytischer Geometrie, zweite Auflage, Grundwissen Mathematik 2, Springer-Verlag, Berlin, Heidelberg, New York, Tokyo, 1985.

[14] P. S. Modenov and A. Parkhomenko: Geometric Transformations, Academic Press, New York, London, 1965.

[15] G. Strang: Introduction to Linear Algebra, 2nd ed., Wellesley-Cambridge Press, 1998.

[16] G. Strang: Linear Algebra and its Applications, 3rd ed., Saunders, Philadelphia, 1988.

[17] G. Strang: The Fundamental Theorem of Linear Algebra, AMS, Vol 100, number 9(1993), 848–855.

[18] S. M. Wilkinson: The Algebraic Eigenvalue Problem, Oxford University Press, New York, 1965.

Other Related Sources

History

[19] M. Kline: Mathematical Thought: From Ancient to Modern Times, Oxford University Press, New York, 1972.

Group

[20] J. J. Rotman: An Introduction to the Theory of Groups, 4th ed., Springer-Verlag, New York, 1995.

Geometry

[21] H. Eves: A Survey of Geometry, Vol 1, 1963, Vol 2, 1965, Allyn and Bacon, Inc., Boston.

Differential Geometry, Lie Group

[22] M. P. do Carmo: Differential Geometry of Curves and Surfaces, Prentice-Hall, Inc., New Jersey, 1976.

[23] C. Chevalley: Theory of Lie Groups, Vol I, Princeton University Press, Princetion, New Jersey, 1946.

[24] S. Helgason: Differential Geometry, Lie Group and Symmetric Spaces, Academic Press, New York, San Francisco, London, 1978.
Note. 佐武一郎: リー群の話, 日本評論所, 東京, 1982. This is a popular book in Japan with its informal, nonrigorous approach to Lie group.

Fractal Geometry

[25] M. Barnsley: Fractals Everywhere, Academic Press, San Diego, 1988.

[26] B. B. Mandelbrot: The Fractal Geometry of Nature, W. H. Freeman, New York, 1983.

Matrix Analysis

[27] R. Bellman: Introduction to Matrix Analysis, Siam, Philadelphia, 1995.

Real Analysis (including Differentiable Manifolds)

[28] L. H. Loomis and S. Sternberg: Advanced Calculus, revised ed., Johnes and Bartlett Publishers, Boston, 1990.
Note. 林義雄 · 林紹雄: 理論分析 (Theoretical Analysis), Vol I (1977, 1978, 1985, 1993); Vol II (1981, 1989); Vol III (1982, 1989), 台北. These three voluminous books contain routine materials in Advanced Calculus plus topics in Metric Space, Functional Analysis, Measure and Integration, and Submanifolds in $\mathbb{R}^n$ up to the Stokes Theorem in Differential Form.

Complex Analysis

[29] L. V. Ahlfors: Complex Analysis, 3rd ed., McGraw-Hill Book Co., New York, 1979.

[30] L. K. Hua(華羅庚): Harmonic Analysis of Functions of Several Complex Variables in the Classical Domains (多複變之典型域上之調和分析), Trans. Math. Monographs, Vol 6, AMS, Providence, R.I., 1963.

[31] O. Lehto and K. I. Virtanen: Quasikonforme Abbildungen, 1965, English Translation: Quasiconformal Mappings in the Plane, Springer-Verlag, Berlin, Heidelberg, New York, 1970.

[32] J. Väisälä: Lectures on n-Dimensional Quasiconformal Mappings, Lecture Notes in Mathematics 229, Springer-Verlag, Berlin, Heidelberg, New York, 1970.

Differential Equations

[33] W. E. Boyce and R. C. Diprema: Elementary Differential Equations, 7th ed., John Wiley & Sons, New York, 2000.

[34] S. J. Farlow: An Introduction to Differential Equations and their Applications, McGraw-Hill, Inc., New York, 1994.

Fourier Analysis

[35] R. E. Edwards: Fourier Series, A Modern Approach, Vols 1 and 2, Pergamon Press, Inc., New York, 1964.

[36] E. C. Titchmarsh: Introduction to the Theory of Fourier Integrals, 2nd ed., Oxford at the Clarendon Press, 1959.

[37] A. Zygmund: Trigonometric Series, Vols I and II, Cambridge University Press, New York, 1959.

Markov Chains

[38] K. L. Chung (鍾開萊): Elementary Probability Theory with Stochastic Processes, UTM, Springer-Verlag, New York, 1974.

[39] K. L. Chung: Markov Chains, 2nd ed., Springer-Verlag, 1967.

[40] J. L. Doob: Stochastic Processes, Wiley & Sons, New York, 1953.

[41] J. G. Kemeny and J. L. Snell: Finite Markov Chains, Springer-Verlag, New York, 1976.

Index of Notations

[1] (etc.)	Reference [1]	
Ex.<A>1 of Sec.2.3 (etc.)	problem 1 in Exercises <A> of Sec.2.3	
Ex.<A> or <B> or <C> or <D>	Exercises <A> or <B> or <C> or <D>	
$A = [a_{ij}]_{m \times n}, [a_{ij}]$ or $A_{m \times n}$	matrix of order $m \times n$: m rows, n columns, where $1 \le i \le m$ and $1 \le j \le n$	700
$A = [a_{ij}]_{n \times n}$ or $A_{n \times n}$	square matrix of order n	700
A^*	transpose of $A = [a_{ij}]_{m \times n}$: $[a_{ji}]_{n \times m}$	124, 702
$\bar{A}$	conjugate matrix of $A = [a_{ij}]$: $[\bar{a}_{ij}]$	702
$\bar{A}^*$	conjugate transpose of A: $(\bar{A})^* = \overline{(A^*)}$	702
$A + B = [a_{ij} + b_{ij}]$	the sum (addition) of the matrices A and B of the same order	702
$\alpha A = [\alpha a_{ij}]$	the scalar multiplication of the scalar α and the matrix A	702
$AB = [c_{ik}]$	the product of $A = [a_{ij}]_{m \times n}$ and $B = [b_{jk}]_{n \times p}$: $c_{ik} = \sum_{j=1}^{n} a_{ij} b_{jk}, 1 \le i \le m, 1 \le k \le p$	705
$A^0 = I_n$	the zero power of a square matrix A is *defined* to be the *identity* matrix I_n	120, 701
A^{-1}	the inverse of an invertible (square) matrix A	47, 55, 336, 711
A^p	the pth power of a square matrix A, where $p \ge 1$ is an integer; negative integer p is permitted if A is invertible	120, 707
$A = B$	equality of two matrices $A_{m \times n}$ and $B_{m \times n}$	700
$A_{\mathcal{B}'}^{\mathcal{B}}$ or $[1_v]_{\mathcal{B}'}^{\mathcal{B}}$	the transition matrix from a basis $\mathcal{B}$ to a basis $\mathcal{B}'$ in a finite-dimensional vector space V; change of coordinates matrix	47, 337, 735

$r(A)$	rank of $A_{m \times n}$, defined to be the dimension dim Im(A), specifically called the *row rank* of A and equal to $r(A^*)$, the *column rank* of A	89, 123, 124, 366, 383, 407, 711, 722, 739
$tr(A)$	trace of a square matrix $A = [a_{ij}]_{n \times n}$: the sum $\sum_{i=1}^{n} a_{ii}$ of diagonal entries of A	122, 407, 713
$det A$; $det[A]_{\mathcal{B}}$	determinant associated to the square matrix $A_{n \times n}$; determinant of linear operator A	47, 87, 121, 704, 727
A_{ij}	the minor of a_{ij} in $A = [a_{ij}]_{n \times n}$	336, 727
$(-1)^{i+j} A_{ij}$	the cofactor of a_{ij} in $A = [a_{ij}]_{n \times n}$	336, 728
$adj A$	adjoint matrix of $A: [b_{ij}]_{n \times n}$ with $b_{ij} = (-1)^{j+i} det A_{ji}, 1 \le i, j \le n$	336, 731
A_{i*}	ith row vector $(a_{i1}, \ldots, a_{in})$, also treated as $1 \times n$ row matrix, of $A = [a_{ij}]_{m \times n}$ for $1 \le i \le m$	375, 408, 700
A_{*j}	jth column vector or $m \times 1$ column matrix $\begin{bmatrix} a_{1j} \\ \vdots \\ a_{mj} \end{bmatrix}$ of $A = [a_{ij}]_{m \times n}$ for $1 \le j \le n$	368, 375, 408, 700
$\vec{x} A = \vec{b}$	$A = [a_{ij}]_{m \times n}$, $\vec{x} = (x_1, \ldots, x_m) \in \mathbb{F}^m$ and $\vec{b} = (b_1, \ldots, b_n) \in \mathbb{F}^n$: $\sum_{i=1}^{m} a_{ij} x_i = b_j$, $1 \le j \le n$, a system of n linear equations in m unknowns $x_1, \ldots, x_m$; homogeneous if $\vec{b} = \vec{0}$, non-homogeneous if $\vec{b} \ne \vec{0}$	87, 150, 367, 415, 424, 442, 711 (etc.)

$\vec{x} A^* = \vec{b}$ or $A\vec{x}^* = \vec{b}^*$	$A = [a_{ij}]_{m \times n}, \vec{x} = (x_1, \dots, x_n) \in$ $\mathbb{F}^n$ and $\vec{b} = (b_1, \dots, b_m) \in \mathbb{F}^m$: $$\sum_{j=1}^{n} a_{ij} x_j = b_i, 1 \le i \le m$$	152, 415, 424, 442, 711, 723, 759 (etc.)
$\left[A \mid \vec{b}^* \right]$ or $\left[\dfrac{A^*}{\vec{b}} \right]$	augmented matrix of *the coefficient matrix A* of $A\vec{x}^* = \vec{b}^*$ or $\vec{x} A^* = \vec{b}$, respectively	150, 152, 156, 377, 416, 425, 724 (etc.)
$\begin{bmatrix} A & B \\ C & D \end{bmatrix}$	block or partitioned matrix	180
A^+	the generalized or pseudo inverse of a real or complex matrix $A_{m \times n}$, which is $A^*(AA^*)^{-1}$ if $r(A) = m$; $(A^*A)^{-1}A^*$ if $r(A) = n$; $C^*(CC^*)^{-1}(B^*B)^{-1}B$ if $A = B_{m \times r} C_{r \times n}$ with $r(B) = r(C) = r$, where $r = r(A)$	175, 177, 419, 429, 462, 469, 766 (etc.)
$\det(A - tI_n)$	the characteristic polynomial $(-1)^n t^n + \alpha_{n-1} t^{n-1} + \cdots + \alpha_1 t + \alpha_0, \alpha_{n-1} = \operatorname{tr}(A)$, $\alpha_0 = \det A$, of a matrix $A_{n \times n}$	106, 399, 407, 491, 719, 791(etc.), 795
$p(A) = \displaystyle\sum_{k=0}^{m} \alpha_k A^k$	polynomial matrix of $A_{n \times n}$ induced by the polynomial $p(t) = \displaystyle\sum_{k=0}^{m} \alpha_k t^k$. Note $A^0 = I_n$	108, 128, 403, 486, 496, 499
$e^A = \displaystyle\sum_{k=0}^{\infty} \frac{1}{k!} A^k$	matrix exponential of A	496, 499, 547, 558
$\rho(A)$	spectral radius of A	496
$\displaystyle\lim_{k \to \infty} A^{(k)}$	the limit matrix (if exists) of $A^{(k)}$, where $A_{n \times n}^{(k)}$, as $k \to \infty$	494
$\vec{a}, \vec{b}, \vec{x}, \vec{y}$ (etc.)	vector	8, 26, 692
$\alpha_1 \vec{a}_1 + \cdots + \alpha_k \vec{a}_k = \displaystyle\sum_{i=1}^{k} \alpha_i \vec{a}_i$	linear combination of vectors $\vec{a}_1, \dots, \vec{a}_k$ with coefficients (scalars) $\alpha_1, \dots, \alpha_k$	31, 324, 695

$\vec{a}_1\,\vec{a}_2$ (etc.)	directed line segment from point $\vec{a}_1$ to point $\vec{a}_2$; line segment with endpoints $\vec{a}_1$ and $\vec{a}_2$	18, 65
$\triangle\,\vec{a}_1\,\vec{a}_2\,\vec{a}_3$	(affine and Euclidean) triangle with vertices at points $\vec{a}_1$, $\vec{a}_2$ and $\vec{a}_3$; base triangle in a barycentric coordinate system for the plane	65, 76
$\bar{\triangle}\,\vec{a}_1\,\vec{a}_2\,\vec{a}_3$	oriented triangle	75
$\triangle\,\vec{a}_1\,\vec{a}_2\,\vec{a}_3\,\vec{a}_4$	a tetrahedron with vertices at points $\vec{a}_1$, $\vec{a}_2$, $\vec{a}_3$ and $\vec{a}_4$; 4-tetrahedron; 4-simplex; base tetrahedron	356, 363, 638, 643
$\triangle\,\vec{a}_0\,\vec{a}_1\cdots\vec{a}_k$	k-tetrahedron or k-simplex, where, $\vec{a}_0,\vec{a}_1,\ldots,\vec{a}_k$ are *affinely independent* points in an affine space	642, 654
$\square\,\vec{a}_0\,\vec{a}_1\cdots\vec{a}_k$	k-parallelogram with $\vec{a}_0$ as vertex and side vectors $\vec{a}_1-\vec{a}_0,\ldots,\vec{a}_k-\vec{a}_0$, where $\vec{a}_0,\vec{a}_1,\ldots,\vec{a}_k$ are affinely independent points in an affine space; k-hyperparallelepiped	61, 446, 638, 661, 667
$\mathcal{B},\mathcal{C},\mathcal{D},\mathcal{N}$ (etc.)	basis for a finite-dimensional vector space V	10, 34, 115, 326, 406, 697 (etc.)
$[P]_{\mathcal{B}}$ or $[\overrightarrow{OP}]_{\mathcal{B}}$ or $[\vec{x}]_{\mathcal{B}}=\alpha$	coordinate vector of a point P in $\mathbb{R}$ or the vector $\overrightarrow{OP}=\vec{x}$ with respect to a basis $\mathcal{B}$	10
$[P]_{\mathcal{B}}$ or $[\overrightarrow{OP}]_{\mathcal{B}}$ or $[\vec{x}]_{\mathcal{B}}=(x_1,x_2)$	coordinate vector of a point P in $\mathbb{R}^2$ or the vector $\overrightarrow{OP}=\vec{x}$ w. r. t. a basis $\mathcal{B}=\{\vec{a}_1,\vec{a}_2\}$, namely, $$\vec{x}=\sum_{i=1}^{2}x_i\,\vec{a}_i$$	34
$[P]_{\mathcal{B}}$ or $[\overrightarrow{OP}]_{\mathcal{B}}$ or $[\vec{x}]_{\mathcal{B}}=(x_1,x_2,x_3)$	coordinate vector of a point P in $\mathbb{R}^3$ or the vector $\overrightarrow{OP}=\vec{x}$ w. r. t. a basis, $\mathcal{B}=\{\vec{a}_1,\vec{a}_2,\vec{a}_3\}$, namely, $$\vec{x}=\sum_{i=1}^{3}x_i\,\vec{a}_i$$	326

$\mathcal{B}^*$	dual basis of a basis $\mathcal{B}$	420, 750
$\mathcal{B} = \{\vec{a}_0, \vec{a}_1, \ldots, \vec{a}_k\}$	affine basis with $\vec{a}_0$ as *base point*, where $\vec{a}_0, \vec{a}_1, \ldots, \vec{a}_k$ are affinely independent points in $\mathbb{R}^n$	19, 71, 362, 640, 641, 642
$(\vec{x})_\mathcal{B} = (\lambda_0, \lambda_1, \ldots, \lambda_k),$ $\lambda_0 + \lambda_1 + \cdots + \lambda_k = 1$	(normalized) barycentric coordinate of the point $\vec{x}$ w. r. t. affine basis $\mathcal{B}$: $$\vec{x} = \sum_{i=0}^{k} \lambda_i \vec{a}_i, \sum_{i=0}^{k} \lambda_i = 1$$	19, 71, 362, 642
$[\vec{x}]_\mathcal{B} = [\vec{x} - \vec{a}_0]_\mathcal{B} = (\lambda_1, \ldots, \lambda_k)$	affine coordinate of the point $\vec{x}$ w. r. t. affine basis $$\mathcal{B} : \vec{x} - \vec{a}_0 = \sum_{i=1}^{k} \lambda_i(\vec{a}_i - \vec{a}_0)$$	19, 72, 363, 641
$(\lambda_1 : \lambda_2 : \lambda_3)$	homogeneous area coordinate or (nonnormalized) barycentric coordinate	76
$\mathbb{C}$	complex field	685
$\mathbb{C}^1$ or $\mathbb{C}$	standard one-dimensional complex vector space	29, 693
$\mathbb{C}^n$	standard n-dimensional complex vector space	42, 693
$C_f(\vec{x})$ or $C(\vec{x})$	f-cycle subspace generated by a vector $\vec{x}: \langle\langle \vec{x}, f(\vec{x}), f^2(\vec{x}), \ldots \rangle\rangle$ where f is a linear operator	210, 212, 570, 802
$\mathrm{diag}[a_{11}, \ldots, a_{nn}]$	diagonal matrix	701
$\dim V$	dimension of a finite dimensional vector space V	43, 698
$\vec{e}_i = (0, \ldots, 0, 1, 0, \ldots, 0)$	ith coordinate vector in the standard basis for $\mathbb{F}^n$ (for example, $\mathbb{R}^n$, $\mathbb{C}^n$, $\mathbb{I}_p^n$), for $n \geq 2$	38, 328, 693, 697
$E_{(i)(j)}$	elementary matrix of type I: interchange of ith row and jth row of I_n, $i \neq j$	150, 160, 442, 720
$E_{\alpha(i)}$	elementary matrix of type II: multiplication of ith row of I_n by scalar $\alpha \neq 0$	150, 160, 442, 720

$\mathbb{R}^3_{\Gamma(O;\ A_1,\ A_2,\ A_3)}$	coordinatized space $\{[P]_{\mathcal{B}} \mid P \in \Gamma\}$ of the space Γ w. r. t. the basis $\mathcal{B} = \{\vec{a}_1, \vec{a}_2, \vec{a}_3\}$, where $\vec{a}_i = \overrightarrow{OA_i}$, $i = 1,\ 2,\ 3$ and $[P]_{\mathcal{B}} = (x_1,\ x_2,\ x_3)$	326
$\mathbb{R}^3$	standard three-dimensional real vector space	327
$\mathbb{R}^n (n \geq 2)$	standard n-dimensional affine, or Euclidean (inner product) or vector space over the real field $\mathbb{R}$	42, 327, 693, 773
$(\mathbb{R}^3)^*$	(first) dual space of $\mathbb{R}^3$	420
$(\mathbb{R}^3)^{**}$	second dual space of $\mathbb{R}^3$	420
S(etc.)	subspace of a vector space V	40, 330, 694
$S_1 \cap S_2$	intersection space of subspaces S_1 and S_2	330, 694
$S_1 + S_2$	sum space of subspaces S_1 and S_2	41, 330, 694
$S_1 \oplus S_2$	direct sum (space) of subspaces S_1 and S_2	41, 330, 694
$\langle\langle S \rangle\rangle$	subspace generated or spanned by a nonempty *subset* S of a vector space V: {finite linear combinations of vectors in S}	41, 330, 695
S°	annihilator $\{f \in V^* \mid f(\vec{x}) = \vec{o} \text{ for all } \vec{x} \in S\}$ of a nonempty subset S of a vector space V, a subspace of V^*; also denoted as $S^{\perp}$ in Sec. B.8	421, 752
$S^{\perp}$	orthogonal complement $\{\vec{y} \in V \mid \langle \vec{y}, \vec{x} \rangle = 0 \text{ for all } \vec{x} \in S\}$ of a nonempty *subset* S in an inner product space $(V, \langle, \rangle)$	129, 412, 778
$S^k = \vec{x}_0 + S$	k-dimensional *affine* subspace of an affine or vector space V, where S is a k-dimensional *subspace* of V and $\vec{x}_0$ is a *point* in V: $\{\vec{x}_0 + \vec{v} \mid \vec{v} \in S\}$	640, 724
$S^r \cap S^k$	intersection (affine) subspace of S^r and S^k	644

$\vec{x}_0 + S$	image of subspace S under $\vec{x} \to \vec{x}_1 + \vec{x}$, an affine subspace	67, 359
$\langle \vec{x}, \vec{y} \rangle$	inner product of $\vec{x}$ and $\vec{y}$	773
$|\vec{x}|$	$\langle \vec{x}, \vec{x} \rangle^{\frac{1}{2}}$, length of $\vec{x}$	360, 774
$\vec{x} \perp \vec{y}$	$\vec{x}$ and $\vec{y}$ is perpendicular or orthogonal to each other: $\langle \vec{x}, \vec{y} \rangle = 0$.	129, 412, 774
V, W(etc.)	vector or linear space over a field	691
V^*	(first) dual space of V	749
$V^{**} = (V^*)^*$	second dual space of V	751
V/S	quotient space of V modulus subspace S	200, 211, 369 (etc.), 695

INDEX